ANNOTATED INSTRUC

BEGINNING ALGEBRA

Fourth Edition

John Tobey

Jeffrey Slater

North Shore Community College
Danvers, Massachusetts

This Annotated Instructor's Edition contains introductory materials written by the author and other mathematics instructors that may assist you in teaching your classes, as well as teaching tips, which appear in the margin of the text pages. In addition, the answers to each exercise, pretest, chapter and cumulative test are displayed in blue next to the exercise or problem. Otherwise this Annotated Instructor's Edition is identical to your students' textbook. When ordering the text for your students, be sure to use the ISBN for the student text,

0-13-743626-2

Prentice Hall, Upper Saddle River, NJ 07458

ANNOTATED INSTRUCTOR'S EDITION

Editor-in-Chief: *Jerome Grant*
Editorial Director: *Tim Bozik*
Acquisitions Editor: *Karin E. Wagner*
Editorial Assistant: *April Thrower/Joanne Wendelken*
Assistant Vice President of Production and Manufacturing: *David W. Riccardi*
Production Editor: *York Production Services*
Managing Editor: *Linda Behrens*
Executive Managing Editor: *Kathleen Schiaparelli*
Marketing Manager: *Jolene Howard*
Marketing Assistant: *Jennifer Pan*
Interior Design: *Amy Rosen/Lisa Jones*
Cover Design: *Amy Rosen*
Creative Director: *Paula Maylahn*
Art Director: *Joseph Sengotta*
Art Manager: *Gus Vibal*
Manufacturing Buyer: *Alan Fischer*
Manufacturing Manager: *Trudy Pisciotti*
Photo Researcher: *Rona Tuccillo*
Photo Editor: *Lori Morris-Nantz*
Supplements Editor: *Audra J. Walsh*
Cover Photo: Burton Pritzker/*Photonica*

Simon & Schuster/A Viacom Company
Upper Saddle River, New Jersey 07458

Photo credits appear on page P–1, which constitutes
a continuation of the copyright page.

Printed in the United States of America

10 9 8 7 6 5 4 3 2

ISBN 0-13-636499-3

Prentice-Hall International (UK) Limited, *London*
Prentice-Hall of Australia Pty. Limited, *Sydney*
Prentice-Hall Canada Inc., *Toronto*
Prentice-Hall Hispanoamericana, S.A., *Mexico City*
Prentice-Hall of India Private Limited, *New Delhi*
Prentice-Hall of Japan, Inc., *Tokyo*
Simon & Schuster Asia Pte. Ltd., *Singapore*
Editora Prentice-Hall do Brasil, Ltda., *Rio de Janeiro*

Contents

CHAPTER 3

CHAPTER 4

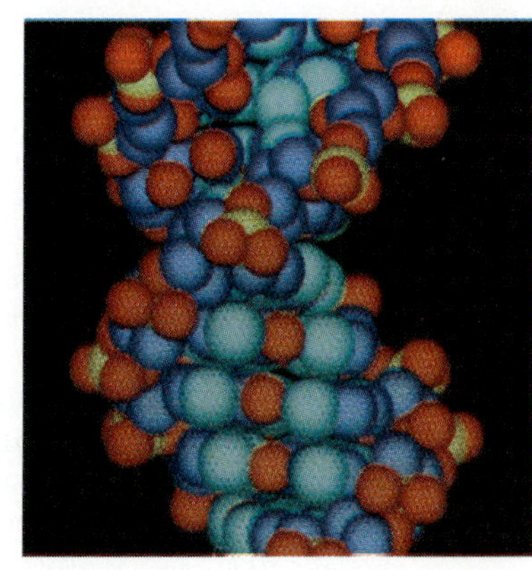

CHAPTER 5

CHAPTER 6

CHAPTER 7

CHAPTER 8

CHAPTER 9

Radicals 503

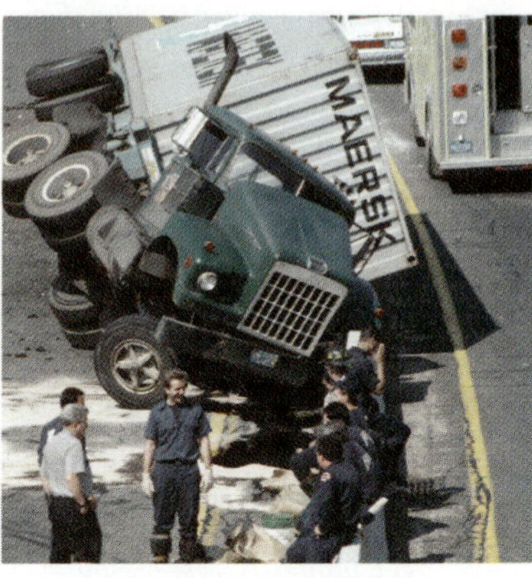

CHAPTER 10

Quadratic Equations 557

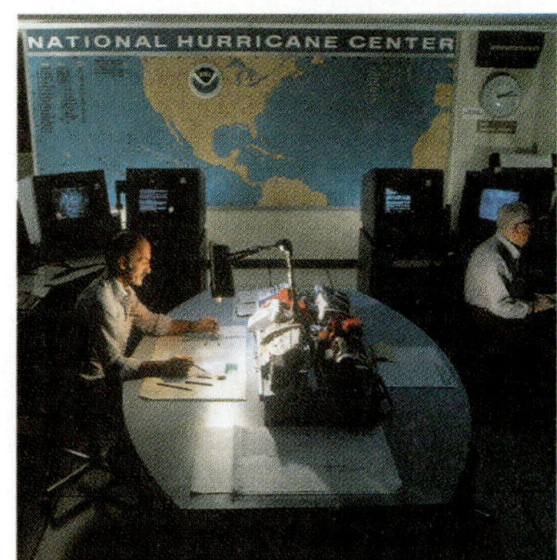

Preface

A Message to the Mathematics Instructor

WHO WAS THIS BOOK WRITTEN FOR?

This text is intended for those students who have not studied algebra, who have been previously unsuccessful in algebra, or who have simply forgotten the algebra they learned. The text is designed to be used in a variety of class settings: lecture-based classes, discussion-oriented classes, self-paced classes, mathematics laboratories, and computer- or audio-visual-supported learning centers. The original manuscript and the first three editions of this book have been class tested by students across the country. The book was developed to help students learn and retain mathematical concepts. Special attention has been given to problem solving in this new edition. This text is organized to help students organize the information in any problem-solving situation, to reduce anxiety, and to provide a guide that enables the student to proceed with the problem-solving process. *Beginning Algebra*, Fourth Edition, is the second in a series of three texts that include the following:

Slater/Tobey	*Basic College Mathematics,* Third Edition
Tobey/Slater	*Beginning Algebra,* Fourth Edition
Tobey/Slater	*Intermediate Algebra,* Third Edition

This series is designed to prepare students for any mathematics course that is required for students to reach their future goals.

WHY DID WE CHANGE THE FOURTH EDITION?

We have listened to teachers across the country, those who are enthusiastic Tobey/Slater users as well as teachers who have not used the text. We have responded to their ideas and suggestions and have incorporated their invaluable insight into this edition. We have retained the same Table of Contents as the Third Edition but have extensively revised the exercises and the application problems.

EMPHASIS: DEVELOPING PROBLEM-SOLVING ABILITIES

Throughout the country there is a renewed interest in improving the critical thinking, reasoning, and problem-solving skills of students. The interest is evident in government, the business community, and the national associations: AMATYC, NCTM, AMS, NADE, and MAA. Faculty have been encouraged to place greater emphasis on the areas of critical thinking, reasoning, and problem solving. In light of this focus, we have specifically designed *Beginning Algebra* to facilitate this objective. It has been carefully structured to improve the student's ability to reason and solve applied problems. It is written to incorporate modeling and connecting to other disciplines. The student is encouraged to read, write, and communicate mathematical ideas. Where scientific calculators, graphing calculators, and the Internet will help student problem solving, provision is made to use these technologies. Students are exposed to new ideas to encourage exploration and mathematical study.

INCREASED NUMBER AND QUALITY OF APPLIED PROBLEMS

Other faculty and students have joined the authors in writing many new applied problems, increasing the quantity, quality, and diversity of the word problems in the book. Now almost every exercise set in the text contains applied problems. The applications come from other academic disciplines, everyday life, and increased emphasis on global issues beyond the borders of the United States. Students will only engage in rich experiences that encourage thinking, analysis, and intellectual investigation if they face realistic and interesting application problems that attract them in the pursuit of knowledge. Over 50% of the applied problems of the new Fourth Edition have been changed in order to help students in this area.

PROVIDING A MATHEMATICAL BLUEPRINT FOR PROBLEM SOLVING

Often students in an elementary algebra course flounder because they "don't know where to begin" in the problem-solving task. The new Fourth Edition has been rewritten with a new emphasis on analyzing an applied problem in arithmetic or in algebra. This unique feature helps students to begin the problem-solving process and to plan the steps to be taken along the way; it provides them with an outline to organize their approach to solving problems. Often the hardest part in problem solving is determining where to begin. Once students fill in the blueprint, they can refer back to their plan as they do what is needed to solve the problem. Because of its flexibility, this feature can be used with single-step problems, multistep problems, applications, and nonroutine problems that require problem-solving strategies. Students will not need to use the blueprint to solve every problem. It is available for students who are faced with a problem with which they are not familiar, to alleviate their anxiety, to show them where to begin, and to assist them in the steps of reasoning.

INTERPRETING GRAPHS, CHARTS, AND TABLES IN THE PROBLEM-SOLVING PROCESS

When students encounter mathematics in real-world settings such as in the daily paper, *USA Today, Time, Newsweek, Sports Illustrated,* or similar published materials, they often encounter data represented in a graph, a chart, or a table and are asked to make a reasonable conclusion based on the data presented. This emphasis on graphical interpretation is a trend that is continuing with the expanding technology of our day. The number of mathematical problems based on charts, graphs, and tables has been significantly increased in the Fourth Edition. Students are asked to make simple interpretations, to solve medium-level problems, and to investigate challenging applied problems based on the data shown in a chart, graph, or table.

EXPANDED USE OF PUTTING YOUR SKILLS TO WORK APPLICATIONS

This widely praised feature in the Third Edition has been expanded with new problems and more in-depth student investigation for the new Fourth Edition. These nonroutine application problems challenge the students to synthesize the knowledge they have gained so far and apply it to a totally new area. These unusual problems encourage critical thinking and a variety of approaches of analysis. Each problem is specifically arranged for **Cooperative Learning Activities and Group Investigation** of mathematical problems that pique student interest. Students are given an opportunity to work together to help one another discover mathematical solutions to extended problems. These investigations feature open-ended questions and extrapolation of data to areas beyond what is normally covered in such a course.

INTERNET CONNECTIONS

As an integral part of each Putting Your Skills to Work Problem students are exposed to an interesting application on the Internet and encouraged to continue their investigation. This use of modern technology inspires students to have confidence in their abilities to successfully use mathematics. It encourages them not to end their use of mathematics at the end of the course. Rather they are inspired to use mathematics and modern technology in the investigation of other disciplines and in their own pursuit of knowledge.

TEXT FEATURES THAT DEVELOP THE MASTERY OF MATHEMATICAL CONCEPTS

Examples and Exercises

The examples and exercises in this text have been carefully chosen to guide students through Beginning Algebra. We have incorporated several different types of exercises and examples to assist your students in retaining the content of this course.

Chapter Pretests

Each chapter opens with a concise pretest to familiarize the student with the learning objectives for that particular chapter.

Practice Problems

These are found throughout the chapter, after the examples, and are designed to provide your students with immediate practice of the skills presented. The complete worked-out solution of each *Practice Problem* is contained in the back of the book in the answer section with yellow trim on the edge of the page.

To Think About

These critical thinking questions follow some of the examples in the text and also appear in the exercise sets. They extend the concept being taught, providing the opportunity for all students to stretch their minds, to look for patterns, and to make conclusions based on their previous experience.

Exercise Sets

Exercise Sets are paired and graded. This design helps to ease the students into the problems, and the answers provide students with immediate feedback.

Cumulative Review Problems

Each exercise set concludes with a section of *Cumulative Review Problems*. These problems review topics previously covered, and are designed to assist students in retaining the material. Many additional applied problems have been added to the *Cumulative Review Sections*.

Graphing and Scientific Calculator Problems

Calculator boxes are placed in the margin of the text to alert students to a scientific calculator application. In the exercise section a scientific calculator icon is used to indicate problems that are designed for solving with a calculator. Those problems where a graphing calculator is helpful are marked. A few optional graphing calculator problems have been added to the exercises.

REVIEWING MATHEMATICAL CONCEPTS

At the end of each chapter we have included problems and tests to provide your students with several different formats to help them review and reinforce the ideas that they have learned. This not only assists them with this chapter, it reviews previously covered topics as well.

Chapter Organizers

The concepts and mathematical procedures covered in each chapter are reviewed at the end of the chapter in a unique *Chapter Organizer*. This device has been extremely popular with faculty and students alike. It not only lists concepts and methods, but provides a completely worked-out example for each type of problem. Students find that preparing a similar chapter organizer on their own in higher-level math courses becomes an invaluable way to master the content of a chapter of material.

Verbal and Writing Skills

Writing exercises provide students with the opportunity to extend a mathematical concept by allowing them to use their own words, to clarify their thinking, and to become familiar with mathematical terms. These exercises have been included at the beginning of exercise sets to set the stage for the practice that follows, or at the end of the practice as a summary.

Review Problems

These problems are grouped by section as a quick refresher at the end of the chapter. These problems can also be used by the student as a quiz of the chapter material.

Tests

Found at the end of the chapter, the *Chapter Test* is a representative review of the material from that particular chapter that simulates an actual testing format. This provides the students with a gauge to their preparedness for the actual examination.

Cumulative Tests

At the end of each chapter is a *Cumulative Test*. One-half of the content of each *Cumulative Test* is based on the math skills learned in previous chapters. By completing these tests for each chapter, the students build confidence that they have mastered not only the contents of the present chapter but the contents of the previous chapters as well.

SUPPLEMENTS

FOR THE INSTRUCTOR

Annotated Instructor's Edition

- Complete student text
- Teaching tips in the margin
- Answers appear next to every exercise.
- Answers to all *Pretests, Review Problems, Tests,* and *Cumulative Tests.*
- Instructor's Materials are in the front of the book. Included is the AMATYC document entitled *Crossroads in Mathematics: Standards for Introductory College Mathematics.*

Instructor's Solutions Manual

Worked solutions to all even-numbered exercises from the text and complete solutions for chapter review problems and chapter tests.

Instructor's Test Item File

Features 9 tests per chapter plus 4 forms of a final examination, prepared and ready to be photocopied. Of these tests, 3 are multiple choice and 6 are free response. Two of the free-response tests are cumulative in nature. The manual also contains 4 forms of a final examination.

TestPro (IBM and Macintosh)

This versatile testing system allows the instructor to easily create up to 99 versions of a customized test. Users may add their own test items and edit existing items in WYSIWYG format. Each objective in the text has at least one multiple choice and free-response algorithm. Free upon adoption.

FOR THE STUDENT

Student Solutions Manual

Worked solutions to all odd-numbered exercises from the text and complete solutions for chapter review problems and chapter tests.

INTERACTIVE COMPUTER LEARNING AND TUTORING

MathPro (IBM and Macintosh)

This tutorial software has been developed exclusively for Prentice-Hall and is text-specific to this Tobey/Slater text. *MathPro* is designed to generate practice exercises based on the exercise sets in the text and will provide the student with interactive help on each exercise. If a student is having difficulty working any exercise, he or she can ask to see a step-by-step example or to get step-by-step help in solving the exercise. At the end of each exercise, results are provided, and when necessary, remediation is provided by referring the student to a specific section of the text for additional skills practice or review of key concepts. Students can also practice taking a test by accessing the Quiz mode. A record-keeping function allows the instructor to track individual or section progress.

INTERACTIVE VIDEO LEARNING AND TUTORING

The Complete Video Series

Every section of this text is explained in detail on a videotape featuring Professors Michael C. Mayne and Professor (Biff) John Pietro of Riverside Community College. These experienced mathematics professors have been successfully teaching developmental mathematics courses for many years. The videotapes are filled with humor and variety, and are designed to put the student at ease. The content of each tape follows the exact format of the textbook. Professors Mayne and Pietro work out the solutions to several even numbered problems in each exercise set.

GUIDE FOR STUDENTS

How to Use Beginning Algebra, Fourth Edition to enrich your class experience and prepare for tests.

CHAPTER 3

Solving Applied Problems

A common activity is the purchase of a new car by taking out an automobile loan. However, often the finance charges for the loan are greater than what the purchaser expected them to be. If you took out a new car loan, would you know how much the loan is really costing you? Turn to page 213 and study the Putting Your Skills to Work in order to find out.

page 179

Each chapter opens with an application. These applications relate to the material you find in the chapter as well as an extended discovery called *Putting Your Skills to Work* which you will encounter later in the chapter.

Putting Your Skills to Work

AUTOMOBILE LOANS AND INSTALLMENT LOANS

The following table provides the amount of the monthly payment of a loan of $1000 at various interest rates for a period of 2 to 5 years. To find the monthly payment for a ...ount of money merely multiply this pay- ...er of thousands of the loan. To borrow ...rest for 3 years you multiply $31.34 by

page 213

Put your new skills to work. These multi-part projects are relevant to chapter concepts as well to problems you encounter from day to day.

Explore on your own...

PROBLEMS FOR INDIVIDUAL STUDY AND INVESTIGATION

Use the table above to solve the following problems.

1. If Ricardo Sanchez wants to buy a new car and borrow \$10,000 at a 9% interest rate for 4 years, what will his monthly payments be? How much interest will he pay over the 4-year period? (How much more than \$10,000 will he pay over 4 years?)

2. If Alicia Wong wants to purchase a new stereo system for her apartment and borrow \$3000 at 12% interest for 2 years, what will her monthly payments be? How much interest will she pay over the 2-year period? (How much more than \$3000 will she pay over the 2 years?)

page 213

PROBLEMS FOR COOPERATIVE GROUP INVESTIGATION

With some other members of your class, determine an answer to the following.

3. If Walter Swenson plans to borrow \$15,000 to purchase a new car at an interest rate of 7%, how much more in interest will he pay if he borrows the money for 5 years as opposed to a period of 4 years?

4. Ray and Shirley Peterson are planning to build a garage and a connecting family room for their house. It will cost \$35,000. They have \$17,000 in savings, and they plan to borrow the rest at 16% interest. They can afford a maximum monthly payment of \$640. What time period for the loan will they have to select to meet that requirement? What will their monthly payment be?

Work cooperatively with fellow students to solve more involved problems. You will find that problems are often solved collaboratively in the workplace.

page 213

INTERNET CONNECTIONS: Go to ''http://www.prenhall.com/tobey'' to be connected

Site: Mortgage Amortization Calculator or a related site

Alternate Site: Amortization Calculator

Kimberly Jones is buying a \$135,000 house. She will make a down payment of 10%. The rest will be financed with a 30-year loan at an interest rate of 8.5%.

5. What will her monthly payment be? How much interest will she pay over the life of the loan?

6. Suppose she decides to pay an extra \$200 every month. How much interest will she save by doing this? In how many years will the loan be paid off?

Surf the 'Net for real data that relates to the *Putting Your Skills to Work* explorations. Extend the Concepts!

page 213

These features have been included in this text to help you make connections to mathematics. Use them to explore, connect, and discover.

PREPARE YOURSELF

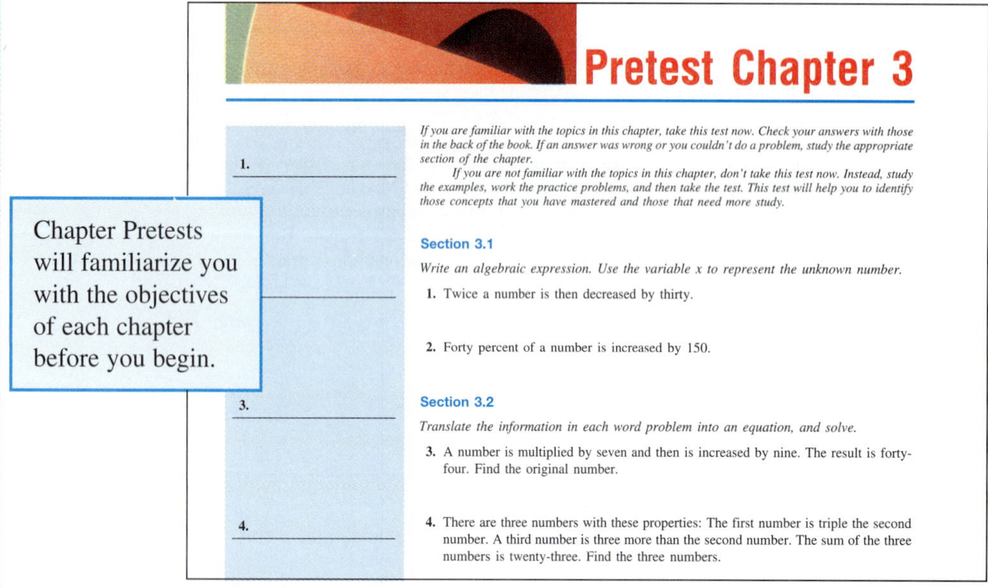

Pretest Chapter 3

If you are familiar with the topics in this chapter, take this test now. Check your answers with those in the back of the book. If an answer was wrong or you couldn't do a problem, study the appropriate section of the chapter.

If you are not familiar with the topics in this chapter, don't take this test now. Instead, study the examples, work the practice problems, and then take the test. This test will help you to identify those concepts that you have mastered and those that need more study.

Section 3.1

Write an algebraic expression. Use the variable x to represent the unknown number.

1. Twice a number is then decreased by thirty.

2. Forty percent of a number is increased by 150.

Section 3.2

Translate the information in each word problem into an equation, and solve.

3. A number is multiplied by seven and then is increased by nine. The result is forty-four. Find the original number.

4. There are three numbers with these properties: The first number is triple the second number. A third number is three more than the second number. The sum of the three numbers is twenty-three. Find the three numbers.

1. ______

3. ______

4. ______

Chapter Pretests will familiarize you with the objectives of each chapter before you begin.

page 180

MATHEMATICS BLUEPRINT FOR PROBLEM SOLVING

Gather the facts.	Label the variables.	Basic equation.	Key points to remember.
\$1250 is invested: part at 8% interest, part at 12% interest. The total interest for the year is \$128.00.	Let x = the amount invested at 8%. $1250 - x$ = the amount invested at 12%. $0.08x$ = the amount of interest for x dollars at 8%. $0.12(1250 - x)$ = the amount of interest for $1250 - x$ dollars at 12%.	Interest earned at 8% + interest earned at 12% = total interest earned during the year which is \$128.	Be careful to write $1250 - x$ for the amount of money invested at 12%. The order is total $- x$ (do not use $x - 1250$).

page 206

1. *Understand the problem.*
   - **(a)** Read the word problem carefully to get an overview.
   - **(b)** Determine what information you will need to solve the problem.
   - **(c)** Draw a sketch. Label it with the known information. Determine what needs to be found.
   - **(d)** Choose a variable to represent one unknown quantity.
   - **(e)** If necessary, represent other unknown quantities in terms of that very same variable.
2. *Write an equation.*
   - **(a)** Look for key words to help you to translate the words into algebraic symbols and expressions.
   - **(b)** Use a given relationship in the problem or an appropriate formula in order to write an equation.
3. *Solve the equation and state the answer.*
4. *Check.*
   - **(a)** Check the solution in the original equation.
   - **(b)** Be sure the solution to the equation answers the question in the word problem. You may need to do some additional calculations if it does not.

page 188

EXAMPLE 2 Five more than six times a number is three hundred and five. Find the number.

1. *Understand the problem.*
   Read the problem carefully. You may not need to draw a sketch.
   Let x = the unknown number.
2. *Write an equation.*
   Five more than six times a number is three hundred and five.
   $$5 + 6x = 305$$
3. *Solve the equation and state the answer.*
   You may want to rewrite the equation to make it easier to solve.
   $$6x + 5 = 305$$
   $$6x + 5 - 5 = 305 - 5 \quad \textit{Subtract 5 from both sides.}$$
   $$6x = 300$$
   $$\frac{6x}{6} = \frac{300}{6} \quad \textit{Divide both sides by 6.}$$
   $$x = 50$$
   The number is 50.
4. *Check.* Is five more than six times 50 three hundred and five?
   $$6(50) + 5 \stackrel{?}{=} 305$$
   $$305 = 305$$

page 189

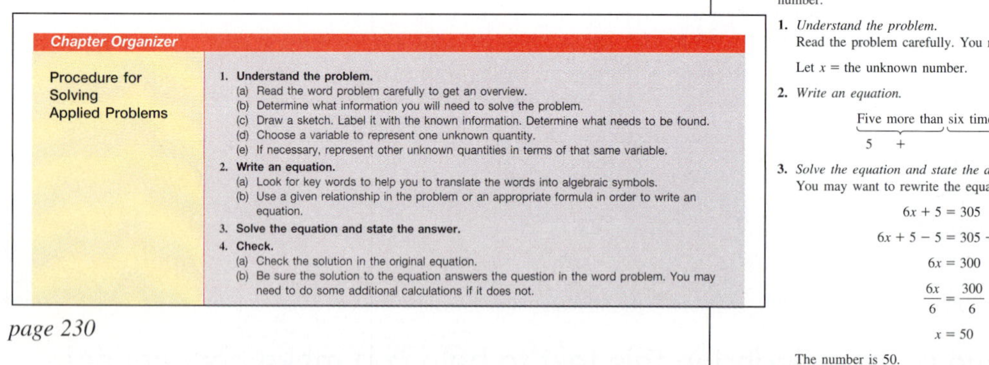

Chapter Organizer

Procedure for Solving Applied Problems	
	1. **Understand the problem.** (a) Read the word problem carefully to get an overview. (b) Determine what information you will need to solve the problem. (c) Draw a sketch. Label it with the known information. Determine what needs to be found. (d) Choose a variable to represent one unknown quantity. (e) If necessary, represent other unknown quantities in terms of that same variable. 2. **Write an equation.** (a) Look for key words to help you to translate the words into algebraic symbols. (b) Use a given relationship in the problem or an appropriate formula in order to write an equation. 3. **Solve the equation and state the answer.** 4. **Check.** (a) Check the solution in the original equation. (b) Be sure the solution to the equation answers the question in the word problem. You may need to do some additional calculations if it does not.

page 230

Have a plan—Develop a problem-solving strategy.

CHECK YOUR UNDERSTANDING

Beginning Algebra, Fourth Edition includes many different types of exercises—Become a Problem Solver!

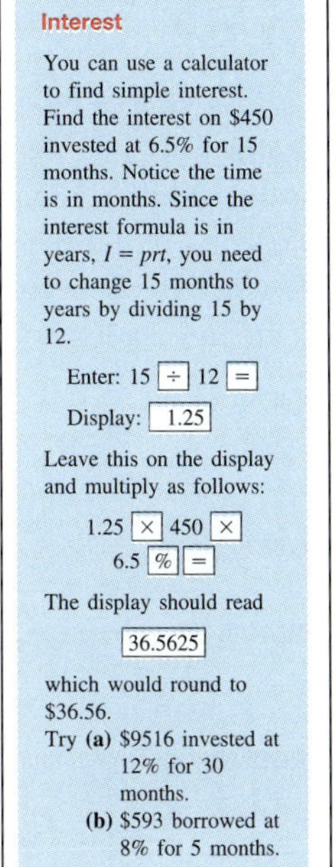

CALCULATOR

Interest

You can use a calculator to find simple interest. Find the interest on \$450 invested at 6.5% for 15 months. Notice the time is in months. Since the interest formula is in years, $I = prt$, you need to change 15 months to years by dividing 15 by 12.

Enter: 15 [÷] 12 [=]

Display: [1.25]

Leave this on the display and multiply as follows:

1.25 [×] 450 [×]
6.5 [%] [=]

The display should read

[36.5625]

which would round to \$36.56.

Try **(a)** \$9516 invested at 12% for 30 months.
(b) \$593 borrowed at 8% for 5 months.

page 205

3.5 Exercises

Verbal and Writing Skills

Write the word(s) to complete each sentence.

1. Perimeter is the __________ a plane figure.

2. __________ is the distance around a circle.

3. Area is a measure of the amount of __________ in a region.

4. A __________ is a four-sided figure with exactly two sides parallel.

5. The sum of the interior angles of any triangle is __________.

page 221

To Think About

42. An isosceles triangle has one side that measures 4 feet and another that measures 7.5 feet. The third side was not measured. Can you find one unique perimeter for this type of triangle? What are the two possible perimeters?

43. Find the total surface area of (a) a sphere of radius 3 inches and (b) a right circular cylinder with radius 2 inches and height 0.5 inch. Which object has a greater surface area?

44. Assume that a very long rope is stretched around the moon supported by poles that are 3 *feet tall.* (Neglect gravitational pull and assume that the rope takes the shape of a large circle, as shown in the figure.) The radius of the moon is approximately 1080 miles. How much longer would you need to make the rope if you wanted the rope to be supported by poles that are 4 *feet tall*?

page 223

Chapter 3 Test

Solve each applied problem.

1. A number is doubled and then decreased by 11. The result is 59. What is the original number?

2. The sum of one-half of a number, one-ninth of a number, and one-twelfth of a number is twenty-five. Find the original number.

3. Triple a number is increased by six. The result is the same as when the original number was diminished by three and then doubled. Find the original number.

4. A triangular region has a perimeter of 66 meters. The first side is two-thirds of the second side. The third side is 14 meters shorter than the second side. What are ... of the tri

1. ____________

2. ____________

page 235

Cumulative Review

Simplify.

45. $2(x - 6) - 3[4 - x(x + 2)]$

46. $-5(x + 3) - 2[4 + 3(x - 1)]$

47. $-\{3 - 2[x - 3(x + 1)]\}$

48. $-2\{x + 3[2 - 4(x - 3)]\}$

page 223

ENHANCE YOUR LEARNING

Beginning Algebra, Fourth Edition is more than a textbook; it is an integrated package of instruction. Ask your professor about these supplements, which are a part of the **Beginning Algebra** suite of learning materials.

Most items are keyed specifically to this text.

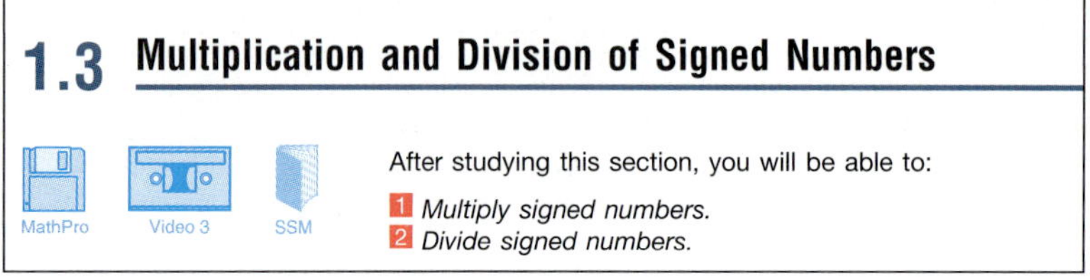

page 86

- Each section of the text begins with a reminder of the additional companion tools that have been designed to enhance your learning experience.

- Contains complete step-by-step solutions for every odd-numbered exercise
- Contains complete step-by-step solutions for all Chapter Review Problems, Chapter Tests, and Cumulative Tests

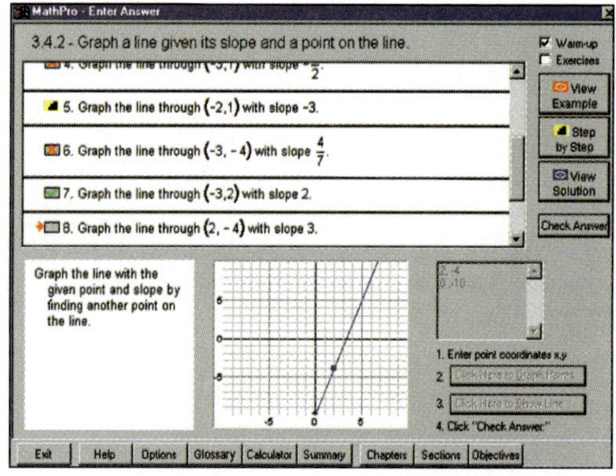

- MathPro Explorer: Interactive and Tutorial Software
- For Windows and Power Macintosh
- Includes preformatted algebra explorations
- Generates unlimited practice exercises

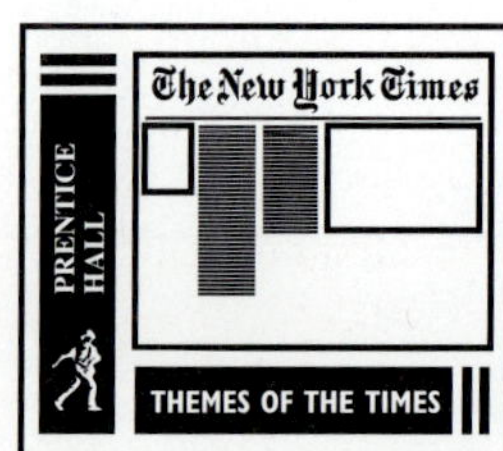

New York Times/Themes of the Times
Newspaper-format supplement–
ask your professor about this free supplement

- Tips include how to prepare for class, how to study for and take tests, and how to improve your grades

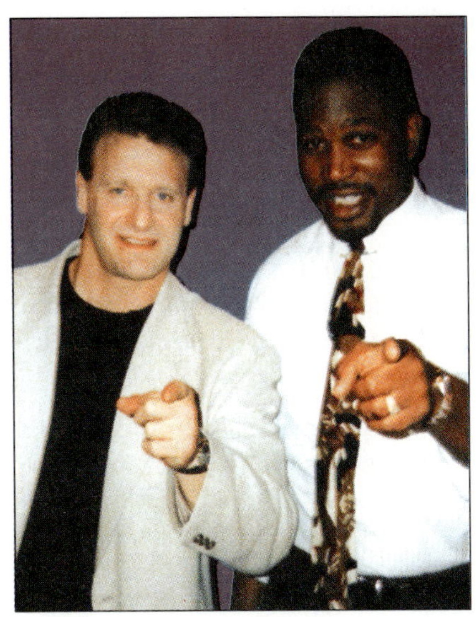

- Team taught video instruction covers each section of the text

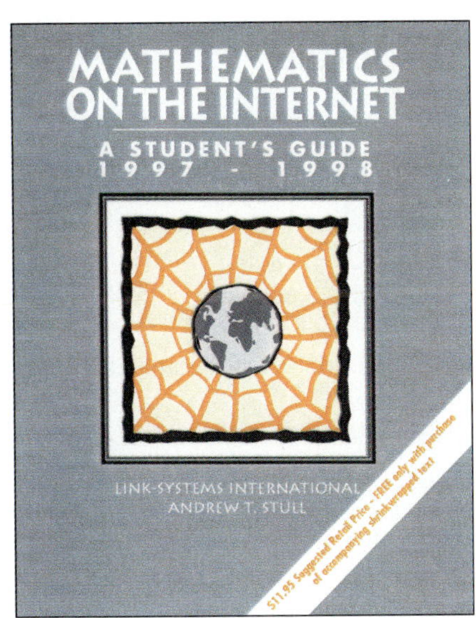

- A guide to navigation strategies through the Internet as well as practice exercises and lists of resources

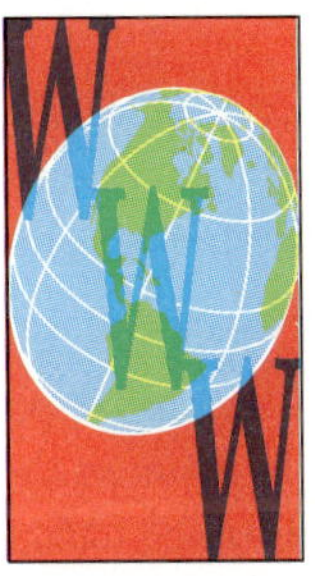

- Visit the companion website for related and extended applications

www.prenhall.com/tobey

- Graphing Utilities Video provides introductory keystroke instruction

CROSSROADS IN MATHEMATICS

Standards for Introductory College Mathematics Before Calculus

Executive Summary

A Document Prepared by the American Mathematical Association of Two-Year Colleges (AMATYC)*

Participating in the preparation of this document were representatives from the American Mathematical Society, the Mathematical Association of America, the National Association of Developmental Education, and the National Council of Teachers of Mathematics. The preparation and dissemination of Crossroads in Mathematics *have been made possible through funding from the National Science Foundation, the Exxon Education Foundation, and Texas Instruments.*

Higher education is situated at the intersection of two major crossroads: A growing societal need exists for a well-educated citizenry and for a workforce adequately prepared in the areas of mathematics, science, engineering, and technology while, at the same time, increasing numbers of academically underprepared students are seeking entrance to postsecondary education.

Mathematics programs at two- and four-year colleges as well as at many universities serve students from diverse personal and academic backgrounds. These students begin their postsecondary education with a wide variety of educational goals and personal aspirations. Some enter college with solid mathematical backgrounds, ready to study calculus. Many others intend to study calculus but lack the preparation to do so. A third group can prepare for their personal and career goals through the study of mathematics below the level of calculus. The students in the last two groups are the ones whose needs are addressed in *Crossroads in Mathematics*. These students make up the majority of the nation's college students who are studying mathematics; they represent over 75 percent of the mathematics students at two-year colleges and 49 percent at four-year colleges and universities (*Statistical Abstract of the Undergraduate Programs in the Mathematical Sciences and Computer Science in the United States, 1990–91 CBMS Survey*).

Mathematics is a vibrant and growing discipline and is being used in more ways by more people than ever before. An alarming situation now exists, however, in the nation's postsecondary institutions. Each year greater numbers of students enter the mathematics "pipeline" at a point below the level of calculus, yet there is no significant gain in the numbers of students studying at higher levels. The failure of many of these students to persist in mathematics not only prevents them from pursuing their chosen careers, it also has a negative impact on our nation's economy, as fewer members of the work force are prepared for jobs in technical fields.

The purpose of *Crossroads in Mathematics* is to address the special circumstances of, establish standards for, and make recommendations about introductory college mathematics. The recommendations are based upon research evidence and the best judgment of the educators who contributed to the document.

Basic Principles

The following principles form the philosophical underpinnings of the document:

1. *All college students should grow in their knowledge of mathematics while attending college.* Students who are not prepared for college-level mathematics upon entering college will obtain the necessary knowledge by studying the Foundation.
2. *The mathematics that students study should be meaningful and relevant.* Mathematics should be introduced in the context of real, understandable problem-solving situations.
3. *Mathematics must be taught as a laboratory discipline.* Effective mathematics instruction should involve active student participation using in-depth projects.
4. *The use of technology is an essential part of an up-to-date curriculum.* Faculty and students will make effective use of appropriate technology.
5. *Students will acquire mathematics through a carefully balanced educational program that emphasizes the content and instructional strategies recommended in the standards along with the viable components of traditional instruction.* These standards emphasize problem solving, technology, intuitive understanding, and collaborative learning strategies. Skill acquisition, mathematical abstraction and rigor, and whole-class instruction, however, are still critical components of mathematics education.
6. *Increased participation in mathematics and in careers using mathematics is a critical goal in our heterogeneous society.* Mathematics education must reach out to all students.

*Inclusion of the Executive Summary of *Crossroads* does not constitute an endorsement of the text by AMATYC, or constitute AMATYC's agreement that the text adheres to the contents of *Crossroads*.

The Standards

The standards provide goals for introductory college mathematics and guidelines for selecting content and instructional strategies for accomplishing the goals.

The *Standards for Intellectual Development* address desired modes of student thinking and represent goals for student outcomes.

Standard I-1: Problem Solving
Students will engage in substantial mathematical problem solving.

Standard I-2: Modeling
Students will learn mathematics through modeling real-world situations.

Standard I-3: Reasoning
Students will expand their mathematical reasoning skills as they develop convincing mathematical arguments.

Standard I-4: Connecting with Other Disciplines
Students will develop the view that mathematics is a growing discipline, interrelated with human culture, and understand its connections to other disciplines.

Standard I-5: Communicating
Students will acquire the ability to read, write, listen to, and speak mathematics.

Standard I-6: Using Technology
Students will use appropriate technology to enhance their mathematical thinking and understanding and to solve mathematical problems and judge the reasonableness of their results.

Standard I-7: Developing Mathematical Power
Students will engage in rich experiences that encourage independent, nontrivial exploration in mathematics, develop and reinforce tenacity and confidence in their abilities to use mathematics, and inspire them to pursue the study of mathematics and related disciplines.

The *Standards for Content* provide guidelines for the selection of content that will be taught at the introductory level.

Standard C-1: Number Sense
Students will perform arithmetic operations, as well as reason and draw conclusions from numerical information.

Standard C-2: Symbolism and Algebra
Students will translate problem situations into their symbolic representations and use those representations to solve problems.

Standard C-3: Geometry
Students will develop a spatial and measurement sense.

Standard C-4: Function
Students will demonstrate understanding of the concept of function by several means (verbally, numerically, graphically, and symbolically) and incorporate it as a central theme into their use of mathematics.

Standard C-5: Discrete Mathematics
Students will use discrete mathematical algorithms and develop combinatorial abilities in order to solve problems of finite character and enumerate sets without direct counting.

Standard C-6: Probability and Statistics
Students will analyze data and use probability and statistical models to make inferences about real-world situations.

Standard C-7: Deductive Proof
Students will appreciate the deductive nature of mathematics as an identifying characteristic of the discipline, recognize the roles of definitions, axioms, and theorems, and identify and construct valid deductive arguments.

The *Standards for Pedagogy* recommend the use of instructional strategies that provide for student activity and interaction and for student-constructed knowledge.

Standard P-1: Teaching with Technology
Mathematics faculty will model the use of appropriate technology in the teaching of mathematics so that students can benefit from the opportunities it presents as a medium of instruction.

Standard P-2: Interactive and Collaborative Learning
Mathematics faculty will foster interactive learning through student writing, reading, speaking, and collaborative activities so that students can learn to work effectively in groups and communicate about mathematics both orally and in writing.

Standard P-3: Connecting with Other Experiences
Mathematics faculty will actively involve students in meaningful mathematics problems that build upon their experiences, focus on broad mathematical themes, and build connections within branches of mathematics and between mathematics and other disciplines so that students will view mathematics as a connected whole relevant to their lives.

Standard P-4: Multiple Approaches
Mathematics faculty will model the use of multiple approaches—numerical, graphical, symbolic, and verbal—to help students learn a variety of techniques for solving problems.

Standard P-5: Experiencing Mathematics
Mathematics faculty will provide learning activities, including projects and apprenticeships, that promote independent thinking and require sustained effort and time so that students will have the confidence to access and use needed mathematics and other technical information independently, to form conjectures from an array of specific examples, and to draw conclusions from general principles.

Interpreting the Standards

The standards reflect many of the same principles found in school reform and calculus reform. They focus, however, on the needs and experiences of college students studying introductory mathematics in various instructional programs. These standards place emphasis on using technology as a tool and as an aid to instruction, developing general strategies for solving real-world problems, and actively involving students in the learning process.

In particular,

- The *Foundation* includes topics traditionally taught in developmental mathematics but also brings in additional topics that all students must understand and be able to use. Courses at this level should not simply be repeats of those offered in high school. Their goal is to prepare college students to study additional mathematics, thus expanding their educational and career options.
- *Technical programs* place strong emphasis on mathematics in the context of real applications. They should prepare students for the immediate needs of employment. At the same time, students should learn to appreciate the usefulness of mathematics and to use mathematics to solve problems in a variety of fields. The content and structure of the mathematics curriculum for technical students must be both rigorous and relevant. The mathematics that technical students study must broaden their options both in careers and in formal education.
- *Mathematics-intensive programs* include the study of calculus and beyond. Consequently, the mathematics that these students study must prepare them to be successful in a wide variety of calculus programs. The study of functions is the heart of precalculus education. While not departing from concerns about mathematical processes and techniques, more emphasis should be placed on developing student understanding of concepts, helping them make connections among concepts, and building their reasoning skills.
- *Liberal arts programs* are designed for bachelor degree–intending students majoring in the humanities and social sciences. These students should gain an appreciation for the roles that mathematics will play in their education, in their careers, and in their personal lives. Each institution has the responsibility of evaluating local needs and resources to determine how best to educate liberal arts majors beyond the Foundation. Options include interdisciplinary modules and introductory statistics.
- *Prospective teacher programs* should shift the emphasis from teaching isolated mathematical knowledge and skills to helping students to apply knowledge and develop in-depth understanding of the subject that goes beyond what they will be expected to teach. Courses for prospective teachers must provide an awareness of what research reveals about how children learn mathematics, models for effective pedagogy, and an understanding of the power and limitations of the use of technology in the classroom.

Implications

Although the standards focus on curriculum and pedagogy, they have wide implications for institutions and their mathematics programs.

- Professional development opportunities must be made available to all faculty members so that they may experience these reform recommendations as learners.
- Colleges must have laboratory and learning center facilities and provide adequate support personnel.
- Appropriate technology must be available for faculty and student use.
- Assessment instruments must measure the full range of what students are expected to know.
- Ongoing program evaluation must be used to make recommendations for improvement while retaining the effective aspects of the program.
- Articulation with high schools, with other colleges and universities, and with employers enables faculty at all levels to work in concert to improve mathematics education.

Implementation

Crossroads in Mathematics provides a framework for the development of improved curriculum and pedagogy. Adoption and implementation of the standards will require a systemic nationwide effort, which must be supported by postsecondary institutions, business and industry, and public and private funding agencies. Faculty, with the support of administrators, must lead the way. National professional organizations and their affiliated groups must promote reform through their conferences and by providing more extensive faculty development opportunities.

Looking to the Future

Introductory college mathematics holds the promise of opening new paths to future learning and fulfilling careers to an often neglected segment of the student population. Mathematics education at this level plays such a critical role in people's lives that its improvement is essential to our nation's vitality. *Crossroads in Mathematics* outlines a standards-based reform effort that will provide all students with a more engaging and valuable learning experience.

Crossroads in Mathematics was prepared by the Task Force and Writing Team of the Standards for Introductory College Mathematics Project with Don Cohen as Editor, Marilyn Mays as Project Director, and Karen Sharp and Dale Ewen as Project Co-Directors.

Individual copies of this document may be obtained by writing to AMATYC, State Technical Institute at Memphis, 5983 Macon Cove, Memphis, TN 38134. A limited number are available free. When the supply is exhausted, copies will be made available at a moderate charge.

Acknowledgments

This book is the product of many years of work and many contributions from faculty and students across the country. We would like to thank the many reviewers and participants in focus groups and special meetings with the authors.

Our deep appreciation to each of the following:

George J. Apostolopoulos, DeVry Institute of Technology
Katherine Barringer, Central Virginia Community College
Jamie Blair, Orange Coast College
Larry Blevins, Tyler Junior College
Robert Christie, Miami-Dade Community College
Mike Contino, California State University at Heyward
Judy Dechene, Fitchburg State University
Floyd L. Downs, Arizona State University
Barbara Edwards, Portland State University
Janice F. Gahan-Rech, University of Nebraska at Omaha
Colin Godfrey, University of Massachusetts, Boston
Carl Mancuso, William Paterson College
Janet McLaughlin, Montclair State College
Gloria Mills, Tarrant County Junior College
Norman Mittman, Northeastern Illinois University
Elizabeth A. Polen, County College of Morris
Ronald Ruemmler, Middlesex County College
Sally Search, Tallahassee Community College
Ara B. Sullenberger, Tarrant County Community College
Michael Trappuzanno, Arizona State University
Jerry Wisnieski, Des Moines Community College

We have been greatly helped by a supportive group of colleagues who not only teach at North Shore Community College but who have provided a number of ideas as well as extensive help on all of our mathematics books. Also, a special word of thanks to Hank Harmeling, Tom Rourke, Wally Hersey, Bob McDonald, Judy Carter, Bob Campbell, Rick Ponticelli, Russ Sullivan, Kathy LeBlanc, Laura Connelly, Sharyn Sharaf, Donna Stefano and Nancy Tufo. Joan Peabody has done an excellent job of typing various materials for the manuscript and her help is gratefully acknowledged.

As a new edition of a book gets finalized new and fresh ideas are always helpful. We want to thank Suellen Robinson for contributing several new exercise problems in every section of the book. We also want to thank Louise Elton for providing several new applied problems and suggested applications. We want to thank Kathy Natale and Cindy Trimbele for diligently error checking the content of this book.

Each textbook is a combination of ideas, writing, and revisions from the authors and wise editorial direction and assistance from the editors. We want to thank our Prentice Hall editor, Karin Wagner, for her administrative support and encouragement, and for her helpful insight and perspective on each phase of the revision of the textbook. Her patience, her willingness to listen, and her flexibility to adapt to changing publishing decisions has been invaluable to the production of this book. We also express our thanks to Jerome Grant and Melissa Acuña who have continued to support our writing projects and given wise direction and focus to our work.

Book writing is impossible for us without the loyal support of our families. Our deepest thanks and love to Nancy, Johnny, Melissa, Marcia, Shelley, Rusty, and Abby. Your understanding, your love and help, and your patience have been a source of great encouragement. Finally, we thank God for the strength and energy to write and the opportunity to help others through this textbook.

We have spent more than 25 years teaching mathematics. Each teaching day we find our greatest joy is helping students learn. We take a personal interest that each student has a good learning experience in taking this course. If you have some personal comments, suggestions, or ideas for future editions of this textbook, please write to us at:

Prof. John Tobey and Prof. Jeff Slater
Prentice Hall Publishing
Office of the College Mathematics Editor
One Lake Street
Upper Saddle River, NJ 07458

We wish you success in this course and in your future life!

John Tobey
Jeffrey Slater

Diagnostic Pretest: Beginning Algebra

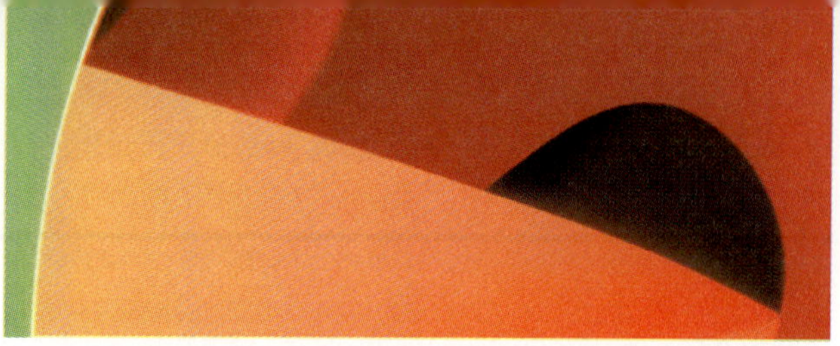

1.	$5\frac{17}{20}$
2.	$3\frac{1}{9}$
3.	10
4.	4.9715
5.	200,000
6.	4480
7.	5
8.	3
9.	$-7x - 14xy$
10.	23
11.	$59 - 30x$
12.	-45
13.	$x = 4$
14.	$x = 0$
15.	$x = 12$
16.	$x = 7$
17.	$\frac{2A + 4f}{3} = p$
18.	$x > 1$
19.	Width = 8 m; Length = 15 m
20.	30 m; 10 m; 23 m
21.	85
22.	\$1300 at 5%; \$4700 at 7%
23.	6 nickels; 9 dimes; 12 quarters
24.	2000 Friday, 1700 Thursday, 2510 Saturday

Follow the directions for each problem. Simplify each answer.

Chapter 0

1. Add. $3\frac{1}{4} + 2\frac{3}{5}$

2. Multiply. $\left(1\frac{1}{6}\right)\left(2\frac{2}{3}\right)$

3. Divide. $\frac{15}{4} \div \frac{3}{8}$

4. Multiply. $(1.63)(3.05)$

5. Divide. $120 \div 0.0006$

6. Find 7% of 64,000.

Chapter 1

7. Add. $(-3) + (-4) + (+12)$

8. Subtract. $-20 - (-23)$

9. Combine. $5x - 6xy - 12x - 8xy$

10. Evaluate. $2x^2 - 3x - 4$ when $x = -3$.

11. Remove the grouping symbols.
$2 - 3\{5 + 2[x - 4(3 - x)]\}$

12. Evaluate.
$-3(2 - 6)^2 + (-12) \div (-4)$

Chapter 2

Solve each equation for x.

13. $40 + 2x = 60 - 3x$

14. $7(3x - 1) = 5 + 4(x - 3)$

15. $\frac{2}{3}x - \frac{3}{4} = \frac{1}{6}x + \frac{21}{4}$

16. $\frac{4}{5}(3x + 4) = 20$

17. Solve for p.
$A = \frac{1}{2}(3p - 4f)$

18. Solve for x and graph the result.
$42 - 18x < 48x - 24$

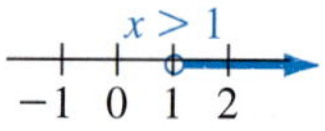

Chapter 3

19. The length of a rectangle is 7 meters longer than the width. The perimeter is 46 meters. Find the dimensions.

20. One side of a triangle is triple the second side. The third side is 3 meters longer than double the second side. Find each side of the triangle if the perimeter of the triangle is 63 meters.

21. Hector has four test scores of 80, 90, 83, and 92. What does he need to score on the fifth test to have an average of 86 on the five tests?

22. Marcia invested \$6000 in two accounts. One earned 5% interest, while the other earned 7% interest. After one year she earned \$394 in interest. How much did she invest in each account?

23. Melissa has 3 more dimes than nickels. She has twice as many quarters as nickels. The value of the coins is \$4.20. How many of each coin does she have?

24. The drama club put on a play for Thursday, Friday, and Saturday nights. The total attendance for the three nights was 6210. Thursday night had 300 fewer people than Friday night. Saturday night had 510 more people than Friday night. How many people came each night?

Chapter 4

25. Multiply. $(-2xy^2)(-4x^3y^4)$

26. Divide. $\dfrac{36x^5y^6}{-18x^3y^{10}}$

27. Raise to the indicated power. $(-2x^3y^4)^5$

28. Evaluate. $(-3)^{-4}$

29. Multiply. $(3x^2 + 2x - 5)(4x - 1)$

30. Divide. $(x^3 + 6x^2 - x - 30) \div (x - 2)$

Chapter 5

Factor completely.

31. $5x^2 - 5$

32. $x^2 - 12x + 32$

33. $8x^2 - 2x - 3$

34. $3ax - 8b - 6a + 4bx$

Solve for x.

35. $x^3 + 7x^2 + 12x = 0$

36. $16x^2 - 24x + 9 = 0$

Chapter 6

37. Simplify. $\dfrac{x^2 + 3x - 18}{2x - 6}$

38. Multiply. $\dfrac{6x^2 - 14x - 12}{6x + 4} \cdot \dfrac{x + 3}{2x^2 - 2x - 12}$

39. Divide and simplify. $\dfrac{x^2}{x^2 - 4} \div \dfrac{x^2 - 3x}{x^2 - 5x + 6}$

40. Add. $\dfrac{3}{x^2 - 7x + 12} + \dfrac{4}{x^2 - 9x + 20}$

41. Solve for x. $2 - \dfrac{5}{2x} = \dfrac{2x}{x + 1}$

42. Simplify. $\dfrac{3 + \dfrac{1}{x}}{\dfrac{9}{x} + \dfrac{3}{x^2}}$

Chapter 7

43. Graph $y = 2x - 4$.

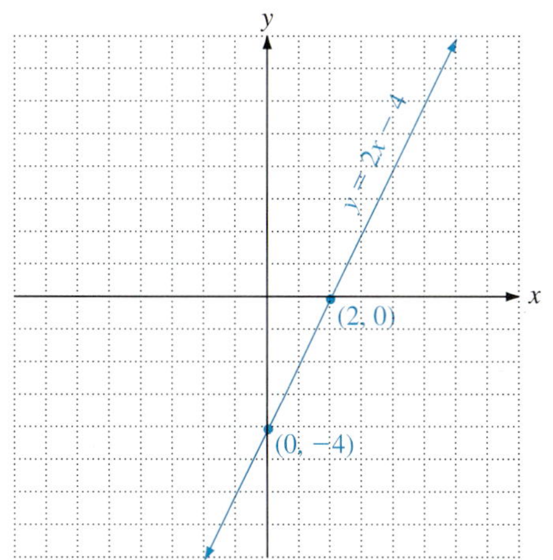

44. Graph $3x + 4y = -12$.

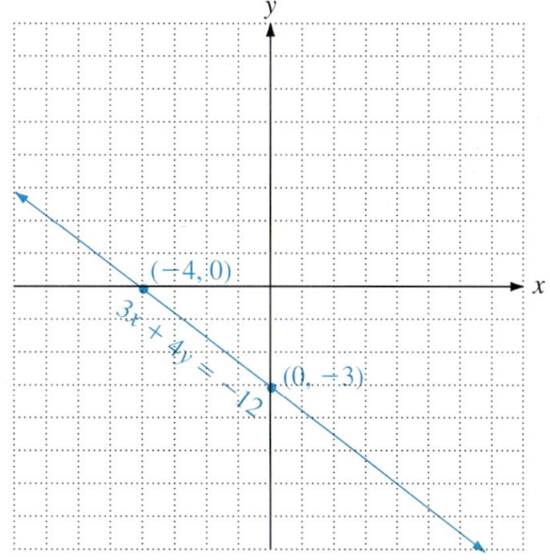

45. What is the slope of a line passing through $(6, -2)$ and $(-3, 4)$?

46. If $f(x) = 2x^2 - 3x + 1$, find $f(3)$.

25. $8x^4y^6$

26. $-\frac{2x^2}{y^4}$

27. $-32x^{15}y^{20}$

28. $\frac{1}{81}$

29. $12x^3 + 5x^2 - 22x + 5$

30. $x^2 + 8x + 15$

31. $5(x + 1)(x - 1)$

32. $(x - 4)(x - 8)$

33. $(2x + 1)(4x - 3)$

34. $(x - 2)(3a + 4b)$

35. $x = 0, x = -3, x = -4$

36. $x = \frac{3}{4}$

37. $\frac{x + 6}{2}$

38. $\frac{x + 3}{2(x + 2)}$

39. $\frac{x}{x + 2}$

40. $\frac{7x - 27}{(x - 3)(x - 4)(x - 5)}$

41. $x = -5$

42. $\frac{x}{3}$

43. See graph.

44. See graph.

45. $m = -\frac{2}{3}$

46. 10

47. See graph.

48. $3x - 5y = -18$

49. $x = 4; y = 13$

50. $x = 0; y = 2$

51. $x = -1$
$y = -2$

52. $x = \frac{4}{3}$
$y = 6$

53. Yes

54. Pair of gloves is $7.00
A scarf is $8.00

55. 11

56. $5xy^2\sqrt{5xy}$

57. $-14 - 2\sqrt{3}$

58. $\frac{\sqrt{30} - 3\sqrt{2}}{6}$

59. $2\sqrt{13}$

60. $y = 77$

47. Graph the region

$y \geq -\frac{1}{3}x + 2$

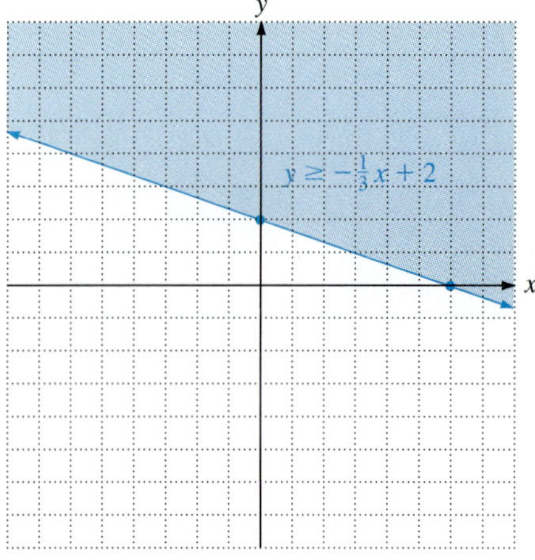

48. Find the equation of a line with a slope of $\frac{3}{5}$ that passes through the point $(-1, 3)$.

Chapter 8

Solve each system by the appropriate method.

49. Substitution method.
$x + y = 17$
$2x - y = -5$

50. Addition method.
$-5x + 4y = 8$
$2x + 3y = 6$

51. Any method.
$2(x - 2) = 3y$
$6x = -3(4 + y)$

52. Any method.
$x + \frac{1}{3}y = \frac{10}{3}$
$\frac{3}{2}x + y = 8$

53. Is $(2, -3)$ a solution for the system
$3x + 5y = -9$
$2x - 3y = 13$?

54. A man bought 3 pairs of gloves and 4 scarves for $53. A woman bought 2 pairs of the same priced gloves and 3 scarves of the same priced scarves for $38. How much did each item cost?

Chapter 9

55. Evaluate. $\sqrt{121}$

56. Simplify. $\sqrt{125x^3y^5}$

57. Multiply and simplify.
$(\sqrt{2} + \sqrt{6})(2\sqrt{2} - 3\sqrt{6})$

58. Rationalize the denominator.
$\frac{\sqrt{5} - \sqrt{3}}{\sqrt{6}}$

59. In the right triangle with sides a, b, and c, find side c if side $a = 4$ and side $b = 6$.

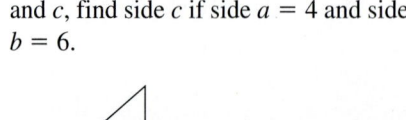

60. y varies directly with x. When $y = 56$, then $x = 8$. Find y when $x = 11$.

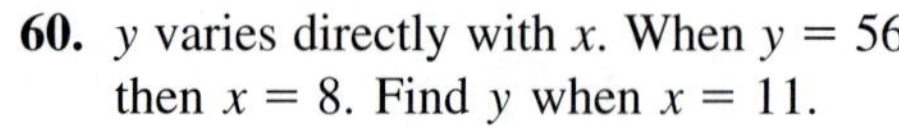

Chapter 10

Solve for x.

61. $14x^2 + 21x = 0$

62. $2x^2 + 1 = 19$

63. $2x^2 - 4x - 5 = 0$

64. $x^2 - x + 8 = 5 + 6x$

65. Graph the equation.
$y = x^2 + 8x + 15$

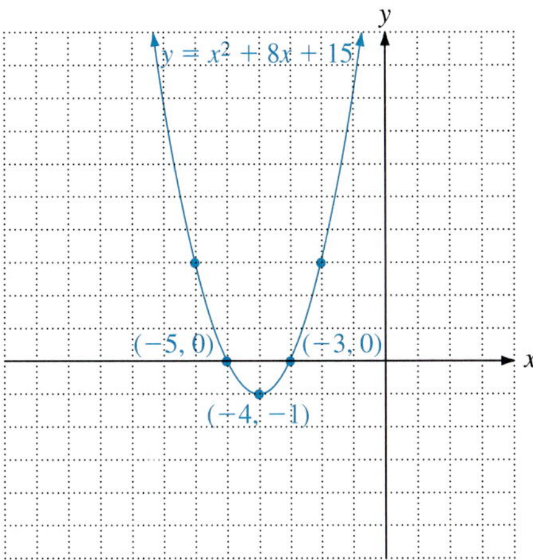

66. A rectangular box is 4 inches longer in length than in width. The area of the box is 96 square inches. Find the length and the width of the box.

61. $x = 0; x = -\frac{3}{2}$

62. $x = \pm 3$

63. $x = \frac{2 \pm \sqrt{14}}{2}$

64. $x = \frac{7 \pm \sqrt{37}}{2}$

65. See graph.

66. Width 8 inches; Length 12 inches

CHAPTER 0
A Brief Review of Arithmetic Skills

Every four years, the athletic competition at the summer Olympics captures the attention of the world. The records of the high jump competition are carefully measured with feet, inches, and fractions of an inch. Could you perform mathematical calculations with these records and make a prediction about future Olympic records? Turn to the Putting your Skills to work on page 21.

Pretest Chapter 0

1. $\frac{7}{9}$	**18.** 19.651
2. $\frac{2}{3}$	**19.** 2.0664
3. $3\frac{3}{4}$	**20.** 0.45
4. $\frac{33}{7}$	**21.** 2.246
5. 6	**22.** 70%
6. 35	**23.** 0.6%
7. 120	**24.** 3.27
8. $\frac{13}{9}$	**25.** 0.02
9. $\frac{5}{12}$	**26.** 407.5
10. $8\frac{5}{12}$	**27.** 7.2%
11. $1\frac{33}{40}$	**28.** 10,000
12. $\frac{10}{9}$ **or** $1\frac{1}{9}$	**29.** 21,000,000
13. $\frac{21}{2}$ **or** $10\frac{1}{2}$	**30.** 20 miles per gallon
14. $\frac{7}{2}$ **or** $3\frac{1}{2}$	**31.** 28.4 miles per gallon
15. $\frac{28}{39}$	
16. 0.625	
17. $0.\overline{5}$	

Take this pretest for Chapter 0 and compare your answers to the solutions for the pretest listed in the Answer Key at the end of the book. If you obtain the correct answer for 25 or more problems, you are ready to begin Chapter 1. If you got more than one wrong in any one section, you should review the necessary sections of Chapter 0 to improve your arithmetic skills. If you take the time to master these arithmetic skills now, you will definitely find Chapter 1 and the remaining chapters of this book easier to complete.

Section 0.1

Simplify each fraction. **1.** $\frac{21}{27}$ **2.** $\frac{26}{39}$

3. Write $\frac{15}{4}$ as a mixed number. **4.** Change $4\frac{5}{7}$ to an improper fraction.

Find the missing number. **5.** $\frac{3}{7} = \frac{?}{14}$ **6.** $\frac{7}{4} = \frac{?}{20}$

Section 0.2

7. Find the LCD, but do not add. $\frac{3}{8}, \frac{5}{6},$ and $\frac{7}{15}$

Perform the calculation indicated. Write the answer in simplest form.

8. $\frac{8}{9} + \frac{5}{9}$ **9.** $\frac{5}{48} + \frac{5}{16}$ **10.** $2\frac{3}{4} + 5\frac{2}{3}$ **11.** $3\frac{1}{5} - 1\frac{3}{8}$

Section 0.3

Perform the calculations indicated. Write the answer in simplest form.

12. $\frac{25}{7} \times \frac{14}{45}$ **13.** $2\frac{4}{5} \times 3\frac{3}{4}$ **14.** $4 \div \frac{8}{7}$ **15.** $2\frac{1}{3} \div 3\frac{1}{4}$

Section 0.4

Change to a decimal. **16.** $\frac{5}{8}$ **17.** $\frac{5}{9}$

Perform the calculations indicated.

18. $15.23 + 3.6 + 0.821$ **19.** 3.28×0.63

20. $3.015 \div 6.7$ **21.** $12.13 - 9.884$

Section 0.5

Change to a percent. **22.** 0.7 **23.** 0.006

Change to a decimal. **24.** 327% **25.** 2%

26. What is 25% of 1630? **27.** What percent of 500 is 36?

Section 0.6

Estimate. **28.** $8456 + 1749$ **29.** 7386×2856

30. Estimate the number of miles per gallon achieved by Marcia's car if she drove for 235.8 miles and used 10.7 gallons of gas.

Section 0.7

Solve. You may want to use the Mathematical Blueprint for Problem Solving.

31. Susan drove 870 miles. She started with a full tank of gas. During the trip, she made the following gasoline purchases: 3.7 gal, 10 gal, 12 gal, and 4.9 gal. How many miles per gallon did her car achieve? (Round to the nearest tenth.)

0.1 Simplifying and Finding Equivalent Fractions

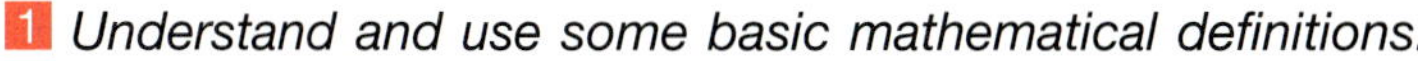

After studying this section, you will be able to:

1 *Understand and use some basic mathematical definitions.*
2 *Simplify fractions to lowest terms.*
3 *Change forms between improper fractions and mixed numbers.*
4 *Change a fraction into an equivalent one with a different denominator.*

Chapter 0 is designed to give you a mental ''warm-up.'' In this chapter you'll be able to step back a bit and tone up your math skills. This brief review of arithmetic will increase your math flexibility and give you a good running start into algebra.

1 *Basic Definitions*

Whole numbers are the set of numbers 0, 1, 2, 3, 4, 5, 6, 7, ... They are used to describe whole objects, or entire quantities.

Fractions are a set of numbers that are used to describe parts of whole quantities. In the object shown on the right there are four equal parts. The *three* of the *four* parts that are shaded are represented by the fraction $\frac{3}{4}$. In the fraction $\frac{3}{4}$ the number 3 is called the **numerator** and the number 4, the **denominator.**

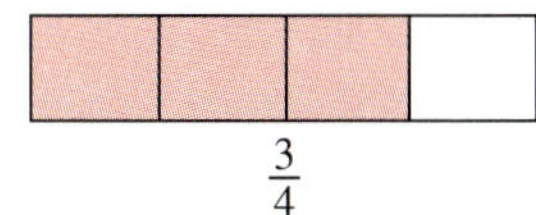

$\frac{3}{4}$

3 ⟵ *Numerator* is on the top
—
4 ⟵ *Denominator* is on the bottom

The *denominator* of a fraction shows the number of equal parts in the whole and the *numerator* shows the number of these parts being talked about or being used.

Numerals are symbols we use to name numbers. There are many different numerals that can be used to describe the same number. We know that $\frac{1}{2} = \frac{2}{4}$. The fractions $\frac{1}{2}$ and $\frac{2}{4}$ both describe the same number.

Usually, we find it more useful to use fractions that are simplified. A fraction is considered to be in simplest form when the numerator (top) and the denominator (bottom) can both be divided exactly by no number other than 1.

TEACHING TIP Throughout mathematics, when referring to rational expressions or to numerical fractions, students are expected to know the terms *numerator* and *denominator*. Emphasize to them which part of the fraction is which. Students who have difficulty keeping them straight appreciate the little rule ''The Denominator is Down on the bottom'' (two D's—Denominator, Down).

$\frac{1}{2}$ is in simplest form.

$\frac{2}{4}$ is *not* in simplest form since the numerator and the denominator can both be divided by 2.

If you get the answer $\frac{2}{4}$ to a problem, you should state it in simplest form, $\frac{1}{2}$. The process of changing $\frac{2}{4}$ to $\frac{1}{2}$ is called simplifying the fraction.

2 *Simplifying Fractions*

Natural numbers or **counting numbers** are the set of whole numbers excluding 0. Thus the natural numbers are the numbers 1, 2, 3, 4, 5, 6, ...

When two or more numbers are multiplied, each number that is multiplied is called a **factor.** For example, when we write $3 \times 7 \times 5$, each of the numbers 3, 7, and 5 is called a factor.

Prime numbers are all natural numbers greater than 1 whose only natural number factors are 1 and itself. The number 5 is prime. The only natural number factors of 5 are 5 and 1.

$$5 = 5 \times 1$$

The number 6 is not prime. The natural number factors of 6 are 3 and 2 or 6 and 1.

$$6 = 3 \times 2 \qquad 6 = 6 \times 1$$

The first 15 prime numbers are

2, 3, 5, 7, 11, 13, 17, 19, 23, 29, 31, 37, 41, 43, 47

Any natural number greater than 1 is either prime or can be written as the product of prime numbers. For example we can take each of the numbers 12, 30, 14, 19, and 29 and either indicate that they are prime or, if they are not prime, write them as the product of prime numbers. We write as follows:

$12 = 2 \times 2 \times 3$ $\quad 30 = 2 \times 3 \times 5$ $\quad$ 29 is a prime number.

$14 = 2 \times 7$ $\quad$ 19 is a prime number.

To reduce a fraction, we use prime numbers to factor numerator and denominator. Write each part of the fraction (numerator and denominator) as a product of prime numbers. Note any *factor* that appears in both the *numerator* (top) and *denominator* (bottom) of the fraction. If we divide numerator and denominator by that value, we will obtain an equivalent fraction in *simplest form.* When the new fraction is simplified, it is said to be in *lowest terms*. Throughout this text, when we say *simplify* a fraction, we always mean to lowest terms.

TEACHING TIP Students will sometimes say, "I don't simplify the fraction $\frac{14}{21}$ that way. I just say 7 goes into 14 twice, and 7 goes into 21 three times; so the fraction simplifies to $\frac{2}{3}$. Can I do it that way?" Stress the importance of making sure they understand the concept. Explain that if they understand "removing the common factor of 7" it will make the simplifying of algebraic fractions easier later. They are welcome to simplify a fraction any way they wish, but the more they understand WHAT is going on and WHY it works, the more mathematics they will be able to learn.

EXAMPLE 1 Simplify each fraction. **(a)** $\frac{14}{21}$ **(b)** $\frac{15}{35}$ **(c)** $\frac{20}{70}$

(a) $\frac{14}{21} = \frac{\cancel{7} \times 2}{\cancel{7} \times 3} = \frac{2}{3}$ *We factor* 14 *and factor* 21. *Then we divide numerator and denominator by* 7.

(b) $\frac{15}{35} = \frac{\cancel{5} \times 3}{\cancel{5} \times 7} = \frac{3}{7}$ *We factor* 15 *and factor* 35. *Then we divide numerator and denominator by* 5.

(c) $\frac{20}{70} = \frac{2 \times \cancel{2} \times \cancel{5}}{7 \times \cancel{2} \times \cancel{5}} = \frac{2}{7}$ *We factor* 20 *and factor* 70. *Then we divide numerator and denominator by both* 2 *and* 5.

Practice Problem 1 Simplify.

(a) $\frac{10}{16}$ **(b)** $\frac{24}{36}$ **(c)** $\frac{42}{36}$ ■

Sometimes when we simplify a fraction, all the prime factors in the top (numerator) are divided out. When this happens, we put a 1 in the numerator. If we did not put the 1 in the numerator, we would not realize that the answer was a fraction.

EXAMPLE 2 Simplify each fraction. **(a)** $\frac{7}{21}$ **(b)** $\frac{15}{105}$

(a) $\frac{7}{21} = \frac{\cancel{7} \times 1}{\cancel{7} \times 3} = \frac{1}{3}$ **(b)** $\frac{15}{105} = \frac{\cancel{5} \times \cancel{3} \times 1}{7 \times \cancel{5} \times \cancel{3}} = \frac{1}{7}$

Practice Problem 2 Simplify.

(a) $\frac{4}{12}$ **(b)** $\frac{25}{125}$ **(c)** $\frac{73}{146}$ ■

If all the prime numbers in the bottom (denominator) are divided out, we do not need to leave a 1 in the denominator, since we do not need to express the answer as a fraction. The answer is then a whole number and is not usually expressed as a fraction.

EXAMPLE 3 Simplify each fraction. **(a)** $\frac{35}{7}$ **(b)** $\frac{70}{10}$

(a) $\frac{35}{7} = \frac{5 \times \cancel{7}}{\cancel{7}} = 5$ **(b)** $\frac{70}{10} = \frac{7 \times \cancel{5} \times \cancel{2}}{\cancel{5} \times \cancel{2}} = 7$

Practice Problem 3 Simplify.

(a) $\frac{18}{6}$ **(b)** $\frac{146}{73}$ **(c)** $\frac{28}{7}$ ■

Sometimes the fraction we use represents how many of a certain thing are successful. For example, if a major league baseball player was at bat 30 times and achieved 12 hits, we could say that he had a hit $\frac{12}{30}$ of the time. If we reduce the fraction, we could say he had a hit $\frac{2}{5}$ of the time.

EXAMPLE 4 Cindy got 48 out of 56 questions correct on a test. Write this as a fraction.

Express as a fraction in simplest form the number of correct responses out of the total number of questions on the test.

$$48 \text{ out of } 56 \longrightarrow \frac{48}{56} = \frac{6 \times \cancel{8}}{7 \times \cancel{8}} = \frac{6}{7}$$

Cindy answered the questions correctly $\frac{6}{7}$ of the time.

Practice Problem 4

The major league pennant winner in 1917 won 56 games out of 154 games played. Express as a fraction in simplest form the number of games won in relation to the number of games played. ■

The number *one* can be expressed as 1, $\frac{1}{1}$, $\frac{2}{2}$, $\frac{6}{6}$, $\frac{8}{8}$, and so on, since

$$1 = \frac{1}{1} = \frac{2}{2} = \frac{6}{6} = \frac{8}{8}$$

We say that these numerals are *equivalent ways* of writing the number *one* because they all express the same quantity even though they appear to be different.

SIDELIGHT When we simplify fractions, we are actually using the fact that we can multiply any number by 1 without changing the value of that number. (Mathematicians call the number 1 the multiplicative identity because it leaves any number it multiplies with the same identical value as before.)

Let's look again at one of the previous problems.

$$\frac{14}{21} = \frac{7 \times 2}{7 \times 3} = \frac{7}{7} \times \frac{2}{3} = 1 \times \frac{2}{3} = \frac{2}{3}$$

So we see that

$$\frac{14}{21} = \frac{2}{3}$$

When we simplify fractions, we are using this property of multiplying by 1.

TEACHING TIP You may need to remind students that there is nothing "improper or incorrect" about an improper fraction. There are times in mathematics when we prefer to leave a fraction in improper form. Some students have been told, "You may never leave your answer as an improper fraction." Remind them that in higher levels of mathematics that statement is not true and that in algebra there is not a "right or wrong" value given to a fraction left in improper form as a final answer. The best rule is to try to leave your answer in an appropriate form. Obviously, no one wants to have to measure out $\frac{15}{4}$ cups of flour or to measure a distance of $\frac{57}{10}$ miles. We would prefer to have the measure of $3\frac{3}{4}$ cups of flour and $5\frac{7}{10}$ miles.

3 *Improper Fractions and Mixed Numbers*

If the numerator is less than the denominator the fraction is a **proper fraction.** A proper fraction is used to describe a quantity smaller than a whole.

Fractions can also be used to describe quantities larger than a whole. The figure below shows two bars that are equal in size. Each bar is divided into 5 equal pieces. The first bar is shaded in completely. The second bar has 2 of the 5 pieces shaded in.

The shaded-in region can be represented by $\frac{7}{5}$ since 7 of the pieces (each of which is $\frac{1}{5}$ of a whole box) are shaded. The fraction $\frac{7}{5}$ is called an improper fraction. An **improper fraction** is one in which the numerator is larger than or equal to the denominator.

The shaded-in region can also be represented by 1 whole added to $\frac{2}{5}$ of a whole, or $1 + \frac{2}{5}$. This is written as $1\frac{2}{5}$. The fraction $1\frac{2}{5}$ is called a mixed number. A **mixed number** consists of a whole number added to a proper fraction (numerator is smaller than the denominator). The addition is understood but not written. When we write $1\frac{2}{5}$, it represents $1 + \frac{2}{5}$. The numbers $1\frac{7}{8}$, $2\frac{3}{4}$, $8\frac{1}{3}$, and $126\frac{1}{10}$ are all mixed numbers. From the figure above it seems clear that $\frac{7}{5} = 1\frac{2}{5}$. This suggests that we can change from one form to the other without changing the value of the fraction.

From a picture it is easy to see how to *change improper fractions to mixed numbers.* For example, if we start with the fraction $\frac{11}{3}$ and represent it by the figure below (where 11 of the pieces that are $\frac{1}{3}$ of a box are shaded), we see that $\frac{11}{3} = 3\frac{2}{3}$, since 3 whole boxes and $\frac{2}{3}$ of a box are shaded.

Changing Improper Fractions to Mixed Numbers

You can do the same procedure without a picture. For example, to change $\frac{11}{3}$ to a mixed number, we can do the following:

$$\frac{11}{3} = \frac{3}{3} + \frac{3}{3} + \frac{3}{3} + \frac{2}{3}$$ *By the rule for adding fractions (which is discussed in detail in Section 0.2).*

$$= 1 + 1 + 1 + \frac{2}{3}$$ *Write 1 in place of $\frac{3}{3}$, since $\frac{3}{3} = 1$.*

$$= 3 + \frac{2}{3}$$ *Write 3 in place of $1 + 1 + 1$.*

$$= 3\frac{2}{3}$$ *Use the notation for mixed numbers.*

Now that you know how to perform the change and why it works, here is a shorter method.

To Change an Improper Fraction to a Mixed Number

1. Divide the denominator into the numerator.
2. The result is the whole-number part of the mixed number.
3. The remainder from the division will be the numerator of the fraction. The denominator of the fraction remains unchanged.

We can write the fraction as a division example and divide. The arrows show how to write the mixed number.

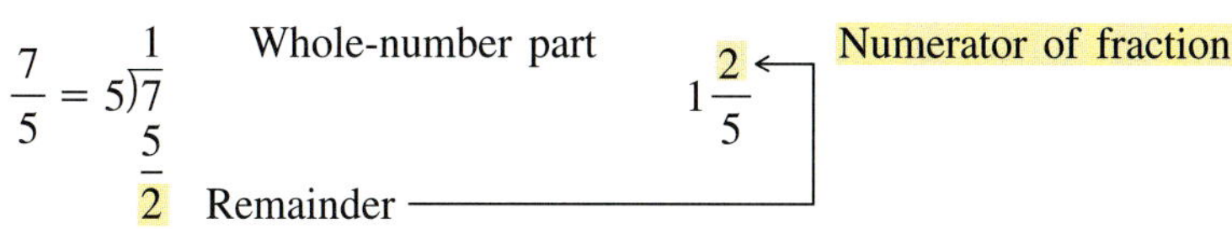

Thus, $\frac{7}{5} = 1\frac{2}{5}$.

$$\frac{11}{3} = 3\overline{)11}\quad\text{quotient } 3,\ 9,\ \text{remainder } 2$$

Whole-number part ⟶ $3\frac{2}{3}$ Numerator of fraction

Remainder

Thus, $\frac{11}{3} = 3\frac{2}{3}$.

Sometimes the remainder is 0. The improper fraction changes to a whole number.

TEACHING TIP You may wish to introduce the following technique to help students to remember which number goes where when writing a fraction as a division example:

The denominator goes on the outside.

EXAMPLE 5 Change to a mixed number or to a whole number.

(a) $\frac{7}{4}$ **(b)** $\frac{15}{3}$

(a) $\frac{7}{4} = 7 \div 4 = 4\overline{)7}$, quotient 1; 4; 3 Remainder

Thus $\frac{7}{4} = 1\frac{3}{4}$.

(b) $\frac{15}{3} = 15 \div 3 = 3\overline{)15}$, quotient 5; 15; 0 Remainder

Thus $\frac{15}{3} = 5$.

Practice Problem 5 Change to a mixed number or to a whole number.

(a) $\frac{12}{7}$ **(b)** $\frac{20}{5}$ ■

Changing Mixed Numbers to Improper Fractions

It is not difficult to see how to change mixed numbers to improper fractions. Suppose that you wanted to write $2\frac{2}{3}$ as an improper fraction:

$$2\frac{2}{3} = 2 + \frac{2}{3} \qquad \textit{The meaning of mixed number notation.}$$

$$= 1 + 1 + \frac{2}{3} \qquad \textit{Since } 1 + 1 = 2.$$

$$= \frac{3}{3} + \frac{3}{3} + \frac{2}{3} \qquad \textit{Since } 1 = \tfrac{3}{3}.$$

When we draw a picture of $\frac{3}{3} + \frac{3}{3} + \frac{2}{3}$, we have this figure:

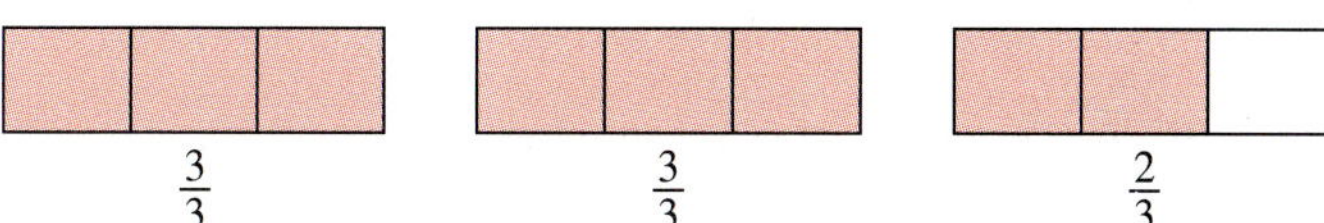

If we count the shaded parts, we see that

$$\frac{3}{3} + \frac{3}{3} + \frac{2}{3} = \frac{8}{3} \quad \text{Thus} \quad 2\frac{2}{3} = \frac{8}{3}$$

Now that you can see how this change can be done, here is a shorter method.

To Change a Mixed Number to an Improper Fraction

1. Multiply the whole number by the denominator.
2. Add this to the numerator. The result is the new numerator. The denominator does not change.

EXAMPLE 6 Change to an improper fraction. (a) $3\frac{1}{7}$ (b) $5\frac{4}{5}$

(a) $3\frac{1}{7} = \frac{(3 \times 7) + 1}{7} = \frac{21 + 1}{7} = \frac{22}{7}$ (b) $5\frac{4}{5} = \frac{(5 \times 5) + 4}{5} = \frac{25 + 4}{5} = \frac{29}{5}$

Practice Problem 6 Change to an improper fraction.

(a) $3\frac{2}{5}$ (b) $1\frac{3}{7}$ (c) $2\frac{6}{11}$ (d) $4\frac{2}{3}$ ■

4 *Changing a Fraction to an Equivalent Fraction with a Different Denominator*

Fractions can be changed to an equivalent fraction with a different denominator by multiplying both numerator and denominator by the same number.

$$\frac{5}{6} = \frac{5 \times 2}{6 \times 2} = \frac{10}{12} \qquad \frac{3}{7} = \frac{3 \times 3}{7 \times 3} = \frac{9}{21}$$

So $\frac{5}{6}$ is equivalent to $\frac{10}{12}$. $\frac{3}{7}$ is equivalent to $\frac{9}{21}$.

We often want to get a *particular denominator.*

TEACHING TIP After discussing Example 6 or a similar example, ask students to find the missing numerator in each of the following as an Additional Class Exercise.

A. $\frac{7}{12} = \frac{?}{36}$ B. $\frac{5}{16} = \frac{?}{64}$

C. $\frac{11}{24} = \frac{?}{144}$ D. $\frac{6}{7} = \frac{?}{49}$

Ans. A. 21 B. 20 C. 66 D. 42

EXAMPLE 7 Find the missing number.

(a) $\frac{3}{5} = \frac{?}{25}$ (b) $\frac{3}{7} = \frac{?}{21}$ (c) $\frac{2}{9} = \frac{?}{36}$

(a) $\frac{3}{5} = \frac{?}{25}$ *Observe that we need to multiply the denominator by 5 to obtain 25. So we multiply the numerator 3 by 5 also.*

$\frac{3 \times 5}{5 \times 5} = \frac{15}{25}$ *The desired numerator is 15.*

(b) $\frac{3}{7} = \frac{?}{21}$ *Observe that $7 \times 3 = 21$. We need to multiply by 3 to get the new numerator.*

$\frac{3 \times 3}{7 \times 3} = \frac{9}{21}$ *The desired numerator is 9.*

(c) $\frac{2}{9} = \frac{?}{36}$ *Observe that $9 \times 4 = 36$. We need to multiply by 4 to get the new numerator.*

$\frac{2 \times 4}{9 \times 4} = \frac{8}{36}$ *The desired numerator is 8.*

Practice Problem 7 Find the missing number.

(a) $\frac{3}{8} = \frac{?}{24}$ (b) $\frac{5}{6} = \frac{?}{30}$ (c) $\frac{12}{13} = \frac{?}{26}$ (d) $\frac{2}{7} = \frac{?}{56}$

(e) $\frac{5}{9} = \frac{?}{27}$ (f) $\frac{3}{10} = \frac{?}{60}$ (g) $\frac{3}{4} = \frac{?}{28}$ (h) $\frac{8}{11} = \frac{?}{55}$ ■

0.1 Exercises

Verbal and Writing Skills

1. In the fraction $\frac{12}{13}$, what number is the numerator? 12
2. In the fraction $\frac{17}{20}$, what number is the denominator? 20
3. What is a factor? Give an example. When two or more numbers are multiplied, each number that is multiplied is called a factor. In 2×3, 2 and 3 are factors.
4. Give some examples of the number 1 written as a fraction. Answers will vary.
5. Draw a diagram to illustrate $2\frac{3}{5}$.

Simplify each fraction.

6. $\frac{18}{24}$ $\frac{3}{4}$ **7.** $\frac{20}{35}$ $\frac{4}{7}$ **8.** $\frac{16}{48}$ $\frac{1}{3}$ **9.** $\frac{12}{60}$ $\frac{1}{5}$ **10.** $\frac{32}{8}$ 4 **11.** $\frac{75}{15}$ 5

Change to a mixed number.

12. $\frac{17}{6}$ $2\frac{5}{6}$ **13.** $\frac{19}{5}$ $3\frac{4}{5}$ **14.** $\frac{21}{9}$ $2\frac{1}{3}$ **15.** $\frac{125}{4}$ $31\frac{1}{4}$ **16.** $\frac{46}{5}$ $9\frac{1}{5}$ **17.** $\frac{89}{6}$ $14\frac{5}{6}$

Change to an improper fraction.

18. $3\frac{1}{5}$ $\frac{16}{5}$ **19.** $2\frac{6}{7}$ $\frac{20}{7}$ **20.** $6\frac{3}{5}$ $\frac{33}{5}$ **21.** $5\frac{3}{8}$ $\frac{43}{8}$ **22.** $8\frac{8}{9}$ $\frac{80}{9}$ **23.** $13\frac{1}{6}$ $\frac{79}{6}$

Find the missing numerator.

24. $\frac{3}{11} = \frac{?}{44}$ $\frac{4 \cdot 3}{4 \cdot 11} = \frac{12}{44}$

25. $\frac{5}{7} = \frac{?}{28}$ $\frac{4 \cdot 5}{4 \cdot 7} = \frac{20}{28}$

26. $\frac{3}{5} = \frac{?}{35}$ $\frac{3 \cdot 7}{5 \cdot 7} = \frac{21}{35}$

27. $\frac{2}{7} = \frac{?}{21}$ $\frac{2 \cdot 3}{7 \cdot 3} = \frac{6}{21}$

28. $\frac{9}{11} = \frac{?}{55}$ $\frac{9 \cdot 5}{11 \cdot 5} = \frac{45}{55}$

29. $\frac{13}{17} = \frac{?}{51}$ $\frac{13 \cdot 3}{17 \cdot 3} = \frac{39}{51}$

Applications

Solve each word problem.

30. While playing in the minor league, Hector Martinez obtained a hit 342 times out of 620 times at bat. Express as a fraction in simplified form how much of the time he obtained a hit.
$\frac{171}{310}$

31. Raintree Estates purchased 105 acres of land. Only 75 acres of land are buildable. Write a fraction in simplified form that indicates the fractional part of the land that is buildable.
$\frac{5}{7}$

32. Charles Barkley of the Phoenix Suns scored 1560 points during 68 games played during the 1994–1995 season. Express as a fraction in simplified form how many points he averaged per game.
$22\frac{16}{17}$

33. Alex Rodriguez, major league baseball's most successful hitter of the 1996 regular season, got 215 hits out of 600 times at bat for the Seattle Mariners. Express as a fraction in simplified form how often he obtained a hit.
$\frac{43}{120}$

34. Last year, my parents had a combined income of $64,000. They paid $13,200 in federal income taxes. What simplified fraction would show how much my parents spent on their federal taxes?
$\frac{33}{160}$

35. REM is making a new CD. 7 out of 8 takes (trys) of their songs are acceptable. The recording engineer tells the band that there is time that day for only 136 takes. How many of the 136 takes will REM need to sing well to maintain the same fractional ratio of acceptable takes.
119

The following chart gives the recipe for a trail mix.

36. What part of the Premium Blend is nuts? $\frac{1}{4}$

37. What part of the High Energy Blend is raisins? $\frac{3}{5}$

38. What part of the Premium Blend is not sunflower seeds? $\frac{1}{2}$

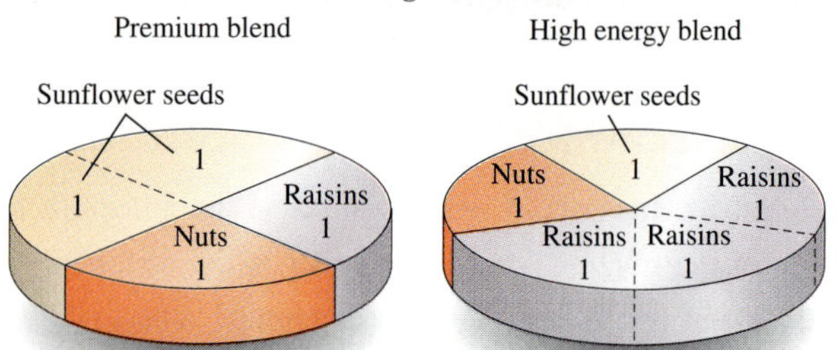

The following chart provides some statistics for the North Andover Knights.

39. Determine how often each player hit the ball based on the number of times at bat. Write each answer as a reduced fraction. Aaron Dunsay $\frac{3}{5}$ Paul Banks $\frac{2}{5}$ Tom Re $\frac{4}{5}$

40. Which player was the best hitter? Tom Re

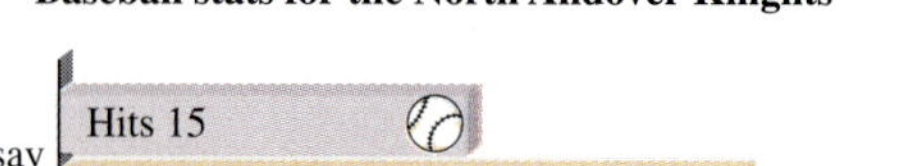

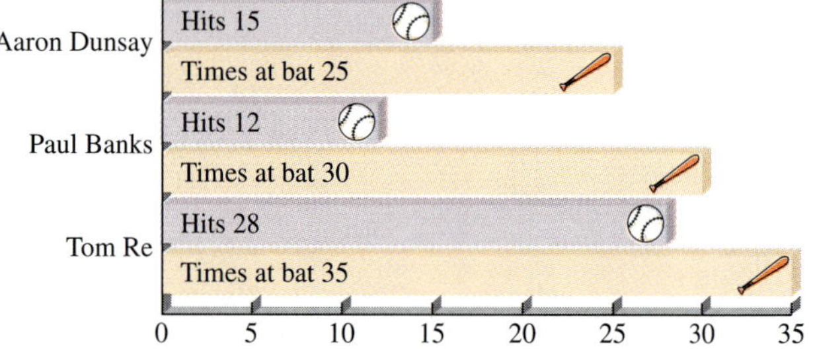

Developing Your Study Skills

WHY IS HOMEWORK NECESSARY?

Mathematics is a set of skills that you learn by doing, not by watching someone else do it. Your instructor may make solving a mathematics problem look very easy, but for you to learn the necessary skills, you must practice them over and over again, just as your instructor once had to do. There is no other way. Learning mathematics is like learning to play a musical instrument, to type, or to play a sport. No matter how much you watch someone else do it, how many books you may read on "how to" do it, or how easy it may seem to be, the key to success is practice on a regular basis.

Homework provides this practice. The amount of practice varies for each individual, but usually students need to do most or all of the exercises provided at the end of each section in the text. The more problems you do, the better you get. Some problems in a set are more difficult than others, and some stress different concepts. Only by working all the problems will you cover the full range of difficulty and coverage.

Make yourself a promise that you will complete each homework assignment in the course for this semester. Stick to this promise, even if it takes a lot of work. If you are able to do this one thing, you will double your chances for successfully completing this course! Make the promise. You will not be sorry.

0.2 Addition and Subtraction of Fractions

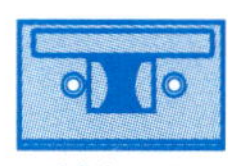

After studying this section, you will be able to:

1. *Add or subtract fractions with a common denominator.*
2. *Find the least common denominator of any two or more fractions.*
3. *Add or subtract fractions that do not have a common denominator.*
4. *Add or subtract mixed numbers.*

1 *Adding or Subtracting Fractions with a Common Denominator*

If fractions have the same denominator, the numerators may be added or subtracted. The denominator remains the same.

TEACHING TIP For some weaker students it is helpful to say the rule WHEN ADDING OR SUBTRACTING FRACTIONS WITH THE SAME DENOMINATOR:

1. Add or subtract the numerators.
2. Keep the common (same) denominator.

To Add or Subtract Two Fractions with the Same Denominator

1. Add or subtract the numerators.
2. Keep the same denominator.
3. Simplify the answer whenever possible.

EXAMPLE 1 Add the fractions. Simplify your answer whenever possible.

(a) $\frac{5}{7}+\frac{1}{7}$ **(b)** $\frac{2}{3}+\frac{1}{3}$ **(c)** $\frac{1}{8}+\frac{3}{8}+\frac{2}{8}$ **(d)** $\frac{3}{5}+\frac{4}{5}$

(a) $\frac{5}{7}+\frac{1}{7}=\frac{5+1}{7}=\frac{6}{7}$ **(b)** $\frac{2}{3}+\frac{1}{3}=\frac{2+1}{3}=\frac{3}{3}=1$

(c) $\frac{1}{8}+\frac{3}{8}+\frac{2}{8}=\frac{1+3+2}{8}=\frac{6}{8}=\frac{3}{4}$ **(d)** $\frac{3}{5}+\frac{4}{5}=\frac{3+4}{5}=\frac{7}{5}=1\frac{2}{5}$

Practice Problem 1 Add.

(a) $\frac{3}{6}+\frac{2}{6}$ **(b)** $\frac{3}{11}+\frac{2}{11}+\frac{6}{11}$ **(c)** $\frac{5}{8}+\frac{2}{8}+\frac{7}{8}$ ■

EXAMPLE 2 Subtract the fractions. Simplify your answer whenever possible.

(a) $\frac{9}{11}-\frac{2}{11}$ **(b)** $\frac{5}{6}-\frac{1}{6}$

(a) $\frac{9}{11}-\frac{2}{11}=\frac{9-2}{11}=\frac{7}{11}$ **(b)** $\frac{5}{6}-\frac{1}{6}=\frac{5-1}{6}=\frac{4}{6}=\frac{2}{3}$

Practice Problem 2 Subtract. **(a)** $\frac{11}{13}-\frac{6}{13}$ **(b)** $\frac{8}{9}-\frac{2}{9}$ ■

Our ability to add and subtract fractions with a common denominator allows us to solve simple problems. Most problems, however, involve fractions that do not have a common denominator. Fractions and mixed numbers such as halves, fourths, and eighths are commonly used. To add or subtract such fractions, we begin by finding a common denominator.

2 *Finding the Least Common Denominator*

Before you can add or subtract, fractions must have the same denominator. To save work, we select the smallest possible common denominator. This is called the **least common denominator** or LCD (also known as the *lowest common denominator*).

> The LCD of two or more fractions is the smallest whole number that is exactly divisible by each denominator of the fractions.

TEACHING TIP Most students are familiar with the concept of LCD from previous mathematics courses. Some students have been taught only to use the *least common multiple* of the numbers in the denominator. You may want to mention that if they are used to finding the LCM then what we are doing is changing fractions to equivalent fractions that have the LCM as the denominator.

EXAMPLE 3 Find the LCD. $\frac{2}{3}$ and $\frac{1}{4}$

The numbers are small enough to find the LCD by inspection. The LCD is 12, since 12 is exactly divisible by 4 and by 3. There is no smaller number that is exactly divisible by 4 and 3.

Practice Problem 3 Find the LCD. $\frac{1}{8}$ and $\frac{5}{12}$ ■

In some cases, the LCD cannot easily be determined by inspection. If we write each denominator as the product of prime factors, we will be able to find the LCD. We will use ($\cdot$) to indicate multiplication. For example, $30 = 2 \cdot 3 \cdot 5$. This means $30 = 2 \times 3 \times 5$.

> ### Procedure to Find the LCD Using Prime Factors
>
> **1.** Write each denominator as the product of prime factors.
>
> **2.** The LCD is the product of each different factor.
>
> **3.** If a factor occurs more than once in any one denominator, the LCD will contain that factor repeated the greatest number of times that it occurs in any one denominator.

EXAMPLE 4 Find the LCD of $\frac{5}{6}$ and $\frac{1}{15}$ by this new procedure.

$$\begin{aligned} 6 &= 2 \cdot 3 \\ 15 &= \quad\ 3 \cdot 5 \\ \text{LCD} &= 2 \cdot 3 \cdot 5 \end{aligned}$$

Write each denominator as the product of prime factors.

$\text{LCD} = 2 \cdot 3 \cdot 5 = 30$ *The LCD is the product of each different prime factor. The different factors are* 2, 3, *and* 5.

Practice Problem 4 Use the prime factors to find the LCD of $\frac{8}{35}$ and $\frac{6}{15}$. ■

Great care should be used to determine the LCD in the case of repeated factors.

EXAMPLE 5 Find the LCD of $\frac{4}{27}$ and $\frac{5}{18}$.

$27 = 3 \cdot 3 \cdot 3$

$18 = \quad 3 \cdot 3 \cdot 2$

$\text{LCD} = 3 \cdot 3 \cdot 3 \cdot 2$

Write each denominator as the product of prime factors. We observe that the factor 3 occurs three times in the factorization of 27.

$\text{LCD} = 3 \cdot 3 \cdot 3 \cdot 2 = 54$

The LCD is the product of each different factor. The factor 3 *occurred most* in the factorization of 27, where it occurred *three* times. Thus the LCD will be the product of *three* 3's and *one* 2.

Practice Problem 5 Find the LCD of $\frac{5}{12}$ and $\frac{7}{30}$. ■

EXAMPLE 6 Find the LCD of $\frac{5}{12}$, $\frac{1}{15}$, and $\frac{7}{30}$.

$12 = 2 \cdot 2 \cdot 3$

$15 = \qquad 3 \cdot 5$

$30 = \quad 2 \cdot 3 \cdot 5$

$\text{LCD} = 2 \cdot 2 \cdot 3 \cdot 5$

Write each denominator as the product of prime factors. Notice that the repeated factor is 2, which occurs twice *in the factorization of* 12.

$\text{LCD} = 2 \cdot 2 \cdot 3 \cdot 5 = 60$

The LCD is the product of each different factor with the factor 2 repeated since it occurred twice in one denominator.

Practice Problem 6 Find the LCD of $\frac{2}{27}$, $\frac{1}{18}$, and $\frac{5}{12}$. ■

3 *Adding or Subtracting Fractions That Do Not Have a Common Denominator*

Before you can add or subtract them, fractions must have the same denominator. Using the LCD will make your work easier. First you must find the LCD. Then change each fraction to a fraction that has the LCD as the denominator. Sometimes one of the fractions will already have the LCD as the denominator. Once all the fractions have the same denominator, you can add or subtract. Be sure to simplify the fraction in your answer if this is possible.

To Add or Subtract Fractions That Do Not Have a Common Denominator

1. Find the LCD of the fractions.
2. Change each fraction to an equivalent fraction with the LCD for a denominator.
3. Add or subtract the fractions.
4. Simplify the answer whenever possible.

Let us return to the two fractions of Example 3. We have previously found that the LCD is 12.

EXAMPLE 7 Add. $\frac{2}{3} + \frac{1}{4}$

We must change $\frac{2}{3}$ and $\frac{1}{4}$ to fractions with the same denominator. We change each fraction to an equivalent fraction with a common denominator of 12, the LCD.

$$\frac{2}{3} = \frac{?}{12} \qquad \frac{2}{3} \times \frac{4}{4} = \frac{8}{12} \quad \text{so} \quad \frac{2}{3} = \frac{8}{12}$$

$$\frac{1}{4} = \frac{?}{12} \qquad \frac{1}{4} \times \frac{3}{3} = \frac{3}{12} \quad \text{so} \quad \frac{1}{4} = \frac{3}{12}$$

Then we rewrite the problem with common denominators and then add.

$$\frac{2}{3} + \frac{1}{4} = \frac{8}{12} + \frac{3}{12} = \frac{8+3}{12} = \frac{11}{12}$$

Practice Problem 7 Add. $\frac{1}{8} + \frac{5}{12}$ ■

Sometimes one of the denominators is the LCD. In such cases the fraction that has the LCD for the denominator will not need to be changed. If every other denominator divides into the largest denominator, the largest denominator is the LCD.

EXAMPLE 8 Find the LCD and then add. $\frac{3}{5} + \frac{7}{20} + \frac{1}{2}$

We can see by inspection that both 5 and 2 divide exactly into 20. Thus 20 is the LCD. Now add:

$$\frac{3}{5} + \frac{7}{20} + \frac{1}{2}$$

We change $\frac{3}{5}$ and $\frac{1}{2}$ to equivalent fractions with a common denominator of 20, the LCD.

$$\frac{3}{5} = \frac{?}{20} \qquad \frac{3}{5} \times \frac{4}{4} = \frac{12}{20} \quad \text{so} \quad \frac{3}{5} = \frac{12}{20}$$

$$\frac{1}{2} = \frac{?}{20} \qquad \frac{1}{2} \times \frac{10}{10} = \frac{10}{20} \quad \text{so} \quad \frac{1}{2} = \frac{10}{20}$$

Then we rewrite the problem with common denominators and add:

$$\frac{3}{5} + \frac{7}{20} + \frac{1}{2} = \frac{12}{20} + \frac{7}{20} + \frac{10}{20} = \frac{12+7+10}{20} = \frac{29}{20} \quad \text{or} \quad 1\frac{9}{20}$$

Practice Problem 8 Find the LCD and add. $\frac{3}{5} + \frac{4}{25} + \frac{1}{10}$ ■

Now we turn to examples where the selection of the LCD is not so obvious. In Examples 9 through 11 we will use the prime factorization method to find the LCD.

EXAMPLE 9 Add. $\frac{7}{18} + \frac{5}{12}$

First we find the LCD.

$$18 = 3 \cdot 3 \cdot 2$$
$$12 = \quad 3 \cdot 2 \cdot 2$$
$$\text{LCD} = 3 \cdot 3 \cdot 2 \cdot 2 = 9 \cdot 4 = 36$$

Now we change $\frac{7}{18}$ and $\frac{5}{12}$ to equivalent fractions that have the LCD.

$$\frac{7}{18} = \frac{?}{36} \qquad \frac{7}{18} \times \frac{2}{2} = \frac{14}{36}$$

$$\frac{5}{12} = \frac{?}{36} \qquad \frac{5}{12} \times \frac{3}{3} = \frac{15}{36}$$

Now we add the fractions.

$$\frac{7}{18} + \frac{5}{12} = \frac{14}{36} + \frac{15}{36} = \frac{29}{36}$$ *This fraction cannot be simplified.*

Practice Problem 9 Add. $\frac{1}{49} + \frac{3}{14}$ ■

EXAMPLE 10 Subtract. $\frac{25}{48} - \frac{5}{36}$

First we find the LCD.

$$48 = 2 \cdot 2 \cdot 2 \cdot 2 \cdot 3$$
$$36 = \quad 2 \cdot 2 \cdot 3 \cdot 3$$
$$\text{LCD} = 2 \cdot 2 \cdot 2 \cdot 2 \cdot 3 \cdot 3 = 16 \cdot 9 = 144$$

Now we change $\frac{25}{48}$ and $\frac{5}{36}$ to equivalent fractions that have the LCD for a denominator.

$$\frac{25}{48} = \frac{?}{144} \qquad \frac{25}{48} \times \frac{3}{3} = \frac{75}{144}$$

$$\frac{5}{36} = \frac{?}{144} \qquad \frac{5}{36} \times \frac{4}{4} = \frac{20}{144}$$

Now we subtract the fractions.

$$\frac{25}{48} - \frac{5}{36} = \frac{75}{144} - \frac{20}{144} = \frac{55}{144}$$ *This fraction cannot be simplified.*

Practice Problem 10 Subtract. $\frac{1}{12} - \frac{1}{30}$ ■

TEACHING TIP Encourage students to line up the factors in the way that is shown in the text. In this way, factors that occur in more than one number will not be repeated.
Additional Class Exercise
Find the LCD: $\frac{11}{40}$, $\frac{7}{15}$, and $\frac{3}{10}$.

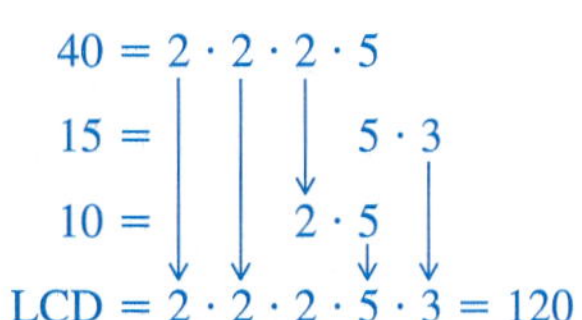

EXAMPLE 11 Add. $\frac{1}{5} + \frac{1}{6} + \frac{3}{10}$

First we find the LCD.

$$5 = 5$$
$$6 = 2 \cdot 3$$
$$10 = 5 \cdot 2$$
$$\text{LCD} = 5 \cdot 2 \cdot 3 = 10 \cdot 3 = 30$$

Now we change $\frac{1}{5}$, $\frac{1}{6}$, and $\frac{3}{10}$ to equivalent fractions that have the LCD for a denominator.

$$\frac{1}{5} = \frac{?}{30} \qquad \frac{1}{5} \times \frac{6}{6} = \frac{6}{30}$$

$$\frac{1}{6} = \frac{?}{30} \qquad \frac{1}{6} \times \frac{5}{5} = \frac{5}{30}$$

$$\frac{3}{10} = \frac{?}{30} \qquad \frac{3}{10} \times \frac{3}{3} = \frac{9}{30}$$

Now we add the three fractions.

$$\frac{1}{5} + \frac{1}{6} + \frac{3}{10} = \frac{6}{30} + \frac{5}{30} + \frac{9}{30} = \frac{20}{30} = \frac{2}{3}$$

Note the important step of simplifying the fraction to obtain the final answer.

TEACHING TIP After discussing Example 12, students may ask if they MUST change to improper fractions first before adding or subtracting. The logical answer is that it is NOT necessary, only suggested. By following this approach, it more logically leads to the methods of adding rational expressions in algebra. You might want to show them an additional example worked both ways:

Add $2\frac{5}{8} + 3\frac{7}{12}$.

METHOD OF CHANGING TO IMPROPER FRACTIONS

$$2\frac{5}{8} = \frac{21}{8} = \frac{63}{24}$$

$$3\frac{7}{12} = \frac{43}{12} = \frac{86}{24}$$

$$\frac{63}{24} + \frac{86}{24} = \frac{149}{24} = 6\frac{5}{24}$$

METHOD OF ADDING INTEGERS AND FRACTIONS SEPARATELY

$$2\frac{5}{8} = 2\frac{15}{24}$$
$$+3\frac{7}{12} = +3\frac{14}{24}$$
$$5\frac{29}{24} = 6\frac{5}{24}$$

Students need not worry about "borrowing" in subtraction if they use the technique illustrated in the text.

Practice Problem 11 Combine. $\frac{2}{3} + \frac{3}{4} - \frac{3}{8}$ ■

4 *Adding or Subtracting Mixed Numbers*

If the problem you are adding or subtracting has mixed numbers, change them to improper fractions first and then combine (addition or subtraction). Often the final answer is changed to a mixed number. Here is a good rule to follow:

> If the original problem contains mixed numbers, express the result as a mixed number rather than as an improper fraction.

EXAMPLE 12 Combine. Simplify your answer whenever possible.

(a) $5\frac{1}{2} + 2\frac{1}{3}$ **(b)** $2\frac{1}{5} - 1\frac{3}{4}$ **(c)** $1\frac{5}{12} + \frac{7}{30}$

(a) First we change the mixed numbers to improper fractions.

$$5\frac{1}{2} = \frac{5 \times 2 + 1}{2} = \frac{11}{2} \qquad 2\frac{1}{3} = \frac{2 \times 3 + 1}{3} = \frac{7}{3}$$

Next we change each fraction to an equivalent form with a common denominator of 6.

$$\frac{11}{2} = \frac{?}{6} \qquad \frac{11}{2} \times \frac{3}{3} = \frac{33}{6}$$

$$\frac{7}{3} = \frac{?}{6} \qquad \frac{7}{3} \times \frac{2}{2} = \frac{14}{6}$$

Finally, we add the two fractions and change our answer to a mixed number.

$$\frac{33}{6} + \frac{14}{6} = \frac{47}{6} = 7\frac{5}{6}$$

Thus $5\frac{1}{2} + 2\frac{1}{3} = 7\frac{5}{6}$.

(b) First we change mixed numbers to improper fractions.

$$2\frac{1}{5} = \frac{2 \times 5 + 1}{5} = \frac{11}{5} \qquad 1\frac{3}{4} = \frac{1 \times 4 + 3}{4} = \frac{7}{4}$$

Next we change each fraction to an equivalent form with a common denominator of 20.

$$\frac{11}{5} = \frac{?}{20} \qquad \frac{11}{5} \times \frac{4}{4} = \frac{44}{20}$$

$$\frac{7}{4} = \frac{?}{20} \qquad \frac{7}{4} \times \frac{5}{5} = \frac{35}{20}$$

Now we subtract the two fractions.

$$\frac{44}{20} - \frac{35}{20} = \frac{9}{20}$$

Thus $2\frac{1}{5} - 1\frac{3}{4} = \frac{9}{20}$.

Note: It is not necessary to use these exact steps to add and subtract mixed numbers. If you know another method and can use it to obtain the correct answers, it is all right to continue to use that method throughout this chapter.

(c) Now we add $1\frac{5}{12} + \frac{7}{30}$.

The LCD of 12 and 30 is 60. Why? Change the mixed number to an improper fraction. Then change each fraction to an equivalent form with a common denominator.

$$1\frac{5}{12} = \frac{17}{12} \times \frac{5}{5} = \frac{85}{60} \qquad \frac{7}{30} \times \frac{2}{2} = \frac{14}{60}$$

Then add the fractions, simplify, and write the answer as a mixed number.

$$\frac{85}{60} + \frac{14}{60} = \frac{99}{60} = \frac{33}{20} = 1\frac{13}{20}$$

Thus $1\frac{5}{12} + \frac{7}{30} = 1\frac{13}{20}$.

Practice Problem 12 Combine. **(a)** $1\frac{2}{3} + 2\frac{4}{5}$ **(b)** $5\frac{1}{4} - 2\frac{2}{3}$ ■

TEACHING TIP Point out the fact that the answer includes the unit of measure. The perimeter of the triangle is $54\frac{3}{8}$ feet. The answer to the word problem is information that Manuel needs to know. An answer of $54\frac{3}{8}$ would not be very useful, and any other unit of measure, for example inches, would be incorrect.

EXAMPLE 13 Manuel is enclosing a triangular-shaped exercise yard for his new dog. He wants to determine how much fencing he will need. The sides of the yard measure $20\frac{3}{4}$ feet, $15\frac{1}{2}$ feet, and $18\frac{1}{8}$ feet. What is the perimeter of (total distance around) the triangle?

Understand the problem. Begin by drawing a picture.

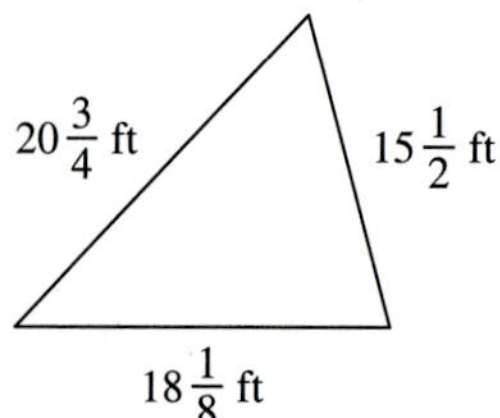

We want to add up the lengths of all three sides of the triangle. This distance around the triangle is called the **perimeter.**

$$20\frac{3}{4} + 15\frac{1}{2} + 18\frac{1}{8} = \frac{83}{4} + \frac{31}{2} + \frac{145}{8}$$

$$= \frac{166}{8} + \frac{124}{8} + \frac{145}{8} = \frac{435}{8} = 54\frac{3}{8} \text{ feet}$$

Practice Problem 13

Find the perimeter of a rectangle with sides of $4\frac{1}{5}$ cm and $6\frac{1}{2}$ cm. Begin by drawing a picture. Label the picture by including the measure of *each* side. ■

Developing Your Study Skills

CLASS ATTENDANCE

A student of mathematics needs to get started in the right direction by choosing to attend class every day, beginning with the first day of class. Statistics show that class attendance and good grades go together. Classroom activities are designed to enhance learning and, therefore, you must be in class to benefit from them. Vital information and explanations are given each day that can help you in understanding concepts. Do not be deceived into thinking that you can just find out from a friend what went on in class. There is no good substitute for firsthand experience. Give yourself a push in the right direction by developing the habit of going to class every day.

CLASS PARTICIPATION

People learn mathematics through active participation, not through observation from the sidelines. If you want to do well in this course, be involved in classroom activities. Sit near the front where you can see and hear well, where your focus is on the instruction process and not on those students around you. Ask questions, be ready to contribute toward solutions, and take part in all classroom activities. Your contributions are valuable to the class and to yourself. Class participation requires an investment of yourself in the learning process, which you will find pays huge dividends.

0.2 Exercises

Verbal and Writing Skills

1. Explain why the denominator 8 is the least common denominator of $\frac{3}{4}$ and $\frac{5}{8}$.
Answers may vary. A sample answer is: 8 is exactly divisible by 4.

2. What must you do before you add or subtract fractions that do not have a common denominator?
You must change the fractions to equivalent fractions with a common denominator.

Find the LCD (least common denominator) of each of the following pairs of fractions. Do not combine the fractions; only find the LCD.

3. $\frac{1}{12}$ and $\frac{5}{9}$
LCD = 36

4. $\frac{11}{12}$ and $\frac{5}{18}$
LCD = 36

5. $\frac{7}{10}$ and $\frac{1}{4}$
LCD = 20

6. $\frac{3}{16}$ and $\frac{1}{24}$
LCD = 48

7. $\frac{5}{63}$ and $\frac{5}{21}$
LCD = 63

8. $\frac{7}{36}$ and $\frac{5}{54}$
LCD = 108

9. $\frac{1}{18}$ and $\frac{13}{30}$
LCD = 90

10. $\frac{3}{14}$ and $\frac{11}{16}$
LCD = 112

Combine. Be sure to simplify your answer whenever possible.

11. $\frac{3}{11} + \frac{2}{11} + \frac{4}{11}$
$\frac{9}{11}$

12. $\frac{9}{13} + \frac{2}{13} + \frac{1}{13}$
$\frac{12}{13}$

13. $\frac{8}{17} - \frac{5}{17}$
$\frac{3}{17}$

14. $\frac{16}{19} - \frac{13}{19}$
$\frac{3}{19}$

15. $\frac{3}{8} + \frac{7}{12}$
$\frac{9 + 14}{24} = \frac{23}{24}$

16. $\frac{1}{12} + \frac{2}{9}$
$\frac{3 + 8}{36} = \frac{11}{36}$

17. $\frac{5}{14} - \frac{1}{4}$
$\frac{10 - 7}{28} = \frac{3}{28}$

18. $\frac{9}{20} - \frac{1}{15}$
$\frac{27 - 4}{60} = \frac{23}{60}$

19. $\frac{2}{9} + \frac{5}{6}$
$\frac{4 + 15}{18} = \frac{19}{18}$ or $1\frac{1}{18}$

20. $\frac{3}{4} + \frac{3}{10}$
$\frac{15 + 6}{20} = \frac{21}{20}$

21. $\frac{4}{5} - \frac{3}{8}$
$\frac{32 - 15}{40} = \frac{17}{40}$

22. $\frac{3}{5} - \frac{1}{3}$
$\frac{9 - 5}{15} = \frac{4}{15}$

23. $\frac{9}{8} + \frac{7}{12}$
$\frac{27 + 14}{24} = \frac{41}{24}$ or $1\frac{17}{24}$

24. $\frac{5}{6} + \frac{7}{15}$
$\frac{25 + 14}{30} = \frac{39}{30} = \frac{13}{10}$ or $1\frac{3}{10}$

25. $\frac{7}{18} + \frac{1}{12}$
$\frac{17}{36}$

26. $\frac{4}{7} + \frac{7}{9}$
$\frac{85}{63}$ or $1\frac{22}{63}$

27. $\frac{2}{3} + \frac{7}{12}$
$\frac{5}{4}$ or $1\frac{1}{4}$

28. $\frac{3}{35} + \frac{4}{7}$
$\frac{23}{35}$

29. $\frac{1}{15} + \frac{7}{12}$
$\frac{13}{20}$

30. $\frac{3}{10} - \frac{3}{25}$
$\frac{15 - 6}{50} = \frac{9}{50}$

31. $\frac{7}{12} - \frac{5}{18}$
$\frac{21 - 10}{36} = \frac{11}{36}$

32. $3\frac{1}{5} + 2\frac{2}{3}$
$5\frac{13}{15}$

33. $5\frac{1}{6} + 3\frac{1}{4}$
$8\frac{5}{12}$

34. $1\frac{5}{24} + \frac{5}{18}$
$1\frac{35}{72}$

35. $6\frac{2}{3} + \frac{3}{4}$
$7\frac{5}{12}$

36. $7\frac{1}{6} - 2\frac{1}{4}$
$4\frac{11}{12}$

37. $7\frac{2}{5} - 3\frac{3}{4}$
$3\frac{13}{20}$

38. $9\frac{4}{5} - 3\frac{1}{2}$
$6\frac{3}{10}$

39. $4\frac{7}{12} - 1\frac{5}{6}$
$2\frac{3}{4}$

40. $2\frac{1}{8} + 3\frac{2}{3}$
$5\frac{19}{24}$

41. $3\frac{1}{7} + 4\frac{1}{3}$
$7\frac{10}{21}$

42. $12\frac{1}{8} - 9\frac{3}{8}$
$2\frac{3}{4}$

43. $14\frac{1}{9} - 12\frac{5}{9}$
$1\frac{5}{9}$

44. $2\frac{1}{8} + 6\frac{3}{4}$
$8\frac{7}{8}$

Applications

Solve each word problem.

45. Jenny and Laura roller-bladed $3\frac{1}{8}$ miles on Monday, $2\frac{2}{3}$ miles on Tuesday, and $4\frac{1}{2}$ miles on Wednesday. What was their total roller-blading distance for those three days?

$3\frac{1}{8} + 2\frac{2}{3} + 4\frac{1}{2} = 10\frac{7}{24}$ miles

46. Keanu has a fish tank holding $9\frac{7}{10}$ gallons of water. While having it cleaned, he drains $3\frac{4}{5}$ gallons. How many gallons of water does Keanu have remaining in the tank?

$9\frac{7}{10} - 3\frac{4}{5} = 5\frac{9}{10}$ gallons

47. Sheryl has $8\frac{1}{2}$ hours this weekend to work on her new video. She estimates that it will take $2\frac{2}{3}$ hours to lip sync the new song and $1\frac{3}{4}$ hours to learn the new dance steps. How much time will she have left over for her MTV interview?

$2\frac{2}{3} + 1\frac{3}{4} = 4\frac{5}{12}$

$8\frac{1}{2} - 4\frac{5}{12} = 4\frac{1}{12}$ hours

48. Gina's landlord is giving her a triangular-shaped space behind her apartment building for her new puppy to play in while she is at work. The sides of the yard measure $10\frac{1}{2}$ feet, $16\frac{1}{4}$ feet, and $24\frac{7}{8}$ feet. How much fencing will she need? How much more or less fencing does she need than Manuel on page 18?

$10\frac{1}{2} + 16\frac{1}{4} + 24\frac{7}{8} = 51\frac{5}{8}$ feet

$54\frac{3}{8} - 51\frac{5}{8} = 2\frac{3}{4}$ feet less fencing than Manuel

Carpenters use fractions in their work. The diagram at the right is a diagram of a spice cabinet. The symbol ″ means inches.

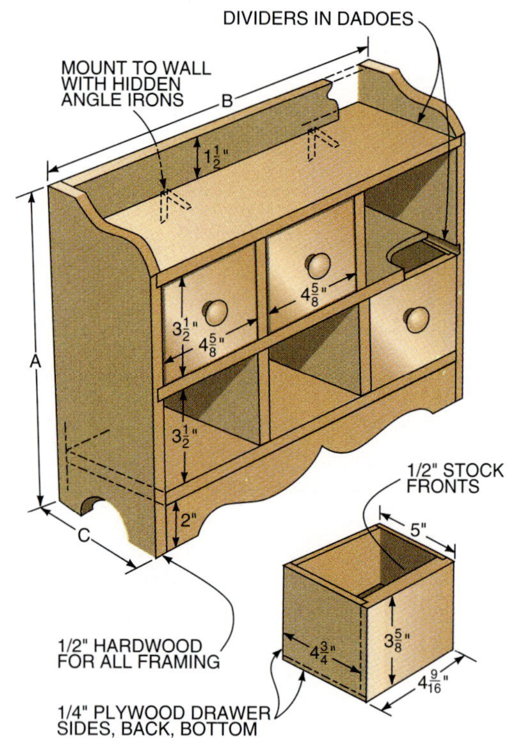

49. To determine where to place the cabinet, calculate the height, A, and the width B. Don't forget to include the $\frac{1}{2}$-inch thickness of the wood where needed.

$A = 12$ inches, $B = 15\frac{7}{8}$ inches

50. Look at the close-up of the drawer. The width is $4\frac{9}{16}''$. In the diagram, the same width is $4\frac{5}{8}''$. What is the difference? $\frac{1}{16}$ inch

Why do you think the drawer is smaller than the opening? so that it can slide in and out

51. The Falmouth Country Club maintains the putting greens with a grass height of $\frac{7}{8}$ inch. The grass on the fairways is maintained at $2\frac{1}{2}$ inches. How much lower must the mower blade be lowered by a person mowing the fairways if that person will be using the same mowing machine on the putting greens?

$1\frac{5}{8}$ inches

52. The Director of Facilities Maintenance at the club discovered that due to slippage in the adjustment lever that the lawn mower actually cuts the grass $\frac{1}{16}$ of an inch too long or too short on some days. What is the longest height that the fairway grass could be after being mowed with this machine? What is the shortest height of the putting greens after being mowed with this machine?

$2\frac{9}{16}$ inches, $\frac{13}{16}$ inch

Cumulative Review Problems

53. Simplify. $\frac{36}{44}$. $\frac{9}{11}$

54. Change to an improper fraction. $26\frac{3}{5}$. $\frac{133}{5}$

55. Change to a mixed number. $\frac{136}{7}$. $19\frac{3}{7}$

56. Find the missing numerator. $\frac{3}{14} = \frac{?}{56}$. 12

Putting Your Skills to Work

THE HIGH JUMP: RAISING THE BAR

In the 1896 Olympics, Ellery Clark of the United States set the world record for the men's high jump by clearing a bar at the height of 5 ft $11\frac{1}{4}$ in. In the 1912 Olympics, Alma Richards of the United States set a new record by clearing a bar at the height of 6 feet 4 in. He had jumped $4\frac{3}{4}$ inches higher than Ellery Clark. An interesting pattern can be observed by examining the record height achieved in the men's high jump in the Olympics over 20-year periods. The bar graph at the right shows the number of inches the jump exceeded the 1896 world record.

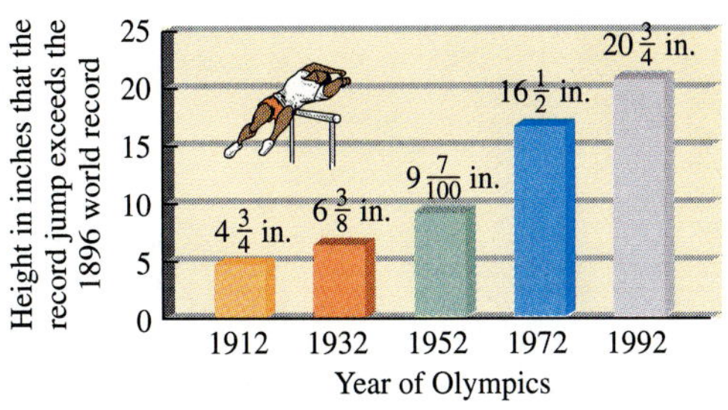

PROBLEMS FOR INDIVIDUAL INVESTIGATION AND ANALYSIS

1. How much higher was the highest jump in the 1972 Olympics than the highest jump in the 1932 Olympics? $10\frac{1}{8}$ in.
2. In what 20-year period did the men's high jump measurement experience the greatest increase? from 1952 to 1972

PROBLEMS FOR GROUP INVESTIGATION AND COOPERATIVE LEARNING

Together with some members of your class see if you can answer the following.

3. During the period from 1912 to 1992 the height jumped increased from $4\frac{3}{4}$ in. above the 1896 record to $20\frac{3}{4}$ in. above the 1896 record. What is the total increase from 1912 to 1992? What is the average increase expected for a 20-year period? What would you expect the average increase to be in a 4-year period? Total increase of 16 in. Average increase of 4 in. per 20 years. Average increase of $\frac{4}{5}$ in. per 4 years.
4. Using your results from (3), what would you predict as the highest jump in the 1996 Olympics? The record height in the 1996 Olympics in Atlanta was a jump of 7 ft 10 in. by Charles Austin of the United States. Was your prediction too high or too low? How much was it in error? At an expected rate of $\frac{4}{5}$ inch in the 4 years since 1992, we would expect a height of 7 ft $8\frac{4}{5}$ in. Our prediction was too low by $1\frac{1}{5}$ in.

INTERNET CONNECTIONS: Go to ''http://www.prenhall.com/tobey'' to be connected

Site: Olympic FAQ—Olympic Winter Games or a related site

This site gives basic information about the Winter Olympic Games each year.

5. Make a table showing the number of male athletes, the number of female athletes, and the total number of athletes for as many years as possible. (Some of these numbers may not be available for all of the years listed.)
6. In 1984, how many more men participated than women?
7. In what 12-year period did the number of female athletes increase the most?
8. Find the total increase in the number of athletes from 1952 to 1984. What is the average increase expected for an eight-year period? Using this result, what would you predict as the number of athletes in 1992? How does your prediction compare with the actual number?

0.3 Multiplication and Division of Fractions

MathPro

Video 1

SSM

After studying this section, you will be able to:

1 *Multiply fractions, whole numbers, and mixed numbers.*
2 *Divide fractions, whole numbers, and mixed numbers.*

1 *Multiplying Fractions, Whole Numbers, and Mixed Numbers*

Multiplication of Fractions

The multiplication rule for fractions states that to multiply two fractions we multiply the two numerators and multiply the two denominators.

> **To Multiply Any Two Fractions**
>
> **1.** Multiply the numerators.
>
> **2.** Multiply the denominators.

Consider these examples.

EXAMPLE 1 Multiply.

(a) $\frac{3}{5} \times \frac{2}{7}$ **(b)** $\frac{1}{3} \times \frac{5}{4}$ **(c)** $\frac{7}{3} \times \frac{1}{5}$ **(d)** $\frac{6}{5} \times \frac{2}{3}$

(a) $\frac{3}{5} \times \frac{2}{7} = \frac{3 \cdot 2}{5 \cdot 7} = \frac{6}{35}$

(b) $\frac{1}{3} \times \frac{5}{4} = \frac{1 \cdot 5}{3 \cdot 4} = \frac{5}{12}$

(c) $\frac{7}{3} \times \frac{1}{5} = \frac{7 \cdot 1}{3 \cdot 5} = \frac{7}{15}$

(d) $\frac{6}{5} \times \frac{2}{3} = \frac{6 \cdot 2}{5 \cdot 3} = \frac{12}{15} = \frac{4}{5}$

Note that we must simplify.

Practice Problem 1 Multiply. **(a)** $\frac{2}{7} \times \frac{5}{11}$ **(b)** $\frac{8}{9} \times \frac{3}{10}$ ■

It is possible to avoid having to simplify a fraction at the last step. In many cases we can divide by a value that appears as a factor in both a numerator and a denominator. Often it is helpful to write a number (as a product of prime factors) in order to do this.

EXAMPLE 2 Multiply.

(a) $\frac{3}{5} \times \frac{5}{7}$ **(b)** $\frac{4}{11} \times \frac{5}{2}$ **(c)** $\frac{15}{8} \times \frac{10}{27}$ **(d)** $\frac{8}{7} \times \frac{5}{12}$

(a) $\frac{3}{5} \times \frac{5}{7} = \frac{3 \cdot 5}{5 \cdot 7} = \frac{3 \cdot \cancel{5}}{7 \cdot \cancel{5}} = \frac{3}{7}$ *Note that here we divided numerator and denominator by 5.*

If we factor each number, we can see the common factors.

(b) $\frac{4}{11} \times \frac{5}{2} = \frac{2 \cdot \cancel{2}}{11} \times \frac{5}{\cancel{2}} = \frac{10}{11}$

(c) $\frac{15}{8} \times \frac{10}{27} = \frac{\cancel{3} \cdot 5}{2 \cdot 2 \cdot \cancel{2}} \times \frac{5 \cdot \cancel{2}}{\cancel{3} \cdot 3 \cdot 3} = \frac{25}{36}$

(d) $\frac{8}{7} \times \frac{5}{12} = \frac{2 \cdot \cancel{2} \cdot \cancel{2}}{7} \times \frac{5}{\cancel{2} \cdot \cancel{2} \cdot 3} = \frac{10}{21}$

By dividing out common factors, the multiplication involves smaller numbers and the answers are in simplified form.

Practice Problem 2 Multiply. **(a)** $\frac{3}{5} \times \frac{4}{3}$ **(b)** $\frac{9}{10} \times \frac{5}{12}$ ■

SIDELIGHT Why does this method of dividing out a value that appears as a factor in both numerator and denominator work? Let's reexamine one of the examples we have solved previously.

$$\frac{3}{5} \times \frac{5}{7} = \frac{3}{\cancel{5}} \times \frac{\cancel{5}}{7} = \frac{3}{7}$$

Consider the following steps and reasons.

$\frac{3}{5} \times \frac{5}{7} = \frac{3 \cdot 5}{5 \cdot 7}$ *Definition of multiplication of fractions.*

$= \frac{5 \cdot 3}{5 \cdot 7}$ *Change the order of the factors in the numerator, since* $3 \cdot 5 = 5 \cdot 3$. *This is called the commutative property of multiplication.*

$= \frac{5}{5} \cdot \frac{3}{7}$ *Definition of multiplication of fractions.*

$= 1 \cdot \frac{3}{7}$ *Write* 1 *in place of* $\frac{5}{5}$ *since* 1 *is another name for* $\frac{5}{5}$.

$= \frac{3}{7}$ $1 \cdot \frac{3}{7} = \frac{3}{7}$ *since any number can be multiplied by* 1 *without changing the value of the number.*

Think about this concept. It is an important one that we will use again when we discuss rational expressions.

Multiplication of a Fraction by a Whole Number

Whole numbers can be named using fractional notation. 3, $\frac{9}{3}$, $\frac{6}{2}$, and $\frac{3}{1}$ are ways of expressing the number *three*. Therefore,

$$3 = \frac{9}{3} = \frac{6}{2} = \frac{3}{1}$$

When we multiply a fraction by a whole number, we merely express the whole number as a fraction whose denominator is 1 and follow the multiplication rule for fractions.

EXAMPLE 3 Multiply. **(a)** $7 \times \frac{3}{5}$ **(b)** $\frac{3}{16} \times 4$

(a) $7 \times \frac{3}{5} = \frac{7}{1} \times \frac{3}{5} = \frac{21}{5} = 4\frac{1}{5}$ **(b)** $\frac{3}{16} \times 4 = \frac{3}{16} \times \frac{4}{1} = \frac{3}{4 \cdot \cancel{4}} \times \frac{\cancel{4}}{1} = \frac{3}{4}$

Notice that in (b) we did not use *prime* factors to factor 16. We recognized that $16 = 4 \cdot 4$. This is a more convenient factorization of 16 for this problem. Choose the factorization that works best for each problem. If you cannot decide what is best, factor into primes.

Practice Problem 3 Multiply. **(a)** $4 \times \frac{2}{7}$ **(b)** $12 \times \frac{3}{4}$ ■

TEACHING TIP This skill becomes very important in Chapter 1 where students have to multiply expressions such as

$$\frac{1}{3}(6x^2 - 2x + 9)$$

and

$$4\left(\frac{1}{2}x^2 - 12x - \frac{3}{4}\right)$$

when they use the distributive property. After presenting Example 3, give students the following Additional Class Exercises.

A. $9 \times \frac{4}{5}$ **B.** $\frac{2}{7} \times 56$

C. $\frac{3}{8} \times 13$ **D.** $16 \times \frac{3}{8}$

Ans. A. $7\frac{1}{5}$ B. 16 C. $4\frac{7}{8}$ D. 6

Multiplication of Mixed Numbers

When multiplying mixed numbers, we first change them to improper fractions and then follow the multiplication rule for fractions.

EXAMPLE 4 Multiply. **(a)** $3\frac{1}{3} \times 2\frac{1}{2}$ **(b)** $1\frac{2}{5} \times 2\frac{1}{3}$

(a) $3\frac{1}{3} \times 2\frac{1}{2} = \frac{10}{3} \times \frac{5}{2} = \frac{\cancel{2} \cdot 5}{3} \times \frac{5}{\cancel{2}} = \frac{25}{3} = 8\frac{1}{3}$

(b) $1\frac{2}{5} \times 2\frac{1}{3} = \frac{7}{5} \times \frac{7}{3} = \frac{49}{15} = 3\frac{4}{15}$

Practice Problem 4 Multiply.

(a) $2\frac{1}{5} \times \frac{3}{7}$ **(b)** $3\frac{1}{3} \times 1\frac{2}{5}$ ■

EXAMPLE 5 Multiply. $2\frac{2}{3} \times \frac{1}{4} \times 6$

$$2\frac{2}{3} \times \frac{1}{4} \times 6 = \frac{8}{3} \times \frac{1}{4} \times \frac{6}{1} = \frac{\cancel{4} \cdot 2}{\cancel{3}} \times \frac{1}{\cancel{4}} \times \frac{2 \cdot \cancel{3}}{1} = \frac{4}{1} = 4$$

Practice Problem 5 Multiply. $3\frac{1}{2} \times \frac{1}{14} \times 4$ ■

2 *Dividing Fractions, Whole Numbers, and Mixed Numbers*

Division of Fractions

To divide two fractions, we invert the second fraction and then multiply the two fractions.

> **To Divide Two Fractions**
>
> **1.** Invert the second fraction.
>
> **2.** Now multiply the two fractions.

EXAMPLE 6 Divide. **(a)** $\frac{1}{3} \div \frac{1}{2}$ **(b)** $\frac{2}{5} \div \frac{3}{10}$ **(c)** $\frac{2}{3} \div \frac{7}{5}$

(a) $\frac{1}{3} \div \frac{1}{2} = \frac{1}{3} \times \frac{2}{1} = \frac{2}{3}$

(b) $\frac{2}{5} \div \frac{3}{10} = \frac{2}{5} \times \frac{10}{3} = \frac{2}{\cancel{5}} \times \frac{\cancel{5} \cdot 2}{3} = \frac{4}{3} = 1\frac{1}{3}$

(c) $\frac{2}{3} \div \frac{7}{5} = \frac{2}{3} \times \frac{5}{7} = \frac{10}{21}$

TEACHING TIP You may want to introduce the concept of a reciprocal. Two numbers whose product is 1 are called *reciprocals.* The reciprocal of $\frac{2}{5}$ is $\frac{5}{2}$ because $\frac{2}{5} \times \frac{5}{2} = 1$. The reciprocal of 5 is $\frac{1}{5}$ because $5 \times \frac{1}{5} = \frac{5}{5} = 1$. Have students look at the problems in Example 6. Point out the reciprocals in each problem and ask students to formulate a rule for division of fractions that uses the word reciprocals.

Practice Problem 6 Divide.

(a) $\frac{2}{5} \div \frac{1}{3}$ **(b)** $\frac{12}{13} \div \frac{4}{3}$ ■

Division of a Fraction and a Whole Number

The process of inverting the second fraction and then multiplying the two fractions should be done very carefully when one of the original values is a whole number. Remember, a whole number such as 2 is equivalent to $\frac{2}{1}$.

EXAMPLE 7 Divide. **(a)** $\dfrac{1}{3} \div 2$ **(b)** $5 \div \dfrac{1}{3}$

(a) $\dfrac{1}{3} \div 2 = \dfrac{1}{3} \div \dfrac{2}{1} = \dfrac{1}{3} \times \dfrac{1}{2} = \dfrac{1}{6}$

(b) $5 \div \dfrac{1}{3} = \dfrac{5}{1} \div \dfrac{1}{3} = \dfrac{5}{1} \times \dfrac{3}{1} = \dfrac{15}{1} = 15$

Practice Problem 7 Divide.

(a) $\dfrac{3}{7} \div 6$ **(b)** $8 \div \dfrac{2}{3}$ ■

SIDELIGHT ***Number Sense*** Look at the answers to the problems in Example 7. In part (a), you will notice that $\frac{1}{6}$ is less than the original number $\frac{1}{3}$. Does this seem reasonable? Let's see. If $\frac{1}{3}$ is divided by 2, it means that $\frac{1}{3}$ will be divided into 2 equal parts. We would expect that each part would be less than $\frac{1}{3}$. $\frac{1}{6}$ is a reasonable answer to this division problem.

In part (b), 15 is greater than the original number 5. Does this seem reasonable? Think of what $5 \div \frac{1}{3}$ means. It means that 5 will be divided into thirds. Let's think of an easier problem. What happens when we divide 1 into thirds? We get *three* thirds. We would expect, therefore, that when we divide 5 by thirds, we would get 5×3 or 15 thirds. 15 is a reasonable answer to this division problem.

Complex Fraction

Sometimes division is written in the form of a complex fraction with one fraction in the numerator and one fraction in the denominator. It is best to write this in standard division notation first; then complete the problem using the rule for division.

EXAMPLE 8 Divide. **(a)** $\dfrac{\frac{3}{7}}{\frac{3}{5}}$ **(b)** $\dfrac{\frac{2}{9}}{\frac{5}{7}}$

(a) $\dfrac{\frac{3}{7}}{\frac{3}{5}} = \dfrac{3}{7} \div \dfrac{3}{5} = \dfrac{\cancel{3}}{7} \times \dfrac{5}{\cancel{3}} = \dfrac{5}{7}$

(b) $\dfrac{\frac{2}{9}}{\frac{5}{7}} = \dfrac{2}{9} \div \dfrac{5}{7} = \dfrac{2}{9} \times \dfrac{7}{5} = \dfrac{14}{45}$

TEACHING TIP After presenting Example 8, ask students to do the following as an Additional Class Exercise.

A. $\dfrac{\frac{2}{3}}{\frac{2}{9}}$ **B.** $\dfrac{\frac{5}{16}}{\frac{15}{8}}$

Ans. A. 3 B. $\dfrac{1}{6}$

Practice Problem 8 Divide.

(a) $\dfrac{\frac{3}{11}}{\frac{5}{7}}$ **(b)** $\dfrac{\frac{12}{5}}{\frac{8}{15}}$ ■

SIDELIGHT Why does the method of "invert and multiply" work? The division rule really depends on the property that any number can be multiplied by 1 without changing the value of the number. Let's look carefully at an example of division of fractions:

$$\frac{2}{5} \div \frac{3}{7} = \frac{\frac{2}{5}}{\frac{3}{7}}$$ *We can write the problem using a complex fraction.*

$$= \frac{\frac{2}{5}}{\frac{3}{7}} \times 1$$ *We can multiply by 1, since any number can be multiplied by 1 without changing the value of the number.*

$$= \frac{\frac{2}{5}}{\frac{3}{7}} \times \frac{\frac{7}{3}}{\frac{7}{3}}$$ *We write 1 in the form $\frac{\frac{7}{3}}{\frac{7}{3}}$, since any nonzero number divided by itself equals 1. We choose this value as a multiplier because it will help simplify the denominator.*

$$= \frac{\frac{2}{5} \times \frac{7}{3}}{\frac{3}{7} \times \frac{7}{3}}$$ *Definition of multiplication of fractions.*

$$= \frac{\frac{2}{5} \times \frac{7}{3}}{1}$$ *The product in the denominator equals 1.*

$$= \frac{2}{5} \times \frac{7}{3}$$

Thus we have shown that $\frac{2}{5} \div \frac{3}{7}$ is equivalent to $\frac{2}{5} \times \frac{7}{3}$ and have shown some justification for the "invert and multiply rule."

Division of Mixed Numbers

This method for division of fractions can be used with mixed numbers. However, we first must change the mixed numbers to improper fractions and then use the rule for dividing fractions.

EXAMPLE 9 Divide. **(a)** $2\frac{1}{3} \div 3\frac{2}{3}$ **(b)** $4\frac{1}{2} \div 1\frac{5}{7}$ **(c)** $\dfrac{2}{3\frac{1}{2}}$

(a) $2\frac{1}{3} \div 3\frac{2}{3} = \frac{7}{3} \div \frac{11}{3} = \frac{7}{\cancel{3}} \times \frac{\cancel{3}}{11} = \frac{7}{11}$

(b) $4\frac{1}{2} \div 1\frac{5}{7} = \frac{9}{2} \div \frac{12}{7} = \frac{9}{2} \times \frac{7}{12} = \frac{3 \cdot \cancel{3}}{2} \times \frac{7}{4 \cdot \cancel{3}} = \frac{21}{8} = 2\frac{5}{8}$

(c) $\dfrac{2}{3\frac{1}{2}} = 2 \div 3\frac{1}{2} = \frac{2}{1} \div \frac{7}{2} = \frac{2}{1} \times \frac{2}{7} = \frac{4}{7}$

Practice Problem 9 Divide.

(a) $1\frac{2}{5} \div 2\frac{1}{3}$ **(b)** $4\frac{2}{3} \div 7$ **(c)** $\dfrac{1\frac{1}{5}}{1\frac{2}{7}}$ ■

EXAMPLE 10 A chemist has 96 fluid ounces of a solution. She pours the solution into test tubes. Each test tube holds $\frac{3}{4}$ fluid ounce. How many test tubes can she fill?

We need to divide the total number of ounces, 96, by the number of ounces in each test tube, $\frac{3}{4}$.

$$96 \div \frac{3}{4} = \frac{96}{1} \div \frac{3}{4} = \frac{96}{1} \times \frac{4}{3} = \frac{\cancel{3} \cdot 32}{1} \times \frac{4}{\cancel{3}} = \frac{128}{1} = 128$$

She will be able to fill 128 test tubes.

Pause for a moment to think about the answer. Does 128 test tubes filled with solution seem like a reasonable answer? Did you perform the correct operation?

TEACHING TIP After explaining Example 10, you might want to show an applied example of division of mixed fractions. A land developer purchased $29\frac{3}{4}$ acres of land in Northern Maine. He wants to divide the land into several equally sized lots that each contains $1\frac{3}{4}$ acres of land. How many lots will he have? The number of lots is obtained by:

$$\begin{pmatrix}\text{Total acres}\\ \text{of land}\end{pmatrix} \div \begin{pmatrix}\text{No. of acres}\\ \text{in each lot}\end{pmatrix}$$

$$29\frac{3}{4} \div 1\frac{3}{4} = \begin{matrix}\text{number}\\ \text{of lots}\end{matrix}$$

$$29\frac{3}{4} \div 1\frac{3}{4} = \frac{119}{4} \div \frac{7}{4}$$

$$= \frac{119}{4} \times \frac{4}{7}$$

$$= \frac{119}{7}$$

$$= 17 \text{ lots}$$

Practice Problem 10

A chemist has 64 fluid ounces of a solution. He wishes to fill several jars each holding $5\frac{1}{3}$ fluid ounces. How many jars can he fill? ■

Sometimes when solving word problems involving fractions, or mixed numbers, it is helpful to solve the problem using simpler numbers. Once you understand what operation is involved, you can go back and solve using the original numbers in the word problem.

EXAMPLE 11 A car traveled 301 miles on $10\frac{3}{4}$ gallons of gas. How many miles per gallon did it get?

Use simpler numbers: 300 miles on 10 gallons of gas. We want to find out how many miles the car traveled on 1 gallon of gas. You may want to draw a picture.

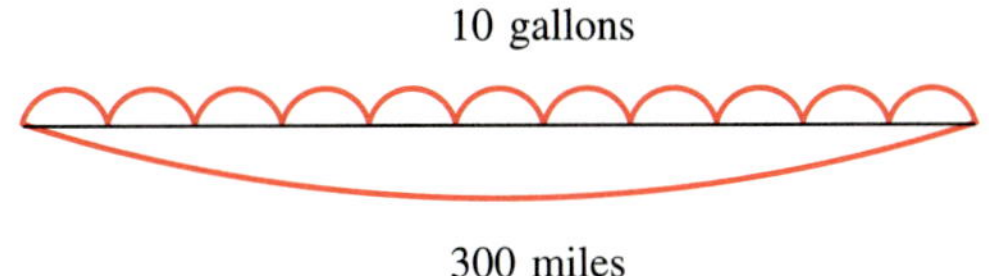

Divide. $300 \div 10 = 30$.

Now use the original numbers given.

$$301 \div 10\frac{3}{4} = \frac{301}{1} \div \frac{43}{4} = \frac{301}{1} \times \frac{4}{43} = \frac{1204}{43} = 28$$

The car got 28 miles to the gallon.

Practice Problem 11

A car can travel $25\frac{1}{2}$ miles on 1 gallon of gas. How many miles can a car travel on $5\frac{1}{4}$ gallons of gas? Check your answer to see if it is reasonable. ■

0.3 Exercises

Multiply. Simplify your answer whenever possible.

1. $\frac{3}{5} \times \frac{2}{11}$ $\frac{6}{55}$

2. $\frac{5}{7} \times \frac{3}{4}$ $\frac{15}{28}$

3. $\frac{21}{5} \times \frac{10}{7}$ 6

4. $\frac{7}{9} \times \frac{18}{5}$ $\frac{14}{5}$ or $2\frac{4}{5}$

5. $8 \times \frac{3}{7}$ $\frac{24}{7}$ or $3\frac{3}{7}$

6. $\frac{7}{13} \times 4$ $\frac{28}{13}$ or $2\frac{2}{13}$

7. $1\frac{1}{3} \times 2\frac{1}{2}$ $\frac{10}{3}$ or $3\frac{1}{3}$

8. $2\frac{3}{4} \times 1\frac{1}{2}$ $\frac{33}{8}$ or $4\frac{1}{8}$

9. $3\frac{1}{3} \times \frac{3}{4} \times 2$ 5

10. $4\frac{1}{5} \times \frac{2}{3} \times 5$ 14

Divide. Simplify your answer whenever possible.

11. $\frac{5}{3} \div \frac{5}{2}$ $\frac{2}{3}$

12. $\frac{7}{8} \div \frac{7}{5}$ $\frac{5}{8}$

13. $\frac{3}{7} \div 3$ $\frac{1}{7}$

14. $\frac{7}{8} \div 4$ $\frac{7}{32}$

15. $10 \div \frac{5}{7}$ 14

16. $18 \div \frac{2}{9}$ 81

17. $\dfrac{\frac{5}{2}}{\frac{2}{3}}$ $\frac{15}{4}$ or $3\frac{3}{4}$

18. $\dfrac{\frac{3}{4}}{\frac{3}{8}}$ 2

19. $1\frac{3}{7} \div 6\frac{1}{4}$ $\frac{8}{35}$

20. $4\frac{1}{2} \div 3\frac{3}{8}$ $\frac{4}{3}$ or $1\frac{1}{3}$

21. $\dfrac{\frac{7}{9}}{1\frac{1}{3}}$ $\frac{7}{12}$

22. $\dfrac{\frac{5}{8}}{1\frac{3}{4}}$ $\frac{5}{14}$

Mixed Practice

Perform the proper calculation. Reduce your answer whenever possible.

23. $\frac{6}{5} \times \frac{10}{12}$ 1

24. $\frac{5}{24} \times \frac{18}{15}$ $\frac{1}{4}$

25. $\frac{5}{16} \div \frac{1}{8}$ $\frac{5}{2}$ or $2\frac{1}{2}$

26. $\frac{2}{11} \div 4$ $\frac{1}{22}$

27. $10\frac{3}{7} \times 5\frac{1}{4}$ $\frac{219}{4}$ or $54\frac{3}{4}$

28. $10\frac{2}{9} \div 2\frac{1}{3}$ $\frac{92}{21}$ or $4\frac{8}{21}$

29. $2\frac{1}{8} \div \frac{1}{4}$ $\frac{17}{2}$ or $8\frac{1}{2}$

30. $4 \div 1\frac{7}{9}$ $\frac{9}{4}$ or $2\frac{1}{4}$

31. $8 \times 3\frac{1}{2}$ 28

32. $\frac{3}{4} \div 6$ $\frac{1}{8}$

33. $2\frac{1}{2} \times \frac{1}{10} \times \frac{3}{4}$ $\frac{3}{16}$

34. $3\frac{1}{3} \times \frac{1}{5} \times \frac{2}{3}$ $\frac{4}{9}$

Applications

Solve each word problem.

35. A denim shirt at the Gap requires $2\frac{3}{4}$ yards of material. How many yards would be needed to make 26 shirts?
$71\frac{1}{2}$ yards

36. Jennifer rode her mountain bike for $4\frac{1}{5}$ miles after work. Two-thirds of the distance was over a mountain bike trail. How long is the mountain bike trail?
$2\frac{4}{5}$ miles

Putting Your Skills to Work

THE STOCK MARKET

The price of stocks bought and sold on the stock exchange is calculated in eighths of a dollar.

NEW YORK STOCK EXCHANGE SELECTED VALUES OCTOBER 11, 1996							
52 WEEK High	52 WEEK Low	Stock	Sales in 100s	Daily High	Daily Low	Last	Change Since Yesterday
$59\frac{1}{4}$	42	Nynex	15293	$43\frac{5}{8}$	$43\frac{1}{8}$	$43\frac{1}{4}$	$+\frac{1}{4}$
$56\frac{1}{8}$	$41\frac{1}{2}$	Raytheon	18283	$47\frac{5}{8}$	$46\frac{1}{4}$	$47\frac{3}{8}$	$+1\frac{1}{4}$
$36\frac{7}{8}$	$24\frac{1}{8}$	Reebok	1123	$33\frac{5}{8}$	$32\frac{7}{8}$	$33\frac{3}{8}$	$+\frac{5}{8}$
$63\frac{1}{4}$	44	Rockwell	2728	$55\frac{3}{4}$	$55\frac{1}{4}$	$55\frac{3}{8}$	$-\frac{3}{8}$
$53\frac{7}{8}$	$32\frac{1}{2}$	Sears	9539	$48\frac{5}{8}$	48	$48\frac{1}{2}$	$+\frac{1}{4}$

Use the abbreviated chart above to solve each of the following questions. Commissions are not taken into account in the following problems.

PROBLEMS FOR INDIVIDUAL INVESTIGATIONS AND STUDY

1. Maria purchased 100 shares of Reebok. How much did she pay for the shares if she purchased them at the low for the day? $\$3287\frac{1}{2}$

2. Joseph sold 50 shares of Sears at the high for the day. How much did he receive for the sale? $\$2431\frac{1}{4}$

PROBLEMS FOR COOPERATIVE GROUP ACTIVITY

Together with other members of your class see if you can complete the following. Remember that only a whole number of shares of stock can be purchased.

3. You have \$10,000 to invest. If you had purchased one of these stocks at the lowest price of the 52 weeks and sold it at the highest price of the 52 weeks, how much profit would you have made? Which stock would have been the best investment under this situation? You could have made a profit of approximately \$6562.13. You should have bought 307 shares of Sears.

INTERNET CONNECTIONS: Go to ``http://www.prenhall.com/tobey'' to be connected

Site: Infoseek Personal Stock Quotes or a related site

Complete this exercise after the stock market closes (4 P.M. Eastern time) or on a weekend so that the price quote you receive will represent a full day. Choose any public company and use Infoseek to find your company's stock symbol. (Not all companies have public stock. If you cannot find the symbol, choose another company.) Write down your company's symbol, high price, low price, last price, and change since yesterday. Then answer the following questions.

4. What is the difference between the high for the day and the low for the day?

5. Use the last price and change information to find yesterday's closing price.

6. If you had bought 300 shares at the daily high price, how much would you have paid?

7. If you had sold 500 shares at the daily low price, how much would you have paid?

0.4 Use of Decimals

MathPro

Video 2

SSM

After studying this section, you will be able to:

1 *Change from a fraction to a decimal.*
2 *Change from a decimal to a fraction.*
3 *Add and subtract decimals.*
4 *Multiply decimals.*
5 *Divide decimals.*
6 *Multiply or divide a number by a multiple of 10.*

The Basic Concept of Decimals

We can express a part of a whole as a fraction or as a decimal. A decimal is another way of writing a fraction whose denominator is 10, 100, 1000, and so on.

$$\frac{3}{10} = 0.3 \qquad \frac{5}{100} = 0.05 \qquad \frac{172}{1000} = 0.172 \qquad \frac{58}{10{,}000} = 0.0058$$

TEACHING TIP Remind students that what we call DECIMALS are really decimal fractions. It is a special kind of fraction with a denominator of 10, 100, 1000, 10000, etc. Decimal fractions are usually referred to simply as decimals.

The period in decimal notation is known as the **decimal point.** The number of digits in a number to the right of the decimal point is known as the number of **decimal places** of the number. The place value of decimals are shown below.

Hundred-thousands	Ten-thousands	Thousands	Hundreds	Tens	Ones	← Decimal point	Tenths	Hundredths	Thousandths	Ten-thousandths	Hundred-thousandths
100,000	10,000	1000	100	10	1	.	$\frac{1}{10}$	$\frac{1}{100}$	$\frac{1}{1000}$	$\frac{1}{10{,}000}$	$\frac{1}{100{,}000}$

EXAMPLE 1 Write the following decimals as a fraction. Give the number of decimal places. Write out in words the way the number would be spoken.

(a) 0.6 **(b)** 0.29 **(c)** 0.527
(d) 1.38 **(e)** 0.00007

TEACHING TIP The skill of being able to read aloud or write in words the decimal form of a number is quite important. It helps the student understand clearly how many places to round to when rounding to the nearest thousandth, etc. After discussing Example 1, ask students to do this Additional Class Exercise. Write down the words to describe:

A. 0.008 **B.** 1328.4
C. 0.0345

Ans.
A. Eight thousandths
B. One thousand three hundred twenty-eight and four-tenths
C. Three hundred forty-five ten-thousandths

	Decimal Form	Fraction Form	Number of Decimal Places	The Words Used to Describe the Number
(a)	0.6	$\frac{6}{10}$	one	six-tenths
(b)	0.29	$\frac{29}{100}$	two	twenty-nine hundredths
(c)	0.527	$\frac{527}{1000}$	three	five hundred twenty-seven thousandths
(d)	1.38	$1\frac{38}{100}$	two	one and thirty-eight hundredths
(e)	0.00007	$\frac{7}{100{,}000}$	five	seven hundred-thousandths

Practice Problem 1 Write each decimal as a fraction and in words.

(a) 1.371 **(b)** 0.09 ■

You have seen that a given fraction can be written in several different but equivalent ways. There are also several different equivalent ways of writing the decimal form of fractions. The decimal 0.18 can be written in the following equivalent ways:

$$\text{Fractional form:}\quad \frac{18}{100} = \frac{180}{1000} = \frac{1800}{10{,}000} = \frac{18{,}000}{100{,}000}$$

$$\text{Decimal form:}\quad 0.18 = 0.180 = 0.1800 = 0.18000$$

Thus we see that any number of *terminal zeros may be added onto the right-hand side* of a decimal without changing its value.

$$0.13 = 0.1300 \qquad 0.162 = 0.162000$$

Similarly, any number of *terminal zeros may be removed* from the right-hand side of a decimal without changing its value.

1 *Changing a Fraction to a Decimal*

A fraction can be changed to a decimal by dividing the denominator into the numerator.

EXAMPLE 2 Write each of the following fractions as a decimal.

(a) $\frac{3}{4}$ **(b)** $\frac{21}{20}$ **(c)** $\frac{1}{8}$ **(d)** $\frac{3}{200}$

(a) $\frac{3}{4} = 0.75$ since $4\overline{)3.00}$ gives 0.75: 28, 20, 20

(b) $\frac{21}{20} = 1.05$ since $20\overline{)21.00}$ gives 1.05: 20, 100, 100

(c) $\frac{1}{8} = 0.125$ since $8\overline{)1.000}$ gives 0.125: 8, 20, 16, 40, 40

(d) $\frac{3}{200} = 0.015$ since $200\overline{)3.000}$ gives 0.015: 200, 1000, 1000

CALCULATOR

Fraction to Decimal

You can use a calculator to change $\frac{3}{5}$ to a decimal.

Enter:

3 [÷] 5 [=]

The display should read:

[0.6]

Try **(a)** $\frac{17}{25}$ **(b)** $\frac{2}{9}$ **(c)** $\frac{13}{10}$ **(d)** $\frac{15}{19}$

Note: 0.7894737 is an approximation for $\frac{15}{19}$.

Practice Problem 2 Write as decimals.

(a) $\frac{3}{8}$ **(b)** $\frac{7}{200}$ **(c)** $\frac{33}{20}$ ■

Sometimes division yields an infinite repeating decimal. We use three dots to indicate that the pattern continues forever. For example:

$$\frac{1}{3} = 0.3333\ldots \qquad 3\overline{)1.000}$$

gives 0.333: 9, 10, 9, 10, 9, 1

An alternative notation is to place a bar over the repeating digits:

$$0.3333\ldots = 0.\overline{3} \qquad 0.575757\ldots = 0.\overline{57}$$

EXAMPLE 3 Write each fraction as a decimal. **(a)** $\frac{2}{11}$ **(b)** $\frac{5}{6}$

(a) $\frac{2}{11} = 0.181818\ldots$ or $0.\overline{18}$ **(b)** $\frac{5}{6} = 0.8333\ldots$ or $0.8\overline{3}$

$$\begin{array}{r} 0.1818 \\ 11\overline{)2.0000} \\ \underline{11} \\ 90 \\ \underline{88} \\ 20 \\ \underline{11} \\ 90 \\ \underline{88} \\ 2 \end{array} \qquad \begin{array}{r} 0.8333 \\ 6\overline{)5.0000} \\ \underline{48} \\ 20 \\ \underline{18} \\ 20 \\ \underline{18} \\ 20 \\ \underline{18} \\ 2 \end{array}$$

Note that the 8 does not repeat. Only the digit 3 is repeating.

Practice Problem 3 Write each fraction as a decimal.

(a) $\frac{1}{6}$ **(b)** $\frac{5}{11}$ ■

Sometimes division must be carried out to many places in order to observe the repeating pattern. This is true in the following example:

$\frac{2}{7} = 0.285714285714285714\ldots$ This can also be written as $\frac{2}{7} = 0.\overline{285714}$.

It can be shown that the denominator determines the maximum number of decimal places that might repeat. So $\frac{2}{7}$ must repeat in the seventh decimal place or sooner.

2 *Changing a Decimal to a Simple Fraction*

To convert from a decimal to a fraction, merely write the decimal as a fraction with a denominator of 10, 100, 1000, 10,000, and so on, and simplify the result when possible.

EXAMPLE 4 Write each decimal as a fraction.

(a) 0.2 **(b)** 0.35 **(c)** 0.516 **(d)** 0.74 **(e)** 0.138 **(f)** 0.008

(a) $0.2 = \frac{2}{10} = \frac{1}{5}$ **(b)** $0.35 = \frac{35}{100} = \frac{7}{20}$

(c) $0.516 = \frac{516}{1000} = \frac{129}{250}$ **(d)** $0.74 = \frac{74}{100} = \frac{37}{50}$

(e) $0.138 = \frac{138}{1000} = \frac{69}{500}$ **(f)** $0.008 = \frac{8}{1000} = \frac{1}{125}$

Practice Problem 4

Write each decimal as a fraction and simplify whenever possible.

(a) 0.8 **(b)** 0.88 **(c)** 0.45 **(d)** 0.148 **(e)** 0.612 **(f)** 0.016 ■

All repeating decimals can also be converted to fractional form. In practice, however, repeating decimals are usually rounded to a few places. It will not be necessary, therefore, to learn how to convert $0.\overline{033}$ to $\frac{11}{333}$ for this course.

3 *Adding and Subtracting Decimals*

Adding or subtracting decimals is similar to adding and subtracting whole numbers except that it is necessary to line up decimal points. To perform the operation 19.8 + 24.7 we line up the numbers in column form and add the digits

$$\begin{array}{r} 19.8 \\ +\ 24.7 \\ \hline 44.5 \end{array}$$

Addition and Subtraction of Decimals

1. Write in column form and line up decimal points.
2. Add or subtract the digits.

EXAMPLE 5 Perform the following operations.

(a) 3.6 + 2.3 **(b)** 127.32 − 38.48
(c) 3.1 + 42.36 + 9.034 **(d)** 5.0006 − 3.1248

(a) $\begin{array}{r} 3.6 \\ +\ 2.3 \\ \hline 5.9 \end{array}$ **(b)** $\begin{array}{r} 127.32 \\ -\ \ 38.48 \\ \hline 88.84 \end{array}$ **(c)** $\begin{array}{r} 3.1 \\ 42.36 \\ +\ \ 9.034 \\ \hline 54.494 \end{array}$ **(d)** $\begin{array}{r} 5.0006 \\ -\ 3.1248 \\ \hline 1.8758 \end{array}$

Practice Problem 5 Add or subtract.

(a) 3.12 + 5.08 + 1.42 **(b)** 152.003 − 136.118
(c) 1.1 + 3.16 + 5.123 **(d)** 1.0052 − 0.1234 ■

SIDELIGHT When we added fractions, we had to have common denominators. Since decimals are really fractions, why can we add them without having common denominators? Actually, we have to have common denominators to add any fractions, whether they are in decimal form or fraction form. However, sometimes the notation does not show this. Let's examine Example 5(c), worked above.

Original Problem

$$\begin{array}{r} 3.1 \\ 42.36 \\ +\ \ 9.034 \\ \hline 54.494 \end{array}$$

We are adding the three numbers:

$3\frac{1}{10} + 42\frac{36}{100} + 9\frac{34}{1000}$

$3\frac{100}{1000} + 42\frac{360}{1000} + 9\frac{34}{1000}$

3.100 + 42.360 + 9.034 *This is the New Problem.*

New Problem	Original Problem
$\begin{array}{r} 3.100 \\ 42.360 \\ +\ 9.034 \\ \hline 54.494 \end{array}$	$\begin{array}{r} 3.1 \\ 42.36 \\ +\ 9.034 \\ \hline 54.494 \end{array}$

We notice that the results are the same. The only difference is the notation. We are using the property that any number of zeros may be added to the right-hand side of a decimal without changing its value.

This shows the convenience of adding and subtracting fractions in the decimal form. Little work is needed to change the decimals to a common denominator. All that is required is to add zeros to the right-hand side of the decimal (and we usually do not even write out that step except when subtracting).

As long as we line up the decimal points, we can add or subtract any decimal fractions.

In the following example we will find it useful to add zeros to the right-hand side of the decimal.

EXAMPLE 6 Perform the following operations.

(a) $1.0003 + 0.02 + 3.4$ **(b)** $12 - 0.057$

We will add zeros so that each number shows the same number of decimal places.

(a)
$$\begin{array}{r} 1.0003 \\ 0.0200 \\ +\ 3.4000 \\ \hline 4.4203 \end{array}$$

(b)
$$\begin{array}{r} 12.000 \\ -\ \ 0.057 \\ \hline 11.943 \end{array}$$

Practice Problem 6 Perform the following operations.

(a) $0.061 + 5.0008 + 1.3$ **(b)** $18 - 0.126$ ■

4 *Multiplying Decimals*

> **Multiplication of Decimals**
>
> To multiply decimals, you first multiply as with whole numbers. To determine the position of the decimal point, you count the total number of decimal places in the two numbers being multiplied. This will determine the number of decimal places that should appear in the answer.

EXAMPLE 7 Multiply. 0.8×0.4

$$\begin{array}{rl} 0.8 & \text{(one decimal place)} \\ \times\ 0.4 & \text{(one decimal place)} \\ \hline 0.32 & \text{(two decimal places)} \end{array}$$

Practice Problem 7 Multiply. 0.5×0.3 ■

TEACHING TIP After showing the students how to do Example 8 or a similar example ask them to do the following Additional Class Exercise.

Multiply.

$$\begin{array}{r} 0.0807 \\ \times\ \ 0.13 \\ \hline \text{Ans.}\ \ 0.010491 \end{array}$$

Note that you will often have to add zeros to the left of the digits obtained in the product so that you obtain the necessary number of decimal places.

EXAMPLE 8 Multiply. 0.123×0.5

$$\begin{array}{rl} 0.123 & \text{(three decimal places)} \\ \times\ \ \ \ 0.5 & \text{(one decimal place)} \\ \hline 0.0615 & \text{(four decimal places)} \end{array}$$

Practice Problem 8 Multiply. 0.12×0.4 ■

Here are some examples that involve more decimal places.

EXAMPLE 9 Multiply. **(a)** 2.56×0.003 **(b)** 0.0036×0.008

(a)

$$\begin{array}{rl} 2.56 & \text{(two decimal places)} \\ \times\ 0.003 & \text{(three decimal places)} \\ \hline 0.00768 & \text{(five decimal places)} \end{array}$$

(b)

$$\begin{array}{rl} 0.0036 & \text{(four decimal places)} \\ \times\ 0.008 & \text{(three decimal places)} \\ \hline 0.0000288 & \text{(seven decimal places)} \end{array}$$

Practice Problem 9 Multiply.

(a) 1.23×0.005 **(b)** 0.003×0.00002 ■

SIDELIGHT Why do we count the number of decimal places? The rule really comes from the properties of fractions. If we write the problem in Example 8 in fraction form, we have

$$(0.123) \times (0.5) = \frac{123}{1000} \times \frac{5}{10} = \frac{615}{10{,}000} = 0.0615$$

5 *Dividing Decimals*

When discussing division of decimals, we frequently refer to the three primary parts of a division problem. Be sure you know the meaning of each term.

The **divisor** is the number you divide into another.

The **dividend** is the number to be divided.

The **quotient** is the result of dividing one number by another.

In the problem $6 \div 2 = 3$ we represent each of these terms as follows:

$$\text{Divisor} \longrightarrow 2\overline{)6} \quad 3 \leftarrow \text{Quotient}, \quad 6 \leftarrow \text{Dividend}$$

$$\begin{array}{r} \text{Quotient} \\ \text{Divisor}\overline{)\text{Dividend}} \end{array}$$

When dividing two decimals, count the *number of decimal places* in the divisor. Then move the *decimal point to the right* that *same number of places* in both *the divisor* and *the dividend.* Mark that position with a caret (‸). Finally, perform the division. Be sure to line up the decimal point in the quotient with the position indicated by the caret in the dividend.

EXAMPLE 10 Divide. $32.68 \div 4$

$$\begin{array}{r} 8.17 \\ 4\overline{)32.68} \\ \underline{32} \\ 6 \\ \underline{4} \\ 28 \\ \underline{28} \end{array}$$

Since there are no decimal places in the divisor, we do not need to move the decimal point. We must be careful, however, to place the decimal point in the quotient directly above the decimal point in the dividend.

Thus $32.68 \div 4 = 8.17$.

Practice Problem 10 Divide. $0.1116 \div 0.18$ ■

TEACHING TIP You may wish to point out that when we move the decimal point to the right we are in fact multiplying by a power of 10. If we move the decimal point one place to the right, we are multiplying by 10. If we move the decimal point two places to the right, we are multiplying by 100. Since we move the decimal point in the divisor and we move the decimal point in the dividend the same number of places, we do not change the value of the expression.

EXAMPLE 11 Divide. $5.75 \div 0.5$

$0.5_\wedge \overline{)\,5.7_\wedge 5}$

One decimal place

*There is **one** decimal place in the divisor, so we move the decimal point **one place** to **the right** in the **divisor** and **dividend** and we mark that new position by a caret (‸).*

```
      11.5
0.5‸)5.7‸5
     5
     -
      7
      5
      -
      25
      25
      --
```

Now we perform the division as with whole numbers. The decimal point in the answer is directly above the caret in the dividend.

Thus $5.75 \div 0.5 = 11.5$

Practice Problem 11 Divide. $0.09 \div 0.3$ ■

Note that sometimes we will need to place extra zeros in the dividend in order to move the decimal point the required number of places.

TEACHING TIP You may want to do an Additional Class Exercise example similar to Example 12 for the class. Divide 1.8864 by 0.036.

```
           52.4
0.036‸)1.886‸4
        180
        ---
         86
         72
         --
         144
         144
         ---
```

Ans. 52.4

EXAMPLE 12 Divide. $16.2 \div 0.027$

$0.027_\wedge \overline{)\,16.200_\wedge}$

Three decimal places

*There are **three** decimal places in the divisor, so we move the decimal point **three places** to **the right** in the divisor and **dividend** and mark the new position by a caret. Note that we must add two zeros to 16.2 in order to do this.*

```
             600.
0.027‸)16.200‸
       16 2
       -----
          000
```

Now perform the division as with whole numbers. The decimal point in the answer is directly above the caret in the dividend.

Thus $16.2 \div 0.027 = 600.$

Practice Problem 12 Divide. $1800 \div 0.06$ ■

Special care must be taken to line up the digits in the quotient. Note that sometimes we will need to place zeros in the quotient after the decimal point.

EXAMPLE 13 Divide. $0.04288 \div 3.2$

$3.2_\wedge \overline{)\,0.0_\wedge 4288}$

*There is **one** decimal place in the divisor, so we move the decimal point **one place** to **the right** in the **divisor** and **dividend** and mark the new position by a caret.*

One decimal place

```
        0.0134
3.2‸)0.0‸4288
          32
          --
          108
           96
          ---
           128
           128
           ---
             0
```

Now perform the division as for whole numbers. The decimal point in the answer is directly above the caret in the dividend. Note the need for the initial zero after the decimal point in the answer.

Thus $0.04288 \div 3.2 = 0.0134$.

Practice Problem 13 Divide. $0.01764 \div 4.9$ ■

SIDELIGHT Why does this method of dividing decimals work? Essentially, we are using the steps we used in Section 0.1 to change a fraction to an equivalent fraction by multiplying both the numerator and denominator by the same number. Let's reexamine Example 13.

$0.04288 \div 3.2 = \dfrac{0.04288}{3.2}$ — *Write the original problem using fraction notation.*

$= \dfrac{0.04288 \times 10}{3.2 \quad \times 10}$ — *Multiply the numerator and denominator by 10. Since this is the same as multiplying by 1, we are not changing the fraction.*

$= \dfrac{0.4288}{32}$ — *Write the result of multiplication by 10.*

$= 0.4288 \div 32$ — *Rewrite the fraction as an equivalent problem with division notation.*

Notice that we have obtained a new problem that is the same as the problem in Example 13 when we moved the decimal one place to the right in the divisor and dividend. We see that the reason we can move the decimal point so many places to the right in divisor and dividend is that the numerator and denominator of a fraction are both being multiplied by 10, 100, 1000, and so on, to obtain an equivalent fraction.

6 Multiplying and Dividing by a Multiple of 10

When multiplying by 10, 100, 1000, and so on, a simple rule may be used to obtain the answer. For every zero in the multiplier, move the decimal point one place to the right.

EXAMPLE 14 Multiply. **(a)** 3.24×10 **(b)** 15.6×100 **(c)** 0.0026×1000

(a) $3.24 \times 10 = 32.4$ — *One zero—move decimal point one place to the right.*

(b) $15.6 \times 100 = 1560$ — *Two zeros—move decimal point two places to the right.*

(c) $0.0026 \times 1000 = 2.6$ — *Three zeros—move decimal point three places to the right.*

TEACHING TIP Point out to students that once you can multiply by 100 and by 1000 using this simple rule you can expand it to other situations. For example, to multiply 3.452×100, we immediately have 345.2 (by moving the decimal point two places to the right). If we wanted to multiply 3.452×200, we could multiply 345.2×2 to obtain 690.4 and then conclude that $3.452 \times 200 = 69{,}040$.

Practice Problem 14 Multiply.

(a) 0.0016×100 **(b)** 2.34×1000 **(c)** $56.75 \times 10{,}000$ ■

The reverse rule is true for division. When dividing by 10, 100, 1000, 10,000, and so on, move the decimal point one place to the left for every zero in the divisor.

EXAMPLE 15 Divide. **(a)** $52.6 \div 10$ **(b)** $0.0038 \div 100$ **(c)** $5936.2 \div 1000$

(a) $\dfrac{52.6}{10} = 5.26$ — *Move one place to the left.*

(b) $\dfrac{0.0038}{100} = 0.000038$ — *Move two places to the left.*

(c) $\dfrac{5936.2}{1000} = 5.9362$ — *Move three places to the left.*

Practice Problem 15 Divide.

(a) $\dfrac{5.82}{10}$ **(b)** $123.4 \div 1000$ **(c)** $\dfrac{0.00614}{10{,}000}$ ■

0.4 Exercises

Verbal and Writing Skills

1. A decimal is another way of writing a fraction whose denominator is 10; 100; 1000; 10,000; and so on.

2. To describe 0.03 in words we write three hundredths.

3. When multiplying 0.059 by 10,000, move the decimal point 4 places to the right.

Write each fraction as a decimal.

4. $\frac{15}{16}$ 0.9375 **5.** $\frac{3}{20}$ 0.15 **6.** $\frac{7}{11}$ $0.\overline{63}$ **7.** $\frac{5}{8}$ 0.625 **8.** $\frac{9}{20}$ 0.45 **9.** $\frac{6}{20}$ 0.3

Write each decimal as a fraction in simplified form.

10. 0.8 $\frac{4}{5}$ **11.** 0.15 $\frac{3}{20}$ **12.** 0.625 $\frac{5}{8}$ **13.** 0.08 $\frac{2}{25}$ **14.** 0.75 $\frac{3}{4}$ **15.** 1.125 $\frac{9}{8}$

Add or subtract.

16. 1.0076 − 0.0982
0.9094

17. 1.2 + 3.9 + 2.62
7.72

18. 0.00381 − 0.00228
0.00153

19. 3.6 + 1.28 + 4.5
9.38

20. 38.02 + 217.0 + 0.036
255.056

21. 33.01 + 0.38 + 175.401
208.791

22. 158.23 − 39.67
118.56

23. 121.98 − 34.78
87.2

Multiply or divide.

24. 7.21 × 0.071
0.51191

25. 7.12 × 2.6
18.512

26. 5.26 × 0.0015
0.00789

27. 2.18 × 1.3
2.834

28. 0.0062 × 0.018
0.0001116

29. 0.17 × 0.0084
0.001428

30. 0.5535 ÷ 0.15
3.69

31. 0.0455 ÷ 0.13
0.35

32. 169,000 × 0.0013
219.7

33. 368,000 × 0.00021
77.28

34. 7.9728 ÷ 3.02
2.64

35. 6.519 ÷ 2.05
3.18

36. 0.5230 ÷ 0.002
261.5

37. 0.031 ÷ 0.005
6.2

38. 186.16 ÷ 5.2
35.8

39. 46.62 ÷ 7.4
6.3

Mixed Practice

Multiply or divide by moving the decimal point.

40. 1.36×1000
1360

41. $0.76 \div 100$
0.0076

42. $164{,}320 \div 10{,}000$
16.432

43. 0.02×100
2

44. $3.52 \div 1000$
0.00352

45. $0.00243 \times 100{,}000$
243

46. $7.36 \times 10{,}000$
73,600

47. $73{,}892 \div 100{,}000$
0.73892

Perform the following calculations.

48. 26.13×0.04
1.0452

49. 1.936×0.003
0.005808

50. $1.62 + 2.005 + 8.1007$
11.7257

51. $28 - 3.64$
24.36

52. $0.05724 \div 0.027$
2.12

53. $77.136 \div 0.003$
25,712

54. $0.9281 \times 100{,}000$
92,810

55. $34.72 \div 10{,}000$
0.003472

Applications

Solve each problem.

56. Fred's car usually gets 32.5 miles per gallon when driven on the highway. The gas tank holds 14.4 gallons. What is the driving range of his car when driven on the highway? $32.5 \times 14.4 = 468$ miles

57. Melanie bought a compact car because of her limited budget. The compact averages 28.5 miles on the highway. The gas tank has a capacity of 23.4 gallons. How far can her car go on a full tank of gas?
$28.5 \times 23.4 = 666.9$ miles

Cumulative Review Problems

Perform each calculation.

58. $3\frac{1}{2} \div 5\frac{1}{4}$
$\frac{2}{3}$

59. $\frac{3}{8} \cdot \frac{12}{27}$
$\frac{1}{6}$

60. $\frac{12}{25} + \frac{9}{20}$
$\frac{48 + 45}{100} = \frac{93}{100}$

61. $1\frac{3}{5} - \frac{1}{2}$
$\frac{16 - 5}{10} = \frac{11}{10}$ or $1\frac{1}{10}$

Developing Your Study Skills

MAKING A FRIEND IN THE CLASS

Attempt to make a friend in your class. You may find that you enjoy sitting together and drawing support and encouragement from one another. Exchange phone numbers so you can call each other whenever you get stuck in your study. Set up convenient times to study together on a regular basis, to do homework, and to review for exams.

You must not depend on a friend or fellow student to tutor you, do your work for you, or in any way be responsible for your learning. However, you will learn from one another as you seek to master the course. Studying with a friend and comparing notes, methods, and solutions can be very helpful. And it can make learning mathematics a lot more fun!

Putting Your Skills to Work

THE MATHEMATICS OF MAJOR WORLD LANGUAGES

You probably know that English is the second most commonly spoken language in the world. However, did you know that Chinese is the most commonly spoken language? Did you have any idea that more people speak Portuguese than French? Use the bar graph chart at the right to answer the following questions.

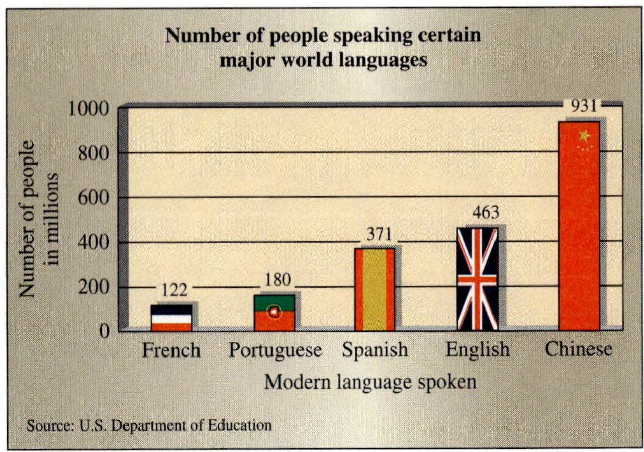

PROBLEMS FOR INDIVIDUAL STUDY AND INVESTIGATION

1. How many more people in the world speak Chinese than English? 468,000,000 people
2. How many more people in the world speak Spanish than French? 249,000,000 people

PROBLEMS FOR GROUP STUDY AND INVESTIGATION

Together with some other members of your class see if you can determine the answers to the following questions. Round your answers to the nearest tenth.

3. If there were approximately 5,734,106,000 people in the world in 1995, what percent of the population of the world spoke English? 8.1%
   What percent of the population of the world spoke Portuguese? 3.1%
4. In 1995, 533,944 college students took modern language courses in Spanish and 272,472 college students took modern language courses in French. How does this relate to the number of people in the world who speak these languages? What can you conclude?
   More students take French rather than Spanish compared to what would be expected if enrollment were based on the number of people in the world who speak each of these languages.

INTERNET CONNECTIONS: Go to ''http://www.prenhall.com/tobey'' to be connected

Site: Language Use Data (U.S. Census Bureau) or a related site

This site includes a table showing languages spoken at home by persons 5 years and over, by state. Use the table to answer the following questions about the state where you live.

5. How many more people speak Greek than Yiddish in your state?
6. What percent of your state's population speaks English at home?
7. What percent of your state's population speaks French at home?
8. Other than English, what is the most commonly spoken language in your state? What percent of your state's population speaks this language at home?

0.5 Use of Percent

MathPro

Video 2

SSM

After studying this section, you will be able to:

1. *Change a decimal to a percent.*
2. *Change a percent to a decimal.*
3. *Find the percent of a number.*
4. *Find the missing percent when given two numbers.*

The Basic Concept of Percents

A **percent** is a fraction that has a denominator of 100. When you say ''sixty-seven percent'' or write 67%, you are just using another way of expressing the fraction $\frac{67}{100}$. The word *percent* is a shortened form of the Latin words *per centum,* which means ''by the hundred.'' In everyday use, percent means per one hundred.

It is important to see that 49% means 49 parts out of 100 parts. It can also be written as a fraction, $\frac{49}{100}$, or as a decimal: 0.49. Understanding the meaning of the notation allows you to change from one notation to another. For example,

$$49\% = 49 \text{ out of } 100 \text{ parts} = \frac{49}{100} = 0.49$$

Similarly, you can express a fraction with denominator 100 as a percent or a decimal.

$$\frac{11}{100} \text{ means 11 parts out of 100 or } 11\%; \qquad \frac{11}{100} \text{ as a decimal is } 0.11$$

So $\frac{11}{100} = 11\% = 0.11$.

1 *Changing a Decimal to a Percent*

Now that we understand the concept, we can use some quick procedures to change from decimals to percent, and vice versa.

Changing a Decimal to Percent

1. Move the decimal point two places to the right.
2. Add the % symbol.

EXAMPLE 1 Change to a percent. **(a)** 0.23 **(b)** 0.461 **(c)** 0.4

We move the decimal point two places to the right and add the % symbol.

(a) 0.23 = 23% **(b)** 0.461 = 46.1% **(c)** 0.4 = 0.40 = 40%

Practice Problem 1 Change to a percent. **(a)** 0.92 **(b)** 0.418 **(c)** 0.7 ■

Be sure to follow the same procedure for percents that are less than 1%. Remember, 0.01 is 1%. Thus we would expect 0.001 to be less than 1%. 0.001 = 0.1% or one-tenth (0.1) of a percent.

EXAMPLE 2 Change to a percent. **(a)** 0.0364 **(b)** 0.0026 **(c)** 0.0008

We move the decimal point two places to the right and add the % symbol.

(a) 0.0364 = 3.64% **(b)** 0.0026 = 0.26% **(c)** 0.0008 = 0.08%

Practice Problem 2 Change to a percent.

(a) 0.0019 **(b)** 0.0736 **(c)** 0.0003 ■

TEACHING TIP A common student error is to write 0.45% = 0.45, which is, of course, not true. After discussing Example 2, have students complete the following Additional Class Exercise.

Change to a percent.

A. 0.5 **B.** 0.05
C. 0.005

Ans. A. 50% B. 5% C. 0.5%

Be sure to follow the same procedure for percents that are greater than 100%. Remember 1 is 100%. Thus we would expect 1.5 to be greater than 100%. 1.5 = 150%

EXAMPLE 3 Change to a percent. **(a)** 1.48 **(b)** 2.938 **(c)** 4.5

We move the decimal point two places to the right and add the percent symbol.

(a) 1.48 = 148% **(b)** 2.938 = 293.8% **(c)** 4.5 = 4.50 = 450%

Practice Problem 3 Change to a percent.

(a) 3.04 **(b)** 5.186 **(c)** 2.1 ■

2 *Changing from a Percent to a Decimal*

In this procedure we move the decimal point to the left and remove the % symbol.

Changing a Percent to a Decimal

1. Move the decimal point two places to the left.
2. Remove the % symbol.

CALCULATOR

Percent to Decimal

You can use a calculator to change 52% to a decimal. If your calculator has a [%] key,

Enter: 52 [%]

The display should read: [0.52]

If your calculator does not have a [%] key, divide the number by 100. Enter:

52 [÷] 100 [=]

Try **(a)** 46% **(b)** 137% **(c)** 9.3% **(d)** 6%

Note: The calculator divides by 100 when the percent key is pressed. If you do not have a [%] key, then you can divide by 100 instead of using the percent key.

EXAMPLE 4 Change to a decimal. **(a)** 16% **(b)** 143%

First we move the decimal point two places to the left. Observe in each case that the decimal point is not written but is understood to be to the right of the last digit in the percent. Then we remove the % symbol.

(a) 16% = 16.% = 0.16

↑ The unwritten decimal point is understood to be here.

(b) 143% = 143.% = 1.43

Practice Problem 4 Change to a decimal. **(a)** 82% **(b)** 347% ■

EXAMPLE 5 Change to a decimal. **(a)** 4% **(b)** 3.2% **(c)** 0.6%

First we move the decimal point two places to the left. Then we remove the % symbol.

(a) 4% = 4.% = 0.04

↑ The unwritten decimal point is understood to be here.

(b) 3.2% = 0.032 **(c)** 0.6% = 0.006

Practice Problem 5 Change to a decimal. **(a)** 7% **(b)** 9.3% **(c)** 0.2% ■

TEACHING TIP Point out that 0.6% is less than 1%. We would expect the decimal form of 0.6% to be less than one one-hundredth or less than 0.01. 0.6% = 0.006

192% is greater than 100%. We would expect the decimal form of 192% to be greater than 1. 192% = 1.92

EXAMPLE 6 Change to a decimal. **(a)** 192% **(b)** 254.8% **(c)** 0.027%

First we move the decimal point two places to the left. Then we remove the % symbol.

(a) 192% = 192.% = 1.92

↑ The unwritten decimal point is understood to be here.

(b) 254.8% = 2.548 **(c)** 0.027% = 0.00027

Practice Problem 6 Change to a decimal.

(a) 131% **(b)** 301.6% **(c)** 0.04% ■

3 *Finding the Percent of a Number*

How do we find 60% of 20? Let us relate it to a problem we did in Section 0.2.

Consider the problem

$$\text{What is } \frac{3}{5} \text{ of } 20?$$

$$\downarrow \quad \downarrow \quad \downarrow \quad \downarrow \quad \downarrow$$

$$\boxed{?} = \frac{3}{5} \times 20$$

$$\boxed{?} = \frac{3}{\cancel{5}} \times \overset{4}{\cancel{20}} = 12 \quad \text{The answer is 12.}$$

TEACHING TIP Some students may find it easier to change 60% to a fraction and multiply. Allow those students who can change easily from a percent to a fraction to use fractions in those situations where fractions work best. Point out, however, that a decimal is always safe to use, and some percents such as 1.3% are best expressed as a decimal for calculation.

Since a percent is really a fraction, a percent problem is solved in a similar way to solving a fraction problem. Since $\frac{3}{5} = \frac{6}{10} = 60\%$, we could write the problem as

$$\text{What is } 60\% \text{ of } 20?$$

$$\downarrow \quad \downarrow \quad \downarrow \quad \downarrow \quad \downarrow$$

$$\boxed{?} = 60\% \times 20$$

$$\boxed{?} = 0.60 \times 20$$

$$\boxed{?} = 12.0 \quad \text{The answer is 12.}$$

Thus we have developed a rule.

> **Finding the Percent of a Number**
>
> To find the percent of a number, merely change the percent to a decimal and multiply by the decimal.

EXAMPLE 7 Find.

(a) 10% of 36 **(b)** 2% of 350 **(c)** 182% of 12 **(d)** 0.3% of 42

(a) 10% of 36 = 0.10 × 36 = 3.6 **(b)** 2% of 350 = 0.02 × 350 = 7

(c) 182% of 12 = 1.82 × 12 = 21.84 **(d)** 0.3% of 42 = 0.003 × 42 = 0.126

Practice Problem 7 Find.

(a) 18% of 50 **(b)** 4% of 64 **(c)** 156% of 35

(d) 0.8% of 60 **(e)** 1.3% of 82 **(f)** 0.002% of 564 ■

There are many real-life applications for finding the percent of a number. When you go shopping in a store, you may find sale merchandise marked 35% off. This means that the sale price is 35% off the regular price. That is, 35% of the regular price is subtracted from the regular price to get the sale price.

CALCULATOR

Using Percents

You can use a calculator to find 12% of 48.

Enter:

12 [%] [×] 48 [=]

The display should read:

[5.76]

If your calculator does not have a [%] key,

Enter: 0.12 [×] 48 [=]

Try What is 54% of 450?

EXAMPLE 8 A store is having a sale of 35% off the retail price of all sofas. Melissa wants to buy a particular sofa that normally sells for \$595.

(a) How much will Melissa save if she buys the sofa on sale?

(b) What is the purchase price if Melissa buys the sofa on sale?

(a) To find 35% of \$595 we will need to multiply 0.35×595.

$$\begin{array}{r} \$595 \\ \times \quad 0.35 \\ \hline 2975 \\ 1785 \\ \hline \$208.25 \end{array}$$

Thus Melissa would save \$208.25 if she buys the sofa on sale.

(b) The purchase price is the difference between the original price and the amount saved.

$$\begin{array}{r} \$595.00 \\ - \quad 208.25 \\ \hline \$386.75 \end{array}$$

Thus Melissa bought the sofa on sale for \$386.75.

Practice Problem 8

John received a 4.2% pay raise at work this year. He previously earned \$18,000 per year.

(a) What is the amount of his pay raise in dollars?

(b) What will his new salary be? ■

4 *Finding the Missing Percent When Given Two Numbers*

Recall that we can write $\frac{3}{4}$ as $\frac{75}{100}$ or 75%. If we were asked the question, ''What percent is 3 of 4?'' we would say 75%. This gives us a procedure for finding what percent one number is of a second number.

Finding the Missing Percent

1. Write a fraction with the two numbers. The number *after* the word ''of'' is always the denominator, and the other number is the numerator.
2. Simplify the fraction (if possible).
3. Change the fraction to a decimal.
4. Express the decimal as a percent.

A very common problem is the following:

EXAMPLE 9 What percent of 24 is 15?

This can be quickly solved as follows:

Step 1 $\frac{15}{24}$ *Write the relationship as a fraction. The number after ''of'' is 24, so the 24 is in the denominator.*

Step 2 $= \frac{5}{8}$ *Simplify the fraction (when possible).*

Step 3 $= 0.625$ *Change the fraction to a decimal.*

Step 4 $= 62.5\%$ *Change the decimal to percent.*

Practice Problem 9 What percent of 148 is 37? ■

The question in Example 9 can also be written as "15 is what percent of 24?" To answer the question, we begin by writing the relationship as $\frac{15}{24}$. Remember "of" 24 means 24 will be in the denominator.

EXAMPLE 10 **(a)** 82 is what percent of 200? **(b)** What percent of 16 is 3.8? **(c)** $150 is what percent of $120?

TEACHING TIP Some calculations are easier than others. Encourage students to do the following mentally.

(a) What is 35.8% of 1000?

(b) 4 is what percent of 400?

(a) 35.8% of $1000 = 0.358 \times 1000 = 358$

To multiply by 1000, you need only move the decimal place three places to the right.

(b) $\frac{4}{400} = \frac{1}{100} = 1\%$

Have the students do the following as Additional Class Exercises.

(a) What is 0.1% of 2000?

(b) What percent of 45 is 15?

Ans. (a) 2 (b) $33\frac{1}{3}\%$

(a) 82 is what percent of 200?

$\frac{82}{200}$ *Write the relationship as a fraction with 200 in the denominator.*

$$\frac{82}{200} = \frac{41}{100} = 0.41 = 41\%$$

(b) What percent of 16 is 3.8?

$\frac{3.8}{16}$ *Write the relationship as a fraction.*

You can divide to change the fraction to a decimal and then change the decimal to a percent.

$$16\overline{)3.8000} = 0.2375 \rightarrow 23.75\%$$

(c) $150 is what percent of $120?

$\frac{150}{120} = \frac{5}{4}$ *Reduce the fraction whenever possible to make the division easier.*

$$4\overline{)5.00} = 1.25 \rightarrow 125\%$$

Practice Problem 10

(a) What percent of 48 is 24?
(b) 4 is what percent of 25? ■

EXAMPLE 11 Marcia had 29 shots on the goal during the last high school field hockey season. She actually scored a goal 8 times. What percent of her total shots were goals? Round your answer to the nearest whole percent.

Marcia scored a goal 8 times out of 29 tries. We want to know what percent of 29 is 8.

Step 1 $\frac{8}{29}$ *Express the relationship as a fraction. The number after the word "of" is 29, so the 29 appears in the denominator.*

Step 2 *Note that this fraction cannot be reduced.*

Step 3 $= 0.2758\ldots$ *The decimal equivalent of the fraction has many digits.*

Step 4 $= 27.58\ldots\%$ $\approx 28\%$ *We change the decimal to a percent, which we round to the nearest whole percent.*

Therefore, Marcia scored a goal approximately 28% of the time she made a shot on the goal.

CALCULATOR

Using Percents

You can use a calculator to find a missing percent.

What percent of 95 is 19?

1. Enter as a fraction.

19 [÷] 95

2. Change to a percent.

19 [÷] 95 [×] 100 [=]

The display should read:

[20]

This means 20%.

Try What percent of 625 is 250?

Practice Problem 11

Roberto scored a basket 430 times out of 1256 attempts during his high school basketball career. What percent of the time did he score a basket? Round your answer to the nearest whole percent. ■

0.5 Exercises

Verbal and Writing Skills

1. When you write 19% what do you really mean? Describe in your own words.
   Answers may vary. Sample answers are below.
   19% means 19 out of 100 parts. Percent means per 100.
   19% is really a fraction with a denominator of 100. In this case it would be $\frac{19}{100}$.

2. When you try to solve a problem like ''What percent of 80 is 30?'' how do you know if you should write the fraction as $\frac{80}{30}$ or as $\frac{30}{80}$? The number directly following the word ''of'' should be the denominator. In this case you should use $\frac{30}{80}$ since 80 follows the word ''of.''

Change to a percent.

3. 0.624 62.4%
4. 0.078 7.8%
5. 0.304 30.4%
6. 0.007 0.7%
7. 1.56 156%
8. 1.6 160%

Change to decimal.

9. 4% 0.04
10. 70% 0.7
11. 0.2% 0.002
12. 0.62% 0.0062
13. 250% 2.50
14. 175% 1.75

Find the following.

15. What is 35.8% of 1000? 358
16. What is 7.7% of 450? 34.65
17. What is 0.8% of 65? 0.52
18. What is 7% of 69? 4.83
19. What is 140% of 212? 296.8
20. What is 176% of 340? 598.4
21. What is 80% of 220? 176
22. What is 0.06% of 410? 0.246
23. What percent of 600 is 30? 5%
24. What percent of 800 is 50? 6.25%
25. 6 is what percent of 120? 5%
26. 33 is what percent of 330? 10%
27. 36 is what percent of 24? 150%
28. 49 is what percent of 28? 175%
29. What percent of 500 is 2? 0.4%
30. 36 is what percent of 300? 12%
31. What percent of 16 is 22? 137.5%
32. 45 is what percent of 18? 250%

Applications

Solve each word problem.

33. One of the Lollapalooza concerts had 30,000 people in the crowd. 47% of the audience were women. How many women were at the concert? $0.47 \times 30{,}000 = 14{,}100$

34. Dave took an exam with 80 questions. He had 12 wrong. What was his grade for the exam? Write the grade as a percent. 85%

35. In a local college survey, it was discovered that 137 out of 180 students had a grandparent not born in the United States. What percent of the students had a grandparent not born in the United States? (Round your answer to the nearest hundredth of a percent.) 76.11%

36. Diana and Russ spent \$32.80 when they went out to dinner. If they want to leave the standard 15% tip for their server, how much will they tip and what will be the new total on the check? \$4.92 tip, \$37.72 new total

37. Music CDs have a failure rate of 1.8% (they skip or get stuck on a song). If 36,000 CDs were manufactured last week, how many of them were defective?
648 CDs were defective.

38. The Gonzalez family has a combined monthly income of \$1850. Their food budget is \$380/month. What percentage of their monthly income is budgeted for food? (Round your answer to the nearest whole percent.)
Approximately 21% of the budget is for food.

39. Eastern Shores charter airline had 2390 flights last year. Of these flights, 560 did not land on time. What percent of the flights landed on time? Round your answer to the nearest whole percent. 77%

40. Jim is earning \$12.50 per hour as a cook. He has earned an 8% pay raise this year. What will his raise be, and what will his new hourly rate be? \$1 an hour raise; new hourly rate will be \$13.50 per hour.

41. Bruce sells medical supplies on the road. He logged 18,600 miles in his car last year. He declared 65% of his mileage as business travel. (a) How many miles did he travel on business last year? (b) If his company reimburses him 31 cents for each mile traveled, how much should he be paid for travel expenses?
(a) 12,090 miles traveled on business.
(b) \$3747.90 for travel expenses.

42. Abdul sells computers for a local computer outlet. He gets paid \$450 per month plus a commission of 3.8% on all the computer hardware he sells. Last year he sold \$780,000 worth of computer hardware. (a) What was his sales commission on the \$780,000 worth of hardware he sold? (b) What was his annual salary if the monthly pay and the commission are added together?
(a) His commission was $0.038 \times 780{,}000 = \$29{,}640$.
(b) His total annual salary was
$12 \times 450 + 29{,}640 = 5400 + 29{,}640 = \$35{,}040$.

Putting Your Skills to Work

ANALYSIS OF CAR SALES IN THE UNITED STATES

The patterns of sales of new cars in the United States show some interesting trends. The following chart designates the number of new cars sold in the United States over a 10-year period based on the size of the car: small, midsize, large, or luxury. Remember this includes only car sales. It does not include pickup trucks (the most popular vehicle based on U.S. sales), nor does it include sport utility vehicles (such as the Jeep Cherokee or the Ford Explorer).

U.S. CAR SALES BY VEHICLE SIZE AND TYPE, 1983–1993
Source: American Automobile Manufacturers Assn.

Year	Small (%)	Midsize (%)	Larger (%)	Luxury (%)	Total U.S. Sales
1993	32.8	43.3	11.1	12.8	8,517,862
1992	32.9	44.5	9.2	13.4	8,214,084
1991	33.0	44.9	8.3	13.9	8,174,656
1990	32.8	44.8	9.4	13.0	9,300,211
1989	36.6	41.9	11.9	11.6	9,772,266
1988	37.6	42.5	10.0	9.9	10,529,730
1987	38.4	42.3	9.1	10.2	10,276,609
1986	37.6	42.5	9.8	10.1	11,459,518
1985	37.9	42.1	9.8	10.2	11,042,287
1984	39.1	39.6	11.6	9.7	10,390,365
1983	38.8	40.6	10.7	9.9	9,182,067

PROBLEMS FOR INDIVIDUAL INVESTIGATIONS AND STUDY

1. In what year were the midsize cars represented by the greatest percentage of the U.S. sales? 1991

2. Between what two years was the greatest difference in percentage of the U.S. car sales of small cars? What was the amount of the difference measured in percent? (Any calendar year not necessarily two consecutive years.) Between 1984 and 1990 or between 1984 and 1993. It differed by 6.3%

PROBLEMS FOR COOPERATIVE GROUP ACTIVITY

Together with other members of your class see if you can complete the following.

3. What was the actual number of large cars sold in 1993 and in 1983? Did the number of large cars sold actually increase or decrease from 1983 to 1993? In 1993 it was 945,483. In 1983 it was 982,481. The percentage increased but the number of cars decreased.

4. Between 1992 and 1993 what was the actual decrease in the number of luxury cars that were sold? (number of cars, not percentage) Decrease of 10,401 cars

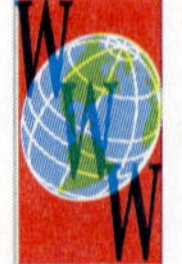

INTERNET CONNECTIONS: Go to ''http://www.prenhall.com/tobey'' to be connected

Site: Car Sales by Series (Volvo 1995) or a related site

This site gives information about Volvo automobiles sold in 1994 and 1995. Answer the following questions.

5. What percent of Volvos sold in 1995 were 850s?

6. Use a percent to compare the number of 940s and 960s sold in 1995.

0.6 Estimation

After studying this section, you will be able to:

1 *Use rounding to estimate.*

MathPro Video 2 SSM

1 *Using Rounding to Estimate*

Estimation is the process of finding an approximate answer. It is not designed to provide an exact answer. Estimation will give you a rough idea of what the answer might be. For any given problem, you may choose to estimate in many different ways.

Estimation by Rounding

1. Round each number to an appropriate place.

2. Perform the calculation with the rounded numbers.

EXAMPLE 1 Find an estimate of the product 5368×2864.

Step 1 Round 5368 to 5000.
Round 2864 to 3000.

Step 2 Multiply.

$$5000 \times 3000 = 15{,}000{,}000$$

An estimate of the product is 15,000,000.

Practice Problem 1

Find an estimate of the product $128{,}621 \times 378$. ■

EXAMPLE 2 The four walls of a college classroom are $22\frac{1}{4}$ feet long and $8\frac{3}{4}$ feet high. A painter needs to know the area of these four walls in square feet. Since paint is sold in gallons, an estimate will do. Estimate the area of the four walls.

Step 1 Round $22\frac{1}{4}$ feet to 20 feet.
Round $8\frac{3}{4}$ feet to 9 feet.

Step 2 Multiply 20×9 to obtain an estimate of the area of one wall.
Multiply $20 \times 9 \times 4$ to obtain an estimate of the area of all four walls.

$$20 \times 9 \times 4 = 720 \text{ square feet}$$

Our estimate for the painter is 720 square feet of wall space.

Practice Problem 2

Mr. and Mrs. Ramirez need to carpet two rooms of their house. One room measures $12\frac{1}{2}$ feet by $9\frac{3}{4}$ feet. The other room measures $14\frac{1}{4}$ feet by $18\frac{1}{2}$ feet. Estimate the number of square feet (square footage) in these two rooms. ■

EXAMPLE 3 Won Lin has a small compact car. He drove 396.8 miles in his car and used 8.4 gallons of gas.

(a) Estimate the number of miles he gets per gallon.

(b) Estimate how much it will cost him for fuel to drive on a cross-country trip of 2764 miles if gasoline usually costs $1.39 9/10 per gallon.

(a) Round 396.8 miles to 400 miles. Round 8.4 gallons to 8 gallons. Now divide.

$$8\overline{)400}\quad = 50$$

Won Lin's car gets about 50 miles per gallon.

(b) We will need to use the information we found in part (a) to determine how many gallons of gasoline Won Lin will use on his trip. Round 2764 miles to 3000 miles and divide 3000 miles by 50 gallons.

$$50\overline{)3000}\quad = 60$$

Won Lin will use about 60 gallons of gas for his cross-country trip.

To estimate the cost, we need to ask ourselves, "What kind of an estimate are we looking for?" It may be sufficient to round $1.39 9/10 to $1.00 and multiply.

$$60 \times \$1.00 = \$60.00$$

Keep in mind that this is a broad estimate. You may want an estimate that will be closer to the exact answer. In that case round $1.39 9/10 to $1.40 and multiply.

$$60 \times \$1.40 = \$84.00$$

Caution: An estimate is only a rough guess. If we are estimating the cost of something, we may want to round the unit cost so that we get closer to the actual amount. It is a good idea to round above the unit cost to make sure that we will have enough money for the expenditure. The actual cost of the cross-country trip to the nearest penny is $81.86.

Practice Problem 3

Roberta drove 422.8 miles in her truck and used 19.3 gallons of gas. Assume that gasoline costs $1.29 9/10 per gallon.

(a) Estimate the number of miles she gets per gallon.

(b) Estimate how much it will cost her to drive to Chicago and back, a distance of 3862 miles. ■

Other words that indicate estimation in word problems are *about* and *approximate*.

EXAMPLE 4 A local manufacturing company releases 1684 pounds of sulfur dioxide per day from its plant. The company operates 294 days per year and has been operating at this level for 32 years. Approximately how many pounds of sulfur dioxide have been released into the air by this company over the last 32 years?

The word *approximately* means we are looking for an estimate. We begin by rounding the given data so that there is only one nonzero digit in each number.

Round 1684 tons to 2000 tons.

Round 294 days to 300 days.

Round 32 years to 30 years.

Now determine the calculation you would perform to find the answer and calculate using the rounded numbers. We multiply

amount released in one day × the number of days in an operating year × years

$$2000 \times 300 \times 30 = 18{,}000{,}000$$

Thus 18,000,000 tons is an approximation of the sulfur dioxide released by the company during 32 years of operation.

Practice Problem 4

A space probe is sent from Earth at 43,300 miles per hour toward the planet Pluto, which has an average distance from earth of 3,580,000,000 miles.

(a) About how many *hours* will it take for the space probe to travel from Earth to Pluto?

(b) About how many *days* will it take for the space probe to travel from Earth to Pluto? ■

EXAMPLE 5 Find an estimate for 2.68% of $54,361.92.

Begin by writing 2.68% as a decimal.

$$2.68\% = 0.0268$$

Next round 0.0268 to the nearest hundredth.

0.0268 is 0.03 rounded to the nearest hundredth.

Finally, round $54,361.92 to $50,000.

Now multiply.

$0.03 \times 50{,}000 = 1500$ *Remember that to find a percent of a number you multiply.*

Thus we estimate the answer to be $1500.

To get a better estimate of 2.68% of $54,361.92, you may wish to round $54,361.92 to the nearest thousand. Try it and compare this estimate to the one above. Remember, an estimate is only an approximation. How close you wish to get to the exact answer depends on what you need the estimate for.

Practice Problem 5

Find an estimate for 56.93% of $293,567.12. ■

0.6 Exercises

Follow the principles of estimation to find an estimated value for each of the following. Do not find the exact value. There are many acceptable answers, depending on your rounding method.

1. 878×203 180,000

2. 329×773 240,000

3. $78{,}295 \times 1872$ 160,000,000

4. $193{,}562 \times 38$ 8,000,000

5. $93 + 87 + 56 + 22 + 34$ 290

6. $127 + 341 + 778 + 567$ 1800

7. $23\overline{)578{,}962}$ 30,000

8. $82\overline{)4{,}320{,}760}$ 50,000

9. $\dfrac{0.002714}{0.0315}$ 0.1

10. $\dfrac{0.5361}{0.00786}$ 60

11. Find 17% of \$21,365.85. \$4000.00

12. Find 4.9% of \$9321.88. \$450

Applications

Determine an estimate of the exact answer. Use estimation by rounding. Do not find the exact value.

13. Estimate the floor area of a ballet studio that measures $36\frac{3}{4}$ feet long and $24\frac{1}{8}$ feet wide. 800 square feet

14. Estimate the floor area of a regulation NBA basketball court measuring 94 feet long and 50 feet wide. 4500 square feet

15. Danielle drove to the Green Mountains of Vermont from New York City for a one-way grand total of 234.8 miles. She used 12.6 gallons of gas. Estimate the number of miles her car gets per gallon. 20 mpg

16. Yoshiko drove for 183.4 miles and used 9.8 gallons of gas. Estimate the number of miles her car gets per gallon. 20 mpg

17. A typical customer at the local Piggly Wiggly supermarket chain in the southern United States spends approximately \$82 at the checkout register. The store keeps four registers open, and each handles 22 customers per hour. Estimate the amount of money the store receives in one hour. \$6400

18. The Westerly Community Credit Union in Rhode Island has found that on Fridays, the average customer withdraws \$85 from the ATM for weekend spending money. Each ATM averages 19 customers per hour. There are five machines. Estimate the amount of money withdrawn by customers in one hour. \$9000

19. Angela and Ramon went to a restaurant and ordered 3 hamburgers at \$3.95 each, 2 salads at \$2.59 each, and 2 soft drinks at \$1.25 each. Estimate the cost of the meal. \$20

20. Betty Jo and Christopher went to a great Tex-Mex restaurant in Sun Valley, Idaho. They ordered 3 tacos at \$2.95 each, 2 burritos at \$3.95 each, and 2 soft drinks at \$1.75 each. Estimate the cost of the meal. \$21

21. Jeff has to pay 6% sales tax on a new John Deere tractor for the farm. Estimate the amount of sales tax to be paid if the tractor costs \$21,385. \$1200

22. The population of Bayside City in Florida 10 years ago was 287,356. Because so many people are retiring there, the population has grown by 18%. Estimate the increase in population over the last 10 years. 60,000

23. The governor of a Midwestern state estimates that only 8.3% of her budget of \$19,364,282,153.18 goes toward education. Estimate the amount of money spent on education. \$1,600,000,000

24. With HMOs (Health Maintenance Organizations) taking the place of many health insurance companies, an industry survey estimates that 6.23% of all operations performed in hospitals are unnecessary. If 28,364,122 operations were performed last year, estimate the number of unnecessary operations. 1,800,000 operations

To Think About

25. Laura's boyfriend James was transferred to San Diego. He averages 43 long-distance calls to her per month. The average cost of his long-distance calls is \$3.24 per call. A new long-distance telephone company has promised James a 23% reduction in the cost of his phone bills. If that is true, estimate how much he will save in *one year*. \$240

26. Eddie usually drives his car to his job in Denver. His car costs 8 cents per mile to drive. He commutes 22 miles round trip everyday, and drives in to work 280 days per year. He can buy an unlimited-use bus pass for one year for \$300. Estimate how much he would save by taking the bus to work. \$180

Developing Your Study Skills

READING THE TEXTBOOK

Your homework time each day should begin with the careful reading of the section(s) assigned in your textbook. Usually, much time and effort have gone into the selection of a particular text, and your instructor has decided that this is the book that will help you to become successful in this mathematics class. Textbooks are expensive, but they can be a wise investment if you take advantage of them by reading them.

Reading a mathematics textbook is unlike reading many other types of books that you may find in your literature, history, psychology, or sociology courses. Mathematics texts are technical books that provide you with exercises to practice on. Reading a mathematics text requires slow and careful reading of each word, which takes time and effort.

Begin reading your textbook with a paper and pencil in hand. As you come across a new definition or concept, underline it in the text and/or write it down in your notebook. Whenever you encounter an unfamiliar term, look it up and make a note of it. When you come to an example, work through it step by step. Be sure to read each word and to follow directions carefully.

Notice the helpful hints that the author provides. They guide you to correct solutions and prevent you from making errors. Take advantage of these pieces of expert advice.

Be sure that you understand what you are reading. Make a note of any of those things that you do not understand and ask your instructor about them. Do not hurry through the material. Learning mathematics takes time.

0.7 Mathematics Blueprint for Problem Solving

MathPro

Video 3

SSM

After studying this section, you will be able to:

1 *Use the mathematical blueprint to solve applied problems.*

1 *Using the Mathematical Blueprint to Solve Applied Problems*

When a builder constructs a new home or office building, he often has a blueprint. This accurate drawing shows the basic form of the building. It also shows the dimensions of the structure to be built. This blueprint serves as a useful reference throughout the construction process.

Similarly, when solving applied problems, it is helpful to have a "mathematical blueprint." This is a simple way to organize the information provided in the word problem, in a chart, or in a graph. You can record the facts you need to use. You can determine what it is you are trying to find and how you can go about actually finding it. You can record other information that you think will be helpful as you work through the problem.

As we solve applied problems, we will utilize three steps.

Step 1 *Understand the problem.* Here we will read through the problem and use the mathematical blueprint as a guide to assist us in thinking through the steps needed to solve the problem.

Step 2 *Solve the problem and state the answer.* We will use arithmetic or algebraic procedures along with problem-solving strategies to find a solution.

Step 3 *Check.* We will use a variety of techniques to see if the answer in step 2 is the solution to the word problem.

EXAMPLE 1 Nancy and John want to install wall-to-wall carpeting in their living room. The floor of the rectangular living room is $11\frac{2}{3}$ feet wide and $19\frac{1}{2}$ feet long. How much will it cost if the carpet is \$18.00 per square yard?

1. *Understand the problem.*
First, read the problem carefully. Drawing a sketch of the living room may help you to see what is required. The carpet will cover the floor of the living room. This is area. Now we fill in the mathematics blueprint.

$19\frac{1}{2}$ ft

$11\frac{2}{3}$ ft $\quad$ $11\frac{2}{3}$ ft

$19\frac{1}{2}$ ft

MATHEMATICS BLUEPRINT FOR PROBLEM SOLVING

Gather the facts.	What am I solving for?	What must I calculate?	Key points to remember.
The living room measures $11\frac{2}{3}$ ft by $19\frac{1}{2}$ ft. The carpet costs \$18.00 per square yard.	(a) The area of the room in square feet (b) The area of the room in square yards (c) The cost of the carpet	(a) Multiply $11\frac{2}{3}$ ft by $19\frac{1}{2}$ ft to get area in square feet. (b) Divide the number of square feet by 9 to get the number of square yards. (c) Multiply the number of square yards by \$18.00.	There are 9 square feet, 3 feet × 3 feet, in 1 square yard; therefore, we must divide the number of square feet by 9 to obtain square yards.

2. *Solve and state the answer.*

(a) To find the area of a rectangle, we multiply the length times the width.

$$11\frac{2}{3} \times 19\frac{1}{2} = \frac{35}{3} \times \frac{39}{2}$$

$$= \frac{455}{2} = 227\frac{1}{2} \text{ sq ft}$$

A minimum of $227\frac{1}{2}$ square feet of carpet will be needed. We say a minimum because some carpet may be wasted in cutting. Carpet is sold by the square yard. We will want to know the amount of carpet needed in square yards.

(b) To determine the area in square yards, we divide $227\frac{1}{2}$ by 9. (9 sq ft = 1 sq yd.)

$$227\frac{1}{2} \div 9 = \frac{455}{2} \div \frac{9}{1}$$

$$= \frac{455}{2} \times \frac{1}{9} = \frac{455}{18} = 25\frac{5}{18} \text{ sq yd}$$

A minimum of $25\frac{5}{18}$ square yards of carpet will be needed.

(c) Since the carpet costs $18.00 per square yard, we will multiply the number of square yards needed by $18.00.

$$25\frac{5}{18} \times 18 = \frac{455}{18} \times \frac{18}{1} = \$455$$

The carpet will cost a minimum of $455.00 for this room.

3. *Check.*

We will estimate to see if our answers are reasonable.

(a) We will estimate by rounding each number to the nearest 10.

$$11\frac{2}{3} \times 19\frac{1}{2} \longrightarrow 10 \times 20 = 200 \text{ sq ft}$$

This is close to our answer of $227\frac{1}{2}$ sq ft. Our answer is reasonable. ✓

(b) We will estimate by rounding to one significant digit.

$$227\frac{1}{2} \div 9 \longrightarrow 200 \div 10 = 20 \text{ sq yd}$$

This is close to our answer of $25\frac{5}{18}$ sq yd. Our answer is reasonable. ✓

(c) We will estimate by rounding each number to the nearest 10.

$$25\frac{5}{18} \times 18 \longrightarrow 30 \times 20 = \$600$$

This is close to our answer of $450. Our answer seems reasonable. ✓

Practice Problem 1

Jeff went to help Abby pick out carpet for her apartment. Her rectangular living room measures $16\frac{1}{2}$ feet by $10\frac{1}{2}$ feet. How much will it cost to carpet the room if the carpet costs $20 per square yard?

MATHEMATICS BLUEPRINT FOR PROBLEM SOLVING

Gather the facts.	What am I solving for?	What must I calculate?	Key points to remember.

To Think About Assume that the carpet in Example 1 comes in a standard width of 12 feet wide. How much wasted carpet will there be if the carpet is laid out for a 12-foot width and a length of $19\frac{1}{2}$ feet? How much wasted carpet would there be if the carpet is laid in two sections side to side that are each $11\frac{2}{3}$ feet long? Assuming you have to pay for wasted carpet, what is the minimum cost to carpet the room?

EXAMPLE 2 Below is a chart showing the 1995 sales of Micropower Computer Software for each of the four regions of the United States. Use the chart to answer the following questions (round all answers to nearest whole percent):

(a) What percent of the sales personnel are assigned to the Northeast?

(b) What percent of the volume of sales is attributed to the Northeast?

(c) What percent of the sales personnel are assigned to the Southeast?

(d) What percent of the volume of sales is attributed to the Southeast?

(e) Which of these two regions of the country has sales personnel that appear to be more effective in terms of the volume of sales?

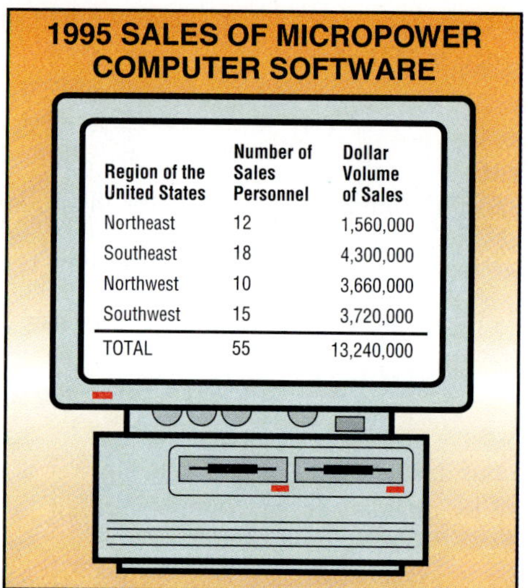

1995 SALES OF MICROPOWER COMPUTER SOFTWARE

Region of the United States	Number of Sales Personnel	Dollar Volume of Sales
Northeast	12	1,560,000
Southeast	18	4,300,000
Northwest	10	3,660,000
Southwest	15	3,720,000
TOTAL	55	13,240,000

1. *Understand the problem*

We will only need to deal with figures from the Northeast region and the Southeast region.

MATHEMATICS BLUEPRINT FOR PROBLEM SOLVING

Gather the facts.	What am I solving for?	What must I calculate?	Key points to remember.
Personnel: 12 Northeast 18 Southeast 55 total Sales Volume: 1,560,000 NE 4,300,000 SE 13,240,000 Total	(a) The percent of the total personnel that is in the Northeast (b) The percent of the total sales made in the Northeast (c) The percent of the total personnel that is in the Southeast (d) The percent of the total sales made in the Southeast (e) Compare the percentages from the two regions	(a) 12 of 55 is what percent? Divide. $12 \div 55$ (b) 1,560,000 of 13,240,000 is what percent? $1{,}560{,}000 \div 13{,}240{,}000$ (c) $18 \div 55$ (d) $4{,}300{,}000 \div 13{,}240{,}000$	We do not need to use the numbers relating to the Northwest or the Southwest in this problem.

2. *Solve and state the answer.*

(a) $\frac{12}{55} = 0.21818\ldots$
$\approx 22\%$

(b) $\frac{1{,}560{,}000}{13{,}240{,}000} = \frac{156}{1324} \approx 0.1178$
$\approx 12\%$

(c) $\frac{18}{55} = 0.32727\ldots$
$\approx 33\%$

(d) $\frac{4{,}300{,}000}{13{,}240{,}000} = \frac{430}{1324} \approx 0.3248$
$\approx 32\%$

(e) We notice that 22% of the sales force in the Northeast made 12% of the sales. The percent of the sales compared to the percent of the sales force is about half (12% of 24% would be half) or 50%. 33% of the sales force in the Southeast made 32% of the sales. The percent of sales compared to the percent of the sales force is close to 100%. We must be cautious here. *If there are no other significant factors,* it would appear that the Southeast sales force is more effective. (There may be other significant factors affecting sales, such as a recession in the Northeast, new and inexperienced sales personnel, or fewer competing companies in the Southeast.)

3. *Check.*
You may want to use a calculator to check the division in step 2, or you may use estimation.

(a) $\frac{12}{55} \rightarrow \frac{10}{60} \approx 0.17$
$= 17\%$ ✓

(b) $\frac{1{,}560{,}000}{13{,}240{,}000} \rightarrow \frac{1.6}{13} \approx 0.12$
$= 12\%$ ✓

(c) $\frac{18}{55} \rightarrow \frac{20}{60} \approx 0.33$
$= 33\%$ ✓

(d) $\frac{4{,}300{,}000}{13{,}240{,}000} \rightarrow \frac{4.3}{13} \approx 0.33$
$= 33\%$ ✓

Practice Problem 2

Using the chart for Example 2, answer the following questions (round all answers to nearest whole percent):

(a) What percent of the sales personnel are assigned to the Northwest?

(b) What percent of the sales volume is attributed to the Northwest?

(c) What percent of the sales personnel are assigned to the Southwest?

(d) What percent of the sales volume is attributed to the Southwest?

(e) Which of these two regions of the country has sales personnel that appear to be more effective in terms of volume of sales?

MATHEMATICS BLUEPRINT FOR PROBLEM SOLVING			
Gather the facts.	What am I solving for?	What must I calculate?	Key points to remember.

To Think About Suppose in 1996 the number of sales personnel (55) increases by 60%. What would the new number of sales personnel be? Suppose in 1997 that the number of sales personnel decreases by 60% from the number of sales personnel in 1996. What would the new number be? Why is this number not 55 since we have increased by 60% and decreased by 60%? Explain.

0.7 Exercises

Use the mathematical blueprint for problem solving to assist you in solving each of the following problems.

1. Carlos and Maria want to install wall-to-wall carpeting in their family room. The floor of the rectangular room is $11\frac{1}{4}$ feet long and $18\frac{1}{2}$ feet wide. How much will it cost to carpet the room if the carpet costs $20.00 per square yard? $462.50

2. Michael and Barbara want to install wall-to-wall carpeting in their bedroom. The floor of the rectangular room is $13\frac{1}{2}$ feet long and $8\frac{3}{4}$ feet wide. How much will it cost to carpet the room if the carpet costs $24.00 per square yard? $315

3. In order to put in a new lawn, the landscaper told Mr. Lopez to add new loam to a depth of $\frac{1}{2}$ foot. Mr. Lopez' lawn is $85\frac{1}{2}$ feet by 60 feet. How many cubic yards of loam does he need? (There are 27 cubic feet in 1 cubic yard.) 95 cubic yards

4. The Brock family removed a built-in swimming pool and decided to fill in the hole with dirt and seed the area. The pool hole is 30 feet by 12 feet by 9 feet deep. How many cubic yards of dirt are needed to fill in the hole? (There are 27 cubic feet in 1 cubic yard.) 120 cubic yards

The following directions are posted on the wall at the gym.

Beginning exercise training schedule
On day 1, each athlete will begin the morning as follows:
Jog .... 1 1/5 miles
Walk for.......... 1 3/4 miles
Rest for........... 2 1/2 minutes
Walk for.......... 1 mile

5. Betty's athletic trainer told her to follow the beginning exercise training schedule on day 1. On day 2 she is to increase all distances and times by $\frac{1}{3}$ of day 1. On day 3 she is to increase all distances and times by $\frac{1}{3}$ of day 2. What will be her training schedule on day 3?

Jog $2\frac{2}{15}$ miles; walk $3\frac{1}{9}$ miles; rest $4\frac{4}{9}$ minutes; walk $1\frac{7}{9}$ miles

6. Melinda's athletic trainer told her to follow the beginning exercise training schedule on day 1. On day 2 she is to increase all distances and times by $\frac{1}{3}$ of day 1. On day 3 she is to once again increase all distances and times by $\frac{1}{3}$ of day 1. What will be her training schedule on day 3?

Jog 2 miles; walk $2\frac{11}{12}$ miles; rest $4\frac{1}{6}$ minutes; walk $1\frac{2}{3}$ miles

To Think About

7. Who will have a more demanding schedule on day 3, Betty or Melinda? Why?

Betty; Melinda increases each activity by $\frac{2}{3}$ by day 3 and Betty increases each activity by $\frac{7}{9}$ by day 3.

8. If Betty kept up the same type of increase day after day, how many miles would she be jogging on day 5?

$3\frac{107}{135}$

9. If Melinda kept up the same type of increase day after day, how many miles would she be jogging on day 7?

$3\frac{3}{5}$

10. Which athletic trainer would appear to have the best plan for training athletes if they used this plan for 14 days? Why? Answers will vary.

11. Roberto decides to visit his grandparents for Thanksgiving break. His college is in El Paso, Texas, and his family is in Minneapolis, Minnesota. His odometer read 45,678.2 when he started his trip and 47,001.2 when he arrived at his destination. He started and ended the trip on a full tank of gas. He made gas purchases of 11 gallons, 12.8 gallons, 13.2 gallons, and 12 gallons for the car during the trip. How many miles per gallon did the car get on the trip? 27 miles per gallon

12. Rochelle wants to go to college in New Orleans. She and her parents take a long weekend to check out the school, and drive from Washington DC. Her odometer read 37,375.3 when she started the trip and 38,473.3 when she arrived at her destination. She started and ended her trip on a full tank of gas. She made gas purchases of 7.9 gallons, 10.3 gallons, and 7 gallons for the car during the trip. How many miles per gallon did the car get on the trip? 26 miles per gallon

13. In 1985, Springfield had a population of 120,000. Between 1985 and 1990, the city increased its population by 10%. Between 1990 and 1995, the city increased its population by 8%. What was the population in 1995? What might be some of the reasons for an increase in population in Springfield? 142,560; Answers will vary. Samples include: improved schools, new job opportunities, availability of affordable housing.

14. In Center City in 1986, 450 newborn infants died in City Hospital within 4 days of their birth. Soon the city began a massive health awareness campaign for pregnant mothers. Between 1986 and 1990, the number of newborn deaths decreased by 50%. Between 1990 and 1994, the number of newborn deaths decreased by 20%. How many newborn infants died within 4 days of their birth in City Hospital in 1994? 180

North Shore Community College students recently conducted a survey of 1000 passengers at Logan Airport. The passengers were classified as follows:

15. What percent of the total travelers were men age 21 or older? What percent of the business travelers were men age 21 or older? Which percentage is greater? Why do you suppose this is the case? 49%; 61%; business travelers; Answers may vary; A sample is: Men 21 or older spend more of their travel time on business than for pleasure.

16. What percent of the business travelers were men age 21 or older? What percent of the business travelers were women age 21 or older? What percent of the business travelers were young people age 20 or younger? Why do you suppose this percentage is so small? 61%; 34%; 5%; Answers may vary. A sample answer is: Parents are usually unable to take their children on business trips.

	Men age 21 or older	Women age 21 or older	Young people age 20 or younger	Total
Business	380	210	30	620
Pleasure	110	150	120	380
Total	490	360	150	1000

Use the following information from a pay check stub to solve Problems 17 through 20.

TOBEY & SLATER, INC.
5000 STILLWELL AVENUE
QUEENS, NY 10001

Check Number	Payroll Period From Date	Payroll Period To Date	Pay Date
495885		11-30-95	12-01-95

Name	Social Security No.	I.D. Number	File No.	Rate/Salary	Department	MS	Dep	Res
Fred J. Gilliani	012-34-5678	01	1379	1150.00	0100	M	5	NY

	Current	Year To Date		Current	Year To Date
GROSS	1,150.00	6,670.00	STATE	67.76	388.45
FEDERAL	138.97	781.07	LOCAL	5.18	30.04
FICA	87.98	510.28	DIS-SUI	.00	.00
W-2 GROSS		6,670.00	NET	790.47	4,960.16

Earnings No.	Type	Hours	Rate	Amount	Dept/Job No.	Deductions/Specials No.	Description	Amount
96	REGULAR			1,150.00	0100	82	Retirement	12.56
						75	Medical	36.28
						56	Union Dues	10.80

Gross pay is the pay an employee receives for his or her services before deductions. The Net pay is the pay the employee actually gets to take home. You may round each amount to the nearest whole percent for Problems 17 through 20.

17. What percent of Fred's gross pay is deducted for federal, state, and local taxes? 18%

18. What percent of Fred's gross pay is deducted for retirement and medical? 4%

19. What percent of Fred's gross pay does he actually get to take home? 69%

20. What percent of Fred's deductions are special deductions? 17%

Chapter Organizer for Fractions

Topic	Procedure	Examples
Simplifying fractions, p. 3.	**1.** Write the **numerator** and **denominator** as a product of prime factors. **2.** Use the basic rule of fractions that $\frac{a \times c}{b \times c} = \frac{a}{b}$ for any factor that appears in both the numerator and the denominator. **3.** Multiply the remaining factors for the numerator and separately for the denominator.	$\frac{15}{25} = \frac{\cancel{5} \cdot 3}{\cancel{5} \cdot 5} = \frac{3}{5}$ $\frac{36}{48} = \frac{\cancel{2} \cdot \cancel{2} \cdot 3 \cdot \cancel{3}}{\cancel{2} \cdot \cancel{2} \cdot 2 \cdot 2 \cdot \cancel{3}} = \frac{3}{4}$ $\frac{26}{39} = \frac{2 \cdot \cancel{13}}{3 \cdot \cancel{13}} = \frac{2}{3}$
Changing improper fractions to mixed numbers, p. 6.	**1.** Divide the denominator into the numerator to obtain the whole-number part of the mixed fraction. **2.** The remainder from the division will be the numerator of the fraction. **3.** The denominator remains unchanged.	$\frac{14}{3} = 4\frac{2}{3}$ $\quad \frac{19}{8} = 2\frac{3}{8}$ since $\begin{array}{r} 4 \\ 3\overline{)14} \\ \underline{12} \\ 2 \end{array}$ $\quad$ since $\begin{array}{r} 2 \\ 8\overline{)19} \\ \underline{16} \\ 3 \end{array}$
Changing mixed numbers to improper fractions, p. 8.	**1.** Multiply the whole number by the denominator and add the result to the numerator. This will yield the new numerator. **2.** The denominator does not change.	$4\frac{5}{6} = \frac{(4 \times 6) + 5}{6} = \frac{24 + 5}{6} = \frac{29}{6}$ $3\frac{1}{7} = \frac{(3 \times 7) + 1}{7} = \frac{21 + 1}{7} = \frac{22}{7}$
Changing fractions to equivalent fractions with a given denominator, p. 14.	**1.** Divide the original denominator into the new denominator. This result is the value that we use for multiplication. **2.** Multiply numerator and denominator of the original fraction by that value.	$\frac{4}{7} = \frac{?}{21}$ $\begin{array}{r} 3 \\ 7\overline{)21} \end{array}$ $\leftarrow$ Use this to multiply $\frac{4 \times 3}{7 \times 3} = \frac{12}{21}$
Addition and subtraction of fractions with the same denominator, p. 11.	**1.** Add or subtract the numerators. **2.** Leave the denominator unchanged. **3.** Simplify the answer if possible.	$\frac{1}{12} + \frac{5}{12} = \frac{6}{12} = \frac{1}{2}$ $\frac{13}{17} - \frac{5}{17} = \frac{8}{17}$
Finding the LCD (least common denominator) of two or more fractions, p. 12.	**1.** Write each denominator as the product of prime factors. **2.** The LCD is the product of each different factor. **3.** If a factor occurs more than once in any one denominator, the LCD will contain that factor repeated the greatest number of times that it occurs in any one denominator.	Find the LCD of $\frac{4}{15}$ and $\frac{3}{35}$. $15 = 5 \cdot 3$ $35 = 5 \cdot 7$ $\text{LCD} = 3 \cdot 5 \cdot 7 = 105$ Find the LCD of $\frac{11}{18}$ and $\frac{7}{45}$. $18 = 3 \cdot 3 \cdot 2$ (factor 3 twice) $45 = 3 \cdot 3 \cdot 5$ (factor 3 twice) $\text{LCD} = 2 \cdot 3 \cdot 3 \cdot 5 = 90$
Adding and subtracting fractions that do not have a common denominator, p. 13.	**1.** Find the LCD. **2.** Change each fraction to an equivalent fraction with the LCD for a denominator. **3.** Add or subtract the fractions, simplifying the answer if possible.	$\frac{3}{8} + \frac{1}{3} = \frac{3 \cdot 3}{8 \cdot 3} + \frac{1 \cdot 8}{3 \cdot 8}$ $= \frac{9}{24} + \frac{8}{24} = \frac{17}{24}$ $\frac{11}{12} - \frac{1}{4} = \frac{11}{12} - \frac{1 \cdot 3}{4 \cdot 3}$ $= \frac{11}{12} - \frac{3}{12} = \frac{8}{12} = \frac{2}{3}$
Adding and subtracting mixed numbers, p. 16.	**1.** Change the mixed numbers to improper fractions. **2.** Follow the rules for adding and subtracting fractions. **3.** If necessary, change your answer to a mixed number.	$1\frac{2}{3} + 1\frac{3}{4} = \frac{5}{3} + \frac{7}{4} = \frac{5 \cdot 4}{3 \cdot 4} + \frac{7 \cdot 3}{4 \cdot 3}$ $= \frac{20}{12} + \frac{21}{12} = \frac{41}{12} = 3\frac{5}{12}$ $2\frac{1}{4} - 1\frac{3}{4} = \frac{9}{4} - \frac{7}{4} = \frac{2}{4} = \frac{1}{2}$

Chapter Organizer for Decimals and Percents

Topic	Procedure	Examples
Multiplying fractions, p. 22.	1. If there are no common factors, multiply the numerators. Then multiply the denominators. 2. If possible, write numerators and denominators as the product of prime factors. Use the basic rule of fractions for any value that appears in both a numerator and a denominator. Multiply the remaining factors in the numerator. Multiply the remaining factors in the denominator.	$\frac{3}{7} \times \frac{2}{13} = \frac{6}{91}$ $\frac{6}{15} \times \frac{35}{91} = \frac{2 \cdot 3}{3 \cdot 5} \times \frac{5 \cdot 7}{7 \cdot 13} = \frac{2}{13}$ $3 \times \frac{5}{8} = \frac{3}{1} \times \frac{5}{8} = \frac{15}{8}$ or $1\frac{7}{8}$
Dividing fractions, p. 24.	1. Change the division sign to multiplication. 2. Invert the second fraction. 3. Multiply the fractions.	$\frac{4}{7} \div \frac{11}{3} = \frac{4}{7} \times \frac{3}{11} = \frac{12}{77}$ $\frac{5}{9} \div \frac{5}{7} = \frac{5}{9} \times \frac{7}{5} = \frac{7}{9}$
Multiplying and dividing mixed numbers, p. 24.	1. Change each mixed number to an improper fraction. 2. Use the rules for multiplying or dividing fractions. 3. Change any improper fractions in the answer to mixed numbers.	$2\frac{1}{4} \times 3\frac{3}{5} = \frac{9}{4} \times \frac{18}{5}$ $= \frac{3 \cdot 3}{2 \cdot 2} \times \frac{2 \cdot 3 \cdot 3}{5}$ $= \frac{3 \cdot 3 \cdot 3 \cdot 3}{2 \cdot 5} = \frac{81}{10} = 8\frac{1}{10}$ $1\frac{1}{4} \div 1\frac{1}{2} = \frac{5}{4} \div \frac{3}{2}$ $= \frac{5}{4} \times \frac{2}{3} = \frac{5}{2 \cdot 2} \times \frac{2}{3} = \frac{5}{6}$
Changing fractional form to decimal form, p. 31.	Divide the **denominator** into the **numerator.**	$\frac{5}{8} = 0.625$ Since $8\overline{)5.000}$ = 0.625 48 20 16 40 40
Changing decimal form to fractional form, p. 32.	1. Write the decimal as a fraction with a denominator of 10, 100, 1000, and so on. 2. Simplify the fraction, if possible.	$0.37 = \frac{37}{100}$ $0.375 = \frac{375}{1000} = \frac{3}{8}$ $0.4 = \frac{4}{10} = \frac{2}{5}$
Adding and subtracting decimals, p. 33.	1. Carefully line up the decimal points as an indicated addition or subtraction. (Extra zeros may be added if desired.) 2. Add or subtract the appropriate digits.	Add. 1.236 + 7.825: 1.236 + 7.825 = 9.061 Subtract. 2 − 1.32: 2.00 − 1.32 = 0.68
Multiplying decimals, p. 34.	1. First multiply the appropriate digits. 2. Count the total number of decimal places in each number being multiplied. 3. Find the sum of the two decimal places. 4. Place the decimal point so that that number of decimal places appears in the answer.	0.9 (one place) × 0.7 (one place) 0.63 (two places) 0.009 (three places) × 0.07 (two places) 0.00063 (five places)
Dividing decimals, p. 35.	1. Count the number of decimal places in the **divisor.** 2. Move the decimal point to the right the same number of places in both the **divisor** and **dividend.** 3. Mark that position with a caret ($_\wedge$). 4. Perform the division. Line up the decimal point in the **quotient** with the position indicated by the caret in the **dividend.**	Divide. 7.5 ÷ 0.6. Move decimal point one place to the right. $0.6_\wedge\overline{)7.5_\wedge0}$ = 12.5 6 15 12 30 30 Therefore, 7.5 ÷ 0.6 = 12.5.

Chapter Organizer for Decimals and Percents

Topic	Procedure	Examples
Multiplying by a multiple of 10, p. 37.	For every zero in the multiplier, move the decimal point one place to the right.	$3.86 \times 100 = 386$ $0.0072 \times 1000 = 7.2$
Dividing by a multiple of 10, p. 37.	For every zero in the **divisor,** move the decimal point one place to the left.	$12.36 \div 10 = 1.236$ $1.97 \div 1000 = 0.00197$
Changing a decimal to percent, p. 41.	**1.** Move the decimal point two places to the right. **2.** Add the % symbol.	$0.46 = 46\%$ $0.002 = 0.2\%$ $1.59 = 159\%$ $0.013 = 1.3\%$ $0.0007 = 0.07\%$
Changing a percent to a decimal, p. 42.	**1.** Remove the % symbol. **2.** Move the decimal point two places to the left.	$49\% = 0.49$ $180\% = 1.8$ $59.8\% = 0.598$ $0.13\% = 0.0013$
Finding a percent of a number, p. 43.	**1.** Convert the **percent** to a decimal. **2.** Multiply the decimal by the number.	Find 12% of 86. $12\% = 0.12$ $\begin{array}{r} 86 \\ \times\ 0.12 \\ \hline 1\ 72 \\ 8\ 6\ \\ \hline 10.32 \end{array}$ Therefore, 12% of 86 = 10.32.
Finding what percent one number is of another number, p. 44.	**1.** Place the number after the word *of* in the denominator. **2.** Place the other number in the **numerator.** **3.** If possible, simplify the fraction. **4.** Change the **fraction** to a decimal. **5.** Express the decimal as a **percent.**	What percent of 8 is 7? $\frac{7}{8} = 0.875 = 87.5\%$ 42 is what percent of 12? $\frac{42}{12} = \frac{7}{2} = 3.5 = 350\%$
Estimation, p. 49.	**1.** Round each number to an appropriate place. **2.** Perform the calculation with the rounded numbers.	Estimate the number of square feet in a room that is 22 feet long and 13 feet wide. Assume that the room is rectangular in shape. **1.** We round 22 to 20. We round 13 to 10. **2.** To find the area of a rectangle, we multiply length times width. We multiply $20 \times 10 = 200.$ We estimate that there are 200 square feet in the room.
Problem Solving, p. 54.	In solving an applied problem, you may find it helpful to complete the following steps. You will not use all of the steps all of the time. Choose the steps that best fit the conditions of the problem. **1.** *Understand the problem.* (a) Read the problem carefully. (b) Draw a picture if this helps you. (c) Use the Mathematics Blueprint for Problem Solving. **2.** *Solve and state the answer.* **3.** *Check.* (a) Estimate to see if your answer is reasonable. (b) Repeat your calculation. (c) Work backwards from your answer. Do you arrive at the original conditions of the problem?	Susan is installing wall to wall carpeting in her $10\frac{1}{2}$ ft by 12 ft bedroom. How much will it cost at $20 a square yard? **1.** *Understand the problem.* (a) We need to find the area of the room in square yards. (b) Then we can find the cost. **2.** *Solve and state the answer.* (a) $10\frac{1}{2} \times 12 = \frac{21}{2} \times \frac{12}{1} = 126$ sq ft (b) $126 \div 9 = 14$ sq yd (c) $14 \times 20 = \$280$ The carpeting will cost $280. **3.** *Check.* (a) $10 \times 12 = 120$ sq ft $120 \div 9 = 13\frac{1}{3}$ sq yd (b) $13 \times 20 = \$260$ Our answer is reasonable. ✓

Chapter 0 Review Problems

0.1 *In Problems 1 through 4, simplify.*

1. $\frac{36}{48}$ $\frac{3}{4}$

2. $\frac{15}{50}$ $\frac{3}{10}$

3. $\frac{24}{72}$ $\frac{1}{3}$

4. $\frac{18}{30}$ $\frac{3}{5}$

5. Write as an improper fraction $2\frac{5}{7}$.

$\frac{19}{7}$

6. Write as a mixed number $\frac{34}{5}$.

$6\frac{4}{5}$

7. Write as a mixed fraction $\frac{27}{4}$.

$6\frac{3}{4}$

Change the given fraction to an equivalent fraction with the specified denominator.

8. $\frac{5}{8} = \frac{?}{24}$ 15

9. $\frac{1}{3} = \frac{?}{45}$ 15

10. $\frac{4}{7} = \frac{?}{21}$ 12

11. $\frac{2}{5} = \frac{?}{55}$ 22

0.2 *Combine.*

12. $\frac{2}{3} + \frac{1}{4}$

$\frac{8+3}{12} = \frac{11}{12}$

13. $\frac{1}{12} + \frac{3}{8}$

$\frac{2+9}{24} = \frac{11}{24}$

14. $\frac{7}{20} - \frac{1}{12}$

$\frac{21-5}{60} = \frac{4}{15}$

15. $\frac{5}{12} - \frac{1}{16}$

$\frac{20-3}{48} = \frac{17}{48}$

16. $3\frac{1}{6} + 2\frac{3}{5}$

$\frac{95+78}{30} = \frac{173}{30}$ or $5\frac{23}{30}$

17. $1\frac{1}{4} + 2\frac{7}{10}$

$\frac{25+54}{20} = \frac{79}{20}$ or $3\frac{19}{20}$

18. $2\frac{1}{5} - 1\frac{2}{3}$

$\frac{33-25}{15} = \frac{8}{15}$

19. $3\frac{1}{15} - 1\frac{3}{20}$

$\frac{184-69}{60} = \frac{23}{12}$ or $1\frac{11}{12}$

0.3 *Multiply.*

20. $\frac{2}{3} \times \frac{9}{10}$

$\frac{3}{5}$

21. $\frac{3}{5} \times \frac{10}{21}$

$\frac{2}{7}$

22. $6 \times \frac{5}{11}$

$\frac{30}{11}$ or $2\frac{8}{11}$

23. $2\frac{1}{3} \times 4\frac{1}{2}$

$\frac{21}{2}$ or $10\frac{1}{2}$

24. $1\frac{1}{8} \times 2\frac{1}{9}$

$\frac{19}{8}$ or $2\frac{3}{8}$

25. $\frac{4}{7} \times 5$

$\frac{20}{7}$ or $2\frac{6}{7}$

Divide.

26. $\frac{1}{3} \div \frac{1}{6}$

2

27. $\frac{1}{2} \div \frac{1}{8}$

4

28. $\frac{2}{5} \div \frac{4}{3}$

$\frac{3}{10}$

29. $\frac{5}{7} \div \frac{15}{14}$

$\frac{2}{3}$

30. $\frac{15}{16} \div 6\frac{1}{4}$

$\frac{3}{20}$

31. $2\frac{6}{7} \div \frac{10}{21}$

6

0.4 *Combine.*

32. 1.324 + 2.008 + 1.130

4.462

33. 24.831 − 17.094

7.737

34. 14.037 − 2.61

11.427

35. 1.6 + 3.21 + 0.004

4.814

36. 1.3 + 1.8 + 2.6 + 7.2 + 0.8

13.7

37. 100.01 + 10.001 + 1.1011

111.1121

Multiply.

38. 0.002×4.31
0.00862

39. 362.341×1000
362,341

40. $2.6 \times 0.03 \times 1.02$
0.07956

41. $1.08 \times 0.06 \times 160$
10.368

Divide.

42. $0.186 \div 100$
0.00186

43. $71.32 \div 1000$
0.07132

44. $0.186 \div 93$
0.002

45. $1.35 \div 0.015$
90

46. $0.147 \div 2.1$
0.07

47. $0.19 \div 0.38$
0.5

48. Write as a percent. $\frac{3}{8}$
37.5%

49. Write as a simplified fraction. 24%
$\frac{6}{25}$

0.5 *In Problems 50 through 53, write the percentages in decimal form.*

50. 1.4% 0.014

51. 36.1% 0.361

52. 0.02% 0.0002

53. 125.3% 1.253

54. What is 65% of 400? 260

55. Find 250% of 36. 90

56. What is 1.8% of 1000?
18

57. What percent of 120 is 15?
12.5%

58. What percent of 1250 is 750?
60%

59. 36% of Americans have blood type B. If there are 270,000,000 Americans, how many of these have blood type B? 97,200,000

60. In a given university, 720 of the 960 freshmen had a math deficiency. What percentage of the class had a math deficiency? 75%

0.6 *Estimate each of the following. Do not find an exact value.*

61. $234{,}897 \times 1{,}936{,}112$
400,000,000,000

62. $357 + 923 + 768 + 417$
2500

63. $780{,}000 - 198{,}000$
600,000

64. $7\frac{1}{3} + 3\frac{5}{6} + 8\frac{3}{7}$
19

65. Find 18% of \$56,297
\$12,000

66. $12{,}482 \div 389$
30

67. Estimate the cost to have a chimney repaired by a stone mason who charges \$18.00 per hour and says the job will take her $3\frac{1}{2}$ days. Assume that she works 8 hours in one day. \$480 or \$640

68. Estimate the monthly cost for each of three roommates who want to share an apartment that costs \$923.50 per month. \$300

0.7 *Solve. You may use the mathematical blueprint for problem solving.*

69. Mr. and Mrs. Carr are installing wall-to-wall carpeting in a room that measures $12\frac{1}{2}$ ft by $9\frac{2}{3}$ ft. How much will it cost if the carpet is \$26.00 per square yard? \$349.07

70. The population of Calais was 34,000 in 1980 and 36,040 in 1995. What was the percent of increase in population? 6%

Chapter 0 Test

In Problems 1 and 2, simplify.

1. $\frac{16}{18}$

2. $\frac{35}{40}$

3. Write as an improper fraction. $6\frac{3}{7}$

4. Write as a mixed number. $\frac{108}{33}$

5. Change the given fraction to an equivalent fraction with the specified denominator. $\frac{11}{16} = \frac{?}{48}$

Perform the operations indicated. Simplify answers whenever possible.

6. $\frac{5}{9} + \frac{1}{6}$

7. $\frac{1}{3} + \frac{1}{4} + \frac{5}{6}$

8. $\frac{2}{3} - \frac{1}{6}$

9. $1\frac{1}{8} + 3\frac{3}{4}$

10. $3\frac{2}{3} - 2\frac{5}{6}$

11. $6 - 4\frac{3}{4}$

12. $\frac{5}{7} \times \frac{28}{15}$

13. $\frac{3}{11} \times \frac{7}{5}$

14. $\frac{7}{4} \div \frac{1}{2}$

15. $2\frac{1}{2} \times 3\frac{1}{4}$

16. $8\frac{5}{6} \div 4\frac{1}{9}$

17. Write as a fraction. 0.72

18. Write as a decimal. $\frac{7}{16}$

1. $\frac{8}{9}$

2. $\frac{7}{8}$

3. $\frac{45}{7}$

4. $3\frac{3}{11}$

5. 33

6. $\frac{13}{18}$

7. $\frac{17}{12}$ or $1\frac{5}{12}$

8. $\frac{1}{2}$

9. $\frac{39}{8}$ or $4\frac{7}{8}$

10. $\frac{5}{6}$

11. $\frac{5}{4}$ or $1\frac{1}{4}$

12. $\frac{4}{3}$ or $1\frac{1}{3}$

13. $\frac{21}{55}$

14. $\frac{7}{2}$ or $3\frac{1}{2}$

15. $\frac{65}{8}$ or $8\frac{1}{8}$

16. $\frac{159}{74}$ or $2\frac{11}{74}$

17. $\frac{18}{25}$

18. 0.4375

19.	14.64
20.	1806.62
21.	1.312
22.	16.32
23.	230
24.	19.3658
25.	7.3%
26.	1.965
27.	63
28.	0.336
29.	12.5%
30.	30%
31.	52%
32.	18
33.	100
34.	99
35.	\$2700
36.	65%

Perform the calculations indicated.

19. $1.6 + 3.24 + 9.8$

20. $2003.42 - 196.8$

21. 32.8×0.04

22. 0.1632×100

23. $12.88 \div 0.056$

24. $19{,}365.8 \div 1000$

25. Write as a percent. 0.073

26. Write as a decimal. 196.5%

27. What is 18% of 350?

28. What is 2% of 16.8?

29. What percent of 120 is 15?

30. What percent of 460 is 138?

31. At Western College, 416 of 800 freshmen are women. What percent are women?

32. A 4-inch stack of computer chips is on the table. Each computer chip is $\frac{2}{9}$ of an inch thick. How many computer chips are in the stack?

Estimate each of the following.

33. $52{,}344\overline{)4{,}678{,}987}$

34. $18\frac{1}{3} + 22\frac{6}{7} + 57\frac{1}{2}$

35. Estimate the amount of money it would cost to buy the following four items: a computer for \$1867.85, a computer stand for \$98.87, a computer printer for \$397.49, and a VGA monitor for \$278.59.

Solve. You may use the mathematical blueprint for problem solving.

36. Allison is paid \$14,000 per year plus a sales commission of 3% of the value of her sales. Last year she sold \$870,000 worth of products. What percent of her total income was her commission?

CHAPTER 1
Real Numbers and Variables

Determining how physically fit you are is not just a subjective measurement of how active you are or how much you weigh. Increasingly, doctors and health fitness professionals are discovering that certain mathematical measurements of your body can be used in a formula to predict the level of fitness of your body. Want to find out if you are physically fit using mathematics? Then turn to the Putting Your Skills to Work on page 113.

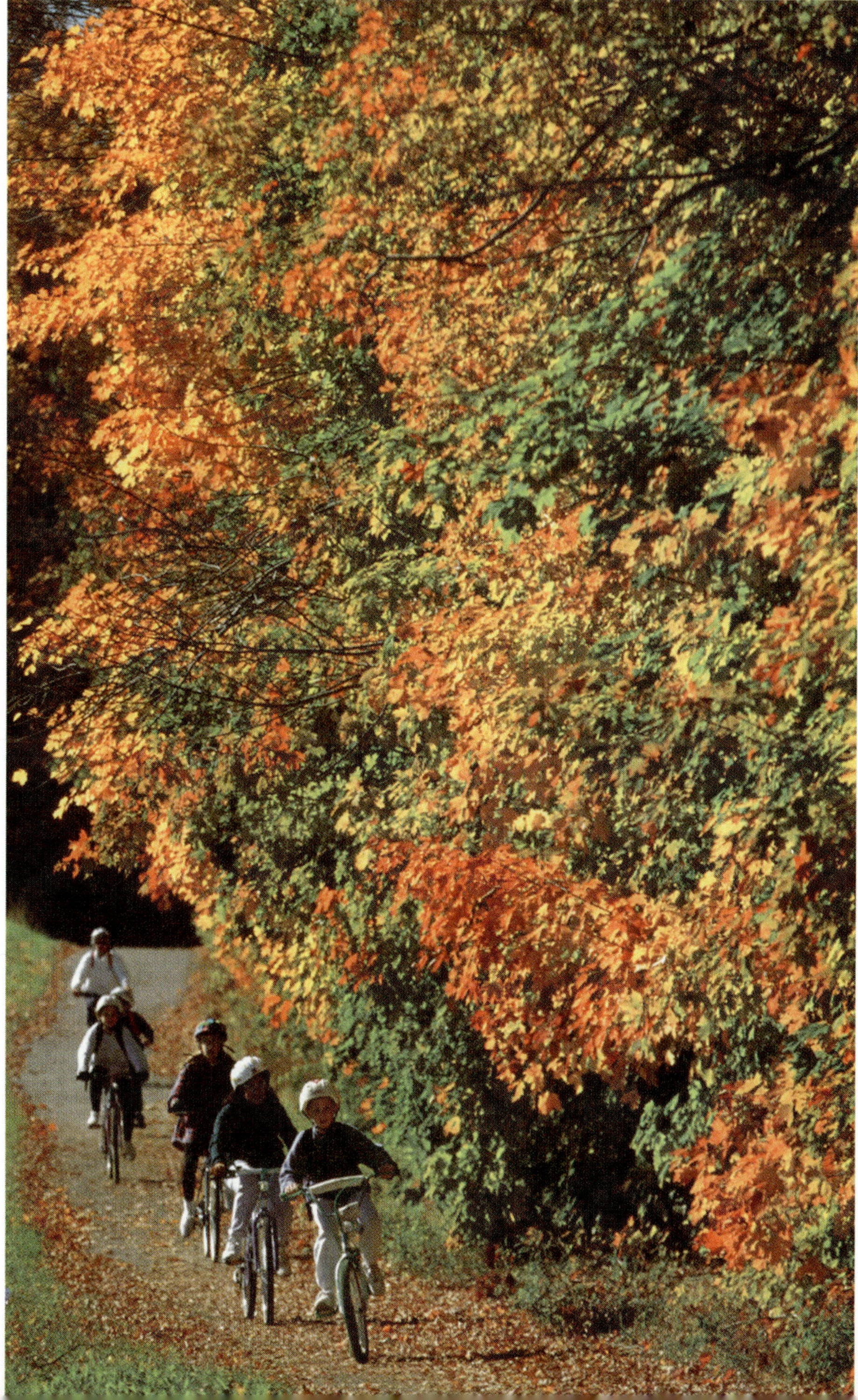

Pretest Chapter 1

If you are familiar with the topics in this chapter, take this test now. Check your answers with those in the back of the book. If you obtained an incorrect answer or could not do a problem, study the appropriate section of the chapter.

If you are not familiar with the topics in this chapter, do not take the test now. Instead, study the examples, work the practice problems, and then take the test.

This test will help you to identify the concepts that you have mastered and the concepts that you must study more.

Perform the necessary operation and simplify your answer.

Sections 1.1–1.2

1. $(-3) - (-6)$ **2.** $-5 + 6 - 3 - 2$

3. $(-7) + (-11)$ **4.** $7 - (-13)$

Section 1.3

5. $(-7)(-2)(+3)(-1)$ **6.** $\dfrac{-\frac{2}{3}}{-\frac{1}{4}}$

7. $(-7)(-4)$ **8.** $20 \div (-5)$

Section 1.4

9. $(-2)^4$ **10.** 4^3 **11.** $\left(-\frac{2}{3}\right)^3$ **12.** -4^4

Section 1.5

13. $-2x(3x - 2xy + z)$ **14.** $-(3x - 4y - 12)$

Section 1.6

Combine like terms.

15. $5x^2 - 3xy - 6x^2y - 8xy$ **16.** $11x - 3y + 7 - 4x + y$

17. $3(2x - 5y) - (x - 8y)$ **18.** $2y(x^2 + y) - 3(x^2y - 5y^2)$

Section 1.7

Simplify.

19. $6 \cdot 2 + 8 \cdot 3 - 4 \cdot 2$ **20.** $3(4)^3 + 18 \div 3 - 1$

21. $-\frac{1}{6}\left(\frac{1}{2}\right) + \frac{3}{4} \div \frac{1}{12}$ **22.** $1.22 - 4.1(1.4) + (-3.3)^2$

Section 1.8

23. Evaluate $3x^2 - 5x - 4$ when $x = -2$.

24. Evaluate $ab + 3a^2 - 5b$ when $a = 3$ and $b = -1$.

25. Determine the Celsius temperature when the Fahrenheit temperature is 77°. Use the formula $C = \frac{5}{9}(F - 32)$.

Section 1.9

Simplify.

26. $3[(x - y) - (2x - y)] - 3(2x + y)$

27. $2x^2 - 3x[2x - (x + 2y)]$

Answers:

1. 3
2. -4
3. -18
4. 20
5. -42
6. $\frac{8}{3}$
7. 28
8. -4
9. 16
10. 64
11. $-\frac{8}{27}$
12. -256
13. $-6x^2 + 4x^2y - 2xz$
14. $-3x + 4y + 12$
15. $5x^2 - 6x^2y - 11xy$
16. $7x - 2y + 7$
17. $5x - 7y$
18. $-x^2y + 17y^2$
19. 28
20. 197
21. $\frac{107}{12}$
22. 6.37
23. 18
24. 29
25. 25°C
26. $-9x - 3y$
27. $-x^2 + 6xy$

1.1 Addition of Signed Numbers

After studying this section, you will be able to:

1. *Understand the names of different types of numbers.*
2. *Recognize real-life situations for signed numbers.*
3. *Add signed numbers with the same sign.*
4. *Add signed numbers with opposite signs.*

MathPro

Video 3

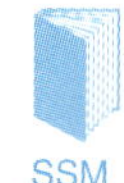
SSM

1 *Different Types of Numbers*

Let's review some of the basic terms we use to talk about numbers.

Integers are numbers such as $\ldots, -3, -2, -1, 0, 1, 2, 3, \ldots$.

Rational numbers are numbers like $\frac{3}{2}$, $\frac{5}{7}$, $-\frac{3}{8}$, $-\frac{4}{13}$, $\frac{6}{1}$, and $-\frac{8}{2}$.

Rational numbers can be written as one integer divided by another integer (as long as the denominator is not zero!). Integers can be written as fractions ($3 = \frac{3}{1}$, for example), so we can see that all integers are rational numbers. Rational numbers can be expressed in decimal form. For example, $\frac{3}{2} = 1.5$, $-\frac{3}{8} = -0.375$, and $\frac{1}{3} = 0.333\ldots$ or $0.\overline{3}$. It is important to note that rational numbers in decimal form are either terminating decimals or repeating decimals.

Irrational numbers are numbers that cannot be expressed as one integer divided by another integer. The numbers π, $\sqrt{2}$, and $\sqrt[3]{7}$ are irrational numbers.

Irrational numbers can be expressed in decimal form. The decimal form of an irrational number is a nonterminating, nonrepeating decimal. For example, $\sqrt{2} = 1.414213\ldots$ can be carried out to an infinite number of decimal places with no repeating of digits.

EXAMPLE 1 Classify each of the following numbers as an integer, as a rational number, or as an irrational number.

(a) 5 **(b)** $-\dfrac{1}{3}$ **(c)** 2.85 **(d)** $\sqrt{2}$ **(e)** $0.777\ldots$

Make a table. Check off the description of the number that applies.

Number	Integer	Rational Number	Irrational Number
a. 5	✓	✓	
b. $-\frac{1}{3}$		✓	
c. 2.85		✓	
d. $\sqrt{2}$			✓
e. $0.777\ldots$		✓	

Practice Problem 1 Classify each of the following numbers.

(a) $-\dfrac{2}{5}$ **(b)** $1.515151\ldots$ **(c)** -8 **(d)** π ■

TEACHING TIP Make sure students are familiar with a number line. Remind them that many of the questions they will have about numbers can be answered through the use of a number line.

Integers and rational numbers can be pictured on a number line.

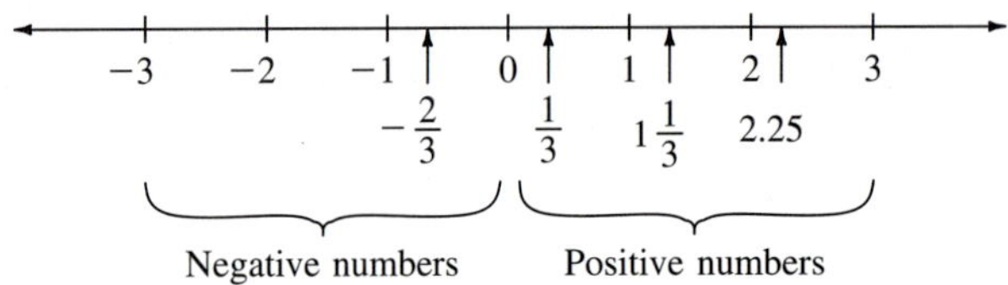

Positive numbers are to the right of 0 on the number line.

Negative numbers are to the left of 0 on the number line.

The **signed numbers** include the positive numbers, the negative numbers, and zero.

2 *Real-life Situations for Signed Numbers*

We often encounter practical examples of number lines that include positive and negative rational numbers. For example, we can tell by reading the thermometer below that the temperature is 20° below 0. From the stock market report, we see that the stock opened at 36, closed at $34\frac{1}{2}$, and the net change for the day was $-1\frac{1}{2}$.

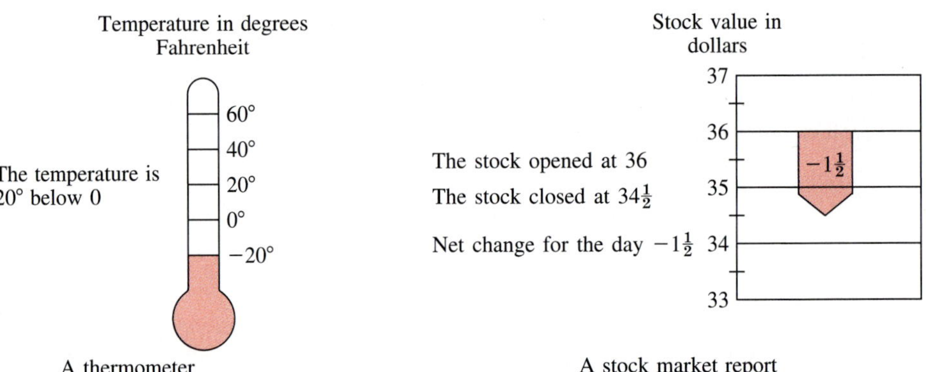

In the following example we use signed numbers to represent real-life situations.

EXAMPLE 2 Use a signed number to represent each situation.

(a) A temperature of 128.6°F below zero is recorded at Vostok, Antarctica.

(b) The Himalayan peak K-2 rises 29,064 feet above sea level.

(c) The Dow gains 10.24 points.

(d) A withdrawal of $316 is made from a savings account.

A key word can help you to decide if a number is positive or negative.

(a) 128.6°F *below* zero is −128.6.

(b) 29,064 feet *above* sea level is +29,064.

(c) A *gain* of 10.24 points is +10.24.

(d) A *withdrawal* of $316 is −316.

Practice Problem 2 Use a signed number to represent each situation.

(a) A population growth of 1,259

(b) A depreciation of $763

(c) A wind chill factor of minus 10 ■

In everyday life we consider positive numbers to be the opposite of negative numbers. For example, a gain of 3 yards in a football game is the opposite of a loss of 3 yards; a check written for \$2.16 on a checking account is the opposite of a deposit of \$2.16.

Each positive number has an opposite negative number. Similarly, each negative number has an opposite positive number. **Opposite numbers** have the same magnitude but different signs and can be represented on the number line.

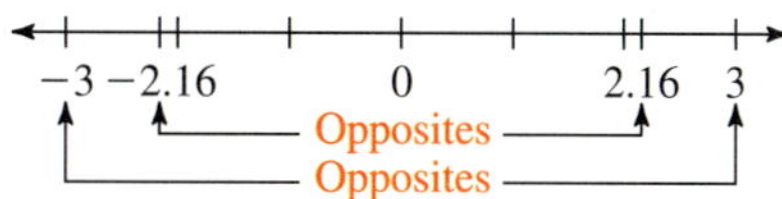

EXAMPLE 3 Find the opposite of each number. **(a)** -7 **(b)** $\frac{1}{4}$

(a) The opposite of -7 is $+7$. **(b)** The opposite of $\frac{1}{4}$ is $-\frac{1}{4}$.

Practice Problem 3 Find the opposite of each number.

(a) $+\frac{2}{5}$ **(b)** -1.92 ■

3 *Adding Signed Numbers with the Same Sign*

To use signed numbers, we need to be clear about the sign part of a signed number. When we write the number three as $+3$, the sign indicates that it is a positive number. The positive sign can be omitted. If someone writes three (3) it is understood that it is a positive three $(+3)$. To write a negative number such as negative three (-3), we must include the sign.

A concept that will help us to add and subtract signed numbers is the idea of absolute value. The **absolute value** of a number is the distance between that number and zero on the number line.

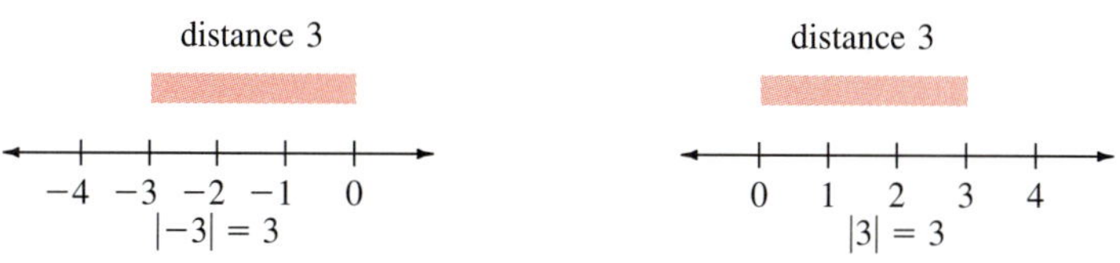

TEACHING TIP The idea of absolute value may be totally new to students. Take some time to explain the notation and to give several different examples. Then ask them to do the following as an Additional Classroom Exercise. Find the following absolute values:

A. $|-3|$
B. $|6|$
C. $|0|$
D. $|-\frac{1}{2}|$
E. $|-2.3|$
F. $|0.008|$

Ans.
A. 3 B. 6 C. 0
D. $\frac{1}{2}$ E. 2.3 F. 0.008

The distance is always a positive number regardless of the direction we travel. This means that the absolute value of any number will be a positive value or zero. We place the symbol "| |" around a number to mean we find the absolute value of the number.

The distance from 0 to 3 is 3, so $|3| = 3$.

This is read "the absolute value of 3 is 3."

The distance from 0 to -3 is 3, so $|-3| = 3$.

This is read "the absolute value of -3 is 3."

Some other examples are

$$|-22| = 22, \qquad |5.6| = 5.6, \qquad |0| = 0$$

Thus, the absolute value of a number can be thought of as the magnitude of the number, without regard to sign.

Now let's look at addition of signed numbers when the two numbers have the same sign. Suppose that you are keeping track of your checking account at a local bank. When you make a deposit of 5 dollars, you record it as $(+5)$. When you write a check for 4 dollars, you record it as (-4), as a debit. Consider two situations.

TEACHING TIP The idea behind these situations is to help students to see logical places in the real world where signed number operations are needed, and the outcome is logical. Feel free to use examples other than banking.

Situation 1

You made a deposit of 20 dollars on one day and a deposit of 17 dollars the next day. You want to know the total value of your deposits.

Your record for situation 1.

$(+20)$	$+$	$(+17)$	$=$	$(+37)$
The amount of the deposit of the first day	added to	the amount of the deposit of the second day	is	the total of deposits made over the two days.

Situation 2

You write a check for 36 dollars to pay one bill and two days later write a check for 5 dollars. You want to know the total value of debits to your account for the two checks.

Your record for situation 2.

(-36)	$+$	(-5)	$=$	(-41)
The value of the first check	added to	the value of the second check	is	the total debit to your account.

In each situation we found that we added the absolute value of each number. (That is, we added the numbers without regarding their sign.) The answer always contained the sign that was common to both numbers.

We will now state these results as a formal rule.

Addition Rule for Two Numbers with the Same Sign

To add two numbers with the same sign, add the absolute values of the numbers and use the common sign in the answer.

EXAMPLE 4 Add. **(a)** $(+14) + (+16)$ **(b)** $(-8) + (-7)$

(a) $(+14) + (+16)$

$14 + 16 = 30$ *Add the absolute values of the numbers.*

$(+14) + (+16) = +30$ *Use the common sign in the answer. Here the common sign is the + sign.*

(b) $(-8) + (-7)$

$8 + 7 = 15$ *Add the absolute values of the numbers.*

$(-8) + (-7) = -15$ *Use the common sign in the answer. Here the common sign is the −.*

Practice Problem 4 Add. $(-23) + (-35)$ ■

EXAMPLE 5 Add. $\left(+\frac{2}{3}\right)+\left(+\frac{1}{7}\right)$

$$\left(+\frac{2}{3}\right)+\left(+\frac{1}{7}\right)$$

$$\left(+\frac{14}{21}\right)+\left(+\frac{3}{21}\right)$$ *Change each fraction to an equivalent fraction with a common denominator of 21.*

$$\frac{14}{21}+\frac{3}{21}=\frac{17}{21}$$ *Add the absolute values of the numbers.*

$$\left(+\frac{14}{21}\right)+\left(+\frac{3}{21}\right)=+\frac{17}{21} \quad or \quad \frac{17}{21}$$ *Use the common sign in the answer. Note that if no sign is written the number is understood to be positive.*

Practice Problem 5 Add. $\left(-\frac{3}{5}\right)+\left(-\frac{4}{7}\right)$ ■

EXAMPLE 6 Add. $(-4.2)+(-3.9)$

$$(-4.2)+(-3.9)$$

$$4.2 + 3.9 = 8.1$$ *Add the absolute values of the numbers.*

$$(-4.2)+(-3.9)=-8.1$$ *Use the common sign in the answer.*

Practice Problem 6 Add. $(-12.7)+(-9.38)$ ■

TEACHING TIP After discussing Example 6 or a similar problem, ask students to do the following as an Additional Class Exercise. Add.

A. $(-20)+(-3)$
B. $(+8.3)+(+9.5)$
C. $(-3/7)+(-2/7)$
D. $(-0.0003)+(-0.0043)$

Ans.
A. -23 B. 17.8
C. $-5/7$ D. -0.0046

The rule for adding two numbers with the same signs can be extended to more than two numbers. If we add more than two numbers with the same sign, the answer will have the sign common to all.

EXAMPLE 7 Add. $(-7)+(-2)+(-5)$

$(-7)+(-2)+(-5)$ *We are adding three signed numbers all with the same sign. We will first add the first two numbers.*

$=(-9)+(-5)$ *Add $(-7)+(-2)=(-9)$.*

$=-14$ *Add $(-9)+(-5)=-14$.*

Of course, this can be shortened by adding the three numbers without regard to sign and then using the common sign for the answer.

Practice Problem 7 Add. $(-7)+(-11)+(-33)$ ■

4 Adding Signed Numbers with Opposite Signs

What if the signs of the numbers you are adding are different? Let's consider our checking account again to see how such a situation might occur.

Situation 3

You made a deposit of 30 dollars on one day. On the next day you write a check for 25 dollars. You want to know the result of your two transactions.

Your record for situation 3.

$(+30)$	$+$	(-25)	$=$	$(+5)$
A positive 30, for the deposit,	added to	a negative 25, for the check, which is a debit,	gives a result of	a net increase of 5 dollars $(+5)$ in the account.

Situation 4

You made a deposit of 10 dollars on one day. The next day you write a check for 40 dollars. You want to know the result of your two transactions.

Your record for situation 4.

$+10$	$+$	(-40)	$=$	(-30)
A positive 10, for the deposit,	added to	a negative 40, for the check, which is a debit,	gives a result of	a net decrease of 30 dollars (-30) in the account

The result is a negative thirty (-30), because the check was larger than the deposit. If you do not have at least 30 dollars in your account at the start of Situation 4, you have overdrawn your account.

What do we observe from Situations 3 and 4? In each case we found the difference of the absolute values of the two numbers. The sign of the result was always the sign of the number with the greater absolute value. Thus, in Situation 3, 30 is larger than 25. The sign of 30 is positive. The sign of the answer $(+5)$ is positive. In Situation 4, 40 is larger than 10. The sign of 40 is negative. The sign of the answer (-30) is negative.

We will now state these results as a formal rule.

Addition Rule for Two Numbers with Different Signs

1. Find the difference between the larger absolute value and the smaller.
2. Give the answer the sign of the number having the larger absolute value.

EXAMPLE 8 Add. $(+8) + (-7)$

$(+8) + (-7)$	*We are to add two signed numbers with opposite signs.*
$8 - 7 = 1$	*Find the difference between the two absolute values, which is 1.*
$(+8) + (-7) = +1$ *or* 1	*The answer will have the sign of the number with the larger absolute value. That number is +8. Its sign is* ***positive,*** *so the answer will be* ***positive*** *1.*

Practice Problem 8 Add. $(-9) + (+3)$ ■

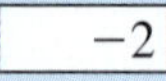

CALCULATOR

Negative Numbers

To enter a negative number on a calculator, find the key marked [+/−]. To enter the number −2, press the key 2 and then the key +/−. The display should read

[−2]

To find $(-32) + (-46)$, enter

32 [+/−] [+] 46 [+/−] [=]

The display should read

[−78]

Try

(a) $(-256) + 184$
(b) $94 + (-51)$
(c) $(-18) - (-24)$
(d) $(-6) + (-10) - (-15)$

Note: The [+/−] key changes the sign of a number from + to − or − to +.

EXAMPLE 9 Add. $(-20) + (+13)$

$(-20) + (+13)$ — *We are adding two signed numbers with different signs.*

$20 \quad - \quad 13 \quad = 7$ — *Take the difference between the two absolute values.*

$(-20) + (+13) = -7$ — *The answer will have the sign of the number with the larger absolute value. That number is -20, which is negative, so the answer will be negative 7.*

Practice Problem 9 Add. $(+37) + (-40)$ ■

It is useful to know the following three properties of signed numbers.

1. *Addition is commutative.* This property states that if two numbers are added, the result is the same if either number is written first. The order of the numbers does not affect the result.

$$(+3) + (+6) = (+6) + (+3) = 9$$

$$(-7) + (-8) = (-8) + (-7) = -15$$

$$(-15) + (+3) = (+3) + (-15) = -12$$

2. *Addition of zero to any given number will result in that given number again.*

$$0 + (+5) = 5$$

$$(-8) + (0) = -8$$

3. *Addition is associative.* This property states that if three numbers are added, it does not matter which two numbers are grouped by parentheses.

$$3 + (5 + 7) = (3 + 5) + 7$$
$$3 + (12) = (8) + 7$$
$$15 = 15$$

First combine numbers inside parentheses; then combine the remaining numbers. The results are the same no matter which numbers are grouped first.

We can use these properties along with the rules we have for adding signed numbers to add three or more numbers. We go from left to right, adding two numbers at a time.

EXAMPLE 10 Add. $\left(+\frac{3}{17}\right) + \left(-\frac{8}{17}\right) + \left(+\frac{4}{17}\right)$

$\left(+\frac{3}{17}\right) + \left(-\frac{8}{17}\right) + \left(+\frac{4}{17}\right)$ — *We observe that all the fractions have a common denominator. We will add the first two numbers first.*

$= \left(-\frac{5}{17}\right) + \left(+\frac{4}{17}\right)$ — *Add $(\frac{3}{17}) + (-\frac{8}{17}) = (-\frac{5}{17})$. The answer is negative since the larger of the two absolute values is negative.*

$= -\frac{1}{17}$ — *Add $(-\frac{5}{17}) + (+\frac{4}{17}) = (-\frac{1}{17})$. The answer is negative since the larger of the two absolute values is negative.*

Practice Problem 10 Add. $\left(-\frac{5}{12}\right) + \left(+\frac{7}{12}\right) + \left(-\frac{11}{12}\right)$ ■

TEACHING TIP After discussing Example 11, you may want to show that the commutative property of addition can be used with this example. Therefore, the problem could just as well be done:

$(-1.8) + (+1.4) + (-2.6)$

$= (-1.8) + (-2.6) + (+1.4)$ by using the commutative property

$= (-4.4) + (+1.4)$

$= -3.0$

The same result is obtained.

Sometimes the numbers being added have the same signs; sometimes they are different. To add three or more numbers, you may need to use rules for both.

EXAMPLE 11 Add. $(-1.8) + (+1.4) + (-2.6)$

$(-1.8) + (1.4) + (-2.6)$ *Add the first two numbers, which are opposite in sign.*

$= (-0.4) + (-2.6)$ *Add* $(-1.8) + (1.4) = (-0.4)$. *We take the difference of* 1.8 *and* 1.4 *and use the sign of the number with the larger absolute value.*

$= -3.0$ *Add* $(-0.4) + (-2.6) = -3.0$. *The signs are the same; we add the absolute values of the numbers and use the common sign.*

Practice Problem 11 Add. $(-6) + (-8) + (+2)$ ■

TEACHING TIP Take some time to show students how Examples 4 through 11 can be written without parentheses and positive signs. Then give them the following as an Additional Class Exercise: Write each of the following problems by avoiding unnecessary use of + signs and parentheses:

A. $(+4.2) + (-3.6)$
B. $(-94) + (+53)$
C. $(-17) + (-15)$
D. $(+1.234) + (+0.287)$

Ans.
A. $4.2 + (-3.6)$
B. $-94 + 53$
C. $-17 + (-15)$
D. $1.234 + 0.287$

Usually students will have some questions, and a short class discussion about the use of the extra parentheses may take place. Remind them that if they want to keep these extra parentheses in while they add the signed numbers, that this is perfectly all right. Later, it will be easier for them to drop the use of these extra symbols.

If many signed numbers are added, it is often easier to add numbers with like signs in a column format. Remember that addition is commutative; therefore, signed numbers can be *added in any order*. You do *not* need to combine the first two numbers as your first step.

EXAMPLE 12 Add. $(-8) + (+3) + (-5) + (-2) + (+6) + (+5)$

$$\begin{array}{r} -8 \\ -5 \\ -2 \\ \hline -15 \end{array} \qquad \begin{array}{r} +3 \\ +6 \\ +5 \\ \hline +14 \end{array}$$

All the signs are the same. Add the three negative numbers to obtain -15.

All the signs are the same. Add the three positive numbers to obtain $+14$.

Add the two results.

$$(-15) + (+14) = -1$$

The answer is negative because the larger absolute value is negative.

Practice Problem 12 Add. $(-6) + (+5) + (-7) + (-2) + (+5) + (+3)$ ■

In actual practice the positive sign and the parentheses are often omitted if it will not change the meaning of the mathematical expression. The only time we really need to show the sign of a number is when the number is negative. For example, -3. The only time we need to show parentheses when we add signed numbers is when we have two different signs preceding a number. For example, $-5 + (-6)$.

EXAMPLE 13 Add each of the following.

(a) $2.8 + (-1.3)$ **(b)** $-\frac{2}{5} + \left(-\frac{3}{5}\right)$ **(c)** $-8 + 4 + (-2)$

(a) $2.8 + (-1.3)$ is equivalent to $(+2.8) + (-1.3) = 1.5$

(b) $-\frac{2}{5} + \left(-\frac{3}{5}\right)$ is equivalent to $\left(-\frac{2}{5}\right) + \left(-\frac{3}{5}\right) = -\frac{5}{5} = -1$

(c) $-8 + 4 + (-2)$ is equivalent to $(-8) + (+4) + (-2) = (-4) + (-2) = -6$

Practice Problem 13 Add each of the following.

(a) $-2.9 + (-5.7)$ **(b)** $\frac{2}{3} + \left(-\frac{1}{4}\right)$ **(c)** $-10 + (-3) + 15 + 4$ ■

1.1 Exercises

Verbal and Writing Skills

Check off the description of the number that applies.

Number	Rational Number	Irrational Number
1. 23	✓	
2. $-\frac{4}{5}$	✓	
3. π		✓
4. 2.34	✓	
5. $-6.666\ldots$	✓	

Number	Rational Number	Irrational Number
6. $-\frac{7}{9}$	✓	
7. $-2.3434\ldots$	✓	
8. 14	✓	
9. $\sqrt{2}$		✓
10. $3.232232223\ldots$		✓

Use a signed number to represent each situation.

11. Aaron writes a check for \$53. -53

12. The house appreciates \$29,000 in value. $+29{,}000$

13. The Dow Jones average is down by $2\frac{3}{8}$. $-2\frac{3}{8}$

14. Jon withdraws \$102 from his account. -102

15. The temperature rises 7°F. $+7$

16. Maya won the game by 12 points. $+12$

Find the opposite of each number.

17. $\frac{3}{4}$ $-\frac{3}{4}$

18. -2 $+2$

19. $-\frac{5}{13}$ $+\frac{5}{13}$

20. 128 -128

Add.

21. $(-9) + (+5)$ -4

22. $(-7) + (+2)$ -5

23. $(-8) + (-5)$ -13

24. $(-12) + (-6)$ -18

25. $\left(-\frac{1}{3}\right) + \left(+\frac{2}{3}\right)$ $\frac{1}{3}$

26. $\left(-\frac{1}{5}\right) + \left(-\frac{3}{5}\right)$ $-\frac{4}{5}$

27. $(+35) + (+20)$ 55

28. $(+18) + (+40)$ 58

29. $(-15) + (-16)$ -31

30. $(-23) + (-13)$ -36

31. $\left(-\frac{2}{13}\right) + \left(-\frac{5}{13}\right)$ $-\frac{7}{13}$

32. $\left(-\frac{5}{14}\right) + \left(+\frac{2}{14}\right)$ $-\frac{3}{14}$

33. $(-1.5) + (-2.3)$ -3.8 **34.** $(-1.8) + (-1.4)$ -3.2 **35.** $(+0.6) + (-0.2)$ 0.4 **36.** $(-0.8) + (+0.5)$ -0.3

37. $(-12) + (-13)$ -25 **38.** $(-17) + (-21)$ -38 **39.** $\left(-\frac{2}{5}\right) + \left(+\frac{3}{7}\right)$ $\frac{1}{35}$ **40.** $\left(-\frac{2}{7}\right) + \left(+\frac{3}{14}\right)$ $-\frac{1}{14}$

41. $(+2) + (-7) + (-6)$ -11 **42.** $-5 + 3 + (-7)$ -9 **43.** $(-3) + (+8) + (+5) + (-7)$ 3

44. $(-2) + (-6) + (+7) + (+3)$ 2 **45.** $(-2) + (+8) + (-3) + (-5)$ -2 **46.** $(+9) + (+5) + (-3) + (-12)$ -1

47. $(+31) + (-16) + (+15) + (-17)$ 13 **48.** $(-22) + (-36) + (+19) + (+23)$ -16 **49.** $(+0.5) + (-3.2) + (-2) + (+1.5)$ -3.2

In Problems 50 to 65 the addition problems have been written without the use of extra parentheses and positive signs. If you find these difficult to do, you may insert the extra parentheses and positive signs before doing the calculation. (See Example 13.)

Add.

50. $2 + (-17)$ -15 **51.** $21 + (-4)$ 17 **52.** $-83 + 42$ -41 **53.** $-114 + 86$ -28

54. $-\frac{3}{4} + \left(-\frac{2}{5}\right)$ $-\frac{23}{20}$ or $-1\frac{3}{20}$ **55.** $-\frac{7}{8} + \frac{1}{2}$ $-\frac{3}{8}$ **56.** $-\frac{1}{10} + \frac{1}{2}$ $\frac{2}{5}$ **57.** $-\frac{2}{3} + \left(-\frac{1}{4}\right)$ $-\frac{11}{12}$

58. $4.3 + (-3.6)$ 0.7 **59.** $5.7 + (-9.1)$ -3.4 **60.** $19 + (-6) + 9$ 22 **61.** $-14 + (-23) + 5$ -32

62. $-27 + 9 + (-54) + 30$ -42 **63.** $18 + (-39) + 25 + (-3)$ 1 **64.** $17.85 + (-2.06) + 0.15$ 15.94 **65.** $23.17 + 5.03 + (-11.81)$ 16.39

Applications

Solve each word problem.

66. Abby has \$25 in her checking account. She writes a check totaling \$32.00 for groceries. What is her final balance? −\$7

67. When we skied at Jackson Hole, Wyoming, yesterday, the temperature at the summit was −12°F. Today when we called the ski report, the temperature had risen 7°F. What is the temperature at the summit today? −5°F

68. Donna is studying rain forest preservation in Central America. She stands on a ridge 126 feet below sea level. Her team hikes down to a gully 43 feet lower. Represent her distance below sea level as a signed number. −169 feet

69. Oceanographers studying effects of light on sea creatures are in a submarine 85 feet below sea level. Then they dive to a point 180 feet lower. Represent the submarine's distance below sea level as a signed number. −265 feet

70. On three successive football running plays, Jon gained 9 yards, lost 11 yards, and gained 5 yards. What was his gain or loss? 3 yard gain

71. The temperature in St. Louis at noon in early May was 75°F. It dropped 18 degrees in the next 5 hours. What was the temperature at 5:00 P.M.? 57°F

72. The population of a particular butterfly species was 8000. Twenty years later there were 3000 fewer. This year, there are 1500 fewer. What is the new population? 3500

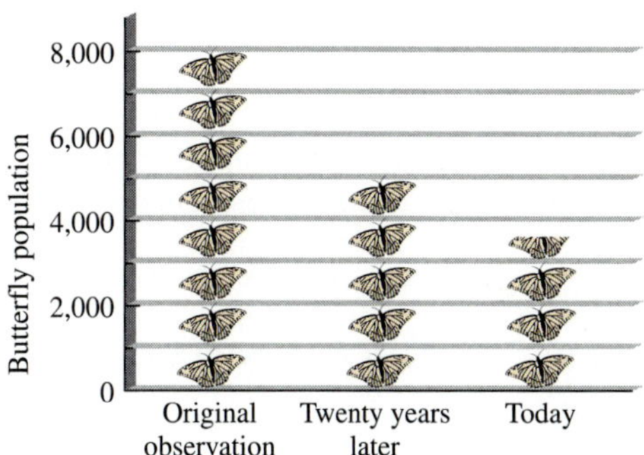

73. Aaron owes \$258 to a credit card company. He makes a purchase of \$32 with the card and then makes a payment of \$150 on the account. How much does he still owe? \$140

To Think About

74. What number must be added to −13 to get 5? 18

75. What number must be added to −18 to get 10? 28

76. Add the following: $(-1.2) + 5.6 + (-2.8) + 1.3 + 2.8 + (-1.3) + 1.2 + (-5.6)$ 0

77. For what values of x and y is $x + y$ negative? Under one of the following cases: (a) $|x| > |y|$ with x negative and y positive, (b) $|y| > |x|$ with x positive and y negative, or (c) x and y both negative.

78. For what values of x and y is $x + (-y) = 0$? $x = y$

79. Glen makes a deposit of \$50 into his checking account and then writes out a check for \$89. What is the least amount of money he needs in his checking account to cover the transactions? \$39.00

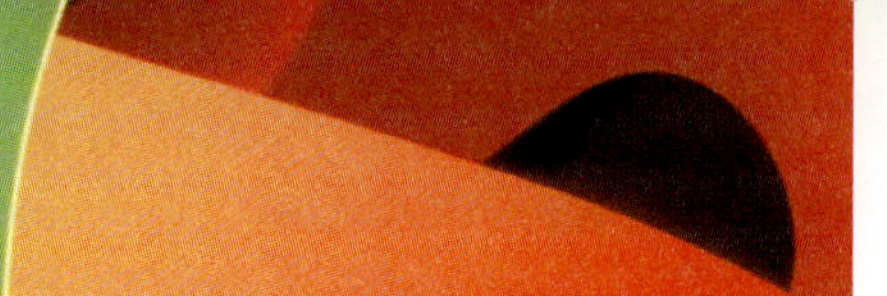

Putting Your Skills to Work

MATHEMATICAL PREDICTION OF FUTURE EXPENSES

The U.S. Bureau of Labor Statistics maintains records that show the cost of living increase for various time periods. From 1970 to 1995 the average increase of the cost of living for each 5-year period is 32.1%. It has been greater than this for some 5-year periods and lower for others as can be seen in the display on the right. This type of a display is helpful for businesses making long-range plans in terms of budgeting for expenses. Use this chart to determine answers to the following questions.

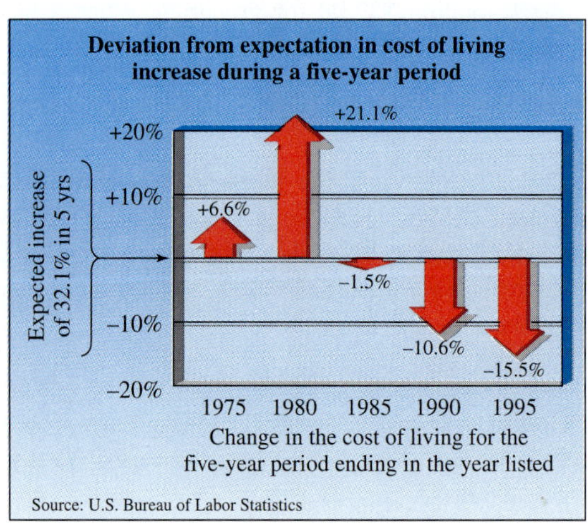

PROBLEMS FOR INDIVIDUAL INVESTIGATION AND ANALYSIS

1. A local department store budgeting for the cost of electricity had established in 1990 a proposed budget of $43,000 for the year 1995. According to this chart, what figure would have been a more accurate budget? $36,335

2. A city taxicab company prepared a budget in 1985 to spend $125,600 for tires in 1990. According to this chart, what figure would have been a more accurate budget? $112,286.40

PROBLEMS FOR GROUP INVESTIGATION AND COOPERATIVE LEARNING

Together with members of your class determine an answer for the following.

3. In 1975 a local elementary school budgeted $45,000 for new textbooks. According to this chart, what would the budget be in 1980? In 1985? $68,940 in 1980; $90,035.64 in 1985

4. In 1985 Russ and Norma Camp established a savings plan for the cost of tuition for their daughter Colette. The college she wanted to attend had a tuition cost of $4500 per semester in 1985. What would the tuition be in 1990? In 1995? $5467.50 in 1990; $6375.11 in 1995

INTERNET CONNECTIONS: Go to ''http://www.prenhall.com/tobey'' to be connected

Site: Historical Income Tables (U.S. Census Bureau) or a related site

5. This site gives historical population and income statistics for Americans. Find the data showing per capita income (all races) in ''current dollars.'' Use this information and the chart above to predict the per capita income for each of the years 1975, 1980, and so on. (For example, you should use the actual per capita income for 1970 to predict the value in 1975.) Compare your predictions with the actual values, and comment on your results.

1.2 Subtraction of Signed Numbers

After studying this section, you will be able to:

1 *Subtract signed numbers.*

1 *Subtracting Signed Numbers*

So far we have developed the rules for adding signed numbers. We can use these rules to subtract signed numbers. Let's look at a checkbook situation to see how.

Situation 5

You have a balance of 20 dollars in your checking account. The bank calls you and says that a deposit of 5 dollars that belongs to another account was erroneously added to your account. They say they will correct the account balance to 15 dollars. The bank tells you that since they cannot take away the erroneous credit, they will add a debit to your account. You want to keep track of what's happening to your account.

Your record for situation 5.

$(+20)$	$-$	$(+5)$	$=$	15
From your present balance	subtract	the deposit	to give	the new balance

This equation shows what needs to be done to your account. The bank tells you that because the error happened in the past they cannot "take it away." However, they can add to your account a debit of 5 dollars. Here is the equivalent addition.

$(+20)$	$+$	(-5)	$=$	15
To your present balance	add	a debit	to give	the new balance

Subtracting a positive 5 has the same effect as adding a negative 5.

> **Subtraction of Signed Numbers**
>
> To subtract signed numbers, add the opposite of the second number to the first.

The rule tells us to do three things when we subtract signed numbers. First, change subtraction to addition. Second, replace the second number by its opposite. Third, add the two numbers using the rules for addition of signed numbers.

EXAMPLE 1 Subtract. $(+6) - (-2)$

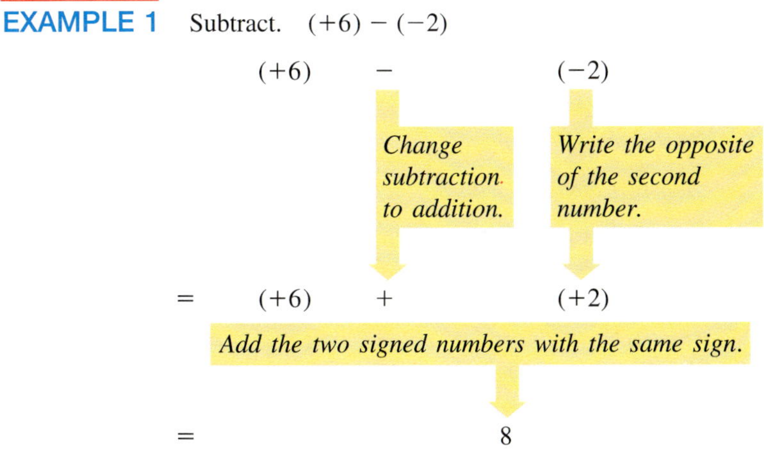

(+6) − (−2)
Change subtraction to addition.
Write the opposite of the second number.
= (+6) + (+2)
Add the two signed numbers with the same sign.
= 8

TEACHING TIP Students often ask, "Why do we have to learn this rule about subtracting signed numbers by adding the opposite of the second number? Why can't we just learn how to subtract signed numbers?" The most satisfactory answer seems to be the following two reasons: (1) By always changing subtracting to adding, we are able to use the commutative property. With subtraction we cannot do this. This will make our work much easier later. (2) Students doing problems this way have fewer errors. Those who try to learn separate rules for how to subtract usually end up very confused by the end of the chapter and make many more errors.

Practice Problem 1 Subtract. $(-9) - (+3)$ ■

EXAMPLE 2 Subtract. $(-8) - (-6)$

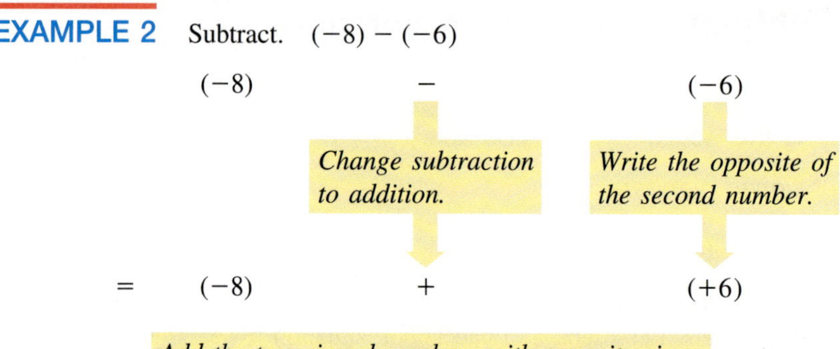
(−8)
−
(−6)
Change subtraction to addition.
Write the opposite of the second number.
= (−8)
+
(+6)

Add the two signed numbers with opposite signs.

$= -2$

Practice Problem 2 Subtract. $(+9) - (+12)$ ■

TEACHING TIP As we get further into the chapter, more of the fraction problems have different denominators. After discussing Example 3, you may want to show the following problem. Subtract.

$$\left(\frac{5}{21}\right) - \left(-\frac{3}{14}\right)$$

Solution. First we change subtracting to adding the opposite.

$$\left(\frac{5}{21}\right) + \left(+\frac{3}{14}\right)$$

Then, since the LCD = 42, we change each fraction to an equivalent fraction with a denominator of 42.

$$= \left(\frac{10}{42}\right) + \left(\frac{9}{42}\right)$$
$$= \frac{19}{42}.$$

EXAMPLE 3 Subtract. **(a)** $\left(+\frac{3}{7}\right) - \left(+\frac{6}{7}\right)$ **(b)** $\left(-\frac{7}{18}\right) - \left(-\frac{1}{9}\right)$

(a) $\left(+\frac{3}{7}\right) + \left(-\frac{6}{7}\right)$ *Change the subtraction problem to one of adding the opposite of the second number. We note that the problem has two fractions with the same denominator.*

$= -\frac{3}{7}$ *Add two numbers with different signs.*

(b) $\left(-\frac{7}{18}\right) + \left(+\frac{1}{9}\right)$ *Change subtracting to adding the opposite.*

$= \left(-\frac{7}{18}\right) + \left(+\frac{2}{18}\right)$ *Change $\frac{1}{9}$ to $\frac{2}{18}$ since LCD = 18.*

$= -\frac{5}{18}$ *Add two numbers with different signs.*

Practice Problem 3 Subtract. $\left(-\frac{1}{5}\right) - \left(+\frac{1}{4}\right)$ ■

TEACHING TIP Some students incorrectly use the word "cancel" here to describe how $(-5.2) + (+5.2) = 0$ in Example 4. It is best to discourage them from using the word "cancel" in this situation. Stress the concept of adding opposites or the idea of an additive inverse.

EXAMPLE 4 Subtract. $(-5.2) - (-5.2)$

$(-5.2) + (+5.2)$ *Change the subtraction problem to one of adding the opposite of the second number.*

$= 0$ *Add two numbers with different signs.*

Example 4 illustrates what is sometimes called the **property of adding opposites.** When you add two real numbers that are opposites of each other, you will obtain zero. Examples of this are the following:

$$(+5) + (-5) = 0 \qquad (-186) + (186) = 0 \qquad \left(-\frac{1}{8}\right) + \left(+\frac{1}{8}\right) = 0$$

Practice Problem 4 Subtract. $\left(-\frac{3}{17}\right) - \left(-\frac{3}{17}\right)$ ■

EXAMPLE 5 Subtract -3 from -20.

TEACHING TIP Emphasize that when you translate the expression "Subtract -3 from -20," the number -20 is written first. Thus we have

$$(-20) - (-3)$$

Note that the -3 is subtracted FROM the -20. Thus we write the -20 first and then subtract -3 from it. Therefore, we have

$$(-20) - (-3) = (-20) + (+3) = -17$$

Practice Problem 5 Subtract -18 from 30. ■

Extra parentheses and extra positive signs within parentheses are often omitted in the writing of subtraction problems in algebra. When such problems are encountered, we want to be sure which number is subtracted from which. If it helps you to insert the optional parentheses and plus signs to make the problem clearer, then you should do so.

EXAMPLE 6 Perform the following.

(a) $-8 - 2$ **(b)** $23 - 28$ **(c)** $5 - (-3)$ **(d)** $\frac{1}{4} - 8$

(a) $-8 - 2$ *Notice that we are subtracting a positive* 2.
$= -8 - (+2)$ *Rewrite using parentheses.*
$= -8 + (-2)$ *Change to addition.*
$= -10$ *Add.*

In a similar fashion we have

(b) $23 - 28 = (+23) - (+28) = (+23) + (-28) = -5$

(c) $5 - (-3) = (+5) - (-3) = (+5) + (+3) = 8$

(d) $\frac{1}{4} - 8 = \left(+\frac{1}{4}\right) - (+8) = \left(+\frac{1}{4}\right) + (-8) = \left(+\frac{1}{4}\right) + \left(-\frac{32}{4}\right) = -\frac{31}{4}$

Practice Problem 6 Perform the following. **(a)** $-21 - 9$ **(b)** $17 - 36$ ■

EXAMPLE 7 A satellite is recording radioactive emissions from nuclear waste buried 3 miles below sea level. The satellite orbits the earth at 98 miles above sea level. How far is the satellite from the nuclear waste?

We want to find the difference between $+98$ miles and -3 miles.

$$98 - (-3)$$
$$= 98 + (+3)$$
$$= 101$$

The satellite is 101 miles from the nuclear waste.

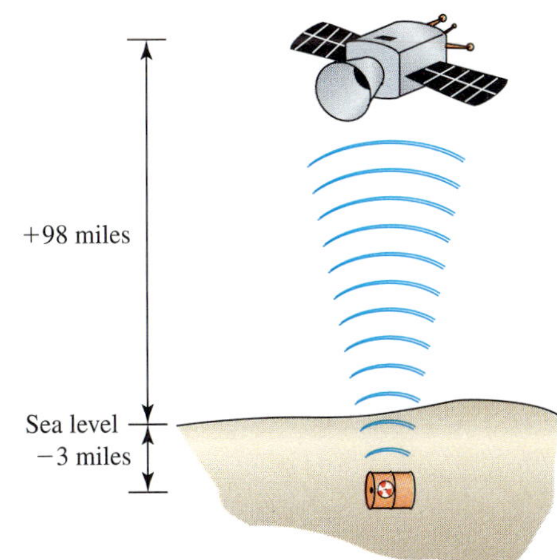

Practice Problem 7 A helicopter is directly over a sunken vessel. The helicopter is 350 feet above sea level. The vessel lies 186 feet below sea level. How far is the helicopter from the sunken vessel? ■

1.2 Exercises

Subtract by adding the opposite.

1. $(+8) - (+5)$ $8 - 5 = 3$
2. $(+12) - (+6)$ $12 - 6 = 6$
3. $(+7) - (-3)$ $7 + 3 = 10$
4. $(-12) - (-8)$ $-12 + 8 = -4$
5. $(-10) - (-4)$ $-10 + 4 = -6$
6. $(-17) - (-6)$ $-17 + 6 = -11$
7. $(+20) - (+46)$ -26
8. $(+15) - (+28)$ -13
9. $(-32) - (-6)$ -26
10. $(+18) - (-8)$ 26
11. $(-52) - (-60)$ 8
12. $(-48) - (-80)$ 32
13. $(0) - (-5)$ 5
14. $(0) - (-7)$ 7
15. $(+15) - (+20)$ -5
16. $(+18) - (+24)$ -6
17. $(-18) - (-18)$ 0
18. $(+24) - (-24)$ 48
19. $(-17) - (-13)$ -4
20. $(-26) - (-7)$ -19
21. $(-0.6) - (+0.3)$ -0.9
22. $(-0.9) - (+0.5)$ -1.4
23. $(+2.64) - (-1.83)$ 4.47
24. $(+0.07) - (-0.09)$ 0.16
25. $(-1.5) - (-3.5)$ 2
26. $\left(+\frac{1}{3}\right) - \left(-\frac{2}{5}\right)$ $\frac{11}{15}$
27. $\left(+\frac{3}{4}\right) - \left(-\frac{3}{5}\right)$ $1\frac{7}{20}$
28. $\left(-\frac{2}{3}\right) - \left(+\frac{1}{4}\right)$ $-\frac{11}{12}$
29. $\left(-\frac{3}{4}\right) - \left(+\frac{5}{6}\right)$ $-1\frac{7}{12}$
30. $\left(-\frac{7}{10}\right) - \left(+\frac{10}{15}\right)$ $-1\frac{11}{30}$

Perform each calculation. You may want to insert the extra parentheses and positive signs before changing the subtraction problem to an equivalent addition problem. (See Example 6.)

31. $34 - 87$ -53
32. $19 - 76$ -57
33. $-67 - 32$ -99
34. $-98 - 34$ -132
35. $2.3 - (-4.8)$ 7.1
36. $8.4 - (-2.7)$ 11.1
37. $3 - \frac{1}{5}$ $\frac{14}{5}$ or $2\frac{4}{5}$
38. $12 - \left(-\frac{2}{3}\right)$ $\frac{38}{3}$ or $12\frac{2}{3}$
39. $\frac{3}{5} - (-8)$ $8\frac{3}{5}$ or $\frac{43}{5}$
40. $\frac{5}{6} - 4$ $-\frac{19}{6}$ or $-3\frac{1}{6}$
41. $-\frac{3}{10} - \frac{3}{4}$ $-\frac{21}{20}$ or $-1\frac{1}{20}$
42. $-\frac{11}{12} - \frac{5}{18}$ $-\frac{43}{36}$ or $-1\frac{7}{36}$
43. $-135 - (-126.5)$ -8.5
44. $-97.6 - (-146)$ 48.4
45. $0.0067 - (-0.0432)$ 0.0499

46. $0.0762 - (-0.0094)$ 0.0856

47. $\frac{1}{5} - 6$ $-5\frac{4}{5}$

48. $\frac{2}{7} - (-3)$ $3\frac{2}{7}$

49. $-\frac{3}{5} - \left(-\frac{3}{10}\right)$ $-\frac{3}{10}$

50. $-1 - 0.0059$ -1.0059

51. Subtract -9 from -2. $-2-(-9) = 7$

52. Subtract -12 from 20. $20-(-12) = 32$

53. Subtract 13 from -35. $-35-(+13) = -48$

Change each subtraction operation to "adding the opposite." Then combine the numbers.

54. $9 + 6 - (-5)$ 20

55. $7 + (-6) - (+3)$ -2

56. $8 + (-4) - (+10)$ -6

57. $-10 + 6 - (-15)$ 11

58. $18 - (-15) - 3$ 30

59. $7 + (-42) - 27$ -62

60. $-37 - (-18) + 5$ -14

61. $-21 - (-36) - 8$ 7

62. $-40 - (-6) + 34 - (-5)$ 5

63. $-84 + 12 - (-45) - (+14)$ -41

64. $-5 - (-30) + 20 + 40 - (-12)$ 97

65. $42 - (-30) - 65 - (-11) + 20$ 38

Applications

Solve each word problem.

66. A rescue helicopter is 300 feet above sea level. The captain has located an ailing submarine directly below it that is 126 feet below sea level. How far is the helicopter from the submarine? 426 feet

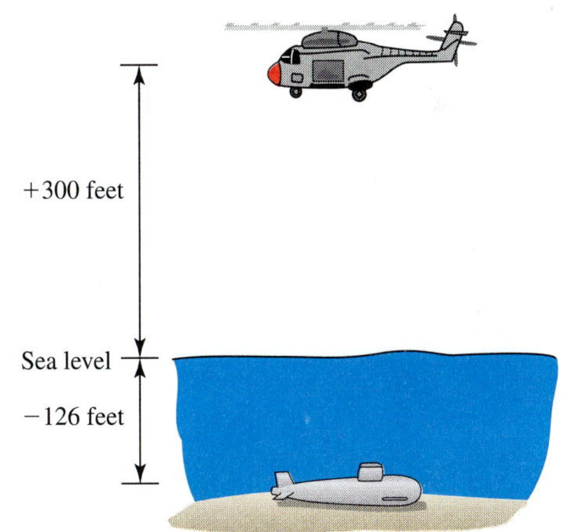

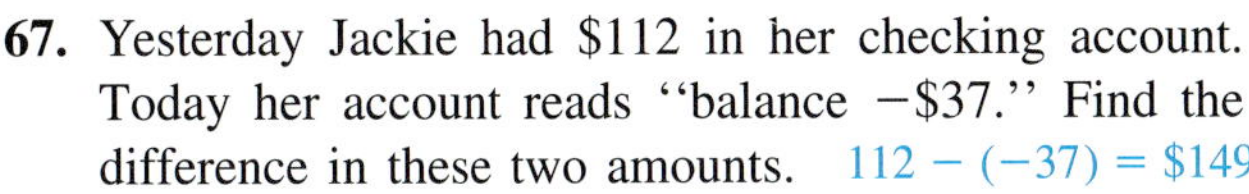

67. Yesterday Jackie had \$112 in her checking account. Today her account reads "balance −\$37." Find the difference in these two amounts. $112 - (-37) = \$149$

68. On January 6, 1971, Hawley Lake, Arizona, had a record low temperature of −23°F. The next day the temperature at the same place was −40°F. What was the change in temperature from January 6 to January 7, 1971? −17°F

69. On January 19, 1937, the temperature at Boca, California, was −29°F. On January 20, the temperature at the same place was −45°F. What was the change in temperature from January 19 to January 20, 1937? −16°F

Cumulative Review Problems

70. $-37 + 16$ -21

71. $-37 + (-14)$ -51

72. $-3 + (-6) + (-10)$ -19

73. What is the temperature after a rise of 13°C from a start of −21°C? −8°C

1.3 Multiplication and Division of Signed Numbers

MathPro

Video 3

SSM

After studying this section, you will be able to:

1. *Multiply signed numbers.*
2. *Divide signed numbers.*

1 *Multiplying Signed Numbers*

We are familiar with the meaning of multiplication for positive numbers. For example, $5 \times 40 = 200$ might mean that you receive five weekly checks of 40 dollars each and you gain \$200. Let's look at a situation that corresponds to $5 \times (-40)$. What might that mean?

TEACHING TIP Point out that we often use parentheses in algebra to mean multiplication. That is, (3)(5) means to *multiply* 3 by 5. (3)(−5) means to *multiply* 3 by −5.

Situation 6

You write a check for five weeks in a row to pay your weekly room rent of 40 dollars. You want to know the total impact on your checking account balance.

Your record for situation 6.

$(+5)$	(-40)	$=$	-200
The number of checks you have written	negative 40, the value of each check that was a debit to your account	gives a result of	negative 200 dollars, a net debit to your account.

Note that a multiplication symbol is not needed between the $(+5)$ and the (-40) because the two sets of parentheses indicate multiplication. In this case the multiplication $(5)(-40)$ is the same as repeated addition of five (-40)'s.

$$\underbrace{(-40) + (-40) + (-40) + (-40) + (-40)}_{\text{repeated addition of five } (-40)\text{'s}} = -200$$

This example seems to show that a positive number multiplied by a negative number is negative.

What if the negative number is the one that is written first? If $(5)(-40) = -200$, then $(-40)(5) = -200$, by the commutative property of multiplication. This is an example showing *that when two numbers with opposite signs* (one positive, one negative) *are multiplied, the result is negative.*

But what if both numbers are negative? Consider the following situation.

Situation 7

Last year at college you rented a room at 40 dollars per week for 36 weeks, which included two semesters and summer school. This year you will not attend the summer session, so you will be renting the room for only 30 weeks. Thus the number of weekly rental checks will be six less than last year. You are making out your budget for this year. You want to know the financial impact of renting the room for six fewer weeks.

Your record for situation 7.

(-6)	(-40)	$=$	240
The number of checks this year compared to last is −6, which is negative to show a decrease,	the value of each check paid out, −40,	gives a result of	+240 dollars. The product is positive, because your financial situation will be 240 dollars better this year.

You could check that the answer is positive by calculation of the total rental expenses.

	Dollars in rent last year	$(36)(40) =$	1440
(subtract)	Dollars in rent this year	$-(30)(40) =$	-1200
	Extra dollars available this year	$=$	$+240$

This agrees with our answer above: $(-6)(-40) = +240$.

In this situation it seems reasonable that a negative number times a negative number yields a positive answer. We have already known from arithmetic that a positive number times a positive number should yield a positive answer. Thus we might see the general rule that *when two numbers with the same sign* (both positive or both negative) *are multiplied, the result is positive.*

We will now state our rule.

TEACHING TIP Sometimes students incorrectly memorize vague rules like "A negative and a negative give a positive." This of course will confuse them later on. If they wish to memorize a short rule like that it should be: "A negative **times** a negative gives a positive." Emphasize to students that they will need to keep all these rules straight when they confront addition, subtraction, multiplication, and division, all in one problem set.

Multiplication of Signed Numbers

To multiply two numbers with the **same sign,** multiply the absolute values. The sign of the result is **positive.**

To multiply two signed numbers with **opposite signs,** multiply the absolute values. The sign of the result is **negative.**

EXAMPLE 1 Multiply.

(a) $(3)(6)$ **(b)** $\left(-\frac{5}{7}\right)\left(-\frac{2}{9}\right)$ **(c)** $(-4)(8)$ **(d)** $\left(\frac{2}{7}\right)(-3)$

(a) $(3)(6) = 18$

(b) $\left(-\frac{5}{7}\right)\left(-\frac{2}{9}\right) = \frac{10}{63}$

When multiplying two numbers with the same sign, the result is a positive number.

(c) $(-4)(8) = -32$

(d) $\left(\frac{2}{7}\right)(-3) = \left(\frac{2}{7}\right)\left(-\frac{3}{1}\right) = -\frac{6}{7}$

When multiplying two numbers with opposite signs, the result is a negative number.

Practice Problem 1 Multiply.

(a) $(-6)(-2)$ **(b)** $(7)(9)$ **(c)** $\left(-\frac{3}{5}\right)\left(\frac{2}{7}\right)$ **(d)** $(40)(-20)$ ■

To multiply more than two numbers, take two numbers at a time.

EXAMPLE 2 Multiply. $(-4)(-3)(-2)$

$= (-4)(-3)(-2)$ *We first multiply the first two numbers.*

$= (+12)(-2)$ *Multiply two numbers, (-4) and (-3). The signs are the same. The answer is positive 12.*

$= -24$ *Multiply two numbers, $(+12)$ and (-2). The signs are different. The answer is negative 24.*

Practice Problem 2 Multiply $(-3)(-4)(-3)$ ■

EXAMPLE 3 Multiply. **(a)** $(-3)(-8)$ **(b)** $\left(-\frac{1}{2}\right)(-1)(-4)$

(c) $(-2)(-2)(-2)(-2)$

Multiply two numbers at a time. See if you find a pattern.

(a) $(-3)(-8) = +24$ or 24

(b) $\left(-\frac{1}{2}\right)(-1)(-4) = +\frac{1}{2}(-4) = -2$

(c) $(-2)(-2)(-2)(-2) = +4(-2)(-2) = -8(-2) = +16$ or 16

What do you think would happen if we multiplied five negative numbers? If you guessed ''negative,'' you probably see the pattern.

Practice Problem 3 Determine the sign of the product. Then multiply to check.

(a) $(-2)(-3)(-4)(-1)$ **(b)** $(-1)(-3)(-2)$ **(c)** $(-4)\left(-\frac{1}{4}\right)(-2)(-6)$ ■

When you multiply two or more signed numbers:

1. The result is always **positive** if there are an **even** number of negative signs.
2. The result is always **negative** if there are an **odd** number of negative signs.

TEACHING TIP After presenting Example 4, ask students to do the following as an Additional Class Exercise. Multiply.

$(-1)(-2)(+2)(-1)(-2)(-1)(2)$

Ans. -16. Emphasize the importance of determining the sign separately in long problems like this, and then multiplying out all the numbers.

EXAMPLE 4 Multiply. $(-3)(+2)(-1)(+6)(-3)$

$(-3)(+2)(-1)(+6)(-3) = -$	*The answer is negative since there are three negative signs and 3 is **odd.***
$3 \times 2 \times 1 \times 6 \times 3 = 108$	*Multiply the numerical values.*
$(-3)(+2)(-1)(+6)(-3) = -108$	*The answer is negative* 108.

Practice Problem 4 Multiply. $(-2)(3)(-4)(-1)(-2)$ ■

For convenience, we will list the properties of multiplication.

1. *Multiplication is commutative.* This property states that if two numbers are multiplied, the order of the numbers does not affect the result. The result is the same if either number is written first.

$$(5)(7) = (7)(5) = 35, \qquad \left(\frac{1}{3}\right)\left(\frac{2}{7}\right) = \left(\frac{2}{7}\right)\left(\frac{1}{3}\right) = \frac{2}{21}$$

2. *Multiplication of any number by zero will result in zero.*

$$(5)(0) = 0, \qquad (-5)(0) = 0, \qquad (0)\left(\frac{3}{8}\right) = 0, \qquad (0)(0) = 0$$

3. *Multiplication of any number by 1 will result in that same number.*

$$(5)(1) = 5, \qquad (1)(-7) = -7, \qquad (1)\left(-\frac{5}{3}\right) = -\frac{5}{3}$$

4. *Multiplication is associative.* This property states that if three numbers are multiplied, it does not matter which two numbers are grouped by parentheses.

$2 \times (3 \times 4) = (2 \times 3) \times 4$	*First multiply numbers in parentheses. Then multiply the remaining numbers.*
$2 \times (12) = (6) \times 4$ $24 = 24$	*The results are the same no matter which numbers are grouped first.*

TEACHING TIP Point out that multiplication in algebra can be written in four different ways.

3×5	Using the symbol ×
(3)(5) 3(5)	Using parentheses
$3 \cdot 5$	Using a centered dot

Each of these expressions indicate ''three times five.'' All these forms of notation will be used in the exercises for this section. The most common notation will be with parentheses.

2 Dividing Signed Numbers

What about division? Any division problem can be rewritten as a multiplication problem.

We know that $(+20) \div (+4) = +5$ because $(+4)(+5) = +20$.

Similarly, $(-20) \div (-4) = +5$ because $(-4)(+5) = -20$.

In both division problems the answer is positive 5. Thus we see that *when you divide two numbers with the same sign* (both positive or both negative) *the answer is positive*. What if the signs are different?

We know that $(-20) \div (+4) = -5$ because $(+4)(-5) = -20$.

Similarly, $(+20) \div (-4) = -5$ because $(-4)(-5) = +20$.

In these two problems the answer is negative 5. So we have reasonable evidence to see that *when you divide two numbers with different signs* (one positive and one negative) *the answer is negative*.

We will now state our rule for division.

Division of Signed Numbers

To divide two numbers with the **same sign,** divide the absolute values. The sign of the result is **positive.**

To divide two numbers with **different signs,** divide the absolute values. The sign of the result is **negative.**

EXAMPLE 5 Divide.

(a) $12 \div 4$ **(b)** $(-25) \div (-5)$ **(c)** $\frac{-36}{18}$ **(d)** $\frac{42}{-7}$

(a) $12 \div 4 = 3$

(b) $(-25) \div (-5) = 5$

When dividing two numbers with the same sign, the result is a positive number.

(c) $\frac{-36}{18} = -2$

(d) $\frac{42}{-7} = -6$

When dividing two numbers with diffferent signs, the result is a negative number.

Practice Problem 5 Divide.

(a) $(-36) \div (-2)$ **(b)** $18 \div 9$ **(c)** $\frac{50}{-10}$ **(d)** $(-49) \div 7$ ■

EXAMPLE 6 Divide. **(a)** $(-36) \div (0.12)$ **(b)** $(-2.4) \div (-0.6)$

(a) $(-36) \div (0.12)$ *Look at the problem to determine the sign. When dividing two numbers with different signs, the result will be a negative number.*

We then divide the absolute values.

$$\begin{array}{r} 3\,00. \\ 0.12_{\wedge})\overline{36.00_{\wedge}} \\ \underline{36} \\ 00 \end{array}$$

Thus $(-36) \div (0.12) = -300$. The answer is a negative number.

(b) $(-2.4) \div (-0.6)$ *Look at the problem to determine the sign. When dividing two numbers with the same sign, the result will be positive.*

We then divide the absolute values.

$$0.6_\wedge\overline{)2.4_\wedge}\quad = 4. \qquad \begin{array}{r} 4. \\ 0.6_\wedge)\overline{2.4_\wedge} \\ \underline{2\,4} \end{array}$$

Thus $(-2.4) \div (-0.6) = 4$. The answer is a positive number.

Practice Problem 6 Divide.

(a) $(-1.242) \div (-1.8)$ **(b)** $(0.235) \div (-0.0025)$ ■

Note how similar the rules for multiplication and division are. When you **multiply** or **divide** two numbers with the **same** sign, you obtain a **positive** number. When you **multiply** or **divide** two numbers with **different** signs, you obtain a **negative** number.

EXAMPLE 7 Divide. $\left(-\frac{12}{5}\right) \div \left(\frac{2}{3}\right)$

$$\left(-\frac{12}{5}\right) \div \left(\frac{2}{3}\right) = \left(-\frac{12}{5}\right)\left(\frac{3}{2}\right)$$ *Divide two fractions. We invert the second fraction and multiply by the first fraction.*

$$\left(-\frac{\overset{6}{\cancel{12}}}{5}\right)\left(\frac{3}{\underset{1}{\cancel{2}}}\right) = -\frac{18}{5} \quad or \quad -3\frac{3}{5}$$ *The answer is negative since the two numbers divided have different signs.*

Practice Problem 7 Divide. $\left(-\frac{5}{16}\right) \div \left(-\frac{10}{13}\right)$ ■

Note that division can be indicated by the symbol $\div$ or by the fraction bar: $-$. $\frac{2}{3}$ means $2 \div 3$.

EXAMPLE 8 Divide. **(a)** $\dfrac{\frac{7}{8}}{-21}$ **(b)** $\dfrac{-\frac{2}{3}}{-\frac{7}{13}}$

TEACHING TIP After discussing Example 8, you may want to expand the horizons of students by showing them a few examples with fractions. Try showing them the following as an Additional Class Exercise. Divide the following.

A. $(-\frac{3}{5})$ divided by $(-\frac{5}{8})$
B. 7 divided by $(-\frac{4}{11})$
C. $(-\frac{7}{9})$ divided by (-14)

Sol.

A. $(-\frac{3}{5})(-\frac{8}{5}) = +\frac{24}{25}$
B. $(7)(-\frac{11}{4}) = -\frac{77}{4}$
C. $(-\frac{7}{9})(-\frac{1}{14}) = +\frac{1}{18}$

(a) $\dfrac{\frac{7}{8}}{-21}$ means $\frac{7}{8} \div (-21)$

$$= \left(\frac{7}{8}\right) \div \left(-\frac{21}{1}\right)$$ *Change -21 to a fraction. $-21 = -\frac{21}{1}$*

$$= \left(\frac{\overset{1}{\cancel{7}}}{8}\right)\left(-\frac{1}{\underset{3}{\cancel{21}}}\right)$$ *Change the division to multiplication. Cancel where possible.*

$$= -\frac{1}{24}$$ *Simplify.*

(b) $\dfrac{-\frac{2}{3}}{-\frac{7}{13}}$ means $\left(-\frac{2}{3}\right) \div \left(-\frac{7}{13}\right) = \left(-\frac{2}{3}\right)\left(-\frac{13}{7}\right) = \frac{26}{21}$ *or* $1\frac{5}{21}$

Practice Problem 8 Divide. **(a)** $\dfrac{-12}{-\frac{4}{5}}$ **(b)** $\dfrac{-\frac{2}{9}}{\frac{8}{13}}$ ■

There are two properties of division that are important to understand.

1. *Division of 0 by any nonzero number gives 0 as a result.*

$$0 \div 5 = 0, \quad 0 \div \frac{2}{3} = 0, \quad \frac{0}{5.6} = 0, \quad \frac{0}{1000} = 0$$

You can divide zero by 5, 3, 5.6, 1000, or any number (except 0).

2. *Division of any number by 0 is not allowed.*

$$7 \div 0 \qquad \frac{64}{0} \qquad \frac{0}{0}$$

↑ ↑ ↑

None of these operations is possible. **You cannot divide by 0!**

TEACHING TIP Students often confuse 0 divided by 3 and 3 divided by 0. Take the time to make sure they realize that division by 0 is never allowed. They must remember this in future problem sets. 0 divided by a nonzero number is always zero.

When combining two numbers, it is important to be sure you know which rule applies. Think about the concepts in the following chart. See if you agree with each example.

Operation	Two Numbers with the Same Sign	Two Numbers with Different Signs
Addition	Result may be positive or negative. $9 + 2 = 11$ $-5 + (-6) = -11$	Result may be positive or negative. $-3 + 7 = 4$ $4 + (-12) = -8$
Subtraction	Result may be positive or negative. $15 - (+6) = 15 + (-6) = 9$ $-12 - (-3) = -12 + (+3) = -9$	Result may be positive or negative. $-12 - (+3) = -12 + (-3) = -15$ $5 - (-6) = 5 + (+6) = 11$
Multiplication	Result is always positive. $(9)(3) = 27$ $(-8)(-5) = 40$	Result is always negative. $(-6)(12) = -72$ $(8)(-3) = -24$
Division	Result is always positive. $150 \div 6 = 25$ $-72 \div (-2) = 36$	Result is always negative. $-60 \div 10 = -6$ $30 \div (-6) = -5$

Developing Your Study Skills

READING THE TEXTBOOK

Begin reading your textbook with a paper and pencil in hand. As you come across a new definition, or concept, underline it in the text and/or write it down in your notebook. Whenever you encounter an unfamiliar term, look it up and make a note of it. When you come to an example, work through it step by step. Be sure to read each word and to follow directions carefully.

Notice the helpful hints the author provides. They guide you to correct solutions and prevent you from making errors. Take advantage of these pieces of expert advice.

Be sure that you understand what you are reading. Make a note of any of those things that you do not understand and ask your instructor about them. Do not hurry through the material. Learning mathematics takes time.

1.3 Exercises

Multiply or divide. Be sure to write your answer in the simplest form.

1. $(3)(-12)$ -36

2. $(5)(-4)$ -20

3. $(-6)(-5)$ 30

4. $(-8)(-7)$ 56

5. $(-5)(-12)$ 60

6. $9(-20)$ -180

7. -8×3 -24

8. -5×11 -55

9. $0(-12)$ 0

10. -3×150 -450

11. 16×1.5 24

12. 24×2.5 60

13. $(-1.32)(-0.2)$ 0.264

14. $(-2.3)(-0.11)$ 0.253

15. $(0.7)(-2.5)$ -1.75

16. $(-6)\left(\frac{3}{10}\right)$ $-1\frac{4}{5}$

17. $\left(\frac{12}{5}\right)(-10)$ -24

18. $\left(\frac{2}{7}\right)\left(\frac{14}{3}\right)$ $1\frac{1}{3}$

19. $\left(-\frac{3}{8}\right)\left(-\frac{5}{6}\right)$ $\frac{5}{16}$

20. $\left(-\frac{5}{21}\right)\left(\frac{3}{10}\right)$ $-\frac{1}{14}$

21. $(12) \div (-4)$ -3

22. $(-36) \div (-9)$ 4

23. $0 \div (-15)$ 0

24. $24 \div (-6)$ -4

25. $-42 \div 7$ -6

26. $(-52) \div (-2)$ 26

27. $-360 \div (-10)$ 36

28. $(-220) \div (-11)$ 20

29. $(240) \div (-15)$ -16

30. $156 \div (-13)$ -12

31. $(-0.6) \div 0.3$ -2

32. $1.2 \div (-0.03)$ -40

33. $(-2.4) \div (-0.6)$ 4

34. $(0.36) \div (-0.9)$ -0.4

35. $(-7.2) \div (8)$ -0.9

36. $\frac{2}{7} \div \left(-\frac{3}{5}\right)$ $-\frac{10}{21}$

37. $\left(-\frac{1}{5}\right) \div \left(\frac{2}{3}\right)$ $-\frac{3}{10}$

38. $\left(-\frac{5}{6}\right) \div \left(-\frac{7}{18}\right)$ $2\frac{1}{7}$

39. $\left(\frac{5}{7}\right) \div \left(-\frac{3}{28}\right)$ $-\frac{20}{3}$ or $-6\frac{2}{3}$

40. $\frac{4}{9} \div \left(-\frac{8}{15}\right)$ $-\frac{5}{6}$

41. $\dfrac{12}{-\frac{2}{5}}$ -30

42. $\dfrac{-\frac{3}{7}}{-6}$ $\frac{1}{14}$

43. $\dfrac{-\frac{3}{8}}{-\frac{2}{3}}$ $\frac{9}{16}$

44. $\dfrac{-\frac{1}{7}}{\frac{3}{14}}$ $-\frac{2}{3}$

45. $\dfrac{\frac{2}{5}}{-\frac{8}{15}}$ $-\frac{3}{4}$

Multiply. You may want to determine the sign of the product before you multiply.

46. $(-6)(2)(-3)(4)$ 144

47. $(-1)(-2)(-3)(4)$ -24

48. $(-2)(-1)(3)(-2)(2)$ -24

49. $(-1)(-3)(-2)(-2)(3)$ 36

50. $(-3)(2)(-4)(0)(-2)$ 0

51. $(-3)(-2)\left(\frac{1}{3}\right)(-4)(2)$ -16

52. $(60)(-0.6)(-20)(0.5)$ 360

53. $(-3)(-0.03)(100)(-2)$ -18

54. $\left(\frac{3}{8}\right)\left(\frac{1}{2}\right)\left(-\frac{5}{6}\right)$ $-\frac{5}{32}$

55. $\left(-\frac{4}{5}\right)\left(-\frac{6}{7}\right)\left(-\frac{1}{3}\right)$ $-\frac{8}{35}$

56. $\left(-\frac{1}{2}\right)\left(\frac{4}{5}\right)\left(-\frac{7}{8}\right)\left(-\frac{2}{3}\right)$ $-\frac{7}{30}$

57. $\left(-\frac{3}{4}\right)\left(-\frac{7}{15}\right)\left(-\frac{8}{21}\right)\left(-\frac{5}{9}\right)$ $\frac{2}{27}$

Mixed Review

Take a minute to review the chart following Example 8. Be sure that you can remember the sign rules for each operation. Then do problems 58 through 67.

58. $(-5) - (-2)$ -3

59. $(-20) \div 5$ -4

60. $(-3)(-9)$ 27

61. $5 + (-7)$ -2

62. $(-32) \div (-4)$ 8

63. $8 - (-9)$ 17

64. $(-6) + (-3)$ -9

65. $6(-12)$ -72

66. $18 \div (-18)$ -1

67. $(-37) \div 37$ -1

Applications

Solve each word problem.

68. Debbie's car loan payment is \$250. If she pays by check, what is the net debit to her account after 6 months?
A debit of \$1500

69. Phil pays a doctor bill for \$100, but forgets to check the balance in his account. When he goes to the ATM, his balance reads $-\$70$. How much was in his account before he wrote the check? \$30

70. The velocity (rate) of a projectile, $r = -30$ meters per second, indicates that it is moving to the left on a number line. Currently, it is at time $t = 0$ and at the zero mark on the number line. Find where it will be (distance) when $t = 3$ seconds. (Use $d = r \cdot t$.) -90 or 90 m left of 0

71. Referring to problem 70, where was the projectile 5 seconds ago ($t = -5$)? 150 or 150 m right of 0

Cumulative Review Problems

72. Add. $(-17.4) + (8.31) + (2.40)$ -6.69

73. Add. $(-13) + (-39) + (-20)$ -72

74. Subtract. $(-3.7) - (-8.33)$ 4.63

75. Subtract. $(-37) - (51)$ -88

Developing Your Study Skills

STEPS TOWARD SUCCESS IN MATHEMATICS

Mathematics is a building process, mastered one step at a time. The foundation of this process is built on a few basic requirements. Those who are successful in mathematics realize the absolute necessity for building a study of mathematics on the firm foundation of these six minimum requirements.

1. Attend class every day.
2. Read the textbook.
3. Take notes in class.
4. Do assigned homework every day.
5. Get help immediately when needed.
6. Review regularly.

1.4 Exponents

MathPro

Video 4

SSM

After studying this section, you will be able to:

1 *Write numbers in exponent form.*
2 *Evaluate numerical expressions that contain exponents.*

1 *Exponent Form*

In mathematics, we use exponents as a way to abbreviate repeated multiplication.

Long Notation	*Exponent Form*
$2 \cdot 2 \cdot 2 \cdot 2 \cdot 2 \cdot 2 =$	2^6

There are two parts to exponent notation: (1) the **base** and (2) the **exponent.** The **base** tells you what number is being multiplied and the **exponent** tells you how many times this number is used as a factor. (A *factor,* you recall, is a number being multiplied.) In the first example above,

$$2 \cdot 2 \cdot 2 \cdot 2 \cdot 2 \cdot 2 = 2^6$$

The *base* is 2 (the number being multiplied)

the *exponent* is 6 (the number of times 2 is used as a factor)

If the base is a *positive integer,* the exponent appears to the right and slightly above the level of the number, for example, 5^6 and 8^3. If the base is a *negative integer,* then parentheses are used around the number and the exponent appears outside the parentheses. For example, $(-2)(-2)(-2) = (-2)^3$.

If we do not know the value of a number, we use a letter to represent the unknown number. We call the letter a **variable.** This is quite useful in the case of exponents. Suppose we do not know the value of a number, but we know the number is multiplied by itself several times. We can represent this with a variable base but a whole number for an exponent. For example when we have an unknown number, represented by the variable x, and this number occurs as a factor four times, we would have

$$(x)(x)(x)(x) = x^4$$

Likewise if an unknown number, represented by the variable w, occurs as a factor five times, we would have

$$(w)(w)(w)(w)(w) = w^5$$

EXAMPLE 1 Write each of the following in exponent form.

(a) $(9)(9)(9)$ **(b)** $(13)(13)(13)(13)$ **(e)** $(x)(x)$

(c) $(-7)(-7)(-7)(-7)(-7)$ **(d)** $(-4)(-4)(-4)(-4)(-4)(-4)$ **(f)** $(y)(y)(y)$

(a) $(9)(9)(9) = 9^3$ **(b)** $(13)(13)(13)(13) = 13^4$

(c) The -7 is used as a factor five times. The answer must contain parentheses. Thus $(-7)(-7)(-7)(-7)(-7) = (-7)^5$

(d) $(-4)(-4)(-4)(-4)(-4)(-4) = (-4)^6$ **(e)** $(x)(x) = x^2$ **(f)** $(y)(y)(y) = y^3$

Practice Problem 1 Write each of the following in exponent form.

(a) $(6)(6)(6)(6)$ **(b)** $(-2)(-2)(-2)(-2)(-2)$

(c) $(108)(108)(108)$ **(d)** $(-11)(-11)(-11)(-11)(-11)(-11)$ ■

If the value has an exponent of 2, we say the value is "squared."

If the value has an exponent of 3, we say the value is "cubed."

If the value has an exponent greater than 3, we say "to the (exponent)-th power."

x^2 is read "x squared." y^3 is read "y cubed."

3^6 is read "three to the sixth power" or simply "three to the sixth."

2 *Evaluating Numerical Expressions That Contain Exponents*

EXAMPLE 2 Evaluate. **(a)** 4^3 **(b)** 2^5 **(c)** 7^4 **(d)** $2^3 + 4^4$

(a) $4^3 = (4)(4)(4) = 64$

(b) $2^5 = (2)(2)(2)(2)(2) = 32$

(c) $7^4 = (7)(7)(7)(7) = 2401$

(d) First we evaluate each power.
$2^3 = (2)(2)(2) = 8$
$4^4 = (4)(4)(4)(4) = 256$
Then we add. $8 + 256 = 264$

Practice Problem 2 Evaluate. **(a)** 5^3 **(b)** 3^5 **(c)** 9^4 ■

If the base is negative, be especially careful in determining the sign. Notice the following:

$$(-3)^2 = (-3)(-3) = +9 \qquad (-3)^3 = (-3)(-3)(-3) = -27$$

$$(-3)^4 = (-3)(-3)(-3)(-3) = +81$$

From Section 1.3 we know that when you multiply two or more signed numbers you multiply their absolute values.

- The result is positive if there are an even number of negative signs.
- The result is negative if there are an odd number of negative signs.

This can be generalized as follows.

Sign Rule for Exponents

Suppose that a number is written in exponent form and the base is negative. The result is **positive** if the exponent is **even.** The result is **negative** if the exponent is **odd.**

Be careful how you read expressions with exponents and negative signs.

$$(-3)^4 \text{ means } (-3)(-3)(-3)(-3) \text{ or } +81$$

$$-3^4 \text{ means } -(3)(3)(3)(3) \text{ or } -81$$

EXAMPLE 3 Evaluate. **(a)** $(-2)^3$ **(b)** $(-4)^6$ **(c)** -3^6

(a) $(-2)^3 = -8$ *The answer is negative since the exponent 3 is odd.*

(b) $(-4)^6 = +4096$ *The answer is positive since the exponent 6 is even.*

(c) $-3^6 = -729$ *The negative sign is not contained in parentheses.* -3^6 *means the negative of 3 raised to the sixth power.*

Practice Problem 3 Evaluate. **(a)** $(-3)^3$ **(b)** -2^4 **(c)** $(-2)^6$ **(d)** $(-1)^{10}$ ■

EXAMPLE 4 Evaluate. **(a)** $\left(\frac{1}{2}\right)^4$ **(b)** $(0.2)^4$ **(c)** $\left(\frac{2}{5}\right)^3$ **(d)** $(3)^3(2)^5$

(a) $\left(\frac{1}{2}\right)^4 = \left(\frac{1}{2}\right)\left(\frac{1}{2}\right)\left(\frac{1}{2}\right)\left(\frac{1}{2}\right) = \frac{1}{16}$

(b) $(0.2)^4 = (0.2)(0.2)(0.2)(0.2) = 0.0016$

(c) $\left(\frac{2}{5}\right)^3 = \left(\frac{2}{5}\right)\left(\frac{2}{5}\right)\left(\frac{2}{5}\right) = \frac{8}{125}$

(d) First we evalute each power.
$3^3 = (3)(3)(3) = 27$
$2^5 = (2)(2)(2)(2)(2) = 32$
Then we multiply. $(27)(32) = 864$

Practice Problem 4 Evaluate. **(a)** $\left(\frac{1}{3}\right)^3$ **(b)** $(0.3)^4$ (c) $\left(\frac{3}{2}\right)^4$ ■

CALCULATOR

Exponents

You can use a calculator to evaluate 3^5. Press the following keys:

[3] [y^x] [5] [=]

The display should read

[243]

Try **(a)** 4^6 **(b)** $(0.2)^5$ **(c)** 18^6 **(d)** 3^{12}

1.4 Exercises

Verbal and Writing Skills

Write an expression in exponent form, given the base and the exponent. Then evaluate the expression.

1. Base = 4, exponent = 3 $4^3 = 64$

2. Base = 7, exponent = 2 $7^2 = 49$

3. Base = −5, exponent = 4 $(-5)^4 = 625$

4. Base = −4, exponent = 4 $(-4)^4 = 256$

5. Base = −7, exponent = 3 $(-7)^3 = -343$

6. Base = −6, exponent = 3 $(-6)^3 = -216$

Evaluate.

7. 2^3 8

8. 2^4 16

9. 3^4 81

10. 3^3 27

11. 7^3 343

12. 5^4 625

13. $(-3)^3$ −27

14. $(-2)^3$ −8

15. $(-2)^6$ 64

16. $(-3)^4$ 81

17. -5^3 −125

18. -4^3 −64

19. $\left(\frac{1}{4}\right)^2$ $\frac{1}{16}$

20. $\left(\frac{1}{2}\right)^3$ $\frac{1}{8}$

21. $\left(\frac{2}{5}\right)^3$ $\frac{8}{125}$

22. $\left(\frac{2}{3}\right)^4$ $\frac{16}{81}$

23. $(0.9)^2$ 0.81

24. $(0.4)^2$ 0.16

25. $(0.5)^3$ 0.125

26. $(0.6)^3$ 0.216

27. $(-8)^4$ 4096

28. $(-7)^4$ 2401

29. -8^4 −4096

30. -7^4 −2401

Write the following in exponent form.

31. (6)(6)(6)(6)(6) 6^5

32. (8)(8)(8)(8)(8)(8) 8^6

33. $(w)(w)$ w^2

34. $(w)(w)(w)$ w^3

35. $(x)(x)(x)(x)$ x^4

36. $(x)(x)(x)(x)(x)$ x^5

37. $(y)(y)(y)(y)(y)$ y^5

38. $(2)(2)(x)(x)(x)$ 2^2x^3

Evaluate each of the following. Approximate when necessary.

39. $2^3 + 3^4$ 89

40. $2^4 + 5^2$ 41

41. $5^3 + 6^2$ 161

42. $7^2 + 6^3$ 265

43. $(-3)^3 - (8)^2$ −91

44. $(-2)^3 - (-5)^4$ −633

45. $7^3 - (-5)^4$ −282

46. $8^2 - (-4)^3$ 128

47. $(-4)^3(-3)^2$ −576

48. $(-7)^3(-2)^4$ −5488

49. $8^2(-2)^3$ −512

50. $9^2(-3)^3$ −2187

51. $(1.43546)^5$ 6.094744099

52. $(1.94657)^6$ 54.40266254

53. 4^{12} 16,777,216

54. 6^{11} 362,797,056

Cumulative Review Problems

Evaluate.

55. $(-11) + (-13) + 6 + (-9) + 8$ −19

56. $\frac{3}{4} \div \left(-\frac{9}{20}\right)$ $-\frac{5}{3}$ or $-1\frac{2}{3}$

57. $-17 - (-9)$ −8

58. $(-2.1)(-1.2)$ 2.52

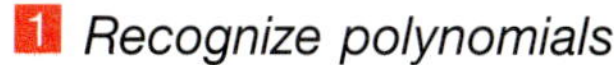

1.5 Use the Distributive Property to Simplify Expressions

After studying this section, you will be able to:

1. *Recognize polynomials.*
2. *Use the distributive property to multiply a polynomial by a monomial.*

MathPro

Video 4

SSM

1 *Polynomials*

If a number is multiplied by a variable we do not need any symbol in between the number and variable. Thus, to indicate $(2)(x)$, we write $2x$. To indicate $3 \cdot y$, we write $3y$. If one variable is multiplied by another variable, we place the variables next to each other. Thus, $(a)(b)$ is written ab. We use exponent form if an unknown number (a variable) is used several times as a factor. Thus, $x \cdot x \cdot x = x^3$. Similarly, $(y)(y)(y)(y) = y^4$. All these ideas are combined in the expression $5ab^2$ below.

Thus, you can combine variables and numbers to write algebraic expressions such as

$$a + b, \qquad 2x - 3, \qquad 5ab^2$$

The kinds of algebraic expressions that we often deal with are called **polynomials.** Polynomials are variable expressions that contain terms with nonnegative integer exponents. The following three expressions are all polynomials.

$$5x^2y + 1, \qquad 3a + 2b^2, \qquad 5a + 6b + 3c$$

There are special names for polynomials with one, two, or three terms. (A **term** is a number, a variable, or a product of numbers and variables.)

A **monomial** has *one* term:

$$5a, \qquad 3x^2yz^6, \qquad 10xy$$

A **binomial** has *two* terms:

$$5x^3 + 2y^2, \qquad 3x + 1, \qquad 10x + y$$

A **trinomial** has *three* terms:

$$5x^2 - 6x + 3, \qquad 2ab^3 - 6ab^2 - 8ab$$

2 *The Distributive Property*

An important property of algebra is the **distributive property.** We can state it in an equation as follows:

$$a(b + c) = ab + ac$$

TEACHING TIP You may want to mention, as you are discussing the distributive property, that it is still called the distributive property in cases such as

$$a(b + c + d) = ab + ac + ad$$

and

$$a(b + c + d + e) = ab + ac + ad + ae$$

A numerical example shows that it does seem reasonable.

$$5(3 + 6) = 5(3) + 5(6)$$

$$5(9) = 15 + 30$$

$$45 = 45$$

We can use the distributive property to multiply any polynomial by a monomial. In Section 1.4, we defined the word *factor*. We can use the word factor with polynomials and with monomials. Two or more algebraic expressions joined by multiplication are called **factors.** Consider the following examples of multiplying a polynomial by a monomial.

EXAMPLE 1 Multiply. **(a)** $5(a + b)$ **(b)** $-1(3x + 2y)$

(a) $5(a + b) = 5a + 5b$ *Multiply the factor* $(a + b)$ *by the factor* 5.

(b) $-1(3x + 2y) = -1(3x) + (-1)(2y) = -3x - 2y$ *Multiply the factor* $(3x + 2y)$ *by the factor* -1.

Practice Problem 1 Multiply.

(a) $-3(x + 2y)$ **(b)** $-a(a - 3b)$ ■

If the parentheses are preceded by a negative sign, we consider this to be the product of (-1) and the polynomial.

EXAMPLE 2 Multiply. **(a)** $-(a - 2b)$ **(b)** $-(3x^2y - 2xy^2 + 6x)$

(a) $-(a - 2b) = (-1)(a - 2b) = (-1)(a) + (-1)(-2b) = -a + 2b$

(b)
$$\begin{aligned} -(3x^2y - 2xy^2 + 6x) &= (-1)(3x^2y - 2xy^2 + 6x) \\ &= (-1)(3x^2y) + (-1)(-2xy^2) + (-1)(6x) \\ &= -3x^2y + 2xy^2 - 6x \end{aligned}$$

Practice Problem 2 Multiply. **(a)** $-(-3x + y)$ **(b)** $-(5r - 3s - 1)$ ■

In general, we see that in all these examples we have multiplied each term of the polynomial in the parentheses by the monomial in front of the parentheses.

EXAMPLE 3 Multiply. **(a)** $\frac{2}{3}(x^2 - 6x + 8)$ **(b)** $1.4(a^2 + 2.5a + 1.8)$

(a)
$$\begin{aligned} \frac{2}{3}(x^2 - 6x + 8) &= \left(\frac{2}{3}\right)(1x^2) + \left(\frac{2}{3}\right)(-6x) + \left(\frac{2}{3}\right)(8) \\ &= \frac{2}{3}x^2 + (-4x) + \frac{16}{3} \\ &= \frac{2}{3}x^2 - 4x + \frac{16}{3} \end{aligned}$$

(b) $1.4(a^2 + 2.5a + 1.8)$
$$\begin{aligned} &= 1.4(1a^2) + (1.4)(2.5a) + (1.4)(1.8) \\ &= 1.4a^2 + 3.5a + 2.52 \end{aligned}$$

Practice Problem 3 Multiply.

(a) $\frac{3}{5}(a^2 - 5a + 25)$ **(b)** $2.5(x^2 - 3.5x + 1.2)$ ■

There are times we multiply a variable by itself and use our exponent notation. Therefore $(x)(x) = x^2$ and $(x)(x)(x) = x^3$. In other cases there will be numbers and variables multiplied at the same time. We will see problems like $(2x)(x) = (2)(x)(x) = 2x^2$. Some expressions will involve the multiplication of more than one variable. We will see problems like $(3x)(xy) = (3)(x)(x)(y) = 3x^2y$. There will be times when we use the distributive property and all of these methods will be used. For example,

$$\begin{aligned} 2x(x - 3y + 2) &= 2x(x) + (2x)(-3y) + (2x)(2) \\ &= 2x^2 + (-6)(xy) + 4(x) \\ &= 2x^2 - 6xy + 4x \end{aligned}$$

We will discuss this type of multiplication of variables with exponents in more detail in Section 4.1. At that point we will expand these examples and other similar examples to develop the general rule for multiplication $(x^a)(x^b) = x^{a+b}$.

EXAMPLE 4 Multiply. **(a)** $5x(x + 4y)$ **(b)** $-2x(3x + y - 4)$

(a) $5x(x + 4y) = 5(x)(x) + 5(x)(4)(y) = 5x^2 + (5)(4)(x)(y) = 5x^2 + 20xy$

(b) $$\begin{aligned} -2x(3x + y - 4) &= -2(x)(3)(x) + (-2)(x)(y) + (-2)(x)(-4) \\ &= -2(3)(x)(x) + (-2)(xy) + (-2)(-4)(x) \\ &= -6x^2 - 2xy + 8x \end{aligned}$$

Practice Problem 4 Multiply. **(a)** $3x(3x + y)$ **(b)** $-4x(x - 2y + 3)$ ■

The distributive property can also be presented with the a on the right.

$$(b + c)a = ba + ca$$

The a is "distributed" over the b and c inside the parentheses.

EXAMPLE 5 Multiply. **(a)** $(2x^2 - x)(-3)$ **(b)** $(5x + 3y - 2)(2y)$

(a) $$\begin{aligned} (2x^2 - x)(-3) &= 2x^2(-3) + (-x)(-3) \\ &= -6x^2 + 3x \end{aligned}$$

(b) $$\begin{aligned} (5x + 3y - 2)(2y) &= 5x(2y) + 3y(2y) + (-2)(2y) \\ &= 10xy + 6y^2 - 4y \end{aligned}$$

Practice Problem 5 Multiply.

(a) $(3x^2 - 2x)(-4)$ **(b)** $(4y - 7x^2 + 1)(4y)$ ■

Developing Your Study Skills

HOW TO DO HOMEWORK

As you begin your homework assignments, read the directions carefully. You need to understand what is being asked for. Concentrate on each exercise or problem, taking time to solve it accurately. Rushing through your work usually causes errors. Check your answers with those given in the back of the textbook. If your answer is incorrect, check to see that you are doing the right problem. Redo the problem, watching for little errors. If it is still wrong, check with a friend. Perhaps the two of you can figure it out.

1.5 Exercises

Verbal and Writing Skills

Complete the sentence by filling in the blank.

1. A ___variable___ is a symbol used to represent an unknown number.

2. A trinomial has ___three___ terms.

3. $3x^2yz^6$ has one term and is called a ___monomial___.

4. Explain why you think the property $a(b + c) = ab + ac$ is called the distributive property. What does distributive mean?

Distribute means to disperse. We distribute the factor to each term in the parentheses.

★ **5.** Does the following distributive property work? $a(b - c) = ab - ac$ Why or why not? Give an example.

Yes $a(b - c)$ can be written as $a[b + (-c)]$.

$3(10 - 2) = (3 \times 10) - (3 \times 2)$

$3 \times 8 = 30 - 6$

$24 = 24$

Multiply. Use the distributive property.

6. $a(a - 2b)$ $a^2 - 2ab$

7. $6(x + 4y)$ $6x + 24y$

8. $-2(4a - 3b)$ $-8a + 6b$

9. $-3(2a - 5b)$ $-6a + 15b$

10. $3(3x - y + 5)$ $9x - 3y + 15$

11. $2(4x + y - 2)$ $8x + 2y - 4$

12. $-(a + 2b - c^2)$ $-a - 2b + c^2$

13. $6x(x + 3y)$ $6x^2 + 18xy$

14. $3x(x - 3y - 7)$ $3x^2 - 9xy - 21x$

15. $2x(4x - y - 6)$ $8x^2 - 2xy - 12x$

16. $-9(9x - 5y + 8)$ $-81x + 45y - 72$

17. $-5(3x + 9 - 7y)$ $-15x - 45 + 35y$

18. $\frac{1}{3}(3x^2 + 2x - 1)$ $x^2 + \frac{2x}{3} - \frac{1}{3}$

19. $\frac{1}{4}(x^2 + 2x - 8)$ $\frac{x^2}{4} + \frac{x}{2} - 2$

20. $\frac{3}{5}(-5x^2 - 10x + 4)$ $-3x^2 - 6x + \frac{12}{5}$

21. $\frac{4}{7}(7x^2 - 21x - 3)$ $4x^2 - 12x - \frac{12}{7}$

22. $\frac{x}{5}(x + 10y - 4)$ $\frac{x^2}{5} + 2xy - \frac{4x}{5}$

23. $\frac{y}{3}(3y - 4x - 6)$ $y^2 - \frac{4xy}{3} - 2y$

24. $5x(x + 2y + z - 1)$ $5x^2 + 10xy + 5xz - 5x$

25. $3a(2a + b - c - 4)$ $6a^2 + 3ab - 3ac - 12a$

26. $(8 - x)(-3)$ $-24 + 3x$

27. $(4 - 2x)(-3)$ $-12 + 6x$

28. $(3x^2 + 5x)(-2)$ $-6x^2 - 10x$

29. $(-5x^3 + x)(-5)$ $25x^3 - 5x$

30. $2x(3x + y - 4)$ $6x^2 + 2xy - 8x$

31. $3x(4x - 5y - 6)$ $12x^2 - 15xy - 18x$

32. $-3x(x + y - 2)$ $-3x^2 - 3xy + 6$

33. $-4x(x + y - 5)$ $-4x^2 - 4xy + 20x$

34. $(3x + 2y - 1)(-xy)$ $-3x^2y - 2xy^2 + xy$

35. $(4a - 2b - 1)(-ab)$ $-4a^2b + 2ab^2 + ab$

36. $(2x + 3y - 2)3xy$ $6x^2y + 9xy^2 - 6xy$

37. $(-2x + y - 3)4xy$ $-8x^2y + 4xy^2 - 12xy$

38. $1.5(2.8x^2 + 3.0x - 2.5)$ $4.2x^2 + 4.5x - 3.75$

39. $2.5(1.5a^2 - 3.5a + 2.0)$ $3.75a^2 - 8.75a + 5$

40. $-0.5x(-0.3y + 1.0xy + 0.2)$ $0.15xy - 0.5x^2y - 0.1x$

41. $-0.4a(0.3a - 0.2b + 0.02b^2)$ $-0.12a^2 + 0.08ab - 0.008ab^2$

42. $0.5x(0.6x + 0.8y - 5)$ $0.30x^2 + 0.40xy - 2.5x$

Cumulative Review Problems

Add.

43. $-18 + (-20) + 36 + (-14)$. -16

Evaluate.

44. $(-2)^6$ 64

45. $(-1)^{200}$ 1

46. $-27 - (-41)$ 14

1.6 Combine Like Terms

After studying this section, you will be able to:

MathPro Video 4 SSM

1 *Identify like terms.*
2 *Combine like terms.*

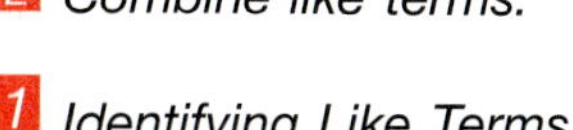

1 *Identifying Like Terms*

We can add or subtract quantities that are like quantities. This is called **combining** like quantities.

$$5 \text{ inches} + 6 \text{ inches} = 11 \text{ inches}$$

$$20 \text{ square inches} - 16 \text{ square inches} = 4 \text{ square inches}$$

However, we cannot combine things that are not the same.

$$16 \text{ square inches} - 4 \text{ inches} \quad \text{(cannot be done!)}$$

Similarly, in algebra we can *combine like terms*. This means to add or subtract like terms. Remember, we cannot combine terms that are not the same. A **term** is a number, a variable, or a product of numbers and variables. **Like terms** are terms that have identical variables and exponents. In other words, like terms must have exactly the same letter parts.

EXAMPLE 1 List the like terms of each expression.

(a) $5x - 2y + 6x$ **(b)** $2x^2 - 3x - 5x^2 - 8x$

(a) $5x$ and $6x$ are like terms. These are the only like terms.

(b) $2x^2$ and $-5x^2$ are like terms.
$-3x$ and $-8x$ are like terms.
Note that x^2 and x are not like terms.

Practice Problem 1 List the like terms of each expression.

(a) $5a + 2b + 8a - 4b$ **(b)** $x^2 + y^2 + 3x - 7y^2$ ■

Do you really understand what a term is? A **term** is a number, a variable, or a product of numbers and variables. Terms are the parts of an algebraic expression separated by plus or minus signs. The sign in front of the product is considered part of the term.

EXAMPLE 2

(a) List the terms of this expression. $3a - 2b + 5a^2 + 6a - 8b - 12a^2$

(b) Identify which terms are like terms.

(a) There are six terms. They are $3a$, $-2b$, $5a^2$, $6a$, $-8b$, and $-12a^2$.

(b) $3a$ and $6a$ are like terms.
$-2b$ and $-8b$ are like terms.
$5a^2$ and $-12a^2$ are like terms.

Practice Problem 2

(a) List the terms of this expression. $5x^2 - 6x + 8 - 4x - 9x^2 + 2$

(b) Identify which terms are like terms. ■

2 Combining Like Terms

It is important to know how to combine like terms. Since

$$4 \text{ inches} + 5 \text{ inches} = 9 \text{ inches}$$

we would expect in algebra that

$$4x + 5x = 9x$$

Why is this true? Let's take a look at the distributive property.

> Like terms may be added or subtracted by using the distributive property:
>
> $$ab + ac = a(b + c) \quad \text{and} \quad ba + ca = (b + c)a$$

For example,

$$-7x + 9x = (-7 + 9)x = 2x$$

$$5x^2 + 12x^2 = (5 + 12)x^2 = 17x^2$$

EXAMPLE 3 Combine like terms. **(a)** $-4x^2 + 8x^2$ **(b)** $5x + 3x + 2x$

(a) Notice that each term contains the factor x^2. Using the distributive property,

$$-4x^2 + 8x^2 = (-4 + 8)x^2 = 4x^2$$

(b) Note that each term contains the factor x. Using the distributive property,

$$5x + 3x + 2x = (5 + 3 + 2)x = 10x$$

Practice Problem 3 Combine like terms.

(a) $5a + 7a + 4a$ **(b)** $16y^3 + 9y^3$ ■

In this section, the direction *simplify* means to combine like terms.

EXAMPLE 4 Simplify. $5a^2 - 2a^2 + 6a^2$

$$5a^2 - 2a^2 + 6a^2 = (5 - 2 + 6)a^2 = 9a^2$$

Practice Problem 4 Simplify. $-8y^2 - 9y^2 + 4y^2$ ■

After doing a few problems, you will find that it is not necessary to write out the step of using the distributive property. We will omit this step for the remaining examples in this section.

EXAMPLE 5 Simplify. **(a)** $5a + 2b + 7a - 6b$

(b) $3x^2y - 2xy^2 + 6x^2y$ **(c)** $2a^2b + 3ab^2 - 6a^2b^2 - 8ab$

(a) $5a + 2b + 7a - 6b = 12a - 4b$ *We combine the a terms and the b terms separately.*

(b) $3x^2y - 2xy^2 + 6x^2y = 9x^2y - 2xy^2$ ***Note:*** *x^2y and xy^2 are not like terms because of different powers.*

(c) $2a^2b + 3ab^2 - 6a^2b^2 - 8ab$ *These terms cannot be combined; there are no like terms in this expression.*

Practice Problem 5 Simplify. **(a)** $-x + 3a - 9x + 2a$

(b) $5ab - 2ab^2 - 3a^2b + 6ab$ **(c)** $7x^2y - 2xy^2 - 3x^2y - 4xy^2 + 5x^2y$ ■

The two skills in this section that a student must practice are identifying like terms and correctly adding or subtracting like terms. If a problem involves many terms, you may find it helpful to rearrange the terms so that like terms are together.

EXAMPLE 6 Simplify. $3a - 2b + 5a^2 + 6a - 8b - 12a^2$

There are three pairs of like terms.

$= 3a + 6a - 2b - 8b + 5a^2 - 12a^2$ — *You can rearrange the terms so that like terms are together, making it easier to combine them.*

(a terms, b terms, a^2 terms)

$= 9a - 10b - 7a^2$ — *Combine like terms.*

The order of terms in an answer in these problems is not significant. These three terms can be rearranged in a different order. $-10b + 9a - 7a^2$ and $-7a^2 + 9a - 10b$ are also correct. However, we usually write polynomials in order of descending powers. $-7a^2 + 9a - 10b$ would be the preferred way to write the answer.

Practice Problem 6 Simplify. $5xy - 2x^2y + 6xy^2 - xy - 3xy^2 - 7x^2y$ ■

EXAMPLE 7 Simplify. $6(2x + 3xy) - 8x(3 - 4y)$

First remove the parentheses; then collect like terms.

$6(2x + 3xy) - 8x(3 - 4y) = 12x + 18xy - 24x + 32xy$ — *Use the distributive property.*

$= -12x + 50xy$ — *Combine like terms.*

Practice Problem 7 Simplify. $5a(2 - 3b) - 4(6a + 2ab)$ ■

Use extra care with fractional coefficients.

EXAMPLE 8 Simplify. $\frac{3}{4}x^2 - 5y - \frac{1}{8}x^2 + \frac{1}{3}y$

We need a least common denominator for the x^2 terms, which is 8. Change $\frac{3}{4}$ to eighths by multiplying the numerator and denominator by 2.

$$\frac{3}{4}x^2 - \frac{1}{8}x^2 = \frac{3 \cdot 2}{4 \cdot 2}x^2 - \frac{1}{8}x^2 = \frac{6}{8}x^2 - \frac{1}{8}x^2 = \frac{5}{8}x^2$$

The least common denominator for the y terms is 3. Change 5 to thirds.

$$-\frac{5}{1}y + \frac{1}{3}y = \frac{-5 \cdot 3}{1 \cdot 3}y + \frac{1}{3}y = \frac{-15}{3}y + \frac{1}{3}y = -\frac{14}{3}y$$

Thus, our solution is $\frac{5}{8}x^2 - \frac{14}{3}y$.

Practice Problem 8 Simplify. $\frac{1}{7}a^2 - \frac{5}{12}b + 2a^2 - \frac{1}{3}b$ ■

TEACHING TIP Students tend to have difficulty with fractional coefficients. After doing Example 8, you may want to do the following as Additional Class Exercise.

Simplify. $\frac{1}{2}a - 3ab - \frac{1}{4}a - \frac{1}{2}ab$

$= \frac{1}{2}a - \frac{1}{4}a - 3ab - \frac{1}{2}ab$

$= \frac{2}{4}a - \frac{1}{4}a - \frac{6}{2}ab - \frac{1}{2}ab$

$= \frac{1}{4}a - \frac{7}{2}ab$

1.6 Exercises

Verbal and Writing Skills

Identify the like terms in each problem. Do not combine.

1. 5 miles + 7 inches + 3 miles
5 miles, 3 miles

2. 2 quarts + 4 gallons + 9 quarts
2 quarts, 9 quarts

3. $5a - 2b - 12a$
$5a, -12a$

4. $3x + 15x - 26$
$3x, 15x$

5. $6x^2 - 3x - 4xy + 11x^2 - 2xy$
$6x^2, 11x^2$, and $-4xy, -2xy$

6. $3xy - 2y^2 + 5x + 3y^2 - 11x$
$-2y^2, 3y^2$, and $5x, -11x$

Simplify.

7. $7x + 3x$ $10x$

8. $2a - 8a$ $-6a$

9. $-12x^3 - 16x^3$ $-28x^3$

10. $-8a^2 + 4a^2$
$-4a^2$

11. $10x^4 + 8x^4 + 7x^2$
$18x^4 + 7x^2$

12. $3xy - 2y + 6x$
$3xy - 2y + 6x$

13. $5a + 2b - 7a^2$
$5a + 2b - 7a^2$

14. $3a^3 - 6a^2 + 5a^3$
$8a^3 - 6a^2$

15. $2ab + 1 - 6ab - 8$ $-4ab - 7$

16. $2x^2 + 3x^2 - 7 - 5x^2$ -7

17. $1.3x - 2.6y + 5.8x - 0.9y$
$7.1x - 3.5y$

18. $3.1ab - 0.2b - 0.8ab + 5.3b$
$2.3ab + 5.1b$

19. $1.6x - 2.8y - 3.6x - 5.9y$
$-2x - 8.7y$

20. $1.9x - 2.4b - 3.8x - 8.2b$
$-1.9x - 10.6b$

21. $\frac{1}{2}x^2 - 3y - \frac{1}{3}y + \frac{1}{4}x^2$
$\frac{3}{4}x^2 - \frac{10}{3}y$

22. $\frac{1}{5}a^2 - 2b - \frac{1}{2}a^2 - 3b$
$\frac{-3}{10}a^2 - 5b$

23. $\frac{1}{3}x - \frac{2}{3}y - \frac{2}{5}x + \frac{4}{7}y$
$-\frac{1}{15}x - \frac{2}{21}y$

24. $\frac{2}{5}s - \frac{3}{8}t - \frac{4}{15}s - \frac{5}{12}t$
$\frac{2}{15}s - \frac{19}{24}t$

25. $3x + 2y - 6 - 8x - 9y - 14$
$-5x - 7y - 20$

26. $-a + 2b + 8 - 7a - b - 1$
$-8a + b + 7$

27. $5x^2y - 10xy^6 + 6xy^2 - 7xy^2$
$5x^2y - 10xy^6 - xy^2$

28. $5bcd - 8cd - 12bcd + cd$
$-7bcd - 7cd$

29. $2ab + 5bc - 6ac - 2ab$
$5bc - 6ac$

30. $5x^2y + 12xy^2 - 8x^2 - 12xy^2$
$5x^2y - 8x^2$

31. $2x^2 - 3x - 5 - 7x + 8 - x^2$
$x^2 - 10x + 3$

32. $5x + 7 - 6x^3 + 6 - 11x + 4x^3$
$-2x^3 - 6x + 13$

33. $2y^2 - 8y + 9 - 12y^2 - 8y + 3$
$-10y^2 - 16y + 12$

34. $5 - 2y^2 + 3y - 8y - 9y^2 - 12$
$-11y^2 - 5y - 7$

35. $ab + 3a - 4ab + 2a - 8b$
$5a - 3ab - 8b$

Simplify. Use the distributive property to remove parentheses; then combine like terms.

36. $5(a - 3b) + 2(-b - 4a)$
$-3a - 17b$

37. $3(x + y) - 5(-2y + 3x)$
$-12x + 13y$

38. $6x(x + y) - 3(x^2 - 2xy)$
$3x^2 + 12xy$

39. $5a(a + 3b) - 2(a^2 - 6ab)$
$3a^2 + 27ab$

40. $-3(x^2 + 3y) + 5(-6y - x^2)$
$-8x^2 - 39y$

41. $-3(7xy - 11y^2) - 2y(-2x + 3y)$
$-17xy + 27y^2$

42. $4(2 - x) - 3(-5 - 12x)$
$32x + 23$

43. $7(3 - x) - 6(8 - 13x)$
$71x - 27$

Cumulative Review Problems

44. $-\frac{1}{3} - \left(-\frac{1}{5}\right)$ $-\frac{2}{15}$

45. $\left(-\frac{5}{3}\right)\left(\frac{1}{2}\right)$ $-\frac{5}{6}$

46. $\frac{4}{5} + \left(-\frac{1}{25}\right) + \left(-\frac{3}{10}\right)$ $\frac{23}{50}$

47. $\left(\frac{5}{7}\right) \div \left(-\frac{14}{3}\right)$
$-\frac{15}{98}$

1.7 Order of Arithmetic Operations

After studying this section, you will be able to:

1 *Use the order of operations to simplify numerical expressions involving addition, subtraction, multiplication, division, and exponents.*

MathPro

Video 4

SSM

1 *Order of Operations*

It is important to know *when* to do certain operations as well as how to do them. For example, for the expression $2 - 4 \cdot 3$, should we subtract first or multiply first? The following list will assist you. It tells which operations to do first: the correct **order of operations.** We call this a *list of priorities.*

TEACHING TIP Students sometimes ask why this order of operations is so important. Remind them that scientific calculators and computers follow the same order of operations. If we want to know how to use either of them, then we must understand which operations these devices will perform in what order. Remind them that the next section, 1.8, will be easier if they master this order of operations now.

Order of Operations for Numbers

Follow this order of operations:

Do first	**1.** Combine numbers inside parentheses.
	2. Raise numbers to a power.
	3. Multiply and divide numbers from left to right.
Do last	**4.** Add and subtract numbers from left to right.

Let's return to the problem: $2 - 4 \cdot 3$. There are no parentheses or numbers raised to a power, so multiplication comes next. We do that first. Then we subtract since this comes last on our list.

$2 - 4 \cdot 3$ *Follow the order of operations by first multiplying* $4 \cdot 3 = 12$.

$= 2 - 12$

$= -10$ *Combine* $2 - 12 = -10$.

TEACHING TIP When discussing Example 1, be sure to mention that multiplication and division have equal priority. As we go from left to right, we do any multiplication or any division problems we encounter. We do *not* go through the problem and do multiplication first and then division. You may want to show students this Additional Class Exercise. Evaluate.

$(-4)^2 \div (-8)(3) - (-2) + 5$

Sol. $16 \div (-8)(3) - (-2) + 5$

$= (-2)(3) - (-2) + 5$

$= -6 - (-2) + 5$

$= -6 + (+2) + 5$

$= -4 + 5 = 1$

EXAMPLE 1 Evaluate. $8 \div 2 \cdot 3 + 4^2$

$= 8 \div 2 \cdot 3 + 16$ *Evaluate* $4^2 = 16$ *because the highest priority in this problem is raising to a power.*

$= 4 \cdot 3 + 16$ *Next multiply and divide from left to right, so* $8 \div 2 = 4$ *and* $4 \cdot 3 = 12$.

$= 12 + 16$ *Finally, add.*

$= 28$

Practice Problem 1 Evaluate. $18 - (-3)^3$ ■

Note: Multiplication and division have equal priority. We do not do multiplication problems first. Rather, we work from left to right, doing any multiplication or division that we encounter. Similarly, addition and subtraction have equal priority.

EXAMPLE 2 Evaluate. $(-3)^3 - 2^4$

The highest priority is to raise the expressions to the appropriate power.

$(-3)^3 - 2^4 = -27 - 16$ *In* $(-3)^3$ *we are cubing the number* -3 *to obtain* -27. *Be careful,* -2^4 *is not* $(-2)^4$! *Raise 2 to the fourth power and take the negative.*

$= -43$ *The last step is to add and subtract from left to right.*

Practice Problem 2 Evaluate. $12 \div (-2)(-3) - 2^4$ ■

CALCULATOR

Order of Operations

Use your calculator to evaluate $3 + 4 \cdot 5$. Enter

3 [+] 4 [×] 5 [=]

If the display is [23], the correct order of operations is built in. If the display is not 23, you will need to modify the way you enter the problem. You should use

4 [×] 5 [+] 3 [=]

Try $6 + 3 \cdot 4 - 8 \div 2$.

TEACHING TIP After discussing Example 3 with your students, ask them to do the following as an Additional Class Exercise. Simplify.

$(4 - 5)^3 + (-2)(-3)^2$

Ans. -19

EXAMPLE 3 Evaluate. $2 \cdot (2 - 3)^3 + 6 \div 3 + (8 - 5)^2$

$2 \cdot (2 - 3)^3 + 6 \div 3 + (8 - 5)^2$ *Combine the numbers inside the parentheses.*

$= 2 \cdot (-1)^3 + 6 \div 3 + 3^2$

$= 2 \cdot (-1) + 6 \div 3 + 9$ *Next, raise to a power. Note that we need parentheses for -1 because of the negative sign, but they are not needed for 3.*

$= -2 + 2 + 9$ *Next, multiply and divide from left to right.*

$= 9$ *Finally, add and subtract from left to right.*

Practice Problem 3 Evaluate. $6 - (8 - 12)^2 + 8 \div 2$ ■

EXAMPLE 4 Evaluate. $\left(-\frac{1}{5}\right)\left(\frac{1}{2}\right) - \left(\frac{3}{2}\right)^2$

The highest priority is to raise $\frac{3}{2}$ to the second power.

$$\left(\frac{3}{2}\right)^2 = \left(\frac{3}{2}\right)\left(\frac{3}{2}\right) = \frac{9}{4}$$

$= \left(-\frac{1}{5}\right)\left(\frac{1}{2}\right) - \frac{9}{4}$ *Next we multiply.*

$= -\frac{1}{10} - \frac{9}{4}$ *We need to write each fraction as an equivalent fraction with the LCD of 20.*

$= -\frac{1 \cdot 2}{10 \cdot 2} - \frac{9 \cdot 5}{4 \cdot 5}$

$= -\frac{2}{20} - \frac{45}{20}$ *Add.*

$= -\frac{47}{20}$

Practice Problem 4 Evaluate. $\left(-\frac{1}{7}\right)\left(\frac{-14}{5}\right) + \left(-\frac{1}{2}\right) \div \left(\frac{3}{4}\right)$ ■

Developing Your Study Skills

PREVIEWING NEW MATERIAL

Part of your study time each day should consist of looking over the sections in your text that are to be covered the following day. You do not necessarily need to study and learn the material on your own, but a survey of the concepts, terminology, diagrams, and examples will help the new ideas seem more familiar as the instructor presents them. You can look for concepts that appear confusing or difficult and be ready to listen carefully for your instructor's explanations. You can be prepared to ask the questions that will increase your understanding. Previewing new material enables you to see what is coming and prepares you to be ready to absorb it.

1.7 Exercises

Verbal and Writing Skills

A friend is counting some money and announces "I have three twos plus four fives."

1. Write this as a number expression. $(3)(2) + (4)(5)$

2. How much money does your friend have? $26

3. What answer would you get to our number expression if
(a) you used left-to-right entry? $50
(b) you used order-of-operation entry? $26

4. Which order best fits this problem? (b)

Evaluate.

5. $4^2 + 2(4)$ 24

6. $2 \div 2 \cdot 3 + 1$ 4

7. $5 + 6 \cdot 2 \div 4 - 1$ 7

8. $(2 - 8)^2 \div 6$ 6

9. $(3 - 5)^2 \cdot 6 \div 4$ 6

10. $2(3 - 5 + 6) + 5$ 13

11. $8 - 2^3 \cdot 5 + 3$
$8 - 40 + 3 = -29$

12. $-14 \div (-7) - 8 \cdot 2 + 3^3$
$2 - 16 + 27 = 13$

13. $4 + 27 \div 3 \cdot 2 - 8$
$4 + 18 - 8 = 14$

14. $3 \cdot 5 + 7 \cdot 3 - 5 \cdot 3$
$15 + 21 - 15 = 21$

15. $5(3)^3 - 20 \div (-2)$
$135 + 10 = 145$

16. $4(2)^4 + 15 \div (-3)$
$64 - 5 = 59$

17. $3(5 - 7)^2 - 6(3)$
$12 - 18 = -6$

18. $-2(3 - 6)^2 - (-2)$
$-18 + 2 = -16$

19. $5 \cdot 6 - (3 - 5)^2 + 8 \cdot 2$
$30 - 4 + 16 = 42$

20. $(-3)^2 \cdot 6 \div 9 + 4 \cdot 2$
$6 + 8 = 14$

21. $\frac{1}{2} \div \frac{2}{3} + 6 \cdot \frac{1}{4}$
$\frac{3}{4} + \frac{6}{4} = \frac{9}{4}$ or $2\frac{1}{4}$

22. $\frac{5}{6} \div \frac{2}{3} - 6 \cdot \left(\frac{1}{2}\right)^2$
$\frac{5}{4} - \frac{6}{4} = -\frac{1}{4}$

23. $0.8 + 0.3(0.6 - 0.2)^2$
0.848

24. $0.05 + 1.4 - (0.5 - 0.7)^3$
$0.05 + 1.4 + 0.008 = 1.458$

25. $\frac{3}{4}\left(-\frac{2}{5}\right) - \left(-\frac{3}{5}\right)$
$-\frac{3}{10} + \frac{6}{10} = \frac{3}{10}$

26. $-\frac{2}{3}\left(\frac{3}{5}\right) + \frac{5}{7} \div \frac{5}{3}$
$\frac{-2}{5} + \frac{5}{7} \cdot \frac{3}{5} = \frac{1}{35}$

27. $-6.3 - (-2.7)(1.1) + (3.3)^2$
$-6.3 + 2.97 + 10.89 = 7.56$

28. $4.35 + 8.06 \div (-2.6) - (2.1)^2$
$4.35 - 3.1 - 4.41 = -3.16$

29. $\left(\frac{1}{2}\right)^3 + \frac{1}{4} - \left(\frac{1}{6} - \frac{1}{12}\right) - \frac{2}{3} \cdot \left(\frac{1}{4}\right)^2$
$\frac{1}{8} + \frac{2}{8} - \frac{1}{12} - \frac{2}{3} \cdot \frac{1}{16} = \frac{1}{4}$

30. $(2.4 \cdot 1.2)^2 - 1.6 \cdot 2.2 \div 4.0 - 3.6$
$8.2944 - 0.88 - 3.6 = 3.8144$

Cumulative Review Problems

Simplify.

31. $(2x - 3y) - (x - 3y)$
$2x - 3y - x + 3y = x$

32. $-\frac{3}{4} - \frac{5}{6}$
$-\frac{19}{12}$ or $-1\frac{7}{12}$

33. -1^{2086}
-1

34. $3\frac{3}{5} \div 6\frac{1}{4}$
$\frac{72}{125}$

1.8 Use Substitution to Evaluate Expressions

MathPro

Video 5

SSM

After studying this section, you will be able to:

1 *Evaluate a variable expression for a specified value.*
2 *Evaluate a formula by substitution.*

1 *Evaluating a Variable Expression*

You will use the order of operations to evaluate variable expressions.

Evaluate $6 + 3x$ for $x = -4$

In general, x represents some unknown number. Here we are told x has the value -4. We can replace x by -4. Use parentheses around -4. Note that we always put replacement values in parentheses.

$$6 + 3(-4) = 6 + (-12) \qquad \textit{Order of operations}$$
$$= -6$$

When we replace a variable by a particular value, we say we have *substituted* the value for the variable. We then evaluate the expression (find a value for it).

EXAMPLE 1 Evaluate. $\frac{2}{3}x - 5$ for $x = -6$

$\frac{2}{3}(-6) - 5$ *Substitute $x = -6$. Be sure to enclose the -6 in parentheses.*

$= -4 \quad -5$ *Multiply $\left(\frac{2}{3}\right)\left(-\frac{6}{1}\right) = -4$.*

$= -9$ *Combine.*

Practice Problem 1 Evaluate. $4 - \frac{1}{2}x$ for $x = -8$ ■

Compare parts (a) and (b) in the next example. The two parts illustrate that you must be careful what value you raise to a power. *Note:* In part (b) we will need parentheses within parentheses. To avoid confusion, we use brackets [ ] to represent the outside parentheses.

EXAMPLE 2 **(a)** Evaluate $2x^2$ for $x = -3$. **(b)** Evaluate $(2x)^2$ for $x = -3$.

(a) Here the value x is squared.

$2x^2 = 2(-3)^2$

$= 2(9)$ *First square -3.*

$= 18$ *Then multiply.*

(b) Here the value $(2x)$ is squared.

$(2x)^2 = [(2)(-3)]^2$

$= (-6)^2$ *First multiply the numbers inside the parentheses.*

$= 36$ *Then square -6.*

Practice Problem 2 Evaluate each expression for $x = -3$.

(a) $-x^4$ **(b)** $(-x)^4$ ■

Carefully study the solutions to Example 2(a) and Example 2(b). You will find that taking the time to see *how* and *why* they are different is a good investment of study time.

EXAMPLE 3 Evaluate $x^2 + x$ for $x = -4$.

$x^2 + x = (-4)^2 + (-4)$ *Replace x by −4 in the original expression.*

$= 16 + (-4)$ *Raise to a power.*

$= 12$ *Finally, add.*

Practice Problem 3 Evaluate $(5x)^3 + 2x$ for $x = -2$. ■

TEACHING TIP After discussing Example 3, it is a good idea to immediately show the class another example, such as: Evaluate $2x^2 + 3x$ for $x = -3$

Sol. $2(-3)^2 + 3(-3)$

$= 2(9) + 3(-3)$

$= 18 + (-9)$

$= 9$

2 *Evaluating Formulas*

We can *evaluate a formula* by substituting values for the variables. For example, the area of a triangle can be found using the formula $A = \frac{1}{2}bh$, where b is the length of the base of a triangle and h is the height (or the altitude) of the triangle (see figure). If we know values for b and h, we can substitute those values into the formula to find the area. The units for area are *square units*.

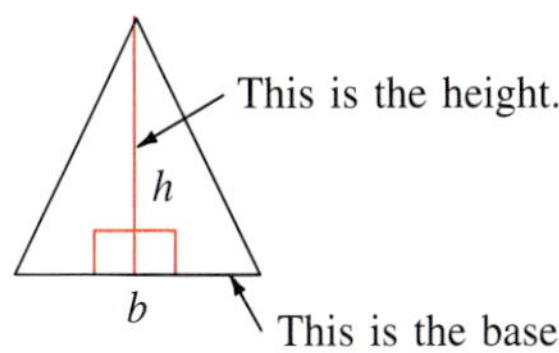

EXAMPLE 4 Find the area of a triangle with a base of 16 centimeters (cm) and a height of 12 centimeters (cm).

Use the formula

$$A = \frac{1}{2}bh$$

Substitute 16 centimeters for b and 12 centimeters for h.

$$A = \frac{1}{2}(16 \text{ centimeters})(12 \text{ centimeters})$$

$$= \frac{1}{2}(16)(12)(\text{cm})(\text{cm})$$ *If you take $\frac{1}{2}$ of 16 first, it will make your calculation easier.*

$$= (8)(12)(\text{cm})^2 = 96 \text{ square centimeters}$$

The area of the triangle is 96 square centimeters.

Practice Problem 4 Find the area of a triangle with a height of 3 meters and a base of 7 meters. ■

TEACHING TIP With some of the famous oil spills off the coast, students find this to be a rather interesting problem. The volume of a right circular cylinder is $V = \pi r^2 h$, where r is the radius of the cylinder and h is the height of the cylinder. Suppose we had a very "short" cylinder with a very large radius, like the shape of a giant coin. This would describe the shape of an ocean oil spill of even thickness if the spill was contained in a circular region (when viewed from overhead).

A recent oil spill of a tanker off the coast of the Mediterranean Sea had a radius of 5000 meters and a uniform thickness of 0.002 meter. How many cubic meters of oil were contained in the spill? (Use $\pi = 3.14$.)

Sol.

$V = \pi r^2 h$

$= \pi(5000 \text{ meters})^2(0.002 \text{ meter})$

$= (3.14)(25000000)(0.002)(\text{m})^3$

$= (78500000)(0.002)(\text{m})^3$

$= 157{,}000$ cubic meters of oil

The area of a circle is given by

$$A = \pi r^2$$

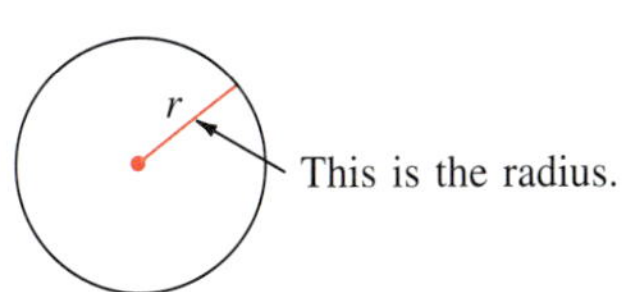

We will use 3.14 as an approximation for the *irrational number* π.

EXAMPLE 5 Find the area of a circle if the radius is 2 inches.

$A = \pi r^2 = (3.14)(2 \text{ inches})^2$ *Write the formula and substitute the given values for the variable.*

$= (3.14)(4)(\text{in.})^2$ *Raise to a power. Then multiply.*

$= 12.56$ square inches

Practice Problem 5 Find the area of a circle if the radius is 3 meters. ■

The formula $C = \frac{5}{9}(F - 32)$ allows us to find the Celsius temperature if we know the Fahrenheit temperature. That is, we can substitute a value for F in degrees Fahrenheit into the formula to obtain a temperature C in degrees Celsius.

EXAMPLE 6 What is the Celsius temperature when the Fahrenheit temperature is $F = -22°$?

Use the formula.

$$C = \frac{5}{9}(F - 32)$$

$$= \frac{5}{9}(-22 - 32) \qquad \textit{Substitute } -22 \textit{ for } F \textit{ in the formula.}$$

$$= \frac{5}{9}(-54) \qquad \textit{Combine the numbers inside the parentheses.}$$

$$= (5)(-6) \qquad \textit{Simplify.}$$

$$= -30 \qquad \textit{Multiply.}$$

The temperature is $-30°$ Celsius.

Practice Problem 6 What is the Celsius temperature when the Fahrenheit temperature is $F = 68°$? Use the formula $C = \frac{5}{9}(F - 32)$. ■

When driving in Canada or Mexico, we must observe speed limits posted in kilometers per hour. A formula that converts r (miles per hour) or k (kilometers per hour) is $k = 1.61r$. Note that this is an approximation.

NOTRE SIGNALISATION ROUTIÈRE
EST MÉTRIQUE
OUR TRAFFIC SIGNS ARE METRIC
55
MAXIMUM
90
km/h

EXAMPLE 7 You are driving on a highway in Mexico. It is posted for a maximum speed of 100 kilometers per hour. You are driving at 61 miles per hour. Are you exceeding the speed limit?

Use the formula.

$$k = 1.61r$$

$$= (1.61)(61) \qquad \textit{Replace } r \textit{ by } 61.$$

$$= 98.21 \qquad \textit{Multiply the numbers.}$$

You are driving at approximately 98 kilometers per hour. You are not exceeding the speed limit.

Practice Problem 7 You are driving behind a heavily loaded truck on a Canadian highway. The highway has a posted minimum speed of 65 kilometers per hour. When you travel at exactly the same speed as the truck ahead of you, you observe that the speedometer reads 35 miles per hour. Assuming that your speedometer is accurate, determine if the truck is violating the minimum speed law. ■

1.8 Exercises

Evaluate.

1. $-2x + 1$ for $x = 3$
$-6 + 1 = -5$

2. $-4x - 2$ for $x = 5$
$-20 - 2 = -22$

3. $\frac{1}{2}x + 2$ for $x = -8$
$-4 + 2 = -2$

4. $\frac{1}{3}x - 6$ for $x = 9$
$3 - 6 = -3$

5. $5x + 10$ for $x = \frac{1}{2}$
$\frac{5}{2} + 10 = 12\frac{1}{2}$

6. $7x + 20$ for $x = -\frac{1}{2}$
$-\frac{7}{2} + 20 = 16\frac{1}{2}$

7. $2 - 4x$ for $x = 7$
$2 - 28 = -26$

8. $3 - 5x$ for $x = 8$
$3 - 40 = -37$

9. $x^2 - 3x$ for $x = -2$
$4 + 6 = 10$

10. $x^2 + 3x$ for $x = 4$
$16 + 12 = 28$

11. $3x^2$ for $x = -1$
3

12. $4x^2$ for $x = -1$
4

13. $-2x^2$ for $x = 5$
-50

14. $-5x^2$ for $x = 3$
-45

15. $9x + 13$ for $x = -2$
$-18 + 13 = -5$

16. $5x + 7$ for $x = -2$
$-10 + 7 = -3$

17. $2x^2 + 3x$ for $x = -3$
$18 - 9 = 9$

18. $18 - 5x$ for $x = -3$
$18 + 15 = 33$

19. $(2x)^2 + x$ for $x = 3$
$6^2 + 3 = 39$

20. $2 - x^2$ for $x = -2$
$2 - (4) = -2$

21. $2 - (-x)^2$ for $x = -2$
$2 - (2)^2 = -2$

22. $2x - 3x^2$ for $x = -4$
$-8 - 3(16) = -56$

23. $7x + (2x)^2$ for $x = -3$
$-21 + 36 = 15$

24. $5x + (3x)^2$ for $x = -2$
$-10 + (-6)^2 = 26$

25. $2x^2 + 3x + 6$ for $x = 2$
$2(4) + 3(2) + 6 = 20$

26. $3x^2 + 2x - 4$ for $x = 3$
$27 + 6 - 4 = 29$

27. $\frac{1}{2}x^2 - 3x + 9$ for $x = -4$
$8 + 12 + 9 = 29$

Evaluate each expression for different values of the variable.

28. $x^2 + 2x + 3$
(a) for $x = 4$ 27
(b) for $x = 0$ 3
(c) for $x = -2$ 3

29. $x^2 - 3x + 4$
(a) for $x = 4$ 8
(b) for $x = 0$ 4
(c) for $x = -2$ 14

30. $5 - 6x - 2x^2$
(a) for $x = 2$ -15
(b) for $x = 0$ 5
(c) for $x = -1$ 9

31. $8 - 3x - 2x^2$
(a) for $x = 3$ -19
(b) for $x = 0$ 8
(c) for $x = -2$ 6

★**32.** $x^2 - 2y + 3y^2$ for $x = -3$ and $y = 4$
$9 - 8 + 48 = 49$

★**33.** $2x^2 - 3xy + 2y$ for $x = 4$ and $y = -1$
$2(16) + 12 - 2 = 42$

★**34.** $a^3 + 2abc - 3c^2$ for $a = 5$, $b = 9$, and $c = -1$
$5^3 + 2(5)(9)(-1) - 3(-1)^2 = 32$

★**35.** $a^2 - 2ab + 2c^2$ for $a = 3$, $b = 2$, and $c = -4$
$3^2 - 2(3)(2) + 2(-4)^2 = 29$

Applications

For Problems 36 and 37, draw and label a diagram. Use the formula $A = \frac{1}{2}bh$.

36. Find the area of a triangular neon gasoline sign with a base of 14 feet and a height of 17 feet. 119 square feet

37. Find the area of a triangular flag with a base of 12 inches and a height of 14 inches. 84 square inches

For Problems 38 and 39, draw and label a diagram. Use the formula $A = \pi r^2$. Use $\pi = 3.14$.

38. The radius of the circular opening of a chemistry flask is 4 cm. What is the area of the opening? 50.24 square centimeters

39. An ancient outdoor sundial has a radius of 5 meters. What is its area? 78.5 square meters

For Problems 40 and 41, use the formula $F = \frac{9}{5}C + 32$ to find the Fahrenheit temperature.

40. Find the Fahrenheit temperature F when the Celsius temperature is 25°C. 77°F

41. It is not uncommon for parts of North Dakota to have a temperature of −10°C. Find the Fahrenheit temperature. 14°F

Solve each word problem. Round to the nearest tenth.

42. Find the cost of covering a circle with thin gold leaf on an outdoor sign at a cost of \$290 per square centimeter if the radius of the circle is 9 cm. (Use $\pi = 3.14$.) \$73,758.60

43. How much would it cost to make a circle of solid aluminum for an outdoor sign at a cost of \$350 per square meter if the radius of the circle is 7 meters. (Use $\pi = 3.14$.) \$53,851

44. Some new computers can be exposed to extreme temperatures (as high as 60°C and as low as −50°C). What is the temperature range in Fahrenheit that these computers can be exposed to? (Use the formula $F = \frac{9}{5}C + 32$.) 140°F to −58°F

45. To deal with extreme temperatures while doing research at the south Pole, scientists have developed accommodations that can comfortably withstand an outside temperature of −60°C with no wind blowing, or −30°C with wind gusts of up to 50 miles per hour. What is the corresponding Fahrenheit temperature range? (Use the formula $F = \frac{9}{5}C + 32$.) −76°F to −22°F

46. Bruce becomes exhausted while on a bicycle trip in Canada. He reads on the map that his present elevation is 2.3 kilometers above sea level. How many miles above sea level is he? Why is he so tired? (Use the formula $k = 1.61\,r$.) 1.4 miles. The air is thin and lacks the usual amount of oxygen at this elevation.

47. While biking down the Pacific coast of Mexico, you see on the map that it is 20 kilometers to the nearest town. Approximately how many miles is it to the nearest town? (Use the formula $k = 1.61r$.) 12.4 miles

Cumulative Review Problems

Simplify.

48. $(-2)^4 - 4 \div 2 - (-2)$ 16

49. $3(x - 2y) - (x^2 - y) - (x - y)$ $-x^2 + 2x - 4y$

50. $-12 - 6 + 8 - (-19) + (-13) - 14 + (-16)$ -34

51. $(-0.3) \times (2.4)$ -0.72

Putting Your Skills to Work

MEASURING YOUR LEVEL OF FITNESS

In recent years there has been a greater awareness of fitness, exercise, and weight control in our country. Often people would like a simple mathematical way to measure if they are physically fit and of the proper weight.

A growing number of doctors and health fitness professionals refer to the Government studies from the National Center for Health Statistics as supporting the BMI, or Body Mass Index as a helpful measure to determine if a person is at the proper weight, is overweight, or is underweight.

To determine your BMI you multiply your weight in pounds by 0.45 to get your weight in kilograms. Next you convert your height to inches. Then you multiply your height in inches by 0.0254 to get your height in meters. Now you multiply that number by itself. Your weight in kilograms is divided by that number. (Your weight in kilograms is divided by the square of your height in meters.)

The result is your BMI. For most people this will be a number that ranges from the 20s to the low 30s. Federal guidelines suggest that you should keep your BMI under 25 in order not to be overweight. However, there is a limit to how low your BMI should be. Many doctors recommend that a person's BMI should not be lower than 18.

PROBLEMS FOR INDIVIDUAL INVESTIGATION AND STUDY

Use the procedure above to answer each of the following questions. Round all measurements of BMI to the nearest tenth.

1. Find the BMI for a woman who weighs 125 pounds and is 5 feet, 4 inches tall. BMI = 21.3

2. Is a man who weighs 185 pounds and is 5 feet, 11 inches tall considered overweight by the government study? (Is his BMI 25 or above?) What is his BMI? Yes, just slightly. BMI = 25.6

PROBLEMS FOR COOPERATIVE GROUP ACTIVITY

Together with other members of your class see if you can complete the following.

3. Is a woman who weighs 95 pounds and is 1.6 meters tall considered underweight by the government study? (Is her BMI under 18?) What is her BMI? Yes, BMI = 16.7

4. Write a formula to find the BMI of a person if the height of the person is measured in meters (m) and the weight of the person is measured in pounds (p).

$$\text{BMI} = \frac{0.45p}{m^2}$$

INTERNET CONNECTIONS: Go to ''http://www.prenhall.com/tobey'' to be connected

Site: BMI Chart (Shape Up America!) or a related site

5. This site gives a ''shortcut method'' for calculating BMI. Write a formula based on this method to find the BMI of a person whose height is measured in inches (i) and weight is measured in pounds (p). Then use your formula to find the BMI of a person who is 4 feet, 10 inches tall and weighs 116 pounds.

1.9 Grouping Symbols

MathPro

Video 5

SSM

After studying this section, you will be able to:

1 *Simplify variable expressions with several grouping symbols.*

1 *Simplifying Variable Expressions*

Many expressions in algebra use **grouping symbols** such as parentheses, brackets, and braces. Sometimes expressions are inside other expressions. Because it can be confusing to have more than one set of parentheses, brackets and braces are also used. How do we know what to do first when we see an expression like $2[5 - 4(a + b)]$?

To simplify the expression, we start with the innermost grouping symbols. Here it is a set of parentheses. We first use the distributive law to multiply

$$2[5 - 4(a + b)] = 2[5 - 4a - 4b]$$

We use the distributive law again.

$$= 10 - 8a - 8b$$

There are no like terms, so this is our final answer.

Notice that we started with two sets of grouping symbols, but our final answer had none. So we can say we *removed* the grouping symbols. Of course, we didn't just take them away: We used the distributive law and the rules for signed numbers to simplify as much as possible. Although we are doing many steps, we sometimes say "remove parentheses" as a shorthand direction. Sometimes we say "simplify."

Remember to remove the innermost parentheses first. Keep working from the inside out.

EXAMPLE 1 Simplify. $3[6 - 2(x + y)]$

We want to remove the innermost parentheses first. Therefore, we first use the distributive property to simplify $-2(x + y)$.

$$3[6 - 2(x + y) = 3[6 - 2x - 2y] \quad \textit{Use the distributive property.}$$

$$= 18 - 6x - 6y \quad \textit{Use the distributive property again.}$$

Practice Problem 1 Simplify. $5[4x - 3(x - 2)]$ ■

You recall that a negative sign in front of parentheses is equivalent to having a coefficient of negative 1. You can write the -1 and then multiply by -1 using the distributive property.

$$-(x + 2y) = -1(x + 2y) = -x - 2y$$

Notice that this has the effect of removing the parentheses. Each term in the result now has its sign changed from that of the original.

Similarly, a positive sign in front of parentheses can be viewed as multiplication by $+1$.

$$+(5x - 6y) = +1(5x - 6y) = 5x - 6y$$

If a grouping symbol has a positive or negative sign in front, we mentally multiply by $+1$ or -1.

EXAMPLE 2 Simplify. $-2[3a - (b + 2c) + (d - 3e)]$

$= -2[3a - b - 2c + d - 3e]$ *Remove the two innermost sets of parentheses. Since one is not inside the other, we remove both sets at once.*

$= -6a + 2b + 4c - 2d + 6e$ *Now we remove the brackets by multiplying each term by -2.*

Practice Problem 2 Simplify. $3ab - [2ab - (2 - a)]$ ■

TEACHING TIP After presenting Example 2, ask students to do the following problem as an Additional Class Exercise.

Simplify.

$-3[2 + 4(x - y) - 3(2w - z)]$

Ans.

$-6 - 12x + 12y + 18w - 9z$

EXAMPLE 3 Simplify. $2[3x - (y + w)] - 3[2x + 2(3y - 2w)]$

$= 2[3x - y - w] - 3[2x + 6y - 4w]$ *In each set of brackets, remove the inner parentheses.*

$= 6x - 2y - 2w - 6x - 18y + 12w$ *Remove each set of brackets by multiplying by the appropriate number.*

$= -20y + 10w$ or $10w - 20y$ *Combine like terms. (Note that $6x - 6x = 0x = 0$.)*

Practice Problem 3 Simplify. $3[4x - 2(1 - x)] - [3x + (x - 2)]$ ■

TEACHING TIP After presenting Example 3, you may want to do the following as an Additional Class Exercise.

Simplify.

$3[2a - (b + 3c)]$
$- 2[-8c - 3(b + 2a)]$

Sol. $3[2a - b - 3c]$
$- 2[-8c - 3b - 6a]$

$= 6a - 3b - 9c + 16c + 6b + 12a$

$= 18a + 3b + 7c$

You can always simplify problems with many sets of grouping symbols by the method shown above. Essentially, you just keep removing one level of grouping symbols at each step. Finally, at the end you add up the like terms if possible.

Sometimes it is possible to collect like terms at each step.

EXAMPLE 4 Simplify. $-3\{7x - 2[x - (2x - 1)]\}$

$= -3\{7x - 2[x - 2x + 1]\}$ *Remove the inner parentheses by multiplying each term within the parentheses by negative 1.*

$= -3\{7x - 2[-x + 1]\}$ *Collect like terms by combining $+x - 2x$.*

$= -3\{7x + 2x - 2\}$ *Remove the brackets by multiplying each term within them by -2.*

$= -3\{9x - 2\}$ *Combine the x terms.*

$= -27x + 6$ *Remove the braces by multiplying each term by -3.*

Practice Problem 4 Simplify. $-2\{5x - 3x[2x - (x^2 - 4x)]\}$ ■

TEACHING TIP After presenting Example 4, you may want to do the following as an Additional Class Exercise.

Simplify.

$-2\{2a^2 - 3[a + 4(a - 2)]\}$

Sol. $-2\{2a^2 - 3[a + 4a - 8]\}$

$= -2\{2a^2 - 3[5a - 8]\}$
$= -2\{2a^2 - 15a + 24\}$
$= -4a^2 + 30a - 48$

Developing Your Study Skills

EXAM TIME: THE NIGHT BEFORE

With adequate preparation, you can spend the night before the exam pulling together the final details.

1. Look over each section to be covered in the exam. Review the steps needed to solve each type of problem.
2. Review your list of terms, rules, and formulas that you are expected to know for the exam.
3. Take the Practice Test at the end of the chapter just as though you were taking the actual exam. Do not look in your text or get help in any way. Time yourself so that you know how long it takes you to complete the test.
4. Check the Practice Test. Redo the problems you missed.
5. Be sure you have ready the necessary supplies for taking your exam.

1.9 Exercises

Verbal and Writing Skills

1. Rewrite the expression $-3x - 2y$ using a negative sign and parentheses. $-(3x + 2y)$

2. Rewrite the expression $-x + 5y$ using a negative sign and parentheses. $-(x - 5y)$

3. To simplify expressions with grouping symbols, we use the ___distributive___ property.

4. When an expression contains many grouping symbols, remove the ___innermost___ parentheses first.

Simplify. Remove grouping symbols and combine like terms.

5. $2y - 3(x + y)$
$2y - 3x - 3y = -3x - y$

6. $y - 3(y + z)$
$y - 3y - 3z = -2y - 3z$

7. $2(a + 3b) - 3(b - a)$
$2a + 6b - 3b + 3a = 5a + 3b$

8. $4(x - y) - 2(3x + y)$
$4x - 4y - 6x - 2y = -2x - 6y$

9. $x^2(x^2 - 3y^2) - 2(x^4 - x^2y^2)$
$x^4 - 3x^2y^2 - 2x^4 + 2x^2y^2 = -x^4 - x^2y^2$

10. $ab(ab - 6b) - 8(a^2b^2 - 2ab^2)$
$a^2b^2 - 6ab^2 - 8a^2b^2 + 16ab^2 = -7a^2b^2 + 10ab^2$

11. $5[3 + 2(x - 26) + 3x]$
$5[3 + 2x - 52 + 3x] = -245 + 25x$

12. $-4[-(x + 3y) - 2(y - x)]$
$-4(x - 5y) = -4x + 20y$

13. $7x^2 + [2(3 - x) - 3x^2]$
$7x^2 + 6 - 2x - 3x^2 = 4x^2 - 2x + 6$

14. $2x[4x^2 - 2(x - 3)]$
$2x[4x^2 - 2x + 6] = 8x^3 - 4x^2 + 12x$

15. $4y[-3y^2 + 2(4 - y)]$
$4y[-3y^2 + 8 - 2y] = -12y^3 - 8y^2 + 32y$

16. $3y^2 + [4(y - 2) - y^2]$
$3y^2 + 4y - 8 - y^2 = 2y^2 + 4y - 8$

17. $2(x - 2y) - [3 - 2(x - y)]$
$2x - 4y - 3 + 2x - 2y = 4x - 6y - 3$

18. $5[a - b(a - b) - b^2]$
$5[a - ab + b^2 - b^2] = 5a - 5ab$

19. $2[a - 3b(a + b) - b^2]$
$2a - 6ab - 8b^2$

20. $3(x + 2y) - [4 - 2(x + y)]$
$3x + 6y - 4 + 2x + 2y = 5x + 8y - 4$

21. $x(x^2 + 2x - 3) - 2(x^3 + 6)$
$-x^3 + 2x^2 - 3x - 12$

22. $5x^2(x + 6) - 2[x - 2(1 + 2x^2)]$
$5x^3 + 38x^2 - 2x + 4$

23. $3a^2 - 4[2b - 3b(b + 2)]$
$3a^2 + 16b + 12b^2$

24. $x - \{2y - 3[x - 2(x + y)]\}$
$-2x - 8y$

25. $5x - \{4y - 2[2x + 3(x - y)]\}$
$5x - \{4y - 4x - 6x + 6y\} = 15x - 10y$

26. $2b^2 - 3[5b + 2b(2 - b)]$
$2b^2 - 27b + 6b^2 = 8b^2 - 27b$

27. $-2\{3(x - 5) + 2[-x^2 - 3(2x + 1)]\}$
$4x^2 + 18x + 42$

28. $-\{-2(5 - y) - 3[2y^2 - 2(3y - 1)]\}$
$6y^2 - 20y + 16$

29. $-4\{3a^2 - 2[4a^2 - (b + a^2)]\}$ $12a^2 - 8b$

30. $-2\{x^2 - 3[x - (x - 2x^2)]\}$ $10x^2$

Cumulative Review Problems

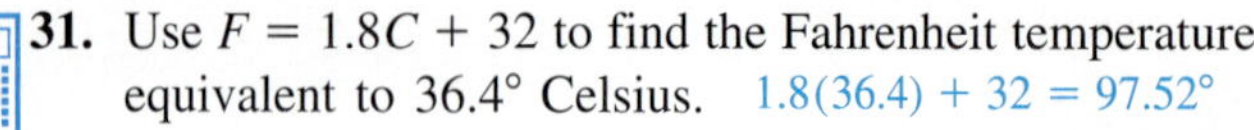

31. Use $F = 1.8C + 32$ to find the Fahrenheit temperature equivalent to 36.4° Celsius. $1.8(36.4) + 32 = 97.52°$

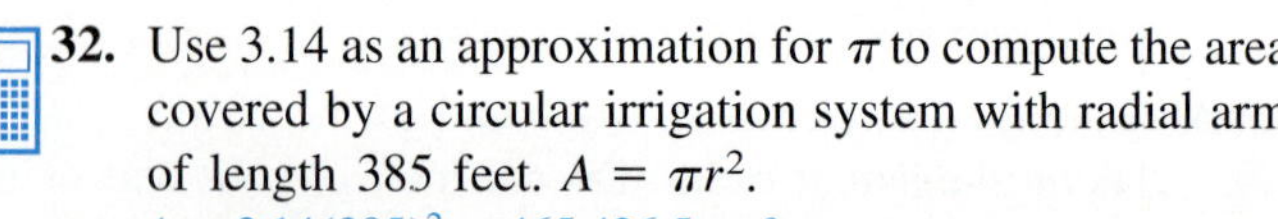

32. Use 3.14 as an approximation for π to compute the area covered by a circular irrigation system with radial arm of length 385 feet. $A = \pi r^2$.
$A = 3.14(385)^2 = 465{,}426.5$ sq ft

33. Simplify. $3 - 4(-2)^3$ $3 - 4(-8) = 35$

34. Evaluate. $4x^2 - x + 11$ for $x = -3$
$4(9) + 3 + 11 = 50$

Putting Your Skills to Work

CONSUMER PRICE INDEX

The federal government produces a statistic that is helpful in determining how much the cost of living is increasing each year and in what areas this increase is most significant. The measure is called the Consumer Price Index (CPI) and it is announced annually by the Bureau of Labor Statistics, U.S. Department of Labor. The CPI measures the average change in the price of goods and services purchased by all urban consumers. Listed below is a table that shows the CPI for certain years.

	1975	1982–1984	1990	1992	1994
Cost of All Items	53.8	100	130.7	140.3	148.2
Food	59.8	100	132.4	137.9	144.3
Housing	50.7	100	128.5	137.5	144.3
Medical Care	47.5	100	162.8	201.4	211.0
Entertainment	62.0	100	132.4	142.3	150.1

The period of 1982–1984 is used as the focus of comparison, and the index for that period is assigned a value of 100. Thus if the cost of food for a family was $100 in 1982–1984, then the cost of similar food in 1994 would be $144.30. However, a similar purchase of food in 1975 would have amounted to $59.80. Use the above table to answer the following questions.

PROBLEMS FOR INDIVIDUAL INVESTIGATION AND STUDY

1. If a family spent $200 for entertainment for a year in the 1982–1984 period, how much would be spent for similar entertainment for a year in 1994? $300.20

2. If an inpatient hospital operation cost $10,000 in the 1982–1984 period, how much would a similar operation cost in 1992? $20,140

PROBLEMS FOR COOPERATIVE GROUP ACTIVITY

Together with other members of your class see if you can complete the following.

3. If Roberta Sanchez had a salary of $22,500 during the 1982–1984 period and her 1994 salary was $31,800, did her salary increase cover the increase of the cost of all items during that period? No. Her salary would have had to increase to $33,345 to cover the increase.

4. If you know the cost of medical care during the 1982–1984 period, what mathematical operation should you do to that value to obtain the cost of similar medical care in 1990? Several methods. One is to multiply the value by 1.628.

INTERNET CONNECTIONS: Go to ``http://www.prenhall.com/tobey'' to be connected

Site: Consumer Price Index (Bureau of Labor Statistics) or a related site

The Bureau of Labor Statistics maintains regional and local price indices. This is because inflation occurs at different rates in different parts of the country. Use this web site to obtain a table showing the CPI for your city or region. (You may find that the basic table format is easiest to use. Note that the annual results are given last.) Then answer the following questions.

5. Suppose you earned $17,300 per year during the 1982–1984 period. If your income increases matched the increase of the CPI for your city or region, how much would you have earned last year?

6. Suppose you spent $5500 on rent last year. If your rent increases matched the CPI increase for your city or region, how much would you have spent for an average year during the 1982–1984 period?

Chapter Organizer

Topic	Procedure	Examples
Absolute value, p. 71.	The absolute value of a number is the distance between that number and zero on the number line. The absolute value of any number will be positive or zero.	$\|3\| = 3$ $\|-2\| = 2$ $\|0\| = 0$ $\left\|-\frac{5}{6}\right\| = \frac{5}{6}$ $\|-1.38\| = 1.38$
Adding signed numbers with the same sign, p. 72.	If the signs are the same, add the absolute values of the numbers. Use the common sign in the answer.	$(-3) + (-7) = -10$
Adding signed numbers with opposite signs, p. 74.	If the signs are different: 1. Find the difference between the larger absolute value and the smaller. 2. Give the answer the sign of the number having the larger absolute value.	$(-7) + (+13) = 6$ $(7) + (-13) = -6$
Adding several signed numbers, p. 76.	When adding several signed numbers, separate them into two groups by common sign. Find the sum of all the positives and all the negatives. Combine these two subtotals by the method described above.	$(-7) + 6 + 8 + (-11) + (-13) + 22$ $-7 \quad +6$ $-11 \quad +8$ $\underline{-13} \quad \underline{+22}$ $-31 \quad +36$ $(-31) + (+36) = +5$ positive since 36 is positive
Subtracting signed numbers, p. 81.	Change the sign of the second number and then add.	$(-3) - (-13) = (-3) + (+13) = 10$
Multiplying and dividing signed numbers, p. 87 and p. 89.	1. If the two numbers have the same sign, multiply (or divide). The result is positive. 2. If the two numbers have different signs, multiply (or divide) as indicated. The result is negative.	$(-5)(-3) = +15$ $(-36) \div (-4) = +9$ $(28) \div (-7) = -4$ $(-6)(3) = -18$
Exponent form, p. 94.	The base tells you what number is being multiplied. The exponent tells you how many times this number is used as a factor.	$2^5 = 2 \cdot 2 \cdot 2 \cdot 2 \cdot 2 = 32$ $4^3 = 4 \cdot 4 \cdot 4 = 64$ $(-3)^4 = (-3)(-3)(-3)(-3) = 81$
Raising a negative number to a power, p. 95.	When the base is negative, the result is positive for even exponents, and negative for odd exponents.	$(-3)^3 = -27$ but $(-2)^4 = 16$
Removing parentheses, p. 98.	Use the distributive law to remove parentheses. $a(b + c) = ab + ac$	$3(5x + 2) = 15x + 6$
Combining like terms, p. 102.	Combine terms that have identical letters and exponents.	$7x^2 - 3x + 4y + 2x^2 - 8x - 9y = 9x^2 - 11x - 5y$
Order of operations, p. 105.	Remember the proper order of operations: 1. Operations inside parentheses 2. Powers 3. Multiplication and division from left to right 4. Addition and subtraction from left to right	$3(5 + 4)^2 - 2^2 \cdot 3 \div (9 - 2^3) = 3 \cdot 9^2 - 4 \cdot 3 \div 1$ $= 3 \cdot 81 - 12$ $= 243 - 12 = 231$
Substituting into variable expressions, p. 108.	1. Replace each letter by the numerical value given. 2. Follow the order of operations in evaluating the expression.	Evaluate $2x^3 + 3xy + 4y^2$ for $x = -3$, $y = 2$. $2(-3)^3 + 3(-3)(2) + 4(2)^2$ $= 2(-27) + 3(-3)(2) + 4(4)$ $= -54 - 18 + 16$ $= -56$

Chapter Organizer

Topic	Procedure	Examples
Using formulas, p. 109.	1. Replace each variable in the formula by the given values. 2. Evaluate the expression. 3. Label units carefully.	Find the area of a circle with radius 4 feet. Use $A = \pi r^2$, with π as approximately 3.14. $A = (3.14)(4 \text{ feet})^2$ $= (3.14)(16 \text{ feet}^2)$ $= 50.24 \text{ feet}^2$ The area of the circle is approximately 50.24 square feet.
Removing grouping symbols, p. 114.	1. Remove innermost grouping symbols first. 2. Then remove remaining innermost grouping symbols. 3. Continue until all grouping symbols are removed. 4. Combine like terms.	$5\{3x - 2[4 + 3(x - 1)]\} = 5\{3x - 2[4 + 3x - 3]\}$ $= 5\{3x - 8 - 6x + 6\}$ $= 15x - 40 - 30x + 30$ $= -15x - 10$

Developing Your Study Skills

PROBLEMS WITH ACCURACY

Strive for accuracy. Mistakes are often made as a result of human error rather than by lack of understanding. Such mistakes are frustrating. A simple arithmetic or sign error can lead to an incorrect answer.

These five steps will help you cut down on errors.

1. Work carefully, and take your time. Do not rush through a problem just to get it done.
2. Concentrate on one problem at a time. Sometimes problems become mechanical, and your mind begins to wander. You become careless and make a mistake.
3. Check your problem. Be sure that you copied it correctly from the book.
4. Check your computations from step to step. Check the solution in the problem. Does it work? Does it make sense?
5. Keep practicing new skills. Remember the old saying, ''Practice makes perfect.'' An increase in practice results in an increase in accuracy. Many errors are due simply to lack of practice.

There is no magic formula for eliminating all errors, but these five steps will be a tremendous help in reducing them.

Chapter 1 Review Problems

1.1 *Add the following.*

1. $(-6) + (-2)$
-8

2. $(-12) + (+7.8)$
-4.2

3. $(+5) + (-2) + (-12)$
-9

4. $(+3.7) + (-1.8)$
1.9

5. $\left(+\frac{1}{2}\right) + \left(-\frac{5}{6}\right)$
$-\frac{1}{3}$

6. $\left(-\frac{3}{11}\right) + \left(-\frac{1}{22}\right)$
$-\frac{7}{22}$

7. $\left(+\frac{3}{4}\right) + \left(-\frac{1}{12}\right) + \left(-\frac{1}{2}\right)$
$\frac{1}{6}$

8. $\left(-\frac{4}{15}\right) + \left(+\frac{12}{5}\right) + \left(-\frac{2}{3}\right)$
$\frac{22}{15}$

9. $(-5) + (-15) + (+6) + (-9) + (+10)$ -13

10. $-8 + 4 + (-3) + (-20) + 18 + 7$
-2

1.2 *Add or subtract the following.*

11. $(+5) - (-3)$
8

12. $(-2) - (-15)$
13

13. $(-30) - (+3)$
-33

14. $(+8) - (-1.2)$
9.2

15. $\left(-\frac{7}{8}\right) + \left(-\frac{3}{4}\right)$
$-\frac{13}{8}$

16. $\left(-\frac{3}{14}\right) + \left(+\frac{5}{7}\right)$
$\frac{1}{2}$

17. $-20.8 - 1.9$
-22.7

18. $-103 - (-76)$
-27

Mixed Review 1.1–1.2 *Perform each operation indicated. Simplify all answers.*

19. $-5 + (-2) - (-3)$
-4

20. $6 - (-4) + (-2) + (8)$
16

21. $(-16) + (-13)$
-29

22. $(-11) - (-12)$
1

1.3

23. $(-7) \times 6$
-42

24. $96 \div (-32)$
-3

25. $(-3)(-2) + (-2)(4)$
-2

26. $\dfrac{-18}{-\frac{2}{3}}$
27

27. $(-20) \div 4$
-5

28. $(-100) \div (-50)$
2

29. $\left(-\frac{1}{2}\right) \div \left(\frac{3}{4}\right)$
$-\frac{2}{3}$

30. $\left(\frac{5}{7}\right) \div \left(-\frac{5}{25}\right)$
$-\frac{25}{7}$

31. $(-5)(-1)(2)$
10

32. $(-6)(3)(4)$
-72

33. $(-1)(-2)(-3)(-5)$
30

34. $(-5)\left(-\frac{1}{2}\right)(4)(-3)$
-30

Mixed Review 1.1–1.3

35. $-\frac{4}{3} + \frac{2}{3} + \frac{1}{6}$
$-\frac{1}{2}$

36. $-\frac{6}{7} + \frac{1}{2} + \left(-\frac{3}{14}\right)$
$-\frac{4}{7}$

37. $(-3)(-2)(-5)$
-30

38. $-6 + (-2) - (-3)$
-5

39. $3.5(-2.6)$
-9.1

40. $(-5.4) \div (-6)$
0.9

41. $5 - (-3.5) + 1.6$
10.1

42. $-8 + 2 - (-4.8)$
-1.2

43. $25 + 4.1 + (-26) + (-3.5)$ -0.4

44. $29 + (-45) + 17.5 + (-3.8)$ -2.3

Solve each word problem.

45. The football team had 3 plays in which they lost 8 yards each time. What was the total yardage lost? 24 yards

46. The low temperature in Anchorage Alaska last night was $-34°F$. During the day the temperature rose $12°F$. What was the temperature during the day? $-22°$

47. A mountain peak is 6895 feet above sea level. A location in Death Valley is 468 feet below sea level. What is the difference in height between these two locations? 7363 feet

48. IBM stock rose $1\frac{1}{2}$ points on Monday, dropped $3\frac{1}{4}$ points on Tuesday, rose 2 points on Wednesday, and dropped $2\frac{1}{2}$ points on Thursday. What was the total gain or loss on the value of the stock over this four-day period? $2\frac{1}{4}$ point loss

1.4 *Evaluate.*

49. $(-3)^4$ 81

50. $(-2)^6$ 64

51. $(-5)^3$ -125

52. $\left(\frac{1}{2}\right)^3$ $\frac{1}{8}$

53. -8^2 -64

54. $(0.7)^2$ 0.49

55. $\left(\frac{5}{6}\right)^2$ $\frac{25}{36}$

56. $\left(\frac{3}{4}\right)^3$ $\frac{27}{64}$

1.5 *Use the distributive property to multiply.*

57. $5(3x - 7y)$ $15x - 35y$

58. $2x(3x - 7y + 4)$ $6x^2 - 14xy + 8x$

59. $-(7x^2 - 3x + 11)$ $-7x^2 + 3x - 11$

60. $(2xy + x - y)(-3y)$ $-6xy^2 - 3xy + 3y^2$

1.6 *Combine like terms.*

61. $3a^2b - 2bc + 6bc^2 - 8a^2b - 6bc^2 + 5bc$ $-5a^2b + 3bc$

62. $3x + 7y + 11x - 9y$ $14x - 2y$

63. $4x^2 - 13x + 7 - 9x^2 - 22x - 16$ $-5x^2 - 35x - 9$

64. $-x + \frac{1}{2} + 14x^2 - 7x - 1 - 4x^2$ $10x^2 - 8x - \frac{1}{2}$

1.7 *Simplify using the order of operations.*

65. $(5)(-4) + (3)(-2)^3$ -44

66. $20 - (-10) - (-6) + (-5) - 1$ 30

67. $(7 - 9)^3 + (-6)(-2) + (-3)$ 1

1.8 *Evaluate each expression for the given value of the variable.*

68. $3x - 4$ for $x = -5$ -19

69. $5 - \frac{2}{3}x$ for $x = 6$ 1

70. $x^2 + 3x - 4$ for $x = -3$ -4

71. $-3x^2 - 4x + 5$ for $x = 2$ -15

72. $-3x^3 - 4x^2 + 2x + 6$ for $x = -2$ 10

73. $vt - \frac{1}{2}at^2$ for $v = 24$, $t = 2$, and $a = 32$ -16

74. $\frac{nRT}{V}$ for $n = 16$, $R = -2$, $T = 4$, and $V = -20$ $\frac{32}{5}$

Solve each word problem.

75. Find the simple interest on a loan of \$6000 at an annual interest rate of 18% per year for $\frac{3}{4}$ of a year. Use $I = prt$. \$810

76. Find the Fahrenheit temperature if a radio announcer in Mexico City says that the high temperature today was 30°C. Use the formula $F = \frac{9C + 160}{5}$. 86°F

77. How much will it cost to paint a circular sign with a radius of 15 meters. The painter charges \$3 per square meter. Use $A = \pi r^2$, where π is approx 3.14. \$2119.50

78. Find the profit P if the initial cost $C = \$1200$, rent $R = \$300$, and sales $S = \$56$. Use the profit formula $P = 180S - R - C$. \$8580.00

1.9 *Simplify.*

79. $3x - 2(x + 4)$ $x - 8$

80. $2(x - 1) - 7(3x + 4)$ $-19x - 30$

81. $2[3 - (4 - 5x)]$ $-2 + 10x$

82. $-3x[x + 3(x - 7)]$ $-12x^2 + 63x$

83. $2xy^3 - 6x^3y - 4x^2y^2 + 3(xy^3 - 2x^2y - 3x^2y^2)$
$5xy^3 - 6x^3y - 13x^2y^2 - 6x^2y$

84. $-2y^2 - 8xy + 3x^2 - 4y^2 + 3xy - 8y^2 - 2x^2 + xy$
$-14y^2 - 4xy + x^2$

85. $-5(x + 2y - 7) + 3x(2 - 5y)$
$x - 10y + 35 - 15xy$

86. $2\{x - 3(y - 2) + 4[x - 2(y + 3)]\}$
$10x - 22y - 36$

87. $8[a - b(3 - 4a)] - 6a[2 - a(3 - b)]$
$-4a - 24b + 32ab + 18a^2 - 6a^2b$

88. $-5\{2a - b[5a - b(3 + 2a)]\}$
$-10a + 25ab - 15b^2 - 10ab^2$

89. $-3\{2x - [x - 3y(x - 2y)]\}$ $-3x - 9xy + 18y^2$

Chapter 1 Test

Simplify each answer.

1. $-3 + (-4) + 9 + 2$

2. $\frac{1}{2} + \left(-\frac{1}{6}\right)$

3. $-0.6 - (-0.8)$

4. $(-8)(-12)$

5. $(-5)(-2)(7)(-1)$

6. $(-12) \div (-3)$

7. $(-1.8) \div (0.6)$

8. $\dfrac{-5}{\frac{3}{2}}$

9. $(-4)^3$

10. $(1.2)^2$

11. -5^3

12. $\left(\frac{2}{3}\right)^4$

1. 4

2. $\frac{1}{3}$

3. 0.2

4. 96

5. -70

6. 4

7. -3

8. $-\frac{10}{3}$ or $-3\frac{1}{3}$

9. -64

10. 1.44

11. -125

12. $\frac{16}{81}$

13. $-5x^2 - 10xy + 35x$

14. $6a^2b^2 + 4ab^3 - 14a^2b^3$

15. $-6xy - 5x^2y - 4xy^2$

16. $2a^2b + \frac{15}{2}ab$

17. $8a^2 + 20ab$

18. $5a + 30$

19. $14x - 16y$

20. 8

21. -25

22. 122

23. 37

24. 1

25. 96.6

26. $-3a - 9ab + 3b^2 - 3ab^2$

27. $-69x + 90y - 51$

13. $-5x(x + 2y - 7)$

14. $-2ab^2(-3a - 2b + 7ab)$

15. $2xy - 5x^2y + 3xy^2 - 8xy - 7xy^2$

16. $6ab - \frac{1}{2}a^2b + \frac{3}{2}ab + \frac{5}{2}a^2b$

17. $12a(a + b) - 4(a^2 - 2ab)$

18. $3(2 - a) - 4(-6 - 2a)$

19. $5(3x - 2y) - (x + 6y)$

20. $7 - (6 - 9)^2 + 5(2)$

21. $3(4 - 6)^3 + 12 \div (-4) + 2$

Evaluate each expression for the value of the variable indicated.

22. $x^3 - 3x^2y + 2y - 5$ for $x = 3$ and $y = -4$

23. $3x^2 - 7x - 11$ for $x = -3$

24. $2a - 3b$ for $a = -4$ and $b = -3$

25. If you are traveling 60 miles per hour in Canada on a highway, how fast are you traveling in kilometers per hour? (Use $k = 1.61r$)

Simplify each expression.

26. $-3\{a + b[3a - b(1 - a)]\}$

27. $3\{[x - (5 - 2y)] - 4[3 + (6x - 7y)]\}$

CHAPTER 2
Equations and Inequalities

By traveling at a constant speed on a modern superhighway, you can maximize the fuel economy of a car when taking trips. However, did you know that the speed at which you travel significantly changes the fuel economy? Do you realize how much more gasoline you use traveling at a higher rate of speed? Turn to the Putting Your Skills to Work section on page 157 to see if your predictions are correct.

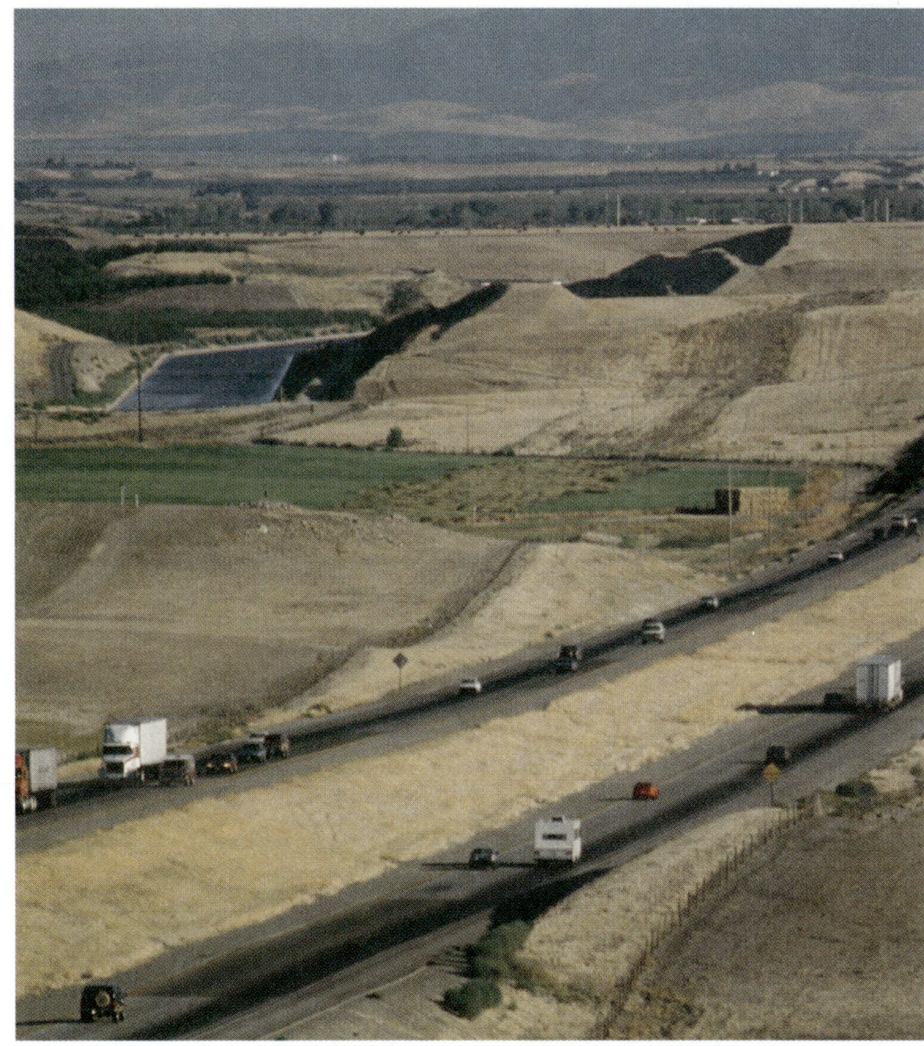

Pretest Chapter 2

1.	$x = 58$
2.	$x = 48$
3.	$x = -4$
4.	$x = 4$
5.	$x = -\frac{7}{2}$
6.	$x = -\frac{1}{7}$
7.	$x = \frac{5}{2}$
8.	$x = \frac{17}{10}$
9. (a)	$F = \frac{9C + 160}{5}$
(b)	$F = 5°$
10. (a)	$r = \frac{I}{Pt}$
(b)	6%
11.	$<$
12.	$>$
13.	$>$
14.	$<$
15.	See graph
16.	See graph

If you are familiar with the topics in this chapter, take this test now. Check your answers with those in the back of the book. If an answer was wrong or you couldn't do a problem, study the appropriate section of the chapter.

If you are not familiar with the topics in this chapter, don't take this test now. Instead, study the examples, work the practice problems, and then take the test.

This test will help you to identify those concepts that you have mastered and those that need more study.

Sections 2.1–2.3

Solve for x.

1. $x - 9 = 49$

2. $\frac{1}{4}x = 12$

3. $-5x = 20$

4. $7x + 2 = 30$

5. $3x - 8 = 7x + 6$

6. $5x + 3 - 6x = 4 - 8x - 2$

Section 2.4

Solve for x.

7. $\frac{2}{5}x + \frac{1}{4} = \frac{1}{2}x$

8. $\frac{2}{3}(x - 2) + \frac{1}{2} = 5 - (3 + x)$

Section 2.5

9. (a) Solve for F. $C = \frac{5}{9}(F - 32)$.

(b) Find the temperature in degrees Fahrenheit if the Celsius temperature is $-15°$.

10. (a) Solve for r. $I = Prt$.

(b) Find the rate r if $P = \$2000$, $t = 2$ years, and $I = \$240$.

Sections 2.6–2.7

Replace the question mark with $<$ or $>$.

11. 1 ? 10 **12.** 2 ? -3 **13.** -12 ? -13 **14.** 0 ? 0.9

Solve and graph.

15. $-2x + 5 \leq 4 - x + 3$

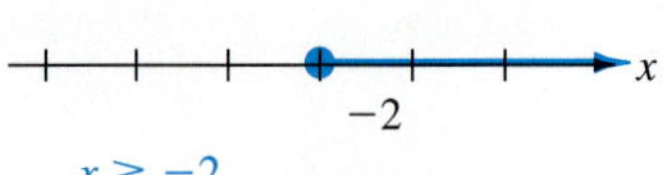

$x \geq -2$

16. $\frac{1}{2}(x + 2) - 1 > \frac{x}{3} - 5 + x$

$x < 6$

2.1 The Addition Principle

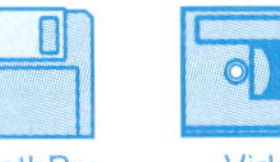

After studying this section, you will be able to:

1 *Solve equations of the form $x + b = c$ using the addition principle.*

1 *Using the Addition Principle*

When we use the equals sign (=), we indicate that two expressions are equal in value. This is called an **equation.** For example, $x + 5 = 23$ is an equation. By choosing certain procedures, you can go step by step from a given equation to the equation $x =$ some number. The number is the solution to the equation.

One of the first procedures used in solving equations has an application in our everyday world. Suppose that we place a 10-kilogram box on one side of a seesaw and a 10-kilogram stone on the other side. If the center of the box is the same distance from the balance point as the center of the stone, we would expect the seesaw to balance. The box and the stone do not look the same, but they have the same value in weight. If we add a 2-kilogram lead weight to the center of weight of each object at the same time, the seesaw should still balance. The results are equal.

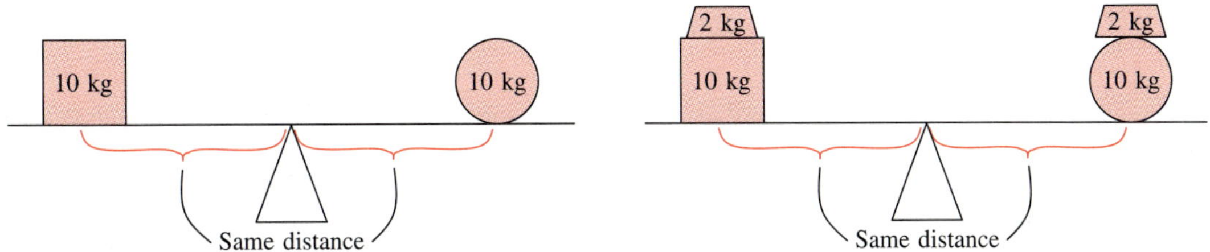

There is a similar principle in mathematics. We can state it in words like this.

> **The Addition Principle**
>
> If the same number is added to both sides of an equation, the results on each side are equal in value.

We can restate it in symbols this way.

> For real numbers a, b, c if $a = b$ then $a + c = b + c$

Here is an example.

$$\text{If } 3 = \frac{6}{2}, \quad \text{then } 3 + 5 = \frac{6}{2} + 5$$

Since we added the same amount 5 to both sides, each side has an equal value.

$$3 + 5 = \frac{6}{2} + 5$$

$$8 = \frac{6}{2} + \frac{10}{2}$$

$$8 = \frac{16}{2}$$

$$8 = 8$$

TEACHING TIP Remind students that when using the addition principle of equality the number added to each side of the equation can be any real number. It may be positive or negative, an integer, a rational number in decimal form, a rational number in fractional form, or even an irrational number.

We can use the addition principle to solve an equation.

EXAMPLE 1 Solve for x. $x + 16 = 20$

$x + 16 + (-16) = 20 + (-16)$ *Use the addition principle to add* -16 *to both sides.*

$x + 0 = 4$ *Simplify.*

$x = 4$ *The value of* x *is* 4.

We have just found the solution of the equation. The **solution** is a value for the variable that makes the equation true. We then say that the value, 4, in our example, **satisfies** the equation. We can easily verify that 4 is a solution by substituting this value in the original equation. This step is called **checking** the solution.

Check. $x + 16 = 20$

$4 + 16 \stackrel{?}{=} 20$

$20 = 20$ ✓

When the same value appears on both sides of the equals sign, we call the equation an **identity.** Because the two sides of the equation in our check have the same value, we know that the original equation has been correctly solved. We have found the solution.

Practice Problem 1 Solve for x and check your solution. $x + 0.3 = 1.2$ ■

When you are trying to solve these types of equations, you notice that you must add a particular number to both sides of the equation. What is the number to choose? Look at the number that is on the same side of the equation with x, that is, the number added to x. Then think of the number that is **opposite in sign.** This is called the **additive inverse** of the number. The additive inverse of 16 is -16. The additive inverse of -3 is 3. The number to add to both sides of the equation is precisely this additive inverse.

It does not matter which side of the equation contains the variable. The x term may be on the right or left. In the next example the x term will be on the right.

TEACHING TIP Mention the definition of ADDITIVE INVERSE a couple of times during the class session. It is an imporant term that will be used frequently in mathematics. Students should understand the concept that, to move a term from one side of an equation to the other, they should add the ADDITIVE INVERSE of that term to both sides of the equation.

EXAMPLE 2 Solve for x. $14 = x - 3$

$14 + 3 = x - 3 + 3$ *Add 3 to both sides, since 3 is the additive inverse of* -3. *This will eliminate the* -3 *on the right and isolate* x.

$17 = x + 0$ *Simplify.*

$17 = x$ *The value of* x *is* 17.

Check. $14 = x - 3$

$14 \stackrel{?}{=} 17 - 3$ *Replace* x *by* 17.

$14 = 14$ ✓ *Simplify. It checks. The solution is* $x = 17$.

Practice Problem 2 Solve for x and check your solution. $17 = x - 5$ ■

Before you add a number to both sides, you should always simplify the equation. The following example shows how combining numbers by addition—separately, on both sides of the equation—simplifies the equation.

TEACHING TIP This concept should be repeated several times at the beginning of Chapter 2. BEFORE YOU ADD A NUMBER TO BOTH SIDES OF THE EQUATION, ALWAYS SIMPLIFY EACH SIDE OF THE EQUATION FIRST. If a student tries to add -6 to both sides of the equation $x + 6 = -12 + 8 - 4$, he is much more likely to make an error. The students who tend to do poorly in learning to solve an equation are most often those who do not take the time to simplify each side of the equation before adding a number to both sides of the equation.

EXAMPLE 3 Solve for x. $15 + 2 = 3 + x + 2$

$17 = x + 5$ *Simplify by adding.*

$17 + (-5) = x + 5 + (-5)$ *Add the value -5 to both sides, since -5 is the additive inverse of 5.*

$12 = x$ *Simplify. The value of x is 12.*

Check. $15 + 2 = 3 + x + 2$

$15 + 2 \stackrel{?}{=} 3 + 12 + 2$ *Replace x by 12 in the original equation.*

$17 = 17$ ✓ *It checks.*

Practice Problem 3 Solve for x and check your solution. $5 - 12 = x - 3$ ■

In Example 3 we added -5 to each side. You could subtract 5 from each side and get the same result. In Chapter 1 we discussed how subtracting a 5 is the same as adding a negative 5. Do you see why?

We can determine if a value is the solution to an equation by following the same steps used to check an answer. Substitute the value to be tested for the variable in the original equation. We will obtain an identity if the value is the solution.

EXAMPLE 4 Is $x = 10$ the solution to the equation $-15 + 2 = x - 3$? If it is not, find the solution.

We substitute 10 for x in the equation and see if we obtain an identity.

$-15 + 2 = x - 3$

$-15 + 2 = 10 - 3$

$-13 \neq 7$ *This is not true. It is not an identity.*

Thus, $x = 10$ is not the solution. Now we take the original equation and solve to find the solution.

$-15 + 2 = x - 3$

$-13 = x - 3$ *Simplify by adding.*

$-13 + 3 = x - 3 + 3$ *Add 3 to both sides. 3 is the additive inverse of -3.*

$-10 = x$

Check to see if $x = -10$ is the solution. The value $x = 10$ was incorrect because of a sign error. We must be especially careful to write the correct sign for each number when solving equations.

Practice Problem 4 Is $x = -2$ the solution to the equation $x + 8 = -22 + 6$? If it is not, find the solution. ■

EXAMPLE 5 Find the value of x that satisfies the equation $\frac{1}{5} + x = -\frac{1}{10} + \frac{1}{2}$.

To combine the fractions, the fractions must have common denominators. The least common denominator (LCD) of the fractions is 10.

$$\frac{1 \cdot 2}{5 \cdot 2} + x = -\frac{1}{10} + \frac{1 \cdot 5}{2 \cdot 5}$$ *Change each fraction to an equivalent fraction with a denominator of* 10.

$$\frac{2}{10} + x = -\frac{1}{10} + \frac{5}{10}$$ *This is an equivalent equation.*

$$\frac{2}{10} + x = \frac{4}{10}$$ *Simplify by adding.*

$$\frac{2}{10} + \left(-\frac{2}{10}\right) + x = \frac{4}{10} + \left(-\frac{2}{10}\right)$$ *Add the additive inverse of* $\frac{2}{10}$ *to each side.*

$$x = \frac{2}{10}$$ *Add the fractions.*

$$x = \frac{1}{5}$$ *Simplify the answer.*

Check. We substitute $\frac{1}{5}$ for x in the original equation and see if we obtain an identity.

$$\frac{1}{5} + x = -\frac{1}{10} + \frac{1}{2}$$

$$\frac{1}{5} + \frac{1}{5} \stackrel{?}{=} -\frac{1}{10} + \frac{1}{2}$$ *Substitute* $\frac{1}{5}$ *for* x.

$$\frac{2}{5} \stackrel{?}{=} -\frac{1}{10} + \frac{5}{10}$$

$$\frac{2}{5} \stackrel{?}{=} \frac{4}{10}$$

$$\frac{2}{5} = \frac{2}{5} \checkmark$$ *It checks.*

Practice Problem 5 Find the value of x that satisfies the following equation.

$$\frac{1}{20} - \frac{1}{2} = x + \frac{3}{5}$$ ■

Developing Your Study Skills

WHY STUDY MATHEMATICS?

In our present-day, technological world, it is easy to see mathematics at work. Many vocational and professional areas—such as the fields of business, statistics, economics, psychology, finance, computer science, chemistry, physics, engineering, electronics, nuclear energy, banking, quality control, and teaching—require a certain level of expertise in mathematics. Those who want to work in these fields must be able to function at a given mathematical level. Those who cannot will not make it. So if your field of study requires you to take higher-level mathematics courses, be sure to master the basics of this course. Then you will be ready for the next one.

2.1 Exercises

Verbal and Writing Skills

Fill in the blank with the appropriate word.

1. When we use the ___equals___ sign, we indicate two expressions are ___equal___ in value.
2. If the ___same___ ___number___ is added to both sides of an equation, the results on each side are equal in value.
3. The ___solution___ of an equation is a value of the variable that makes the equation true.
4. What is the additive inverse of -20? 20
5. Why do we add the additive inverse of a to each side of $x + a = b$ to solve for x?
Answers may vary. A sample answer is to isolate the variable.

Find the value of x that satisfies each equation. Check your answers for Problems 6 through 25.

6. $x + 5 = 9$
$x = 4$
7. $x + 3 = 6$
$x = 3$
8. $4 = 7 + x$
$-3 = x$
9. $6 = x + 7$
$x = -1$
10. $x - 16 = 5$
$x = 21$
11. $x - 18 = 2$
$x = 20$
12. $0 = x + 5$
$x = -5$
13. $0 = x - 7$
$x = 7$
14. $3 + 5 = x - 7$
$x = 15$
15. $8 - 2 = x + 5$
$x = 1$
16. $7 + 3 + x = 5 + 5$
$x = 0$
17. $18 - 2 + 3 = x + 19$
$x = 0$
18. $x - 6 = -19$
$x = -13$
19. $x - 11 = -13$
$x = -2$
20. $-12 + x = 50$
$x = 62$
21. $-18 + x = 48$
$x = 66$
22. $18 - 11 = x - 5$
$x = 12$
23. $23 - 8 = x - 12$
$x = 27$
24. $5 - 19 + 3 = x + 1$
$x = -12$
25. $x + 7 = 15 - 26 + 4$
$x = -14$

In Problems 26 through 33, determine if the given solution is correct. If it is not, find the solution.

26. Is $x = 8$ the solution to $-12 + x = 4$?
no, $x = 16$
27. Is $x = 12$ the solution to $-19 + x = 7$?
no, $x = 26$
28. Is $x = -4$ the solution to $-12 + 5 = 3 + x$?
no, $x = -10$
29. Is $x = -3$ the solution to $-10 - 3 = x - 20$?
no, $x = 7$

30. Is $x = -33$ the solution to $x - 23 = -56$?
yes

31. Is $x = -8$ the solution to $-39 = x - 47$?
no, $x = 8$

32. Is $x = 35$ the solution to $15 - 3 + 20 = x - 3$?
yes

33. Is $x = -12$ the solution to $x + 8 = 12 - 19 + 3$?
yes

Find the value of x that satisfies each equation.

34. $-3 = x - 8$
$5 = x$

35. $-11 + x = -7$
$x = 4$

36. $-16 = x + 25$
$x = -41$

37. $27 = 5 + x$
$x = 22$

38. $1.3 + x + 1.8 = 0.2$
$x + 3.1 = 0.2$
$x = -2.9$

39. $3.6 + 1.2 = x + 1.3$
$4.8 = x + 1.3$
$3.5 = x$

40. $-1.5 + x = -3.8$
$x = -2.3$

41. $-0.6 + x = -1.8$
$x = -1.2$

42. $x - \frac{1}{4} = \frac{3}{4}$
$x = 1$

43. $x + \frac{1}{3} = \frac{2}{3}$
$x = \frac{1}{3}$

44. $\frac{1}{2} + x = \frac{1}{4} + \frac{1}{3}$
$x = \frac{1}{12}$

45. $x - \frac{1}{5} = \frac{1}{2} + \frac{1}{10}$
$x = \frac{4}{5}$

46. $\frac{1}{18} - \frac{5}{9} = x - \frac{1}{2}$
$x = 0$

47. $\frac{7}{12} - \frac{2}{3} = x - \frac{5}{4}$
$x = \frac{7}{6}$ or $1\frac{1}{6}$

48. $5\frac{1}{6} + x = 8$
$x = \frac{17}{6}$ or $2\frac{5}{6}$

49. $7\frac{1}{8} = -20 + x$
$x = 27\frac{1}{8}$

★**50.** $\frac{1}{2} + 3x = \frac{1}{4} + 2x - 3 - \frac{1}{2}$
$3x + \frac{1}{2} = 2x - \frac{13}{14}$
$x = \frac{15}{4}$ or $-3\frac{3}{4}$

51. $1.6 - 5x - 3.2 = -2x + 5.6 + 4x - 8x$
$-5x - 1.6 = -6x + 5.6$
$x = 7.2$

52. $x - 0.8613 = 2.1754$
$x = 3.0367$

53. $x + 9.3715 = -18.1261$
$x = -27.4976$

Cumulative Review Problems

Simplify by adding like terms.

54. $5x - 8 + 2x - 12$ $7x - 20$

55. $x + 3y - 5x - 7y + 2x$ $-2x - 4y$

56. $-3x^2 - 6x + 1 - 4x^2 + 2x - 8$ $-7x^2 - 4x - 7$

57. $y^2 + y - 12 - 3y^2 - 5y + 16$ $-2y^2 - 4y + 4$

58. A telemarketer has been given the task of selling credit cards to people who subscribe to a certain medical journal. After having made 85 calls, she has succeeded in selling 20 credit cards. What percent of the people did not choose to buy a credit card?
(Round your answer to nearest tenth of a percent.) 76.5% did not choose to buy a credit card.

2.2 The Multiplication Principle

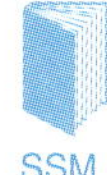

After studying this section, you will be able to:

1. *Solve equations of the form* $\frac{1}{a}x = b$.
2. *Solve equations of the form* $ax = b$.

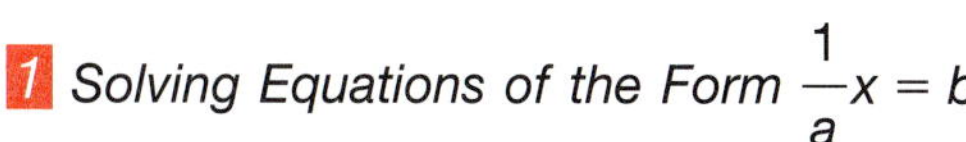

1 *Solving Equations of the Form* $\frac{1}{a}x = b$

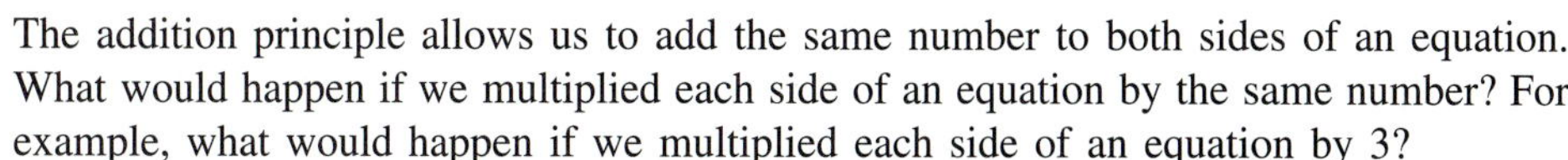

The addition principle allows us to add the same number to both sides of an equation. What would happen if we multiplied each side of an equation by the same number? For example, what would happen if we multiplied each side of an equation by 3?

To answer this question, let's return to our simple example of the box and the stone on a balanced seesaw. If we triple the number of weights on each side (we are multiplying each side by 3), the seesaw should still balance. The "weight value" of each side remains equal.

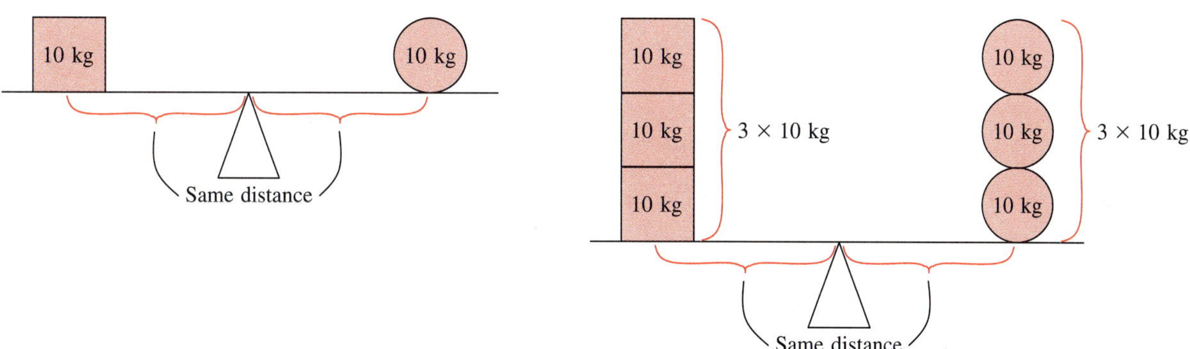

In words we can state this principle thus.

> **Multiplication Principle**
>
> If both sides of an equation are multiplied by the same number, the results on each side are equal in value.

In symbols we can restate the multiplication principle this way.

> For real numbers a, b, c with $c \neq 0$ if $a = b$ then $ca = cb$

Let us look at an equation where it would be helpful to multiply each side of the equation by 3.

EXAMPLE 1 Solve for x. $\frac{1}{3}x = -15$

We know that $(3)(\frac{1}{3}) = 1$. We will multiply each side of the equation by 3, because we want to isolate the variable x.

$$3\left(\frac{1}{3}x\right) = 3(-15) \quad \textit{Multiply each side of the equation by 3 since } (3)(\tfrac{1}{3}) = 1.$$

$$\left(\frac{3}{1}\right)\left(\frac{1}{3}\right)(x) = -45$$

$$1x = -45 \quad \textit{Simplify.}$$

$$x = -45$$

Check. $\frac{1}{3}(-45) \stackrel{?}{=} -15$ *Substitute* -45 *for* x *in the original equation.*

$-15 = -15$ ✓ *It checks.*

Practice Problem 1 Solve for x. $\frac{1}{8}x = -2$ ■

Note that $\frac{1}{5}x$ can be written as $\frac{x}{5}$. To solve the equation $\frac{x}{5} = 3$, we could multiply each side of the equation by 5. Try it. Then check your solution.

2 *Solving Equations of the Form* $ax = b$

We can see that using the multiplication principle to multiply each side of an equation by $\frac{1}{2}$ is the same as dividing each side of the equation by 2. Thus, it would seem that the multiplication principle would allow us to divide each side of the equation by any nonzero real number. Is there a real-world example of this idea?

Let's return to our simple example of the box and the stone on a balanced seesaw. Suppose that we were to cut the two objects in half (so that the amount of weight of each was divided by 2). We then return the objects to the same places on the seesaw. The seesaw would still balance. The "weight value" of each side remains equal.

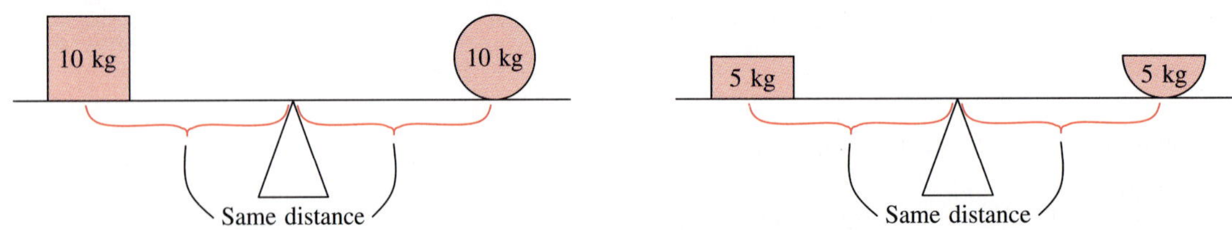

In words we can state this principle thus:

TEACHING TIP Emphasize the fact that the division principle of equality, if $a = b$, then $\frac{a}{c} = \frac{b}{c}$ is only valid if $c \neq 0$. We may never divide by zero. This concept requires continual emphasis.

Division Principle

If both sides of an equation are divided by the same nonzero number, the results on each side are equal in value.

Note: We put a restriction on the number by which we are dividing. We *cannot divide* by zero. We say that expressions like $\frac{2}{0}$ are not defined. Thus we restrict our divisor to *nonzero* numbers. We can restate the division principle this way.

For real numbers a, b, c where $c \neq 0$ if $a = b$ then $\frac{a}{c} = \frac{b}{c}$

EXAMPLE 2 Solve for x. $5x = 125$

$$\frac{5x}{5} = \frac{125}{5} \quad \textit{Divide both sides by 5.}$$

$$x = 25 \quad \textit{Simplify. The solution is 25.}$$

Check. $5x = 125$

$5(25) \stackrel{?}{=} 125$ *Replace x* by 25.

$125 = 125$ ✓ *It checks.*

Practice Problem 2 Solve for x. $9x = 72$ ■

For equations of the form $ax = b$ (a number multiplied by x equals another number), we solve the equation by choosing to divide both sides by a particular number. What is the number to choose? We look at the side of the equation that contains x. We notice the number that is multiplied by x. We divide by that number. The division principle tells us that we can still have a true equation provided that we divide by that number *on both sides* of the equation.

The solution to an equation may be a proper fraction or an improper fraction.

EXAMPLE 3 Solve for x. $4x = 38$

$$\frac{4x}{4} = \frac{38}{4} \quad \textit{Divide both sides by 4.}$$

$$x = \frac{19}{2} \quad \textit{Simplify. The solution is } \tfrac{19}{2}.$$

If you leave the solution as a fraction, it will be easier to check that solution in the original equation.

Check: $4x = 38$ *Replace x by* $\frac{19}{2}$.

$$\overset{2}{\cancel{4}}\left(\frac{19}{\cancel{2}}\right) \stackrel{?}{=} 38$$

$38 = 38$ ✓ *It checks.*

Practice Problem 3 Solve for x. $6x = 50$ ■

In Examples 2 and 3 we *divided by the number multiplied by x (the coefficient of x).* This procedure is followed regardless of whether the sign of that number is positive or negative.

EXAMPLE 4 Solve for x. $-3x = 48$

$$\frac{-3x}{-3} = \frac{48}{-3} \qquad \textit{Divide both sides by } -3.$$

$$x = -16 \qquad \textit{The solution is } -16.$$

Check. Can you check this solution?

Practice Problem 4 Solve for x. $-27x = 54$ ■

The coefficient of x may be 1 or -1. You may have to rewrite the equation so that the coefficient of 1 or -1 is obvious. With practice you may be able to "see" the coefficient without actually rewriting the equation.

EXAMPLE 5 Solve for x. $-x = -24$.

$$-1x = -24 \qquad \textit{Rewrite the equation. } -1x \textit{ is the same as } -x.$$
Now the coefficient of -1 *is obvious.*

$$\frac{-1x}{-1} = \frac{-24}{-1} \qquad \textit{Divide both sides by } -1.$$

$$x = 24 \qquad \textit{The solution is } 24.$$

Check. Can you check this solution?

Practice Problem 5 Solve for x. $-x = 36$ ■

The variable can be on either side of the equation. The equation $-78 = -3x$ can be solved in exactly the same way as $-3x = -78$.

EXAMPLE 6 Solve for x. $-78 = -3x$

$$\frac{-78}{-3} = \frac{-3x}{-3} \qquad \textit{Divide both sides by } -3.$$

$$26 = x \qquad \textit{The solution is } 26.$$

Check: $-78 = -3x$

$-78 \stackrel{?}{=} -3(26)$ *Replace x by* 26.

$-78 = -78$ ✓ *It checks.*

Practice Problem 6 Solve for x. $-51 = -6x$ ■

TEACHING TIP After doing Example 7, ask students to do the following problem as an Additional Class Exercise.
Solve for x.

$$-0.4x + 3.0x = -18.2$$

Sol. $x = -7.0$

EXAMPLE 7 Solve for x. $31.2 = 6.0x - 0.8x$

$31.2 = 6.0x - 0.8x$ *There are like terms on the right side.*

$31.2 = 5.2x$ *Collect like terms.*

$$\frac{31.2}{5.2} = \frac{5.2x}{5.2} \qquad \textit{Divide both sides by } 5.2.$$

$$6 = x \qquad \textit{The solution is } 6.$$

Note. Be sure to place the decimal point in the quotient directly above the caret (‸) when performing the division.

$$5.2_{\wedge}\overline{)31.2_{\wedge}} = 6., \quad 31\,2,\ \text{remainder } 0$$

Check. The check is up to you.

Practice Problem 7 Solve for x. $21 = 4.2x$ ■

2.2 Exercises

Verbal and Writing Skills

1. To solve the equation $-5x = -30$, divide each side of the equation by $\underline{-5}$.

2. To solve the equation $20 = -10x$, divide each side of the equation by $\underline{-10}$.

3. To solve the equation $\frac{1}{7}x = -2$, multiply each side of the equation by $\underline{7}$.

4. To solve the equation $\frac{1}{9}x = 5$, multiply each side of the equation by $\underline{9}$.

In Problems 5 through 14, solve for x. Be sure to reduce your answers. Check your solution.

5. $\frac{1}{3}x = 21$
$x = 63$

6. $\frac{1}{5}x = 30$
$x = 150$

7. $\frac{1}{5}x = -3$
$x = -15$

8. $\frac{1}{7}x = -4$
$x = -28$

9. $\frac{x}{5} = 16$
$x = 80$

10. $\frac{x}{10} = 8$
$x = 80$

11. $-4 = \frac{x}{12}$
$x = -48$

12. $-5 = \frac{x}{15}$
$x = -75$

13. $8x = 56$
$x = 7$

14. $7x = 42$
$x = 6$

15. $-16 = 6x$
$x = -\frac{8}{3}$

16. $-35 = 21x$
$x = -\frac{5}{3}$

17. $1.5x = 75$
$x = 50$

18. $2x = 0.36$
$x = 0.18$

19. $-15 = -x$
$x = 15$

20. $32 = -x$
$x = -32$

21. $-84 = 12x$
$x = -7$

22. $-72 = -9x$
$x = 8$

23. $0.5x = 0.20$
$x = \frac{2}{5}$

24. $0.31x = 9.3$
$x = 30$

Determine if the given solution is correct. If it is not, find the correct solution.

25. Is $x = 7$ the solution for $-3x = 21$?
no, $x = -7$

26. Is $x = 8$ the solution for $5x = -40$?
no, $x = -8$

27. Is $x = -6$ the solution for $-11x = 66$?
yes

28. Is $x = -20$ the solution for $-x = 20$?
yes

Find the value of the variable that satisfies each equation.

29. $-3y = 2.4$ $y = -0.8$

30. $5z = -1.8$ $z = -0.36$

31. $-27 = -12z$ $z = \frac{9}{4}$

32. $63 = -28y$ $y = -\frac{9}{4}$

33. $4.6y = -3.22$ $y = -0.7$

34. $-2.8y = -3.08$ $y = 1.1$

35. $4x + 3x = 21$ $x = 3$

36. $5x + 4x = 36$ $x = 4$

37. $2x - 7x = 20$ $x = -4$

38. $3x - 9x = 18$ $x = -3$

39. $-12y + y = -15$ $y = \frac{15}{11}$

40. $-14y + y = -17$ $y = \frac{17}{13}$

★**41.** $-\frac{2}{3} = -\frac{4}{7}x$ $x = \frac{7}{6}$

★**42.** $5.6 = -2.7x$ $x = -\frac{56}{27}$

43. $3.6172x = -19.026472$ $x = -5.26$

44. $-4.0518x = 14.505444$ $x = -3.58$

Cumulative Review Problems

Evaluate using the correct order of operations. (Be careful to avoid sign errors.)

45. $5 - 3(2)$ $5 - 6 = -1$

46. $(-6)(-8) + (-3)(2)$ $48 - 6 = 42$

47. $(-3)^3 + (-20) \div 2$ $-27 - 10 = -37$

48. $5 + (2 - 6)^2$ $5 + 16 = 21$

49. An off-price clothing store chain specializes in last year's merchandise. Their contact in Hong Kong has purchased 12,000 famous designer men's sport coats for the stores. When the shipment is unloaded, 800 sport coats have no left sleeve. What percent of the shipment is acceptable?

$93\frac{1}{3}\%$ is acceptable.

Developing Your Study Skills

GETTING HELP

Getting the right kind of help at the right time can be a key ingredient in being successful in mathematics. When you have gone to class on a regular basis, taken careful notes, methodically read your textbook, and diligently done your homework—all of which means making every effort possible to learn the mathematics—you may find that you are still having difficulty. If this is the case, then you need to seek help. Make an appointment with your instructor to find out what help is available to you. The instructor, tutoring services, a mathematics lab, video tapes, and computer software may be among resources you can draw on.

Once you discover the resources available in your school, you need to take advantage of them. Do not put it off, or you will find yourself getting behind. You cannot afford that. When studying mathematics, you must keep up with your work.

2.3 Use the Addition and Multiplication Principles Together

After studying this section, you will be able to:

1 *Solve equations of the form $ax + b = c$.*
2 *Solve equations in which the variable appears on both sides of the equation.*
3 *Solve equations with parentheses.*

MathPro

Video 6

SSM

1 *Solving Equations of the Form $ax + b = c$*

To solve many equations, we must use both the addition principle and the multiplication principle.

EXAMPLE 1 Solve for x and check your solution.

$$5x + 3 = 18$$

$$5x + 3 + (-3) = 18 + (-3)$$ *Add -3 to both sides, using the addition principle.*

$$5x = 15$$ *Simplify.*

$$\frac{5x}{5} = \frac{15}{5}$$ *Divide both sides by 5, using the division principle.*

$$x = 3$$ *The solution is 3.*

Check. $5(3) + 3 \stackrel{?}{=} 18$

$$15 + 3 \stackrel{?}{=} 18$$

$$18 = 18 \quad ✓$$ *It checks.*

Practice Problem 1 Solve for x and check your solution. $9x + 2 = 38$ ■

2 *A Variable on Both Sides of the Equation*

In some cases the variable appears on both sides of the equation. We would like to rewrite the equation so that all the terms containing the variable appear on one side. To do this, we apply the addition principle to the variable term.

EXAMPLE 2 Solve for x. $9x = 6x + 15$

$$9x + (-6x) = 6x + (-6x) + 15$$ *Add $-6x$ to both sides. Notice $6x + (-6x)$ eliminates the variable on the right side.*

$$3x = 15$$ *Collect like terms.*

$$\frac{3x}{3} = \frac{15}{3}$$ *Divide both sides by 3.*

$$x = 5$$ *The solution is 5.*

Check. Left to the student.

Practice Problem 2 Solve for x. $13x = 2x - 66$ ■

Many problems have variable terms and constant terms on both sides of the equation. You will want to get all the variable terms on one side and all the constant terms on the other side.

TEACHING TIP Some students prefer to use vertical form when solving equations. If they are more comfortable with vertical form, allow them to use it. The steps for solving Example 3 in vertical form are as follows:

$$\begin{array}{rcr} 9x + 3 & = & 7x - 2 \\ \underline{-7x \quad\;\;} & & \underline{-7x \quad\;\;} \\ 2x + 3 & = & -2 \\ \underline{\quad\;\; -3} & & \underline{-3} \\ 2x \quad\;\; & = & -5 \end{array}$$

Some students may be able to combine the above procedure into one step.

$$\begin{array}{rcr} 9x + 3 & = & 7x - 2 \\ \underline{-7x - 3} & & \underline{-7x - 3} \\ 2x \quad\;\; & = & -5 \end{array}$$

Encourage students to cross out the terms that zero out so that they can better keep track of the resulting terms.

EXAMPLE 3 Solve for x and check your solution. $9x + 3 = 7x - 2$.

$9x + (-7x) + 3 = 7x + (-7x) - 2$ *Add $-7x$ to both sides of the equation.*

$2x + 3 = -2$ *Combine like terms.*

$2x + 3 + (-3) = -2 + (-3)$ *Add -3 to both sides.*

$2x = -5$ *Simplify.*

$\frac{2x}{2} = \frac{-5}{2}$ *Divide both sides by 2.*

$x = -\frac{5}{2}$ *The solution is $-\frac{5}{2}$.*

Check. $9x + 3 = 7x - 2$

$9\left(-\frac{5}{2}\right) + 3 \stackrel{?}{=} 7\left(-\frac{5}{2}\right) - 2$ *Replace x by $-\frac{5}{2}$.*

$-\frac{45}{2} + 3 \stackrel{?}{=} -\frac{35}{2} - 2$ *Simplify.*

$-\frac{45}{2} + \frac{6}{2} \stackrel{?}{=} -\frac{35}{2} - \frac{4}{2}$ *Change to equivalent fractions with a common denominator.*

$-\frac{39}{2} = -\frac{39}{2}$ ✓ *It checks. $x = -\frac{5}{2}$ is the solution.*

TEACHING TIP This is a good time to stress the importance of checking your solution. After discussing Example 4, you may want to show students the following Additional Class Exercise.

Solve. $3x - 7 = 5x - 20$

Check your solution.

Sol. $3x - 5x - 7 = 5x - 5x - 20$

$-2x - 7 = -20$

$-2x - 7 + 7 = -20 + 7$

$-2x = -13$

$x = \frac{13}{2}$

Check. $3\left(\frac{13}{2}\right) - 7 \stackrel{?}{=} 5\left(\frac{13}{2}\right) - 20$

$\frac{39}{2} - 7 \stackrel{?}{=} \frac{65}{2} - 20$

$\frac{39}{2} - \frac{14}{2} \stackrel{?}{=} \frac{65}{2} - \frac{40}{2}$

$\frac{25}{2} = \frac{25}{2}$ ✓

Practice Problem 3 Solve for x and check your solution. $3x + 2 = 5x + 2$ ■

In our next example we will study equations that need simplifying before any other steps are taken. Where it is possible, you should first collect like terms on one or both sides of the equation. The variable terms can be collected on the right side or the left side. In this example we will collect all the x terms on the right side.

EXAMPLE 4 Solve for x. $5x + 26 - 6 = 9x + 12x$

$5x + 20 = 21x$ *Combine like terms.*

$5x + (-5x) + 20 = 21x + (-5x)$ *Add $-5x$ to both sides.*

$20 = 16x$ *Combine like terms.*

$\frac{20}{16} = \frac{16x}{16}$ *Divide both sides by 16.*

$\frac{5}{4} = x$ *(Don't forget to reduce the resulting fraction.)*

Check. Left to the student.

Practice Problem 4 Solve for z. $-z + 8 - z = 3z + 10 - 3$ ■

Do you really need all these steps? No. As you become more proficient you will be able to combine or eliminate some of these steps. However, it is best to write each step in its entirety until you are consistently obtaining the correct solution. It is much better to show every step than to take a lot of shortcuts but obtain a wrong "solution." This is a section of the algebra course where working neatly and accurately will help you—both now and as you progress through the course.

All the equations we have been studying so far are called first-degree equations. This means the variable terms are not squared (such as x^2 or y^2) or some higher power. It is possible to solve equations with x^2 and y^2 terms by the same methods we have used so far. If x^2 or y^2 terms appear, try to collect them on one side of the equation. If the squared term drops out, you may solve it as a first-degree equation using the methods discussed in this section.

EXAMPLE 5 Solve for y. $5y^2 + 6y - 2 = -y + 5y^2 + 12$

$5y^2 - 5y^2 + 6y - 2 = -y + 5y^2 - 5y^2 + 12$	*Subtract $5y^2$ from both sides.*
$6y - 2 = -y + 12$	*Combine, since $5y^2 - 5y^2 = 0$.*
$6y + y - 2 = -y + y + 12$	*Add y to each side.*
$7y - 2 = 12$	*Simplify.*
$7y - 2 + 2 = 12 + 2$	*Add 2 to each side.*
$7y = 14$	*Simplify.*
$\frac{7y}{7} = \frac{14}{7}$	*Divide each side by 7.*
$y = 2$	*Simplify. The solution is 2.*

Check. Left to the student.

TEACHING TIP After showing students Example 5, give them the following problem as an Additional Class Exercise.
Solve for x.

$x^2 + 2x - 8x + 16$
$= x^2 + 22 - 3x + 15$

Sol. $x = -7$

Practice Problem 5 Solve for x. $2x^2 - 6x + 3 = -4x - 7 + 2x^2$ ■

3 *Solving Equations with Parentheses*

The equations that you just solved are simpler versions of equations that we will now discuss. These equations contain parentheses. If the parentheses are first removed, the problems then become just like those encountered previously. We use the distributive property to remove the parentheses.

EXAMPLE 6 Solve for x and check your solution. $4(x + 1) - 3(x - 3) = 25$

$4(x + 1) - 3(x - 3) = 25$

$4x + 4 - 3x + 9 = 25$	*Multiply by 4 and −3 to remove parentheses. Be careful of the signs. Remember $(-3)(-3) = 9$.*

After removing the parentheses, it is important to collect like terms on each side of the equation. Do this before going on to isolate the variable.

$x + 13 = 25$	*Collect like terms.*
$x + 13 - 13 = 25 - 13$	*Add −13 to both sides to isolate the variable.*
$x = 12$	*The solution is 12.*

Check.	$4(12 + 1) - 3(12 - 3) \stackrel{?}{=} 25$	*Replace x by 12.*
	$4(13) - 3(9) \stackrel{?}{=} 25$	*Combine numbers inside parentheses.*
	$52 - 27 \stackrel{?}{=} 25$	*Multiply.*
	$25 = 25$ ✓	*Simplify. It checks.*

TEACHING TIP After showing students Example 6, ask them to do the following problem as an Additional Class Exercise.
Solve. $4(x - 3) - 6(x + 1) = 0$

Sol. $x = -9$

Practice Problem 6 Solve for x and check your solution.

$4x - (x + 3) = 12 - 3(x - 2)$ ■

EXAMPLE 7 Solve for x. $3(-x - 7) = -2(2x + 5)$

$-3x - 21 = -4x - 10$	*Remove parentheses. Watch the signs carefully.*
$-3x + 4x - 21 = -4x + 4x - 10$	*Add 4x to both sides.*
$x - 21 = -10$	*Simplify.*
$x - 21 + 21 = -10 + 21$	*Add 21 to both sides.*
$x = 11$	*The solution is 11.*

Check. Left to the student.

Practice Problem 7 Solve for x. $4(-2x - 3) = -5(x - 2) + 2$ ■

TEACHING TIP After showing students how to do Example 8, ask them to do the following problem as an Additional Class Exercise. Solve for x.

$$0.3x - 0.4(2 + x) = 0.5$$

Sol. $x = -13$

In problems that involve decimals, great care should be taken. In some steps you will be multiplying decimal quantities, and in other steps you will be adding them.

EXAMPLE 8 Solve for x. $0.3(1.2x - 3.6) = 4.2x - 16.44$

$0.36x - 1.08 = 4.2x - 16.44$	*Remove parentheses.*
$0.36x - 0.36x - 1.08 = 4.2x - 0.36x - 16.44$	*Subtract 0.36x from both sides.*
$-1.08 = 3.84x - 16.44$	*Collect like terms.*
$-1.08 + 16.44 = 3.84x - 16.44 + 16.44$	*Add 16.44 to both sides.*
$15.36 = 3.84x$	*Simplify.*
$\dfrac{15.36}{3.84} = \dfrac{3.84x}{3.84}$	*Divide both sides by 3.84.*
$4 = x$	*The solution is $x = 4$.*

Check. Left to the student.

Practice Problem 8 Solve for x. $0.3x - 2(x + 0.1) = 0.4(x - 3) - 1.1$ ■

TEACHING TIP After showing students how to do Example 9, ask them to do the following as an Additional Class Exercise. Solve and check.

$$-4(1 - y) + 3(2y + 3) = 7y - 10$$

Sol. $y = -5$
Check: $-45 = -45$

EXAMPLE 9 Solve for z and check. $2(3z - 5) + 2 = 4z - 3(2z + 8)$

$6z - 10 + 2 = 4z - 6z - 24$	*Remove parentheses.*
$6z - 8 = -2z - 24$	*Collect like terms.*
$6z - 8 + 2z = -2z + 2z - 24$	*Add 2z to each side.*
$8z - 8 = -24$	*Simplify.*
$8z - 8 + 8 = -24 + 8$	*Add 8 to each side.*
$8z = -16$	*Simplify.*
$\dfrac{8z}{8} = \dfrac{-16}{8}$	*Divide each side by 8.*
$z = -2$	*Simplify. The solution is −2.*

Check. $2[3(-2) - 5] + 2 \stackrel{?}{=} 4(-2) - 3[2(-2) + 8]$	*Replace z by −2.*
$2[-6 - 5] + 2 \stackrel{?}{=} -8 - 3[-4 + 8]$	*Multiply.*
$2[-11] + 2 \stackrel{?}{=} -8 - 3[4]$	*Simplify.*
$-22 + 2 \stackrel{?}{=} -8 - 12$	
$-20 = -20$ ✓	*It checks.*

Practice Problem 9 Solve for z and check. $5(2z - 1) + 7 = 7z - 4(z + 3)$ ■

2.3 Exercises

Find the value of the variable that satisfies each equation. Check your solution.

1. $7x + 9 = 51$
$7x = 42$
$x = 6$

2. $3x + 8 = 50$
$3x = 42$
$x = 14$

3. $6x - 4 = 62$
$6x = 66$
$x = 11$

4. $5x - 9 = 71$
$5x = 80$
$x = 16$

5. $4x - 13 = -81$
$4x = -68$
$x = -17$

6. $9x - 13 = -76$
$9x = -63$
$x = -7$

7. $-4x + 17 = -35$
$x = 13$

8. $-6x + 25 = -83$
$-6x = -108$
$x = 18$

9. $-15 + 2x = 15$
$2x = 30$
$x = 15$

10. $-8 + 4x = 8$
$4x = 16$
$x = 4$

11. $\frac{1}{2}x - 3 = 11$
$\frac{1}{2}x = 14$
$x = 28$

12. $\frac{1}{3}x - 2 = 13$
$\frac{1}{3}x = 15$
$x = 45$

13. $\frac{1}{6}x + 2 = -4$
$\frac{1}{6}x = -6$
$x = -36$

14. $\frac{1}{5}x + 6 = -24$
$\frac{1}{5}x = -30$
$x = -150$

15. $8x = 48 + 2x$
$6x = 48$
$x = 8$

16. $5x = 22 + 3x$
$2x = 22$
$x = 11$

17. $-6x = -27 + 3x$
$-9x = -27$
$x = 3$

18. $-7x = -26 + 6x$
$-13x = -26$
$x = 2$

19. $30 - x = 5x$
$30 = 6x$
$5 = x$

20. $42 - x = 6x$
$42 = 7x$
$6 = x$

21. $54 - 2x = -8x$
$54 = -6x$
$-9 = x$

To Think About

22. Is $y = 2$ the solution for $2y + 3y = 12 - y$?
yes

23. Is $y = 4$ the solution for $5y + 2 = 6y - 6 + y$?
yes

24. Is $x = 11$ a solution for $7x + 6 - 3x = 2x - 5 + x$?
no, $x = -11$

25. Is $x = -12$ a solution for $9x + 2 - 5x = -8 + 5x - 2$?
no, $x = 12$

Solve for y by getting all the y terms on the left. Then solve for y by getting all the y terms on the right. Which approach is better?

26. $-3 + 10y + 6 = 15 + 12y - 18$
$10y + 3 = 12y - 3$ $\quad$ $10y + 3 = 12y - 3$
$6 = 2y$ $\quad$ $-2y = -6$
$3 = y$ $\quad$ $y = 3$

27. $7y + 21 - 5y = 5y - 7 + y$
$2y + 21 = 6y - 7$ $\quad$ $2y + 21 = 6y - 7$
$-4y = -28$ $\quad$ $28 = 4y$
$y = 7$ $\quad$ $7 = y$

Solve for the variable. You may move the variable terms to the right or to the left.

28. $15 - 3x = -x - 5$

$20 = 2x$
$x = 10$

29. $5x + 6 = -7x - 4$

$12x = -10$
$x = -\frac{5}{6}$

30. $x - 6 = 8 - x$

$2x = 14$
$x = 7$

31. $2x + 5 = 4x - 5$

$10 = 2x$
$5 = x$

32. $6y - 5 = 8y - 7$

$2 = 2y$
$1 = y$

33. $11y - 8 = 9y - 16$

$2y = -8$
$y = -4$

34. $5x - 9 + 2x = 3x + 23 - 4x$

$7x - 9 = -x + 23$
$8x = 32$
$x = 4$

35. $9x - 5 + 4x = 7x + 43 - 2x$

$13x - 5 = 5x + 43$
$8x = 48$
$x = 6$

Remove the parentheses. Solve for the variable. Check your solution.

36. $4(x + 15) = 16$

$4x + 60 = 16$
$4x = -44$
$x = -11$

37. $3(2y - 4) = 12$

$6y - 12 = 12$
$6y = 24$
$y = 4$

38. $6(3x + 2) - 8 = -2$

$18x + 4 = -2$
$x = -\frac{1}{3}$

39. $4(2x + 1) - 7 = 6 - 5$

$8x - 3 = 1$
$x = \frac{1}{2}$

40. $7x - 3(5 - x) = 10$

$7x - 15 + 3x = 10$
$10x = 25$
$x = 2.5$

41. $6(3 - 4x) + 17 = 8x - 3(2 - 3x)$

$18 - 24x + 17 = 8x - 6 + 9x$
$41 = 41x$
$1 = x$

42. $0.3x - 0.2(x + 1) = 0.1$

$0.3x - 0.2x - 0.2 = 0.1$
$0.1x = 0.3$
$x = 3$

43. $3(x + 0.2) = 2(x - 0.3) + 5.2$

$3x + 0.6 = 2x - 0.6 + 5.2$
$x = 4$

44. $2(x - 3) + 5 = 3(x + 2)$

$2x - 6 + 5 = 3x + 6$
$-7 = x$

45. $3(x - 2) + 2 = 2(x - 4)$

$3x - 6 + 2 = 2x - 8$
$x = -4$

46. $0.2(x + 3) - (x - 1.5) = 0.3(x + 2) - 2.9$

$0.2x + 0.6 - x + 1.5 = 0.3x + 0.6 - 2.9$
$-0.8x + 2.1 = 0.3x - 2.3$
$4 = x$

47. $3(x + 0.2) - (2x + 0.5) = 2(x + 0.3) - 0.5$

$3x + 0.6 - 2x - 0.5 = 2x + 0.6 - 0.5$
$x = 2x$
$0 = x$

48. $-3(y - 3y) + 4 = -4(3y - y) + 6 + 13y$

$6y + 4 = 5y + 6$
$y = 2$

49. $2(4x - x) + 6 = 2(2x + x) + 8 - x$

$6x + 6 = 5x + 8$
$x = 2$

50. $3(x - 1) + 2x + 8 - 4x - 2 = 3[(2x - 1) - 2(x + 3)] + 6 + 2x + 2$

$3x - 3 + 2x + 8 - 4x - 2$
$= 3[2x - 1 - 2x - 6] + 6 + 2x + 2$
$x + 3 = 2x - 13$
$16 = x$

Mixed Practice

Solve for the variable.

51. $1.2x + 4 = 3.2x - 8$
$12 = 2.0x$
$6 = x$

52. $9x + 1.9 = -7x - 3.1$
$16x = -5.0$
$x = -\frac{5}{16}$ or -0.3125

53. $5z + 7 - 2z = 32 - 2z$
$5z = 25$
$z = 5$

54. $8 - 7z + 2z = 20 + 5z$
$8 - 5z = 20 + 5z$
$-12 = 10z$
$-1.2 = z$

55. $-4w - 28 = -7 - w$
$-21 = 3w$
$-7 = w$

56. $-6w - 7 = -3 - 8w$
$2w = 4$
$w = 2$

57. $6x + 8 - 3x = 11 - 12x - 13$
$3x + 8 = -2 - 12x$
$15x = -10$
$x = -\frac{2}{3}$

58. $4 - 7x - 13 = 8x - 3 - 5x$
$-7x - 9 = 3x - 3$
$-10x = 6$
$x = -\frac{3}{5}$

59. $2x^2 - 3x - 8 = 2x^2 + 5x - 6$
$-2 = 8x$
$-\frac{1}{4} = x$

60. $3x^2 + 4x - 7 = 3x^2 - 5x + 2$
$9x = 9$
$x = 1$

61. $-3.5x + 1.3 = -2.7x + 1.5$
$-0.2 = 0.8x$
-0.25 *or* $-\frac{1}{4} = x$

62. $2.8x - 0.9 = 5.2x - 3.3$
$2.4 = 2.4x$
$1 = x$

63. $5(4 + x) = 3(3x - 1) - 9$
$20 + 5x = 9x - 3 - 9$
$8 = x$

64. $x - 0.8x + 4 = 2.6$
$0.2x + 4 = 2.6$
$x = -7$

65. $17(y + 3) - 4(y - 10) = 13$
$17y + 51 - 4y + 40 = 13$
$13y = -78$
$y = -6$

66. $3x + 2 - 1.7x = 0.6x + 31.4$
$1.3x + 2 = 0.6x + 31.4$
$0.7x = 29.4$
$x = 42$

67. $10(x - 5) - 2(x - 9) = -28$
$10x - 50 - 2x + 18 = -28$
$x = \frac{1}{2}$

68. $2(5z + 4) + 19 = 4z - 3(2z + 11)$
$10z + 8 + 19 = 4z - 6z - 33$
$10z + 27 = -2x - 33$
$12z = -60$
$z = -5$

Solve for x. Round your answer to the nearest hundredth.

69. $1.63x - 9.23 = 5.71x + 8.04$ $x = -4.23$

70. $-2.21x + 8.65 = 3.69x - 7.78$ $x = 2.78$

Cumulative Review Problems

Simplify.

71. $2x(3x - y) + 4(2x^2 - 3xy)$ $14x^2 - 14xy$

72. $5[3(x + y) - 2(x - y)]$ $5x + 25y$

73. $2\{3 - 2[x - 4(2x + 3)]\}$ $28x + 54$

74. $2\{x - 3[4 + 2(3 + x)]\}$ $-10x - 60$

2.4 Equations with Fractions

MathPro

Video 6

SSM

After studying this section, you will be able to:

1 *Solve equations with fractions.*

1 *Solving Equations with Fractions*

Equations with fractions can be rather difficult to solve. This difficulty is simply due to the extra care we usually have to use when computing with fractions. The actual equation-solving procedures are the same, with fractions or without. To avoid unnecessary work, we transform the given equation with fractions to an equivalent equation that does not contain fractions. How do we do this? We multiply each side of the equation by the lowest common denominator of all the fractions contained in the equation. We then use the distributive property so that the LCD is multiplied by each term of the equation.

EXAMPLE 1 Solve for x. $\frac{1}{4}x - \frac{2}{3} = \frac{5}{12}x$

First we find that the LCD = 12.

$$12\left(\frac{1}{4}x - \frac{2}{3}\right) = 12\left(\frac{5}{12}x\right) \quad \textit{Multiply each side by 12.}$$

$$\left(\frac{12}{1}\right)\left(\frac{1}{4}\right)(x) - \left(\frac{12}{1}\right)\left(\frac{2}{3}\right) = \left(\frac{12}{1}\right)\left(\frac{5}{12}\right)(x) \quad \textit{Use the distributive property.}$$

$$3x - 8 = 5x \quad \textit{Simplify.}$$

$$3x + (-3x) - 8 = 5x + (-3x) \quad \textit{Add } -3x \textit{ to each side.}$$

$$-8 = 2x \quad \textit{Simplify.}$$

$$-\frac{8}{2} = \frac{2x}{2} \quad \textit{Divide each side by 2.}$$

$$-4 = x \quad \textit{Simplify.}$$

Check. $\frac{1}{4}(-4) - \frac{2}{3} \stackrel{?}{=} \frac{5}{12}(-4)$

$$-1 - \frac{2}{3} \stackrel{?}{=} -\frac{5}{3}$$

$$-\frac{3}{3} - \frac{2}{3} \stackrel{?}{=} -\frac{5}{3}$$

$$-\frac{5}{3} = -\frac{5}{3} \quad \checkmark \quad \textit{It checks.}$$

Practice Problem 1 Solve for x. $\frac{3}{8}x - \frac{3}{2} = \frac{1}{4}x$ ■

In Example 1 we multiplied each side of the equation by the LCD. It is common practice to immediately go to the second step and multiply each term by the LCD, rather than to write out a separate step using the distributive property.

EXAMPLE 2 Solve for x and check your solution. $\frac{x}{3} + 3 = \frac{x}{5} - \frac{1}{3}$

TEACHING TIP Point out that it is important to multiply each term on both sides of the equation by the lowest common denominator. A common student mistake is to multiply the LCD only by the fractions. Be sure students multiply the LCD by every term.

$$15\left(\frac{x}{3}\right) + 15(3) = 15\left(\frac{x}{5}\right) - 15\left(\frac{1}{3}\right)$$ *The LCD is 15. Use the multiplication principle to multiply each term by 15.*

$$5x + 45 = 3x - 5$$ *Simplify.*

$$5x - 3x + 45 = 3x - 3x - 5$$ *Add $-3x$ to both sides.*

$$2x + 45 = -5$$ *Collect like terms.*

$$2x + 45 - 45 = -5 - 45$$ *Add -45 to both sides.*

$$2x = -50$$ *Simplify.*

$$\frac{2x}{2} = \frac{-50}{2}$$ *Divide both sides by 2.*

$$x = -25$$ *The solution is -25.*

Check. $\frac{-25}{3} + 3 \stackrel{?}{=} \frac{-25}{5} - \frac{1}{3}$

$$-\frac{25}{3} + \frac{9}{3} \stackrel{?}{=} -\frac{5}{1} - \frac{1}{3}$$

$$-\frac{16}{3} \stackrel{?}{=} -\frac{15}{3} - \frac{1}{3}$$

$$-\frac{16}{3} = -\frac{16}{3} \quad \checkmark$$

TEACHING TIP This section is more difficult for students than most other parts of Chapter 2. In addition to Example 2, you may want to show them the following as an Additional Class Exercise. Solve for x.

$$2x - \frac{1}{3} + \frac{2}{3}x = \frac{4}{9} + \frac{1}{3}x$$

Sol. $x = \frac{1}{3}$

Practice Problem 2 Solve for x and check your solution. $\frac{5x}{4} - 1 = \frac{3x}{4} + \frac{1}{2}$ ■

EXAMPLE 3 Solve for x. $\frac{x+5}{7} = \frac{x}{4} + \frac{1}{2}$

$$\frac{x}{7} + \frac{5}{7} = \frac{x}{4} + \frac{1}{2}$$ *First we write as separate fractions.*

$$28\left(\frac{x}{7}\right) + 28\left(\frac{5}{7}\right) = 28\left(\frac{x}{4}\right) + 28\left(\frac{1}{2}\right)$$ *We observe that the LCD is 28, so we multiply each term by 28.*

$$4x + 20 = 7x + 14$$ *Simplify.*

$$4x - 4x + 20 = 7x - 4x + 14$$ *Add $-4x$ to both sides.*

$$20 = 3x + 14$$ *Collect like terms.*

$$20 - 14 = 3x + 14 - 14$$ *Add -14 to both sides.*

$$6 = 3x$$ *Collect like terms.*

$$\frac{6}{3} = \frac{3x}{3}$$ *Divide both sides by 3.*

$$2 = x$$ *The solution is $x = 2$.*

Check. Left to the student.

Practice Problem 3 Solve for x. $\frac{5x}{6} - \frac{5}{8} = \frac{3x}{4} - \frac{1}{3}$ ■

If a problem contains both parentheses and fractions, it is best to remove the parentheses first. Many students find it is helpful to have a written procedure to follow in solving these more involved equations.

Procedure to Solve Linear Equations

1. Remove any parentheses.
2. If fractions exist, multiply all terms on both sides by the lowest common denominator of all the fractions.
3. Collect like terms if possible. Simplify numerical work if possible.
4. Add or subtract terms on both sides of the equation to get all terms with the variable on one side of the equation.
5. Add or subtract a value on both sides of the equation to get all terms not containing the variable on the other side of the equation.
6. Divide both sides of the equation by the coefficient of the variable.
7. Simplify the solution (if possible).
8. Check your solution.

Let's use each step in solving this example.

TEACHING TIP Fractional notation is encountered with this type of problem in two ways. Explain to students that both forms of notation are correct and that either may be used. In the first step of solving Example 4 we may write

$$\frac{x}{3} - \frac{2}{3} = \frac{x}{5} + \frac{4}{5} + 2$$

or an equivalent form for the same equation is

$$\frac{1}{3}x - \frac{2}{3} = \frac{1}{5}x + \frac{4}{5} + 2$$

Make sure students understand that both forms are equivalent.

TEACHING TIP With fractional equations that contain parentheses, some students prefer to remove parentheses first, while others prefer to clear the equation of fractions first. If the work is done carefully, either method is correct. It seems that for weaker students it is advisable to always remove the parentheses first. It tends to minimize errors in solving the equation.

EXAMPLE 4 Solve for x and check your solution. $\frac{1}{3}(x - 2) = \frac{1}{5}(x + 4) + 2$

Step 1 $\frac{x}{3} - \frac{2}{3} = \frac{x}{5} + \frac{4}{5} + 2$ *Remove parentheses.*

Step 2 $15\left(\frac{x}{3}\right) - 15\left(\frac{2}{3}\right) = 15\left(\frac{x}{5}\right) + 15\left(\frac{4}{5}\right) + 15(2)$ *Multiply by the LCD, 15.*

Step 3 $5x - 10 = 3x + 12 + 30$ *Simplify.*

$5x - 10 = 3x + 42$ *Simplify.*

Step 4 $5x - 3x - 10 = 3x - 3x + 42$ *Add $-3x$ to both sides.*

$2x - 10 = 42$ *Simplify.*

Step 5 $2x - 10 + 10 = 42 + 10$ *Add 10 to both sides.*

$2x = 52$ *Simplify.*

Step 6 $\frac{2x}{2} = \frac{52}{2}$ *Divide both sides by 2.*

Step 7 $x = 26$ *Simplify the solution.*

Step 8 *Check.* $\frac{1}{3}(26 - 2) \stackrel{?}{=} \frac{1}{5}(26 + 4) + 2$ *Replace x by 26.*

$\frac{1}{3}(24) \stackrel{?}{=} \frac{1}{5}(30) + 2$ *Combine values within parentheses.*

$8 \stackrel{?}{=} 6 + 2$ *Simplify.*

$8 = 8$ ✓ *$x = 26$ is the solution.*

Practice Problem 4 Solve for x and check your solution.

$\frac{1}{3}(x - 2) = \frac{1}{4}(x + 5) - \frac{5}{3}$ ■

It should be remembered that not every step will be needed in each problem. You can combine some steps as well, *as long as you are consistently obtaining the correct solution.* However, you are encouraged to write out every step as a way of helping you to avoid careless errors.

It is important to remember that when we write decimals these numbers are really fractions written in a special way. Thus, $0.3 = \frac{3}{10}$ and $0.07 = \frac{7}{100}$. It is possible to take a linear equation containing decimals and to multiply each term by the appropriate value to obtain integer coefficients.

EXAMPLE 5 Solve for x. $0.2(1 - 8x) + 1.1 = -5(0.4x - 0.3)$

$$0.2 - 1.6x + 1.1 = -2.0x + 1.5$$ *Remove parentheses.*

$$10(0.2) + 10(1.6x) + 10(1.1) = 10(-2.0x) + 10(1.5)$$ *Multiply each term by 10.*

$$2 - 16x + 11 = -20x + 15$$ *Multiplying by 10 moves the decimal point one place to the right.*

$$-16x + 13 = -20x + 15$$ *Simplify.*

$$-16x + 20x + 13 = -20x + 20x + 15$$ *Add 20x to each side.*

$$4x + 13 = 15$$ *Simplify.*

$$4x + 13 + (-13) = 15 + (-13)$$ *Add −13 to each side.*

$$4x = 2$$ *Simplify.*

$$\frac{4x}{4} = \frac{2}{4}$$ *Divide each side by 4.*

$$x = \frac{1}{2} \quad \text{or} \quad 0.5$$ *Simplify.*

Check. $0.2[1 - 8(0.5)] + 1.1 \stackrel{?}{=} -5[0.4(0.5) - 0.3]$

$$0.2[1 - 4] + 1.1 \stackrel{?}{=} -5[0.2 - 0.3]$$

$$0.2[-3] + 1.1 \stackrel{?}{=} -5[-0.1]$$

$$-0.6 + 1.1 \stackrel{?}{=} 0.5$$

$$0.5 = 0.5 \quad ✓$$

Practice Problem 5 Solve for x. $2.8 = 0.3(x - 2) + 2(0.1x - 0.3)$ ■

Developing Your Study Skills

TAKING NOTES IN CLASS

An important part of studying mathematics is taking notes. To take meaningful notes, you must be an active listener. Keep your mind on what the instructor is saying, and be ready with questions whenever you do not understand something.

If you have previewed the lesson material, you will be prepared to take good notes. The important concepts will seem somewhat familiar. If you frantically try to write all that the instructor says or copy all the examples done in class, you may find your notes to be nearly worthless when you are home alone. Write down *important* ideas and examples as the instructor lectures, making sure that you are listening and following the logic. Include any helpful hints or suggestions that your instructor gives you or refers to in your text.

2.4 Exercises

In Problems 1 through 16, solve for the variable and check your answer.

1. $\frac{1}{5}x + \frac{1}{10} = \frac{1}{2}$
$2x + 1 = 5$
$2x = 4$
$x = 2$

2. $\frac{1}{3}x - \frac{1}{9} = \frac{8}{9}$
$3x - 1 = 8$
$3x = 9$
$x = 3$

3. $\frac{3}{4}x = \frac{1}{2}x + \frac{5}{8}$
$6x = 4x + 5$
$2x = 5$
$x = \frac{5}{2}$ or $2\frac{1}{2}$

4. $\frac{1}{2} = \frac{3}{10} - \frac{2}{5}x$
$5 = 3 - 4x$
$2 = -4x$
$-\frac{1}{2} = x$

5. $\frac{y}{2} + \frac{y}{3} = \frac{5}{6}$
$3y + 2y = 5$
$y = 1$

6. $\frac{x}{4} - 1 = \frac{x}{5}$
$5x - 20 = 4x$
$x = 20$

7. $20 - \frac{1}{3}x = \frac{1}{2}x$
$120 - 2x = 3x$
$24 = x$

8. $\frac{1}{5}x - \frac{1}{2} = \frac{1}{6}x$
$6x - 15 = 5x$
$15 = x$

9. $2 + \frac{y}{2} = \frac{3y}{4} - 3$
$8 + 2y = 3y - 12$
$20 = y$

10. $\frac{x}{3} - 1 = -\frac{1}{2} - x$
$2x - 6 = -3 - 6x$
$x = \frac{3}{8}$

11. $\frac{y - 1}{2} = 4 - \frac{y}{7}$
$7y - 7 = 56 - 2y$
$y = 7$

12. $\frac{x - 7}{6} = -\frac{1}{2}$
$x - 7 = -3$
$x = 4$

13. $2.8 - 0.4x = 8$
$28 - 4x = 80$
$x = -13$

14. $0.3x - 2.2 = 3.2$
$3x - 22 = 32$
$x = 18$

15. $5.9 = 2.5x - 1.6$
$59 = 25x - 16$
$3 = x$

16. $0.09x + 2 = 6.5$
$9x + 200 = 650$
$x = 50$

17. Is $y = 4$ a solution to $\frac{1}{2}(y - 2) + 2 = \frac{3}{8}(3y - 4)$?
yes

18. Is $y = 2$ a solution to $\frac{1}{5}(y + 2) = \frac{1}{10}y + \frac{3}{5}$?
yes

19. Is $y = \frac{5}{8}$ a solution to $\frac{1}{2}\left(y - \frac{1}{5}\right) = \frac{1}{5}(y + 2)$?
no

20. Is $y = \frac{13}{3}$ a solution to $\frac{y}{2} - \frac{7}{9} = \frac{y}{6} + \frac{2}{3}$?
yes

Solve for the variable.

21. $4(3x - 2) = \frac{1}{3}(x - 1) + 4$
$36x - 24 = x - 1 + 12$
$35x = 35$
$x = 1$

22. $\frac{2}{3}(x + 2) = -2(1 - 3x)$
$2x + 4 = -6 + 18x$
$10 = 16x$
$\frac{5}{8} = x$

23. $0.5 - 2.1x = 6.4 + 0.3x - 5.9$
$5 - 21x = 64 + 3x - 59$
$-24x = 0$
$x = 0$

24. $0.7x - 3.3 = 2.5 - 0.2x - 5.8$
$7x - 33 = 25 - 2x - 58$
$9x = 0$
$x = 0$

25. $-0.3(x - 1) + 2 = -0.4(2x + 3)$
$-0.3x + 0.3 + 2 = -0.8x - 1.2$
$5x = -35$
$x = -7$

26. $0.7 - 0.3(6 - 2x) = -0.1 + 0.4x$
$0.7 - 1.8 + 0.6x = -0.1 + 0.4x$
$0.2x = 1$
$x = 5$

27. $-5(0.2x + 0.1) - 0.6 = 1.9$
$-x - 0.5 - 0.6 = 1.9$
$-x = 3$
$x = -3$

28. $0.3x + 1.7 = 0.2x - 0.4(5x + 1)$
$0.3x + 1.7 = 0.2x - 2x - 0.4$
$21x = -21$
$x = -1$

Mixed Practice

Solve.

29. $\frac{1}{3}(y+2) = 3y - 5(y-2)$

$y + 2 = 9y - 15y + 30$
$7y = 28$
$y = 4$

30. $\frac{2}{5}(y+3) - \frac{1}{2} = \frac{1}{3}(y-2) + \frac{1}{2}$

$12y + 36 - 15 = 10y - 20 + 15$
$2y = -26$
$y = -13$

31. $\frac{1+2x}{5} + \frac{4-x}{3} = \frac{1}{15}$

$3 + 6x + 20 - 5x = 1$
$x = -22$

32. $\frac{x+3}{4} = 4x - 2(x-3)$

$x + 3 = 16x - 8x + 24$
$-3 = x$

33. $\frac{x}{5} - \frac{2}{3}x + \frac{16}{15} = \frac{1}{3}(x-4)$

$3x - 10x + 16 = 5x - 20$
$36 = 12x$
$3 = x$

34. $3 + \frac{1}{2}(x-1) = 2 - \frac{1}{3}x + 3$

$18 + 3x - 3 = 12 - 2x + 18$
$5x = 15$
$x = 3$

35. $\frac{1}{3}(x-2) = 3x - 2(x-1) + \frac{16}{3}$

$x - 2 = 9x - 6x + 6 + 16$
$-24 = 2x$
$-12 = x$

36. $\frac{3}{4}(x-2) + \frac{3}{5} = \frac{1}{5}(x+1)$

$15x - 30 + 12 = 4x + 4$
$11x = 22$
$x = 2$

37. $\frac{3}{2}x + \frac{1}{3} = \frac{2x-3}{4}$

$18x + 4 = 6x - 9$
$12x = -13$
$x = -\frac{13}{12}$ or $-1\frac{1}{12}$

38. $\frac{5}{3} - \frac{1}{6}x = \frac{3x+5}{4}$

$20 - 2x = 9x + 15$
$-11x = -5$
$x = \frac{5}{11}$

39. $0.8(x-3) = -5(2.1x + 0.4)$

$0.8x - 2.4 = -10.5x - 2$
$8x - 24 = -105x - 20$
$113x = 4$
$x = \frac{4}{113}$

40. $0.7(2x+3) = -4(6.1x + 0.2)$

$1.4x + 2.1 = -24.4x - 0.8$
$14x + 21 = -244x - 8$
$x = -\frac{29}{258}$

41. $\frac{1}{5}x + \frac{2}{3} + \frac{1}{15} = \frac{4}{3}x - \frac{7}{15}$

$3x + 10 + 1 = 20x - 7$
$18 = 17x$
$\frac{18}{17} = x$

42. $\frac{1}{20}x - \frac{1}{4} + \frac{3}{5} = -\frac{1}{2}x + \frac{1}{4}x$

$x - 5 + 12 = -10x + 5x$
$6x = -7$
$x = -\frac{7}{6}$

Solve for x in the following equations. Work carefully.

★ **43.** $7x + 2[(x+2)+3] + 2(x+2) = 2[2(x+2) + 2x] + 8$

$7x + 2x + 4 + 6 + 2x + 4 = 4x + 8 + 4x + 8$
$3x = 2$
$x = \frac{2}{3}$

★ **44.** $\frac{7}{2} - \frac{2x-3}{2} - (3x+5) = \frac{3}{2} - \frac{5x}{2} - x$

$7 - 2x + 3 - 6x - 10 = 3 - 5x - 2x$
$-8x = 3 - 7x$
$-3 = x$

Cumulative Review Problems

45. Add. $\frac{3}{7} + 1\frac{5}{10}$

$\frac{27}{14}$ or $1\frac{13}{14}$

46. Subtract. $3\frac{1}{5} - 2\frac{1}{4}$

$\frac{19}{20}$

47. Multiply. $\left(2\frac{1}{5}\right)\left(6\frac{1}{8}\right)$

$\frac{539}{40}$ or $13\frac{19}{40}$

48. Divide. $5\frac{1}{2} \div 1\frac{1}{4}$

$\frac{22}{5}$ or $4\frac{2}{5}$

2.5 Formulas

MathPro

Video 7

SSM

After studying this section, you will be able to:

1 *Solve formulas for a specified variable.*

1 *Solving for a Specified Variable in a Formula*

Formulas are equations with one or more variables that are used to describe real-world situations. The formula describes the relationship that exists among the variables. For example, in the formula $d = rt$, distance (d) is related to the rate of speed (r) and to time (t). We can use this formula to find distance if we know the rate and time. Sometimes, however, we are given the distance and the rate, and we are asked to find the time.

EXAMPLE 1 Joseph drove a distance of 156 miles at an average speed of 52 miles per hour. How long did it take Joseph to make the trip?

$d = rt$ *Use the distance formula.*

$156 = 52t$ *Substitute the known values for the variables.*

$\frac{156}{52} = \frac{52}{52}t$ *Divide both sides of the equation by 52 to solve for t.*

$3 = t$ *We have solved for t.*

It took Joseph 3 hours to drive 156 miles at 52 miles per hour.

Practice Problem 1 Find the rate for $d = 240$ miles and $t = 5$ hours. ■

If we have many problems that ask us to find the time given the distance and rate, it may be worthwhile to rewrite the formula in terms of time.

EXAMPLE 2 Solve for t. $d = rt$

$\frac{d}{r} = \frac{rt}{r}$ *We want to isolate t. Therefore, we are dividing both sides of the equation by the coefficient of t, which is r.*

$\frac{d}{r} = t$ *You have solved for the variable indicated.*

Practice Problem 2 Solve for m. $E = mc^2$ ■

A simple first degree equation with two variables can be thought of as the equation of a line. It is often useful to solve for y in order to make graphing the line easier.

EXAMPLE 3 Solve for y. $3x - 2y = 6$

$-2y = 6 - 3x$ *We want to isolate the term containing y, so we subtract 3x from both sides.*

$\frac{-2y}{-2} = \frac{6 - 3x}{-2}$ *Divide both sides by the coefficient of y.*

$y = \frac{6}{-2} + \frac{-3x}{-2}$ *Rewrite the fraction.*

$y = \frac{3}{2}x - 3$ *Simplify and regroup.*

This is known as the slope–intercept form of the equation of a line.

Practice Problem 3 Solve for y. $8 - 2y + 3x = 0$ ■

Our procedure for solving a first-degree equation can be rewritten to give us a procedure for solving a formula for a specified variable.

Procedure to Solve a Formula for a Specified Variable

1. Remove any parentheses.
2. If fractions exist, multiply all terms on both sides by the LCD of all the fractions.
3. Collect like terms or simplify if possible.
4. Add or subtract terms on both sides of the equation to get all terms with the desired variable on one side of the equation.
5. Add or subtract the appropriate quantity to get all terms that do *not* have the desired variable on the other side of the equation.
6. Divide both sides of the equation by the coefficient of the desired variable.
7. Simplify if possible.

EXAMPLE 4 A trapezoid is a four-sided figure with two parallel sides. If the parallel sides are a and b and the altitude is h, the area is given by

$$A = \frac{h}{2}(a + b)$$

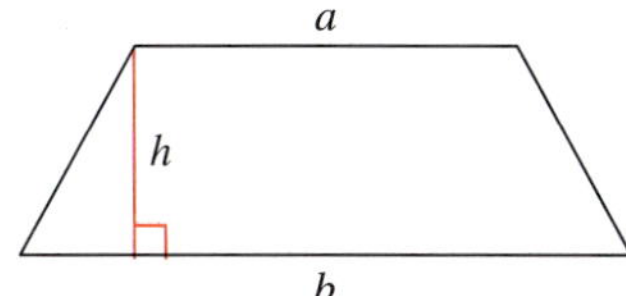

Solve this equation for a.

$$A = \frac{h}{2}(a + b)$$

$$A = \frac{ha}{2} + \frac{hb}{2} \qquad \textit{Remove the parentheses.}$$

$$2(A) = 2\left(\frac{ha}{2}\right) + 2\left(\frac{hb}{2}\right) \qquad \textit{Multiply all terms by LCD of 2.}$$

$$2A = ha + hb \qquad \textit{Simplify.}$$

$$2A - hb = ha$$ *We want to isolate the term containing a. Therefore, we subtract hb from both sides.*

$$\frac{2A - hb}{h} = \frac{ha}{h} \qquad \textit{Divide both sides by h (the coefficient of a).}$$

$$\frac{2A - hb}{h} = a \qquad \textit{The solution is obtained.}$$

Note: Although the solution is in simple form, it could be written in an alternative way. Since

$$\frac{2A - hb}{h} = \frac{2A}{h} - \frac{hb}{h} = \frac{2A}{h} - b$$

we could have $\frac{2A}{h} - b = a$ as an alternative way of writing the answer.

Practice Problem 4 Solve for d. $c = \pi d$ ■

2.5 Exercises

Applications

1. Use the formula $d = rt$ to find the rate of speed of a car that covers 450 miles in 9 hours.

$450 = 9r$
$r = 50$ miles per hour

2. The formula for calculating the temperature in degrees Fahrenheit when you know the temperature in degrees Celsius is $F = \frac{9}{5}C + 32$. Use this formula to find the Celsius temperature when the Fahrenheit temperature is $-49°$.

$$-49 = \frac{9}{5}C + 32$$
$$C = -45°$$

3. The formula for the area of a triangle is $A = \frac{1}{2}bh$
where b is the *base* of the triangle
h is the *height* or *altitude* of the triangle

(a) Use this formula to find the base of a triangle that has an area of 60 square meters and a height of 12 meters.

$$60 = \frac{1}{2}b(12)$$
$$b = 10 \text{ meters}$$

(b) Use this formula to find the altitude of a triangle that has an area of 88 square meters and a base of 11 meters.

$$88 = \frac{1}{2}(11)h$$
$$h = 16 \text{ meters}$$

4. The formula for calculating simple interest is $I = Prt$
where P is the *principal* (amount of money invested)
r is the *rate* at which the money is invested
t is the *time*

(a) Use this formula to find how long it would take to earn \$720 in interest on an investment of \$3000 at the rate of 6%.

$720 = 3000(0.06)t$
$t = 4$ years

(b) Use this formula to find the rate of interest if \$5000 earns \$400 interest in 2 years.

$400 = 5000(2)r$
$r = 0.04$ or 4%

(c) Use this formula to find the amount of money invested if the interest earned was \$120 and the rate of interest was 5% over 3 years.

$120 = 0.05(3)P$
$P = \$800$

5. The standard form of a line is $3x - 5y = 15$.

(a) Solve for the variable y.

$$y = \frac{3}{5}x - 3$$

(b) Use this result to find y with $x = -5$. $y = -6$

In each formula or equation, solve for the variable indicated.

Area of a triangle

6. $A = \frac{1}{2}bh$ Solve for b. $b = \dfrac{2A}{h}$

7. $A = \frac{1}{2}bh$ Solve for h. $h = \dfrac{2A}{b}$

Simple interest formula

8. $I = Prt$ Solve for P. $P = \dfrac{I}{rt}$

9. $I = Prt$ Solve for r. $r = \dfrac{I}{Pt}$

Slope–intercept form of a line

10. $y = mx + b$ Solve for m. $m = \dfrac{y - b}{x}$

11. $y = mx + b$ Solve for b. $b = y - mx$

Simple interest formula

12. $A = P(1 + rt)$ Solve for t. $t = \dfrac{A - P}{Pr}$

Area of a trapezoid

13. $A = \frac{1}{2}a(b_1 + b_2)$ Solve for b_1. $\dfrac{2A - ab_2}{a} = b_1$

Standard form of a line

14. $5x - 6y = 6$ Solve for y. $y = \dfrac{5}{6}x - 1$

15. $5x + 9y = -18$ Solve for y. $y = -\dfrac{5}{9}x - 2$

Slope–intercept form of a line

16. $y = -\frac{3}{4}x + 9$ Solve for x. $x = -\dfrac{4}{3}y + 12$

17. $y = \frac{6}{7}x - 12$ Solve for x. $x = \dfrac{7}{6}y + 14$

Standard form of a line

18. $ax + by = c$ Solve for y. $y = \dfrac{c - ax}{b}$

19. $ax + by = c$ Solve for x. $x = \dfrac{c - by}{a}$

Area of a circle

20. $A = \pi r^2$ Solve for r^2. $r^2 = \dfrac{A}{\pi}$

Surface area of a sphere

21. $s = 4\pi r^2$ Solve for r^2. $r^2 = \dfrac{s}{4\pi}$

Energy equation

22. $E = mc^2$ Solve for c^2. $c^2 = \dfrac{E}{m}$

23. $E = mc^2$ Solve for m. $m = \dfrac{E}{c^2}$

Distance of falling object

24. $S = \frac{1}{2}gt^2$ Solve for g. $g = \dfrac{2S}{t^2}$

25. $S = \frac{1}{2}gt^2$ Solve for t^2. $t^2 = \dfrac{2S}{g}$

Electronics equation

26. $E = IR$ Solve for R. $R = \dfrac{E}{I}$

27. $E = IR$ Solve for I. $I = \dfrac{E}{R}$

Volume of a rectangular prism

28. $V = LWH$ Solve for L. $L = \dfrac{V}{WH}$

29. $V = LWH$ Solve for H. $H = \dfrac{V}{LW}$

Volume of a cone

30. $V = \frac{1}{3}\pi r^2 h$ Solve for r^2. $r^2 = \dfrac{3V}{\pi h}$

31. $V = \frac{1}{3}\pi r^2 h$ Solve for h. $h = \dfrac{3V}{\pi r^2}$

Perimeter of a rectangle

32. $P = 2L + 2W$ Solve for W.
$W = \dfrac{P - 2L}{2}$

33. $P = 2L + 2W$ Solve for L.
$L = \dfrac{P - 2W}{2}$

Pythagorean theorem

34. $c^2 = a^2 + b^2$ Solve for a^2.
$a^2 = c^2 - b^2$

35. $c^2 = a^2 + b^2$ Solve for b^2.
$b^2 = c^2 - a^2$

Temperature conversion formulas

36. $F = \frac{9}{5}C + 32$ Solve for C.
$C = \dfrac{5(F - 32)}{9}$

37. $C = \frac{5}{9}(F - 32)$ Solve for F.
$F = \dfrac{9}{5}C + 32$

Boyle's Law for gases

38. $P = k\left(\dfrac{T}{V}\right)$ Solve for T.
$T = \dfrac{PV}{k}$

39. $P = k\left(\dfrac{T}{V}\right)$ Solve for V.
$V = \dfrac{kT}{P}$

Area of a sector of a circle

40. $A = \dfrac{\pi r^2 S}{360}$

(a) Solve for S (angle measure).
$S = \dfrac{360A}{\pi r^2}$

(b) Solve for r^2 (radius).
$r^2 = \dfrac{360A}{\pi S}$

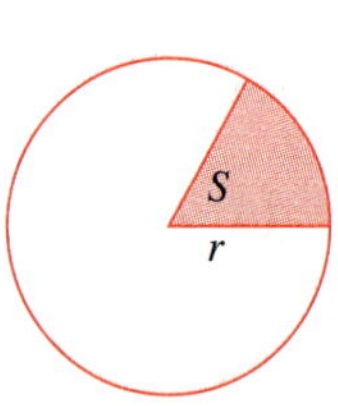

To Think About

41. In $I = Prt$, if t doubles, what is the effect on I?
I doubles

42. In $I = Prt$, if both r and t double, what is the effect on I? I increases four times

43. In $A = \pi r^2$, if r doubles, what is the effect on A?
A increases four times

44. In $A = \pi r^2$, if r is halved, what is the effect on A?
A becomes $\dfrac{1}{4}A$

Cumulative Review Problems

45. Find 12% of 260. 31.2

46. What is 0.2% of 48? 0.096

47. One day last year, 12 out of 30 people were able to climb Mt. Rainier in Washington. What percent were successful? 40%

48. A major auto company received a shipment of car stereos. Four stereos out of 160 are defective. What percent of the shipment is defective? 2.5%

Putting Your Skills to Work

FUEL ECONOMY OF A CAR AT VARIOUS SPEEDS

It has been reported that Dr. Robert Oman, a former professor at North Shore Community College and a former race car driver collected the following data. His 1996 Honda Accord with a four-cylinder engine averages 32 miles per gallon of gasoline when traveling at a speed of 50 miles per hour. For every 5 miles per hour increase in speed there is a decrease of 3 miles per gallon. He observed that this relationship seems quite accurate for his car most of the time under actual driving conditions.

PROBLEMS FOR INDIVIDUAL INVESTIGATION AND THOUGHT

Using the relationship described by Dr. Oman answer the following.

1. How many miles per gallon would he obtain if he were traveling at 60 miles per hour? 26 mpg

2. How many miles per gallon would he obtain if he were traveling at 70 miles per hour? What speed would you expect he was traveling if he obtained 23 miles per gallon?
20 mpg if he were traveling at 70 miles per hour.
He would be traveling at an estimated 65 miles per hour if he obtained 23 mpg.

PROBLEMS FOR COOPERATIVE GROUP INVESTIGATION

Together with some of the members of your class determine the following.

3. Dr. Oman did a trip on an interstate highway in Montana where the only speed limit is to travel at ''a reasonable and prudent speed.'' On a clear, sunny day on a deserted road he traveled 85 miles per hour and found his Honda only obtained 8 miles per gallon of gasoline. Is this what should be expected at this speed? No, it is 3 miles per gallon less than what is predicted by the relationship.

4. Dr. Oman's son is planning a trip from Chicago to Seattle. If he travels 1800 miles using his dad's Honda Accord and regular unleaded gasoline has an average cost of $1.35 per gallon, how much more will it cost him to drive at an average speed of 65 miles per hour instead of an average speed of 55 miles per hour? It will cost $21.86 more for gasoline to drive at 65 mph.

INTERNET CONNECTIONS: Go to ''http://www.prenhall.com/tobey'' to be connected
Site: Fuel Economy by Speed, 1973 and 1984 (Center for Transportation Analysis) or a related site

This site gives the typical gas mileage at various speeds in 1973 and 1984.

5. In 1984, Heidi Stephens traveled 618 miles from Birmingham to Cleveland. She used 23 gallons of gasoline for her trip. How many miles per gallon did she get? If her car got typical gas mileage for 1984, how fast do you think she was going?

6. In 1973, Jim Garcia planned to travel 403 miles from Los Angeles to San Francisco. If his car got typical gas mileage for 1973, how much more would it cost him to drive at an average speed of 70 miles per hour instead of 50 miles per hour? Assume that he paid $0.57 per gallon of gasoline.

7. Discuss the differences between the gas mileage values in 1973 and 1984. Why do you think the fuel economy changed during this time period?

2.6 Write and Graph Inequalities

MathPro

Video 7

SSM

After studying this section, you will be able to:

1 *Interpret an inequality statement.*
2 *Graph an inequality on a number line.*

1 *Inequality Statements*

We frequently speak of one value being greater than or less than another value. We say that ''5 is less than 7'' or ''9 is greater than 4.'' These relationships are called **inequalities.** We can write inequalities in mathematics by using symbols. We use the symbol $<$ to represent the words ''is less than.'' We use the symbol $>$ to represent the words ''is greater than.''

Statement in Words	***Statement in Algebra***
5 is less than 7.	$5 < 7$
9 is greater than 4.	$9 > 4$

TEACHING TIP As a sidelight, you may want to point out that it is possible to use an inequality to express relationships between irrational and rational numbers. Draw attention to the decimal approximation of π in Section 1.8. Although π is a nonterminating, nonrepeating decimal, we can say that $\pi > 3.14$ and represent this relationship on the number line.

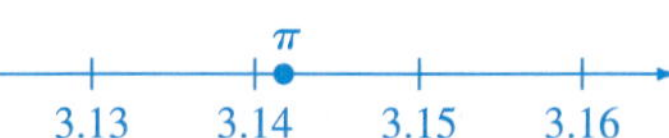

We will study irrational numbers extensively in Chapter 8. In this section we limit our study of inequalities to inequalities between rational numbers.

Note. ''5 is less than 7'' and ''7 is greater than 5'' have the same meaning. Similarly, $5 < 7$ and $7 > 5$ have the same meaning. They represent two equivalent ways of describing the same relationship between the two numbers 5 and 7.

We can illustrate the concept of inequality graphically if we examine a number line.

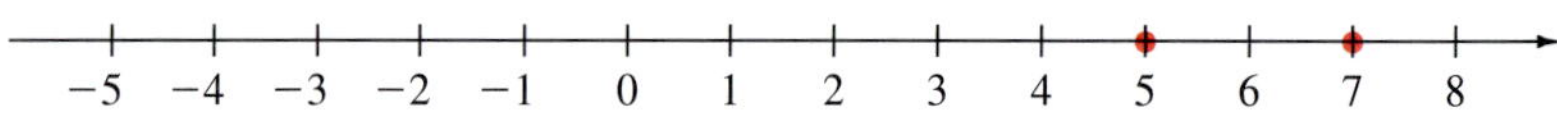

We say that **one number is greater than another** if it is to the right of the other on the number line. Thus $7 > 5$, since 7 is to the right of 5.

What about negative numbers? We can say ''-1 is greater than -3'' and write it in symbols $-1 > -3$ because we know that -1 lies to the right of -3 on the number line.

EXAMPLE 1 Replace the question mark with the symbol $<$ or $>$ in each statement.

(a) $3 ? -1$ **(b)** $-2 ? 1$ **(c)** $-3 ? -4$ **(d)** $0 ? 3$ **(e)** $-3 ? 0$

(a) $3 > -1$ *Use $>$, since 3 is to the right of -1.*
(b) $-2 < 1$ *Use $<$, since -2 is to the left of 1.*
(Or equivalently, we could say that 1 is to the right of -2.)
(c) $-3 > -4$ *Since -3 is to the right of -4.*
(d) $0 < 3$
(e) $-3 < 0$

Practice Problem 1 Replace the question mark with the symbol $<$ or $>$ in each statement.

(a) $7 ? 2$ **(b)** $-3 ? -4$ **(c)** $-1 ? 2$

(d) $-8 ? -5$ **(e)** $0 ? -2$ **(f)** $\frac{2}{5} ? \frac{3}{8}$ ■

2 *Graphing an Inequality on a Number Line*

Sometimes we will use an inequality to express the relationship between a variable and a number. $x > 3$ means that x could have the value of *any number* greater than 3. This can be pictured on the number line in a graph as follows:

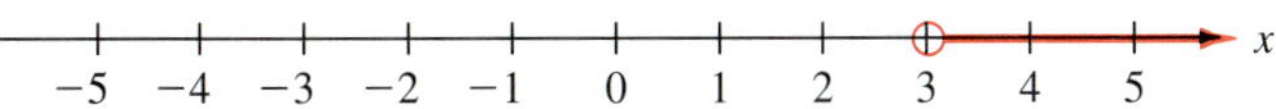

Note that the open circle at 3 suggests that we do not include the point for the number 3.

Similarly, we could represent graphically $x < -2$ as follows:

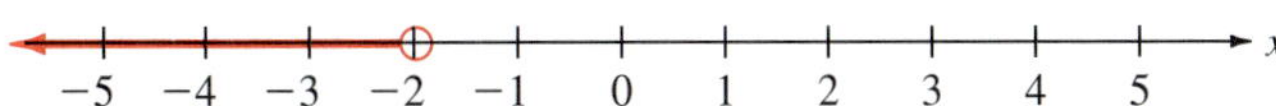

Sometimes a variable will be either greater than or equal to a certain number. In the statement "x is greater than or equal to 3," we are implying that x could have the value of 3 or any number greater than 3. We write this as $x \geq 3$. We represent it graphically as follows:

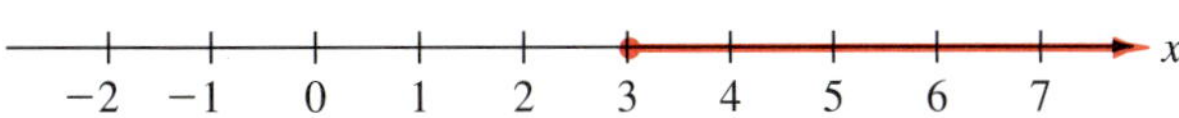

Note that the closed circle at 3 suggests that we *do* include the point for the number 3.

Similarly, we could represent graphically $x \leq -2$ as follows:

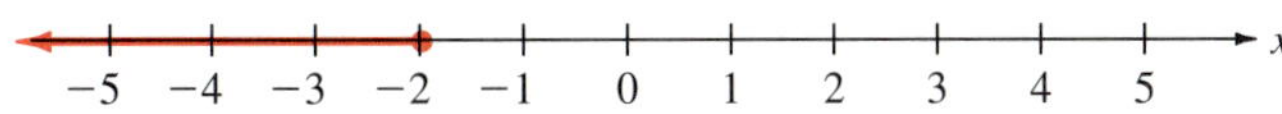

TEACHING TIP Although almost all elementary algebra books use this notation, it is not the only one in practice. There are several variations. In upper-level math courses the interval notations are usually employed. You may want to mention to your students that they may use an alternative notation if it is clear and mathematically acceptable.

$x \geq 2$ may be graphed

or

$x < -3$ may be graphed

or

EXAMPLE 2 For parts (a) through (d), state each mathematical relationship in words and then illustrate it graphically.

(a) $x < -2$ **(b)** $-3 < x$ **(c)** $x \geq -2$ **(d)** $x \leq -6$

(a) We state that "x is less than -2."

$x < -2$

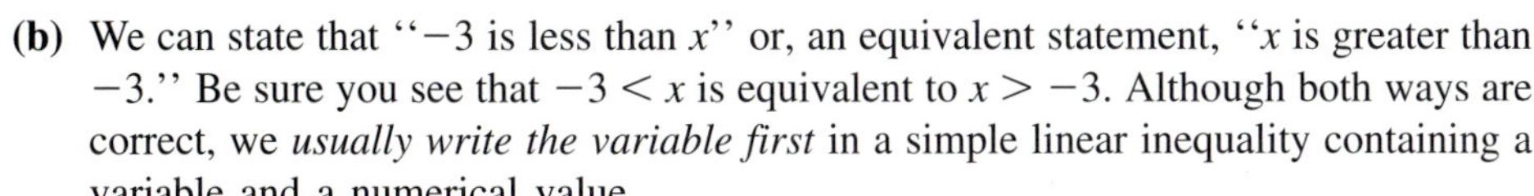

(b) We can state that "-3 is less than x" or, an equivalent statement, "x is greater than -3." Be sure you see that $-3 < x$ is equivalent to $x > -3$. Although both ways are correct, we *usually write the variable first* in a simple linear inequality containing a variable and a numerical value.

$x > -3$

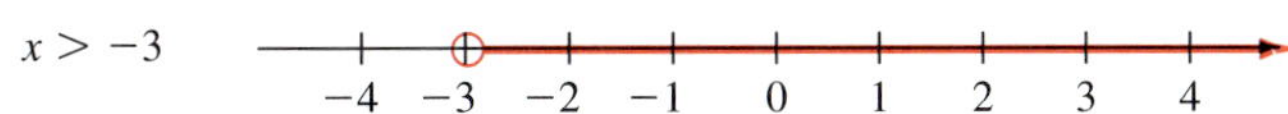

(c) We state that "x is greater than or equal to -2."

$x \geq -2$

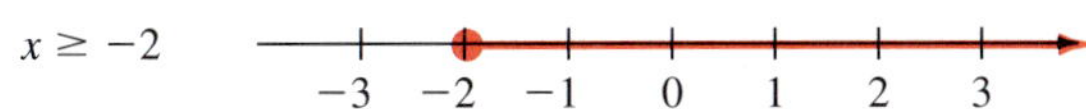

(d) We state that "x is less than or equal to -6."

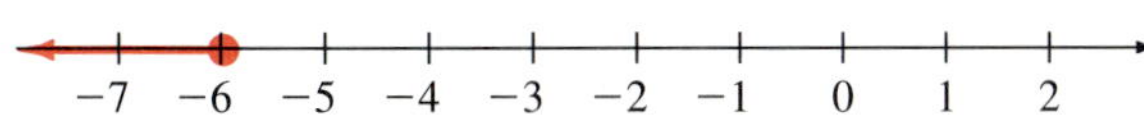

Practice Problem 2 State each mathematical relationship in words and then illustrate it graphically on the number lines in the margin.

(a) $x > 5$ **(b)** $x \leq -2$ **(c)** $3 > x$ **(d)** $x \geq -\frac{3}{2}$ ■

PRACTICE PROBLEM 2

(a)

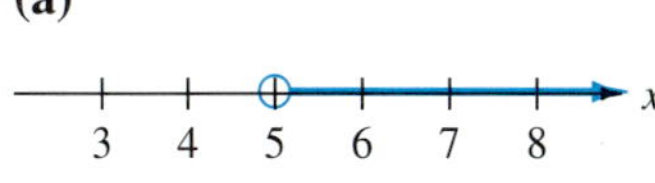

(b)

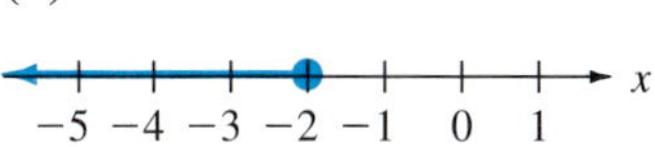

(c)

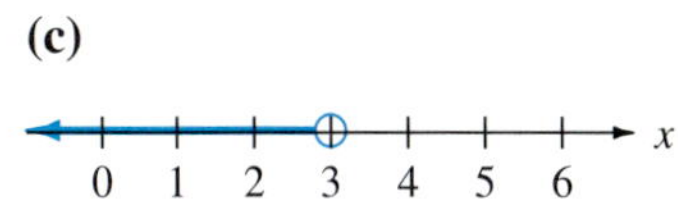

(d)

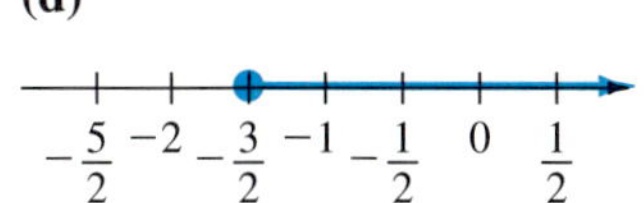

There are many everyday situations involving an unknown value and an inequality. We can translate these situations into algebraic statements. This is the first step in solving word problems using inequalities.

EXAMPLE 3 Translate each English statement into an algebraic statement.

(a) The police on the scene said that the car was traveling greater than 80 miles per hour (use the variable s for speed).

(b) The owner of the trucking company said that the payload of a truck must never exceed 4500 pounds (use the variable p for payload).

(a) Since the speed must be greater than 80 we have $s > 80$.

(b) If the payload of the truck can never exceed 4500 pounds, then the payload must be always less than or equal to 4500 pounds. Thus we write $p \leq 4500$.

Practice Problem 3 Translate the English statement into an inequality.

(a) During the drying cycle, the temperature inside the clothes dryer must never exceed 180 degrees Fahrenheit (use the variable t for temperature).

(b) The bank loan officer said that the total consumer debt incurred by Wally and Mary must be less than \$15,000 if they want to qualify for a mortgage to buy their first home (use the variable d for debt). ■

Developing Your Study Skills

KEEP TRYING

You may be one of those students who have had much difficulty with mathematics in the past and who are sure that you cannot do well in this course. Perhaps you are thinking, "I have never been any good at mathematics," or "I have always hated mathematics," or "Math always scares me," or "I have not had any math for so long that I have forgotten it all." You may even have picked up on the label "math anxiety" and attached it to yourself. That is most unfortunate, and it is time for you to reprogram your thinking. Replace those negative thoughts with more positive ones. You need to say things like, "I will give this math class my best shot," or "I can learn mathematics if I work at it," or "I will try to do better than I have done in previous math classes." You will be pleasantly surprised at the difference this more positive attitude makes!

We live in a highly technical world, and you cannot afford to give up on the study of mathematics. Dropping mathematics may prevent you from entering certain career fields that you may find interesting. You may not have to take math courses as high-level as calculus, but such courses as intermediate algebra, finite math, college algebra, and trigonometry may be necessary. Learning mathematics can open new doors for you.

Learning mathematics is a process that takes time and effort. You will find that regular study and daily practice are necessary to strengthen your skills and to help you to grow academically. This process will lead you toward success in mathematics. Then, as you become more successful, your confidence in your ability to do mathematics will grow.

2.6 Exercises

Verbal and Writing Skills

1. Is the statement $5 > -6$ equivalent to the statement $-6 < 5$? Why? yes
Both statements imply 5 is to the right of -6 on the number line.

2. Is the statement $-8 < -3$ equivalent to the statement $-3 > -8$? Why? yes
Both statements imply -3 is to the right of -8 on the number line.

Replace the ? by < or >.

3. $9 \ ? \ -3$ $>$

4. $-2 \ ? \ 5$ $<$

5. $-4 \ ? \ -2$ $<$

6. $-3 \ ? \ -6$ $>$

7. $\frac{2}{3} \ ? \ \frac{3}{4}$ $<$

8. $\frac{5}{7} \ ? \ \frac{6}{8}$ $<$

9. $-1.2 \ ? \ +2.1$ $<$

10. $-3.6 \ ? \ +2.4$ $<$

11. $-\frac{13}{3} \ ? \ -4$ $<$

12. $-3 \ ? \ -\frac{15}{4}$ $>$

13. $-\frac{5}{8} \ ? \ -\frac{3}{5}$ $<$

14. $-\frac{2}{3} \ ? \ -\frac{3}{4}$ $>$

Which is greater?

15. $\frac{123}{4986}$ or 0.0247? 0.0247

16. $\frac{997}{6384}$ or 0.15613? $\frac{997}{6384}$

Give a graphical representation on the number line.

17. $x \geq -6$

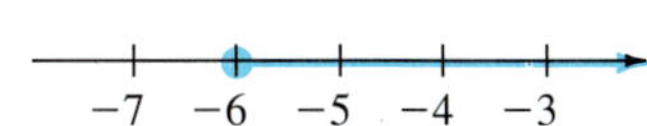

18. $x \leq -2$

19. $x < 3$

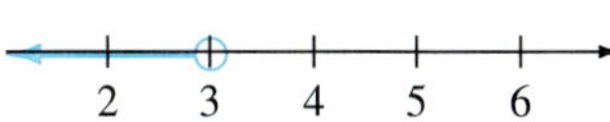

20. $x > 5$

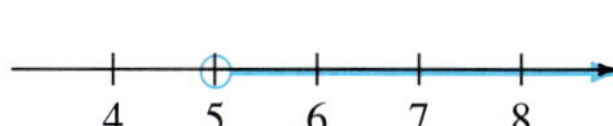

21. $x > \frac{3}{4}$

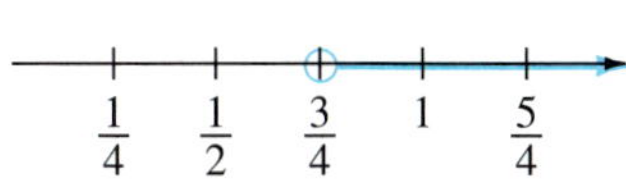

22. $x \geq -\frac{5}{2}$

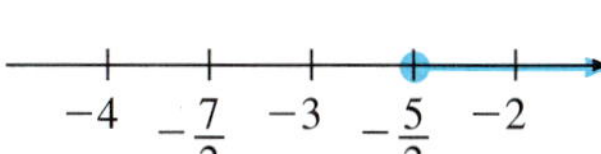

23. $x \leq -3.6$

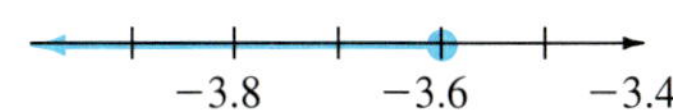

24. $x < -2.2$

25. $25 < x$

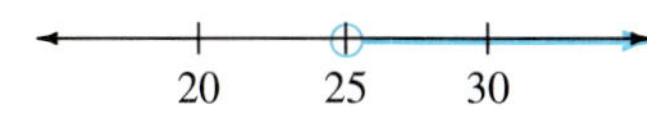

26. $35 \geq x$

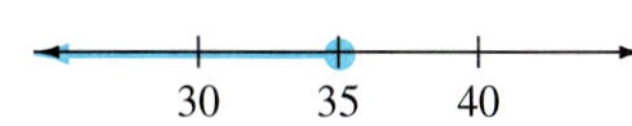

Translate the graphical representation to an inequality using the variable x.

27.

$x \geq -1$

28. $-\frac{3}{2}$ -1 $-\frac{1}{2}$ 0 $\frac{1}{2}$ $x > -\frac{3}{2}$

29.

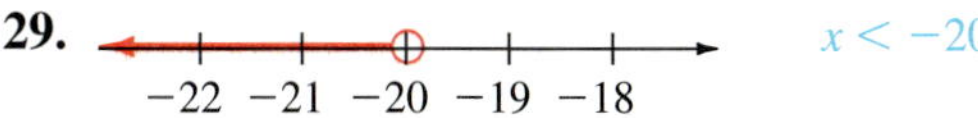

$x < -20$

30. −37 −36.5 −36 −35.5 −35 $x < -36$

31.

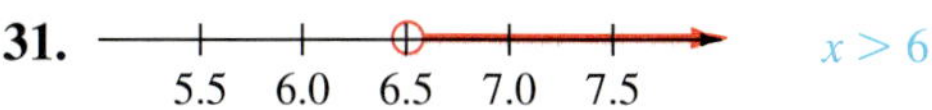

$x > 6.5$

32. 20 30 40 50 60 $x \leq 50$

Translate the English statement into an inequality.

33. The cost of the hiking boots must be less than \$56. (Use the variable c for cost.) $c < 56$

34. The speed of the rocket was greater than 580 kilometers per hour. (Use the variable V for speed.) $V > 580$

35. The number of hours for a full-time position at this company cannot be less than 37, in order to receive full-time benefits. (Use the variable h for hours.) $h \geq 37$

36. The number of nurses on duty on the floor can never exceed 6. (Use the variable n for the number of nurses.) $n \leq 6$

To Think About

37. Suppose that the variable x must satisfy *all* of these conditions.

(a) $x \leq 2$ **(b)** $x > -3$ **(c)** $x < \frac{5}{2}$ **(d)** $x \geq -\frac{5}{2}$

Graph on a number line the region that satisfies all conditions.

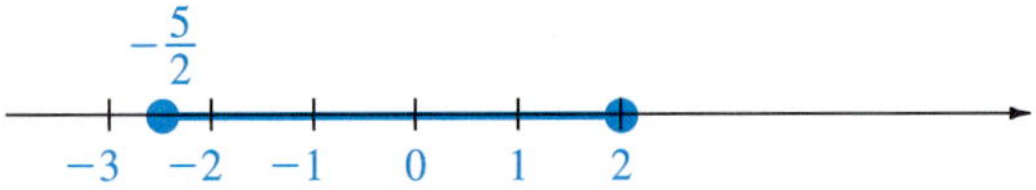

38. Suppose that the variable x must satisfy *all* of these conditions.

(a) $x < 4$ **(b)** $x > -4$ **(c)** $x \leq \frac{7}{2}$ **(d)** $x \geq -\frac{9}{2}$

Graph on a number line the region that satisfies all conditions.

Cumulative Review Problems

39. Find 16% of 38. 6.08

40. Find 1.3% of 1250. 16.25

41. For the most coveted graduate study positions, only 16 out of 800 students are accepted. What percent are accepted? 2%

42. Write the fraction $\frac{3}{8}$ as a percent. 37.5%

Developing Your Study Skills

EXAM TIME: GETTING ORGANIZED

Studying adequately for an exam requires careful preparation. Begin early so that you will be able to spread your review over several days. Even though you may still be learning new material at this time, you can be reviewing concepts previously learned in the chapter. Giving yourself plenty of time for review will take the pressure off. You need this time to process what you have learned and to tie concepts together.

Adequate preparation enables you to feel confident and to think clearly with less tension and anxiety.

2.7 Solve Inequalities

After studying this section, you will be able to:

1 *Solve an inequality.*

MathPro  Video 7  SSM

1 *Solving Inequalities*

The possible values that make an inequality true are called its **solutions.** Thus, when we **solve an inequality,** we are finding *all* the values that make it true. To solve an inequality, we simplify it to the point where we can clearly see the possible values for the variable. We've solved equations by adding, subtracting, multiplying, and dividing a particular value on both sides of the equation. Here we do similar operations with inequalities, with one important exception. We'll show some examples so that you can see the operations we can do with inequalities just as with equations.

We will first examine the pattern that takes place when we perform a given operation on both sides of an inequality.

EXAMPLE 1

	Original Inequality				*New Inequality*
(a)	$3 < 5$	$\longrightarrow$	Multiply both sides by 2.	$\longrightarrow$	$6 < 10$
(b)	$-2 < -1$	$\longrightarrow$	Add -3 to both sides.	$\longrightarrow$	$-5 < -4$
(c)	$0 > -4$	$\longrightarrow$	Divide both sides by 2.	$\longrightarrow$	$0 > -2$
(d)	$8 > 4$	$\longrightarrow$	Subtract 6 from both sides.	$\longrightarrow$	$2 > -2$

Note that *we avoided multiplying or dividing by a negative number!*

Practice Problem 1 Perform the given operation and write a new inequality.

(a) $9 > 6$ Divide both sides by 3.

(b) $2 < 5$ Multiply both sides by 2.

(c) $-8 < -3$ Add 4 to each side.

(d) $-3 < 7$ Add -7 to each side. ■

Now let us examine what would happen if we did multiply or divide by a negative number. We start with an original, true inequality. We want to get a new, also true inequality.

Original Inequality				*New Inequality*
$3 < 5$	$\longrightarrow$	Multiply by -2.	$\longrightarrow$	$-6\ ?\ -10$

What is the correct inequality sign? Since -6 is to the right of -10, we know the new inequality should be $-6 > -10$, if we wish the statement to remain true. Notice how we reverse the direction of the inequality from $<$ (less than) to $>$ (greater than). We would thus obtain the new inequality $-6 > -10$. Thus

$$3 < 5 \quad \longrightarrow \quad \text{Multiply by } -2. \quad \longrightarrow \quad -6 > -10$$

The $<$ sign we started with $(3 < 5)$ is reversed to $>$ $(-6 > -10)$. A similar reversal takes place in the following example.

EXAMPLE 2

	Original Inequality				*New Inequality*
(a)	$-2 < -1$	$\longrightarrow$	Multiply by -3.	$\longrightarrow$	$6 > 3$
(b)	$0 > -4$	$\longrightarrow$	Divide both sides by -2.	$\longrightarrow$	$0 < 2$
(c)	$8 > 4$	$\longrightarrow$	Divide both sides by -4.	$\longrightarrow$	$-2 < -1$

Notice that we do the arithmetic with signed numbers just as we always do. But the new inequality has its inequality sign reversed (from that of the original inequality). *Whenever both sides of an inequality are multiplied or divided by a negative quantity, the direction of the inequality is reversed.*

Practice Problem 2

(a) $7 > 2$ Multiply each side by -2.

(b) $-3 < -1$ Multiply each side by -1.

(c) $-10 > -20$ Divide each side by -10.

(d) $-15 < -5$ Divide each side by -5. ■

Procedure for Solving Inequalities

You may use the same procedures to solve inequalities that you did to solve equations *except* that the direction of an inequality is *reversed* if you *multiply* or *divide* both sides *by a negative number.*

EXAMPLE 3 Solve and graph $3x + 7 \geq 13$.

$3x + 7 - 7 \geq 13 - 7$ *Subtract 7 from both sides.*

$3x \geq 6$ *Simplify.*

$\frac{3x}{3} \geq \frac{6}{3}$ *Divide both sides by 3.*

$x \geq 2$ *Simplify. Note the direction of the inequality is not changed, since we have divided by a positive number.*

The graphical representation is 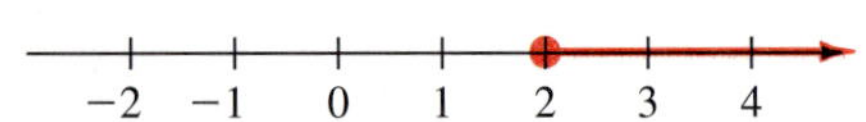

PRACTICE PROBLEM 3

$\frac{3}{8}$ $\frac{1}{2}$ $\frac{5}{8}$ $\frac{3}{4}$ $\frac{7}{8}$ 1 $1\frac{1}{8}$

Practice Problem 3 Solve and graph $8x - 2 < 3$. ■

EXAMPLE 4 Solve and graph $5 - 3x > 7$.

$5 - 5 - 3x > 7 - 5$ *Subtract 5 from both sides.*

$-3x > 2$ *Simplify.*

$\frac{-3x}{-3} < \frac{2}{-3}$ *Divide by -3, and* ***reverse the inequality,*** *since both sides are divided by negative 3.*

$x < -\frac{2}{3}$ *Note the direction of the inequality.*

The graphical representation is 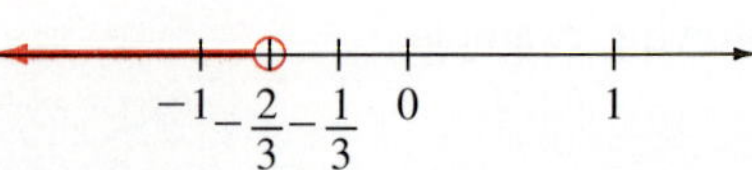

PRACTICE PROBLEM 4

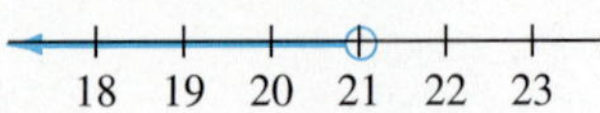

Practice Problem 4 Solve and graph $7 + x < 28$. ■

Just like equations, some inequalities contain parentheses and fractions. The initial steps to solve these inequalities will be the same as those used to solve equations with parentheses and fractions. When the variable appears on both sides of the inequality, it is advisable to collect the x terms on the left side of the inequality symbol.

EXAMPLE 5 Solve and graph $-\frac{13x}{2} \leq \frac{x}{2} - \frac{15}{8}$.

$8\left(\frac{-13x}{2}\right) \leq 8\left(\frac{x}{2}\right) - 8\left(\frac{15}{8}\right)$	*Multiply all terms by LCD = 8. We do **not** reverse the direction of the inequality symbol since we are multiplying by a positive number, 8.*
$-52x \leq 4x - 15$	*Simplify.*
$-52x - 4x \leq 4x - 15 - 4x$	*Add $-4x$ to both sides.*
$-56x \leq -15$	*Combine like terms.*
$\frac{-56x}{-56} \geq \frac{-15}{-56}$	*Divide both sides by -56. We **reverse** the direction of the inequality when we divide both sides by a negative number.*
$x \geq \frac{15}{56}$	

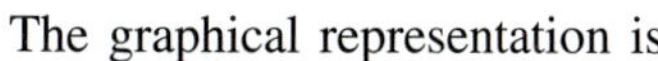
The graphical representation is

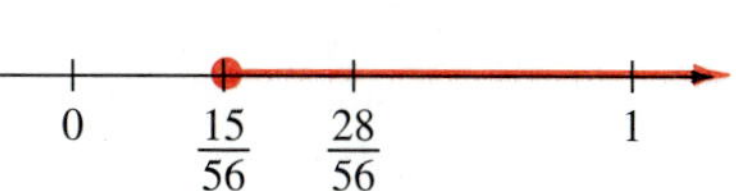

PRACTICE PROBLEM 5

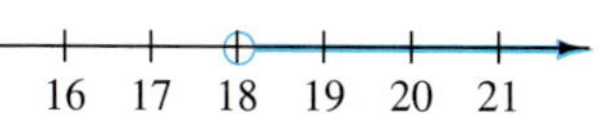

Practice Problem 5 Solve and graph $\frac{1}{2}x + 3 < \frac{2}{3}x$. ■

EXAMPLE 6 Solve and graph $\frac{1}{3}(3 - 2x) \leq -4(x + 1)$.

$1 - \frac{2x}{3} \leq -4x - 4$	*Remove parentheses.*
$3(1) - 3\left(\frac{2x}{3}\right) \leq 3(-4x) - 3(4)$	*Multiply all terms by LCD = 3.*
$3 - 2x \leq -12x - 12$	*Simplify.*
$3 - 2x + 12x \leq -12x + 12x - 12$	*Add $12x$ to both sides.*
$3 + 10x \leq -12$	*Combine like terms.*
$3 - 3 + 10x \leq -12 - 3$	*Subtract 3 from both sides.*
$10x \leq -15$	*Simplify.*
$\frac{10x}{10} \leq \frac{-15}{10}$	*Divide both sides by 10. Since we are dividing by a **positive** number, the inequality is **not** reversed.*
$x \leq -\frac{3}{2}$	

The graphical representation is

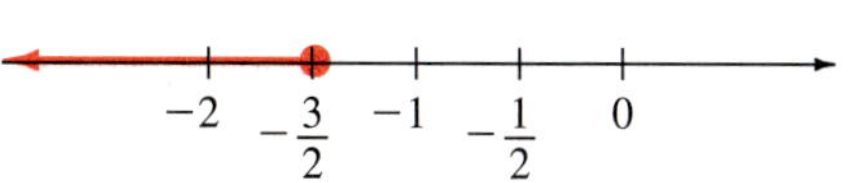

TEACHING TIP After solving Example 6, ask students to do the following problem as an Additional Class Exercise. Solve and graph.

$$\frac{1}{4}(x + 5) + \frac{1}{4} \leq \left(\frac{1}{8}\right)x + \frac{3}{4}$$

Sol. $x \leq -6$

PRACTICE PROBLEM 6

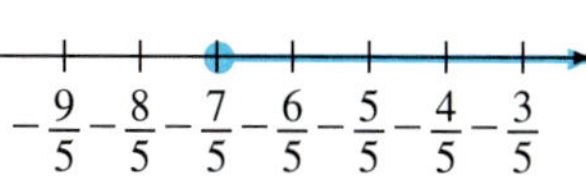

Practice Problem 6 Solve and graph $\frac{1}{2}(3 - x) \leq 2x + 5$. ■

The most common error students make in solving inequalities is forgetting to reverse the direction of the inequality symbol when multiplying or dividing by a negative number.

CAUTION When each side of an inequality is multiplied or divided by a negative number, the direction of the inequality symbol is reversed.

EXAMPLE 7 A hospital director has determined that the costs of operating one floor of the hospital for an 8-hour shift must never exceed \$2370. An expression for the cost of operating one floor of the hospital is $130n + 1200$, where n is the number of nurses. This expression is determined by \$1200 in fixed costs and \$130 per nurse for an 8-hour shift. Set up and solve an inequality to determine the number of nurses that may be on duty on this floor for an 8-hour shift if this cost factor is to be followed.

The cost expression must never exceed \$2370. This means the cost expression is always less than or equal to \$2370.

$$130n + 1200 \leq 2370 \qquad \textit{Write the inequality.}$$

$$130n + 1200 + (-1200) \leq 2370 + (-1200) \qquad \textit{Add } (-1200) \textit{ to each side.}$$

$$130n \leq 1170 \qquad \textit{Simplify.}$$

$$\frac{130n}{130} \leq \frac{1170}{130} \qquad \textit{Divide each side by } 130.$$

$$n \leq 9$$

The number of nurses on duty on this floor for an 8-hour shift must always be less than or equal to 9.

Practice Problem 7 The company president of Staywell, Inc., wants the monthly profits to never be less than \$2,500,000. He has determined that an expression for monthly profit for the company is $2000n - 700{,}000$. In the expression, n is the number of exercise machines manufactured each month. The profit on each machine is \$2000 and the $-\$700{,}000$ represents fixed costs to run the manufacturing division. How many machines must be made and sold each month to satisfy these financial goals? ■

2.7 Exercises

Solve each inequality. Graph the results.

1. $x - 5 > 3$ $x > 8$

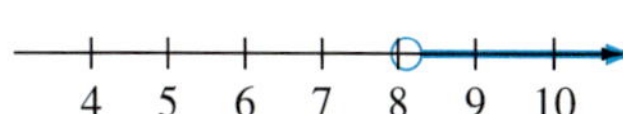

2. $x + 9 \le 2$ $x \le -7$

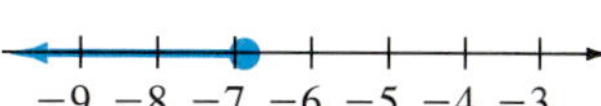

3. $-2x < 18$ $x > -9$

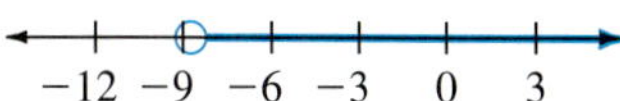

4. $5x \le 25$ $x \le 5$

5. $\frac{1}{2}x \ge 4$ $x \ge 8$

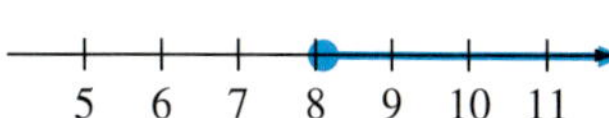

6. $-\frac{1}{5}x < 10$ $x > -50$

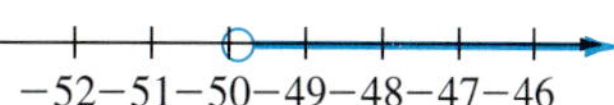

7. $2x - 3 < 4$ $x < \frac{7}{2}$

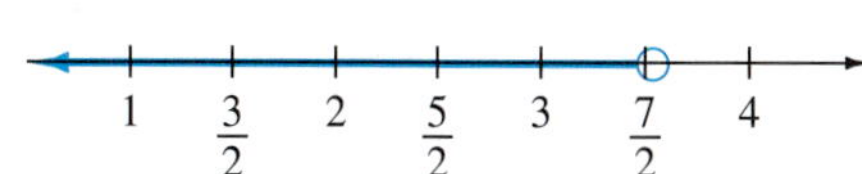

8. $3 - 3x > 12$ $x < -3$

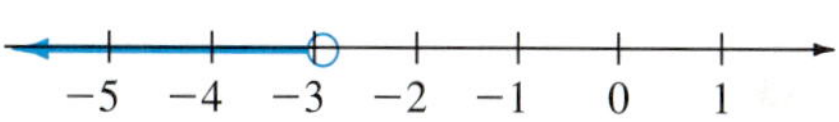

9. $-4 - 14x < 6 - 6x$ $x > -\frac{5}{4}$

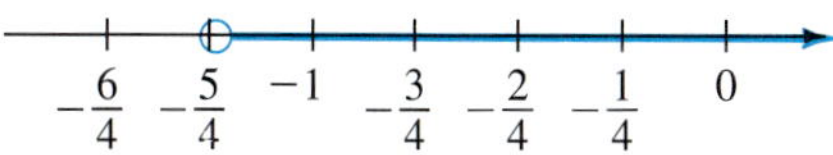

10. $7 - 8x \le -6x - 5$ $x \ge 6$

11. $\frac{5x}{6} - 5 > \frac{x}{6} - 9$ $4x > -24$
$x > -6$

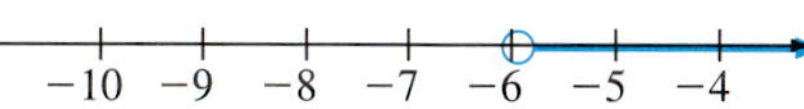

12. $\frac{x}{4} - 2 < \frac{3x}{4} + 5$ $-2x < 28$
$x > -14$

13. $3(x + 2) \le 2x + 4$ $3x + 6 \le 2x + 4$
$x \le -2$

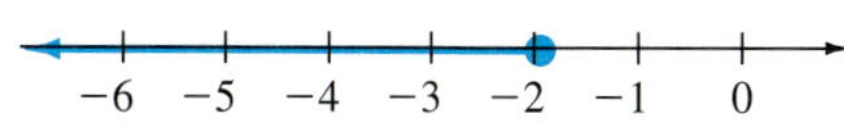

14. $4(x + 5) \ge 3x + 7$ $4x + 20 \ge 3x + 7$
$x \ge -13$

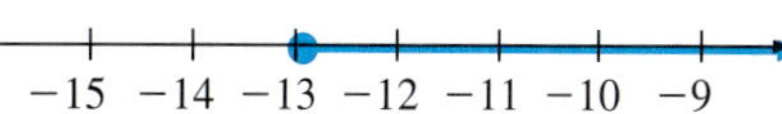

Verbal and Writing Skills

15. Add -2 to both sides of the inequality $5 > 3$. What is the result? $3 > 1$
Why was the direction of the inequality not reversed?
Adding *any* number to both sides of an inequality does not reverse the direction.

16. Divide -3 into both sides of the inequality $-21 > -29$. What is the result?
Why was the direction of the inequality reversed?

$7 < \frac{29}{3}$ Whenever you divide both sides of an inequality by a negative number, the direction of the inequality is reversed.

Mixed Practice

Solve each inequality. Collect the variable terms on the left side of the inequality.

17. $2x - 5 < 5x - 11$
$-3x < -6$
$x > 2$

18. $4x - 7 > 9x - 2$
$-5x > 5$
$x < -1$

19. $6x - 2 \geq 4x + 6$
$2x \geq 8$
$x \geq 4$

20. $5x - 5 \leq 2x + 10$
$3x \leq 15$
$x \leq 5$

21. $0.3(x - 1) < 0.1x - 0.5$
$0.2x < -0.2$
$2x < -2$
$x < -1$

22. $0.2(3 - x) + 0.1 > 0.1(x - 2)$
$-0.3x > -0.9$
$-3x > -9$
$x < 3$

23. $5 - 2(3 - x) \leq 2(2x + 5) + 1$
$5 - 6 + 2x \leq 4x + 10 + 1$
$-2x \leq 12$
$x \geq -6$

24. $5(x + 1) + 2 \geq x - 3(2x + 1)$
$5x + 5 + 2 \geq x - 6x - 3$
$10x \geq -10$
$x \geq -1$

25. $\dfrac{x + 6}{7} - \dfrac{6}{14} > \dfrac{x + 3}{2}$
$2x + 12 - 6 > 7x + 21$
$-5x > 15$
$x < -3$

26. $\dfrac{3x + 5}{4} + \dfrac{7}{12} > -\dfrac{x}{6}$
$9x + 15 + 7 > -2x$
$22 > -11x$
$x > -2$

★**27.** $3(0.3 + 0.1x) + 0.1 < 0.5(x + 2)$
$0.9 + 0.3x + 0.1 < 0.5x + 1$
$-0.2x < 0$
$-2x < 0$
$x > 0$

★**28.** $0.3 + 0.4(2 - x) \leq 6(-0.2 + 0.1x) + 0.3$
$0.3 + 0.8 - 0.4x \leq -1.2 + 0.6x + 0.3$
$-x \leq -2$
$x \geq 2$

★**29.** $\dfrac{1}{6} - \dfrac{1}{2}(3x + 2) < \dfrac{1}{3}\left(x - \dfrac{1}{2}\right)$
$1 - 9x - 6 < 2x - 1$
$x > -\dfrac{4}{11}$

★**30.** $\dfrac{2}{3}(2x - 5) + 3 \geq \dfrac{1}{4}(3x + 1) - 5$
$16x - 40 + 36 \geq 9x + 3 - 60$
$x \geq -\dfrac{53}{7}$

Solve for x. Round your answer to the nearest hundredth.

31. $1.96x - 2.58 < 9.36x + 8.21$ $x > -1.46$

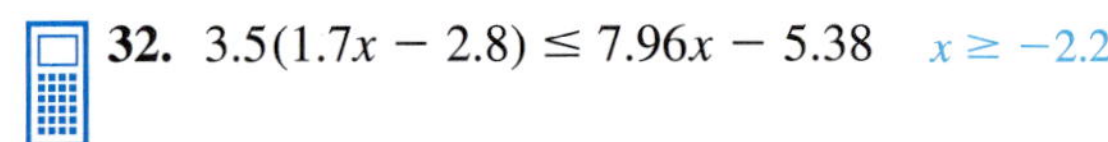
32. $3.5(1.7x - 2.8) \leq 7.96x - 5.38$ $x \geq -2.20$

Applications

33. To pass a course with a B grade, a student must have an average of 80 or greater. A student's grades on three tests are 75, 83, and 86. What score must the student get on the next test to get a B average or better?
A score of 76 or greater.

34. Sharon sells very expensive European sports cars. She may choose to receive \$10,000.00 or 8% of her sales as payment for her work. How much does she need to sell to make the 8% offer a better deal?
She needs to sell more than \$125,000 worth of cars.

35. Tess supervises a computer chip manufacturing facility. She has determined that her monthly profit factor is given by the expression $12.5n - 300{,}000$. Here n represents the number of chips manufactured each month. Each finished chip produces a profit of \$12.50. The fixed costs (overhead) of the factory are \$300,000 per month.
(a) How many chips must be manufactured to break even? 24,000 chips
(b) How many chips must be manufactured monthly to ensure a profit of \$650,000? A profit of \$470,000?
76,000 chips for a profit of \$650,000; 61,600 chips for a profit of \$470,000

Putting Your Skills to Work

OVERSEAS TRAVEL

Karin lives in Dallas and works for an American computer company. Her employer often sends her to Europe on business trips. She keeps the following charts handy.

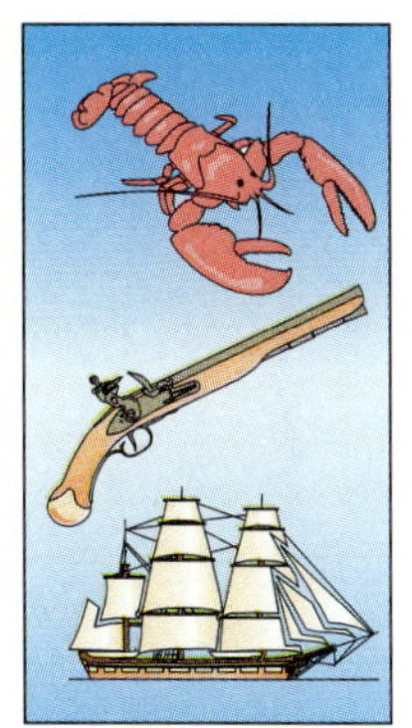

CORPORATE RATES FLYING OVERSEAS AIRLINES		
Round-trip Fares Boston to:	Stay at Least One Week Including a Saturday	Not Staying One Week Including a Saturday
London	$460	$680
Paris	$490	$710
Geneva	$500	$730

CORPORATE RATES FLYING DOMESTIC AIRLINES		
Round-trip Fares between Boston and Dallas on:	Spend One Week Including a Saturday	Not Staying One Week Including a Saturday
First class	$640	$900
Business class	$500	$680
Regular coach class	$450	$570
14-Day advance coach class	$310	$460

PROBLEMS FOR INDIVIDUAL INVESTIGATION AND STUDY

1. In three weeks she must travel from Dallas to Paris. She will be in Paris for 2 weeks. What is the least amount that her airfare could be? $800

2. Her company has notified her that she must fly from Dallas to London in 3 days. They want her to travel first class on the domestic part of the trip in order to prepare for a company presentation. She will only be in London for 4 days. What is the least amount that her airfare could be? $1580

PROBLEMS FOR COOPERATIVE GROUP ACTIVITY

Together with other members of your class see if you can complete the following.

3. Soon her company will send her to Geneva for 7 days. She does not know how soon she will be asked to leave. What is the least amount that her airfare could be? If she flies regular coach class, what is the greatest amount that her airfare could be? $\$810 \leq \text{airfare} \leq \$950$

4. Karin has a travel budget of $3000. Her company has authorized her to travel overseas on first class. First class tickets cost 50% more than business class tickets for overseas travel. If she has one trip from Dallas to London for 14 days and one trip from Dallas to Geneva for 2 days, will she be able to travel first class on all flights? Explain why or why not. No. This total of $3325 exceeds the budget.

INTERNET CONNECTIONS: Go to ``http://www.prenhall.com/tobey'' to be connected

Site: American Airlines Fares or a related site

Frederick lives in Denver, Colorado. His employer has told him that he needs to go to a conference in Amsterdam, the Netherlands. He is to leave within the next week. Note that the fares may vary from day to day.

5. Find the lowest round-trip fare he can get on this airline. What day or days can he leave to get this fare?

6. If he travels first class on the same day, how much more will he pay for the round trip?

Chapter Organizer

Topic	Procedure	Examples
Solving equations without parentheses or fractions, p. 139.	1. On each side of the equation, collect like terms if possible. 2. Add or subtract terms on both sides of the equation in order to get all terms with the variable on one side of the equation. 3. Add or subtract a value on both sides of the equation to get all terms not containing the variable on the other side of the equation. 4. Divide both sides of the equation by the coefficient of the variable. 5. If possible, simplify solution. 6. Check your solution by substituting the obtained value into the original equation.	$5x + 2 + 2x = -10 + 4x + 3$ $7x + 2 = -7 + 4x$ $7x + (-4x) + 2 = -7 + 4x + (-4x)$ $3x + 2 = -7$ $3x + 2 + (-2) = -7 + (-2)$ $3x = -9$ $\frac{3x}{3} = \frac{-9}{3}$ $x = -3$ *Check:* Is $x = -3$ the solution of $5x + 2 + 2x = -10 + 4x + 3$? $5(-3) + 2 + 2(-3) \stackrel{?}{=} -10 + 4(-3) + 3$ $-15 + 2 - 6 \stackrel{?}{=} -10 + (-12) + 3$ $-13 - 6 \stackrel{?}{=} -22 + 3$ $-19 = -19$
Solving equations with parentheses and/or fractions, p. 141 and p. 148.	1. Remove any parentheses. 2. Simplify, if possible. 3. If fractions exist, multiply all terms on both sides by the lowest common denominator of all the fractions. 4. Now follow the remaining steps of solving an equation without parentheses or fractions.	Solve for y: $5(3y - 4) = \frac{1}{4}(6y + 4) - 48$ $15y - 20 = \frac{3}{2}y + 1 - 48$ $15y - 20 = \frac{3}{2}y - 47$ $2(15y) - 2(20) = 2\left(\frac{3}{2}y\right) - 2(47)$ $30y - 40 = 3y - 94$ $30y - 3y - 40 = 3y - 3y - 94$ $27y - 40 = -94$ $27y - 40 + 40 = -94 + 40$ $27y = -54$ $\frac{27y}{27} = \frac{-54}{27}$ $y = -2$

Chapter Organizer

Topic	Procedure	Examples
Solving formulas, p. 152.	1. Remove any parentheses and simplify if possible. 2. If fractions exist, multiply all terms on both sides by the LCD, which may be a variable. 3. Add or subtract terms on both sides of the equation in order to get all terms containing the *desired variable* on one side of the equation. 4. Add or subtract terms on both sides of the equation in order to get all other terms on the opposite side of the equation. 5. Divide both sides of the equation by the coefficient of the desired variable. This division may involve other variables. 6. Simplify, if possible. 7. (Optional) Check your solution by substituting the obtained expression into the original equation.	Solve for z: $B = \frac{1}{3}(hx + hz)$. First we remove parentheses. $B = \frac{1}{3}hx + \frac{1}{3}hz$ Now we multiply each term by 3. $3(B) = 3\left(\frac{1}{3}hx\right) + 3\left(\frac{1}{3}hz\right)$ $3B = hx + hz$ $3B - hx = hx - hx + hz$ $3B - hx = hz$ The coefficient of z is h, so we divide each side by h. $\frac{3B - hx}{h} = z$
Solving inequalities, p. 163.	1. Follow the steps for solving a first-degree equation up until the division step. 2. If you divide both sides of the inequality by a positive number, the direction of the inequality is not reversed. 3. If you divide both sides of the inequality by a *negative number,* the direction of the inequality is reversed.	Solve for x and graph your solution: $\frac{1}{2}(3x - 2) \leq -5 + 5x - 3$ First remove parentheses and simplify. $\frac{3}{2}x - 1 \leq -8 + 5x$ Now multiply each term by 2. $2\left(\frac{3}{2}x\right) - 2(1) \leq 2(-8) + 2(5x)$ $3x - 2 \leq -16 + 10x$ $3x - 10x - 2 \leq -16 + 10x - 10x$ $-7x - 2 \leq -16$ $-7x - 2 + 2 \leq -16 + 2$ $-7x \leq -14$ When we divide both sides by a negative number, the inequality is reversed. $\frac{-7x}{-7} \geq \frac{-14}{-7}$ $x \geq 2$ Graphical solution: 0 1 2 3 4

Chapter 2 Review Problems

2.1–2.3 *Solve for x.*

1. $5x + 20 = 3x$ -10

2. $7x + 3 = 4x$ -1

3. $7(x - 4) = x + 2$ 5

4. $2(17 - x) = 6x + 2$ 4

5. $4x - 3(x + 2) = 4$ 10

6. $1 - 2(6 - x) = 3x + 2$ -13

7. $x - (0.5x + 2.6) = 17.6$ 40.4

8. $-0.2(x + 1) = 0.3(x + 11)$ -7

9. $3(x - 2) = -4(5 + x)$ -2

10. $\frac{1}{4}x = -16$ -64

11. $x + 37 = 26$ -11

12. $6(8x + 3) = 5(9x + 8)$ $\frac{22}{3}$ or $7\frac{1}{3}$

13. $3(x - 3) = 13x + 21$ -3

14. $9x + 10 = 3x + 4$ -1

15. $24 - 3x = 4(x - 1)$ 4

16. $12 - x + 2 = 3x - 10 + 4x$ 3

17. $36 = 9x - (3x - 18)$ 3

18. $8 - 3x = -5x - 4$ -6

19. $2(3 - x) = 1 - (x - 2)$ 3

20. $4(x + 5) - 7 = 2(x + 3)$ $-\frac{7}{2}$ or $-3\frac{1}{2}$

21. $0.9x + 3 = 0.4x + 1.5$ -3

22. $7x - 3.4 = 11.3$ 2.1

23. $8(3x + 5) - 10 = 9(x - 2) + 13$ $-\frac{7}{3}$ or $-2\frac{1}{3}$

24. $8 - 3x + 5 = 13 + 4x + 2$ $-\frac{2}{7}$

25. $3 = 2x + 5 - 3(x - 1)$ 5

26. $-2(x - 3) = -4x + 3(3x + 2)$ 0

27. $2(5x - 1) - 7 = 3(x - 1) + 5 - 4x$ 1

2.4 *Solve for the variable.*

28. $\frac{3}{4}x - 3 = \frac{1}{2}x + 2$ 20

29. $1 = \frac{5x}{6} + \frac{2x}{3}$ $\frac{2}{3}$

30. $\frac{7x}{5} = 5 + \frac{2x}{5}$ 5

31. $\frac{7x - 3}{2} - 4 = \frac{5x + 1}{3}$ $\frac{35}{11}$

32. $\frac{3x - 2}{2} + \frac{x}{4} = 2 + x$ 4

33. $\frac{-3}{2}(x + 5) = 1 - x$ -17

34. $\frac{-4}{3}(2x + 1) = -x - 2$ $\frac{2}{5}$

35. $\frac{1}{3}(x - 2) = \frac{x}{4} + 2$ 32

36. $\frac{1}{5}(x - 3) = 2 - \frac{x}{2}$ $\frac{26}{7}$

37. $\frac{4}{5} + \frac{1}{2}x = \frac{1}{5}x + \frac{1}{2}$ -1

38. $3x + \frac{6}{5} - x = \frac{6}{5}x - \frac{4}{5}$ $-\frac{5}{2}$ or $-2\frac{1}{2}$

39. $\frac{10}{3} - \frac{5}{3}x + x = \frac{2}{9} + \frac{1}{9}x$ 4

40. $-\frac{8}{3}x - 8 + 2x - 5 = -\frac{5}{3}$ -17

41. $\frac{1}{2} + \frac{5}{4}x = \frac{2}{5}x - \frac{1}{10} + 4$ 4

42. $5 + 3x = \frac{7}{3}x + 5$ 0

43. $\frac{1}{6}x - \frac{2}{3} = \frac{1}{3}(x - 4)$ 4

44. $\frac{1}{2}(x - 3) = \frac{1}{4}(3x - 1)$ -5

45. $\frac{7}{12}(x - 3) = -\frac{1}{3}x + 1$ 3

46. $\frac{5}{6} + \frac{1}{3}(x - 3) = \frac{1}{2}(x + 9)$ -28

47. $\frac{1}{7}(x + 5) - \frac{6}{14} = \frac{1}{2}(x + 3)$ $-\frac{17}{5}$

48. $\frac{1}{6}(8x + 3) = \frac{1}{2}(2x + 7)$ 9

49. $-\frac{1}{3}(2x - 6) = \frac{2}{3}(3 + x)$ 0

50. $\frac{7}{9}x + \frac{2}{3} = 2 + \frac{1}{3}x$ 3

Solve for the variable indicated.

51. Solve for y. $3x - y = 10$ $y = 3x - 10$

52. Solve for y. $5x + 2y + 7 = 0$ $y = \frac{-5x - 7}{2}$

53. Solve for r. $A = P(1 + rt)$ $r = \frac{A - P}{Pt}$

54. Solve for h. $A = 4\pi r^2 + 2\pi rh$ $h = \frac{A - 4\pi r^2}{2\pi r}$

55. Solve for p. $H = \frac{1}{3}(a + 2p + 3)$ $p = \frac{3H - a - 3}{2}$

56. **(a)** Solve for T. $C = \frac{WRT}{1000}$. $T = \frac{1000\,C}{WR}$

(b) Use your result to find T if $C = 0.36$, $W = 30$, and $R = 0.002$. $T = 6000$

57. **(a)** Solve for y. $5x - 3y = 12$ $y = \frac{5}{3}x - 4$

(b) Use your result to find y if $x = 9$. $y = 11$

58. **(a)** Solve for R. $I = \frac{E}{R}$ $R = \frac{E}{I}$

(b) Use your result to find R if $E = 100$ and $I = 20$. $R = 5$

2.6–2.7 *Solve the inequalities and graph the result.*

59. $7 - 2x \geq 4x - 5$ $\quad$ −2 −1 0 1 2 3 $\quad x \leq 2$

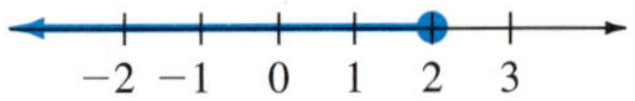

60. $2 - 3x \leq -5 + 4x$ $\quad$ −2 −1 0 1 2 3 $\quad x \geq 1$

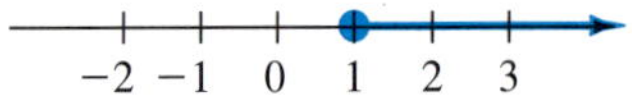

61. $2x - 3 + x > 5(x + 1)$ $\quad$ −5 −4 −3 −2 −1 0 $\quad x < -4$

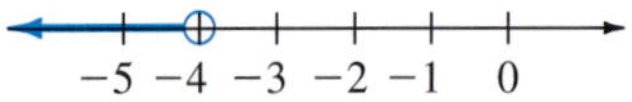

62. $-x + 4 < 3x + 16$ $\quad$ −5 −4 −3 −2 −1 0 $\quad x > -3$

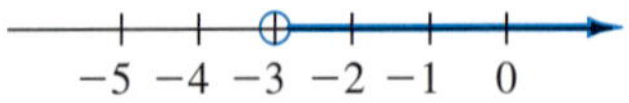

63. $4x \geq 2(12 - 2x)$ $\quad$ −1 0 1 2 3 4 $\quad x \geq 3$

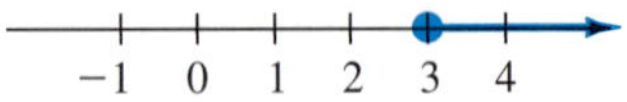

64. $5 - \frac{1}{2}x > 4$ $\quad$ −1 0 1 2 3 4 $\quad x < 2$

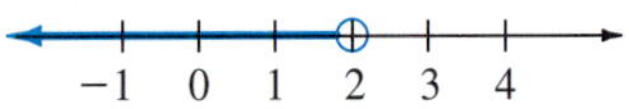

65. $2(x - 1) \geq 3(2 + x)$ $\quad$ −10 −9 −8 −7 −6 −5 $\quad x \leq -8$

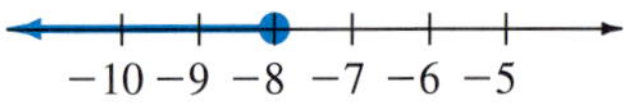

66. $3x + 5 - 7x \leq -2x - 1$ $\quad$ 0 1 2 3 4 5 $\quad x \geq 3$

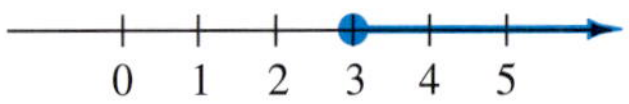

67. $-4x - 14 < 4 - 2(3x - 1)$ $\quad$ 7 8 9 10 11 12 $\quad x < 10$

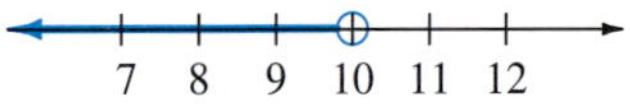

68. $3(x - 2) + 8 < 7x + 14$ $\quad$ −5 −4 −3 −2 −1 0 $\quad x > -3$

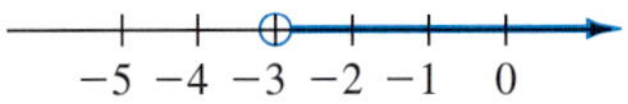

69. $\frac{1}{2}(2x + 3) > 10$ $\quad$ $\frac{15}{2}$ 8 $\frac{17}{2}$ 9 $\frac{19}{2}$ 10

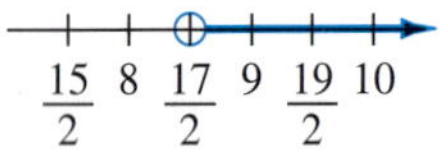

$x > \frac{17}{2}$

70. $\frac{1}{3}(x + 2) \leq \frac{1}{2}(3x - 5)$ $\quad$ $\frac{15}{7}$ $\frac{17}{7}$ $\frac{19}{7}$

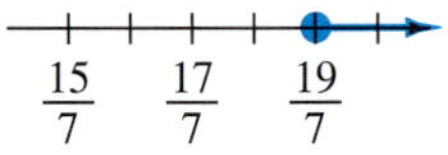

$x \geq \frac{19}{7}$

71. $4(2 - x) - (-5x + 1) \geq -8$ $\quad$ −15 −13 −11

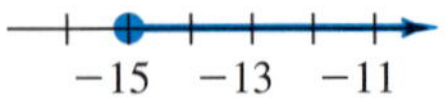

$x \geq -15$

72. $5(1 - x) < 3(x - 1) - 2(3 - x)$

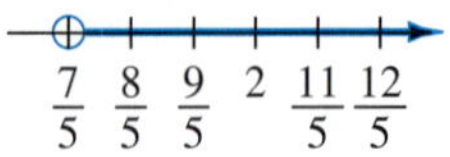

$\frac{7}{5}$ $\frac{8}{5}$ $\frac{9}{5}$ 2 $\frac{11}{5}$ $\frac{12}{5}$ $\quad x > \frac{7}{5}$

Use inequalities to solve the following problems.

73. The cost of a substitute teacher for one day at Central High school is \$70. Let n = the number of substitute teachers. Set up an inequality to determine how many times a substitute teacher may be hired if the monthly budget for substitute teachers is \$3220. What is the maximum number of substitute teachers that may be hired during the month?

$70n \leq 3220$

$n \leq 46$

74. The cost of hiring a temporary secretary for the day is \$85. Let n = the number of temporary secretaries. Set up an inequality to determine how may times a temporary secretary may be hired if the company budget for temporary secretaries is \$1445 per month. What is the maximum number of days a temporary secretary may be hired during the month?

$85n \leq 1445$

$n \leq 17$

Chapter 2 Test

Solve for the variable.

1. $2x + 0.8 = 5.0$

2. $9x - 3 = -6x + 27$

3. $2(2y - 3) = 4(2y + 2)$

4. $\frac{1}{2}y + 5 = \frac{1}{3}y$

5. $2(6 - x) = 3x + 2$

6. $0.8x + 0.18 - 0.4x = 0.3(x + 0.2)$

7. $\frac{2y}{3} + \frac{1}{5} - \frac{3y}{5} + \frac{1}{3} = 1$

8. $3 - 2y = 2(3y - 2) - 5y$

Solve for x.

9. $5(20 - x) + 10x = 165$

10. $2(x + 75) + 5x = 1025$

11. $-2(2 - 3x) = 76 - 2x$

12. $20 - (2x + 6) = 5(2 - x) + 2x$

1. $x = 2.1$

2. $x = 2$

3. $y = -\frac{7}{2}$

4. $y = -30$

5. $x = 2$

6. $x = -1.2$

7. $y = 7$

8. $y = \frac{7}{3}$

9. $x = 13$

10. $x = 125$

11. $x = 10$

12. $x = -4$

Solve for x.

13. $2x - 3 = 12 - 6x + 3(2x + 3)$

13. $x = 12$

14. $\frac{1}{3}x - \frac{3}{4}x = \frac{1}{12}$

14. $x = -\frac{1}{5}$

15. $\frac{3}{5}x + \frac{7}{10} = \frac{1}{3}x + \frac{3}{2}$

15. $x = 3$

16. $\frac{15x - 2}{28} = \frac{5x - 3}{7}$

16. $x = 2$

17. $\frac{1}{3}(7x - 1) + \frac{1}{4}(2 - 5x) = \frac{1}{3}(5 + 3x)$

17. $x = 18$

18. Solve for w. $A = 3w + 2P$

18. $w = \frac{A - 2P}{3}$

19. Solve for w. $\frac{2w}{3} = 4 - \frac{1}{2}(x + 6)$

19. $w = \frac{6 - 3x}{4}$

20. Solve for a. $A = \frac{1}{2}h(a + b)$

20. $a = \frac{2A - hb}{h}$

21. Solve for y. $5ax(2 - y) = 3axy + 5$

21. $y = \frac{10ax - 5}{8ax}$

Solve and graph the inequalities.

22. $4(x - 1) \geq 12x$

$-\frac{3}{2}$ $-\frac{1}{2}$

$x \leq -\frac{1}{2}$

23. $2 - 7(x + 1) - 5(x + 2) < 0$

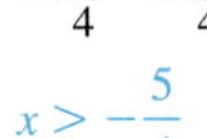

$-\frac{7}{4}$ $-\frac{5}{4}$ $-\frac{3}{4}$

$x > -\frac{5}{4}$

24. $5 + 8x - 4 < 2x + 13$

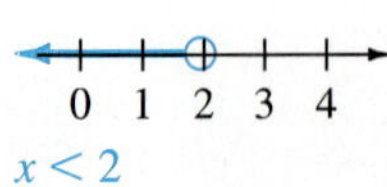

$x < 2$

25. $\frac{1}{4}x + \frac{1}{16} \leq \frac{1}{8}(7x - 2)$

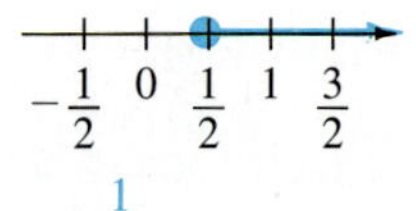

$x \geq \frac{1}{2}$

22. See graph

23. See graph

24. See graph

25. See graph

Cumulative Test for Chapters 0–2

Approximately one-half of this test covers the content of Chapters 0 and 1. The remainder covers the content of Chapter 2.

1. $\frac{6}{7} - \frac{2}{3}$

2. $1\frac{3}{4} + 2\frac{1}{5}$

3. $3\frac{1}{5} \div 1\frac{1}{2}$

4. Multiply. $\begin{array}{r} 1.23 \\ \times\ 0.56 \\ \hline \end{array}$

5. Divide. $0.144 \div 1.2$

6. What is 18% of 340?

7. Multiply. $(-3)(-5)(-1)(2)(-1)$

8. Collect like terms. $5ab - 7ab^2 - 3ab - 12ab^2 + 10ab - 9ab^2$

9. Simplify. $(5x)^2$

10. Simplify. $2\{x + y[3 - 2x(1 - 4y)]\}$

11. Solve for x. $3(5 - x) = 2x - 10$

12. Solve for x. $\frac{1}{3}(x + 5) = 2x - 5$

13. Solve for y. $\frac{2y}{3} - \frac{1}{4} = \frac{1}{6} + \frac{y}{4}$

1. $\frac{4}{21}$

2. $\frac{79}{20}$ or $3\frac{19}{20}$

3. $\frac{32}{15}$ or $2\frac{2}{15}$

4. 0.6888

5. 0.12

6. 61.2

7. 30

8. $12ab - 28ab^2$

9. $25x^2$

10. $2x + 6y - 4xy + 16xy^2$

11. $x = 5$

12. $x = 4$

13. $y = 1$

14. $y = \frac{3x + 2}{7}$

15. $b = \frac{3H - 8a}{2}$

16. $t = \frac{I}{Pr}$

17. $a = \frac{2A}{h} - b$

18. See graph

19. See graph

20. See graph

21. See graph

22. See graph

23. 92

14. Solve for y. $3x - 7y + 2 = 0$

15. Solve for b. $H = \frac{2}{3}(b + 4a)$

16. Solve for t. $I = Prt$

17. Solve for a. $A = \frac{ha}{2} + \frac{hb}{2}$

In Problems 18 through 22, solve and graph the inequalities.

18. $-6x - 3 < 2x - 10x + 7$

2 3 4 5

$x < 5$

19. $\frac{1}{2}(x - 5) \geq x - 4$

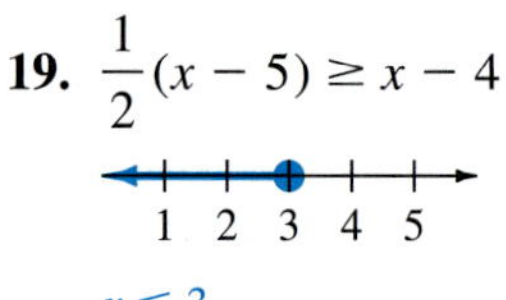

$x \leq 3$

20. $4(2 - x) > 1 - 5x - 8$

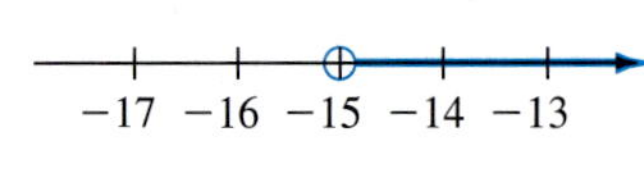

$x > -15$

21. $x + \frac{5}{9} \leq \frac{1}{3} + \frac{7}{9}x$

−2 −1 0 1 2

$x \leq -1$

22. $4 - 16x \leq 6 - 5(3x - 2)$

−14 −13 −12 −11 −10

$x \geq -12$

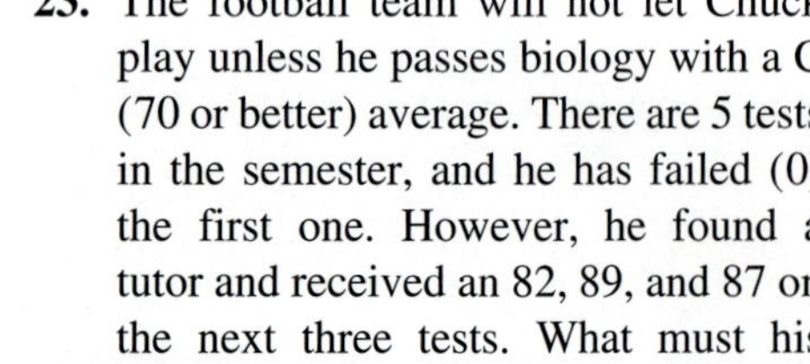

23. The football team will not let Chuck play unless he passes biology with a C (70 or better) average. There are 5 tests in the semester, and he has failed (0) the first one. However, he found a tutor and received an 82, 89, and 87 on the next three tests. What must his minimum score be on the last test to pass the course and play football?
92 or higher

CHAPTER 3

Solving Applied Problems

A common activity is the purchase of a new car by taking out an automobile loan. However, often the finance charges for the loan are greater than what the purchaser expected them to be. If you took out a new car loan, would you know how much the loan is really costing you? Turn to page 213 and study the Putting Your Skills to Work in order to find out.

Pretest Chapter 3

If you are familiar with the topics in this chapter, take this test now. Check your answers with those in the back of the book. If an answer was wrong or you couldn't do a problem, study the appropriate section of the chapter.

If you are not familiar with the topics in this chapter, don't take this test now. Instead, study the examples, work the practice problems, and then take the test. This test will help you to identify those concepts that you have mastered and those that need more study.

Section 3.1

Write an algebraic expression. Use the variable x to represent the unknown number.

1. Twice a number is then decreased by thirty.

2. Forty percent of a number is increased by 150.

Section 3.2

Translate the information in each word problem into an equation, and solve.

3. A number is multiplied by seven and then is increased by nine. The result is forty-four. Find the original number.

4. There are three numbers with these properties: The first number is triple the second number. A third number is three more than the second number. The sum of the three numbers is twenty-three. Find the three numbers.

Section 3.3

5. One side of a triangular piece of land is 3 meters less than double the second side. The third side is 4 meters more in length than the first side. The perimeter of the triangle is 41 meters. Find the length of each side of the triangle.

6. Charlene took three packages to the post office to mail. The second package was $3\frac{1}{2}$ pounds less than the first package. The third package was 2 pounds less than the first package. The total weight of the three packages was 17 pounds. How much did each package weigh?

7. The length of a rectangle is 13 meters longer than twice the width. The perimeter of the rectangle is 74 meters. Find the dimensions of the rectangle.

Section 3.4

8. Two investments were made totaling $1000. The investments gain interest once each year. The first investment yielded 12% interest after one year. The second investment yielded 9% interest. The total interest for both investments was $102. How much was invested at each interest rate?

Answers

1. $2x - 30$

2. $0.40x + 150$

3. 5

4. 1st number = 12
2nd number = 4
3rd number = 7

5. $x = 8.6$ m
$2x - 3 = 14.2$ m
$2x + 1 = 18.2$ m

6. $x = 7.5$ pounds
$x - 3.5 = 4$ pounds
$x - 2 = 5.5$ pounds

7. width = 8 m
length = 29 m

8. $400 at 12%
$600 at 9%

9. The population of Springville has grown 11% from the population level of five years ago. The town now has population of 24,420 people. What was the population five years ago?

10. Enrique has $3.35 in change in his pocket. He has only nickels, dimes, and quarters. He has four more dimes than nickels. He has three more quarters than dimes. How many coins of each type does he have?

Section 3.5

11. Find the area of a triangle whose base is 16 inches and whose altitude is 9 inches.

12. Find the volume of a cylinder whose height is 12 meters and whose radius is 3 meters. (Use 3.14 for π.)

13. Find the circumference of a circle whose radius is 4 centimeters. (Use 3.14 for π.)

14. A stainless-steel plate in the shape of a trapezoid is to be manufactured according to the following diagram. The cost to manufacture the plate is $1.85 per square centimeter. What is the cost to manufacture this steel plate?

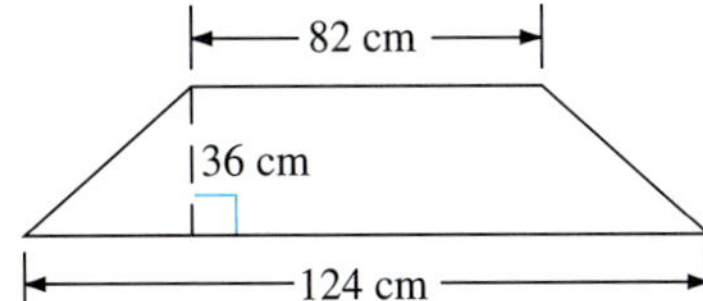

Section 3.6

Translate each phrase into an inequality.

15. The height must be no more than 72 inches.

16. The rent will be at least $450 a month.

17. The average number must be less than 15.

18. The cost of the car is more than $8900.

Solve.

19. Bob makes 3% on the amount of his sales. If he wants to earn more than $900, how much will he need to sell?

20. Lauren is making a rectangular pen. The width is 11 feet. The perimeter cannot be more than 54 feet. What is the length?

9. 22,000 people

10. 3 nickels
7 dimes
10 quarters

11. 72 square inches

12. 339.12 cubic meters

13. 25.12 centimeters

14. $6859.80

15. $h \leq 72$

16. $r \geq 450$

17. $a < 15$

18. $c > 8{,}900$

19. More than $30,000

20. Length $\leq$ 16 feet

3.1 Translate English Phrases into Algebraic Expressions

MathPro

Video 8

SSM

After studying this section, you will be able to:

1. *Write an algebraic expression for a simple English phrase.*
2. *Write an algebraic expression for two or more quantities that are being compared.*

1 *Translating English Phrases into Algebra*

One of the most useful applications of algebra is solving word problems. One of the first steps in solving word problems is translating the conditions of the problem into algebra. In this section we show you how to translate common English phrases into algebraic symbols. This process is similar to translating between languages like Spanish and French.

There are several English phrases to describe the operation of addition. If we represent an unknown number by the variable x, all of the following phrases can be translated into algebra as $x + 3$.

ENGLISH PHRASES DESCRIBING ADDITION	ALGEBRAIC EXPRESSION	DIAGRAM
Three *more than* a number The *sum of* a number and three A number *increased by* three Three is *added to* a number. Three *greater than* a number A number *plus* three	$\mathbf{x + 3}$	x, 3; $x + 3$

In a similar way we can use algebra to express English phrases that describe the operations of subtraction, multiplication, and division.

ENGLISH PHRASES DESCRIBING SUBTRACTION	ALGEBRAIC EXPRESSION	DIAGRAM
A number *decreased by* four Four *less than* a number Four is *subtracted from* a number. Four *smaller than* a number Four *fewer than* a number A number *diminished by* four A number *minus* 4 The *difference between* a number and four	$\mathbf{x - 4}$	x; 4; $x - 4$

ENGLISH PHRASES DESCRIBING MULTIPLICATION	ALGEBRAIC EXPRESSION	DIAGRAM
Double a number *Twice* a number The *product* of two and a number Two *of* a number Two *times* a number	$\mathbf{2x}$	x, x; $2x$

ENGLISH PHRASES DESCRIBING DIVISION	ALGEBRAIC EXPRESSION	DIAGRAM
A number *divided by* five One-*fifth* of a number	$\mathbf{\frac{x}{5}}$	x; $\frac{x}{5}$

It is important to review these phrases and to see how they are represented as algebraic expressions. Often other words are used in English instead of the word "number." We can use a variable, such as x, here also.

EXAMPLE 1

ENGLISH PHRASE	ALGEBRAIC EXPRESSION
(a) A *quantity* is increased by five.	$x + 5$
(b) Double the *value*	$2x$
(c) One-third of the *weight*	$\frac{x}{3}$ or $\frac{1}{3}x$
(d) Twelve *more than* a number	$x + 12$
(e) Seven *less than* a number	$x - 7$

Note that the algebraic expression for "seven less than a number" does not follow the order of the words in the English phrase. The variable or expression that follows the words *less than* always comes first.

seven less than x

$x - 7$

The variable or expression that follows the words *more than* always comes first.

TEACHING TIP Students who have had algebra before very often have very bad memories of "Word Problems." This is a time to tell them that they can learn to do word problems without anxiety if they master a few basic steps. Taking the time to translate English phrases into algebraic expressions will prove valuable in solving word problems.

Practice Problem 1 Write each English phrase as an algebraic expression.

(a) Four more than a number **(b)** Triple a value

(c) Eight less than a number **(d)** One-fourth of a height ■

Finally, we see that more than one operation can be described in an English phrase. Sometimes parentheses must be used to make clear which operation is done first.

EXAMPLE 2

ENGLISH PHRASE	ALGEBRAIC EXPRESSION
(a) Seven more than double a number	$2x + 7$
	*Note that these are **not** the same.*
(b) The value of the number is increased by seven and then doubled.	$2(x + 7)$

Practice Problem 2 Write each English phrase as an algebraic expression.

(a) Eight more than triple a number

(b) A number is increased by eight and then it is tripled. ■

TEACHING TIP After presenting Examples 1, 2, and 3, ask students to do the following as an Additional Class Exercise. Write an algebraic expression for each of the following (use the letter x to represent the number):

A. Eight more than a number.

B. Two fewer than a number.

C. Triple a number.

D. One-fourth of a number.

E. Three less than double a number.

Sol. A. $x + 8$ B. $x - 2$ C. $3x$ D. $(\frac{1}{4})x$ E. $2x - 3$

EXAMPLE 3

Write the following English phrase as an algebraic expression.

A number increased by five is subtracted from three times the same number.

We let x represent the number. We can visualize it this way:

A number	increased by	5
$(x$	$+$	$5)$

Therefore, "a number increased by 5" is $(x + 5)$.

Now we must subtract the entire quantity $(x + 5)$ from three times the number.

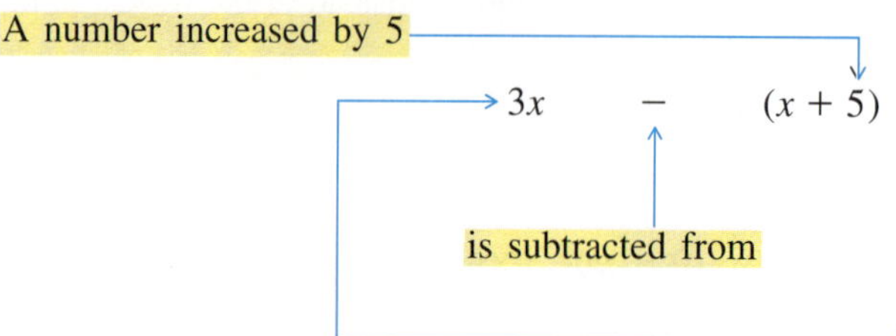

Compare the English phrase with the algebraic expression. Notice that the part that follows the word *from* in the English phrase comes first in the algebraic expression. It is also important to note the use of parentheses here. $3x - x + 5$ would be wrong.

Practice Problem 3 Write each English phrase as an algebraic expression.

(a) Four less than three times a number

(b) Two-thirds of the sum of a number and five ■

2 *When Two or More Quantities Are Compared*

Often in a word problem two or more quantities are described in terms of another. We will want to use a variable to represent one quantity and then write an algebraic expression using *the same variable* to represent the other quantity. Which quantity should we let the variable represent? We usually let the variable represent the quantity that is the basis of comparison: the quantity that the others are being *compared to.*

EXAMPLE 4 Use a variable and an algebraic expression to describe the two quantities in the English phrase ''Mike's salary is \$2000 more than Fred's salary.''

The two quantities that are being compared are Mike's salary and Fred's salary. Since Mike's salary is being *compared to* Fred's salary, we let the variable represent Fred's salary. The choice of the letter f helps us to remember that the variable represents Fred's salary.

Let f = Fred's salary

Then $f + \$2000$ = Mike's salary *Since Mike's salary is* \$2000 *more than Fred's.*

Practice Problem 4 Use a variable and an algebraic expression to describe the two quantities in the English phrase ''Marie works 17 hours per week less than Ann.'' ■

EXAMPLE 5 The length of a rectangle is 3 meters shorter than twice the width. Use a variable and an algebraic expression to describe the length and the width. Draw a picture.

The length of the rectangle is being *compared to* the width. Use the letter w for width.

Let w = the width

3 meters shorter than twice the width

Then $2w - 3$ = the length

Draw a picture and use the algebraic expressions to identify the length and the width.

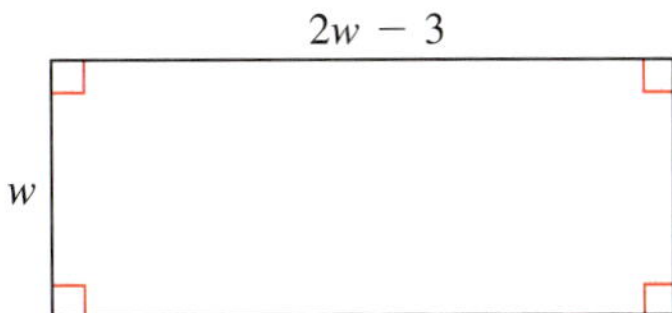

Practice Problem 5 The length of a rectangle is 5 meters longer than double the width. Use a variable and an algebraic expression to describe the length and the width. Draw a picture. ■

EXAMPLE 6 The first angle of a triangle is triple the second angle. The third angle of a triangle is 12° more than the second angle. Describe each angle algebraically. Draw a diagram.

Since the first and third angles are described in terms of the second angle, we let the variable represent the number of degrees in the second angle.

Let s = the number of degrees in the second angle

$3s$ = the number of degrees in the first angle

$s + 12$ = the number of degrees in the third angle

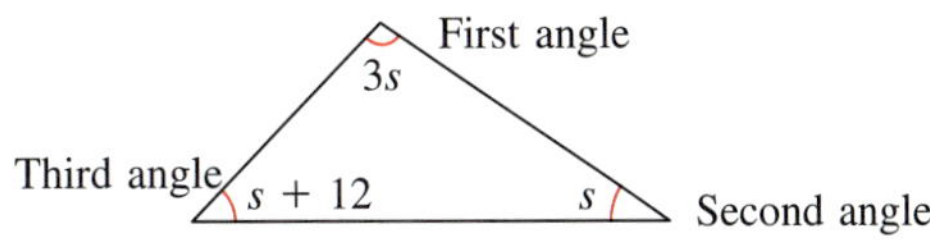

TEACHING TIP After presenting Example 6, ask students to do the following as an Additional Class Exercise. Describe each salary in terms of the letter b.

Walter's salary is \$500 more than Bob's salary.

Jim's salary is \$1700 less than Bob's salary.

Write an expression for Walter's salary, Jim's salary, and Bob's salary.

Sol. Let b = Bob's salary
Let $b + 500$ = Walter's salary
Let $b - 1700$ = Jim's salary

Practice Problem 6 The first angle of a triangle is 16° less than the second angle. The third angle is double the second angle. Describe each angle algebraically. Draw a diagram. ■

Some comparisons will involve fractions.

EXAMPLE 7 A theater manager was examining the records of attendance for last year. The number of people attending the theater in January was one-half of the number of people attending the theater in February. The number of people attending the theater in March was three-fifths of the number of people attending the theater in February. Use algebra to describe the attendance each month.

What are we looking for? The *number of people* who attended the theater *each month*. The basis of comparison is February. That is where we begin.

Let f = the number of people who attended in February

$\frac{1}{2}f$ = the number of people who attended in January

$\frac{3}{5}f$ = the number of people who attended in March

Practice Problem 7 The college dean noticed that in the spring the number of students on campus was two-thirds of the number of students on campus in the fall. She also noticed that in the summer the number of students on campus was one-fifth the number of students on campus in the fall. Use algebra to describe the number of students on campus in each of these three time periods. ■

3.1 Exercises

Verbal and Writing skills

Write an algebraic expression for each quantity. Let x represent the unknown value.

1. A quantity increased by 5
$x + 5$

2. Nine greater than a number
$x + 9$

3. Five less than a value
$x - 5$

4. Twelve smaller than a quantity
$x - 12$

5. One-third of a number
$\frac{x}{3}$

6. Three-fourths of a number
$\frac{3x}{4}$

7. Triple a number
$3x$

8. Double a number
$2x$

9. One-eighth of a quantity
$\frac{x}{8}$

10. The product of twelve and a number
$12x$

11. Five more than double a number
$2x + 5$

12. Eight less than triple a number
$3x - 8$

13. Two-sevenths of a value decreased by three
$\frac{2}{7}x - 3$

14. Double a quantity increased by four
$2x + 4$

15. Five times the sum of a number and twelve
$5(x + 12)$

16. One-fifth of the sum of a number and eight
$\frac{1}{5}(x + 8)$

17. Five less than half of a number
$\frac{1}{2}x - 5$

18. The sum of four and one-third of a number
$4 + \frac{1}{3}x$

19. Three times a number added to one-half of the same number
$3x + \frac{1}{2}x$

20. Twice a number added to one-third of the same number
$2x + \frac{1}{3}x$

21. Four times a number decreased by one-half of the same number
$4x - \frac{1}{2}x$

Write an algebraic expression for each of the quantities that is being compared.

22. Wally's salary is \$1400 more than George's salary.
$g + 1400$ = Wally's salary
g = George's salary

23. Alicia works 15 hours per week more than Barbara.
$b + 15$ = no. of hours Alicia works
b = no. of hours Barbara works

24. The value of a share of IBM stock on that day was \$74.50 more than the value of a share of AT&T stock.
x = value of a share of AT&T stock
$x + 74.50$ = value of a share of IBM stock

25. The annual income from Dr. Smith's mutual fund was \$833 less than the annual income from her retirement fund. x = income from retirement fund
$x - 833$ = income from mutual fund

26. The length of the rectangle is 7 inches more than double the width.
$2w + 7$ = length of rectangle
w = width of rectangle

27. The length of the rectangle is 3 meters more than triple the width.
$3w + 3$ = length of rectangle
w = width of rectangle

28. The attendance on Monday was 1185 people more than on Tuesday. The attendance on Wednesday was 365 people fewer than on Tuesday.

$t + 1185$ = attendance on Monday
t = attendance on Tuesday
$t - 365$ = attendance on Wednesday

29. The attendance on Friday was 1600 people more than on Saturday. The attendance on Thursday was 783 people fewer than on Saturday.

s = attendance on Saturday
$s + 1600$ = attendance on Friday
$s - 783$ = attendance on Thursday

30. The first angle of a triangle is 16 degrees less than the second angle. The third angle of a triangle is double the second angle.

1st angle $= s - 16$; 2nd angle $= s$; 3rd angle $= 2s$

31. The first angle of a triangle is 19 degrees more than the third angle. The second angle is triple the third angle.

1st angle $= t + 19$; 2nd angle $= 3t$; 3rd angle $= t$

32. The attendance on Tuesday was 350 people more than double the attendance on Monday.

$2m + 350$ = attendance on Tuesday
m = attendance on Monday

33. The attendance on Friday was 200 people less than triple the attendance on Thursday.

$3t - 200$ = attendance on Friday
t = attendance on Thursday

34. The first angle of a triangle is triple the second angle. The third angle of a triangle is 14 degrees less than the second angle.

1st angle $= 3x$; 2nd angle $= x$; 3rd angle $= x - 14$

35. The cost of Hiro's biology book was \$13 more than the cost of his history book. The cost of his English book was \$27 less than the cost of his history book.

x = the cost of his history book in dollars
$x + 13$ = the cost of his biology book in dollars
$x - 27$ = the cost of his English book in dollars

Applications

Use algebra to describe the situation in each word problem.

36. A local car dealer observed that car sales in August were one-half of the car sales in July. She also observed that the car sales in September were two-thirds of the car sales in July. Describe the number of cars sold in each of these three months.

July $= x$; August $= \frac{1}{2}x$; September $= \frac{2}{3}x$

37. Computer City found that the sales of computers in April were three-fourths of the sales of computers in March. The sales of computers in February were five-sixths of the sales of computers in March. Describe the number of computers sold in each of these three months.

February $= \frac{5}{6}x$; March $= x$; April $= \frac{3}{4}x$

38. The mass of Jupiter is 32.5 more than triple the mass of Saturn. Write an expression for the mass of each planet.

x = the mass of Saturn
$3x + 32.5$ = the mass of Jupiter

39. The atomic weight of lead is 51.2 greater than triple the atomic weight of chromium. Write an expression for the atomic weight of each element.

x = the atomic weight of chromium
$3x + 51.2$ = the atomic weight of lead

Cumulative Review Problems

Solve for the variable in each equation.

40. $x + 2(x + 2) = 55$ $\quad x = 17$

41. $2x + 2(3x - 4) = 72$ $\quad x = 10$

42. $5(x - 8) = 13 + x - 5$ $\quad x = 12$

43. $6(w - 1) - 3(2 + w) = 9$ $\quad w = 7$

3.2 Use Equations to Solve Word Problems

MathPro

Video 8

SSM

After studying this section, you will be able to:

1 *Solve number problems.*
2 *Solve applied word problems.*
3 *Use a formula to solve a word problem.*

The skills you have just developed in translating English phrases into algebraic expressions will enable you to solve a variety of word problems. Solving word problems can be challenging. You need to be organized. The following steps may help.

1. *Understand the problem.*
   (a) Read the word problem carefully to get an overview.
   (b) Determine what information you will need to solve the problem.
   (c) Draw a sketch. Label it with the known information. Determine what needs to be found.
   (d) Choose a variable to represent one unknown quantity.
   (e) If necessary, represent other unknown quantities in terms of that very same variable.
2. *Write an equation.*
   (a) Look for key words to help you to translate the words into algebraic symbols and expressions.
   (b) Use a given relationship in the problem or an appropriate formula in order to write an equation.
3. *Solve the equation and state the answer.*
4. *Check.*
   (a) Check the solution in the original equation.
   (b) Be sure the solution to the equation answers the question in the word problem. You may need to do some additional calculations if it does not.

1 *Number Problems*

EXAMPLE 1 Two-thirds of a number is eighty-four. What is the number?

1. *Understand the problem.*
   Draw a sketch.

   Let x = the unknown number.

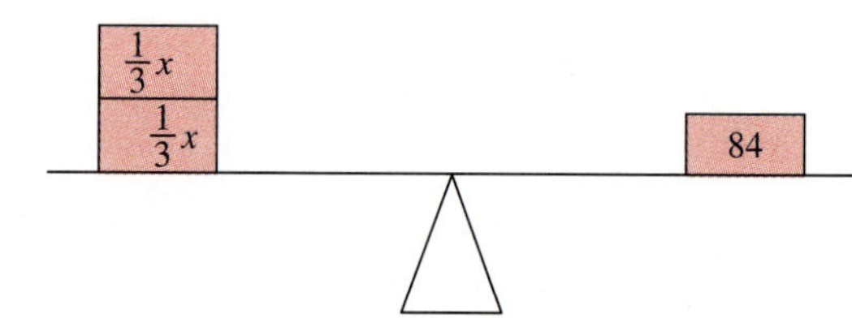

2. *Write an equation.*

Two-thirds of a number is eighty-four.

$$\frac{2}{3}x = 84$$

3. *Solve the equation and state the answer.*

$$\frac{2}{3}x = 84$$

$$3\left(\frac{2}{3}x\right) = 3(84) \qquad \textit{Multiply both sides of the equation by 3.}$$

$$2x = 252$$

$$\frac{2x}{2} = \frac{252}{2} \qquad \textit{Divide both sides by 2.}$$

$$x = 126$$

The number is 126.

4. *Check.* Is two-thirds of 126 eighty-four?

$$\frac{2}{3}(126) \stackrel{?}{=} 84$$

$$84 = 84 \checkmark$$

Practice Problem 1 Three-fourths of a number is negative eighty-one. What is the number? ■

EXAMPLE 2 Five more than six times a number is three hundred and five. Find the number.

1. *Understand the problem.*
Read the problem carefully. You may not need to draw a sketch.

Let $x =$ the unknown number.

2. *Write an equation.*

$$\underbrace{\text{Five more than}}_{5\ +}\ \underbrace{\text{six times a number}}_{6x}\ \underbrace{\text{is}}_{=}\ \underbrace{\text{three hundred and five.}}_{305}$$

3. *Solve the equation and state the answer.*
You may want to rewrite the equation to make it easier to solve.

$$6x + 5 = 305$$

$$6x + 5 - 5 = 305 - 5 \qquad \textit{Subtract 5 from both sides.}$$

$$6x = 300$$

$$\frac{6x}{6} = \frac{300}{6} \qquad \textit{Divide both sides by 6.}$$

$$x = 50$$

The number is 50.

4. *Check.* Is five more than six times 50 three hundred and five?

$$6(50) + 5 \stackrel{?}{=} 305$$

$$305 = 305$$

Practice Problem 2

Two less than triple a number is forty-nine. Find the number. ■

TEACHING TIP After presenting Examples 2 and 3 to students, ask them to do the following problem as an Additional Class Exercise. The smaller of two numbers is 3 less than one-half of the larger number. The sum of the numbers is -15. What is each number?

Sol. Let $x =$ the larger number. Let $0.5x - 3 =$ the smaller number. (They may use $\frac{1}{2}$ instead of 0.5.)

$$x + 0.5x - 3 = -15$$
$$1.5x - 3 = -15$$
$$1.5x = -12$$
$$x = -8$$
$$0.5x - 3 = -4 - 3 = -7$$

The larger number is -8 and the smaller number is -7.

EXAMPLE 3 The larger of two numbers is three more than twice the smaller. The sum of the numbers is thirty-nine. Find each number.

1. *Understand the problem.*
Read the problem carefully. The problem refers to *two* numbers. We must write an algebraic expression for *each number* before writing the equation. The larger number is being compared to the smaller number.

Let $s =$ the smaller number

$$\underbrace{2s + 3}_{\text{three more than twice the smaller number}} = \text{the larger number}$$

2. *Write an equation.*
Write an equation for the sum of the two numbers.

The sum of the numbers is thirty-nine.

$$s + (2s + 3) \quad = \quad 39$$

3. *Solve the equation.*

$$s + (2s + 3) = 39$$
$$3s + 3 = 39 \qquad \textit{Collect like terms.}$$
$$3s = 36 \qquad \textit{Subtract 3 from each side.}$$
$$s = 12 \qquad \textit{Divide both sides by 3.}$$

4. *Check.*

$$12 + [2(12) + 3] \stackrel{?}{=} 39$$
$$39 = 39 \checkmark$$

The solution checks, but have we solved the word problem? We need to find *each* number. 12 is the smaller number. Substitute 12 into the expression $2s + 3$ to find the larger number.

$$2s + 3 = ?$$
$$2(12) + 3 = 27$$

The smaller number is 12. The larger number is 27.

Practice Problem 3 Consider two numbers. The second number is twelve less than triple the first number. The sum of the two numbers is twenty-four. Find each number. ■

To facilitate understanding the problem we will use a ''Mathematical Blueprint'' similar to the one we used in Section 0.7. This blueprint format is a simple way to organize facts, determine what to set variables equal to, and select a method or approach that will assist you in finding the desired quantity. You will find using this form helpful, particularly in those cases when you read through a word problem and mentally say to yourself, ''Now where do I begin?'' You begin by responding to the headings of the Blueprint. Soon a pattern for solving the problem will emerge.

MATHEMATICS BLUEPRINT FOR PROBLEM SOLVING

Gather the facts.	Label the variable.	Basic formula or equation.	Key points to remember.

EXAMPLE 4 The mean annual snowfall in Juneau, Alaska, is 105.8 inches. This is 20.2 inches less than three times the annual snowfall in Boston. What is the annual snowfall in Boston?

MATHEMATICS BLUEPRINT FOR PROBLEM SOLVING

Gather the facts.	Label the variable.	Basic formula or equation.	Key points to remember.
Snowfall in Juneau is 105.8 inches. This is 20.2 inches less than three times the snowfall in Boston.	We do not know the snowfall in Boston. Let b = annual snowfall in Boston. Then $3b - 20.2$ = annual snowfall in Juneau.	Set $3b - 20.2$ equal to 105.8, which is the snowfall in Juneau.	All measurements of snowfall are recorded in inches.

Juneau's snowfall is 20.2 less than three times Boston's snowfall.

$$105.8 = 3b - 20.2$$

Solve the equation and state the answer.
You may want to rewrite the equation to make it easier to solve.

$$3b - 20.2 = 105.8$$

$$3b = 126 \quad \textit{Add 20.2 to both sides.}$$

$$b = 42 \quad \textit{Divide both sides by 3.}$$

The annual snowfall in Boston is 42 inches.

Check. Reread the word problem. Work backward.

Three times 42 is 126.

126 less 20.2 is 105.8.

Is this the annual snowfall in Juneau? ✓

Practice Problem 4 The maximum 24-hour rainfall for Utah occurred at Bug Point on 9/5/70. It was 2 inches less than that recorded at Elk Point, South Dakota, on 9/10/70. If the combined rainfall was 14 inches, how much rainfall was recorded for each location? ■

Some word problems require a simple translation of the facts. Others require a little more detective work. You will not always need to use the Mathematics Blueprint to solve every word problem. As you gain confidence in problem solving, you will no doubt leave out some of the steps. We suggest that you use the procedure when you find yourself on unfamiliar ground. It is a powerful organizational tool.

3 *Formulas*

Sometimes the relationship between two quantities is so well understood that we have developed a formula to describe that relationship. We have already done some work with formulas in Section 2.5. The following examples show how you can use a formula to solve a word problem.

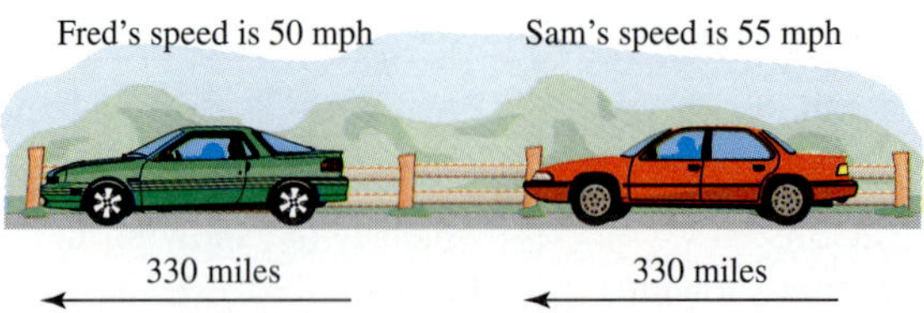

EXAMPLE 5 Two people travel in separate cars. They each travel a distance of 330 miles on an interstate highway. To maximize fuel economy, Fred travels at exactly 50 mph. Sam travels at exactly 55 mph. How much time did the trip take each person?

MATHEMATICS BLUEPRINT FOR PROBLEM SOLVING

Gather the facts.	Label the variable.	Basic formula or equation.	Key points to remember.
Each person drives 330 miles. Fred drives at 50 mph. Sam drives at 55 mph.	Time is the unknown quantity for each driver. Use subscripts to denote different values of t. t_f = Fred's time t_s = Sam's time	distance = (rate) × (time) or $d = rt$	The time is expressed in hours.

Substitute the known values into the formula and solve for t.

$$d = rt \qquad\qquad d = rt$$

$$330 = 50t_f \qquad\qquad 330 = 55t_s$$

$$6.6 = t_f \qquad\qquad 6 = t_s$$

It took Fred 6.6 hours to drive 330 miles. It took Sam 6 hours to drive 330 miles.

Check. Is this reasonable? Yes, you would expect Fred to take longer to drive the same distance because Fred is driving at a lower rate of speed.

Note: You may wish to express 6.6 hours in hours and minutes. To change 0.6 hours to minutes, proceed as follows:

$$0.6 \cancel{\text{hour}} \times \frac{60 \text{ minutes}}{1 \cancel{\text{hour}}} = (0.6)(60) \text{ minutes} = 36 \text{ minutes}$$

Thus, Fred drove for 6 hours and 36 minutes.

Practice Problem 5 Sarah left the city to visit her aunt and uncle, who live in a rural area north of the city. She traveled the 220-mile trip in 4 hours. When she went home she took a slightly longer route, which measured 225 miles on the car odometer. The return trip took 4.5 hours.

(a) What was her average speed on the trip leaving the city?

(b) What was her average speed on the return trip?

(c) On which trip did she travel faster and by how much? ■

EXAMPLE 6 A teacher told Melinda that she had a course average of 78 based on her six math tests. When she got home Melinda found five of her tests. She had scores of 87, 63, 79, 71, and 96 on the five tests. She could not find her sixth test. What score did she obtain on that test?

MATHEMATICS BLUEPRINT FOR PROBLEM SOLVING

Gather the facts.	Label the variable.	Basic formula or equation.	Key points to remember.
Her five test scores are 87, 63, 79, 71, and 96. Her course average is 78.	We do not know the score Melinda received on her sixth test. Let x = the score on the sixth test.	$\text{Average} = \dfrac{\text{Sum of scores}}{\text{Number of scores}}$	Since there are six test scores we will need to divide the sum by 6.

We now write the equation you would use to find the average and then solve for x.

$$\frac{87 + 63 + 79 + 71 + 96 + x}{6} = 78$$

$$\frac{396 + x}{6} = 78$$ *Add the numbers in the numerator.*

$$6\left(\frac{396 + x}{6}\right) = 6(78)$$ *Multiply both sides of the equation by 6 to remove the fraction.*

$$396 + x = 468$$ *Simplify.*

$$x = 72$$ *Subtract* 396 *from both sides to find x.*

Melinda's score on the sixth test was 72.

Check. To verify that this is correct, we see if the average of the 6 tests is 78.

$$\frac{87 + 63 + 79 + 71 + 96 + 72}{6} \stackrel{?}{=} 78$$

$$\frac{468}{6} \stackrel{?}{=} 78$$

$$78 = 78 \checkmark$$

The problem checks. We know that the score on the sixth test was 72.

Practice Problem 6 Barbara's math course has four tests and one final exam. The final exam counts as much as two tests. Barbara has test scores of 78, 80, 100, and 96 on her four tests. What grade does she need on the final exam if she wants to have a 90 average for the course? ■

3.2 Exercises

Solve each word problem. Check your solution.

1. What number added to 125 is 170?

$x + 125 = 170$
$x = 45$

2. Nine times what number is 756?

$9x = 756$
$x = 84$

3. A number divided by eight is 296. What is the number?

$\frac{x}{8} = 296$
$x = 2368$

4. Eighteen less than a number is 23. What is the number?

$x - 18 = 23$
$x = 41$

5. Fifteen greater than a number is 305. Find the number.

$x + 15 = 305$
$x = 290$

6. Twice a number is one. What is the number?

$2x = 1$
$x = \frac{1}{2}$

7. A number is doubled and then increased by three. The result is eighty-one. What is the original number?

$2x + 3 = 81$
$x = 39$

8. Nine less than twelve times a number is one hundred seventeen. Find the original number.

$12x - 9 = 117$
$x = 10\frac{1}{2}$

9. When six is subtracted from one-third of a number, the result is 11. What is the original number?

$\frac{x}{3} - 6 = 11$
$x = 51$

10. A number is tripled and then increased by seven. The result is seventy-nine. What is the original number?

$3x + 7 = 79$
$x = 24$

11. Four less than twice a number is the same as four times the number. Find the number.

$2x - 4 = 4x$
$-4 = 2x$
$-2 = x$

12. Triple a number added to eight gives the same result as the number divided by three. What is the number?

$3x + 8 = \frac{x}{3}$
$9x + 24 = x$
$24 = -8x$
$-3 = x$

13. One number is 12 more than another number. The sum of the two numbers is 156. Find each number.

$x + (x + 12) = 156$
$x = 72;\ x + 12 = 84$

14. One number is 15 less than another number. The sum of the two numbers is 177. Find each number.

$x + (x - 15) = 177$
$x = 96;\ x - 15 = 81$

15. The sum of two numbers is -2. One number is eight more than the other number. Find each number.

$8 + x + x = -2$
$2x = -10$
$x = -5$
One number is 3. The other number is -5.

16. A number is two thirds of another number. Their sum is twenty. Find each number.

$x + \frac{2}{3}x = 20$
$3x + 2x = 60$
$5x = 50$
$x = 12$
One number is 8. The other number is 12.

17. One-fourth of a number, one-eighth of that number, and one-fifth of that number are added. The result is forty-six. What is the original number?

$\frac{x}{4} + \frac{x}{8} + \frac{x}{5} = 46$
$10x + 5x + 8x = 40(46)$
$x = 80$

18. One-eighth of a number, one-third of that number, and one-sixth of that number are added. The result is sixty. What is the original number?

$\frac{x}{8} + \frac{x}{3} + \frac{x}{6} = 60$
$3x + 8x + 4x = 60(24)$
$15x = 1440$
$x = 96$

19. One number is 41.8632 greater than another. The numbers have a sum of 196.0578. Find each number.

$x + x + 41.8632 = 196.0578$
$x = 77.0973$
$x + 41.8632 = 118.9605$

20. The sum of three numbers is 88.9635. The first number is double the second number. The third number is 6.3254 greater than the second number. Find each number.

1st number = 41.31905
2nd number = 20.659525
3rd number = 26.984925

Applications

Solve each word problem. Check to see if your answer is reasonable.

21. A Harley Davidson motorcycle shop maintains an inventory of four times as many new bikes as used bikes. If there are 60 new bikes, how many used bikes are now in stock?

$4x = 60$
$x = 15$
There are 15 used bikes in stock.

22. In 1992 the number of U.S. citizens living in the Netherlands was half of the number of U.S. citizens living in Greece. At that time the number of U.S. citizens living in the Netherlands was 35,000. How many U.S. citizens were living in Greece at that time?

$\frac{1}{2}x = 35{,}000$
$x = 70{,}000$
There were 70,000 U.S. citizens living in Greece.

23. At Center City Ford, the monthly rental on a Ford Aspire is \$320. The cost of a new car is \$8000. In how many months will the rental equal the cost of the car?

$320x = 8000$
$x = 25$ months

24. The cost of renting a subcompact car at Sudbay Chrysler is \$20 a day plus 25 cents per mile. How far can Jeff drive in one day if he has only \$75?

$.25x + 20 = 75$
$x = 220$ miles

25. The sale price of a new Panasonic compact disc player is \$218 at a local discount store. At the store where this sale is going on each new CD is on sale for \$11 each. If Kyle purchases a player and some CDs for \$284, how many CDs did he purchase?

$218 + 11x = 284$
$11x = 66$
He purchased 6 CDs.

26. Suellen subscribes to an online computer service that charges \$9.50 per month for 5 hours online and \$1.50 for each hour on line in excess of 5 hours. Last month her bill was \$20. How many extra hours was she charged for?

$9.50 + 1.50x = 20$ $\quad x = 7$
7 extra hours

27. Rob is on a diet and can have 500 calories for lunch. A 3-ounce hamburger on whole-wheat bread has 315 calories and 12 fluid ounces of soda has 145 calories. How many French fries can he eat if there are 10 calories in one French fry?

$315 + 145 + 10x = 500$
$x = 4$ French fries

28. Lauren Lee makes \$24 per hour for a 40-hour week and time and a half for every hour over 40 hours. If Lauren made \$1140 last week, how many overtime hours did she work?

$(24)(40) + 36x = 1140$
$x = 5$ hours

29. Mona drove on U.S. 1, (the Pacific Coast Highway), for a distance of 270 miles. She averaged 45 miles per hour on the trip. How many hours did the trip take?

$270 = 45x$ $\quad x = 6$
The trip took 6 hours.

30. Mike traveled on the Ohio turnpike for a distance of 260 miles. The trip took 4 hours. What was his average rate of speed?

$260 = 4x$
$x = 65$ miles per hour

31. Alice jogs 3 miles in 30 minutes. What is her average rate of speed?

$3 = 0.5x$
$x = 6$ miles per hour

32. Paul drove 75 miles at 50 miles per hour. How long did the trip take?

$75 = 50x$
$x = 1\frac{1}{2}$ hours

33. Two roller-bladers, Nell and Kristin, start from the same point and "blade" in the same direction. Nell skates at 12 miles per hour and Kristin skates at 14 miles per hour. If they can keep up that pace for 2.5 hours, how far apart will they be at that time?

5 miles apart

34. Two trains leave a train station at the same time. One train travels east at 50 miles per hour. The other train travels west at 55 miles per hour. In how many hours will the two trains be 315 miles apart?

$315 = 55t + 50t$
$315 = 105t$
$t = 3$ hours

★ **35.** Sammy drove from Albuquerque, New Mexico, to the "Garden of the Gods" rock formation in Colorado Springs. He took 6 hours over the mountain road to travel 312 miles. He came home on the highway. On the highway he took 5 hours to travel 320 miles. How fast did he travel using the highway route? How fast did he travel using the mountain route? How much faster (in miles per hour) did he travel using the highway route?

He traveled 64 miles per hour on the highway.
He traveled 52 miles per hour on the mountain road.
It was 12 mph faster on the highway route.

★ **36.** Allison drives 30 miles per hour through the city and 55 miles per hour on the New Jersey Turnpike. She drove 90 miles from Battery Park to the Jersey Shore. How much of the time was city driving if she spent 1.2 hours on the turnpike?

$(55)(1.2) + 30x = 90$
0.8 hours or 48 minutes was city driving.

37. Juanita will have eight quizzes in her chemistry class. Her grades for her first seven quizzes are 75, 90, 85, 88, 92, 73, and 81. What score does she need to obtain on her last quiz to raise her quiz average to an 85?

$$\frac{75 + 90 + 85 + 88 + 92 + 73 + 81 + x}{8} = 85$$

$$x = 96$$

38. In Won Sun's math class he will take four quizzes and two one-hour exams. Each exam will count as much as two quizzes. His four quiz grades are 91, 80, 86, and 83. He scored a 93 on his first one-hour exam. What score does he need on his last hour exam to have a 90 average for the course?

$$\frac{91 + 80 + 86 + 83 + 93 + 93 + x + x}{8} = 90$$

$$x = 97$$

★ **39.** The Ramirez family has five licensed drivers and three cars. The smallest family car is a subcompact and gets excellent fuel mileage of 38 miles per gallon (mpg) in city driving. The second car is a compact and gets 21 mpg in city driving. The third car is an ancient heavy station wagon. Dad Ramirez does not like to admit the mileage rating of the "old beast." Rita calculated that if all three cars were driven the same number of miles in city driving each year, the *average* mpg rating of the three Ramirez cars would be $22\frac{2}{3}$ mpg in city driving. What is the mpg rating of the old station wagon?

$$\frac{38 + 21 + x}{3} = \frac{68}{3}$$

$$x = 9 \text{ mpg}$$

★ **40.** A local hospital revealed that the five chief officers of the hospital had an average salary of \$125,000 per year. Three of the annual salaries of these officers were revealed. They were \$50,000, \$60,000, and \$65,000. The annual salaries for the executive vice president and the president were not revealed. It was disclosed, however, that the president makes double the salary of the executive vice president. Based on this information can you find the salaries of the president and the executive vice president?

Let x = salary of executive vice president.

$$\frac{(2x) + (x) + 50{,}000 + 60{,}000 + 65{,}000}{5} = 125{,}000$$

Executive vice president earns \$150,000/year. President earns \$300,000 per year.

To Think About

41. In warmer climates, approximate temperature predictions can be made by counting the number of chirps of a cricket during a minute. The Fahrenheit temperature decreased by forty is equivalent to one-fourth of the number of cricket chirps.

(a) Write an equation for this relationship.

(b) Approximately how many chirps should be recorded if the temperature is 90°F?

(c) If a person recorded 148 cricket chirps in a minute, what would be the Fahrenheit temperature according to this formula?

$$F - 40 = \frac{x}{4}$$

$$90 - 40 = \frac{x}{4}$$

$$x = 200 \text{ chirps}$$

$$F - 40 = \frac{148}{4}$$

$$F = 77°$$

Cumulative Review Problems

Simplify.

42. $5x(2x^2 - 6x - 3)$ $10x^3 - 30x^2 - 15x$

43. $-2a(ab - 3b + 5a)$ $-2a^2b + 6ab - 10a^2$

Collect like terms.

44. $7x - 3y - 12x - 8y + 5y$ $-5x - 6y$

45. $5x^2y - 7xy^2 - 8xy - 9x^2y$ $-4x^2y - 7xy^2 - 8xy$

46. The local Apple Factory produce market sells premium MacIntosh apples at the price of 4 apples for \$3.60. They purchase the apples from the apple orchards for \$5.40 per dozen. How many apples will they have to sell to make a profit of \$1350? 3000 apples

3.3 Solve Word Problems: Comparisons

After studying this section, you will be able to:

1 *Solve word problems involving comparisons.*

MathPro

Video 8

SSM

1 *Comparisons*

Many real-life problems involve comparisons. We often compare quantities such as length, height, or income. Sometimes not all the information is known about the quantities that are being compared. You need to identify each quantity and write an algebraic expression that describes the situation in the word problem.

EXAMPLE 1 Cindy and Aaron Dunsay have a combined income of \$32,500. If Cindy makes \$4000 more than Aaron, what is Aaron's income?

1. *Understand the problem.*

Read the problem: What information is given?
The combined income is \$32,500.
What is being compared?
Cindy's income is being compared to Aaron's income.

Express this algebraically:

$$\text{Let } a = \text{Aaron's income}$$
$$a + 4000 = \text{Cindy's income}$$

2. *Write an equation.*

Reword the problem and translate:

Cindy's income combined with Aaron's income is \$32,500.

$$(a + 4000) \quad + \quad a \quad = 32{,}500$$

3. *Solve the equation and state the answer.*

$$(a + 4000) + a = 32{,}500$$

$$2a + 4000 = 32{,}500 \qquad \textit{Combine like terms.}$$

$$2a + 4000 - 4000 = 32{,}500 - 4000 \qquad \textit{Subtract 4000 from both sides.}$$

$$2a = 28{,}500$$

$$\frac{2a}{2} = \frac{28{,}500}{2} \qquad \textit{Divide both sides by 2.}$$

$$a = 14{,}250$$

Aaron's income is \$14,250.

4. *Check.*

$$(14{,}250 + 4000) + 14{,}250 \stackrel{?}{=} 32{,}500 \qquad \textit{Substitute 14,250 for a in the equation.}$$

$$18{,}250 + 14{,}250 \stackrel{?}{=} 32{,}500$$

$$32{,}500 = 32{,}500 \quad \checkmark$$

Practice Problem 1 A deck hand on a fishing boat is working with a rope that measures 89 feet. He needs to cut it into two pieces. The long piece must be 17 feet longer than the short piece. Find the length of each piece of rope. ■

If the word problem contains three unknown quantities, determine what is the basis of comparison for two of the quantities.

TEACHING TIP After discussing Example 2 with students, ask them to solve the following problem as an Additional Class Exercise. The first angle of a triangle is double the second angle. The third angle of the triangle is 20 degrees larger than the second angle. If the sum of the angles of a triangle is 180 degrees, find each angle.

Sol. The first angle is 80 degrees. The second angle is 40 degrees. The third angle is 60 degrees.

EXAMPLE 2 An airport filed a report showing the number of plane departures that took off from the airport during each month last year. The number of departures in March was 50 more than the number of departures in January. In July the number of departures was 150 less than triple the number of departures in January. In those three months, the airport had 2250 departures. How many departures were recorded for each month?

1. *Understand the problem.*
   What is the basis of comparison?

   The number of departures in March is compared to January.
   The number of departures in July is compared to January.

   Express this algebraically. It may help to underline the key phrases.

   Let j = the departures in January
   $\underbrace{\text{March was 50 more than January}}$
   $j + 50$ = the departures in March
   $\underbrace{\text{July was 150 less than triple January}}$
   $3j - 150$ = the departures in July

2. *Write an equation.*

Number of departures in January	+	Number of departures in March	+	Number of departures in July	=	3 months' total departures
j	+	$(j + 50)$	+	$(3j - 150)$	=	2250

3. *Solve the equation and state the answer.*

$$j + (j + 50) + (3j - 150) = 2250$$
$$5j - 100 = 2250 \quad \textit{Collect like terms.}$$
$$5j = 2350 \quad \textit{Add 100 to each side.}$$
$$j = 470 \quad \textit{Divide both sides by 5.}$$

   Now, if $j = 470$, then

$$j + 50 = 470 + 50 = 520$$

   and

$$3j - 150 = 3(470) - 150 = 1410 - 150 = 1260$$

   The number of departures in January was 470; the number of departures in March was 520; the number of departures in July was 1260.

4. *Check.* Do these answers seem reasonable? Yes. Do these answers verify all the statements in the word problem?

   Is the number of departures in March 50 more than those in January?

$$520 \stackrel{?}{=} 50 + 470$$
$$520 = 520 \checkmark$$

   Is the number of departures in July 150 less than triple those in January?

$$1260 \stackrel{?}{=} 3(470) - 150$$
$$1260 \stackrel{?}{=} 1410 - 150$$
$$1260 = 1260 \checkmark$$

Is the total number of departures in the 3 months equal to 2250?

$$470 + 520 + 1260 \stackrel{?}{=} 2250$$

$$2250 = 2250 \quad ✓$$

Yes, all conditions are satisfied. The three answers are correct.

Practice Problem 2 A social services worker was comparing the cost incurred by three families in heating their homes for the year. The first family has an annual heating bill of \$360 more than that of the second family. The third family has a heating bill of \$200 less than double the heating bill of the second family. The total annual heating bill for the three families was \$3960. What was the annual heating bill for each family? ■

EXAMPLE 3 A small plot of land is in the shape of a rectangle. The length is 7 meters longer than the width. The perimeter of the rectangle is 86 meters. Find the dimensions of the rectangle.

1. *Understand the problem.*

Read the problem: What information is given?

The perimeter of a rectangle is 86 meters.

What is being compared?

The length is being compared to the width.

Express this algebraically, and draw a picture.

Let w = the width

$w + 7$ = the length

$w + 7$

w $\quad$ w

$w + 7$

Reread the problem: What are you being asked to do?

Find the dimensions of the rectangle. The dimensions of the rectangle are the length and the width of the rectangle.

2. *Write an equation.*

The perimeter is the total distance around the rectangle.

$$w + (w + 7) + w + (w + 7) = 86$$

3. *Solve the equation and state the answer.*

$$w + (w + 7) + w + (w + 7) = 86$$

$4w + 14 = 86$ *Combine like terms.*

$4w = 72$ *Subtract 14 from both sides.*

$w = 18$ *Divide both sides by 4.*

The width of the rectangle is 18 meters. What is the length?

$$w + 7 = \text{the length}$$

$$18 + 7 = 25$$

The length of the rectangle is 25 meters.

4. *Check.* Put in the actual dimensions in your drawing and add the length of the sides. Is the sum 86 meters? ✓

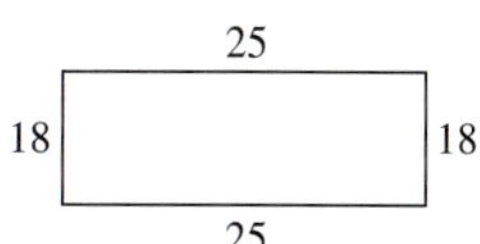

Practice Problem 3 A farmer purchased 720 meters of wire fencing to enclose a pasture. The pasture is in the shape of a triangle. One side of the triangle is 30 meters less than the longest side. The shortest side is one-half as long as the longest side. Find the dimensions of the triangle. ■

TEACHING TIP Remind students that when doing perimeter problems they must take the time to determine what geometric figure is involved. Students often confuse the perimeter of a rectangle and the perimeter of a triangle. After explaining Example 3, ask them to do the following problem as an Additional Class Exercise: The perimeter of a triangle is 240 meters. The second side of the triangle is 50 meters less than double the first side. The third side is 10 meters longer than the first side. How long is each side?

Sol. The first side is 70 meters, the second side is 90 meters, and the third side is 80 meters.

3.3 Exercises

Applications

Solve each word problem. Check to see if your answer is reasonable. Have you answered the question that was asked?

1. Russ Camp purchased a computer three years ago that had a small hard drive. His new computer has a hard drive that is 2.46 gigabytes larger than his older computer. The combined hard drive space of his two computers is 3.14 gigabytes. What is the hard drive space of each of his computers?

Older computer has 0.34 gigabyte.
New computer has 2.8 gigabytes.

2. Charlie walked 86 miles more than Robert last year in a local YMCA "walk your way to health" program. The two boys together walked a total of 1022 miles last year. How many miles did each person walk?

$R = R$
$C = R + 86$
$R + R + 86 = 1022$
$R = 468$

Robert walked 468 miles.
Charlie walked 554 miles.

3. For their Homecoming parade, the students of Wheaton College have created a colorful banner, 47 meters in length, that is made of two pieces of parachute material. The short piece is 17 meters shorter than the long piece. Find the length of each piece.

The long piece is 32 meters long.
The short piece is 15 meters long.

4. A copper conducting wire measures 84 centimeters in length. It is cut into two pieces. The shorter piece is 5 centimeters shorter than the long piece. Find the length of each piece.

long piece = 44.5 centimeters
short piece = 39.5 centimeters

5. The main span of the Verrazano Narrows bridge is 60 feet longer than the main span of the Golden Gate bridge. The main span of the George Washington bridge is 700 feet shorter than the main span of the Golden Gate bridge. The combined length of the main spans of these three famous suspension bridges is 11,960 feet. Find the length of the main span of each bridge.

The main span of the Verrazano Narrows bridge is 4260 feet.
The main span of the George Washington bridge is 3500 feet.
The main span of the Golden Gate bridge is 4200 feet.

6. Alice bought a new car that cost $1368 more than one purchased by Michael. Bettina's car cost $852 less than the new car Michael bought. The three cars cost $27,588. Can you determine how much each car cost?

$B = M - 852$
$A = M + 1368$
$M = M$
$3M + 516 = 27{,}588$
$M = 9024$

Alice's car = $10,392
Bettina's car = $8172
Michael's car = $9024

7. Three people carrying more weight than they needed went to a weight-reducing class for a year. The total weight lost by all three people was 156 pounds. Jan lost two-thirds the number of pounds that Margaret lost. Tony lost 20 pounds less than Margaret. How many pounds did each lose?

Margaret lost 66 lb.
Tony lost 46 lb.
Jan lost 44 lb.

8. Larry bought an Acura that cost $3880 more than the Chrysler purchased by Brett. Curtis bought a Saturn that cost $1410 more than Brett's Chrysler. The three cars together cost $49,030. What did each of them pay for their cars?

Brett's Chrysler cost $14,580.
Larry's Acura cost $18,460.
Curtis' Saturn cost $15,990.

9. The total flying time for three flights is 14.25 hours. The flight time of the first flight is half of the second flight. The flying time of the third flight is one and one-half times longer than the second. How long is each flight?

The first flight is 2.375 hours.
The second flight is 4.75 hours.
The third flight is 7.125 hours.

10. Sue has three cups of liquid that total 117 ml. The amount of liquid in the first cup is $\frac{1}{3}$ the amount in the second cup. The amount of liquid in the third cup is 3 times the amount in the second cup. How much liquid is in each of the cups?

The second cup contains 27 ml.
The first cup contains 9 ml.
The third cup contains 81 ml.

11. The length of a rectangular piece of tin is 3 centimeters more than double the width. The perimeter of the rectangle is 42 centimeters. Find the dimensions in centimeters of the width and length of the piece of tin.

width = 6 cm
length = 15 cm

12. A state park in Colorado has a perimeter of 92 miles. The park is in the shape of a rectangle. The length of the park is 30 miles less than triple the width of the park. What are the dimensions of the park?

width = 19 miles
length = 27 miles

13. A giant rectangular chocolate bar was made for a special promotion. The length is 6 meters more than half the width. The perimeter of the chocolate bar is 24 meters. Find the length and width of the giant chocolate bar.

width = 4 meters
length = 8 meters

14. A solid gold jewelry box was found in the underwater palace of Cleopatra just off the shore of Alexandria. The length of the rectangular box is 35 centimeters less than triple the width. The perimeter of the box is 190 centimeters. Find the length and width of Cleopatra's solid gold jewelry box.

width = 32.5 cm
length = 62.5 cm

15. A sculptor is joining together different pieces of marble for her new sculpture. A triangular piece of marble has a perimeter of 86 centimeters. The third side is 4 centimeters less than the first side. The second side is one-half of the first side in length. Find the length in centimeters of each side of the triangular marble piece.

First side = 36 cm, second side = 18 cm, third side = 32 cm

16. The symbol of Concordia College has an otter inside a blue triangle. When the maintenance department drew it in the middle of the football field, the perimeter of the triangle measured 58 meters. The second side is one-third of the first side in length. The third side is 5 meters less than the first side. Find the length in meters of each side of the college's symbol.

The first side is 27 meters.
The second side is 9 meters.
The third side is 22 meters.

17. A small triangular piece of metal is welded onto the hull of a Gloucester whale watch boat. The perimeter of the metal piece is 46 inches. The shortest side of the triangle is four inches longer than one-half of the longest side. The second longest side is 3 inches shorter than the longest side. Find the length of each side.

Longest side $= x = 18$ in.

Shortest side $= \frac{1}{2}x + 4 = 13$ in.

Third side $= x - 3 = 15$ in.

$$x + \left(\frac{1}{2}x + 4\right) + (x - 3) = 46$$
$$x = 18$$

18. Carmelina's uncle owns a triangular piece of land in Maryland. The perimeter fence that surrounds the land measures 378 yards. The shortest side is 30 yards longer than one-half of the longest side. The second longest side is 2 yards shorter than the longest side. Find the length of each side.

Longest side $= x = 140$ yards

Shortest side $= \frac{1}{2}x + 30 = 100$ yards

Third side $= x - 2 = 138$ yards

$$x + \left(\frac{1}{2}x + 30\right) + (x - 2) = 378$$
$$x = 140$$

19. The perimeter of a rectangle is 49.240 centimeters. The length of the rectangle is 1.334 centimeters larger than double the width. Find the length and width.

Length = 16.858 cm Width = 7.762 cm

20. A triangle has a perimeter of 8.07844 meters. The first side is 0.20508 meter longer than the second side. The third side is double the second side. Find the length of each side.

1st side = 2.17342 m

2nd side = 1.96834 m 3rd side = 3.93668 m

To Think About

21. A small square is constructed. Then a new square is made by increasing each side by 2 meters. The perimeter of the new square is 3 meters shorter than five times the length of one side of the original square. Find the dimensions of the original square.

x = side of original square $4x + 8 = 5x - 3$

$x + 2$ = side of new square $11 = x$

Original square was 11 m × 11 m

22. A rectangle is constructed. The length of this rectangle is double the width. Then a new rectangle is made by increasing each side by 3 meters. The perimeter of the new rectangle is 2 meters greater than 4 times the length of the old rectangle. Find the dimensions of the original rectangle.

width of old rectangle $= x$ width of new rectangle $= x + 3$

length of old rectangle $= 2x$ length of new rectangle $= 2x + 3$

$$2(2x + 3) + 2(x + 3) = 4(2x) + 2$$
$$x = 5$$
$$2x = 10$$

Width of original rectangle = 5 meters

Length of original rectangle = 10 meters

Cumulative Review Problems

Simplify.

23. $-4x(2x^2 - 3x + 8)$ $-8x^3 + 12x^2 - 32x$

24. $5a(ab + 6b - 2a)$ $5a^2b + 30ab - 10a^2$

Collect like terms.

25. $-7x + 10y - 12x - 8y - 2$ $-19x + 2y - 2$

26. $3x^2y - 6xy^2 + 7xy + 6x^2y$ $9x^2y - 6xy^2 + 7xy$

3.4 Solve Word Problems: The Value of Money and Percents

After studying this section, you will be able to:

1. *Solve applied problems with periodic rate charges.*
2. *Solve percent problems.*
3. *Solve investment problems involving simple interest.*
4. *Solve coin problems.*

MathPro

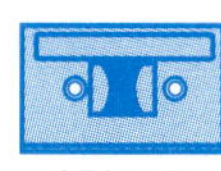
Video 9

SSM

The problems we now present are frequently encountered in business. They deal with money: buying, selling, and renting items; earning and borrowing money; the value of collections of stamps or coins. Many applications require an understanding of the use of percents and decimals. Review Sections 0.4 and 0.5 if you are weak in these skills.

1 *Periodic Rate Charges*

EXAMPLE 1 A business executive rented a car. The Car Rental Agency charged \$39 per day and \$0.28 per mile. The executive rented the car for two days and the total rental cost was computed to be \$176. How many miles did the executive drive the rented car?

TEACHING TIP Sometimes in the effort to make problems simple and understandable they become easy enough for a student to do without algebra. Some students may comment that they can solve Example 1 and Homework Problems 1 and 2 in Exercise 3.4 without the use of algebra.

Encourage them to do the problem using an algebraic equation. The more experience they have in using equations, the better they will be in solving word problems. In the ultimate sense, however, it is best to allow students the freedom to solve any problem in Chapter 3 by the methods of arithmetic if they are able to do so. It only discourages them if you absolutely prohibit them from solving a problem in a valid mathematical way (even if it does not use algebra). Most mathematics teachers conclude it is better to encourage, but not demand, that every single problem be done with the use of an algebraic equation.

1. *Understand the problem.* How do you calculate the cost of renting a car?
Total cost = Per day cost + Mileage cost
What is known?
It cost \$176 to rent the car for two days.
What do you need to find?
The number of miles the car was driven.

Choose a variable:

Let m = the number of miles driven in the rented car

2. *Write an equation.*
Use the relationship for calculating the total cost.

Per day cost + Mileage cost = Total cost

$$(39)(2) \quad + \quad (0.28)m \quad = \quad 176$$

3. *Solve the equation and state the answer.*

$$(39)(2) + (0.28)(m) = 176$$

$$78 + 0.28m = 176 \qquad \textit{Simplify the equation.}$$

$$0.28m = 98 \qquad \textit{Subtract 78 from both sides.}$$

$$\frac{0.28m}{0.28} = \frac{98}{0.28} \qquad \textit{Divide both sides by } 0.28.$$

$$m = 350 \qquad \textit{Simplify.}$$

The executive drove 350 miles.

4. *Check.* Does this seem reasonable? If he drove the car 350 miles in two days, would it cost \$176?

(Cost of \$39 per day for 2 days) + (cost of \$0.28 per mile for 350 miles)

$$\stackrel{?}{=} \text{total cost of } \$176$$

$$(\$39)(2) + (350)(\$0.28) \stackrel{?}{=} \$176$$

$$\$78 + \$98 \stackrel{?}{=} \$176$$

$$\$176 = \$176 \checkmark$$

Practice Problem 1 Alfredo wanted to rent a truck to move to Florida. He determined that the cheapest rental rates for a truck of the correct size are from a local company that will charge him \$22 per day and \$0.18 per mile. He has not yet completed an estimate of the mileage of the trip, but he knows that he will need the truck for three days. He has allowed \$363 in his moving budget for the truck. How far can he travel for a rental cost of exactly \$363? ■

2 *Percents*

Many applied situations require finding a percent of an unknown number. If we want to find 23% of \$400, we multiply $0.23(400) = 92$. If we want to find 23% of an unknown number, we can express this using algebra by writing $0.23n$, where n represents the unknown number.

EXAMPLE 2 A sofa was marked with the following sign: ''The price of this sofa has been reduced by 23%. You can save \$138 if you buy now.'' What was the original price of the sofa?

Understand the problem.

Let $s =$ the original price of the sofa

Then $0.23s =$ the amount of the price reduction, which is \$138

Write an equation and solve.

$$0.23s = 138 \quad \textit{Write the equation.}$$

$$\frac{0.23s}{0.23} = \frac{138}{0.23} \quad \textit{Divide each side of the equation by } 0.23.$$

$$s = 600 \quad \textit{Simplify.}$$

The original price of the sofa was \$600.

Check: Is \$600 a reasonable answer? ✓ Is 23% of \$600 = \$138? ✓

Practice Problem 2 John earns a commission of 38% of the cost of every set of encyclopedias that he sells. Last year he earned \$4560 in commissions. What was the cost of the encyclopedias that he sold last year? ■

EXAMPLE 3 Hector received a pay raise this year. The raise was 6% of last year's salary. This year he earns \$15,900 per year. What was his salary last year before the raise?

Understand the problem. What do we need to find?
Hector's salary last year
What do we know?
Hector received a 6% pay raise and now earns \$15,900.
What does this mean?

Reword the problem: *This year's salary of* \$15,900 *is* 6% *more than last year's salary.*

Choose a variable:

Let $x =$ Hector's salary last year

Then $0.06x =$ the amount of the raise

Write an equation and solve.

Last year's salary	+	the amount of his raise	=	this year's salary

$$x + 0.06x = 15{,}900 \quad \textit{Write the equation.}$$

$$1.00x + 0.06x = 15{,}900 \quad \textit{Rewrite } x \textit{ as } 1.00x.$$

$$1.06x = 15{,}900 \quad \textit{Collect like terms.}$$

$$x = \frac{15{,}900}{1.06} \quad \textit{Divide by } 1.06.$$

$$x = 15{,}000 \quad \textit{Simplify.}$$

Thus Hector's salary was $15,000 last year before the raise.

Check. Does it seem reasonable that Hector's salary last year was $15,000? The check is up to you.

TEACHING TIP After explaining how to do Examples 2 and 3 in this section, ask students to do the following problem as an Additional Class Exercise:

The present population of Oakwood is 34% greater than it was 10 years ago. The population now is 46,900. How many people lived in Oakwood 10 years ago?

Sol. 35,000 people lived in Oakwood 10 years ago.

Practice Problem 3 The price of Betsy's new car is 7% more than the price of a similar model last year. She paid $13,910 for her car this year. What would a similar model have cost last year? ■

3 *Interest Problems*

Interest rates affect our lives. They affect the national economy and they affect a consumer's ability to borrow money for big purchases. For these reasons, a student of mathematics should be able to solve problems involving interest. There are two basic types of interest: simple interest and compound interest.

Simple Interest

Simple interest is a charge for borrowing money or an income from investing money. It is computed by multiplying the amount of money borrowed or invested (which is called the *principal*) times the rate of interest times the period of time it is borrowed or invested (usually measured in years unless otherwise stated).

$$\text{Interest} = \text{principal} \times \text{rate} \times \text{time}$$

$$I = prt$$

Compound Interest

You often hear of banks offering a certain interest rate *compounded* quarterly, monthly, weekly, or daily. In **compound interest** the amount of interest is added to the amount of the original principal at the end of each time period, so future interest is based on the sum of both principal and previous interest. Most financial institutions use compound interest in their transactions.

Problems involving compound interest may be solved by:

1. Repeated calculations using the simple interest formula
2. Using a compound interest table
3. Using exponential functions, a topic that is usually covered in a higher-level college algebra course

All examples and exercises in this chapter will involve **simple interest.**

CALCULATOR

Interest

You can use a calculator to find simple interest. Find the interest on $450 invested at 6.5% for 15 months. Notice the time is in months. Since the interest formula is in years, $I = prt$, you need to change 15 months to years by dividing 15 by 12.

Enter: 15 [÷] 12 [=]

Display: [1.25]

Leave this on the display and multiply as follows:

1.25 [×] 450 [×] 6.5 [%] [=]

The display should read

[36.5625]

which would round to $36.56.

Try **(a)** $9516 invested at 12% for 30 months.

(b) $593 borrowed at 8% for 5 months.

EXAMPLE 4 Find the interest on \$3000 borrowed at an interest rate of 18% for one year.

$I = prt$ *The simple interest formula.*

$I = (3000)(0.18)(1)$ *Substitute the value of the principal = 3000; the rate = 18% = 0.18; the time = one year.*

$I = 540$

Thus the interest charge for borrowing \$3000 for one year at a simple interest rate of 18% is \$540.

Practice Problem 4 Find the interest on \$7000 borrowed at an interest rate of 12% for one year. ■

Now we apply this concept to a word problem about investments.

EXAMPLE 5 A woman invested an amount of money in two places for one year. She invested some at 8% interest and the rest at 12% interest. Her total amount invested was \$1250. At the end of the year she had earned \$128.00 in interest. How much money did she invest in each place?

MATHEMATICS BLUEPRINT FOR PROBLEM SOLVING

Gather the facts.	Label the variables.	Basic equation.	Key points to remember.
\$1250 is invested: part at 8% interest, part at 12% interest. The total interest for the year is \$128.00.	Let x = the amount invested at 8%. $1250 - x$ = the amount invested at 12%. $0.08x$ = the amount of interest for x dollars at 8%. $0.12(1250 - x)$ = the amount of interest for $1250 - x$ dollars at 12%.	Interest earned at 8% + interest earned at 12% = total interest earned during the year which is \$128.	Be careful to write $1250 - x$ for the amount of money invested at 12%. The order is total $- x$ (do not use $x - 1250$).

interest earned at 8%	+	interest earned at 12%	=	total interest earned during the year
$0.08x$	+	$0.12(1250 - x)$	=	128

NOTE: Be sure you write $(1250 - x)$ for the amount of money invested at 12%. Students often write it backwards by mistake. It is *not* correct to use $(x - 1250)$ instead of $(1250 - x)$. The order of the terms is very important.

Solve the equation and state the answer.

$0.08x + 150 - 0.12x = 128$ — *Remove parentheses.*

$-0.04x + 150 = 128$ — *Collect like terms.*

$-0.04x = -22$ — *Subtract* 150 *from both sides.*

$\frac{-0.04x}{-0.04} = \frac{-22}{-0.04}$ — *Divide both sides by* -0.04

$x = 550$ — *The amount invested at* 8% *interest is* $550.

$1250 - x = 1250 - 550 = 700$ — *The amount invested at* 12% *interest is* $700.

Check. Are these values reasonable? Yes. Do the amounts equal $1250?

$$\$550 + \$700 \stackrel{?}{=} \$1250$$

$$\$1250 = \$1250 \ ✓$$

Would these amounts earn $128 interest in one year invested at the specified rates?

$$0.08(\$550) + 0.12(\$700) \stackrel{?}{=} \$128$$

$$\$44 + \$84 \stackrel{?}{=} \$128$$

$$\$128 = \$128 \ ✓$$

TEACHING TIP The most common error students encounter in solving problems like Example 5 is that they write $x - 1250$ instead of the correct expression $1250 - x$. Emphasize that the expression should be the whole minus the part. The expression will always be of the form (Total $- x$).
After explaining Example 5, ask students to do the following as an Additional Class Exercise: Hillary invested $4000 for one year. She invested part at 5% interest and part at 9% interest. At the end of one year she earned $256 in interest. How much did she invest at each interest rate?

Sol. $2600 at 5% and $1400 at 9%

Practice Problem 5 A woman invested her savings of $8000 in two accounts that each calculate interest only once per year. She placed one amount in a special notice account that yields 15% annual interest. The remainder she placed in a tax-free All-Savers account that yields 12% annual interest. At the end of the year she earned $1155 in interest from the two accounts together. How much did she invest in each account? ■

4 *Coin Problems*

Coin problems provide an unmatched opportunity to use the concept of *value.* We must make a distinction between how many coins and the *value* of the coins.

Consider this example. Here we know *the value* of some coins, but do not know *how many* we have.

EXAMPLE 6 When Bob got out of math class, he had to make a long-distance call. He had exactly enough dimes and quarters to make a phone call that would cost $2.55. He had one less quarter than he had dimes. How many coins of each type did he have?

Let d = the number of dimes

Then $d - 1$ = the number of quarters

The total value of the coins is $2.55. How can we represent the value of the dimes and the value of the quarters? Think.

Each dime is worth $0.10.	Each quarter is worth $0.25.
5 dimes are worth $(5)(0.10) = 0.50$	8 quarters are worth $(8)(0.25) = 2.00$
d dimes are worth $(d)(0.10) = 0.10d$	$(d - 1)$ quarters are worth $(d - 1)(0.25) = 0.25(d - 1)$

Now we can write an equation for the total value.

$$\text{(value of dimes)} + \text{(value of quarters)} = \$2.55$$

$$0.10d + 0.25(d-1) = 2.55$$

$$0.10d + 0.25d - 0.25 = 2.55 \quad \textit{Remove parentheses.}$$

$$0.35d - 0.25 = 2.55 \quad \textit{Collect like terms.}$$

$$0.35d = 2.80 \quad \textit{Add 0.25 to both sides.}$$

$$\frac{0.35d}{0.35} = \frac{2.80}{0.35} \quad \textit{Divide both sides by 0.35.}$$

$$d = 8 \quad \textit{Simplify.}$$

$$d - 1 = 7$$

Thus Bob had 8 dimes and 7 quarters.

Check. Is this answer reasonable? Yes. Does Bob have one less quarter than he has dimes?

$$8 - 7 \stackrel{?}{=} 1$$

$$1 = 1 \checkmark$$

Are 8 dimes and 7 quarters worth \$2.55?

$$8(\$0.10) + 7(\$0.25) \stackrel{?}{=} \$2.55$$

$$\$0.80 + \$1.75 \stackrel{?}{=} \$2.55$$

$$\$2.55 = \$2.55 \checkmark$$

Practice Problem 6 Ginger has five more quarters than dimes. She has \$5.10 in change. If she has only quarters and dimes, how many coins of each type does she have? ■

EXAMPLE 7 Michele returned from the store with \$2.80 in change. She had twice as many quarters as nickels. She had two more dimes than nickels. How many nickels, dimes, and quarters did she have?

MATHEMATICS BLUEPRINT FOR PROBLEM SOLVING

Gather the facts.	Label the variables.	Basic equation.	Key points to remember.
Michele had \$2.80 in change. She had twice as many quarters as nickels. She had two more dimes than nickels.	Let x = the number of nickels. $2x$ = the number of quarters. $x + 2$ = the number of dimes. $0.05x$ = the value of the nickels $0.25(2x)$ = the value of the quarters. $0.10(x + 2)$ = the value of the dimes.	The value of the nickels + the value of the dimes + the value of the quarters = \$2.80.	Don't add the number of coins to get \$2.80. You must add the value of the coins!

$$\text{(value of nickels)} + \text{(value of dimes)} + \text{(value of quarters)} = \$2.80$$

$$0.05x + 0.10(x+2) + 0.25(2x) = 2.80$$

Solve the equation and check.

$$0.05x + 0.10x + 0.20 + 0.50x = 2.80 \quad \textit{Remove parentheses.}$$

$$0.65x + 0.20 = 2.80 \quad \textit{Collect like terms.}$$

$$0.65x = 2.60 \quad \textit{Subtract 0.20 from both sides.}$$

$$\frac{0.65x}{0.65} = \frac{2.60}{0.65} \quad \textit{Divide both sides by 0.65.}$$

$$x = 4 \quad \textit{Simplify. Michele had 4 nickels.}$$

$$2x = 8 \quad \textit{She had 8 quarters.}$$

$$x + 2 = 6 \quad \textit{She had 6 dimes.}$$

Check. Is the answer reasonable? Yes. Did Michele have twice as many quarters as nickels?

$$(4)(2) \stackrel{?}{=} 8 \qquad 8 = 8 \checkmark$$

Did she have two more dimes than nickels?

$$4 + 2 \stackrel{?}{=} 6 \qquad 6 = 6 \checkmark$$

Do 4 nickels, 8 quarters, and 6 dimes have a value of \$2.80?

$$4(\$0.05) + 8(\$0.25) + 6(\$0.10) \stackrel{?}{=} \$2.80$$

$$\$0.20 + \$2.00 + \$0.60 \stackrel{?}{=} \$2.80$$

$$\$2.80 = \$2.80 \checkmark$$

Practice Problem 7 A young boy told his friend that he had twice as many nickels as dimes in his pocket. He also said that he had four more quarters than dimes. He said that he had \$2.35 in change in his pocket. Can you determine how many nickels, dimes, and quarters he had? ■

Developing Your Study Skills

APPLICATIONS OR WORD PROBLEMS

Applications or word problems are the very life of mathematics! They are the reason for doing mathematics because they teach you how to put into use the mathematical skills you have developed. Learning mathematics without ever doing word problems is similar to learning all the skills of a sport without ever playing a game or learning all the notes on an instrument without ever playing a song.

The key to success is practice. Make yourself do as many problems as you can. If you need help organizing your facts, use the "Mathematics Blueprint." You may not be able to do them all correctly at first, but keep trying. Do not give up whenever you reach a difficult one. If you cannot solve it, just try another one. Then come back and try it again later.

A misconception among students when they begin studying word problems is that each problem is different. At first the problems may seem this way, but as you practice more and more, you will begin to see the similarities, the different "types." You will see patterns in solving problems, which will enable you to solve problems of a given type more easily.

3.4 Exercises

Applications

Solve each word problem.

1. Sierra rented some computer time at a computer center because her PC was in for repair. She was charged \$20 per day. She was charged \$0.15 per page for everything she printed out. She rented the computer time for three days. Her total bill was \$68.40. How many pages did she print out?
$3(20) + 0.15x = 68.90$
$x = 56$ She printed 56 pages.

2. Jack and Helen hired a clown for their son's birthday party. The clown charged \$75 plus \$0.20 per balloon toy made for the children. The bill at the end of the afternoon came to \$96.40. How many balloon toys did the clown make?
$75 + 0.20x = 96.40$
$x = 107$ 107 balloon toys

3. Jane is a tour guide for the local chamber of commerce. She earns \$35 per day plus \$9 for every tour she gives. Last week she worked 4 days and earned \$302. How many tours did she give?
$(35)(4) + 9x = 302$
$x = 18$ She gave 18 tours.

4. Marybelle is contemplating a job as a waitress. She would be paid \$2 per hour plus tips. The other waitresses have told her that an average tip at that restaurant is \$3 per table served. If she works 20 hours per week, how many tables would she have to serve in order to make \$151 during the week?
$2(20) + 3x = 151$
$x = 37$ tables

5. Ramon has a summer job to earn his college tuition. He gets paid \$6.00 per hour for the first 40 hours and \$9.00 per hour for each hour in the week worked above the 40 hours. This summer his goal is to earn at least \$303.00 per week. How many hours of overtime per week will he need to achieve his goal?
$9x + (6)(40) = 303$
$x = 7$ hours

6. Maria has a summer job to earn her college tuition. She gets paid \$6.50 per hour for the first 40 hours and \$9.75 for each hour in the week worked above the 40 hours. This summer her goal is to earn at least \$338.00 per week. How many hours of overtime per week will she need to achieve her goal?
$9.75x + 260 = 338$
$x = 8$ hours

7. The camera Melissa wanted for her birthday is on sale for 28% lower than usual. The amount of the discount is \$100.80. What was the original price of the camera?
$0.28x = 100.80$ $x = 360$
Original price was \$360.

8. The number of women working full-time in Springfield has risen 7% this year. This means 126 more women have full-time jobs. What was the number of women working full-time last year?
$0.07x = 126$ $x = 1800$
1800 women

9. The crime rate in Center City has increased by 7% from last year to this year. There were 56 more reported crimes this year in Center City than there were last year. How many reported crimes were there last year in Center City?
$0.07x = 56$
$x = 800$ crimes

10. Due to a serious budget shortfall the state has been forced to lay off 9% of its employees. The number of employees laid off was 6030. How many employees did the state have before these employees were laid off?
$0.09x = 6030$
$x = 67{,}000$ employees

11. When Harvey got promoted at work, he received a 12% pay raise. He now earns \$16,240 per year. What did he earn annually before the pay raise?
$1.12x = 16{,}240$
$x = \$14{,}500$

12. Shirley borrowed some money from Easy Finance Company. They charged 18% interest on the loan. At the end of one year Shirley paid off the debt by paying back the amount of the original loan plus the 18% interest. She paid the company \$2006. How much did she borrow originally?
$1.18x = 2006$
$x = \$1700$

13. Paul Frydrych invested some money at 14% interest. At the end of the year the total amount of his original principal and the interest was \$5700. How much did he originally invest?

$1.14x = 5700$
$x = \$5000$

14. The cost of living last year went up 13%. Fred Whitney was fortunate to get a 13% increase in salary. He now makes \$15,820 per year. How much did he make last year?

$1.13x = 15{,}820$
$x = \$14{,}000$

15. Linda Couselco has a job as a loan counselor in a bank. She advised a customer to invest part of her money in a 12% money market fund and the rest in a 14% investment fund. The customer has \$4000 to invest. If she earned \$508 in interest after one year, how much did she invest in each fund?

$0.12x + 0.14(4000 - x) = 508$ \$2600 at 12%
$12x + 56{,}000 - 14x = 50{,}800$ \$1400 at 14%
$x = 2600$

16. A retired couple earned \$656 last year from investments made in tax-free bonds. They invested some of the money at 8% and the rest at 10%. The total amount they invested was \$7200. If the couple invested the \$7200 for exactly one year, how much did they invest at each interest rate?

$0.08x + 0.10(7200 - x) = 656$ \$3200 at 8%
$6400 = 2x$ \$4000 at 10%
$3200 = x$

17. Castle Cinema Productions invested \$320,000 in two types of mutual funds to help replace sets that were destroyed by a recent fire. Part was invested in a conservative mutual fund that pays 8% interest. The remainder was invested in an aggressive fund that pays 12% interest. At the end of one year the production company received \$30,400 in interest. How much was invested at each amount?

$0.08x + 0.12(320{,}000 - x) = \$30{,}400$
\$200,000 at 8%, \$120,000 at 12%

18. Eva invested \$5000 for one year. She invested part of it at 10% and part of it at 12%. At the end of the year she earned \$566. How much did she invest at each rate?

$0.12x + 0.10(5000 - x) = 566$ \$3300 at 12%
$2x = 6600$ \$1700 at 10%
$x = 3300$

19. Ted and Linda invested \$9000 for one year. They invested part of it at 14% and the remainder at 9%. At the end of the year they had earned \$1135. How much did they invest at each rate?

$0.14x + 0.09(9000 - x) = 1135$ \$6500 at 14%
$5x = 32{,}500$ \$2500 at 9%

20. Prior to going to college Carlos had saved up \$3000. He invested it for one year. Part of it was invested at 7% and the remainder at 9%. At the end of one year he had earned \$242 in interest. How much did he invest at each rate?

$0.07x + 0.09(3000 - x) = 242$ \$1400 at 7%
$2800 = 2x$ \$1600 at 9%
$1400 = x$

21. Little Melinda has nickels and quarters in her bank. She has four fewer nickels than quarters. She has \$3.70 in the bank. How many coins of each type does she have?

$5(x - 4) + 25x = 370$ 13 quarters
$30x = 390$ 9 nickels
$x = 13$

22. Fred's younger brother had several coins when he returned from his paper route. He had, in dimes and quarters, a total of \$5.35. He said he had six more quarters than he had dimes. How many of each coin did he have?

$25(x + 6) + 10x = 535$ 11 dimes
$35x = 385$ 17 quarters
$x = 11$

23. A newspaper carrier has \$3.75 in change. He has three more quarters than dimes but twice as many nickels as quarters. How many coins of each type does he have?

$25(x + 3) + 10x + 5(2x + 6) = 375$ 18 nickels
$45x = 270$ 6 dimes
$x = 6$ 9 quarters

24. Alice collected a number of nickels, dimes, and quarters from her room. She found three more dimes than nickels. She also found twice as many quarters as dimes. The value of her coins was \$5.05. How many coins of each type did she have?

$5x + 10(x + 3) + 25(2x + 6) = 505$ 5 nickels
$65x = 325$ 8 dimes
$x = 5$ 16 quarters

25. Charlie Saulnier cashed his paycheck and came home from the bank with \$100 bills, \$20 bills, and \$10 bills. He has twice as many \$20 bills as he has \$10 bills. He has three more \$100 bills than he has \$10 bills. He is carrying \$1500 in bills. How many of each denomination does he have?
$10x + 20(2x) + 100(x + 3) = 1500$
eight \$10 bills, sixteen \$20 bills, eleven \$100 bills

26. Roberta Burgess came home with \$325 in tips from two nights on her job as a waitress. She had \$20 bills, \$10 bills, and \$5 bills. She discovered she had three times as many \$5 as she had \$10 bills. She also found that she had 4 fewer \$20 bills than she had \$10 bills. How many of each denomination did she have?
$5(3x) + 10(x) + 20(x - 4) = 325$
nine \$10 bills, twenty-seven \$5 bills, five \$20 bills

★**27.** The sum of the measures of the three interior angles of a triangle is always 180°. In a certain triangle the measure of one angle is double the measure of the second angle but is 5 degrees less than the measure of the third angle. What is the measure of each angle?
First angle measures 70°
Second angle measures 35°
Third angle measures 75°

★**28.** The length of a spring increases 1.32 centimeters for each kilogram it supports. The spring is 12.00 centimeters long when it is supporting 3.5 kilograms. Find the original length of the spring. 7.38 cm

29. Walter invested \$2969 for one year and earned \$273.47 in interest. He invested part of the \$2969 at 7% and the remainder at 11%. How much did he invest at each rate?
$0.07x + 0.11(2969 - x) = 273.47$
\$1328 at 7%, \$1641 at 11%

30. Last year the town of Waterbury paid an interest payment of \$52,396.08 for a one-year note. This represented 11% of the amount the town borrowed. How much did they borrow? \$476,328

To Think About

31. The West Suburban Car Rental Agency will rent a compact car for \$35 per day and an additional charge of \$0.24 per mile. The Golden Gate Car Rental Agency only charges \$0.16 per mile but charges \$41 per day. If a salesperson wanted to rent a car for three days, how many miles would that person have to drive to make the Golden Gate Car Rental Agency car a better bargain? 225 miles

32. A Peabody pumping station pumps 2000 gallons per hour into a reservoir tank for a town's drinking water. The tank is initially empty. The station pumps for 3 hours. Then a leak in the reservoir tank is created by a large crack. Some water flows out of the reservoir tank at a constant rate. The pumping station continues pumping for 6 more hours while the leak is undetected. At the end of 9 hours the reservoir has 17,640 gallons. During the last 6 hours, how many gallons per hour were leaking from the reservoir? 60 gals/hr

Cumulative Review Problems

Do each operation in the proper order.

33. $5(3) + 6 \div (-2)$ $15 - 3 = 12$

34. $2 - (8 - 10)^3 + 5$ $2 + 8 + 5 = 15$

Evaluate when $x = -2$ and $y = 3$.

35. $2x^2 + 3xy - 2y^2$ $8 - 18 - 18 = -28$

36. $x^3 - 5x^2 + 3y - 6$ $-8 - 20 + 9 - 6 = -25$

Putting Your Skills to Work

AUTOMOBILE LOANS AND INSTALLMENT LOANS

The following table provides the amount of the monthly payment of a loan of $1000 at various interest rates for a period of 2 to 5 years. To find the monthly payment for a loan of a larger amount of money merely multiply this payment by the number of thousands of the loan. To borrow $14,000 at 8% interest for 3 years you multiply $31.34 by 14 to obtain a payment of $438.76.

LOAN AMORTIZATION TABLE FOR REPRESENTATIVE INTEREST RATES
Monthly Payment per $1000 to Pay Principal and Interest

Period of Loan in Years	7%	8%	9%	10%	12%	16%
2	44.77	45.23	45.68	46.14	47.07	48.96
3	30.88	31.34	31.80	32.27	33.21	35.16
4	23.95	24.41	24.89	25.36	26.33	28.34
5	19.80	20.28	20.76	21.25	22.24	24.32

PROBLEMS FOR INDIVIDUAL STUDY AND INVESTIGATION

Use the table above to solve the following problems.

1. If Ricardo Sanchez wants to buy a new car and borrow $10,000 at a 9% interest rate for 4 years, what will his monthly payments be? How much interest will he pay over the 4-year period? (How much more than $10,000 will he pay over 4 years?) Monthly payments will be $248.90. He will pay $1947.20 in interest.

2. If Alicia Wong wants to purchase a new stereo system for her apartment and borrow $3000 at 12% interest for 2 years, what will her monthly payments be? How much interest will she pay over the 2-year period? (How much more than $3000 will she pay over the 2 years?) Monthly payments will be $141.21. She will pay $389.04 in interest.

PROBLEMS FOR COOPERATIVE GROUP INVESTIGATION

With some other members of your class, determine an answer to the following.

3. If Walter Swenson plans to borrow $15,000 to purchase a new car at an interest rate of 7%, how much more in interest will he pay if he borrows the money for 5 years as opposed to a period of 4 years? He will pay $576 more in interest.

4. Ray and Shirley Peterson are planning to build a garage and a connecting family room for their house. It will cost $35,000. They have $17,000 in savings, and they plan to borrow the rest at 16% interest. They can afford a maximum monthly payment of $640. What time period for the loan will they have to select to meet that requirement? What will their monthly payment be? They will pick a time period of 3 years or more. The monthly payment for this loan for 3 years is $632.88.

INTERNET CONNECTIONS: Go to ``http://www.prenhall.com/tobey'' to be connected

Site: Mortgage Amortization Calculator or a related site

Alternate Site: Amortization Calculator

Kimberly Jones is buying a $135,000 house. She will make a down payment of 10%. The rest will be financed with a 30-year loan at an interest rate of 8.5%.

5. What will her monthly payment be? How much interest will she pay over the life of the loan?

6. Suppose she decides to pay an extra $200 every month. How much interest will she save by doing this? In how many years will the loan be paid off?

3.5 Solve Word Problems Using Geometric Formulas

MathPro

Video 9

SSM

After studying this section, you will be able to:

1. *Find the area and the perimeter of two-dimensional objects. Find the missing dimension given the area or perimeter and one dimension.*
2. *Find the volume and the surface area of three-dimensional objects. Find a missing dimension.*
3. *Solve more involved geometric problems.*

1 *Area and Perimeter*

We can use our knowledge of geometry to solve word problems. To do this, we first review some geometric facts and formulas. **Perimeter** is the distance around a plane figure. Perimeter is measured in linear units (inches, feet, centimeters, miles). **Area** is a measure of the amount of surface in a region. Area is measured in square units (square inches, square feet, square centimeters).

In our sketches we will show angles of 90° by using a small square (∟ ⌟). This indicates that the two lines are at right angles. All angles that measure 90° are called *right angles.* An *altitude* is perpendicular to the base of a figure. That is, the altitude forms right angles with the base. The small corner square in a sketch helps us to identify the altitude of the figure.

The following box provides a handy guide to some facts and formulas you will need to know. Use it as a reference when solving word problems involving geometric figures.

Geometric Formulas: Two-dimensional Figures

A **parallelogram** is a four-sided figure with opposite sides parallel. In a parallelogram, opposite sides are equal and opposite angles are equal.

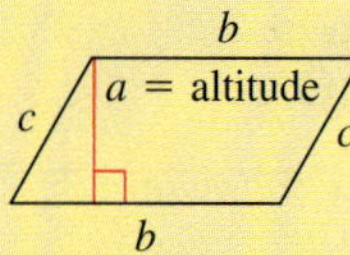

Perimeter = the sum of all four sides

Area $= ab$

A **rectangle** is a parallelogram with all interior angles measuring 90°.

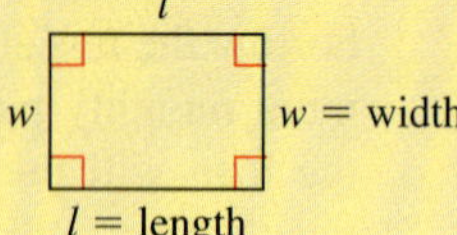

Perimeter $= 2l + 2w$

Area $= lw$

A **square** is a rectangle with all four sides equal.

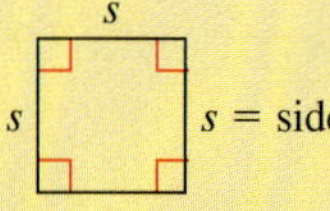

Perimeter $= 4s$

Area $= s^2$

Geometric Formulas: Two-dimensional Figures (continued)

A **trapezoid** is a four-sided figure with two sides parallel. The parallel sides are called the bases of the trapezoid.

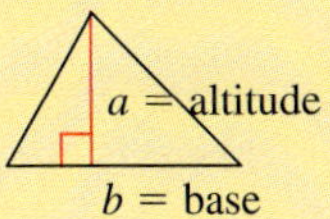

Perimeter = the sum of all four sides

$$\text{Area} = \frac{1}{2}a(b_1 + b_2)$$

A **triangle** is a closed plane figure with three sides.

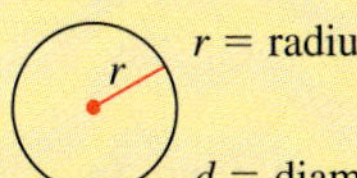

Perimeter = the sum of the three sides

$$\text{Area} = \frac{1}{2}ab$$

A **circle** is a plane curve consisting of all points at an equal distance from a given point called the center.

Circumference is the distance around a circle.

r = radius

d = diameter = $2r$

$$\text{Circumference} = 2\pi r$$

$$\text{Area} = \pi r^2$$

π is a constant associated with circles. It is an irrational number that is approximately 3.141592654. We usually use 3.14 as a sufficiently accurate approximation. Thus we write $\pi \approx 3.14$ for most of our calculations involving π.

We frequently encounter triangles in word problems. There are four important facts about triangles, which we list for convenient reference.

Triangle Facts

1. The sum of the interior angles of any triangle is 180°. That is,

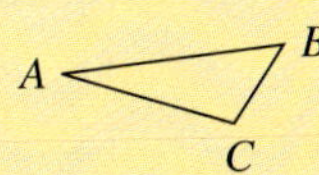

$$\text{measure of } \angle A + \text{measure of } \angle B + \text{measure of } \angle C = 180°$$

2. An **equilateral** triangle is a triangle with three sides equal in length and three angles that measure 60° each.

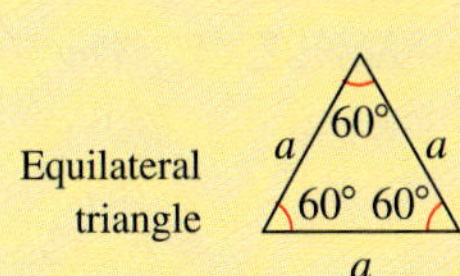

3. An **isosceles** triangle is a triangle with two equal sides. The two angles opposite the equal sides are also equal.

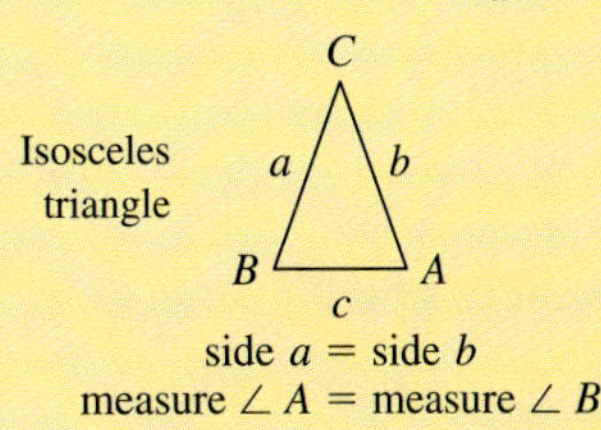

4. A **right** triangle is a triangle with one angle that measures 90°.

Right triangle

90°

EXAMPLE 1 Find the area of a triangle whose base is 5 centimeters (cm) and whose altitude is 12 cm.

We can use the formula for the area of a triangle. Substitute the known values in the formula and solve for the unknown.

$$A = \frac{1}{2}ab \qquad \textit{Write the formula for the area of a triangle.}$$

$$= \frac{1}{2}(12\text{ cm})(5\text{ cm}) \qquad \textit{Substitute the known values in the formula.}$$

$$= \frac{1}{2}(12)(5)(\text{cm})(\text{cm}) \qquad \textit{Simplify.}$$

$$A = 30\text{ cm}^2 \quad \text{or} \quad 30\text{ square centimeters}$$

The area of the triangle is 30 square centimeters.

Practice Problem 1 Find the area of a triangle whose base is 14 inches and whose altitude is 20 inches. ■

EXAMPLE 2 The area of a rectangular field is 72 square meters (m^2). Find the width of the field if the length is 9 meters (m).

Draw a diagram.

9 meters

72 m^2

w

Write the formula for the area of a rectangle and solve for the unknown value.

$$A = lw$$

$$72(\text{m})^2 = (9\text{ m})(w) \qquad \textit{Substitute the known values into the formula.}$$

$$\frac{72(\text{m})(\text{m})}{9\text{ m}} = \frac{9\text{ m}}{9\text{ m}}w \qquad \textit{Divide both sides by 9 m.}$$

$$8\text{ m} = w$$

The width of the rectangular field is 8 meters.

Practice Problem 2 The area of a rectangular field is 120 square yards. If the width of the field is 8 yards, what is the length? ■

EXAMPLE 3 The area of a trapezoid is 80 square inches (in.^2). The altitude is 20 inches and one of the bases is 3 inches. Find the length of the other base.

$$A = \frac{1}{2}a(b_1 + b_2) \qquad \textit{Write the formula for the area of a trapezoid.}$$

$$80(\text{in.}^2) = \frac{1}{2}(20\text{ in.})(3\text{ in.} + b_2) \qquad \textit{Substitute the known values.}$$

$$80(\text{in.})(\text{in.}) = 10\text{ in.}(3\text{ in.} + b_2) \qquad \textit{Simplify.}$$

$$80(\text{in.})(\text{in.}) = 30(\text{in.})(\text{in.}) + (10\text{ in.})b_2 \qquad \textit{Remove parentheses.}$$

$$50(\text{in.})(\text{in.}) = (10\text{ in.})b_2 \qquad \textit{Subtract 30 in.}^2\textit{ from both sides.}$$

$$5\text{ in.} = b_2 \qquad \textit{Divide both sides by 10 in.}$$

The other base is 5 inches long.

Practice Problem 3 The area of a trapezoid is 256 square feet. The bases are 12 feet and 20 feet, respectively. Find the length of the altitude. ■

TEACHING TIP Students often confuse r^2 with $2r$. Have students find the circumference and the area of a circle with radius 3 inches.

$$C = 2\pi r = 2(3.14)(3) = 18.84 \text{ inches}$$

$$A = \pi r^2 = (3.14)(3)(3) = 28.26 \text{ square inches}$$

Point out the difference in the answers.

EXAMPLE 4 Find the area of a circle whose diameter is 14 inches (use $\pi = 3.14$). Round to the nearest square inch.

$$A = \pi r^2 \quad \textit{Write the formula for the area of a circle.}$$

Note that the length of the diameter is given in the word problem. We need to know the length of the radius to use the formula. Since $d = 2r$, then 14 in. $= 2r$ and $r = 7$ in.

$A = \pi(7 \text{ in.})(7 \text{ in.})$ *Substitute known values into the formula.*

$= (3.14)(7 \text{ in.})(7 \text{ in.})$

$= (3.14)(49 \text{ in.}^2)$

$= 153.86 \text{ in.}^2$

Rounded to the nearest square inch, the area of the circle is approximately 159 square inches.

In this example, you were asked to find the area of the circle. Do not confuse the formula for area with the formula for circumference. $A = \pi r^2$, while $C = 2\pi r$. Remember that area involves square units, so it is only natural that in the area formula you would square the radius, $A = \pi r^2$.

Practice Problem 4 Find the circumference of a circle whose radius is 15 meters (use $\pi = 3.14$). Round your answer to the nearest meter. ■

EXAMPLE 5 Find the perimeter of a parallelogram whose longest side is 4 feet and whose shortest side is 2.6 feet.

Draw a picture.

4 feet

2.6 feet

The perimeter is the distance around a figure. To find the perimeter, add the lengths of the sides. Since the opposite sides of a parallelogram are equal, we can write

$P = 2(4 \text{ feet}) + 2(2.6 \text{ feet})$

$= 8 \text{ feet} + 5.2 \text{ feet}$

$= 13.2 \text{ feet}$

The perimeter of the parallelogram is 13.2 feet.

Practice Problem 5 Find the perimeter of an equilateral triangle with a side that measures 15 centimeters. ■

2 *Volume and Surface Area*

Now let's examine three-dimensional figures. **Surface area** is the total area of the faces of the figure. You can find surface area by calculating the area of each face and then finding the sum. **Volume** is the measure of the amount of space inside the figure. Some formulas for the surface area and the volume of regular figures can be found in the table below.

Geometric Formulas: Three-dimensional Figures

Rectangular prism

h = height

w = width

l = length

Surface area $= 2lw + 2wh + 2lh$

Volume $= lwh$

Notice that all of the faces are rectangles.

Sphere

r = radius

Surface area $= 4\pi r^2$

Volume $= \dfrac{4}{3}\pi r^3$

Right circular cylinder

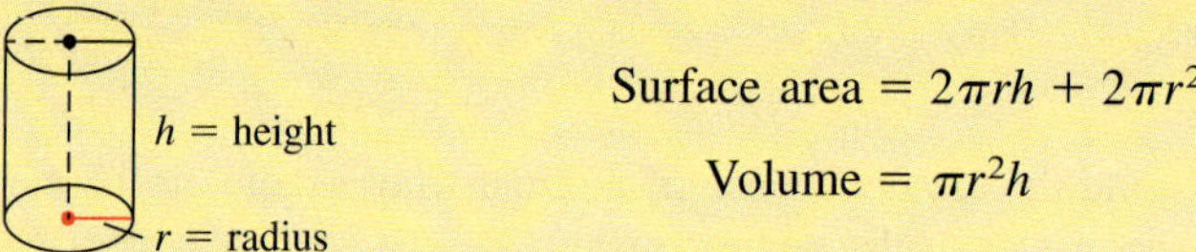

Surface area $= 2\pi rh + 2\pi r^2$

Volume $= \pi r^2 h$

EXAMPLE 6 Find the volume of a sphere with radius 4 centimeters (use $\pi = 3.14$). Round your answer to the nearest cubic centimeter.

$V = \dfrac{4}{3}\pi r^3$ *Write the formula for the volume of a sphere.*

$= \dfrac{4}{3}(3.14)(4 \text{ cm})^3$ *Substitute the known values into the formula.*

$= \dfrac{4}{3}(3.14)(64)\text{cm}^3$

$= 267.946667 \text{ cm}^3$

Rounded to the nearest whole number, the volume of the sphere is 268 cubic centimeters.

Practice Problem 6 Find the surface area of a sphere with radius 5 meters. Round your answer to the nearest square meter. ■

EXAMPLE 7 A can is made of aluminum. It has a flat top and a flat bottom. The height is 5 inches and the radius is 2 inches. How much aluminum is needed to make the can? How much aluminum is needed to make 10,000 cans?

Understand the problem.
What do we need to find? Reword the problem.

We need to find the total surface area of a cylinder.

Draw a picture.

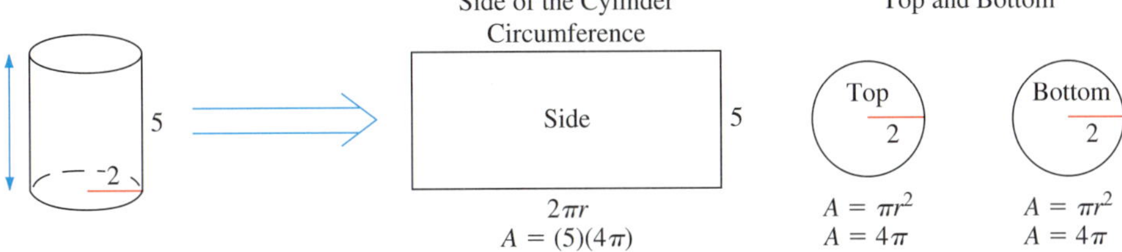

You may calculate the area of each piece and find the sum or you may use the formula.

$$\begin{aligned}\text{Surface area} &= 2\pi rh + 2\pi r^2\\ &= 2(3.14)(2\text{ in.})(5\text{ in.}) + 2(3.14)(2\text{ in.})^2\\ &= 62.8\text{ in.}^2 + 25.12\text{ in.}^2\\ &= 87.92\text{ in.}^2\end{aligned}$$

87.92 square inches of aluminum is needed to make each can.

Have we answered all the questions in the word problem? Reread the problem. How much aluminum is needed to make 10,000 cans?

$$(\text{aluminum for 1 can})(10{,}000)$$

$$(87.92\text{ in.}^2)(10{,}000) = 879{,}200 \text{ square inches}$$

It would take 879,200 square inches of aluminum to make 10,000 cans.

Practice Problem 7 Sand is stored in a cylindrical drum that is 4 feet high and has a radius of 3 feet. How much sand can be stored in the drum? ■

3 *More Involved Geometric Problems*

Problems that at first seem complicated can sometimes be reduced to a combination of several simple problems.

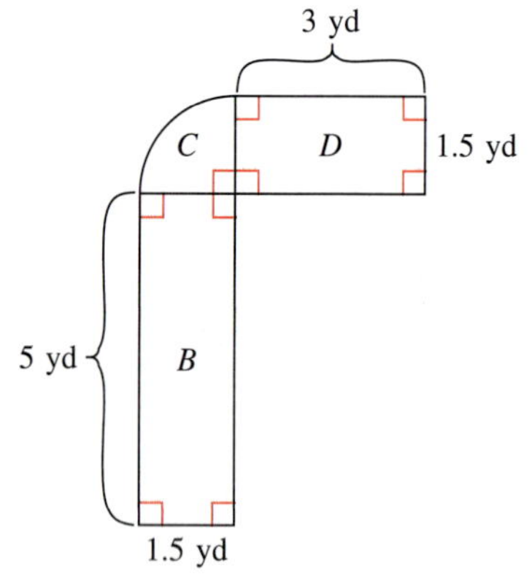

EXAMPLE 8 A quarter-circle (of radius 1.5 yards) is connected to two rectangles with dimensions labeled on the sketch in the margin. You need to lay a strip of carpet in your house according to this sketch. How many square yards of carpeting will be needed on the floor? (Use $\pi \approx 3.14$. Round your final answer to the nearest tenth.)

The desired area is the sum of three areas, which we will call B, C, and D. Area B and area D are rectangular in shape. It is relatively easy to find the areas of these shapes.

$$A_B = (5\text{ yd})(1.5\text{ yd}) = 7.5\text{ yd}^2 \qquad A_D = (3\text{ yd})(1.5\text{ yd}) = 4.5\text{ yd}^2$$

Area C is one-fourth of a circle. The radius of the circle is 1.5 yd.

$$A_C = \pi\frac{r^2}{4} = \frac{(3.14)(1.5\text{ yd})^2}{4} = \frac{(3.14)(2.25)\text{yd}^2}{4} = 1.76625\text{ yd}^2 \approx 1.8\text{ yd}^2$$

The total area is $A_B + A_D + A_C = 7.5\text{ yd}^2 + 4.5\text{ yd}^2 + 1.8\text{ yd}^2 = 13.8\text{ yd}^2$

13.8 square yards of carpeting will be needed to cover the floor.

TEACHING TIP Discuss the answer to Example 8. Point out that this is the amount of carpeting to be placed on the floor. There would undoubtedly be some waste when the carpeting is cut to these dimensions. Ask, what would be the amount of waste if area C were cut from a square piece of carpeting 1.5 yd on a side?

$$A_{\text{square}} = (1.5\text{ yd})(1.5\text{ yd}) = 2.25\text{ yd}^2$$

$$A_{\text{square}} - A_C = 2.25\text{ yd}^2 - 1.8\text{ yd}^2 = 0.45\text{ yd}^2$$

Practice Problem 8 John has a swimming pool that measures 8 feet by 12 feet. He plans to make a concrete walkway around the pool that is 3 feet wide. What will be the total cost of the walkway at $12 a square foot? ■

EXAMPLE 9 Find the weight of the water in a full cylinder containing water if the height of the cylinder is 5 feet and the radius is 2 feet. 1 cubic foot of water weighs 62.4 pounds. Round your answer to the nearest ten pounds.

Understand the problem.

Draw a diagram.

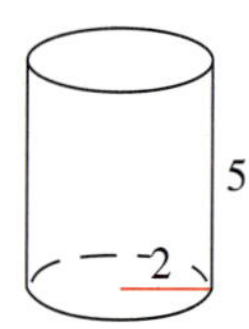

Write the formula for the volume of a cylinder.

$$\begin{aligned} V &= \pi r^2 h \\ &= (3.14)(2\text{ ft})^2(5\text{ ft}) \\ &= (3.14)(4\text{ ft}^2)(5\text{ ft}) \\ &= 62.8\text{ ft}^3 \end{aligned}$$

The volume of the cylinder is 62.8 cubic feet.

Each cubic foot of water weighs 62.4 pounds. That is, water weighs 62.4 pounds per cubic foot. This can be written as 62.4 lb/ft³. Since the volume of the cylinder is 62.8 ft³, we multiply.

$$\begin{aligned} \text{Weight of the water} &= 62.8\,\cancel{\text{ft}^3} \times \frac{62.4\text{ lb}}{1\,\cancel{\text{ft}^3}} \\ &= (62.8)(62.4)\text{ lb} \\ &= 3918.72\text{ lb} \end{aligned}$$

The weight of the water in the cylinder is 3920 rounded to the nearest ten pounds.

Practice Problem 9 Find the weight of the water in a rectangular watertight container that is 5 feet wide, 6 feet long, and 8 feet high. Remember, 1 cubic foot of water weighs 62.4 pounds. Round your answer to the nearest 100 pounds. ■

3.5 Exercises

Verbal and Writing Skills

Write the word(s) to complete each sentence.

1. Perimeter is the distance around a plane figure.
2. Circumference is the distance around a circle.
3. Area is a measure of the amount of surface in a region.
4. A trapezoid is a four-sided figure with exactly two sides parallel.
5. The sum of the interior angles of any triangle is 180°.

Applications

6. Find the area of a triangle whose altitude is 24 inches and whose base is 13 inches.
$A = \frac{1}{2}(24)(13) = 156 \text{ in.}^2$

7. Find the area of a parallelogram whose altitude is 14 inches and whose base is 7 inches.
$A = (14)(7) = 98 \text{ in.}^2$

8. The area of a parallelogram is 345 square meters. The altitude is 15 meters. What is the base?
$345 = (15)(b)$
$b = 23$ m

9. The area of a triangle is 80 square feet. Find the altitude if the base is 16 feet.
$80 = \frac{1}{2}(a)(16) = \frac{1}{2}(16)(a)$
$80 = 8a$
$a = 10$ feet

10. Find the area of a rectangle whose length is 12 centimeters and whose width is 7 centimeters.
$A = (12)(7) = 84 \text{ cm}^2$

11. Find the length of a rectangle whose area is 225 square inches and whose width is 15 inches. What kind of a rectangle is this? Find the perimeter.
$225 = 15l$
$l = 15$ cm
Square $P = 15 + 15 + 15 + 15 = 60$ cm

12. Find the area of a circular sign whose radius is 7.00 feet. (Use $\pi = 3.14$ as an approximate value.)
153.86 ft^2

13. Find the area of a circular flower bed whose diameter is 6 meters. (Use $\pi = 3.14$ as an approximate value.)
28.26 m^2

14. Find the area of a trapezoid with an altitude of 12 meters and whose bases are 5 meters and 7 meters, respectively.
$A = \frac{1}{2}(12)(5 + 7) = 72 \text{ m}^2$

15. The area of a trapezoid is 600 square inches and the bases are 20 inches and 30 inches, respectively. Find the altitude.
$600 = \frac{1}{2}a(30 + 20)$
$a = 24$ inches

16. Find the perimeter of a rectangular driveway whose width is 13 feet and whose length is 18.5 feet.
$P = 2(13) + 2(18.5) = 63$ feet

17. Find the perimeter of a triangular-shaped platform whose sides are 15 meters, 27 meters, and 31 meters, respectively.
$P = 15 + 27 + 31 = 73$ meters

18. The side of a barn is in the shape of a rectangle. The perimeter of the rectangle is 88 meters. Find the length of the rectangle if the width is 20 meters.
$88 = 2x + 2(20)$
$x = 24$ meters

19. An exihibit area has a floor base that is circular. Find the circumference of the circle if the diameter is 20 meters. (Use $\pi = 3.14$ as an approximate value.)
$C = (3.14)(20) = 62.8$ meters

20. The circumference of a circle is 31.4 centimeters. Find the radius. (Use $\pi = 3.14$ as an approximate value.)
$31.4 = (3.14)(2r)$
$r = 5$ centimeters

21. The perimeter of an equilateral triangle is 27 inches. Find the length of each side of the triangle.
$27 = 3x$
$x = 9$ inches

22. What is the length of one side of a square whose perimeter is 148 centimeters?
$148 = 4s$
$s = 37$ centimeters

23. The perimeter of an isosceles triangle is 32 feet. The base is $\frac{2}{3}$ as long as one of the equal sides. How long is each side of the triangle?
equal sides $= x = 12$ feet
base $= \frac{2}{3}x = 8$ feet
$32 = 2x + \frac{2}{3}x$
$x = 12$

24. Two angles of a triangle measure 47 degrees and 59 degrees, respectively. What is the measure of the third angle?
$x + 47 + 59 = 180$
$x = 74°$

25. A right triangle has one angle that measures 77 degrees. What does the other angle measure?
$x + 77 = 90$
$x = 13°$

26. Each of the equal angles of an isosceles triangle is 4 times as large as the vertex angle. What is the measure of each angle?
Vertex angle $= x = 20°$
Base angles $= 4x = 80°$
$4x + 4x + x = 180$
$x = 20$

★ **27.** In a triangle, the measure of the first angle is twice the measure of the second angle. The measure of the third angle is 20 degrees less than the second angle. What is the measure of each angle?
Second angle $= x = 50°$
First angle $= 2x = 100°$
Third angle $= x - 20 = 30°$
$x + 2x + (x - 20) = 180$

28. What is the volume of a rectangular solid whose height is 12 centimeters, whose width is 5 centimeters, and whose length is 21 centimeters?
$V = (12)(5)(21) = 1260 \text{ cm}^3$

29. What is the volume of a cylinder whose height is 8 inches and whose radius is 10 inches. (Use $\pi = 3.14$ as an approximate value.)
$V = (3.14)(8)(10)(10) = 2512 \text{ in.}^3$

30. A storage box holds furniture. The volume of this rectangular solid is 1320 cubic meters. Find the length if the height is 11 meters and the width is 10 meters.
$1320 = (11)(10)x$
$x = 12$ meters

31. A cylinder holds propane gas. The volume of the cylinder is 235.5 cubic feet. Find the height if the radius is 5 feet. (Use $\pi = 3.14$ as an approximate value.)
$235.5 = (3.14)(5)(5)x$
$x = 3$ feet

32. A beach ball is inflated to the shape of a sphere. Find the surface area of this sphere with radius 3 feet. (Use $\pi = 3.14$ as an approximate value.)
$S = 4(3.14)(3)(3) = 113.04 \text{ ft}^2$

33. A cabinet in the garage is shaped like a rectangular solid. The volume of the rectangular solid is 864 cubic inches. (a) Find the height of the solid if the length is 12 inches and the width is 9 inches. (b) What is the surface area?
$864 = (12)(9)x$ $S = 2(9 \times 12) + 2(8 \times 9) + 2(8 \times 12)$
$x = 8$ inches $S = 552$ square inches

34. A plastic cylinder made to hold milk is constructed with a solid top and bottom. The radius is 6.00 centimeters and the height is 4.00 centimeters. (a) Find the volume of the cylinder. (b) Find the total surface area of the cylinder.
(Use $\pi = 3.14$.)
$V = (3.14)(6)(6)(4) = 452.16 \text{ cm}^3$
$S = 2(3.14)(6)(4) + 2(3.14)(6)(6) = 376.8 \text{ cm}^2$

35. A Pyrex glass sphere is made to hold liquids in a science lab. The radius of the sphere is 3.00 centimeters. (a) Find the volume of the sphere. (b) Find the total surface area of the sphere.
(Use $\pi = 3.14$.) $V = \frac{4}{3}(3.14)(3)(3)(3) = 113.04 \text{ cm}^3$
$S = 4(3.14)(3)(3) = 113.04 \text{ cm}^2$

36. Inside the dashboard of a Dodge, Caravan is an unusually shaped piece of insulation. Find the area of the object shown in the figure. The length of the rectangle is 4 inches. The width of the rectangle and the diameter of the small half-circle are both 2 inches. The radius of the large quarter-circle is 2 inches. $A = 12.71 \text{ in.}^2$

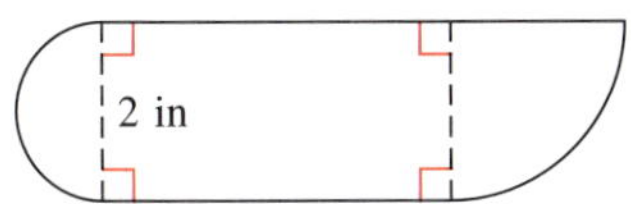

37. An oddly shaped field in front of Gordon College consists of a triangular region and a quarter-circle. This field must be fertilized. Green Lawn, Inc., will do the job for \$0.40 per square yard. How much will the job cost? $A = 59.625$ $C = \$23.85$

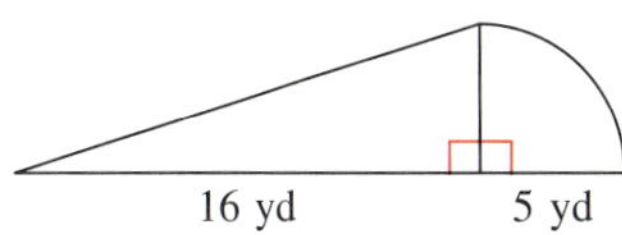

38. A cement walkway is to be poured. It consists of the two rectangles with dimensions as shown in the diagram and a quarter of a circle with radius 1.5 yards. (a) How many square yards will the walkway have? (b) If a painter paints it for \$2.50 per square yard, how much will the painting cost? 22.77 sq.yd.; \$56.93

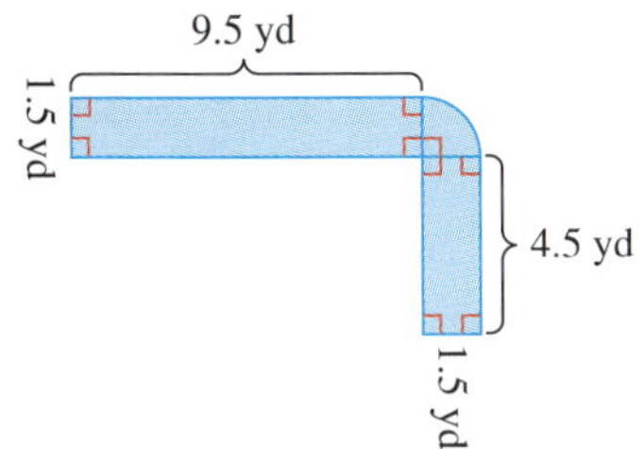

39. An aluminum plate is made according to the following dimensions. The outer radius is 10 inches and the inner radius is 3 inches. (a) How many square inches in area is the aluminum plate (the shaded region in the sketch)? (b) If the cost to construct the plate is \$7.50 per square inch, what is the construction cost?
285.74 sq.in.; \$2143.05

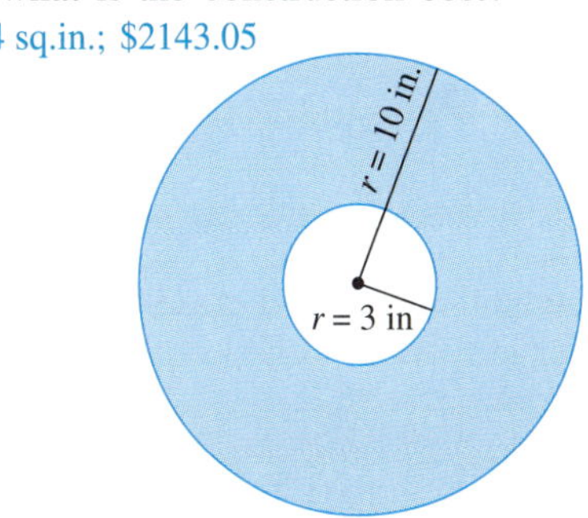

40. It cost \$120.93 to frame a painting with a flat aluminum strip. This was calculated at \$1.39 per foot for materials and labor. Find the dimensions of the painting if its length is 3 feet less than four times its width.
$x = W = 9.3$ ft $\quad [2(x) + 2(4x - 3)]1.39 = 120.93$
$4x - 3 = L = 34.2$ ft $\quad (10x - 6)1.39 = 120.93$
$x = 9.3$

41. A concrete driveway measures 18 yards by 4 yards. The driveway will be 0.5 yards thick. (a) How many cubic yards of concrete will be needed? (b) How much will it cost if concrete is \$3.50 per cubic yard?
36 cubic yd.; \$126

To Think About

42. An isosceles triangle has one side that measures 4 feet and another that measures 7.5 feet. The third side was not measured. Can you find one unique perimeter for this type of triangle? What are the two possible perimeters? No. The perimeter may be 19 feet or it may be 15.5 feet.

43. Find the total surface area of (a) a sphere of radius 3 inches and (b) a right circular cylinder with radius 2 inches and height 0.5 inch. Which object has a greater surface area? The sphere.
$A = 4\pi(3)^2 = 36\pi \text{ in.}^2$
$A = 2\pi(2)(.5) + 2\pi(2)^2$
$= 2\pi + 8\pi = 10\pi \text{ in.}^2$

44. Assume that a very long rope is stretched around the moon supported by poles that are 3 *feet tall.* (Neglect gravitational pull and assume that the rope takes the shape of a large circle, as shown in the figure.) The radius of the moon is approximately 1080 miles. How much longer would you need to make the rope if you wanted the rope to be supported by poles that are 4 *feet tall*?
$2\pi(1080 \cdot 5280 + 4) - 2\pi(1080 \cdot 5280 + 3)$
$= 2\pi = 6.28$ ft

Cumulative Review

Simplify.

45. $2(x - 6) - 3[4 - x(x + 2)]$ $\quad 3x^2 + 8x - 24$

46. $-5(x + 3) - 2[4 + 3(x - 1)]$ $\quad -11x - 17$

47. $-\{3 - 2[x - 3(x + 1)]\}$ $\quad -4x - 9$

48. $-2\{x + 3[2 - 4(x - 3)]\}$ $\quad 22x - 84$

Putting Your Skills to Work

A MATHEMATICAL PREDICTION OF RAINFALL

You have probably seen pictures of Seattle in the rain and talked to friends who have visited Seattle in the rain. Perhaps you have been there yourself in the rain. According to government records kept over the last 50 years, Seattle has an average of 154 days a year when it rains or snows for at least part of the day. However, some periods of the year have more precipitation than others, as you can see from the circle graph shown on the right.

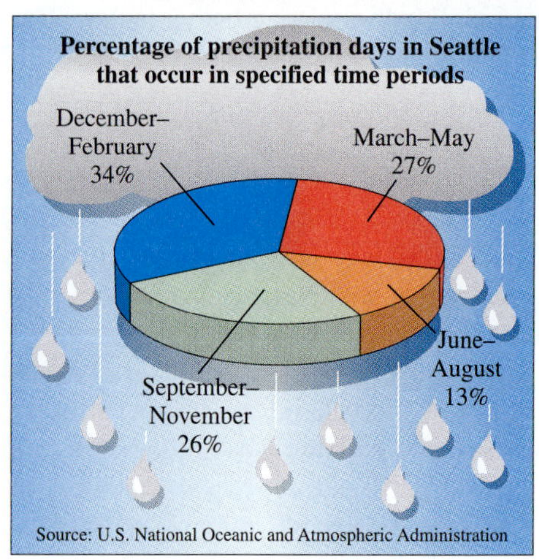

PROBLEMS FOR INDIVIDUAL INVESTIGATION

1. What percent of the days of precipitation per year in Seattle occur in the time period of September to February? 60%

2. How many of the 154 days of precipitation per year in Seattle occur in the time period of June to November? 60 days

PROBLEMS FOR JOINT INVESTIGATION AND COOPERATIVE GROUP ACTIVITY

With some other members of your class determine the answer to the following.

3. Write a Precipitation Equation of the form $P = ad$, where d is the number of days that you visit Seattle during the period September through November and P is the number of days that you are likely to see precipitation. What is the value of the number a in this case? (You may express a as a fraction or as a decimal rounded to the nearest hundredth.)
$P = (40/91)d$ or $P = 0.44d$ $a = 40/91$ or $a = 0.44$

4. Use your equation obtained in (3) in order to find how many days of precipitation someone would expect to have if they visited Seattle over a period of several years for 46 days during the months of September, October, and November. Round your answer to the nearest whole number. 20 days

INTERNET CONNECTIONS: Go to ``http://www.prenhall.com/tobey'' to be connected

Site: Climate of San Francisco Graphics (National Weather Service) or a related site

Use the graph of Average Rainfall and Rainy Days, which shows the average number of rainy days in San Francisco for each month of the year. Answer the following questions.

5. On the average, how many rainy days occur in the time period of November to January? What is the total number of days during this time period?

6. Write a Precipitation Equation of the form $P = ad$, where d is the number of days that you visit San Francisco during the period of November to January and P is the number of rainy days. (You may express a as a fraction or as a decimal rounded to the nearest thousandth.)

7. Suppose you are planning to be in San Francisco for 20 days during November, December, and January. How many rainy days do you expect to see? Round your answer to the nearest whole number.

3.6 Use Inequalities to Solve Word Problems

After studying this section, you will be able to:

1 *Translate English sentences into inequalities.*

2 *Solve applied problems using inequalities.*

1 *Translating English Sentences into Inequalities*

Real-life situations often involve inequalities. For example, if a student wishes to maintain a certain grade point average (GPA), you may hear him or her say, "I have to get a grade of at least an 87 on the final exam." What does this mean? What grade does the student need to achieve? 87 would fit the condition, *at least an 87*. What about an 88? The student would be happy with an 88. In fact the student would be happy with any grade that *is greater than or equal to* an 87. We can write this as

$$\text{grade} \geq 87$$

Notice that the English phrase *at least an 87* translates mathematically as *greater than or equal to an 87*. Be careful when translating English phrases that involve inequalities into algebra. If you are not sure which symbol to use, try a few numbers to see if they fit the conditions of the problem and note the direction they take. Choose the mathematical symbol accordingly.

EXAMPLE 1 Translate each phrase into an inequality.

(a) The perimeter of the rectangle must be no more than 70 inches.

(b) The average number of errors must be less than 2 mistakes per page.

(c) His income must be at least \$950 per week.

(d) The new car will cost at most \$12,070.

(a) Perimeter must be *no more than* 70.

$$\text{Perimeter} \leq 70$$

Try 80. *Does* 80 *fit the condition, no more than* 70? *No. Do* 60 *and* 50 *fit? Yes.* Use the less than or equal to symbol.

(b) Average number must be *less than* 2.

$$\text{Average} < 2$$

(c) Income must be *at least* \$950.

$$\text{Income} \geq 950$$

"At least" means, "not less than." The income must be \$950 *or more. Use* $\geq$.

(d) The new car will cost *at most* \$12,070.

$$\text{Car} \leq 12{,}070$$

At most \$12,070 *means it is the top price for the car. The cost of the car cannot be more than* \$12,070. *Use* $\leq$.

Practice Problem 1 Translate each phrase into an inequality.

(a) The height of the person must be no greater than 6 feet.

(b) The speed of the car must have been greater than 65 miles per hour.

(c) The area of the room must be greater than or equal to 560 square feet.

(d) The profit margin cannot be less than \$50 per television set. ■

2 Solving Applied Problems Using Inequalities

We can use the same problem-solving techniques we used previously to solve word problems involving inequalities.

EXAMPLE 2 A manufacturing company makes 60-watt light bulbs. For every 1000 bulbs manufactured, a random sample of 5 bulbs is selected and tested. To meet quality control standards, the average number of hours these five bulbs last must be at least 900 hours. The test engineer tested 5 bulbs. Four of them lasted 850 hours, 1050 hours, 1000 hours, and 950 hours, respectively. How many hours must the fifth bulb last in order to meet quality control standards?

1. *Understand the problem.*

Let n = the number of hours the fifth bulb must last

The average that the five bulbs last *must be at least* 900 hours. We could write

$$\text{Average} \geq 900 \text{ hours}$$

2. *Write an inequality.*

$$\text{Average of five bulbs} \geq 900 \text{ hours}$$

$$\frac{850 + 1050 + 1000 + 950 + n}{5} \geq 900$$

3. *Solve the inequality and state the answer.*

$$\frac{850 + 1050 + 1000 + 950 + n}{5} \geq 900$$

$$\frac{3850 + n}{5} \geq 900 \qquad \textit{Simplify the numerator.}$$

$$3850 + n \geq 4500 \qquad \textit{Multiply both sides by 5.}$$

$$n \geq 650 \qquad \textit{Subtract 3850 from each side.}$$

If the fifth bulb burns for 650 hours or more, the quality control standards will be met.

4. *Check.* Any value of 650 or larger should satisfy the original inequality. Suppose we pick $n = 650$ and substitute it into the inequality.

$$\frac{850 + 1050 + 1000 + 950 + 650}{5} \stackrel{?}{\geq} 900$$

$$\frac{4500}{5} \stackrel{?}{\geq} 900$$

$$900 \geq 900 \checkmark$$

If we try $n = 700$, the left side of the inequality will be 910. $910 \geq 900$. In fact, if we try any number greater than 650 for n, the left side of the inequality will be greater than 900. The inequality will be valid for $n = 650$ or any larger value.

Practice Problem 2 A manufacturing company makes coils of $\frac{1}{2}$-inch maritime rope. For every 300 coils of rope made, a random sample of four coils of rope is selected and tested. To meet quality control standards, the average number of pounds that each coil will hold must be at least 1100 pounds. The test engineer tested four coils of rope. The first three held 1050 pounds, 1250 pounds, and 950 pounds, respectively. How many pounds must the last coil hold if the rope is to meet quality control standards? ■

EXAMPLE 3 Juan is selling supplies for restaurants in the Southwest. He gets paid \$700 per month plus 8% of all the supplies he sells to restaurants. If he wants to earn more than \$3100 per month, what value of products will he need to sell?

1. *Understand the problem.*
   Juan earns a fixed salary and a salary that depends on how much he sells (commission salary) each month.

   Let x = the value of the supplies he sells each month

   Then $0.08x$ = the amount of the commission salary he earns each month

   He wants the total income to be more than \$3100 per month; thus

   $$\text{Total income} > 3100$$

2. *Write the inequality.*

The fixed income	added to	the commission income	must be greater than	3100
↓	↓	↓	↓	↓
700	+	$0.08x$	>	3100

3. *Solve the inequality and state the answer.*

   $$700 + 0.08x > 3100$$

   $$0.08x > 2400 \qquad \textit{Subtract 700 from each side.}$$

   $$x > 30{,}000 \qquad \textit{Divide each side by 0.08.}$$

   Juan must sell more than \$30,000 worth of supplies each month.

4. *Check.* Any value greater than 30,000 should satisfy the original inequality. Suppose we pick $x = 30{,}001$ and substitute it into the inequality.

$$700 + 0.08(30{,}001) \stackrel{?}{>} 3100$$

$$700 + 2400.08 \stackrel{?}{>} 3100$$

$$3100.08 > 3100 \checkmark$$

Practice Problem 3 Rita is selling commercial fire and theft alarm systems to local companies. She gets paid \$1400 per month plus 2% of the cost of all the systems she sells in one month. She wants to earn more than \$2200 per month. What value of products will she need to sell each month? ■

3.6 Exercises

Verbal and Writing Skills

Translate to an inequality.

1. The cost is greater than \$67,000. $x > 67{,}000$

2. The clearance of the bridge is less than 12 feet. $x < 12$

3. The number of people is not more than 120. $x \leq 120$

4. The tax increase is more than \$7890. $x > 7890$

5. The perimeter is less than 34 miles. $x < 34$

6. The area is not greater than 550 square yards. $x \leq 550$

7. His earnings cannot be less than \$3500 per month. $x \geq 3500$

8. The number of passengers in the plane is not less than 26 people. $x \geq 26$

9. When you bake the cake, the temperature cannot exceed 350 degrees Fahrenheit. $x \leq 350$

10. The A student will never exceed four unexcused absences per semester. $x \leq 4$

Applications

Solve each of the following applied problems by using an inequality.

11. The fencing around a farmer's triangular field needs to be installed. Two sides are 180 feet and 135 feet respectively, and the farmer has 539 feet of fencing available. What are the possible values for the length of the third side of the field?
$180 + 135 + x \leq 539$ $\quad x \leq 224$
It must be less than or equal to 224 feet.

12. A triangular-shaped window on David Kirk's new Searay boat has two sides that measure 87 centimeters and 64 centimeters, respectively. The perimeter of the triangle must not exceed 291 centimeters. What are the possible values for the length of the third side of the window?
$87 + 64 + x \leq 291$
It must be less than or equal to 140 cm.

13. Sophia has scored 85, 77, and 68 on three tests in her biology course. She has one final test to take and wants to obtain an average of 80 or more in the course. What possible values can she obtain on her final test and still achieve her goal?
$\dfrac{85 + 77 + 68 + x}{4} \geq 80$ $\quad$ She must score 90 or more.

14. A video rental store needs to average 48 or more customers per day for six days each week in order to make a profit. On the first five days the following number of customers came into the store: 60, 36, 40, 52, and 59. What possible number of customers can come into the rental store on the last day to achieve the customer goal for the week?
$\dfrac{60 + 36 + 40 + 52 + 59 + x}{6} \geq 48$
$x \geq 41$

15. A computer software salesman has a limit of \$655 per month in travel and entertainment expenses. He spends \$230 monthly for travel. He wants to take as many clients to lunch as possible. If the average lunch for two people is \$25, how many potential clients can he feed?
$230 + 25x \leq 655$
$x \leq 17$
He must take 17 potential clients or fewer.

16. Tim drives a cab part-time. He gets paid \$100 per week plus tips. His average tip is \$3 per trip. If he needs to earn an average of at least \$217 per week to meet his college expenses, how many taxi trips must he take per week to meet his expense requirements?
$100 + 3x \geq 217$
$x \geq 39$
He must have 39 cab trips or more.

17. Dick and Ann Wright finally saved up enough money to build a summer lake house. The living room will be 12 feet wide. What are the possible lengths of the living room if the area of the room must be greater than or equal to 204 square feet.
$12x \geq 204$ $\quad x > 17$
The length must be 17 feet or longer.

18. Martel Construction is constructing a house with a master bedroom that is 16 feet wide. What are the possible lengths of the bedroom if the area of the room must be greater than or equal to 224 square feet?
$16x \geq 224$ $\quad x \geq 14$ feet
The length must be 14 feet or longer.

19. John is selling cars at Centerville Chrysler. He receives \$850 per month plus \$130 for every car he sells during the month. He wants to earn at least \$2800 per month. How many cars will he need to sell each month?
$850 + 130x \geq 2800$
$x \geq 15$ cars

20. Marcia is selling cars at the Beverly Jeep Center. She receives \$920 per month plus \$120 for every car or wagon she sells during the month. She wants to earn at least \$2600 per month. How many cars and wagons will she need to sell to meet that goal?
$920 + 120x \geq 2600$
$x \geq 14$ cars and wagons

21. Melissa is selling supplies for MidCentral Medical Service. She gets paid \$900 per month plus 3% of the sales price of all the medical supplies she sells. If she wants to earn more than \$2700 per month, what value of products will she need to sell?
$900 + 0.03x > 2700$
$x > \$60{,}000$

22. Jerome is selling supplies for Commercial Photography Centers. He gets paid \$600 per month plus 4% of the sales price of all the supplies he sells. If he wants to earn more than \$2500 per month, what value of products will he need to sell?
$600 + 0.04x > 2500$
$x > \$47{,}500$

23. The coolant in Tom's car is effective for radiator temperatures of less than 110 degrees Celsius. What would be the corresponding temperature range in degrees Fahrenheit? (Use $C = \frac{5}{9}(F - 32)$ in your inequality.)
$\frac{5}{9}(F - 32) < 110$
$F < 230°$

24. A person's IQ is found by multiplying mental age (m), as indicated by standard tests, by 100 and dividing this result by the chronological age (c). This gives the formula IQ $= 100m/c$. If a group of 14-year-old children has an IQ of not less than 80, what is the mental age range of this group?
$\frac{100m}{14} \geq 80$
$m \geq 11.2$

25. Dr. Slater is purchasing computer tables and chairs for the tutoring center at Westmont College. The dean has told him that he cannot spend more than \$850. Computer tables will cost him \$55 and chairs will cost him \$28. He has decided to purchase nine computer tables. What are the different possible numbers of chairs that he can buy?
$9(55) + 28x \leq 850$
$x \leq 12$ chairs

26. Franklin and Roberto are planning a pizza party for their college friends. They have decided they cannot spend more than \$88 for food. Each large pizza costs \$11.50 at their favorite pizza parlor. They purchase six-packs of cold soda for \$3.50 each. They have decided to buy six large pizzas. What are the different possible numbers of six-packs of cold soda that they can buy?
$6(11.50) + 3.50x \leq 88$
$x \leq 5$ six-packs

★ **27.** A compact disc recording company has a cost equation for the week of cost $= 5000 + 7n$, where n is the number of compact discs manufactured in one week. The income equation for the amount of income produced from selling these n discs is income $= 18n$. How many discs need to be manufactured and sold in one week for the income to be greater than the cost?
$18n > 5000 + 7n$
$n > 454$ discs

★ **28.** A company that manufactures VCRs has a cost equation for the week of cost $= 12{,}000 + 100n$, where n is the number of VCRs manufactured in one week. The income equation for the amount of income produced from selling these n VCRs is income $= 210n$. How many VCRs need to be manufactured and sold in one week for the income to be greater than the cost?
$210n > 12{,}000 + 100n$
$n > 109$ VCRs

Cumulative Review Problems

Solve each inequality.

29. $10 - 3x > 14 - 2x$ $\quad x < -4$

30. $2x - 3 \geq 7x - 18$ $\quad x \leq 3$

31. $30 - 2(x + 1) \leq 4x$ $\quad x \geq 4\frac{2}{3}$

32. $2(x + 3) - 22 < 4(x - 2)$ $\quad x > -4$

Chapter Organizer

Procedure for Solving Applied Problems

1. **Understand the problem.**
   (a) Read the word problem carefully to get an overview.
   (b) Determine what information you will need to solve the problem.
   (c) Draw a sketch. Label it with the known information. Determine what needs to be found.
   (d) Choose a variable to represent one unknown quantity.
   (e) If necessary, represent other unknown quantities in terms of that same variable.
2. **Write an equation.**
   (a) Look for key words to help you to translate the words into algebraic symbols.
   (b) Use a given relationship in the problem or an appropriate formula in order to write an equation.
3. **Solve the equation and state the answer.**
4. **Check.**
   (a) Check the solution in the original equation.
   (b) Be sure the solution to the equation answers the question in the word problem. You may need to do some additional calculations if it does not.

Example A

The perimeter of a rectangle is 126 meters. The length of the rectangle is 6 meters less than double the width. Find the dimensions of the rectangle.

Length / Width / Width / Length

1. *Understand the problem.* We want to find the length and the width of a rectangle whose perimeter is 126 meters.

   The length is *compared to* the width, so we start with the width. — Let w = width

   The length is 6 meters less than double the width. — Let $l = 2w - 6$
2. *Write an equation.* The perimeter of a rectangle is $P = 2w + 2l$. — $126 = 2w + 2(2w - 6)$
3. *Solve the equation.*

   $126 = 2w + 4w - 12$

   $126 = 6w - 12$

   $138 = 6w$

   $23 = w$ The width is 23 meters.

   $2w - 6 = 2(23) - 6 = 46 - 6 = 40$

   The length is 40 meters.
4. *Check:* Is this reasonable? Yes. A rectangle 23 meters wide and 40 meters long seems to be about right for the perimeter to be 126 meters.

   40 / 23 / 23 / 40

   Is the perimeter exactly 126 meters?

   $2(23) + 2(40) \stackrel{?}{=} 126$

   $46 + 80 \stackrel{?}{=} 126$

   Yes $126 = 126$ ✓

   Is the length exactly 6 meters less than double the width?

   $40 \stackrel{?}{=} 2(23) - 6$

   $40 \stackrel{?}{=} 46 - 6$

   Yes. $40 = 40$ ✓

Example B

Gina saved some money for college. For one year she invested $2400 and earned $225 in interest. She invested part of it at 12% and the rest of it at 9%. How much did she invest at each rate?

1. *Understand the problem.* We want to find each amount that was invested.

   Interest earned at 12% + interest earned at 9% = total interest of $225

   We let x represent one quantity of money. — Let x = amount of money invested at 12%

   We started with $2400. If we invest x at 12%, we still have $(2400 - x)$ left. — Let $2400 - x$ = the amount of money invested at 9%

   Interest = prt

   Interest at 12% — $I_1 = 0.12x$

   Interest at 9% — $I_2 = 0.09(2400 - x)$
2. *Now write the equation.* $I_1 + I_2 = 225$. — $0.12x + 0.09(2400 - x) = 225$
3. *Solve the equation.*

   $0.12x + 216 - 0.09x = 225$

   $0.03x + 216 = 225$

   $0.03x = 9$

   $x = \frac{9.00}{0.03}$

   $x = 300$

   $300 was invested at 12%

   $2400 - x = 2400 - 300 = 2100$

   $2100 was invested at 9%
4. *Check:* Is this reasonable? Yes.

   Are the conditions of the problem satisfied? Does the total amount invested equal $2400?

   $2100 + 300 \stackrel{?}{=} 2400$

   Yes. $2400 = 2400$ ✓

   Will $2100 at 9% and $300 at 12% yield $225 in interest?

   $0.09(2100) + 0.12(300) \stackrel{?}{=} 225$

   $189 + 36 \stackrel{?}{=} 225$

   Yes. $225 = 225$ ✓

Topic	Procedure	Examples
Geometric formulas, p. 214.	Know the definitions of and formulas related to basic geometric shapes such as the triangle, rectangle, parallelogram, trapezoid, and circle. Review these formulas from the beginning of Section 3.5 as needed. Formulas for the volume and total surface area of three solids are also given there.	The volume of a right circular cylinder found in a four-cylinder car engine is determined by the formula $V = \pi r^2 h$. Find V if $r = 6$ centimeters and $h = 5$ centimeters. Use $\pi = 3.14$. $V = 3.14(6 \text{ cm})^2(5 \text{ cm})$ $= (3.14)(36)(5) \text{ cm}^3 = (113.04)(5) \text{ cm}^3$ $= 565.2$ cc (or cm^3) Four such cylinders would make this a 2260-cc engine.
Using inequalities to solve word problems, p. 225.	Be sure that the inequality you use fits the conditions of the word problem. 1. At least 87 means ≥ 87 2. No less than \$150 means $\geq$\$150 3. At most \$500 means $\leq$\$500 4. No more than 70 means ≤ 70 5. Less than 2 means <2 6. More than 10 means >10	Julie earns 2% on sales over \$1000. How much must Julie make in sales to earn at least \$300? Let x = amount of sales $(2\%)(x - \$1000) \geq \$300$ $0.02(x - 1000) \geq 300$ $x - 1000 \geq 15{,}000$ $x \geq 16{,}000$ Julie must make sales of \$16,000 or more (at least \$16,000).

Chapter 3 Review Problems

3.1 *Write an algebraic expression. Use the variable x to represent the unknown value.*

1. 19 more than a number $x + 19$

2. Two-thirds of a number $\frac{2}{3}x$

3. 56 less than a number $x - 56$

4. Triple a number $3x$

5. Double a number increased by seven $2x + 7$

6. Twice a number decreased by three $2x - 3$

7. Five times a number decreased by one-third of that same number $5x - \frac{x}{3}$

8. One-fourth of the sum of a number and six $\frac{x + 6}{4}$

Write an expression for each of the quantities compared. Use the letter specified.

9. The speed of the jet was 370 miles per hour faster than the speed of the helicopter. (Use the letter h.)

$h + 370$ = the speed of the jet
h = the speed of the helicopter

10. The cost of the truck was \$2000 less than the cost of the garage. (Use the letter g.)

$g - 2000$ = the cost of the truck
g = the cost of the garage

11. The length of the rectangle is 5 meters more than triple the width. (Use the letter w.)

$3w + 5$ = the length of the rectangle
w = the width of the rectangle

12. The number of working people is four times the number of retired people. The number of unemployed people is one-half the number of retired people. (Use the letter r.)

r = the number of retired people
$4r$ = the number of working people
$0.5r$ = the number of unemployed people

13. A triangle has three angles A, B, and C. The number of degrees in angle A is double the number of degrees in angle B. The number of degrees in angle C is 17 degrees less than the number in angle B. (Use the letter b.)

b = the number of degrees in angle B
$2b$ = the number of degrees in angle A
$b - 17$ = the number of degrees in angle C

14. There are 29 more students in biology class than in algebra. There are one-half as many students in geology class as in algebra. (Use the letter a.)

a = the number of students in algebra
$a + 29$ = the number of students in biology
$0.5a$ = the number of students in geology

3.2 *Solve each word problem.*

15. Twice a number is increased by five. The result is 29. What is the original number? $x = 12$

16. Triple a number is decreased by seven. The result is 47. What is the original number? $x = 18$

17. When one-fourth of a number is diminished by two the result is nine. Find the original number. $x = 44$

18. Seven times a number diminished by three is equal to four times the original number. Find the original number. $x = 1$

19. Jon is twice as old as his cousin David. If Jon is 32, how old is David? $x = 16$ years old

20. Six chairs and a table cost \$450. If the table cost \$210, how much does one chair cost? $x = \$40$

21. Two cars left Center City and drove to the shore 330 miles away. One car was driven at 50 miles an hour. The other car was driven at 55 miles an hour. How long did it take each car to make the trip?
$t_1 = 6.6$ hours; $t_2 = 6$ hours

22. A student took five math tests. She obtained scores of 85, 78, 65, 92, and 70. She needs to complete one more test. If she wants to get an average of 80 on all six tests, what score does she need on her last test? 90

3.3

23. A rectangle has a perimeter of 66 meters. The length of the rectangle is 3 meters longer than double the width. Find the dimensions of the rectangle.
$w = 10$ m; $l = 23$ m

24. The length of a rectangle is $2\frac{1}{2}$ times its width. The rectangle has a perimeter of 84 feet. Find the length and width of the rectangle. $w = 12$ ft; $l = 30$ ft

25. The perimeter of a triangle is 40 yards. The second side is 1 yard shorter than double the length of the first side. The third side is 9 yards longer than the first side. Find the length of each side.
1st side = 8 yd; 2nd side = 15 yd; 3rd side = 17 yd

26. The measure of the second angle of a triangle is three times the measure of the first angle. The measure of the third angle is 12 degrees less than twice the measure of the first. Find the measure of each of the three angles.
1st angle = 32°; 2nd angle = 96°; 3rd angle = 52°

27. A piece of rope 50 yards long is cut into two pieces. One piece is three-fifths as long as the other. Find the length of each piece. 31.25 yd and 18.75 yd

28. Jon and Lauren have a combined income of \$48,000. Lauren earns \$3000 more than half of what Jon earns. How much do each earn?
Jon = \$30,000
Lauren = \$18,000

3.4

29. Abel rented a car from Sunshine Car Rentals. He was charged \$39 per day and \$0.25 per mile. He rented the car for 3 days and paid a rental cost of \$187. How many miles did he drive the car? 280 miles

30. The electric bill at Jane's house this month was \$71.50. The charge is based on a flat rate of \$25 per month plus a charge of \$0.15 per kilowatt-hour of electricity used. How many kilowatt-hours of electricity were used?
310 kilowatt-hours

31. Jamie bought a new tape deck for her car. When she bought it, she found that the price had been decreased by 18%. She was able to buy the tape deck for \$36 less than if the item had not been on sale. What was the original price? \$200.00

32. Alfred has worked for his company for 10 years as a store manager. Last year he received a 9% raise in salary. He now earns \$24,525 per year. What was his salary last year before the raise? \$22,500

33. Peter and Shelly invested \$9000 for one year. Part of it was invested at 12% and the remainder was invested at 8%. At the end of one year the couple had earned exactly \$1000 in interest. How much did they invest at each rate?
\$7000 at 12% and \$2000 at 8%

34. A man invests \$5000 in a savings bank. He places part of it in a checking account that earns 4.5% and the rest in a regular savings account that earns 6%. His total annual income from this investment is \$270. How much was invested in each account?
\$2000 at $4\frac{1}{2}$% and \$3000 at 6%

35. Mary has \$3.75 in nickels, dimes, and quarters. She has three more quarters than dimes. She has twice as many nickels as quarters. How many of each coin does she have? 18 nickels; 6 dimes; 9 quarters

36. Tim brought in \$3.65 in coins. He had two more quarters than he had dimes. He had one fewer nickel than the number of dimes. How many coins of each type did he have? 7 nickels; 8 dimes; 10 quarters

3.5 *Solve each word problem.*

37. Find the diameter of a circle whose circumference is 25.12 meters. (Use $\pi \approx 3.14$.) 8 m

38. What is the perimeter of a rectangle whose length is 8 inches and whose width is $\frac{1}{4}$ inch? $16\frac{1}{2}$ in.

39. Find the third angle of a triangle if one angle measures 62 degrees and a second angle measures 47 degrees. 71°

40. Find the area of a triangle whose base is 8 miles and whose altitude is 10.5 miles. 42 sq. mi.

41. Find the perimeter of a triangle whose smallest side is 8 inches, whose second side is double the first, and whose third side is double the second. 56 in.

42. Find the area of a parallelogram whose altitude is 8 centimeters and whose base is 21 centimeters. $A = 168$ square centimeters

43. Find the volume of a rectangular solid whose width is 8 feet, whose length is 12 feet, and whose height is 20 feet. 1920 ft^3

44. Find the volume of a sphere if the radius is 3 centimeters. (Use $\pi \approx 3.14$.) 113.04 cm^3

45. The perimeter of an isosceles triangle is 46 inches. The two equal sides are each 17 inches long. How long is the other side? 12 in.

46. Find the area of a circle whose diameter is 18.00 centimeters. (Use $\pi \approx 3.14$.) 254.34 cm^2

47. Find the cost to build a cylinder out of aluminum if the radius is 5 meters and the height is 14 meters and the cylinder will have a flat top and flat bottom. The cost factor has been determined to be $40.00 per square meter. (*Hint:* First find the total surface area.) $23,864

48. Fred wants to have the exterior of his house painted. Two sides of the house have a height of 18 feet and a length of 22 feet, and the other two sides of the house have a height of 18 feet and a length of 19 feet. The painter has quoted a price of $2.50 per square foot. The windows and doors of the house take up 100 square feet. How much will the painter charge to paint the walls? $3440

3.6 *Solve.*

49. Bruce scored 92, 85, 78, and 86 on his math tests. What must he score on his last test to obtain an 85 or more for the course? 84

50. Diane has more than $10 in dimes and quarters. She has 10 times as many dimes as quarters. At least how many of each kind of coin does she have? 80 dimes, 8 quarters

51. Chelsea rents a car for $35 a day and 15 cents a mile. How far can she drive if she wants to rent the car for 2 days and does not want to spend more than $100? 200 miles

52. Philip earns $12,000 per year in salary and 4% commission on his sales. How much were his sales if his annual income was more than $24,000? more than $300,000

53. Michael plans to spend less than $70 on shirts and ties. He bought three shirts for $17.95 each. At most, how much can he spend on ties? $16.15

54. Two cars start from the same point going in opposite directions. One car travels at 50 miles per hour. The other car travels at 55 miles per hour. How long must they travel to be at least 315 miles apart? 3 hours

Miscellaneous

Solve the following miscellaneous applied problems.

55. Fred timed a popular TV show that was scheduled for 30 minutes. The entertainment portion of the show lasted 4 minutes longer than four times the amount of time that was devoted to commercials. How many minutes long was the actual entertainment portion of the show? $24\frac{4}{5}$ min

56. Faye conducted an experiment. To verify the weight of four identical steel balls, she placed them on a balance. She placed three of the balls on one side of the balance. On the other side she placed an 8-ounce weight and one of the balls. She discovered that these two sides exactly balanced on the balance beam. How much does each steel ball weigh? 4 ounces

57. Susan borrowed $333 from her uncle to help purchase a car. She promised to pay him back with an initial payment of $45 and then $18 per month until the loan is paid back. How many payments will it take to pay off the loan? 16

58. The college baseball team has won seven of their first nine games. They will play an additional 16 games. The coach desires that they win at least 84% of their games in order to be eligible for the regional playoffs. How many of the 16 games will they need to win? 14 games

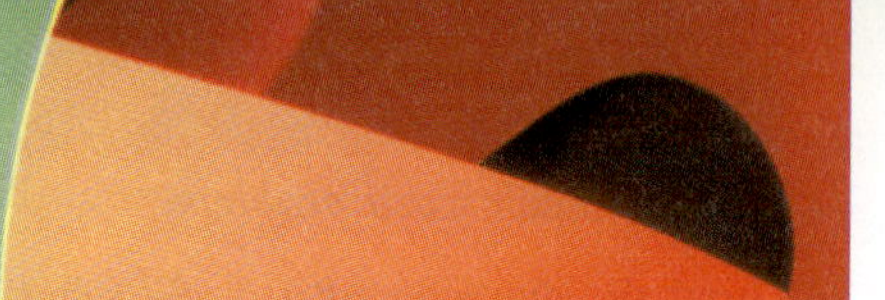

Putting Your Skills to Work

USING MATHEMATICS FOR SEARCH AND RESCUE

The Coast Guard employs modern computer methods to determine the time to reach the area where help is needed. However, the important questions are determined by mathematical relationships. To answer valuable questions regarding rescue operations requires either a person to solve equations by hand or a computer to perform similar calculations.

PROBLEMS FOR INDIVIDUAL STUDY AND INVESTIGATION

A sinking boat reports that it is 100 miles from the Coast Guard rescue station. The boat is sinking and will only stay afloat for one more hour. It is slowly heading toward the rescue station at 5 miles per hour. A rescue helicopter flying at 120 miles per hour leaves the station 5 minutes after the call comes in.

1. Let t = the time elapsed from the time of the call to the time of the helicopter reaching the ship. Write two equations using $d = rt$ to describe the distance traveled by the sinking boat and the distance traveled by the helicopter. $d = 5t \qquad d = 120(t - \frac{1}{12})$

2. Write an equation using the two distance expressions in (1) and the total distance of 100 miles. Solve this equation for t. Will the helicopter reach the boat before it sinks?
$5t + 120(t - \frac{1}{12}) = 100$
$t = \frac{22}{25}$ Time is about 52.8 minutes. Yes.

PROBLEMS FOR COOPERATIVE GROUP INVESTIGATION

Together with other members of your class, determine an answer to the following.

3. A sinking boat is 76 miles from a Coast Guard station. The boat is traveling toward the Coast Guard station battling heavy seas. It is only able to move toward the station at 2 miles per hour. At the same time a rescue helicopter leaves the Coast Guard station and flies directly toward the boat at 150 miles per hour. If the boat does not sink, how long will it take the helicopter to reach the boat?
$\frac{1}{2}$ hour or 30 minutes

4. When the helicopter arrives, the boat has already sunk. The Coast Guard orders a search by eight rescue helicopters. The search area is defined by the trapezoid shown. (shaded region)
How many square miles are in the search area? If a helicopter can search 15 square miles in an hour, how long will it take the eight helicopters to search the entire area?
18,850 square miles, 157 hours or 6.5 days

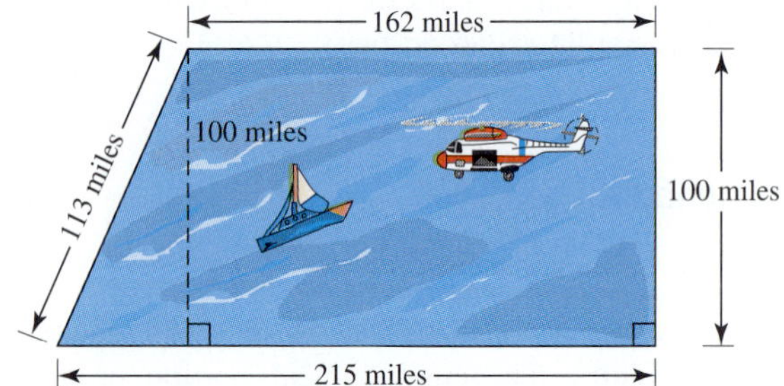

INTERNET CONNECTIONS: Go to ''http://www.prenhall.com/tobey'' to be connected

Site: U.S. Coast Guard Air Station, San Francisco or a related site

This site gives general information about the Dolphin helicopters used by the U.S. Coast Guard.

5. Suppose the crew of the Dolphin needs to rescue someone who is 150 miles away from the air station. Assume that they need to stay at the rescue location for 30 minutes. How long will the complete rescue operation take, including round-trip travel?

6. Explain why it might not be practical for the Dolphin crew to rescue someone who is 200 nautical miles away, even though the range of the helicopter is sufficient to travel 400 nautical miles.

Chapter 3 Test

Solve each applied problem.

1. A number is doubled and then decreased by 11. The result is 59. What is the original number?

2. The sum of one-half of a number, one-ninth of a number, and one-twelfth of a number is twenty-five. Find the original number.

3. Triple a number is increased by six. The result is the same as when the original number was diminished by three and then doubled. Find the original number.

4. A triangular region has a perimeter of 66 meters. The first side is two-thirds of the second side. The third side is 14 meters shorter than the second side. What are the lengths of the three sides of the triangular region?

5. A rectangle has a length 7 meters longer than double the width. The perimeter is 134 meters. Find the dimensions of the rectangle.

6. Three harmful pollutants were measured by a consumer group in the city. The total sample contained 15 parts per million of three harmful pollutants. The amount of the first pollutant was double the second. The amount of the third pollutant was 75% of the second. How many parts per million of each pollutant were found?

7. Raymond has a budget of $1940 to rent a computer for his company office. The computer company he wants to rent from charges $200 for installation and service as a one-time fee. Then they charge $116 per month rental for the computer. How many months will Raymond be able to rent a computer with this budget?

8. Last year the yearly tuition at King's College went up 8%. This year's charge for tuition for the year is $9180. What was it last year before the increase went into effect?

1. 35

2. 36

3. −12

4. first side = 20 m; second side = 30 m; third side = 16 m

5. width = 20 m; length = 47 m

6. 1st pollutant = 8 ppm; 2nd pollutant = 4 ppm; 3rd pollutant = 3 ppm

7. 15 months

8. $8500

9. $1400 at 14%; $2600 at 11%

10. 16 nickels; 7 dimes; 8 quarters

11. 138.16 in.

12. 192 square inches

13. 4187 in.^{3}

14. 70 cm^2

15. $450

16. at least an 82

17. more than $110,000

9. Franco invested $4000 in money market funds. Part was invested at 14% interest, the rest at 11% interest. At the end of each year the fund company pays interest. After one year he earned $482 in interest. How much was invested at each interest rate?

10. Mary has $3.50 in change. She has twice as many nickels as quarters. She has one less dime than she has quarters. How many of each coin does she have?

11. Find the circumference of a circle with radius of 22.0 inches. Use $\pi = 3.14$ as an approximation.

12. Find the area of a trapezoid if the two bases are 10 inches and 14 inches, respectively, and the altitude is 16 inches.

13. Find the volume of a sphere with radius 10 inches. Use $\pi = 3.14$ as an approximation and round your answer to the nearest cubic inch.

14. Find the area of a parallelogram with a base of 10 centimeters and an altitude of 7 centimeters.

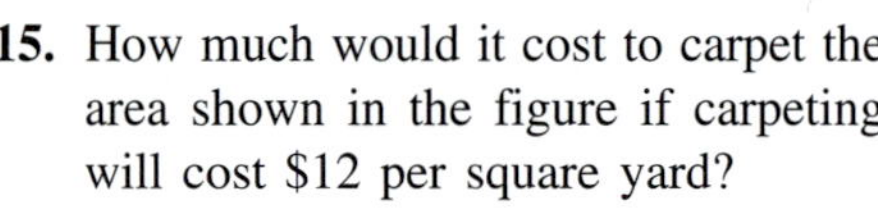

15. How much would it cost to carpet the area shown in the figure if carpeting will cost $12 per square yard?

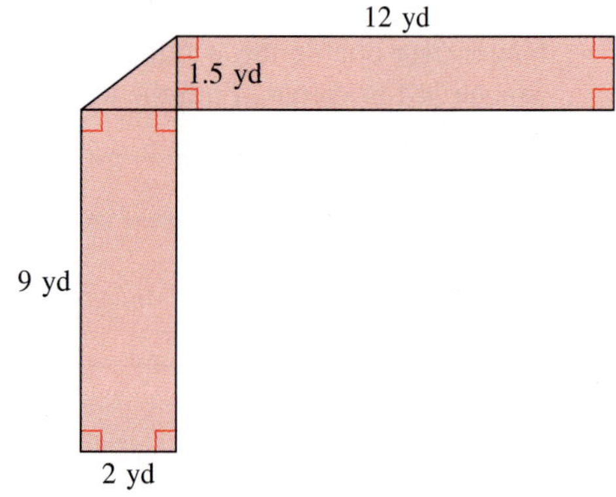

16. Cathy scored 76, 84, and 78 on three tests. How much must she score on the next test to have at least an 80 average?

17. Carol earns $15,000 plus 5% commission on sales over $10,000. How much must she make in sales to earn more than $20,000?

Cumulative Test for Chapters 0–3

Approximately one-half of this test covers the content of Chapters 0–2. The remainder covers the content of Chapter 3.

1. Divide. $2.4\overline{)8.856}$

2. Add. $\dfrac{3}{8} + \dfrac{5}{12} + \dfrac{1}{2}$

3. Simplify. $3(2x - 3y + 6) + 2(-3x - 7y)$

4. Evaluate. $5 - 2(3 - 7) + 15$

5. Solve for b. $H = \dfrac{1}{2}(3a + 5b)$

6. Solve for x and graph your solution. $5x - 3 \leq 2(4x + 1) + 4$

7. A number is first tripled and then decreased by 17. The result is 103. Find the original number.

8. The number of students in Psychology class is 34 fewer than the number of students in World History. The total enrollment for the two classes is 134 students. How many students are in each class?

1. 3.69

2. $\frac{31}{24}$

3. $-23y + 18$

4. 28

5. $\dfrac{2H - 3a}{5}$

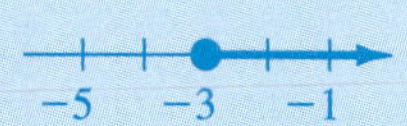

6. $x \geq -3$; See graph

7. 40

8. Psych students = 50; Wrld Hist. students = 84

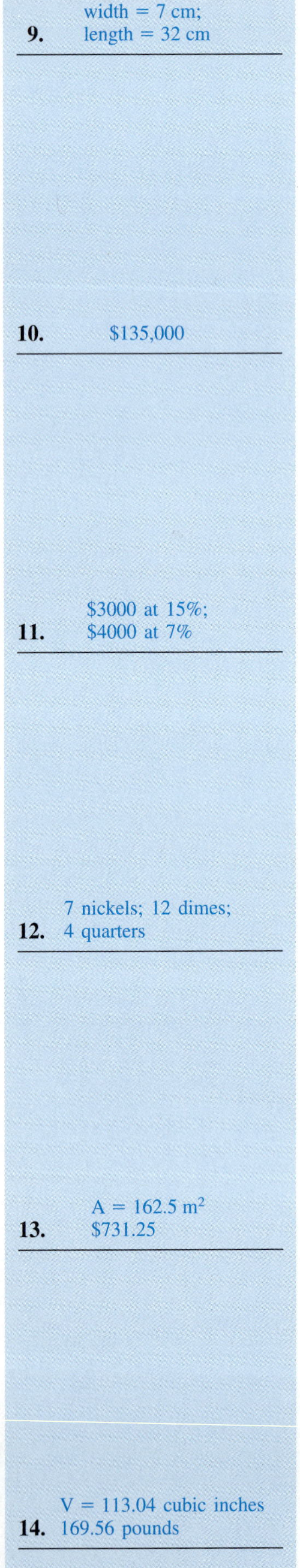

9. width = 7 cm; length = 32 cm
10. $135,000
11. $3000 at 15%; $4000 at 7%
12. 7 nickels; 12 dimes; 4 quarters
13. A = 162.5 m^2 $731.25
14. V = 113.04 cubic inches 169.56 pounds

9. A rectangle has a perimeter of 78 centimeters. The length of the rectangle is 11 centimeters longer than triple the width. Find the dimensions of the rectangle.

10. This year the sales for Homegrown Video are 35% higher than last year. This year the sales are $182,250. What were the sales last year?

11. Hassan invested $7000 for one year. He invested part of it in a high-risk fund that pays 15% interest. He placed the rest in a safe, low-risk fund that pays 7% interest. At the end of one year he earned $730. How much did he invest in each of the two funds?

12. Linda has some dimes, nickels, and quarters in her purse. The total value of these coins is $2.55. She has three times as many dimes as quarters. She has three more nickels than quarters. How many coins of each type does she have?

13. Find the area of a triangle with an altitude of 13 meters and a base of 25 meters. What is the cost to construct such a triangle out of sheet metal that costs $4.50 per square meter?

14. Find the volume of a sphere that has a radius of 3.00 inches. How much will the contents of the sphere weigh if it is filled with a liquid that weighs 1.50 pounds per cubic inch?

CHAPTER 4

Exponents and Polynomials

Much of the latest research in biology involves a study of DNA. The genetic material that organisms inherit from their parents consists of DNA. The careful study of DNA is leading to a better understanding of diseases. A DNA molecule is composed of two strands bonded together and twisted into a helix. This helix is very long and very thin. Turn to the Putting Your Skills to Work on page 275 to see if you can determine the relationships involved in the DNA molecule.

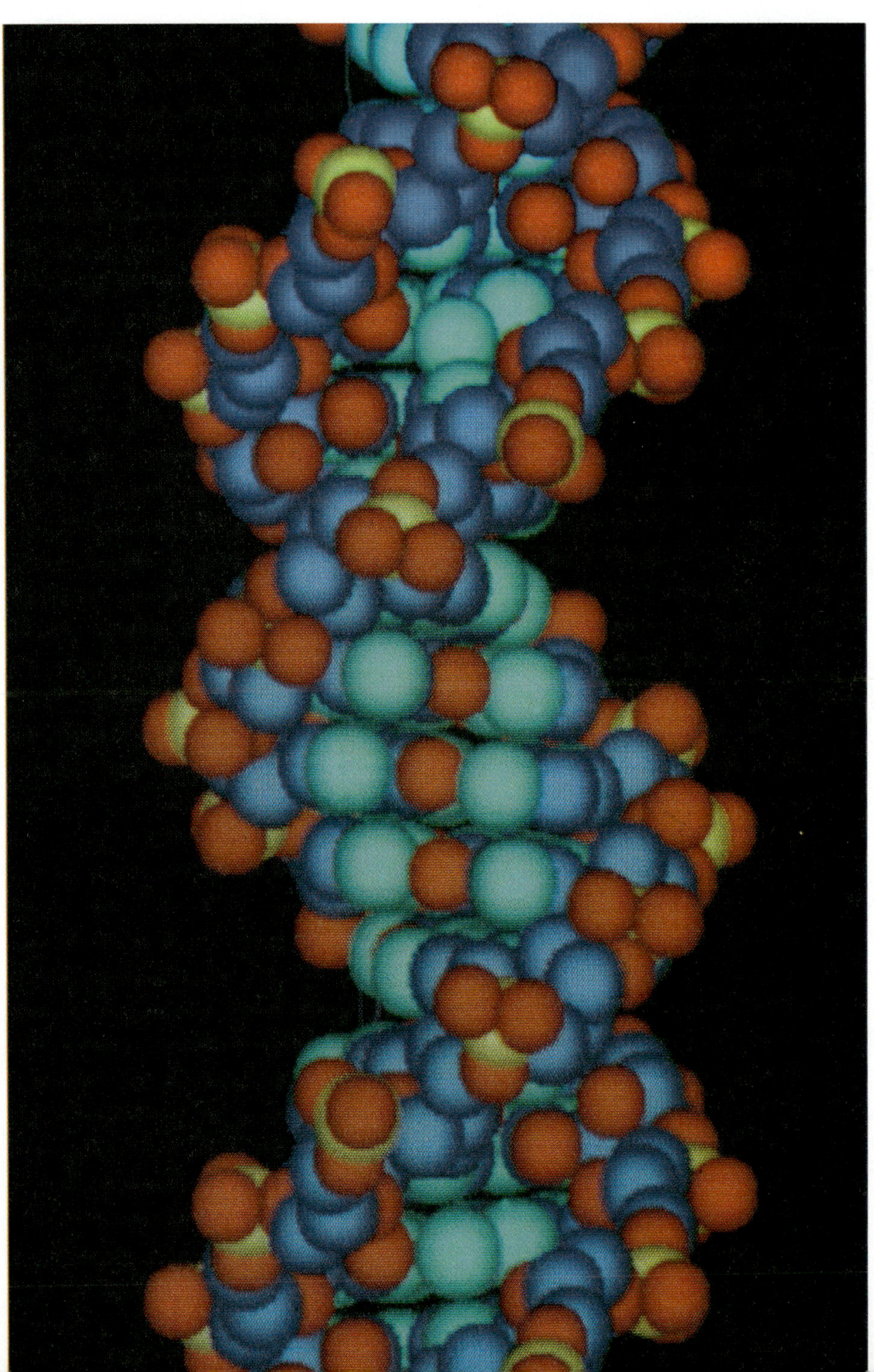

Pretest Chapter 4

If you are familiar with the topics in this chapter, take this test now. Check your answers with those in the back of the book. If an answer was wrong or you couldn't do a problem, study the appropriate section of the chapter.

If you are not familiar with the topics in this chapter, don't take this test now. Instead, study the examples, work the practice problems, and then take the test.

This test will help you to identify those concepts that you have mastered and those that need more study.

Section 4.1

Multiply and leave your answer in exponent form.

1. $(3^6)(3^{10})$ **2.** $(-5x^3)(2x^4)$ **3.** $(-4ab^2)(2a^3b)(3ab)$

Simplify each fraction. Assume that all variables are nonzero.

4. $\dfrac{x^{26}}{x^3}$ **5.** $\dfrac{12xy^2}{-6x^3y^4}$ **6.** $\dfrac{-25a^0b^2c^3}{-15abc^2}$

Simplify. Assume that all variables are nonzero.

7. $(x^5)^{10}$ **8.** $(-2x^2y)^3$ **9.** $\left(\dfrac{3ab^2}{c^3}\right)^2$

Section 4.2

Express with positive exponents.

10. $3x^2y^{-3}z^{-4}$ **11.** $\dfrac{-2a^{-3}}{b^{-2}c^4}$

12. Evaluate. $(-3)^{-2}$

13. Write in scientific notation. 0.000638 **14.** Write in decimal notation. 1.894×10^{12}

Section 4.3

Combine the following polynomials.

15. $(3x^2 - 6x - 9) + (-5x^2 + 13x - 20)$ **16.** $(2x^3 - 5x^2 + 7x) - (x^3 - 4x + 12)$

Section 4.4

Multiply.

17. $-3x(2 - 7x)$ **18.** $2x^2(x^2 + 4x - 1)$ **19.** $(x^3 - 3xy + y^2)(5xy^2)$

20. $(x + 5)(x + 9)$ **21.** $(3x - 2y)(4x + 3y)$ **22.** $(2x^2 - y^2)(x^2 - 3y^2)$

Section 4.5

Multiply.

23. $(8x - 11y)(8x + 11y)$ **24.** $(6x + 3y)^2$ **25.** $(5ab - 6)^2$

26. $(x^2 - 3x + 2)(4x - 1)$ **27.** $(2x - 3)(2x + 3)(4x^2 + 9)$

Section 4.6

Divide.

28. $(42x^4 - 36x^3 + 66x^2) \div (6x)$ **29.** $(15x^3 - 28x^2 + 18x - 7) \div (3x - 2)$

1. 3^{16}
2. $-10x^7$
3. $-24a^5b^4$
4. x^{23}
5. $-\dfrac{2}{x^2y^2}$
6. $\dfrac{5bc}{3a}$
7. x^{50}
8. $-8x^6y^3$
9. $\dfrac{9a^2b^4}{c^6}$
10. $\dfrac{3x^2}{y^3z^4}$
11. $\dfrac{-2b^2}{a^3c^4}$
12. $\frac{1}{9}$
13. 6.38×10^{-4}
14. 1,894,000,000,000
15. $-2x^2 + 7x - 29$
16. $x^3 - 5x^2 + 11x - 12$
17. $-6x + 21x^2$
18. $2x^4 + 8x^3 - 2x^2$
19. $5x^4y^2 - 15x^2y^3 + 5xy^4$
20. $x^2 + 14x + 45$
21. $12x^2 + xy - 6y^2$
22. $2x^4 - 7x^2y^2 + 3y^4$
23. $64x^2 - 121y^2$
24. $36x^2 + 36xy + 9y^2$
25. $25a^2b^2 - 60ab + 36$
26. $4x^3 - 13x^2 + 11x - 2$
27. $16x^4 - 81$
28. $7x^3 - 6x^2 + 11x$
29. $5x^2 - 6x + 2 - \dfrac{3}{3x - 2}$

4.1 The Rules of Exponents

After studying this section, you will be able to:

1. *Multiply exponential expressions with like bases.*
2. *Divide exponential expressions with like bases.*
3. *Raise exponential expressions to a power.*

1 *The Product Rule for Exponents*

Recall that x^2 means $x \cdot x$. That is, x appears as a factor two times. What happens when we multiply $x^2 \cdot x^2$? Is there a pattern that will help us to form a general rule?

Notice that

$$(2^2)(2^3) = \overbrace{(2 \cdot 2)(2 \cdot 2 \cdot 2)}^{5 \text{ twos}} = 2^5$$ *The exponent means 2 occurs 5 times as a factor.*

$$(3^3)(3^4) = \overbrace{(3 \cdot 3 \cdot 3)(3 \cdot 3 \cdot 3 \cdot 3)}^{7 \text{ threes}} = 3^7$$ *Notice that* $3 + 4 = 7$.

$$(x^3)(x^5) = \overbrace{(x \cdot x \cdot x)(x \cdot x \cdot x \cdot x \cdot x)}^{8\ x\text{'s}} = x^8$$ *The sum of the exponents* $3 + 5 = 8$.

$$(y^4)(y^2) = \overbrace{(y \cdot y \cdot y \cdot y)(y \cdot y)}^{6\ y\text{'s}} = y^6$$ *The sum of the exponents* $4 + 2 = 6$.

We can state the pattern in words and then use variables.

The Product Rule

To multiply two numbers or variables in exponent form that have the same base, *add the exponents* but keep the base unchanged.

$$x^a \cdot x^b = x^{a+b}$$

Be sure to notice that this rule applies only to numbers or variables that have the *same base*. Here x represents the base (the same in both cases), while the letters a and b represent the exponents that are added.

It is important that you apply this rule when the exponent is 1. Every variable that does not have a written exponent is understood to have an exponent of 1. Thus $x^1 = x$, $y^1 = y$, and so on.

EXAMPLE 1 Multiply. **(a)** $x^3 \cdot x^6$ **(b)** $x \cdot x^5$

(a) $x^3 \cdot x^6 = x^{3+6} = x^9$

(b) $x \cdot x^5 = x^{1+5} = x^6$ *Note that the exponent of the first* x *is* 1.

TEACHING TIP After completing Example 1, you may want to mention that any letter may be used for the variable. Thus, we have $w^8 \cdot w^2 = w^{10}$ and also $p^8 \cdot p = p^9$

Practice Problem 1 Multiply. **(a)** $a^7 \cdot a^5$ **(b)** $w^{10} \cdot w$ ■

EXAMPLE 2 Simplify. **(a)** $y^5 \cdot y^{11}$ **(b)** $2^3 \cdot 2^5$ **(c)** $x^6 \cdot y^8$

(a) $y^5 \cdot y^{11} = y^{5+11} = y^{16}$

(b) $2^3 \cdot 2^5 = 2^{3+5} = 2^8$ *Note that the base does not change! Only the exponent changes.*

(c) $x^6 \cdot y^8$ *The rule for multiplying numbers with exponents does not apply since the bases are not the same. This cannot be simplified.*

Practice Problem 2 Simplify if possible.

(a) $x^3 \cdot x^9$ **(b)** $3^7 \cdot 3^4$ **(c)** $a^3 \cdot b^2$ ■

We can now look at multiplying expressions such as

$$(2x^5)(3x^6)$$

The number 2 in $2x^5$ is called the **numerical coefficient.** A numerical coefficient is a number that is multiplied by a variable. What is the numerical coefficient in $3x^6$? When we multiply two expressions such as $2x^5$ and $3x^6$, we first multiply the numerical coefficients; we multiply the variables with their exponents separately.

EXAMPLE 3 Multiply. **(a)** $(2x^5)(3x^6)$ **(b)** $(5x^3)(x^6)$ **(c)** $(-6x)(-4x^5)$

(a) $(2x^5)(3x^6) = (2 \cdot 3)(x^5 \cdot x^6)$ *Multiply the numerical coefficients.*

$= 6(x^5 \cdot x^6)$ *Use the rule for multiplying expressions with exponents. Add the exponents.*

$= 6x^{11}$

(b) Every variable that does not have a visible numerical coefficient is understood to have a numerical coefficient of 1. Thus x^6 has a numerical coefficient of 1.

$$(5x^3)(x^6) = (5 \cdot 1)(x^3 \cdot x^6) = 5x^9$$

(c) $(-6x)(-4x^5) = (-6)(-4)(x^1 \cdot x^5) = 24x^6$ *Remember that x has an exponent of 1.*

Practice Problem 3 Multiply.

(a) $(-a^8)(a^4)$ **(b)** $(3y^2)(-2y^3)$ **(c)** $(-4x^3)(-5x^2)$ ■

Problems of this type may involve more than one variable or more than two factors.

EXAMPLE 4 Multiply. **(a)** $(-4x^2y^3)(-2x^5y^3)$ **(b)** $(5ab)(-\frac{1}{3}a)(9b^2)$

(a) $(-4x^2y^3)(-2x^5y^3) = (-4)(-2)(x^2 \cdot x^5)(y^3 \cdot y^3)$

$= 8x^7y^6$

(b) $(5ab)(-\frac{1}{3}a)(9b^2) = (5)(-\frac{1}{3})(9)(a \cdot a)(b \cdot b^2)$

$= -15a^2b^3$

As you do the following problems, keep in mind the rule for multiplying numbers with exponents and the rules for multiplying signed numbers.

Practice Problem 4 Multiply.

(a) $(-5a^3b)(-2ab^4)$ **(b)** $(2xy)(-\frac{1}{4}x^2y)(6xy^3)$ ■

2 The Quotient Rule for Exponents

Frequently, we must divide expressions like the ones we have just learned to multiply. Since division by zero is undefined, in all problems in this chapter we assume that the denominator of any variable expression is not zero. We'll look at division in three separate parts.

Suppose that we want to divide $x^5 \div x^2$. We could do the division in a long fashion.

$$\frac{x^5}{x^2} = \frac{(x)(x)(x)(\cancel{x})(\cancel{x})}{(\cancel{x})(\cancel{x})} = x^3$$

Here we are using the arithmetical property of reducing fractions (see Section 0.1). When the same factor appears in both numerator and denominator, that factor can be removed.

A simpler way is to *subtract the exponents.* Notice that the base remains the same.

The Quotient Rule (Part 1)

$$\frac{x^a}{x^b} = x^{a-b} \quad \text{if the larger exponent is in the numerator and } x \neq 0$$

EXAMPLE 5 Divide. **(a)** $\frac{2^{16}}{2^{11}}$ **(b)** $\frac{x^5}{x^3}$ **(c)** $\frac{y^{16}}{y^7}$

(a) $\frac{2^{16}}{2^{11}} = 2^{16-11} = 2^5$ Note that the base does *not* change.

(b) $\frac{x^5}{x^3} = x^{5-3} = x^2$ **(c)** $\frac{y^{16}}{y^7} = y^{16-7} = y^9$

Practice Problem 5 Divide. **(a)** $\frac{10^{13}}{10^7}$ **(b)** $\frac{x^{11}}{x}$ ■

Now we consider the situation where the larger exponent is in the denominator. Suppose that we want to divide $x^2 \div x^5$.

$$\frac{x^2}{x^5} = \frac{(\cancel{x})(\cancel{x})}{(\cancel{x})(\cancel{x})(x)(x)(x)} = \frac{1}{x^3}$$

The Quotient Rule (Part 2)

$$\frac{x^a}{x^b} = \frac{1}{x^{b-a}} \quad \text{if the larger exponent is in the denominator and } x \neq 0$$

EXAMPLE 6 Divide. **(a)** $\frac{12^{17}}{12^{20}}$ **(b)** $\frac{b^7}{b^9}$ **(c)** $\frac{x^{20}}{x^{24}}$

(a) $\frac{12^{17}}{12^{20}} = \frac{1}{12^{20-17}} = \frac{1}{12^3}$ Note that the base does *not* change.

(b) $\frac{b^7}{b^9} = \frac{1}{b^{9-7}} = \frac{1}{b^2}$ **(c)** $\frac{x^{20}}{x^{24}} = \frac{1}{x^{24-20}} = \frac{1}{x^4}$

Practice Problem 6 Simplify. **(a)** $\frac{c^3}{c^4}$ **(b)** $\frac{10^{31}}{10^{56}}$ ■

When there are numerical coefficients, use the rules for dividing signed numbers to reduce fractions to lowest terms.

TEACHING TIP After explaining Example 7, give students the following problems as an Additional Class Exercise. Divide.

A. $\dfrac{36x^3}{12x^5}$ B. $\dfrac{-15x^8}{25x^{10}}$

C. $\dfrac{-125x^6}{-5x^3}$

Ans.

A. $\dfrac{3}{x^2}$ B. $-\dfrac{3}{5x^2}$ C. $25x^3$

EXAMPLE 7 Divide.

(a) $\dfrac{5x^5}{25x^7}$ **(b)** $\dfrac{-12x^8}{4x^3}$ **(c)** $\dfrac{2x^6}{6x^4}$ **(d)** $\dfrac{-16x^7}{-24x^8}$

(a) $\dfrac{5x^5}{25x^7} = \dfrac{1}{5x^{7-5}} = \dfrac{1}{5x^2}$ **(b)** $\dfrac{-12x^8}{4x^3} = -3x^{8-3} = -3x^5$

(c) $\dfrac{2x^6}{6x^4} = \dfrac{x^{6-4}}{3} = \dfrac{x^2}{3}$ **(d)** $\dfrac{-16x^7}{-24x^8} = \dfrac{2}{3x^{8-7}} = \dfrac{2}{3x}$

Practice Problem 7 Simplify. **(a)** $\dfrac{-7x^7}{-21x^9}$ **(b)** $\dfrac{15x^{11}}{-3x^4}$ ■

You have to work very carefully if two or more variables are involved. Treat the coefficients and each variable separately.

TEACHING TIP In addition to Example 8, you may wish to do the following as an Additional Class Exercise. Divide.

$$\frac{-5x^4yz^3}{-15x^3y^2z^8} = \frac{1x}{3yz^5}$$

$$= \frac{x}{3yz^5} \quad \text{Ans.}$$

EXAMPLE 8 Divide. **(a)** $\dfrac{x^3y^2}{5xy^6}$ **(b)** $\dfrac{x^5y^3}{x^6w}$ **(c)** $\dfrac{-3x^2y^5}{12x^6y^8}$

(a) $\dfrac{x^3y^2}{5xy^6} = \dfrac{x^2}{5y^4}$ **(b)** $\dfrac{x^5y^3}{x^6w} = \dfrac{y^3}{xw}$ **(c)** $\dfrac{-3x^2y^5}{12x^6y^8} = -\dfrac{1}{4x^4y^3}$

Practice Problem 8 Simplify. **(a)** $\dfrac{x^7y^9}{y^{10}}$ **(b)** $\dfrac{12x^5y^6}{-24x^3y^8}$ ■

Suppose that, for a given base, the exponent is the same in numerator and denominator. In these cases we can use the fact that *any nonzero number divided by itself is* 1.

EXAMPLE 9 Divide. **(a)** $\dfrac{x^6}{x^6}$ **(b)** $\dfrac{3x^5}{x^5}$

(a) $\dfrac{x^6}{x^6} = 1$ **(b)** $\dfrac{3x^5}{x^5} = 3\left(\dfrac{x^5}{x^5}\right) = 3(1) = 3$

Practice Problem 9 Divide. **(a)** $\dfrac{10^7}{10^7}$ **(b)** $\dfrac{12a^4}{15a^4}$ ■

Do you see that if we had subtracted exponents for $\frac{x^6}{x^6}$ we would have obtained x^0 in Example 9? So we can surmise that any number (except 0) to the 0 power equals 1. We can write this fact as a separate rule.

TEACHING TIP Point out that 0^0 is undefined.

The Quotient Rule (Part 3)

$$\frac{x^a}{x^a} = x^0 = 1 \quad \text{if } x \neq 0 \quad (0^0 \text{ remains undefined})$$

EXAMPLE 10 Divide. **(a)** $\dfrac{4x^0y^2}{8^0y^5z^3}$ **(b)** $\dfrac{5x^2y}{10x^2y^3}$

(a) $\dfrac{4x^0y^2}{8^0y^5z^3} = \dfrac{4(1)y^2}{(1)y^5z^3} = \dfrac{4y^2}{y^5z^3} = \dfrac{4}{y^3z^3}$ **(b)** $\dfrac{5x^2y}{10x^2y^3} = \dfrac{1x^0}{2y^2} = \dfrac{(1)(1)}{2y^2} = \dfrac{1}{2y^2}$

TEACHING TIP In addition to Example 10, you may wish to point out to students that since $x^0 = 1$, whenever $x \neq 0$, it is also true that x could be an entire expression. Thus

$$\left(\frac{ab^2}{c}\right)^0 = 1 \text{ as long as } a \neq 0,\ b \neq 0,\ c \neq 0$$

Practice Problem 10 Simplify. $\dfrac{-20a^3b^8c^4}{28a^3b^7c^5}$ ■

We can combine all three parts of the rule we have developed.

> **The Quotient Rule**
>
> $\dfrac{x^a}{x^b} = x^{a-b}$ if the larger exponent is in the numerator and $x \neq 0$
>
> $\dfrac{x^a}{x^b} = \dfrac{1}{x^{b-a}}$ if the larger exponent is in the denominator and $x \neq 0$
>
> $\dfrac{x^a}{x^a} = x^0 = 1$ if there is the same exponent in the numerator and denominator and $x \neq 0$

We can combine the product rule and the quotient rule to simplify algebraic expressions that involve both multiplication and division.

EXAMPLE 11 Simplify. $\dfrac{(8x^2y)(-3x^3y^2)}{-6x^4y^3}$

$$\frac{(8x^2y)(-3x^3y^2)}{-6x^4y^3} = \frac{-24x^5y^3}{-6x^4y^3} \quad \textit{Simplify the numerator. Then divide.}$$

$$= 4x$$

Practice Problem 11 Simplify. $\dfrac{(-6ab^5)(3a^2b^4)}{16a^5b^7}$ ■

3 *Raising a Power to a Power*

How do we simplify an expression such as $(x^4)^3$? $(x^4)^3$ is x^4 raised to the third power. For this type of problem we say that we are raising a power to a power. A problem such as $(x^4)^3$ could be done by writing

$$(x^4)^3 = x^4 \cdot x^4 \cdot x^4 \quad \text{By definition.}$$

$$= x^{12} \quad \text{By adding exponents.}$$

Notice that when we add the exponents we get $4 + 4 + 4 = 12$. This is the same as multiplying 4 by 3. That is, $4 \cdot 3 = 12$. This process can be summarized by the following rule.

> **Raising a Power to a Power**
>
> To raise a power to a power, keep the same base and multiply the exponents.
>
> $$(x^a)^b = x^{ab}$$

TEACHING TIP Take a little extra time to make sure that students see where this rule comes from. They will remember it a lot longer if they can see some examples of how and why the rule works. Essentially, we are appealing to the concept that raising to a positive integer power is repeated multiplication of the same factor. $(x^a)^b = x^a \cdot x^a \cdot x^a \cdots x^a$. The x^a occurs as a factor b times.

Recall what happens when you raise a negative number to a power. $(-1)^2 = 1$. $(-1)^3 = -1$. In general,

$$(-1)^n = \begin{cases} +1, & n \text{ is even} \\ -1, & n \text{ is odd} \end{cases}$$

If an expression can be raised to a power, we sometimes give the direction *simplify*.

EXAMPLE 12 Simplify. **(a)** $(x^3)^5$ **(b)** $(2^7)^3$

(a) $(x^3)^5 = x^{3 \cdot 5} = x^{15}$ **(b)** $(2^7)^3 = 2^{7 \cdot 3} = 2^{21}$

Note that in both parts (a) and (b) the base does not change.

Practice Problem 12 Raise each expression to the power indicated.

(a) $(a^4)^3$ **(b)** $(10^5)^2$ ■

Here are two similar rules involving products and quotients that are very useful. We'll illustrate each with an example.

If a product in parentheses is raised to a power, the parentheses indicate that *each factor* must be raised to that power.

$$(xy)^2 = x^2y^2 \qquad (xy)^3 = x^3y^3$$

$$(xy)^a = x^ay^a$$

EXAMPLE 13 Simplify. **(a)** $(ab)^8$ **(b)** $(3x)^4$ **(c)** $(-2x^2)^3$

(a) $(ab)^8 = a^8b^8$ **(b)** $(3x)^4 = (3)^4x^4 = 81x^4$ **(c)** $(-2)^3 \cdot (x^2)^3 = -8x^6$

Practice Problem 13 Raise each expression to the power indicated.

(a) $(3xy)^3$ **(b)** $(yz)^{37}$ ■

If $\left(\frac{x}{y}\right)$ is raised to the second power, the parentheses indicate that numerator and denominator are each raised to that power.

$$\left(\frac{x}{y}\right)^2 = \frac{x^2}{y^2} \quad \text{if } y \neq 0$$

$$\left(\frac{x}{y}\right)^a = \frac{x^a}{y^a} \quad \text{if } y \neq 0$$

EXAMPLE 14 Simplify. **(a)** $\left(\frac{x}{y}\right)^5$ **(b)** $\left(\frac{7}{w}\right)^4$

(a) $\left(\frac{x}{y}\right)^5 = \frac{x^5}{y^5}$ **(b)** $\left(\frac{7}{w}\right)^4 = \frac{7^4}{w^4}$

Practice Problem 14 Raise to the power indicated. $\left(\frac{4a}{b}\right)^6$ ■

Many problems can be simplified by using the previous rules involving exponents. Be sure to take particular care to determine the correct sign, especially if the numerical coefficient of the term is a negative number.

EXAMPLE 15 Simplify. **(a)** $\left(\frac{-3x^2z^0}{y^3}\right)^4$ **(b)** $\left(\frac{-2xw^4}{x^3w}\right)^3$

(a) $\left(\frac{-3x^2z^0}{y^3}\right)^4 = \left(\frac{-3x^2}{y^3}\right)^4$ *Since $z^0 = 1$. Simplify inside the parentheses first.*

$= \frac{(-3)^4x^8}{y^{12}}$ *Apply the rules for raising a power to a power. Notice that we wrote $(-3)^4$ and not -3^4. We are raising -3 to the fourth power.*

$= \frac{81x^8}{y^{12}}$ *Simplify the coefficient. $(-3)^4 = +81$*

(b) $\left(\frac{-2xw^4}{x^3w}\right)^3 = \left(\frac{-2w^3}{x^2}\right)^3$ *Simplify the expression inside the parentheses first.*

$= \frac{(-2)^3w^9}{x^6}$ *Apply the rules for raising a power to a power.*

$= \frac{-8w^9}{x^6}$ *Simplify the coefficient $(-2)^3 = -8$.*

TEACHING TIP After explaining Example 15, ask students to do the following as an Additional Class Exercise. Simplify:

A. $\left(\frac{-2x^3}{y^4}\right)^3$ **B.** $\left(\frac{-3xy^0z^4}{x^4y^3}\right)^4$

Ans. A. $\frac{-8x^9}{y^{12}}$ B. $\frac{81z^{16}}{x^{12}y^{12}}$

Practice Problem 15 Simplify. **(a)** $\left(\frac{-2x^3y^0z}{4xz^2}\right)^5$ **(b)** $\left(\frac{4x^8y^5}{-3x^4z}\right)^3$ ■

Developing Your Study Skills

WHY IS REVIEW NECESSARY?

You master a course in mathematics by learning the concepts one step at a time. Thus the study of mathematics is built step by step upon this foundation, each step supporting the next. The process is a carefully designed procedure, so no steps can be skipped. A student of mathematics needs to realize the importance of this building process to succeed.

Because learning new concepts depends on those previously learned, students often need to take time to review. The reviewing process will strengthen the understanding and application of concepts that are weak due to a lack of mastery or the passage of time. Review at the right time on the right concepts can strengthen previously learned skills and make progress possible.

Timely, periodic review of previously learned mathematical concepts is absolutely necessary in order to master new concepts. You may have forgotten a concept or grown a bit rusty in applying it. Reviewing is the answer. Make use of the Cumulative Review Problems in your textbook, whether they are assigned or not. Look back to previous chapters whenever you have forgotten how to do something. Review the Chapter Organizers from previous chapters. Study the examples and practice some exercises to refresh your understanding.

Be sure that you understand and can perform the computations of each new concept. This will enable you to be able to move successfully on to the next ones.

Remember, mathematics is a step-by-step building process. Learn one concept at a time, skipping none, and reinforce and strengthen with review whenever necessary.

4.1 Exercises

Verbal and Writing Skills

1. Write in your own words the product rule for exponents. Answers may vary. Check student's response.

2. To be able to use the rules of exponents, what must be true of the bases? The bases must be the same.

3. If the larger exponent is in the denominator, the quotient rule states that $\frac{x^a}{x^b} = \frac{1}{x^{b-a}}$. Provide an example to show why this is true. A sample example is $\frac{2^2}{2^3} \stackrel{?}{=} \frac{1}{2^{3-2}}$

$$\frac{4}{8} = \frac{1}{2}$$

Identify the numerical coefficient, the base(s), and the exponent(s).

4. $-8x^5y^2$ -8; x, y; 5, and 2

5. $6x^{11}y$ 6; x, y; 11, and 1

6. Evaluate. (a) $3x^0$ (b) $(3x)^0$ Why are the results different? (a) $3x^0 = 3(1) = 3$ (b) $(3x)^0 = 1$;
In (a) the coefficient is multiplied by x^0 or 1.
In (b) the coefficient is in the parentheses and is part of the expression that is raised to the zero power which is 1.

Write in simplest exponent form.

7. $3 \cdot x \cdot x \cdot x \cdot y \cdot y$ $3x^3y^2$

8. $4 \cdot a \cdot a \cdot a \cdot a \cdot b \cdot b \cdot b$ $4a^4b^3$

9. $(-3)(a)(a)(b)(c)(b)(c)(c)$ $-3a^2b^2c^3$

10. $(-7)(x)(y)(z)(y)(x)$ $-7x^2y^2z$

Multiply and leave your answer in exponent form.

11. $(3^8)(3^7)$ 3^{15}

12. $(2^5)(2^8)$ 2^{13}

13. $(5^{10})(5^{16})$ 5^{26}

14. $(5^3)(2^6)$ $5^3 \cdot 2^6$

15. $(3^5)(8^2)$ $3^5 \cdot 8^2$

Multiply.

16. $-9x^2 \cdot 8x^3$ $-72x^5$

17. $-12x^2 \cdot 5x$ $-60x^3$

18. $(5x)(10x^2)$ $50x^3$

19. $(-4x^2)(-3x^3)$ $12x^5$

20. $(-2a^5)(-5a^3)$ $10a^8$

21. $(4x^8)(-3x^3)$ $-12x^{11}$

22. $(-2ab)(3a^2b^3)$ $-6a^3b^4$

23. $(5xy^2)(-2x^3y)$ $-10x^4y^3$

24. $\left(\frac{1}{2}x^2y\right)\left(\frac{3}{4}xy^2\right)$ $\frac{3}{8}x^3y^3$

25. $\left(\frac{2}{3}x^3y\right)\left(\frac{5}{7}xy^4\right)$ $\frac{10}{21}x^4y^5$

26. $(1.1x^2z)(-2.5xy)$ $-2.75x^3yz$

27. $(2.3x^4w)(-3.5xy^4)$ $-8.05x^5wy^4$

28. $(-3x)(2y)(5x^3y)$ $-30x^4y^2$

29. $(8a)(2a^3b)(0)$ 0

30. $(-5ab)(2a^2)(0)$ 0

31. $(-4x)(2x^2y)(3x)$ $-24x^4y$

32. $(-20xy^3)(-4x^3y^2)$ $80x^4y^5$

33. $(-6xy^3)(-12x^5y^8)$ $72x^6y^{11}$

34. $(-3x^5)(20y^8)$ $-60x^5y^8$

35. $(14a^5)(-2b^6)$ $-28a^5b^6$

36. $(5a^3b^2)(-ab)$ $-5a^4b^3$

37. $(-x^2y)(3x)$ $-3x^3y$

38. $(-3x^2y)(0)(-5xy^3)$ 0

39. $(-a^2b^2)(2ab^3)(-6a^4b)$ $12a^7b^6$

40. $(3x^4y^3)(-3x^2y)(-4xy^5)$ $36x^7y^9$

41. $(4x^3y)(0)(-12x^4y^2)$ 0

42. $(5x^3y)(-2w^4z)$ $-10w^4x^3yz$

43. $(6w^5z^6)(-4xy)$ $-24w^5xyz^6$

44. $(3ab)(5a^2c)(-2b^2c^3)$ $-30a^3b^3c^4$

Divide. Leave your answer in exponent form. Assume that all variables in any denominator are nonzero.

45. $\frac{x^8}{x^{10}}$ $\frac{1}{x^2}$

46. $\frac{a^{15}}{a^{13}}$ a^2

47. $\frac{y^{12}}{y^5}$ y^7

48. $\frac{b^{20}}{b^{23}}$ $\frac{1}{b^3}$

49. $\frac{3^{45}}{3^{17}}$ 3^{28}

50. $\frac{7^{20}}{7^{33}}$ $\frac{1}{7^{13}}$

51. $\frac{13^{20}}{13^{30}}$ $\frac{1}{13^{10}}$

52. $\frac{5^{19}}{5^{11}}$ 5^8

53. $\frac{-3^{18}}{-3^{14}}$ 3^4

54. $\frac{-5^{16}}{-5^{12}}$ 5^4

55. $\frac{a^{13}}{4a^5}$ $\frac{a^8}{4}$

56. $\frac{4b^{16}}{b^{13}}$ $4b^3$

57. $\frac{x^7}{y^9}$ $\frac{x^7}{y^9}$

58. $\frac{x^{20}}{y^3}$ $\frac{x^{20}}{y^3}$

59. $\frac{-10a^5b^2}{-20a^5b^3}$ $\frac{1}{2b}$

60. $\frac{12x^5y^6}{-3xy}$ $-4x^4y^5$

61. $\frac{-36x^3y^6}{-18xy^8}$ $\frac{2x^2}{y^2}$

62. $\frac{-20a^3b^2}{4a^5b^2}$ $-\frac{5}{a^2}$

63. $\frac{12ab}{-24a^3b^2}$ $-\frac{1}{2a^2b}$

64. $\frac{-13x^5y^6}{13x^5y^6}$ -1

65. $\frac{a^5b^6}{a^5b^6}$ 1

66. $\frac{-36x^3y^7}{72x^5y}$ $-\frac{y^6}{2x^2}$

67. $\frac{48x^7y}{-24x^3y^6}$ $-\frac{2x^4}{y^5}$

68. $\frac{42m^6n^3}{4.2mn^6}$ $\frac{10m^5}{n^3}$

69. $\frac{2.3x^8y^{10}}{23x^{12}y^5}$ $\frac{y^5}{10x^4}$

70. $\frac{-51x^6y^8z^{12}}{17x^3y^8z^7}$ $-3x^3z^5$

71. $\frac{30x^5y^4}{5x^3y^4}$ $6x^2$

72. $\frac{27x^3y^2}{3xy^2}$ $9x^2$

73. $\frac{8^0x^2y^3}{16x^5y}$ $\frac{y^2}{16x^3}$

74. $\frac{3^2x^3y^7}{3^0x^5y^2}$ $\frac{9y^5}{x^2}$

75. $\frac{18a^6b^3c^0}{24a^5b^3}$ $\frac{3a}{4}$

76. $\frac{12a^7b^8}{16a^3b^8c^0}$ $\frac{3a^4}{4}$

77. $\frac{25x^6}{35y^8}$ $\frac{5x^6}{7y^8}$

78. $\frac{85x^5y^3z}{-5x^2y}$ $-17x^3y^2z$

79. $\frac{-4x^3y^5}{68x^8y^7z}$ $-\frac{1}{17x^5y^2z}$

80. $\frac{24y^5}{16x^3}$ $\frac{3y^5}{2x^3}$

Multiply the numerator **first.** *Then divide the two expressions.*

81. $\frac{(3x^2)(2x^3)}{3x^4}$ $\frac{6x^5}{3x^4} = 2x$

82. $\frac{(4a^5)(3a^6)}{2a^3}$ $6a^8$

83. $\frac{(9a^2b)(2a^3b^6)}{-27a^8b^7}$ $\frac{18a^5b^7}{-27a^8b^7} = -\frac{2}{3a^3}$

To Think About

84. What expression can be multiplied by $(-3x^3yz)$ to obtain $81x^8y^2z^4$? $-27x^5yz^3$

85. $63a^5b^6$ is divided by an expression and the result is $-9a^4b$. What is this expression? $-7ab^5$

Find a value for the variable to show that the two expressions are **not** *equivalent.*

86. $(x^2)(x^3)$; x^6
$(2^2)(2^3) = (4)(8)$; $2^6 = 64$
$= 32$

87. $\frac{y^8}{y^4}$; y^2
$\frac{2^8}{2^4} = \frac{256}{16}$; $2^2 = 4$
$= 16$

88. $\frac{z^3}{z^6}$; z^3
$\frac{2^3}{2^6} = \frac{8}{64}$; $2^3 = 8$
$= \frac{1}{8}$

Simplify.

89. $(a^{2b})(a^{3b})$ a^{5b}

90. $\frac{b^{5x}}{b^{2x}}$ b^{3x}

91. $\frac{(c^{3y})(c^4)}{c^{2y+2}}$ c^{y+2}

Raise each expression to the power indicated.

92. $(x^2)^6$ x^{12}

93. $(w^5)^8$ w^{40}

94. $(xy^2)^7$ x^7y^{14}

95. $(a^3b)^4$ $a^{12}b^4$

96. $(a^5b^2)^6$ $a^{30}b^{12}$

97. $(a^3b^2c)^8$ $a^{24}b^{16}c^8$

98. $(2ab^2c^7)^4$ $16a^4b^8c^{28}$

99. $(3^2xy^2)^4$ $3^8x^4y^8$

100. $(-3a^4)^2$ $9a^8$

101. $(-2a^5)^4$ $16a^{20}$

102. $\left(\dfrac{7}{w^2}\right)^8$ $\dfrac{7^8}{w^{16}}$

103. $\left(\dfrac{12x}{y^2}\right)^5$ $\dfrac{12^5x^5}{y^{10}}$

104. $\left(\dfrac{3a}{2b}\right)^4$ $\dfrac{81a^4}{16b^4}$

105. $\left(\dfrac{6x}{5y^3}\right)^2$ $\dfrac{36x^2}{25y^6}$

106. $(-3a^2b^3c^0)^4$ $81a^8b^{12}$

107. $(-2a^5b^2c)^5$ $-32a^{25}b^{10}c^5$

108. $(-2x^3yz)^3$ $-8x^9y^3z^3$

109. $(-4xy^0z^4)^3$ $-64x^3z^{12}$

110. $\dfrac{(2x)^4}{(2x)^5}$ $\dfrac{1}{2x}$

111. $\dfrac{(4a^2b)^2}{(4ab^2)^3}$ $\dfrac{a}{4b^4}$

112. $(3ab^2)^3(ab)$ $27a^4b^7$

113. $(-2a^2b^3)^3(ab^2)$ $-8a^7b^{11}$

114. $\left(\dfrac{2x^2y^3}{3z^0}\right)^3$ $\dfrac{8x^6y^9}{27}$

115. $\left(\dfrac{4x^0y^4}{3z^3}\right)^2$ $\dfrac{16y^8}{9z^6}$

116. $\left(\dfrac{x^5}{3}\right)^3$ $\dfrac{x^{15}}{27}$

117. $\left(\dfrac{y^6}{2}\right)^3$ $\dfrac{y^{18}}{8}$

118. $\left(\dfrac{ab^2}{c^3d^4}\right)^4$ $\dfrac{a^4b^8}{c^{12}d^{16}}$

119. $\left(\dfrac{a^3b}{c^5d}\right)^5$ $\dfrac{a^{15}b^5}{c^{25}d^5}$

120. $(5xy^2)^3(xy^2)$ $125x^4y^8$

121. $(4x^3y)^2(x^3y)$ $16x^9y^3$

122. $\left(\dfrac{a^2b^6}{b^2c^5}\right)^3$ $\dfrac{a^6b^{12}}{c^{15}}$

123. $\left(\dfrac{a^4b^7}{a^8c^3}\right)^5$ $\dfrac{b^{35}}{a^{20}c^{15}}$

To Think About

124. What expression raised to the third power is $-27x^9y^{12}z^{21}$? $-3x^3y^4z^7$

125. What expression raised to the fourth power is $16x^{20}y^{16}z^{28}$? $\pm 2x^5y^4z^7$

Find a value of the variable that shows that the two expressions are **not** *equivalent.*

126. $5x^3$; $(5x)^3$
$5(2)^3 = 5 \cdot 8;\ (5 \cdot 2)^3 = 10^3$
$= 40 \qquad = 1000$

127. $(y + 2)^2$; $y^2 + 2^2$
$(3 + 2)^2 = 5^2;\ 3^2 + 2^2 = 9 + 4$
$= 25; \qquad = 13$

128. $(z - 3)^2$; $z^2 - 3^2$
$(4 - 3)^2 = 1;\ 4^2 - 3^2 = 16 - 9$
$= 7$

129. Write $4^2 \cdot 8^3 \cdot 16$ as a power of 2. 2^{17}

130. Write $9^3 \cdot 27 \cdot 81$ as a power of 3. 3^{13}

Cumulative Review Problems

Simplify.

131. $-3 - 8$ -11

132. $-17 + (-32) + (-24) + (27)$ -46

133. $\left(\dfrac{2}{3}\right)\left(-\dfrac{21}{8}\right)$ $-\dfrac{7}{4}$

134. $\dfrac{-3}{4} \div \dfrac{-12}{80}$ 5

135. The population of Mexico City in 1991 was 20.9 million. What will the population be in the year 2000 if it is predicted to increase by 33.4% over this 9-year period? Approx. 27.9 million

4.2 Negative Exponents and Scientific Notation

After studying this section, you will be able to:

1. *Use negative exponents.*
2. *Write numbers in scientific notation.*

MathPro

Video 10

SSM

1 *Negative Exponents*

If n is a positive integer and $x \neq 0$, then x^{-n} is defined as follows:

Definition of a Negative Exponent

$$x^{-n} = \frac{1}{x^n}$$

EXAMPLE 1 Write with positive exponents.

(a) y^{-3} **(b)** z^{-6} **(c)** w^{-1} **(d)** x^{-12}

(a) $y^{-3} = \frac{1}{y^3}$ **(b)** $z^{-6} = \frac{1}{z^6}$

(c) $w^{-1} = \frac{1}{w^1} = \frac{1}{w}$ **(d)** $x^{-12} = \frac{1}{x^{12}}$

Practice Problem 1 Write with positive exponents.

(a) x^{-12} **(b)** w^{-5} **(c)** z^{-2} **(d)** y^{-7} ■

To evaluate a numerical expression with a negative exponent, first write the expression with a positive exponent. Then simplify.

EXAMPLE 2 Evaluate. **(a)** 2^{-5} **(b)** 3^{-4}

(a) $2^{-5} = \frac{1}{2^5} = \frac{1}{32}$ **(b)** $3^{-4} = \frac{1}{3^4} = \frac{1}{81}$

Practice Problem 2 Evaluate. **(a)** 4^{-3} **(b)** 6^{-2} ■

All the previously studied laws of exponents can be applied when the exponent is any integer. These laws are the summarized in the following box.

Laws of Exponents

The Product Rule

$$x^a \cdot x^b = x^{a+b}$$

The Quotient Rule

$$\frac{x^a}{x^b} = x^{a-b} \quad \text{if } a > b, \qquad \frac{x^a}{x^b} = \frac{1}{x^{b-a}} \quad \text{if } a < b$$

Raising a Power to a Power

$$(xy)^a = x^a y^a, \qquad (x^a)^b = x^{ab}, \qquad \left(\frac{x}{y}\right)^a = \frac{x^a}{y^a}$$

By using the definition of a negative exponent and the properties of fractions, we can derive two more helpful properties of exponents.

Properties of Negative Exponents

$$\frac{1}{x^{-n}} = x^n \qquad \frac{x^{-m}}{y^{-n}} = \frac{y^n}{x^m}$$

EXAMPLE 3 Simplify. Write the expression so that no negative exponents appear.

(a) $\frac{1}{x^{-6}}$ **(b)** $\frac{x^{-3}y^{-2}}{z^{-4}}$ **(c)** $x^{-2}y^3$

(a) $\frac{1}{x^{-6}} = x^6$ **(b)** $\frac{x^{-3}y^{-2}}{z^{-4}} = \frac{z^4}{x^3y^2}$ **(c)** $x^{-2}y^3 = \frac{y^3}{x^2}$

Practice Problem 3 Simplify by writing with no negative exponents.

(a) $\frac{3}{w^{-4}}$ **(b)** $\frac{x^{-6}y^4}{z^{-2}}$ **(c)** $x^{-6}y^{-5}$ ■

EXAMPLE 4 Simplify. Write each expression so that no negative exponents appear.

(a) $(3x^{-4}y^2)^{-3}$ **(b)** $\frac{x^2y^{-4}}{x^{-5}y^3}$

(a) $(3x^{-4}y^2)^{-3} = 3^{-3}x^{12}y^{-6} = \frac{x^{12}}{3^3y^6} = \frac{x^{12}}{27y^6}$

(b) $\frac{x^2y^{-4}}{x^{-5}y^3} = \frac{x^2x^5}{y^4y^3} = \frac{x^7}{y^7}$ *First rewrite the expression so that only positive exponents appear. Then simplify using the product rule.*

Practice Problem 4 Simplify by writing with no negative exponents.

(a) $(2x^4y^{-5})^{-2}$ **(b)** $\frac{y^{-3}z^{-4}}{y^2z^{-6}}$ ■

2 *Scientific Notation*

One common use of negative exponents is in writing numbers in scientific notation. Scientific notation is most useful in expressing very large and very small numbers.

Scientific Notation

A positive number is written in scientific notation if it is in the form $a \times 10^n$, where $1 \le a < 10$ and n is an integer.

EXAMPLE 5 Write in scientific notation. (a) 4567 (b) 157,000,000

(a) $4567 = 4.567 \times 1000$ *To change 4567 to a number that is greater than 1 but less than 10, we move the decimal point three places to the left. We must then multiply the number by a power of 10 so that we do not change the value of the number. Use 1000.*

$= 4.567 \times 10^3$

(b) $157{,}000{,}000 = \underleftarrow{1.57000000}_{\text{8 places}} \times \underbrace{100000000}_{\text{8 zeros}}$

$= 1.57 \times 10^8$

Practice Problem 5 Write in scientific notation.

(a) 78,200 **(b)** 4,786,000 ■

Numbers that are smaller than 1 will have a negative power of 10 if they are written in scientific notation.

EXAMPLE 6 Write in scientific notation.

(a) 0.061 **(b)** 0.0034 **(c)** 0.000052

(a) We need to write 0.061 as a number that is greater than 1 but less than 10. In which direction do we move the decimal point?

$0.061 = 6.1 \times 10^{-2}$ *Move the decimal point 2 places to the right. Then multiply by* 10^{-2}.

(b) $0.0034 = 3.4 \times 10^{-3}$ *Move the decimal point 3 places to the right. Then multiply by* 10^{-3}.

(c) $0.000052 = 5.2 \times 10^{-5}$ *Why?*

Practice Problem 6 Write in scientific notation.

(a) 0.98 **(b)** 0.000763 **(c)** 0.000092 ■

The reverse procedure is used to go from scientific notation to ordinary decimal notation.

EXAMPLE 7 Write in decimal notation. (a) 1.568×10^2 (b) 7.432×10^{-3}

(a) $1.568 \times 10^2 = 1.568 \times 100$

$= 156.8$

Alternative method.

$1.568 \times 10^2 = 156.8$ *The exponent 2 tells us to move the decimal point 2 places to the right.*

(b) $7.432 \times 10^{-3} = 7.432 \times \frac{1}{1000}$

$= 0.007432$

Alternative method.

$7.432 \times 10^{-3} = 0.007432$ *The exponent −3 tells us to move the decimal point 3 places to the left.*

Practice Problem 7 Write in decimal notation.

(a) 2.96×10^3 **(b)** 1.93×10^6 **(c)** 5.43×10^{-2} **(d)** 8.562×10^{-5} ■

CALCULATOR

Scientific Notation

Most calculators can display only eight digits at one time. Numbers with more than eight digits are usually shown in scientific notation. 1.12 E 08 or 1.12 8 means 1.12×10^8. You can use the calculator to compute large numbers by entering the numbers using scientific notation. For example,

$(7.48 \times 10^{24}) \times (3.5 \times 10^8)$

Enter 7.48 [EXP] 24 [×] 3.5 [EXP] 8 [=]

Display: [2.618 E 33]

Note: Some calculators have an [EE] key instead of [EXP].

Compute on a calculator.

1. 35,000,000,000 + 77,000,000,000
2. $(6.23 \times 10^{12}) \times (4.9 \times 10^5)$
3. $(2.5 \times 10^7)^5$
4. $3.3284 \times 10^{32} \div (6.28 \times 10^{24})$
5. How many seconds are there in 1000 years?

TEACHING TIP Some students may find using a place value chart helpful when changing from scientific notation to decimal notation. To change 5.326×10^{-4} to decimal notation, have students begin by placing the 5 under 10^{-4}. Then have them fill in the remaining places.

	10^{-1}	10^{-2}	10^{-3}	10^{-4}	10^{-5}	10^{-6}	10^{-7}
.	0	0	0	5	3	2	6

The distance light travels in one year is called a *light-year.* A light-year is a convenient unit of measure to use when investigating the distances between stars.

EXAMPLE 8 A light-year is a distance of 9,460,000,000,000,000 meters. Write this in scientific notation.

$$9{,}460{,}000{,}000{,}000{,}000 = 9.46 \times 10^{15} \text{ meters}$$

Practice Problem 8 Astronomers measure distances to faraway galaxies in parsecs. A parsec is a distance of 30,900,000,000,000,000 meters. Write this in scientific notation. ■

To perform a calculation involving very large or very small numbers, it is usually helpful to write the numbers in scientific notation and then to use the laws of exponents to do the calculation.

EXAMPLE 9 Use scientific notation and the laws of exponents to find these answers. Leave your answer in scientific notation.

(a) (32,000,000)(1,500,000,000,000) **(b)** $\dfrac{0.00063}{0.021}$

(a) (32,000,000)(1,500,000,000,000)

$= (3.2 \times 10^7)(1.5 \times 10^{12})$ *Change each number to scientific notation.*

$= 3.2 \times 1.5 \times 10^7 \times 10^{12}$ *Rearrange the order. Remember multiplication is commutative.*

$= 4.8 \times 10^{19}$ *Multiply* 3.2×1.5. *Multiply* $10^7 \times 10^{12}$.

(b) $\dfrac{0.00063}{0.021} = \dfrac{6.3 \times 10^{-4}}{2.1 \times 10^{-2}}$ *Change each number to scientific notation.*

$= \dfrac{6.3}{2.1} \times \dfrac{10^{-4}}{10^{-2}}$ *Rearrange the order. We are actually using the definition of multiplication of fractions.*

$= \dfrac{6.3}{2.1} \times \dfrac{10^{2}}{10^{4}}$ *Rewrite with positive exponents.*

$= 3.0 \times 10^{-2}$

Practice Problem 9 Use scientific notation to find the following. Leave your answer in scientific notation.

(a) (56,000)(1,400,000,000) **(b)** $\dfrac{0.000111}{0.00000037}$ ■

EXAMPLE 10 The approximate distance from Earth to the star Polaris is 208 parsecs. A parsec is a distance of approximately 3.09×10^{13} kilometers. How long would it take a space probe traveling at 40,000 kilometers per hour to reach the star? Round your answer to three significant digits.

1. *Understand the problem.*
Recall that the distance formula is

$$\text{distance} = \text{rate} \times \text{time}$$

We are given the distance and the rate. We need to find the time.

Let's take a look at the distance. The distance is given in parsecs, but the rate is given in kilometers per hour. We need to change the distance to kilometers. We are told that a parsec is approximately 3.09×10^{13} kilometers. That is, there are 3.09×10^{13} kilometers per parsec. We use this information to change 208 parsecs to kilometers.

$$208 \text{ parsecs} = \frac{(208 \text{ parsecs})(3.09 \times 10^{13} \text{ kilometers})}{1 \text{ parsec}} = 642.72 \times 10^{13} \text{ kilometers}$$

2. *Write an equation.*
Use the distance formula.

$$d = r \times t$$

3. *Solve the equation and state the answer.*
Substitute the known values into the formula and solve for the unknown, time.

$$642.72 \times 10^{13} \text{ km} = \frac{40{,}000 \text{ km}}{1 \text{ hour}} \times t$$

$$6.4272 \times 10^{15} \text{ km} = \frac{4 \times 10^4 \text{ km}}{1 \text{ hr}} \times t$$ *Change the numbers to scientific notation.*

$$\frac{6.4272 \times 10^{15} \text{ km}}{\dfrac{4 \times 10^4 \text{ km}}{1 \text{ hr}}} = t$$ *Divide both sides by* $\dfrac{4 \times 10^4 \text{ km}}{1 \text{ hr}}$.

$$\frac{(6.4272 \times 10^{15} \cancel{\text{km}})(1 \text{ hr})}{4 \times 10^4 \cancel{\text{km}}} = t$$

$$1.6068 \times 10^{11} \text{ hr} = t$$

1.6068×10^{11} is 160.68×10^9 or 160.68 billion hours. The space probe will take approximately 160.68 billion hours to reach the star.

Reread the problem. Are we finished? What is left to do? We need to round the answer to three significant digits. Rounding to three significant digits, we have

$$160.68 \times 10^9 = 161 \times 10^9$$

This is approximately 161 billion hours or a little more than 18 million years.

4. *Check.* Unless you have had a great deal of experience working in astronomy, it would be difficult to determine if this is a reasonable answer. You may wish to reread your analysis and redo your calculations as a check.

Practice Problem 10 The average distance from Earth to the star Betelgeuse is 159 parsecs. How many hours would it take a space probe to travel from Earth to Betelgeuse at a speed of 50,000 kilometers per hour? Round your answer to three significant digits. ■

4.2 Exercises

Simplify each term and express your answer with positive exponents. Assume that all variables are nonzero.

1. $3x^{-2}$ $\frac{3}{x^2}$

2. $4xy^{-4}$ $\frac{4x}{y^4}$

3. $(2xy)^{-1}$ $\frac{1}{2xy}$

4. $(3x^2y^3)^{-2}$ $\frac{1}{9x^4y^6}$

5. $\frac{3xy^{-2}}{z^{-3}}$ $\frac{3xz^3}{y^2}$

6. $\frac{4x^{-2}y^{-3}z^0}{y^4}$ $\frac{4}{x^2y^7}$

7. $\frac{(3x)^{-2}}{(3x)^{-3}}$ $3x$

8. $\frac{(2ab^2)^{-3}}{(2ab^2)^{-4}}$ $2ab^2$

9. $x^6y^{-2}z^{-3}w^{-10}$ $\frac{x^6}{y^2z^3w^{10}}$

10. $2x^2y^{-3}z^{-4}w^5$ $\frac{2x^2w^5}{y^3z^4}$

11. $(8^{-2})(2^3)$ $\frac{1}{8}$

12. $(9^2)(3^{-3})$ 3

13. $\left(\frac{3xy^2}{z^4}\right)^{-2}$ $\frac{z^8}{9x^2y^4}$

14. $\left(\frac{2a^3b^0}{c^2}\right)^{-3}$ $\frac{c^6}{8a^9}$

15. $\frac{x^{-2}y^{-3}}{x^4y^{-2}}$ $\frac{1}{x^6y}$

16. $\frac{a^{-6}b^3}{a^{-2}b^{-5}}$ $\frac{b^8}{a^4}$

17. $(-2x^3y^{-2})^{-3}$ $-\frac{y^6}{8x^9}$

18. $(-3x^{-4}y^2)^{-4}$ $\frac{x^{16}}{81y^8}$

Write in scientific notation.

19. 123,780 1.2378×10^5

20. 0.063 6.3×10^{-2}

21. 0.000742 7.42×10^{-4}

22. 889,610,000,000 8.8961×10^{11}

23. 7,652,000,000 7.652×10^9

24. 0.00000001963 1.963×10^{-8}

Write in decimal notation.

25. 5.63×10^4 56,300

26. 1.776×10^8 177,600,000

27. 3.3×10^{-5} 0.000033

28. 1.99×10^{-1} 0.199

29. 9.83×10^5 983,000

30. 3.5×10^{-8} 0.000000035

31. The speed of light is 3.00×10^8 meters per second. Write this in decimal notation. 300,000,000 meters per second

32. The mass of a proton is

    0.00000000000000000000000000167 kilogram

    Write this in scientific notation. 1.67×10^{-27} kilogram

33. The average volume of an atom of gold is 0.00000000000000000000001695 cubic centimeters. Write this in scientific notation. 1.695×10^{-23} cubic centimeters

34. A single human red blood cell is about 7×10^{-6} meters in diameter. Write this in decimal notation. 0.000007 m

Evaluate by using scientific notation and the laws of exponents. Leave your answer in scientific notation.

35. $\frac{(5{,}000{,}000)(16{,}000)}{8{,}000{,}000{,}000}$ 1.0×10^1

36. $(0.0075)(0.0000002)(0.001)$ 1.5×10^{-12}

37. $(0.0002)^5$ 3.2×10^{-19}

38. $(40{,}000{,}000)^3$ 6.4×10^{22}

39. $(150{,}000{,}000)(0.00005)(0.002)(30{,}000)$ 4.5×10^5

40. $\frac{(1{,}600{,}000)(0.00003)}{2400}$ 2×10^{-2}

Applications

For fiscal year 1995 the national debt was determined to be approximately 4.693×10^{12} *dollars.*

41. In 1994 the census bureau estimates that the entire population of the United States was 2.60×10^8 people. If the national debt were evenly divided among every person in the country, how much debt would be assigned to each individual? Round your answer to three significant digits. 1.81×10^4 dollars

42. In 1994 the census bureau estimates that the number of people in the United States who were over the age of 18 was approximately 1.92×10^8 people. If the national debt were evenly divided among every person over the age of 18 in the country, how much debt would be assigned to each individual? Round your answer to three significant digits. 2.44×10^4 dollars

A parsec is a distance of approximately 3.09×10^{13} *kilometers.*

43. How long would it take a space probe to travel from Earth to the star Rigel, which is 276 parsecs from Earth? Assume that the space probe travels at 45,000 kilometers per hour. Round your answer to three significant digits. 1.9×10^{11} hours

44. How long would it take a space probe to travel from earth to the star Hadar, which is 150 parsecs from Earth? Assume that the space probe travels at 55,000 kilometers per hour. Round your answer to three significant digits. 8.43×10^{10} hours

45. Antares, a supergiant star, is one of the brightest stars seen by the naked eye. It is 130 parsecs from Earth. How long would it take an experimental robotic probe to reach Antares if we assume that the probe travels at 35,000 kilometers per hour? Round your answer to three significant digits. 1.15×10^{11} hours

46. The mass of an electron is approximately 9.11×10^{-28} gram. Find the mass of 125,000,000 electrons. 1.13875×10^{-19} gram

47. The mass of a proton is approximately 1.67×10^{-24} gram. Find the mass of 250,000,000 protons. 4.175×10^{-16} gram

48. A 26-gram sample of water containing tritium emits 1.50×10^3 beta particles per second. How many beta particles are emitted in one week? 9.072×10^8 particles

49. The sun radiates energy into space at the rate of 3.9×10^{26} joules per second. How many joules are emitted in two weeks? 4.717×10^{26} joules

Cumulative Review Problems

Simplify.

50. $-2.7 - (-1.9)$ -0.8

51. $(-1)^{2935}$ -1

52. $\frac{-3}{4} + \frac{5}{7}$ $-\frac{1}{28}$

Putting Your Skills to Work

THE MATHEMATICS OF FORESTS

One-third of the world's area that was originally covered by forests and woodlands no longer has either. The welfare of people and animals in future generations on earth may well depend on the ability of people to stop the destruction of forests worldwide. Some countries have a very significant amount of the present total of 9.587×10^9 acres of forests in the world.

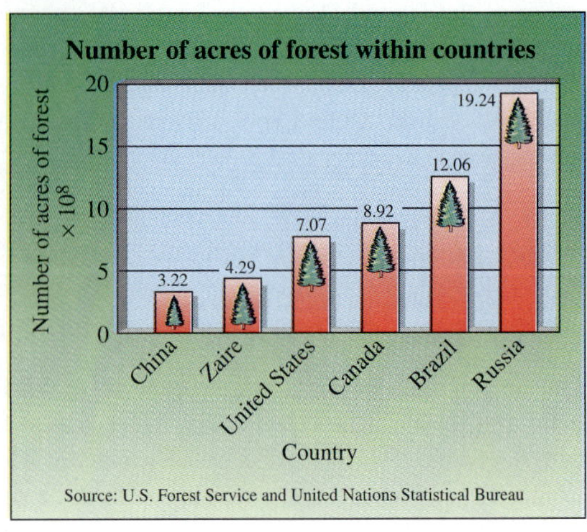

PROBLEMS FOR INDIVIDUAL INVESTIGATION AND ANALYSIS

1. How many more acres of forests are in Brazil than are in Canada? 3.14×10^8 acres

2. What is combined acreage of forests in Russia and China? 2.246×10^9 acres

PROBLEMS FOR GROUP INVESTIGATION AND COOPERATIVE GROUP ACTIVITY

Together with some members of your class see if you can determine the answers to the following.

3. If the current population of China is 1.203×10^9 people, how many people are there for each acre of forest in China? Round your answer to the nearest whole number. There are approximately 4 people per acre of forest.

4. Rainforests for quite some time have been destroyed at the rate of 40 million acres of forest per year. If we assume that this rate has been the same for the last 25 years, approximately how many acres of forests were in the world 25 years ago if we make adjustments for this one factor? 10.587×10^9 acres of forests

INTERNET CONNECTIONS: Go to ''http://www.prenhall.com/tobey'' to be connected

Site: Rates of Rainforest Loss (Rainforest Action Network) or a related site

This site presents information about rates of rainforest destruction. Use the information to answer the following questions. (There is some difference of opinion among scientists as to the rate of destruction.)

5. About how many acres of rainforest are destroyed in one hour?

6. About how many acres of Brazilian rainforest were destroyed during the last 12 years?

7. The current indigenous population of Brazil's forests is about what percent of the indigenous population in 1500?

4.3 Addition and Subtraction of Polynomials

After studying this section, you will be able to:

MathPro Video 11 SSM

1 *Add polynomials.*
2 *Subtract polynomials.*

1 *Adding Polynomials*

Recall that a polynomial in x is the sum of a finite number of terms of the form ax^n, where a is any real number and n is a whole number. Usually these polynomials are written in descending powers of the variable, such as

$$5x^3 + 3x^2 - 2x - 5 \quad \text{and} \quad 3.2x^2 - 1.4x + 5.6$$

You can add, subtract, multiply, and divide polynomials.

Let us take a look at addition. To add two polynomials, we add the respective like terms.

EXAMPLE 1 Add. $(5x^2 - 6x - 12) + (-3x^2 - 9x + 5)$

To begin we need to group like terms.

$$\begin{aligned}(5x^2 - 6x - 12) + (-3x^2 - 9x + 5) &= [5x^2 + (-3x^2)] + [-6x + (-9x)] + [-12 + 5]\\ &= [(5 - 3)x^2] + [(-6 - 9)x] + [-12 + 5]\\ &= 2x^2 + (-15x) + (-7)\\ &= 2x^2 - 15x - 7\end{aligned}$$

Alternative Method

We could write the problem in vertical form by lining up the like terms.

$$\begin{array}{r} 5x^2 - 6x - 12\\ + \; -3x^2 - 9x + 5\\ \hline 2x^2 - 15x - 7\end{array}$$

You may use either horizontal form or vertical form to group the like terms, whichever form you find easier to use.

Practice Problem 1 Add.

$(-8x^3 + 3x^2 + 6) + (2x^3 - 7x^2 - 3)$ ■

The numerical coefficients of the polynomials may be any real number. Thus the polynomials may have numerical coefficients that are decimals or fractions.

EXAMPLE 2 Add. $(\frac{1}{2}x^2 - 6x + \frac{1}{3}) + (\frac{1}{5}x^2 - 2x - \frac{1}{2})$

$$\begin{aligned}(\tfrac{1}{2}x^2 - 6x + \tfrac{1}{3}) + (\tfrac{1}{5}x^2 - 2x - \tfrac{1}{2}) &= [\tfrac{1}{2}x^2 + \tfrac{1}{5}x^2] + [-6x + (-2x)] + [\tfrac{1}{3} + (-\tfrac{1}{2})]\\ &= [(\tfrac{1}{2} + \tfrac{1}{5})x^2] + [(-6 - 2)x] + [\tfrac{1}{3} + (-\tfrac{1}{2})]\\ &= [(\tfrac{5}{10} + \tfrac{2}{10})x^2] + [-8x] + [\tfrac{2}{6} - \tfrac{3}{6}]\\ &= \tfrac{7}{10}x^2 - 8x - \tfrac{1}{6}\end{aligned}$$

Practice Problem 2 Add. $(-\frac{1}{3}x^2 - 6x - \frac{1}{12}) + (\frac{1}{4}x^2 + 5x - \frac{1}{3})$ ■

EXAMPLE 3 Add. $(1.2x^3 - 5.6x^2 + 5) + (-3.4x^3 - 1.2x^2 + 4.5x - 7)$

Group like terms.

$$(1.2x^3 - 5.6x^2 + 5) + (-3.4x^3 - 1.2x^2 + 4.5x - 7) = (1.2 - 3.4)x^3 + (-5.6 - 1.2)x^2 + 4.5x + (5 - 7)$$
$$= -2.2x^3 - 6.8x^2 + 4.5x - 2$$

Practice Problem 3

Add. $(3.5x^3 - 0.02x^2 + 1.56x - 3.5) + (-0.08x^2 - 1.98x + 4)$ ■

Polynomials may involve more than one variable.

EXAMPLE 4 Add. $(-5x^2 + 2xy - 15y^2) + (-30x^2 - 5xy + 2y^2)$

Group like terms.

$$(-5x^2 + 2xy - 15y^2) + (-30x^2 - 5xy + 2y^2) = (-5 - 30)x^2 + (2 - 5)xy + (-15 + 2)y^2$$
$$= -35x^2 - 3xy - 13y^2$$

Practice Problem 4 Add. $(-3a^3 + 2ab + 7b) + (-4a^3 - 6ab + 3b^3)$ ■

2 *Subtracting Polynomials*

Recall that subtraction of real numbers can be defined as adding the opposite of the second number. Thus $(a) - (b) = (a) + (-b)$. That is, $3 - 5 = 3 + (-5)$. A similar method is used to subtract two polynomials.

To subtract two polynomials, change all the signs of the second polynomial and it becomes an addition problem.

EXAMPLE 5 Subtract. $(7x^2 - 6x + 3) - (5x^2 - 8x - 12)$

We change all the signs of the second polynomial and then add.

$$(7x^2 - 6x + 3) - (5x^2 - 8x - 12) = (7x^2 - 6x + 3) + (-5x^2 + 8x + 12)$$
$$= (7 - 5)x^2 + (-6 + 8)x + (3 + 12)$$
$$= 2x^2 + 2x + 15$$

Practice Problem 5

Subtract. $(5x^3 - 15x^2 + 6x - 3) - (-4x^3 - 10x^2 + 5x + 13)$ ■

EXAMPLE 6 Subtract. $(-6x^2y - 3xy + 7xy^2) - (5x^2y - 8xy - 15x^2y^2)$

Change the sign of each term in the second polynomial. Then look for like terms.

$$(-6x^2y - 3xy + 7xy^2) + (-5x^2y + 8xy + 15x^2y^2) = (-6 - 5)x^2y + (-3 + 8)xy + 7xy^2 + 15x^2y^2$$
$$= -11x^2y + 5xy + 7xy^2 + 15x^2y^2$$

Nothing further can be done to combine these four terms.

Practice Problem 6 Subtract. $(x^3 - 7x^2y + 3xy^2 - 2y^3) - (2x^3 + 4xy - 6y^3)$ ■

4.3 Exercises

Add the following polynomials.

1. $(7x + 3) + (-5x - 20)$ $2x - 17$

2. $(-8x - 4) + (3x + 12)$ $-5x + 8$

3. $(2x^2 - 8x + 7) + (-5x^2 - 2x + 3)$ $-3x^2 - 10x + 10$

4. $(x^2 - 6x - 3) + (-4x^2 + x - 4)$ $-3x^2 - 5x - 7$

5. $(x^3 - 3x^2 + 7) + (x^2 + 5x - 8)$ $x^3 - 2x^2 + 5x - 1$

6. $(x^2 - 7x - 8) + (x^3 + 3x^2 + 10)$ $x^3 + 4x^2 - 7x + 2$

7. $(\frac{1}{2}x^2 + \frac{1}{3}x - 4) + (\frac{1}{3}x^2 + \frac{1}{6}x - 5)$ $\frac{5}{6}x^2 + \frac{1}{2}x - 9$

8. $(\frac{1}{4}x^2 - \frac{2}{3}x - 10) + (-\frac{1}{3}x^2 + \frac{1}{9}x + 2)$ $-\frac{1}{12}x^2 - \frac{5}{9}x - 8$

9. $(3.4x^3 - 5.6x^2 - 7.1x + 3.4) + (-1.7x^3 + 2.2x^2 - 6.1x - 8.8)$ $1.7x^3 - 3.4x^2 - 13.2x - 5.4$

10. $(-4.6x^3 - 2.9x^2 + 5.6x - 0.3) + (-2.6x^3 + 9.8x^2 + 4.5x - 1.7)$ $-7.2x^3 + 6.9x^2 + 10.1x - 2$

Subtract the following polynomials.

11. $(-8x + 2) - (3x + 4)$ $-11x - 2$

12. $(12x + 9) - (8x - 6)$ $4x + 15$

13. $(3x^4 - 2x^2) - (x^4 - 3x^2)$ $2x^4 + x^2$

14. $(5x^3 + 3x) - (6x^3 - 2x)$ $-x^3 + 5x$

15. $(\frac{1}{2}x^2 + 3x - \frac{1}{5}) - (\frac{2}{3}x^2 - 4x + \frac{1}{10})$ $-\frac{1}{6}x^2 + 7x - \frac{3}{10}$

16. $(\frac{1}{8}x^2 - 5x + \frac{2}{7}) - (\frac{1}{4}x^2 - 6x + \frac{1}{14})$ $-\frac{1}{8}x^2 + x + \frac{3}{14}$

17. $(-3x^2 + 5x) - (2x^3 - 3x^2 + 10)$
$-2x^3 + 5x - 10$

18. $(7x^3 - 3x^2 - 5x + 4) - (x^3 - 7x + 3)$
$6x^3 - 3x^2 + 2x + 1$

19. $(0.5x^4 - 0.7x^2 + 8.3) - (5.2x^4 + 1.6x + 7.9)$
$-4.7x^4 - 0.7x^2 - 1.6x + 0.4$

20. $(1.3x^4 - 3.1x^3 + 6.3x) - (x^4 - 5.2x^2 + 6.5x)$
$0.3x^4 - 3.1x^3 + 5.2x^2 - 0.2x$

Perform the indicated operations.

21. $(7x - 8) - (5x - 6) + (8x + 12)$
$10x + 10$

22. $(-8x + 3) + (-6x - 20) - (-3x + 9)$
$-11x - 26$

23. $(5x^2y - 6xy^2 + 2) + (-8x^2y + 12xy^2 - 6)$
$-3x^2y + 6xy^2 - 4$

24. $(7x^2y^2 - 6xy + 5) + (-15x^2y^2 - 6xy + 18)$
$-8x^2y^2 - 12xy + 23$

25. $(-7a^4 + 3a^2 - 2) - (4a^4 + 5a^3 + 3a^2)$
$-11a^4 - 5a^3 - 2$

26. $(-9a^4 - 6a^3 + 2a^2) - (a^4 + 6a^2 - 3a)$
$-10a^4 - 6a^3 - 4a^2 + 3a$

27. $(3a^3 - 5a^2b - ab^2 + 4b^2) + (-2a^3 - 3a^2b + 6ab^2 - 11b^2)$ $a^3 - 8a^2b + 5ab^2 - 7b^2$

28. $(x^3 - 4x^2y - 6xy^2 - 10y^3) + (-5x^3 - 4x^2y + xy^2 - y^3)$ $-4x^3 - 8x^2y - 5xy^2 - 11y^3$

To Think About

29. When $5x^2 - 7xy - 9y^2$ is added to $-63x^2 + 34xy + y^2$, what is the result? $-58x^2 + 27xy - 8y^2$

30. When $7x^2y^2 - 16xy - 45$ is added to $-8x^2y^2 + 23xy + 45$, what is the result? $-x^2y^2 + 7xy$

31. When $5x^3 - 7x^2 - 28$ is subtracted from $12x^3 + 34x^2 - 6$, what is the result? $7x^3 + 41x^2 + 22$

32. When $7x^3 + 4x - 18$ is subtracted from $-18x^3 - 16x + 3$, what is the result? $-25x^3 - 20x + 21$

Cumulative Review Problems

33. Solve for x. $3y - 8x = 2$
$x = \dfrac{3y - 2}{8}$

34. Solve for B. $3A = 2BCD$
$B = \dfrac{3A}{2CD}$

35. Solve for b. $A = \frac{1}{2}h(b + c)$
$\dfrac{2A}{h} - c = b$

36. Solve for d. $B = \dfrac{5xy}{d}$
$d = \dfrac{5xy}{B}$

37. The total expenditure for health care in the United States increased by 26.9% over the years from 1990 to 1993. In 1993 a total of 884.2 billion dollars was spent. How much was spent in 1990? Approx. 696.8 billion dollars

4.4 Multiplication of Polynomials

After studying this section, you will be able to:

1 *Multiply a monomial by a polynomial.*
2 *Multiply two binomials.*

1 *Multiplying a Monomial by a Polynomial*

We use the distributive property to multiply a monomial by a polynomial. Remember, the distributive property states that for real numbers a, b, c it is true that

$$a(b + c) = ab + ac$$

EXAMPLE 1 Multiply. $3x^2(5x - 2)$

$$3x^2(5x - 2) = 3x^2(5x) + 3x^2(-2) \qquad \textit{Use the distributive property.}$$
$$= (3 \cdot 5)(x^2 \cdot x) + (3)(-2)x^2$$
$$= 15x^3 - 6x^2$$

Practice Problem 1 Multiply. $4x^3(-2x^2 + 3x)$ ■

Try to do as much of the multiplication mentally as you can.

EXAMPLE 2 Multiply.

(a) $3(2 - y)$ **(b)** $2x(x^2 + 3x - 1)$ **(c)** $-2xy^2(x^2 - 2xy - 3y^2)$

(a) $3(2 - y) = 6 - 3y$

(b) $2x(x^2 + 3x - 1) = 2x^3 + 6x^2 - 2x$

(c) $-2xy^2(x^2 - 2xy - 3y^2) = -2x^3y^2 + 4x^2y^3 + 6xy^4$
Notice in part (c) that you are multiplying each term by the negative number $-2xy^2$. This will affect the sign of each term in the product.

Practice Problem 2 Multiply.

(a) $7(3 - 2x)$ **(b)** $-3x(x^2 + 2x - 4)$ **(c)** $6xy(x^3 + 2x^2y - y^2)$ ■

When we multiply by a monomial, the monomial may be on the right side.

EXAMPLE 3 Multiply. **(a)** $(x + 2y + z)3x$ **(b)** $(x^2 - 2x + 6)(-2xy)$

(a) $(x + 2y + z)3x = 3x^2 + 6xy + 3xz$

(b) $(x^2 - 2x + 6)(-2xy) = -2x^3y + 4x^2y - 12xy$

Practice Problem 3 Multiply.

(a) $(3x^2 - 7x + 1)4x$ **(b)** $(-6x^3 + 4x^2 - 2x)(-3xy)$ ■

2 *Multiplying Two Binomials*

We can build on our knowledge of the distributive property and our experience with multiplying monomials to learn how to multiply two binomials. Let's suppose that we want to multiply $(x + 2)(3x + 1)$. We can use the distributive property. Since a can represent any number, let $a = x + 2$. Then let $b = 3x$ and $c = 1$. We now have

$$\begin{aligned} a(b + c) &= \quad ab \quad + \quad ac \\ (x + 2)(3x + 1) &= (x + 2)(3x) + (x + 2)(1) \\ &= 3x^2 + 6x + x + 2 \\ &= 3x^2 + 7x + 2 \end{aligned}$$

Let's take another look at the original problem, $(x + 2)(3x + 1)$. This time we will assign a letter to each term in the binomials. That is, let $a = x$, $b = 2$, $c = 3x$, and $d = 1$. Using substitution, we have

$$\begin{aligned} &(x + 2)(3x + 1) \\ &(a + b)(c + d) = (a + b)c + (a + b)d \\ &\qquad = ac + bc + ad + bd \\ &\qquad = (x)(3x) + (2)(3x) + (x)(1) + (2)(1) \qquad \textit{By substitution.} \\ &\qquad = 3x^2 + 6x + x + 2 \\ &\qquad = 3x^2 + 7x + 2 \end{aligned}$$

How does this compare with the result above?

The distributive property shows us *how* the problem can be done and *why* it can be done. In actual practice there is a somewhat easier approach to obtain the answer. It is often referred to as the FOIL method. The letters FOIL stand for

F	multiply the *F*irst terms
O	multiply the *O*uter terms
I	multiply the *I*nner terms
L	multiply the *L*ast terms

The FOIL letters are simply a way to remember the four terms in the final product and how they are obtained. Let's return to our original problem.

$(x + 2)(3x + 1)$	F	Multiply *first* terms and obtain $3x^2$.
$(x + 2)(3x + 1)$	O	Multiply the *outer* terms and obtain x.
$(x + 2)(3x + 1)$	I	Multiply the two *inner* terms and obtain $6x$.
$(x + 2)(3x + 1)$	L	Multiply the *last* two terms and obtain 2.

The result so far is $3x^2 + x + 6x + 2$. These four terms are the same four terms that we obtained when we multiplied using the distributive property. We can combine the like terms to obtain the final answer: $3x^2 + 7x + 2$. Our result is again the same as when we used the distributive property. Now let's study the use of the FOIL method in a few examples.

EXAMPLE 4 Multiply. $(2x - 1)(3x + 2)$

First Last

$(2x - 1)(3x + 2)$ First + Outer + Inner + Last

Inner

Outer

$\begin{matrix} F & O & I & L \end{matrix}$

$= 6x^2 + 4x - 3x - 2$

$= 6x^2 + x - 2$ *Collect like terms.*

TEACHING TIP Some students prefer to multiply the four products in a different order, particularly if they have a learning disability. This is OK and students should be given this freedom.

Practice Problem 4 Multiply. $(5x - 1)(x - 2)$ ■

EXAMPLE 5 Multiply. $(3a + 2b)(4a - b)$.

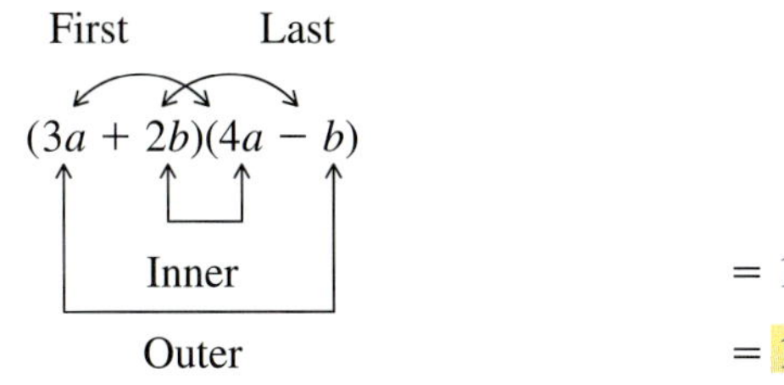

$= 12a^2 - 3ab + 8ab - 2b^2$

$= 12a^2 + 5ab - 2b^2$

Practice Problem 5 Multiply. $(8a - 5b)(3c - d)$ ■

After you have done several problems you may be able to combine the outer and inner products mentally.

EXAMPLE 6 Multiply. $(5x + 1)(x - 2)$

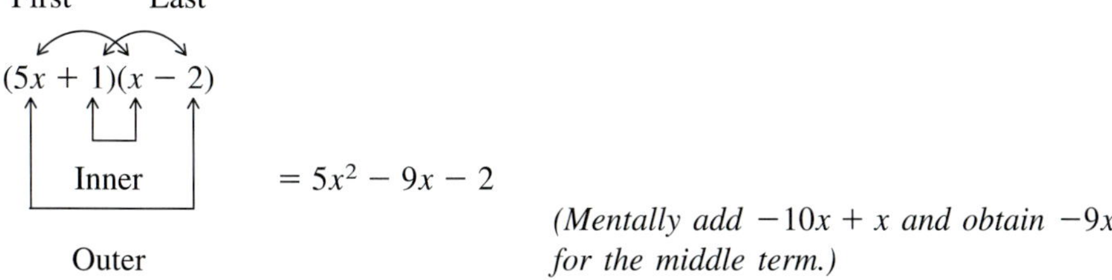

$= 5x^2 - 9x - 2$

(Mentally add $-10x + x$ and obtain $-9x$ for the middle term.)

Practice Problem 6 Multiply. $(3x + 2)(x - 1)$ ■

In some problems the inner and outer products cannot be combined.

EXAMPLE 7 Multiply. $(3x + 2y)(5x - 3z)$

First Last

$(3x + 2y)(5x - 3z)$

Inner

Outer

$= 15x^2 - 9xz + 10xy - 6yz$

Since there are no like terms, we cannot combine any terms.

TEACHING TIP After presenting Examples 6 and 7 ask students as an Additional Class Exercise to multiply out the following binomials and express their answers in simplest form.

A. $(3x - 2)(4x - 1)$

B. $(7x - 6)(x + 3)$

C. $(5x + 2)(2x + 9)$

Sol.

A. $12x^2 - 11x + 2$

B. $7x^2 + 15x - 18$

C. $10x^2 + 49x + 18$

Practice Problem 7 Multiply. $(3a + 2b)(2a - 3b)$ ■

TEACHING TIP Students have a bit more trouble with two binomials when both the first and last terms contain a variable. In addition to Example 8, you may want to do the following as an Additional Class Exercise.
Multiply. $(9a + 5b)^2$
Sol. $81a^2 + 90ab + 25b^2$

EXAMPLE 8 Multiply. $(7x - 2y)^2$

$(7x - 2y)(7x - 2y)$ *When we square a binomial, it is the same as multiplying the binomial by itself.*

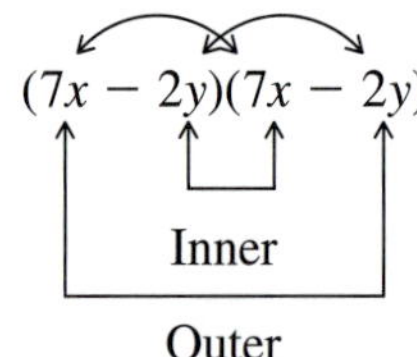

$$= 49x^2 - 14xy - 14xy + 4y^2$$
$$= 49x^2 - 28xy + 4y^2$$

Practice Problem 8 Multiply. $(3x - 2y)^2$ ■

The exponents of the variable may be higher than 1. We may multiply binomials containing x^2 or y^3, and so on.

TEACHING TIP If the degree of the binomials is two or higher, students have more difficulty. After doing example 9, you may wish to show the students the following Additional Class Exercise.
Multiply. $(7x^3 - 3y^3)(2x^3 - 8y^3)$
Sol. Multiplying each product yields
$14x^6 - 56x^3y^3 - 6x^3y^3 + 24y^6$
$= 14x^6 - 62x^3y^3 + 24y^6$

EXAMPLE 9 Multiply. $(3x^2 + 4y^3)(2x^2 + 5y^3)$

$$(3x^2 + 4y^3)(2x^2 + 5y^3) = 6x^4 + 15x^2y^3 + 8x^2y^3 + 20y^6$$
$$= 6x^4 + 23x^2y^3 + 20y^6$$

Practice Problem 9 Multiply. $(2x^2 + 3y^2)(5x^2 + 6y^2)$ ■

EXAMPLE 10 The width of a living-room is $(x + 4)$ feet. The length of the room is $(3x + 5)$ feet. What is the area of the room in square feet?

$$A = \text{(length)(width)} = (3x + 5)(x + 4)$$
$$= 3x^2 + 12x + 5x + 20$$
$$= 3x^2 + 17x + 20$$

There are $(3x^2 + 17x + 20)$ square feet in the room.

Practice Problem 10 What is the area in square feet of a room that is $(2x - 1)$ feet wide and $(7x + 3)$ feet long? ■

Developing Your Study Skills

EXAM TIME: HOW TO REVIEW

Reviewing adequately for an exam enables you to bring together the concepts you have learned over several sections. For your review, you will need to:

1. Reread your textbook. Make a list of any terms, rules, or formulas you need to know for the exam. Be sure you understand them all.
2. Reread your notes. Go over returned homework and quizzes. Redo the problems you missed.
3. Practice some of each type of problem covered in the chapter(s) you are to be tested on.
4. Use the end-of-chapter materials provided in your textbook. Read carefully through the Chapter Organizer. Do the Review Problems. Take the Chapter Test. When you are finished, check your answers. Redo any problems you missed.
5. Get help if any concepts give you difficulty.

4.4 Exercises

Multiply.

1. $3x(5x^2 - x)$
$15x^3 - 3x^2$

2. $-6x(2x^2 + 3x)$
$-12x^3 - 18x^2$

3. $-5x(3x^2 - 2x + 5)$
$-15x^3 + 10x^2 - 25x$

4. $3x(-2x^2 + 7x - 11)$
$-6x^3 + 21x^2 - 33x$

5. $2x(x^3 - 3x^2 + x - 6)$
$2x^4 - 6x^3 + 2x^2 - 12x$

6. $-3x(x^3 + 5x^2 - 6x - 7)$
$-3x^4 - 15x^3 + 18x^2 + 21x$

7. $-5xy(x^2y^2 - 3xy + 5)$
$-5x^3y^3 + 15x^2y^2 - 25xy$

8. $2xy(3x^2y^2 + 9xy - 8)$
$6x^3y^3 + 18x^2y^2 - 16xy$

9. $3x^3(x^4 + 2x^2 - 3x + 7)$
$3x^7 + 6x^5 - 9x^4 + 21x^3$

10. $-7x^3(x^5 - 2x^3 + 3x^2 - 4)$
$-7x^8 + 14x^6 - 21x^5 + 28x^3$

11. $2ab^2(3 - 2ab - 5b^3)$
$6ab^2 - 4a^2b^3 - 10ab^5$

12. $-2x^2y(x^2 - 8x + 2y^2)$
$-2x^4y + 16x^3y - 4x^2y^3$

13. $(5x^3 - 2x^2 + 6x)(-3xy^2)$
$-15x^4y^2 + 6x^3y^2 - 18x^2y^2$

14. $(3b^2 - 6b + 8ab)(5b^3)$
$15b^5 - 30b^4 + 40ab^4$

15. $(2b^2 + 3b - 4)(2b^2)$
$4b^4 + 6b^3 - 8b^2$

16. $(2x^3 + x^2 - 6x)(2xy)$
$4x^4y + 2x^3y - 12x^2y$

17. $(x^3 - 3x^2 + 5x - 2)(3x)$
$3x^4 - 9x^3 + 15x^2 - 6x$

18. $(-2x^3 + 4x^2 - 7x + 3)(2x)$
$-4x^4 + 8x^3 - 14x^2 + 6x$

19. $(x^2y^2 - 6xy + 8)(-2xy)$
$-2x^3y^3 + 12x^2y^2 - 16xy$

20. $(x^2y^2 + 5xy - 9)(-3xy)$
$-3x^3y^3 - 15x^2y^2 + 27xy$

21. $(-7x^3 + 3x^2 + 2x - 1)(4x^2y)$
$-28x^5y + 12x^4y + 8x^3y - 4x^2y$

22. $(-5x^3 - 6x^2 + x - 1)(5xy^2)$
$-25x^4y^2 - 30x^3y^2 + 5x^2y^2 - 5xy^2$

23. $(5a^3 + a^2 - 7)(-3ab^2)$
$-15a^4b^2 - 3a^3b^2 + 21ab^2$

24. $(6a^3 - 5a + 2)(-4a^2b)$
$-24a^5b + 20a^3b - 8a^2b$

25. $6x^3(2x^4 - x^2 + 3x + 9)$
$12x^7 - 6x^5 + 18x^4 + 54x^3$

26. $8x^3(-2x^4 + 3x^2 - 5x - 14)$
$-16x^7 + 24x^5 - 40x^4 - 112x^3$

27. $-4x^3(6x^4 - 3x^2 + 2)$
$-24x^7 + 12x^5 - 8x^3$

28. $-3x^5(-5x^4 + 2x^2 - 3)$
$15x^9 - 6x^7 + 9x^5$

Multiply. Try to do most of the problems mentally without writing down intermediate steps.

29. $(x + 10)(x + 3)$
$x^2 + 13x + 30$

30. $(x + 3)(x + 4)$
$x^2 + 7x + 12$

31. $(x + 6)(x + 2)$
$x^2 + 8x + 12$

32. $(x - 8)(x + 2)$
$x^2 - 6x - 16$

33. $(x + 3)(x - 6)$
$x^2 - 3x - 18$

34. $(x - 5)(x - 4)$
$x^2 - 9x + 20$

35. $(x - 6)(x - 5)$
$x^2 - 11x + 30$

36. $(3x + 1)(-x - 4)$
$-3x^2 - 13x - 4$

37. $(5x + 1)(-x - 3)$
$-5x^2 - 16x - 3$

38. $(7x - 4)(x + 2y)$
$7x^2 + 14xy - 4x - 8y$

39. $(x - 9)(3x + 4y)$
$3x^2 + 4xy - 27x - 36y$

40. $(2y + 3)(5y - 2)$
$10y^2 + 11y - 6$

41. $(3y + 2)(4y - 3)$
$12y^2 - y - 6$

42. $(7y - 2)(3y - 1)$
$21y^2 - 13y + 2$

43. $(5y - 3)(4y - 2)$
$20y^2 - 22y + 6$

To Think About

44. What is *wrong* with this multiplication? $(x - 2)(-3) = 3x - 6$
The signs are incorrect. The result should be $-3x + 6$.

45. What is *wrong* with this answer? $-(3x - 7) = -3x - 7$
The last term is incorrect. The result should be $-3x + 7$.

46. What is the missing term? $(3x + 2)(3x + 2) = 9x^2 + \underline{\quad 12x \quad} + 4$.

47. After multiplying these two binomials, write a brief description of what is special about your answer. $(7x + 3)(7x - 3)$
$49x^2 - 9$, difference of 2 perfect squares

Multiply.

48. $(x + y)(-5x - 6y)$
$-5x^2 - 11xy - 6y^2$

49. $(4x - 2y)(-7x - 3y)$
$-28x^2 + 2xy + 6y^2$

50. $(5x - 3y)(-8x - 9y)$
$-40x^2 - 21xy + 27y^2$

51. $(3a - 2b^2)(5a - 6b^2)$
$15a^2 - 28ab^2 + 12b^4$

52. $(8a - 3b^2)(2a - 5b^2)$
$16a^2 - 46ab^2 + 15b^4$

53. $(2x^2 - 5y^2)(2x^2 + 5y^2)$
$4x^4 - 25y^4$

54. $(8x^2 - 3y^2)(8x^2 - 3y^2)$
$64x^4 - 48x^2y^2 + 9y^4$

55. $(8x - 2)^2$
$64x^2 - 32x + 4$

56. $(5x - 3)^2$
$25x^2 - 30x + 9$

57. $(5x^2 + 2y^2)^2$
$25x^4 + 20x^2y^2 + 4y^4$

58. $(3x^2 - 7y^2)^2$
$9x^4 - 42x^2y^2 + 49y^4$

59. $(7m - 5g)(3m + 2g)$
$21m^2 - 1gm - 10g^2$

60. $(5m - 8g)(9m + 3g)$
$45m^2 - 57gm - 24g^2$

61. $(3ab - 5d)(2ab - 7d)$
$6a^2b^2 - 31abd + 35d^2$

62. $(6xy - 2z)(5xy + 4z)$
$30x^2y^2 + 14xyz - 8z^2$

63. $(5x + 8y)(6x - y)$
$30x^2 + 43xy - 8y^2$

64. $(3x + 8y)(5x - 2y)$
$15x^2 + 34xy - 16y^2$

65. $(7x - 2y)(5x + 3z)$
$35x^2 + 21xz - 10xy - 6yz$

66. $(4a - 5b)(a + 3f)$
$4a^2 + 12af - 5ab - 15bf$

67. $(a - 8b)(4c - 3b)$
$4ac - 3ab - 32bc + 24b^2$

68. $(x - 3y)(z - 8y)$
$xz - 8xy - 3yz + 24y^2$

★**69.** $(\frac{1}{3}x^3y^4 - \frac{1}{7}xy^6)(\frac{1}{2}x^3y^4 + \frac{1}{3}xy^6)$
$\frac{1}{6}x^6y^8 + \frac{5}{126}x^4y^{10} - \frac{1}{21}x^2y^{12}$

★**70.** $(3.2a^3b - 0.9ab^3)(5.3a^3b - 1.1ab^3)$
$16.96a^6b^2 - 8.29a^4b^4 + 0.99a^2b^6$

Cumulative Review Problems

71. Solve for x. $3(x - 6) = -2(x + 4) + 6x$
$x = -10$

72. Solve for w. $3(w - 7) - (4 - w) = 11w$
$-\frac{25}{7} = w$

73. La Tanya has three more quarters than dimes, giving her \$3.55. How many of each coin does she have? (Use d for the number of dimes.) 8 dimes; 11 quarters

74. A number decreased by three is seven more than three times the number. Find the number. -5

75. A total of 10% less than half the population of China owns a color television. In India 28% less of the population owns a color television. In the United States, the percent of the people who own a color television is 17% greater than double the percent of the population of China who own them. What percent of the people in each of these three countries own a color television?
40% of people in China, 97% of people in the United States, 12% of people in India

76. Heather returned from the bank with \$375. She had one more \$20 bill than the number of \$10 bills. The number of \$5 bills she had was one less than triple the number of \$10 bills. How many of each denomination did she have? Nine 20's, eight 10's, and twenty-three 5's

4.5 Multiplication: Special Cases

MathPro Video 11 SSM

After studying this section, you will be able to:

1. *Multiply binomials of the type* $(a + b)(a - b)$.
2. *Multiply binomials of the type* $(a + b)^2$ and $(a - b)^2$.
3. *Multiply polynomials with more than two terms.*

1 *Multiplying Binomials of the Type* $(a + b)(a - b)$

The case when you multiply $(x + y)(x - y)$ is interesting and deserves special consideration. Using the FOIL method, we find

$$(x + y)(x - y) = x^2 - xy + xy - y^2 = x^2 - y^2$$

Notice that the sum of the inner product and the outer product is zero. We see that

$$(x + y)(x - y) = x^2 - y^2$$

This works in all cases when the binomials have the same two terms. In one factor the terms are added, while in the other factor the terms are subtracted.

$$(5a + 2b)(5a - 2b) = 25a^2 - 10ab + 10ab - 4b^2$$
$$= 25a^2 - 4b^2$$

The product is the difference of squares. That is, $(5a)^2 - (2b)^2$ or $25a^2 - 4b^2$.

Many students find it helpful to memorize this equation.

> **Multiplying Binomials: A Sum and a Difference**
>
> $$(a + b)(a - b) = a^2 - b^2$$

You may use this relationship to find the product quickly in cases where it applies. The terms must be the same and there must be a sum and a difference.

EXAMPLE 1 Multiply. $(7x + 2)(7x - 2)$

$$(7x + 2)(7x - 2) = (7x)^2 - (2)^2 = 49x^2 - 4$$

Practice Problem 1 Multiply. $(6x + 7)(6x - 7)$ ■

TEACHING TIP The formula $(a + b)(a - b) = a^2 - b^2$ is useful now for multiplying and later in the chapter for factoring. Therefore, it is worthwhile to memorize it. Emphasize to your students the future value of the formula.

EXAMPLE 2 Multiply. $(5x - 8y)(5x + 8y)$

$$(5x - 8y)(5x + 8y) = (5x)^2 - (8y)^2 = 25x^2 - 64y^2$$

Practice Problem 2 Multiply. $(3x + 5y)(3x - 5y)$ ■

TEACHING TIP After presenting Example 3 to your students, ask them to multiply the following binomials as an Additional Class Exercise.

A. $(6x + 11)(6x - 11)$
B. $(5x - 8y)(5x + 8y)$
C. $(3x^2 + 7y^2)(3x^2 - 7y^2)$

Sol.

A. $36x^2 - 121$
B. $25x^2 - 64y^2$
C. $9x^4 - 49y^4$

EXAMPLE 3 Multiply. $(2x^2 - 9)(2x^2 + 9)$

$$(2x^2 - 9)(2x^2 + 9) = (2x^2)^2 - (9)^2 = 4x^4 - 81$$

Practice Problem 3 Multiply. $(4x^2 - 1)(4x^2 + 1)$ ■

2 *Multiplying Binomials of the Type $(a + b)^2$ and $(a - b)^2$*

A second case that is worth special consideration is when a binomial is squared. Consider the problem

$$\begin{aligned}(3x + 2)^2 &= (3x + 2)(3x + 2)\\ &= 9x^2 + 6x + 6x + 4\\ &= 9x^2 + 12x + 4\end{aligned}$$

If you complete enough problems of this type, you will notice a pattern. The answer always contains the square of the first term added to double the product of the first and last terms added to the square of the last term. That is,

$3x$ is the first term	2 is the last term		Square the first term: $(3x)^2$		Double the product of the first and last terms: $2(3x)(2)$		Square the last term: $(2)^2$
↓	↓		↓		↓		↓
$(3x$ +	$2)^2$	=	$9x^2$	+	$12x$	+	4

We show the same steps using variables instead of words.

$$(a + b)^2 = a^2 + 2ab + b^2$$

This is also true for the square of a difference.

$$\begin{aligned}(2x - 3)^2 &= (2x)^2 - 2(2x)(3) + (3)^2\\ &= 4x^2 - 12x + 9\end{aligned}$$

You may wish to multiply this out using FOIL to verify. That is, multiply $(2x - 3)(2x - 3)$.

$$(a - b)^2 = a^2 - 2ab + b^2$$

The two types of multiplications, the square of a sum and the square of a difference, can be summarized as follows:

A Binomial Squared

$$\boldsymbol{(a + b)^2 = a^2 + 2ab + b^2}$$

$$\boldsymbol{(a - b)^2 = a^2 - 2ab + b^2}$$

EXAMPLE 4 Multiply. **(a)** $(5y - 2)^2$ **(b)** $(8x + 9y)^2$

(a) $(5y - 2)^2 = (5y)^2 - (2)(5y)(2) + (2)^2$

$= 25y^2 - 20y + 4$

(b) $(8x + 9y)^2 = (8x)^2 + (2)(8x)(9y) + (9y)^2$

$= 64x^2 + 144xy + 81y^2$

Practice Problem 4 Multiply. **(a)** $(5x + 4)^2$ **(b)** $(4a - 9b)^2$ ■

Warning

$(a+b)^2 \neq a^2+b^2$! The two sides are not equal! Squaring a sum $(a+b)$ does not give a^2+b^2! Beginning algebra students often make this error. Make sure you remember that when you square a binomial there is always a *middle term.*

$$(a+b)^2 = a^2+2ab+b^2$$

Sometimes a numerical example helps you to see this.

$$(3+4)^2 \neq 3^2+4^2$$

$$7^2 \neq 9+16$$

$$49 \neq 25$$

Notice that what is missing on the right is the $2ab = 2 \cdot 3 \cdot 4 = 24$.

TEACHING TIP This is beyond a doubt the most common student error found in all of elementary algebra! Be sure to warn your students often to avoid this error!

3 *Multiplying Polynomials with More Than Two Terms*

We used the distributive property to multiply two binomials $(a+b)(c+d)$, and we obtained $ac+ad+bc+bd$. We could also use the distributive property to multiply the polynomials $(a+b)\ (c+d+e)$ and we would obtain $ac+ad+ae+bc+bd+be$. Let us see if we can find a direct way to multiply polynomials such as $(3x-2)(x^2-2x+3)$. It can be quickly done using an approach similar to that used in arithmetic for multiplying whole numbers. Consider the following arithmetic problem.

$$\begin{array}{rl} 128 & \\ \times\ \ 43 & \\ \hline 384 & \longleftarrow \text{The product of 128 and 3.} \\ 512\ \ & \longleftarrow \text{The product of 128 and 4 moved one space to the left.} \\ \hline 5504 & \longleftarrow \text{The sum of the two products.} \end{array}$$

Let us follow a similar format to multiply the two polynomials. For example, multiply $(x^2-2x+3)(3x-2)$.

$$\begin{array}{rl} x^2 - \ \ 2x + 3 & \text{We place the polynomial with the greatest number of} \\ 3x - 2 & \text{terms above the other polynomial.} \\ \hline -2x^2 + \ \ 4x - 6 & \longleftarrow \text{The product } (x^2-2x+3)(-2). \\ 3x^3 - 6x^2 + \ \ 9x\ \ \ \ \ \ & \longleftarrow \text{The product } (x^2-2x+3)(3x) \text{ moved one space to the} \\ \hline & \quad\ \ \text{left so that like terms are underneath each other.} \\ 3x^3 - 8x^2 + 13x - 6 & \longleftarrow \text{The sum of the two products.} \end{array}$$

EXAMPLE 5 Multiply *vertically*. $(3x^3+2x^2+x)(x^2-2x-4)$

$$\begin{array}{rl} 3x^3 + 2x^2 + x & \text{We place one polynomial over the other.} \\ x^2 - 2x - 4 & \text{This is often called vertical multiplication.} \\ \hline -12x^3 - 8x^2 - 4x & \longleftarrow \text{The product } (3x^3+2x^2+x)(-4). \\ -6x^4 - 4x^3 - 2x^2\ \ \ \ \ \ & \longleftarrow \text{The product } (3x^3+2x^2+x)(-2x). \\ 3x^5 + 2x^4 + x^3\ \ \ \ \ \ \ \ \ \ \ \ \ \ & \longleftarrow \text{The product } (3x^3+2x^2+x)(x^2). \\ \hline 3x^5 - 4x^4 - 15x^3 - 10x^2 - 4x & \longleftarrow \text{The sum of the three products.} \end{array}$$

Note that the answers for each product are placed so that like terms are underneath each other.

Practice Problem 5 Multiply vertically the following polynomials.

$(3x^2-2xy+4y^2)(x-2y)$ ■

Alternative Method

Some students prefer to do this type of multiplication using a horizontal format similar to the FOIL method. The following example illustrates this approach.

EXAMPLE 6 Multiply horizontally. $(x^2 + 3x + 5)(x^2 - 2x - 6)$

We will use the distributive property repeatedly.

$$\begin{aligned}(x^2 + 3x + 5)(x^2 - 2x - 6) &= x^2(x^2 - 2x - 6) + 3x(x^2 - 2x - 6) + 5(x^2 - 2x - 6)\\ &= x^4 - 2x^3 - 6x^2 + 3x^3 - 6x^2 - 18x + 5x^2 - 10x - 30\\ &= x^4 + x^3 - 7x^2 - 28x - 30\end{aligned}$$

Practice Problem 6 Multiply horizontally. $(2x^2 - 5x + 3)(x^2 + 3x - 4)$ ■

Some problems may need to be done in two or more separate steps.

TEACHING TIP Problems like Example 7 are quite important. As soon as you have discussed Example 7, have students do the following problem as an Additional Class Exercise. Multiply.

$$(3x - 2)(4x + 1)(x - 2)$$

Sol. By multiplying the first two binomials we have

$$(12x^2 - 5x - 2)(x - 2)$$

These are multiplied to obtain

$$12x^3 - 29x^2 + 8x + 4$$

EXAMPLE 7 Multiply. $(2x - 3y)(x + 2y)(x + y)$

We first need to multiply any two binomials. Let us select the first pair.

$$\underbrace{(2x - 3y)(x + 2y)}(x + y)$$

Do this product first.

$$\begin{aligned}(2x - 3y)(x + 2y) &= 2x^2 + 4xy - 3xy - 6y^2\\ &= 2x^2 + xy - 6y^2\end{aligned}$$

Now we replace the first two sets of parentheses with the result obtained above.

$$\underbrace{(2x^2 + xy - 6y^2)}(x + y)$$

Result of first product

We then multiply.

$$\begin{aligned}(2x^2 + xy - 6y^2)(x + y) &= (2x^2 + xy - 6y^2)x + (2x^2 + xy - 6y^2)y\\ &= 2x^3 + x^2y - 6xy^2 + 2x^2y + xy^2 - 6y^3\\ &= 2x^3 + 3x^2y - 5xy^2 - 6y^3\end{aligned}$$

The vertical format of Example 5 is an alternative method for this type of problem.

$$\begin{array}{r}2x^2 + xy - 6y^2\\ x + y\\ \hline 2x^2y + xy^2 - 6y^3\\ 2x^3 + x^2y - 6xy^2 \quad\\ \hline 2x^3 + 3x^2y - 5xy^2 - 6y^3\end{array}$$

Be sure to use special care in writing the exponents correctly in problems with two different variables.

Thus we have

$$(2x - 3y)(x + 2y)(x + y) = 2x^3 + 3x^2y - 5xy^2 - 6y^3$$

Note that it does not matter which two binomials are multiplied first. For example, you could first multiply $(2x - 3y)(x + y)$ to obtain $2x^2 - xy - 3y^2$ and then multiply that result by $(x + 2y)$ and still obtain the same result.

Practice Problem 7 Multiply. $(3x - 2)(2x + 3)(3x + 2)$ (*Hint:* Rearrange.) ■

4.5 Exercises

Verbal and Writing Skills

1. In the special case of $(a + b)(a - b)$, a binomial times a binomial is a binomial.

2. Identify which of the following could be the answer to a problem using the formula for $(a + b)(a - b)$.
(a) $9x^2 - 16$ **(b)** $4x^2 + 25$ **(c)** $9x^2 + 12x + 4$ **(d)** $x^4 - 1$ (a) and (d)

3. A student evaluated $(4x - 7)^2$ as $16x^2 + 49$. Is this correct? Why? Why not? What is missing? State the correct answer. The answer is not correct, the middle term is missing. The answer should be $16x^2 - 56x + 49$.

4. The square of a binomial, $(a - b)^2$, always produces a (choose):
(a) Binomial **(b)** Trinomial **(c)** Four-term polynomial
(b) Trinomial

Use the relationship $(a + b)(a - b) = a^2 - b^2$ to multiply each of the following.

5. $(x + 4)(x - 4)$
$x^2 - 16$

6. $(y - 8)(y + 8)$
$y^2 - 64$

7. $(x - 9)(x + 9)$
$x^2 - 81$

8. $(x + 6)(x - 6)$
$x^2 - 36$

9. $(8x + 3)(8x - 3)$
$64x^2 - 9$

10. $(6x + 5)(6x - 5)$
$36x^2 - 25$

11. $(2x - 7)(2x + 7)$
$4x^2 - 49$

12. $(3x - 10)(3x + 10)$
$9x^2 - 100$

13. $(2x - 5y)(2x + 5y)$
$4x^2 - 25y^2$

14. $(9w - 4z)(9w + 4z)$
$81w^2 - 16z^2$

15. $(7st + 1)(7st - 1)$
$49s^2t^2 - 1$

16. $(8r - 5s)(8r + 5s)$
$64r^2 - 25s^2$

17. $(12x^2 + 7)(12x^2 - 7)$
$144x^4 - 49$

18. $(1 - 8ab^2)(1 + 8ab^2)$
$1 - 64a^2b^4$

19. $(3x^2 - 4y^3)(3x^2 + 4y^3)$
$9x^4 - 16y^6$

Use the relationship for a binomial squared to multiply each of the following.

20. $(3y + 1)^2$
$9y^2 + 6y + 1$

21. $(4x - 1)^2$
$16x^2 - 8x + 1$

22. $(7x - 2)^2$
$49x^2 - 28x + 4$

23. $(8x + 3)^2$
$64x^2 + 48x + 9$

24. $(9x + 5)^2$
$81x^2 + 90x + 25$

25. $(5x - 7)^2$
$25x^2 - 70x + 49$

26. $(3x - 7)^2$
$9x^2 - 42x + 49$

27. $(2x + 3y)^2$
$4x^2 + 12xy + 9y^2$

28. $(3x + 4y)^2$
$9x^2 + 24xy + 16y^2$

29. $(8x - 5y)^2$
$64x^2 - 80xy + 25y^2$

30. $(7 - 3y)^2$
$49 - 42y + 9y^2$

31. $(11 + 2y)^2$
$121 + 44y + 4y^2$

32. $(4a + 5b)^2$
$16a^2 + 40ab + 25b^2$

33. $(6w + 5z)^2$
$36w^2 + 60wz + 25z^2$

34. $(5xy - 6z)^2$
$25x^2y^2 - 60xyz + 36z^2$

Multiply. Use the special relationship that applies.

35. $(7x + 3y)(7x - 3y)$ $49x^2 - 9y^2$

36. $(12a - 5b)(12a + 5b)$ $144a^2 - 25b^2$

37. $(4x^2 - 7y)^2$ $16x^4 - 56x^2y + 49y^2$

38. $(3a - 10b^2)^2$ $9a^2 - 60ab^2 + 100b^4$

Multiply out the polynomials.

39. $(x^2 + 3x - 2)(x - 3)$ $x^3 - 11x + 6$

40. $(x^2 + 3x - 1)(x + 3)$ $x^3 + 6x^2 + 8x - 3$

41. $(3y + 2z)(3y^2 - yz - 2z^2)$ $9y^3 + 3y^2z - 8yz^2 - 4z^3$

42. $(4w - 2z)(w^2 + 2wz + z^2)$ $4w^3 + 6w^2z - 2z^3$

43. $(4x + 1)(x^3 - 2x^2 + x - 1)$ $4x^4 - 7x^3 + 2x^2 - 3x - 1$

44. $(3x - 1)(x^3 + x^2 - 4x - 2)$ $3x^4 + 2x^3 - 13x^2 - 2x + 2$

45. $(x + 2)(x - 3)(2x + 5)$ $2x^3 + 3x^2 - 17x - 30$

46. $(x + 6)(x - 1)(3x - 2)$ $3x^3 + 13x^2 - 28x + 12$

47. $(3x + 5)(x - 2)(x - 4)$ $3x^3 - 13x^2 - 6x + 40$

48. $(2x - 7)(x + 1)(x - 2)$ $2x^3 - 9x^2 + 3x + 14$

49. $(x + 4)(2x - 7)(x - 4)$ $2x^3 - 7x^2 - 32x + 112$

50. $(x - 2)(5x + 3)(x - 2)$ $5x^3 - 18x^2 + 12x + 8$

51. $(a^2 - 3a + 2)(a^2 + 4a - 3)$ $a^4 + a^3 - 13a^2 + 17a - 6$

52. $(x^2 + 4x - 5)(x^2 - 3x + 4)$ $x^4 + x^3 - 13x^2 + 31x - 20$

★ **53.** $(x^2 + 3y)(x^2 - 3y)(2x^2 + 5xy - 6y^2)$
$2x^6 + 5x^5y - 6x^4y^2 - 18x^2y^2 - 45xy^3 + 54y^4$

★ **54.** $(x^3 + 3x^2 - 4x + 2)(2x^3 - 4x^2 + x + 5)$
$2x^6 + 2x^5 - 19x^4 + 28x^3 + 3x^2 - 18x + 10$

Cumulative Review Problems

55. An executive invested \$18,000 in two accounts, one cash-on-reserve account that yielded 7% simple interest, the other a long-term certificate paying 11% simple interest. At the end of one year, the two accounts had earned her a total of \$1540. How much did she invest in each account? \$11,000 at 7%; \$7,000 at 11%

56. The perimeter of a rectangular room measures 34 meters. The width is 2 meters more than half the length. Find the dimensions of the room.
width = 7 m; length = 10 m

57. Ammonia has a boiling point of −33.4°C. What is the boiling point of ammonia measured in degrees Fahrenheit? (Use $F = 1.8C + 32$). Approx. −28.12°F

58. A man with 5.5 liters of blood would have approximately 2.75×10^{10} red blood cells. Each red blood cell is a small disk that measures 7.5×10^{-6} meters in diameter. If all of the red blood cells were arranged in a long straight line, how long would that line be?
2.0625×10^5 meters

Putting Your Skills to Work

THE MATHEMATICS OF DNA

The genetic material that organisms inherit from their parents consists of DNA. Each time a cell reproduces itself by dividing, its DNA is copied and passed on from one generation to the next. The DNA molecule consists of two strands twisted in the shape of a double helix. Each strand is composed of fundamental units called nucleotides. Each nucleotide on one strand is bonded (base paired) to a complementary nucleotide on the opposing strand. (**Source:** Dr. Russ Camp, Biology Department, Gordon College)

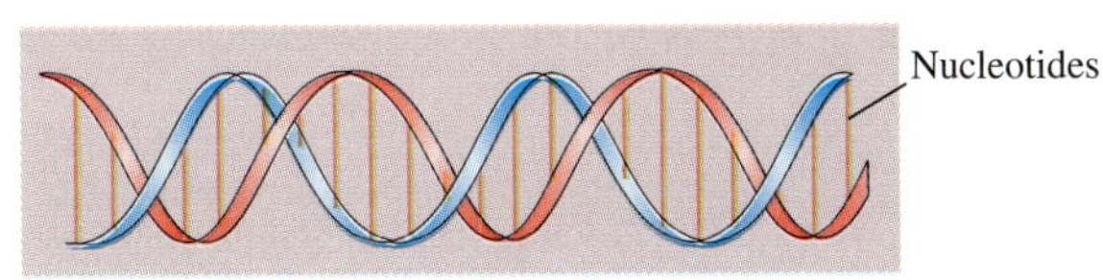

PROBLEMS FOR INDIVIDUAL INVESTIGATION AND ANALYSIS

1. *E. coli,* the most popular bacterial cell used in molecular biology research, has a single chromosome made up of one double-stranded DNA molecule. The linear distance from one nucleotide pair to the next is 0.34 nanometer (one nanometer $= 1 \times 10^{-9}$ meter). The entire *E. coli* chromosome contains 4.5×10^6 nucleotide pairs (or base pairs). How long is the chromosome in nanometers? 1.53×10^6 nanometers

2. How long is the chromosome in millimeters? 1.53 millimeters

PROBLEMS FOR COOPERATIVE GROUP INVESTIGATION

Together with other members of your class, see if you can determine the following.

3. If the human cell contains 46 chromosomes and if all the DNA in each chromosome is added together, how long would the DNA be in meters if there are 8.0×10^9 nucleotide pairs? 2.72 meters

4. How many times larger is the human DNA in a cell compared to the DNA in the bacterial cell above? 1777 times larger

INTERNET CONNECTIONS: Go to ''http://www.prenhall.com/tobey'' to be connected

Introduction to Primer on Molecular Genetics (Department of Energy) or a related site

This site provides some basic information about genes and DNA. Use the information to answer the following questions.

5. Estimate the volume of the DNA in a human genome by assuming that, when unwound and tied together, the strands of DNA form a right circular cylinder.

6. Tell what percent of a human genome is made up of intron sequences and other noncoding regions.

4.6 Division of Polynomials

MathPro

Video 12

SSM

After studying this section, you will be able to:

1 *Divide a polynomial by a monomial.*
2 *Divide a polynomial by a binomial.*

1 *Dividing a Polynomial by a Monomial*

To divide a polynomial by a monomial, divide each term of the numerator by the denominator; then write the sum of the results. We are using the property of fractions that

$$\frac{a+b}{c} = \frac{a}{c} + \frac{b}{c}$$

Dividing a Polynomial by a Monomial

1. Divide each term of the polynomial by the monomial.
2. When dividing variables, use the property $\frac{x^a}{x^b} = x^{a-b}$.

EXAMPLE 1 Divide. $\frac{25x^2 - 20x}{5x}$

$$\frac{25x^2 - 20x}{5x} = \frac{25x^2}{5x} - \frac{20x}{5x} = 5x - 4$$

Practice Problem 1 Divide. $\frac{14y^2 - 49y}{7y}$ ■

EXAMPLE 2 Divide. $\frac{8y^6 - 8y^4 + 24y^2}{8y^2}$

$$\frac{8y^6 - 8y^4 + 24y^2}{8y^2} = \frac{8y^6}{8y^2} - \frac{8y^4}{8y^2} + \frac{24y^2}{8y^2} = y^4 - y^2 + 3$$

Practice Problem 2 Divide. $\frac{15y^4 - 27y^3 - 21y^2}{3y^2}$ ■

2 *Dividing a Polynomial by a Binomial*

Division of a polynomial by a binomial is similar to long division in arithmetic. Notice the similarity in the following division problems.

Division of a Three-digit by a Two-digit Number	***Division of a Polynomial by a Binomial***
$\begin{array}{r} 32 \\ 21\overline{)672} \\ \underline{63} \\ 42 \\ \underline{42} \end{array}$	$\begin{array}{r} 3x + 2 \\ 2x + 1\overline{)6x^2 + 7x + 2} \\ \underline{6x^2 + 3x} \\ 4x + 2 \\ \underline{4x + 2} \end{array}$
To check, multiply. $(21)(32) = 672$	*To check, multiply.* $(2x + 1)(3x + 2) = 6x^2 + 7x + 2$

Dividing a Polynomial by a Binomial

1. Place the terms of the polynomial and binomial in descending order. Insert a 0 for any missing term.
2. Divide the first term of the polynomial by the first term of the binomial. The result is the first term of the answer.
3. Multiply the first term of the answer by the binomial and subtract the results from the first two terms of the polynomial. Bring down the next term to obtain a new polynomial.
4. Divide the new polynomial by the binomial using the process described in step 2.
5. Continue dividing, multiplying, and subtracting until the remainder is at a lower power than the variable in the first term of the binomial divisor.

EXAMPLE 3 Divide. $(x^3 + 5x^2 + 11x + 14) \div (x + 2)$

Step 1. The terms are arranged in descending order. No terms are missing.

Step 2. Divide the first term of the polynomial by the first term of the binomial. In this case, divide x^3 by x to get x^2.

$$\begin{array}{r} x^2 \\ x + 2\overline{)x^3 + 5x^2 + 11x + 14} \end{array}$$

Step 3. Multiply x^2 by $x + 2$ and subtract the result from the first two terms of the polynomial: $x^3 + 5x^2$, in this case.

$$\begin{array}{rl} x^2 & \\ x + 2\overline{)x^3 + 5x^2 + 11x + 14} & \\ \underline{x^3 + 2x^2} \downarrow & \\ 3x^2 + 11x & \quad \textit{Bringing down the next term.} \end{array}$$

Step 4. Continue to use the step 2 process. Divide $3x^2$ by x. Write the resulting $3x$ as the next term of the answer.

$$\begin{array}{r} x^2 + 3x \\ x + 2\overline{)x^3 + 5x^2 + 11x + 14} \\ \underline{x^3 + 2x^2} \\ 3x^2 + 11x \end{array}$$

Step 5. Continue multiplying, dividing, and subtracting until the remainder is at a power lower than the variable in the binomial's first term. In this case, we stop when the remainder does not have an x.

$$\begin{array}{rl} x^2 + 3x + 5 & \\ x + 2\overline{)x^3 + 5x^2 + 11x + 14} & \\ \underline{x^3 + 2x^2} & \\ 3x^2 + 11x & \\ \underline{3x^2 + 6x} & \\ 5x + 14 & \\ \underline{5x + 10} & \\ 4 & \longleftarrow \text{The remainder is 4.} \end{array}$$

Answer is $x^2 + 3x + 5 + \dfrac{4}{x + 2}$.

To check the answer, we multiply $(x + 2)(x^2 + 3x + 5)$ and add the remainder 4.

$$(x + 2)(x^2 + 3x + 5) = x^3 + 5x^2 + 11x + 10$$

Now add 4. We get $x^3 + 5x^2 + 11x + 14$. This is the original polynomial. It checks.

Practice Problem 3 Divide. $(x^3 + 10x^2 + 31x + 35) \div (x + 4)$ ■

Take great care with the subtraction step when negative numbers are involved.

EXAMPLE 4 Divide. $(5x^3 - 24x^2 + 9) \div (5x + 1)$

We must first put in a $0x$ to represent the missing term. Then we divide $5x^3 - 24x^2$ by $5x + 1$.

$$\begin{array}{r@{}l}
 & x^2 \\
5x + 1 \overline{)} & 5x^3 - 24x^2 + 0x + 9 \\
 & \underline{5x^3 + x^2} \\
 & - 25x^2
\end{array}$$

Note that we are subtracting:
$-24x^2 - (+1x^2)$
$= -24x^2 - 1x^2 = -25x^2$

Next we divide $-25x^2 + 0x$ by $5x + 1$.

$$\begin{array}{r@{}l}
 & x^2 - 5x \\
5x + 1 \overline{)} & 5x^3 - 24x^2 + 0x + 9 \\
 & \underline{5x^3 + x^2} \\
 & - 25x^2 + 0x \\
 & \underline{- 25x^2 - 5x} \\
 & 5x
\end{array}$$

Note we are subtracting:
$0x - (-5x)$
$= 0x + (5x) = 5x$

Finally, we divide $5x + 9$ by $5x + 1$.

$$\begin{array}{r@{}l}
 & x^2 - 5x + 1 \\
5x + 1 \overline{)} & 5x^3 - 24x^2 + 0x + 9 \\
 & \underline{5x^3 + x^2} \\
 & - 25x^2 + 0x \\
 & \underline{- 25x^2 - 5x} \\
 & 5x + 9 \\
 & \underline{5x + 1} \\
 & 8 \longleftarrow \text{The remainder is 8.}
\end{array}$$

The answer is $x^2 - 5x + 1 + \dfrac{8}{5x + 1}$.

To check, multiply $(5x + 1)(x^2 - 5x + 1)$ and add the remainder 8.

$$(5x + 1)(x^2 - 5x + 1) = 5x^3 - 24x^2 + 1$$

Now add the remainder. $5x^3 - 24x^2 + 1 + 8 = 5x^3 - 24x^2 + 9$. This is the original polynomial. Our answer is correct.

Practice Problem 4 Divide. $(2x^3 - x^2 + 1) \div (x - 1)$ ■

Now we will perform the division by writing a minimum of steps. See if you can follow each step.

EXAMPLE 5 Divide and check. $(12x^3 - 11x^2 + 8x - 4) \div (3x - 2)$

$$\begin{array}{r} 4x^2 - x + 2 \\ 3x - 2 \overline{)12x^3 - 11x^2 + 8x - 4} \\ \underline{12x^3 - 8x^2} \\ - 3x^2 + 8x \\ \underline{- 3x^2 + 2x} \\ 6x - 4 \\ \underline{6x - 4} \end{array}$$

Check. $(3x - 2)(4x^2 - x + 2) = 12x^3 - 3x^2 + 6x - 8x^2 + 2x - 4$

$= 12x^3 - 11x^2 + 8x - 4$ *Our answer is correct.*

Practice Problem 5 Divide and check. $(20x^3 - 11x^2 - 11x + 6) \div (4x - 3)$ ■

Developing Your Study Skills

EXAM TIME: TAKING THE EXAM

Allow yourself plenty of time to get to your exam. You may even find it helpful to arrive a little early in order to collect your thoughts and ready yourself. This will help you feel more relaxed.

After you get your exam, you will find it helpful to do the following.

1. Take two or three moderately deep breaths. Inhale, then exhale slowly. You will feel your entire body begin to relax.
2. Write down on the back of the exam any formulas or ideas that you need to remember.
3. Look over the entire test quickly in order to pace yourself and use your time wisely. Notice how many points each problem is worth. Spend more time on items of greater worth.
4. Read directions carefully, and be sure to answer all questions clearly. Keep your work neat and easy to read.
5. Ask your instructor about anything that is not clear to you.
6. Work the problems and answer the questions that are easiest for you first. Then come back to the more difficult ones.
7. Do not get bogged down on one problem too long because it may jeopardize your chances of finishing other problems. Leave the tough problem and come back to it when you have time later.
8. Check your work. This will help you to catch minor errors.
9. Stay calm if others leave before you do. You are entitled to use the full amount of allotted time.

4.6 Exercises

Divide.

1. $\dfrac{8x^3 - 20x^2 + 4x}{2x}$

$\dfrac{8x^3}{2x} - \dfrac{20x^2}{2x} + \dfrac{4x}{2x} = 4x^2 - 10x + 2$

2. $\dfrac{21x^3 - 6x^2 + 9x}{3x}$

$\dfrac{21x^3}{3x} - \dfrac{6x^2}{3x} + \dfrac{9x}{3x} = 7x^2 - 2x + 3$

3. $\dfrac{8y^4 - 12y^3 - 4y^2}{4y^2}$

$\dfrac{8y^4}{4y^2} - \dfrac{12y^3}{4y^2} - \dfrac{4y^2}{4y^2} = 2y^2 - 3y - 1$

4. $\dfrac{10y^4 - 35y^3 + 5y^2}{5y^2}$

$\dfrac{10y^4}{5y^2} - \dfrac{35y^3}{5y^2} + \dfrac{5y^2}{5y^2} = 2y^2 - 7y + 1$

5. $\dfrac{49x^6 - 21x^4 + 56x^2}{7x^2}$

$\dfrac{49x^6}{7x^2} - \dfrac{21x^4}{7x^2} + \dfrac{56x^2}{7x^2} = 7x^4 - 3x^2 + 8$

6. $\dfrac{36x^6 + 54x^4 - 6x^2}{6x^2}$

$\dfrac{36x^6}{6x^2} + \dfrac{54x^4}{6x^2} - \dfrac{6x^2}{6x^2} = 6x^4 + 9x^2 - 1$

7. $(48x^7 - 54x^4 + 36x^3) \div 6x^3$

$8x^4 - 9x + 6$

8. $(72x^8 - 56x^5 - 40x^3) \div 8x^3$

$9x^5 - 7x^2 - 5$

9. $(30x^4 - 21x^3 - 3x^2 + 15x) \div (3x)$

$\dfrac{30x^4}{3x} - \dfrac{21x^3}{3x} - \dfrac{3x^2}{3x} + \dfrac{15x}{3x} = 10x^3 - 7x^2 - x + 5$

10. $(100x^4 - 16x^3 - 28x^2 + 56x) \div (2x)$

$\dfrac{100x^4}{2x} - \dfrac{16x^3}{2x} - \dfrac{28x^2}{2x} + \dfrac{56x}{2x} = 50x^3 - 8x^2 - 14x + 28$

Divide. Check your answers for Exercises 11 through 18 by multiplication.

11. $\dfrac{6x^2 + 13x + 5}{2x + 1}$

$$\begin{array}{r} 3x + 5 \\ 2x + 1\overline{)6x^2 + 13x + 5} \\ \underline{6x^2 + 3x} \\ 10x + 5 \\ \underline{10x + 5} \end{array}$$

12. $\dfrac{12x^2 + 19x + 5}{3x + 1}$

$$\begin{array}{r} 4x + 5 \\ 3x + 1\overline{)12x^2 + 19x + 5} \\ \underline{12x^2 + 4x} \\ 15x + 5 \\ \underline{15x + 5} \end{array}$$

13. $\dfrac{x^2 - 9x - 6}{x - 7}$

$$\begin{array}{r} x - 2 \\ x - 7\overline{)x^2 - 9x - 6} \\ \underline{x^2 - 7x} \\ -2x - 6 \\ \underline{-2x + 14} \\ -20 \end{array}$$

14. $\dfrac{x^2 - 8x - 4}{x - 6}$

$$\begin{array}{r} x - 2 \\ x - 6\overline{)x^2 - 8x - 4} \\ \underline{x^2 - 6x} \\ -2x - 4 \\ \underline{-2x + 12} \\ -16 \end{array}$$

15. $\dfrac{3x^3 - x^2 + 4x - 2}{x + 1}$

$$\begin{array}{r} 3x^2 - 4x + 8 \\ x + 1\overline{)3x^3 - x^2 + 4x - 2} \\ \underline{3x^3 + 3x^2} \\ -4x^2 + 4x \\ \underline{-4x^2 - 4x} \\ 8x - 2 \\ \underline{8x + 8} \\ -10 \end{array}$$

16. $\dfrac{2x^3 - 3x^2 - 3x + 6}{x - 1}$

$$\begin{array}{r} 2x^2 - x - 4 \\ x - 1\overline{)2x^3 - 3x^2 - 3x + 6} \\ \underline{2x^3 - 2x^2} \\ -x^2 - 3x \\ \underline{-x^2 + x} \\ -4x + 6 \\ \underline{-4x + 4} \\ +2 \end{array}$$

17. $\dfrac{2x^3 - x^2 + 3x + 2}{2x + 1}$

$$\begin{array}{r} x^2 - x + 2 \\ 2x + 1\overline{)2x^3 - x^2 + 3x + 2} \\ \underline{2x^3 + x^2} \\ -2x^2 + 3x \\ \underline{-2x^2 - x} \\ 4x + 2 \\ \underline{4x + 2} \end{array}$$

18. $\dfrac{2x^3 - 3x^2 + 4x + 4}{2x + 1}$

$$\begin{array}{r} x^2 - 2x + 3 \\ 2x + 1\overline{)2x^3 - 3x^2 + 4x + 4} \\ \underline{2x^3 + x^2} \\ -4x^2 + 4x \\ \underline{-4x^2 - 2x} \\ 6x + 4 \\ \underline{6x + 3} \\ 1 \end{array}$$

19. $\dfrac{6x^3 + x^2 - 7x + 2}{3x - 1}$

$$\begin{array}{r} 2x^2 + x - 2 \\ 3x - 1\overline{)6x^3 + x^2 - 7x + 2} \\ \underline{6x^3 - 2x^2} \\ 3x^2 - 7x \\ \underline{3x^2 - x} \\ -6x + 2 \\ \underline{-6x + 2} \end{array}$$

20. $\dfrac{3x^3 + 8x^2 - 6x + 2}{3x - 1}$

$$\begin{array}{r} x^2 + 3x - 1 \\ 3x - 1\overline{)3x^3 + 8x^2 - 6x + 2} \\ \underline{3x^3 - x^2} \\ 9x^2 - 6x \\ \underline{9x^2 - 3x} \\ -3x + 2 \\ \underline{-3x + 1} \\ 1 \end{array}$$

21. $\dfrac{12y^3 - 12y^2 - 25y + 31}{2y - 3}$

$$\begin{array}{r} 6y^2 + 3y - 8 \\ 2y - 3\overline{)12y^3 - 12y^2 - 25y + 31} \\ \underline{12y^3 - 18y^2} \\ 6y^2 - 25y \\ \underline{6y^2 - 9y} \\ -16y + 31 \\ \underline{-16y + 24} \\ 7 \end{array}$$

22. $\dfrac{9y^3 - 30y^2 + 31y - 4}{3y - 5}$

$$\begin{array}{r} 3y^2 - 5y + 2 \\ 3y - 5\overline{)9y^3 - 30y^2 + 31y - 4} \\ \underline{9y^3 - 15y^2} \\ -15y^2 + 31y \\ \underline{-15y^2 + 25y} \\ 6y - 4 \\ \underline{6y - 10} \\ 6 \end{array}$$

23. $(y^3 - y^2 - 13y - 12) \div (y + 3)$

$$\begin{array}{r}
y^2 - 4y - 1 \\
y+3\overline{)y^3 - y^2 - 13y - 12} \\
\underline{y^3 + 3y^2} \\
-4y^2 - 13y \\
\underline{-4y^2 - 12y} \\
-y - 12 \\
\underline{-y - 3} \\
-9
\end{array}$$

24. $(4y^3 - 17y^2 + 7y + 10) \div (4y - 5)$

$$\begin{array}{r}
y^2 - 3y - 2 \\
4y-5\overline{)4y^3 - 17y^2 + 7y + 10} \\
\underline{4y^3 - 5y^2} \\
-12y^2 + 7y \\
\underline{-12y^2 + 15y} \\
-8y + 10 \\
\underline{-8y + 10}
\end{array}$$

25. $(4y^3 - 1 + 3y) \div (2y - 1)$

$$\begin{array}{r}
2y^2 + y + 2 \\
2y-1\overline{)4y^3 + 0y^2 + 3y - 1} \\
\underline{4y^3 - 2y^2} \\
2y^2 + 3y \\
\underline{2y^2 - y} \\
4y - 1 \\
\underline{4y - 2} \\
1
\end{array}$$

26. $(y^3 - 7y^2 - 8y - 9) \div (y + 3)$

$$\begin{array}{r}
y^2 - 10y + 22 \\
y+3\overline{)y^3 - 7y^2 - 8y - 9} \\
\underline{y^3 + 3y^2} \\
-10y^2 - 8y \\
\underline{-10y^2 - 30y} \\
22y - 9 \\
\underline{22y + 66} \\
-75
\end{array}$$

27. $(y^4 - 3y^2 + 2y - 1) \div (y + 3)$

$$\begin{array}{r}
y^3 - 3y^2 + 6y - 16 \\
y+3\overline{)y^4 + 0y^3 - 3y^2 + 2y - 1} \\
\underline{y^4 + 3y^3} \\
-3y^3 - 3y^2 \\
\underline{-3y^3 - 9y^2} \\
6y^2 + 2y \\
\underline{6y^2 + 18y} \\
-16y - 1 \\
\underline{-16y - 48} \\
47
\end{array}$$

28. $(y^4 - 2y^3 + y^2 - 1) \div (y - 2)$

$$\begin{array}{r}
y^3 + 0y^2 + y + 2 \\
y-2\overline{)y^4 - 2y^3 + y^2 + 0y - 1} \\
\underline{y^4 - 2y^3} \\
0y^3 + y^2 \\
\underline{0y^3 + 0y^2} \\
y^2 + 0y \\
\underline{y^2 - 2y} \\
2y - 1 \\
\underline{2y - 4} \\
3
\end{array}$$

29. $(y^4 - 9y^2 - 5) \div (y - 2)$

$$\begin{array}{r}
y^3 + 2y^2 - 5y - 10 \\
y-2\overline{)y^4 - 0y^3 - 9y^2 + 0y - 5} \\
\underline{y^4 - 2y^3} \\
2y^3 - 9y^2 \\
\underline{2y^3 - 4y^2} \\
-5y^2 + 0y \\
\underline{-5y^2 + 10y} \\
-10y - 5 \\
\underline{-10y + 20} \\
-25
\end{array}$$

30. $(2y^4 + 3y^2 - 5) \div (y - 2)$

$$\begin{array}{r}
2y^3 + 4y^2 + 11y + 22 \\
y-2\overline{)2y^4 + 0y^3 + 3y^2 + 0y - 5} \\
\underline{2y^4 - 4y^3} \\
4y^3 + 3y^2 \\
\underline{4y^3 - 8y^2} \\
11y^2 + 0y \\
\underline{11y^2 - 22y} \\
22y - 5 \\
\underline{22y - 44} \\
39
\end{array}$$

Cumulative Review Problems

31. The average home in the United States uses 110,000 gallons of water per year. Most environmental groups feel that with improved efficiency of water-using devices (washers, toilets, faucets, shower heads), this could be reduced by 30%. How many gallons of water per year would be used in the average home if this reduction were accomplished? 77,000 gallons

32. From 1900 to 1990 the population of New York City increased by 115% to a new total of 7.3 million. What was the population of the city in 1900? Approx. 3.4 million

Chapter Organizer

Topic	Procedure	Examples
Multiplying monomials, p. 241.	$x^a \cdot x^b = x^{a+b}$ 1. Multiply the numerical coefficients. 2. Add the exponents of a given base.	$3^{12} \cdot 3^{15} = 3^{27}$ $x^3 \cdot x^4 = x^7$ $(-3x^2)(6x^3) = -18x^5$ $(2ab)(4a^2b^3) = 8a^3b^4$
Dividing monomials, p. 243.	$\frac{x^a}{x^b} = \begin{cases} x^{a-b} & \text{if } a \text{ is greater than } b \\ \frac{1}{x^{b-a}} & \text{if } b \text{ is greater than } a \end{cases}$ 1. Divide or reduce the fraction created by the quotient of the numerical coefficients. 2. Subtract the exponents of a given base.	$\frac{16x^7}{8x^3} = 2x^4$ $\frac{5x^3}{25x^5} = \frac{1}{5x^2}$ $\frac{-12x^5y^7}{18x^3y^{10}} = -\frac{2x^2}{3y^3}$
Exponent of zero, p. 244.	$x^0 = 1 \quad \text{if } x \neq 0$	$5^0 = 1 \quad \frac{x^6}{x^6} = 1$ $w^0 = 1 \quad 3x^0y = 3y$
Raising a power to a power, p. 245.	$(x^a)^b = x^{ab}$ $(xy)^a = x^ay^a$ $\left(\frac{x}{y}\right)^a = \frac{x^a}{y^a} \quad (y \neq 0)$ 1. Raise the numerical coefficient to the power outside the parentheses. 2. Multiply the exponent outside the parentheses times the exponent inside the parentheses.	$(x^9)^3 = x^{27}$ $(3x^2)^3 = 27x^6$ $\left(\frac{2x^2}{y^3}\right)^3 = \frac{8x^6}{y^9}$ $(-3x^4y^5)^4 = 81x^{16}y^{20}$ $(-5ab)^3 = -125a^3b^3$
Negative exponents, p. 251.	If $x \neq 0$, $y \neq 0$ Then $x^{-n} = \frac{1}{x^n}$ $\frac{1}{x^{-n}} = x^n$ $\frac{x^{-m}}{y^{-n}} = \frac{y^n}{x^m}$	Write with positive exponents. $3^{-4} = \frac{1}{3^4} = \frac{1}{81}$ $x^{-6} = \frac{1}{x^6}$ $\frac{1}{w^{-3}} = w^3$ $\frac{w^{-12}}{z^{-5}} = \frac{z^5}{w^{12}}$ $(2x^2)^{-3} = 2^{-3}x^{-6} = \frac{1}{2^3x^6}$ $= \frac{1}{8x^6}$
Scientific notation, p. 252.	A positive number is written in scientific notation if it is in the form $a \times 10^n$, where $1 \leq a < 10$ and n is an integer.	$128 = 1.28 \times 10^2$ $2{,}568{,}000 = 2.568 \times 10^6$ $13{,}200{,}000{,}000 = 1.32 \times 10^{10}$ $0.16 = 1.6 \times 10^{-1}$ $0.00079 = 7.9 \times 10^{-4}$ $0.0000034 = 3.4 \times 10^{-6}$
Add polynomials, p. 259.	To add two polynomials, we add the respective like terms.	$(-7x^3 + 2x^2 + 5) + (x^3 + 3x^2 + x)$ $= -6x^3 + 5x^2 + x + 5$
Subtracting polynomials, p. 260.	To subtract the polynomials, change all signs of the second polynomial and add the result to the first polynomial. $(a) - (b) = (a) + (-b)$	$(5x^2 - 6) - (-3x^2 + 2)$ $= (5x^2 - 6) + (+3x^2 - 2)$ $= 8x^2 - 8$
Multiplying a monomial by a polynomial, p. 263.	Use the distributive property. $a(b + c) = ab + ac$ $(b + c)a = ba + ca$	Multiply. $-5x(2x^2 + 3x - 4) = -10x^3 - 15x^2 + 20x$ $(6x^3 - 5xy - 2y^2)(3xy) = 18x^4y - 15x^2y^2 - 6xy^3$

Chapter Organizer

Topic	Procedure	Examples
Multiplying two binomials, p. 264.	**1.** The product of the sum and difference of the same two values yields the difference of their squares. $(a + b)(a - b) = a^2 - b^2$ **2.** The square of a binomial yields a trinomial: the square of the first, plus twice the product of first and second, plus the square of the second. $(a + b)^2 = a^2 + 2ab + b^2$ $(a - b)^2 = a^2 - 2ab + b^2$ **3.** Use FOIL for other binomial multiplication. The middle terms can often be combined, giving a trinomial answer.	$(3x + 7y)(3x - 7y) = 9x^2 - 49y^2$ $(3x + 7y)^2 = 9x^2 + 42xy + 49y^2$ $(3x - 7y)^2 = 9x^2 - 42xy + 49y^2$ $(3x - 5)(2x + 7) = 6x^2 + 21x - 10x - 35$ $= 6x^2 + 11x - 35$
Multiplying two polynomials, p. 271.	To multiply two polynomials, multiply each term of one by each term of the other. This method is similar to the multiplication of many-digit numbers.	Vertical method $(3x^2 - 7x + 4)$ $\times \quad (3x - 1)$ $-3x^2 + 7x - 4$ $9x^3 - 21x^2 + 12x$ $9x^3 - 24x^2 + 19x - 4$ Horizontal method $(5x + 2)(2x^2 - x + 3)$ $= 10x^3 - 5x^2 + 15x + 4x^2 - 2x + 6$ $= 10x^3 - x^2 + 13x + 6$
Multiplying three or more polynomials, p. 272.	**1.** Multiply any two polynomials. **2.** Multiply the result by any remaining polynomials.	$(2x + 1)(x - 3)(x + 4) = (2x^2 - 5x - 3)(x + 4)$ $2x^2 - 5x - 3$ $x + 4$ $8x^2 - 20x - 12$ $2x^3 - 5x^2 - 3x$ $2x^3 + 3x^2 - 23x - 12$
Dividing a polynomial by a monomial, p. 276.	**1.** Divide each term of the polynomial by the monomial. **2.** When dividing variables, use the property $\frac{x^a}{x^b} = x^{a-b}$	Divide. $(15x^3 + 20x^2 - 30x) \div (5x)$ $= \frac{15x^3}{5x} + \frac{20x^2}{5x} + \frac{-30x}{5x}$ $= 3x^2 + 4x - 6$
Dividing a polynomial by a binomial, p. 277.	**1.** Place the terms of the polynomial and binomial in descending order. Insert a 0 for any missing term. **2.** Divide the first term of the polynomial by the first term of the binomial. **3.** Multiply the partial answer by the binomial, and subtract the results from the first two terms of the polynomial. Bring down the next term to obtain a new polynomial. **4.** Divide the new polynomial by the binomial using the process described in step 2. **5.** Continue dividing, multiplying, and subtracting until the remainder is at a lower power than the variable in the first term of the binomial divisor.	Divide. $(8x^3 - 13x + 2x^2 + 7) \div (4x - 1)$ We rearrange the terms. $2x^2 + x - 3$ $4x - 1 \overline{)8x^3 + 2x^2 - 13x + 7}$ $8x^3 - 2x^2$ $4x^2 - 13x$ $4x^2 - x$ $-12x + 7$ $-12x + 3$ 4 The answer is $2x^2 + x - 3 + \frac{4}{4x - 1}$

Chapter 4 Review Problems

4.1 *Simplify and leave your answer in exponent form.*

1. $(-6a^2)(3a^5)$
$-18a^7$

2. $(5^{10})(5^{13})$
5^{23}

3. $(3xy^2)(2x^3y^4)$
$6x^4y^6$

4. $5xy(-2x^3)(-4xy^4)$
$40x^5y^5$

5. $\dfrac{8^{20}}{8^3}$
8^{17}

6. $\dfrac{7^{15}}{7^{27}}$
$\dfrac{1}{7^{12}}$

7. $\dfrac{x^8}{x^{12}}$
$\dfrac{1}{x^4}$

8. $\dfrac{w^{20}}{w^{15}}$
w^5

9. $\dfrac{3x^8y^0}{9x^4}$
$\dfrac{x^4}{3}$

10. $\dfrac{-15xy^2}{25x^6y^6}$
$-\dfrac{3}{5x^5y^4}$

11. $\dfrac{-12a^3b^6}{18a^2b^{12}}$
$-\dfrac{2a}{3b^6}$

12. $\dfrac{-20a^8b^6c^2}{-12a^5b^5c^3}$
$\dfrac{5a^3b}{3c}$

13. $(x^3)^8$
x^{24}

14. $(5xy^2)^3$
$125x^3y^6$

15. $(-3a^3b^2)^2$
$9a^6b^4$

16. $[3(ab)^2]^3$
$27a^6b^6$

17. $\dfrac{2x^4}{3y^2}$
$\dfrac{2x^4}{3y^2}$

18. $\left(\dfrac{5ab^2}{c^3}\right)^2$
$\dfrac{25a^2b^4}{c^6}$

19. $\dfrac{(2xy^2)^3}{(2xy^2)^4}$
$\dfrac{1}{2xy^2}$

20. $\left(\dfrac{x^0y^3}{4w^5z^2}\right)^3$
$\dfrac{y^9}{64w^{15}z^6}$

4.2 *Simplify. Write with positive exponents.*

21. x^{-3}
$\dfrac{1}{x^3}$

22. $x^{-6}y^{-8}$
$\dfrac{1}{x^6y^8}$

23. $\dfrac{2x^{-6}}{y^{-3}}$
$\dfrac{2y^3}{x^6}$

24. $2^{-1}x^5y^{-6}$
$\dfrac{x^5}{2y^6}$

25. $(2x^3)^{-2}$
$\dfrac{1}{4x^6}$

26. $\dfrac{3x^{-3}}{y^{-2}}$
$\dfrac{3y^2}{x^3}$

27. $\dfrac{4x^{-5}y^{-6}}{w^{-2}z^8}$
$\dfrac{4w^2}{x^5y^6z^8}$

28. $\dfrac{3^{-3}a^{-2}b^5}{c^{-3}d^{-4}}$
$\dfrac{b^5c^3d^4}{27a^2}$

Write in scientific notation.

29. 156,340,200,000
1.563402×10^{11}

30. 88,367
8.8367×10^4

31. 179,632
1.79632×10^5

32. 0.0078
7.8×10^{-3}

33. 0.00006173
6.173×10^{-5}

34. 0.13
1.3×10^{-1}

Write in decimal notation.

35. 1.2×10^5
120,000

36. 8.367×10^{10}
83,670,000,000

37. 3×10^6
3,000,000

38. 2.5×10^{-1}
0.25

39. 5.708×10^{-8}
0.00000005708

40. 6×10^{-9}
0.000000006

Perform the calculation and leave the answer in scientific notation.

41. $\dfrac{(28{,}000{,}000)(5{,}000{,}000{,}000)}{7000}$
2.0×10^{13}

42. $(3.12 \times 10^5)(2.0 \times 10^6)(1.5 \times 10^8)$
9.36×10^{19}

43. $(1.6 \times 10^{-3})(3.0 \times 10^{-5})(2.0 \times 10^{-2})$
9.6×10^{-10}

44. $\dfrac{(0.00078)(0.000005)(0.00004)}{0.002}$
7.8×10^{-11}

45. If a space probe travels at 40,000 kilometers per hour for 1 year, how far will it travel? (Assume that 1 year = 365 days.) 3.504×10^8 kilometers

46. An atomic clock is based on the fact that cesium emits 9,192,631,770 cycles of radiation in one second. How many of these cycles occur in one day? Round your answer to 3 significant digits. 7.94×10^{14} cycles

47. Today's fastest modern computers can perform one operation in 1×10^{-8} second. How many operations can such a computer perform in 1 minute? 6×10^9

4.3 *Combine the following polynomials.*

48. $(2x^2 - 3x + 5) + (-7x^2 - 8x - 23)$
$-5x^2 - 11x - 18$

49. $(-7x^2 - 6x + 3) + (2x^2 - 10x - 15)$
$-5x^2 - 16x - 12$

50. $(x^3 + x^2 - 6x + 2) - (2x^3 - x^2 - 5x - 6)$
$-x^3 + 2x^2 - x + 8$

51. $(4x^3 - x^2 - x + 3) - (-3x^3 + 2x^2 + 5x - 1)$
$7x^3 - 3x^2 - 6x + 4$

52. $(-6x^2 - 3xy + y^2) + (-5x^2 + 10xy - 21y^2)$
$-11x^2 + 7xy - 20y^2$

53. $(8x^2 + 2y + 3xy) + (-5x^2 - 8y + 17xy)$
$3x^2 - 6y + 20xy$

54. $(9x^3y^3 + 3xy - 4) - (4x^3y^3 + 2x^2y^2 - 7xy)$
$5x^3y^3 - 2x^2y^2 + 10xy - 4$

55. $(4x^2y^2 + 6xy - 2y) - (-7x^2y^2 + 5xy + 3y^2)$
$11x^2y^2 + xy - 2y - 3y^2$

56. $(5x^2 + 3x) + (-6x^2 + 2) - (5x - 8)$
$-x^2 - 2x + 10$

57. $(2x^2 - 7) - (3x^2 - 4) + (-5x^2 - 6x)$
$-6x^2 - 6x - 3$

4.4 *Multiply.*

58. $(3x + 1)(5x - 1)$
$15x^2 + 2x - 1$

59. $(8x - 2)(4x - 3)$
$32x^2 - 32x + 6$

60. $(2x + 3)(10x + 9)$
$20x^2 + 48x + 27$

61. $5x(2x^2 - 6x + 3)$
$10x^3 - 30x^2 + 15x$

62. $(3x^2y^2 - 5xy + 6)(-2xy)$
$-6x^3y^3 + 10x^2y^2 - 12xy$

63. $(x^3 - 3x^2 + 5x - 2)(4x)$
$4x^4 - 12x^3 + 20x^2 - 8x$

64. $(5a + 7b)(a - 3b)$
$5a^2 - 8ab - 21b^2$

65. $(2x^2 - 3)(4x^2 - 5y)$
$8x^4 - 12x^2 - 10x^2y + 15y$

66. $-3x^2y(5x^4y + 3x^2 - 2)$
$-15x^6y^2 - 9x^4y + 6x^2y$

4.5 *Multiply.*

67. $(3x - 2)^2$
$9x^2 - 12x + 4$

68. $(5x + 3)(5x - 3)$
$25x^2 - 9$

69. $(7x + 6y)(7x - 6y)$
$49x^2 - 36y^2$

70. $(5a - 2b)^2$
$25a^2 - 20ab + 4b^2$

71. $(8x + 9y)^2$
$64x^2 + 144xy + 81y^2$

72. $(x^2 + 7x + 3)(4x - 1)$
$4x^3 + 27x^2 + 5x - 3$

73. $(x^3 + 2x^2 - x - 4)(2x - 3)$
$2x^4 + x^3 - 8x^2 - 5x + 12$

74. $(x - 6)(2x - 3)(x + 4)$
$2x^3 - 7x^2 - 42x + 72$

75. $(2x + 3)^3$
$8x^3 + 36x^2 + 54x + 27$

4.6 *Divide.*

76. $(12y^3 + 18y^2 + 24y) \div (6y)$
$2y^2 + 3y + 4$

77. $(30x^5 + 35x^4 - 90x^3) \div (5x^2)$
$6x^3 + 7x^2 - 18x$

78. $(16x^3y^2 - 24x^2y + 32xy^2) \div (4xy)$
$4x^2y - 6x + 8y$

79. $(106x^6 - 24x^5 + 38x^4 + 26x^3) \div (2x^3)$
$53x^3 - 12x^2 + 19x + 13$

80. $(15x^2 + 41x + 14) \div (5x + 2)$
$3x + 7$

81. $(16x^2 - 8x - 3) \div (4x - 3)$
$4x + 1$

82. $(2x^3 - x^2 + 3x - 1) \div (x + 2)$
$2x^2 - 5x + 13 + \dfrac{-27}{x + 2}$

83. $(6x^3 + x^2 + 6x + 5) \div (2x - 1)$
$3x^2 + 2x + 4 + \dfrac{9}{2x - 1}$

84. $(8x^2 - 6x + 6) \div (2x + 1)$
$4x - 5 + \dfrac{11}{2x + 1}$

85. $(12x^2 + 11x + 2) \div (3x + 2)$
$4x + 1$

86. $(x^3 - x - 24) \div (x - 3)$
$x^2 + 3x + 8$

87. $(2x^3 - 3x + 1) \div (x - 2)$
$2x^2 + 4x + 5 + \dfrac{11}{x - 2}$

88. $(3x^4 + 2x^3 - 4x - 5) \div (x + 1)$
$3x^3 - x^2 + x - 5$

89. $(12x^3 + 11x^2 - 2x - 1) \div (3x - 1)$
$4x^2 + 5x + 1$

Chapter 4 Test

Simplify each problem. Leave your answer in exponent form.

1. $(3^{10})(3^{24})$ **2.** $\dfrac{3^{35}}{3^{7}}$ **3.** $(8^4)^6$

Simplify.

4. $(-2x^2y)(-3x^5y^6)$

5. $\dfrac{-35x^8y^{10}}{25x^5y^{10}}$

6. $(-5xy^6)^3$

7. $\left(\dfrac{7a^7b^2}{3c^0}\right)^2$

8. $\dfrac{4a^5b^6}{16a^{10}b^{12}}$

9. Evaluate. 5^{-3}

Write with only positive exponents.

10. $6a^{-4}b^{-3}c^5$

11. $\dfrac{2x^{-3}y^{-4}}{w^{-6}z^8}$

12. Write in scientific notation. 0.0000001765

13. Write in decimal notation. 5.82×10^8

14. Multiply and leave your answer in scientific notation.
$(4.0 \times 10^{-3})(3.0 \times 10^{-8})(2.0 \times 10^{4})$

1. 3^{34}

2. 3^{28}

3. 8^{24}

4. $6x^7y^7$

5. $\dfrac{-7x^3}{5}$

6. $-125x^3y^{18}$

7. $\dfrac{49a^{14}b^4}{9}$

8. $\dfrac{1}{4a^5b^6}$

9. $\dfrac{1}{125}$

10. $\dfrac{6c^5}{a^4b^3}$

11. $\dfrac{2w^6}{x^3y^4z^8}$

12. 1.765×10^{-7}

13. 582,000,000

14. 2.4×10^{-6}

15. $-2x^2 + 5x$

16. $3x^2 + 3xy - 6y - 7y^2$

17. $-21x^5 + 28x^4 - 42x^3 + 14x^2$

18. $15x^4y^3 - 18x^3y^2 + 6x^2y$

19. $10a^2 + 7ab - 12b^2$

20. $6x^3 - 11x^2 - 19x - 6$

21. $49x^4 + 28x^2y^2 + 4y^4$

22. $81x^2 - 4y^2$

23. $12x^4 - 14x^3 + 25x^2 - 29x + 10$

24. $3x^4 + 4x^3y - 15x^2y^2$

25. $3x^3 - x + 5$

26. $2y^2 + 3y - 1$

Combine the following polynomials.

15. $(2x^2 - 3x - 6) + (-4x^2 + 8x + 6)$

16. $(5x^2 + 6xy - 7y^2) - (2x^2 + 3xy + 6y)$

Multiply.

17. $-7x^2(3x^3 - 4x^2 + 6x - 2)$

18. $(5x^2y^2 - 6xy + 2)(3x^2y)$

19. $(5a - 4b)(2a + 3b)$

20. $(3x + 2)(2x + 1)(x - 3)$

21. $(7x^2 + 2y^2)^2$

22. $(9x - 2y)(9x + 2y)$

23. $(3x - 2)(4x^3 - 2x^2 + 7x - 5)$

24. $(3x^2 - 5xy)(x^2 + 3xy)$

Divide.

25. $15x^6 - 5x^4 + 25x^3 \div 5x^3$

26. $(4y^4 + 6y^3 + 3y - 1) \div (2y^2 + 1)$

Cumulative Test for Chapters 0–4

Approximately one-half of this test covers the content of Chapters 0 through 3. The remainder covers the content of Chapter 4.

Simplify.

1. $\dfrac{5}{12} - \dfrac{7}{8}$

2. $(-3.7) \times (0.2)$

3. $\left(-4\dfrac{1}{2}\right) \div \left(5\dfrac{1}{4}\right)$

4. 7% of \$128.00

5. $7x(3x - 4) - 5x(2x - 3) - (3x)^2$

6. Evaluate $2x^2 - 3xy + y^2$ when $x = -2$ and $y = 3$.

Solve.

7. $7x - 3(4 - 2x) = 14x - (3 - x)$

8. $\dfrac{x - 5}{3} = \dfrac{1}{4}x + 2$

9. $4 - 7x < 11$

10. Solve for f: $B = \frac{1}{2}a(c + 3f)$

11. A national walkout of 11,904 employees of the VBM Corp. occurred last month. This was 96% of the number of employees. How many employees does VBM have?

12. How much interest would \$3320 earn in one year if invested at 6%?

1. $-\frac{11}{24}$

2. -0.74

3. $-\frac{6}{7}$

4. \$8.96

5. $2x^2 - 13x$

6. 35

7. $x = -\frac{9}{2}$

8. $x = 44$

9. $x > -1$

10. $f = \dfrac{2B}{3a} - \dfrac{c}{3}$

11. 12,400 employees

12. \$199.20

13. Multiply. $(3x - 7)(5x - 4)$

14. Multiply. $(3x - 5)^2$

15. $(2x + 1)^3$ *Hint:* Write it in the form $(2x + 1)(2x + 1)(2x + 1)$ and then multiply.

Simplify.

16. $(-4x^4y^5)(5xy^3)$

17. $\dfrac{14x^8y^3}{-21x^5y^{12}}$

18. $(-3xy^4z^2)^3$

19. Write with only positive exponents. $\dfrac{9x^{-3}y^{-4}}{w^2z^{-8}}$

20. Write in scientific notation. 1,360,000,000,000,000

21. Perform the calculation and leave your answer in scientific notation.
$$\frac{(2.0 \times 10^{-12})(8.0 \times 10^{-20})}{4.0 \times 10^3}$$

22. Subtract the polynomials. $(x^3 - 3x^2 - 5x + 20) - (-4x^3 - 10x^2 + x - 30)$

Multiply.

23. $-6xy^2(6x^2 - 3xy + 8y^2)$

24. $(x^2 - 6x + 3)(2x^2 - 3x + 4)$

13. $15x^2 - 47x + 28$

14. $9x^2 - 30x + 25$

15. $8x^3 + 12x^2 + 6x + 1$

16. $-20x^5y^8$

17. $-\dfrac{2x^3}{3y^9}$

18. $-27x^3y^{12}z^6$

19. $\dfrac{9z^8}{w^2x^3y^4}$

20. 1.36×10^{15}

21. 4.0×10^{-35}

22. $5x^3 + 7x^2 - 6x + 50$

23. $-36x^3y^2 + 18x^2y^3 - 48xy^4$

24. $2x^4 - 15x^3 + 28x^2 - 33x + 12$

CHAPTER 5

Factoring

Often we think of our annual income as being rather small and not quite what we would like. Have you ever thought about how much you might make in your lifetime? Do you realize how much money you might make over the next twenty years? Turn to the Putting Your Skills to Work on page 330 for a mathematical formula that allows us to quickly make these kind of calculations.

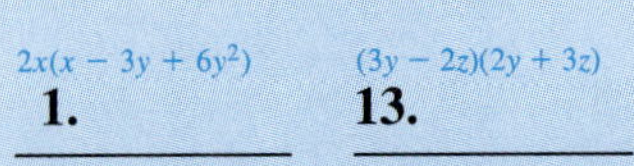

Pretest Chapter 5

1. $2x(x - 3y + 6y^2)$
2. $(3x + 4y)(a - 2b)$
3. $18ab(2b - 1)$
4. $(a - 2b)(5 - 3x)$
5. $(x + y)(3x - 4)$
6. $(3x - 2)(7x - 3)$
7. $(x - 24)(x + 2)$
8. $(x - 5)(x - 3)$
9. $(x + 5)(x + 1)$
10. $2(x + 6)(x - 2)$
11. $3(x - 9)(x + 7)$
12. $(5x - 2)(3x - 2)$
13. $(3y - 2z)(2y + 3z)$
14. $4(3x + 5)(x + 2)$
15. $(9x^2 + 4)(3x + 2)(3x - 2)$
16. $(7x - 2y)^2$
17. $(5x + 8)^2$
18. $3x(2x - 1)(x + 3)$
19. $2y^2(4x - 3)^2$
20. Cannot be factored
21. $x = 1, x = -\frac{3}{2}$
22. $x = 5, x = 6$
23. $x = -\frac{2}{3}, x = 3$

If you are familiar with the topics in this chapter, take this test now. Check your answers with those in the back of the book. If an answer was wrong or you couldn't do a problem, study the appropriate section of the chapter.

If you are not familiar with the topics in this chapter, don't take this test now. Instead, study the examples, work the practice problems, and then take the test.

This test will help you to identify those concepts that you have mastered and those that need more study.

Section 5.1

Remove the greatest common factor.

1. $2x^2 - 6xy + 12xy^2$ **2.** $3x(a - 2b) + 4y(a - 2b)$ **3.** $36ab^2 - 18ab$

Section 5.2

Factor.

4. $5a - 10b - 3ax + 6xb$ **5.** $3x^2 - 4y + 3xy - 4x$ **6.** $21x^2 - 14x - 9x + 6$

Section 5.3

Factor.

7. $x^2 - 22x - 48$ **8.** $x^2 - 8x + 15$ **9.** $x^2 + 6x + 5$

10. $2x^2 + 8x - 24$ **11.** $3x^2 - 6x - 189$

Section 5.4

Factor.

12. $15x^2 - 16x + 4$ **13.** $6y^2 + 5yz - 6z^2$ **14.** $12x^2 + 44x + 40$

Section 5.5

Factor completely.

15. $81x^4 - 16$ **16.** $49x^2 - 28xy + 4y^2$ **17.** $25x^2 + 80x + 64$

Section 5.6

Factor completely. If not possible, so state.

18. $6x^3 + 15x^2 - 9x$ **19.** $32x^2y^2 - 48xy^2 + 18y^2$ **20.** $36x^2 + 49$

Section 5.7

Solve for the roots of each quadratic equation.

21. $2x^2 + x - 3 = 0$ **22.** $x^2 + 30 = 11x$ **23.** $\dfrac{3x^2 - 7x}{2} = 3$

5.1 Introduction to Factoring

After studying this section, you will be able to:

1 *Factor polynomials containing a common factor in each term.*

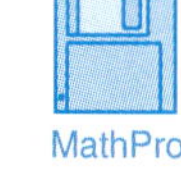
MathPro

Video 12

SSM

1 *Common Factors*

Recall that when two or more numbers, variables, or algebraic expressions are multiplied, each is called a **factor.**

$$\underbrace{3}_{\text{factor}} \cdot \underbrace{2}_{\text{factor}} \qquad \underbrace{3x^2}_{\text{factor}} \cdot \underbrace{5x^3}_{\text{factor}} \qquad \underbrace{(2x-3)}_{\text{factor}}\underbrace{(x+4)}_{\text{factor}}$$

When you are asked to factor a number or an algebraic expression, you are being asked, ''What, when multiplied, will give that number or expression?'' For example, you can factor 6 as $3 \cdot 2$ since $3 \cdot 2 = 6$. You can factor $15x^5$ as $3x^2 \cdot 5x^3$ since $3x^2 \cdot 5x^3 = 15x^5$. Factoring is simply the reverse of multiplying. 6 and $15x^5$ are simple expressions to factor and can be factored in different ways. The factors of the polynomial $2x^2 + x - 12$ are not so easy to recognize. In this chapter we will be learning techniques to find the factors of a polynomial. We will begin with common factors.

EXAMPLE 1 Factor. **(a)** $3x - 6y$ **(b)** $9x + 2xy$

Begin by looking for a common factor, the factor that both terms have in common. Then rewrite the expression as a product.

(a) $3x - 6y = 3(x - 2y)$ *This is true because* $3(x - 2y) = 3x - 6y$.

(b) $9x + 2xy = x(9 + 2y)$ *This is true because* $x(9 + 2y) = 9x + 2xy$.

Some people find it helpful to think of factoring as the distributive property in reverse. When we write $3x - 6y = 3(x - 2y)$, we are ''undistributing'' the 3.

Practice Problem 1 Factor. **(a)** $21a - 7b$ **(b)** $p + prt$ ■

When we factor, we are looking for the largest common factor. For example, in the polynomial $48x - 16y$, a common factor is 2. We could factor $48 - 16y$ as $2(24 - 8y)$. However, this is not complete. To factor $48 - 16y$ completely, we look for the greatest common factor of 48 and of 16.

$$48x - 16y = 16(3x - y)$$

TEACHING TIP You will want to stress that removing the largest common factor may involve removing a number, a variable, or both. After presenting Example 2, ask students to do the following as an Additional Class Exercise. Remove the greatest common factor.

A. $12x^2y^2 - 36x^3y$
B. $21a^4 - 14a^3 + 35a^2$
C. $96x - 144y + 48z$

Sol.

A. $12x^2y(y - 3x)$
B. $7a^2(3a^2 - 2a + 5)$
C. $48(2x - 3y + z)$

EXAMPLE 2 Factor. $24xy + 12x^2 + 36x^3$

Find the greatest common factor of 24, 12, and 36. You may want to factor each number or you may notice that 12 is a common factor. 12 is the greatest common factor. Notice also that x is a factor of each term.

$$24xy + 12x^2 + 36x^3 = 12x(2y + x + 3x^2)$$

Practice Problem 2 Factor. $12a^2 + 16ab^2 - 12a^2b$ ■

Common Factors of a Polynomial

1. You can determine the numerical common factor by asking, ''What is the largest integer that will divide into the coefficient of each term?''
2. You can determine the variable common factor by asking, ''What variables are common to each term, and what is the largest exponent of those variables that is common?''

TEACHING TIP Explain to your students that to see if they have factored the greatest common factor they must do two things:

1. See if any further common factor exists within the terms in the parentheses. (If so, you did not correctly remove the GREATEST common factor.)
2. Multiply back the factors. See if the result matches exactly the original problem.

EXAMPLE 3 Factor. **(a)** $12x^2 + 18y^2$ **(b)** $x^2y^2 + 3xy^2 + y^3$

(a) Note that the largest integer that is common to both terms is 6 (not 3 or 2).

$$12x^2 + 18y^2 = 6(2x^2 + 3y^2)$$

(b) Although y is common to all terms, we factor out y^2 since 2 is the largest exponent of y that is common to all terms. We do not factor out x since x is not common to all terms.

$$x^2y^2 + 3xy^2 + y^3 = y^2(x^2 + 3x + y)$$

Practice Problem 3 Factor. **(a)** $16a^3 - 24b^3$ **(b)** $r^3s^2 - 4r^4s + 7r^5$ ■

Checking

You can check any factoring problem by multiplying the factors you obtain. The result should be the same as the original polynomial.

EXAMPLE 4 Factor. $8x^3y + 16x^2y^2 + 24x^3y^3$

We see that 8 is the largest integer that will divide evenly into the three numerical coefficients. We can factor an x^2 out of each term (not just x). For the other variable we factor y out of each term.

$$8x^3y + 16x^2y^2 + 24x^3y^3 = 8x^2y(x + 2y + 3xy^2)$$

Check

$$8x^2y(x + 2y + 3xy^2) = 8x^3y + 16x^2y^2 + 24x^3y^3 \checkmark$$

Practice Problem 4 Factor. $18a^3b^2c - 27ab^3c^2 - 45a^2b^2c^2$ ■

EXAMPLE 5 Factor. $9a^3b^2 + 9a^2b^2$

We observe that both terms contain a common factor of 9. We can remove a^2 and b^2 from each term.

$$9a^3b^2 + 9a^2b^2 = 9a^2b^2(a + 1)$$

WARNING Don't forget to include the 1 inside the parentheses in Example 5. The solution is wrong without it. You will see why if you try to check a result written without the 1.

Practice Problem 5 Factor and check. $30x^3y^2 - 24x^2y^2 + 6xy^2$ ■

EXAMPLE 6 Factor. $3x(x - 4y) + 2(x - 4y)$

Be sure you understand what are terms and what are factors in this example. There are two *terms.* The expression $3x(x - 4y)$ is one term. The expression $2(x - 4y)$ is the second term. Each term is made up of two factors. Observe that the binomial $(x - 4y)$ is a common *factor* in each term. A common factor may be a set of parentheses containing any type of polynomial. Thus we can remove the common factor $(x - 4y)$.

$$3x(x - 4y) + 2(x - 4y) = (x - 4y)(3x + 2)$$

Practice Problem 6 Factor. $3(a + b) + x(a + b)$ ■

TEACHING TIP Students find the type of factoring in Example 6 more difficult. You may wish to do the following as an Additional Class Exercise after you have finished explaining Example 6.
Factor. $a(5b - 6) - 3b(5b - 6)$

Sol. The common factor is the parentheses containing $5b - 6$. If we remove that, the terms remaining are $a - 3b$. Thus

$$a(5b - 6) - 3b(5b - 6) = (5b - 6)(a - 3b)$$

EXAMPLE 7 Factor. $7x^2(2x - 3y) - (2x - 3y)$

The common factor for each term is $(2x - 3y)$. What happens when we factor out $(2x - 3y)$? What are we left with in the second term? Recall that $(2x - 3y) = 1(2x - 3y)$. Thus

$7x^2(2x - 3y) - (2x - 3y) = 7x^2(2x - 3y) - 1(2x - 3y)$ *Rewrite the original expression.*

$= (7x^2 - 1)(2x - 3y)$ *Factor out* $(2x - 3y)$.

Practice Problem 7 Factor. $8y(9y^2 - 2) - (9y^2 - 2)$ ■

EXAMPLE 8 A computer programmer is writing a program to find the area of 4 circles. She uses the formula $A = \pi r^2$. The radii of the circles are a, b, c, and d, respectively. She wants the final answer to be in factored form with the value of π occurring only once, in order to minimize round-off area. Write the total area with a formula that has π occurring only once.

For each circle $A = \pi r^2$, where $r = a, b, c,$ or d.

The total area is $\pi a^2 + \pi b^2 + \pi c^2 + \pi d^2$.

In factored form the total area $= \pi(a^2 + b^2 + c^2 + d^2)$

Practice Problem 8 Use $A = \pi r^2$ to find the shaded area. The radius of the larger circle is b. The radius of the smaller circle is a. Write the total area formula in factored form so that π only appears once.

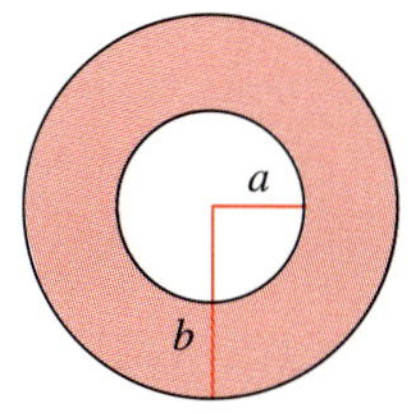

■

5.1 Exercises

Verbal and Writing Skills

Write a word or words to complete each sentence.

1. In the expression $3x^2 \cdot 5x^3$, $3x^2$ and $5x^3$ are called ______factors______.
2. In the expression $3x^2 + 5x^3$, $3x^2$ and $5x^3$ are called ______terms______.
3. We can factor $30a^4 + 15a^3 - 45a^2$ as $5a(6a^3 + 3a^2 - 9a)$. Is the factoring complete? Why or why not?
   No. $6a^3 + 3a^2 - 9a$ has a common factor of $3a$.

Remove the largest possible common factor. Check your answers for Problems 4 through 15 by multiplication.

4. $6x^2 - 6x$ $\quad 6x(x - 1)$
5. $5ab - 5b$ $\quad 5b(a - 1)$
6. $21abc - 14ab^2$
   $7ab(3c - 2b)$
7. $18wz - 27w^2z$
   $9wz(2 - 3w)$
8. $5x^3 + 25x^2 - 15x$
   $5x(x^2 + 5x - 3)$
9. $8x^3 - 10x^2 - 14x$
   $2x(4x^2 - 5x - 7)$
10. $12ab - 28bc + 20ac$
    $4(3ab - 7bc + 5ac)$
11. $12xy - 18yz - 36xz$
    $6(2xy - 3yz - 6xz)$
12. $ax^2 - 2bx^2 - cx^2 + dx$
    $x(ax - 2bx - cx + d)$
13. $a^4y - a^3y^2 + a^2y^3 - 2a$
    $a(a^3y - a^2y^2 + ay^3 - 2)$
14. $60x^3 - 50x^2 + 25x$
    $5x(12x^2 - 10x + 5)$
15. $6x^9 - 8x^7 + 4x^5$
    $2x^5(3x^4 - 4x^2 + 2)$
16. $2\pi rh + 2\pi r^2$
    $2\pi r(h + r)$
17. $9a^2b^2 - 36ab$
    $9ab(ab - 4)$
18. $14x^2y - 35xy - 63x$
    $7x(2xy - 5y - 9)$
19. $40a^2 - 16ab - 24a$
    $8a(5a - 2b - 3)$
20. $39x^2 - 26xy - 13x$
    $13x(3x - 2y - 1)$
21. $34x^2 - 51xy - 17x$
    $17x(2x - 3y - 1)$
22. $20a^2b - 4ab^2 + 12ab + 8a^2b^2$
    $4ab(5a - b + 3 + 2ab)$
23. $18xy^2 + 6x^2y + 24xy - 12x^2y^2$
    $6xy(3y + x + 4 - 2xy)$

Hint: In Problems 24 through 37, refer to Examples 6 and 7.

24. $2(a + b) + 3x(a + b)$
    $(a + b)(2 + 3x)$
25. $5(x - 2y) - z(x - 2y)$
    $(x - 2y)(5 - z)$
26. $3x(x - 4) - 2(x - 4)$
    $(x - 4)(3x - 2)$
27. $5x(x - 7) + 3(x - 7)$
    $(x - 7)(5x + 3)$
28. $6b(2a - 3c) - 5d(2a - 3c)$
    $(2a - 3c)(6b - 5d)$
29. $7x(3y + 5z) - 6t(3y + 5z)$
    $(3y + 5z)(7x - 6t)$

30. $3(x^2 + 1) + 2y(x^2 + 1) + w(x^2 + 1)$
$(x^2 + 1)(3 + 2y + w)$

31. $5a(bc - 1) + b(bc - 1) + c(bc - 1)$
$(bc - 1)(5a + b + c)$

32. $2a(xy - 3) - 4(xy - 3) - z(xy - 3)$
$(xy - 3)(2a - 4 - z)$

33. $3c(bc - 3a) - 2(bc - 3a) - 6b(bc - 3a)$
$(bc - 3a)(3c - 2 - 6b)$

34. $4y(x + 2y) + (x + 2y)$
$(x + 2y)(4y + 1)$

35. $3x^2(x - 2y) - (x - 2y)$
$(x - 2y)(3x^2 - 1)$

36. $9a(5a - b) - (5a - b)$
$(5a - b)(9a - 1)$

37. $2ab(9x + 5y) + (9x + 5y)$
$(9x + 5y)(2ab + 1)$

Factor. Be sure to remove the greatest common factor.

★ 38. $100x^6 - 25x^5 + 36x^4 - 72x^3 - 68x^2 + 120x$ $x(100x^5 - 25x^4 + 36x^3 - 72x^2 - 68x + 120)$

★ 39. $3x^2y(5a - 3b + 1) - 19xy(5a - 3b + 1) - y(5a - 3b + 1)$ $y(5a - 3b + 1)(3x^2 - 19x - 1)$

40. Remove a common factor of 1.6845 from $5.0535x^2 - 1.6845w^2 + 15.1605z^2$.
$1.6845(3x^2 - w^2 + 9z^2)$

41. Remove a common factor of 57,982 from $173{,}946x^2 - 347{,}892y^2 - 637{,}802z^2$.
$57{,}982(3x^2 - 6y^2 - 11z^2)$

Applications

42. Find a formula for the area of 4 rectangles of width 2.786 inches. The length of each rectangle is a, b, c, or d inches. Write the formula in factored form.
$A = 2.786(a + b + c + d)$ sq in.

43. Find a formula for the total cost of all purchases by four people. Each person went to the local wholesale warehouse and spent $29.95 per item. Harry bought a items, Richard bought b items, Lyle bought c items, and Selena bought d items. Write the formula in factored form. $C = 25.95(a + b + c + d)$

Cumulative Review Problems

Multiply the following:

44. $(2x - 3)(2x + 3)$ $4x^2 - 9$

45. $(4a - b)^2$ $16a^2 - 8ab + b^2$

46. The sum of three consecutive odd integers is 40 more than the smallest of these integers. Find each of these three consecutive odd integers. (*Hint:* consecutive odd integers are numbers included in the following pattern: $-3, -1, 1, 3, 5, 7, 9, \ldots, x, x + 2, \ldots$).
The three consecutive odd integers are 17, 19, 21.

47. Tuition, room, and board at a college in Colorado cost Lisa $27,040. This was a 4% increase over the cost of these items last year. What did Lisa pay last year for tuition, room, and board? $26,000

48. The Federal Budget Deficit for 1996 was $117 billion. This was a decrease of 29% from the deficit for 1995. What was the deficit for 1995? (Round your answer to the nearest billion dollars.) $165 billion

49. A videotape plays 6.72 feet per minute for 2 hours at standard speed. It plays for 6 hours at extended long play speed. What is the rate of the tape at the slower speed? 2.24 feet per minute

5.2 Factor by Grouping

MathPro

Video 12

SSM

After studying this section, you will be able to:

1 *Factor problems with four terms by grouping.*

1 *Factoring by Grouping*

A common factor can be a number, a variable, or an algebraic expression. Sometimes the polynomial is written so that it is easy to recognize the common factor. This is especially true when the common factor is enclosed by parentheses.

EXAMPLE 1 Factor. $x(x - 3) + 2(x - 3)$

Observe each term:

$$\underbrace{x(x-3)}_{\text{first term}} + \underbrace{2(x-3)}_{\text{second term}}$$

The common factor for both the first and second terms is the quantity $(x - 3)$, so we have

$$x(\underline{x-3}) + 2(\underline{x-3}) = (\underline{x-3})(x+2)$$

Practice Problem 1 Factor. $3y(2x - 7) - 8(2x - 7)$ ■

In many cases these types of problems do not have any parentheses and we are asked to factor the polynomial. In such cases we remove a common factor from the first two terms and a different common factor from the second two terms. If the two parentheses contain the same expressions at this step, the problem may then be completed as we did in Example 1. This procedure of factoring is often called *factoring by grouping.*

EXAMPLE 2 Factor. $2x^2 + 3x + 6x + 9$

$2x^2 + 3x$	+	$6x + 9$
Remove a common factor of x from the first two terms.		Remove a common factor of 3 from the second two terms.
$x(2x + 3)$		$3(2x + 3)$

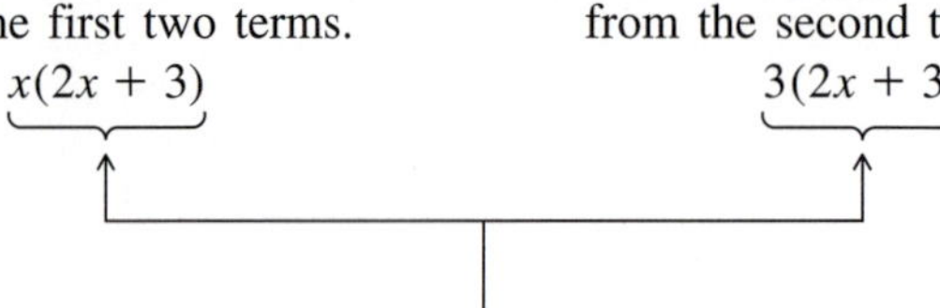

Note that the two sets of parentheses contain the same expression at this step. The expression in parentheses is now a common factor.

Now we finish the factoring.

$$x(2x + 3) + 3(2x + 3) = (2x + 3)(x + 3)$$

Practice Problem 2 Factor. $6x^2 - 15x + 4x - 10$ ■

EXAMPLE 3 Factor. $4x + 8y + ax + 2ay$

Remove a common factor of 4 from the first two terms.

$$4x + 8y + ax + 2ay = 4(x + 2y) + a(x + 2y)$$

Remove a common factor of a from the second two terms.

$4(\underline{x + 2y}) + a(\underline{x + 2y}) = (\underline{x + 2y})(4 + a)$ *Both sets of parentheses contain the same expression. The common factor of each term is the expression in parentheses,* $(x + 2y)$.

TEACHING TIP After you have explained Examples 2 and 3, give students the following as an Additional Class Exercise.
Factor. $3a^2 - 15ab + 7ad - 35bd$
Sol. $(a - 5b)(3a + 7d)$

Practice Problem 3 Factor by grouping. $ax + 2a + 4bx + 8b$ ■

In practice, these problems are done in just two steps.

EXAMPLE 4 Factor. $cx + cy + 4x + 4y$

$cx + cy + 4x + 4y = c(x + y) + 4(x + y)$ *Remove the common factor of c from the first two terms. Remove the common factor of 4 from the second two terms.*

$= (x + y)(c + 4)$ *Since the two parentheses above are the same, we complete the factoring.*

Practice Problem 4 Factor. $12x + 4y + 21ax + 7ay$ ■

In some problems the terms are out of order. We have to rearrange the order of the terms first so that the first two terms have a common factor.

EXAMPLE 5 Factor. $bx + 4y + 4b + xy$

$= bx + 4b + xy + 4y$ *Rearrange the terms so that the first terms have a common factor.*

$= b(x + 4) + y(x + 4)$ *Remove the common factor of b from the first two terms.*
Remove the common factor of y from the second two terms.

$= (x + 4)(b + y)$ *Since the two sets of parentheses above contain the same expression, we can complete the factoring.*

Practice Problem 5 Factor. $6a^2 + 5bc + 10ab + 3ac$ ■

Sometimes you will need to remove a negative common factor from the second two terms to obtain two sets of parentheses that contain the same expression.

TEACHING TIP The idea of removing a common factor that is negative gives students some difficulty. After you have shown them how to do Example 6, you may want to expand the idea by showing the following: The expression $-3a - 15b$ can be factored two ways.
We can write
$-3a - 15b = -3(a + 5b)$,
or we can write
$-3a - 15b = 3(-a - 5b)$.
Show students how to factor
$2a^2 + 10ab - 3a - 15b$.
We factor as follows:

$2a(a + 5b) - 3(a + 5b)$
$= (a + 5b)(2a - 3)$

EXAMPLE 6 Factor. $2x^2 + 5x - 4x - 10$

$= x(2x + 5) - 2(2x + 5)$ — *Remove the common factor of x from the first two terms and a common factor of −2 from the second two terms.*

$= (2x + 5)(x - 2)$ — *Since the two sets of parentheses above contain the same expression, we can complete the factoring.*

Notice that if you removed a common factor of +2 in the first step, the two sets of parentheses would not contain the same expression. If the expressions inside the two sets of parentheses are not exactly the same, you cannot express the polynomial as a product of two factors!

Practice Problem 6 Factor. $6xy + 14x - 15y - 35$ ■

TEACHING TIP After explaining Example 7, ask students to do the following problem as an Additional Class Exercise.
Factor. $10ab - 3c + 5ac - 6b$
Sol. $(2b + c)(5a - 3)$
When discussing the solution, ask students what order they put their terms in before factoring. Usually you will get two or three different suggestions. By putting them all on the blackboard and factoring each way, you will be able to convince students that any order is OK as long as the first two terms have a common factor.

EXAMPLE 7 Factor. $2ax - a - 2bx + b$

$= a(2x - 1) - b(2x - 1)$ — *Remove the common factor of a from the first two terms.*
Remove the common factor of −b from the second two terms.

$= (2x - 1)(a - b)$ — *Since the two sets of parentheses contain the same expression, we can complete the factoring.*

Practice Problem 7 Factor. $3x - 10ay + 6y - 5ax$ ■

WARNING Many students find that they make a factoring error in the first step of problems like Example 7. Multiplying back the results of the first step (even if you do it in your head) will usually help you to detect any error you may have made.

EXAMPLE 8 Factor and check your answer. $8ad + 21bc - 6bd - 28ac$

We observe that the first two terms do not have a common factor.

$= 8ad - 6bd - 28ac + 21bc$ — *Rearrange the order using the commutative and associative properties of addition.*

$= 2d(4a - 3b) - 7c(4a - 3b)$ — *Remove the common factor of 2d from first terms.*
Remove the common factor of −7c from last terms.

$= (4a - 3b)(2d - 7c)$ — *Remove the common factor of (4a − 3b).*

To check, we multiply out the two binomials using the FOIL procedure.

$(4a - 3b)(2d - 7c) = 8ad - 28ac - 6bd + 21bc$

$= 8ad + 21bc - 6bd - 28ac$ ✓ — *Rearrange the order of the terms. This is the original problem. Thus it checks.*

Practice Problem 8 Factor and check your answer.
$10ad + 27bc - 6bd - 45ac$ ■

5.2 Exercises

Factor by grouping. Check your answers for Problems 1 through 12.

1. $3x - 6 + xy - 2y$
$(x - 2)(3 + y)$

2. $2x - 10 + xy - 5y$
$(x - 5)(2 + y)$

3. $2ax + 6bx - ay - 3by$
$(a + 3b)(2x - y)$

4. $4x + 8y - 3wx - 6wy$
$(x + 2y)(4 - 3w)$

5. $x^3 - 4x^2 + 3x - 12$
$(x - 4)(x^2 + 3)$

6. $x^3 - 6x^2 + 2x - 12$
$(x - 6)(x^2 + 2)$

7. $3ax + bx - 6a - 2b$
$(3a + b)(x - 2)$

8. $4ax + bx - 28a - 7b$
$(4a + b)(x - 7)$

9. $5a + 12bc + 10b + 6ac$
$(a + 2b)(5 + 6c)$

10. $2x + 15yz + 6y + 5xz$
$(x + 3y)(2 + 5z)$

11. $5a - 5b - 2ax + 2xb$
$(a - b)(5 - 2x)$

12. $xy - 4x - 3y + 12$
$(y - 4)(x - 3)$

13. $y^2 - 2y - 3y + 6$
$(y - 2)(y - 3)$

14. $x^2 + 4x - 5x - 20$
$(x + 4)(x - 5)$

15. $3ax - y + 3ay - x$
$(x + y)(3a - 1)$

16. $xa + 2bx - a - 2b$
$(a + 2b)(x - 1)$

17. $6ax - y + 2ay - 3x$
$(3x + y)(2a - 1)$

18. $6tx - 3t - 2rx + r$
$(2x - 1)(3t - r)$

19. $x^2 + 6x - 8x - 48$
$(x + 6)(x - 8)$

20. $x^2 - 7x - 9x + 63$
$(x - 7)(x - 9)$

21. $28x^2 + 8xy^2 + 21xw + 6y^2w$
$(4x + 3w)(7x + 2y^2)$

22. $8xw + 10x^2 + 35xy^2 + 28y^2w$
$(5x + 4w)(2x + 7y^2)$

23. $8a^2 - 6ab + 20ae - 15be$
$(4a - 3b)(2a + 5e)$

24. $21a^2 - 6ab - 28ae + 8be$
$(3a - 4e)(7a - 2b)$

★ **25.** $44x^3 - 63y^2w - 36xy^2 + 77x^2w$
$(4x + 7w)(11x^2 - 9y^2)$

★ **26.** $50x^7 - 40x^3y^5 - 15x^4w^2 + 12y^5w^2$
$(10x^3 - 3w^2)(5x^4 - 4y^5)$

Verbal and Writing Skills

27. Although $6a^2 - 12bd - 8ad + 9ab = 6(a^2 - 2bd) - a(8d - 9b)$ is true, it is not correct as the solution to the problem "Factor $6a^2 - 12bd - 8ad + 9ab$." Explain. Can this problem be factored? We must rearrange the terms in a different order so that the expression in the parentheses is the same in each case. We use the order $6a^2 - 8ad + 9ab - 12bd$ to factor $2a(3a - 4d) + 3b(3a - 4d) = (3a - 4d)(2a + 3b)$

Cumulative Review Problems

28. Multiply. $(11x - 7y)(11x + 7y)$. $121x^2 - 49y^2$

29. A polynomial of three terms is called a ___trinomial___.

30. A train was traveling at 73 miles per hour when the engineer spotted a stalled truck on the railroad crossing ahead. He jammed on the brakes but was unable to stop the train before it collided with the stalled truck. For every second the engineer applied the brakes, the train slowed down by 4 miles per hour. The accident reconstruction team found that the train was still traveling at 41 miles per hour at the time of impact. For how many seconds were the brakes applied? 8 seconds

31. Tim Brown made an average profit of \$320 per car for each car he sold at Donahue Motors over the last 5 months. In January and February he averaged only \$200 per car. In March he averaged \$300 per car, but in April he averaged \$480 per car. Assuming that he sells around the same number of cars each month, what was his average profit per car for the month of May? He made an average profit of \$420 per car during the month of May.

5.3 Factoring Trinomials of the Form $x^2 + bx + c$

MathPro

Video 13

SSM

After studying this section, you will be able to:

1 *Factor polynomials of the form $x^2 + bx + c$.*
2 *Factor polynomials that have a common factor and a factor of the form $x^2 + bx + c$.*

1 *Factoring Polynomials of the Form $x^2 + bx + c$*

Suppose that you wanted to factor $x^2 + 5x + 6$. After some trial and error you *might* obtain $(x + 2)(x + 3)$ or you might get discouraged and not get an answer. If you did get these factors, you could check this answer by the FOIL method.

$$(x + 2)(x + 3) = x^2 + 3x + 2x + 6$$
$$= x^2 + 5x + 6$$

But trial and error can be a long process. There is another way.

Look at the original polynomial $x^2 + 5x + 6$. We know immediately that the answer will be of the form ( )( ). Since the first term of the polynomial is x^2, we can place an x in each of the parentheses, $(x \quad)(x \quad)$. Next look at the sign of the last term. The sign is positive. This means that the sign of each factor must *be the same.* Since the sign of the middle term is positive, the sign of each factor must be positive, and we can write $(x + \quad)(x + \quad)$. Look again at the last term. Think of factors of 6. Try 2 and 3 since $2 + 3 = 5$, the coefficient of the middle term. We write $(x + 2)(x + 3)$ and our factoring is complete. We can check this using FOIL as shown above. To summarize:

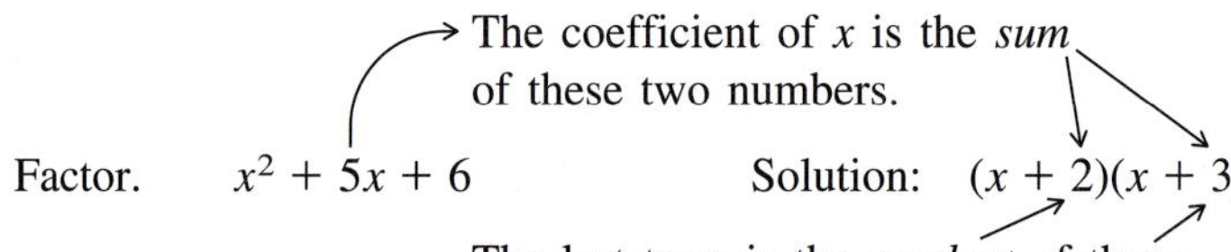

Let's write the procedure we have observed and try a few examples.

TEACHING TIP Students should be told that factoring trinomials of the form $x^2 + bx + c$ is the most common form of factoring. Students should work at a total mastery of this type of problem (including speed).

Factoring Trinomials of the Form $x^2 + bx + c$

1. The answer will be of the form $(x + \_\_)(x + \_\_)$.
2. The two numbers at the end of each set of parentheses are numbers such that:
   (a) When you multiply them, you get the last term, which is c.
   (b) When you add them, you get the coefficient of x, which is b.

TEACHING TIP To factor problems of the type $x^2 + bx + c$, we want to find two numbers that we can multiply to get c but add up to get b. You may want to try the following exercise.
Write this problem on the board: $x^2 + 12x + 32$.
What two numbers can we multiply to get 32 but add up to get 12? (8 and 4)
Then we can factor as follows:

$$x^2 + 12x + 32 = (x + 8)(x + 4)$$

EXAMPLE 1 Factor. $x^2 + 7x + 12$

The answer is of the form $(x + \_\_)(x + \_\_)$. We want to find the two numbers you can multiply to get 12 but add to get 7. The numbers are 3 and 4.

$$x^2 + 7x + 12 = (x + 3)(x + 4)$$

Practice Problem 1 Factor. $x^2 + 8x + 12$ ■

EXAMPLE 2 Factor. $x^2 + 12x + 20$

We want two numbers that have a product of 20 but a sum of 12. The numbers are 10 and 2.

$$x^2 + 12x + 20 = (x + \underline{10})(x + \underline{2})$$

Note: If you cannot think of the numbers in your head, write down the possible factors whose product is 20.

Product	***Sum***
$1 \cdot 20 = 20$	$1 + 20 = 21$
$2 \cdot 10 = 20$	$2 + 10 = 12$ ←
$4 \cdot 5 = 20$	$4 + 5 = 9$

Then select the pair whose sum is 12. Select this pair.

Practice Problem 2 Factor. $x^2 + 17x + 30$ You may find that it is helpful to list all the factors whose product is 30 first. ■

EXAMPLE 3 Factor. $x^2 + 11x + 10$

Two numbers whose product is 10 but whose sum is 11 are the numbers 10 and 1.

$$x^2 + 11x + 10 = (x + 10)(x + 1)$$

Note: The order of the parenthetical factors is not important. You can write the answer as $(x + 1)(x + 10)$. This concept applies to all factoring problems. Remember, multiplication is commutative.

Practice Problem 3 Factor. $x^2 + 7x + 6$ ■

So far we have only factored trinomials of the form $x^2 + bx + c$, where b and c are positive numbers. The same procedure applies if b is a negative number and c is positive. Remember, if c is positive, the sign of both factors will *be the same.*

EXAMPLE 4 Factor. $x^2 - 8x + 15$

We want two numbers that have a product of $+15$ but a sum of -8. They must be negative numbers since the sign of the middle term is negative.

the sum of $-5 + (-3)$

$x^2 - 8x + 15 = (x - 5)(x - 3)$ *Think:* $(-5)(-3) = +15$ *and* $-5 + (-3) = -8$.

the product of $(-5)(-3)$

Multiply using FOIL to check.

Practice Problem 4 Factor. $x^2 - 11x + 18$ ■

EXAMPLE 5 Factor. $x^2 - 9x + 14$

We want two numbers whose product is 14 but whose sum is -9. The numbers are -7 and -2. So

$$x^2 - 9x + 14 = (x - 7)(x - 2) \quad \text{or} \quad (x - 2)(x - 7)$$

Practice Problem 5 Factor. $y^2 - 11y + 24$ ■

TEACHING TIP After explaining Example 5, ask students to factor the following four problems as an Additional Class Exercise.

A. Factor $x^2 - 13x + 36$
B. Factor $x^2 - 16x + 60$
C. Factor $y^2 - 16y + 63$
D. Factor $y^2 - 15y + 44$

Sol.

A. $(x - 4)(x - 9)$
B. $(x - 6)(x - 10)$
C. $(y - 7)(y - 9)$
D. $(y - 4)(y - 11)$

All the examples so far have had a positive last term. What happens when the last term is negative? If the last term is negative, we will need one positive number and one negative number. Why? The product of a positive and a negative number is negative.

EXAMPLE 6 Factor. $x^2 + 2x - 8$

We want two numbers whose product is -8 but whose sum is $+2$.

$x^2 + 2x - 8 = (x + 4)(x - 2)$ *Think:* $(+4)(-2) = -8$ *and* $+4 + (-2) = +2$.

Practice Problem 6 Factor. $x^2 + 5x - 6$ ■

EXAMPLE 7 Factor. $x^2 - 3x - 10$

We want two numbers whose product is -10 but whose sum is -3. The two numbers are -5 and $+2$.

$$x^2 - 3x - 10 = (x - 5)(x + 2)$$

Practice Problem 7 Factor. $a^2 - 5a - 24$ ■

What if we made a sign error and *incorrectly* factored the trinomial $x^2 - 3x - 10$ as $(x + 5)(x - 2)$? We could detect the error immediately since the sum of $+5$ and $-2 = 3$. We need a sum of -3!

EXAMPLE 8 Factor. $x^2 + 10x - 24$ Check your answer.

The two numbers whose product is -24 but whose sum is $+10$ are the numbers $+12$ and -2.

$$x^2 + 10x - 24 = (x + 12)(x - 2)$$

WARNING It is very easy to make a sign error in these problems. Make sure that you mentally multiply your answer back to obtain the original expression. Check each sign carefully.

Check: $(x + 12)(x - 2) = x^2 - 2x + 12x - 24 = x^2 + 10x - 24$ ✓

Practice Problem 8 Factor. $x^2 + 17x - 60$ Multiply your answer to check. ■

TEACHING TIP Remind students that if the last term is negative the two numbers they pick for the last term in each parentheses will be opposite in sign.

EXAMPLE 9 Factor. $x^2 - 16x - 36$

We want two numbers whose product is -36 and whose sum is -16.

List all the possible factors of 36 (without regard to sign). Find the pair that has a difference of 16. We are looking for a difference because the signs of the factors are different.

Factors of 36	*The Difference between the Factors*
36 and 1	35
18 and 2	16 ⟵ This is the value we want.
12 and 3	9
9 and 4	5
6 and 6	0

Once we have picked the pair of numbers (18 and 2), it is easy to find the signs. For the coefficient of the middle term to be -16, we will have to add the numbers $\underline{-18}$ and $\underline{+2}$.

$$x^2 - 16x - 36 = (x - 18)(x + 2)$$

TEACHING TIP Point out that the largest factor (the largest absolute value) gets the sign of the middle term in the trinomial.

Practice Problem 9 Factor. $x^2 - 7x - 60$ You may find it helpful to list the pairs of numbers whose product is 60. ■

At this point you should work several problems to develop your factoring skill. This is one section where you really need to drill by doing many problems.

Feel a little confused about the signs? If you do, you may find these facts helpful.

Facts about Factoring Trinomials of the Form $x^2 + bx + c$

The *two numbers* at the end of the parentheses will have the *same sign* if the last term is *positive.* $\quad x^2 + bx + c = (x \quad)(x \quad)$

1. They will both be *positive* if the *coefficient* of the *second* term is *positive.* $\quad x^2 + 5x + 6 = (x + 2)(x + 3)$
2. They will both be *negative* if the *coefficient* of the *second* term is *negative.* $\quad x^2 - 5x + 6 = (x - 2)(x - 3)$

The two numbers at the end of the parentheses will have *opposite signs* if the last term is *negative.*

1. The *larger* of the two numbers *without regard to sign* will be given a plus sign if the coefficient of the *second term* is *positive.* $\quad x^2 + 6x - 7 = (x + 7)(x - 1)$
2. The larger of the two numbers *without regard to sign* will be given a negative sign if the coefficient of the *second term* is *negative.* $\quad x^2 - 6x - 7 = (x - 7)(x + 1)$

Do not memorize these facts; rather, try to understand the pattern.

Sometimes the exponent of the first term will be greater than 2. If the exponent is an even power, it is a square. For example, $x^4 = (x^2)(x^2)$. Likewise, $x^6 = (x^3)(x^3)$.

TEACHING TIP After presenting Example 10 to your students, ask them to factor each of the following problems as an Additional Class Exercise.

A. Factor $x^2 + 14x + 48$

B. Factor $y^2 - 18y + 77$

C. Factor $x^4 - 6x^2 - 7$

Sol.

A. $(x + 6)(x + 8)$

B. $(y - 11)(y - 7)$

C. $(x^2 - 7)(x^2 + 1)$

EXAMPLE 10 Factor. $y^4 - 2y^2 - 35$

Think: $y^4 = (y^2)(y^2)$

$(y^2 \quad)(y^2 \quad)$ *Place y^2 in the parentheses. Next determine*

$(y^2 + \quad)(y^2 - \quad)$ *the signs. The last term of the polynomial is negative. The signs will be different. Now think of factors of 35 whose difference is 2.*

$(y^2 + 5)(y^2 - 7)$ *Check by multiplying.*

Practice Problem 10 Factor. $a^4 + a^2 - 42$ ■

2 *Factoring Polynomials That Have a Common Factor*

Some factoring problems require two steps. Often we must first remove a common factor from each term of the polynomial. Once this is done, we may find that the other factor is a trinomial that can be factored using the methods previously discussed in this section.

EXAMPLE 11 Factor. $2x^2 + 36x + 160$

$2(x^2 + 18x + 80)$ *First remove the common factor of 2 from each term of the polynomial.*

$2(x + 8)(x + 10)$ *Then factor the remaining polynomial.*

The final answer is $2(x + 8)(x + 10)$. *Be sure to list all parts of the answer.*

Check: $2(x + 8)(x + 10) = 2(x^2 + 18x + 80) = 2x^2 + 36x + 160$ ✓

Thus we are sure that the answer is $2(x + 8)(x + 10)$

Practice Problem 11 Factor. $3x^2 + 45x + 150$ ■

EXAMPLE 12 Factor. $3x^2 + 9x - 162$

$3(x^2 + 3x - 54)$ *First remove the common factor of 3 from each term of the polynomial.*

$3(x - 6)(x + 9)$ *Then factor the remaining polynomial.*

The final answer is $3(x - 6)(x + 9)$.

Check: $3(x - 6)(x + 9) = 3(x^2 + 3x - 54) = 3x^2 + 9x - 162$ ✓

Thus we are sure that the answer is $3(x - 6)(x + 9)$

Practice Problem 12 Factor. $4x^2 - 8x - 140$ ■

It is quite easy to forget to remove the greatest common factor as the first step of factoring a trinomial. Therefore, it is a good idea to look at your final answer in any factoring problem and ask yourself the question, ''Can I remove a common factor from any binomial contained inside a parentheses?'' Often you will be able to see a common factor at that point if you missed it in doing the first step of the problem.

5.3 Exercises

Verbal and Writing Skills

Fill in the blanks.

1. To factor $x^2 + 5x + 6$, find two numbers whose ___product___ is 6 but whose ___sum___ is 5.

2. To factor $x^2 + 5x - 6$, find two numbers whose ___product___ is -6 but whose ___sum___ is 5.

Factor.

3. $x^2 + 9x + 8$
$(x + 1)(x + 8)$

4. $x^2 + 8x + 15$
$(x + 5)(x + 3)$

5. $x^2 + 12x + 35$
$(x + 7)(x + 5)$

6. $x^2 + 11x + 24$
$(x + 8)(x + 3)$

7. $x^2 - 4x + 3$
$(x - 3)(x - 1)$

8. $x^2 - 6x + 8$
$(x - 2)(x - 4)$

9. $x^2 - 13x + 22$
$(x - 11)(x - 2)$

10. $x^2 - 15x + 14$
$(x - 14)(x - 1)$

11. $x^2 + x - 12$
$(x + 4)(x - 3)$

12. $x^2 + 2x - 8$
$(x + 4)(x - 2)$

13. $x^2 - 13x - 14$
$(x - 14)(x + 1)$

14. $x^2 - 6x - 16$
$(x - 8)(x + 2)$

15. $x^2 + x - 20$
$(x + 5)(x - 4)$

16. $x^2 + 7x - 18$
$(x + 9)(x - 2)$

17. $x^2 - 5x - 24$
$(x - 8)(x + 3)$

18. $x^2 - 11x - 26$
$(x - 13)(x + 2)$

Look over your answers to problems 3 through 18 carefully. Be sure that you are clear on your sign rules. Problems 19 through 50 contain a mixture of all the types of problems in this section. Make sure you can do them all. Check your answers by multiplication.

19. $x^2 - 7x + 10$
$(x - 2)(x - 5)$

20. $x^2 + 3x - 10$
$(x + 5)(x - 2)$

21. $x^2 - 5x - 14$
$(x - 7)(x + 2)$

22. $x^2 + 10x + 21$
$(x + 3)(x + 7)$

23. $x^2 + 5x - 14$
$(x + 7)(x - 2)$

24. $x^2 - 2x - 15$
$(x - 5)(x + 3)$

25. $x^2 - 9x + 20$
$(x - 4)(x - 5)$

26. $x^2 - 8x + 12$
$(x - 6)(x - 2)$

27. $x^2 + 17x + 30$
$(x + 15)(x + 2)$

28. $x^2 - 3x - 28$
$(x - 7)(x + 4)$

29. $y^2 - 4y - 5$
$(y - 5)(y + 1)$

30. $y^2 - 8y + 7$
$(y - 1)(y - 7)$

31. $a^2 + 6a - 16$
$(a + 8)(a - 2)$

32. $a^2 - 13a + 30$
$(a - 3)(a - 10)$

33. $x^2 - 12x + 32$
$(x - 4)(x - 8)$

34. $x^2 - 6x - 27$
$(x + 3)(x - 9)$

35. $x^2 + 4x - 21$
$(x + 7)(x - 3)$

36. $x^2 - 11x + 18$
$(x - 9)(x - 2)$

37. $x^2 + 13x + 40$
$(x + 5)(x + 8)$

38. $x^2 + 15x + 50$
$(x + 5)(x + 10)$

39. $x^2 - 21x - 22$
$(x - 22)(x + 1)$

40. $x^2 + 12x - 28$
$(x + 14)(x - 2)$

41. $x^2 + 9x - 36$
$(x + 12)(x - 3)$

42. $x^2 - 13x + 36$
$(x - 4)(x - 9)$

43. $x^2 - xy - 42y^2$
$(x - 7y)(x + 6y)$

44. $x^2 - xy - 56y^2$
$(x - 8y)(x + 7y)$

45. $x^2 - 16xy + 63y^2$
$(x - 7y)(x - 9y)$

46. $x^2 + 19xy + 48y^2$
$(x + 3y)(x + 16y)$

47. $x^4 - 3x^2 - 40$
$(x^2 - 8)(x^2 + 5)$

48. $x^4 + 11x^2 + 28$
$(x^2 + 4)(x^2 + 7)$

49. $x^4 - 12x^2 - 45$
$(x^2 - 15)(x^2 + 3)$

50. $x^4 - 13x^2 - 30$
$(x^2 - 15)(x^2 + 2)$

In Problems 51 through 62, first remove the greatest common factor from each term. Then factor the remaining polynomial. Refer to Examples 11 and 12.

51. $2x^2 - 12x + 16$
$2(x - 2)(x - 4)$

52. $2x^2 - 14x + 24$
$2(x - 4)(x - 3)$

53. $3x^2 - 6x - 72$
$3(x + 4)(x - 6)$

54. $3x^2 - 12x - 63$
$3(x + 3)(x - 7)$

55. $4x^2 + 24x + 20$
$4(x + 5)(x + 1)$

56. $4x^2 + 28x + 40$
$4(x + 2)(x + 5)$

57. $7x^2 + 21x - 70$
$7(x + 5)(x - 2)$

58. $7x^2 + 7x - 84$
$7(x + 4)(x - 3)$

59. $6x^2 + 18x + 12$
$6(x + 1)(x + 2)$

60. $6x^2 + 24x + 18$
$6(x + 1)(x + 3)$

61. $2x^2 - 32x + 126$
$2(x - 7)(x - 9)$

62. $2x^2 - 30x + 112$
$2(x - 8)(x - 7)$

Cumulative Review Problems

63. Solve for t. $A = P + Prt$
$t = \dfrac{A - P}{Pr}$

64. Solve for x. $2 - 3x \leq 7$ Give a graphical representation on the number line.

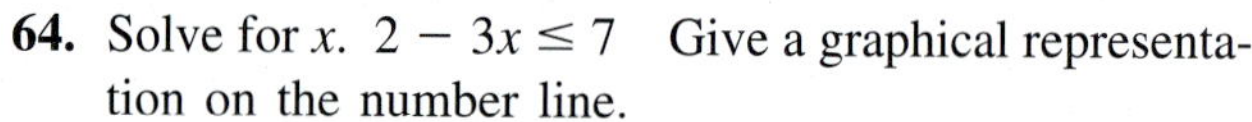

$x \geq -\dfrac{5}{3}$

65. A car travels from Watch Hill, Rhode Island, to Greenwich, Connecticut, in 2 hours, maintaining a constant speed. A train, traveling 20 mph faster, makes the trip in $1\frac{1}{2}$ hours. How far is it from Watch Hill to Greenwich? (*Hint:* Let $c =$ the car's speed. First find the speed of the car and the speed of the train. Then be sure to answer the question.) 120 miles

66. Kerri works as a radio advertising sales rep. She earns a guaranteed minimum salary of \$500 per month plus 3% commission on her sales. She wants to earn \$4400 or more this month. At least how much must she generate in sales? She must generate at least \$130,000 in sales.

The equation $T = 19 + 2M$ has been used by some meteorologists to predict the monthly average temperature for the small island of Menorca off the coast of Spain during the first 6 months of the year. The variable T represents the average monthly temperature measured in degrees Celsius. The variable M represents the number of months since January. Use this formula to answer the following questions.

67. What is the average temperature of Menorca during the month of April? 25°C

68. During what month will the average temperature be 29°C? June

5.4 Factoring Trinomials of the Form $ax^2 + bx + c$

After studying this section, you will be able to:

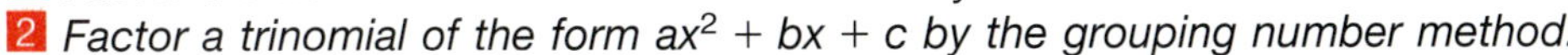

1. *Factor a trinomial of the form $ax^2 + bx + c$ by the trial and error method.*
2. *Factor a trinomial of the form $ax^2 + bx + c$ by the grouping number method.*
3. *Factor a trinomial of the form $ax^2 + bx + c$ after a common factor has been removed for each term.*

MathPro

Video 13

SSM

1 *Using the Trial and Error Method to Factor*

When the coefficient of the x^2 term is not 1, the trinomial is more difficult to factor. Several possibilities must be considered.

EXAMPLE 1 Factor. $2x^2 + 5x + 3$

To get the coefficient of the first term to be 2, the factors would be 2 and 1. $2x^2 + 5x + 3 = (2x \quad)(x \quad)$.

To get the last term to be 3, the factors would be 3 and 1.

Since all signs are positive, we know that each set of parentheses will contain only positive signs. However, we still have two possibilities. They are

$$(2x + 3)(x + 1)$$

$$(2x + 1)(x + 3)$$

We check them by multiplying by the FOIL method:

$(2x + 1)(x + 3) = 2x^2 + 7x + 3$ *Wrong middle term.*

$(2x + 3)(x + 1) = 2x^2 + 5x + 3$ *Right middle term.*

Thus the correct answer is

$$(2x + 3)(x + 1) \quad \text{or} \quad (x + 1)(2x + 3)$$

Practice Problem 1 Factor. $2x^2 - 7x + 5$ ■

TEACHING TIP Except for those students who remember quite well how to factor trinomials of the general form $ax^2 + bx + c$, most students greatly prefer to use the grouping method, which begins with Example 5. A few students will still prefer the old trial and error method. Here is a suggested way to approach the lesson:

1. Explain how to factor Example 1 by the trial and error method.
2. Ask students to factor $4x^2 - 7x - 15$ by the trial and error method. (Give the class about 90 seconds to try to do it.)
3. Show them the correct answer $(4x + 5)(x - 3)$.
4. Ask how many students obtained the correct answer? (Usually about 20% of the class or less.)
5. Ask the class how many students would like to see a method that most students feel is shorter and easier? (Usually 80% of the class will answer yes.)
6. Go directly to Examples 5, 6, and 7 and show the class how to do the grouping method.

Some problems have many more possibilities.

EXAMPLE 2 Factor. $4x^2 - 13x + 3$

The Different Factors of 4 Are:	*The Factors of 3 Are:*
2 and 2	1 and 3
1 and 4	

Let us list the possible factoring combinations and compute the middle term by the FOIL method. Note that both signs will be negative. Why?

Possible Factors	*Middle Term*	*Correct?*
$(2x - 3)(2x - 1)$	$-8x$	No
$(4x - 3)(x - 1)$	$-7x$	No
$(4x - 1)(x - 3)$	$-13x$	Yes

The correct answer is $(4x - 1)(x - 3)$ or $(x - 3)(4x - 1)$

This method is called the **trial-and-error method.**

Practice Problem 2 Factor. $9x^2 - 64x + 7$ ■

TEACHING TIP Remind students that, when factoring problems of the type $ax^2 + bx + c$, they should ALWAYS CHECK by multiplying the factors out to see if the original trinomial is obtained.

EXAMPLE 3 Factor. $6x^2 + x - 5$

Let's consider the last number in each set of parentheses. Because -5 is negative, the two numbers will have opposite signs. Since the last number can be either positive or negative, we have more possibilities. In this case:

Factors of 6	*Factors of 5*
3 and 2	1 and 5
6 and 1	

Possible Factors	*Middle Term*	*Correct?*
$(3x + 1)(2x - 5)$	$-13x$	No
$(3x - 1)(2x + 5)$	$+13x$	No
$(3x + 5)(2x - 1)$	$+7x$	No
$(3x - 5)(2x + 1)$	$-7x$	No
$(6x + 1)(x - 5)$	$-29x$	No
$(6x - 1)(x + 5)$	$+29x$	No
$(6x + 5)(x - 1)$	$-x$	No
$(6x - 5)(x + 1)$	$+x$	Yes

The correct answer is

$$(6x - 5)(x + 1) \quad \text{or} \quad (x + 1)(6x - 5)$$

Practice Problem 3 Factor. $3x^2 + 4x - 4$ ■

Look back to Example 3. Let's see if we can shorten our work. Notice that there are four pairs of possible factors. The only difference between the two parentheses in the pair is that the last signs in each set of parentheses are reversed. This suggests that we list only half as many possibilities. Then, if the coefficient of the middle term is the opposite in sign of what we want, we can just reverse signs. Let's try this shorter method.

EXAMPLE 4 Factor. $3x^2 - 2x - 8$

Factors of 3	*Factors of 8*
3 and 1	8 and 1
	4 and 2

Let us list only one-half of the possibilities. The last term of that first set of parentheses will always be positive.

Possible Factors	*Middle Term*	*Correct Factors?*
$(x + 8)(3x - 1)$	$+23x$	No
$(x + 1)(3x - 8)$	$-5x$	No
$(x + 4)(3x - 2)$	$+10x$	No
$(x + 2)(3x - 4)$	$+2x$	No (but it is exactly opposite in sign from what is needed)

So we just *reverse* the signs of the last term in each set of parentheses.

	Middle Term	***Correct Factor?***
$(x - 2)(3x + 4)$	$-2x$	Yes

The correct answer is

$$(x - 2)(3x + 4) \quad \text{or} \quad (3x + 4)(x - 2)$$

Practice Problem 4 Factor. $3x^2 - x - 14$ ■

It takes a good deal of practice to readily factor problems of this type. The more problems you do, the more proficient you will become. The following method will help you factor more quickly.

2 *Using the Grouping Number Method for Factoring*

One way to factor a trinomial of the form $ax^2 + bx + c$ is to write it with four terms and factor by grouping, as we did in Section 5.2. For example, the trinomial $2x^2 + 13x + 20$ can be written as $2x^2 + 5x + 8x + 20$. Using the methods of Section 5.2, we factor

$$\begin{aligned} 2x^2 + 5x + 8x + 20 &= x(2x + 5) + 4(2x + 5) \\ &= (2x + 5)(x + 4) \end{aligned}$$

We can factor all factorable trinomials of the form $ax^2 + bx + c$ in this way. We will use the following procedure:

> **Grouping Number Method for Factoring Trinomials of the Form $ax^2 + bx + c$**
>
> **1.** Obtain the grouping number ac.
>
> **2.** Find the factors of the grouping number whose sum is b.
>
> **3.** Use those factors to write bx as the sum of two terms.
>
> **4.** Factor by grouping.
>
> **5.** Multiply to check.

Let's try the same problem we did in Example 1.

EXAMPLE 5 Factor by grouping. $2x^2 + 5x + 3$

1. The grouping number is $(2)(3) = 6$.

2. The factors of 6 are $6 \cdot 1$ and $3 \cdot 2$. *We chose the factors whose sum is 5.*

3. We write $5x$ as the sum $3x + 2x$.

4. Factor by grouping.

$$\begin{aligned} 2x^2 + 5x + 3 &= 2x^2 + 2x + 3x + 3 \\ &= 2x(x + 1) + 3(x + 1) \\ &= (x + 1)(2x + 3) \end{aligned}$$

5. Multiply to check.

$$\begin{aligned} (x + 1)(2x + 3) &= 2x^2 + 3x + 2x + 3 \\ &= 2x^2 + 5x + 3 \ \checkmark \end{aligned}$$

Practice Problem 5 Factor by grouping. $2x^2 - 7x + 5$ ■

EXAMPLE 6 Factor by grouping. $4x^2 - 13x + 3$

1. The grouping number is $(4)(3) = 12$.
2. The factors of 12 are $(12)(1)$ or $(4)(3)$ or $(6)(2)$. We choose the factors whose sum is 13. Both factors receive the sign of the middle term in the trinomial.
3. We write $-13x$ as the sum $-12x + (-1x)$.
4. Factor by grouping.

$$4x^2 - 13x + 3 = 4x^2 - 12x - 1x + 3$$

$$= 4x(x - 3) - 1(x - 3)$$ *Remember to factor out a -1 from the last two terms so that both parentheses contain the same expression.*

$$= (x - 3)(4x - 1)$$

Practice Problem 6 Factor by grouping. $9x^2 - 64x + 7$ ■

EXAMPLE 7 Factor by grouping. $3x^2 - 2x - 8$

1. The grouping number is $(3)(-8) = -24$.
2. Since the last sign is negative we want two numbers whose product is 24 but whose difference is 2. They are 6, 4. The largest gets the middle sign; so we have -6 and $+4$.
3. We write $-2x$ as a sum $-6x + 4x$.
4. Factor by grouping.

$$3x^2 - 6x + 4x - 8 = 3x(x - 2) + 4(x - 2)$$

$$= (x - 2)(3x + 4)$$

Practice Problem 7 Factor by grouping. $3x^2 + 4x - 4$ ■

Use the method that works best for you, either trial and error or grouping.

3 *Factoring Polynomials That Have a Common Factor*

Some problems require first removing a common factor and then factoring the trinomial by one of the two methods of this section.

EXAMPLE 8 Factor. $9x^2 + 3x - 30$

$$9x^2 + 3x - 30 = 3(3x^2 + 1x - 10)$$ *We first remove the common factor of 3 from each term of the trinomial.*

$$= 3(3x - 5)(x + 2)$$ *We then factor the trinomial by the grouping method or by the trial and error method.*

Practice Problem 8 Factor. $8x^2 - 8x - 6$ ■

EXAMPLE 9 Factor. $32x^2 - 40x + 12$

$$32x^2 - 40x + 12 = 4(8x^2 - 10x + 3)$$ *We first remove the greatest common factor of 4 from each term of the trinomial.*

$$= 4(2x - 1)(4x - 3)$$ *We then factor the trinomial by the grouping method or by the trial and error method.*

Practice Problem 9 Factor. $24x^2 - 38x + 10$ ■

5.4 Exercises

Factor. Check your answers for Problems 1 through 20 using the FOIL method.

1. $3x^2 + 7x + 2$
$(3x + 1)(x + 2)$

2. $2x^2 + 7x + 3$
$(2x + 1)(x + 3)$

3. $2x^2 - 5x + 2$
$(2x - 1)(x - 2)$

4. $3x^2 - 8x + 4$
$(3x - 2)(x - 2)$

5. $3x^2 - 4x - 7$
$(3x - 7)(x + 1)$

6. $5x^2 + 7x - 6$
$(5x - 3)(x + 2)$

7. $2x^2 - 5x - 3$
$(2x + 1)(x - 3)$

8. $2x^2 - x - 6$
$(2x + 3)(x - 2)$

9. $5x^2 + 3x - 2$
$(5x - 2)(x + 1)$

10. $6x^2 + x - 2$
$(3x + 2)(2x - 1)$

11. $6x^2 - 13x + 6$
$(3x - 2)(2x - 3)$

12. $6x^2 - 7x + 2$
$(2x - 1)(3x - 2)$

13. $2x^2 + 3x - 20$
$(2x - 5)(x + 4)$

14. $6x^2 + 11x - 10$
$(3x - 2)(2x + 5)$

15. $9x^2 + 9x + 2$
$(3x + 1)(3x + 2)$

16. $4x^2 + 11x + 6$
$(4x + 3)(x + 2)$

17. $6x^2 - 5x - 6$
$(3x + 2)(2x - 3)$

18. $3x^2 - 13x - 10$
$(3x + 2)(x - 5)$

19. $6x^2 - 19x + 10$
$(3x - 2)(2x - 5)$

20. $10x^2 - 19x + 6$
$(2x - 3)(5x - 2)$

21. $4x^2 + 16x - 9$
$(2x - 1)(2x + 9)$

22. $8x^2 + x - 9$
$(8x + 9)(x - 1)$

23. $9y^2 - 13y + 4$
$(9y - 4)(y - 1)$

24. $5y^2 - 11y + 2$
$(5y - 1)(y - 2)$

25. $5a^2 - 13a - 6$
$(5a + 2)(a - 3)$

26. $3a^2 - 10a - 8$
$(3a + 2)(a - 4)$

27. $12x^2 - 20x + 3$
$(6x - 1)(2x - 3)$

28. $9x^2 + 5x - 4$
$(9x - 4)(x + 1)$

29. $15x^2 + 4x - 4$
$(5x - 2)(3x + 2)$

30. $8x^2 - 11x + 3$
$(8x - 3)(x - 1)$

31. $10x^2 + 21x + 9$
$(5x + 3)(2x + 3)$

32. $12x^2 + 11x + 2$
$(4x + 1)(3x + 2)$

33. $12x^2 - 16x - 3$
$(6x + 1)(2x - 3)$

34. $12x^2 + x - 6$
$(4x + 3)(3x - 2)$

35. $2x^4 + 15x^2 - 8$
$(2x^2 - 1)(x^2 + 8)$

36. $4x^4 - 11x^2 - 3$
$(4x^2 + 1)(x^2 - 3)$

37. $4x^2 + 8xy - 5y^2$
$(2x - y)(2x + 5y)$

38. $3x^2 + 8xy + 4y^2$
$(3x + 2y)(x + 2y)$

39. $5x^2 + 16xy - 16y^2$
$(5x - 4y)(x + 4y)$

40. $12x^2 + 11xy - 5y^2$
$(3x - y)(4x + 5y)$

Factor each of the following by first removing the greatest common factor. See Examples 8 and 9.

41. $10x^2 + 22x + 12$
$2(5x + 6)(x + 1)$

42. $4x^2 + 34x + 42$
$2(2x + 3)(x + 7)$

43. $12x^2 - 24x + 9$
$3(2x - 1)(2x - 3)$

44. $8x^2 - 26x + 6$
$2(4x - 1)(x - 3)$

45. $18x^2 + 39x - 15$
$3(3x - 1)(2x + 5)$

46. $12x^2 + 12x - 45$
$3(2x - 3)(2x + 5)$

47. $6x^3 - 16x^2 - 6x$
$2x(3x + 1)(x - 3)$

48. $6x^3 + 9x^2 - 60x$
$3x(2x - 5)(x + 4)$

Factor the following.

★49. $12x^2 + 16x - 35$
$(2x + 5)(6x - 7)$

★50. $20x^2 - 53x + 18$
$(4x - 9)(5x - 2)$

★51. $20x^2 - 27x + 9$
$(4x - 3)(5x - 3)$

★52. $12x^2 - 29x + 15$
$(3x - 5)(4x - 3)$

Cumulative Review Problems

53. Solve. $7x - 3(4 - 2x) = 2(x - 3) - (5 - x)$
$x = \frac{1}{10}$

54. Factor. $x^2 + 5x - 36$ $(x + 9)(x - 4)$

55. David Damato used the Bay State East calling plan on his telephone this year. His bill for local calls in February was \$59.12. Under this plan you pay \$30 per month for 120 minutes of local calls. Then you are charged \$0.052 per minute for each additional minute beyond the 120 minutes. How many minutes of local calls did David make in February? 680 minutes

56. In the school year 1992–1993 there were 31.1 million children enrolled in the United States in kindergarten through grade 8. This represented an increase of 65% in the number of enrolled children since the school year 1939–1940. How many children were enrolled in grades K–8 during 1939–1940? (Round your answer to the nearest tenth of a million.) 18.8 million children

5.5 Special Cases of Factoring

MathPro

Video 13

SSM

After studying this section, you will be able to:

1. *Recognize and factor problems of the type $a^2 - b^2$.*
2. *Recognize and factor problems of the type $a^2 + 2ab + b^2$.*
3. *Factor problems that require removing a common factor and then using the special case formula.*

As we proceed in this section you will be able to reduce the time it takes you to factor a problem by quickly recognizing and factoring two special types of factoring problems. By becoming very familiar with the appropriate formula, you will be able to factor the difference of two squares and perfect square trinomials.

1 *Factoring the Difference of Two Squares*

Recall the formula from Section 4.5:

$$(a + b)(a - b) = a^2 - b^2$$

In reverse form we can use it for factoring.

Difference of Two Squares

$$a^2 - b^2 = (a + b)(a - b)$$

We can state it in words in this way: ''The difference of two squares can be factored into the sum and difference of those values that were squared.''

EXAMPLE 1 Factor. $9x^2 - 1$

We see that the problem is in the form of the difference of two squares. $9x^2$ is a square and 1 is a square. So using the formula we can write

$$9x^2 - 1 = (3x + 1)(3x - 1) \qquad \textit{Because } 9x^2 = (3x)^2 \textit{ and } 1 = (1)^2.$$

Practice Problem 1 Factor. $1 - 64x^2$ ■

EXAMPLE 2 Factor. $25x^2 - 16$

We see that by using the formula for the difference of squares, we can write

$$25x^2 - 16 = (5x + 4)(5x - 4) \qquad \textit{Because } 25x^2 = (5x)^2 \textit{ and } 16 = (4)^2.$$

Practice Problem 2 Factor. $36x^2 - 49$ ■

Sometimes the problem will contain two variables.

EXAMPLE 3 Factor. $4x^2 - 49y^2$

We see that

$$4x^2 - 49y^2 = (2x + 7y)(2x - 7y)$$

Practice Problem 3 Factor. $100x^2 - 81y^2$ ■

TEACHING TIP After you have shown students how to factor Examples 2 and 3, ask them to factor the following four problems as an Additional Class Exercise. Factor.

A. $36x^2 - 1$
B. $1 - 100y^2$
C. $25a^2 - 64b^2$
D. $121y^2 - 4$

Sol.

A. $(6x + 1)(6x - 1)$
B. $(1 + 10y)(1 - 10y)$
C. $(5a + 8b)(5a - 8b)$
D. $(11y + 2)(11y - 2)$

Some problems may involve more than one step.

EXAMPLE 4 Factor. $81x^4 - 1$

We see that

$$81x^4 - 1 = (9x^2 + 1)(9x^2 - 1) \qquad \textit{Because } 81x^4 = (9x^2)^2 \textit{ and } 1 = (1)^2.$$

Is the factoring complete? We can factor $9x^2 - 1$.

$$81x^4 - 1 = (9x^2 + 1)(3x - 1)(3x + 1) \qquad \textit{Because } (9x^2 - 1) = (3x - 1)(3x + 1).$$

Practice Problem 4 Factor. $x^8 - 1$ ■

2 *Factoring Perfect Square Trinomials*

There is a formula that will help us to factor very quickly certain trinomials, called *perfect square trinomials.* Recall the formula for a binomial squared from Section 4.5:

$$(a + b)^2 = a^2 + 2ab + b^2$$

$$(a - b)^2 = a^2 - 2ab + b^2$$

In reverse form we can use these two equations for factoring.

Perfect Square Trinomials

$$a^2 + 2ab + b^2 = (a + b)^2$$

$$a^2 - 2ab + b^2 = (a - b)^2$$

TEACHING TIP The key to recognizing the perfect square trinomial is determining if the first and last terms of a given trinomial are perfect squares. Spend some time emphasizing to the class the definition of perfect square and the list of such values.

A perfect square trinomial is a trinomial that is the result of squaring a binomial. How can we recognize a perfect square trinomial?

1. The first and last terms are *perfect squares.*
2. The middle term is twice the product of the values whose squares are the first and last terms.

EXAMPLE 5 Factor. $x^2 + 6x + 9$

This *is* a perfect square trinomial.

1. The first and last terms are perfect squares because $x^2 = (x)^2$ and $9 = (3)^2$.
2. The middle term is twice the product of x and 3.

Since $x^2 + 6x + 9$ is a perfect square trinomial, we can use the formula

$$a^2 + 2ab + b^2 = (a + b)^2$$

and we have

$$x^2 + 6x + 9 = (x + 3)^2$$

Practice Problem 5 Factor. $16x^2 + 8x + 1$ ■

EXAMPLE 6 Factor. $4x^2 - 20x + 25$

This *is* a perfect square trinomial. $20x = 2(2x \cdot 5)$ Note the negative sign. We have

$$4x^2 - 20x + 25 = (2x - 5)^2 \qquad \textit{Since } a^2 - 2ab + b^2 = (a - b)^2.$$

Practice Problem 6 Factor. $25x^2 - 30x + 9$ ■

More than one variable may be involved and the exponents may be higher than 2. The same principles apply.

EXAMPLE 7 Factor. **(a)** $49x^2 + 42xy + 9y^2$ **(b)** $36x^4 - 12x^2 + 1$

(a) This is a perfect square trinomial. Why?

$49x^2 + 42xy + 9y^2 = (7x + 3y)^2$ *Because* $49x^2 = (7x)^2$ *and* $9y^2 = (3y)^2$. *Also,* $42xy = 2(7x \cdot 3y)$.

(b) This is a perfect square trinomial. Why?

$36x^4 - 12x^2 + 1 = (6x^2 - 1)^2$ *Because* $36x^4 = (6x^2)^2$ *and* $1 = (1)^2$. *Also,* $12x^2 = 2(6x^2 \cdot 1)$.

Practice Problem 7 Factor. **(a)** $25x^2 - 60xy + 36y^2$ **(b)** $64x^6 - 48x^3 + 9$ ■

Some problems appear to be perfect square trinomials but are not. They were factored as other trinomials in Section 5.4.

EXAMPLE 8 Factor. $49x^2 + 35x + 4$

This is *not* a perfect square trinomial! Although the first and last terms are perfect squares since $(7x)^2 = 49x^2$ and $(2)^2 = 4$, the middle term is not double the product of 2 and $7x$! $35x \neq 28x$! So we must factor by trial and error or by grouping to obtain

$$49x^2 + 35x + 4 = (7x + 4)(7x + 1)$$

Practice Problem 8 Factor. $9x^2 - 15x + 4$ ■

3 *Factoring Polynomials by First Removing a Common Factor and Then Using One of the Special Factoring Formulas*

In some cases, we will first remove a common factor. Then we will find an opportunity to use the difference of two squares formula or one of the perfect square trinomial formulas.

EXAMPLE 9 Factor. $12x^2 - 48$

$12x^2 - 48 = 12(x^2 - 4)$ *First we remove the greatest common factor* 12.

$= 12(x + 2)(x - 2)$ *Then we use the difference of two squares formula:* $a^2 - b^2 = (a + b)(a - b)$.

Practice Problem 9 Factor. $20x^2 - 45$ ■

EXAMPLE 10 Factor. $24x^2 + 72x + 54$

$24x^2 + 72x + 54 = 6(4x^2 + 12x + 9)$ *First we remove the greatest common factor* 6.

$= 6(2x + 3)^2$ *Then we use the perfect square trinomial formula:* $a^2 + 2ab + b^2 = (a + b)^2$.

Practice Problem 10 Factor. $75x^2 - 60x + 12$ ■

5.5 Exercises

Factor.

1. $9x^2 - 16$
$(3x + 4)(3x - 4)$

2. $36x^2 - 1$
$(6x + 1)(6x - 1)$

3. $16 - 9x^2$
$(4 - 3x)(4 + 3x)$

4. $49 - 25x^2$
$(7 - 5x)(7 + 5x)$

5. $25x^2 - 16$
$(5x - 4)(5x + 4)$

6. $81x^2 - 1$
$(9x - 1)(9x + 1)$

7. $4x^2 - 25$
$(2x - 5)(2x + 5)$

8. $16x^2 - 25$
$(4x - 5)(4x + 5)$

9. $36x^2 - 25$
$(6x - 5)(6x + 5)$

10. $1 - 25x^2$
$(1 - 5x)(1 + 5x)$

11. $1 - 49x^2$
$(1 - 7x)(1 + 7x)$

12. $1 - 36x^2$
$(1 - 6x)(1 + 6x)$

13. $81x^2 - 49y^2$
$(9x - 7y)(9x + 7y)$

14. $16x^4 - y^4$
$(4x^2 + y^2)(2x + y)(2x - y)$

15. $25 - 121x^2$
$(5 + 11x)(5 - 11x)$

16. $9x^2 - 49$
$(3x + 7)(3x - 7)$

17. $81x^2 - 100y^2$
$(9x + 10y)(9x - 10y)$

18. $25x^4 - 16y^4$
$(5x^2 + 4y^2)(5x^2 - 4y^2)$

19. $25a^2 - 1$
$(5a + 1)(5a - 1)$

20. $36a^2 - 49$
$(6a + 7)(6a - 7)$

21. $81x^4 - 4y^2$
$(9x^2 - 2y)(9x^2 + 2y)$

22. $4x^2 - 49y^2$
$(2x + 7y)(2x - 7y)$

23. $49x^2 + 14x + 1$
$(7x + 1)^2$

24. $36x^2 + 12x + 1$
$(6x + 1)^2$

25. $y^2 - 6y + 9$
$(y - 3)^2$

26. $y^2 - 8y + 16$
$(y - 4)^2$

27. $9x^2 - 24x + 16$
$(3x - 4)^2$

28. $4x^2 + 20x + 25$
$(2x + 5)^2$

29. $49x^2 + 28x + 4$
$(7x + 2)^2$

30. $25x^2 + 30x + 9$
$(5x + 3)^2$

31. $x^2 + 10x + 25$
$(x + 5)^2$

32. $x^2 + 12x + 36$
$(x + 6)^2$

33. $49 - 70x + 25x^2$
$(7 - 5x)^2$

34. $36x^2 - 60x + 25$
$(6x - 5)^2$

35. $25x^2 - 40x + 16$
$(5x - 4)^2$

36. $49x^2 - 42x + 9$
$(7x - 3)^2$

37. $81x^2 + 36xy + 4y^2$
$(9x + 2y)^2$

38. $36x^2 + 60xy + 25y^2$
$(6x + 5y)^2$

39. $9x^2 - 42xy + 49y^2$
$(3x - 7y)^2$

40. $49x^2 - 70xy + 25y^2$
$(7x - 5y)^2$

41. $16a^2 + 72ab + 81b^2$
$(4a + 9b)^2$

42. $169a^2 + 26ab + b^2$
$(13a + b)^2$

43. $9x^4 - 6x^2y + y^2$
$(3x^2 - y)^2$

44. $y^4 - 22y^2 + 121$
$(y^2 - 11)^2$

45. $49x^2 + 70x + 9$
$(7x + 1)(7x + 9)$

46. $25x^2 - 50x + 16$
$(5x - 2)(5x - 8)$

47. $16x^4 - 1$
$(4x^2 + 1)(2x + 1)(2x - 1)$

48. $x^8 - 1$
$(x^4 + 1)(x^2 + 1)(x + 1)(x - 1)$

49. $x^{10} - 36y^{10}$
$(x^5 + 6y^5)(x^5 - 6y^5)$

50. $x^4 - 49y^6$
$(x^2 + 7y^3)(x^2 - 7y^3)$

51. $4x^8 + 12x^4 + 9$
$(2x^4 + 3)^2$

52. $4x^6 + 4x^3 + 1$
$(2x^3 + 1)^2$

To Think About

53. When you factor Example 4 as $81x^4 - 1 = (9x^2 + 1)(9x^2 - 1)$, you can then factor $9x^2 - 1 = (3x + 1)(3x - 1)$. Show why you cannot factor $9x^2 + 1$. Because no matter what combination you try, you cannot multiply two binomials to equal $9x^2 + 1$.

54. What two numbers could replace the b in $25x^2 + bx + 16$ so that the resulting trinomial would be a perfect square? (*Hint:* One number is negative.) ± 40

55. What value could you give to c so that $16y^2 - 56y + c$ would become a perfect square trinomial? Is there only one answer or more than one? There is one answer. It is 49.

56. Jerome says that he can find two values of b so that $100x^2 + bx - 9$ will be a perfect square. Kesha says there is only one that fits, and Larry says there are none. Who is correct, and why? Larry; the last term must be positive in order to have a perfect square.

Factor by first removing the greatest common factor. See Examples 9 and 10.

57. $27x^2 - 3$
$3(3x - 1)(3x + 1)$

58. $32x^2 - 2$
$2(4x - 1)(4x + 1)$

59. $147x^2 - 3y^2$
$3(7x - y)(7x + y)$

60. $16y^2 - 100x^2$
$4(2y - 5x)(2y + 5x)$

61. $12x^2 - 36x + 27$
$3(2x - 3)^2$

62. $125x^2 - 100x + 20$
$5(5x - 2)^2$

63. $98x^2 + 84x + 18$
$2(7x + 3)^2$

64. $128x^2 + 96x + 18$
$2(8x + 3)^2$

Mixed Practice

Factor. Be sure to remove common factors first.

65. $x^2 - 9x + 14$
$(x - 2)(x - 7)$

66. $x^2 - 9x - 36$
$(x - 12)(x + 3)$

67. $2x^2 + 5x - 3$
$(2x - 1)(x + 3)$

68. $15x^2 - 11x + 2$
$(3x - 1)(5x - 2)$

69. $16x^2 - 121$
$(4x - 11)(4x + 11)$

70. $9x^2 - 100y^2$
$(3x - 10y)(3x + 10y)$

71. $9x^2 + 42x + 49$
$(3x + 7)^2$

72. $9x^2 + 30x + 25$
$(3x + 5)^2$

73. $3x^2 + 6x - 45$
$3(x + 5)(x - 3)$

74. $4x^2 + 24x + 32$
$4(x + 4)(x + 2)$

75. $7x^2 - 63$
$7(x - 3)(x + 3)$

76. $11x^2 - 44$
$11(x - 2)(x + 2)$

77. $5x^2 + 20x + 20$
$5(x + 2)^2$

78. $8x^2 + 48x + 72$
$8(x + 3)^2$

79. $2x^2 - 32x + 126$
$2(x - 9)(x - 7)$

80. $2x^2 - 32x + 110$
$2(x - 11)(x - 5)$

Cumulative Review Problems

81. Multiply. $(x + 4)(2x + 3)(x - 2)$
$2x^3 + 7x^2 - 10x - 24$

82. Solve for x. $7 - 11x < -(13 + 3x)$
$x > 2.5$

83. Divide. $(x^3 + x^2 - 2x - 11) \div (x - 2)$
$x^2 + 3x + 4 + \dfrac{-3}{x - 2}$

84. Divide. $(6x^3 + 11x^2 - 11x - 20) \div (3x + 4)$
$2x^2 + x - 5$

The peak of Mount Washington is at an altitude of 6288 feet above sea level. The altitude in feet (A) of a car driving down the mountain road a distance measured in miles (M) is given by $A = 6288 - 700M$.

85. What is the altitude of a car that has driven from the mountain peak down a distance of 3.5 miles?
3838 feet above sea level

86 A car drives from the peak down the mountain road to a point where the altitude is 2788 feet above sea level. How many miles down the road has the car driven?
5 miles

5.6 A Brief Review of Factoring

After studying this section, you will be able to:

1 *Identify and factor any polynomial that can be factored.*
2 *Determine if a polynomial is prime.*

1 *Review of Factoring*

Often the various types of factoring problems are all mixed together. We need to be able to identify each type of factoring problem quickly. The following table may be helpful in summarizing your knowledge of factoring.

FACTORING ORGANIZER

Number of Terms	Identifying Name and/or Formula	Example
A. Any number of terms	**Common factor** Each term has a common factor consisting of a number, a variable, or both.	$2x^2 - 16x = 2x(x - 8)$ $3x^2 + 9y - 12 = 3(x^2 + 3y - 4)$ $4x^2y + 2xy^2 - wxy + xyz = xy(4x + 2y - w + z)$
B. Two terms	**Difference of two squares** First and last terms are perfect squares. $a^2 - b^2 = (a + b)(a - b)$	$16x^2 - 1 = (4x + 1)(4x - 1)$ $25y^2 - 9x^2 = (5y + 3x)(5y - 3x)$
C. Three terms	**Perfect square trinomial** First and last terms are perfect squares. $a^2 + 2ab + b^2 = (a + b)^2$ $a^2 - 2ab + b^2 = (a - b)^2$	$25x^2 - 10x + 1 = (5x - 1)^2$ $16x^2 + 24x + 9 = (4x + 3)^2$
D. Three terms	**Trinomial of the form $x^2 + bx + c$** It starts with x^2. The two numbers at the end of each parentheses are numbers whose product is c but whose sum is b.	$x^2 - 7x + 12 = (x - 3)(x - 4)$ $x^2 + 11x - 26 = (x + 13)(x - 2)$ $x^2 - 8x - 20 = (x - 10)(x + 2)$
E. Three terms	**Trinomial of the form $ax^2 + bx + c$** It starts with ax^2, where a is any number but 1.	Use trial and error or the grouping number method. Factor. $12x^2 - 5x - 2$ 1. The grouping number is -24. 2. Find two factors whose sum is -5. -8 and 3 3. $12x^2 - 5x - 2 = 12x^2 + 3x - 8x - 2$ $= 3x(4x + 1) - 2(4x + 1)$ $= (4x + 1)(3x - 2)$
F. Four terms	**Factor by grouping** Rearrange the order if the first two terms do not have a common factor.	Factor. $wx - 6yz + 2wy - 3xz$ $= wx + 2wy - 3xz - 6yz$ $= w(x + 2y) - 3z(x + 2y)$ $= (x + 2y)(w - 3z)$

TEACHING TIP After discussing Example 1, ask students to do the following problems as an Additional Class Exercise.
Factor completely. (BE SURE TO REMOVE THE COMMON FACTOR FIRST!)

A. $3ax + 9bx - 12aw - 36bw$

B. $x^3 + 2x^2y - 15xy^2$

C. $8 + 7x - x^2$

Sol.

A. $3[ax + 3bx - 4aw - 12bw]$
$= 3[x(a + 3b) - 4w(a + 3b)]$
$= 3(a + 3b)(x - 4w)$

B. $x(x^2 + 2xy - 15y^2)$
$= x(x + 5y)(x - 3y)$

C. $-1(x^2 - 7x - 8)$
$= -1(x - 8)(x + 1)$ or the answer may be $(8 - x)(x + 1)$

Many problems involve more than one step when factoring. When you are asked to factor a polynomial, it is expected that you will factor it completely. Usually, the first step of factoring is removing the common factor, then the next step will be more apparent.

EXAMPLE 1 Factor.

(a) $25x^3 - 10x^2 + x$ **(b)** $20x^2y^2 - 45y^2$

(c) $2ax + 4ay + 4x + 8y$ **(d)** $15x^2 - 3x^3 + 18x$

(a) $25x^3 - 10x^2 + x = x(25x^2 - 10x + 1)$ *Remove the common factor of x. The other factor is a perfect square trinomial.*

$= x(5x - 1)^2$

(b) $20x^2y^2 - 45y^2 = 5y^2(4x^2 - 9)$ *Remove the common factor of $5y^2$. The other factor is the difference of squares.*

$= 5y^2(2x - 3)(2x + 3)$

(c) $2ax + 4ay + 4x + 8y = 2[ax + 2ay + 2x + 4y]$ *Remove the common factor of 2.*

$= 2[a(x + 2y) + 2(x + 2y)]$ *Factor the terms inside the bracket by the grouping method.*

$= 2[(x + 2y)(a + 2)]$ *Remove the common factor of $(x + 2y)$.*

(d) $15x^2 - 3x^3 + 18x = -3x^3 + 15x^2 + 18x$ *Rearrange the terms in descending order of powers of x.*

$= -3x(x^2 - 5x - 6)$ *Remove the common factor of $-3x$.*

$= -3x(x - 6)(x + 1)$ *Factor the trinomial.*

Practice Problem 1 Factor. Be careful. These practice problems are mixed.

(a) $9x^4y^2 - 9y^2$ **(b)** $12x - 9 - 4x^2$ **(c)** $3x^2 - 36x + 108$

(d) $5x^3 - 15x^2y + 10x^2 - 30xy$ ■

2 *Determining If a Polynomial Is Prime*

Not all polynomials can be factored using the methods in this chapter. If we cannot factor a polynomial by elementary methods, we will identify it as a **prime** polynomial. When a polynomial cannot be factored by our previously learned techniques, you will be comfortable in saying "It cannot be done so it is prime," rather than "I can't do it—I give up!"

TEACHING TIP Prime problems need some special emphasis. After explaining Examples 2 and 3, put the following three problems on the board as an Additional Class Exercise. Ask students why each of them is PRIME. Usually they will have some good discussion and learn quite a bit as they listen to their classmates on each problem. Factor if possible.

A. $x^2 + 36$ (The sum of two squares cannot be factored, only the difference of two squares.)

B. $x^2 - 7x + 14$ (There are no factors of 14 whose sum is -7.)

C. $2x^2 + x + 6$ (Using the grouping number method, the grouping number is 12. There are no factors of 12 whose sum is $+1$.)

EXAMPLE 2 Factor, if possible. $x^2 + 6x + 12$

Since we are only using rational numbers, we find this polynomial is prime. The rational factors of 12 are

$$(1)(12) \quad \text{or} \quad (2)(6) \quad \text{or} \quad (3)(4)$$

None of these pairs will add up to 6, the coefficient of the middle term. Thus the problem cannot be factored by the methods of this chapter. It is **prime.**

Practice Problem 2 Factor. $x^2 - 9x - 8$ ■

EXAMPLE 3 Factor, if possible. $25x^2 + 4$

We have a formula to factor the difference of two squares. There is no way to factor the sum of two squares. That is, $a^2 + b^2$ cannot be factored. Thus

$$25x^2 + 4 \text{ is prime}$$

Practice Problem 3 Factor, if possible. $25x^2 + 82x + 4$ ■

5.6 Exercises

Review the six basic types of factoring in the Factoring Organizer on page 319. Each of the six types is included in the first 12 exercises. Be sure you can find two of each type.

Factor. Check your answer by multiplying.

1. $9x - 6y + 15$
   $3(3x - 2y + 5)$
2. $x^2 + 5xy - 7x$
   $x(x + 5y - 7)$
3. $36x^2 - 9y^2$
   $9(2x + y)(2x - y)$
4. $100x^2 - 1$
   $(10x - 1)(10x + 1)$
5. $9x^2 - 12xy + 4y^2$
   $(3x - 2y)^2$
6. $16x^2 + 24xy + 9y^2$
   $(4x + 3y)^2$
7. $x^2 + 8x + 15$
   $(x + 5)(x + 3)$
8. $x^2 + 15x + 54$
   $(x + 9)(x + 6)$
9. $5x^2 - 13x + 6$
   $(5x - 3)(x - 2)$
10. $21x^2 + 25x - 4$
    $(3x + 4)(7x - 1)$
11. $ax - 3ay - 6by + 2bx$
    $(x - 3y)(a + 2b)$
12. $ax - 20y - 5x + 4ay$
    $(x + 4y)(a - 5)$

Factor, if possible. Be sure to factor completely. Always remove the greatest common factor first if one exists.

13. $3x^4 - 12$
    $3(x^4 - 4)$
    $3(x^2 + 2)(x^2 - 2)$
14. $y^2 + 16y + 64$
    $(y + 8)^2$
15. $4x^2 - 12x + 9$
    $(2x - 3)^2$
16. $108x^2 - 3$
    $3(36x^2 - 1)$
    $3(6x - 1)(6x + 1)$
17. $2x^2 - 11x + 12$
    $(2x - 3)(x - 4)$
18. $2xy^2 - 50x$
    $2x(y^2 - 25)$
    $2x(y + 5)(y - 5)$
19. $x^2 - 3xy - 70y^2$
    $(x - 10y)(x + 7y)$
20. $2x^3 - 7x^2 + 4x - 14$
    $x^2(2x - 7) + 2(2x - 7)$
    $(2x - 7)(x^2 + 2)$
21. $ax + 20 - 5a - 4x$
    $x(a - 4) - 5(a - 4)$
    $(a - 4)(x - 5)$
22. $3x^2 - 9x + 15$
    $3(x^2 - 3x + 5)$
23. $45x - 5x^3$
    $5x(9 - x^2)$
    $5x(3 - x)(3 + x)$
24. $18y^2 + 3y - 6$
    $3(6y^2 + y - 2)$
    $3(2y - 1)(3y + 2)$
25. $5x^3y^3 - 10x^2y^3 + 5xy^3$
    $5xy^3(x^2 - 2x + 1)$
    $5xy^3(x - 1)^2$
26. $12x^2 - 36x + 27$
    $3(4x^2 - 12x + 9)$
    $3(2x - 3)^2$
27. $27xyz^2 - 12xy$
    $3xy(9z^2 - 4)$
    $3xy(3z + 2)(3z - 2)$
28. $12x^2 - 2x - 18x^3$
    $-2x(9x^2 - 6x + 1)$
    $-2x(3x - 1)^2$
29. $3x^2 + 6x - 105$
    $3(x^2 + 2x - 35)$
    $3(x + 7)(x - 5)$
30. $4x^2 - 28x - 72$
    $4(x^2 - 7x - 18)$
    $4(x - 9)(x + 2)$
31. $5x^2 - 30x + 40$
    $5(x^2 - 6x + 8)$
    $5(x - 2)(x - 4)$
32. $7x^2 + 3x - 2$
    prime
33. $7x^2 - 2x^4 + 4$
    $-1(2x^4 - 7x^2 - 4)$
    $-1(2x^2 + 1)(x + 2)(x - 2)$
34. $2x^4 - 9x^2 - 5$
    $(2x^2 + 1)(x^2 - 5)$
35. $6x^2 - 3x + 2$
    prime
36. $4x^3 + 8x^2 - 60x$
    $4x(x^2 + 2x - 15)$
    $4x(x + 5)(x - 3)$
37. $3x^2 - 15xy + 18y^2$
    $3(x^2 - 5xy + 6y^2)$
    $3(x - 2y)(x - 3y)$
38. $3x^2 - 24xy + 30y^2$
    $3(x^2 - 8xy + 10y^2)$
39. $30x^3 + 3x^2y - 6xy^2$
    $3x(10x^2 + xy - 2y^2)$
    $3x(2x + y)(5x - 2y)$
40. $56x^2 - 14xy - 7y^2$
    $7(8x^2 - 2xy - y^2)$
    $7(4x + y)(2x - y)$
41. $8x^2 + 28x - 16$
    $4(2x^2 + 7x - 4)$
    $4(2x - 1)(x + 4)$
42. $12x^2 - 30x + 12$
    $6(2x^2 - 5x + 2)$
    $6(2x - 1)(x - 2)$
43. $99x^4 - 11x^2$
    $11x^2(9x^2 - 1)$
    $11x^2(3x - 1)(3x + 1)$
44. $6x^4 - 150x^2$
    $6x^2(x^2 - 25)$
    $6x^2(x - 5)(x + 5)$
45. $a^2x + 2a^2y - 3ax - 6ay$
    $a(ax + 2ay - 3x - 6y)$
    $a(x + 2y)(a - 3)$
46. $2ax^2 + 2axy - 5x^2 - 5xy$
    $x(2ax + 2ay - 5x - 5y)$
    $x(x + y)(2a - 5)$

To Think About

47. A polynomial that cannot be factored by the methods of this chapter is called ___prime___.

48. A binomial of the form $x^2 - d$ can be quickly factored or identified as prime. If it can be factored, what is true of the number d? The number d must be a perfect square such as 1, 4, 9, 16, 25, 36, 49, 64, 81, 100, . . .

Cumulative Review Problems

49. When Dave Barry decided to leave the company and work as an independent contractor, he took a pay cut of 14%. He earned \$24,080 this year. What did he earn in his previous job? \$28,000

50. A major pharmaceutical company is testing a new powerful antiviral drug. It kills 13 strains of virus every hour. If there are presently 294 live strains of virus in the test container, how many live strains were there 6 hours ago? 372 live strains 6 hours ago

5.7 Solving Quadratic Equations by Factoring

MathPro

Video 14

SSM

After studying this section, you will be able to:

1 *Solve a quadratic equation by factoring.*
2 *Solve a variety of applied problems by using a quadratic equation.*

1 *Solving a Quadratic Equation by Factoring*

In Chapter 2 we learned how to solve linear equations such as $3x + 5 = 0$ to find the root (or value of x) that satisfied the equation. Now we turn to the question of how to solve equations like $3x^2 + 5x + 2 = 0$. Such equations are called ***quadratic equations.*** A quadratic equation is a polynomial equation that contains the power 2 of a variable as the highest power in the equation.

The *standard form* of a quadratic equation is $ax^2 + bx + c = 0$, where a, b, and c are real numbers and $a \neq 0$.

In this section, we will study quadratic equations in standard form, where a, b, and c are integers.

Usually we can find two real roots that satisfy a quadratic equation. But how can we find them? The most direct approach is the factoring method. This method depends on a very powerful property.

Zero Factor Property

If $a \cdot b = 0$, then $a = 0$ or $b = 0$.

When we make a statement in mathematics using the word **or,** we intend it to mean one or the other or both. Therefore, *both* a and b can equal zero if the product is zero. We can use this principle to solve a quadratic equation. Before you start, make sure that the equation is in standard form.

1. Factor, if possible, the quadratic expression that equals 0.
2. Set each factor equal to 0.
3. Solve the resulting equations to find each root.
4. Check each root.

EXAMPLE 1 Find the two roots. $3x^2 + 5x + 2 = 0$

$3x^2 + 5x + 2 = 0$ *The equation is in standard form.*

$(3x + 2)(x + 1) = 0$ *Factor the quadratic expression.*

$3x + 2 = 0 \qquad x + 1 = 0$ *Set each factor equal to 0.*

$3x = -2 \qquad x = -1$ *Solve each equation to find the two roots.*

$x = -\dfrac{2}{3}$

The two roots are $-\frac{2}{3}$ and -1.

Check. We can determine if the two roots $-\frac{2}{3}$ and -1 are each a solution to the equation. Substitute $-\frac{2}{3}$ for x in the *original equation.* If an identity results, $-\frac{2}{3}$ is a solution. Do the same for -1.

$$3x^2 + 5x + 2 = 0$$
$$3\left(-\frac{2}{3}\right)^2 + 5\left(-\frac{2}{3}\right) + 2 \stackrel{?}{=} 0$$
$$3\left(\frac{4}{9}\right) + 5\left(-\frac{2}{3}\right) + 2 \stackrel{?}{=} 0$$
$$\frac{4}{3} - \frac{10}{3} + 2 \stackrel{?}{=} 0$$
$$\frac{4}{3} - \frac{10}{3} + \frac{6}{3} \stackrel{?}{=} 0$$
$$0 = 0 \quad \checkmark$$

$$3x^2 + 5x + 2 = 0$$
$$3(-1)^2 + 5(-1) + 2 \stackrel{?}{=} 0$$
$$3(1) + 5(-1) + 2 \stackrel{?}{=} 0$$
$$3 - 5 + 2 \stackrel{?}{=} 0$$
$$-2 + 2 \stackrel{?}{=} 0$$
$$0 = 0 \quad \checkmark$$

Thus $x = -\frac{2}{3}$ and $x = -1$ are both roots for the equation $3x^2 + 5x + 2 = 0$.

Practice Problem 1 Find the roots by factoring and check. $10x^2 - x - 2 = 0$ ■

EXAMPLE 2 Find the two roots. $2x^2 + 13x - 7 = 0$

$2x^2 + 13x - 7 = 0$		*The equation is in standard form.*
$(2x - 1)(x + 7) = 0$		*Factor.*
$2x - 1 = 0$	$x + 7 = 0$	*Set each factor equal to 0.*
$2x = 1$	$x = -7$	*Solve each equation to find the two roots.*
$x = \frac{1}{2}$		

The two roots are $\frac{1}{2}$ and -7.
Check. If $x = \frac{1}{2}$, then

$$2\left(\frac{1}{2}\right)^2 + 13\left(\frac{1}{2}\right) - 7 = 2\left(\frac{1}{4}\right) + 13\left(\frac{1}{2}\right) - 7$$
$$= \frac{1}{2} + \frac{13}{2} - \frac{14}{2} = 0 \quad \checkmark$$

If $x = -7$, then

$$2(-7)^2 + 13(-7) - 7 = 2(49) + 13(-7) - 7$$
$$= 98 - 91 - 7 = 0 \quad \checkmark$$

Thus $x = \frac{1}{2}$ and $x = -7$ are both roots for the equation $2x^2 + 13x - 7 = 0$.

Practice Problem 2 Solve. $3x^2 - 5x + 2 = 0$ ■

TEACHING TIP It is critical that students feel that they can factor trinomials of the form $ax^2 + bx + c = 0$. After explaining Examples 1 and 2, ask students to do the following as an Additional Class Exercise. Find the two roots of the equation $8x^2 - 22x + 5 = 0$.

Sol. $x = \frac{1}{4}$ and $x = \frac{5}{2}$

If the quadratic equation $ax^2 + bx + c = 0$ has no visible constant term, then $c = 0$. All such quadratic equations can be factored by removing a common factor to obtain two solutions that are real numbers.

EXAMPLE 3 Find the two roots. $7x^2 - 3x = 0$

$7x^2 - 3x = 0$ *The equation is in standard form. Here $c = 0$.*

$x(7x - 3) = 0$ *Factor by removing the common factor.*

$x = 0 \qquad 7x - 3 = 0$ *Set each factor equal to 0 by the zero factor property.*

$7x = 3$

$x = \frac{3}{7}$ *Solve each equation to find the two roots.*

The two roots are $x = 0$ and $x = \frac{3}{7}$.
Check. Verify that $x = 0$ and $x = \frac{3}{7}$ are the roots of $7x^2 - 3x = 0$.

Practice Problem 3 Solve. $7x^2 + 11x = 0$ ■

If the quadratic equation is not in standard form, we use the same basic algebraic methods we studied in Sections 2.1 to 2.3 to set the equation equal to zero so that we can use the zero factor property.

TEACHING TIP Students should always be in the habit of collecting like terms and putting the equation in standard form before factoring. After doing Example 4, you may want to emphasize this as you do the following Additional Class Exercise. Solve.

$1 + 2x^2 + 13x = 8x + 1$

Sol. $2x^2 + 13x = 8x$
$2x^2 + 5x = 0$
$x(2x + 5) = 0$
$x = 0,\ x = -\frac{5}{2}$

EXAMPLE 4 Find the two roots. $x^2 = 12 - x$

$x^2 = 12 - x$ *The equation is not in standard form.*

$x^2 + x - 12 = 0$ *Add x and -12 to both sides of the equation so that the left side is equal to zero; we can now factor.*

$(x - 3)(x + 4) = 0$ *Factor.*

$x - 3 = 0 \qquad x + 4 = 0$ *Set each factor equal to 0 by the zero factor property.*

$x = 3 \qquad x = -4$ *Solve each equation for x.*

Check. If $x = 3$: $(3)^2 \stackrel{?}{=} 12 - 3$ If $x = -4$: $(-4)^2 \stackrel{?}{=} 12 - (-4)$

$9 \stackrel{?}{=} 12 - 3 \qquad 16 \stackrel{?}{=} 12 + 4$

$9 = 9$ ✓ $\qquad 16 = 16$ ✓

Both roots check.

Practice Problem 4 Solve. $x^2 - 6x + 4 = -8 + x$ ■

EXAMPLE 5 Solve. $(x + 3)(x - 2) = 50$

Be careful here! You must have a zero on the right hand side to use the zero factor property.

$(x + 3)(x - 2) = 50$	
$x^2 + 3x - 2x - 6 = 50$	*Remove the parentheses.*
$x^2 + x - 56 = 0$	*Place in standard form.*
$(x + 8)(x - 7) = 0$	*Factor.*
$x + 8 = 0 \qquad x - 7 = 0$	*Set each factor equal to zero.*
$x = -8 \qquad x = 7$	*Solve each equation for x.*

The check is left to the student.

Practice Problem 5 Solve. $x(3x - 7) = 20$ ■

EXAMPLE 6 Solve. $\dfrac{x^2 - x}{2} = 6$

We must first clear the fractions from the equation.

$2\left(\dfrac{x^2 - x}{2}\right) = 2(6)$	*Multiply each side by 2.*
$x^2 - x = 12$	*Simplify.*
$x^2 - x - 12 = 0$	*Place in standard form.*
$(x - 4)(x + 3) = 0$	*Factor.*
$x - 4 = 0 \qquad x + 3 = 0$	*Set each factor equal to zero.*
$x = 4 \qquad x = -3$	*Solve each equation for x.*

The check is left to the student.

Practice Problem 6 Solve. $\dfrac{2x^2 - 7x}{3} = 5$ ■

2 *Using a Quadratic Equation to Solve Applied Problems*

Certain types of word problems lead to quadratic equations. These problems may be number problems, problems about geometric figures, or applied problems. We'll show the solution to some model problems in this section.

It is particularly important to check the apparent solutions to the quadratic equation with conditions stated in the word problem. Often a particular solution to the quadratic equation will be eliminated by the conditions of the word problem.

EXAMPLE 7 The length of a rectangle is 3 meters longer than twice the width. The area of the rectangle is 44 square meters. Find the length and width of the rectangle.

1. *Understand the problem.*
Draw a picture.

Let w = the width in meters

Then $2w + 3$ = the length in meters

2w + 3
w
w
2w + 3

2. *Write an equation.*

$$\text{area} = (\text{width})(\text{length})$$

$$44 = w(2w + 3)$$

3. *Solve the equation and state the answer.*

$$44 = w(2w + 3)$$

$$44 = 2w^2 + 3w \quad \textit{Remove parentheses.}$$

$$0 = 2w^2 + 3w - 44 \quad \textit{Subtract 44 from both sides.}$$

$$0 = (2w + 11)(w - 4) \quad \textit{Factor.}$$

$$2w + 11 = 0 \qquad w - 4 = 0 \quad \textit{Set each factor equal to 0.}$$

$$2w = -11 \qquad w = 4 \quad \textit{Simplify and solve.}$$

$$w = -5\frac{1}{2}$$

Note that $w = -5\frac{1}{2}$ is not a valid solution. It would not make sense to have a rectangle with a negative number as a width.

Since $w = 4$, the width of the rectangle is 4 meters. The length is $2w + 3$, so we have $2(4) + 3 = 8 + 3 = 11$. Thus the length of the rectangle is 11 meters.

4. *Check.* Is the length 3 meters more than twice the width?

$$11 \stackrel{?}{=} 3 + 2(4) \qquad 11 = 3 + 8 \ \checkmark$$

Is the area of the rectangle 44 square meters?

$$4 \times 11 \stackrel{?}{=} 44 \qquad 44 = 44 \ \checkmark$$

Practice Problem 7 The length of a rectangle is 2 meters longer than triple the width. The area of the rectangle is 85 square meters. Find the length and width of the rectangle. ■

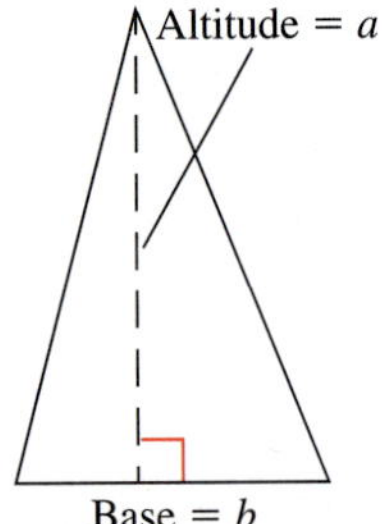

EXAMPLE 8 The area of a triangle is 49 square centimeters. The altitude of the triangle is 7 centimeters longer than the base. Find the altitude and the base of the triangle.

Let b = the length of the base in centimeters

$b + 7$ = the length of the altitude in centimeters

To find the area of a triangle, we use

$$\text{area} = \frac{1}{2}(\text{altitude})(\text{base}) = \frac{1}{2}ab = \frac{ab}{2}$$

$$\frac{ab}{2} = 49 \quad \textit{Write an equation.}$$

$$\frac{(b + 7)(b)}{2} = 49 \quad \textit{Substitute the expressions for altitude and base.}$$

$$\frac{b^2 + 7b}{2} = 49 \quad \textit{Simplify.}$$

$$b^2 + 7b = 98 \quad \textit{Multiply each side of the equation by 2.}$$

$$b^2 + 7b - 98 = 0 \quad \textit{Place the quadratic equation in standard form.}$$

$$(b - 7)(b + 14) = 0 \quad \textit{Factor.}$$

$$b - 7 = 0 \qquad b + 14 = 0 \quad \textit{Set each factor equal to zero.}$$

$$b = 7 \qquad b = -14 \quad \textit{Solve each equation for b.}$$

We cannot have a base of -14 centimeters, so we reject the negative answer. The only possible answer is $b = 7$. So the base is 7 centimeters. The altitude is $b + 7 = 7 + 7 = 14$. The altitude is 14 centimeters.

Check. When you do the check, answer the following two questions.

1. Is the altitude 7 centimeters longer than the base?
2. Is the area of a triangle with a base of 7 centimeters and an altitude of 14 centimeters actually 49 square centimeters?

Practice Problem 8 A triangle has an area of 35 square centimeters. The altitude of the triangle is 3 centimeters shorter than the base. Find the altitude and the base of the triangle. ■

Many examples in the sciences require the use of quadratic equations. You will study these in more detail if you take a course in physics or calculus in college. Often a quadratic equation is given as part of the problem.

When an object is thrown upward, its height (S) in meters is given, approximately, by the quadratic equation

$$S = -5t^2 + vt + h$$

The letter h represents the initial height in meters. The letter v represents the initial velocity of the object thrown. The letter t represents the time in seconds starting from the time the object is thrown.

EXAMPLE 9 Suppose that the initial height above the ground is 6 meters. A tennis ball is thrown upward with an initial velocity of 29 meters/second. At what time t will the ball hit the ground?

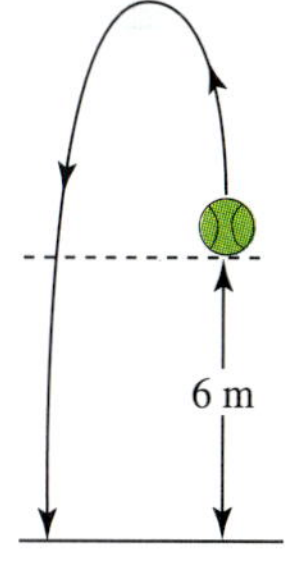

In this case $S = 0$ since the ball will hit the ground. The initial upward velocity is $v = 29$ meters/second. The initial height is 6 meters, so $h = 6$.

$S = -5t^2 + vt + h$ *Write an equation.*

$0 = -5t^2 + 29t + 6$ *Substitute all values into the equation.*

$5t^2 - 29t - 6 = 0$ *Transpose to left side. (Most students can factor more readily if the squared variable is positive.)*

$(5t + 1)(t - 6) = 0$ *Factor.*

$5t + 1 = 0 \qquad t - 6 = 0$ *Set each factor = 0.*

$5t = -1 \qquad t = 6$ *Solve for t.*

$t = -\frac{1}{5}$

We want a positive time t in seconds; thus we do not use $t = -\frac{1}{5}$. Thus the ball will strike the ground 6 seconds after you throw it.

Check. Verify the solution.

Practice Problem 9 A diver does a handstand dive from the 45-meter platform. This constitutes free fall, so the initial $v = 0$, and there is no upward spring, so $h = 45$ meters. How long will it be until she breaks the water's surface? ■

5.7 Exercises

Using the factoring method, solve for the roots of each quadratic equation. Be sure to place the equation in standard form before factoring. Check your answers.

1. $x^2 - 3x - 10 = 0$
$(x - 5)(x + 2) = 0$
$x = 5, -2$

2. $x^2 + 2x - 24 = 0$
$(x + 6)(x - 4) = 0$
$x = -6, 4$

3. $x^2 + 13x + 40 = 0$
$(x + 5)(x + 8) = 0$
$x = -5, -8$

4. $x^2 + 16x + 63 = 0$
$(x + 9)(x + 7) = 0$
$x = -9, -7$

5. $2x^2 - 5x - 3 = 0$
$(2x + 1)(x - 3) = 0$
$x = -\frac{1}{2}, 3$

6. $3x^2 - 5x - 2 = 0$
$(3x + 1)(x - 2) = 0$
$x = -\frac{1}{3}, 2$

7. $2x^2 - 7x + 6 = 0$
$(2x - 3)(x - 2) = 0$
$x = \frac{3}{2}, 2$

8. $2x^2 - 11x + 12 = 0$
$(2x - 3)(x - 4) = 0$
$x = \frac{3}{2}, 4$

9. $x^2 - 6x = 16$
$(x - 8)(x + 2) = 0$
$x = 8, -2$

10. $x^2 - 21 = 4x$
$(x - 7)(x + 3) = 0$
$x = 7, -3$

11. $10 = 7x - x^2$
$(x - 5)(x - 2) = 0$
$x = 2, 5$

12. $6 = 7x - x^2$
$(x - 1)(x - 6) = 0$
$x = 1, 6$

13. $6x^2 - 13x = -6$
$(3x - 2)(2x - 3) = 0$
$x = \frac{2}{3}, \frac{3}{2}$

14. $10x^2 + 19x = 15$
$(5x - 3)(2x + 5) = 0$
$x = \frac{3}{5}, -\frac{5}{2}$

15. $x^2 - 8x = 0$
$x(x - 8) = 0$
$x = 0, 8$

16. $5x^2 - x = 0$
$x(5x - 1) = 0$
$x = 0, \frac{1}{5}$

17. $8x^2 = 72$
$8x^2 - 72 = 0$
$8(x + 3)(x - 3) = 0$
$x = 3, -3$

18. $9x^2 = 81$
$9x^2 - 81 = 0$
$9(x + 3)(x - 3) = 0$
$x = 3, -3$

19. $5x^2 + 3x = 8x$
$5(x^2 - x) = 0$
$5x(x - 1) = 0$ $x = 0, 1$

20. $6x^2 - 4x = 3x$
$x(6x - 7) = 0$
$x = 0, \frac{7}{6}$

21. $x(9 + x) = 4(2x + 5)$
$x^2 + x - 20 = 0$
$(x + 5)(x - 4) = 0$
$x = -5, 4$

22. $(x - 2)(x - 3) = 6(x - 3)$
$x^2 - 11x + 24 = 0$
$(x - 3)(x - 8) = 0$
$x = 3, 8$

23. $x^2 - 12x + 36 = 0$
$(x - 6)^2 = 0$
$x = 6$

24. $x^2 - 14x + 49 = 0$
$(x - 7)^2 = 0$
$x = 7$

25. $4x^2 - 3x + 1 = -7x$
$(2x + 1)^2 = 0$
$x = -\frac{1}{2}$

26. $9x^2 - 2x + 4 = 10x$
$(3x - 2)^2 = 0$
$x = \frac{2}{3}$

27. $\frac{x^2}{2} - 8 + x = -8$
$x(x + 2) = 0$
$x = 0, -2$

28. $4 + \frac{x^2}{3} = 2x + 4$
$x^2 - 6x = 0$
$x(x - 6) = 0$
$x = 0, 6$

29. $\frac{x^2 - 3x}{2} = 27$
$x^2 - 3x - 54 = 0$
$(x - 9)(x + 6) = 0$
$x = 9, -6$

30. $\frac{x^2 - 9x}{2} = -9$
$x^2 - 9x + 18 = 0$
$(x - 3)(x - 6) = 0$
$x = 3, 6$

31. $\frac{9x^2 - 12x}{3} = 15$
$9x^2 - 12x - 45 = 0$
$3(3x + 5)(x - 3) = 0$
$x = -\frac{5}{3}, 3$

32. $\frac{4x^2 - 10x}{2} = 12$
$4x^2 - 10x - 24 = 0$
$2(2x + 3)(x - 4) = 0$
$x = -\frac{3}{2}, 4$

To Think About

33. Why can an equation in standard form with $c = 0$ (leaving $ax^2 + bx = 0$) always be solved? You can always factor out x.

34. Martha solved $(x + 3)(x - 2) = 14$ as follows:

$$x + 3 = 14 \quad \text{or} \quad x - 2 = 14$$
$$x = 11 \quad \text{or} \quad x = 16$$

Josette said this had to be wrong because these values did not check. Explain what is wrong with Martha's method. You must have the product of two or more factors equal to zero, not 14. Remember the zero factor property.

Applications

35. The area of a rectangle is 140 square meters. The width is 3 meters longer than one-half of the length. Find the length and the width of the rectangle.

$$x\left(\frac{x}{2}+3\right) = 140$$
$$x^2 + 6x - 280 = 0$$
$$(x + 20)(x - 14) = 0$$
$$L = 14 \text{ m}$$
$$W = 10 \text{ m}$$

36. The area of a triangle is 33 square centimeters. The base of the triangle is 1 centimeter less than double the altitude. Find the altitude and the base of the triangle.

$$\frac{1}{2}x(2x - 1) = 33$$
$$2x^2 - x - 66 = 0$$
$$(2x + 11)(x - 6) = 0$$
$$\text{alt} = 6 \text{ cm}$$
$$\text{base} = 11 \text{ cm}$$

37. Find two positive numbers whose difference is 10 and whose product is 56.

$$x(x + 10) = 56$$
$$(x + 14)(x - 4) = 0$$
$$x = 4$$

4 and 14

38. The sum of two numbers is -17. The product of these two numbers is 66. Find the numbers.

$$x(-17 - x) = 66$$
$$x^2 + 17x + 66 = 0$$
$$(x + 6)(x + 11) = 0$$

-6 and -11

Use the following information for Problems 39 and 40. When an object is thrown upward, its height (S), in meters, is given (approximately) by the quadratic equation

$$S = -5t^2 + vt + h$$

where v = the upward initial velocity in meters/second

t = the time of flight in seconds

h = the height above level ground from which the object is thrown

39. Johnnie is standing on a platform 6 meters high and throws a ball straight up as high as he can at a velocity of 13 meters/second. At what time t will the ball hit the ground? How far from the ground is the ball after 2 seconds have elapsed from the time of the throw? (Assume that the ball is 6 meters from the ground when it leaves his hand.)

$$-5t^2 + 13t + 6 = 0$$
$(5t + 2)(t - 3) = 0$ It will hit the ground after 3 seconds.
$$S = -5(2)^2 + 13(2) + 6$$
$= 12$ meters above ground after 2 sec

40. You are standing on the edge of a cliff near Acapulco, overlooking the ocean. The place where you stand is 180 meters from the ocean. You drop a pebble into the water. (''Dropping'' the pebble implies that there is no initial velocity, so $v = 0$.) How many seconds will it take to hit the water? How far has the pebble dropped in 3 seconds?

$$0 = -5t^2 + 180$$
$$t^2 - 36 = 0$$
$t = 6$ It will hit the water in 6 seconds.
$$S = -5(3)^2 + 180$$
$= 135$ m $\quad 180 - 135 = 45$ meters
After 3 secs it has dropped 45 m

41. Paying overtime to employees can be very expensive, but if profits are realized, it is worth putting a shift on overtime. In a high-tech helicopter company, the additional hourly cost in dollars for producing x additional helicopters is given by the cost equation $C = 2x^2 - 7x$. If the additional hourly cost is \$15, how many additional helicopters were produced?

$$15 = 2x^2 - 7x$$
$$2x^2 - 7x - 15 = 0$$
$$(2x + 3)(x - 5) = 0$$
$x = 5$ 5 additional helicopters

42. A boat generator on a Gloucester fishing boat is required to produce 64 watts of power. The amount of current I measured in amperes needed to produce the power for this generator is given by the equation $P = 40I - 4I^2$. What is the *minimum* number of amperes required to produce the necessary power?

$$4(I^2 - 10I + 16) = 0$$
$$4(I - 2)(I - 8) = 0$$
The minimum number of amperes is 2.

Cumulative Review Problems

Simplify.

43. $(2x^2y^3)(-6xy^4)$
$-12x^3y^7$

44. $(3a^4b^5)(4a^6b^8)$
$12a^{10}b^{13}$

45. $\dfrac{21a^5b^{10}}{-14ab^{12}}$
$-\dfrac{3a^4}{2b^2}$

46. $\dfrac{18x^3y^6}{54x^8y^{10}}$
$\dfrac{1}{3x^5y^4}$

Putting Your Skills to Work

PREDICTING TOTAL EARNINGS BY A MATHEMATICAL SERIES

It is often helpful to look at the total income of a company or a person over a long period of time. If the income tends to go up at a fairly constant rate, then the sum of this income can be determined by a formula for a mathematical series.

Angela Sanchez has been employed by West Coast Cable Vision for 9 years. She started at an annual salary of \$18,000. Her income has increased about \$1200 per year for the last 9 years. She started earning \$18,000 and now is earning \$27,600. To find her total earnings for 9 years we can use the formula in factored form

$T = 0.5n(s + e)$

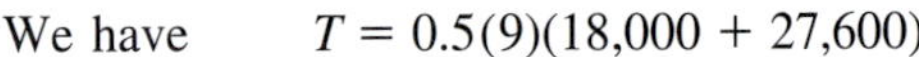

where n = the **number** of years she has worked

s = her **starting** salary

e = her salary at the **end** of the time period

We have $\quad T = 0.5(9)(18{,}000 + 27{,}600)$

$T = 0.5(9)(45{,}600)$

$T = \$205{,}200$

Thus Angela has earned \$205,200 in the last 9 years.

Use the above formula to determine the following.

PROBLEMS FOR INDIVIDUAL STUDY AND INVESTIGATION

1. Monique Bussell has worked for the last 20 years at a salary that has increased about \$1500 per year. Her starting salary was \$24,000. How much has she earned in that 20-year period? \$765,000

2. The Software Service Center has been in operation for 8 years. During the first year they made a profit of \$150,000. Each year the profit has increased by \$70,000. What is their total profit for 8 years? \$3,160,000

PROBLEMS FOR COOPERATIVE GROUP INVESTIGATION

3. James Kerr has opened his own travel business. Last year he earned \$40,000. This year he earned \$56,000. If his business continues to increase in earnings for each year at the same rate, how many years will it take until he makes one million dollars in combined total income? 10 years

4. A national chain of department stores made a profit of \$780,000,000 ten years ago. Each year the profit has been decreasing at the rate of \$15,000,000 per year. What will be the total earnings of the store for the next 15 years (starting with this present year)? \$7,875,000,000

INTERNET CONNECTIONS: Go to ''http://www.prenhall.com/tobey'' to be connected

Site: DataMasters Salary Survey or a related site

This site gives salary information for executives and professionals in the computer industry.

5. Karl is an experienced software engineer whose salary has kept up with the regional median for the west coast since 1990. Use the DataMasters site to determine Karl's salary for 1990 and for the current year. Then use the formula $T = 0.5n(s + e)$ to find his total salary for all years from 1990 to the current year, assuming that his income has increased by a constant amount each year.

Chapter Organizer

Topic	Procedure	Examples
A. Common factor, p. 294.	Remove the largest common factor from each term.	$2x^2 - 2x = 2x(x - 1)$ $3a^2 + 3ab - 12a = 3a(a + b - 4)$ $8x^4y - 24x^3 = 8x^3(xy - 3)$
Special cases, B. Difference of squares, p. 314. C. Perfect square trinomials, p. 315.	If you recognize the special cases, you will be able to factor quickly. $a^2 - b^2 = (a + b)(a - b)$ $a^2 + 2ab + b^2 = (a + b)^2$ $a^2 - 2ab + b^2 = (a - b)^2$	$25x^2 - 36y^2 = (5x + 6y)(5x - 6y)$ $16x^4 - 1 = (4x^2 + 1)(2x + 1)(2x - 1)$ $25x^2 + 10x + 1 = (5x + 1)^2$ $49x^2 - 42xy + 9y^2 = (7x - 3y)^2$
D. Trinomial that starts with x^2, p. 302.	Factor trinomials of the form $x^2 + bx + c$ by asking what two numbers have a product of c but a sum of b. If each term of the trinomial has a common factor, remove the common factor as the first step.	$x^2 - 18x + 77 = (x - 7)(x - 11)$ $x^2 + 7x - 18 = (x + 9)(x - 2)$ $5x^2 - 10x - 40 = 5(x^2 - 2x - 8)$ $= 5(x - 4)(x + 2)$
E. Trinomial that starts with ax^2, where $a \neq 1$, p. 309.	Factor trinomials of the form $ax^2 + bx + c$ by the grouping number method or by trial and error.	$6x^2 + 11x - 10$ Grouping number $= -60$ Two numbers whose product is -60 and whose sum is $+11$ are $+15$ and -4 $6x^2 + 15x - 4x - 10 = 3x(2x + 5) - 2(2x + 5)$ $= (2x + 5)(3x - 2)$
F. Four terms. Factor by grouping, p. 298.	Rearrange the terms if necessary so that the terms have a common factor. Then remove the common factors. $ax + ay - bx - by = a(x + y) - b(x + y)$ $= (a - b)(x + y)$	$2ax^2 + 21 + 3a + 14x^2$ $= 2ax^2 + 14x^2 + 3a + 21$ $= 2x^2(a + 7) + 3(a + 7)$ $= (a + 7)(2x^2 + 3)$
Prime polynomials, p. 320.	A polynomial that is not factorable is called *prime*.	$x^2 + y^2$ is *prime*. $x^2 + 5x + 7$ is *prime*.
Multistep factoring, p. 320.	Many problems require two or three steps of factoring. Always try to remove a common factor as the first step.	$3x^2 - 21x + 36 = 3(x^2 - 7x + 12)$ $= 3(x - 4)(x - 3)$ $2x^3 - x^2 - 6x = x(2x^2 - x - 6)$ $= x(2x + 3)(x - 2)$ $25x^3 - 49x = x(25x^2 - 49)$ $= x(5x + 7)(5x - 7)$ $8x^2 - 24x + 18 = 2(4x^2 - 12x + 9)$ $= 2(2x - 3)^2$
Solving quadratic equations by factoring, p. 322.	**1.** Write as $ax^2 + bx + c = 0$. **2.** Factor. **3.** Set each factor equal to 0. **4.** Solve the resulting equations.	Solve: $3x^2 + 5x = 2$ $3x^2 + 5x - 2 = 0$ $(3x - 1)(x + 2) = 0$ $3x - 1 = 0$ or $x + 2 = 0$ $x = \frac{1}{3}$ or $x = -2$
Using quadratic equations to solve applied problems, p. 325.	Word problems involving the product of two numbers, area, and formulas with a squared variable can be solved using the factoring methods we have shown.	The length of a rectangle is 4 less than three times the width. Find the length and width if the area is 55 square inches. Let w = width $3w - 4$ = length $55 = w(3w - 4)$ $55 = 3w^2 - 4w$ $0 = 3w^2 - 4w - 55$ $0 = (3w + 11)(w - 5)$ $w = -\frac{11}{3}$ or $w = 5$ $-\frac{11}{3}$ is not a valid solution. width = 5 inches length = 11 inches

Chapter 5 Review Problems

5.1 *Factor out the greatest common factor from each term.*

1. $15x^3y - 9x^2y^2$ $3x^2y(5x - 3y)$

2. $21x^2 - 14x$ $7x(3x - 2)$

3. $7x^2y - 14xy^2 - 21x^3y^3$ $7xy(x - 2y - 3x^2y^2)$

4. $50a^4b^5 - 25a^4b^4 + 75a^5b^5$ $25a^4b^4(2b - 1 + 3ab)$

5. $27x^3 - 9x^2$ $9x^2(3x - 1)$

6. $2x - 4y + 6z + 12$ $2(x - 2y + 3z + 6)$

7. $2a(a + 3b) - 5(a + 3b)$ $(a + 3b)(2a - 5)$

8. $15x^3y + 6xy^2 + 3xy$ $3xy(5x^2 + 2y + 1)$

5.2 *Factor by grouping.*

9. $3ax - 7a - 6x + 14$ $(3x - 7)(a - 2)$

10. $a^2 + 5ab - 4a - 20b$ $(a + 5b)(a - 4)$

11. $x^2y + 3y - 2x^2 - 6$ $(x^2 + 3)(y - 2)$

12. $4ad - 25c - 5d + 20ac$ $(4a - 5)(d + 5c)$

13. $15x^2 - 3x + 10x - 2$ $(5x - 1)(3x + 2)$

14. $30w^2 - 18w + 5wz - 3z$ $(5w - 3)(6w + z)$

5.3 *Factor completely. Be sure to remove any common factors as your first step.*

15. $x^2 - 3x - 18$ $(x - 6)(x + 3)$

16. $x^2 - 10x + 24$ $(x - 4)(x - 6)$

17. $x^2 + 14x + 48$ $(x + 6)(x + 8)$

18. $x^2 + 8xy + 15y^2$ $(x + 3y)(x + 5y)$

19. $x^4 + 13x^2 + 42$ $(x^2 + 7)(x^2 + 6)$

20. $x^2 - 10x - 39$ $(x + 3)(x - 13)$

21. $5x^2 + 20x + 15$ $5(x + 1)(x + 3)$

22. $3x^2 + 39x + 36$ $3(x + 1)(x + 12)$

23. $2x^2 - 28x + 96$ $2(x - 6)(x - 8)$

24. $4x^2 - 44x + 120$ $4(x - 5)(x - 6)$

5.4 *Factor completely. Be sure to remove any common factors as your first step.*

25. $4x^2 + 7x - 15$ $(4x - 5)(x + 3)$

26. $12x^2 + 11x - 5$ $(3x - 1)(4x + 5)$

27. $15x^2 + 7x - 4$ $(5x + 4)(3x - 1)$

28. $6x^2 - 13x + 6$ $(3x - 2)(2x - 3)$

29. $2x^2 - x - 3$ $(2x - 3)(x + 1)$

30. $3x^2 + 14x - 5$ $(3x - 1)(x + 5)$

31. $20x^2 + 48x - 5$ $(10x - 1)(2x + 5)$

32. $20x^2 + 21x - 5$ $(5x - 1)(4x + 5)$

33. $4a^2 - 11a - 3$ $(4a + 1)(a - 3)$

34. $4a^2 - a - 3$ $(4a + 3)(a - 1)$

35. $6x^2 + 4x - 10$ $2(x - 1)(3x + 5)$

36. $6x^2 - 4x - 10$ $2(x + 1)(3x - 5)$

37. $4x^2 - 26x + 30$ $2(2x - 3)(x - 5)$

38. $4x^2 - 20x - 144$ $4(x - 9)(x + 4)$

39. $10x^2 - 22x + 12$ $2(x - 1)(5x - 6)$

40. $2x^2 + 3x - 44$ $(2x + 11)(x - 4)$

41. $12x^2 + 5x - 3$ $(3x - 1)(4x + 3)$

42. $16x^2 + 14x - 30$ $2(x - 1)(8x + 15)$

43. $6x^2 - 19xy + 10y^2$ $(3x - 2y)(2x - 5y)$

44. $6x^2 - 32xy + 10y^2$ $2(3x - y)(x - 5y)$

45. $3x^4 - 5x^2 - 8$ $(3x^2 - 8)(x^2 + 1)$

46. $3x^4 + 10x^2 - 8$ $(3x^2 - 2)(x^2 + 4)$

5.5 *Factor these special cases. Be sure to remove any common factors.*

47. $49x^2 - y^2$ $(7x + y)(7x - y)$

48. $16x^2 - 36y^2$ $4(2x - 3y)(2x + 3y)$

49. $9x^2 - 12x + 4$ $(3x - 2)^2$

50. $16x^2 - 1$ $(4x - 1)(4x + 1)$

51. $25x^2 - 36$ $(5x - 6)(5x + 6)$

52. $100x^2 - 9$ $(10x - 3)(10x + 3)$

53. $1 - 49x^2$ $(1 - 7x)(1 + 7x)$

54. $4 - 49x^2$ $(2 - 7x)(2 + 7x)$

55. $36x^2 + 12x + 1$ $(6x + 1)^2$

56. $25x^2 - 20x + 4$ $(5x - 2)^2$

57. $16x^2 - 24xy + 9y^2$ $(4x - 3y)^2$

58. $49x^2 - 28xy + 4y^2$ $(7x - 2y)^2$

59. $2x^2 - 18$ $2(x - 3)(x + 3)$

60. $3x^2 - 75$ $3(x - 5)(x + 5)$

61. $8x^2 + 40x + 50$ $2(2x + 5)^2$

62. $50x^2 + 120x + 72$ $2(5x + 6)^2$

5.6 *If possible, factor each problem completely. If it cannot be factored, state "prime."*

63. $4x^2 - 9y^2$
$(2x + 3y)(2x - 3y)$

64. $x^2 + 6x + 9$
$(x + 3)^2$

65. $x^2 - 9x + 18$
$(x - 3)(x - 6)$

66. $x^2 + 13x - 30$
$(x + 15)(x - 2)$

67. $6x^2 + x - 7$
$(x - 1)(6x + 7)$

68. $10x^2 + x - 2$
$(5x - 2)(2x + 1)$

69. $12x + 16$
$4(3x + 4)$

70. $8x^2y^2 - 4xy$
$4xy(2xy - 1)$

71. $50x^3y^2 + 30x^2y^2 - 10x^2y^2$
$10x^2y^2(5x + 2)$

72. $26a^3b - 13ab^3 + 52a^2b^4$
$13ab(2a^2 - b^2 + 4ab^3)$

73. $x^3 - 16x^2 + 64x$
$x(x - 8)^2$

74. $2x^2 + 40x + 200$
$2(x + 10)^2$

75. $3x^2 - 18x + 27$
$3(x - 3)^2$

76. $25x^3 - 60x^2 + 36x$
$x(5x - 6)^2$

77. $7x^2 + 3x - 10$
$(7x + 10)(x - 1)$

78. $5x^2 - 8x - 4$
$(5x + 2)(x - 2)$

79. $9x^3y - 4xy^3$
$xy(3x + 2y)(3x - 2y)$

80. $3x^3a^3 - 11x^4a^2 - 20x^5a$
$x^3a(3a + 4x)(a - 5x)$

81. $12a^2 + 14ab - 10b^2$
$2(3a + 5b)(2a - b)$

82. $16a^2 - 40ab + 25b^2$
$(4a - 5b)^2$

83. $7a - 7 - ab + b$
$(a - 1)(7 - b)$

84. $3d - 4 - 3cd + 4c$
$(3d - 4)(1 - c)$

85. $2x - 1 + 2bx - b$
$(2x - 1)(1 + b)$

86. $5xb - 35x + 4by - 28y$
$(b - 7)(5x + 4y)$

87. $2a^2x - 15ax + 7x$
$x(2a - 1)(a - 7)$

88. $x^5 - 17x^3 + 16x$
$x(x + 4)(x - 4)(x + 1)(x - 1)$

89. $x^4 - 81y^{12}$
$(x^2 + 9y^6)(x + 3y^3)(x - 3y^3)$

90. $6x^4 - x^2 - 15$
$(3x^2 - 5)(2x^2 + 3)$

91. $28yz - 16xyz + x^2yz$
$yz(14 - x)(2 - x)$

92. $12x^3 + 17x^2 + 6x$
$x(3x + 2)(4x + 3)$

93. $16w^2 - 2w - 5$
$(2w + 1)(8w - 5)$

94. $12w^2 - 12w + 3$
$3(2w - 1)^2$

95. $4y^3 + 10y^2 - 6y$
$2y(2y - 1)(y + 3)$

96. $10y^2 + 33y - 7$
$(5y - 1)(2y + 7)$

97. $8y^{10} - 16y^8$
$8y^8(y^2 - 2)$

98. $49x^4 - 49$
$49(x^2 + 1)(x + 1)(x - 1)$

99. $x^2 + 13x + 54$
prime

100. $8x^2 - 19x - 6$
prime

101. $8y^5 + 4y^3 - 60y$
$4y(2y^2 - 5)(y^2 + 3)$

102. $9xy^2 + 3xy - 42x$
$3x(3y + 7)(y - 2)$

103. $16x^4y^2 - 56x^2y + 49$
$(4x^2y - 7)^2$

104. $128x^3y - 2xy$
$2xy(8x + 1)(8x - 1)$

105. $2ax + 5a - 10b - 4bx$
$(2x + 5)(a - 2b)$

106. $2x^3 - 9 + x^2 - 18x$
$(2x + 1)(x + 3)(x - 3)$

5.7 *Solve these quadratic equations by factoring.*

107. $x^2 - 3x - 18 = 0$
$x = -3, x = 6$

108. $x^2 + 3x - 10 = 0$
$x = 2, x = -5$

109. $5x^2 = 2x - 7x^2$
$x = 0, x = \frac{1}{6}$

110. $8x^2 + 5x = 2x^2 - 6x$
$x = 0, x = -\frac{11}{6}$

111. $2x^2 + 9x - 5 = 0$
$x = -5, x = \frac{1}{2}$

112. $x^2 + 11x + 24 = 0$
$x = -8, x = -3$

113. $x^2 + 14x + 45 = 0$
$x = -5, x = -9$

114. $5x^2 = 7x + 6$
$x = -\frac{3}{5}, x = 2$

115. $3x^2 + 6x = 2x^2 - 9$
$x = -3$

116. $4x^2 + 9x - 9 = 0$
$x = -3, x = \frac{3}{4}$

117. $5x^2 - 11x + 2 = 0$
$x = \frac{1}{5}, x = 2$

Solve the following applied problems.

118. The area of a triangle is 35 square centimeters. The base is 3 centimeters longer than the altitude of the triangle. Find the length of the base and the altitude.
base = 10 cm altitude = 7 cm

119. The area of a rectangle is 105 square feet. The length of the rectangle is 1 foot longer than double the width. Find the length and width of the rectangle.
width = 7 feet, length = 15 feet

120. The height in feet that a model rocket attains is given by height $= -16t^2 + 80t + 96$, where t is the time measured in seconds. How many seconds will it take until the rocket finally reaches the ground? (*Hint:* At ground level the height = 0.) 6 seconds

121. An electronic technician is working with a 100-volt electric generator. The output power of the generator is given by the equation, power $= -5x^2 + 100x$, where x is the amount of current measured in amperes. The technician wants to find the value for x when the power is 480 watts. Can you find the two answers?
$x = 8$ amperes, $x = 12$ amperes

Putting Your Skills to Work

THE MATHEMATICS OF COLLEGE ENROLLMENT

Probably you know that the number of students enrolled in the public two-year and four-year colleges in the United States is increasing, but did you realize how much this number is increasing? Look at the bar graph shown at the right and then answer the following questions.

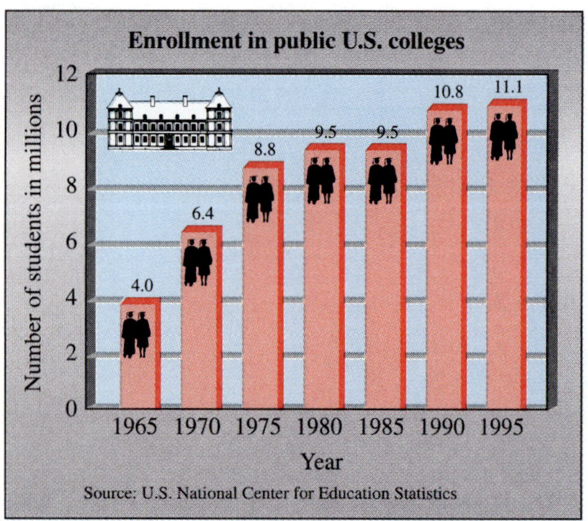

PROBLEMS FOR INDIVIDUAL INVESTIGATION AND ANALYSIS

1. How many more students were enrolled in public colleges in 1975 than were enrolled in 1970?
2.4 million students

2. What was the percent of increase in the number of students enrolled in public colleges from 1990 to 1995? Round to the nearest tenth. Approximately 2.8%

PROBLEMS FOR GROUP INVESTIGATION AND COOPERATIVE GROUP ACTIVITY

Together with some of the members of your class answer the following.

3. In 1995 there were 11.1 million students in public colleges. In the year 2005 it has been estimated that there will be 12.6 million students. Write an equation of the form $P = ax + 11.1$, where x is the number of years since 1995 and P is the number of students in public colleges. See if your equation will obtain the value $P = 11.1$ when $x = 0$ and the value $P = 12.6$ when $x = 10$. What is the value of a in your equation?
$P = 0.15x + 11.1$ $a = 0.15$

4. Factor the equation you obtained in (3) above by removing a common factor of 1/100 for each term of the right hand side of the equation. Now use the equation to find a prediction for the number of students in public colleges in the year 2010.
$P = \frac{1}{100}(15x + 1110)$
Predicted population is approximately 13.35 million students.

INTERNET CONNECTIONS: Go to ''http://www.prenhall.com/tobey'' to be connected

Site: Enrollment By College/School: Graduate (Auburn University) or a related site

This site gives graduate enrollment statistics for Auburn University in Alabama. Use the information to answer the following questions.

5. How many more graduate students were enrolled in 1994 than in 1987?

6. What was the percent increase in the number of graduate students from 1989 to 1992? Round to the nearest tenth.

7. Use the data for 1990 and 1995 to write an equation of the form $P = ax + 2544$, where x is the number of years since 1990 and P is the number of graduate students at Auburn University.

8. What value does your equation predict for the most recent year for which the data is shown on the web page? Is the prediction accurate?

Chapter 5 Test

If possible, factor each problem completely. If a problem cannot be factored, state "prime."

1. $x^2 + 12x - 28$

2. $9x^2 - 100y^2$

3. $10x^2 + 27x + 5$

4. $9a^2 - 30ab + 25b^2$

5. $7x - 9x^2 + 14xy$

6. $3x^2 - 4wy - 2wx + 6xy$

7. $6x^3 - 20x^2 + 16x$

8. $5a^2c - 11abc + 2b^2c$

9. $100x^4 - 16y^4$

10. $9x^2 - 15xy + 4y^2$

11. $7x^2 - 42x$

12. $a^2 + 1$

1. $(x + 14)(x - 2)$

2. $(3x + 10y)(3x - 10y)$

3. $(5x + 1)(2x + 5)$

4. $(3a - 5b)^2$

5. $x(7 - 9x + 14y)$

6. $(x + 2y)(3x - 2w)$

7. $2x(3x - 4)(x - 2)$

8. $c(5a - b)(a - 2b)$

9. $4(5x^2 + 2y^2)(5x^2 - 2y^2)$

10. $(3x - y)(3x - 4y)$

11. $7x(x - 6)$

12. prime

13. $3x^2 + 5x + 1$

14. $60xy^2 - 20x^2y - 45y^3$

15. $16x^2 - 1$

16. $x^{16} - 1$

17. $2ax + 6a - 5x - 15$

18. $aw^2 - 8b + 2bw^2 - 4a$

19. $3x^2 - 3x - 90$

20. $2x^3 - x^2 - 15x$

Solve the following quadratic equations.

21. $x^2 + 14x + 45 = 0$

22. $14 + 3x(x + 2) = -7x$

23. $2x^2 + x - 10 = 0$

Solve the following problem by using a quadratic equation.

24. The park service is studying a rectangular piece of land that contains 91 square miles. The length of this piece of land is 1 mile shorter than double the width. Find the length and width of this rectangular piece of land.

Answers

13. prime

14. $-5y(2x - 3y)^2$

15. $(4x + 1)(4x - 1)$

16. $(x^8 + 1)(x^4 + 1)(x^2 + 1)(x + 1)(x - 1)$

17. $(x + 3)(2a - 5)$

18. $(a + 2b)(w + 2)(w - 2)$

19. $3(x - 6)(x + 5)$

20. $x(2x + 5)(x - 3)$

21. $x = -5, x = -9$

22. $x = -\frac{7}{3}, x = -2$

23. $x = -\frac{5}{2}, x = 2$

24. width = 7 miles, length = 13 miles

Cumulative Test for Chapters 0–5

Approximately one-half of this test covers the content of Chapters 0–4. The remainder covers the content of Chapter 5.

1. What % of 480 is 72?

2. What is 13% of 3.8?

Simplify.

3. $-3.2 - 6.4 + 0.24 - 1.8 + 0.8$

4. $(-2x^3y^4)(-4xy^6)$

5. -2^6

6. $(7x - 3)(2x + 1)$

7. $(2x^2 - 6x + 1)(x - 3)$

Solve.

8. $3x - 4 \geq 6x + 5$

9. $3x - (7 - 5x) = 3(4x - 5)$

10. $\frac{1}{2}x - 3 = \frac{1}{4}(3x + 3)$

11. Solve for t: $s = \frac{1}{2}(2a + 3t)$.

1. 15%

2. 0.494

3. -10.36

4. $8x^4y^{10}$

5. -64

6. $14x^2 + x - 3$

7. $2x^3 - 12x^2 + 19x - 3$

8. $x \leq -3$

9. $x = 2$

10. $x = -15$

11. $t = \frac{2s - 2a}{3}$

12.	$(3x-1)(2x-1)$
13.	$(6x+1)(x-1)$
14.	$(3x+2)(3x-1)$
15.	$(11x+8y)(11x-8y)$
16.	$-4(5x+6)(4x-5)$
17.	prime
18.	$x(4x+5)^2$
19.	$(9x^2+4b^2)(3x+2b)(3x-2b)$
20.	$(2x+3)(a-2b)$
21.	$(x^2+5)(x^2+3)$
22.	$x=-8, x=3$
23.	$x=\frac{5}{3}, x=2$
24.	base = 6 miles, altitude = 19 miles

Factor each problem completely. If a problem cannot be factored, state "prime."

12. $6x^2 - 5x + 1$

13. $6x^2 - 5x - 1$

14. $9x^2 + 3x - 2$

15. $121x^2 - 64y^2$

16. $4x + 120 - 80x^2$

17. $x^2 + 1$

18. $16x^3 + 40x^2 + 25x$

19. $81x^4 - 16b^4$

20. $2ax - 4bx + 3a - 6b$

21. $x^4 + 8x^2 + 15$

Solve the following quadratic equations.

22. $x^2 + 5x - 24 = 0$

23. $3x^2 - 11x + 10 = 0$

Solve the following problem by using a quadratic equation.

24. The park service is studying a triangular piece of land that contains 57 square miles. The altitude of this triangle is 7 miles longer than double the base. Find the altitude and the base of this triangular piece of land.

CHAPTER 6

Rational Expressions and Equations

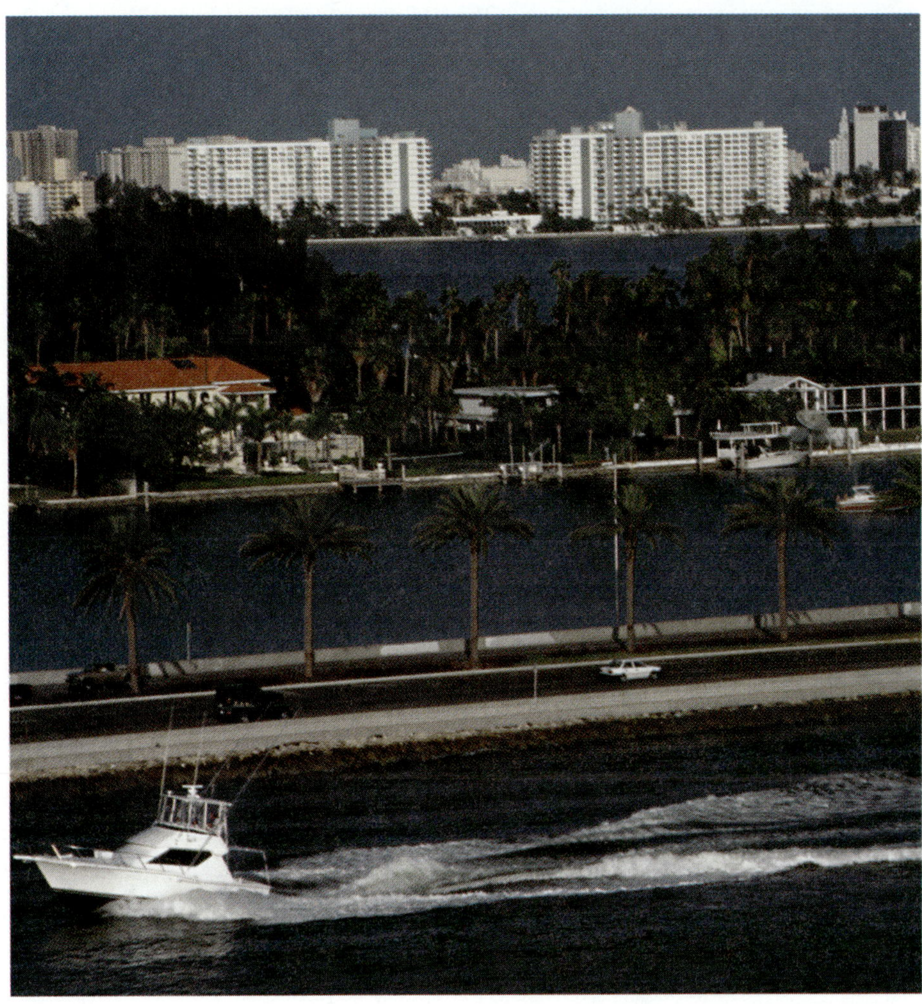

Boating can be a lot of fun, but did you realize how much mathematics can be involved? What speed should you travel at to achieve the maximum range on a boat trip? How much does the sound level increase as you accelerate a boat? Do you have any idea how many gallons of fuel per hour are used by a 23-foot power boat? If you would like to address these questions turn to page 363 and look at the Putting Your Skills to Work.

Pretest Chapter 6

1.	-2
2.	$\frac{3x-2}{4x+3}$
3.	$\frac{b}{2a-b}$
4.	$\frac{2a-b}{6}$
5.	$\frac{x}{x+5}$
6.	$\frac{1}{x+3}$
7.	$\frac{y}{x-1}$
8.	3
9.	$\frac{-2y-7}{(y-1)(2y+3)}$
10.	$\frac{1}{x+5}$
11.	$\frac{5x^2+6x-9}{3x(x-3)^2}$
12.	$\frac{2a-3}{5a^2+a}$
13.	$\frac{a^2-2a-2}{3a^2(a+1)}$
14.	$\frac{-x^2-y^2}{x^2+2xy-y^2}$
15.	$x = -5$
16.	$x = 7$
17.	9.3
18.	\$263.50

If you are familiar with the topics in this chapter, take this test now. Check your answers with those in the back of the book. If an answer was wrong or you couldn't do a problem, study the appropriate section of the chapter.

If you are not familiar with the topics in this chapter, don't take this test now. Instead, study the examples, work the practice problems, and then take the test.

This test will help you to identify those concepts that you have mastered and those that need more study.

Simplify the algebraic expressions in Problems 1 through 14.

Section 6.1

1. $\dfrac{6a-4b}{2b-3a}$ **2.** $\dfrac{3x^2-14x+8}{4x^2-13x-12}$ **3.** $\dfrac{a^2b+2ab^2}{2a^3+3a^2b-2ab^2}$

Section 6.2

4. $\dfrac{4a^2-b^2}{6a-6b}\cdot\dfrac{3a-3b}{6a+3b}$ **5.** $\dfrac{x^2-6x+9}{x^2-x-6}\div\dfrac{x^2+2x-15}{x^2+2x}$

6. $\dfrac{x^2+5x+6}{3x^2+8x+4}\cdot\dfrac{1}{x+3}\cdot\dfrac{6x^2-11x-10}{2x^2+x-15}$ **7.** $\dfrac{xy+3y}{x^2-x}\div\dfrac{x+3}{x}$

Section 6.3

8. $\dfrac{3y+1}{y+2}+\dfrac{5}{y+2}$ **9.** $\dfrac{2y-1}{2y^2+y-3}-\dfrac{2}{y-1}$

10. $\dfrac{4x+10}{x^2-25}-\dfrac{3}{x-5}$ **11.** $\dfrac{x+3}{x^2-6x+9}+\dfrac{2x+3}{3x^2-9x}$

Section 6.4

12. $\dfrac{\dfrac{2}{a}-\dfrac{3}{a^2}}{5+\dfrac{1}{a}}$ **13.** $\dfrac{\dfrac{a}{a+1}-\dfrac{2}{a}}{3a}$ **14.** $\dfrac{\dfrac{x-y}{x}-\dfrac{x+y}{y}}{\dfrac{x-y}{x}+\dfrac{x+y}{y}}$

Section 6.5

Solve for x. If the equation has no solution, so state.

15. $\dfrac{5}{2x}=2-\dfrac{2x}{x+1}$ **16.** $\dfrac{24}{x^2-3x-10}=\dfrac{2}{x-5}+\dfrac{3}{x+2}$

Section 6.6

17. Solve for x and round your answer to the nearest tenth. $\dfrac{5}{x}=\dfrac{7}{13}$

18. When Marcia went to England, she found that \$3.10 in U.S. money was needed at the current exchange rate to receive 2 British pounds. How much U.S. money was needed to receive 170 British pounds?

6.1 Simplify Rational Expressions

After studying this section, you will be able to:

1 *Simplify an algebraic fraction by factoring.*

Recall that a rational number is a number that can be written as one integer divided by another integer such as $3 \div 4$ or $\frac{3}{4}$. We usually use the word fraction to mean $\frac{3}{4}$. We can extend this idea to algebraic expressions. A rational expression is an algebraic expression divided by another algebraic expression such as

$$(3x + 2) \div (x + 4) \quad \text{or} \quad \frac{3x + 2}{x + 4}$$

The last fraction is sometimes called a **fractional algebraic expression.** There is a special restriction for all fractions, including fractional algebraic expressions. The denominator of the fraction cannot be 0. For example, in the expression

$$\frac{3x + 2}{x + 4}$$

the denominator cannot be 0. Therefore, the value of x cannot be -4. The following important restriction will apply throughout this chapter. We state it here to avoid having to mention it repeatedly throughout this chapter.

Restriction

The denominator of an algebraic fraction cannot be zero. Any value of the variable that would make the denominator zero is not allowed.

We have discovered that fractions can be simplified (or reduced) in the following way.

$$\frac{15}{25} = \frac{3 \cdot 5}{5 \cdot 5} = \frac{3}{5}$$

This is sometimes referred to as the *basic rule of fractions* and can be stated as follows:

Basic Rule of Fractions

For any polynomials a, b, c (where $b \neq 0$ and $c \neq 0$),

$$\frac{ac}{bc} = \frac{a}{b}$$

We will examine several examples where a, b, c are real numbers, as well as more involved examples where a, b, and c are polynomials. In either case we shall make extensive use of our factoring skills in this section.

There is one essential property that is revealed by the basic rule of fractions. It is this: If the numerator and denominator of a given fraction are multiplied by the same quantity, an equivalent fraction is obtained. The rule can be used two ways. You can start with $\frac{ac}{bc}$ and end with the equivalent fraction $\frac{a}{b}$. Or, you can start with $\frac{a}{b}$ and end with the equivalent fraction $\frac{ac}{bc}$.

TEACHING TIP Taking the time to review simple rules of building and reducing fractions with numerical values will help to establish a clearer picture for students of how to simplify algebraic fractions. Taking the time to connect the rules of arithmetic with the rules of algebra is always profitable for the students taking a Beginning Algebra course.

EXAMPLE 1

(a) Write a fraction equivalent to $\frac{3}{5}$ with a denominator of 10.

(b) Reduce $\frac{21}{39}$.

(a) $\frac{3}{5} = \frac{3 \cdot 2}{5 \cdot 2} = \frac{6}{10}$ *Use the rule* $\frac{a}{b} = \frac{ac}{bc}$. *Let* $c = 2$ *since* $5 \cdot 2 = 10$.

(b) $\frac{21}{39} = \frac{7 \cdot 3}{13 \cdot 3} = \frac{7}{13}$ *Use the rule* $\frac{ac}{bc} = \frac{a}{b}$. *Let* $c = 3$ *because* 3 *is the greatest common factor of* 21 *and* 39.

Practice Problem 1

(a) Find a fraction equivalent to $\frac{7}{3}$ with a denominator of 18.

(b) Reduce $\frac{28}{63}$. ■

1 *Simplifying Algebraic Fractions*

The process of reducing the fraction shown above is sometimes called *dividing out* common factors. Remember, only factors of both the numerator and the denominator can be divided out. Now, to apply this rule, it is usually necessary that the numerator and denominator of the fraction be completely factored. You will need to use your factoring skills from Chapter 5 to accomplish this step. When you are applying this rule you are *simplifying the fraction.*

TEACHING TIP After going through the steps of Example 2, you may also want to show an Additional Class Exercise in which the factor common to numerator and denominator is a numerical value.

Simplify. $\frac{8x - 24}{16x - 40}$

Sol. We factor both numerator and denominator and then apply the basic rule of fractions.

$\frac{8(x - 3)}{8(2x - 5)} = \frac{x - 3}{2x - 5}$

EXAMPLE 2 Simplify. $\frac{4x + 12}{5x + 15}$

$$\frac{4x + 12}{5x + 15} = \frac{4(x + 3)}{5(x + 3)}$$ *Factor 4 from the numerator. Factor 5 from the denominator.*

$$= \frac{4\,(x + 3)}{5\,(x + 3)}$$ *Apply the basic rule of fractions.*

$$= \frac{4}{5}$$

Practice Problem 2 Simplify. $\frac{12x - 6}{14x - 7}$ ■

EXAMPLE 3 Simplify. $\dfrac{x^2 + 9x + 14}{x^2 - 4}$

$$= \frac{(x + 7)(x + 2)}{(x - 2)(x + 2)}$$ *Factor the numerator.* *Factor the denominator.*

$$= \frac{(x + 7)\,(x + 2)}{(x - 2)\,(x + 2)}$$ *Apply the basic rule of fractions.*

$$= \frac{x + 7}{x - 2}$$

Practice Problem 3 Simplify. $\dfrac{4x - 6}{2x^2 - x - 3}$ ■

EXAMPLE 4 Simplify. $\dfrac{ab + ac}{ax - 3ay}$

$$= \frac{a(b + c)}{a(x - 3y)}$$ *Factor the numerator.* *Factor the denominator.*

$$= \frac{a\,(b + c)}{a\,(x - 3y)}$$ *Apply the basic rule of fractions.*

$$= \frac{b + c}{x - 3y}$$

Practice Problem 4 Simplify. $\dfrac{7x^2 - 14xy}{14x^2 - 28xy}$ ■

Some problems may involve more than one step of factoring. Always remember to remove any common factors as the first step, if it is possible to do so.

EXAMPLE 5 Simplify. $\dfrac{9x - x^3}{x^3 + x^2 - 6x}$

$$= \frac{x\,(9 - x^2)}{x\,(x^2 + x - 6)}$$ *Remove a common factor from the polynomial in the numerator and in the denominator.*

$$= \frac{x(3 + x)\,(3 - x)}{x(x + 3)\,(x - 2)}$$ *Factor each polynomial and apply the basic rule of fractions.* *Note that $(3 + x)$ is equivalent to $(x + 3)$ since addition is commutative.*

$$= \frac{3 - x}{x - 2}$$

TEACHING TIP After discussing Example 5, ask students to simplify the following fraction as an Additional Class Exercise. Remind them to remove a common factor first before performing any other factoring step for each polynomial.

Simplify. $\dfrac{2x^3 + 5x^2 - 12x}{6x^3 - 11x^2 + 3x}$

Sol. $\dfrac{x + 4}{3x - 1}$

Practice Problem 5 Simplify. $\dfrac{x^3 + 11x^2 + 30x}{3x^3 + 17x^2 - 6x}$ ■

Be on the lookout for a special simplifying situation. Watch for a situation where *each term in the denominator is opposite in sign* from each term in the numerator. In such cases you should factor -1 or another negative number from one polynomial so that the expression in each set of parentheses is equivalent. Look carefully at the following two examples.

EXAMPLE 6 Simplify. $\dfrac{5x - 15}{6 - 2x}$

Notice that the variable term in the numerator, $5x$, and the variable term in the denominator, $-2x$, *are opposite in sign.* Likewise, the numerical terms -15 and 6 *are opposite in sign.* Factor out a negative number from the denominator.

$$= \frac{5(x-3)}{-2(-3+x)}$$ *Factor 5 from the numerator. Factor -2 from the denominator.*

Note that $(x - 3)$ and $(-3 + x)$ are equivalent (since $+x - 3 = -3 + x$).

$$= \frac{5(x-3)}{-2(-3+x)}$$ *Apply the basic rule of fractions.*

$$= -\frac{5}{2}$$

Note that $\frac{5}{-2}$ is not considered to be in simple form. We usually avoid leaving a negative number in the denominator. Therefore, to simplify, give the result as $-\frac{5}{2}$ or $\frac{-5}{2}$.

Practice Problem 6 Simplify. $\dfrac{2x - 5}{5 - 2x}$ ■

EXAMPLE 7 Simplify. $\dfrac{2x^2 - 11x + 12}{16 - x^2}$

$$= \frac{(x-4)(2x-3)}{(4-x)(4+x)}$$ *Factor numerator and denominator. Observe that $(x - 4)$ and $(4 - x)$ are opposite in sign.*

$$= \frac{(x-4)(2x-3)}{-1(-4+x)(4+x)}$$ *Factor -1 out of $(+4 - x)$ to obtain $-1(-4 + x)$.*

$$= \frac{(x-4)(2x-3)}{-1(-4+x)(4+x)}$$ *Apply the basic rule of fractions, since $(x - 4)$ and $(-4 + x)$ are equivalent.*

$$= \frac{2x-3}{-1(4+x)}$$

$$= -\frac{2x-3}{4+x}$$

Practice Problem 7 Simplify. $\dfrac{4x^2 + 3x - 10}{25 - 16x^2}$ ■

After doing Examples 6 and 7, you will notice a pattern. Whenever the factor in the numerator and the factor in the denominator are exactly opposite in sign, the value -1 results. We could actually make this a definition of that property.

For all monomials A and B where $A \neq B$, it is true that

$$\frac{A - B}{B - A} = -1$$

You may use this definition in reducing fractions if it is helpful to you. Otherwise, you may use the factoring method discussed in Examples 6 and 7.

Some problems will involve two or more variables. In such cases, you will need to factor carefully and make sure that each set of parentheses contains the correct letters.

EXAMPLE 8 Simplify. $\dfrac{x^2 - 7xy + 12y^2}{2x^2 - 7xy - 4y^2}$

$= \dfrac{(x - 4y)(x - 3y)}{(2x + y)(x - 4y)}$ *Factor the numerator. Factor the denominator.*

$= \dfrac{(x - 4y)(x - 3y)}{(2x + y)(x - 4y)}$ *Apply the basic rule of fractions.*

$= \dfrac{x - 3y}{2x + y}$

Practice Problem 8 Simplify. $\dfrac{4x^2 - 9y^2}{4x^2 + 12xy + 9y^2}$ ■

EXAMPLE 9 Simplify. $\dfrac{6a^2 + ab - 7b^2}{36a^2 - 49b^2}$

$= \dfrac{(6a + 7b)(a - b)}{(6a + 7b)(6a - 7b)}$ *Factor the numerator. Factor the denominator.*

$= \dfrac{(6a + 7b)(a - b)}{(6a + 7b)(6a - 7b)}$ *Apply the basic rule of fractions.*

$= \dfrac{a - b}{6a - 7b}$

Practice Problem 9 Simplify. $\dfrac{12x^3 - 48x}{6x - 3x^2}$ ■

TEACHING TIP After discussing Example 9, some students may comment that the factoring seems more difficult if there is a variable in both the first and the last terms of the polynomial. In addition to Example 9, you may wish to do an Additional Class Exercise for them to study.

Simplify. $\dfrac{2x^2 - 15xy + 18y^2}{x^2 - 4xy - 12y^2}$

Sol. $\dfrac{(x - 6y)(2x - 3y)}{(x + 2y)(x - 6y)}$

$= \dfrac{2x - 3y}{x + 2y}$

6.1 Exercises

Simplify.

1. $\dfrac{6x - 3y}{2x - y}$

$\dfrac{3(2x - y)}{2x - y} = 3$

2. $\dfrac{7a + 2b}{14a + 4b}$

$\dfrac{7a + 2b}{2(7a + 2b)} = \dfrac{1}{2}$

3. $\dfrac{6x + 18}{x^2 + 3x}$

$\dfrac{6(x + 3)}{x(x + 3)} = \dfrac{6}{x}$

4. $\dfrac{8x - 12}{2x^3 - 3x^2}$

$\dfrac{4(2x - 3)}{x^2(2x - 3)} = \dfrac{4}{x^2}$

5. $\dfrac{2x - 8}{x^2 - 8x + 16}$

$\dfrac{2(x - 4)}{(x - 4)^2} = \dfrac{2}{x - 4}$

6. $\dfrac{4x^2 + 4x + 1}{1 - 4x^2}$

$\dfrac{(2x + 1)^2}{(1 + 2x)(1 - 2x)} = \dfrac{2x + 1}{1 - 2x}$

7. $\dfrac{x^2 - 9y^2}{x + 3y}$

$\dfrac{(x + 3y)(x - 3y)}{x + 3y} = x - 3y$

8. $\dfrac{xy(x + y^2)}{x^2y^2}$

$\dfrac{x + y^2}{xy}$

9. $\dfrac{6x^2}{2x(x - 3y)}$

$\dfrac{3x}{x - 3y}$

10. $\dfrac{x^2 + 3x - 4}{x^2 - 1}$

$\dfrac{(x + 4)(x - 1)}{(x + 1)(x - 1)} = \dfrac{x + 4}{x + 1}$

11. $\dfrac{x^2 + x - 2}{x^2 - x}$

$\dfrac{(x + 2)(x - 1)}{x(x - 1)} = \dfrac{x + 2}{x}$

12. $\dfrac{x^2 - x - 6}{x^2 - 5x + 6}$

$\dfrac{(x - 3)(x + 2)}{(x - 3)(x - 2)} = \dfrac{x + 2}{x - 2}$

13. $\dfrac{x^2 + 2x - 3}{4x^2 - 5x + 1}$

$\dfrac{(x + 3)(x - 1)}{(4x - 1)(x - 1)} = \dfrac{x + 3}{4x - 1}$

14. $\dfrac{4x^2 - 10x + 6}{2x^2 + x - 3}$

$\dfrac{2(2x - 3)(x - 1)}{(2x + 3)(x - 1)} = \dfrac{2(2x - 3)}{2x + 3}$

15. $\dfrac{x^3 - 8x^2 + 16x}{x^2 + 2x - 24}$

$\dfrac{x(x - 4)^2}{(x + 6)(x - 4)} = \dfrac{x(x - 4)}{x + 6}$

16. $\dfrac{x^2 - 10x - 24}{x^3 + 9x^2 + 14x}$

$\dfrac{(x - 12)(x + 2)}{x(x + 2)(x + 7)} = \dfrac{x - 12}{x(x + 7)}$

17. $\dfrac{3x^2 + 7x - 6}{x^2 + 7x + 12}$

$\dfrac{(3x - 2)(x + 3)}{(x + 4)(x + 3)} = \dfrac{3x - 2}{x + 4}$

18. $\dfrac{x^2 - 5x - 14}{2x^2 - x - 10}$

$\dfrac{(x - 7)(x + 2)}{(2x - 5)(x + 2)} = \dfrac{x - 7}{2x - 5}$

19. $\dfrac{3x^2 - 8x + 5}{4x^2 - 5x + 1}$

$\dfrac{(3x - 5)(x - 1)}{(4x - 1)(x - 1)} = \dfrac{3x - 5}{4x - 1}$

20. $\dfrac{5y^2 + 13y - 6}{5y^2 + 18y - 8}$

$\dfrac{(5y - 2)(y + 3)}{(5y - 2)(y + 4)} = \dfrac{y + 3}{y + 4}$

21. $\dfrac{2x^2 - 5x - 12}{2x^2 - x - 6}$

$\dfrac{(2x + 3)(x - 4)}{(2x + 3)(x - 2)} = \dfrac{x - 4}{x - 2}$

22. $\dfrac{2x^2 + 2x - 12}{x^2 + 3x - 4}$

$\dfrac{2(x + 3)(x - 2)}{(x + 4)(x - 1)}$

23. $\dfrac{2x^3 + 11x^2 + 5x}{2x^3 - 5x^2 - 3x}$

$\dfrac{x(2x + 1)(x + 5)}{x(2x + 1)(x - 3)} = \dfrac{x + 5}{x - 3}$

24. $\dfrac{6x^3 + 5x^2 - 6x}{6x^3 - x^2 - 2x}$

$\dfrac{x(3x - 2)(2x + 3)}{x(3x - 2)(2x + 1)} = \dfrac{2x + 3}{2x + 1}$

25. $\dfrac{6-3x}{2x-4}$
$\dfrac{-3(x-2)}{2(x-2)}=\dfrac{-3}{2}$

26. $\dfrac{5-ay}{ax^2y-5x^2}$
$\dfrac{1(5-ay)}{-x^2(5-ay)}=-\dfrac{1}{x^2}$

27. $\dfrac{2x^2-7x-15}{25-x^2}$
$\dfrac{(2x+3)(x-5)}{(-1)(x+5)(x-5)}=\dfrac{-2x-3}{x+5}$

28. $\dfrac{49-x^2}{2x^2-9x-35}$
$\dfrac{-1(x+7)(x-7)}{(2x+5)(x-7)}=\dfrac{-x-7}{2x+5}$

29. $\dfrac{(4x+5)^2}{8x^2+6x-5}$
$\dfrac{(4x+5)(4x+5)}{(4x+5)(2x-1)}=\dfrac{4x+5}{2x-1}$

30. $\dfrac{6x^2-13x-8}{(3x-8)^2}$
$\dfrac{2x+1}{3x-8}$

31. $\dfrac{2y^2-5y-12}{8+2y-y^2}$
$\dfrac{(2y+3)(y-4)}{-1(y+2)(y-4)}=\dfrac{-2y-3}{y+2}$

32. $\dfrac{6+13y-5y^2}{2y^2-5y-3}$
$\dfrac{(-1)(5y+2)(y-3)}{(2y+1)(y-3)}=\dfrac{-5y-2}{2y+1}$

33. $\dfrac{a^2+2ab-3b^2}{2a^2+5ab-3b^2}$
$\dfrac{(a+3b)(a-b)}{(2a-b)(a+3b)}=\dfrac{a-b}{2a-b}$

34. $\dfrac{a^2+3ab-10b^2}{3a^2-7ab+2b^2}$
$\dfrac{(a+5b)(a-2b)}{(3a-b)(a-2b)}=\dfrac{a+5b}{3a-b}$

35. $\dfrac{9x^2-4y^2}{9x^2+12xy+4y^2}$
$\dfrac{3x-2y}{3x+2y}$

36. $\dfrac{16x^2-24xy+9y^2}{16x^2-9y^2}$
$\dfrac{(4x-3y)(4x-3y)}{(4x-3y)(4x+3y)}=\dfrac{4x-3y}{4x+3y}$

★ 37. $\dfrac{6x^4-9x^3-6x^2}{12x^3+42x^2+18x}$
$\dfrac{x(x-2)}{2(x+3)}$

★ 38. $\dfrac{xa-yb-ya+xb}{xa-ya+2bx-2by}$
$\dfrac{a+b}{a+2b}$

Cumulative Review Problems

39. Multiply. $(3x-7)^2$
$9x^2-42x+49$

40. $(10x+9y)(10x-9y)$ $100x^2-81y^2$

41. $(2x+3)(x-4)(x-2)$ $2x^3-9x^2-2x+24$

42. $(x^2-3x-5)(2x^2+x+3)$
$2x^4-5x^3-10x^2-14x-15$

43. Walter and Ann Daumier wish to divide $4\frac{7}{8}$ acres of farmland into three equal-sized house lots. What will be the acreage of each lot?
$1\dfrac{5}{8}$ acre per lot

44. Ron and Mary Larson are planning to cook a $17\frac{1}{2}$-pound turkey. The directions suggest a cooking time of 22 minutes per pound for turkeys that weigh between 16 and 20 pounds. How many hours and minutes should they allow for an approximate cooking time?
6 hours, 25 minutes

6.2 Multiplication and Division of Rational Expressions

MathPro

Video 15

SSM

After studying this section, you will be able to:

1. *Multiply algebraic fractions and write the answer in simplest form.*
2. *Divide algebraic fractions and write the answer in simplest form.*

1 *Multiplying Algebraic Fractions*

To multiply two fractional expressions, we multiply the numerators and we multiply the denominators. As before, the denominators cannot equal zero.

For any two fractions with polynomials a, b, c, d where $b \neq 0$, $d \neq 0$

$$\frac{a}{b} \text{ and } \frac{c}{d},$$

the definition of multiplication of these two fractions is $\frac{a}{b} \cdot \frac{c}{d} = \frac{ac}{bd}$.

Simplifying or reducing fractions *prior to multiplying them* usually makes the problem easier to do. To do the problem the long way would certainly increase the chance for error! This long approach should be avoided.

As an example, let's do the same problem two ways to see which way seems easier. First let's multiply by the "long way":

$$\frac{5}{7} \times \frac{49}{125}$$

$\frac{5}{7} \times \frac{49}{125} = \frac{245}{875}$ *Multiply the numerators and multiply the denominators.*

$= \frac{7}{25}$ *Reduce the fraction.* (Note: *It takes a bit of trial and error to discover how to reduce it.)*

Compare this with the method of reducing the fractions prior to multiplying them. Let's multiply by simplifying *before* multiplication:

$$\frac{5}{7} \times \frac{49}{125}$$

$\frac{5}{7} \times \frac{7 \cdot 7}{5 \cdot 5 \cdot 5}$ Step 1. *It is easier to factor first. We factor the numerator and denominator of the second fraction.*

$\frac{5 \cdot 7 \cdot 7}{7 \cdot 5 \cdot 5 \cdot 5}$ Step 2. *We express the product as one fraction (by the definition of multiplication of fractions).*

$\frac{5 \cdot 7 \cdot 7}{7 \cdot 5 \cdot 5 \cdot 5} = \frac{7}{25}$ Step 3. *Then we apply the basic rule of fractions to divide common factors of 5 and 7 that appear in the denominator and in the numerator.*

A similar approach can be used with the multiplication of rational algebraic expressions. We first factor the numerator and denominator of each fraction wherever possible. Then we cancel any factor that is common to a numerator and a denominator. Finally, we multiply the remaining numerators and the remaining denominators.

EXAMPLE 1 Multiply. $\dfrac{x^2 - x - 12}{x^2 - 16} \cdot \dfrac{2x^2 + 7x - 4}{x^2 - 4x - 21}$

$$\frac{(x-4)(x+3)}{(x-4)(x+4)} \cdot \frac{(x+4)(2x-1)}{(x+3)(x-7)}$$

Factor wherever possible is always the first step.

$$= \frac{(x-4)(x+3)}{(x-4)(x+4)} \cdot \frac{(x+4)(2x-1)}{(x+3)(x-7)}$$

Apply the basic rule of fractions. (Three pairs of factors divide out.)

$$= \frac{2x-1}{x-7}$$

The final answer.

Practice Problem 1 Multiply. $\dfrac{10x}{x^2 - 7x + 10} \cdot \dfrac{x^2 + 3x - 10}{25x}$ ■

TEACHING TIP To further illustrate the reduction of each fraction, you may want to show the following Additional Class Exercise after you have presented Example 1 to the class.

Simplify.

$$\frac{x^2 - 12x + 20}{18x} \cdot \frac{27x}{2x^2 - 19x - 10}$$

Sol.

$$\frac{(x-10)(x-2)}{(9x)(2)} \cdot \frac{(9x)(3)}{(x-10)(2x+1)}$$

$$= \frac{3(x-2)}{2(2x+1)} \text{ or } \frac{3x-6}{4x+2}$$

The answer may or may not be left in factored form.

In some cases, a given numerator can be factored more than once. You should always be sure that you remove the *common factor first* wherever possible.

EXAMPLE 2 Multiply. $\dfrac{x^4 - 16}{x^3 + 4x} \cdot \dfrac{2x^2 - 8x}{4x^2 + 2x - 12}$

$$\frac{(x^2+4)(x^2-4)}{x(x^2+4)} \cdot \frac{2x(x-4)}{2(2x^2+x-6)}$$

Factor each numerator and denominator. Factoring out the common factor first is very important.

$$= \frac{(x^2+4)(x+2)(x-2)}{x(x^2+4)} \cdot \frac{2x(x-4)}{2(x+2)(2x-3)}$$

Factor again where possible.

$$= \frac{(x^2+4)(x+2)(x-2)}{x(x^2+4)} \cdot \frac{2x(x-4)}{2(x+2)(2x-3)}$$

Remove factors that appear in both numerator and denominator. (There are four such pairs to be removed.)

$$= \frac{(x-2)(x-4)}{(2x-3)} \text{ or } \frac{x^2 - 6x + 8}{2x - 3}$$

Write the answer as one fraction. (Usually, if there is more than one set of parentheses in a numerator, the answer is left in factored form.)

Practice Problem 2 Multiply. $\dfrac{2y^2 - 6y - 8}{y^2 - y - 2} \cdot \dfrac{y^2 - 5y + 6}{2y^2 - 4y - 6}$. ■

2 *Dividing Algebraic Fractions*

For any two fractions $\frac{a}{b}$ and $\frac{c}{d}$ the operation of division can be performed by inverting the second fraction and multiplying it by the first fraction.

The definition for division of fractions is

$$\frac{a}{b} \div \frac{c}{d} = \frac{a}{b} \cdot \frac{d}{c}$$

This property holds whether a, b, c, and d are polynomials or numerical values. (It is assumed, of course, that no denominator is zero.)

In the first step for dividing two algebraic fractions, you should invert the second fraction and write the problem as a multiplication. Then you follow the procedure for multiplying algebraic fractions.

TEACHING TIP After presenting Example 3 to the class, ask them to divide the following two fractions as an Additional Class Exercise.

$$\frac{6x^2 + 5x + 1}{4x^3 - x} \div \frac{3x^2 - 2x - 1}{2x^2 - 3x + 1}$$

Sol. $\frac{1}{x}$

EXAMPLE 3 Divide. $\dfrac{6x + 12y}{2x - 6y} \div \dfrac{9x^2 - 36y^2}{4x^2 - 36y^2}$

$$\frac{6x + 12y}{2x - 6y} \cdot \frac{4x^2 - 36y^2}{9x^2 - 36y^2}$$ *Invert the second fraction and write the problem as the product of two fractions.*

$$= \frac{6(x + 2y)}{2(x - 3y)} \cdot \frac{4(x^2 - 9y^2)}{9(x^2 - 4y^2)}$$ *Factor each numerator and denominator.*

$$= \frac{(3)(2)(x + 2y)}{2(x - 3y)} \cdot \frac{(2)(2)(x + 3y)(x - 3y)}{(3)(3)(x + 2y)(x - 2y)}$$ *Factor again where possible.*

$$= \frac{(3)(2)(x + 2y)}{2(x - 3y)} \cdot \frac{(2)(2)(x + 3y)(x - 3y)}{(3)(3)(x + 2y)(x - 2y)}$$ *Remove factors that appear in both numerator and denominator.*

$$= \frac{(2)(2)(x + 3y)}{3(x - 2y)}$$ *Write the result as one fraction.*

$$= \frac{4(x + 3y)}{3(x - 2y)}$$ *Simplify. (Usually, answers are left in this form.)*

Although it is correct to write this answer as $\dfrac{4x + 12y}{3x - 6y}$ it is customary to leave the answer in factored form in order to ensure that the final answer is simplified.

Practice Problem 3 Divide. $\dfrac{x^2 + 5x + 6}{x^2 + 8x} \div \dfrac{2x^2 + 5x + 2}{2x^2 + x}$ ■

A polynomial that is not in fraction form can be written as a fraction if you write a denominator of 1.

EXAMPLE 4 Divide. $\dfrac{15-3x}{x+6} \div (x^2-9x+20)$

TEACHING TIP After presenting Example 4 to the class, ask them to divide the following two fractions as an Additional Class Exercise.

$$\frac{x^2-25}{6x+30} \div \frac{x-5}{x}$$

Sol. $\dfrac{x}{6}$

Note that $x^2-9x+20$ can be written as $\dfrac{x^2-9x+20}{1}$.

$$\frac{15-3x}{x+6} \cdot \frac{1}{x^2-9x+20}$$ *Invert and multiply.*

$$= \frac{-3(-5+x)}{x+6} \cdot \frac{1}{(x-5)(x-4)}$$ *Factor where possible. Note we had to factor -3 from the first numerator so that we would have a common factor with the second denominator.*

$$= \frac{-3(-5+x)}{x+6} \cdot \frac{1}{(x-5)(x-4)}$$ *Divide out the common factor. $(-5+x)$ is equivalent to $(x-5)$.*

$$= \frac{-3}{(x+6)(x-4)}$$ *The final answer. Note that the answer can be written in several equivalent forms.*

$$\text{or} \quad -\frac{3}{(x+6)(x-4)}$$

$$\text{or} \quad \frac{3}{(x+6)(4-x)}$$ *Take a minute to study each. Each is equivalent to the others.*

Practice Problem 4 Divide. $\dfrac{x+3}{x-3} \div (9-x^2)$ ■

A word of caution:

It is logical to assume that the types of problems encountered in Section 6.2 have at least one common factor that can be divided out. Therefore if after factoring you do not observe any common factors, you should be somewhat suspicious. In such cases, it would be wise to double check your factoring steps to see if an error has been made.

6.2 Exercises

Verbal and Writing Skills

1. Before we do any multiplication in the multiplication of algebraic fractions, we always first try to factor completely the numerator and denominator and try to simplify.

2. Division of two algebraic functions is done by keeping the first fraction unchanged and then multiplying by the reciprocal of the second fraction.

Perform the operation indicated.

3. $\dfrac{x+3}{x+7} \cdot \dfrac{x^2+3x-10}{x^2+x-6}$

$\dfrac{x+3}{x+7} \cdot \dfrac{(x+5)(x-2)}{(x+3)(x-2)} = \dfrac{x+5}{x+7}$

4. $\dfrac{5x^2-5}{3x+3} \cdot \dfrac{3x+6}{5x^2-15x+10}$

$\dfrac{5(x+1)(x-1)}{3(x+1)} \cdot \dfrac{3(x+2)}{5(x-1)(x-2)} = \dfrac{x+2}{x-2}$

5. $\dfrac{x^2+2x}{6x} \cdot \dfrac{3x^2}{x^2-4}$

$\dfrac{x(x+2)}{(3)(2)(x)} \cdot \dfrac{3x^2}{(x+2)(x-2)} = \dfrac{x^2}{2(x-2)}$

6. $\dfrac{3x+12}{8x^3} \cdot \dfrac{16x^2}{9x+36}$

$\dfrac{3(x+4)}{(8)(x)(x)(x)} \cdot \dfrac{(2)(8)(x)(x)}{(3)(3)(x+4)} = \dfrac{2}{3x}$

7. $\dfrac{x^2+3x-10}{x^2+x-20} \cdot \dfrac{x^2-3x-4}{x^2+4x+3}$

$\dfrac{(x-2)(x+5)}{(x-4)(x+5)} \cdot \dfrac{(x-4)(x+1)}{(x+3)(x+1)} = \dfrac{x-2}{x+3}$

8. $\dfrac{x^2-x-20}{x^2-3x-10} \cdot \dfrac{x^2+7x+10}{x^2+4x-5}$

$\dfrac{(x-5)(x+4)}{(x-5)(x+2)} \cdot \dfrac{(x+2)(x+5)}{(x+5)(x-1)} = \dfrac{x+4}{x-1}$

9. $\dfrac{x^2+7x-8}{2x^2-18} \cdot \dfrac{2x^2+20x+42}{7x^2-7x}$

$\dfrac{(x+8)(x-1)}{2(x+3)(x-3)} \cdot \dfrac{2(x+3)(x+7)}{7x(x-1)} = \dfrac{(x+8)(x+7)}{7x(x-3)}$

10. $\dfrac{3x^2-3x-18}{x^2+2x-15} \cdot \dfrac{2x^2+6x-20}{2x^2-12x+16}$

$\dfrac{3(x-3)(x+2)}{(x-3)(x+5)} \cdot \dfrac{2(x+5)(x-2)}{2(x-4)(x-2)} = \dfrac{3(x+2)}{x-4}$

11. $\dfrac{5x^2+6x+1}{x^2+5x+6} \div (5x+1)$

$\dfrac{(5x+1)(x+1)}{(x+2)(x+3)} \cdot \dfrac{1}{5x+1} = \dfrac{x+1}{(x+2)(x+3)}$

12. $(3x-2) \div \dfrac{9x^2-4}{6x+4}$

$\dfrac{(3x-2)}{1} \cdot \dfrac{2(3x+2)}{(3x+2)(3x-2)} = 2$

13. $\dfrac{3x^2+12xy+12y^2}{x^2+4xy+3y^2} \div \dfrac{4x+8y}{x+y}$

$\dfrac{3(x+2y)(x+2y)}{(x+3y)(x+y)} \cdot \dfrac{(x+y)}{4(x+2y)} = \dfrac{3(x+2y)}{4(x+3y)}$

14. $\dfrac{5x^2+10xy+5y^2}{x^2+5xy+6y^2} \div \dfrac{3x+3y}{x+2y}$

$\dfrac{5(x+y)(x+y)}{(x+3y)(x+2y)} \cdot \dfrac{(x+2y)}{3(x+y)} = \dfrac{5(x+y)}{3(x+3y)}$

15. $\dfrac{xy - y^2}{x^2 + 2x + 1} \div \dfrac{2x^2 + xy - 3y^2}{2x^2 + 5xy + 3y^2}$ $\dfrac{y(x-y)}{(x+1)^2} \cdot \dfrac{(2x+3y)(x+y)}{(2x+3y)(x-y)} = \dfrac{y(x+y)}{(x+1)^2}$

16. $\dfrac{4a^2 - b^2}{a^2 + 4ab + 4b^2} \div \dfrac{4a - 2b}{3a + 6b}$ $\dfrac{(2a+b)(2a-b)}{(a+2b)^2} \cdot \dfrac{3(a+2b)}{2(2a-b)} = \dfrac{3(2a+b)}{2(a+2b)}$

17. $\dfrac{x^2 + 5x - 14}{x - 5} \div \dfrac{x^2 + 12x + 35}{15 - 3x}$ $\dfrac{(x+7)(x-2)}{(x-5)} \cdot \dfrac{-3(x-5)}{(x+5)(x+7)} = \dfrac{-3(x-2)}{(x+5)}$

18. $\dfrac{3x^2 + 13x + 4}{16 - x^2} \div \dfrac{3x^2 - 5x - 2}{3x - 12}$ $\dfrac{(3x+1)(x+4)}{(-1)(x+4)(x-4)} \cdot \dfrac{3(x-4)}{(3x+1)(x-2)} = \dfrac{-3}{x-2}$

19. $\dfrac{(x+5)^2}{3x^2 - 7x + 2} \cdot \dfrac{x^2 - 4x + 4}{x + 5}$ $\dfrac{(x+5)(x-2)}{(3x-1)}$

20. $\dfrac{3x^2 - 10x - 8}{(4x+5)^2} \cdot \dfrac{4x + 5}{(x-4)^2}$ $\dfrac{3x+2}{(4x+5)(x-4)}$

21. $\dfrac{y^2 - y - 2}{y^2 + 4y + 3} \cdot \dfrac{y^2 + y - 6}{y^2 - 4y + 4}$ $\dfrac{(y-2)(y+1)}{(y+3)(y+1)} \cdot \dfrac{(y+3)(y-2)}{(y-2)(y-2)} = 1$

22. $\dfrac{y^2 - y - 6}{y^2 + 5y - 24} \cdot \dfrac{y^2 + 6y - 16}{y^2 - 4}$ $\dfrac{(y-3)(y+2)}{(y+8)(y-3)} \cdot \dfrac{(y+8)(y-2)}{(y+2)(y-2)} = 1$

23. $\dfrac{3x^2 - 7x + 2}{3x^2 + 2x - 1} \cdot \dfrac{2x^2 - 9x - 5}{x^2 + x - 6} \cdot \dfrac{4x^2 + 11x - 3}{x^2 - 11x + 30}$ $\dfrac{(3x-1)(x-2)}{(3x-1)(x+1)} \cdot \dfrac{(2x+1)(x-5)}{(x+3)(x-2)} \cdot \dfrac{(4x-1)(x+3)}{(x-5)(x-6)}$ $= \dfrac{(2x+1)(4x-1)}{(x+1)(x-6)}$

24. $\dfrac{6a^2 - a - 1}{4a^2 - 1} \cdot \dfrac{10a^2 + a - 2}{3a^2 - 17a - 6} \cdot \dfrac{a^2 - 3a - 18}{5a^2 + 3a - 2}$ $\dfrac{(3a+1)(2a-1)(2a+1)(5a-2)(a-6)(a+3)}{(2a+1)(2a-1)(3a+1)(a-6)(5a-2)(a+1)}$ $= \dfrac{a+3}{a+1}$

Cumulative Review Problems

25. Solve. $5x^2 - 7x + 11 = 5x^2 - x + 2$ $x = \dfrac{3}{2}$

26. Multiply. $(7x^2 - x - 1)(x - 3)$ $7x^3 - 22x^2 + 2x + 3$

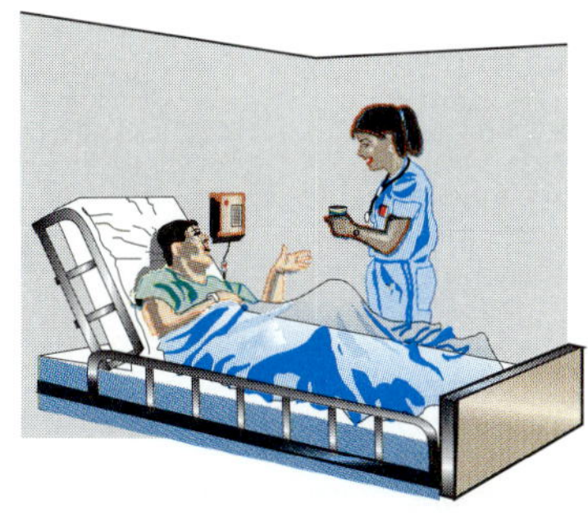

27. Suzanne Hartling is a nurse at Meadowbrook Rehabilitation Center. She must give an 80-kilogram patient a medication based on his weight. The dosage is $\frac{1}{16}$ milligram of medication for each kilogram of patient weight. Find the amount of medication she should give this patient. 5 milligrams of medication

28. Denise Abrahamson purchased 4 gallons of milk. The price of 4 gallons of milk is the same as the price of 2 gallons of milk plus some other groceries that cost \$4.76. What is the price of a gallon of milk? \$2.38 for a gallon of milk

6.3 Addition and Subtraction of Rational Expressions

MathPro

Video 16

SSM

After studying this section, you will be able to:

1. *Add and subtract algebraic fractions with the same denominator.*
2. *Determine the LCD for two or more algebraic fractions with different denominators.*
3. *Add or subtract algebraic fractions with different denominators.*

1 *Adding and Subtracting Algebraic Fractions with the Same Denominator*

If fractional algebraic expressions have the same denominator, they can be combined in a fashion similar to that used in arithmetic. The numerators are added or subtracted and the denominator remains the same.

TEACHING TIP A comprehensive example with both numerical and algebraic fractions helps students to tie in previous knowledge of fractions and bridges the gap from not needing a common denominator (when multiplying and dividing) to needing one (when adding or subtracting). Presentation to class: ''Just as we add numerical fractions with a common denominator, so we can add algebraic fractions with a common denominator.''

$$\frac{3}{7}+\frac{2}{7}=\frac{5}{7}$$

$$\frac{x+3}{(x+2)(x+5)}+\frac{2x+2}{(x+2)(x+5)}$$
$$=\frac{3x+5}{(x+2)(x+5)}$$

''Just as we can subtract numerical fractions with a common denominator, we can subtract algebraic fractions with a common denominator.''

$$\frac{9}{13}-\frac{5}{13}=\frac{4}{13}$$

$$\frac{9}{17x^2y}-\frac{3}{17x^2y}=\frac{6}{17x^2y}$$

Adding Fractional Algebraic Expressions

For any polynomials a, b, c,

$$\frac{a}{b}+\frac{c}{b}=\frac{a+c}{b} \quad \text{where } b \neq 0$$

EXAMPLE 1 Add. $\dfrac{5a}{a+2b}+\dfrac{6a}{a+2b}$

$$\frac{5a}{a+2b}+\frac{6a}{a+2b}=\frac{5a+6a}{a+2b}=\frac{11a}{a+2b}$$

Note the denominators are the same. Add the numerators.

Practice Problem 1 Add. $\dfrac{2s+t}{2s-t}+\dfrac{s-t}{2s-t}$ ■

Subtracting Fractional Algebraic Expressions

For any polynomials a, b, c,

$$\frac{a}{b}-\frac{c}{b}=\frac{a-c}{b} \quad \text{where } b \neq 0$$

EXAMPLE 2 Subtract. $\dfrac{3x}{(x+y)(x-2y)}-\dfrac{8x}{(x+y)(x-2y)}$

$$\frac{3x}{(x+y)(x-2y)}-\frac{8x}{(x+y)(x-2y)}=\frac{3x-8x}{(x+y)(x-2y)}$$ *Write as one fraction.*

$$=\frac{-5x}{(x+y)(x-2y)}$$ *Simplify.*

Practice Problem 2 Subtract. $\dfrac{b}{(a-2b)(a+b)}-\dfrac{2b}{(a-2b)(a+b)}$ ■

2 *Determining the LCD*

How do we add or subtract fractional expressions when the denominators are not the same? First we must find the **least common denominator** (LCD). You need to be clear on how to find a least common denominator and how to add and subtract fractions from arithmetic before you attempt this section. Review Sections 0.1 and 0.2 if you have any weaknesses on this topic.

How to Find the LCD of Two or More Algebraic Fractions

1. Factor each denominator completely.
2. The LCD is the product of each *different factor.*
3. If a factor occurs more than once in any one denominator, the LCD will contain that factor repeated the greatest number of times that it occurs in any one denominator.

TEACHING TIP After explaining the method to find the LCD and going over a couple of examples, ask students to find the LCD for each of the following fractions as an Additional Class Exercise.
Find the LCD only of

A. $\frac{3}{x+6}, \frac{5}{x^2-36}$

B. $\frac{7}{a^3b}, \frac{3}{ab^2}$

C. $\frac{1}{x^2-12x+20}, \frac{3}{2x^2-19x-10}$

Sol.

A. LCD $= (x+6)(x-6)$

B. LCD $= a^3b^2$

C. LCD $= (x-2)(x-10)(2x+1)$

EXAMPLE 3 Find the LCD of the following algebraic fractions. $\frac{5}{2x-4}, \frac{6}{3x-6}$

Factor each denominator.

$$2x - 4 = 2(x-2)$$
$$3x - 6 = 3(x-2)$$

The different factors are 2, 3, and $(x-2)$. Since no factor is repeated more than once in any one denominator, the LCD is the product of these three factors.

$$\text{LCD} = (2)(3)(x-2) = 6(x-2)$$

Practice Problem 3 Find the LCD of the following. $\frac{7}{6x+21}, \frac{13}{10x+35}$ ■

EXAMPLE 4 Find the LCD. **(a)** $\frac{5}{12ab^2c}, \frac{13}{18a^3bc^4}$

(b) $\frac{7}{x^2-2x}, \frac{9}{x^2-4x+4}$ **(c)** $\frac{8}{x^2-5x+4}, \frac{12}{x^2+2x-3}$

If a factor occurs more than once in any one denominator, the LCD will contain that factor repeated the greatest number of times that it occurs in any one denominator.

(a)
$$\begin{aligned} 12ab^2c &= 2\cdot 2\cdot 3\cdot \quad a \quad \cdot\; b\cdot b\cdot c \\ 18a^3bc^4 &= \quad 2\cdot 3\cdot 3\cdot a\cdot a\cdot a\cdot b\cdot \quad c\cdot c\cdot c\cdot c \\ &\quad \downarrow\;\downarrow\;\downarrow\;\downarrow\;\downarrow\;\downarrow\;\downarrow\;\downarrow\;\downarrow\;\downarrow\;\downarrow\;\downarrow\;\downarrow \\ &\quad 2\cdot 2\cdot 3\cdot 3\cdot a\cdot a\cdot a\cdot b\cdot b\cdot c\cdot c\cdot c\cdot c \end{aligned}$$

$$\begin{aligned} \text{LCD} &= 2^2\cdot 3^2\cdot a^3\cdot b^2\cdot c^4 \\ &= 36a^3b^2c^4 \end{aligned}$$

(b)
$$\begin{aligned} x^2 - 2x &= x(x-2) \\ x^2 - 4x + 4 &= (x-2)(x-2) \\ \text{LCD} &= x(x-2)(x-2) \end{aligned}$$

(c)
$$\begin{aligned} x^2 - 5x + 4 &= (x-4)(x-1) \\ x^2 + 2x - 3 &= (x-1)(x+3) \\ \text{LCD} &= (x-4)(x-1)(x+3) \end{aligned}$$

Practice Problem 4 Find the LCD. **(a)** $\frac{3}{50xy^2z}, \frac{19}{40x^3yz}$

(b) $\frac{12}{x^2+5x}, \frac{16}{x^2+10x+25}$ **(c)** $\frac{2}{x^2+5x+6}, \frac{6}{3x^2+5x-2}$ ■

3 *Adding and Subtracting Fractions with Different Denominators*

If two fractions have different denominators, we first change them to equivalent fractions with the least common denominator. Then we add or subtract the numerators and keep the common denominator.

EXAMPLE 5 Add. $\frac{5}{xy} + \frac{2}{y}$

The denominators are different. Find the LCD. The two factors are x and y. We observe that the LCD is xy.

$$\frac{5}{xy} + \frac{2}{y} = \frac{5}{xy} + \frac{2}{y} \cdot \frac{x}{x}$$ *Multiply the second fraction by* $\frac{x}{x}$.

$$= \frac{5}{xy} + \frac{2x}{xy}$$ *Now each fraction has a common denominator of* xy.

$$= \frac{5 + 2x}{xy}$$ *Write the sum as one fraction.*

Practice Problem 5 Add. $\frac{7}{a} + \frac{3}{abc}$ ■

EXAMPLE 6 Add. $\frac{3x}{(x + y)(x - y)} + \frac{5}{x + y}$

The two factors are $(x + y)$ and $(x - y)$. We observe that the LCD $= (x + y)(x - y)$.

$$\frac{3x}{(x + y)(x - y)} + \frac{5}{(x + y)} \cdot \frac{x - y}{x - y}$$ *Multiply the second fraction by* $\frac{x - y}{x - y}$.

$$= \frac{3x}{(x + y)(x - y)} + \frac{5x - 5y}{(x + y)(x - y)}$$ *Now each fraction has a common denominator of* $(x + y)(x - y)$.

$$= \frac{3x + 5x - 5y}{(x + y)(x - y)}$$ *Write the sum of the numerators over one common denominator.*

$$= \frac{8x - 5y}{(x + y)(x - y)}$$ *Collect like terms.*

Practice Problem 6 Add. $\frac{2a - b}{(a + 2b)(a - 2b)} + \frac{2}{(a + 2b)}$ ■

It is important to remember that the LCD is the smallest algebraic expression into which each denominator can be divided. For algebraic expressions the LCD must contain *each factor* that appears in any denominator. If the factor is repeated, the LCD must contain each factor the greatest number of times that it appears in any one denominator.

EXAMPLE 7 Add. $\frac{5}{xy^2} + \frac{3}{x^2y} + \frac{2}{y^3}$

The x factor is squared in one fraction. The y factor is cubed in one fraction. Therefore, the LCD is x^2y^3.

$$\frac{5}{xy^2} \cdot \frac{xy}{xy} + \frac{3}{x^2y} \cdot \frac{y^2}{y^2} + \frac{2}{y^3} \cdot \frac{x^2}{x^2}$$

Multiply each fraction by the appropriate value to obtain x^2y^3 in each denominator.

*(*Note: *Here we* always *multiply the numerator and denominator of any one fraction by the* same value *to obtain an equivalent fraction.*

$$= \frac{5xy}{x^2y^3} + \frac{3y^2}{x^2y^3} + \frac{2x^2}{x^2y^3}$$

Now all fractions have a common denominator.

$$= \frac{5xy + 3y^2 + 2x^2}{x^2y^3}$$

Write the sum as one fraction.

Practice Problem 7 Add. $\frac{3}{x^2} + \frac{4x}{xy} + \frac{7}{y}$ ■

In many cases, the denominators in an addition or subtraction problem are not in factored form. You must factor each denominator in order to determine the LCD. Collect like terms in the numerator; then look to see if that final numerator can be factored. If so, the fraction can sometimes be simplified.

EXAMPLE 8 Add. $\frac{5}{x^2 - y^2} + \frac{3x}{x^3 + x^2y}$

$$\frac{5}{x^2 - y^2} + \frac{3x}{x^3 + x^2y} = \frac{5}{(x + y)(x - y)} + \frac{3x}{x^2(x + y)}$$

Factor the two denominators. Observe that the LCD is $x^2(x + y)(x - y)$.

$$= \frac{5}{(x + y)(x - y)} \cdot \frac{x^2}{x^2} + \frac{3x}{x^2(x + y)} \cdot \frac{x - y}{x - y}$$

Multiply each fraction by the appropriate value to obtain a common denominator of $x^2(x + y)(x - y)$.

$$= \frac{5x^2}{x^2(x + y)(x - y)} + \frac{3x^2 - 3xy}{x^2(x + y)(x - y)}$$

$$= \frac{5x^2 + 3x^2 - 3xy}{x^2(x + y)(x - y)}$$

Write the sum of the numerators over one common denominator.

$$= \frac{8x^2 - 3xy}{x^2(x + y)(x - y)}$$

Collect like terms.

$$= \frac{x(8x - 3y)}{x^2(x + y)(x - y)}$$

Remove the common factor x in the numerator and denominator and simplify.

$$= \frac{8x - 3y}{x(x + y)(x - y)}$$

Practice Problem 8 Add. $\frac{7a}{a^2 + 2ab + b^2} + \frac{4}{a^2 + ab}$ ■

It is very easy to make a mistake in sign when subtracting two fractions. You will find it helpful to place parentheses around the numerator of the second fraction so that you will not forget to subtract the entire numerator.

EXAMPLE 9 Subtract. $\dfrac{3x+4}{x-2} - \dfrac{x-3}{2x-4}$

Factor the second denominator.

$$\frac{3x+4}{x-2} - \frac{x-3}{2(x-2)}$$

Observe that the LCD is $2(x-2)$.

$$= \frac{2}{2} \cdot \frac{(3x+4)}{x-2} - \frac{x-3}{2(x-2)}$$

Multiply the first fraction by $\frac{2}{2}$ so that the resulting fraction will have the common denominator.

$$= \frac{2(3x+4)-(x-3)}{2(x-2)}$$

Write the indicated subtraction as one fraction. Note the parentheses around $(x-3)$.

$$= \frac{6x+8-x+3}{2(x-2)}$$

Remove the parentheses in the numerator.

$$= \frac{5x+11}{2(x-2)}$$

Collect like terms.

Practice Problem 9 Subtract. $\dfrac{x+7}{3x-9} - \dfrac{x-6}{x-3}$ ■

To avoid making errors when subtracting two fractions, some students find it helpful to change subtraction to addition of the opposite of the second fraction.

TEACHING TIP Students will often incorrectly divide out common factors before adding or subtracting fractions. In Example 10 and in similar problems, students will be tempted to divide out the $(x+4)$ quantities when they arrive at the expression

$$\frac{8x+(-4)(x+4)}{(x+4)(x-4)}$$

Be sure to point out that dividing out such quantities at this step would not be correct. The *terms* in the numerator must be combined first. Stress the idea that common factors should only be removed at the very last step of the problem. Remind students that the value they remove must be a *factor* of the entire numerator and the entire denominator. They cannot divide out terms that are separated by a plus or a minus sign.

EXAMPLE 10 Subtract. $\dfrac{8x}{x^2-16} - \dfrac{4}{x-4}$

$$\frac{8x}{x^2-16} - \frac{4}{x-4} = \frac{8x}{(x+4)(x-4)} + \frac{-4}{x-4}$$

Factor the first denominator. Use the property that

$$\frac{a}{b} - \frac{c}{b} = \frac{a}{b} + \frac{-c}{b}.$$

$$= \frac{8x}{(x+4)(x-4)} + \frac{-4}{x-4} \cdot \frac{x+4}{x+4}$$

Multiply the second fraction by $\dfrac{x+4}{x+4}$.

$$= \frac{8x+(-4)(x+4)}{(x+4)(x-4)}$$

Write the addition of the numerators over one common denominator.

$$= \frac{8x-4x-16}{(x+4)(x-4)}$$

Remove parentheses.

$$= \frac{4x-16}{(x+4)(x-4)}$$

Collect like terms. Note that the numerator can be factored.

$$= \frac{4(x-4)}{(x+4)(x-4)}$$

Divide out. Since $(x-4)$ is a factor *of the numerator* and *the denominator, we may divide out the common factor.*

$$= \frac{4}{x+4}$$

Practice Problem 10 Subtract and simplify your answer. $\dfrac{x-2}{x^2-4} - \dfrac{x+1}{2x^2+4x}$ ■

6.3 Exercises

Perform the operation indicated. Be sure to simplify your answer.

1. $\dfrac{7}{12x^3} + \dfrac{1}{12x^3}$ $\dfrac{2}{3x^3}$

2. $\dfrac{9}{21ab^2} + \dfrac{5}{21ab^2}$ $\dfrac{2}{3ab^2}$

3. $\dfrac{5}{x+2} + \dfrac{x+3}{2+x}$ $\dfrac{x+8}{x+2}$

4. $\dfrac{4x+5}{x+6} + \dfrac{8}{6+x}$ $\dfrac{4x+13}{x+6}$

5. $\dfrac{x}{x-4} - \dfrac{x+1}{x-4}$ $\dfrac{-1}{x-4}$

6. $\dfrac{x+3}{x+5} - \dfrac{x-8}{x+5}$ $\dfrac{11}{x+5}$

7. $\dfrac{8x+3}{5x+7} - \dfrac{6x+10}{5x+7}$ $\dfrac{2x-7}{5x+7}$

8. $\dfrac{7x-1}{3x-5} - \dfrac{4x-6}{3x-5}$ $\dfrac{3x+5}{3x-5}$

9. $\dfrac{8x-1}{2x-7} - \dfrac{6x-8}{2x-7}$ $\dfrac{2x+7}{2x-7}$

Find the LCD *of each of the following fractions. Find the* LCD *only. Do not combine fractions.*

10. $\dfrac{13}{3ab}, \dfrac{7}{a^2b^2}$ LCD $= 3a^2b^2$

11. $\dfrac{12}{5a^2}, \dfrac{9}{a^3}$ $5a^3$

12. $\dfrac{3}{16xy^4}, \dfrac{5}{24x^3y^2}$ LCD $= 48x^3y^4$

13. $\dfrac{7}{15x^3y^5}, \dfrac{13}{40x^4y}$ LCD $= 120x^4y^5$

14. $\dfrac{8}{x+3}, \dfrac{15}{x^2-9}$ LCD $= x^2 - 9$

15. $\dfrac{13}{x^2-16}, \dfrac{7}{x-4}$ LCD $= x^2 - 16$

16. $\dfrac{6}{2x^2+x-3}, \dfrac{5}{4x^2+12x+9}$ LCD $= (x-1)(2x+3)^2$

17. $\dfrac{3}{16x^2-8x+1}, \dfrac{6}{4x^2+7x-2}$ LCD $= (x+2)(4x-1)^2$

Add or subtract the following fractions.

18. $\dfrac{2}{x^2y} + \dfrac{7}{xy}$ $\dfrac{2}{x^2y} + \dfrac{7}{xy}\left(\dfrac{x}{x}\right) = \dfrac{2+7x}{x^2y}$

19. $\dfrac{8}{xy} + \dfrac{2}{xy^2}$ $\dfrac{8}{xy}\left(\dfrac{y}{y}\right) + \dfrac{2}{xy^2} = \dfrac{8y+2}{xy^2}$

20. $\dfrac{2}{x+2} + \dfrac{3}{x^2-4}$
$\dfrac{2}{(x+2)}\left(\dfrac{x-2}{x-2}\right) + \dfrac{3}{(x+2)(x-2)} = \dfrac{2x-1}{(x+2)(x-2)}$

21. $\dfrac{5}{x^2+5x+6} + \dfrac{2}{x+3}$
$\dfrac{5}{(x+2)(x+3)} + \dfrac{2}{(x+3)}\left(\dfrac{x+2}{x+2}\right) = \dfrac{2x+9}{(x+3)(x+2)}$

22. $\dfrac{3y}{y+2} + \dfrac{y}{y-2}$ $\dfrac{3y}{(y+2)}\left(\dfrac{y-2}{y-2}\right) + \dfrac{y}{(y-2)}\left(\dfrac{y+2}{y+2}\right)$
$= \dfrac{4y^2-4y}{(y+2)(y-2)}$

23. $\dfrac{2}{y-1} + \dfrac{2}{y+1}$
$\dfrac{2}{(y-1)}\left(\dfrac{y+1}{y+1}\right) + \dfrac{2}{(y+1)}\left(\dfrac{y-1}{y-1}\right)$
$= \dfrac{4y}{(y+1)(y-1)}$

24. $\dfrac{4}{a+3}+\dfrac{2}{3a}$ $\quad \dfrac{4}{a+3}\left(\dfrac{3a}{3a}\right)+\dfrac{2}{3a}\left(\dfrac{a+3}{a+3}\right)$

$= \dfrac{14a+6}{3a(a+3)}$

25. $\dfrac{5}{2ab}+\dfrac{1}{2a+b}$ $\quad \dfrac{5}{2ab}\left(\dfrac{2a+b}{2a+b}\right)+\dfrac{1}{(2a+b)}\left(\dfrac{2ab}{2ab}\right)$

$= \dfrac{10a+5b+2ab}{2ab(2a+b)}$

26. $\dfrac{2}{3xy}+\dfrac{1}{6yz}$ $\quad \dfrac{2}{3xy}\left(\dfrac{2z}{2z}\right)+\dfrac{1}{6yz}\left(\dfrac{x}{x}\right)$

$= \dfrac{4z+x}{6xyz}$

27. $\dfrac{x-3}{4x}+\dfrac{6}{x^2}$ $\quad \dfrac{x-3}{4x}\left(\dfrac{x}{x}\right)+\dfrac{6}{x^2}\left(\dfrac{4}{4}\right)$

$= \dfrac{x^2-3x+24}{4x^2}$

28. $\dfrac{1}{x^2+3x+2}+\dfrac{1}{x^2+4x+3}$

$\dfrac{1}{(x+1)(x+2)}\left(\dfrac{x+3}{x+3}\right)+\dfrac{1}{(x+1)(x+3)}\left(\dfrac{x+2}{x+2}\right)$

$= \dfrac{2x+5}{(x+1)(x+2)(x+3)}$

29. $\dfrac{5}{x^2-25}+\dfrac{1}{x^2+10x+25}$

$\dfrac{5}{(x+5)(x-5)}\left(\dfrac{x+5}{x+5}\right)+\dfrac{1}{(x+5)^2}\left(\dfrac{x-5}{x-5}\right)$

$= \dfrac{6x+20}{(x+5)^2(x-5)}$

30. $\dfrac{3x-8}{x^2-5x+6}+\dfrac{x+2}{x^2-6x+8}$

$\dfrac{3x-8}{(x-2)(x-3)}\left(\dfrac{x-4}{x-4}\right)+\dfrac{x+2}{(x-2)(x-4)}\left(\dfrac{x-3}{x-3}\right)$

$= \dfrac{4x^2-21x+26}{(x-2)(x-4)(x-3)} = \dfrac{4x-13}{(x-4)(x-3)}$

31. $\dfrac{3x+5}{x^2+4x+3}+\dfrac{-x+5}{x^2+2x-3}$

$\dfrac{3x+5}{(x+1)(x+3)}\left(\dfrac{x-1}{x-1}\right)+\dfrac{(-x+5)}{(x+3)(x-1)}\cdot\left(\dfrac{x+1}{x+1}\right)$

$= \dfrac{2x}{(x+1)(x-1)}$

32. $\dfrac{a+b}{3}-\dfrac{a-2b}{4}$

$\dfrac{a+b}{3}\left(\dfrac{4}{4}\right)+\dfrac{-a+2b}{4}\left(\dfrac{3}{3}\right)$

$= \dfrac{a+10b}{12}$

33. $\dfrac{a+1}{2}-\dfrac{a-1}{3}$

$\dfrac{a+1}{2}\left(\dfrac{3}{3}\right)+\dfrac{-a+1}{3}\left(\dfrac{2}{2}\right)$

$= \dfrac{a+5}{6}$

34. $\dfrac{3+y}{7y}-\dfrac{2y+5}{2y}$

$\dfrac{3+y}{7y}\left(\dfrac{2}{2}\right)-\dfrac{2y+5}{2y}\left(\dfrac{7}{7}\right)$

$= \dfrac{-12y-29}{14y}$

35. $\dfrac{2z+1}{6z}-\dfrac{3z+2}{5z}$

$\dfrac{2z+1}{6z}\left(\dfrac{5}{5}\right)-\dfrac{3z+2}{5z}\left(\dfrac{6}{6}\right)$

$= \dfrac{-8z-7}{30z}$

36. $\dfrac{8}{2x-3}-\dfrac{6}{x+2}$

$\dfrac{8}{2x-3}\left(\dfrac{x+2}{x+2}\right)+\dfrac{-6}{(x+2)}\left(\dfrac{2x-3}{2x-3}\right)$

$= \dfrac{-4x+34}{(2x-3)(x+2)}$

37. $\dfrac{6}{3x-4}-\dfrac{5}{4x-3}$

$\dfrac{6}{3x-4}\left(\dfrac{4x-3}{4x-3}\right)+\dfrac{-5}{4x-3}\left(\dfrac{3x-4}{3x-4}\right)$

$= \dfrac{9x+2}{(3x-4)(4x-3)}$

38. $\dfrac{x}{x^2 + 2x - 3} - \dfrac{x}{x^2 - 5x + 4}$

$\dfrac{x}{(x+3)(x-1)}\left(\dfrac{x-4}{x-4}\right) + \dfrac{-x}{(x-1)(x-4)}\left(\dfrac{x+3}{x+3}\right)$

$= \dfrac{-7x}{(x+3)(x-1)(x-4)}$

39. $\dfrac{1}{x^2 - 2x} - \dfrac{5}{x^2 - 4x + 4}$

$\dfrac{1}{x(x-2)}\left(\dfrac{x-2}{x-2}\right) + \dfrac{-5}{(x-2)(x-2)}\left(\dfrac{x}{x}\right)$

$= \dfrac{-4x-2}{x(x-2)^2}$

40. $\dfrac{3y}{8y^2 + 2y - 1} - \dfrac{5y}{2y^2 - 9y - 5}$

$\dfrac{3y}{(4y-1)(2y+1)}\left(\dfrac{y-5}{y-5}\right) + \dfrac{-5y}{(2y+1)(y-5)}\left(\dfrac{4y-1}{4y-1}\right)$

$= \dfrac{-17y^2 - 10y}{(4y-1)(2y+1)(y-5)}$

41. $\dfrac{2x}{x^2 + 5x + 6} - \dfrac{x+1}{x^2 + 2x - 3}$

$\dfrac{2x}{(x+2)(x+3)}\left(\dfrac{x-1}{x-1}\right) + \dfrac{(-x-1)}{(x+3)(x-1)}\left(\dfrac{x+2}{x+2}\right)$

$= \dfrac{x^2 - 5x - 2}{(x+2)(x+3)(x-1)}$

42. $\dfrac{5}{x} + \dfrac{x}{x-5} + \dfrac{3}{5x}$

$\dfrac{5}{x}\left[\dfrac{5(x-5)}{5(x-5)}\right] + \dfrac{x}{x-5}\left[\dfrac{5x}{5x}\right] + \dfrac{3}{5x}\left[\dfrac{x-5}{x-5}\right]$

$= \dfrac{5x^2 + 28x - 140}{5x(x-5)}$

43. $\dfrac{3}{ab} + \dfrac{2}{a-b} + \dfrac{a}{b}$

$\dfrac{3}{ab}\left[\dfrac{a-b}{a-b}\right] + \dfrac{2}{a-b}\left[\dfrac{ab}{ab}\right] + \dfrac{a}{b}\left[\dfrac{a(a-b)}{a(a-b)}\right]$

$= \dfrac{3a - 3b + 2ab + a^3 - a^2b}{ab(a-b)}$

44. $\dfrac{6a}{a-7} - \dfrac{3a}{7-a}$

$\dfrac{6a}{a-7} - \dfrac{3a}{7-a}\left(\dfrac{-1}{-1}\right)$

$= \dfrac{9a}{a-7}$

45. $\dfrac{7y}{y-3} - \dfrac{2y}{3-y}$

$\dfrac{7y}{y-3} - \dfrac{2y}{3-y}\left(\dfrac{-1}{-1}\right)$

$= \dfrac{9y}{y-3}$

46. $\dfrac{4y}{y^2 + 4y + 3} + \dfrac{2}{y+1}$

$\dfrac{4y}{(y+1)(y+3)} + \dfrac{2}{y+1}\left[\dfrac{y+3}{y+3}\right] = \dfrac{6}{y+3}$

47. $\dfrac{y-23}{y^2 - y - 20} + \dfrac{2}{y-5}$

$\dfrac{y-23}{(y-5)(y+4)} + \dfrac{2}{y-5}\left[\dfrac{y+4}{y+4}\right] = \dfrac{3}{y+4}$

48. $\dfrac{x}{x-1} + \dfrac{x}{x+1} + \dfrac{2}{x^2-1}$

$\dfrac{x}{x-1}\left[\dfrac{x+1}{x+1}\right] + \dfrac{x}{x+1}\left[\dfrac{x-1}{x-1}\right] + \dfrac{2}{(x-1)(x+1)}$

$= \dfrac{2x^2 + 2}{(x+1)(x-1)}$

49. $\dfrac{2}{x^2 - 9} + \dfrac{x}{x+3} + \dfrac{2x}{x-3}$

$\dfrac{2}{(x+3)(x-3)} + \dfrac{x}{(x+3)}\left[\dfrac{x-3}{x-3}\right] + \dfrac{2x}{x-3}\left[\dfrac{x+3}{x+3}\right]$

$= \dfrac{3x^2 + 3x + 2}{(x+3)(x-3)}$

50. $\dfrac{3x}{x^2 + 3x - 10} + \dfrac{5}{4-2x}$

$\dfrac{3x}{(x+5)(x-2)}\left[\dfrac{2}{2}\right] + \dfrac{-5}{2(x-2)}\left[\dfrac{x+5}{x+5}\right]$

$= \dfrac{x-25}{2(x+5)(x-2)}$

51. $\dfrac{2y}{3y^2 - 8y - 3} + \dfrac{1}{6y - 2y^2}$

$\dfrac{2y}{(3y+1)(y-3)}\left[\dfrac{2y}{2y}\right] + \dfrac{-1}{2y(y-3)}\left[\dfrac{3y+1}{3y+1}\right]$

$= \dfrac{(4y+1)(y-1)}{2y(3y+1)(y-3)}$

52. $\dfrac{5}{3x-6}-\dfrac{2}{4x-8}$

$\dfrac{5}{3(x-2)}\left[\dfrac{2}{2}\right]+\dfrac{-1}{2(x-2)}\left[\dfrac{3}{3}\right]$

$=\dfrac{7}{6(x-2)}$

53. $\dfrac{12}{15x-3}+\dfrac{1}{10x-2}$

$\dfrac{4}{(5x-1)}\left[\dfrac{2}{2}\right]+\dfrac{1}{2(5x-1)}=\dfrac{9}{2(5x-1)}$

54. $\dfrac{1}{x-3y}+\dfrac{1}{x-4y}+\dfrac{y}{x^2-7xy+12y^2}$

$\dfrac{1}{x-3y}\left[\dfrac{x-4y}{x-4y}\right]+\dfrac{1}{x-4y}\left[\dfrac{x-3y}{x-3y}\right]+\dfrac{y}{(x-4y)(x-3y)}$

$=\dfrac{2}{x-4y}$

55. $\dfrac{42x}{y^2-49}+\dfrac{3x}{y-7}+\dfrac{3x}{y+7}$

$\dfrac{42x}{(y+7)(y-7)}+\dfrac{3x}{y-7}\left[\dfrac{y+7}{y+7}\right]+\dfrac{3x}{y+7}\left[\dfrac{y-7}{y-7}\right]$

$=\dfrac{6x}{y-7}$

56. $\dfrac{4}{2x+3}+\dfrac{7}{2x-5}+\dfrac{32}{4x^2-4x-15}$

$\dfrac{4}{2x+3}\left[\dfrac{2x-5}{2x-5}\right]+\dfrac{7}{2x-5}\left[\dfrac{2x+3}{2x+3}\right]+\dfrac{32}{(2x+3)(2x-5)}$

$=\dfrac{11}{2x-5}$

57. $\dfrac{2}{3a+1}+\dfrac{5}{3a-2}+\dfrac{6}{9a^2-3a-2}$

$\dfrac{2}{3a+1}\left[\dfrac{3a-2}{3a-2}\right]+\dfrac{5}{3a-2}\left[\dfrac{3a+1}{3a+1}\right]+\dfrac{6}{(3a+1)(3a-2)}$

$=\dfrac{7}{3a-2}$

Cumulative Review Problems

58. Solve. $\dfrac{1}{3}(x-2)+\dfrac{1}{2}(x+3)=\dfrac{1}{4}(3x+1)$

$4x-8+6x+18=9x+3$

$x=-7$

59. Solve for y. $5ax=2(ay-3bc)$ $\quad y=\dfrac{5ax+6bc}{2a}$

60. Solve for x. $\dfrac{1}{2}x<\dfrac{1}{3}x+\dfrac{1}{4}$ $\quad x<\dfrac{3}{2}$

61. Simplify. $(3x^3y^4)^4$ $\quad 81x^{12}y^{16}$

62. A subway token costs $1.50. A monthly unlimited ride subway pass costs $50. How many days would you have to use the subway to go to work (assume one subway token to get to work and one subway token to get back home) in order for it to be cheaper to buy a monthly subway pass? At least 17 days per month

63. In Finland the unemployment rate went from 3.5% in 1990 to 10.9% in 1993. If the population of Finland was 5,100,000 during this period, how many more people were unemployed in 1993 than in 1990? Approximately 377,400 more people were unemployed.

Putting Your Skills to Work

THE MATHEMATICS OF BOATING

Power boating is a pleasurable pastime, but that does not mean you can take a vacation from mathematics. In fact a number of decisions on a boat are made based on an analysis of mathematical facts.

Dr. Donald Abt made a detailed study of the following test results on his 23-foot-long Caravelle. His boat is powered by a 7.4-liter stern drive engine. The boat has a 50-gallon fuel tank.

Engine Revolutions per Minute	Miles per Hour	Gallons of Fuel Used per Hour	Range of Boat in Nautical Miles	Sound Level in Decibels
1000	6.4	3.9	64	72
2000	23.0	8.3	108	81
3000	41.9	13.2	124	92
4000	56.5	24.7	89	99
4400	59.0	26.2	88	103

Using the results of the test above answer the following questions. Round all answers to the nearest tenth.

PROBLEMS FOR INDIVIDUAL STUDY AND INVESTIGATION

1. What is the maximum distance in miles that this boat can travel at a constant 23 miles per hour? approximately 138.6 miles

2. Between what two engine readings (measured in revolutions per minute) does the greatest increase in speed occur? between 2000 rpm and 3000 rpm

PROBLEMS FOR COOPERATIVE GROUP INVESTIGATION AND ANALYSIS

Together with some members from your class see if you can determine the following.

3. What is the percent of increase in the sound level (measured in decibels) as you increase the engine from 2000 rpm to 3000 rpm? 13.6%

4. A nautical mile is approximately 6076 feet, a distance substantially longer than the usual measurement of a mile (a statute mile) of 5280 feet. How many statute miles can this boat travel at a constant engine speed of 4000 rpm? 102.4 statute miles

5. To avoid running out of fuel while traveling on the water, the ''Range of Boat'' figures represent a certain percent of the possible maximum distance that can be traveled with the boat traveling at that constant rate of speed. Using your answer from Problem 4 determine what that percent is. (Round your answer to the nearest whole percent.) 90%

INTERNET CONNECTIONS: Go to ''http://www.prenhall.com/tobey'' to be connected

Site: Wizard Yachts, Ltd.—Sailing Test of SC52 or a related site

6. This site gives a review of the Santa Cruz 52 yacht. The article includes the cruising speed and the fuel consumption for this boat. (A *knot* represents one nautical mile per hour, or about 1.15 statute miles per hour.) The fuel tank holds 86 gallons of fuel. How many statute miles can the Santa Cruz 52 go at cruising speed on a full tank of gas?

6.4 Simplify Complex Rational Expressions

MathPro

Video 16

SSM

After studying this section, you will be able to:

1. *Simplify complex fractions by adding or subtracting in the numerator and denominator.*
2. *Simplify complex fractions by multiplying by the LCD of all the denominators.*

1 *Simplifying Complex Fractions by Adding or Subtracting in the Numerator and the Denominator*

A complex fraction has a fraction in the numerator or in the denominator, or both.

$$\frac{3 + \frac{2}{x}}{\frac{x}{7} + 2} \qquad \frac{\frac{x}{y} + 1}{2} \qquad \frac{\frac{a+b}{3}}{\frac{x-2y}{4}}$$

The bar in a complex fraction is both a grouping symbol and a symbol for division.

$$\frac{\frac{a+b}{3}}{\frac{x-2y}{4}} \quad \text{is equivalent to} \quad \left(\frac{a+b}{3}\right) \div \left(\frac{x-2y}{4}\right)$$

We need a procedure for simplifying complex fractions.

Procedure to Simplify a Complex Fraction: Adding and Subtracting

1. Add or subtract so that you have a single fraction in the numerator and in the denominator.
2. Divide the fraction in the numerator by the fraction in the denominator. This is done by inverting the fraction in the denominator and multiplying it by the numerator.

TEACHING TIP Before presenting Example 1 to the class, you might want to ask the class "How would you simplify the expression $\frac{\frac{1}{7}}{\frac{2}{15} + \frac{1}{3}}$?" Usually some class members will suggest finding an LCD and adding the two fractions in the denominator. Continue to work out the problem

$$\frac{\frac{1}{7}}{\frac{2}{15} + \frac{1}{3}} = \frac{\frac{1}{7}}{\frac{2}{15} + \frac{5}{15}} = \frac{\frac{1}{7}}{\frac{7}{15}}$$

so

$$\frac{1}{7} \div \frac{7}{15} = \frac{1}{7} \cdot \frac{15}{7} = \frac{15}{49}$$

Then bridge the gap by saying, "In a similar fashion we can simplify a complex fraction that contains algebraic fractions." Then go on to show students how to do Example 1.

EXAMPLE 1 Simplify. $\dfrac{\frac{1}{x}}{\frac{2}{y^2} + \frac{1}{y}}$

Step 1 Add the two fractions in the denominator.

$$\frac{\frac{1}{x}}{\frac{2}{y^2} + \frac{1}{y} \cdot \frac{y}{y}} = \frac{\frac{1}{x}}{\frac{2+y}{y^2}}$$

Step 2 Divide the fraction in the numerator by the fraction in the denominator.

$$\frac{1}{x} \div \frac{2+y}{y^2} = \frac{1}{x} \cdot \frac{y^2}{2+y} = \frac{y^2}{x(2+y)}$$

Practice Problem 1 Simplify. $\dfrac{\frac{1}{a} + \frac{1}{b}}{\frac{2}{ab^2}}$ ■

In some problems there may be two or more fractions in the numerator and the denominator.

EXAMPLE 2 Simplify. $\dfrac{\dfrac{1}{x}+\dfrac{1}{y}}{\dfrac{3}{a}-\dfrac{2}{b}}$

TEACHING TIP Students find the type of problem containing two fractions in the numerator and two fractions in the denominator to be a bit more difficult. In addition to Example 2, you may want to present the following as an Additional Class Exercise.
Simplify.

$$\dfrac{\dfrac{1}{3x}-\dfrac{2}{x}}{\dfrac{5}{6x}+\dfrac{1}{3}}=\dfrac{\dfrac{1}{3x}-\dfrac{6}{3x}}{\dfrac{5}{6x}+\dfrac{2x}{6x}}=\dfrac{\dfrac{-5}{3x}}{\dfrac{5+2x}{6x}}$$

So

$$\frac{-5}{3x}\div\frac{5+2x}{6x}=\frac{-5}{3x}\cdot\frac{6x}{5+2x}=\frac{-10}{5+2x}$$

We observe that the LCD of the fractions in the numerator is xy. The LCD of the fractions in the denominator is ab.

$$\dfrac{\dfrac{1}{x}\cdot\dfrac{y}{y}+\dfrac{1}{y}\cdot\dfrac{x}{x}}{\dfrac{3}{a}\cdot\dfrac{b}{b}-\dfrac{2}{b}\cdot\dfrac{a}{a}}$$

Multiply each fraction by the appropriate value to obtain common denominators.

$$=\dfrac{\dfrac{y+x}{xy}}{\dfrac{3b-2a}{ab}}$$

Add the two fractions in the numerator.

Subtract the two fractions in the denominator.

$$=\frac{y+x}{xy}\cdot\frac{ab}{3b-2a}$$

Invert the fraction in the denominator and multiply it by the numerator.

$$=\frac{ab(y+x)}{xy(3b-2a)}$$

Write the answer as one fraction.

Practice Problem 2 Simplify. $\dfrac{\dfrac{1}{a}+\dfrac{1}{b}}{\dfrac{1}{a}-\dfrac{1}{b}}$ ■

In some problems, factoring may be necessary to determine the LCD and to combine fractions.

EXAMPLE 3 Simplify. $\dfrac{\dfrac{1}{x^2-1}+\dfrac{2}{x+1}}{x}$

TEACHING TIP Many weaker students may have some difficulty understanding Example 3. You may wish to show the following as an Additional Class Exercise after you have discussed Example 3.
Simplify.

$$\dfrac{\dfrac{x}{x^2-7x+12}+\dfrac{3}{x-4}}{x}=\dfrac{\dfrac{x}{(x-3)(x-4)}+\dfrac{3(x-3)}{(x-3)(x-4)}}{x}$$

So

$$\frac{4x-9}{(x-3)(x-4)}\div\frac{x}{1}=\frac{4x-9}{x(x-3)(x-4)}$$

We observe the need for factoring x^2-1.

$$=\dfrac{\dfrac{1}{(x+1)(x-1)}+\dfrac{2}{(x+1)}\cdot\dfrac{x-1}{x-1}}{x}$$

The LCD for the fractions in the numerator is $(x+1)(x-1)$.

$$=\dfrac{\dfrac{1+2x-2}{(x+1)(x-1)}}{x}$$

Add the two fractions in the numerator.

$$=\frac{2x-1}{(x+1)(x-1)}\cdot\frac{1}{x}$$

Simplify the numerator. Invert the fraction in the denominator and multiply.

$$=\frac{2x-1}{x(x+1)(x-1)}$$

Write the answer as one fraction.

Practice Problem 3 Simplify. $\dfrac{\dfrac{x}{x^2+4x+3}+\dfrac{2}{x+1}}{x+1}$ ■

In simplifying complex fractions, always be alert to see if the final fraction can be reduced or simplified.

EXAMPLE 4 Simplify. $\dfrac{\dfrac{3}{a+b} - \dfrac{3}{a-b}}{\dfrac{5}{a^2 - b^2}}$

The LCD of the two fractions in the numerator is $(a + b)(a - b)$.

$$\frac{\dfrac{3}{a+b} \cdot \dfrac{a-b}{a-b} - \dfrac{3}{a-b} \cdot \dfrac{a+b}{a+b}}{\dfrac{5}{a^2-b^2}} = \frac{\dfrac{3a-3b}{(a+b)(a-b)} - \dfrac{3a+3b}{(a+b)(a-b)}}{\dfrac{5}{a^2-b^2}}$$

Study carefully how we combine the two fractions in the numerator. Do you see how we obtain $-6b$?

$$= \frac{\dfrac{-6b}{(a+b)(a-b)}}{\dfrac{5}{(a+b)(a-b)}} \qquad \textit{Factor } a^2 - b^2 \textit{ as } (a+b)(a-b).$$

$$= \frac{-6b}{(a+b)(a-b)} \cdot \frac{(a+b)(a-b)}{5}$$

Note that since $(a + b)(a - b)$ are factors in both numerator and denominator they may be divided out.

$$= \frac{-6b}{5}$$

Practice Problem 4 Simplify. $\dfrac{\dfrac{6}{x^2 - y^2}}{\dfrac{1}{x-y} + \dfrac{3}{x+y}}$ ■

2 *Simplifying Complex Fractions by Multiplying by the LCD of All the Denominators*

There is another way to simplify complex fractions: Multiply the numerator and denominator of the complex fraction by the least common denominator of all the denominators appearing in the complex fraction.

Procedure to Simplify a Complex Fraction: Multiplying by the LCD

1. Determine the LCD of all individual denominators occurring in the numerator and denominator of the complex fraction.
2. Multiply both the numerator and the denominator of the complex fraction by the LCD.
3. Simplify if possible.

EXAMPLE 5 Simplify by multiplying by the LCD. $\dfrac{\dfrac{5}{ab^2} - \dfrac{2}{ab}}{3 - \dfrac{5}{2a^2b}}$

The LCD of all the denominators in the complex fraction is $2a^2b^2$.

$$\frac{2a^2b^2\left(\frac{5}{ab^2} - \frac{2}{ab}\right)}{2a^2b^2\left(3 - \frac{5}{2a^2b}\right)} = \frac{2a^2b^2\left(\frac{5}{ab^2}\right) - 2a^2b^2\left(\frac{2}{ab}\right)}{2a^2b^2(3) - 2a^2b^2\left(\frac{5}{2a^2b}\right)}$$ *Multiply each term by* $2a^2b^2$.

$$= \frac{10a - 4ab}{6a^2b^2 - 5b}$$ *Simplify.*

Practice Problem 5 Simplify by using the alternative method. $\dfrac{\dfrac{2}{3x^2} - \dfrac{3}{y}}{\dfrac{5}{xy} - 4}$ ■

So that you can compare the two methods, we will redo Example 4 by the alternative method.

EXAMPLE 6 Simplify. $\dfrac{\dfrac{3}{a+b} - \dfrac{3}{a-b}}{\dfrac{5}{a^2 - b^2}}$ Use alternative method.

The LCD of all individual fractions contained in the complex fraction is $(a + b)(a - b)$.

$$\frac{(a+b)(a-b)\left(\frac{3}{a+b}\right) - (a+b)(a-b)\left(\frac{3}{a-b}\right)}{(a+b)(a-b)\left(\frac{5}{(a+b)(a-b)}\right)}$$ *Multiply each term by the LCD.*

$$= \frac{3(a-b) - 3(a+b)}{5}$$ *Simplify.*

$$= \frac{3a - 3b - 3a - 3b}{5}$$ *Remove parentheses.*

$$= -\frac{6b}{5}$$ *Simplify.*

Practice Problem 6 Simplify by using the alternative method. $\dfrac{\dfrac{6}{x^2 - y^2}}{\dfrac{7}{x-y} + \dfrac{3}{x+y}}$ ■

6.4 Exercises

Simplify.

1. $\dfrac{\frac{3}{x}}{\frac{2}{y}}$ $\quad \dfrac{3y}{2x}$

2. $\dfrac{\frac{3}{ab}}{\frac{b}{a}}$ $\quad \dfrac{3}{b^2}$

3. $\dfrac{\frac{1}{x}+\frac{1}{y}}{\frac{1}{xy}}$ $\quad y + x$

4. $\dfrac{\frac{1}{x}+1}{\frac{1}{x}}$ $\quad 1 + x$

5. $\dfrac{\frac{1}{x}+y}{\frac{1}{y}+x}$ $\quad \dfrac{y}{x}$

6. $\dfrac{\frac{1}{x}+\frac{1}{y}}{x+y}$ $\quad \dfrac{1}{xy}$

7. $\dfrac{1-\frac{9}{x^2}}{\frac{3}{x}+1}$ $\quad \dfrac{x-3}{x}$

8. $\dfrac{\frac{4}{x}+1}{1-\frac{16}{x^2}}$ $\quad \dfrac{x}{x-4}$

9. $\dfrac{\frac{2}{x+1}-2}{3}$ $\quad \dfrac{-2x}{3(x+1)}$

10. $\dfrac{1-\frac{3}{xy}}{x+2y}$ $\quad \dfrac{xy-3}{xy(x+2y)}$

11. $\dfrac{a+\frac{3}{a}}{\frac{a^2+2}{3a}}$ $\quad \dfrac{3a^2+9}{a^2+2}$

12. $\dfrac{a+\frac{1}{a}}{\frac{3}{a}-a}$ $\quad \dfrac{a^2+1}{3-a^2}$

13. $\dfrac{a+1-\frac{12}{a-2}}{\frac{-2}{a-2}+a-1}$ $\quad \dfrac{a^2-a-14}{a^2-3a}$

14. $\dfrac{y-\frac{4}{y}}{1+\frac{3}{2y+1}}$ $\quad \dfrac{(2y+1)(y-2)}{2y}$

15. $\dfrac{\frac{x}{4}-\frac{1}{2}}{\frac{x}{4}+\frac{1}{x}}$ $\quad \dfrac{x(x-2)}{x^2+4}$

16. $\dfrac{\frac{1}{x}-\frac{1}{2x}}{\frac{1}{2}+\frac{1}{2x}}$ $\quad \dfrac{1}{x+1}$

17. $\dfrac{\frac{1}{x^2-9}+\frac{2}{x+3}}{\frac{3}{x-3}}$ $\quad \dfrac{2x-5}{3(x+3)}$

18. $\dfrac{\frac{5}{x+4}}{\frac{1}{x-4}-\frac{2}{x^2-16}}$ $\quad \dfrac{5(x-4)}{x+2}$

19. $\dfrac{\frac{2}{y-1}+2}{\frac{2}{y+1}-2}$ $\quad \dfrac{y+1}{-y+1}$

20. $\dfrac{\frac{y}{y+1}+1}{\frac{2y+1}{y-1}}$ $\quad \dfrac{y-1}{y+1}$

★**21.** $\dfrac{\frac{y}{y+1}-\frac{y-1}{y}}{\frac{y+1}{y}+\frac{y}{y-1}}$ $\quad \dfrac{y-1}{2y^3+2y^2-y-1}$

★**22.** $\dfrac{y+2-\frac{6}{2y+3}}{\frac{8y}{2y-1}+y}$ $\quad \dfrac{2y-1}{2y+3}$

Cumulative Review Problems

23. Solve for w. $P = 2(l + w)$ $\quad w = \dfrac{P-2l}{2}$

24. Solve and graph. $7 + x < 11 + 5x$
$x > -1$

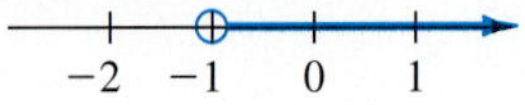

6.5 Equations Involving Rational Expressions

After studying this section, you will be able to:

1 *Solve equations involving algebraic fractions that have solutions.*
2 *Determine if an equation involving algebraic fractions has no solution.*

1 *Solving Equations Involving Algebraic Fractions*

In Section 2.4 we developed procedures to solve linear equations containing fractions whose denominators were numerical values. In this section we use a similar approach to solve equations containing fractions whose denominators are polynomials. It would be wise for you to review Section 2.4 briefly *before you begin this section.* It will be especially helpful to carefully study Examples 1 and 2.

To Solve a Rational Equation

1. Determine the LCD of all the denominators.
2. Multiply each term of the equation by the LCD.
3. Solve the resulting equation.
4. Check your solution. Exclude from your solution any value that would make the LCD equal to zero. If such a value is obtained, there is *no solution.*

TEACHING TIP You may want to show the connection more precisely between Section 2.4, which they have previously studied, and this present section. You might find this approach helpful.

"In Section 2.4 you did the following homework problem."

$$\frac{x}{4} - 1 = \frac{x}{5}$$

LCD = (4)(5) = 20

$$20\left(\frac{x}{4}\right) - 20(1) = 20\left(\frac{x}{5}\right)$$

$$5x - 20 = 4x$$

$$5x - 4x = 20$$

$$x = 20$$

"In this section you will be doing very similar problems. This time, however, the denominators will contain variables."

EXAMPLE 1 Solve for x and check your solution. $\frac{5}{x} + \frac{2}{3} = 2 - \frac{2}{x} - \frac{1}{6}$

We observe that the LCD is $6x$.

$6x\left(\frac{5}{x}\right) + 6x\left(\frac{2}{3}\right) = 6x(2) - 6x\left(\frac{2}{x}\right) - 6x\left(\frac{1}{6}\right)$ *Multiply each term by $6x$.*

$30 + 4x = 12x - 12 - x$ *Simplify. Do you see how each term is obtained?*

$30 + 4x = 11x - 12$ *Collect like terms.*

$30 = 7x - 12$ *Subtract $4x$ from both sides.*

$42 = 7x$ *Add 12 to both sides.*

$6 = x$ *Divide both sides by 7.*

Check. $\frac{5}{6} + \frac{2}{3} \stackrel{?}{=} 2 - \frac{2}{6} - \frac{1}{6}$ *Replace each x by 6.*

$\frac{5}{6} + \frac{4}{6} \stackrel{?}{=} \frac{12}{6} - \frac{2}{6} - \frac{1}{6}$

$\frac{9}{6} = \frac{9}{6}$ ✓ *It checks.*

Practice Problem 1 Solve for x and check your solution.

$\frac{2}{x} + \frac{4}{x} = 3 - \frac{1}{x} - \frac{17}{8}$ ■

EXAMPLE 2 Solve and check. $\dfrac{6}{x+3} = \dfrac{3}{x}$

We observe that the LCD $= x(x + 3)$.

$$x(x+3)\left(\frac{6}{x+3}\right) = x(x+3)\left(\frac{3}{x}\right) \quad \textit{Multiply both sides by } x(x+3).$$

$$6x = 3(x + 3) \quad \textit{Simplify. Do you see how this is done?}$$

$$6x = 3x + 9 \quad \textit{Remove parentheses.}$$

$$3x = 9 \quad \textit{Subtract } 3x \textit{ from both sides.}$$

$$x = 3 \quad \textit{Divide both sides by } 3.$$

Check. $\dfrac{6}{3+3} \stackrel{?}{=} \dfrac{3}{3}$ *Replace each x by 3.*

$\dfrac{6}{6} = \dfrac{3}{3}$ ✓ *It checks.*

Practice Problem 2 Solve and check. $\dfrac{4}{2x+1} = \dfrac{6}{2x-1}$ ■

It is sometimes necessary to factor denominators before the correct LCD can be determined.

TEACHING TIP After you have presented Example 3 you may wish to do the following as an Additional Class Exercise of this type of problem.
Solve for y.

$$\frac{y}{y+5} = \frac{10}{y^2-25} + 1$$

Sol.
LCD $= (y + 5)(y - 5)$. We multiply each term by the LCD.

$$(y+5)(y-5)\left[\frac{y}{y+5}\right]$$
$$= (y+5)(y-5)\left[\frac{10}{(y+5)(y-5)}\right] + (y+5)(y-5)(1)$$
$$y(y-5) = 10 + (y+5)(y-5)$$
$$y^2 - 5y = 10 + y^2 - 25$$
$$-5y = 10 - 25$$
$$-5y = -15$$
$$y = 3$$

EXAMPLE 3 Solve and check your solution. $\dfrac{3}{x+5} - 1 = \dfrac{4-x}{2x+10}$

We observe a need to factor the $2x + 10$.

$$\frac{3}{x+5} - 1 = \frac{4-x}{2(x+5)} \quad \textit{Factor. We determine that the LCD is } 2(x+5).$$

$$2(x+5)\left(\frac{3}{x+5}\right) - 2(x+5)(1) = 2(x+5)\left[\frac{4-x}{2(x+5)}\right] \quad \textit{Multiply each term by the LCD.}$$

$$2(3) - 2(x + 5) = 4 - x \quad \textit{Simplify.}$$

$$6 - 2x - 10 = 4 - x \quad \textit{Remove parentheses.}$$

$$-2x - 4 = 4 - x \quad \textit{Collect like terms.}$$

$$-4 = 4 + x \quad \textit{Add } 2x \textit{ to both sides.}$$

$$-8 = x \quad \textit{Subtract 4 from both sides.}$$

Check. $\dfrac{3}{-8+5} - 1 \stackrel{?}{=} \dfrac{4-(-8)}{2(-8)+10}$ *Replace each x in the original equation by −8.*

$$\frac{3}{-3} - 1 \stackrel{?}{=} \frac{4+8}{-16+10}$$

$$-1 - 1 \stackrel{?}{=} \frac{12}{-6}$$

$-2 = -2$ ✓ *It checks. $x = -8$ is the solution.*

Practice Problem 3 Solve and check. $\dfrac{x-1}{x^2-4} = \dfrac{2}{x+2} + \dfrac{4}{x-2}$ ■

2 *Determining If an Equation Involving Algebraic Fractions Has No Solution*

Linear equations containing fractions that have variables in denominators sometimes appear to have solutions when in fact they do not. By this we mean that the "solutions" we get by using completely correct methods are, in actuality, not solutions.

In the case where a value makes a denominator in the equation equal to zero, we say there is **no solution** to the equation. Such a value is called an **extraneous solution.** An extraneous solution is an apparent solution that does *not* satisfy the original equation. It is important that you check the number in the original equation.

EXAMPLE 4 Solve and check. $\dfrac{y}{y-2} - 4 = \dfrac{2}{y-2}$

TEACHING TIP After you have discussed Example 4 and the extraneous solution, present the students with the following question. Suppose in the solution of

$$\frac{12}{x^2-9} + \frac{3}{x+3} = \frac{2}{x-3}$$

I obtained an answer of $x = 3$. How would I find out if this is an extraneous solution? Will $x = 3$ satisfy the original equation? By the end of this little discussion the students will have a clearer understanding of how to quickly determine if any given solution is extraneous or not by determining if the obtained solution ever causes a denominator to be zero.

We observe that the LCD is $y - 2$.

$$(y-2)\left(\frac{y}{y-2}\right) - (y-2)(4) = (y-2)\left(\frac{2}{y-2}\right) \quad \textit{Multiply each term by } (y-2).$$

$$y - 4(y-2) = 2 \quad \textit{Simplify. Do you see how this is done?}$$

$$y - 4y + 8 = 2 \quad \textit{Remove parentheses.}$$

$$-3y + 8 = 2 \quad \textit{Collect like terms.}$$

$$-3y = -6 \quad \textit{Subtract 8 from both sides.}$$

$$\frac{-3y}{-3} = \frac{-6}{-3} \quad \textit{Divide both sides by negative 3.}$$

$$y = 2 \quad \textit{y = 2 is only an apparent solution.}$$

There is no solution to this problem.

Why? We can see immediately that $y = 2$ is not a possible solution for the original equation. The use of the value $y = 2$ in a denominator would make the denominator equal to zero, and the expression is undefined.

Check. $\dfrac{y}{y-2} - 4 = \dfrac{2}{y-2}$ *Suppose that you try to check the apparent solution by substituting $y = 2$.*

$$\frac{2}{2-2} - 4 \stackrel{?}{=} \frac{2}{2-2}$$

$$\frac{2}{0} - 4 = \frac{2}{0} \quad \textit{This does not check since you do not obtain a real number when you divide by zero.}$$

↑ ↑

These expressions are not defined.

There is no such number as $2 \div 0$. We see that $y = 2$ does *not* check. There is **no solution** to this problem.

Practice Problem 4 Solve and check. $\dfrac{2x}{x+1} = \dfrac{-2}{x+1} + 1$ ■

6.5 Exercises

Solve and check your solution.

1. $\frac{x+1}{2x} = \frac{2}{3}$
$3x + 3 = 4x$
$x = 3$

2. $\frac{9}{2x-1} = \frac{3}{x}$
$9x = 6x - 3$
$x = -1$

3. $\frac{1}{x} = \frac{1}{4-x}$
$x = 4 - x$
$x = 2$

4. $\frac{x-2}{4x} = \frac{1}{6}$
$3x - 6 = 2x$
$x = 6$

5. $\frac{5}{3x-4} = \frac{3}{x-3}$
$5x - 15 = 9x - 12$
$x = -\frac{3}{4}$

6. $\frac{5}{x-1} = \frac{3}{x+1}$
$3x - 3 = 5x + 5$
$x = -4$

7. $\frac{2}{x} + \frac{x}{x+1} = 1$
$2x + 2 + x^2 = x^2 + x$
$x = -2$

8. $\frac{5}{2} = 3 + \frac{2x+7}{x+6}$
$5x + 30 = 6x + 36 + 4x + 14$
$-20 = 5x$
$-4 = x$

9. $\frac{x+1}{x} = 1 + \frac{x-2}{2x}$
$2x + 2 = 2x + x - 2$
$4 = x$

10. $\frac{7x-4}{5x} = \frac{9}{5} - \frac{4}{x}$
$7x - 4 = 9x - 20$
$x = 8$

11. $\frac{x+1}{x} = \frac{1}{2} - \frac{4}{3x}$
$6x + 6 = 3x - 8 \quad x = \frac{-14}{3}$

12. $\frac{2x-1}{3x} + \frac{1}{9} = \frac{x+2}{x} + \frac{1}{9x}$
$6x - 3 + x = 9x + 18 + 1$
$x = -11$

13. $\frac{x}{x-2} - 2 = \frac{2}{x-2}$
$x - 2x + 4 = 2$
$2 = x$ There is no solution.

14. $4 - \frac{8}{x+1} = \frac{8x}{x+1}$
$4x + 4 - 8 = 8x$
$x = -1$ There is no solution.

15. $\frac{2}{x+1} - \frac{1}{x-1} = \frac{2x}{x^2-1}$
$2x - 2 - x - 1 = 2x$
$-3 = x$

16. $\frac{8x}{4x^2-1} = \frac{3}{2x+1} + \frac{3}{2x-1}$
$8x = 6x - 3 + 6x + 3$
$x = 0$

17. $\frac{y+1}{y^2+2y-3} = \frac{1}{y+3} - \frac{1}{y-1}$
$y + 1 = y - 1 - y - 3$
$y = -5$

18. $\frac{2y}{y+1} + \frac{1}{3y-2} = 2$
$6y^2 - 4y + y + 1 = 6y^2 + 2y - 4$
$y = 1$

19. $\frac{79-x}{x} = 5 + \frac{7}{x}$
$79 - x = 5x + 7$
$x = 12$

20. $\frac{2x-14}{x^2+3x-28} + \frac{x-2}{x-4} = \frac{x+3}{x+7}$
$2x - 14 + x^2 + 5x - 14 = x^2 - x - 12$
$8x = 16$
$x = 2$

21. $\frac{6}{x-3} = \frac{-5}{x-2} + \frac{-5}{x^2-5x+6}$
$6(x-2) = -5(x-3) - 5$
$x = 2$
There is no solution.

22. $\frac{2}{3x+6} + \frac{1}{2x+4} = \frac{1}{6}$
$4 + 3 = x + 2$
$x = 5$

23. $\frac{x+11}{x^2-5x+4} + \frac{3}{x-1} = \frac{5}{x-4}$
$x + 11 + 3x - 12 = 5x - 5$
$x = 4$ There is no solution.

24. $\frac{52-x}{x} = 9 + \frac{2}{x}$
$52 - x = 9x + 2$
$x = 5$

25. $\frac{2x}{x+4} - \frac{8}{x-4} = \frac{2x^2+32}{x^2-16}$
$2x^2 - 8x - 32 - 8x = 2x^2 + 32$
$-64 = 16x$
$-4 = x$
There is no solution.

26. $\frac{4x}{x+3} - \frac{12}{x-3} = \frac{4x^2+36}{x^2-9}$
$4x^2 - 12x - 12x - 36 = 4x^2 + 36$
$-72 = 24x$
$-3 = x$
There is no solution.

27. $\frac{23}{5} - \frac{3}{5x - 10} = \frac{4}{x - 2}$

$23x - 46 - 3 = 20$
$23x = 69$
$x = 3$

28. $\frac{x - 3}{x - 2} + \frac{x + 1}{x + 3} = \frac{2x^2 - 15x}{x^2 + x - 6}$

$x^2 - 9 + x^2 - x - 2 = 2x^2 - 15x$
$-11 = -14x$
$\frac{11}{14} = x$

29. $\frac{2}{x + 7} + \frac{4x - 1}{x^2 + 5x - 14} = \frac{1}{x - 2}$

$2x - 4 + 4x - 1 = x + 7$
$5x = 12$
$x = \frac{12}{5}$

30. $\frac{2}{x - 2} + \frac{3}{x} = \frac{4}{-2x + x^2}$

$2x + 3x - 6 = 4$
$5x = 10$
$x = 2$
There is no solution.

★ 31. $\frac{10}{x - 2} - \frac{40}{x^2 + x - 6} = \frac{12}{x + 3}$

$10x + 30 - 40 = 12x - 24$
$14 = 2x$
$7 = x$

★ 32. $\frac{10 - 2y^2}{y^2 + 3y - 10} + \frac{4y + 2}{y - 2} = \frac{2y + 2}{y + 5}$

$10 - 2y^2 + 4y^2 + 22y + 10 = 2y^2 - 2y - 4$
$24y = -24$
$y = -1$

To Think About

In each of the following equations, determine the extraneous solutions. Do not solve the equation. (Hint: *The denominator can never be zero.)*

33. $\frac{3x}{x - 2} - \frac{4x}{x - 4} = \frac{3}{x^2 - 6x + 8}$

$x = 2, x = 4$

34. $\frac{2x}{x + 2} - \frac{x}{x + 3} = \frac{-3}{x^2 + 5x + 6}$

$x = -2, x = -3$

Cumulative Review Problems

35. Factor. $6x^2 - x - 12$ $(3x + 4)(2x - 3)$

36. Solve. $4 - (3 - x) = 7(x + 3)$ $x = \frac{-10}{3}$

37. The perimeter of a rectangle is 54 meters. The length is 1 meter less than triple the width. Find the dimensions of the rectangle.

$2x + 2(3x - 1) = 54$ width = 7 m length = 20 m
$x = 7$

38. Wally and Adele Panzas plan to purchase a new home and borrow \$115,000. They plan to take out a 25-year mortgage at an annual interest rate of 8.75%. The bank will charge them \$8.23 per month for each \$1000 of mortgage. What will their monthly payments be?

Monthly payment is \$946.45

6.6 Ratio, Proportion, and Other Applied Problems

MathPro

Video 17

SSM

After studying this section, you will be able to:

1 *Solve problems involving ratio and proportion.*
2 *Solve problems involving similar triangles.*
3 *Solve certain distance problems.*
4 *Solve work problems.*

1 *Solving Problems Involving Ratio and Proportion*

A **ratio** is a comparison of two quantities. You may be familiar with ratios that compare miles to hours or miles to gallons. A ratio is often written as a quotient in the form of a fraction. For example, the ratio of 7 to 9 can be written as $\frac{7}{9}$.

A **proportion** is an equation that states that two ratios are equal. For example,

$$\frac{7}{9} = \frac{21}{27}, \qquad \frac{2}{3} = \frac{10}{15}, \qquad \frac{a}{b} = \frac{c}{d}$$

Let's take a closer look at the last proportion. We can see that the LCD of the fractional equation is bd.

$(bd)\frac{a}{b} = (bd)\frac{c}{d}$ *Multiply each side by the LCD.*

$da = bc$

$ad = bc$ *Since multiplication is commutative, $da = ad$.*

Thus we have proved the following:

The Proportion Equation

If

$$\frac{a}{b} = \frac{c}{d} \quad \text{then} \quad ad = bc$$

for all real numbers a, b, c, and d, where $b \neq 0$, $d \neq 0$.

This is sometimes called **cross-multiplying.** It can be applied only if you have *one* fraction and nothing else on each side of the equation.

We can use a proportion and the technique of cross-multiplying to solve a variety of applied problems involving two ratios.

EXAMPLE 1 Michael took 5 hours to drive 245 miles on the turnpike. If he continues at the same rate, how many hours will it take him to drive a distance of 392 miles?

1. *Understand the problem.*
   Let x = the number of hours it will take to drive 392 miles. If 5 hours are needed to drive 245 miles, then x hours is needed to drive 392.
2. *Write an equation.*
   We can write this as a proportion. Compare time to distance in each ratio.

$$\begin{array}{lcl} \text{Time} \longrightarrow & \dfrac{5 \text{ hours}}{245 \text{ miles}} = \dfrac{x \text{ hours}}{392 \text{ miles}} & \longleftarrow \text{Time} \\ \text{Distance} \longrightarrow & & \longleftarrow \text{Distance} \end{array}$$

3. *Solve the equation and state the answer.*

$$5(392) = 245x \qquad \textit{Cross-multiply.}$$

$$\frac{1960}{245} = x \qquad \textit{Divide both sides by 245.}$$

$$8 = x$$

It would take Michael 8 hours to drive 392 miles.

4. *Check.* Is $\frac{5}{245} = \frac{8}{392}$? Do the computation and see.

Practice Problem 1 It took Brenda 8 hours to drive 420 miles. At the same rate, how long would it take her to drive 315 miles? ■

EXAMPLE 2 If $\frac{3}{4}$ inch on a map represents an actual distance of 20 miles, how long is the distance represented by $4\frac{1}{8}$ inches on the same map?

Let x = the distance represented by $4\frac{1}{8}$ inches.

Initial measurement on map ⟶ $\frac{3}{4}$; $4\frac{1}{8}$ ⟵ Second measurement on the map

Initial distance ⟶ 20 ; x ⟵ Second distance

$$\frac{\frac{3}{4}}{20} = \frac{4\frac{1}{8}}{x}$$

$$\left(\frac{3}{4}\right)(x) = (20)\left(4\frac{1}{8}\right) \qquad \textit{Cross-multiply.}$$

$$\left(\frac{3}{4}\right)(x) = (\overset{5}{\cancel{20}})\left(\frac{33}{\underset{2}{\cancel{8}}}\right) \qquad \textit{Write } 4\tfrac{1}{8} \textit{ as } \tfrac{33}{8} \textit{ and simplify.}$$

$$\frac{3x}{4} = \frac{165}{2} \qquad \textit{Multiplication of fractions.}$$

$$\cancel{4}\left(\frac{3x}{\cancel{4}}\right) = \overset{2}{\cancel{4}}\left(\frac{165}{\cancel{2}}\right) \qquad \textit{Multiply each side by 4.}$$

$$3x = 330 \qquad \textit{Simplify.}$$

$$x = 110 \qquad \textit{Divide both sides by 3.}$$

$4\frac{1}{8}$ inches on the map represents an actual distance of 110 miles.

Practice Problem 2 If $\frac{5}{8}$ inch on a map represents 30 miles, how long is the distance represented by $2\frac{1}{2}$ inches on the same map? ■

2 *Solving Problems Involving Similar Triangles*

Similar triangles are triangles that have the same shape, but may be a different size. For example, if you draw a triangle on a sheet of paper and place the paper in a photocopy machine and make a copy that is reduced by 25%, you would create a triangle that was similar to the original triangle. The two triangles would have the same shape. The corresponding sides of the triangles would be proportional.

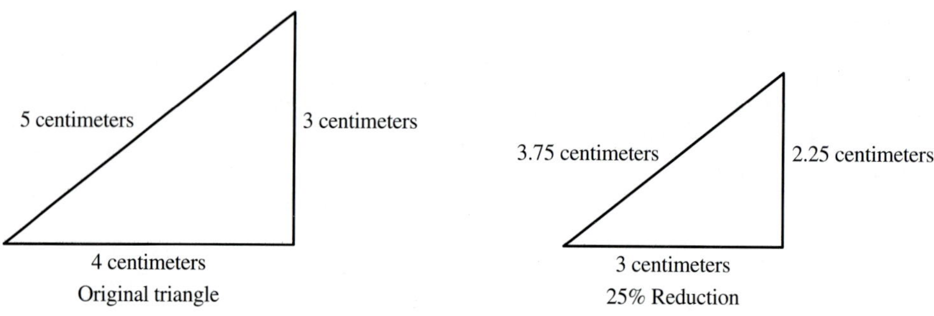

You can use the proportion equation to show that the corresponding sides of the above triangles are proportional. In fact, you can use the proportion equation to find the length of a side of a triangle if it is not known.

EXAMPLE 3 Find the length of side x in the two similar triangles pictured below.

32 meters, 15 meters — Triangle A

9 meters, x — Triangle B

Triangle A, longest side ⟶ 32 (numerator); 15 (numerator) ⟵ Shortest side, triangle A

Triangle B, longest side ⟶ 9 (denominator); x (denominator) ⟵ Shortest side, triangle B

$$\frac{32}{9} = \frac{15}{x}$$

$$32x = (9)(15) \qquad \textit{Cross-multiply.}$$

$$32x = 135 \qquad \textit{Multiply.}$$

$$x = \frac{135}{32} \qquad \textit{Divide both sides by 32.}$$

$$\text{or} \quad x = 4\frac{7}{32} \text{ meters}$$

Practice Problem 3 Triangle C is similar to triangle D. Find the length of side x. Leave your answer as a fraction.

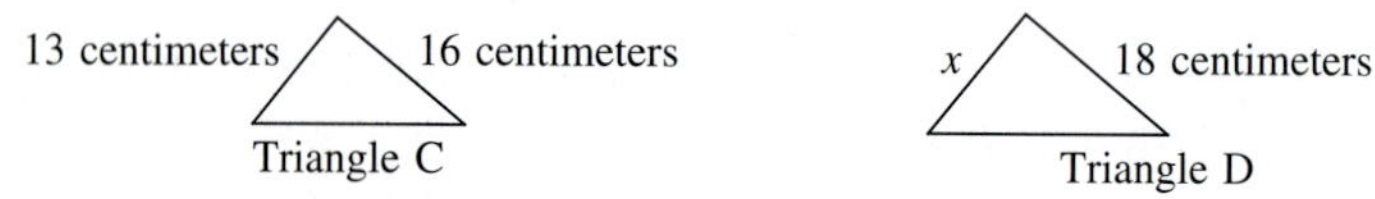

We can also use similar triangles for indirect measurement, for instance, to find the measure of an object that is too tall to measure using standard measuring devices. When the sun shines on two vertical objects at the same time, the shadows and the objects form similar triangles.

EXAMPLE 4 A woman who is 5 feet tall casts a shadow that is 8 feet long. At the same time of day, a building casts a shadow that is 72 feet long. How tall is the building?

1. *Understand the problem.*
First we draw a sketch.

We do not know the height of the building, so we call it x.

2. *Write an equation and solve.*

Height of woman ⟶ $\dfrac{5}{8} = \dfrac{x}{72}$ ⟵ Height of building
Length of woman's shadow ⟶ (8) (72) ⟵ Length of building's shadow

$$\frac{5}{8} = \frac{x}{72}$$

$$(5)(72) = 8x \qquad \textit{Cross-multiply.}$$

$$360 = 8x$$

$$45 = x$$

The height of the building is 45 feet.

Practice Problem 4 A man who is 6 feet tall casts a shadow that is 7 feet long. At the same time of day, a large flagpole casts a shadow that is 38.5 feet long. How tall is the flagpole? ■

In problems such as Example 4 we are assuming that the building and the person are standing exactly perpendicular to the ground. In other words, each triangle is assumed to be a right triangle. In some similar problems of this type, if the triangle is not a right triangle you must be careful to be sure that the angle between the object and the ground is the same in each case.

3 Solving Distance Problems Involving Algebraic Fractions

Some distance problems are solved using rational equations. We will need the formula Distance = Rate × Time, $D = RT$, which we can write in the form $T = \frac{D}{R}$.

EXAMPLE 5 Plane A flies at a speed that is 50 kilometers per hour faster than plane B. Plane A flies 500 kilometers in the amount of time that plane B flies 400 kilometers. Find the speed of each plane.

1. *Understand the problem.*
   Let s = the speed of plane B in kilometers per hour.
   Let $s + 50$ = the speed of plane A in kilometers per hour.
   Make a simple table for D, R, and T.

	D	R	$T = \frac{D}{R}$
Plane A	500	$s + 50$	?
Plane B	400	s	?

   Since $T = \frac{D}{R}$, we divide the expression for D by the expression for R and write it in the table in the column for time.

	D	R	$T = \frac{D}{R}$
Plane A	500	$s + 50$	$\frac{500}{s + 50}$
Plane B	400	s	$\frac{400}{s}$

2. *Write an equation and solve.*
   Each plane flies for the same amount of time. That is, the time for plane A equals the time for plane B.

$$\frac{500}{s + 50} = \frac{400}{s}$$

   You can solve this equation using the methods in Section 6.5 or you may cross-multiply. Here we will cross-multiply.

$$500s = (s + 50)(400) \qquad \textit{Cross-multiply.}$$

$$500s = 400s + 20{,}000 \qquad \textit{Remove parentheses.}$$

$$100s = 20{,}000 \qquad \textit{Subtract } 400s \textit{ from each side.}$$

$$s = 200 \qquad \textit{Divide each side by } 100.$$

   Plane B travels 200 kilometers per hour. Since

$$s + 50 = 200 + 50 = 250$$

   plane A travels 250 kilometers per hour.

Practice Problem 5 Two European freight trains traveled toward Paris for the same amount of time. Train A traveled 180 kilometers, while train B traveled 150 kilometers. Train A travels 10 kilometers per hour faster than train B. What is the speed of each train? ■

4 Solving Work Problems

Some applied problems involve the length of time needed to do a job. These problems are often referred to as work problems.

EXAMPLE 6 Reynaldo can sort a huge stack of mail on an old sorting machine in 9 hours. His brother Carlos can sort the same amount of mail using a newer sorting machine in 8 hours. How long would it take them to do the job working together? Express your answer in hours and minutes. Round to the nearest minute.

1. *Understand the problem.*

Let's do a little reasoning.

If Reynaldo can do the job in 9 hours, then in *1 hour* he could do $\frac{1}{9}$ of the job.
If Carlos can do the job in 8 hours, then in *1 hour* he could do $\frac{1}{8}$ of the job.
Let x = the number of hours it takes Reynaldo and Carlos to do the job together.

In *1 hour* together they could do $\frac{1}{x}$ of the job.

2. *Write an equation and solve.*

The amount of work Reynaldo can do in 1 hour plus the amount of work Carlos can do in 1 hour must be equal to the amount of work they could do together in 1 hour.

Amount of work done by Reynaldo	+	Amount of work done by Carlos	=	Amount of work done together
$\frac{1}{9}$	+	$\frac{1}{8}$	=	$\frac{1}{x}$

We observe the LCD is $72x$.

$$72x\left(\frac{1}{9}\right) + 72x\left(\frac{1}{8}\right) = 72x\left(\frac{1}{x}\right) \quad \textit{Multiply each term by the LCD.}$$

$$8x + 9x = 72 \quad \textit{Simplify.}$$

$$17x = 72 \quad \textit{Collect like terms.}$$

$$x = \frac{72}{17} \quad \textit{Divide each side by 17.}$$

$$x = 4\frac{4}{17}$$

To change $\frac{4}{17}$ of an hour to minutes we multiply.

$$\frac{4}{17}\,\cancel{\text{hour}} \times \frac{60 \text{ minutes}}{1\,\cancel{\text{hour}}} = \frac{240}{17} \text{ minutes,}$$ which is approximately 14.118 minutes

To the nearest minute this is 14 minutes. Thus doing the job together will take 4 hours and 14 minutes.

Practice Problem 6 John and Dave obtained night custodian jobs at a local factory while going to college part time. Using the buffer machine, John can buff all the floors in the building in 6 hours. Dave takes a little longer and can do all the floors in the building in 7 hours. Their supervisor bought another buffer machine. How long will it take John and Dave to do all the floors in the building working together, each with his own machine? Express your answer in hours and minutes. Round your answer to the nearest minute. ■

6.6 Exercises

Solve each proportion.

1. $\frac{7}{5} = \frac{21}{x}$ $x = 15$

2. $\frac{3}{8} = \frac{x}{14}$ $x = 5\frac{1}{4}$

3. $\frac{x}{17} = \frac{12}{5}$ $x = 40\frac{4}{5}$

4. $\frac{16}{x} = \frac{3}{4}$ $x = \frac{64}{3}$

5. $\frac{5}{3} = \frac{x}{8}$ $x = \frac{40}{3}$

6. $\frac{150}{70} = \frac{9}{x}$ $x = \frac{630}{150} = 4.2$

7. $\frac{7}{x} = \frac{40}{130}$ $x = \frac{910}{40} = 22.75$

8. $\frac{x}{18} = \frac{13}{2}$ $x = 117$

Applications

Use a proportion to answer each question.

9. If 18 books fit into a packing case, how many books can be placed into 7 packing cases?
books/cases $\frac{18}{1} = \frac{x}{7}$ $x = 126$ books

10. A paint store sold 512 gallons of paint in the past 8 months. At the same rate, how many gallons will the store sell over the next 3 months?
gallons/months $\frac{512}{8} = \frac{x}{3}$ $x = 192$ gallons

11. Ben's engine is using a mixture of 4 pints of oil for every 7 gallons of gas. He has 11 pints of oil on hand to prepare a mixture. How many gallons of gas should he use?
$\frac{4}{7} = \frac{11}{x}$ $x = 19.25$ gallons

12. Nella Coastes' recipe for Shoofly Pie contains $\frac{3}{4}$ cup of unsulfured molasses. This recipe makes a small pie that serves 8 people. If she makes the larger pie that serves 12 people and the ratio of molasses to people remains the same, how much molasses will she need for the larger pie?
$1\frac{1}{8}$ cup of molasses

13. James LeBlanc spent a summer semester in England. When he arrived in England he coverted \$800 to pounds. The posted exchange rate that summer was that 1 British pound was worth \$1.53 in American currency.
(a) How many pounds did he receive for his \$800?
(b) At the end of the summer he had 140 pounds left and he changed it back into dollars. How many dollars did he receive?
(a) 522.88 British pounds; (b) \$214.20

14. Christine Maney spent a summer studying in Paris as an exchange student. The exchange rate for the summer was 5.56 francs per American dollar. When she arrived she brought \$900.
(a) How many French francs was her \$900 worth?
(b) At the end of the summer she had 350 francs left and exchanged them for American dollars. How many dollars did she receive?
(a) 5004 francs; (b) \$62.95

15. Alfonse and Melinda took a drive in Mexico. They know that if you drive 100 kilometers per hour that this is about 62 miles per hour. They are now driving on a Mexican road that has a speed limit of 90 kilometers per hour. How many miles per hour is the speed limit? Round your answer to the nearest mile per hour.
56 miles per hour

16. Dick and Anne took a trip to France. They know that 50 kilograms is equivalent to 110 pounds. Their suitcases were weighed at the airport and the weight recorded was 39 kilograms. How many pounds did their suitcases weigh? Round your answer to the nearest pound.
86 pounds

17. On the highway, Marcia and Melissa found that their car uses 8 gallons of gas to travel 170 miles. They need to travel on the highway from Boston to Chicago, a distance of 970 miles. Approximately how many gallons of gas will they use? Round your answer to the nearest gallon. 46 gallons

18. Hector uses 11 ounces of enchilada sauce for his famous beef taco recipe. The recipe makes 14 tacos. He wants to make 30 tacos. How many ounces of enchilada sauce will he need? Round your answer to the nearest tenth of an ounce. 23.6 ounces

19. On a map the distance between two mountains is $3\frac{1}{2}$ inches. The actual distance between the mountains is 136 miles. Russ is camped at a location that on the map is $\frac{3}{4}$ inch from the base of the mountain. How many miles is he from the base of the mountain? Round your answer to the nearest mile. 29 miles

20. Maria is adding a porch to her house that is 18 feet long. On the drawing done by the carpenter the length is shown as 11 inches. The drawing shows that the width of the porch is 8 inches. How many feet wide will the porch be? Round your answer to the nearest foot. 13 feet

Triangles A and B are similar. Use this in answering questions 21–24. Leave your answers as fractions.

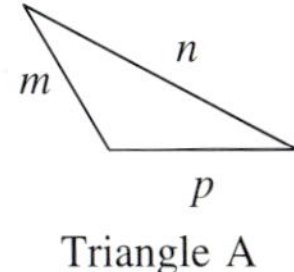

Triangle A

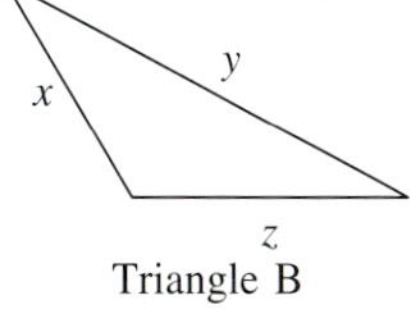

Triangle B

21. If $x = 20$ in., $y = 29$ in., and $m = 13$ in., find the length of side n.

$\frac{x}{y} = \frac{m}{n}$ $\frac{20}{29} = \frac{13}{n}$

$n = \frac{377}{20}$ $n = 18\frac{17}{20}$ inches

22. If $p = 14$ in., $m = 17$ in. and $z = 23$ in., find the length of side x.

$\frac{p}{m} = \frac{z}{x}$ $\frac{14}{17} = \frac{23}{x}$

$14x = 391$ $x = \frac{391}{14}$ $x = 27\frac{13}{14}$ inches

23. If $n = 40$ meters, $m = 35$ meters, and $x = 100$ meters, find the length of side y.

$\frac{n}{m} = \frac{y}{x}$ $\frac{40}{35} = \frac{y}{100}$

$4000 = 35y$ $\frac{800}{7} = y$ $y = 114\frac{2}{7}$ meters

24. If $z = 18$ cm, $y = 25$ cm, and $n = 9$ cm, find the length of side p.

$\frac{z}{y} = \frac{p}{n}$ $\frac{18}{25} = \frac{p}{9}$

$162 = 25p$ $\frac{162}{25} = p$ $p = 6\frac{12}{25}$ cm

Use a proportion to solve each problem.

25. A rectangle that is pleasing to the eye is called a "golden rectangle." The ratio of the width to the length is approximately 5 to 8. Using this ratio, what should be the length of a rectangular picture if its width is to be 30 inches? 48 inches long

26. A 5-foot-tall woman casts a shadow of 4 feet. At the same time, a tree casts a shadow of 31 feet. How tall is the tree? Round your answer to the nearest foot. 39 feet

27. A kite is held out on a line that is almost perfectly straight. When the kite is held out on 7 meters of line, it is 5 meters high off the ground. Assuming the same relationship, how high would the kite be if it is held out on 120 meters of line? Round your answer to the nearest meter. 86 meters

28. A wire line helps to secure a radio transmission tower. The wire measures 23 meters from the tower to the ground anchor pin. The wire is secured 14 meters up on the tower. Assuming the same relationship, if a wire is secured 130 meters up on the tower, how long would it need to be to reach from the tower to an anchor pin on the ground? Round your answer to the nearest meter. 214 meters

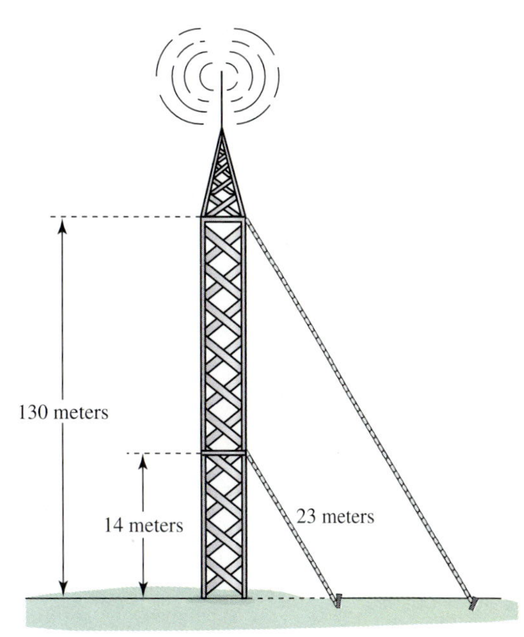

29. Ben Hale is driving his new Toyota Camry on Interstate 90 at 45 miles per hour. He accelerates at the rate of 3 miles per hour every 2 seconds. How fast will he be traveling after accelerating for 11 seconds?
61.5 miles per hour

30. Tim Newitt is driving a U-Haul truck to Chicago. He is driving at 55 miles per hour and has to hit the brakes because of heavy traffic. His truck slows at the rate of 2 miles per hour for every 3 seconds. How fast will he be traveling 10 seconds after he hits the brakes?
approx. 48.3 miles per hour

31. A Montreal commuter airline travels 40 kilometers per hour faster than the television news helicopter over the city. The commuter airline travels 1250 kilometers during the same time that the television news helicopter travels only 1050 kilometers to film footage for a special news story. How fast does the commuter airline fly? How fast does the television news helicopter fly? The commuter airline flies at 250 kilometers pr hour. The helicopter flies at 210 kilometers per hour.

32. Melissa drove to Dallas, while Marcia drove to Houston in the same amount of time. Melissa drove 360 kilometers, while Marcia drove 280 kilometers. Melissa averaged 20 kilometers per hour faster than Marcia on her trip. What is the average speed in kilometers per hour for each woman? Marcia averaged 70 kilometers per hour. Melissa averaged 90 kilometers per hour.

33. It takes a person using a large rotary mower 4 hours to mow all the lawns at the town park. It takes a person using a small rotary mower 5 hours to mow these same lawns. If both mowers are used, how long should it take two people to mow these lawns? Round your answer to the nearest minute.
$2\frac{2}{9}$ hours or 2 hours 13 minutes

34. It takes 7 hours to write the addresses on envelopes for a business mailing list with a secretary doing the work by typewriter. It takes 3 hours to have the computer print out the addresses and have the secretary put the computer labels on the envelopes. The company boss is in a big rush and wants one secretary to do part of the job with the computer and another secretary to do the rest of the list with a typewriter. If they work efficiently, how long should it take with the two secretaries working together? Round your answer to the nearest minute.
$2\frac{1}{10}$ hours or 2 hours 6 minutes

35. The larger pump at the town pumping station can fill the town water tank in 15 hours. The smaller pump can do the same job in 20 hours. How long will it take the pumps working together? Round your answer to the nearest minute.
$8\frac{4}{7}$ hours or 8 hours 34 minutes

36. It took 400 hours for highway team A to clean all the streets of the city. It took 300 hours for highway team B to do the same job. The mayor of the city is in a big hurry to get the streets clean for visiting dignitaries. If team A and team B both work together, how many hours should it take to clean all the streets of the city? Round your answer to the nearest hour. 171 hours

Cumulative Review Problems

37. Write in scientific notation. 0.0000006316 6.316×10^{-7}

38. Write in decimal notation. 5.82×10^8 582,000,000

39. Write with positive exponents. $\dfrac{x^{-3}y^{-2}}{z^4w^{-8}}$ $\dfrac{w^8}{x^3y^2z^4}$

40. Evaluate. $\left(\dfrac{2}{3}\right)^{-3}$ $\dfrac{27}{8}$ or $3\dfrac{3}{8}$

Putting Your Skills to Work

MATHEMATICAL MEASUREMENT OF PLANET ORBIT TIME

The closer a planet is to the sun, the fewer the number of days it takes to complete one orbit around the sun. The Earth takes approximately 365 days to complete one orbit. Only two of the nine planets have a shorter orbit time. The orbit times of the four planets closest to the sun are shown in the bar graph to the right. Use this bar graph to answer the following questions. Round all answers to the nearest whole number.

PROBLEMS FOR INDIVIDUAL INVESTIGATION

1. How many more days does it take Mars than Venus to complete one rotation around the sun? 462 days more

2. In 5 years, the Earth will rotate around the sun five times. In that time period how many orbits will Mercury make around the sun? approx. 21 orbits

PROBLEMS FOR GROUP INVESTIGATION AND COOPERATIVE LEARNING

Together with some other members of your class see if you can answer the following.

The **synodic period** of a planet is the number of days it takes for a planet closer to the sun to gain one orbit on a planet farther from the sun. The synodic period (S) can be found by the formula

$$\frac{1}{S} = \frac{1}{a} - \frac{1}{b}$$

where a = the number of days it takes the planet closer to the sun to complete one orbit

b = the number of days it takes the planet farther from the sun to complete one orbit

3. What is the synodic period for Mercury to gain one orbit on Venus? approx. 145 days

4. What is the synodic period for the Earth to gain one orbit on Jupiter if the number of days it takes Jupiter to orbit the sun is approximately 4333 days? approx. 399 days

INTERNET CONNECTIONS: Go to ''`http://www.prenhall.com/tobey`'' to be connected

Site: The Nine Planets or a related site

This site provides a wealth of information about our solar system. Note that many of the numbers here are given in an abbreviated form of scientific notation; for example, 1.35e23 means 1.35×10^{23}.

5. Tell how many days it takes each of the planets Saturn, Uranus, Neptune, and Pluto to complete one rotation about the sun. This information is listed as ''O–Period'' in the Solar System Data appendix of the web site.

6. How many more days does it take Neptune than Saturn to complete one orbit about the sun?

7. Use the formula to find the synodic period for Uranus to gain one orbit on Pluto.

Chapter Organizer

Topic	Procedure	Examples
Simplifying rational expressions, p. 342.	1. Factor numerator and denominator. 2. Divide out any factor common to both numerator and denominator.	$\frac{36x^2 - 16y^2}{18x^2 + 24xy + 8y^2} = \frac{\overset{2}{4}(3x - 2y)(3x + 2y)}{2(3x + 2y)(3x + 2y)}$ $= \frac{2(3x - 2y)}{3x + 2y}$
Multiplying rational expressions, p. 348.	1. Factor all numerators and denominators, writing the problem as one fraction. 2. Proceed as in simplifying, above.	$\frac{x^2 - y^2}{x^2 + 2xy + y^2} \cdot \frac{x^2 + 4xy + 3y^2}{x^2 - 4xy + 3y^2}$ $= \frac{(x + y)(x - y)(x + y)(x + 3y)}{(x + y)(x + y)(x - y)(x - 3y)}$ $= \frac{x + 3y}{x - 3y}$
Dividing rational expressions, p. 350.	1. Invert the second fraction and change the problem to multiplication. 2. Proceed as for multiplication.	$\frac{14x^2 + 17x - 6}{x^2 - 25} \div \frac{4x^2 - 8x - 21}{x^2 + 10x + 25}$ $= \frac{(2x + 3)(7x - 2)}{(x + 5)(x - 5)} \cdot \frac{(x + 5)(x + 5)}{(2x - 7)(2x + 3)}$ $= \frac{(7x - 2)(x + 5)}{(x - 5)(2x - 7)}$
Adding rational expressions, p. 354.	1. If the denominators differ, factor them and determine the least common denominator (LCD). 2. Change, by multiplication, each fraction into an equivalent one with the LCD. 3. Add (add numerators; put answer over the LCD). 4. Simplify as needed.	$\frac{x - 1}{x^2 - 4} + \frac{x - 1}{3x + 6}$ $(x + 2)(x - 2) \quad 3(x + 2)$ LCD $= 3(x + 2)(x - 2)$ $\frac{x - 1}{(x + 2)(x - 2)} = \frac{?}{3(x + 2)(x - 2)}$ Need to multiply by $\frac{3}{3}$. $\frac{(x - 1)}{3(x + 2)} = \frac{?}{3(x + 2)(x - 2)}$ Need to multiply by $\frac{x - 2}{x - 2}$. $= \frac{(x - 1) \cdot 3}{3(x + 2)(x - 2)} + \frac{(x - 1)(x - 2)}{3(x + 2)(x - 2)}$ $= \frac{3x - 3 + x^2 - 3x + 2}{3(x + 2)(x - 2)}$ $= \frac{x^2 - 1}{3(x + 2)(x - 2)}$
Subtracting rational expressions, p. 354.	Move a subtraction sign to the numerator of the following fraction. Add. Simplify if possible. $\frac{a}{b} - \frac{c}{b} = \frac{a}{b} + \frac{-c}{b}$	$\frac{5x}{x - 2} - \frac{3x + 4}{x - 2}$ $= \frac{5x}{x - 2} + \frac{-(3x + 4)}{x - 2}$ $= \frac{5x - 3x - 4}{x - 2} = \frac{2x - 4}{x - 2}$ $= \frac{2(x - 2)}{x - 2} = 2$

Chapter Organizer

<table>
<tr><th>Topic</th><th>Procedure</th><th>Examples</th></tr>
<tr><td>Simplifying complex fractions, p. 364.</td><td>1. Add together the two fractions in the numerator.
2. Add together the two fractions in the denominator.
3. Divide the fraction in the numerator by the fraction in the denominator. This is done by inverting the fraction in the denominator and multiplying by the numerator.
4. Simplify.</td><td>$$\frac{\frac{x}{x^2-4}+\frac{1}{x+2}}{\frac{3}{x+2}-\frac{4}{x-2}}$$ $$=\frac{\frac{x}{(x+2)(x-2)}+\frac{1(x-2)}{(x+2)(x-2)}}{\frac{3(x-2)}{(x+2)(x-2)}+\frac{-4(x+2)}{(x+2)(x-2)}}$$ $$=\frac{\frac{x+x-2}{(x+2)(x-2)}}{\frac{3x-6-4x-8}{(x+2)(x-2)}}$$ $$=\frac{2x-2}{(x+2)(x-2)}\div\frac{-x-14}{(x+2)(x-2)}$$ $$=\frac{2(x-1)}{(x+2)(x-2)}\cdot\frac{(x+2)(x-2)}{-x-14}$$ $$=\frac{2x-2}{-x-14}\quad\text{or}\quad\frac{-2x+2}{x+14}$$</td></tr>
<tr><td>Solving equations involving algebraic fractions, p. 369.</td><td>1. Determine the LCD of all denominators.
2. Note what values will make the LCD equal to 0. These are excluded from your solutions.
3. Multiply each side by the LCD, distributing as needed.
4. Solve the resulting polynomial equation.
5. Check. Be sure to exclude those values found in step 2.</td><td>$$\frac{3}{x-2}=\frac{4}{x+2}$$ LCD $=(x-2)(x+2)$. Since LCD $\neq 0$, $x \neq 2, -2$. $$(x-2)(x+2)\frac{3}{x-2}=\frac{4}{x+2}(x-2)(x+2)$$ $$3(x+2)=4(x-2)$$ $$3x+6=4x-8$$ $$-x=-14$$ $$x=14$$ (Since $x \neq 2, -2$ this should check unless an error has been made.)
Check: $\frac{3}{14-2}\stackrel{?}{=}\frac{4}{14+2}$ $$\frac{3}{12}\stackrel{?}{=}\frac{4}{16}$$ $$\frac{1}{4}=\frac{1}{4}\quad ✓$$</td></tr>
<tr><td>Solving applied problems with proportions, p. 374.</td><td>1. Organize the data.
2. Write a proportion equating the respective parts, letting x represent the value that is not known.
3. Solve the proportion.</td><td>Renee can make 5 cherry pies with 3 cups of flour. How many cups of flour does she need to make 8 cherry pies? $$\frac{5\text{ cherry pies}}{3\text{ cups flour}}=\frac{8\text{ cherry pies}}{x\text{ cups flour}}$$ $$\frac{5}{3}=\frac{8}{x}$$ $$5x=24$$ $$x=\frac{24}{5}$$ $$x=4\frac{4}{5}$$ $4\frac{4}{5}$ cups of flour are needed for 8 cherry pies.</td></tr>
</table>

Chapter 6 Review Problems

6.1 *Simplify.*

1. $\dfrac{4x-4y}{5y-5x}$ $-\dfrac{4}{5}$

2. $\dfrac{bx}{bx-by}$ $\dfrac{x}{x-y}$

3. $\dfrac{2x^2+5x-3}{2x^2-9x+4}$ $\dfrac{x+3}{x-4}$

4. $\dfrac{6x^2+7x+2}{3x^2+5x+2}$ $\dfrac{2x+1}{x+1}$

5. $\dfrac{x^2-9}{x^2-10x+21}$ $\dfrac{x+3}{x-7}$

6. $\dfrac{2x^2+18x+40}{3x+15}$ $\dfrac{2(x+4)}{3}$

7. $\dfrac{4x^2+4x-3}{4x^2-2x}$ $\dfrac{2x+3}{2x}$

8. $\dfrac{x^3+3x^2}{x^3-2x^2-15x}$ $\dfrac{x}{x-5}$

9. $\dfrac{2x^2-2xy-24y^2}{2x^2+5xy-3y^2}$ $\dfrac{2(x-4y)}{2x-y}$

10. $\dfrac{4-y^2}{3y^2+5y-2}$ $\dfrac{2-y}{3y-1}$

11. $\dfrac{5x^3-10x^2}{25x^4+5x^3-30x^2}$ $\dfrac{x-2}{(5x+6)(x-1)}$

12. $\dfrac{16x^2-4y^2}{4x-2y}$ $4x+2y$

6.2 *Multiply or divide.*

13. $\dfrac{3x^2-13x-10}{3x^2+2x}\cdot\dfrac{x^2-25x}{x^2-25}$ $\dfrac{x-25}{x+5}$

14. $\dfrac{x^2+7x+12}{4x+12}\cdot\dfrac{x^2-16}{x^2+8x+16}$ $\dfrac{x-4}{4}$

15. $\dfrac{2x^2-18}{3y^2+3y}\div\dfrac{y^2+6y+9}{y^2+4y+3}$ $\dfrac{2(x+3)(x-3)}{3y(y+3)}$

16. $\dfrac{2y^2+3y-2}{2y^2+y-1}\div\dfrac{2y^2+y-1}{2y^2-3y-2}$ $\dfrac{(y+2)(2y+1)(y-2)}{(y+1)(2y-1)(y+1)}$

17. $\dfrac{x^3-36x}{12x^2+2x}\cdot\dfrac{36x+6}{x^3-6x^2}\cdot\dfrac{x^2+2x}{x^2+11x+30}$ $\dfrac{3(x+2)}{x(x+5)}$

18. $\dfrac{10x^2}{x^2-7x+10}\cdot\dfrac{x^2+3x-10}{25x^2}\cdot\dfrac{2x-8}{x^2+x-20}$ $\dfrac{4}{5(x-5)}$

19. $\dfrac{6y^2+13y-5}{9y^2+3y}\div\dfrac{4y^2+20y+25}{12y^2}$ $\dfrac{4y(3y-1)}{(3y+1)(2y+5)}$

20. $\dfrac{3xy^2+12y^2}{2x^2-11x+5}\div\dfrac{2xy+8y}{8x^2+2x-3}$ $\dfrac{3y(4x+3)}{2(x-5)}$

21. $\dfrac{11}{x-2}\cdot\dfrac{2x^2-8}{44}$ $\dfrac{(x+2)}{2}$

22. $\dfrac{2x^2+10x+2}{8x-8}\cdot\dfrac{3x-3}{4x^2+20x+4}$ $\dfrac{3}{16}$

23. $\dfrac{5x^7}{16x^2-9}\cdot\dfrac{8x+6}{10x^8}$ $\dfrac{1}{x(4x-3)}$

24. $\dfrac{x^2-5xy-24y^2}{2x^2-2xy-24y^2}\cdot\dfrac{4x^2+4xy-24y^2}{x^2-10xy+16y^2}$ $\dfrac{2(x+3y)}{x-4y}$

6.3 *Add or subtract.*

25. $\dfrac{7}{x+1}+\dfrac{4}{2x}$ $\dfrac{9x+2}{x(x+1)}$

26. $5+\dfrac{1}{x}+\dfrac{1}{x+1}$ $\dfrac{5x^2+7x+1}{x(x+1)}$

27. $\dfrac{2}{x^2-9}+\dfrac{x}{x+3}$ $\dfrac{(x-1)(x-2)}{(x+3)(x-3)}$

28. $\dfrac{7}{x+2}+\dfrac{3}{x-4}$ $\dfrac{10x-22}{(x+2)(x-4)}$

29. $\dfrac{x}{y}+\dfrac{3}{2y}+\dfrac{1}{y+2}$ $\dfrac{2xy+4x+5y+6}{2y(y+2)}$

30. $\dfrac{4}{a}+\dfrac{2}{b}+\dfrac{3}{a+b}$ $\dfrac{(2a+b)(a+4b)}{ab(a+b)}$

31. $\dfrac{3x+1}{3x}-\dfrac{1}{x}$ $\dfrac{3x-2}{3x}$

32. $\dfrac{x+4}{x+2}-\dfrac{1}{2x}$ $\dfrac{2x^2+7x-2}{2x(x+2)}$

33. $\dfrac{1}{x^2+7x+10}-\dfrac{x}{x+5}$ $\dfrac{1-2x-x^2}{(x+5)(x+2)}$

34. $\dfrac{7}{x^2-9x+20}-\dfrac{3x}{x-4}$ $\dfrac{7+15x-3x^2}{(x-4)(x-5)}$

35. $\dfrac{2x+1}{x^2+5x+6}+\dfrac{4x}{x^2-9}$ $\dfrac{3(2x-1)(x+1)}{(x+2)(x+3)(x-3)}$

36. $\dfrac{3x}{x^2+x-2}-\dfrac{2x+1}{x^2+6x+8}$ $\dfrac{x^2+13x+1}{(x+2)(x-1)(x+4)}$

6.4 *Simplify.*

37. $\dfrac{\dfrac{3}{2y} - \dfrac{1}{y}}{\dfrac{4}{y} + \dfrac{3}{2y}}$ $\dfrac{1}{11}$

38. $\dfrac{\dfrac{2}{x} + \dfrac{1}{2x}}{x + \dfrac{x}{2}}$ $\dfrac{5}{3x^2}$

39. $\dfrac{w - \dfrac{4}{w}}{1 + \dfrac{2}{w}}$ $w - 2$

40. $\dfrac{1 - \dfrac{w}{w-1}}{1 + \dfrac{w}{1-w}}$ 1

41. $\dfrac{1 + \dfrac{1}{y^2 - 1}}{\dfrac{1}{y+1} - \dfrac{1}{y-1}}$ $-\dfrac{y^2}{2}$

42. $\dfrac{\dfrac{1}{y} + \dfrac{1}{x+y}}{1 + \dfrac{2}{x+y}}$ $\dfrac{x + 2y}{y(x + y + 2)}$

43. $\dfrac{\dfrac{1}{a+b} - \dfrac{1}{a}}{b}$ $\dfrac{-1}{a(a+b)}$

44. $\dfrac{\dfrac{2}{a+b} - \dfrac{3}{b}}{\dfrac{1}{a+b}}$ $\dfrac{-3a - b}{b}$

45. $\left(\dfrac{1}{x+2y} - \dfrac{1}{x-y}\right) \div \dfrac{2x - 4y}{x^2 - 3xy + 2y^2}$ $\dfrac{-3y}{2(x + 2y)}$

46. $\dfrac{x + 5y}{x - 6y} \div \left(\dfrac{1}{5y} - \dfrac{1}{x + 5y}\right)$ $\dfrac{5y(x + 5y)^2}{x(x - 6y)}$

6.5 *Solve for the variable. If no solution is possible, say so.*

47. $\dfrac{8}{a-3} = \dfrac{12}{a+3}$ $a = 15$

48. $\dfrac{8a - 1}{6a + 8} = \dfrac{3}{4}$ $a = 2$

49. $\dfrac{2x - 1}{x} - \dfrac{1}{2} = -2$ $x = \dfrac{2}{7}$

50. $\dfrac{5 - x}{x} - \dfrac{7}{x} = -\dfrac{3}{4}$ $x = -8$

51. $\dfrac{5}{2} - \dfrac{2y + 7}{y + 6} = 3$ $y = -4$

52. $\dfrac{5}{4} - \dfrac{1}{2x} = \dfrac{1}{x} + 2$ $x = -2$

53. $\dfrac{1}{3x} + 2 = \dfrac{5}{6x} - \dfrac{1}{2}$ $x = \dfrac{1}{5}$

54. $\dfrac{7}{8x} - \dfrac{3}{4} = \dfrac{1}{4x} + \dfrac{1}{2}$ $x = \dfrac{1}{2}$

55. $\dfrac{x - 8}{x - 2} = \dfrac{2x}{x + 2} - 2$ $x = -4, x = 6$

56. $\dfrac{3}{y - 3} = \dfrac{3}{2} + \dfrac{y}{y - 3}$ no solution

57. $\dfrac{3y - 1}{3y} - \dfrac{6}{5y} = \dfrac{1}{y} - \dfrac{4}{15}$ $y = 2$

58. $\dfrac{9}{2} - \dfrac{7y - 4}{y + 2} = -\dfrac{1}{4}$ $y = 6$

59. $\dfrac{4}{x^2 - 1} = \dfrac{2}{x - 1} + \dfrac{2}{x + 1}$ no solution

60. $\dfrac{y + 18}{y^2 - 16} = \dfrac{y}{y + 4} - \dfrac{y}{y - 4}$ $y = -2$

61. $\dfrac{9y - 3}{y^2 + 2y} - \dfrac{5}{y + 2} = \dfrac{3}{y}$ $y = 9$

62. $\dfrac{2}{3 - 3y} + \dfrac{2}{2y - 1} = \dfrac{4}{3y - 3}$ $y = 0$

6.6 *Solve each proportion. Round your answer to the nearest tenth.*

63. $\frac{8}{5} = \frac{2}{x}$ $x = 1.3$

64. $\frac{x}{4} = \frac{12}{17}$ $x = 2.8$

65. $\frac{33}{10} = \frac{x}{8}$ $x = 26.4$

66. $\frac{5}{x} = \frac{22}{9}$ $x = 2$

Use a proportion to answer each question.

67. It took 9 hours to register 450 students. At the same rate, how long will it take to register 550 students? Round your answer to the nearest minute. 11 hours

68. Fred is driving in Canada. He knows that 100 kilometers per hour is about the same as 62 miles per hour. The speed limit on this back road is 65 kilometers per hour. How many miles per hour is the speed limit? Round your answer to the nearest mile per hour. 40 miles per hour

69. A 5-gallon can of paint will cover 240 square feet. How many gallons of paint will be needed to cover 400 square feet? Round your answer to the nearest tenth of a gallon. 8.3 gallons

70. Aunt Lexie uses 3 pounds of sugar to make 100 cookies. How many cookies can she make with 5 pounds of sugar? Round to nearest whole number. 167 cookies

71. Ron found that his car used 7 gallons of gas to travel 200 miles. He plans to drive 1300 miles from his home to Denver, Colorado. How many gallons of gas will he use if his car continues to use gas at the same rate? Round your answer to the nearest whole number. 46 gallons

72. The distance on a map between two cities is 4 inches. The actual distance between these cities is 122 miles. Two lakes are 3 inches apart on the same map. How many miles is the actual distance between these two lakes? 91.5 miles

73. A train travels 180 miles in the same time that a car travels 120 miles. The speed of the train is 20 miles per hour faster than the speed of the car. Find the speed of the train and the speed of the car. train: 60 miles per hour; car: 40 miles per hour

74. A professional painter can paint the interior of the Jackson's house in 5 hours. John Jackson can do the same job in 8 hours. How long will it take the two people working together on the painting job? Round your answer to the nearest minute. 3 hours 5 minutes

75. A flagpole that is 8 feet tall casts a shadow of 3 feet. At the same time of day, a tall office building in the city casts a shadow of 450 feet. How tall is the office building? Round your answer to the nearest foot. 1200 feet tall

76. Mary took a walk across a canyon in New Mexico. She stands 5.75 feet tall and her shadow is 3 feet long. At the same time the shadow from the peak of the canyon wall casts a shadow that is 95 feet long. How tall is the peak of the canyon? Round your answer to the nearest foot. 182 feet tall

Chapter 6 Test

Perform the operation indicated. Simplify your answers.

1. $\dfrac{2ac + 2ad}{3a^2c + 3a^2d}$

2. $\dfrac{8x^2 - 2x^2y^2}{y^2 + 4y + 4}$

3. $\dfrac{x^2 + 2x}{2x - 1} \cdot \dfrac{10x^2 - 5x}{12x^3 + 24x^2}$

4. $\dfrac{a^2 + ab}{b} \cdot \dfrac{a^2 - 2ab + b^2}{a^4 - b^4} \cdot \dfrac{a^2 + b^2}{3a^2b - 3ab^2}$

5. $\dfrac{2a^2 - 3a - 2}{4a^2 + a - 14} \div \dfrac{2a^2 + 5a + 2}{16a^2 - 49}$

6. $\dfrac{9}{x - 2} + \dfrac{2}{x - 3} + \dfrac{1}{2x - 6}$

7. $\dfrac{x - y}{xy} - \dfrac{a - y}{ay}$

8. $\dfrac{3x}{x^2 - 3x - 18} - \dfrac{x - 4}{x - 6}$

9. $\dfrac{\dfrac{x}{3y} - \dfrac{1}{2}}{\dfrac{4}{3y} - \dfrac{2}{x}}$

10. $\dfrac{\dfrac{2}{x + 3} + \dfrac{1}{x}}{3x + 9}$

11. $\dfrac{2x^2 + 3xy - 9y^2}{4x^2 + 13xy + 3y^2}$

12. $\dfrac{1}{x + 4} - \dfrac{2}{x^2 + 6x + 8}$

Answers:

1. $\dfrac{2}{3a}$
2. $\dfrac{2x^2(2 - y)}{(y + 2)}$
3. $\dfrac{5}{12}$
4. $\dfrac{1}{3b^2}$
5. $\dfrac{(a - 2)(4a + 7)}{(a + 2)^2}$
6. $\dfrac{23x - 64}{2(x - 2)(x - 3)}$
7. $\dfrac{x - a}{ax}$
8. $-\dfrac{x + 2}{x + 3}$
9. $\dfrac{x}{4}$
10. $\dfrac{x + 1}{x(x + 3)^2}$
11. $\dfrac{2x - 3y}{4x + y}$
12. $\dfrac{x}{(x + 2)(x + 4)}$

13. $x = 3$

14. $x = 4$

15. no solution

16. $x = \frac{47}{6}$

17. $x = \frac{45}{13}$

18. 6 hours, 25 minutes

19. \$368

20. 102 feet

Solve for x. Check your answers.

13. $\frac{15}{x} + \frac{9x - 7}{x + 2} = 9$

14. $\frac{x - 3}{x - 2} = \frac{2x^2 - 15}{x^2 + x - 6} - \frac{x + 1}{x + 3}$

15. $3 - \frac{7}{x + 3} = \frac{x - 4}{x + 3}$

16. $\frac{3}{3x - 5} = \frac{7}{5x + 4}$

17. $\frac{9}{x} = \frac{13}{5}$

18. Katie is typing a term paper for her English class. She typed 3 pages of the paper in 55 minutes. If she continues at the same rate, how long will it take her to type the entire 21-page paper? Express your answer in hours and minutes.

19. In northern Michigan the Gunderson family heats their home with firewood. They used \$100 worth of wood in 25 days. Mr. Gunderson estimates that he needs to burn wood at that rate for about 92 days during the winter. If that is so, how much will the 92-day supply of wood cost?

20. A hiking club is trying to construct a rope bridge across a canyon. A 6-foot construction pole held upright casts a 7-foot shadow. At the same time of day, a tree at the edge of the canyon casts a shadow that exactly covers the distance that is needed for the rope bridge. The tree is exactly 87 feet tall. How long should the rope bridge be? Round your answer to the nearest foot.

Cumulative Test for Chapters 0–6

Approximately one-half of this test covers the content of Chapters 0–5. The remainder covers the content of Chapter 6.

1. To the nearest thousandth, how much error can be tolerated in the length of a wire that is supposed to be 2.57 centimeters long? Specifications allow an error of no more than 0.25%.

2. Fifteen percent of what number is \$37.50?

3. After the quake of 1989, California raised its sales tax by $\frac{1}{4}$% (one-quarter of 1%) to aid earthquake relief. How much extra sales tax will be paid on a new \$19,990 car?

4. Solve.

$5(x - 3) - 2(4 - 2x) = 7(x - 1) - (x - 2)$

5. Solve for h.

$A = \pi r^2 h$

6. Solve and graph on a number line. $4(2 - x) < 3$

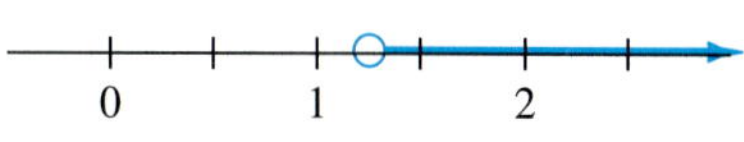

7. Solve for x.

$\frac{1}{4}x + \frac{3}{4} < \frac{2}{3}x - \frac{4}{3}$

8. Factor. $3ax + 3bx - 2ay - 2by$

9. Factor. $8a^3 - 38a^2b - 10ab^2$

10. Simplify. $(-3x^2y)(4x^3y^6)$

11. Simplify. $\dfrac{4x^2 - 25}{2x^2 + 9x - 35}$

1. 0.006 cm

2. \$250

3. \$49.98

4. $x = 6$

5. $h = \frac{A}{\pi r^2}$

6. $x > 1.25$ See graph

7. $x > 5$

8. $(a + b)(3x - 2y)$

9. $2a(4a + b)(a - 5b)$

10. $-12x^5y^7$

11. $\frac{2x + 5}{x + 7}$

12.	$3x + 1$
13.	$\frac{11x - 3}{2(x + 2)(x - 3)}$
14.	$\frac{x^2 + 2x + 6}{(x + 4)(x - 3)(x + 2)}$
15.	$x = -2$
16.	$x = \frac{-9}{2}$
17.	$\frac{x + 8}{6x(x - 3)}$
18.	$\frac{3ab^2 + 2a^2b}{5b^2 - 2a^2}$
19.	$x = \frac{98}{5}$
20.	208 miles
21.	484 phone calls

Perform the indicated operations.

12. $\frac{x^2 - 4}{x^2 - 25} \cdot \frac{3x^2 - 14x - 5}{3x^2 + 6x} \div \frac{x - 2}{3x^2 + 15x}$

13. $\frac{5}{2x + 4} + \frac{3}{x - 3}$

14. $\frac{2x + 1}{x^2 + x - 12} - \frac{x - 1}{x^2 - x - 6}$

Solve for x.

15. $\frac{3x - 2}{3x + 2} = 2$

16. $\frac{x - 3}{x} = \frac{x + 2}{x + 3}$

Simplify.

17. $\dfrac{\frac{1}{x - 3} + \frac{5}{x^2 - 9}}{\frac{6x}{x + 3}}$

18. $\dfrac{\frac{3}{a} + \frac{2}{b}}{\frac{5}{a^2} - \frac{2}{b^2}}$

19. Solve for x. $\frac{5}{7} = \frac{14}{x}$

20. Jane is looking at a road map. The distance between two cities is 130 miles. This distance is represented by $2\frac{1}{2}$ inches on the map. She sees that the distance she has to drive today is a totally straight interstate highway represented by 4 inches on the map. How many miles does she have to drive? Round your answer to the nearest mile.

21. Roberto is working as a telemarketing salesperson for a large corporation. In the last 22 telephone calls he has made, he was able to make a sale 5 times. His goal is to make 110 sales this month. How many phone calls must he make if his rate for making a sale continues as in the past?

CHAPTER 7
Graphing and Functions

Suppose you are walking through the streets of Paris and see a wonderful painting that you want to buy. It costs 850 French francs. Do you have any idea of how many U.S. dollars you will need to buy it? If you continue shopping in France, do you know that there is a simple graph that can help you estimate how much something in another country will cost you in U.S. dollars? If this sounds helpful to you, turn to the Putting Your Skills to Work on page 446.

Pretest Chapter 7

1. See graph

2. A $(-1, -2)$; B $(3, -2)$; C $(-1, 2)$

3. (a) $(-1, 10)$

(b) $(2, 1)$

4. See graph

5. See graph

6. $\frac{2}{7}$

7. $\frac{4}{3}$

8. (a) $y = 4x - 5$

(b) $4x - y = 5$

If you are familiar with the topics in this chapter, take this test now. Check your answers with those in the back of the book. If an answer was wrong or you couldn't do a problem, study the appropriate section of the chapter.

If you are not familiar with the topics in this chapter, don't take this test now. Instead, study the examples, work the Practice Problems, and then take the test.

This test will help you to identify those concepts that you have mastered and those that need more study.

Section 7.1

1. Graph the points.
$(-3, 1)$, $(4, -4)$, $(-2, -6)$

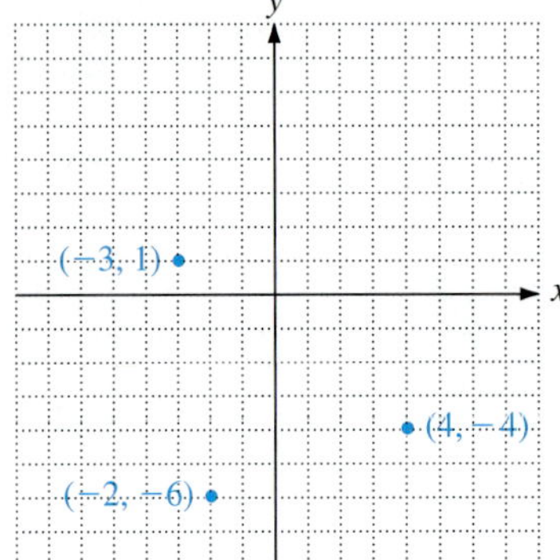

2. Write the coordinates of each point below.

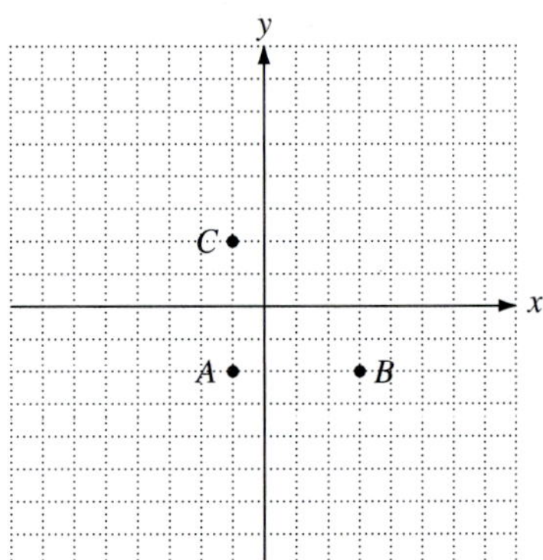

3. Complete each ordered pair so that it is a solution to the equation $y = -3x + 7$.
(a) $(-1, \quad)$ **(b)** $(\quad , 1)$

Section 7.2

4. Graph $5x - 2y = -10$.

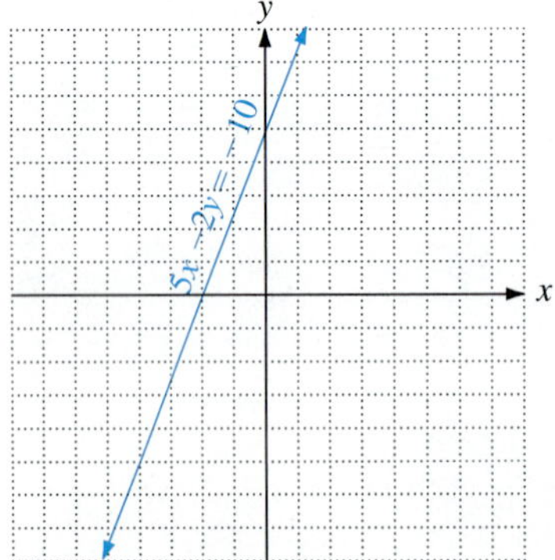

5. Graph $y = 4$.

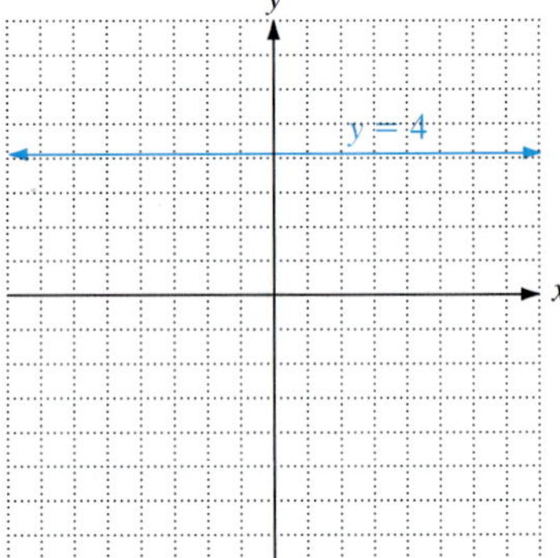

Section 7.3

6. What is the slope of the line passing through $(4, -2)$ and $(-3, -4)$?

7. Find the slope of the line $4x - 3y - 7 = 0$.

8. Find the equation of the line with slope 4 and y-intercept -5.
(a) Write the equation in slope–intercept form.
(b) Write the equation in $Ax + By = C$ form.

Section 7.4

9. Graph the equation $y = 2x + 1$.

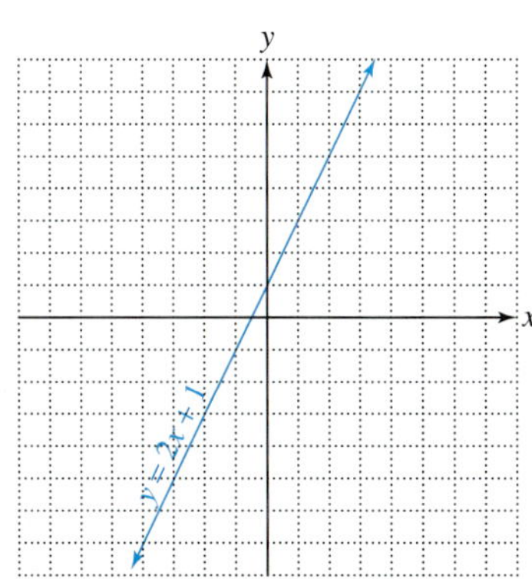

10. Write the equation of the graph.

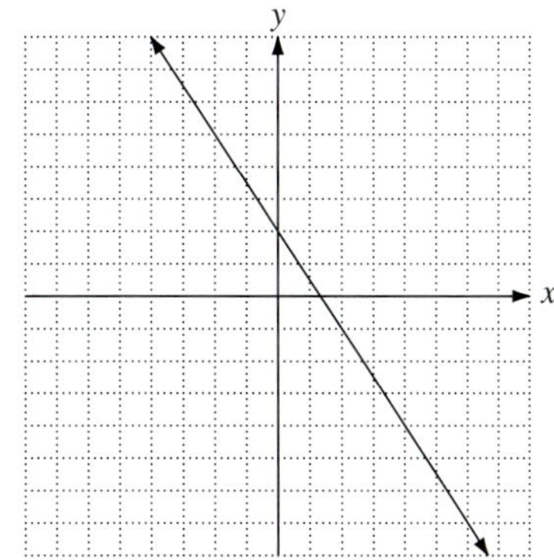

11. Line p has a slope of $-\frac{1}{4}$.

(a) What is the slope of a line parallel to line p?

(b) What is the slope of a line perpendicular to line p?

12. Find the equation of a line with slope $\frac{2}{3}$ that passes through the point $(-3, -3)$.

13. Find the equation of a line that passes through $(-1, 6)$ and $(2, 3)$.

Section 7.5

14. Graph $y \geq 3x - 2$.

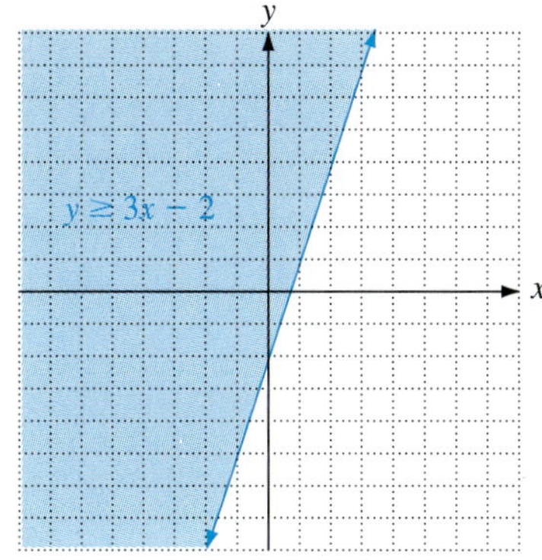

15. Graph $4x + 2y < -12$.

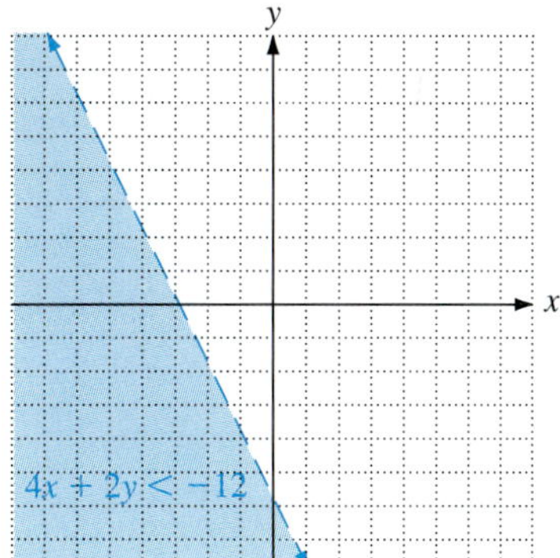

Section 7.6

Determine whether each of the following is a function.

16. Circle

r	1	5	10	15
C	6.28	31.4	62.8	94.2

17.

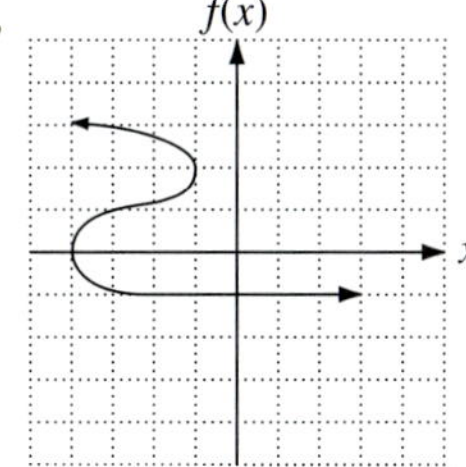

Find the indicated values for each function.

18. $f(x) = 3x - 7$

(a) $f(-2)$ **(b)** $f(4)$

19. $g(x) = 2x^2$

(a) $g(-6)$ **(b)** $g(\frac{1}{2})$

9. See graph

10. $y = -\frac{3}{2}x + 2$

11. (a) $-\frac{1}{4}$

(b) 4

12. $y = \frac{2}{3}x - 1$

13. $y = -x + 5$

14. See graph

15. See graph

16. yes

17. no

18. (a) -13

(b) 5

19. (a) 72

(b) $\frac{1}{2}$

7.1 Rectangular Coordinate System

MathPro

Video 17

SSM

After studying this section, you will be able to:

1 *Plot a point, given the coordinates.*
2 *Name the coordinates of a plotted point.*
3 *Find ordered pairs for a given linear equation.*

1 *Plotting a Point*

Many things in everyday life are clearer if we see a picture. Similarly, in mathematics, we can picture algebraic relationships by drawing a graph. To draw a graph, we need a frame of reference.

In Chapter 1 we showed that any real number can be represented on a number line. Look at the number line below. The arrow indicates the positive direction.

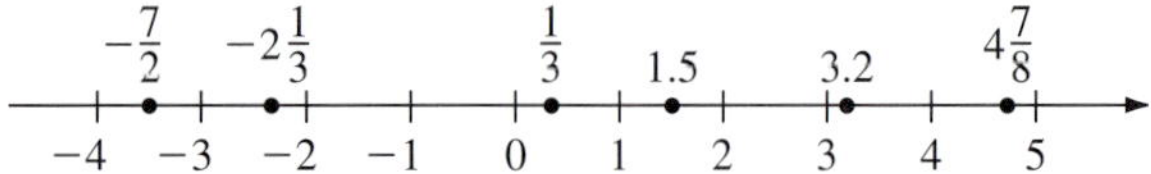

To form a rectangular coordinate system, we draw a second number line vertically. We construct it so that the 0 point on each number line is exactly at the same place. We refer to this location as the **origin.** The horizontal number line is often called the ***x*-axis.** The vertical number line is often called the ***y*-axis.** Arrows show the positive direction for each axis.

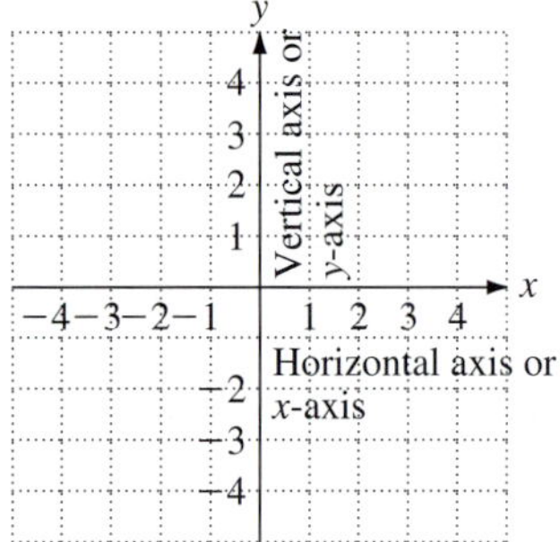

We can represent a point in this rectangular coordinate system by using an **ordered pair** of numbers. The first number of the pair represents the distance from the origin measured along the horizontal or x-axis. The second number of the pair represents the distance measured on the y-axis or on a line parallel to the y-axis. For example (5, 2) is an ordered pair that represents a point in the rectangular coordinate system.

The ordered pair of numbers that represent a point are often referred to as the **coordinates** of a point. The first value is called the x-coordinate. The second value is called the y-coordinate. If the x-coordinate is positive, we count the proper number of squares to the right (in the positive direction). If the x-coordinate is negative, we count to the left. If the y-coordinate is positive, we count the proper number of squares upward (in the positive direction). If the y-coordinate is negative, we count downward.

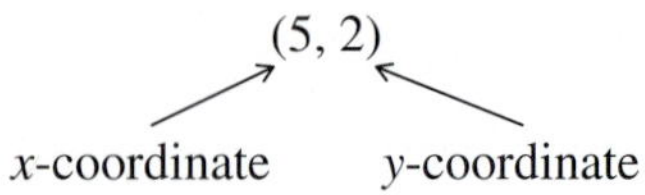

EXAMPLE 1 Plot the point (5, 2) on a rectangular coordinate system. Label this as point A.

Since the x-coordinate is 5, we first count 5 units to the right on the x-axis. Then, because the y-coordinate is 2, we count 2 units up from the point where we stopped on the x-axis. This locates the point corresponding to (5, 2). We label this point A.

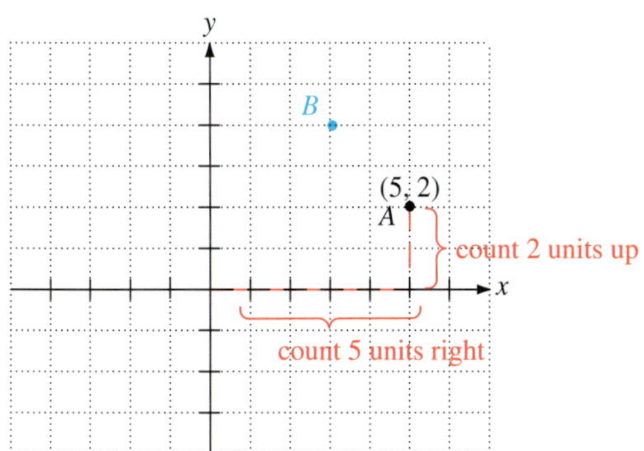

TEACHING TIP The most common mistake students make in the beginning stages of graphing is to mix up the x-axis and the y-axis or else to reverse the (x, y) coordinates. Remind students of the importance of remembering that the first number of the ordered pair (x, y) is measured horizontally (in the x direction), and the second number of the pair is measured vertically (in the y direction).

Practice Problem 1 Plot the point (3, 4) on the coordinate system above. Label this as point B. ■

It is important to remember that the first number in an ordered pair is the x-coordinate, and the second number is the y-coordinate. The points represented by (5, 2) and (2, 5) are totally different, as we shall see in the next example.

EXAMPLE 2 Plot the point (2, 5) on a rectangular coordinate system.

We count 2 units to the right for the x value of the point. Then we count 5 units upward for the y value of the point. We draw a dot and label it C at the location of the ordered pair (2, 5).

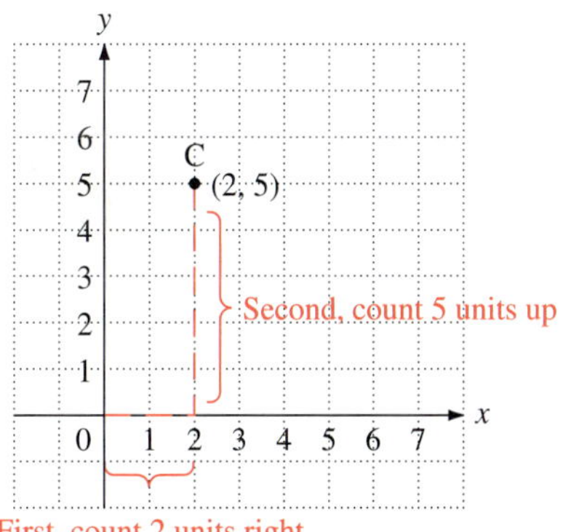

We observe that point C (2, 5) is in a very different location from point A (5, 2) in Example 1.

Practice Problem 2 Plot the point (1, 4) and the point (4, 1) on the coordinate system in the margin. Label the points D and E, respectively. ■

To Think About Describe how the locations of points D and E differ.

PRACTICE PROBLEM 2

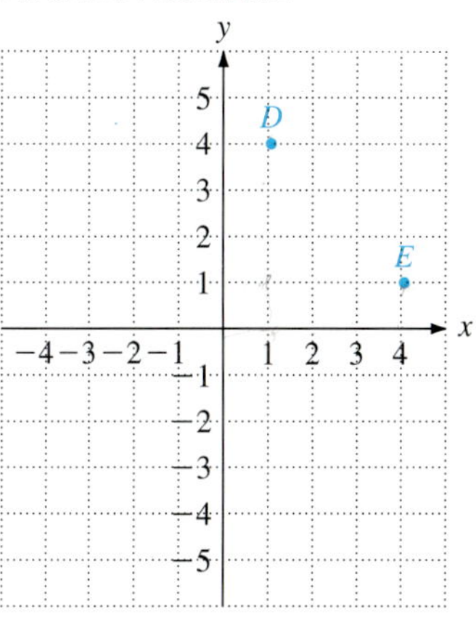

EXAMPLE 3 Plot each of the following points on a rectangular coordinate system. Label the points F, G, and H, respectively.

(a) $(-5, 3)$ **(b)** $(2, -6)$ **(c)** $(-4, -5)$

(a) Notice that the x-coordinate, -5, is negative. On the coordinate grid the x-coordinate is negative to the left of the origin. Thus, we will begin by counting 5 squares to the left starting at the origin. Since the y-coordinate, 3, is positive, we will count 3 units up from the point where you stopped on the x-axis.

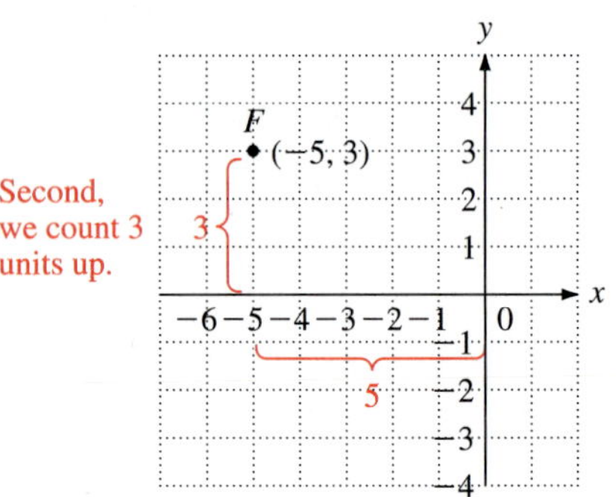

(b) The x-coordinate is positive. Begin by counting squares to the right of the origin. Then count down because the y-coordinate is negative.

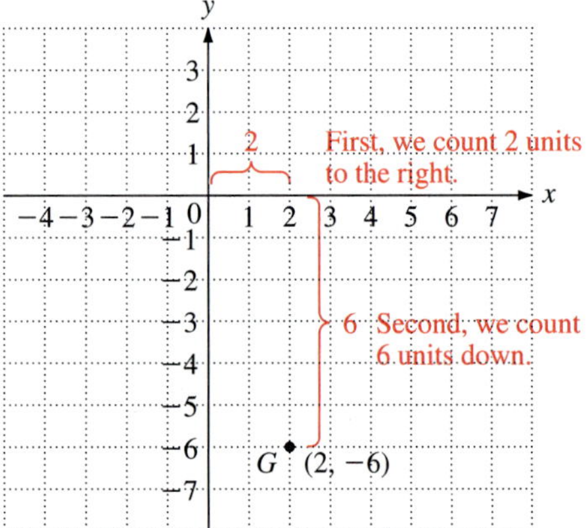

(c) The x-coordinate is negative. Begin by counting squares to the left of the origin. Then count down because the y-coordinate is negative.

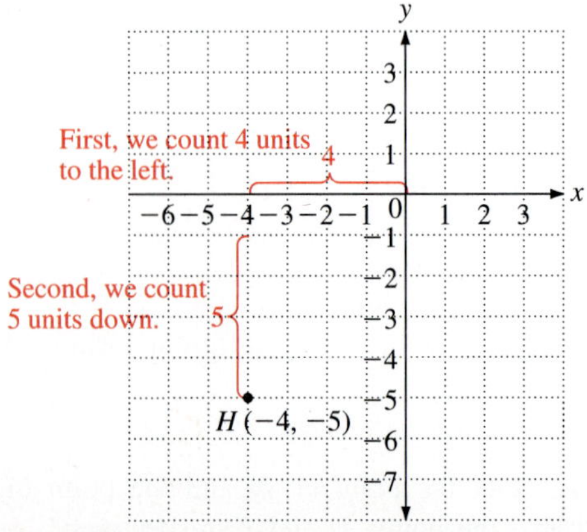

PRACTICE PROBLEM 3

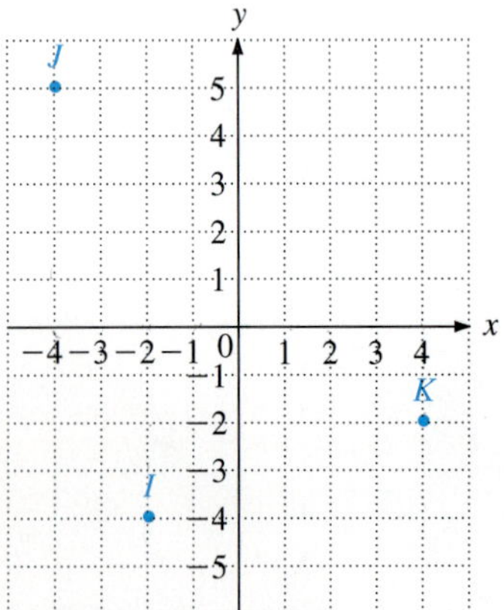

Practice Problem 3 Use the rectangular coordinate system in the margin to plot the following points. Label the points I, J, and K, respectively.

(a) $(-2, -4)$ **(b)** $(-4, 5)$ **(c)** $(4, -2)$ ■

EXAMPLE 4 Plot the following points.

F: $(0, 5)$ G: $(3, \frac{3}{2})$ H: $(-6, 4)$ I: $(-3, -4)$

J: $(-4, 0)$ K: $(2, -3)$ L: $(6.5, -7.2)$

These points are plotted in the figure.

TEACHING TIP Students need a little encouragement in Example 4 when plotting points like $(3, \frac{3}{2})$ or $(6.5, -7.2)$. Remind them that it is only an approximation. However, there are many times we have to approximate the location of a point when graphing an equation.

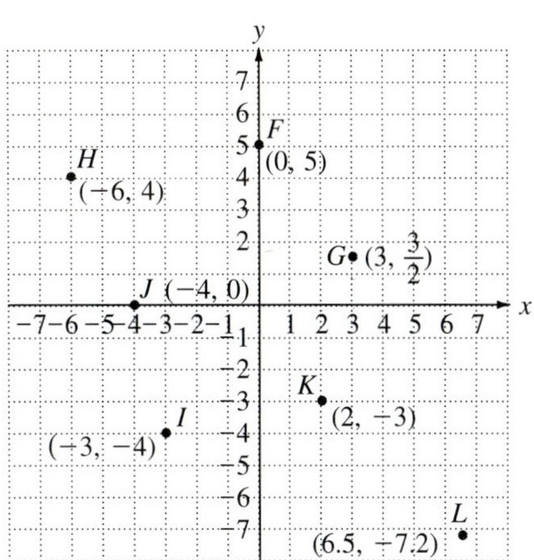

Note: When you are plotting decimal values like $(6.5, -7.2)$, put the location of the plotted point halfway between 6 and 7 for the 6.5 and at your best approximation in the y-direction for the -7.2.

PRACTICE PROBLEM 4

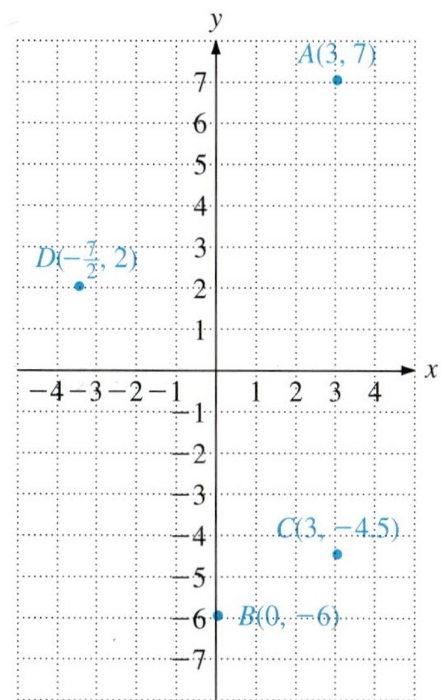

Practice Problem 4 Plot the following points. Label each point with both the letter and the ordered pair. Use the coordinate system provided in the margin.

A: $(3, 7)$ B: $(0, -6)$ C: $(3, -4.5)$ D: $(-\frac{7}{2}, 2)$ ■

2 *Determining the Coordinates of a Plotted Point*

Sometimes, we need to find the coordinates of a point that has been plotted. First, we count the units we need on the x-axis to get as close as possible to that point. Next we count the units up or down we need to go from the x-axis to finally reach that point.

EXAMPLE 5 What ordered pair of numbers represents the point A in the figure on the right?

We see that the point A is represented by the ordered pair $(5, 4)$. We first counted 5 units to the right of the origin on the x-axis. Thus we obtain 5 as the first number of the ordered pair. Then we counted 4 units upward on a line parallel to the y-axis. Thus we obtain 4 as the second number on the ordered pair.

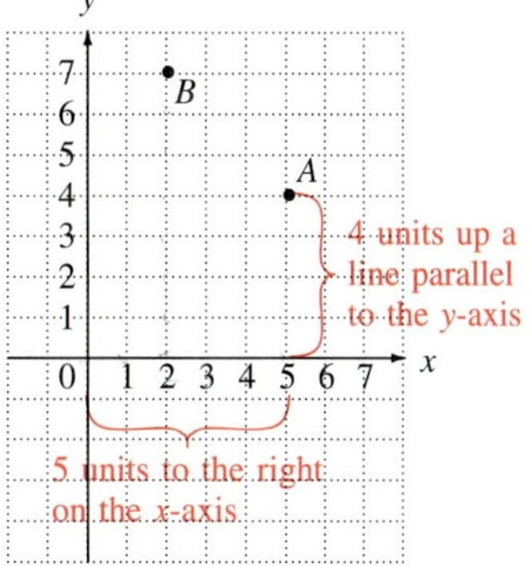

5, 4
2, 7

Practice Problem 5 Give the ordered pair of numbers that represents the point B in the graph on the right. ■

EXAMPLE 6 Write the coordinates of each point plotted below.

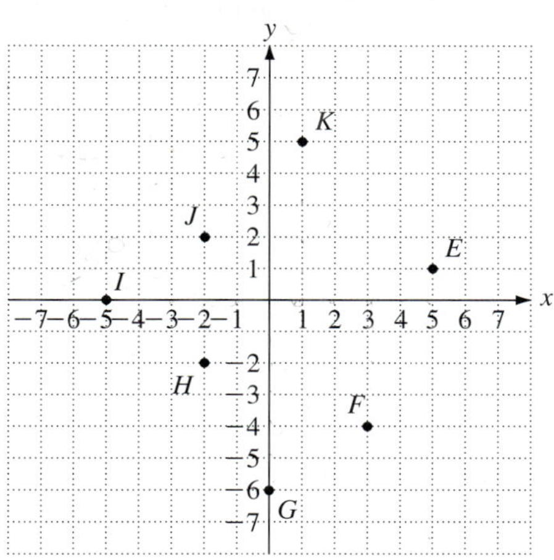

The coordinates of each point are:

$E = (5, 1)$ $\quad$ $I = (-5, 0)$

$F = (3, -4)$ $\quad$ $J = (-2, 2)$

$G = (0, -6)$ $\quad$ $K = (1, 5)$

$H = (-2, -2)$

Be very careful that you put the x-coordinate first and the y-coordinate second. Be careful that each sign is correct.

Practice Problem 6 Give the coordinates of each point in the graph in the margin. ■

PRACTICE PROBLEM 6

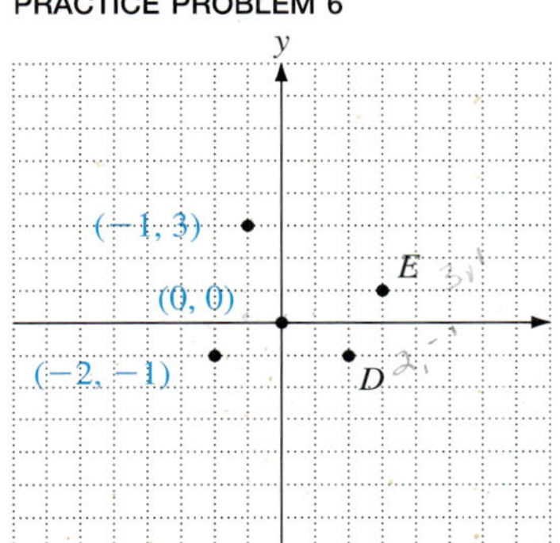

3 *Finding Ordered Pairs for a Given Linear Equation*

Equations such as $3x + 2y = 5$ and $6x + y = 3$ are called linear equations in two variables. A **linear equation in two variables** is an equation that can be written in the form $Ax + By = C$ where A, B, and C are real numbers but A and B cannot *both* be zero. Replacement values for x and y that make *true mathematical statements* of the equation are called *truth values* and an ordered pair of these truth values is called a **solution.**

Consider the equation $3x + 2y = 5$. The ordered pair (1, 1) is a solution to the equation. When we replace x by 1 and y by 1 we obtain a true statement.

$$3(1) + 2(1) = 5 \text{ or } 3 + 2 = 5$$

There are an infinite number of solutions for any given linear equation in two variables.

We obtain ordered pairs from examining such a linear equation. If one value of an ordered pair is known, the other can be quickly obtained. To do so, we replace one variable in the linear equation by the known value. Then using the methods learned in Chapter 2 we solve the resulting equation for the other variable.

EXAMPLE 7 Find the missing coordinate to complete the following ordered pairs for the equation $2x + 3y = 15$.

(a) (0, ?) $\quad$ **(b)** (? , 1)

(a) In the ordered pair (0, ?) we know that $x = 0$. Replace x by 0 in the equation.

$$2x + 3y = 15$$
$$2(0) + 3y = 15$$
$$0 + 3y = 15$$
$$y = 5$$

Thus we have the ordered pair (0, 5).

(b) In the ordered pair (? , 1), we *do not know* the value of x. However, we do know that $y = 1$. So we start by replacing the variable y by 1. We will end up with an equation with one variable x. We can then solve for x.

$$2x + 3y = 15$$
$$2x + 3(1) = 15$$
$$2x + 3 = 15$$
$$2x = 12$$
$$x = 6$$

Thus we have the ordered pair (6, 1).

Practice Problem 7 Find the missing coordinate to complete the following ordered pairs for the equation $3x - 4y = 12$.

(a) (0, ?) **(b)** (? , 3) **(c)** (? , −6) ■

The linear equations that we work with are not always written in the form $Ax + By = C$ but are sometimes solved for y as in $y = -6x + 3$. Consider the equation $y = -6x + 3$. The ordered pair $(2, -9)$ is a solution to the equation. When we replace x by 2 and y by -9 we obtain a true mathematical statement:

$$(-9) = -6(2) + 3 \text{ or } -9 = -12 + 3.$$

EXAMPLE 8 Find the missing coordinate to complete the following ordered pairs for the equation $y = -3x + 4$

(a) (2, ?) **(b)** (−3, ?)

(a) For the ordered pair (2, ?) we know that x is 2, so we replace x by 2 in the equation and solve for y as in Chapter 2.

$y = -3x + 4$	Check your answers.
$y = -3(2) + 4$	$y = -3x + 4$
$y = -6 + 4$	$(-2) = -3(2) + 4$
$y = -2$	$-2 = -6 + 4$ ✓

Thus, the solution is the complete ordered pair (2, −2).

(b) For the ordered pair (−3, ?) we replace x by -3 in the equation.

$y = -3x + 4$	Check. $y = -3x + 4$
$y = -3(-3) + 4$	$13 = -3(-3) + 4$
$y = 9 + 4$	$13 = 9 + 4$ ✓
$y = 13$	

Thus, the solution is the ordered pair (−3, 13).

Practice Problem 8 Find the missing coordinate to complete the following ordered pairs for the equation $y = -2x - 5$.

(a) (3, ?) **(b)** (−4, ?) ■

7.1 Exercises

1. Plot the following points.
A: $(0, -3)$ C: $(-3, -7)$
B: $(5, 8)$ D: $(-3, 7)$

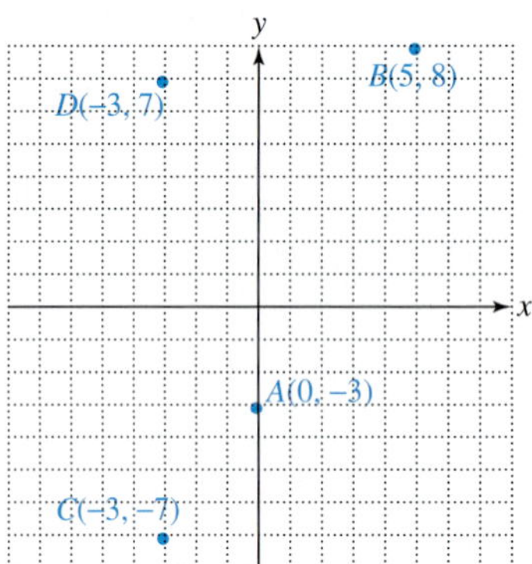

2. Plot the following points.
E: $(-6, 2)$ F: $(3, 5)$
G: $(-6, -1)$ H: $(0, -5)$

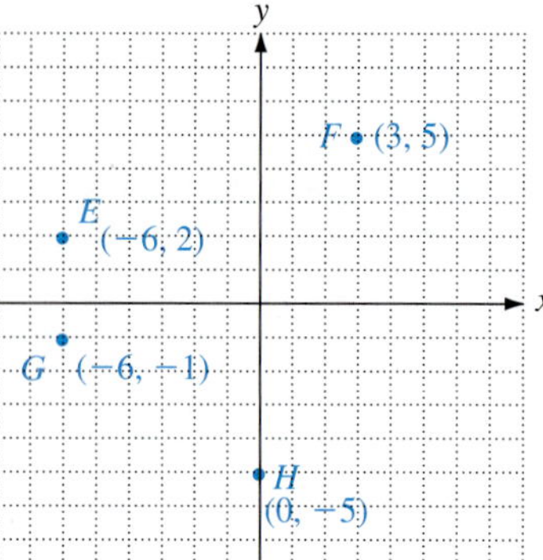

3. Plot the following points.
J: $(5, 0)$ K: $(-5, 0)$
L: $\left(-3\frac{1}{2}, 7\right)$ M: $(-2, -4.5)$

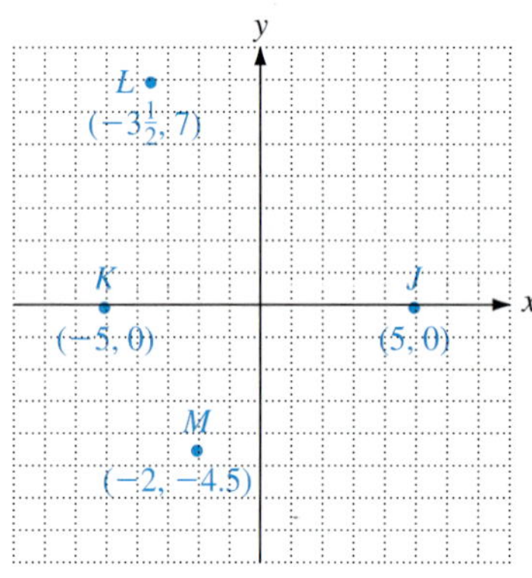

4. Plot the following points.
N: $(0, 4)$ O: $(-4, 0)$
P: $\left(4\frac{1}{2}, -3\right)$ Q: $(-2, 3.5)$

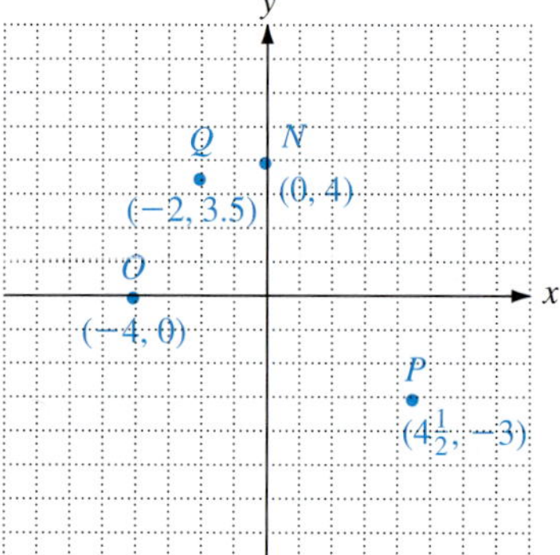

Consider the points plotted in the graph on the right.

5. Give the coordinates for the points R, S, X, and Y.
R: $(-3, -5)$
S: $\left(-4\frac{1}{2}, 0\right)$
X: $(3, -5)$
Y: $\left(2\frac{1}{2}, 6\right)$

6. Give the coordinates for points T, V, W, and Z.
T: $(-6, 4)$, V: $\left(3\frac{1}{2}, -2\right)$, W: $\left(6\frac{1}{2}, 2\right)$, Z: $\left(0, -3\frac{1}{2}\right)$

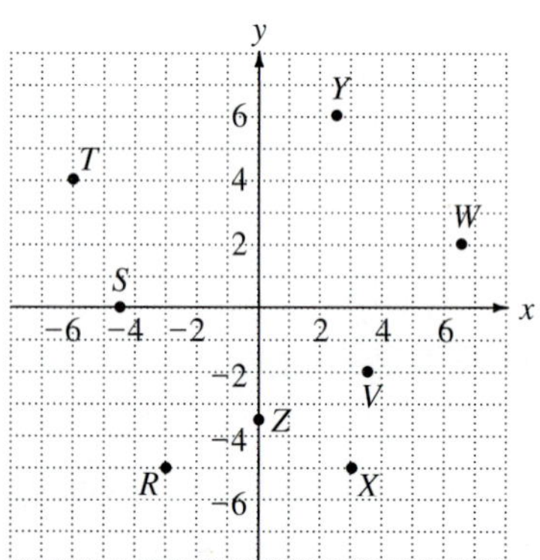

Verbal and Writing Skills

7. What is the x-coordinate of the origin? 0

8. What is the y-coordinate of the origin? 0

9. Explain why (5, 1) is referred to as an ordered pair of numbers. The order in which you write the numbers matters. The graph of (5, 1) is not the same as the graph of (1, 5).

10. Ten points are plotted in the figure shown. List all the ordered pairs needed to represent these points.

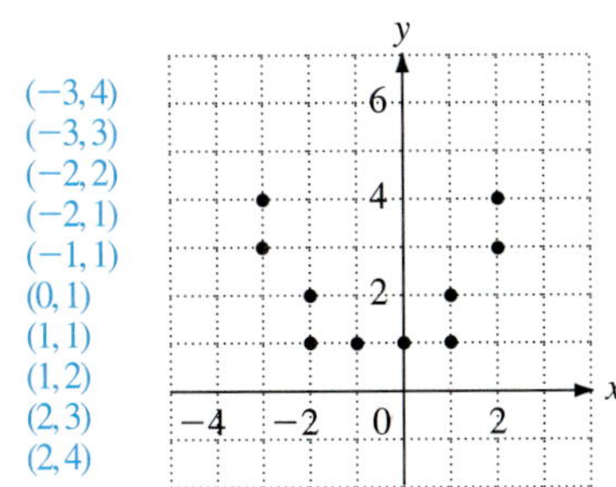

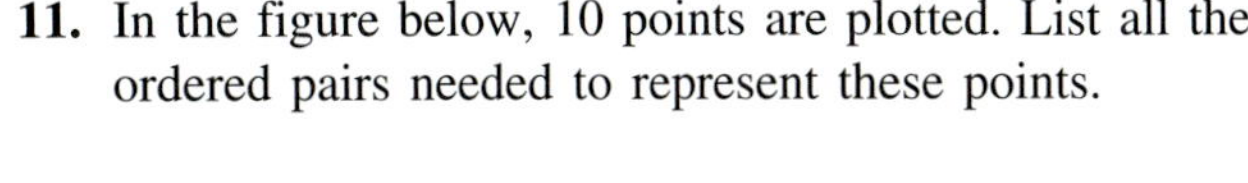

11. In the figure below, 10 points are plotted. List all the ordered pairs needed to represent these points.

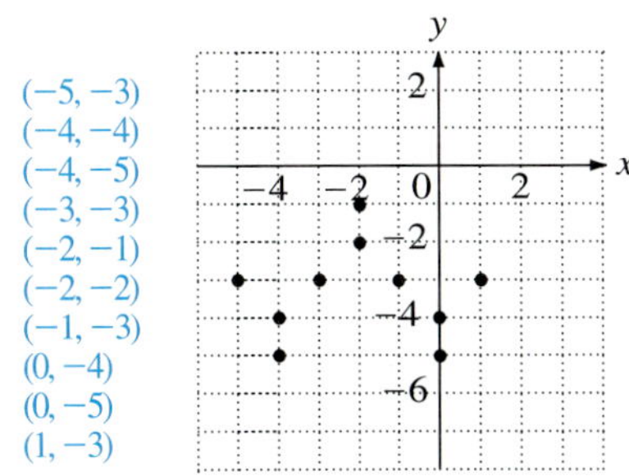

Complete the ordered pairs so that each is a solution to the given linear equation.

12. $y = 2x + 5$
(a) (0,) (0, 5) **(b)** (3,) (3, 11)

13. $y = 3x + 8$
(a) (0,) (0, 8) **(b)** (4,) (4, 20)

14. $y = -4x + 2$
(a) (−5,) (−5, 22) **(b)** (4,) (4, −14)

15. $y = -2x + 3$
(a) (−6,) (−6, 15) **(b)** (3,) (3, −3)

16. $6x - 2y = 24$
(a) (1,) (1, −9) **(b)** (, −3) (3, −3)

17. $2x - 8y = 16$
(a) (8,) (8, 0) **(b)** (, −5) (−12, −5)

18. $2y + 3x = -6$
(a) (−2,) (−2, 0) **(b)** (, 3) (−4, 3)

19. $8x + 5y = -41$
(a) (−7,) (−7, 3) **(b)** (, −5) (−2, −5)

20. $-10x + 11y = 30$
(a) $\left(-\frac{4}{5},\ \right)$ $\left(-\frac{4}{5}, 2\right)$ **(b)** (, 3) $\left(\frac{3}{10}, 3\right)$

Cumulative Review Problems

21. Simplify. $\dfrac{8x}{x+y} + \dfrac{8y}{x+y}$ $\dfrac{8x+8y}{x+y} = 8$

22. Simplify. $\dfrac{x^2 - y^2}{x^2 + 4xy + 3y^2} \cdot \dfrac{x^2 - 9y^2}{x^2 - 6xy + 9y^2}$

$\dfrac{(x+y)(x-y)}{(x+y)(x+3y)} \cdot \dfrac{(x+3y)(x-3y)}{(x-3y)(x-3y)} = \dfrac{x-y}{x-3y}$

23. The circular pool at the hotel we stayed at in Orlando, Florida has a radius of 19 yards. What is the area of the pool? (Use $\pi \approx 3.14$). Area ≈ 1133.54 sq yd

24. The membership at my gym has increased by 16% over the last 2 years. We used to have 795 members. How many members do we have now? 922 members

7.2 Graphing Linear Equations

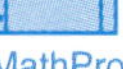

MathPro Video 18 SSM

After studying this section, you will be able to:

1 *Graph a straight line by finding three ordered pairs that are solutions to the linear equation.*
2 *Graph a straight line by finding its x- and y-intercepts.*
3 *Graph horizontal and vertical lines.*

1 *Graphing a Linear Equation by Plotting Three Ordered Pairs*

We have seen that the graph of an ordered pair is a point. An ordered pair can also be a solution to a linear equation in two variables. Since there are an infinite number of solutions, there are an infinite number of ordered pairs and an infinite number of points. If we can plot these points, we will be graphing the equation. What will this graph look like? Let's look at the equation $y = -3x + 4$. To look for a solution to the equation, we can choose any value for x. For convenience we will choose $x = 0$. That is, the first coordinate of the ordered pair will be 0. To complete the ordered pair (0,), we substitute 0 for x in the equation:

$$\begin{aligned} y &= -3x + 4 \\ &= -3(0) + 4 \\ &= 0 + 4 \\ &= 4 \end{aligned}$$

Thus the ordered pair is (0, 4).

To find another ordered pair that is a solution to the equation $y = -3x + 4$, let $x = 1$.

$$\begin{aligned} y &= -3x + 4 \\ &= -3(1) + 4 \\ &= -3 + 4 \\ &= 1 \end{aligned}$$

Thus the ordered pair (1, 1) is another solution to the equation.

TEACHING TIP Some students do not realize that when we draw the equation of a straight line we draw arrow tips at each end. Other students will draw the arrow tips but not know what they mean. The purpose of the arrowhead is to indicate that the straight line continues forever in that direction. Later in this course students will graph some lines that do not continue forever, and they will not have arrowheads at each end.

The graph of these two ordered pairs or solutions is shown at the right. From geometry we know that two points determine a line. Thus we can say that the line that contains the two points is the graph of the equation $y = -3x + 4$. In fact, *the graph of any linear equation in two variables is a straight line.*

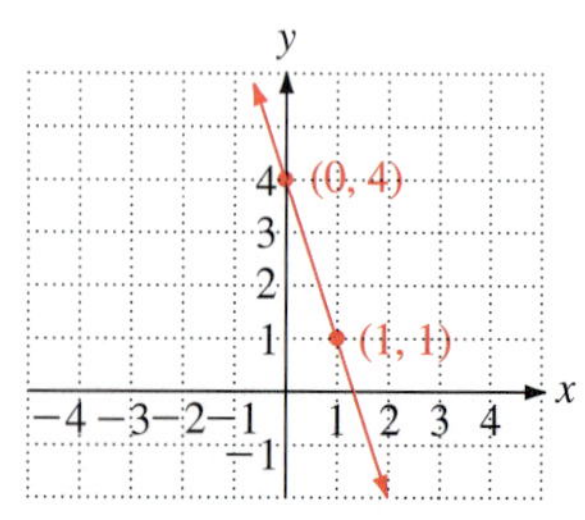

While you only need two points to determine a line, we recommend that you use three points to graph an equation. Two points to determine the line and a third point to verify. For ease in plotting the points it is better that the ordered pairs contain integers.

To Graph a Linear Equation

1. Look for three ordered pairs that are solutions to the equation.
2. Plot the points.
3. Draw a line through the points.

EXAMPLE 1 Find three ordered pairs that satisfy $2x + y = 4$. Then graph the resulting straight line.

Since we can choose any value for x, choose those numbers that are convenient. To organize the results, we will make a table of values. We will let $x = 0$, $x = 1$, and $x = 3$, respectively. We write these numbers under x in our table of values. For each of these x values, we find the corresponding y value in the equation $2x + y = 4$.

TABLE OF VALUES

x	y
0	4
1	2
3	−2

$2x + y = 4$	$2x + y = 4$	$2x + y = 4$
$2(0) + y = 4$	$2(1) + y = 4$	$2(3) + y = 4$
$0 + y = 4$	$2 + y = 4$	$6 + y = 4$
$y = 4$	$y = 2$	$y = -2$

We record these results by placing each y value in the table next to its corresponding x value. Keep in mind that these values represent ordered pairs each of which is a solution to the equation. Since there are an infinite number of ordered pairs, we will prefer those with integers whenever possible. If we plot these ordered pairs and connect the three points, we get a straight line that is the graph of the equation $2x + y = 4$. The graph of the equation is shown in the figure on the right.

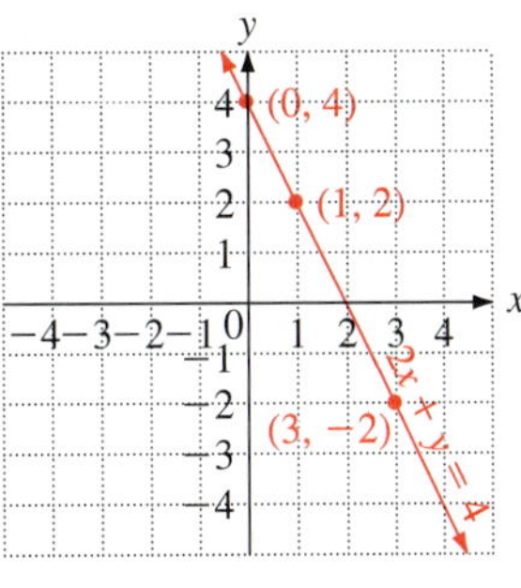

Practice Problem 1 Graph $x + y = 10$ on the coordinate system below. ■

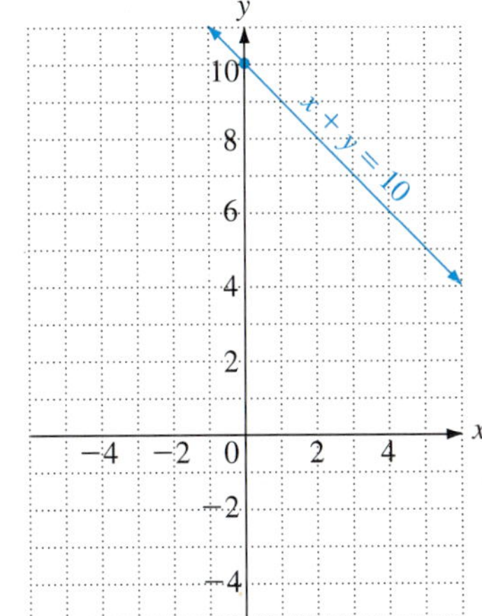

EXAMPLE 2 Graph $5x - 4y + 2 = 2$.

First, we simplify the equation by adding -2 to each side.

$$5x - 4y + 2 = 2$$
$$5x - 4y + 2 - 2 = 2 - 2$$
$$5x - 4y = 0$$

Since we are free to choose any value of x, $x = 0$ is a natural choice. Calculate the value of y when $x = 0$.

$$5x - 4y = 0$$
$$5(0) - 4y = 0$$
$$-4y = 0$$
$$y = 0$$

Remember: Any number times 0 is 0. Since $-4y = 0$, y must equal 0.

Now let's see what happens when $x = 1$.

$$5x - 4y = 0$$
$$5(1) - 4y = 0$$
$$5 - 4y = 0$$
$$-4y = -5$$
$$y = \frac{-5}{-4} \text{ or } \frac{5}{4}$$

This is not an easy number to graph.

A better choice for a replacement of x is a number that is divisible by 4. Let's see why. Let $x = 4$ and let $x = -4$.

$5x - 4y = 0$	$5x - 4y = 0$
$5(4) - 4y = 0$	$5(-4) - 4y = 0$
$20 - 4y = 0$	$-20 - 4y = 0$
$-4y = -20$	$-4y = 20$
$y = \frac{-20}{-4}$ or 5	$y = \frac{20}{-4}$ or -5

PRACTICE PROBLEM 2

Now we can put these numbers in our table of values and graph.

TABLE OF VALUES	
x	y
0	0
4	5
−4	−5

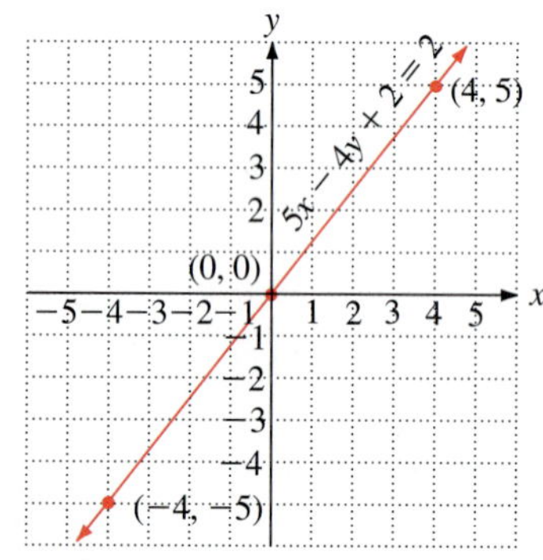

Practice Problem 2 Graph $7x + 3 = -2y + 3$ on the coordinate system in the margin. ■

To Think About In Example 2, we picked values of x and found the corresponding values for y. An alternative approach is to *first* solve the equation for the variable y. Thus

$$5x - 4y = 0$$

$$-4y = -5x \qquad \textit{Add } -5x \textit{ to each side.}$$

$$\frac{-4y}{-4} = \frac{-5x}{-4} \qquad \textit{Divide each side by } -4.$$

$$y = \frac{5}{4}x$$

Now let $x = -4$, $x = 0$, and $x = 4$, and find the corresponding values of y. Explain why you would choose multiples of 4 as replacements of x in this equation. Graph the equation and compare it to the graph in Example 2.

In the previous two examples we began by picking values for x. We could just as easily have chosen values for y.

EXAMPLE 3 Graph $3x - 4y = 12$.

We first find three ordered pairs by arbitrarily picking three values for y and in each case solving for x. We will choose $y = 0$, $y = 3$, and $y = -3$. Can you see why?

TEACHING TIP Some instructors prefer that students always solve the linear equation for y and then substitute in the x values. If that is your own viewpoint, then before finding points for the equation in Example 3, first rewrite the equation in the form $y = \frac{3}{4}x - 3$.

$3x - 4y = 12$	$3x - 4y = 12$	$3x - 4y = 12$
$3x - 4(0) = 12$	$3x - 4(3) = 12$	$3x - 4(-3) = 12$
$3x - 0 = 12$	$3x - 12 = 12$	$3x + 12 = 12$
$3x = 12$	$3x = 24$	$3x = 0$
$x = 4$	$x = 8$	$x = 0$

Our table of values is:

x	y
4	0
8	3
0	−3

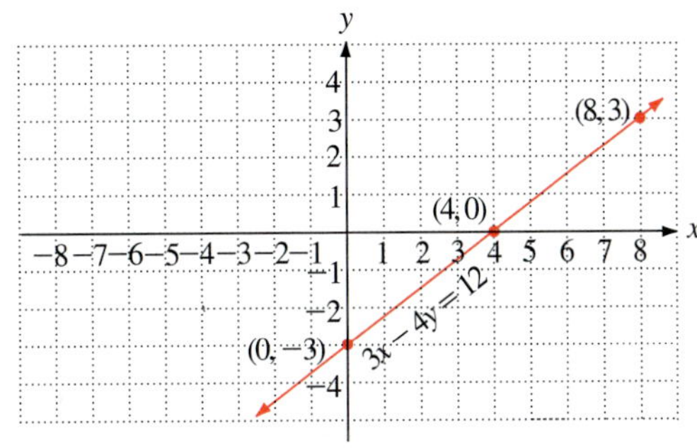

Practice Problem 3 Graph $3x - 2y = 6$ by arbitrarily picking three values for y and in each case solving for x. Use the coordinate system below. ■

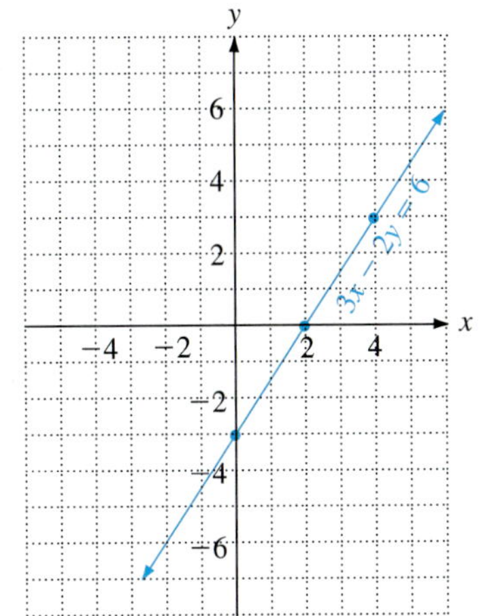

2 *Graphing a Straight Line by Plotting Its Intercepts*

What values should we pick for x and y? Which points should we use for plotting? For many straight lines it is easiest to pick the two *intercepts.* A few lines have only one intercept. We will discuss these separately.

The number a is the *x-intercept* of a line if the line crosses the x-axis at $(a, 0)$. The number b is the *y-intercept* of a line if the line crosses the y-axis at the point $(0, b)$.

Intercept Method of Graphing

To graph an equation by using intercepts, we:

1. Find the x-intercept by letting $y = 0$ and solving for x.
2. Find the y-intercept by letting $x = 0$ and solving for y.
3. Find one additional ordered pair so that we have three points with which to plot the line.

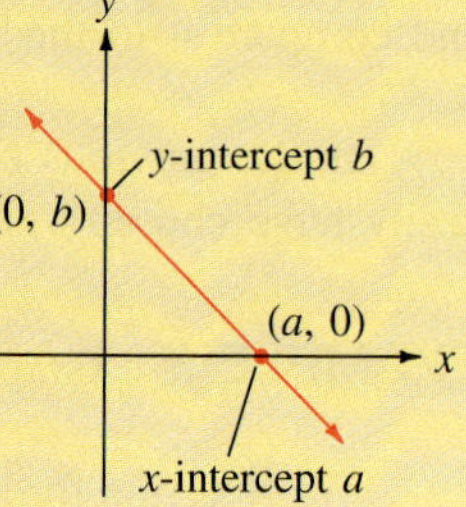

EXAMPLE 4 Graph $5y - 3x = 15$ by the intercept method.

Let $y = 0$.

$5(0) - 3x = 15$ *Replace y by 0.*

$-3x = 15$ *Divide both sides by −3.*

$x = -5$ *The x-intercept is −5.*

The ordered pair is $(-5, 0)$ is the x-intercept point.

$5y - 3x = 15$

Let $x = 0$.

$5y - 3(0) = 15$ *Replace x by 0.*

$5y = 15$ *Divide both sides by 5.*

$y = 3$ *The y-intercept is 3.*

The ordered pair is $(0, 3)$ is the y-intercept point.

We find another pair to have a third point.

Let $y = 6$.

$5(6) - 3x = 15$ *Replace y by 6.*

$30 - 3x = 15$ *Simplify.*

$-3x = -15$ *Subtract 30 from both sides.*

$$x = \frac{-15}{-3} \text{ or } 5$$

The ordered pair is $(5, 6)$.

Our table of values is

x	y
0	3
−5	0
5	6

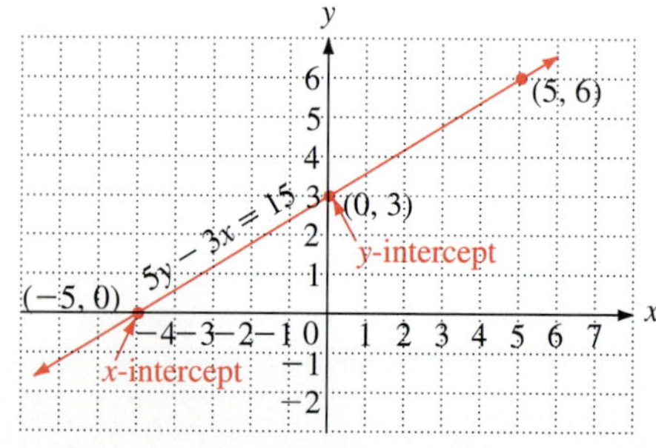

Practice Problem 4 Use the intercept method to graph $2y - x = 6$. Use the coordinate system in the margin. ■

PRACTICE PROBLEM 4

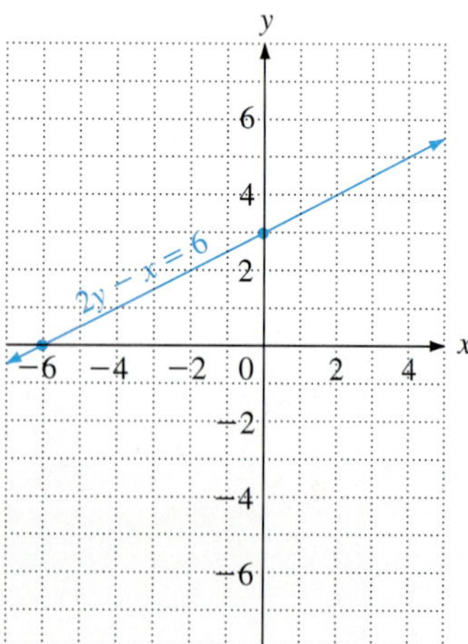

SIDELIGHT Can you draw all straight lines by the intercept method? Not really. Some straight lines may go through the origin and have only one intercept. If a line goes through the origin, it will have an equation of the form $Ax + By = 0$, where $A \neq 0$ or $B \neq 0$ or both. In such cases you should plot two additional points besides the origin. Be sure to simplify each equation before attempting to graph it.

TEACHING TIP You may wish to show the general condition which determines that the graph of an equation of the form $Ax + By = C$ goes through the origin. That is, point out that when $Ax + By = 0$ the graph of the equation goes through the origin. Ask students to find the x- and y-intercepts, respectively. Remind students that $A \neq 0$ or $B \neq 0$ or both.

x-intercept	*y-intercept*
Let $y = 0$	Let $x = 0$
$Ax + By = 0$	$Ax + By = 0$
$Ax + B(0) = 0$	$A(0) + By = 0$
$Ax = 0$	$By = 0$
$x = 0$	$y = 0$

Thus the x-intercept is (0, 0) and the y-intercept is (0, 0).

You may wish to provide a specific example such as $3x + 2y = 0$. Since the intercept is at the origin (0, 0), have students look for two other points to graph.

3 *Graphing Horizontal and Vertical Lines*

You will notice that the x-axis is a horizontal line. It is the line $y = 0$, since for any x, y is 0. Try a few points. (1, 0), (3, 0), (−2, 0) all lie on the x-axis. Any horizontal line will be parallel to the x-axis. Lines such as $y = 5$ and $y = -2$ are horizontal lines. What does $y = 5$ mean? It means that, for any x, y is 5. Likewise $y = -2$ means that, for any x, $y = -2$.

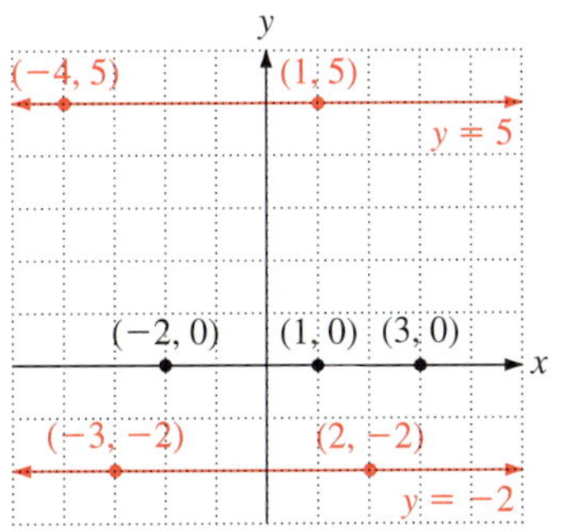

How can we recognize the equation of a line that is horizontal, that is, parallel to the x-axis.

> If an equation is a straight line that is parallel to the x-axis (that is, a horizontal line), it will be of the form $y = b$, where b is some real number.

TEACHING TIP It is important for students to recognize and graph the equation of a horizontal line. Remind them that the straight line that is graphed in Example 5 can have the form $y = -3$, or the form $5y = -15$, or the form $4y + 11 = -1$. Both of the latter equations can be simplified to the form $y = -3$. This type of simplification should be done prior to graphing.

EXAMPLE 5 Graph $y = -3$.

You could write the equation as $0x + y = -3$. Then it is clear that for any value of x that you substitute you will always obtain $y = -3$. Thus, as shown in the figure, $(4, -3)$, $(0, -3)$, and $(-3, -3)$ are all ordered pairs that satisfy the equation $y = -3$. Since the y-coordinate of every point on this line is −3, it is easy to see that the horizontal line will be 3 units below the x-axis.

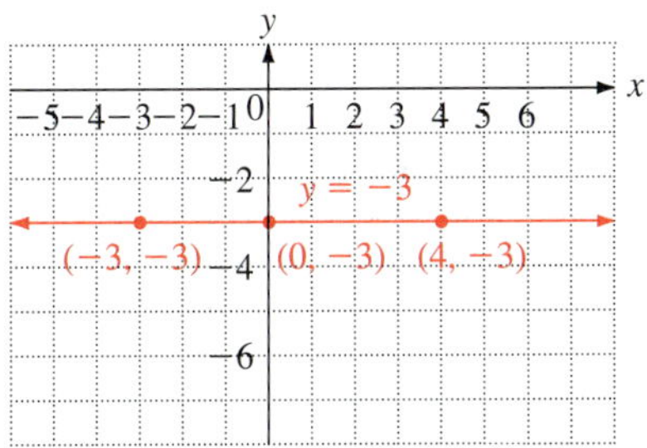

Practice Problem 5 Graph $2y - 3 = 0$ on the coordinate system in the margin. ■

PRACTICE PROBLEM 5

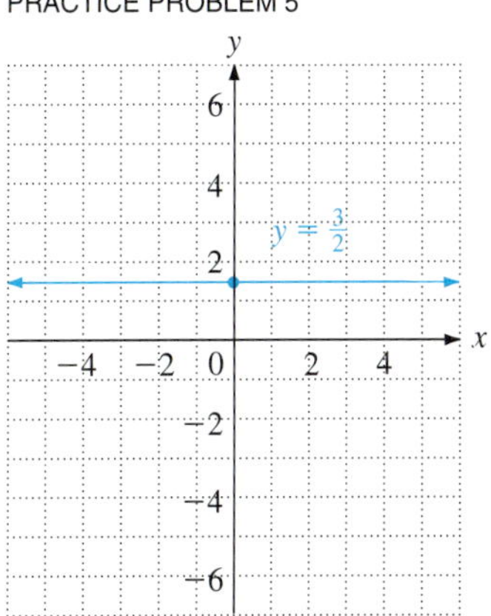

Notice that the y-axis is a vertical line. This is the line $x = 0$, since, for any y, x is 0. Try a few points. $(0, 2)$, $(0, -3)$, $(0, \frac{1}{2})$ all lie on the y-axis. Any vertical line will be parallel to the y-axis. Lines such as $x = 2$ and $x = -3$ are vertical lines. Think of what $x = 2$ means. It means that, for any value of y, x is 2. $x = 2$ is a vertical line two units to the right of the y-axis.

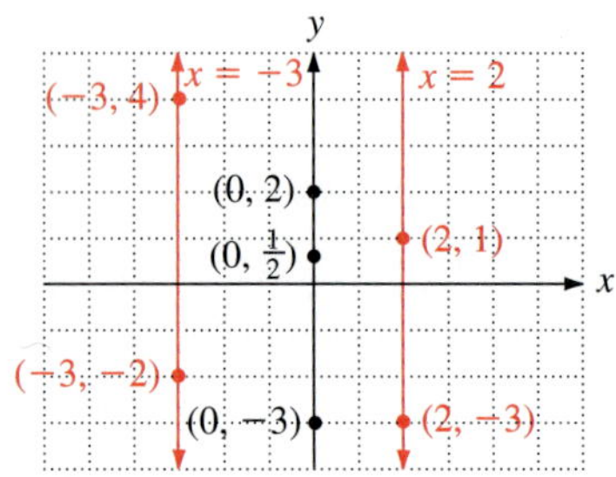

How can we recognize the equation of a line that is vertical, that is, parallel to the y-axis?

> If the graph of an equation is a straight line that is parallel to the y-axis, that is, a vertical line, its equation will be of the form $x = a$, where a is some real number.

EXAMPLE 6 Graph $x = 5$.

This can be done immediately by drawing a vertical line 5 units to the right of the origin. The x-coordinate of every point on this line is 5.

The equation $x - 5 = 0$ can be rewritten as $x = 5$ and graphed as shown.

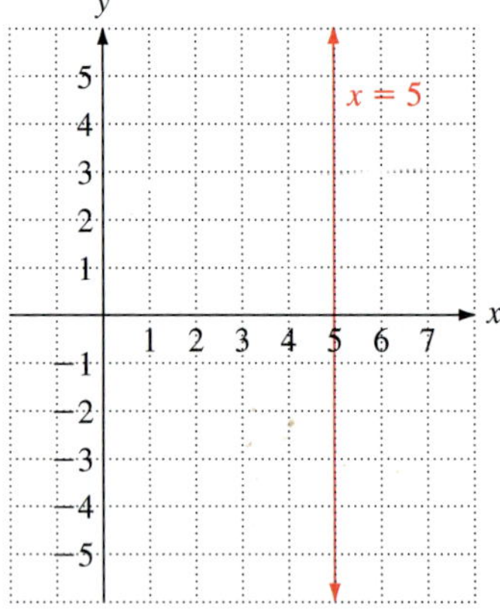

TEACHING TIP After you have discussed the vertical line and have reviewed the graph of Example 6, ask students to do the following as an Additional Class Exercise. Graph $2(x + 2y) = -4(2 - y)$.

Sol. The equation simplifies to $x = -4$.

Practice Problem 6 Graph $x + 3 = 0$ on the coordinate system below. ■

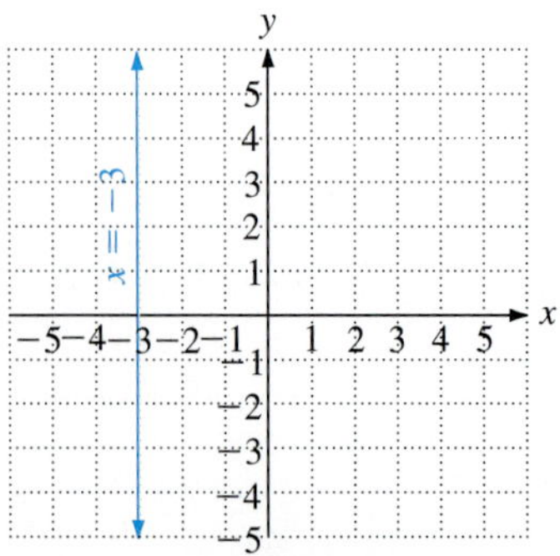

7.2 Exercises

Verbal and Writing Skills

1. Is the point $(-2, 5)$ a solution to the equation $2x + 5y = 0$? Why or why not? No. Replacing x by -2 and y by 5 in the equation does not result in a true statement.

2. The graph of a linear equation in two variables is a straight line.

3. The x-intercept of a line is the point where the line crosses the x-axis.

4. The graph of the equation $y = b$ is a horizontal line.

Complete the ordered pairs so that each is a solution of the given linear equation. Then plot each solution and graph the equation by connecting the points by a straight line.

5. $y = -2x + 1$
$(0, 1)$
$(-2, 5)$
$(1, -1)$

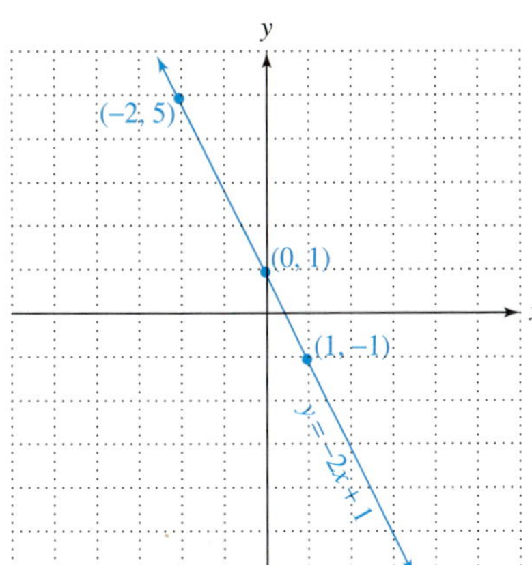

6. $y = -3x - 4$
$(-2, 2)$
$(-1, -1)$
$(0, -4)$

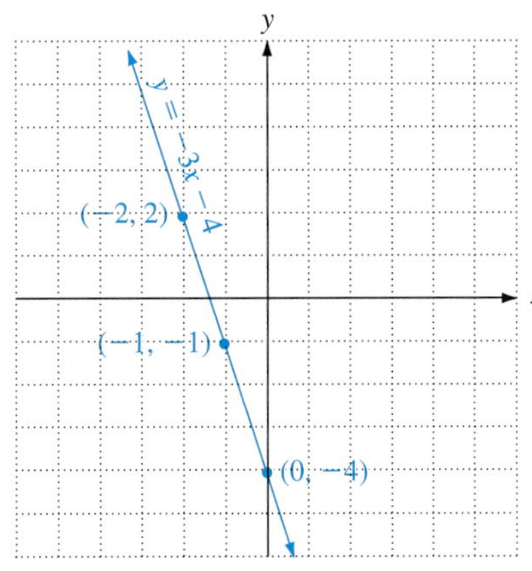

7. $y = x - 4$
$(0, -4)$
$(2, -2)$
$(4, 0)$

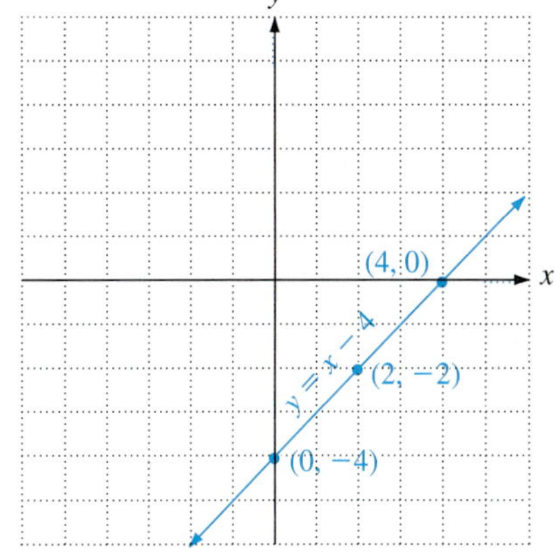

Graph each equation by plotting three points and connecting them.

8. $y = 2x - 3$

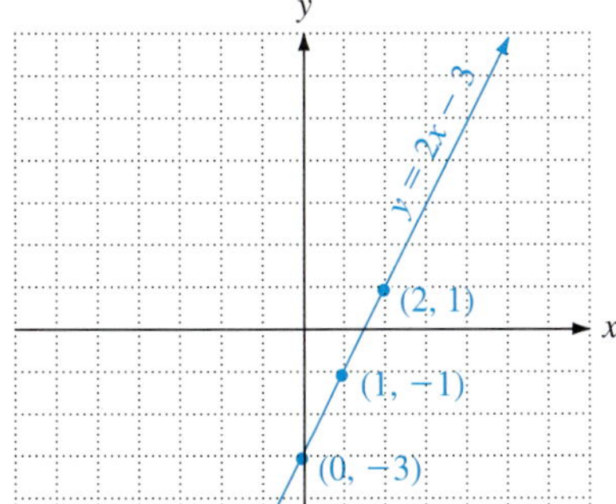

9. $y = 3x - 1$

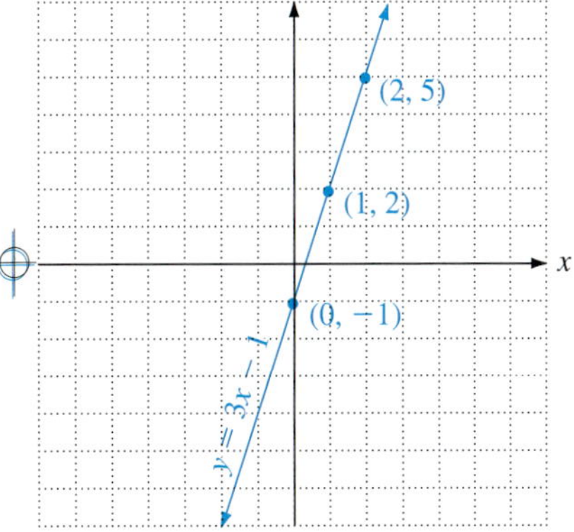

10. $y = 3x + 2$

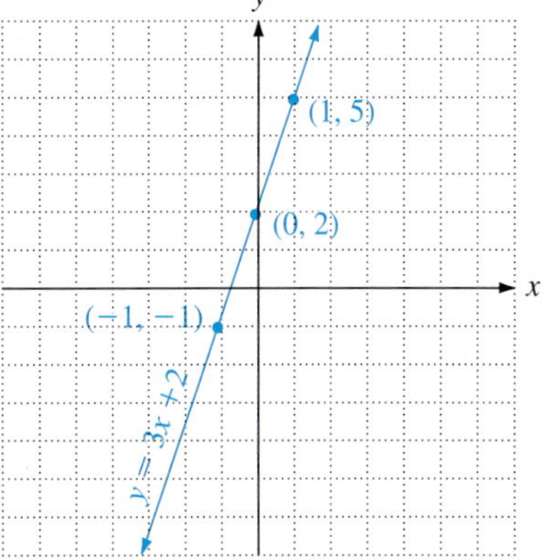

11. $y = 2x + 1$

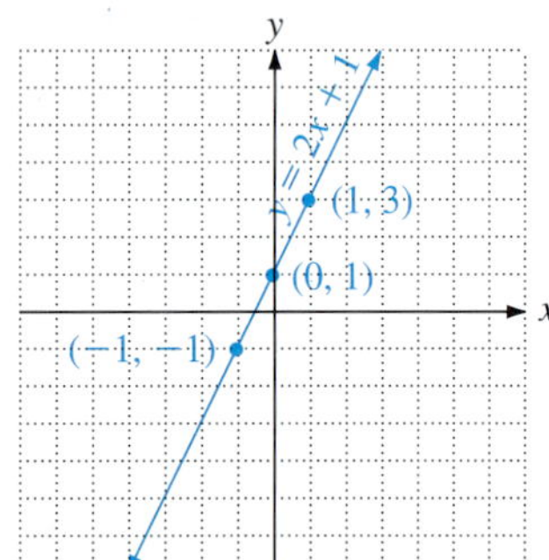

12. $3x - 5y = 0$

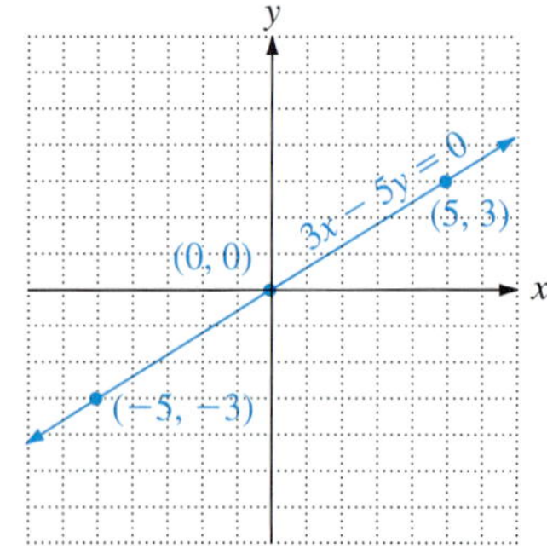

13. $2y - 4x = 0$

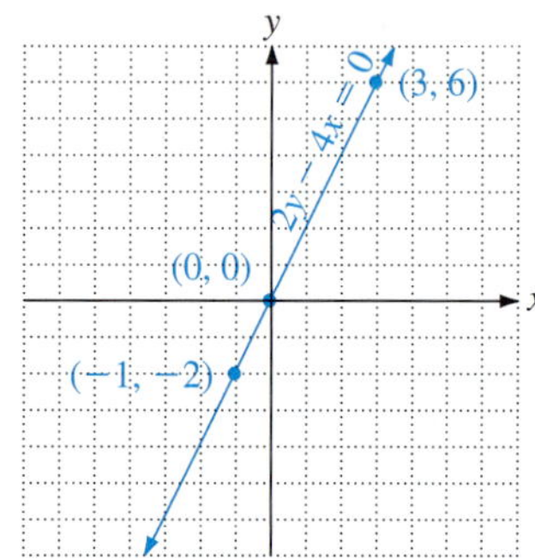

Graph each equation by plotting the intercepts and one other point.

14. $y = -\frac{1}{2}x + 3$

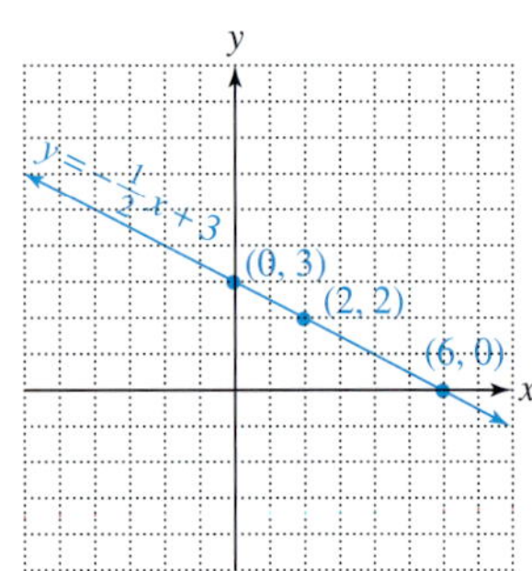

15. $y = -\frac{2}{3}x + 2$

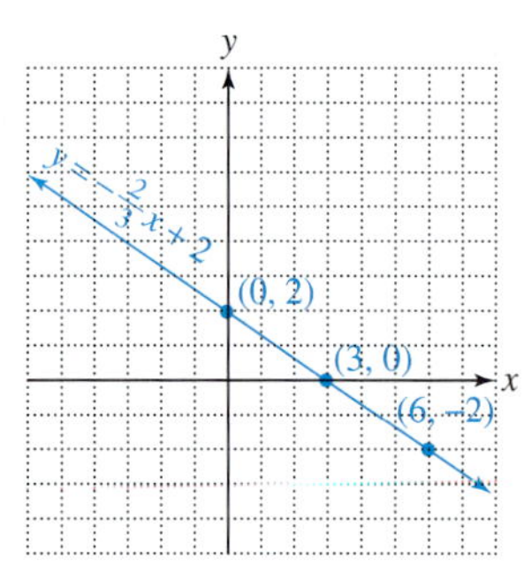

16. $4x + 3y = 12$

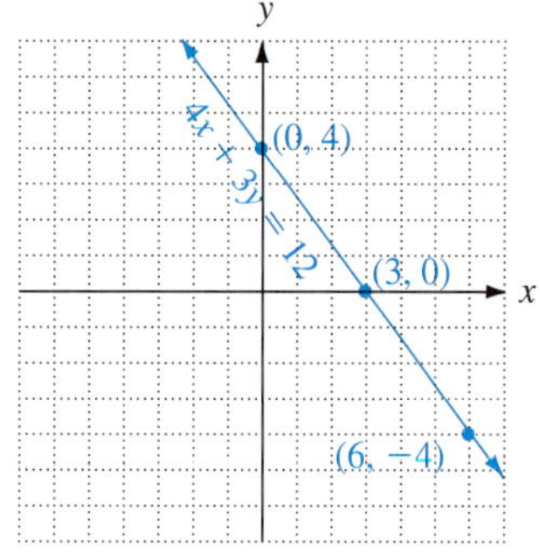

17. $y = 6 - 2x$

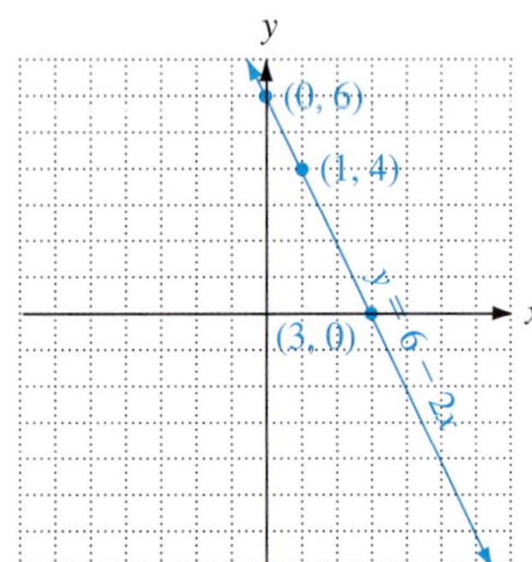

18. $y = 4 - 2x$

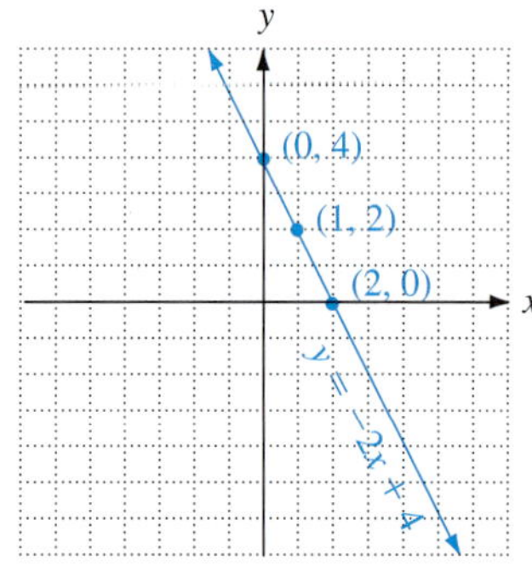

19. $y = -\frac{3}{4}x + 1$

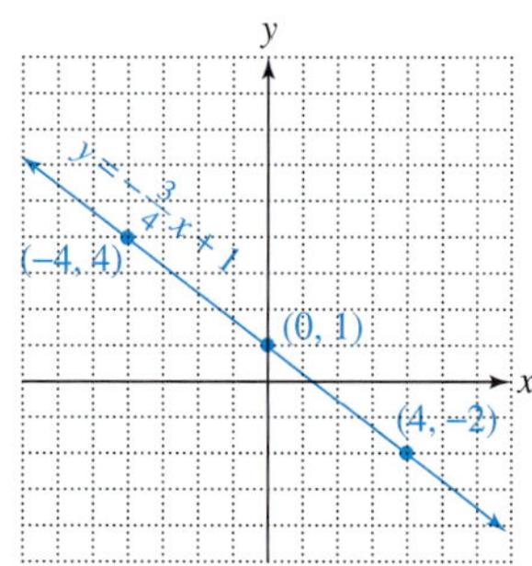

20. $4x = 8$

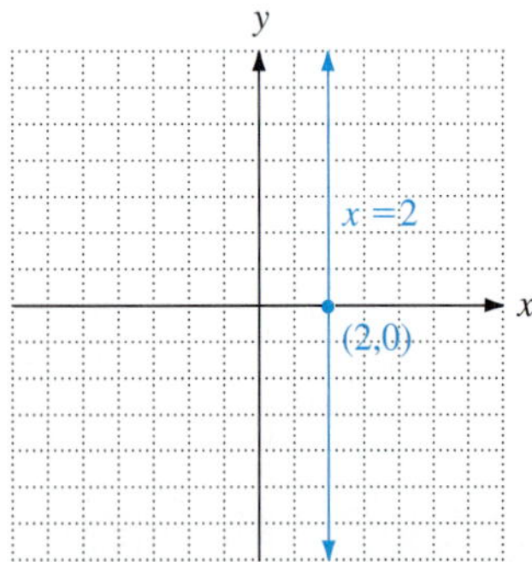

21. $x - 4 = 2y$

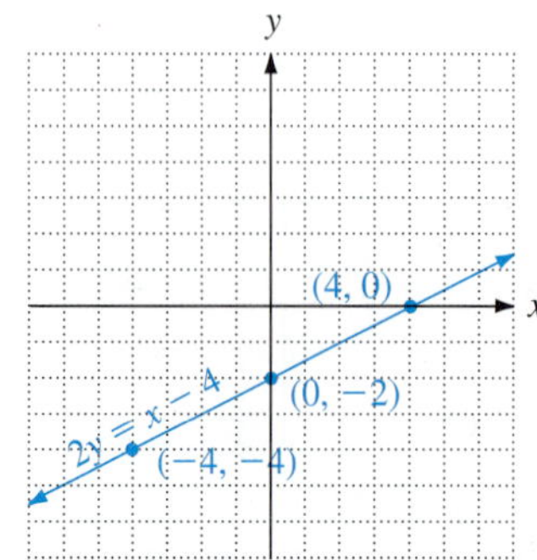

22. $x - 2 = 4y$

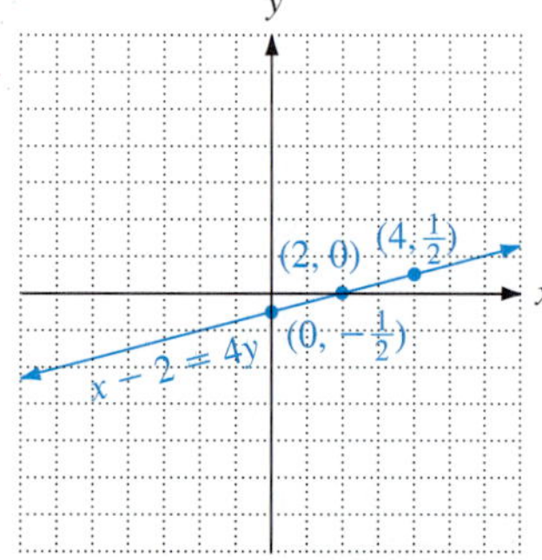

Graph each equation. Be sure to simplify your equation before graphing it.

23. $3x + 2y = 6$

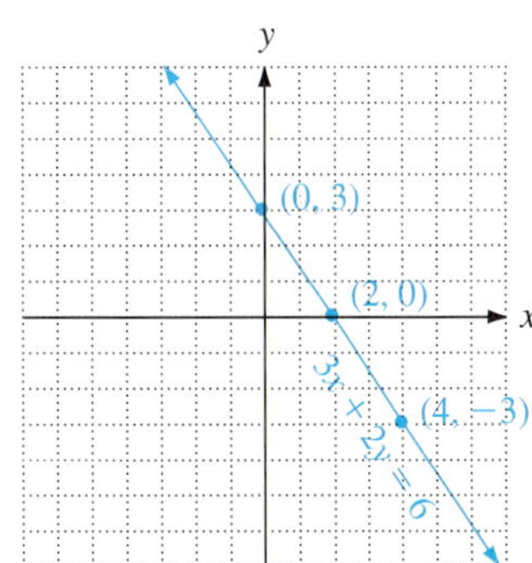

24. $y - 2 = 3y$

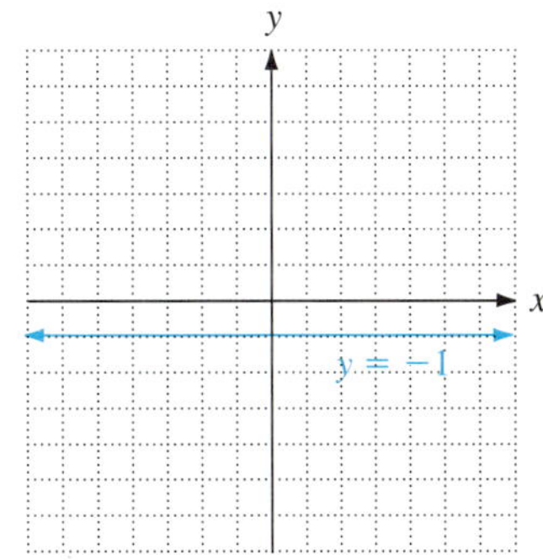

25. $2x + 5y - 2 = -12$

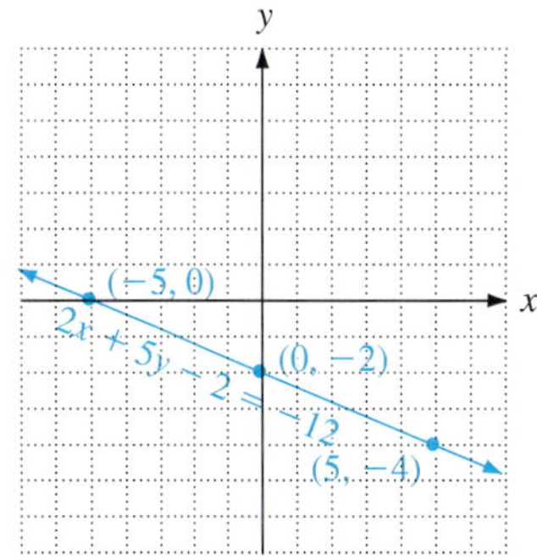

26. $3x - 4y - 5 = -17$

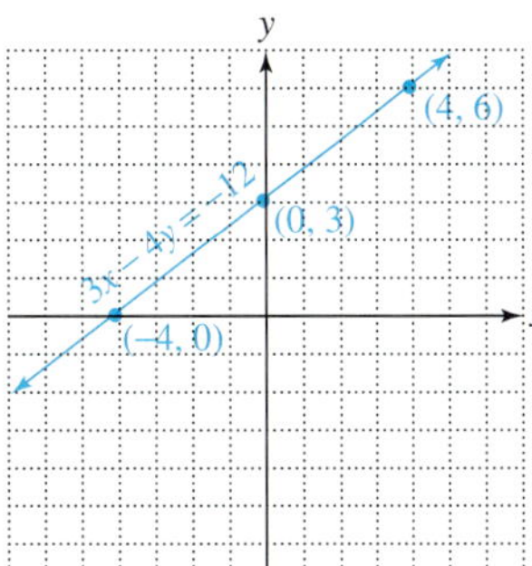

27. $3y + 1 = 7$

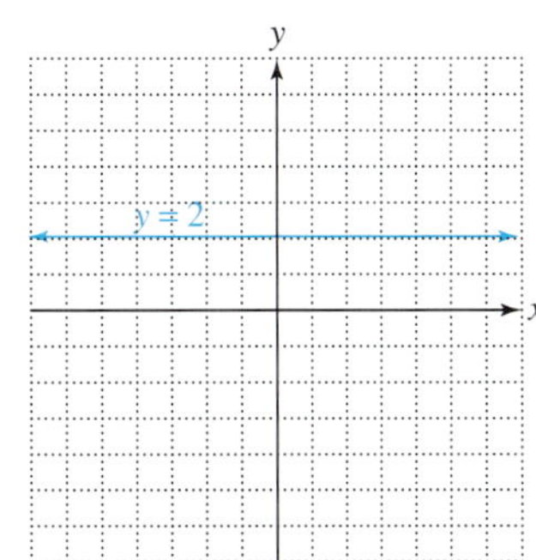

28. $3x - 4 = -13$

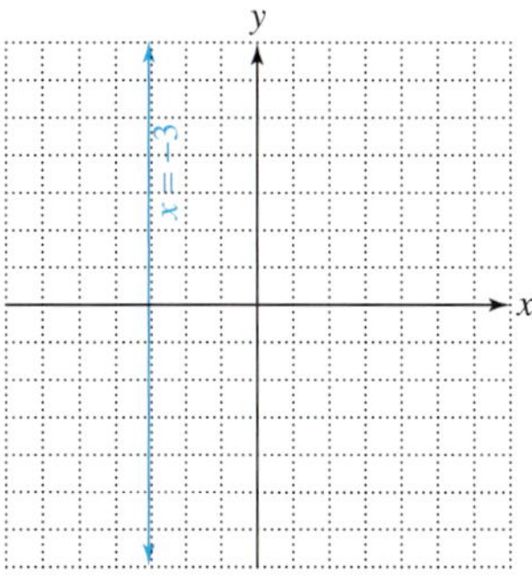

Applications

29. The maximum amount of carbon dioxide in the air in Mauna Loa, Hawaii, is approximated by the equation $C = 1.6t + 343$ where C is the concentration of CO_2 measured in parts per million and t is the number of years since 1982. Graph the equation for $t = 0$, 4, and 8. (**Source:** *Environmental Science,* Fifth Edition, B. Nebel and R. Wright, Upper Saddle River, NJ, Prentice-Hall.)

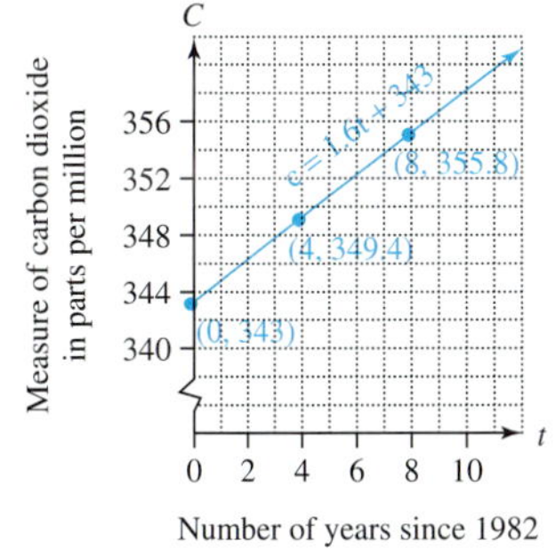

30. The approximate population of the United States is given by the equation $P = 2.3t + 179$ where P is the population in millions and t is the number of years since 1960. Graph the equation for $t = 0$, 10, 30. (**Source:** Bureau of the Census, U.S. Department of Commerce.)

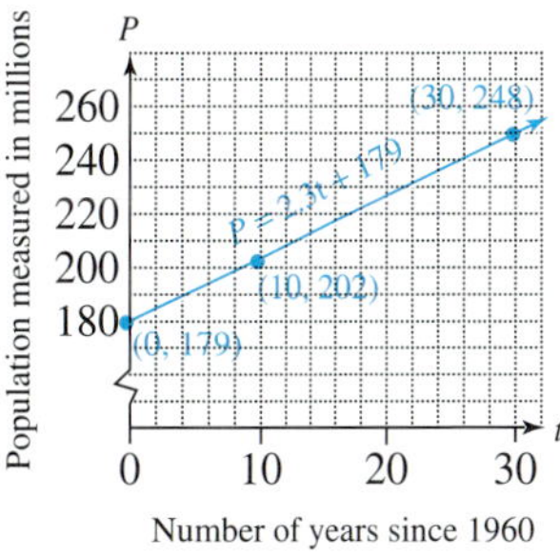

Cumulative Review Problems

31. Solve. $\dfrac{x}{x+4} + \dfrac{x-1}{x-4} = 2$

$x^2 - 4x + x^2 + 3x - 4 = 2x^2 - 32$

$x = 28$

32. Solve and graph on a number line. $4 - 3x \leq 18$

$x \geq -\dfrac{14}{3}$

−6 −5 $-\frac{14}{3}$ −4

33. A rectangle's width is 1 meter more than half of its length. Find the dimensions if the perimeter measures 53 meters.

$W = 9.5$ meters
$L = 17$ meters

34. There are 29 students for every 2 faculty members at Skyline University. Last semester the school had 3074 students. If this ratio is true, how many faculty members are there?

212 faculty members

7.3 Slope of a Line

MathPro

Video 18

SSM

TEACHING TIP The slope can be thought of as a move from one point to another.

$$m = \frac{\text{vertical move}}{\text{horizontal move}}$$

After studying this section, you will be able to:

1. *Find the slope of a straight line given two points on the line.*
2. *Find the slope and y-intercept of a straight line, given the equation of the line.*
3. *Write the equation of a line, given the slope and the y-intercept.*
4. *Graph a line using the slope and y-intercept.*
5. *Find the slopes of lines that are parallel or perpendicular.*

1 *Find the Slope of a Straight Line, Given Two Points on the Line*

We often use the word slope to describe the incline (the steepness) of a hill. A carpenter or a builder will refer to the *pitch* or *slope* of a roof. The slope is the change in the vertical distance compared to the change in the horizontal distance as you go from one point to another point along the roof. If the change in the vertical distance is greater than the change in the horizontal distance, the slope will be steep. If the change in the horizontal distance is greater than the change in the vertical distance, the slope will be gentle.

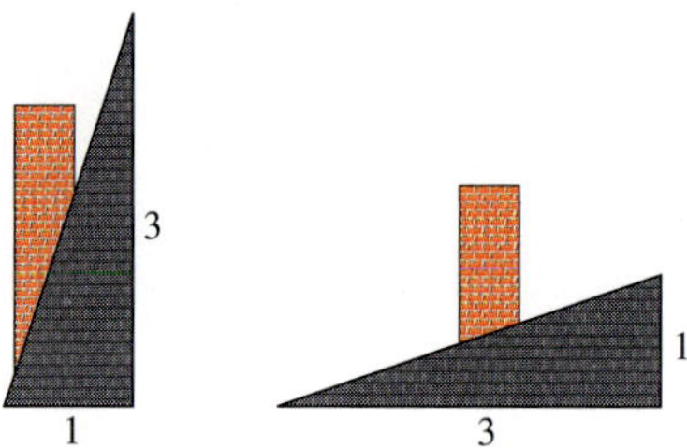

In a coordinate plane, the **slope** of a straight line is defined by the change in y divided by the change in x.

$$\text{Slope} = \frac{\text{change in } y}{\text{change in } x}$$

Consider the line drawn through points A and B in the figure. If we measure the change from point A to point B in the x-direction and the y-direction, we will have an idea of the steepness (or the slope) of the line. From point A to point B the change in y values is from 2 to 4, a *change of* 2. From point A to point B the change in x values is from 1 to 5, a *change of* 4. Thus

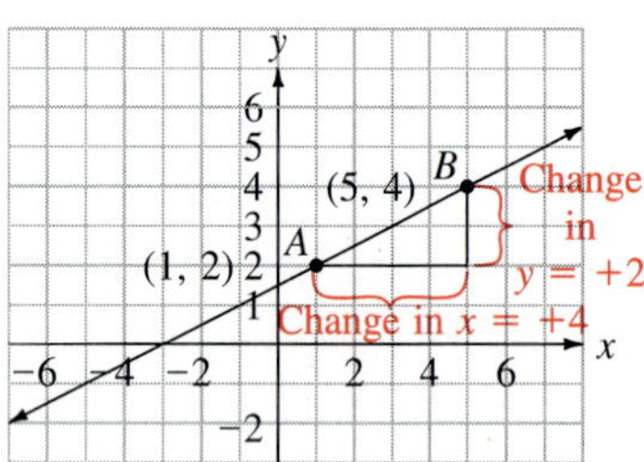

$$\text{slope} = \frac{\text{change in } y}{\text{change in } x} = \frac{2}{4} = \frac{1}{2}$$

Informally, we can describe this move as the rise over the run: **slope** $= \dfrac{\text{rise}}{\text{run}}$.

We now state a more formal (and more frequently used) definition.

Definition of Slope of a Line

The *slope of any nonvertical straight line* that contains the points with coordinates (x_1, y_1) and (x_2, y_2) has a slope defined by the difference ratio

$$\text{slope} = m = \frac{y_2 - y_1}{x_2 - x_1} \qquad \text{where } x_2 \neq x_1$$

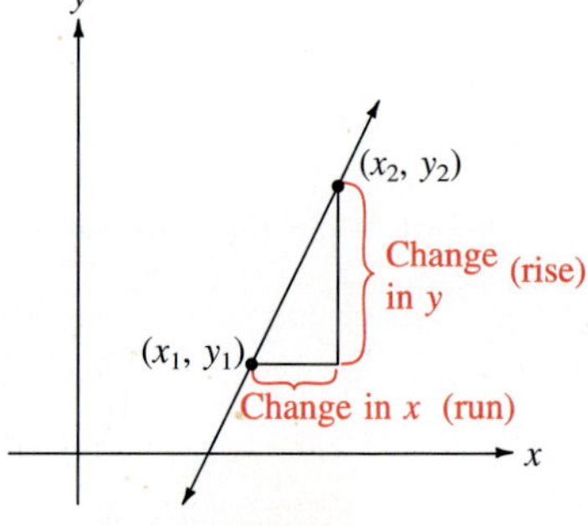

The use of subscripted terms such as x_1, x_2, and so on, is just a way of indicating that the first x value is x_1 and the second x value is x_2. Thus (x_1, y_1) are the coordinates of the first point and (x_2, y_2) are the coordinates of the second point. The letter m is commonly used for the slope.

EXAMPLE 1 Find the slope of the line that passes through (2, 0) and (4, 2).

Let (2, 0) be the first point (x_1, y_1)

(4, 2) be the second point (x_2, y_2)

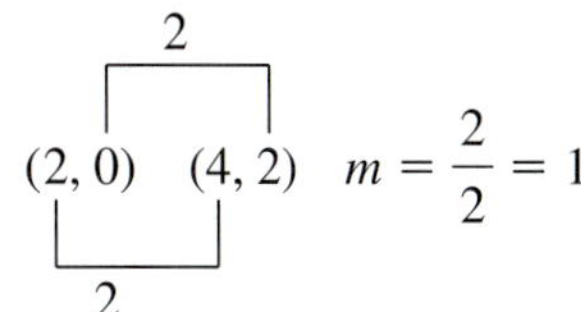

$$m = \frac{2}{2} = 1$$

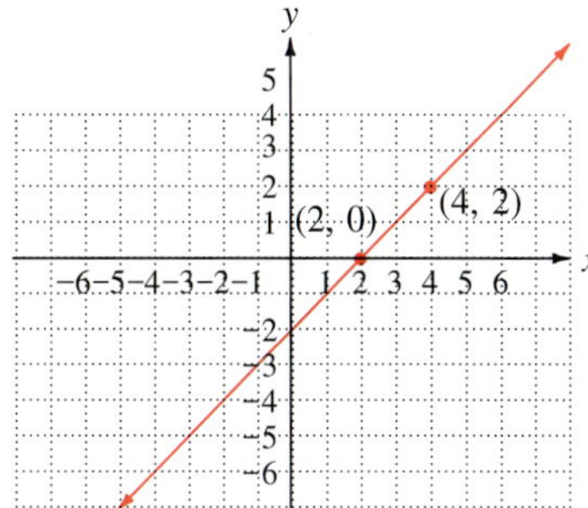

By the formula

$$\text{slope} = m = \frac{y_2 - y_1}{x_2 - x_1} = \frac{2 - 0}{4 - 2} = \frac{2}{2} = 1$$

The sketch of the line is shown in the figure above.

Note that the slope of the line will be the same if we let (4, 2) be the first point (x_1, y_1) and (2, 0) be the second point (x_2, y_2).

−2
(4, 2) (2, 0)
−2

$$m = \frac{y_2 - y_1}{x_2 - x_1} = \frac{0 - 2}{2 - 4} = \frac{-2}{-2} = 1$$

Thus, given two points, it does not matter which you call (x_1, y_1) and which you call (x_2, y_2).

WARNING Be careful, however, not to put the x's in one order and the y's in another order when finding the slope, given two points on a line.

Practice Problem 1 Find the slope of the line through (6, 1) and (−4, −1). ■

TEACHING TIP Students may find it easier to use the formula informally and diagram the difference ratio. They may do so for Example 2 in the following way.

$$\overbrace{(2,0)\quad(4,4)}^{4} \qquad m = \frac{4}{2} = 2$$
$$\underbrace{\qquad\qquad\qquad}_{2}$$

EXAMPLE 2 Find the slope of the line that passes through (2, 0) and (4, 4).

Let (2, 0) be the first point (x_1, y_1)

(4, 4) be the second point (x_2, y_2)

$$m = \frac{y_2 - y_1}{x_2 - x_1} = \frac{4 - 0}{4 - 2} = \frac{4}{2} = 2$$

The slope of the line is 2. Notice that the slope (rate of change) of this line is twice that of the slope of the line in Example 1.

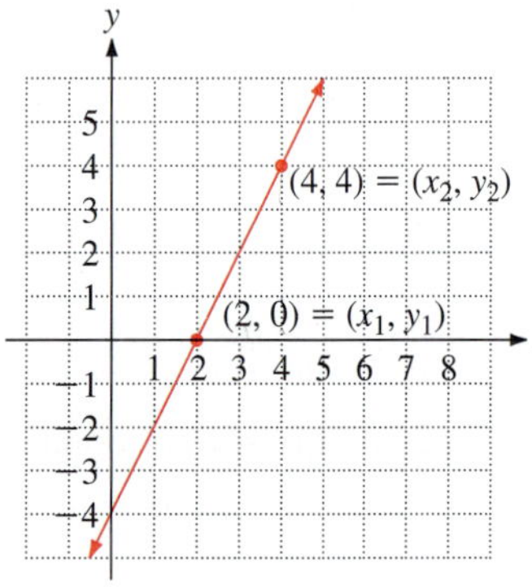

Practice Problem 2 Find the slope of the line through (−1, −4) and (1, 6). ■

To Think About Compare the slope of the line in Practice Problem 1 with the slope of the line in Practice Problem 2. Which line is steeper? Why? Draw the lines to verify.

It is important to see how positive and negative slopes affect the graphs of the line.

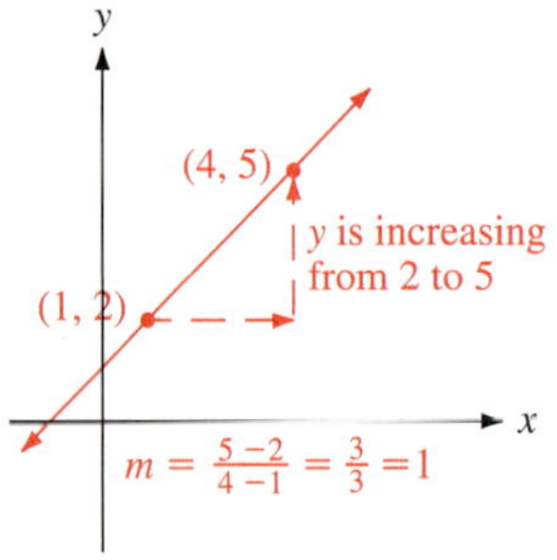

1. Line with positive slope

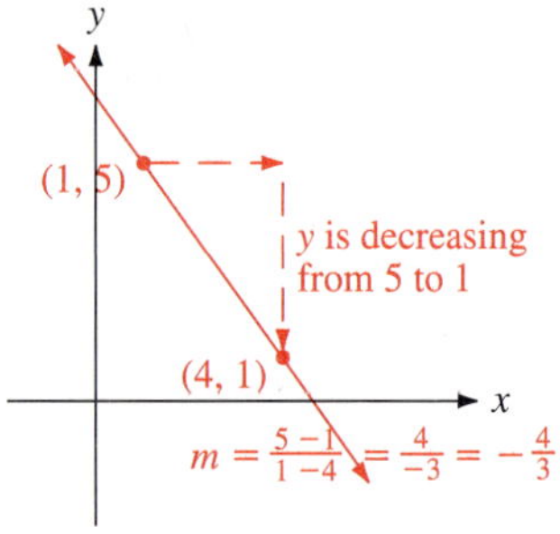

2. Line with negative slope

1. If the y values increase as you go from left to right, the slope of the line is positive.
2. If the y values decrease as you go from left to right, the slope of the line is negative.

TEACHING TIP It is easy for students to make sign errors when finding the slope. After explaining Example 3, ask students to find the slope of the line through (4, −7) and (−3, −2) as an Additional Class Exercise.
Sol. The slope $m = -\frac{5}{7}$.

EXAMPLE 3 Find the slope of the line through (−3, 2) and (2, −4).

Let (−3, 2) be (x_1, y_1)

(2, −4) be (x_2, y_2)

$$m = \frac{y_2 - y_1}{x_2 - x_1} = \frac{-4 - 2}{2 - (-3)} = \frac{-4 - 2}{2 + 3} = \frac{-6}{5} = -\frac{6}{5}$$

The slope of this line is negative. We would expect this since the y value decreased from 2 to −4. What does the graph of this line look like? Plot the points and draw the line to verify.

Practice Problem 3 Find the slope of the line through (2, 0) and (−1, 1). ■

To Think About Describe the line in Practice Problem 3 by looking at its slope. Then verify by drawing the graph.

EXAMPLE 4 Find the slope of the line that passes through the given points.

(a) (0, 2) and (5, 2) **(b)** (−4, 0) and (−4, −4)

(a) Take a moment to look at the y values. What do you notice? What does this tell you about the line? Now calculate the slope.

$$m = \frac{2-2}{5-0} = \frac{0}{5} = 0$$

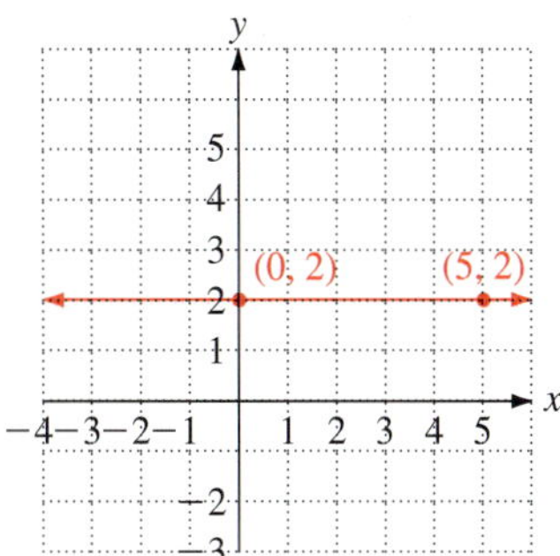

A horizontal line has a slope of 0.

(b) Take a moment to look at the x values. What do you notice? What does this tell you about the line? Now calculate the slope.

$$m = \frac{-4-0}{-4-(-4)} = \frac{-4}{0}$$

Recall that division by 0 is undefined.

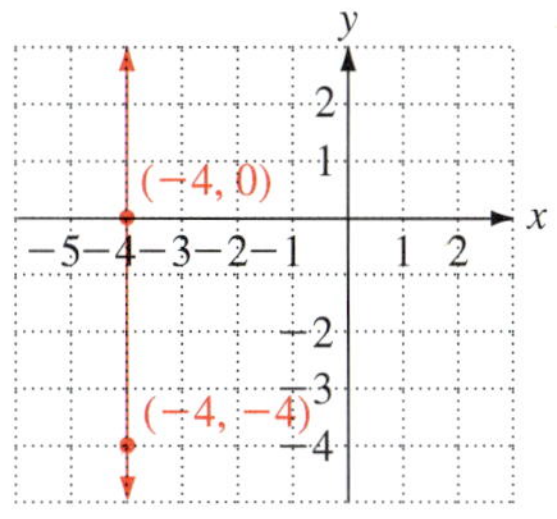

The slope of a vertical line is undefined. We say that this line has **no slope.**

Notice in our definition of slope that $x_2 \neq x_1$. Thus it is not appropriate to use the formula for slope for the points in (b). We do so to illustrate what would happen if $x_2 = x_1$. We get an impossible situation, $\frac{y_2 - y_1}{0}$. Now you can see why we include the restriction $x_2 \neq x_1$ in our definition.

Practice Problem 4 Find the slope of the line that passes through the given points.

(a) (−4, 0) and (−4, 4) **(b)** (−7, −11) and (3, −11) ■

TEACHING TIP To show that three points lie on the same line, have students find the slope between pairs of points. For example, show that (1, 3), (3, 7), (5, 11) are on the same line.

(1, 3) (3, 7), $m = \frac{4}{2} = 2$ (difference in y: 4; difference in x: 2)

(3, 7) (5, 11), $m = \frac{4}{2} = 2$ (difference in y: 4; difference in x: 2)

(1, 3) (5, 11), $m = \frac{8}{4} = 2$ (difference in y: 8; difference in x: 4)

The points have a *common* difference ratio (slope). They must be on the *same* line.

TEACHING TIP It is important for students to know that a horizontal line has a slope of zero but that a vertical line has **no slope.** Emphasize this to students when discussing horizontal and vertical lines.

Slope of a Straight Line

Positive slope
Line goes upward to the right

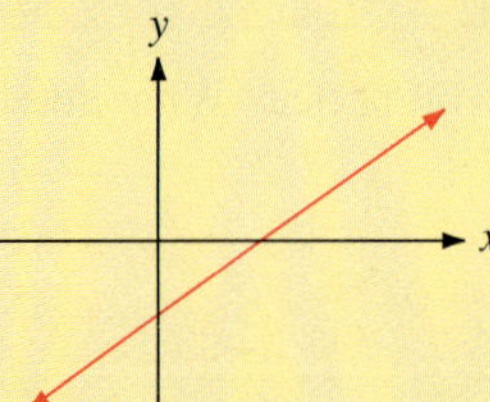

1. *Lines with positive slopes go upward as x increases* (that is, as you go from left to right).

Negative slope
Line goes downward to the right

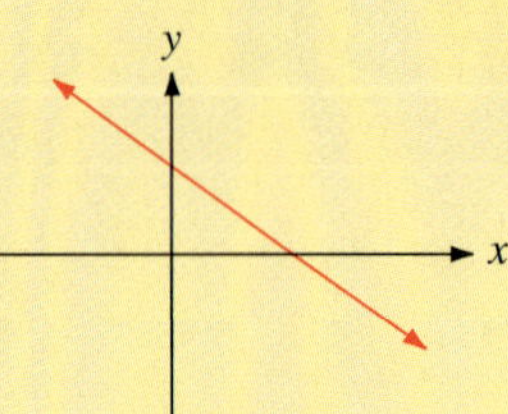

2. *Lines with negative slopes go downward as x increases* (that is, as you go from left to right).

Zero slope
Horizontal line

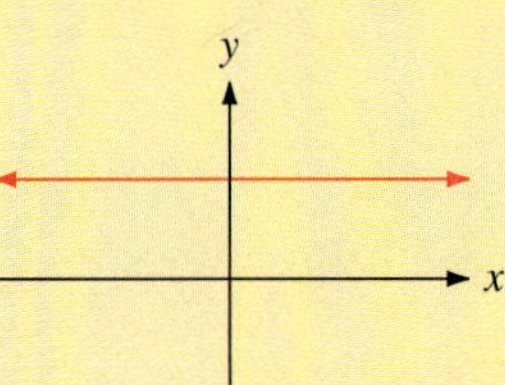

3. Horizontal lines have a slope of 0.

Undefined slope
Vertical line

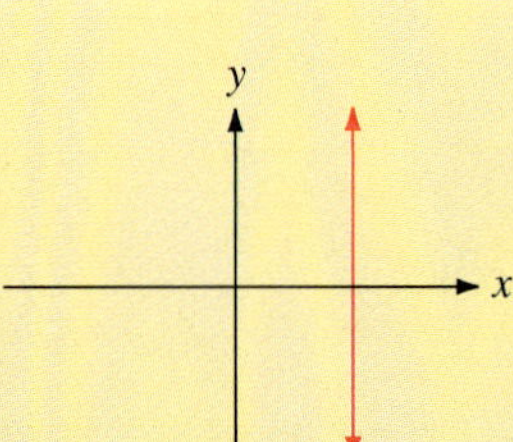

4. A vertical line is said to have undefined slope. The slope of a vertical line is not defined. In other words, a vertical line has no slope.

2 *Finding the Slope and y-Intercept of a Line, Given the Equation of a Line*

Recall that the equation of a a line is a linear equation in two variables. This equation can be written in several different ways. A very useful form of the equation of a straight line is the slope–intercept form. The form can be derived in the following way. Suppose that a straight line, with slope m, crosses the y-axis at a point $(0, b)$. Consider any other point on the line and label the point (x, y).

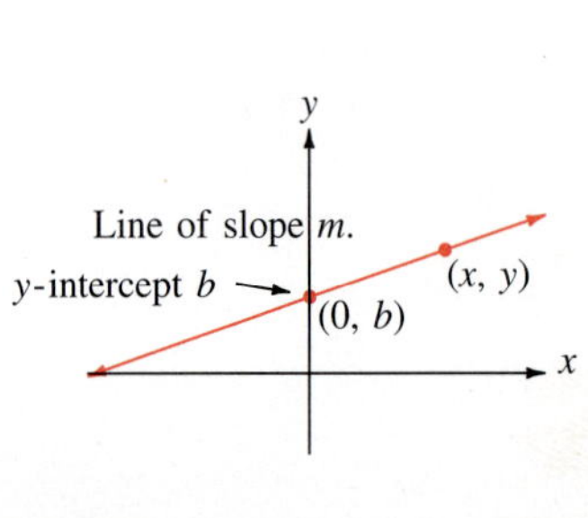

$$\frac{y_2 - y_1}{x_2 - x_1} = m \qquad \textit{Definition of slope.}$$

$$\frac{y - b}{x - 0} = m \qquad \textit{Substitute } (0, b) \textit{ for } (x_1, y_1) \textit{ and } (x, y) \textit{ for } (x_2, y_2).$$

$$\frac{y - b}{x} = m \qquad \textit{Simplify.}$$

$$y - b = mx \qquad \textit{Multiply both sides by } x.$$

$$y = mx + b \qquad \textit{Add } b \textit{ to both sides.}$$

This form of a linear equation immediately reveals the slope of the line, m, and where the line intercepts (crosses) the y-axis, b.

Slope–Intercept Form of a Line

The slope–intercept form of the equation of the line that has slope m and the y-intercept b at point $(0, b)$ is given by

$$y = mx + b$$

EXAMPLE 5 What is the slope and the y-intercept of the line whose equation is $y = 4x - 5$?

The equation is in the form $y = mx + b$.

$$y = 4x + (-5)$$

slope $= 4$ y-intercept $= -5$

The slope is 4 and the y-intercept is -5.

Practice Problem 5 What is the slope and the y-intercept of $y = -2x + 6$? ■

By using algebraic operations, we can write any linear equation in slope–intercept form and use this form to identify the slope and the y-intercept of the line.

EXAMPLE 6 What is the slope and the y-intercept of the line $5x + 3y = 2$?

TEACHING TIP After explaining how to do Example 6, ask students to find the slope and the y-intercept of the line $7x - 8y = -9$ as an Additional Class Exercise.

Sol. The slope is $m = \frac{7}{8}$. The y-intercept is $b = \frac{9}{8}$.

We want to solve for y and get the equation in the form $y = mx + b$. We need to isolate the y variable.

$5x + 3y = 2$

$3y = -5x + 2$ *Subtract 5x from both sides.*

$y = \dfrac{-5x + 2}{3}$ *Divide both sides by 3.*

$y = -\dfrac{5}{3}x + \dfrac{2}{3}$ *Write the right-hand side as two fractions, using the property* $\dfrac{a + b}{c} = \dfrac{a}{c} + \dfrac{b}{c}$.

The *slope* is $-\dfrac{5}{3}$. The *y-intercept* is $\dfrac{2}{3}$.

Practice Problem 6 Find the slope and y-intercept of the line $4x - 2y = -5$. ■

3 Writing the Equation of a Line, Given the Slope and the y-Intercept

If we know the slope of a line m, and the y-intercept, b, we can write the equation of the line, $y = mx + b$.

EXAMPLE 7 Find the equation of the line with slope $\frac{2}{5}$ and y-intercept -3.

(a) Write the equation in slope–intercept form, $y = mx + b$.

(b) Write the equation in $Ax + By = C$ form.

(a) We are given that $m = \frac{2}{5}$ and $b = -3$. Since

$$y = mx + b$$

$$y = \frac{2}{5}x + (-3)$$

$$y = \frac{2}{5}x - 3$$

(b) Recall, for the form $Ax + By = C$, that A, B, and C are integers. We first clear the equation of fractions. Then we move the x term to the left side.

$$5y = 5\frac{2x}{5} - 5(3) \qquad \textit{Multiply each term by 5.}$$

$$5y = 2x - 15 \qquad \textit{Simplify.}$$

$$-2x + 5y = -15 \qquad \textit{Subtract 2x from each side.}$$

$$2x - 5y = 15$$ *Multiply each term by -1. The form $Ax + By = C$ is usually written with A as a positive integer.*

Practice Problem 7 Find the equation of the line with slope $-\frac{3}{7}$ and y-intercept $\frac{2}{7}$.

(a) Write the equation in slope–intercept form.

(b) Write the equation in the form $Ax + By = C$. ■

4 Graphing a Line Using the Slope and the y-Intercept

If we know the slope of a line and the y-intercept, we can draw the graph of the line.

EXAMPLE 8 Graph a line with slope $m = \frac{2}{3}$ and y-intercept of -3. Use the coordinate system below.

Recall that the y-intercept is the point where the line crosses the y-axis. The x-coordinate of this point is 0. Thus the coordinates of the y-intercept for this line are $(0, -3)$. We plot the point.

Recall that slope $= \frac{\text{rise}}{\text{run}}$. Since the slope for this line is $\frac{2}{3}$, we will go up (rise) 2 units and go over (run) to the right 3 units from the point $(0, -3)$. Look at the figure on the right. This is the point $(3, -1)$. Plot the point. Draw a line that connects the two points $(0, -3)$ and $(3, -1)$.

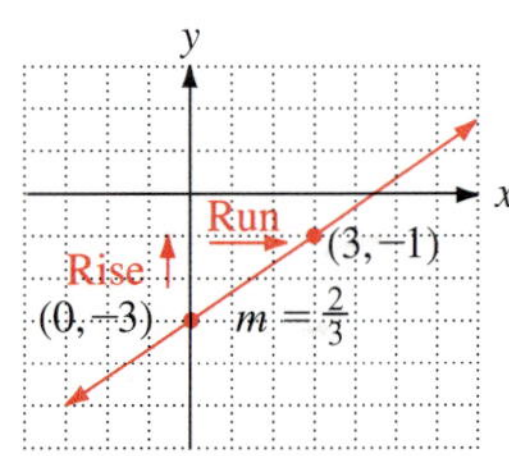

This is the graph of the line with slope $\frac{2}{3}$ and y-intercept of -3.

PRACTICE PROBLEM 8

y
x
$m = \frac{3}{4}$
(4, 2)
(0, −1)

Practice Problem 8 Graph a line with slope $= \frac{3}{4}$ and a y-intercept of -1. Use the coordinate system in the margin. ■

EXAMPLE 9 Graph the equation $y = -\frac{1}{2}x + 4$. Use the coordinate system below.

Begin with the y-intercept. Since the y-intercept is 4, plot the point $(0, 4)$. Now look at the slope $-\frac{1}{2}$. This can be written as $\frac{-1}{2}$. Begin at $(0, 4)$ and go *down* 1 unit and to the right 2 units. This is the point $(2, 3)$. Plot the point. Draw the line that connects the points $(0, 4)$ and $(2, 3)$.

This is the graph of the equation $y = -\frac{1}{2}x + 4$.

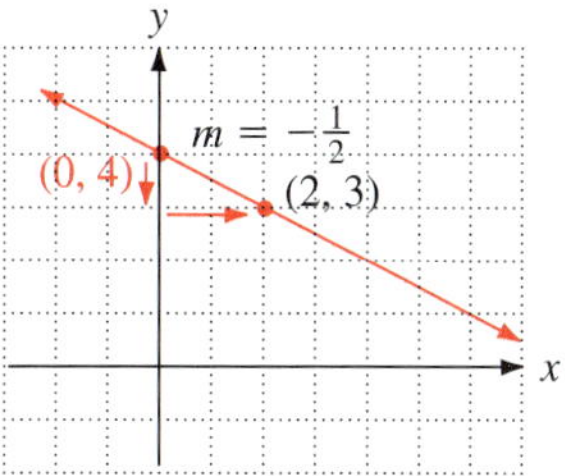

Practice Problem 9 Graph the equation $y = -\frac{2}{3}x + 5$ on the coordinate system in the margin. ■

PRACTICE PROBLEM 9

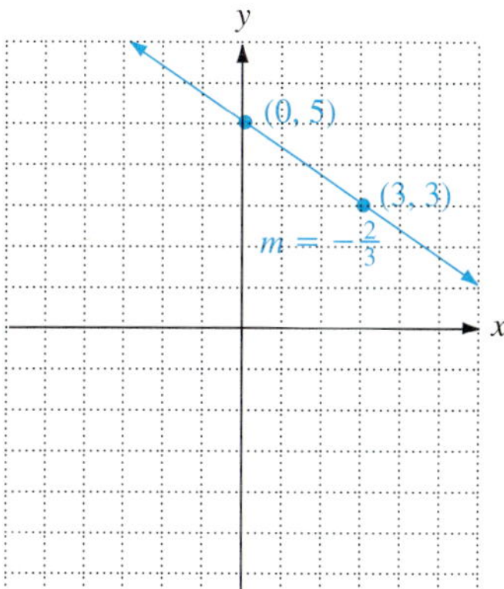

To Think About Explain why you count down 1 unit and move to the right 2 units to represent the slope $-\frac{1}{2}$. Could you have done this in another way? Try it. Verify that this is the same line.

5 *Finding the Slopes of Lines That Are Parallel or Perpendicular*

Parallel lines are two straight lines that never touch. Look at the parallel lines in the figure at the right. Notice that the slope of line a is -3 and the slope of line b is also -3. Why do you think the slopes must be equal? What would happen if the slope of line b was -1. Try it.

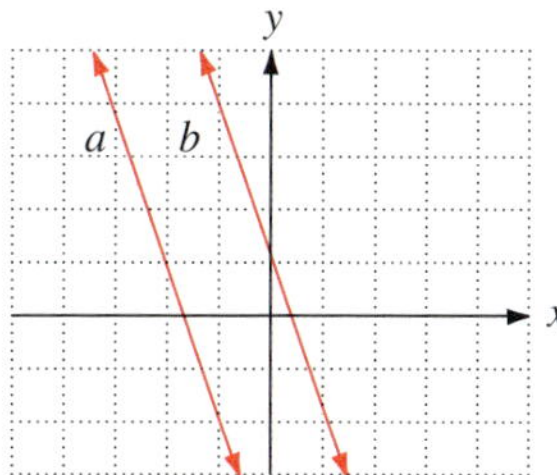

Parallel Lines

Parallel lines are two straight lines that never touch.

Parallel lines have the same slope.

$$m_1 = m_2$$

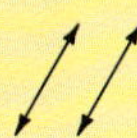

Parallel lines

Perpendicular lines are two lines that meet at a 90° angle. Look at the perpendicular lines in the figure on the right. The slope of line c is -3. The slope of line d is $\frac{1}{3}$. Notice that

$$(-3)\left(\frac{1}{3}\right) = \left(-\frac{3}{1}\right)\left(\frac{1}{3}\right) = -1$$

You may wish to draw several pairs of perpendicular lines to determine if this will always happen.

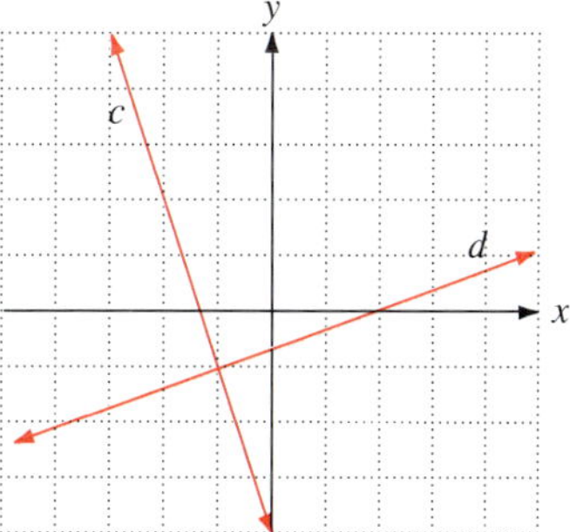

Perpendicular Lines

Perpendicular lines are two lines that meet at a 90° angle.

Perpendicular lines have slopes whose product is -1. If m_1 and m_2 are slopes of perpendicular lines, then

$$m_1 m_2 = -1 \quad \text{or} \quad m_1 = -\frac{1}{m_2}$$

Perpendicular lines

GRAPHING CALCULATOR

Graphing Lines

You can graph a line given in the form $y = mx + b$ using a graphing calculator. For example, to graph $y = 2x + 4$ enter the right-hand side of the equation in the Y = editor of your calculator and graph. Choose an appropriate window to show all the intercepts. The following window is -10 to 10 by -10 to 10.

Display:

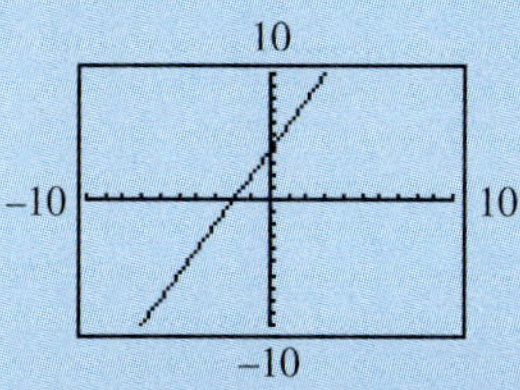

Try graphing other equations given in slope-intercept form.

EXAMPLE 10 Line e has a slope of $-\frac{2}{3}$.

(a) If line f is parallel to line e, what is its slope?

(b) If line g is perpendicular to line e, what is its slope?

(a) Parallel lines have the same slope.
Line f will have a slope of $-\frac{2}{3}$.

(b) Perpendicular lines have slopes whose product is -1.

$$m_1 m_2 = -1$$

$$-\frac{2}{3}m_2 = -1 \qquad \textit{Substitute } -\tfrac{2}{3} \textit{ for } m_1.$$

$$\left(-\frac{3}{2}\right)\left(-\frac{2}{3}\right)m_2 = -1\left(-\frac{3}{2}\right) \qquad \textit{Multiply both sides by } -\tfrac{3}{2}.$$

$$m_2 = \frac{3}{2}$$

Practice Problem 10 Line h has a slope of $\frac{1}{4}$.

(a) If line j is parallel to line h, what is its slope?

(b) If line k is perperpendicular to line h, what is its slope? ■

EXAMPLE 11 The equation of line l is $y = -2x + 3$.

(a) What is the slope of a line that parallel to line l?

(b) What is the slope of a line that is perpendicular to line l?

(a) Looking at the equation, we can see that the slope of line l is -2.

$$y = -2x + 3$$

$$m_1 = -2$$

The slope of a line that is parallel to line l is -2.

(b) Perpendicular lines have slopes whose product is -1.

$$m_1 m_2 = -1$$

$$(-2)m_2 = -1 \qquad \textit{Substitute } -2 \textit{ for } m_1.$$

$$m_2 = \frac{1}{2} \qquad \textit{Because } (-2)\left(\frac{1}{2}\right) = -1.$$

The slope of a line that is perpendicular to line l is $\frac{1}{2}$.

Practice Problem 11 The equation of line n is $y = \frac{1}{4}x - 1$.

(a) What is the slope of a line that is parallel to line n?

(b) What is the slope of a line that is perpendicular to line n? ■

7.3 Exercises

Find the slope of a straight line that passes through the given pair of points.

1. (7, 3) and (6, 9) -6

2. (5, 2) and (8, 5) 1

3. (−2, 1) and (3, 4) $\frac{3}{5}$

4. (5, 6) and (−3, 1) $\frac{5}{8}$

5. (−7, −4) and (3, −8) $-\frac{2}{5}$

6. (−4, −6) and (5, −9) $-\frac{1}{3}$

7. (3, −4) and (−6, −4) 0

8. (7, −2) and (−12, −2) 0

9. (−3,0) and (0, −4) $-\frac{4}{3}$

10. (0, 5) and (5, 3) $-\frac{2}{5}$

11. $(\frac{1}{3}, -2)$ and $(\frac{2}{3}, -5)$ -9

12. $(4, -3)$ and $(\frac{5}{2}, 5)$ $-\frac{16}{3}$

Verbal and Writing Skills

13. Can you find the slope of the line passing through (5, −12) and (5, −6)? Why? Why not?
no, division by zero is impossible ⇒ undefined slope

14. Can you find the slope of the line passing through (6, −2) and (−8, −2)? Why? Why not?
yes, $\frac{0}{-14} = 0$; the slope is 0.

Find the slope and the y-intercept of each line.

15. $y = 8x + 9$
$m = 8; b = 9$

16. $y = 2x + 10$
$m = 2; b = 10$

17. $y = 3x - 7$
$m = 3; b = -7$

18. $y = 5x - 8$
$m = 5; b = -8$

19. $y = \frac{5}{6}x - \frac{2}{9}$
$m = \frac{5}{6}; b = -\frac{2}{9}$

20. $y = -\frac{3}{4}x + \frac{5}{6}$
$m = -\frac{3}{4}; b = \frac{5}{6}$

21. $y = -6x$
$m = -6; b = 0$

22. $y = -2$
$m = 0; b = -2$

23. $4x + y = -\frac{1}{3}$
$m = -4; b = -\frac{1}{3}$

24. $3x + y = \frac{2}{5}$
$m = -3; b = \frac{2}{5}$

25. $3x - y = -2$
$m = 3; b = 2$

26. $4x - y = 5$
$m = 4; b = -5$

27. $5x + 2y = 3$
$m = -\frac{5}{2}; b = \frac{3}{2}$

28. $7x + 3y = 4$
$m = -\frac{7}{3}; b = \frac{4}{3}$

29. $7x - 3y = 4$
$m = \frac{7}{3}; b = -\frac{4}{3}$

30. $9x - 4y = 18$
$m = \frac{9}{4}; b = -\frac{9}{2}$

Write the equation of the line (a) in the slope–intercept form and (b) in the form $Ax + By = C$.

31. $m = \frac{3}{4}$, $b = 2$
(a) $y = \frac{3}{4}x + 2$
(b) $3x - 4y = -8$

32. $m = \frac{4}{5}$, $b = 3$
(a) $y = \frac{4}{5}x + 3$
(b) $4x - 5y = -15$

33. $m = 6$, $b = -3$
(a) $y = 6x - 3$
(b) $6x - y = 3$

34. $m = 5$, $b = -6$
(a) $y = 5x - 6$
(b) $5x - y = 6$

35. $m = \frac{-5}{2}$, $b = 6$
(a) $y = \frac{-5}{2}x + 6$
(b) $5x + 2y = 12$

36. $m = \frac{-2}{3}$, $b = 7$
(a) $y = \frac{-2}{3}x + 7$
(b) $2x + 3y = 21$

37. $m = -2$, $b = \frac{1}{2}$
(a) $y = -2x + \frac{1}{2}$
(b) $4x + 2y = 1$

38. $m = -3$, $b = \frac{1}{4}$
(a) $y = -3x + \frac{1}{4}$
(b) $12x + 4y = 1$

Graph the line.

39. $m = \frac{1}{2}$, $b = -3$

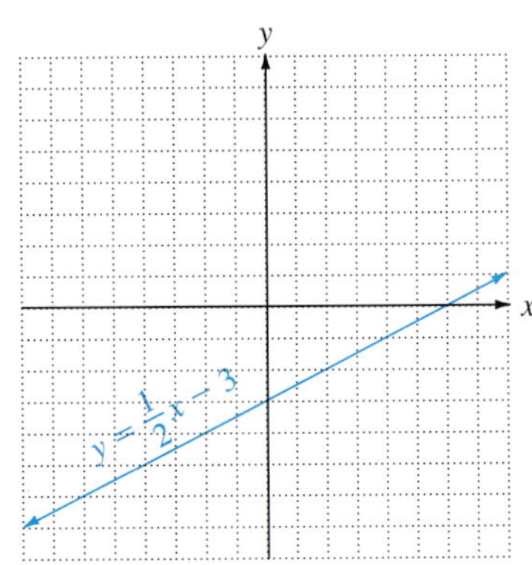

40. $m = \frac{2}{3}$, $b = -4$

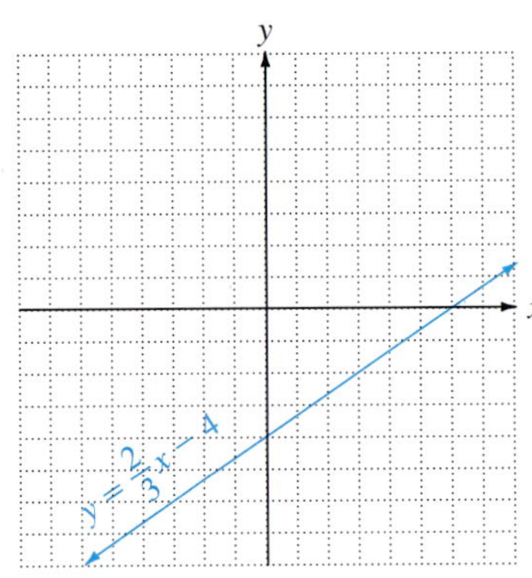

41. $m = -\frac{5}{3}$, $b = 2$

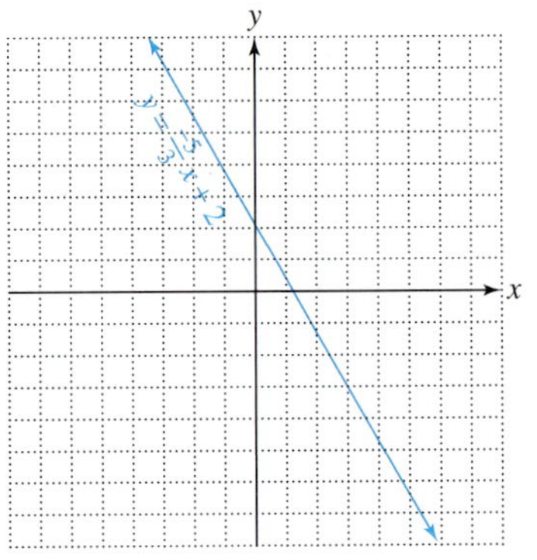

42. $m = -\frac{3}{2}$, $b = 4$

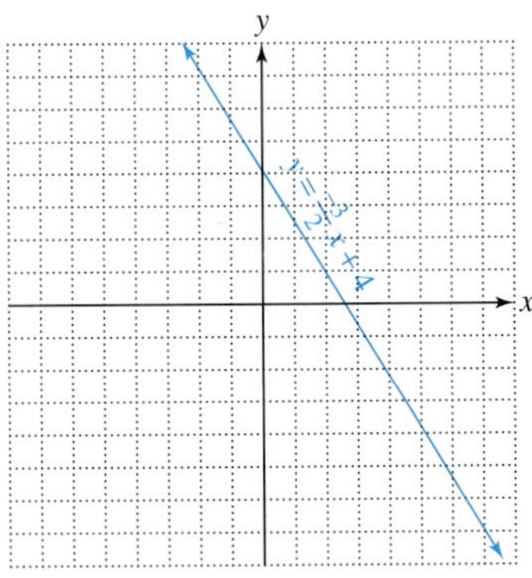

43. $y = \frac{3}{4}x - 5$

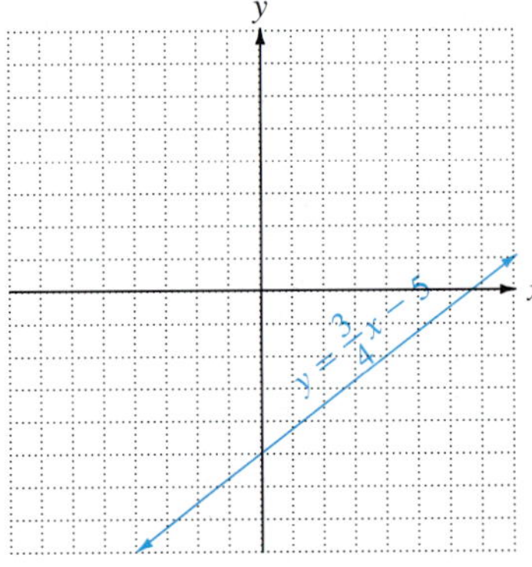

44. $y = \frac{2}{3}x - 6$

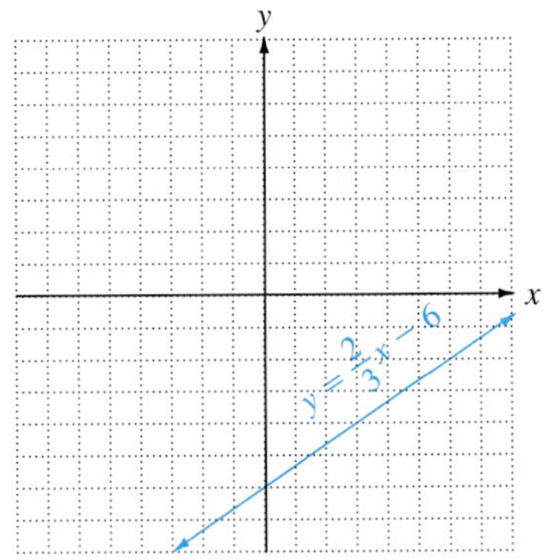

45. $y + 3x = 2$

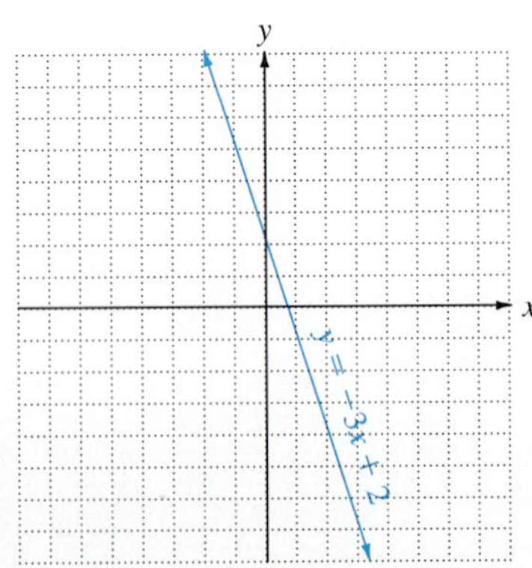

46. $y + 2x = 4$

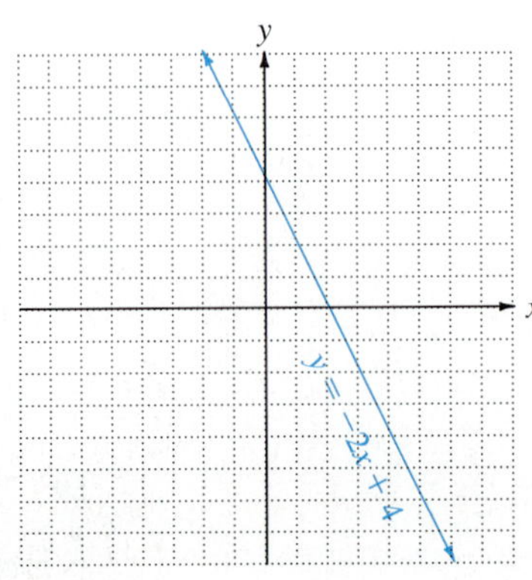

47. $y = 2x$

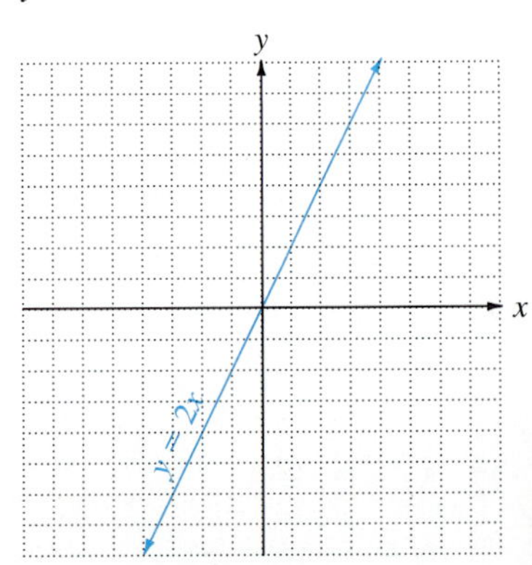

48. A line has a slope of $\frac{2}{3}$.

(a) What is the slope of a line parallel to it? $\frac{2}{3}$

(b) What is the slope of a line perpendicular to it? $-\frac{3}{2}$

49. A line has a slope of $\frac{13}{5}$.

(a) What is the slope of a line parallel to it? $\frac{13}{5}$

(b) What is the slope of a line perpendicular to it? $-\frac{5}{13}$

50. A line has a slope of -4.

(a) What is the slope of a line parallel to it? -4

(b) What is the slope of a line perpendicular to it? $\frac{1}{4}$

51. A line has a slope of $\frac{1}{8}$.

(a) What is the slope of a line parallel to it? $\frac{1}{8}$

(b) What is the slope of a line perpendicular to it? -8

52. The equation of a line is $3y = -9x - 5$.

(a) What is the slope of a line parallel to it? -3

(b) What is the slope of a line perpendicular to it? $\frac{1}{3}$

53. The equation of a line is $2y = 4x - 1$.

(a) What is the slope of a line parallel to it? 2

(b) What is the slope of a line perpendicular to it? $-\frac{1}{2}$

54. The equation of a line is $y = \frac{1}{3}x + 2$.

(a) What is the slope of a line parallel to it? $\frac{1}{3}$

(b) What is the slope of a line perpendicular to it? -3

55. The equation of a line is $y = -\frac{1}{2}x - 3$.

(a) What is the slope of a line parallel to it? $-\frac{1}{2}$

(b) What is the slope of a line perpendicular to it? 2

Cumulative Review Problems

Solve for x and graph your solution.

56. $3x + 8 > 2x + 12$ $x > 4$

57. $\frac{1}{4}x + 3 > \frac{2}{3}x + 2$ $3x + 36 > 8x + 24$

$12 > 5x$

$x < \frac{12}{5}$

$\frac{11}{5}$ $\frac{12}{5}$ $\frac{13}{5}$ $\frac{14}{5}$ 3

58. $\frac{1}{2}(x + 2) \le \frac{1}{3}x + 5$ $3x + 6 \le 2x + 30$

$x \le 24$

22 23 24 25 26

59. $7x - 2(x + 3) \le 4(x - 7)$ $7x - 2x - 6 \le 4x - 28$

$x \le -22$

-25 -24 -23 -22 -21

60. Captain Charles Allen has a 35-foot charter offshore fishing boat in Falmouth, Massachusetts. His boat has a cruising range of 390 miles when traveling at 25 miles per hour. However the cruising range is decreased by 62% when cruising at his maximum speed of 45 miles per hour. What is his cruising range at maximum speed? (Round to the nearest mile.)
148 miles at maximum speed

61. New England Camp Cherith tries to maintain a ratio of 3 counselors for every 25 campers. One week last summer there were 189 campers. What is the minimum number of counselors who should have been there that week?
There should be a minimum of 23 counselors.

Putting Your Skills to Work

UNDERWATER PRESSURE

The pressure on a submarine or other submersible object deep under the ocean's surface is very significant. In 1960 a record was set for a manned submersible by the *Trieste,* which descended to a depth of more than 35,800 feet. The pressure on the hull of the *Trieste* at that depth was over 8 tons per square inch!

The underwater pressure in pounds per square inch on an object submerged in the ocean is given by the equation $p = \frac{5}{11}d + 15$, where d is the depth in feet below the surface.

PROBLEMS FOR INDIVIDUAL STUDY AND INVESTIGATION

1. Find the underwater pressure at a depth of 22 feet.
25 pounds per square inch

2. The underwater pressure was found to be 35 pounds per square inch. What was the depth at which this measurement was taken? 44 feet

3. Using the coordinates of the two points obtained in problems 1 and 2, calculate the slope of the line.
$\frac{5}{11}$

4. From the linear equation, determine the coordinates of a third point. Plot the point and the points obtained in Problems 1 and 2. Draw a line.

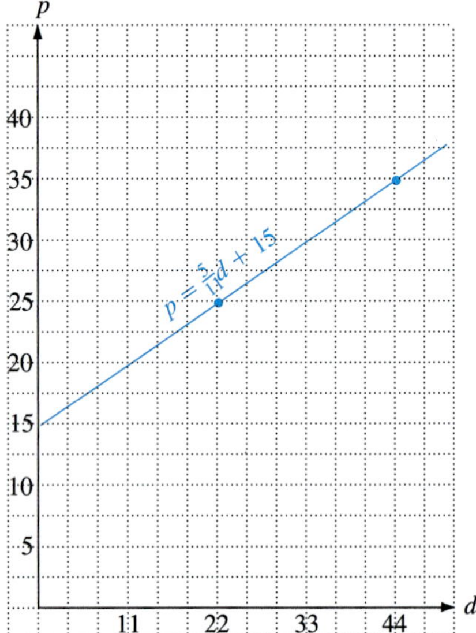

5. From the graph, determine the slope of the line. Does this agree with the slope in problem 3? Does this agree with the equation? Why? Why not?
$\frac{5}{11}$; yes. You can determine the slope of a line using the coordinates of two points, from the equation in slope–intercept form, from the graph by counting.

6. Complete the following table of values.

d	0	11	22	33	44
p	15	20	25	30	35

(a) What is the difference between successive p values? 5

(b) What is the difference between successive d values? 11

(c) How is this related to the slope? Why? Slope is the difference between the p values over the difference between the d values.

PROBLEMS FOR COOPERATIVE GROUP ACTIVITY AND ANALYSIS

Together with members of your class, see if you can determine the following.

Scientists are working to find a pressure equation similar to $p = \frac{5}{11}d + 15$ for a liquid that has different properties than water. They have compiled the following table of values in which p is the pressure in pounds per square inch and d is the depth in feet below the surface of this liquid:

d	0	5	12
p	18	22	27.6

7. Using as coordinates for (d, p) the values $(0, 18)$ and $(5, 22)$, find the slope of the line.
$\frac{4}{5}$

8. Using as coordinates for (d, p) the values $(5, 22)$ and $(12, 27.6)$, find the slope of the line. How does this compare with your answer from problem 7? What does this imply? $\frac{5.6}{7}$, (students can determine the decimal equivalent of $\frac{4}{5}$ and of $\frac{5.6}{7}$ to see that the slopes are the same.) There is a constant rate of change. The graph is a line.

9. Write the equation of the pressure for this new liquid in the form $p = md + b$ by determining the slope m and the p-intercept b.
$p = \frac{4}{5}d + 18$

10. What would the pressure be if an object is submerged in this liquid at a depth of 44 feet? How does this compare with the pressure of an object submerged in water at a depth of 44 feet? 53.2 pounds per square inch; the pressure is greater.

11. Is this new liquid more dense than water or less dense than water? What evidence do you have to support your conclusions? The new liquid is more dense because the pressure is greater at the same depth.

INTERNET CONNECTIONS: Go to ``http://www.prenhall.com/tobey'' to be connected

Site: Boat Dives (Kona Coast Divers) or a related site

This site provides information about diving locations in Hawaii. Use the information provided to find the underwater pressure that you would experience in a dive to the maximum possible depth at each of the following locations.

12. Robert's Reef

13. Garden Eel Cove

14. Fish Rock: (Ka'iwi Point)

15. Long Lava Tube

16. Amphitheatre

7.4 Obtaining the Equation of a Line

MathPro

Video 18

SSM

After studying this section, you will be able to:

1 *Write the equation of a line, given a point and a slope.*
2 *Write the equation of a line, given two points.*
3 *Write the equation of a line, given a graph of the line.*

1 *Writing the Equation of a Line, Given a Point and a Slope*

If we know the slope of a line and the y-intercept, we can write the equation of the line in slope–intercept form. Sometimes we are given the slope and a point on the line. We use the information to find the y-intercept. Then we can write the equation of the line.

EXAMPLE 1 Find the equation of a line with a slope of 4 that passes through the point (3, 1).

We are given the following values: $m = 4$, $x = 3$, $y = 1$.

$y = mx + b$

$y = 4x + b$ *We are given that the slope of the line is 4.*

$1 = 4(3) + b$ *Since (3, 1) is a point on the line, it satisfies the equation. Substitute $x = 3$ and $y = 1$ into the equation.*

$1 = 12 + b$ *Solve for b.*

$-11 = b$

Thus the y-intercept is -11. We can now write the equation of the line.

$$y = 4x - 11$$

Practice Problem 1 Find the equation of the line through $(-2, 4)$ with slope $\frac{1}{2}$. ■

It may be helpful to summarize our approach.

To Find the Equation of a Line Given a Point and a Slope

1. Substitute the given values of x, y, and m into the equation $y = mx + b$.
2. Solve for b.
3. Use the values of b and m to write the equation in the form $y = mx + b$.

TEACHING TIP Remind students that this type of problem is a key problem. Ask them to find the equation of a line passing through $(6, -1)$ having a slope of 3 as an Additional Class Exercise.

Sol. $y = 3x - 19$

EXAMPLE 2 Find the equation of a line with slope $-\frac{2}{3}$ that passes through the point $(-3, 6)$.

We write out the values $m = -\frac{2}{3}$, $x = -3$, $y = 6$.

$y = mx + b$

$6 = \left(-\frac{2}{3}\right)(-3) + b$ *Substitute known values.*

$6 = 2 + b$

$4 = b$

The equation of the line is $y = -\frac{2}{3}x + 4$.

Practice Problem 2 Find the equation of a line with slope $-\frac{3}{4}$ that passes through $(-8, 12)$. ■

2 *Finding the Equation of a Line, Given Two Points*

Our procedure can be extended to the case for which two points are given.

TEACHING TIP Ask students to find the equation of a straight line that passes through $(1, -3)$ and $(-5, -2)$ as an Additional Class Exercise.

Sol. $x + 6y = -17$ or $y = -\frac{1}{6}x - \frac{17}{6}$

EXAMPLE 3 Find the equation of a line that passes through (2, 5) and (6, 3).

We first find the slope of the line. Then we proceed as in Example 2.

$$m = \frac{y_2 - y_1}{x_2 - x_1}$$

$$m = \frac{3 - 5}{6 - 2} \qquad \textit{Substitute } (x_1, y_1) = (2, 5) \textit{ and } (x_2, y_2) = (6, 3) \textit{ into the formula.}$$

$$m = \frac{-2}{4} = -\frac{1}{2}$$

Choose either point, say (2, 5), and proceed as in Example 2.

$$y = mx + b$$

$$5 = -\frac{1}{2}(2) + b$$

$$5 = -1 + b$$

$$6 = b$$

The equation of the line is $y = -\frac{1}{2}x + 6$.

Note: After finding the slope $m = -\frac{1}{2}$, we could have used the other point (6, 3) and would have arrived at the same answer. Try it.

Practice Problem 3 Find the equation of the line through (3, 5) and (−1, 1). ■

3 *Finding the Equation of a Line from the Graph*

EXAMPLE 4 What is the equation of the line in the figure on the right?

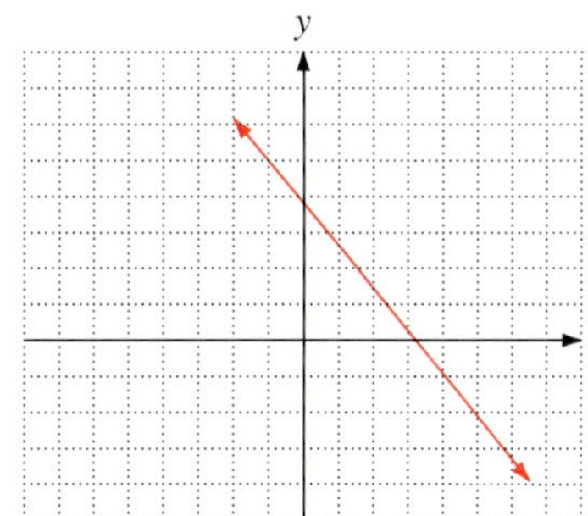

First, look for the y-intercept. The line crosses the y-axis at (0, 4). Thus $b = 4$.

Second, find the slope.

$$m = \frac{\text{change in } y}{\text{change in } x}$$

TEACHING TIP You may want to point out that we chose $(5, -2)$ because it is easy to determine the coordinates of this point since it lies on the intersection of the grid lines.

Look for another point on the line. We chose (5, −2). Count the number of vertical units from 4 to −2 (rise). Count the number of horizontal units from 0 to 5 (run).

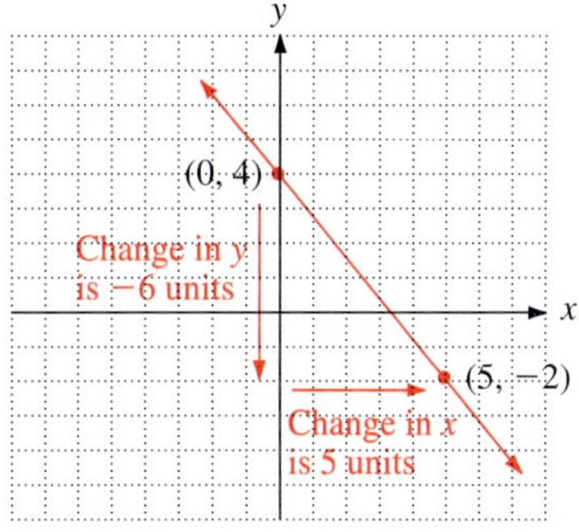

$$m = \frac{-6}{5}$$

Now we can write the equation of the line, $m = -\frac{6}{5}$ and $b = 4$.

$$y = mx + b$$

$$= -\frac{6}{5}x + 4$$

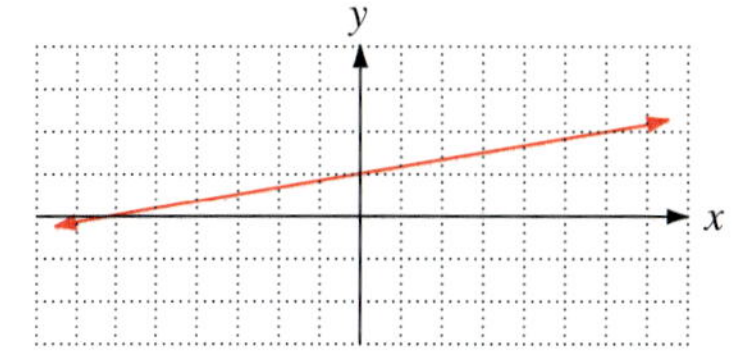

Practice Problem 4 What is the equation of the line in the figure on the right? ■

7.4 Exercises

Find the equation of the line that has the given slope and passes through the given point.

1. $m = 6, (-3, 2)$
$y = 6x + 20$

2. $m = 5, (4, -1)$
$y = 5x - 21$

3. $m = -2, (4, 3)$
$y = -2x + 11$

4. $m = -4, (5, 7)$
$y = -4x + 27$

5. $m = \frac{2}{3}, (-3, -2)$
$y = \frac{2}{3}x$

6. $m = \frac{2}{5}, (-4, -2)$
$y = \frac{2}{5}x - \frac{2}{5}$

7. $m = -3, \left(\frac{1}{2}, 2\right)$
$y = -3x + \frac{7}{2}$

8. $m = -2, \left(3, \frac{1}{3}\right)$
$y = -2x + \frac{19}{3}$

9. $m = -\frac{2}{5}, (5, -3)$
$y = -\frac{2}{5}x - 1$

10. $m = \frac{2}{3}, (3, -2)$
$y = \frac{2}{3}x - 4$

11. $m = 0, (7, -4)$
$y = -4$

12. $m = 0, (-3, 2)$
$y = 2$

Write the equation of a line passing through the two points given.

13. $(1, 5)$ and $(4, 11)$
$y = 2x + 3$

14. $(-1, -5)$ and $(2, 7)$
$y = 4x - 1$

15. $(1, -2)$ and $(3, -8)$
$y = -3x + 1$

16. $(2, -7)$ and $(-1, 8)$
$y = -5x + 3$

17. $(2, -3)$ and $(-1, 6)$
$y = -3x + 3$

18. $(1, -8)$ and $(2, -14)$
$y = -6x - 2$

19. $(3, 5)$ and $(-1, -15)$
$y = 5x - 10$

20. $(-1, -19)$ and $(2, 2)$
$y = 7x - 12$

21. $(1, -3)$ and $(2, 6)$
$y = 9x - 12$

22. $(0, 0)$ and $(1, \frac{1}{2})$
$y = \frac{1}{2}x$

★ **23.** $(1, \frac{5}{6})$ and $(3, \frac{3}{2})$
$y = \frac{1}{3}x + \frac{1}{2}$

★ **24.** $(2, 0)$ and $(\frac{3}{2}, \frac{1}{2})$
$y = -x + 2$

Write the equation of each line.

25.

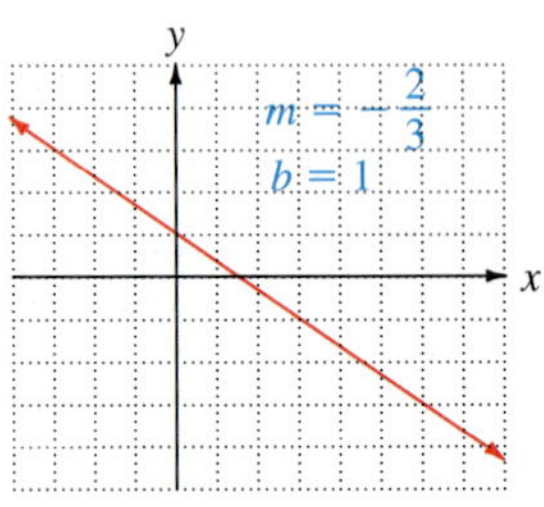

$y = -\frac{2}{3}x + 1$

26.

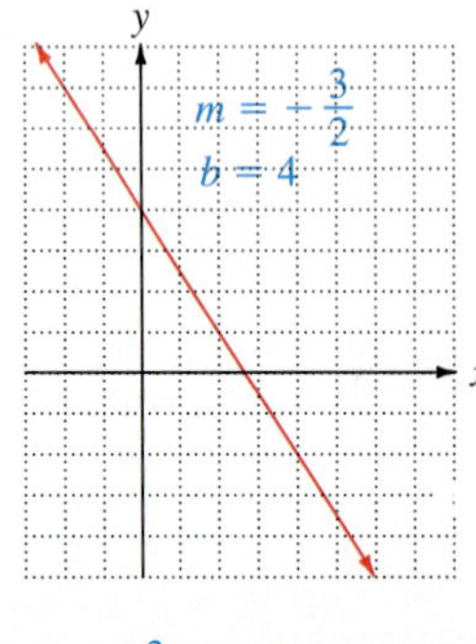

$y = -\frac{3}{2}x + 4$

27.

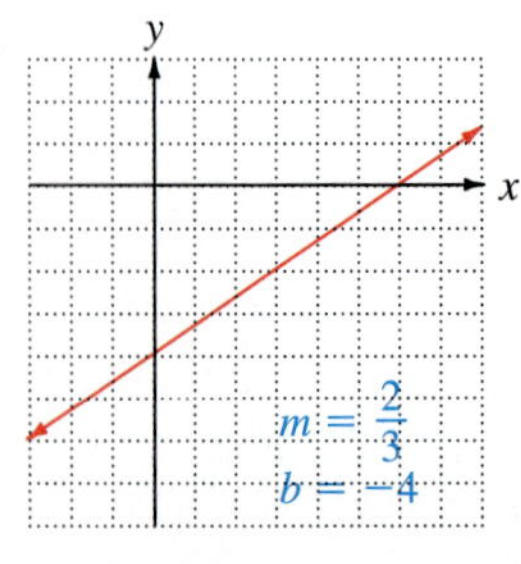

$y = \frac{2}{3}x - 4$

28.

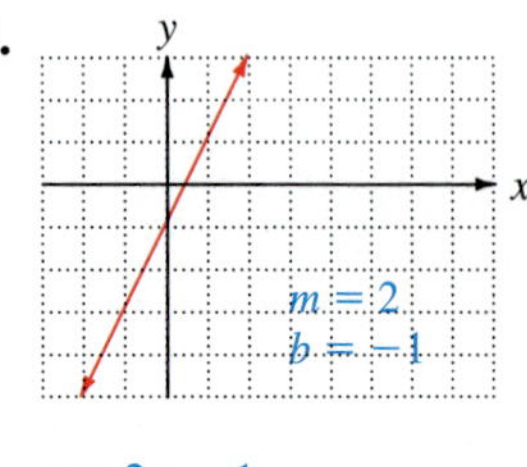

$y = 2x - 1$

29.

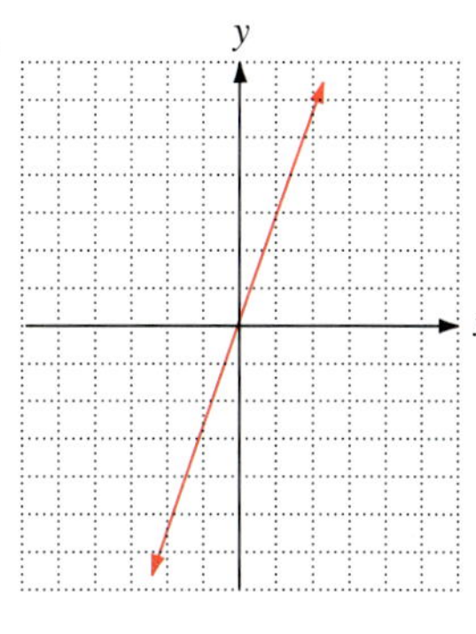

$y = 3x$

30.

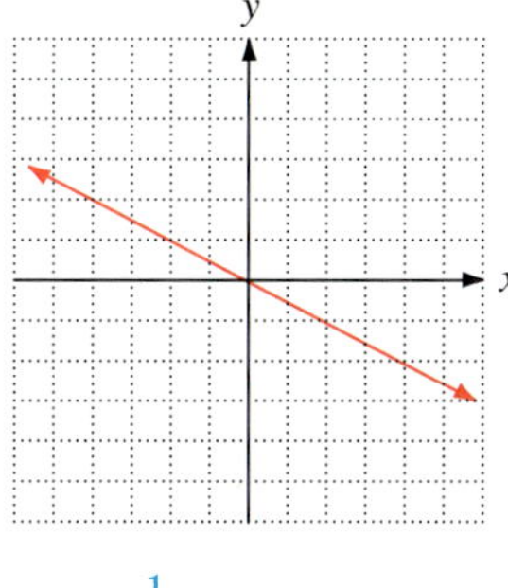
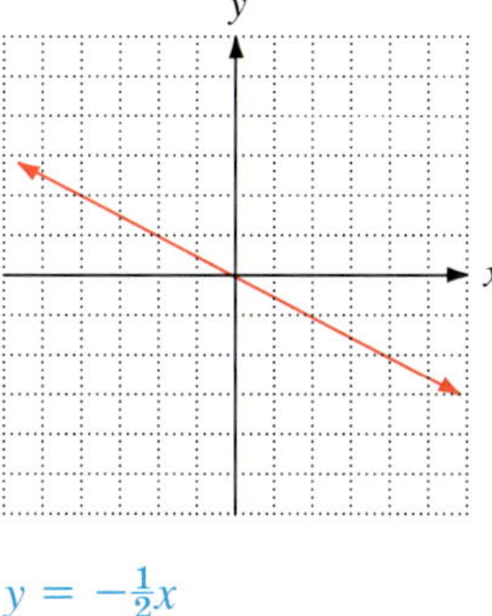

$y = -\frac{1}{2}x$

31.

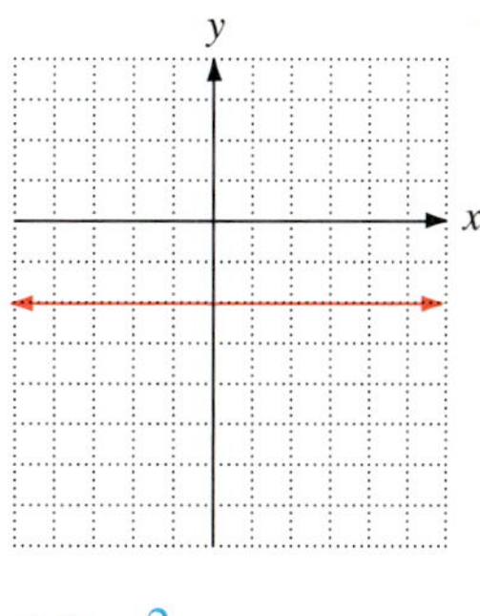

$y = -2$

★ **32.**

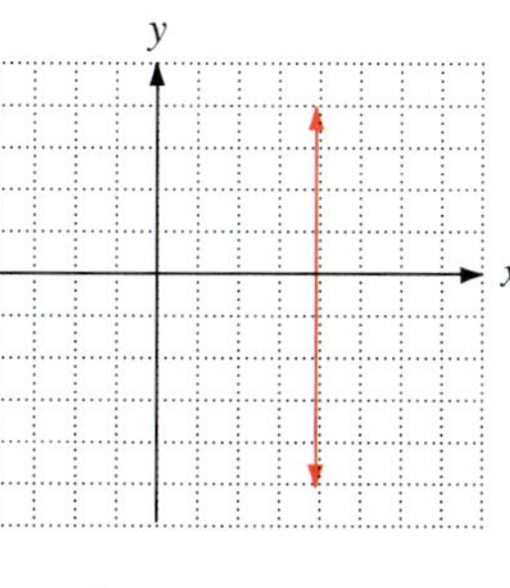

$x = 4$

To Think About

Find the equation of the line passing through the following.

33. $(6, -1)$ with undefined slope.
$x = 6$

34. $(2, 5)$ with slope 0.
$y = 5$

35. $(-5, 7)$ that is parallel to the y-axis.
$x = -5$

36. $(-3, -3)$ that is parallel to the x-axis.
$y = -3$

37. $(0, 2)$ that is perpendicular to the y-axis.
$y = 2$

38. $(-1, 4)$ that is perpendicular to the x-axis.
$x = -1$

39. $(0, 3)$ that is parallel to $y = 2x + 1$.
$y = 2x + 3$

40. $(0, 3)$ that is perpendicular to $y = 2x + 1$.
$y = -\dfrac{1}{2}x + 3$

41. $(2, 3)$ that is perpendicular to $y = 2x - 9$.
$y = -\dfrac{1}{2}x + 4$

42. $(2, 9)$ that is parallel to $y = 5x - 3$.
$y = 5x - 1$

Cumulative Review Problems

43. Solve. $\dfrac{1}{8} + \dfrac{1}{t} = \dfrac{1}{3}$

$3t + 24 = 8t$

$t = \dfrac{24}{5}$

44. Simplify. $\dfrac{4x^2 - 25}{4x^2 - 20x + 25} \cdot \dfrac{x^2 - 9}{2x^2 + x - 15} \div (2x^2 - x - 15)$.

$\dfrac{(2x + 5)(2x - 5)}{(2x - 5)(2x - 5)} \cdot \dfrac{(x + 3)(x - 3)}{(2x - 5)(x + 3)} \cdot \dfrac{1}{(2x + 5)(x - 3)} = \dfrac{1}{(2x - 5)^2}$

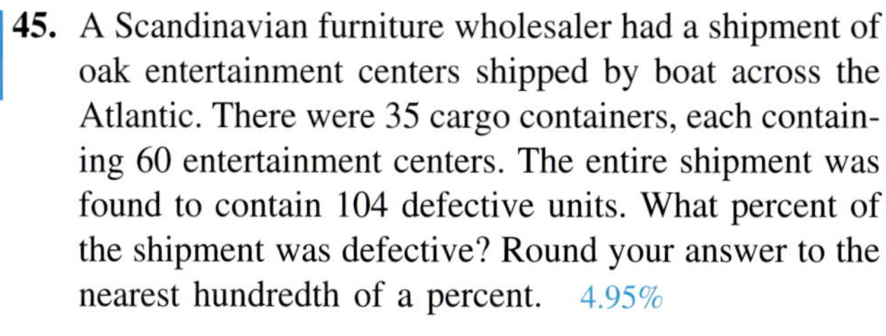

45. A Scandinavian furniture wholesaler had a shipment of oak entertainment centers shipped by boat across the Atlantic. There were 35 cargo containers, each containing 60 entertainment centers. The entire shipment was found to contain 104 defective units. What percent of the shipment was defective? Round your answer to the nearest hundredth of a percent. 4.95%

46. Jackie earned \$2122.40 in interest after investing part of her \$24,000 in savings in an account earning 7% simple interest. She took a risk on some stock options with the rest of the savings, which yielded her 10.5%. How much was invested in the stock options? \$12,640

7.5 Graphing Linear Inequalities

MathPro

Video 19

SSM

After studying this section, you will be able to:

1 *Graph linear inequalities in two variables.*

In Section 2.7 we discussed inequalities in one variable. Look at the inequality $x < -2$ (x is less than -2). Some of the solutions to the inequality are -3, -5, and $-5\frac{1}{2}$. In fact all numbers to the left of -2 on the number line are solutions. The graph of the inequality is given below. Notice that the open circle at 2 indicates that 2 is *not* a solution.

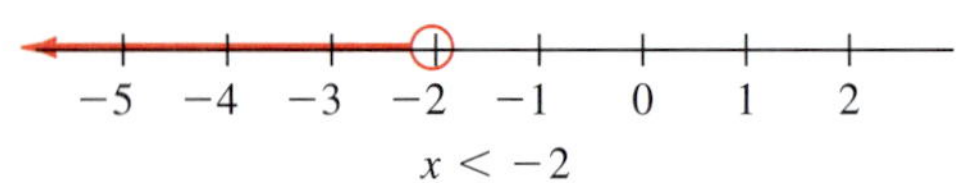

$x < -2$

1 *Graphing Linear Inequalities in Two Variables*

Consider the inequality $y \geq x$. The solution of the inequality is the set of all possible ordered pairs that when substituted into the inequality will yield a true statement. Which ordered pairs will make the statement $y \geq x$ true? Let's try some.

(0, 6)	(−2, 1)	(1, −2)	(3, 5)	(4, 4)
$6 \geq 0$, true	$1 \geq -2$, true	$-2 \geq 1$, false	$5 \geq 3$, true	$4 \geq 4$, true

(0, 6), (−2, 1), (3, 5), and (4, 4) are solutions to the inequality $y \geq x$. In fact, every point at which the y-coordinate is greater than or equal to the x-coordinate is a solution to the inequality. This is shown by the shaded region in the graph at the right. Notice that the solution set includes the points on the line $y = x$.

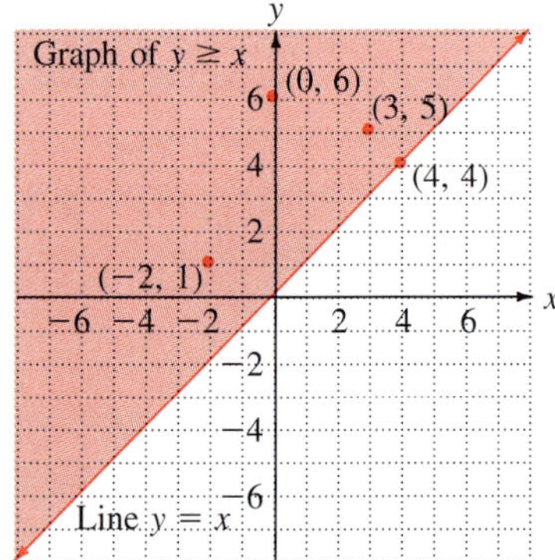

Did we need to test so many points? The solution set for an inequality in two variables will be the region above the line or the region below the line. It is sufficient to test one point. If the point is a solution to the inequality, shade the region that includes the point. If the point is not a solution, shade the region on the other side of the line.

EXAMPLE 1 Graph $5x + 3y > 15$. Use the coordinate system below.

We begin by graphing the line $5x + 3y = 15$. You may use any method discussed previously to graph the line. Since there is no equal sign in the inequality, we will draw a dashed line to indicate that the line is *not* part of the solution set.

Look for a test point. The easiest point to test is (0, 0). Substitute (0, 0) for (x, y) in the inequality.

$$5x + 3y > 15$$
$$5(0) + 3(0) > 15$$
$$0 > 15 \quad \text{false}$$

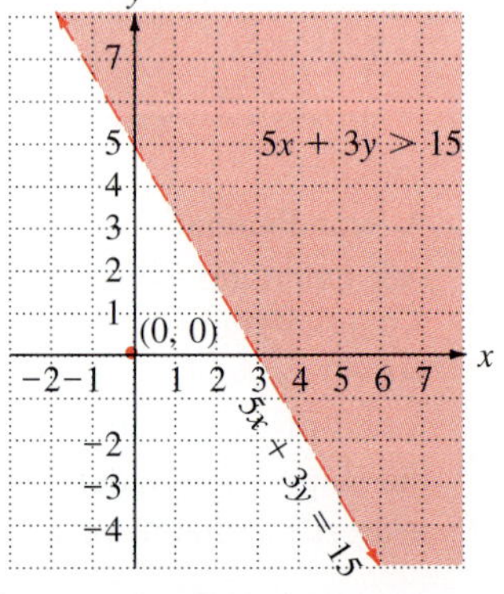

(0, 0) is not a solution. Shade the side of the line that does *not* include (0, 0).

PRACTICE PROBLEM 1

Practice Problem 1 Graph $x - y \geq -10$ on the coordinate system in the margin.

Graphing Linear Inequalities

1. Replace the inequality symbol by an equality symbol. Graph the line.
   (a) The line will be solid if the inequality is $\geq$ or $\leq$.
   (b) The line will be dashed if the inequality is $>$ or $<$.
2. Test the point (0, 0) in the inequality.
   (a) If the inequality is true, shade the side of the line that includes (0, 0).
   (b) If the inequality is false, shade the side of the line that does not include (0, 0).
3. If the point (0, 0) is a point on the line, choose another test point and proceed accordingly.

EXAMPLE 2 Graph $2y \leq -3x$.

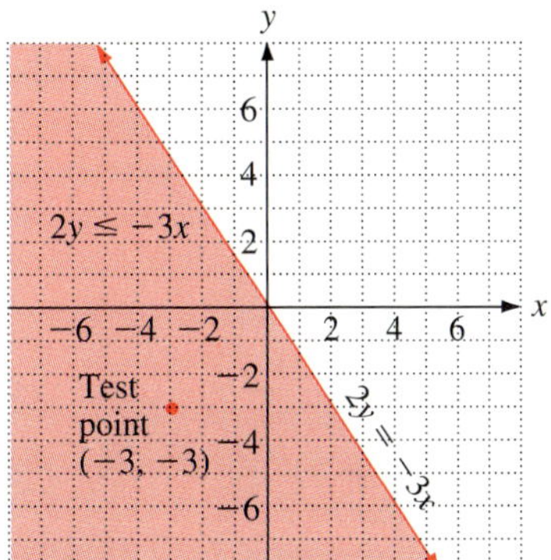

Step 1 Graph $2y = -3x$. Since $\leq$ is used, the line should be a solid line.
Step 2 We see that the line passes through (0, 0).
Step 3 Choose another test point. We will choose $(-3, -3)$.

$$2y \leq -3x$$
$$2(-3) \leq -3(-3)$$
$$-6 \leq 9 \quad \text{true}$$

Shade the region that includes $(-3, -3)$.
Shade the region below the line.

PRACTICE PROBLEM 2

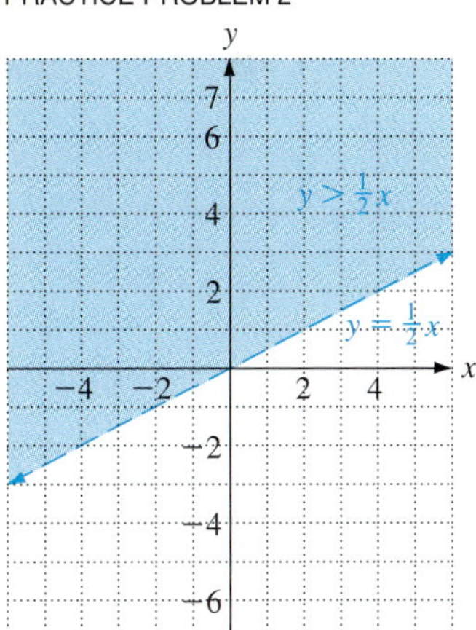

Practice Problem 2 Graph $y > \frac{1}{2}x$ on the coordinate system in the margin. ■

If we are graphing the inequality $x < -2$ on the coordinate plane, the solution will be a region. Notice that this is very different from the solution $x < -2$ on the number line discussed earlier.

EXAMPLE 3 Graph $x < -2$ on the coordinate plane.

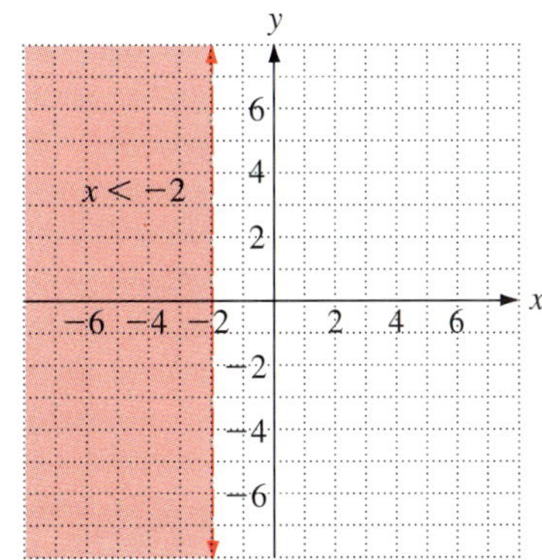

Step 1 Graph $x = -2$. Since $<$ is used, the line should be dashed.
Step 2 Test (0, 0) in the inequality.

$$x < -2$$
$$0 < -2 \quad \text{false}$$

Shade the region that does not include (0, 0). Shade the region to the left of the line $x = -2$.

PRACTICE PROBLEM 3

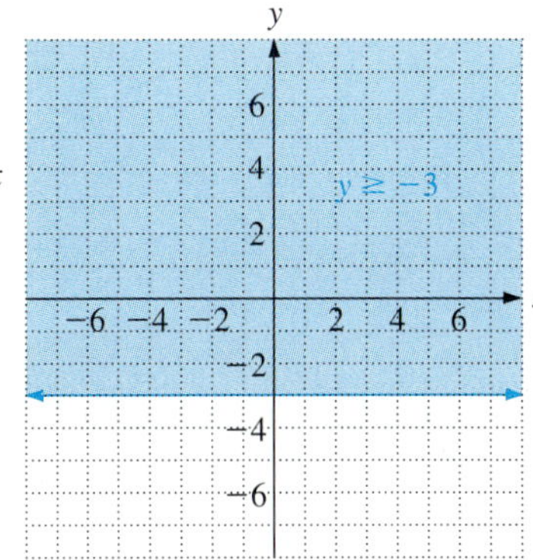

Practice Problem 3 Graph $y \geq -3$ on the coordinate system in the margin. ■

7.5 Exercises

Verbal and Writing Skills

1. Does it matter what point you use as your test point? Justify your response. Any convenient point suffices as long as the test point is not on the boundary line. If a true inequality results, shade the region that includes the test point. If a false inequality results, shade the region on the opposite side of the line from the test point.

2. Explain when to use a solid line or a dashed line when graphing linear inequalities in two variables. Use a solid line for inequalities containing $\leq$ or $\geq$. Use a dashed line for inequalities containing $<$ or $>$.

Graph the region described by the inequality.

3. $y > 2 - 3x$

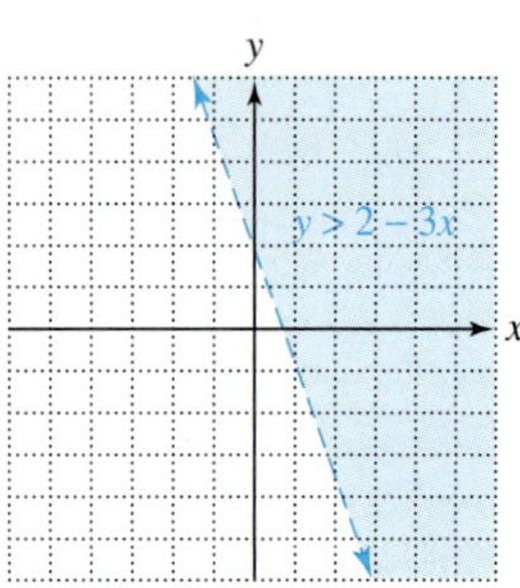

4. $y > 3x - 1$

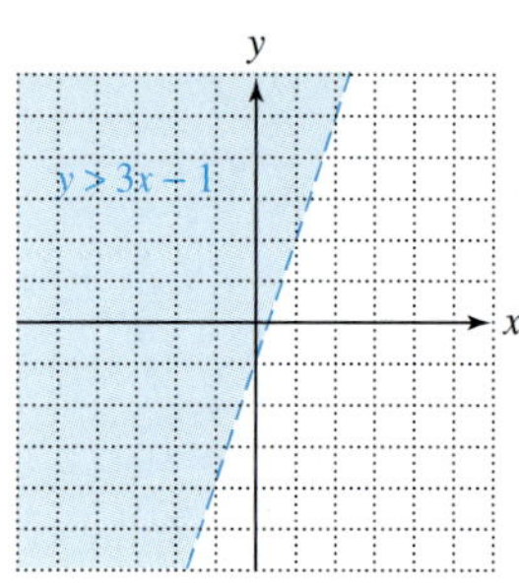

5. $2x - 3y < 6$

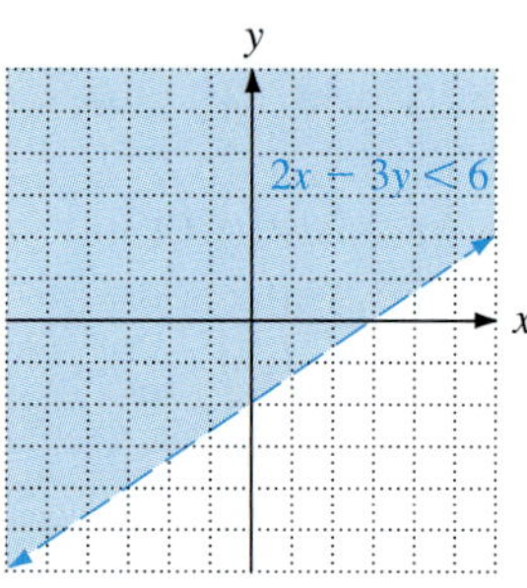

6. $3x + 2y < -6$

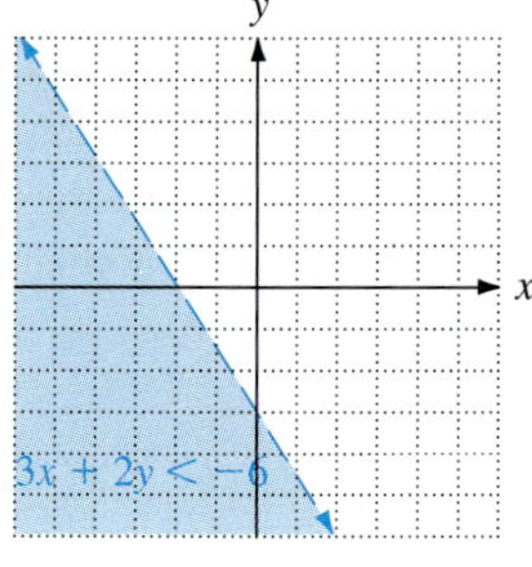

7. $x - y \geq 2$

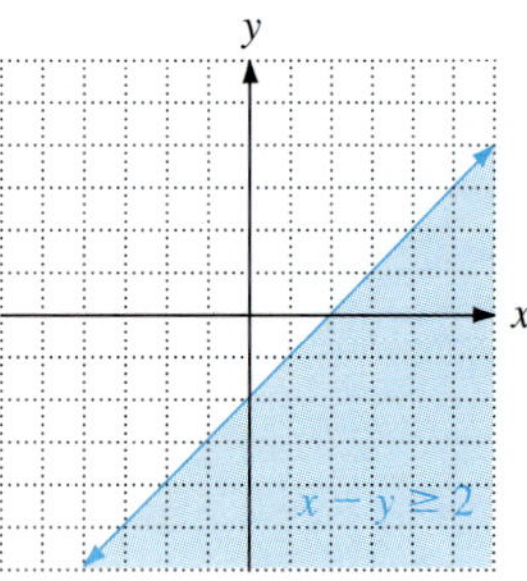

8. $-y + x \geq 3$

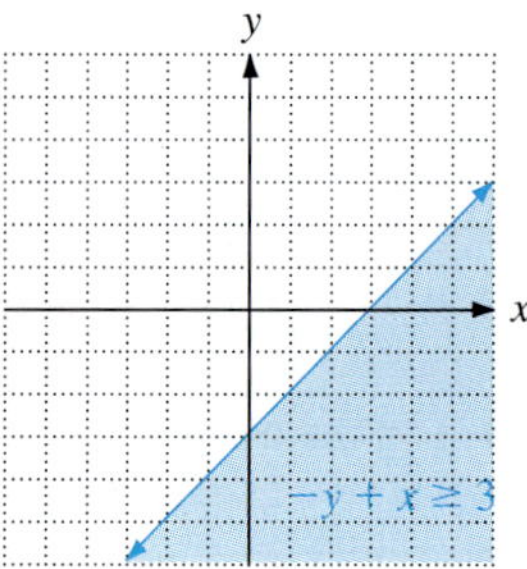

9. $y \geq 3x$

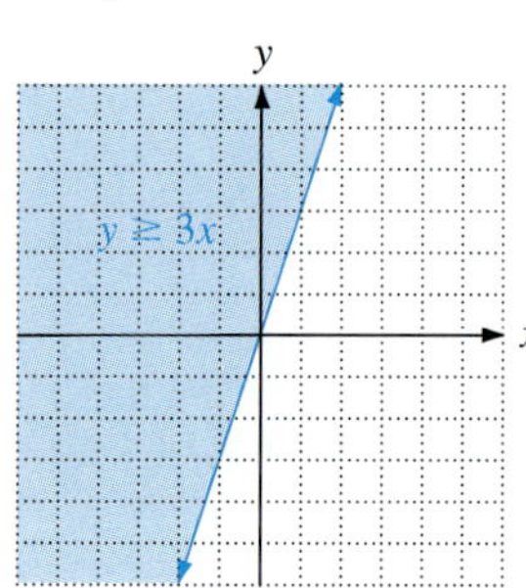

10. $y \geq -4x$

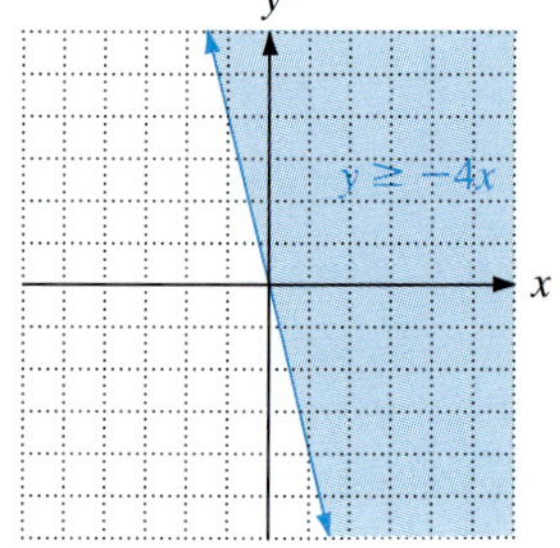

11. $y < -\frac{1}{2}x$

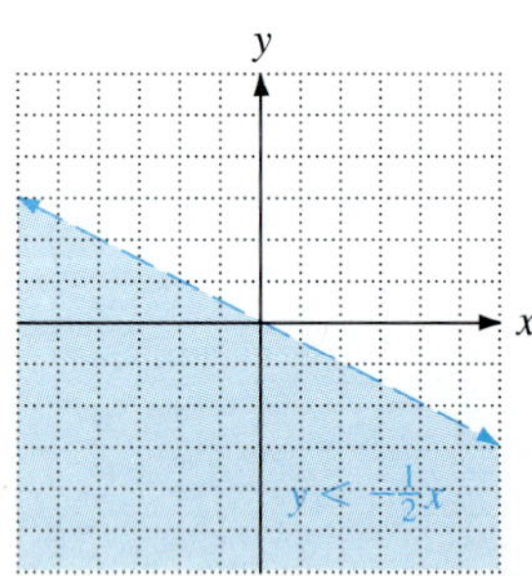

12. $y > \frac{1}{5}x$

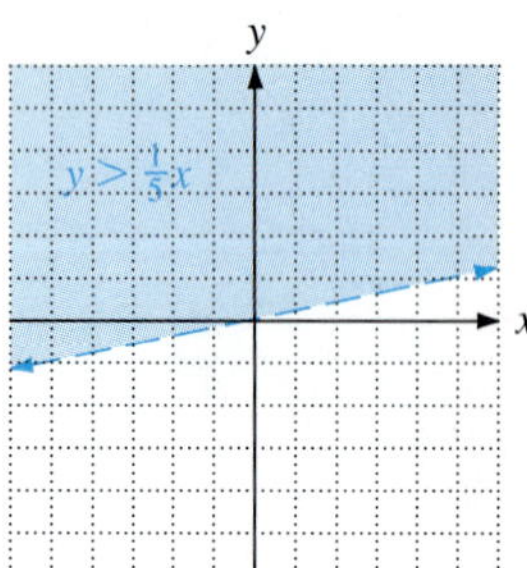

13. $x \leq -3$

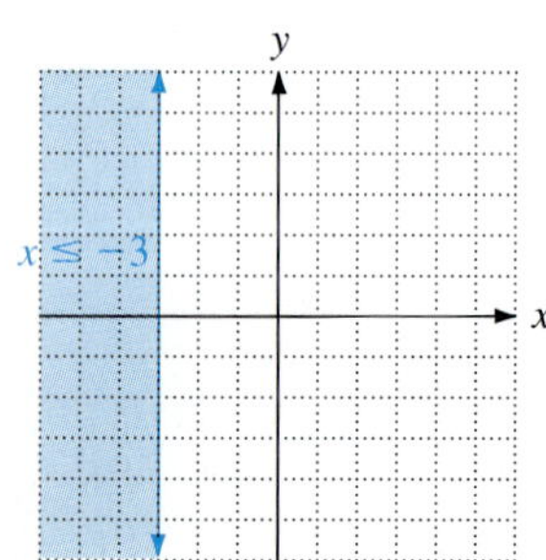

14. $y \geq -4$

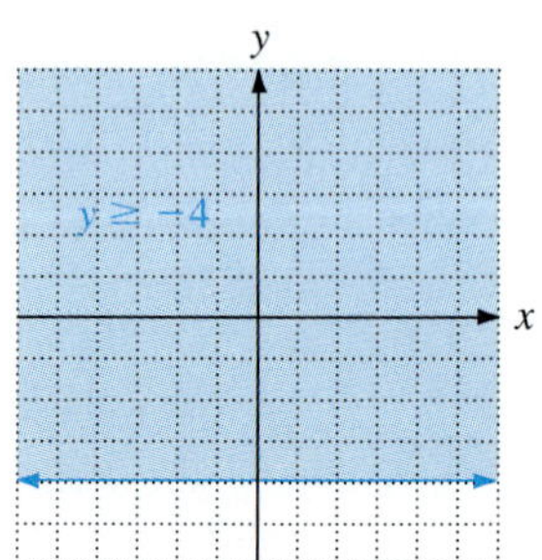

15. $3x - y + 1 \geq 0$

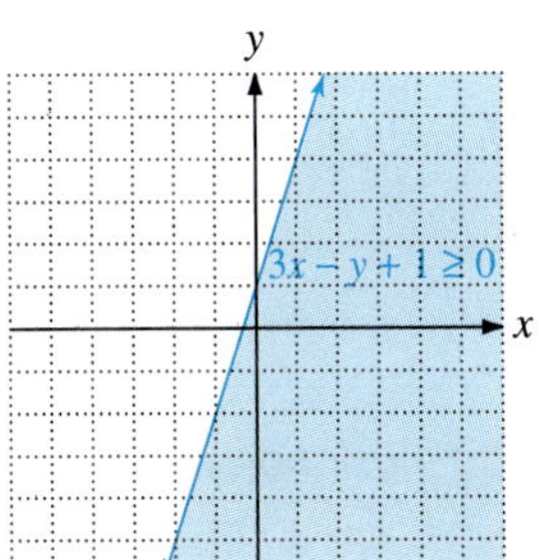

16. $2x + y - 5 \leq 0$

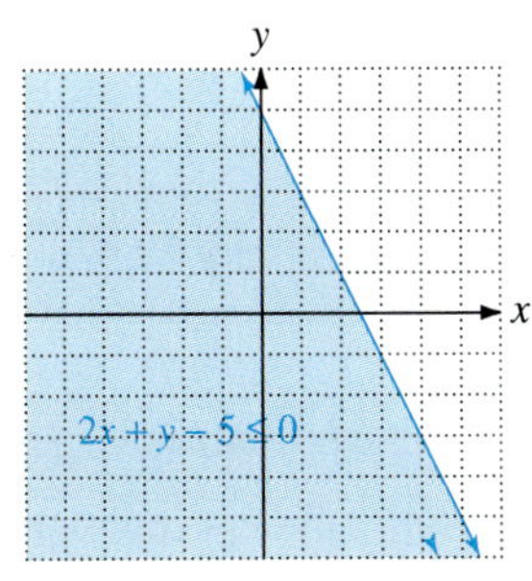

17. $3y \geq -2x$

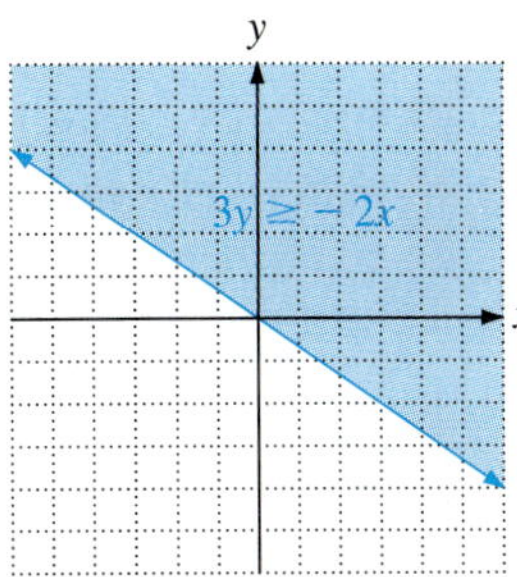

18. $2y < 5x$

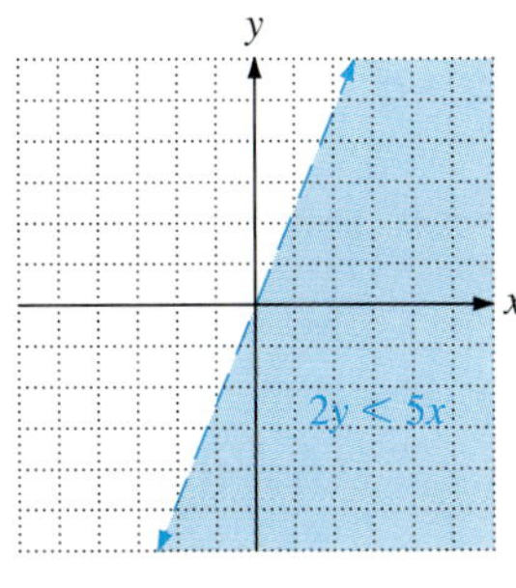

★**19.** $2x > 3 - y$

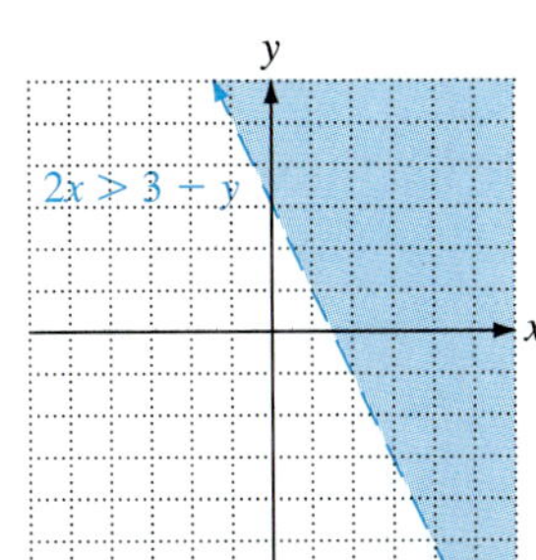

★**20.** $3x > 1 + y$

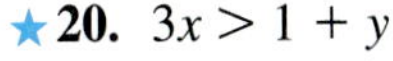

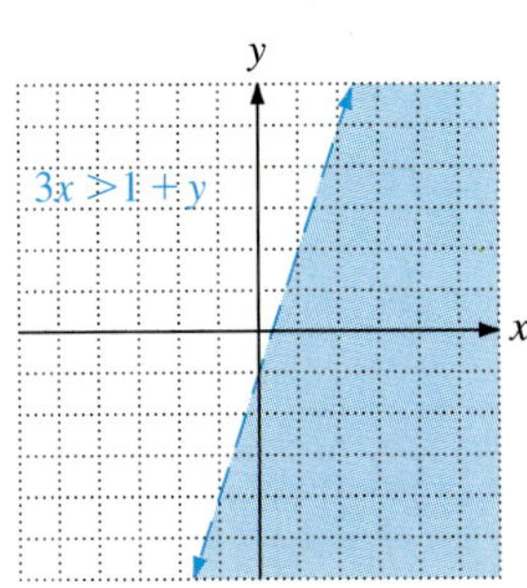

★**21.** $x > -2y$

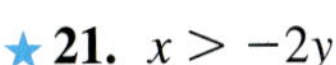

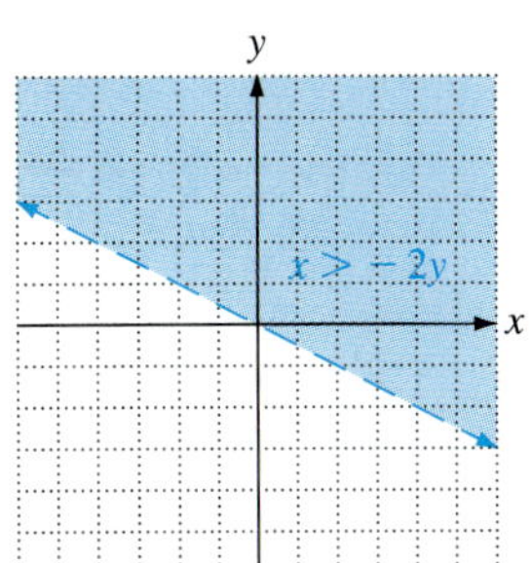

★**22.** $x < -3y$

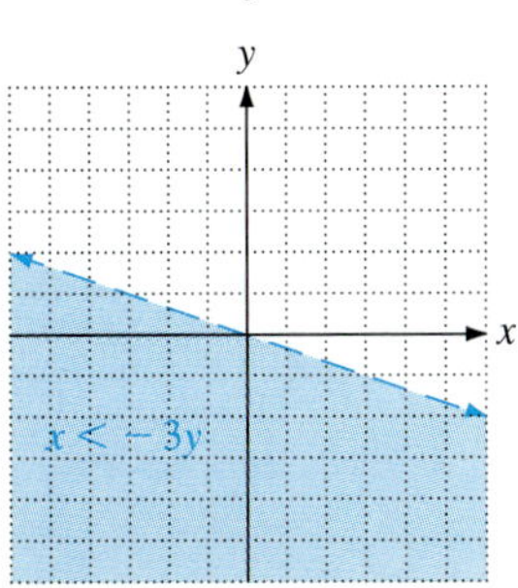

To Think About

23. Graph the inequality $3x + 3 > 5x - 3$ by graphing the lines $y = 3x + 3$ and $y = 5x - 3$ on the same coordinate plane.

(a) For which values of x does the graph of $y = 3x + 3$ lie above the graph of $y = 5x - 3$? $x < 3$

(b) For which values of x is $3x + 3 > 5x - 3$ true? $x < 3$

(c) Explain how the answer to (a) can help you to find the answer to (b). $3x + 3$ will be greater than $5x - 3$ for those values of x where the graph of $y = 3x + 3$ lies *above* the graph of $y = 5x - 3$.

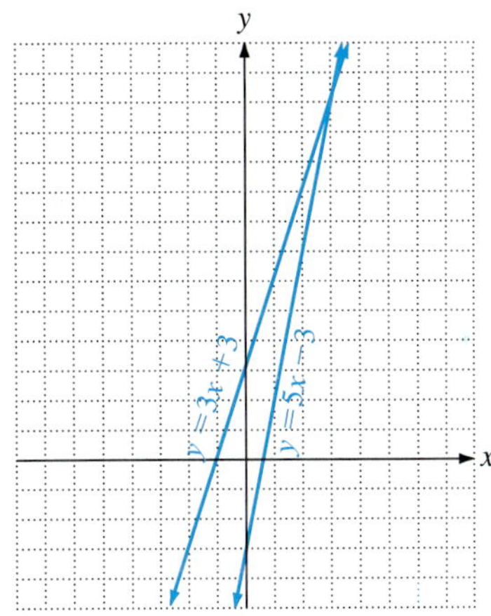

Cumulative Review Problems

24. What is the slope–intercept form of the equation for the line through $(3, 0)$ and $(6, -7)$?

$y = -\frac{7}{3}x + 7$

25. Solve for x. $2x^2 + 7x + 6 = 0$

$x = -\frac{3}{2}, x = -2$

26. Samantha is a great telemarketer. She has great success in selling subscriptions to a travel club. She has an average of 0.492 after having made 1000 telephone calls. (She sells a subscription 49.2% of the time.) Assuming that she makes a sale with each of her next telephone calls, what is the minimum number of calls she needs to make in order to raise her average to 0.500? If she makes a sale with each call, she will only need 16 more calls.

27. Angela has a new sales job that allows her to lease a car. Her lease budget is \$8400 for a 3-year period. She wants to lease a new Chrysler Cirrus at \$195 per month for the next 36 months. However, all mileage in excess of 36,000 miles driven over this 3-year period will be charged at \$0.15 per mile. How many miles above the 36,000-mile limit can Angela drive and still not exceed the leasing budget? She can drive an extra 9200 miles above the limit during the 3 years.

7.6 Functions

MathPro

Video 19

SSM

After studying this section, you will be able to:

1. *Understand and use the definition of a relation and a function.*
2. *Graph simple nonlinear equations.*
3. *Determine if a graph represents a function.*
4. *Use function notation.*

1 *Understanding and Using the Definitions of Relation and Function*

Thus far you have studied linear equations in two variables. You have seen that such an equation can be represented by a table of values, by the algebraic equation itself, and by a graph.

x	0	1	3
y	4	1	-5

$y = -3x + 4$

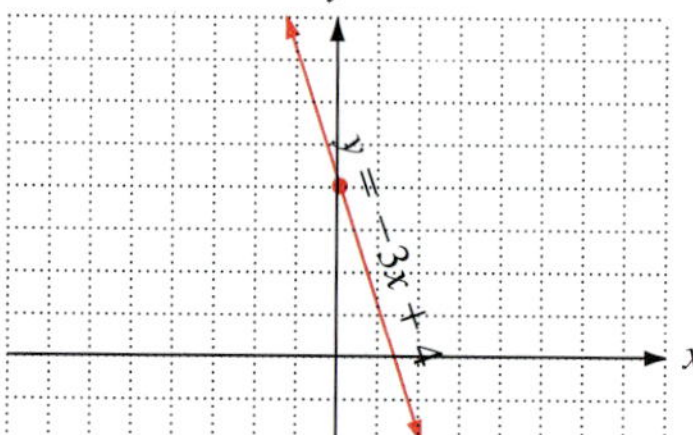

The solutions to the linear equation are all the ordered pairs that satisfy the equation (make the equation true). They are all the points that lie on the graph of the line. These ordered pairs can be represented in a table of values. Notice the relationship between the ordered pairs. We can choose any value for x. But once we have chosen a value for x, the value of y is determined. For the above equation, if x is 0, then y must be 4. We say that x is the *independent variable* and that y is the *dependent variable.*

Mathematicians call such a pairing of two values a *relation.*

Definition of a Relation

A **relation** is any set of ordered pairs.

All the first coordinates in each ordered pair make up the **domain** of the relation. All the second coordinates in each ordered pair make up the **range** of the relation. Notice that the definition of a relation is very broad. Some relations cannot be described by an algebraic expression. These relations may simply be a set of discrete ordered pairs.

EXAMPLE 1 State the domain and range of the following relation.

$$\{(5, 7), (9, 11), (10, 7), (12, 14)\}$$

The domain consists of all the first coordinates in the ordered pairs.

Domain

$$\{(5, 7), (9, 11), (10, 7), (12, 14)\}$$

Range

The range consists of all of the second coordinates in the ordered pairs. Thus

The domain is $\{5, 9, 10, 12\}$

The range is $\{7, 11, 14\}$ *We only list 7 once.*

Practice Problem 1 State the domain and range of the following relation.

$$\{(-3, -5), (3, 5), (0, -5), (20, 5)\}$$ ■

Some relations have a special property and are called *functions*. The relation $y = -3x + 4$ is a function. If you look at its table of values, no two different ordered pairs have the same first coordinate. If you look at its graph, you will notice that any vertical line intersects the graph only once. Sometimes we get a better idea if we look at some relations that are *not* functions. Look at $y^2 = x$.

x	1	4	9
y	± 1	± 2	± 3

$y^2 = x$

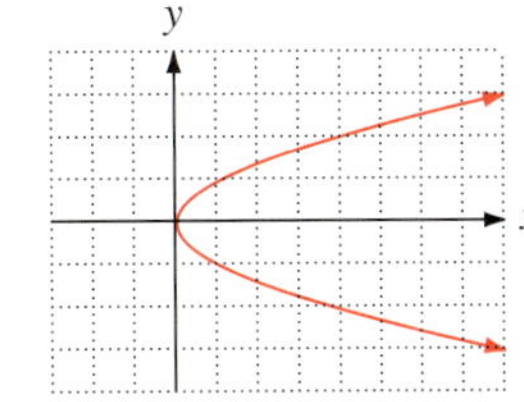

$\{(1, -1), (1, 1), (4, -2), (4, 2), (9, -3), (9, 3)\}$

Notice that, if x is 1, y could be 1 since $1^2 = 1$ or y could be -1 since $(-1)^2 = 1$. That is, the ordered pairs $(1, -1)$ and $(1, 1)$ both have the same first coordinate. $y^2 = x$ is not a function. What do you notice about the graph?

Definition of a Function

A **function** is a relation in which no two different ordered pairs have the same first coordinate.

EXAMPLE 2 Determine if the relation is a function or not a function.

(a) $\{(3, 9), (4, 16), (5, 9), (6, 36)\}$ **(b)** $\{(7, 8), (9, 10), (12, 13), (7, 14)\}$

(a) Look at the ordered pairs. No two ordered pairs have the same first coordinate. Thus (a) is a function. Note that the ordered pairs $(3, 9)$ and $(5, 9)$ have the same second coordinate, but this does not keep the relation from being a function.

(b) Look at the ordered pairs. Two different ordered pairs, $(7, 8)$ and $(7, 14)$, have the same first coordinate. Thus this relation is *not* a function.

Practice Problem 2 Determine if the relation is a function or not a function.

(a) $\{(-5, -6), (9, 30), (-3, -3), (8, 30)\}$

(b) $\{(60, 30), (40, 20), (20, 10), (60, 120)\}$ ■

Functions are what we hope to find when we analyze two sets of data. Look at the table of values that compares Celsius temperature with Fahrenheit temperature. Is there a relationship between degrees Fahrenheit and degrees Celsius? Is the relation a function?

Temperature

°F	23	32	41	50
°C	−5	0	5	10

Since every Fahrenheit temperature produces a unique Celsius temperature, we would expect this to be a function. We can verify our assumption by looking at the formula $C = \frac{5}{9}(F - 32)$ and its graph. The formula is a linear equation, and its graph is a line with a slope $\frac{5}{9}$ and y-intercept at about -17.8. The relation is a function. Notice that the *dependent variable* is C, since the value of C *depends* on the value of F. We say that F is the *independent variable*. The *domain* can be described as the set of possible values of the independent variable. The *range* is the set of corresponding values of the dependent variable. Scientists believe that the coldest temperature possible is approximately −273°C. They call this temperature **absolute zero.** Thus,

Domain = {all possible Fahrenheit temperatures from absolute zero to infinity}

Range = {all corresponding Celsius temperatures from −273°C to infinity}

EXAMPLE 3 Determine if the relation is a function or not a function. If it is a function, identify the domain and range.

(a) Circle

Radius	1	2	3	4	5
Area	3.14	12.56	28.26	50.24	78.5

(b) \$4000 loan at 8%

Time (yr)	1	2	3	4	5
Interest	\$320	\$665.60	\$1038.85	\$1441.96	\$1877.31

(a) Looking at the table, we see that no two different ordered pairs have the same first coordinate. The area of a circle is a function of the length of the radius.

Next we need to identify the independent variable in order to determine the domain. Sometimes it is easier to identify the dependent variable. Here we notice that the area of the circle depends on the length of the radius. Thus radius is the independent variable. Since a negative length does not make sense, the radius cannot be a negative number.

Domain = {all nonnegative real numbers}

Range = {all nonnegative real numbers}

(b) No two different ordered pairs have the same first coordinate. Interest is a function of time.

Since the amount of interest paid on a loan depends on the number of years (term of the loan), interest is the dependent variable and time is the independent variable. Negative numbers do not apply in this situation.

Domain = {all positive real numbers greater than 1}

Range = {all positive real numbers greater than \$320}

Practice Problem 3 Determine if the relation is a function or not a function. If it is a function, identify the domain and the range.

(a) 28 mpg at \$1.20 per gallon

Distance	0	28	42	56	70
Cost	0	1.20	1.80	2.40	3.00

(b) Store's inventory of shirts

Number of shirts	5	10	5	2	8
Price of shirt	\$20	\$25	\$30	\$45	\$50

■

To Think About Look at the following bus schedule. Determine if the relation is a function. Which is the independent variable? Explain your choice.

Bus Schedule

Bus Stop	Main St.	8th Ave.	42nd St.	Sunset Blvd.	Cedar Lane
Time	7:00	7:10	7:15	7:30	7:39

2 *Graphing Simple Nonlinear Equations*

Thus far in this chapter we have graphed linear equations in two variables. We now turn to graphing a few nonlinear equations. We will need to plot more than three points to get a good idea of what the graph will look like.

EXAMPLE 4 Graph $y = x^2$.

Begin by constructing a table of values. We select values for x and then determine by the equation the corresponding values of y. We will include negative values for x as well as positive values. We then plot the ordered pairs and connect the points with a smooth curve.

x	$y = x^2$	y
-2	$y = (-2)^2 = 4$	4
-1	$y = (-1)^2 = 1$	1
0	$y = (0)^2 = 0$	0
1	$y = (1)^2 = 1$	1
2	$y = (2)^2 = 4$	4

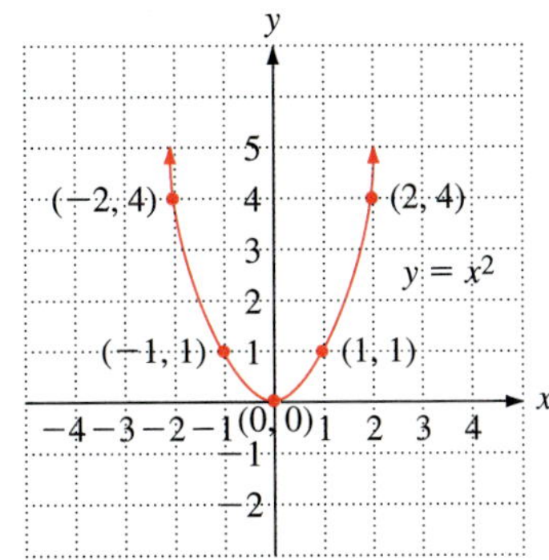

This type of curve is called a *parabola*. We will study the graph of these types of curves more extensively in Chapter 10.

Practice Problem 4 Graph $y = x^2 - 2$ on the coordinate system in the margin. ■

PRACTICE PROBLEM 4

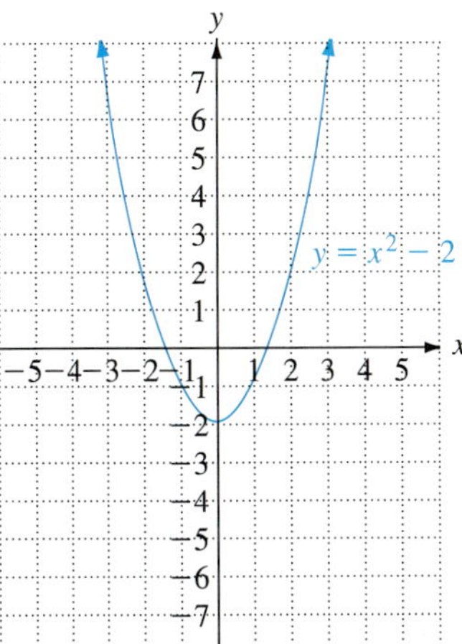

To Think About Compare the graph for Example 4 with the graph for Practice Problem 4. How are the graphs the same? How are the graphs different? Now compare the equations. How does the -2 in the equation $y = x^2 - 2$ affect the graph? What do you think the graph of $y = x^2 + 3$ will look like? Verify by drawing the graph.

Some equations are solved for x. Usually, in those cases we pick values of y and then obtain the corresponding values of x from the equation.

GRAPHING CALCULATOR

Graphing Nonlinear Equations

You can graph nonlinear equations solved for y using a graphing calculator. For example, graph $y = x^2 - 2$ on a graphing calculator using an appropriate window. Display:

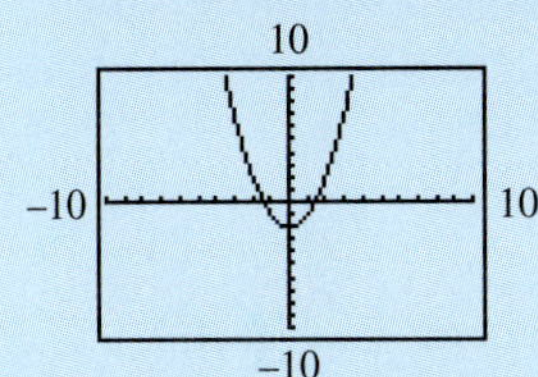

Try graphing $y = \frac{5}{x}$.

EXAMPLE 5 Graph $x = y^2 + 2$.

We will select a value of y and then substitute it into the equation to obtain x. For convenience in graphing, we will repeat the y column at the end so that it is easy to write the ordered pairs (x, y).

y	$x = y^2 + 2$	x	y
-2	$x = (-2)^2 + 2 = 4 + 2 = 6$	6	-2
-1	$x = (-1)^2 + 2 = 1 + 2 = 3$	3	-1
0	$x = (0)^2 + 2 = 0 + 2 = 2$	2	0
1	$x = (1)^2 + 2 = 1 + 2 = 3$	3	1
2	$x = (2)^2 + 2 = 4 + 2 = 6$	6	2

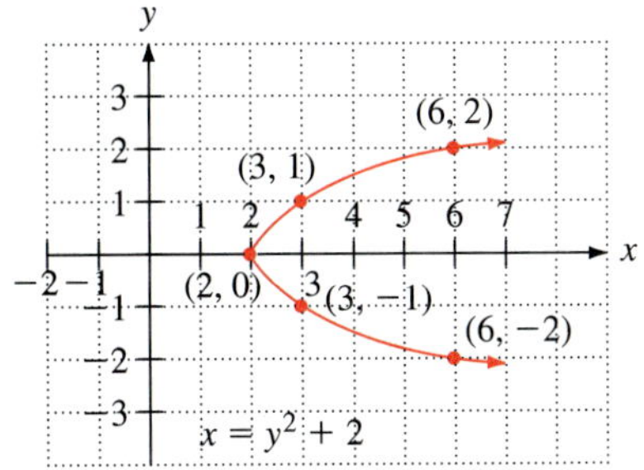

Practice Problem 5 Graph $x = y^2 - 1$ on the coordinate system in the margin. ■

PRACTICE PROBLEM 5

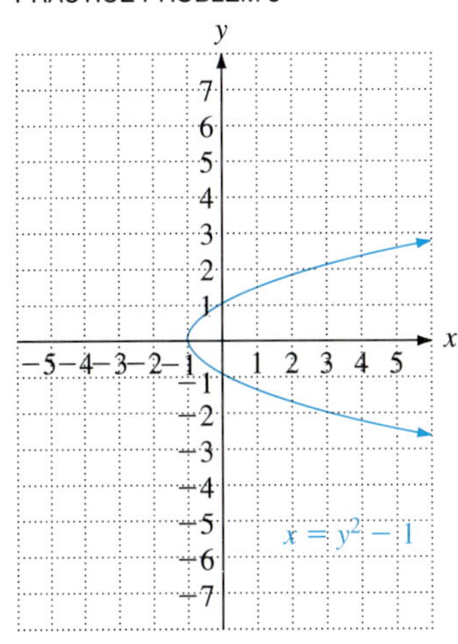

If the equation involves fractions with variables in the denominator, we must use extra caution. You may never divide by zero.

EXAMPLE 6 Graph $y = \frac{4}{x}$.

It is important to note that x cannot be zero because division by zero is not defined. $y = \frac{4}{0}$ is not allowed! Observe that when we draw the graph we get two separate branches of the curve that do not touch.

x	$y = \frac{4}{x}$	y
-4	$y = \frac{4}{-4} = -1$	-1
-2	$y = \frac{4}{-2} = -2$	-2
-1	$y = \frac{4}{-1} = -4$	-4
0	We cannot divide by zero.	There is no value.
1	$y = \frac{4}{1} = 4$	4
2	$y = \frac{4}{2} = 2$	2
4	$y = \frac{4}{4} = 1$	1

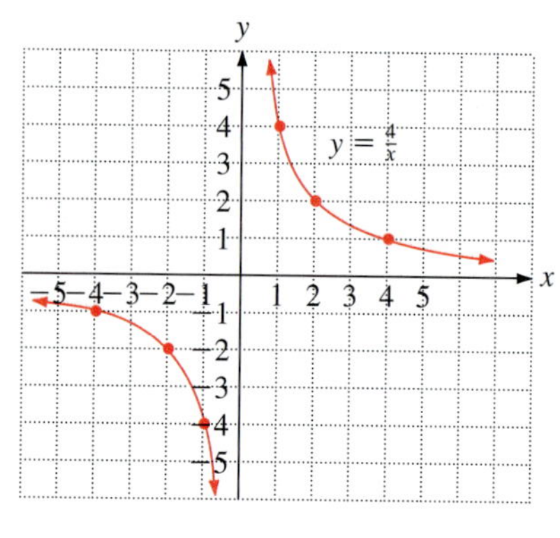

PRACTICE PROBLEM 6

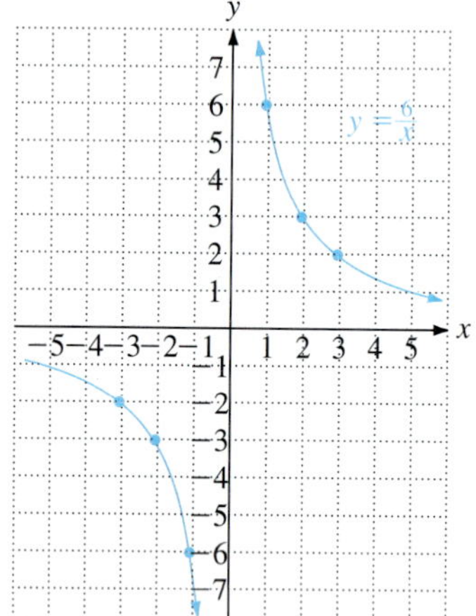

Practice Problem 6 Graph $y = \frac{6}{x}$ on the coordinate system in the margin. ■

3 *Determining If a Graph Represents a Function*

Can we tell from a graph if it is the graph of a function? Recall that a function must have the property that no two different ordered pairs have the same first coordinate. That is, each value of x must have a separate unique value of y. Look at the graph of $y = x^2$ in Example 4. Each x value has a unique y value. Look at the graph of $x = y^2 + 2$ in Example 5. At $x = 3$ there are two y values, 1 and -1. In fact, for every x value greater than 2 there are two y values. $x = y^2 + 2$ is not a function.

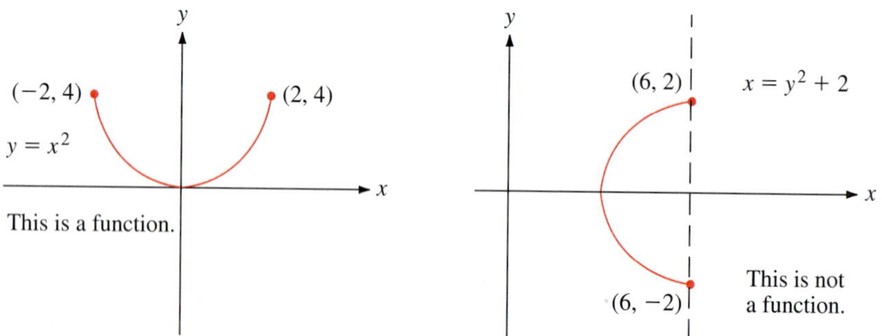

Observe that we can draw a vertical line through (6, 2) and (6, -2). Any graph that is not a function will have at least one region in which a vertical line will cross the curve more than once.

Vertical Line Test

If a vertical line can intersect the graph of a relation more than once, the relation is not a function.

EXAMPLE 7 Determine if each of the following is the graph of a function.

(a)

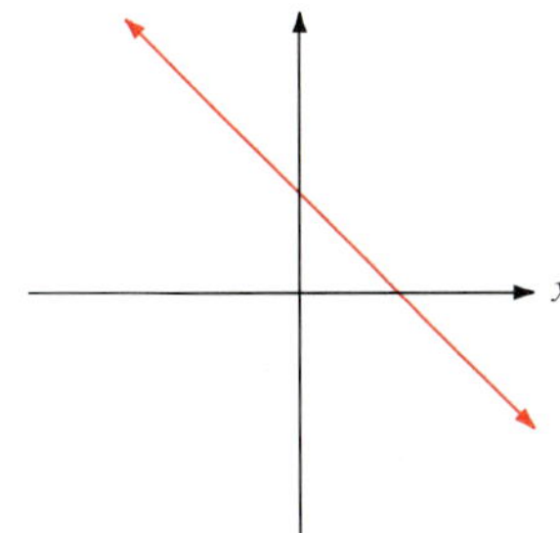

(b)

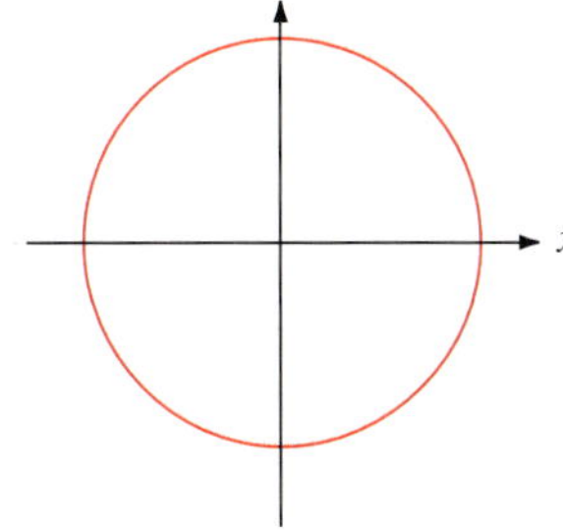

(c)

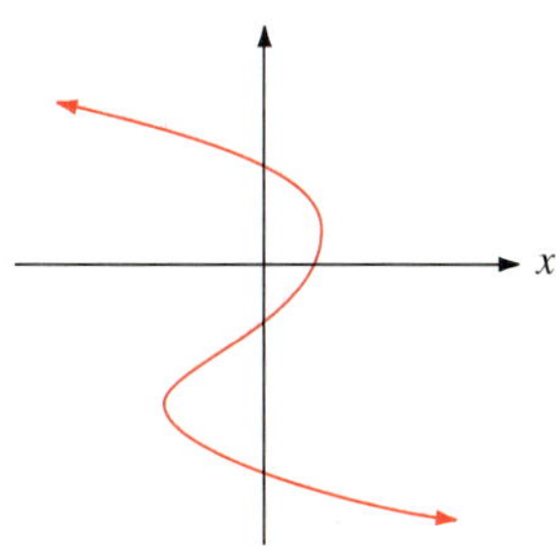

(a) The graph of the straight line is a function. A vertical line can only cross this straight line in one location.

(b) and **(c)** Each of these graphs is not the graph of a function. In each case a vertical line can cross the curve in more than one place.

(a)

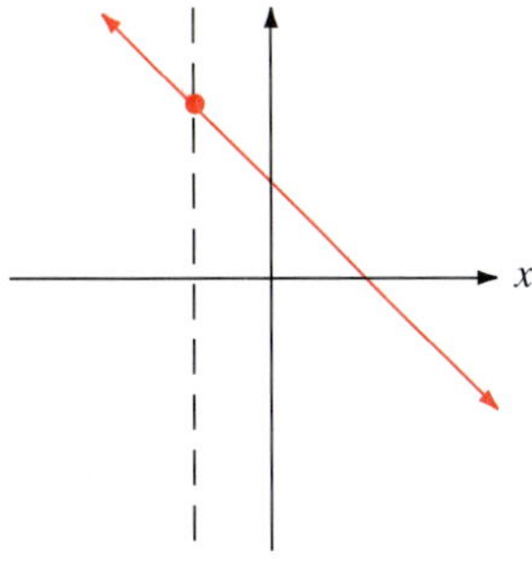

A function

(b)

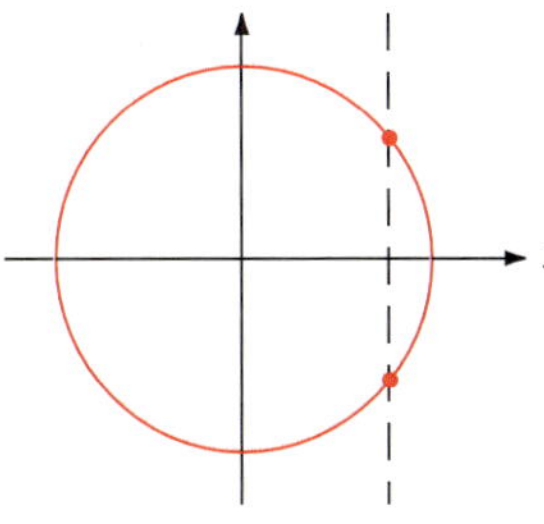

Not a function

(c)

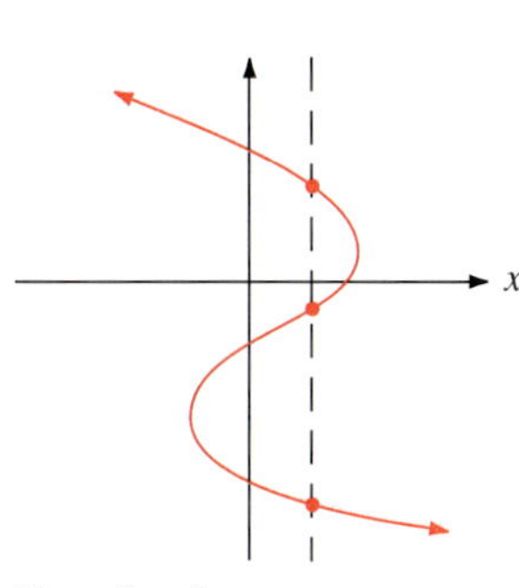

Not a function

Practice Problem 7 Determine if each of the following is the graph of a function.

(a)

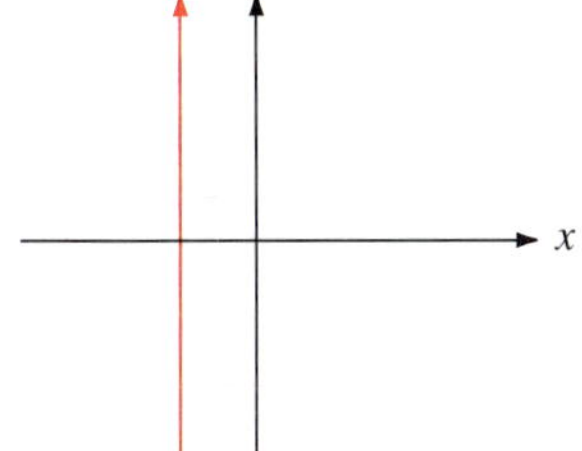

(b)

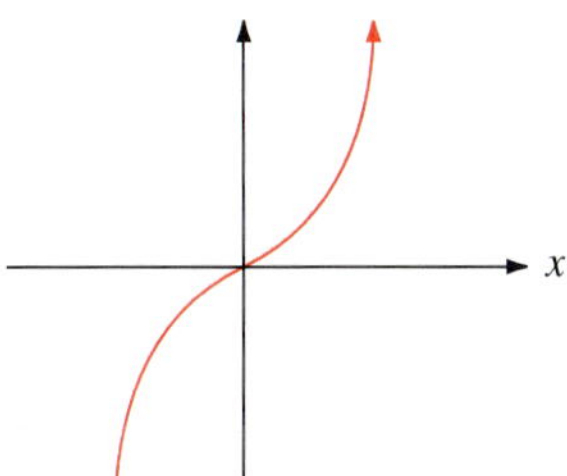

(c)

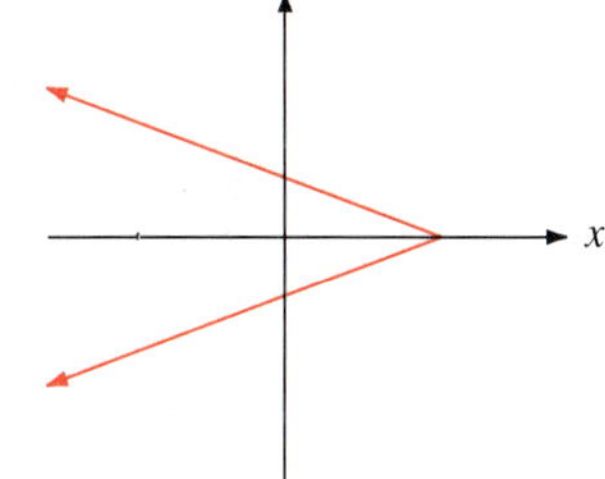

4 *Using Function Notation*

We have seen that an equation like $y = 2x + 7$ is a function. For each value of x, the equation assigns a unique value to y. We could say "y is a function of x." This statement can be symbolized by using the function notation $y = f(x)$. Thus you can avoid the y variable completely and describe an expression like $2x + 7$ as a function of x by writing $f(x) = 2x + 7$.

WARNING Be careful. The notation $f(x)$ does not mean f multiplied by x.

EXAMPLE 8 If $f(x) = 2x + 7$, then find $f(3)$.

$f(3)$ means that we want to find the value of the function $f(x)$ when $x = 3$. In this example $f(x)$ is $2x + 7$. Thus we substitute 3 for x wherever x occurs in the equation and evaluate.

$$f(x) = 2x + 7$$

$$f(3) = 2(3) + 7 = 6 + 7 = 13$$

$$= 13$$

Practice Problem 8 If $f(x) = -3x + 4$, then find $f(5)$. ■

EXAMPLE 9 If $f(x) = 3x^2 - 4x + 5$, find:

(a) $f(2)$ **(b)** $f(-2)$ **(c)** $f(4)$ **(d)** $f(0)$

(a) $f(2) = 3(2)^2 - 4(2) + 5 = 3(4) - 4(2) + 5 = 12 - 8 + 5 = 9$

(b) $f(-2) = 3(-2)^2 - 4(-2) + 5 = 3(4) - 4(-2) + 5 = 12 + 8 + 5 = 25$

(c) $f(4) = 3(4)^2 - 4(4) + 5 = 3(16) - 4(4) + 5 = 48 - 16 + 5 = 37$

(d) $f(0) = 3(0)^2 - 4(0) + 5 = 3(0) - 4(0) + 5 = 0 - 0 + 5 = 5$

Practice Problem 9 If $f(x) = -2x^2 + 3x - 8$, find:

(a) $f(-1)$ **(b)** $f(2)$ **(c)** $f(-3)$ **(d)** $f(0)$ ■

When evaluating a function it is helpful to place parentheses around the value that is being substituted for x. Taking the time to do this will minimize sign errors in your work.

7.6 Exercises

Verbal and Writing Skills

1. What are the three ways you can use to describe a function?
You can describe a function using a table of values, an algebraic equation, or a graph.

2. What is the difference between a function and a relation?
A relation is any set of ordered pairs. A function is a relation in which no two different ordered pairs have the same first coordinate.

3. The domain of a function is the set of __possible__ __values__ of the __independent__ variable.

4. The range of a function is the set of __corresponding__ __values__ of the __dependent__ variable.

5. How can we tell if a graph is the graph of a function? If a vertical line can intersect the graph more than once, the relation is not a function.

(a) *Find the domain and range of each relation.* **(b)** *Determine if each relation is a function.*

6. $\{(9, 3), (4, 5), (7, 8), (1, 3)\}$
Domain = $\{1, 4, 7, 9\}$
Range = $\{3, 5, 8\}$
Function

7. $\{(1, 8), (3, -2), (5, 8), (7, 7)\}$
Domain = $\{1, 3, 5, 7\}$
Range = $\{-2, 7, 8\}$
Function

8. $\{(\frac{1}{2}, \frac{1}{2}), (-10, \frac{1}{2}), (7, \frac{1}{4}), (\frac{1}{2}, \frac{1}{4})\}$
Domain = $\{-10, \frac{1}{2}, 7\}$
Range = $\{\frac{1}{4}, \frac{1}{2}\}$
Not a function

9. $\{(\frac{1}{2}, 5), (\frac{1}{4}, 10), (\frac{3}{4}, 6), (\frac{1}{2}, 6)\}$
Domain = $\{\frac{1}{4}, \frac{1}{2}, \frac{3}{4}\}$
Range = $\{5, 6, 10\}$
Not a function

10. $\{(8.6, 3), (9.2, 6), (8.6, -3)\}$
Domain = $\{8.6, 9.2\}$
Range = $\{-3, 3, 6\}$
Not a function

11. $\{(10, 1), (1, 10), (8, 3), (3, 8)\}$
Domain = $\{1, 3, 8, 10\}$
Range = $\{1, 3, 8, 10\}$
Function

12. $\{(12, 1), (14, 3), (1, 12), (9, 12)\}$
Domain = $\{1, 9, 12, 14\}$
Range = $\{1, 3, 12\}$
Function

13. $\{(5.6, 8), (5.8, 6), (6, 5.8), (5, 6)\}$
Domain = $\{5, 5.6, 5.8, 6\}$
Range = $\{5.8, 6, 8\}$
Function

14. $\{(1, 20), (2, 20), (3, 20), (1, 40)\}$
Domain = $\{1, 2, 3\}$
Range = $\{20, 40\}$
Not a function

15. $\{(40, 2), (50, 30), (30, 60), (40, 80)\}$
Domain = $\{30, 40, 50\}$
Range = $\{2, 30, 60, 80\}$
Not a function

Graph the equation. Compare the graph with its algebraic equation. Is the graph what you might expect? If not, check your work.

16. $y = x^2 + 3$

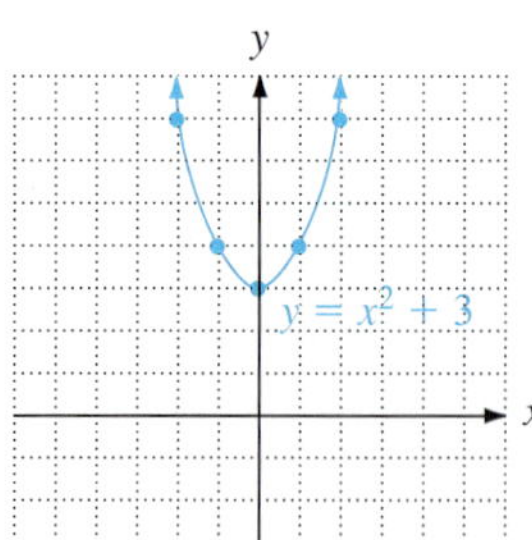

17. $y = x^2 - 1$

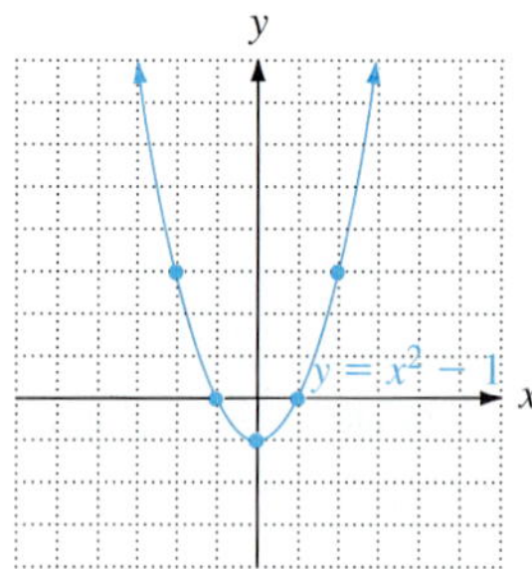

18. $y = 2x^2$

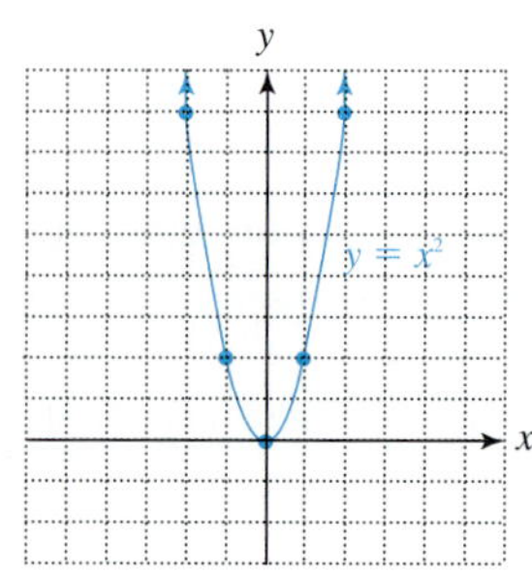

19. $y = \frac{1}{2}x^2$

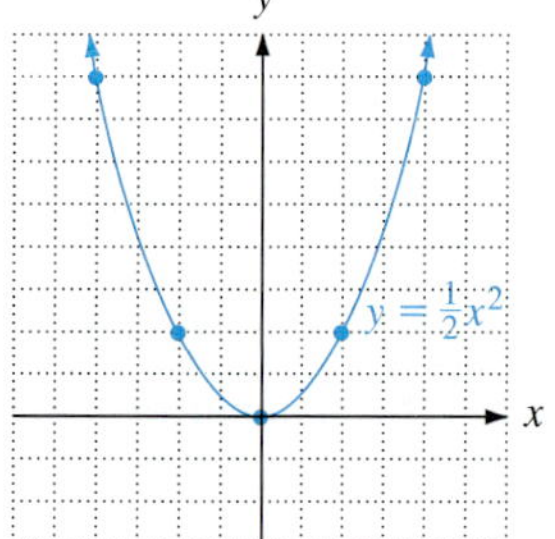

20. $x = y^2$

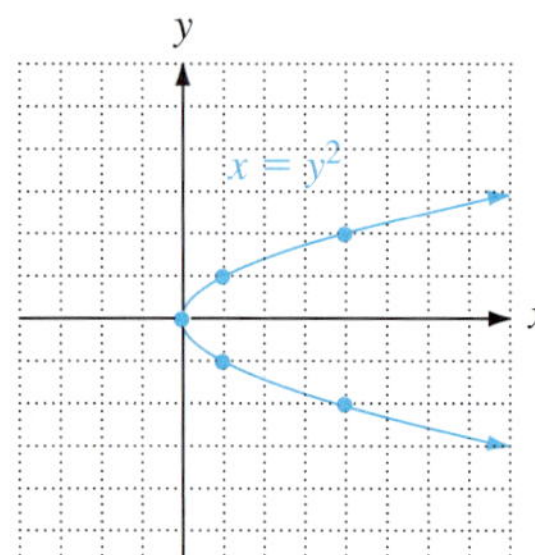

21. $x = y^2 - 3$

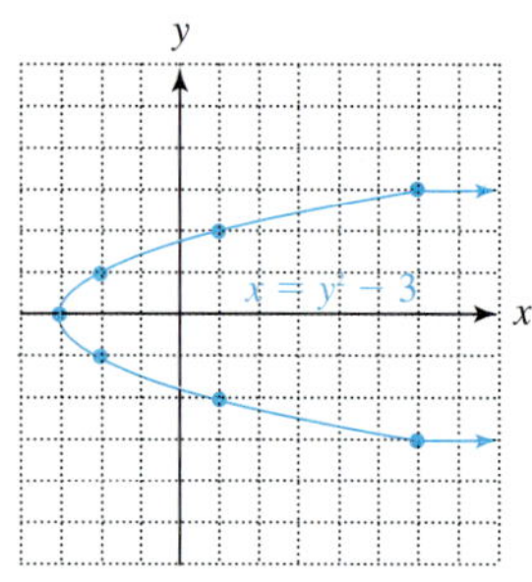

22. $x = y^2 + 1$

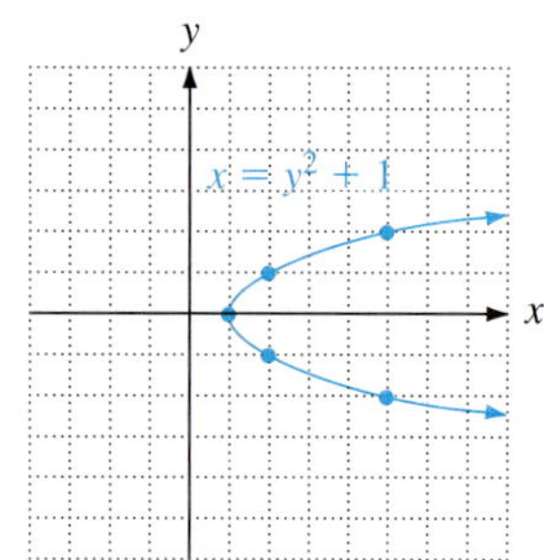

23. $x = 2y^2$

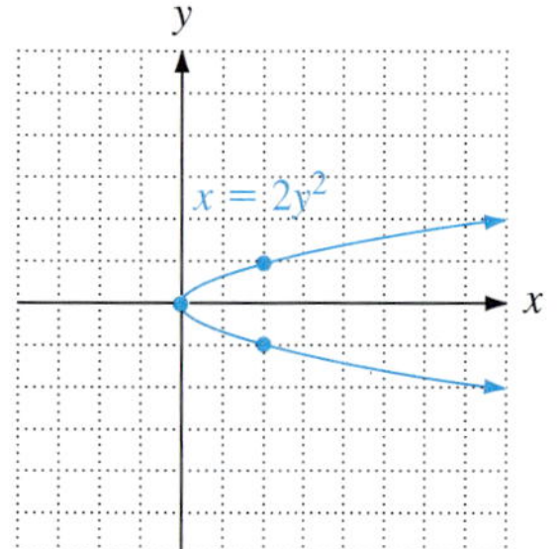

24. $y = \frac{2}{x}$

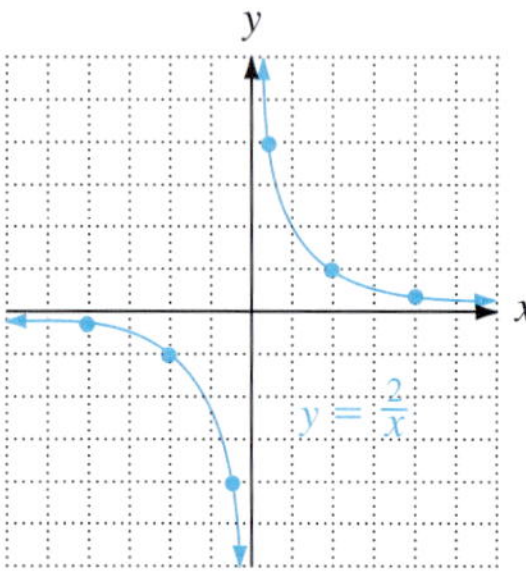

25. $y = -\frac{2}{x}$

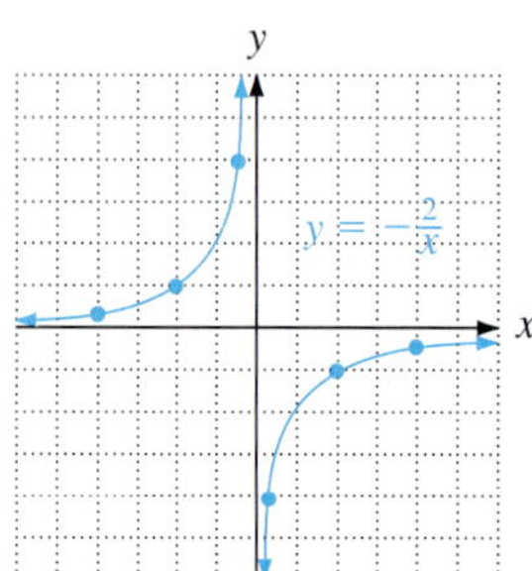

26. $y = \frac{1}{x}$

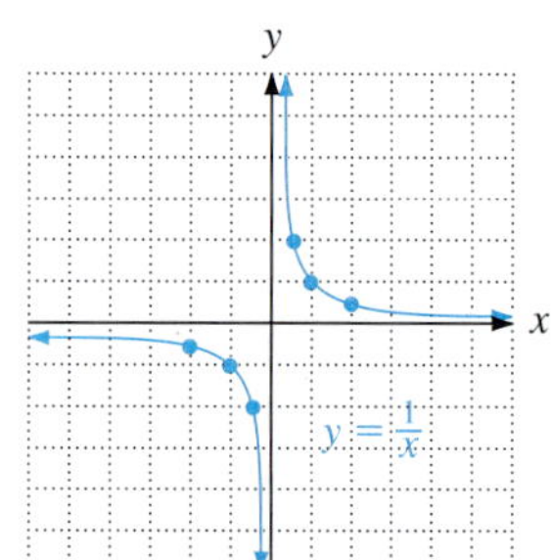

27. $y = -\frac{1}{x}$

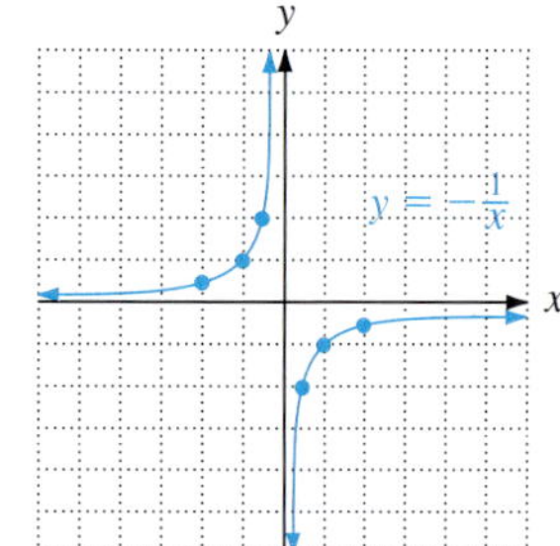

★**28.** $y = (x + 1)^2$

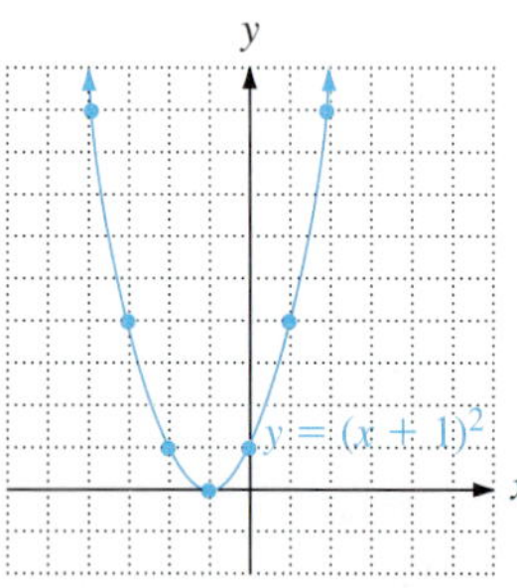

★**29.** $x = (y - 2)^2$

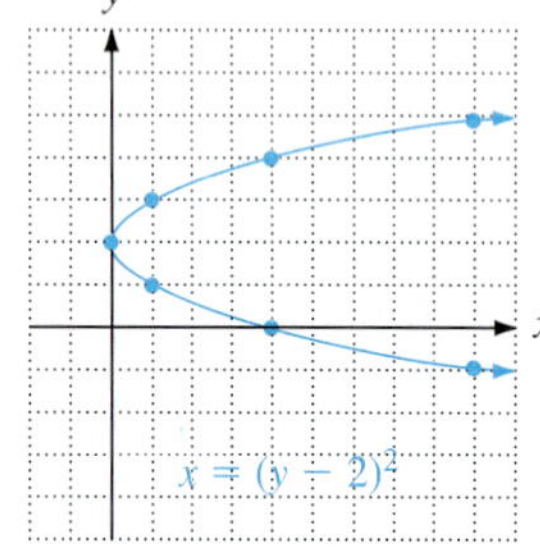

★**30.** $y = \dfrac{4}{x - 2}$

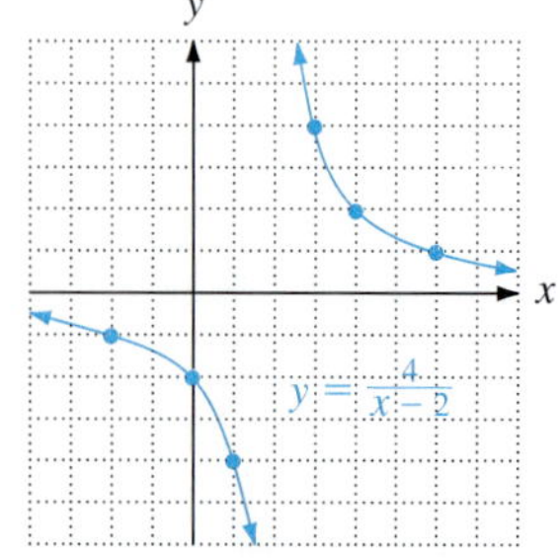

★**31.** $x = \dfrac{2}{y + 1}$

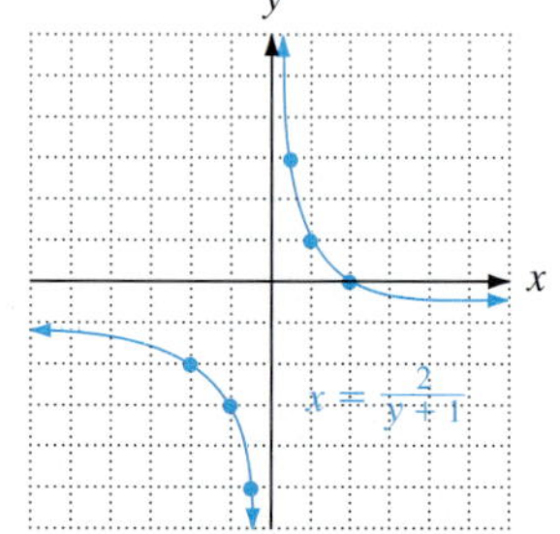

Determine if each relation is a function.

32. Function

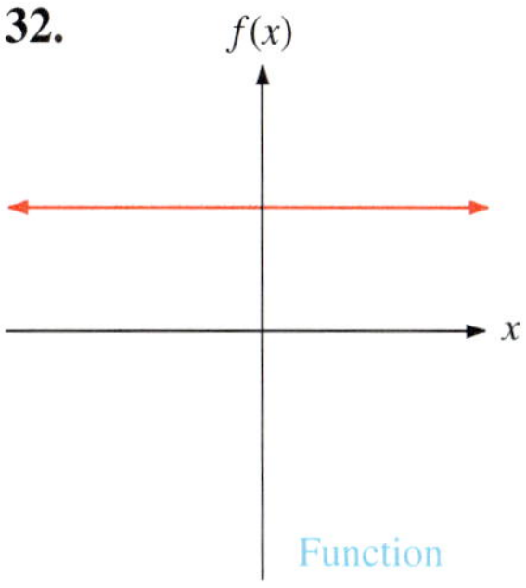

33. Function

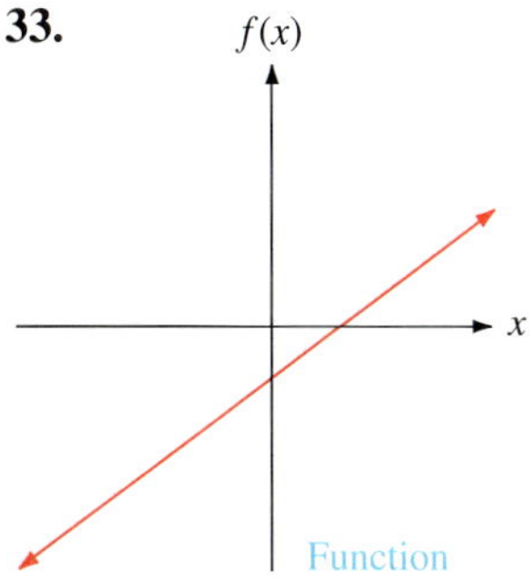

34. Not a function

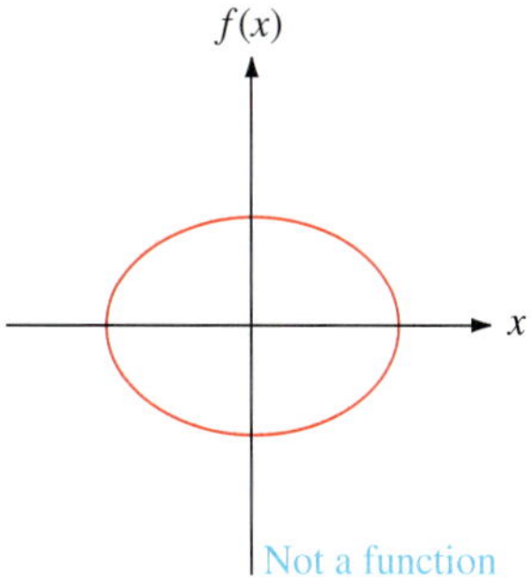

35. Not a function

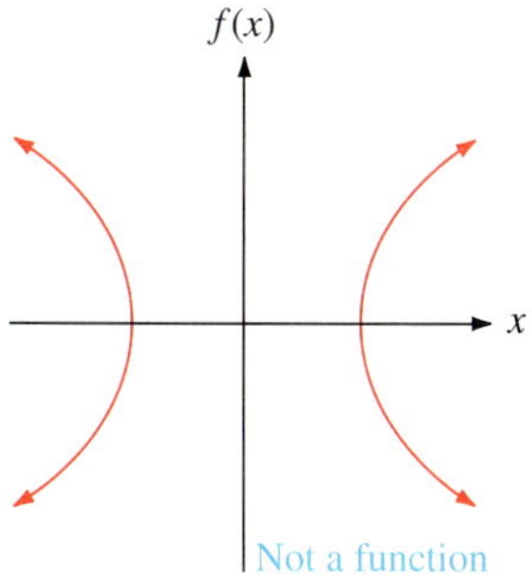

36. Function

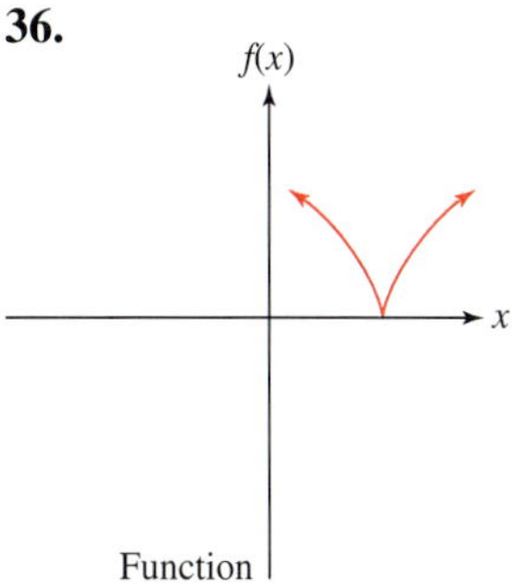

37. Function

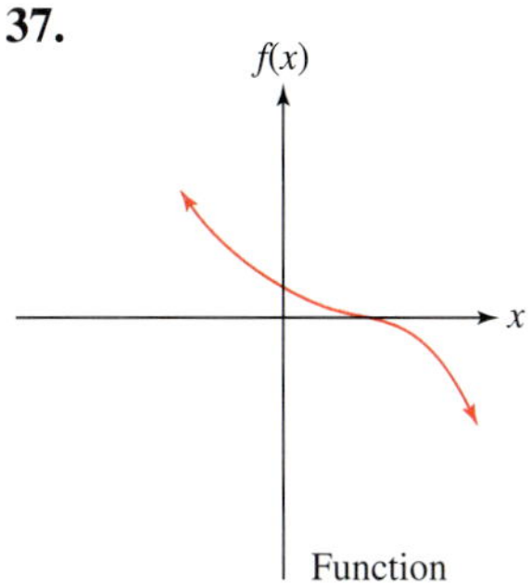

38. Not a function

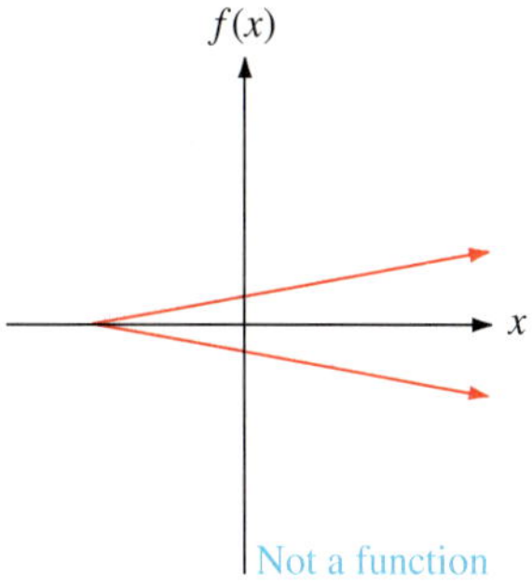

39. Not a function

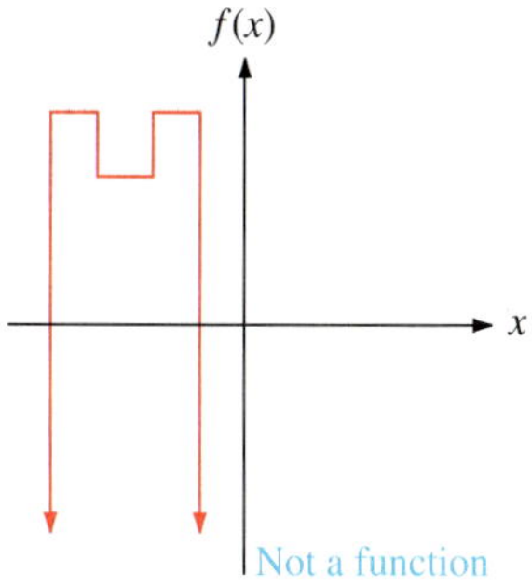

Given the following functions, find the indicated values.

40. $f(x) = 2x - 6$ **(a)** $f(-1)$ -8 **(b)** $f(2)$ -2 **(c)** $f(3)$ 0

41. $f(x) = 5x + 4$ **(a)** $f(-3)$ -11 **(b)** $f(0)$ 4 **(c)** $f(4)$ 24

42. $f(x) = 4 - 8x$ **(a)** $f(2)$ -12 **(b)** $f(0.5)$ 0 **(c)** $f(-4)$ 36

43. $f(x) = 1 - 10x$ **(a)** $f(0.5)$ -4 **(b)** $f(-0.1)$ 2 **(c)** $f(3)$ -29

44. $f(x) = x^2 + 3x + 4$ **(a)** $f(2)$ 14 **(b)** $f(0)$ 4 **(c)** $f(-3)$ 4

45. $f(x) = x^2 + 2x - 3$ **(a)** $f(-1)$ -4 **(b)** $f(0)$ -3 **(c)** $f(3)$ 12

46. $g(x) = 2 - 6x^2$ **(a)** $g(0)$ 2 **(b)** $g(4)$ -94 **(c)** $g(-2)$ -22

47. $g(x) = 3 - 4x^2$ **(a)** $g(0)$ 3 **(b)** $g(2)$ -13 **(c)** $g(-3)$ -33

48. $f(x) = 2x + 3 - \frac{12}{x}$ **(a)** $f(-2)$ 5 **(b)** $f(1)$ -7 **(c)** $f(4)$ 8

49. $f(x) = x - 2 + \frac{8}{x}$ **(a)** $f(-2)$ -8 **(b)** $f(1)$ 7 **(c)** $f(4)$ 4

Cumulative Review Problems

Simplify each of the following

50. $(5x^2 - 6x + 2) + (-7x^2 - 8x + 4)$
$-2x^2 - 14x + 6$

51. $(-2x^2 - 3x + 8) - (x^2 + 4x - 3)$
$-3x^2 - 7x + 11$

52. $(x^2 + 2x + 1) + (x^2 - 3x + 2)$
$2x^2 - x + 3$

53. $(6x^2 + 10x - 5) \div (3x - 1)$
$2x + 4 + \frac{-1}{3x - 1}$

Putting Your Skills to Work

CURRENCY EXCHANGE GRAPH

If you were planning a trip to Paris you would be interested in the exchange rate from U.S. to French currency. As of Nov. 19, 1996, the exchange rate was 100 U.S. dollars = 508 French francs. (This rate varies somewhat. For example on Nov. 19, 1995, the exchange rate was 100 U.S. dollars = 458 French francs, but on Nov. 19, 1994, the exchange rate was 100 U.S. dollars = 534 French francs.)

We can graph the equation $f = cd$ where c = the daily conversion constant of how much one U.S. dollar is worth that day in French francs, d = the number of U.S. dollars and f = the number of French francs. On November 19, 1996, the equation would be $f = 5.08d$.

If you investigate the graph you will find that it allows you to make a quick estimate.

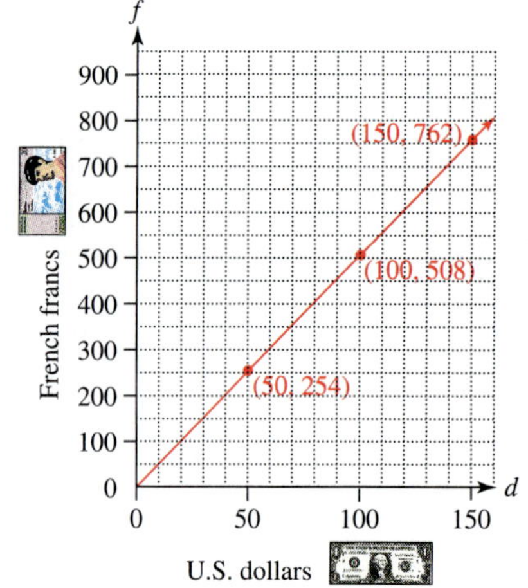

PROBLEMS FOR INDIVIDUAL INVESTIGATION AND STUDY

Use the graph above to answer the following questions.

1. You see a painting by a sidewalk artist that you want to purchase for 700 French francs. What is the approximate equivalent in U.S. dollars?
any answer close to 138 dollars

2. You have 60 dollars and are eating in a very nice but somewhat expensive restaurant just outside of Paris. Your friends have warned you that you may pay from 200 to 250 French francs to eat there. Do you have enough money? What is the most that the meal would cost you in U.S. dollars if your friends have made an accurate prediction? yes; about 50 dollars

PROBLEMS FOR COOPERATIVE INVESTIGATION AND GROUP STUDY

Together with some members of your class answer the following questions.

Use an appropriate Web site, a reference book in the library, or the exchange rate published in major daily papers such as *The New York Times, The Wall Street Journal,* or *The Boston Globe* to assist you in completing the following.

3. Find the exchange rate for one dollar in terms of British pounds. Create an equation to graph the relationship of dollars to pounds. Graph the equation. Use your answer to find the range in British pounds it will cost if you buy a sweater in England in the range of 60 to 90 U.S. dollars.
Answers will vary. As of Nov. 1996, equation is $p = 0.6d$.

4. Find the exchange rate for one dollar in terms of German marks. Create an equation to graph the relationship of dollars to marks. Graph the equation. Use your answer to find the range in German marks it will cost if you purchase a German camera in the range of 320 to 380 U.S. dollars.
Answers will vary. As of Nov. 1996, equation is $m = 1.5d$.

INTERNET CONNECTIONS: Go to ``http://www.prenhall.com/tobey'' to be connected

Site: Olsen & Associates Currency Converter or a related site

5. Use the Currency Converter to find the dollar value of 15,000 Japanese yen on January 1 of each year beginning with 1992. Choose an appropriate scale for plotting the values. Use the year (t) as the horizontal axis and the value (v) as the vertical axis. Connect the points by a smooth curve.

6. Use your graph to estimate the dollar value of ¥15,000 on July 1, 1995. Then use the Currency Converter to find out if your guess was accurate.

Chapter Organizer

Topic	Procedure	Examples
Plotting points, p. 396.	To plot (x, y): **1.** Begin at the origin. **2.** If x is positive, move to the right along the x-axis. If x is negative, move to the left along the x-axis. **3.** If y is positive, move up. If y is negative, move down.	To plot $(-2, 3)$: $(-2, 3)$ 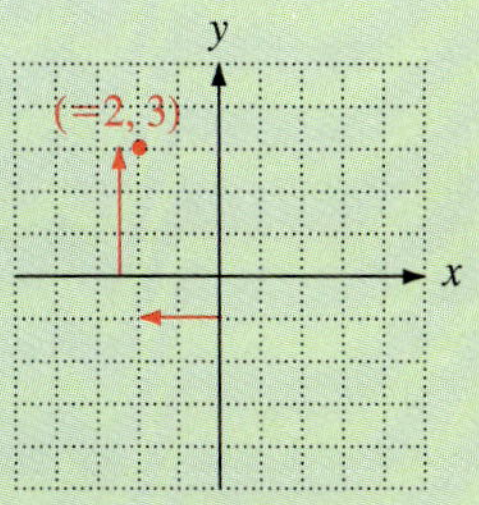
Graphing straight lines, p. 404.	An equation of the form $Ax + By = C$ has a graph that is a straight line. To graph such an equation, plot any three points; two give the line and the third checks it. (Where possible, use the x- and y-intercepts.)	Graph $3x + 2y = 6$. x | y: 0 | 3; 4 | −3; 2 | 0 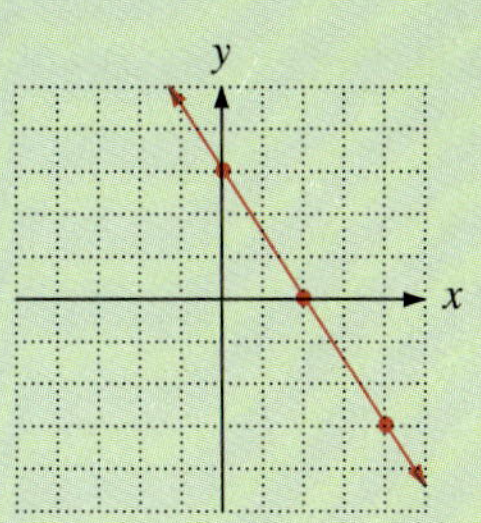
Finding the slope given two points, p. 414.	Nonvertical lines passing through distinct points (x_1, y_1) and (x_2, y_2) have slope $m = \dfrac{y_2 - y_1}{x_2 - x_1}$ The slope of a horizontal line is 0. The slope of a vertical line is undefined.	What is the slope of the line through (2, 8) and (5, 1)? $m = \dfrac{1 - 8}{5 - 2} = -\dfrac{7}{3}$
Finding the slope and y-intercept of a line given the equation, p. 418.	**1.** Place the equation in the form $y = mx + b$. **2.** The slope is m. **3.** The y-intercept is b.	Find the slope and y-intercept of $3x - 4y = 8$ $-4y = -3x + 8$ $y = \frac{3}{4}x - 2$ The slope is $\frac{3}{4}$. The y-intercept is -2.
Finding the equation of a line given the slope and y-intercept, p. 420.	The slope–intercept form of the equation of a line is $y = mx + b$ The slope is m and the y-intercept is at $(0, b)$.	Find the equation of the line through (0, 7) with slope $m = 3$. $y = 3x + 7$
Graphing a line using slope and y-intercept, p. 420.	**1.** Plot the y-intercept. **2.** Starting from $(0, b)$, plot a second point using the slope. slope $= \dfrac{\text{rise}}{\text{run}}$ **3.** Connect the two points.	Graph $y = -4x + 1$. Slope $= -4$ or $\frac{-4}{1}$ 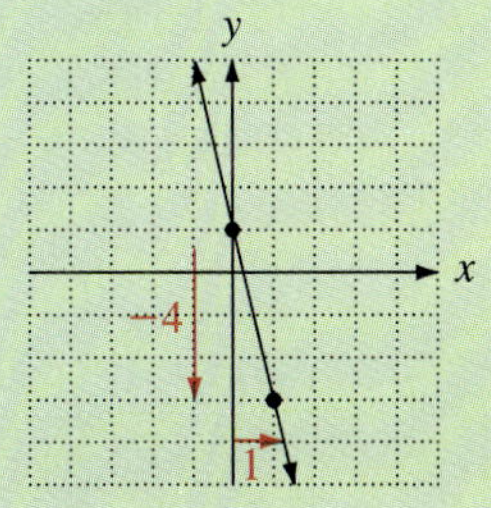 −4 1

Chapter Organizer

Topic	Procedure	Examples
Finding the slope of parallel and perpendicular lines, p. 421.	Parallel lines have the same slope. Perpendicular lines have slopes whose product is -1.	Line q has a slope of 2. The slope of a line parallel to q is 2. The slope of a line perpendicular to q is $-\frac{1}{2}$.
Finding the equation of a line through a point with a given slope, p. 428.	**1.** Substitute the known values in the equation $y = mx \pm b$. **2.** Solve for b. **3.** Use the values of m and b to write the general equation.	Find the equation of the line through (3, 2) with slope $m = \frac{4}{5}$. $y = mx + b$ $2 = \frac{4}{5}(3) + b$ $2 = \frac{12}{5} + b$ $-\frac{2}{5} = b$ The equation is $y = \frac{4}{5}x - \frac{2}{5}$.
Finding the equation of a line through two points, p. 429.	**1.** Find the slope. **2.** Use the procedure when given a point and the slope.	Find the equation of the line through (3, 2) and (13, 10). $m = \frac{y_2 - y_1}{x_2 - x_1} = \frac{10 - 2}{13 - 3} = \frac{8}{10} = \frac{4}{5}$ We choose to use the point (3, 2). $y = mx + b$ $2 = \frac{4}{5}(3) + b$ $2 = \frac{12}{5} + b$ $-\frac{2}{5} = b$ The equation is $y = \frac{4}{5}x - \frac{2}{5}$.
Writing the equation of a line given the graph, p. 429.	**1.** Identify the y-intercept. **2.** Find the slope. $\text{slope} = \frac{\text{change in } y}{\text{change in } x}$	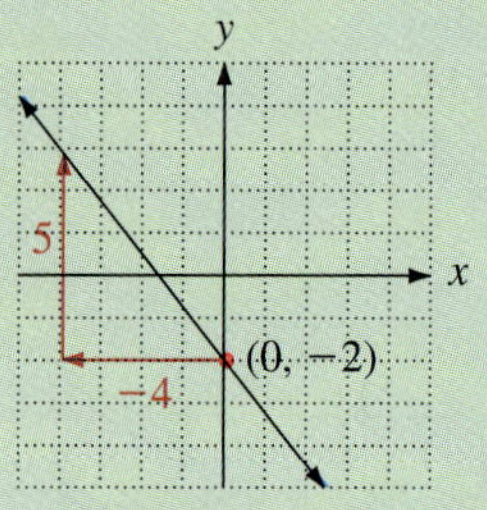 y x 5 -4 (0, -2) The equation is $y = -\frac{5}{4}x - 2$
Graphing linear inequalities, p. 432.	**1.** Graph as if it were an equation. If the inequality symbol is $>$ or $<$, use a dashed line. If the inequality symbol is $\geq$ or $\leq$, use a solid line. **2.** Look for a test point. The easiest test point is (0, 0). **3.** Substitute the coordinates of the test point into the inequality. **4.** If it is a true statement, shade the side of the line containing the test point. If it is a false statement, shade the side of the line that does *not* contain the test point.	Graph $y \geq 3x + 2$. Graph the line $y = 3x + 2$. Use a solid line. Test (0, 0). $0 \geq 3(0) + 2$ $0 \geq 2$ false Shade the side of the line that does not contain (0, 0). 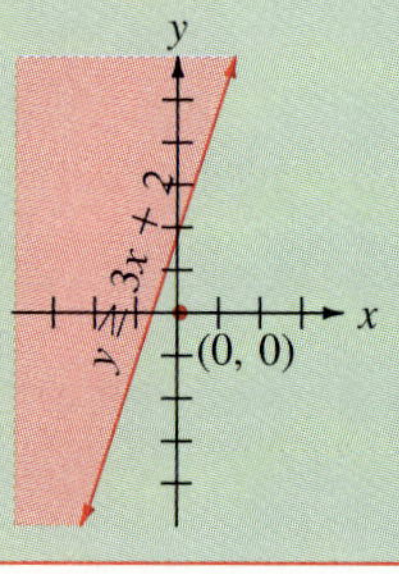 y x $y \geq 3x + 2$ (0, 0)

Chapter Organizer

Topic	Procedure	Examples
Determining if a relation is a function, p. 436.	A function is a relation in which no two different ordered pairs have the same first coordinates.	Is this relation a function? $\{(5, 7), (3, 8), (5, 10)\}$ It is *not* a function since in (5, 7) and (5, 10) we have two different ordered pairs with an x-coordinate of 5.
Determining if a graph represents a function, p. 440.	If a vertical line can intersect the graph of a relation more than once, the relation is not a function.	Is this graph a function? Yes. A vertical line can only intersect it once.

Chapter 7 Review Problems

7.1

1. Plot the following points.
   A: $(2, -3)$ B: $(-1, 0)$ C: $(3, 2)$ D: $(-2, -3)$

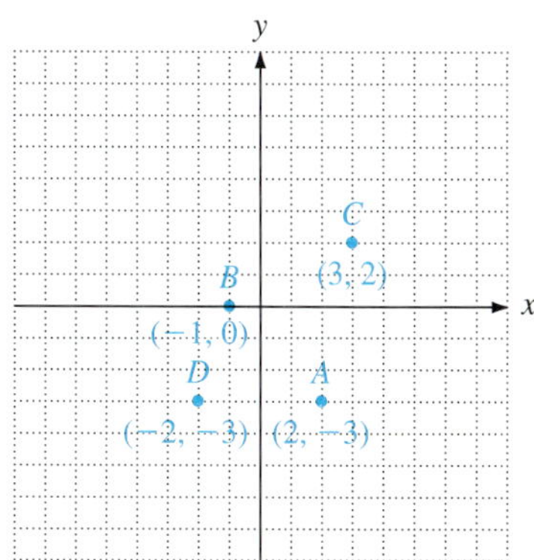

2. Give the coordinates of each point.

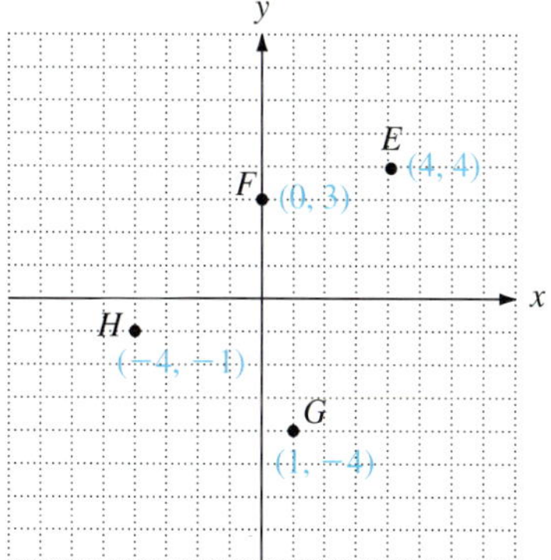

Complete the ordered pairs so that each is a solution to the given equation.

3. $y = 3x - 5$
   (a) $(0, \ -5\ )$ **(b)** $(3, \ 4\ )$

4. $2x + 5y = 12$
   (a) $(1, \ 2\ )$ **(b)** $(\ -4\ , 4)$

5. $x = 6$
   (a) $(\ 6\ , -1)$ **(b)** $(\ 6\ , 3)$

7.2

6. Graph $3y = 2x + 6$.

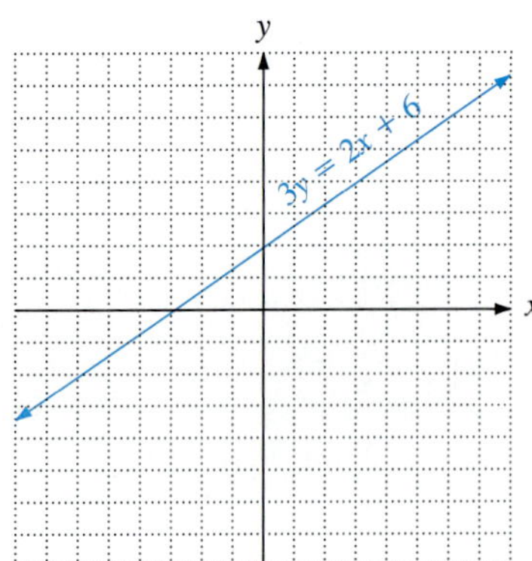

7. Graph $5y + x = -15$.

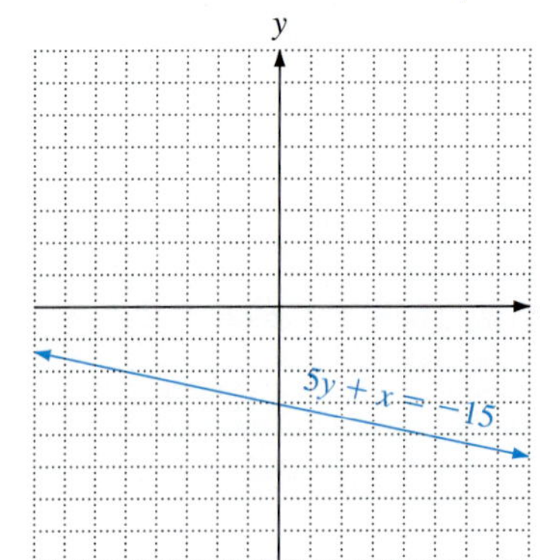

8. Graph $3x + 5y = 15 + 3x$.

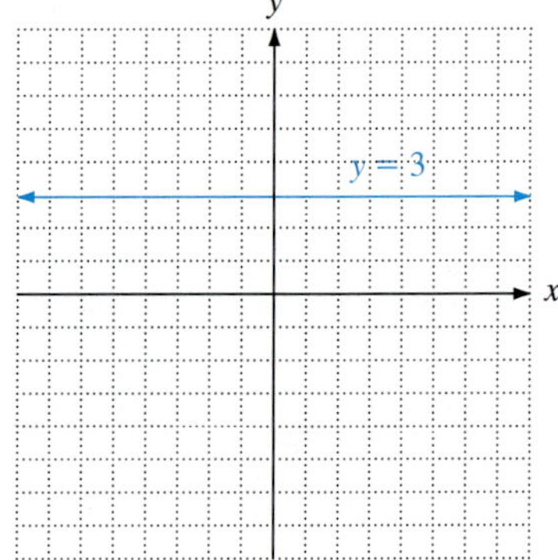

7.3

9. Find the slope of a line passing through $(5, -3)$ and $(2, -\frac{1}{2})$.

$m = -\dfrac{5}{6}$

10. Find the slope and y-intercept of the line $5x - 8y + 12 = 0$.

$m = \dfrac{5}{8}$, $b = \dfrac{3}{2}$

11. Write an equation of the line with slope $-\frac{1}{2}$ and y-intercept of 3.

$y = -\dfrac{1}{2}x + 3$

12. Write an equation of the line with slope 0 and y-intercept of 7. $y = 7$

13. A line has a slope of $-\frac{2}{3}$. What is the slope of a line perpendicular to that line?

$\dfrac{3}{2}$

14. A line has a slope of $\frac{1}{7}$. What is the slope of a line parallel to that line?

$\dfrac{1}{7}$

15. Graph $y = -\frac{1}{2}x + 3$.

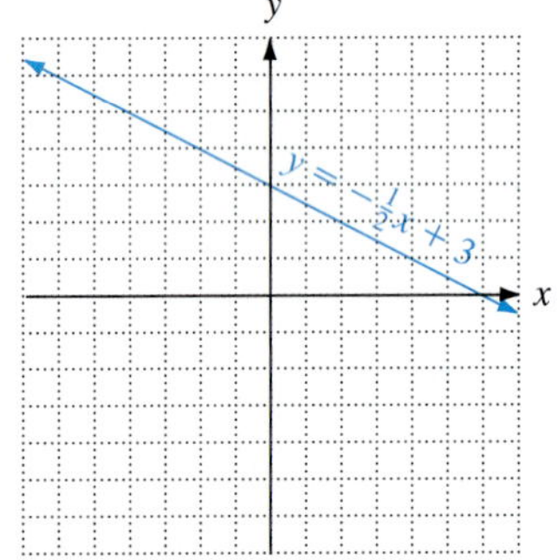

16. Graph $2x - 3y = -12$.

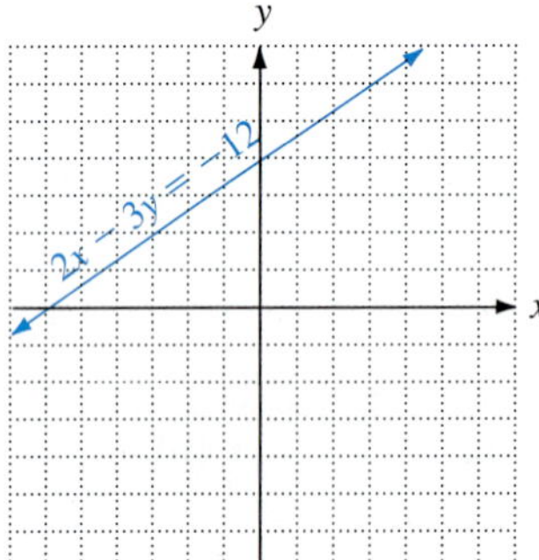

17. Graph $5x + 2y = 20 + 2y$.

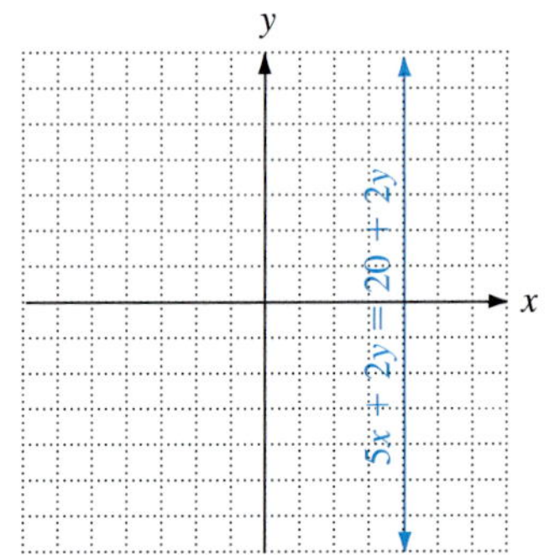

7.4

18. Write an equation of the line passing through $(5, 6)$ having a slope of 2. $y = 2x - 4$

19. Write an equation of the line passing through $(3, -4)$ having a slope of -6. $y = -6x + 14$

20. Write an equation of the line passing through $(6, -3)$ having a slope of $\frac{1}{3}$.

$y = \dfrac{1}{3}x - 5$

21. Write an equation of the line passing through (5, 2) and (−3, 6). $y = -\frac{1}{2}x + \frac{9}{2}$

22. Write an equation of the line passing through (6, 1) and (2, −3). $y = x - 5$

23. Write an equation of the line passing through (0, 5) and (−1, 5). $y = 5$

Write the equation of each graph.

24.

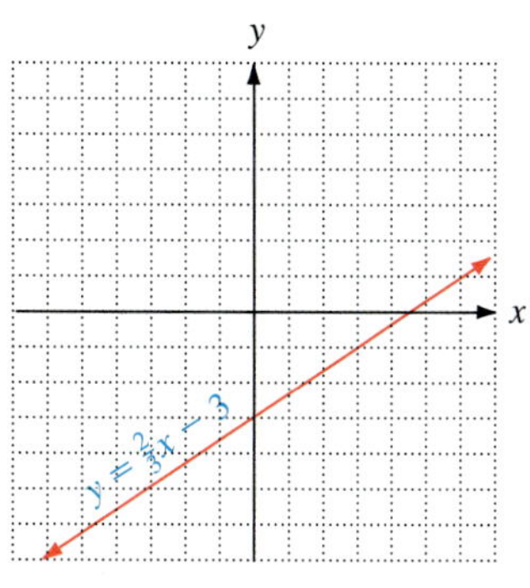

25.

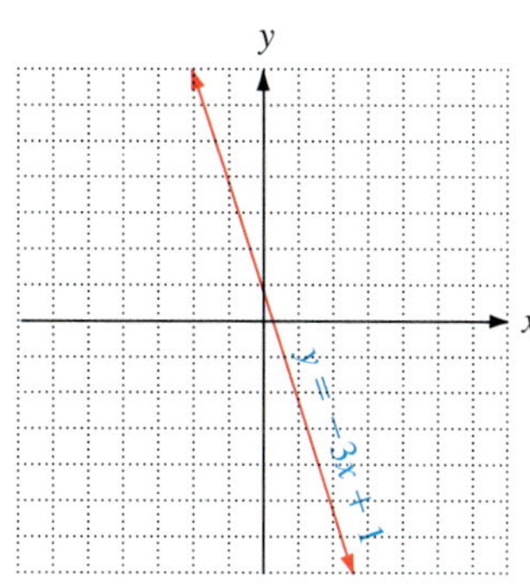

26.

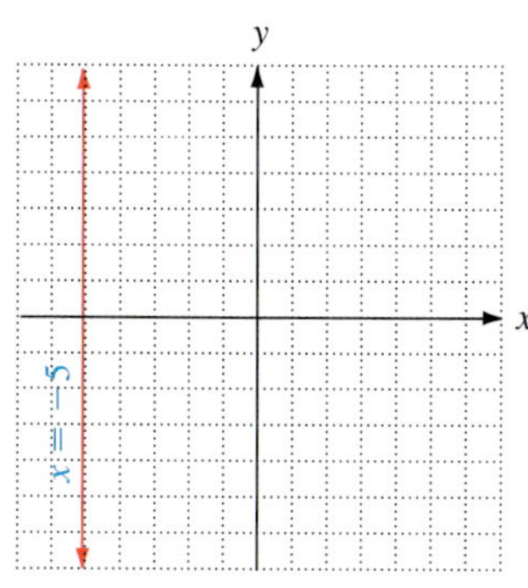

7.5

27. Graph $y < \frac{1}{3}x + 2$.

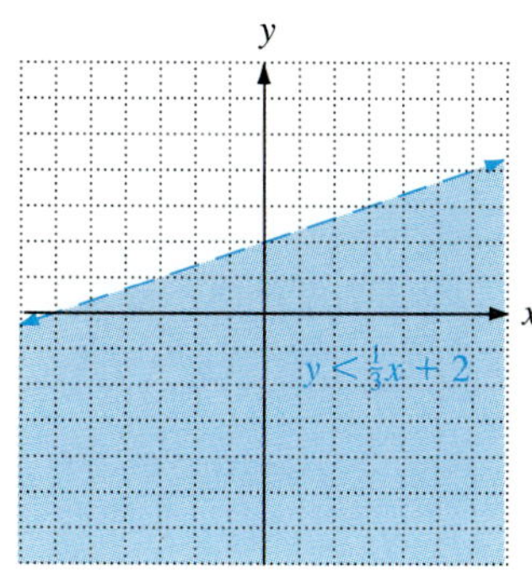

28. Graph $3y + 2x \geq 12$.

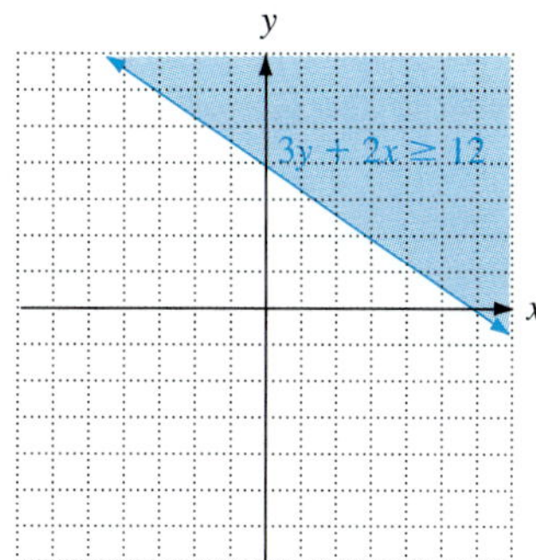

29. Graph $y \geq -2$.

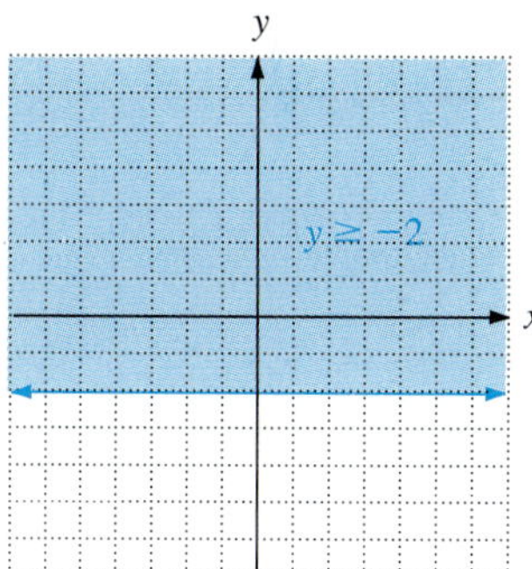

7.6

Determine the domain and range of each relation. Determine if each relation is a function.

30. {(−3, −2), (0, 4), (1, 7), (6, −2)}
Domain: {−3, 0, 1, 6}, Range: {−2, 4, 7}, function

31. {(0, 6), (1, 4), (2, 8), (0, 8)}
Domain: {0, 1, 2}, Range: {6, 4, 8}, not a function

32. {(5, −6), (−6, 5), (−5, 5), (−6, −6)}
Domain: {5, −6, −5}, Range: {−6, 5}, not a function

33. {(3, −7), (−7, 3), (−3, 7), (7, −3)}
Domain {3, −7, −3, 7}, Range: {−7, −3, 3, 7}, function

Which of the following graphs represent a function?

34.

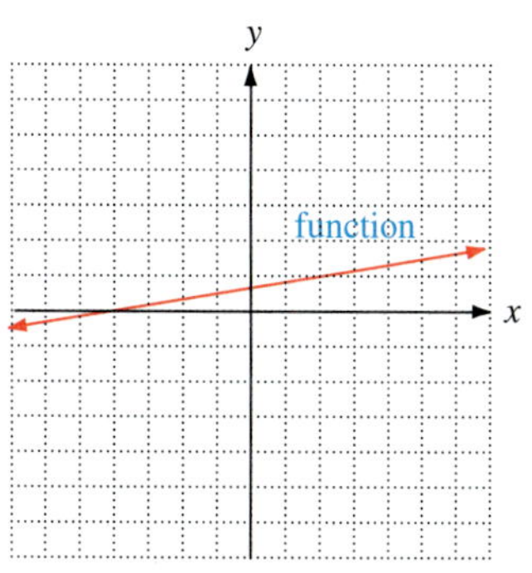

35.

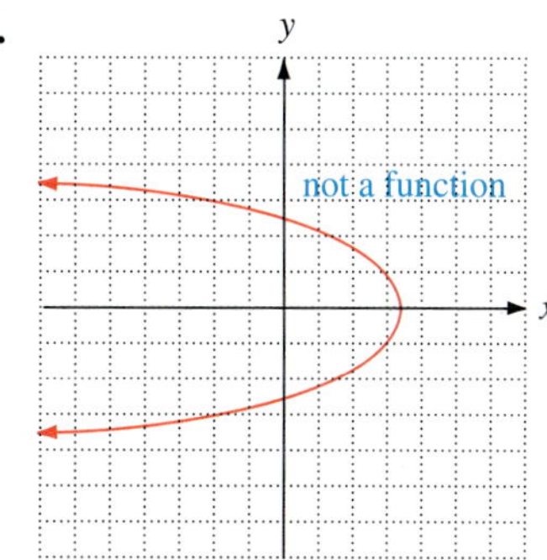

36.

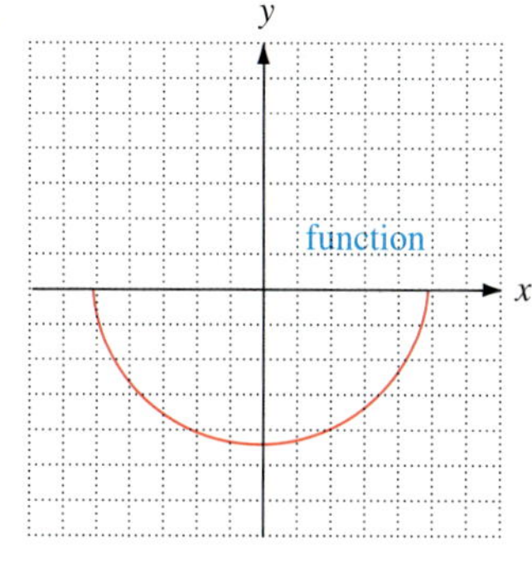

37. Graph $y = x^2 - 5$.

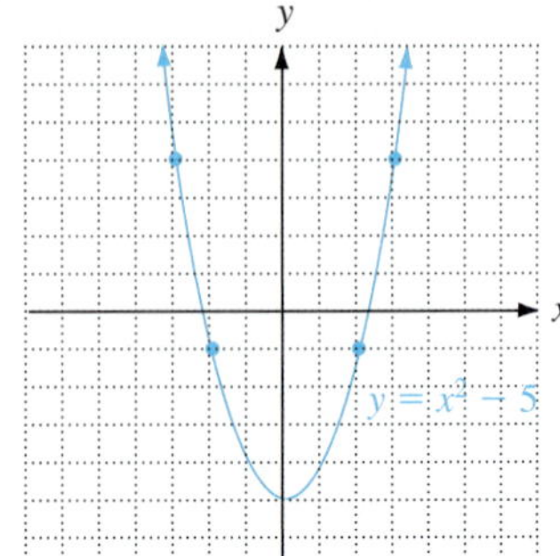

38. Graph $x = y^2 + 3$.

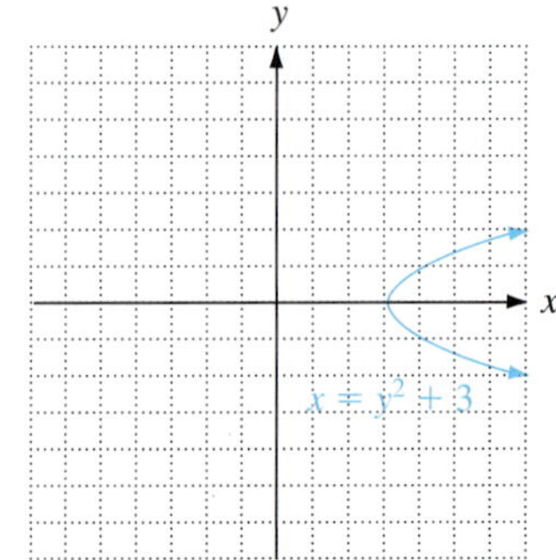

39. Graph $y = (x - 3)^2$.

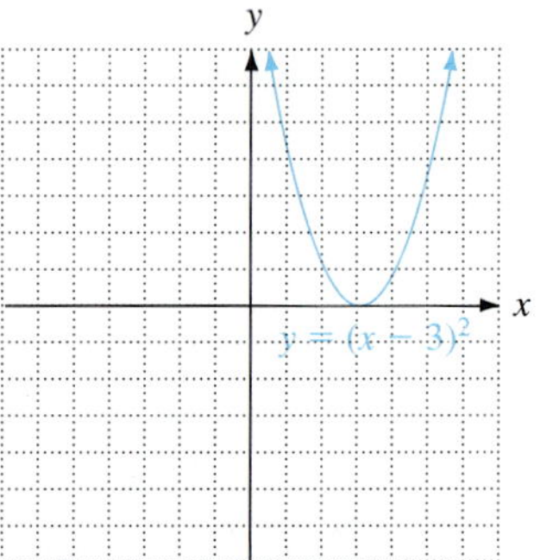

Given the following functions, find the indicated values.

40. $f(x) = 7 - 6x$ **(a)** $f(0)$ 7 **(b)** $f(-2)$ 19

41. $f(x) = 2x^2 - x + 3$ **(a)** $f(2)$ 9 **(b)** $f(-3)$ 24

42. $g(x) = -2x^2 + 3x + 4$ **(a)** $g(-1)$ -1 **(b)** $g(2)$ 2

43. $h(x) = -5 + 4x^2$ **(a)** $h(-3)$ 31 **(b)** $h(4)$ 59

44. $p(x) = \frac{-3}{x}$ **(a)** $p(-3)$ 1 **(b)** $p(0.5)$ -6

45. $f(x) = \frac{2}{x + 4}$ **(a)** $f(-2)$ 1 **(b)** $f(6)$ $\frac{1}{5}$

Chapter 7 Test

1. Plot and label the following points:
C: $(-4, -3)$ B: $(6, 1)$
D: $(-3, 0)$ E: $(5, -2)$

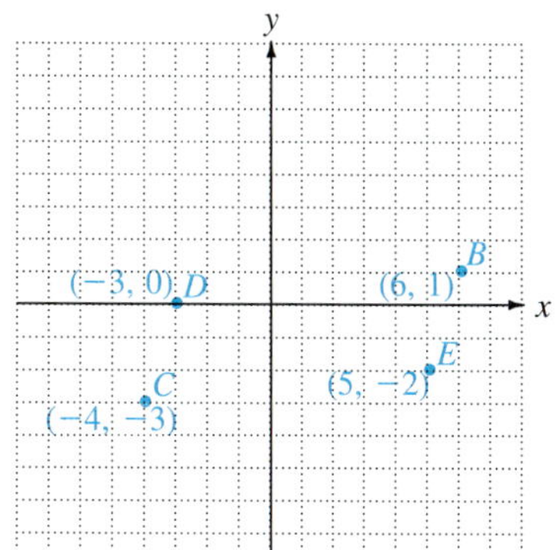

2. Graph the line $6x - 3 = 5x - 2y$.

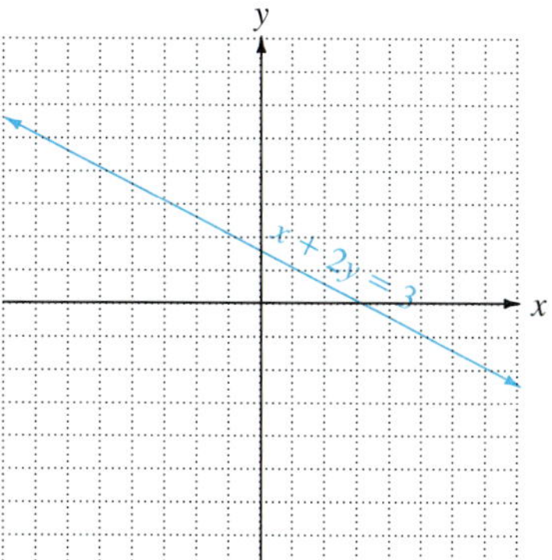

3. Graph the line $12x - 3y = 6$.

4. Graph $y = \frac{2}{3}x - 4$.

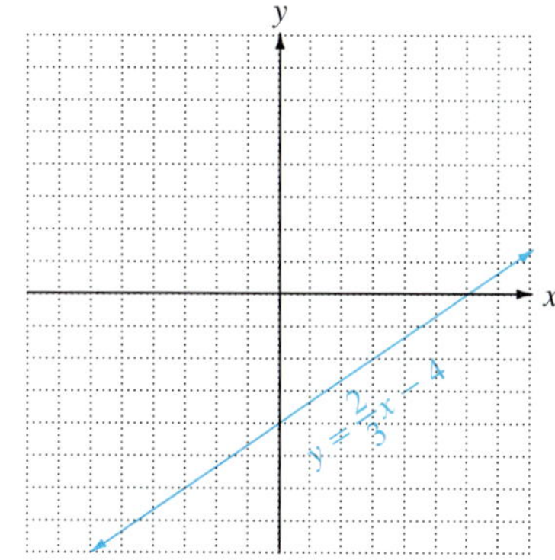

5. What is the slope and the y-intercept of the line $3x + 2y - 5 = 0$?

6. Find the slope of the line that passes through $(5, -2)$ and $(-3, -6)$.

7. Write an equation for the line that passes through $(4, -2)$ and has a slope of $\frac{1}{2}$.

8. Find the slope of the line through $(-3, 11)$ and $(6, 11)$.

9. Find an equation for the line passing through $(2, 5)$ and $(8, 3)$.

10. Write an equation for the line through $(2, 7)$ and $(2, -2)$. What is the slope of this line?

1. See graph
2. See graph
3. See graph
4. See graph
5. $m = -\frac{3}{2}, b = \frac{5}{2}$
6. $m = \frac{1}{2}$
7. $x - 2y = 8$
8. $m = 0$
9. $x + 3y = 17$
10. $x = 2$, no slope

11. See graph

12. See graph

13. Yes. No two different ordered pairs have the same first coordinate.

14. Yes. A vertical line passes through every point on the graph once.

15. See graph

16. (a) -4

(b) -2

17. (a) -3

(b) $-\frac{3}{4}$

11. Graph the region described by $4y \leq 3x$.

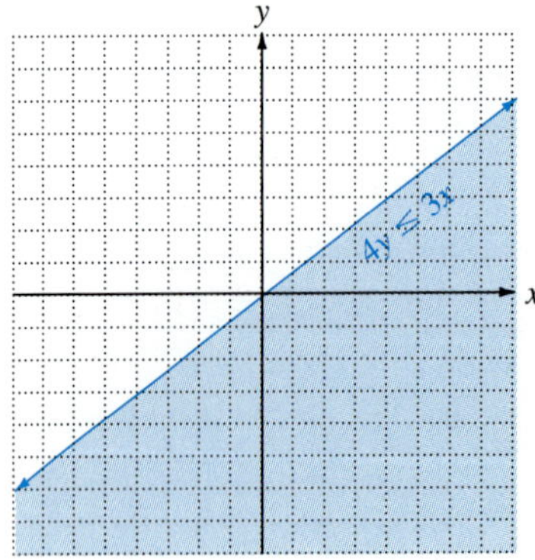

12. Graph the region described by $-3x - 2y > 10$.

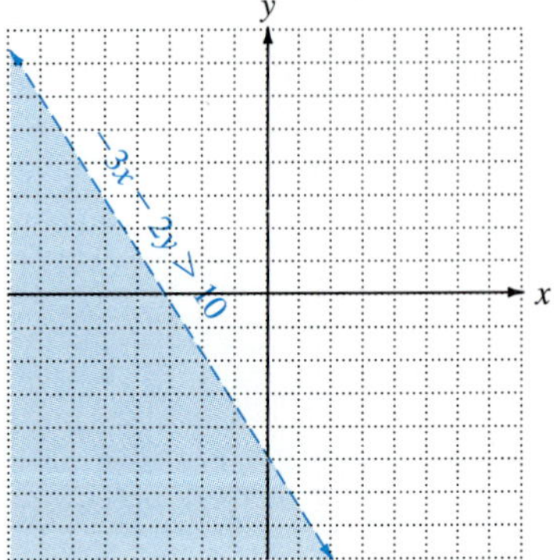

13. Is this relation a function? $\{(-8, 2), (3, 2), (5, -2)\}$ Why?

14. Is this relation a function? Why?

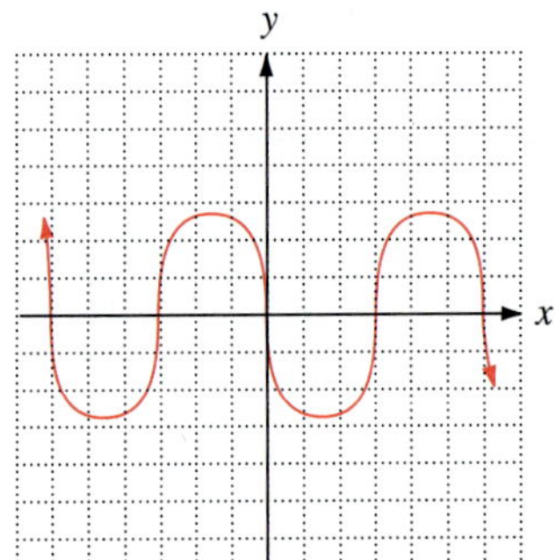

15. Graph $y = 2x^2 - 3$.

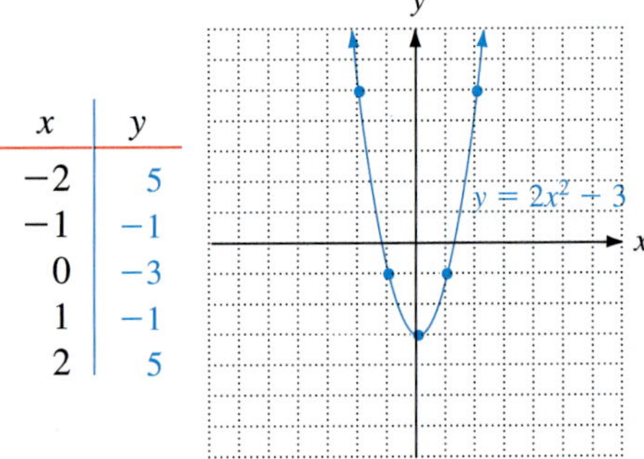

x	y
-2	5
-1	-1
0	-3
1	-1
2	5

16. For $f(x) = 2x^2 + 3x - 4$:
(a) Find $f(0)$.
(b) Find $f(-2)$.

17. For $g(x) = \dfrac{3}{x - 4}$:
(a) Find $g(3)$.
(b) Find $g(0)$.

Cumulative Test for Chapters 0–7

Approximately one-half of this test covers the content of Chapters 0–6. The remainder covers the content of Chapter 7.

1. Simplify. $(-3x^3y^4z)^3$

2. Simplify. $-\dfrac{18a^3b^4}{24ab^6}$

Solve.

3. $2 - 3(4 - x) = x - (3 - x)$

4. $\dfrac{600}{R} = \dfrac{600}{R + 50} + \dfrac{500(R + 20)}{R^2 + 50R}$

5. $x = 2^3 - 6 \div 3 - 1$

6. $A = lw + lh$, for w

7. Factor. $50a^2 - 98b^2$

8. Factor. $2x^2 + 13x + 20$

9. Simplify. $\dfrac{x - \dfrac{1}{x}}{\dfrac{1}{2} + \dfrac{1}{2x}}$

10. Find an equation of the line through (6, 8) and (7, 11).

1. $-27x^9y^{12}z^3$

2. $-\dfrac{3a^2}{4b^2}$

3. $x = 7$

4. $R = 40$

5. $x = 5$

6. $\dfrac{A}{l} - h = w$ or $\dfrac{A - lh}{l} = w$

7. $2(5a + 7b)(5a - 7b)$

8. $(2x + 5)(x + 4)$

9. $2(x - 1)$

10. $3x - y = 10$

11. $x = 7$

12. $x - 3y = -11$

13. 0

14. $\frac{3}{7}$

15. See graph

16. See graph

17. See graph

18. See graph. No

19. Yes

20. (a) 5

(b) 3

Give the equations of the lines.

11. Through $(7, -4)$ and vertical.

12. Through $(-2, 3)$ with slope of $\frac{1}{3}$.

13. Find the slope of the line through $(-8, -3)$ and $(11, -3)$.

14. What is the slope of the line $3x - 7y = -2$?

15. Graph $y = \frac{2}{3}x - 4$.

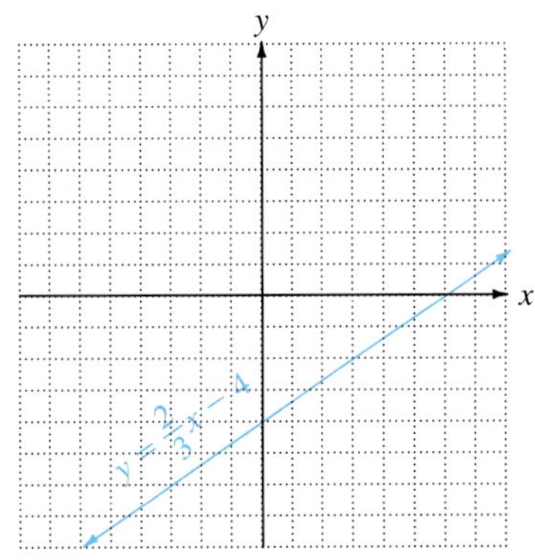

16. Graph $3x + 8 = 5x$.

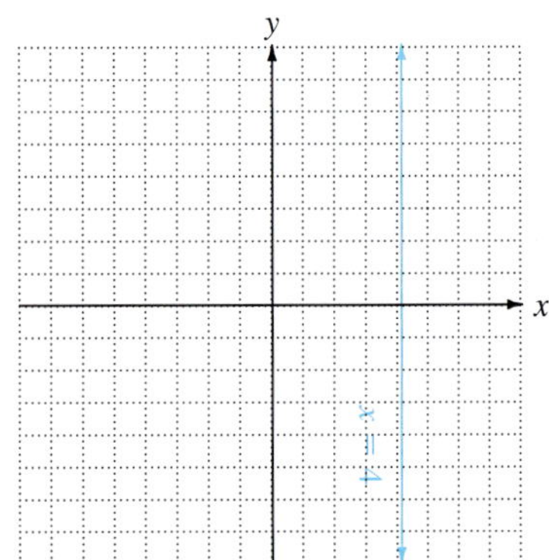

17. Graph the region $2x + 5y \leq -10$.

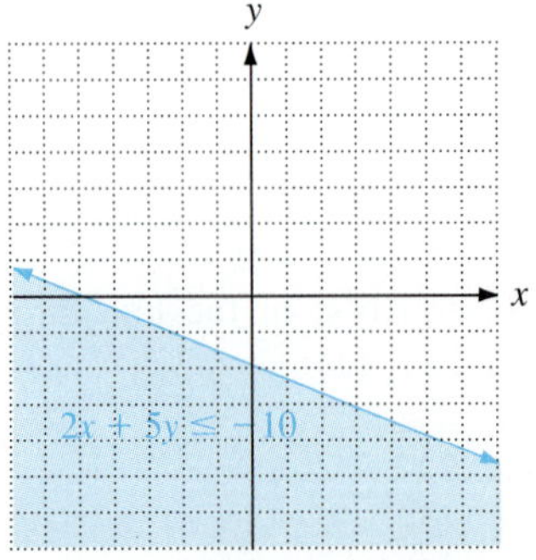

18. Graph $x = y^2 + 3$.
Is this a function?

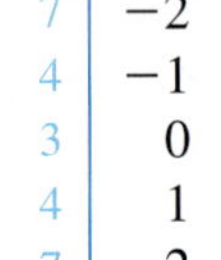

x	y
7	−2
4	−1
3	0
4	1
7	2

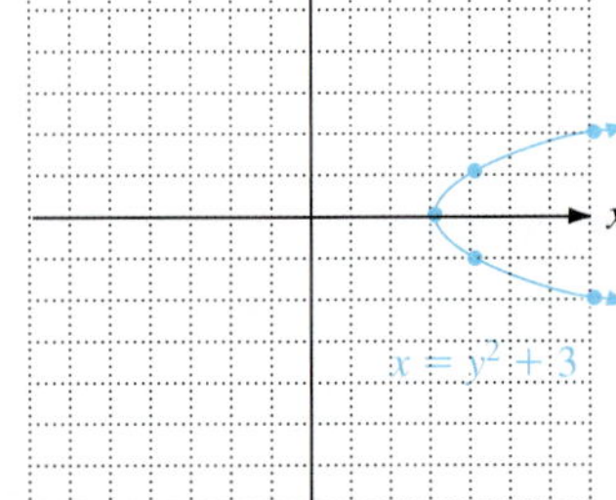

19. Is this relation a function?
$\{(3, 10), (10, -3), (0, -3), (5, 10)\}$

20. For $f(x) = -2x^2 - 3x + 5$:
(a) Find $f(0)$.
(b) Find $f(-2)$.

CHAPTER 8

Systems of Equations

If you work with a computer on a daily basis you probably know that it can do millions of calculations in less than a second. But did you know that new computers are being developed that can do billions of calculations in less than a second? Maybe even trillions of calculations in a second? In fact, some scientists think they have already built one that will accomplish this speed and more. To find out how fast it really is, turn to the Putting Your Skills to Work on page 478.

Pretest Chapter 8

If you are familiar with the topics in this chapter, take this test now. Check your answers with those in the back of the book. If an answer was wrong or you couldn't do a problem, study the appropriate section of the chapter.

If you are not familiar with the topics in this chapter, don't take this test now. Instead, study the examples, work the practice problems, and then take the test.

This test will help you identify those concepts that you have mastered and those that need more study.

Section 8.1

Solve each system by the graphing method.

1. $4x + 2y = -8$
$-2x + 3y = 12$

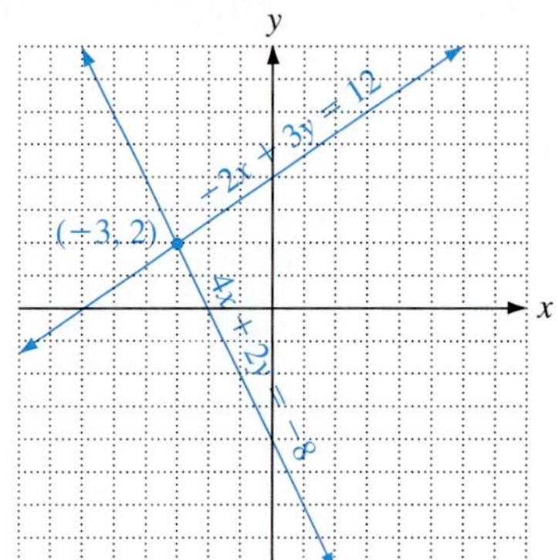

2. $x - 2y = -2$
$3x - 6y = 12$

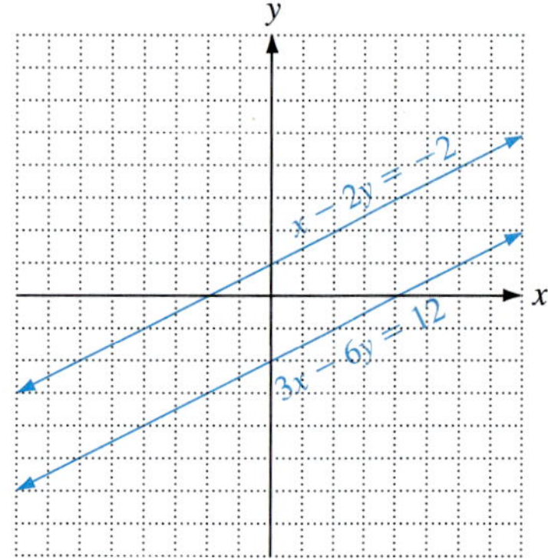

Section 8.2

Solve each system by the substitution method.

3. $3x + y = -11$
$4x - 3y = -6$

4. $\frac{2}{3}x + y = 1$
$3x + 2y = 3$

Section 8.3

Solve each system by the addition method.

5. $5x - 3y = 10$
$4x + 7y = 8$

6. $\frac{1}{2}x - \frac{3}{4}y = \frac{3}{2}$
$x + y = 8$

7. $0.8x - 0.3y = 0.7$
$1.2x + 0.6y = 4.2$

8. If a system has one solution, the graphs of the equations will be lines that ______________.

9. If a system has no solutions, the graphs of the equations will be lines that ______________.

10. If a system has an infinite number of solutions, the graphs of the equations will be lines that ______________.

Section 8.4

Solve by any method.

11. $x + 8y = 7$
$5x - 2y = 14$

12. $3x - y = 5$
$-5x + 2y = -10$

13. $2x - 3y = 7$
$3x + 5y = 1$

Section 8.5

14. James and Michael work sorting letters for the post office. Yesterday James sorted letters for 3 hours and Michael for 2 hours. Together they sorted 2250 letters. Today James sorted letters for 2 hours and Michael for 3 hours. Together they sorted 2650 letters. How many letters per hour can be sorted by James? by Michael?

1.  $x = -3, y = 2$

2. no solution

3. $x = -3, y = -2$

4. $x = \frac{3}{5}, y = \frac{3}{5}$

5. $x = 2, y = 0$

6. $x = 6, y = 2$

7. $x = 2, y = 3$

8. intersect

9. are parallel

10. coincide

11. $x = 3, y = \frac{1}{2}$

12. $x = 0, y = -5$

13. $x = 2, y = -1$

14. James = 290
Michael = 690

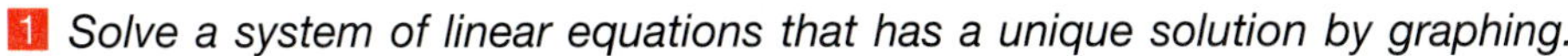

8.1 Solve a System of Equations by Graphing

After studying this section, you will be able to:

1. *Solve a system of linear equations that has a unique solution by graphing.*
2. *Graph a linear system of equations to determine the type of solution.*

MathPro

Video 19

SSM

In Chapter 7 we examined linear equations and inequalities in two variables. In this chapter we will examine systems of linear equations and inequalities. Recall that the graph of a linear equation is a straight line. If we have two linear equations on one coordinate plane, what are the possibilities with regard to their graphs?

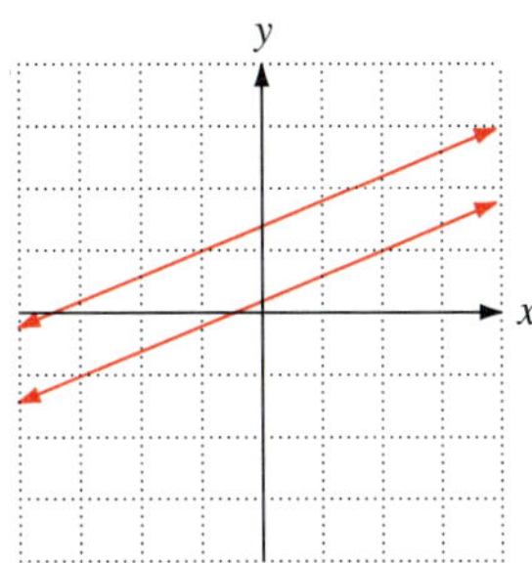

The lines may be parallel.

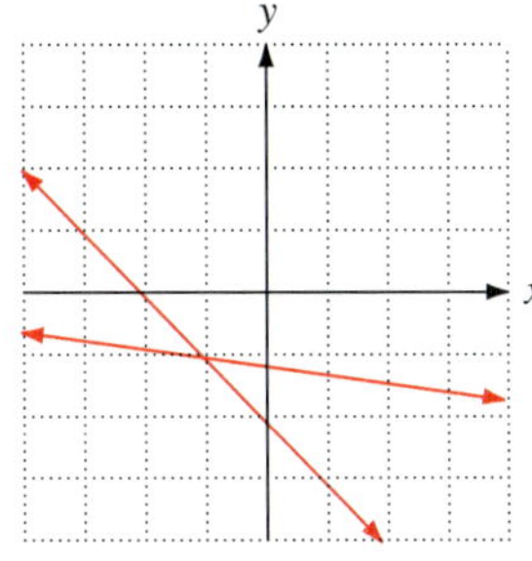

The lines may intersect.

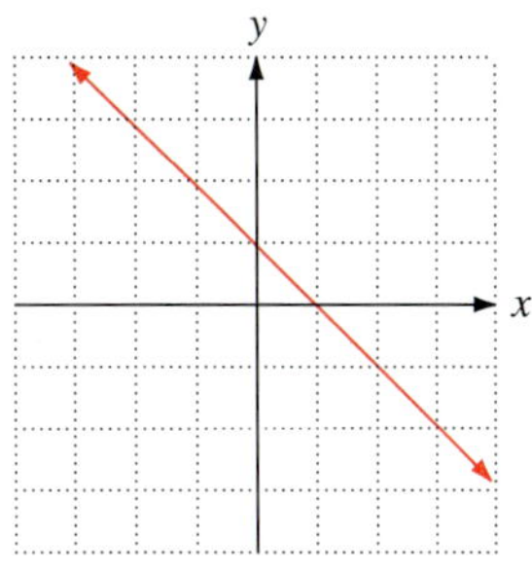

The lines may coincide.

The solution to the system of equations is those points that both lines have in common. Note that parallel lines have no points in common. The system has no solution. Intersecting lines intersect at one point. The system has one solution. Lines that coincide have an infinite number of points in common. The system has an infinite number of solutions. You can determine the solution to a system of equations by graphing.

TEACHING TIP After covering the content of Example 1, you may wish to give this problem as an Additional Class Exercise that involves a system of equations that is fairly easy to graph. Ask students to graph the system and find the solution for

$$x + y = 5$$
$$x - 4y = 5$$

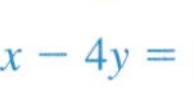

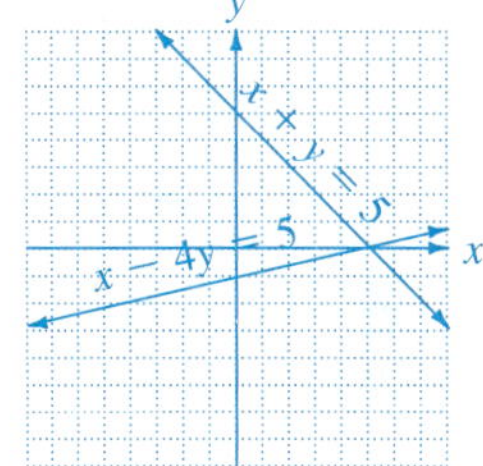

The solution is $x = 5$, $y = 0$ or $(5, 0)$

1 Solving a System of Linear Equations That Has a Unique Solution by Graphing

To solve a system of equations by graphing, graph each equation on the same coordinate system. If the lines intersect, the system has a unique solution. The coordinates of the point of intersection is the solution to the system of equations.

EXAMPLE 1 Solve the following system of equations by graphing.

$$-2x + 3y = 6$$
$$2x + 3y = 18$$

Graph each equation on the same coordinate plane and determine the point of intersection. The graph of the above system is shown in the figure on the right. The lines intersect at the point (3, 4). Thus the unique solution to the system of equations is $x = 3$ and $y = 4$. Always check.

$-2(3) + 3(4) \stackrel{?}{=} 6 \qquad 2(3) + 3(4) \stackrel{?}{=} 18$

$6 = 6 \qquad 18 = 18$ ✓

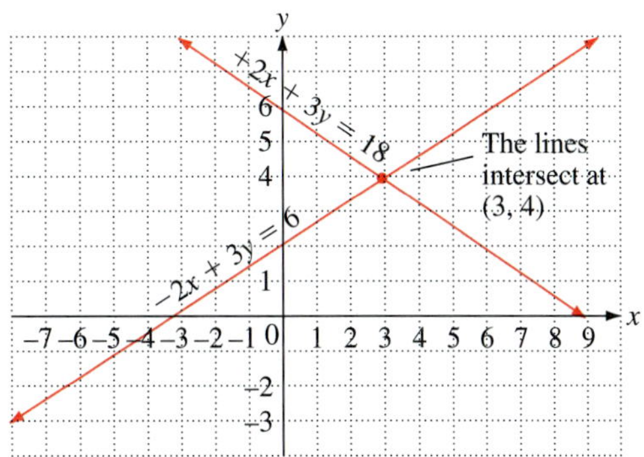

A linear system of equations that has one solution is said to be *consistent.*

Practice Problem 1 Solve by graphing. $x + y = 12$

$$-x + y = 4$$

PRACTICE PROBLEM 1

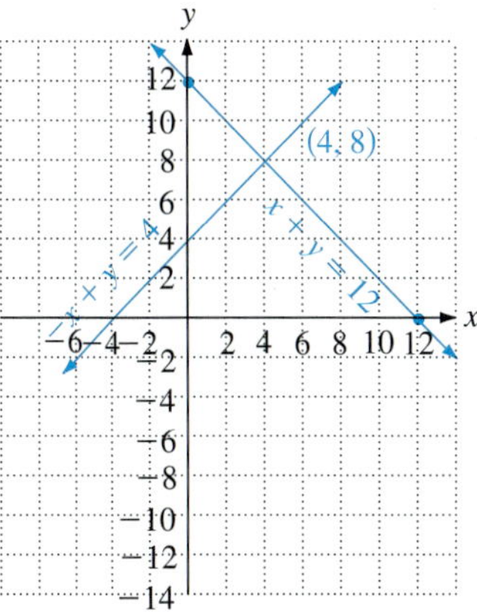

2 Graphing a Linear System of Equations to Determine the Type of Solution

Two lines in a plane may intersect, may never intersect (be parallel lines), or may be the same line. The corresponding system of equations will have one solution, no solution, or an infinite number of solutions. Thus far we have focused on systems of linear equations that have one unique solution. We now draw your attention to the other two cases.

EXAMPLE 2 Solve the following system of equations by graphing.

$$3x - y = -1$$
$$3x - y = -7$$

We graph each equation on the same coordinate plane. The graph is on the right. Notice that the lines are parallel. They do not intersect. Hence there is no solution to this system of equations.

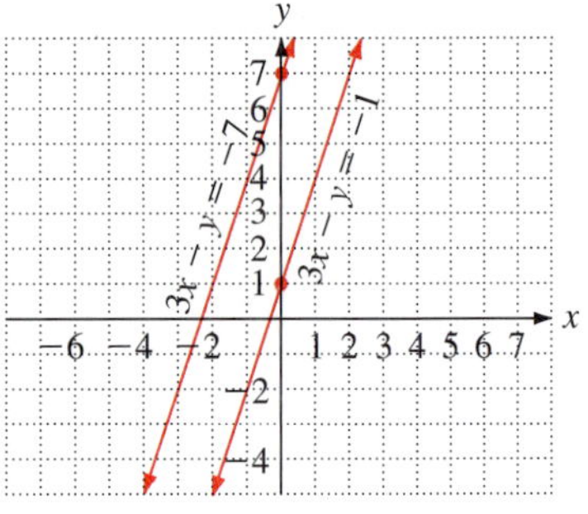

A linear system of equations that has no solution is called *inconsistent.*

PRACTICE PROBLEM 2

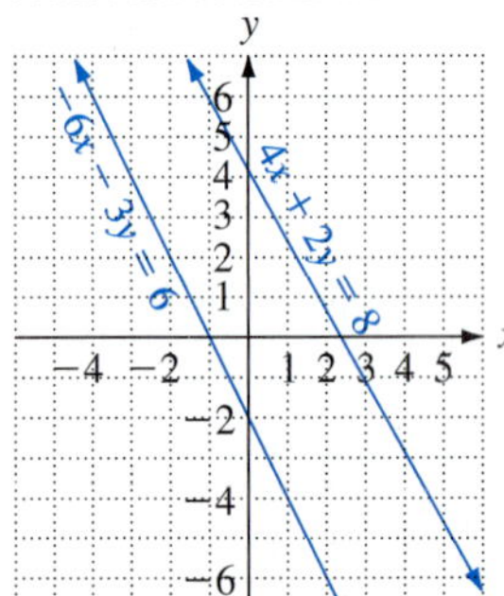

Practice Problem 2 Solve the following system of equations by graphing.

$$4x + 2y = 8$$
$$-6x - 3y = 6 \blacksquare$$

To Think About Without graphing, how could you tell that the system in Example 2 represented parallel lines? Without graphing, determine which of the following systems is inconsistent (has no solution) or is consistent (has a solution).

(a) $2x - 3y = 6$
$6y = 4x + 1$

(b) $3x + y = 2$
$2x + y = 3$

(c) $x - 2y = 1$
$3x - 6y = 3$

EXAMPLE 3 Solve the following system of equations by graphing.

$$x + y = 4$$
$$3x + 3y = 12$$

We graph each equation on the same coordinate plane on the right. Notice that both equations represent the same line (coincide). The coordinates of every point on the line will satisfy both equations. That is, every point on the line represents a solution to $x + y = 4$ and to $3x + 3y = 12$. Thus there is *an infinite number of solutions* to this system.

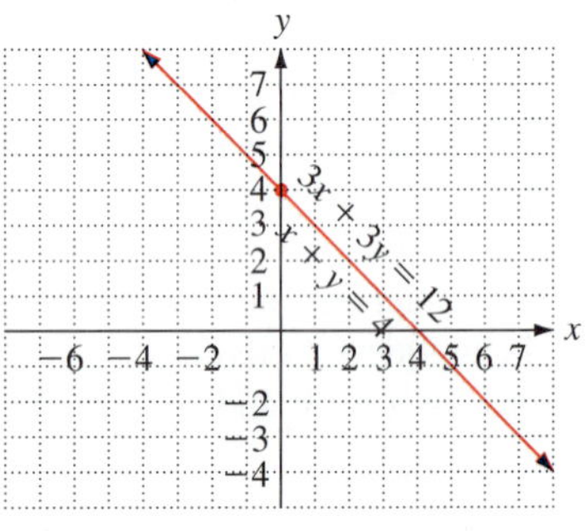

These equations are said to be *dependent.*

PRACTICE PROBLEM 3

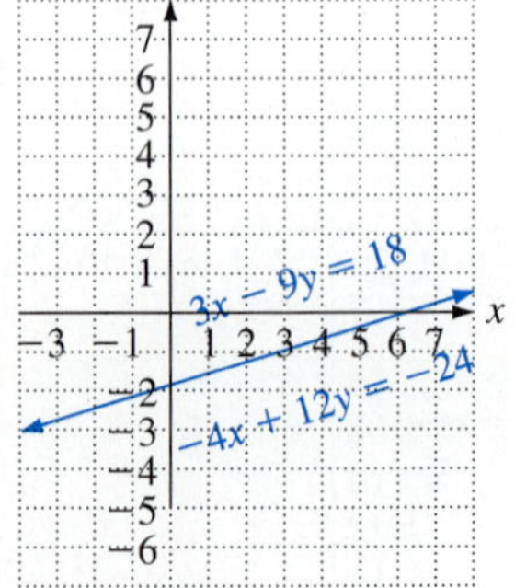

Practice Problem 3 Solve the following system of equations by graphing.

$$3x - 9y = 18$$
$$-4x + 12y = -24 \blacksquare$$

8.1 Exercises

Solve each system by graphing. If there is not one solution to a system, state the reason.

1. $x + y = 8$
$x - y = 2$

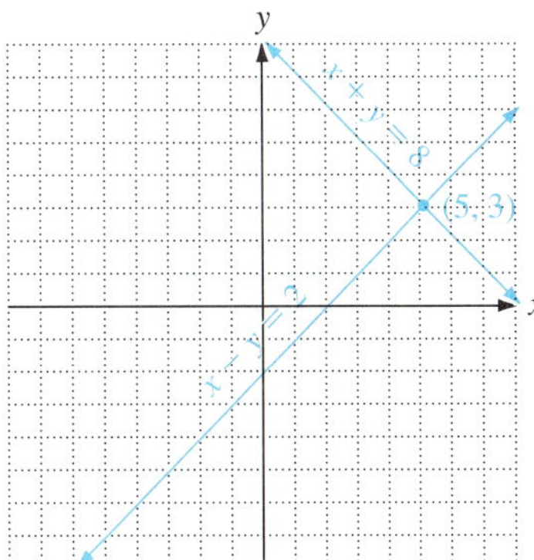

2. $x + y = 6$
$-x + y = -2$

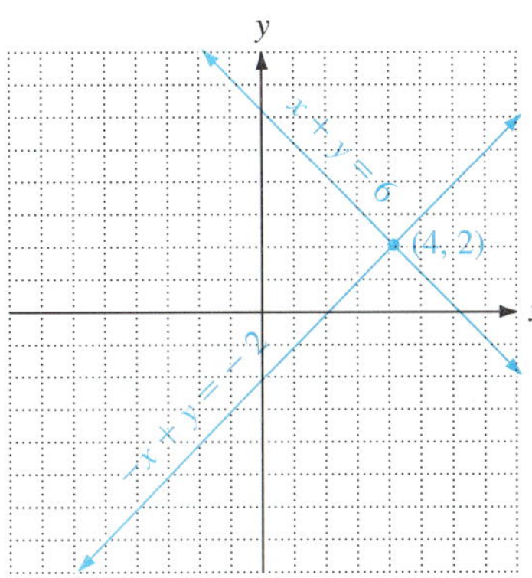

3. $x + 2y = 5$
$-3x + 2y = 17$

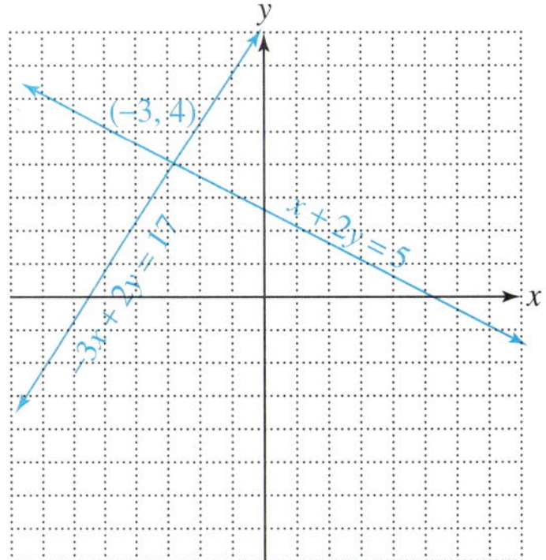

4. $2x + y = 0$
$-3x + y = 5$

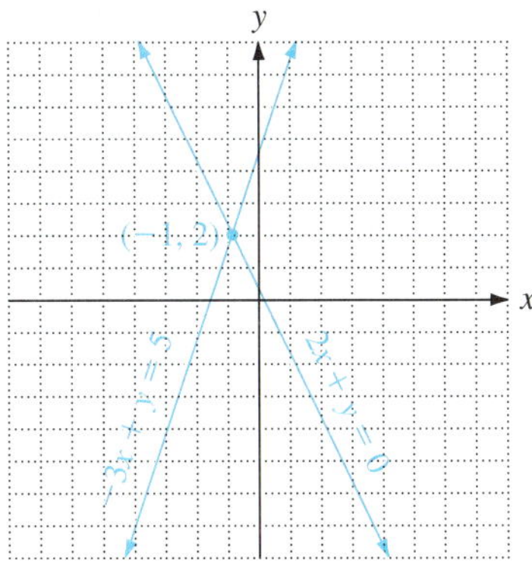

5. $2x + y = 6$
$-2x + y = 2$

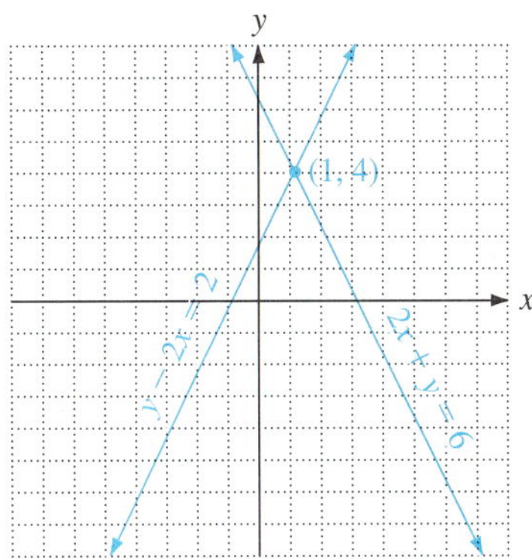

6. $x - 3y = 6$
$4x + 3y = 9$

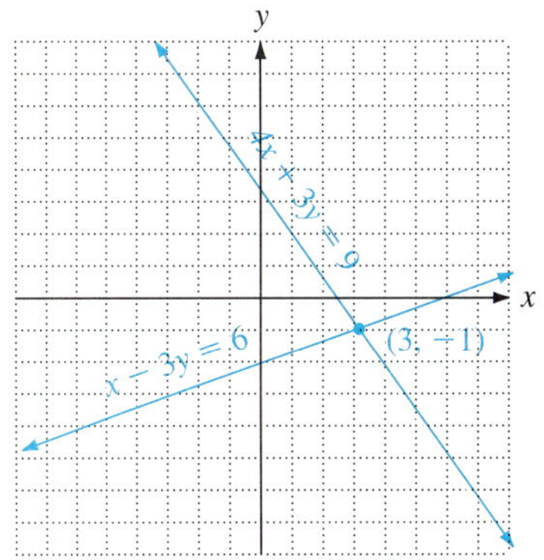

7. $-2x + y - 3 = 0$
$4x + y + 3 = 0$

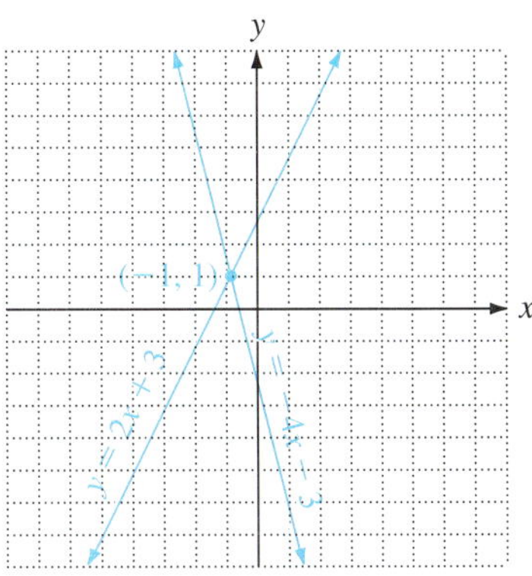

8. $x - 2y - 10 = 0$
$2x + 3y - 6 = 0$

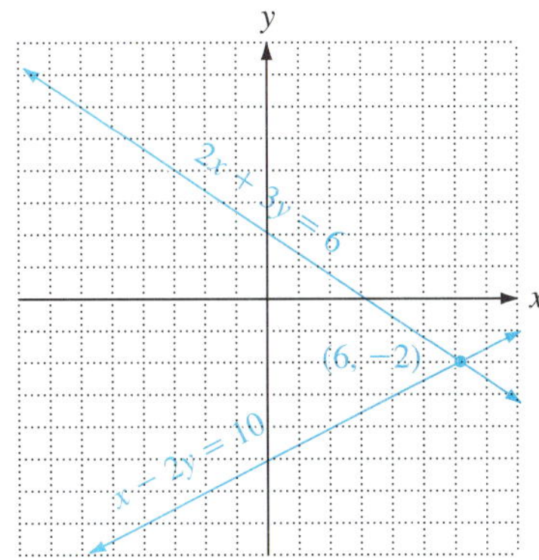

9. $3x - 2y = -18$
$2x + 3y = 14$

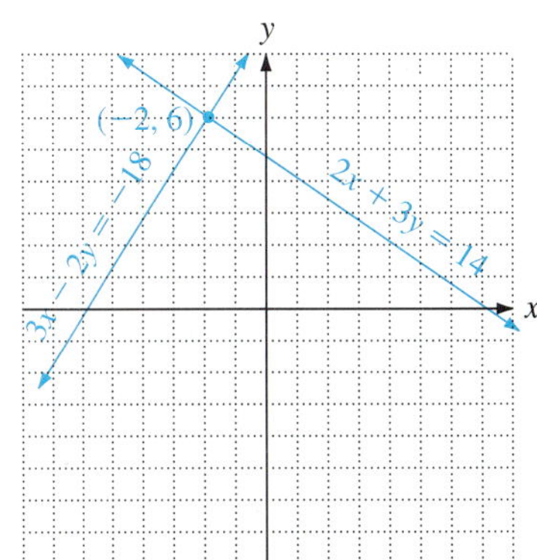

10. $3x + 2y = -10$

$-2x + 3y = 24$

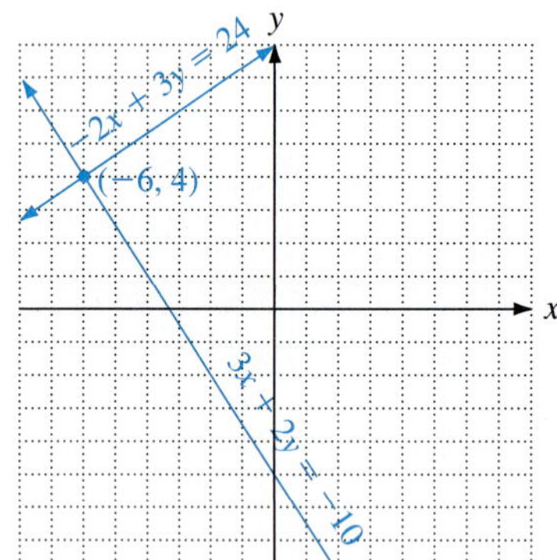

11. $y = \frac{5}{7}x - 2$

$y = \frac{1}{3}x + \frac{2}{3}$

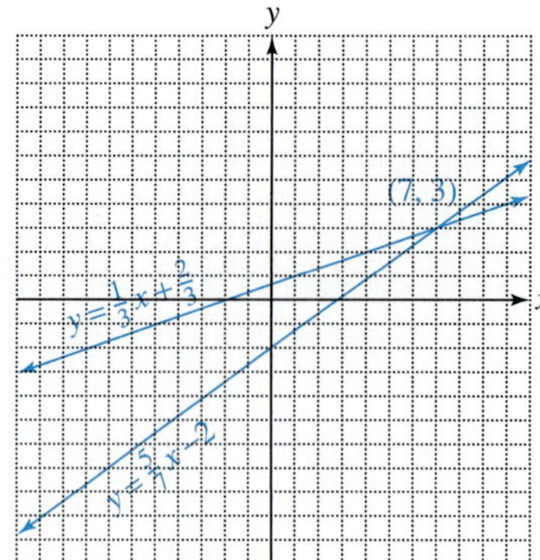

12. $y = \frac{3}{4}x + 7$

$y = -\frac{1}{2}x + 2$

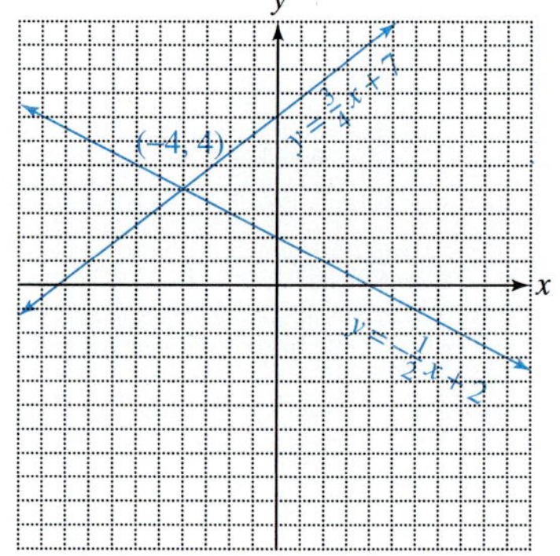

13. $4x - 6y = 8$

$-2x + 3y = -4$

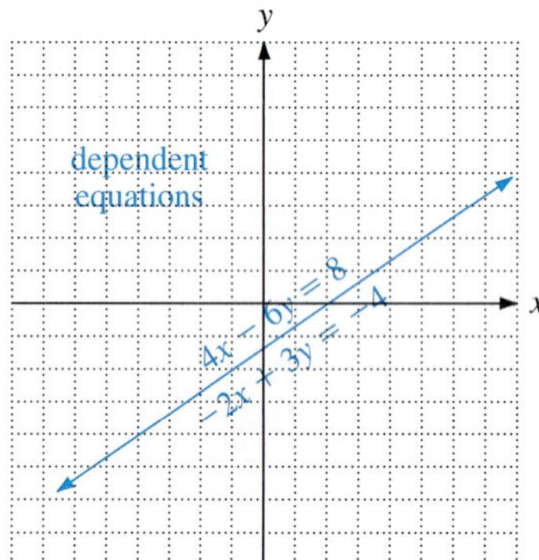

14. $3x - 2y = -4$

$-9x + 6y = -8$

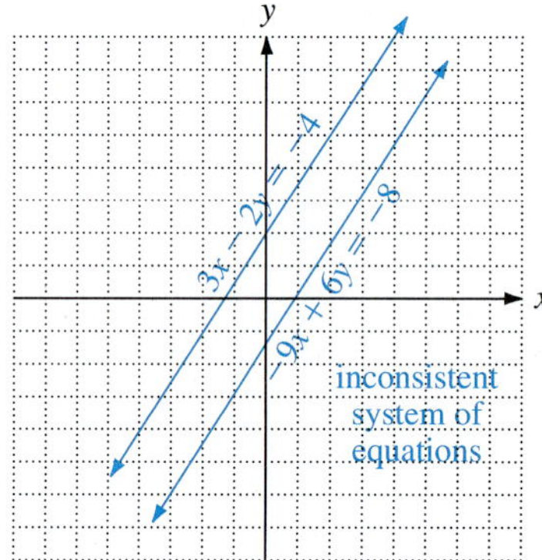

15. $2y + x - 6 = 0$

$y + \frac{1}{2}x = 4$

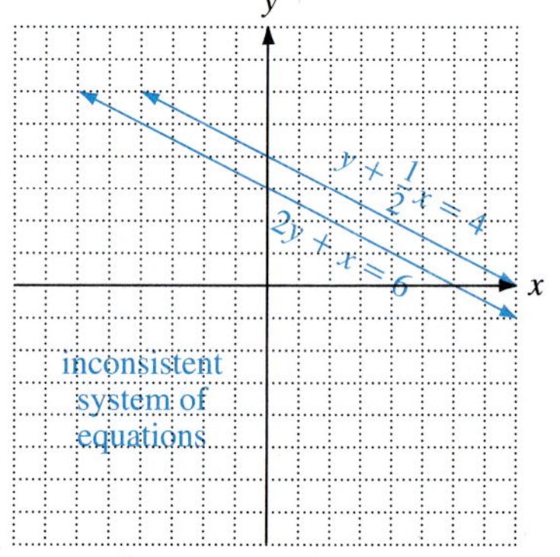

16. $y - 2x - 6 = 0$

$\frac{1}{2}y - 3 = x$

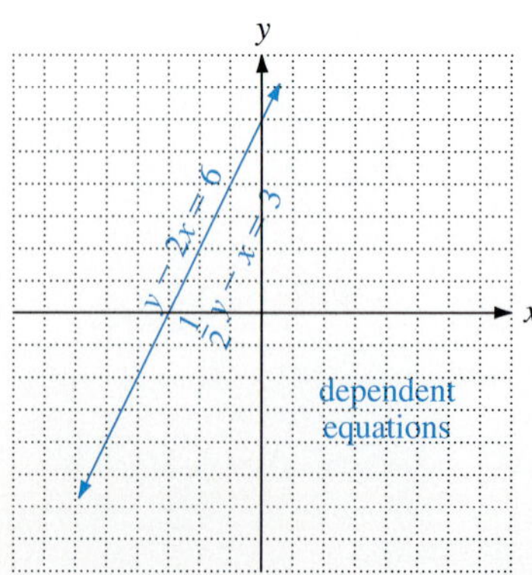

★ **17.** $\frac{1}{7}x - \frac{3}{7}y = 1$

$\frac{1}{2}x - \frac{1}{4}y = -1$

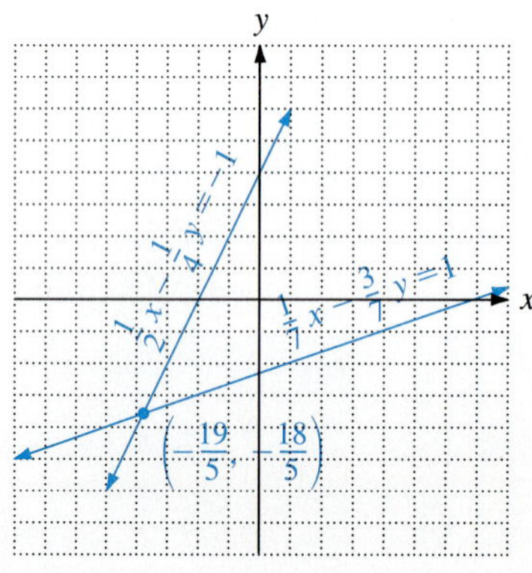

★ **18.** $\frac{1}{4}x - \frac{5}{12}y = -\frac{5}{4}$

$-\frac{1}{2}x + \frac{5}{6}y = \frac{5}{2}$

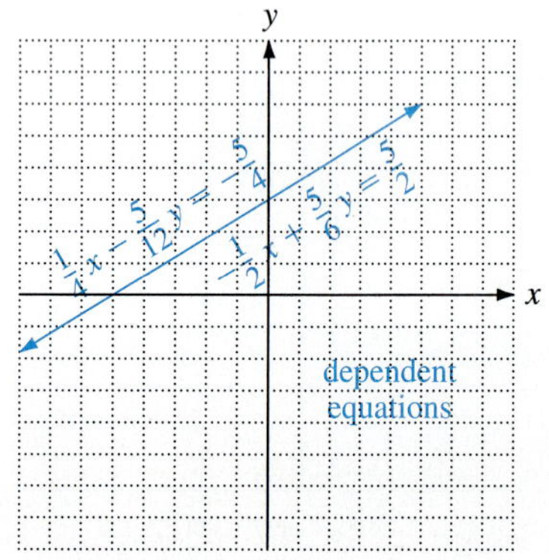

Verbal and Writing Skills

19. In the system $y = 2x - 5$ and $y = 2x + 6$, the lines have the same slope but different y-intercepts. What does this tell you about the graph of this system? The lines are parallel. One line intersects the y-axis at -5. The other line interects the y-axis at 6. There is no solution. The system is inconsistent.

20. In the system $y = -3x + 4$ and $y = -3x + 4$, the lines have the same slope and the same y-intercepts. What does this tell you about the graph of the system? The two lines coincide. They are the same line. There are an infinite number of solutions. The equations are dependent.

21. Before you graph a system of equations, if you notice that the lines have different slopes, you can conclude they ___intersect___ and the system has ___one___ solution(s).

22. If lines have the same slope, you cannot conclude that the lines must be parallel. Why? How else might the lines look when graphed? They may coincide.

23. Two lines have different slopes but the same y-intercept. What does this tell you about the graphs of the lines and about the solution of the system? The lines intersect at the y-intercept. The y-intercept is the solution of the system.

Graphing Calculator Exercises (Optional)

If you have a graphing calculator, solve each system by graphing the equations and estimating the point of intersection to the nearest hundredth. You will need to adjust the window for the appropriate x and y values in order to see the region where the straight lines intersect.

24. $y = 56x + 1808$
$y = -62x - 2086$
$(-33, -40)$

25. $y = 47x - 960$
$y = -36x + 783$
$(21, 27)$

26. $88x + 57y = 683.10$
$95x - 48y = 7460.64$
$(47.52, -61.38)$

27. $64x + 99y = -8975.52$
$58x - 73y = 3503.58$
$(-29.61, -71.52)$

Cumulative Review Problems

28. Solve. $7 - 3(4 - x) \leq 11x - 6$
$7 - 12 + 3x \leq 11x - 6$
$\frac{1}{8} \leq x$

29. Solve for d. $3dx - 5y = 2(dx + 8)$
$d = \dfrac{5y + 16}{x}$

30. Multiply. $(3x - 7)^2$
$9x^2 - 42x + 49$

31. Factor. $25x^2 - 60x + 36$
$(5x - 6)^2$

8.2 Solve a System of Equations by the Substitution Method

MathPro

Video 20

SSM

TEACHING TIP It is important for students to understand the concept that the solution must be verified in *both* of the linear equations before we can be certain that it is a solution. After explaining the idea of a solution of a system of equations, present students with the following question.

"Here is a system of equations:

$$4x - 5y = 22$$
$$x + 2y = 1$$

Is $x = 3$ and $y = -2$ a solution to the system?" (Then give the students time to determine the answer to this question.)

The answer is *no*. The values $x = 3$ and $y = -2$ check in the first equation, but not in the second equation. Many students will only check in the first equation and make an erroneous conclusion. However, if you stress how important it is to check both equations based on this illustration, they will not likely make that mistake again.

TEACHING TIP This method should only be used if at least one of the coefficients in the original system is ± 1.

After studying this section, you will be able to:

1. *Solve a system of two linear equations with integer coefficients by the substitution method.*
2. *Solve a system of two linear equations with fractional coefficients by the substitution method.*

1 *Solving a System of Two Linear Equations with Integer Coefficients by the Substitution Method*

Since the solution to a system of linear equations may be a point where two lines intersect, we would expect the solution to be an ordered pair. Since the point must lie on both lines, the ordered pair (x, y) must satisfy both equations. For example, the solution to the system $3x + 2y = 6$ and $x + y = 3$ is $(0, 3)$. Let's check.

$3x + 2y = 6$	$x + y = 3$
$3(0) + 2(3) \stackrel{?}{=} 6$	$0 + 3 \stackrel{?}{=} 3$
$6 = 6$ ✓	$3 = 3$ ✓

Given a system of linear equations, we can develop an algebraic method to find the solution. We know how to solve an equation with one variable. The following methods reduce the system to one equation in one variable. Once we have solved for the variable, we can substitute this known value into any one of the original equations and solve for the second variable.

EXAMPLE 1 Find the solution to the following system of equations.

$$y = 3x - 1$$
$$3x - 2y = -4$$

We observe that the first equation is solved for the variable y in terms of x. This means that y is equivalent to the expression $3x - 1$. Let us substitute $3x - 1$ for y in the second equation. The second equation will then have only one variable, and we will be able to solve for that variable.

$3x - 2y = -4$	*This is the original second equation.*
$3x - 2(3x - 1) = -4$	*Replace the variable y by 3x − 1. Note that we need parentheses here.*
$3x - 6x + 2 = -4$	*Remove the parentheses.*
$-3x + 2 = -4$	*Collect like terms.*
$-3x = -6$	*Subtract 2 from both sides.*
$x = 2$	*Divide both sides by −3.*

We have now found the value for x. Now we substitute this value for x into one of the original equations and solve for y.

$y = 3x - 1$	*This is the original first equation.*
$= 3(2) - 1$	*Replace the variable x by 2.*
$= 6 - 1$	*Simplify.*
$= 5$	

Thus we see that the solution to the system is $x = 2$, $y = 5$.

Check. $y = 3x - 1$ $\qquad$ $3x - 2y = -4$

$5 \stackrel{?}{=} 3(2) - 1$ $\qquad$ $3(2) - 2(5) \stackrel{?}{=} -4$

$5 \stackrel{?}{=} 6 - 1$ $\qquad$ $6 - 10 \stackrel{?}{=} -4$

$5 = 5$ ✓ $\qquad$ $-4 = -4$ ✓

Practice Problem 1 Find the solution to the following system of equations.

$$y = 3x + 10$$
$$2x - 3y = -16$$ ■

Sometimes neither equation is solved for a variable. In this case we need to begin by taking one equation (the choice is up to you) and solve for one variable. We have broken the procedure to follow into five steps.

EXAMPLE 2 Find the solution to the following system of equations.

$$x - 2y = 7$$
$$-5x + 4y = -5$$

TEACHING TIP After you have presented Example 2, ask students to solve the following system by the substitution method as an Additional Class Exercise.

$$2x + 5y = 13$$
$$-3x + y = -11$$

Sol. $x = 4$, $y = 1$

Step 1 Solve for one variable.

$x - 2y = 7$ $\qquad$ *The first equation is the easiest one in which to isolate a variable.*

$x = 7 + 2y$ $\qquad$ *Add 2y to both sides to solve for x.*

Step 2 Substitute the expression into the other equation.

$-5x + 4y = -5$ $\qquad$ *This is the original second equation.*

$-5(7 + 2y) + 4y = -5$ $\qquad$ *Substitute the value 7 + 2y for x in this equation.*

Step 3 Solve this equation.

$-35 - 10y + 4y = -5$ $\qquad$ *Remove the parentheses.*

$-35 - 6y = -5$ $\qquad$ *Collect like terms.*

$-6y = 30$ $\qquad$ *Add 35 to both sides.*

$y = -5$ $\qquad$ *Divide both sides by −6.*

Step 4 Obtain the value of the second variable. We will now use the value $y = -5$ to find the value for x in one of the equations that contains both variables.

$x = 7 + 2y$ $\qquad$ *The easiest equation to use is the one that is already solved for x.*

$= 7 + 2(-5)$ $\qquad$ *Replace y by −5.*

$= 7 - 10$

$= -3$

The solution is $x = -3$ and $y = -5$.

Step 5 Check. To be sure that we have the correct solution, we will need to check that the obtained values of x and y can be substituted into *both original* equations to obtain true mathematical statements.

$x - 2y = 7$ $\qquad$ $-5x + 4y = -5$

$-3 - 2(-5) \stackrel{?}{=} 7$ $\qquad$ $-5(-3) + 4(-5) \stackrel{?}{=} -5$

$-3 + 10 \stackrel{?}{=} 7$ $\qquad$ $15 - 20 \stackrel{?}{=} -5$

$7 = 7$ ✓ $\qquad$ $-5 = -5$ ✓

Procedure for Solving a System of Equations by the Substitution Method

1. Solve for one variable in one of the two equations.
2. Substitute the expression you obtain in step 1 for this variable in the *other* equation.
3. You now have one equation with one variable. Solve this equation to find the value for that one variable.
4. Substitute this value for the variable into one of the original equations to obtain a value for the second variable.
5. Check each solution against each equation to verify your results.

Practice Problem 2 Solve the system.

$$5x + 3y = 19$$
$$2x - y = 12$$ ■

TEACHING TIP Some students may have difficulty following all the steps of Example 3. After they have had some time to study it and to ask questions, you may want to work the following Additional Class Exercise. Find a solution to the following system by the substitution method:

$$2x - y = 1$$
$$\frac{2}{5}x - y = -\frac{1}{5}$$

Multiply the second equation by 5 to obtain $2x - 5y = -1$. Solve the first equation for y.

$$y = -1 + 2x$$

Substitute that expression for y into the second equation.

$$2x - 5(-1 + 2x) = -1$$
$$-8x + 5 = -1$$
$$x = \frac{3}{4}$$

Now substitute $x = \frac{3}{4}$ into the equation $y = -1 + 2x$.

$$y = -1 + 2\left(\frac{3}{4}\right)$$
$$= \frac{1}{2}$$

The solution is $\left(\frac{3}{4}, \frac{1}{2}\right)$.

2 *Solving a System of Two Linear Equations with Fractional Coefficients by the Substitution Method*

If a system of equations contains fractions, clear the equations of fractions *before* doing any other steps.

EXAMPLE 3 Find the solution to the following system of equations.

$$\frac{3}{2}x + y = \frac{5}{2}$$
$$-y + 2x = -1$$

We want to clear the first equation of the fractions. We observe that the LCD of the fractions is 2.

$$2\left(\frac{3}{2}x\right) + 2(y) = 2\left(\frac{5}{2}\right)$$ *Multiply each term of the equation by the LCD.*

$$3x + 2y = 5$$ *This is equivalent to* $\frac{3}{2}x + y = \frac{5}{2}$

Now follow the five-step procedure.

Step 1 Solve for one variable. We will use the second equation.

$$-y + 2x = -1$$
$$-y = -1 - 2x$$ *Add* $-2x$ *to both sides.*
$$y = 1 + 2x$$ *Multiply each term by* -1.

Step 2 Substitute the expression into the other equation.

$$3x + 2(1 + 2x) = 5$$

$$3x + 2 + 4x = 5 \qquad \textit{Remove parentheses.}$$

$$7x + 2 = 5$$

Step 3 Solve this equation for the variable.

$$7x + 2 = 5$$

$$7x = 3$$

$$x = \frac{3}{7}$$

Step 4 Find the value of the second variable. Always use one of the original equations. We will use the second equation.

$$-y + 2x = -1$$

$$-y + 2\left(\frac{3}{7}\right) = -1$$

$$-y + \frac{6}{7} = -1$$

$$y - \frac{6}{7} = 1$$

$$y = 1 + \frac{6}{7} = \frac{13}{7}$$

The solution to the system is $x = \frac{3}{7}$ and $y = \frac{13}{7}$.

Step 5 Check

$$\frac{3}{2}x + y = \frac{5}{2} \qquad\qquad -y + 2x = -1$$

$$\frac{3}{2}\left(\frac{3}{7}\right) + \left(\frac{13}{7}\right) \stackrel{?}{=} \frac{5}{2} \qquad\qquad -\frac{13}{7} + 2\left(\frac{3}{7}\right) \stackrel{?}{=} -1$$

$$\frac{9}{14} + \frac{13}{7} \stackrel{?}{=} \frac{5}{2} \qquad\qquad -\frac{13}{7} + \frac{6}{7} \stackrel{?}{=} -\frac{7}{7}$$

$$\frac{9}{14} + \frac{26}{14} \stackrel{?}{=} \frac{35}{14} \qquad\qquad -\frac{7}{7} = -\frac{7}{7} \checkmark$$

$$\frac{35}{14} = \frac{35}{14} \checkmark$$

Practice Problem 3 Solve the system of equations. Be sure to check your answer.

$$\frac{x}{3} - \frac{y}{2} = 1$$

$$x + 4y = -8 \;\blacksquare$$

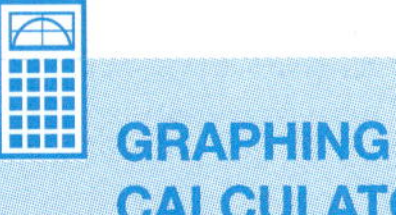

GRAPHING CALCULATOR

Solving Systems of Equations

We can solve systems of equations graphically by using a graphing calculator. For example, to solve the system of equations in Example 3, first rewrite each equation in slope-intercept form.

$$y = -\tfrac{3}{2}x + \tfrac{5}{2}$$

$$y = 2x + 1$$

Then graph $y_1 = -\frac{3}{2}x + \frac{5}{2}$ and $y_2 = 2x + 1$ on the same screen.
Display:

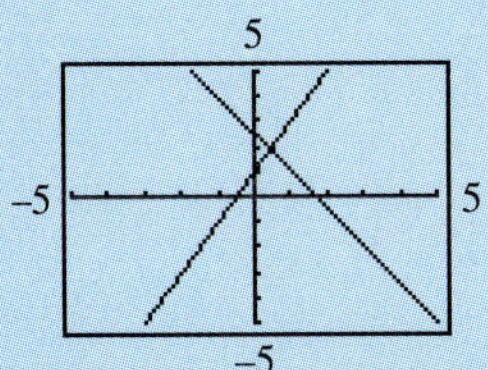

Next you can use the Trace and Zoom features to find the intersection of the two lines. Some graphing calculators have a command to find and calculate the intersection.
Display:

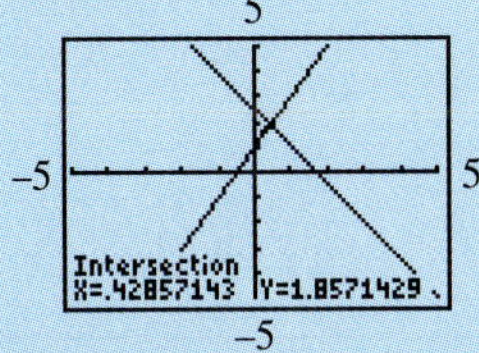

Rounded to four decimal places the solution is $x = 0.4286$ and $y = 1.8571$. Observe that $\frac{3}{7} = 0.4285714\ldots$ and $\frac{13}{7} = 1.8571428\ldots$ so the answer agrees with the answer found in Example 3.

8.2 Exercises

Find the solution for each system of equations by the substitution method. Check your answers.

1. $y = 4x - 3$
$3x - y = 2$
$3x - (4x - 3) = 2$
$x = 1;\ y = 1$

2. $x = 6 + 2y$
$3x - 2y = 12$
$3(6 + 2y) - 2y = 12$
$y = -\frac{3}{2};\ x = 3$

3. $4x + 3y = 9$
$x - 3y = 6$
$4(3y + 6) + 3y = 9$
$y = -1;\ x = 3$

4. $x + 4y = 4$
$-x + 2y = 2$
$x = 4 - 4y$
$-(4 - 4y) + 2y = 2$
$y = 1;\ x = 0$

5. $2x + y = 4$
$2x - y = 0$
$2x = y$
$2x + 2x = 4$
$x = 1;\ y = 2$

6. $7x - 3y = -10$
$x + 3y = 2$
$x = 2 - 3y$
$7(2 - 3y) - 3y = -10$
$14 - 21y - 3y = -10$
$y = 1;\ x = -1$

7. $3x - 4y = -8$
$y = 2x$
$3x - 4(2x) = -8$
$x = \frac{8}{5};\ y = \frac{16}{5}$

8. $2x - 7y = -3$
$x = 3y$
$2(3y) - 7y = -3$
$y = 3;\ x = 9$

9. $5x + 2y = 5$
$3x + y = 4$
$y = 4 - 3x$
$5x + 2(4 - 3x) = 5$
$x = 3;\ y = -5$

10. $4x + 2y = 4$
$3x + y = 4$
$y = 4 - 3x$
$4x + 2(4 - 3x) = 4$
$x = 2;\ y = -2$

11. $-x + 2y = -3$
$-2x + 3y = -6$
$x = 2y + 3$
$-2(2y + 3) + 3y = -6$
$y = 0;\ x = 3$

12. $6x - y = 6$
$-x + 3y = -1$
$y = 6x - 6$
$-x + 3(6x - 6) = -1$
$x = 1;\ y = 0$

13. $3a - 5b = 2$
$3a + b = 32$
$b = 32 - 3a$
$3a - 5(32 - 3a) = 2$
$b = 5;\ a = 9$

14. $a - 5b = 21$
$3a + 4b = -13$
$a = 21 + 5b$
$3(21 + 5b) + 4b = -13$
$a = 1;\ b = -4$

15. $3x - 4y = 6$
$-2x - y = -4$
$y = -2x + 4$
$3x - 4(-2x + 4) = 6$
$x = 2;\ y = 0$

16. $-2x - 3y = -1$
$3x - y = 7$
$y = 3x - 7$
$-2x - 3(3x - 7) = -1$
$x = 2;\ y = -1$

17. $p + 2q - 4 = 0$
$7p - q - 3 = 0$
$q = 7p - 3$
$p + 2(7p - 3) - 4 = 0$
$p = \frac{2}{3};\ q = \frac{5}{3}$

18. $7s + 2t - 10 = 0$
$5s - t + 5 = 0$
$t = 5s + 5$
$7s + 2(5s + 5) - 10 = 0$
$s = 0;\ t = 5$

19. $3x - y - 9 = 0$
$8x + 5y - 1 = 0$
$y = 3x - 9$
$8x + 5(3x - 9) - 1 = 0$
$x = 2;\ y = -3$

20. $8x + 2y - 7 = 0$
$-2x - y + 2 = 0$
$y = -2x + 2$
$8x + 2(-2x + 2) - 7 = 0$
$x = \frac{3}{4};\ y = \frac{1}{2}$

21. $\frac{5}{3}x + \frac{1}{3}y = -3$
$-2x + 3y = 24$
$5x + y = -9$
$y = -5x - 9$
$-2x + 3(-5x - 9) = 24$
$x = -3;\ y = 6$

22. $-x + y = -4$
$\frac{3x}{7} + \frac{2y}{3} = 5$
$y = x - 4$
$9x + 14y = 105$
$9x + 14(x - 4) = 105$
$x = 7;\ y = 3$

23. $3x - 2y = -11$
$x - \frac{1}{3}y = \frac{-10}{3}$
$x = \frac{1}{3}y - \frac{10}{3}$
$3\left(\frac{1}{3}y - \frac{10}{3}\right) - 2y = -11$
$y - 10 - 2y = -11$
$x = -3;\ y = 1$

24. $\frac{1}{6}x - \frac{1}{9}y = -1$
$x + 4y = 8$
$3x - 2y = -18$
$x = -4y + 8$
$3(-4y + 8) - 2y = -18$
$x = -4;\ y = 3$

25. $\frac{3}{5}x - \frac{2}{5}y = 5$
$x + 4y = -1$
$3x - 2y = 25$
$x = -4y - 1$
$3(-4y - 1) - 2y = 25$
$x = 7;\ y = -2$

26. $-\frac{1}{3}x + \frac{1}{2}y = 1$
$2x - y = 2$
$-2x + 3y = 6$
$y = 2x - 2$
$-2x + 3(2x - 2) = 6$
$x = 3;\ y = 4$

27. $-\frac{1}{2}x + \frac{1}{6}y = 2$
$3x - 2y = -15$
$-3x + y = 12$
$y = 3x + 12$
$3x - 2(3x + 12) = -15$
$x = -3;\ y = 3$

28. $\frac{1}{4}x + \frac{1}{8}y = 1$
$3x - 2y = 5$
$2x + y = 8$
$y = -2x + 8$
$3x - 2(-2x + 8) = 5$
$x = 3;\ y = 2$

★ **29.** $3(x + y) = 10 - y$
$x - 2 = 3(y - 1)$
$3x + 4y = 10$
$x - 3y = -1 \Rightarrow x = 3y - 1$
$3(3y - 1) + 4y = 10$
$x = 2;\ y = 1$

★ **30.** $2(x - 8) = 1 - y$
$2(y - 2x) = -6$
$2x + y = 17 \Rightarrow y = -2x + 17$
$2y - 4x = -6$
$2(-2x + 17) - 4x = -6$
$x = 5;\ y = 7$

★ **31.** $\frac{1}{4}x - 3y = 0$
$\frac{3}{4}x + 6y = \frac{15}{4}$
$x - 12y = 0 \Rightarrow x = 12y$
$3x + 24y = 15$
$3(12y) + 24y = 15$
$x = 3;\ y = \frac{1}{4}$

★ **32.** $\frac{1}{8}x + \frac{1}{6}y = 4$
$-\frac{1}{4}x + \frac{1}{2}y = 2$
$-x + 2y = 8 \Rightarrow x = 2y - 8$
$3x + 4y = 96$
$3(2y - 8) + 4y = 96$
$x = 16;\ y = 12$

Solve for x and y by the substitution method. Round your answer to four decimal places.

33. $2.9864x + y = 9.8632$
$1.8926x + 4.7618y = 10.7891$
$x = 2.9346;\ y = 1.0993$

34. $2.0153x + 3.1087y = -13.1286$
$x - 2.2154y = 4.1387$
$x = -2.1415;\ y = -2.8348$

Verbal and Writing Skills

35. How many equations do you think you would need in order to solve a system with three unknowns? With seven unknowns? Explain your reasoning. 3; 7; Use substitution to reduce the system to one equation with one unknown. Solve. Then work backward to solve for each of the other variables.

36. The point(s) where the graphs of the linear equations intersect is the ___solution(s)___ to the system.

37. The solution to a system must satisfy ___both___ equations.

38. How many solutions will a system of equations have if the graphs of the lines intersect? Justify your answer. One. Since the solution must satisfy both equations, the point must lie on both lines. This is the point of intersection.

39. How many solution(s) will a system of equations have if the graphs are parallel lines? Why? No solution. Parallel lines have no points in common.

Cumulative Review Problems

40. Find the slope and y-intercept of $7x + 11y = 19$.

$m = -\frac{7}{11};\ b = \frac{19}{11}$

41. Graph $3x - y < 2$.

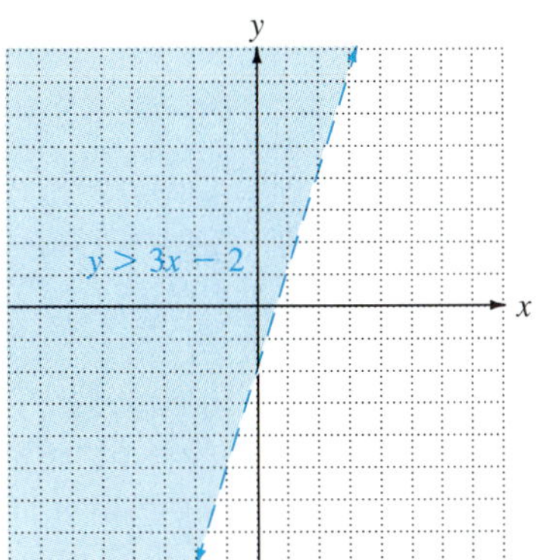

42. Find the running speed r of a 85-kilogram runner who is using up energy at the rate of 35 kilocalories per minute. Use the equation $85(dr - g) = 35$ where d is 0.0011, g is 0.026, and r is the running speed measured in meters per minute. Round your answer to the nearest whole number. approximately 398 meters per minute

43. West Coast Scholastic Construction has been given a contract to build the new wing of the Benson Memorial Middle School for \$3.8 million. To avoid opening the school at an inconvenient time, the construction company was offered a 3% bonus if the work was completed 3 months ahead of schedule. What will be the total cost of the construction if the work is completed 3 months ahead of schedule? \$3,914,000

8.3 Solve a System of Equations by the Addition Method

MathPro

Video 20

SSM

After studying this section, you will be able to:

1. *Use the addition method to solve a system of two linear equations with integer coefficients.*
2. *Use the addition method to solve a system of two linear equations with fractional coefficients.*
3. *Use the addition method to solve a system of two linear equations with decimal coefficients.*

1 *Using the Addition Method to Solve a System of Two Linear Equations with Integer Coefficients*

The substitution method is useful when the coefficient of one variable is 1. In the following system, the variable x in the first equation has a coefficient of 1, and we can easily solve for x.

$$x - 2y = 7 \Rightarrow x = 7 + 2y$$
$$-5x + 4y = -5$$

This makes the substitution method for solving a system a natural choice. We may not always have 1 as a coefficient of any variable. We need a method of solving systems of equations that will work for integer coefficients, fractional coefficients, and decimal coefficients. One such method is the addition method.

EXAMPLE 1 Find the solution to the following system of equations by the *addition method.*

$2x + 5y = -2$	*Add the two equations. Since $5y - 5y = 0$, we have*
$\underline{-7x - 5y = 32}$	*eliminated the y variable.*
$-5x = 30$	
$\dfrac{-5x}{-5} = \dfrac{30}{-5}$	*Divide both sides by -5.*
$x = -6$	*Simplify.*
$2(-6) + 5y = -2$	*Substitute $x = -6$ into one of the original equations.*
$-12 + 5y = -2$	*Remove parentheses.*
$5y = 10$	*Add 12 to both sides.*
$y = 2$	*Divide both sides by 5.*

The solution is $x = -6$ and $y = 2$.

Notice that when we add the two equations together, one variable is eliminated. The addition method is therefore often called the *elimination* method.

Practice Problem 1 Find the solution to the system of equations by the addition method.

$$3x + 5y = 1$$
$$y - 3x = 11$$

(Be careful to note the position of the variables in this problem. You may change the position of the variables if needed.) ■

Most systems of equations cannot be solved this easily by the addition method. If you look back at Example 1, you will notice that there was a $5y$ in the first equation and a $-5y$ in the second equation. When these equations were added, the y terms became 0 and were eliminated. Our goal in the following examples will be to eliminate one of the variables.

EXAMPLE 2 Use the addition method to solve this system.

$$5x + 2y = 7 \qquad \textbf{(1)}$$
$$3x - y = 13 \qquad \textbf{(2)}$$

We would like the coefficients of the y terms to be equal in value and opposite in sign. One should be $+2y$ and the other $-2y$. Therefore, we multiply each term of equation **(2)** by 2.

$2(3x) - 2(y) = 2(13)$	*Multiply each term of equation* **(2)** *by* 2.
$5x + 2y = 7$	*We now have an equivalent system of equations.*
$6x - 2y = 26$	*Add the two equations. This will eliminate the*
$11x \quad = 33$	*y variable.*
$x = 3$	*Divide both sides by* 11.
$5(3) + 2y = 7$	*Substitute* $x = 3$ *into one of the original equations. Arbitrarily, we pick equation* **(1)**.
$15 + 2y = 7$	*Remove parentheses.*
$2y = -8$	*Subtract* 15 *from both sides.*
$y = -4$	*Divide both sides by* 2.

The solution is $x = 3$ and $y = -4$.

Check: Replace x by 3 and y by -4 in *both* original equations.

$5x + 2y = 7$	$3x - y = 13$
$5(3) + 2(-4) \stackrel{?}{=} 7$	$3(3) - (-4) \stackrel{?}{=} 13$
$15 - 8 \stackrel{?}{=} 7$	$9 + 4 \stackrel{?}{=} 13$
$7 = 7$ ✓	$13 = 13$ ✓

TEACHING TIP After discussing Example 2, present students with the following problem. Show the students this system of equations:

$$12x + 5y = -9$$
$$2x - 7y = -25$$

Ask the students, "By what number should you multiply one equation in order that either the x variables or y variables will drop out?" Allow them to pause and think, and contribute some suggestions.

Soon, students will reach an agreement that we should multiply the second equation by -6 so that the x variables will drop out when the equations are added. Then finish the problem as an Additional Class Exercise.

Sol.
$$12x + 5y = -9$$
$$-12x + 42y = 150$$
$$47y = 141$$
$$y = 3$$
$$2x - 7(3) = -25$$
$$x = -2$$

Practice Problem 2 Solve by addition.

$$3x + y = 7$$
$$5x - 2y = 8$$

For convenience, we will make a list of the steps we are using to solve a system of equations by the addition method.

Procedure for Solving a System of Equations by the Addition Method

1. Multiply each term of one or both equations by some nonzero integer so that the coefficients of one of the variables are opposites.
2. Add the equations of this new system.
3. Solve the resulting equation for the remaining variable.
4. Substitute the value found in step 3 into either of the original equations to find the value of the second variable.
5. Check your solution in both of the original equations.

TEACHING TIP After you have discussed Example 3 with students, present them with the following systems of equations as an Additional Class Exercise. In each case ask them by what should they multiply the top equation and by what should they multiply the bottom equation to most easily eliminate the x variable?

A. $2x + 5y = 3$
$-5x + 3y = -23$

B. $5x + 4y = 5$
$-7x + 6y = -36$

C. $9x + 5y = 14$
$4x - 7y = -3$

Sol.

A. Multiply top by 5, bottom by 2

B. Multiply top by 7, bottom by 5

C. Multiply top by 4, bottom by -9 *or* multiply top by -4, bottom by 9

EXAMPLE 3 Use the addition method to solve.

$$3x + 4y = 7 \quad \textbf{(1)}$$
$$2x + 7y = 9 \quad \textbf{(2)}$$

Suppose we want the x terms to have opposite coefficients. One equation could have $+6x$ and the other $-6x$. This would happen if we multiply equation **(1)** by $+2$ and equation **(2)** by -3. We will do so.

$(2)(3x) + (2)(4y) = (2)(7)$ *Multiply each term of equation* **(1)** *by* 2.

$(-3)(2x) + (-3)(7y) = (-3)(9)$ *Multiply each term of equation* **(2)** *by* -3.

(Often this step of multiplying each equation by 2 or -3 can be done mentally.)

$6x + 8y = 14$ *We now have an equivalent system of equations.*

$-6x - 21y = -27$ *The coefficients of the x terms are opposite.*

$-13y = -13$ *Add the two equations to eliminate the variable x.*

$y = 1$ *Divide both sides by* -13.

Substitute $y = 1$ into one of the original equations.

$$3x + 4(1) = 7$$

$3x + 4 = 7$ *Solve for x.*

$$3x = 3$$
$$x = 1$$

The solution is $x = 1$ and $y = 1$. Check the solution in both original equations.

Alternative solution: You could also have eliminated the y variable. For example, if you multiply the first equation by 7 and the second equation by -4, you would obtain the equivalent system.

$$21x + 28y = 49$$
$$-8x - 28y = -36$$

You can add these equations to eliminate the y variable. Since the numbers involved in this approach are somewhat larger, it is probably wiser to eliminate the x variable in this example.

Note: A common error occurs when students forget that we want to obtain two terms with *opposite signs* that when added will equal zero. If all the coefficients in the equations are positive, such as Example 3, it will be necessary to multiply *one* of the equations by a negative number.

Practice Problem 3 Solve by addition.

$$4x + 5y = 7$$
$$3x + 7y = 2$$

2 Using the Addition Method to Solve a System of Two Linear Equations with Fractional Coefficients

If the system of equations has fractional coefficients, you should first clear each equation of the fractions. To do so, you will need to multiply each term in the equation by the LCD of the fractions.

TEACHING TIP After you have presented Example 4, ask students to do the following similar problem as an Additional Class Exercise.

Solve the system:

$$\frac{4}{7}x + \frac{3}{7}y = -1$$

$$-\frac{1}{5}x + \quad y = \frac{19}{5}$$

Sol. $x = -4$, $y = 3$

EXAMPLE 4 Solve.

$$x - \frac{5}{2}y = \frac{5}{2} \qquad \textbf{(1)}$$

$$\frac{4}{3}x + \quad y = \frac{23}{3} \qquad \textbf{(2)}$$

$2(x) - 2\left(\frac{5}{2}y\right) = 2\left(\frac{5}{2}\right)$ *Multiply each term of equation* **(1)** *by* 2.

$2x - 5y = 5$

$3\left(\frac{4}{3}x\right) + 3(y) = 3\left(\frac{23}{3}\right)$ *Multiply each term of equation* **(2)** *by* 3.

$4x + 3y = 23$

We now have an equivalent system of equations that does not contain fractions.

$$2x - 5y = 5 \qquad \textbf{(3)}$$

$$4x + 3y = 23 \qquad \textbf{(4)}$$

Let us eliminate the x variable. We want the coefficients of x to be opposites.

$(-2)(2x) - (-2)5y = (-2)(5)$ *Multiply each term of equation* **(3)** *by* -2.

$-4x + 10y = -10$ *We now have an equivalent system of equations.*

$4x + 3y = 23$ *The coefficients of the x terms are opposites.*

$13y = 13$ *Add the two equations to eliminate the variable x.*

$y = 1$ *Divide both sides by* 13.

$4x + 3(1) = 23$ *Substitute* $y = 1$ *into equation* **(4)**.

$4x + 3 = 23$

$4x = 20$

$x = 5$

The solution is $x = 5$ and $y = 1$.

Check: Check the solution in both of the *original* equations.

Practice Problem 4 Solve.

$$\frac{2}{3}x - \frac{3}{4}y = 3$$

$$-2x + \quad y = 6$$ ■

3 *Using the Addition Method to Solve a System of Two Linear Equations with Decimal Coefficients*

Some linear equations will have decimal coefficients. It will be easier to work with the equations if we change the decimal coefficients to integer coefficients. To do so, we will multiply each term of the equation by a power of 10.

EXAMPLE 5 Solve.

$$0.12x + 0.05y = -0.02 \quad \textbf{(1)}$$
$$0.08x - 0.03y = -0.14 \quad \textbf{(2)}$$

Since the decimals are hundredths, we will multiply each term of both equations by 100.

$$100(0.12x) + 100(0.05y) = 100(-0.02)$$
$$100(0.08x) + 100(-0.03y) = 100(-0.14)$$
$$12x + 5y = -2 \quad \textbf{(3)}$$
$$8x - 3y = -14 \quad \textbf{(4)}$$

We now have an equivalent system of equations that has integer coefficients.

We will eliminate the variable y. We want the coefficients of the y terms to be opposites.

$$36x + 15y = -6$$
$$40x - 15y = -70$$
Multiply equation **(3)** *by* 3 *and equation* **(4)** *by* 5.

$$76x \qquad = -76$$
Add the equations.

$$x = -1$$
Solve for x.

$$12(-1) + 5y = -2$$
Replace x by -1 *in equation* **(3)**, *and solve for y.*

$$-12 + 5y = -2$$
$$5y = -2 + 12$$
$$5y = 10$$
$$y = 2$$

The solution is $x = -1$ and $y = 2$.

Check: We substitute $x = -1$ and $y = 2$ into the original equations. I'm sure most students would rather not use equation **(1)** and equation **(2)**! However, *we must check the solutions in the original equations* if we want to be sure of our answers. It is possible to make an error going from equations **(1)** and **(2)** to equations **(3)** and **(4)**. The solutions would satisfy equations **(3)** and **(4)**, but would not satisfy equations **(1)** and **(2)**.

$$0.12x + 0.05y = -0.02 \qquad\qquad 0.08x - 0.03y = -0.14$$
$$0.12(-1) + 0.05(2) \stackrel{?}{=} -0.02 \qquad\qquad 0.08(-1) - 0.03(2) \stackrel{?}{=} -0.14$$
$$-0.12 + 0.10 \stackrel{?}{=} -0.02 \qquad\qquad -0.08 - 0.06 \stackrel{?}{=} -0.14$$
$$-0.02 = -0.02 \checkmark \qquad\qquad -0.14 = -0.14 \checkmark$$

TEACHING TIP Before students start the exercise set, you may wish to do one system of equations as an example that requires some algebraic operations to place in a standard form for a system of two linear equations. The students will encounter this type of problem in Exercises 8.3, Problems 40–43.

Solve the system:

$$7 + 3(a + b) = b + 8$$
$$3(b - 4) = -2a - 3$$

Sol. Removing parentheses, we have

$$3a + 2b = 1$$
$$2a + 3b = 9$$
$$a = -3;\ b = 5$$

Practice Problem 5 Solve. $0.2x + 0.3y = -0.1$

$0.5x - 0.1y = -1.1$ ■

Warning

A common error is to find the value for x and then stop. You have not solved a system in x and y until you find both the x and the y values that make both equations true.

8.3 Exercises

Verbal and Writing Skills

1. Look at the following systems of equations. Decide which variable to eliminate in each system, and explain how you would eliminate that variable.

(a) $3x + 2y = 5$
$5x - y = 3$

(b) $2x - 9y = 1$
$2x + 3y = 2$

(c) $4x - 3y = 10$
$5x + 4y = 0$

(a) Multiply equation (2) by 2. This will eliminate y.
(b) Multiply equation (1) or equation (2) by -1. This will eliminate x.
(c) Multiply equation (1) by 4. Multiply equation (2) by 3. This will eliminate y.

Find the solution to each system of equations by the addition method. *Check your answers.*

2. $2x + y = -1$
$3x - y = -9$
$5x = -10$
$x = -2;\ y = 3$

3. $2x + y = 2$
$4x - y = -8$
$6x = -6$
$x = -1;\ y = 4$

4. $2x + 3y = 1$
$x - 2y = 4$
$2x + 3y = 1$
$-2x + 4y = -8$
$x = 2;\ y = -1$

5. $2x + y = 4$
$3x - 2y = -1$
$4x + 2y = 8$
$3x - 2y = -1$
$x = 1;\ y = 2$

6. $x + 5y = 2$
$2x + 3y = -3$
$-2x - 10y = -4$
$2x + 3y = -3$
$x = -3;\ y = 1$

7. $x + y = 5$
$2x - y = -5$
$3x = 0$
$x = 0;\ y = 5$

8. $3x - 5y = 1$
$2x - y = -4$
$3x - 5y = 1$
$-10x + 5y = 20$
$x = -3;\ y = -2$

9. $3x - 2y = -6$
$x - 3y = 5$
$3x - 2y = -6$
$-3x + 9y = -15$
$y = -3;\ x = -4$

10. $2x + 3y = 1$
$3x + 2y = -6$
$4x + 6y = 2$
$-9x - 6y = 18$
$x = -4;\ y = 3$

11. $2x + 5y = 11$
$3x + 8y = 16$
$-6x - 15y = -33$
$6x + 16y = 32$
$x = 8;\ y = -1$

12. $a - 2b = 0$
$6a - 13b = 5$
$-6a + 12b = 0$
$6a - 13b = 5$
$a = -10;\ b = -5$

13. $a + 3b = 0$
$-3a - 10b = -2$
$3a + 9b = 0$
$b = 2;\ a = -6$

14. $8x + 6y = -2$
$10x - 9y = -8$
$24x + 18y = -6$
$20x - 18y = -16$
$y = \frac{1}{3};\ x = -\frac{1}{2}$

15. $4x + 9y = 0$
$8x - 5y = -23$
$-8x - 18y = 0$
$-23y = -23$
$y = 1;\ x = -\frac{9}{4}$

16. $2x + 3y = -8$
$5x + 4y = -34$
$10x + 15y = -40$
$-10x - 8y = 68$
$x = -10,\ y = 4$

17. $9x - 2y = 14$
$4x + 3y = 14$
$27x - 6y = 42$
$8x + 6y = 28$
$x = 2;\ y = 2$

18. $5x - 2y = 6$
$6x + 7y = 26$
$35x - 14y = 42$
$12x + 14y = 52$
$x = 2;\ y = 2$

19. $-5x + 4y = -13$
$11x + 6y = -1$
$-55x + 44y = -143$
$55x + 30y = -5$
$x = 1;\ y = -2$

20. $3x + 7y = 15$
$-5x + 2y = 16$
$15x + 35y = 75$
$-15x + 6y = 48$
$x = -2;\ y = 3$

21. $5x - 2y = 24$
$4x - 5y = 9$
$25x - 10y = 120$
$-8x + 10y = -18$
$x = 6;\ y = 3$

Verbal and Writing Skills

22. Before using the addition method to solve each of the following systems, you will need to simplify the equation(s). Explain how you would change the fractional coefficients to integer coefficients in each system.

(a) $\frac{1}{4}x - \frac{3}{4}y = 2$
$2x + 3y = 1$

(b) $5x + 2y = 9$
$\frac{4}{9}x - \frac{2}{3}y = 4$

(c) $\frac{1}{2}x + \frac{2}{3}y = \frac{1}{3}$
$\frac{3}{4}x - \frac{4}{5}y = 2$

(a) Multiply equation (1) by 4.
(b) Multiply equation (2) by 9.
(c) Multiply equation (1) by 6. Multiply equation (2) by 20.

Find the solution to each system of equations. Check your answers.

23. $x + \frac{5}{4}y = \frac{9}{4}$
$\frac{2}{5}x - y = \frac{3}{5}$

$4x + 5y = 9$
$2x - 5y = 3$
$x = 2;\ y = \frac{1}{5}$

24. $\frac{2}{3}x + y = 2$
$x + \frac{1}{2}y = 7$

$2x + 3y = 6$, $2x + y = 14$ $\Rightarrow$ $2x + 3y = 6$, $-2x - y = -14$
$x = 9;\ y = -4$

25. $\frac{4}{3}x + \frac{4}{5}y = 0$
$-2x + 6y = 36$

$20x + 12y = 0$, $-2x + 6y = 36$ $\Rightarrow$ $20x + 12y = 0$, $-20x + 60y = 360$
$x = -3;\ y = 5$

26. $\frac{5}{6}x - 5y = \frac{15}{2}$
$6x - 2y = 20$

26. $5x - 30y = 45$, $6x - 2y = 20$ $\Rightarrow$ $5x - 30y = 45$, $-90x + 30y = -300$
$x = 3;\ y = -1$

27. $\frac{5}{6}x + y = -\frac{1}{3}$
$-8x + 9y = 28$

$5x + 6y = -2$, $-8x + 9y = 28$ $\Rightarrow$ $15x + 18y = -6$, $16x - 18y = -56$
$x = -2;\ y = \frac{4}{3}$

28. $\frac{2}{9}x - \frac{1}{3}y = 1$
$4x + 9y = -2$

$2x - 3y = 9$, $4x + 9y = -2$ $\Rightarrow$ $6x - 9y = 27$, $4x + 9y = -2$
$x = \frac{5}{2};\ y = -\frac{4}{3}$

29. $\frac{1}{2}x - \frac{3}{4}y = -\frac{1}{2}$
$\frac{1}{6}x + \frac{1}{2}y = -\frac{5}{3}$

$2x - 3y = -2$
$x + 3y = -10$
$x = -4;\ y = -2$

30. $\frac{x}{5} - \frac{y}{3} = \frac{4}{5}$
$-\frac{x}{7} + \frac{y}{3} = -\frac{8}{7}$

$3x - 5y = 12$
$-3x + 7y = -24$
$y = -6;\ x = -6$

Verbal and Writing Skills

31. Explain how you would change each of the following systems to an equivalent system with integer coefficients. Write the equivalent system.

(a) $0.5x - 0.3y = 0.1$
$5x + 3y = 6$

(b) $0.08x + y = 0.05$
$2x - 0.1y = 3$

(c) $4x + 0.5y = 9$
$0.2x - 0.05y = 1$

(a) Multiply equation (1) by 10.
$5x - 3y = 1;$
$5x + 3y = 6$

(b) Multiply equation (1) by 100. Multiply equation (2) by 10.
$8x + 100y = 5;$
$20x - y = 30$

(c) Multiply equation (1) by 10. Multiply equation (2) by 100.
$40x + 5y = 90;$
$20x - 5y = 100$

Solve for x and y using the addition method.

32. $0.2x - 0.3y = 0.4$
$0.3x - 0.4y = 0.9$

$2x - 3y = 4 \Rightarrow 6x - 9y = 12$
$3x - 4y = 9 \Rightarrow -6x + 8y = -18$
$x = 11;\ y = 6$

33. $0.5x - 0.2y = 0.5$
$0.4x + 0.7y = 0.4$

$5x - 2y = 5 \Rightarrow 35x - 14y = 35$
$4x + 7y = 4 \Rightarrow 8x + 14y = 8$
$x = 1;\ y = 0$

34. $0.02x - 0.04y = 0.26$
$0.07x - 0.09y = 0.66$

$2x - 4y = 26 \Rightarrow 14x - 28y = 182$
$7x - 9y = 66 \Rightarrow -14x + 18y = -132$
$y = -5;\ x = 3$

35. $0.04x - 0.03y = 0.05$
$0.05x + 0.08y = -0.76$
$4x - 3y = 5 \Rightarrow 20x - 15y = 25$
$5x + 8y = -76 \Rightarrow -20x - 32y = 304$
$-47y = 329$
$x = -4;\ y = -7$

36. $0.4x - 5y = -1.2$
$-0.03x + 0.5y = 0.14$
$4x - 50y = -12$
$-3x + 50y = 14$
$x = 2;\ y = \frac{2}{5}$ or 0.4

37. $-0.6x - 0.08y = -4$
$3x + 2y = 4$
$-60x - 8y = -400 \Rightarrow -60x - 8y = -400$
$30x + 20y = 40 \Rightarrow 60x + 40y = 80$
$x = 8;\ y = -10$

38. $2x + 3y = 13$
$351x + 274y = 2226$
(Round answer to four decimal places)
$x = 6.1703;\ y = 0.2198$

39. $3x + 5y = 11$
$291x + 243y = 1551$
$x = 7;\ y = -2$

To Think About

Find the solution to each system of equations.

40. $y + 1 = 2(x - 6)$
$4(-x - 2) = 9y - 1$
$x = 5;\ y = -3$

41. $5(x - y) = -7 - 3y$
$3(2 + x) = y + 1$
$x = -3;\ y = -4$

42. $5(x + 5) = 2(y + 13)$
$3x + 11 = 2(17 - y)$
$x = 3;\ y = 7$

43. $3(x - 6) = -2(1 + y)$
$7(x + 3) = 2y + 25$
$x = 2;\ y = 5$

Let a and b represent nonzero constants. Find the solution to each system of equations by the addition method.

44. $ax + by = b^2 + a^2$
$-bx + ay = b^2 + a^2$
$abx + b^2y = b^3 + a^2b$
$-abx + a^2y = ab^2 + a^3$
$x = a - b;\ y = a + b$

45. $-\frac{x}{a} + \frac{y}{b} = -2$
$\frac{x}{a} + \frac{y}{b} = 6$
$-bx + ay = -2ab$
$bx + ay = 6ab$
$x = 4a;\ y = 2b$

Cumulative Review Problems

46. Carlos and Bobby are professional race car drivers. They are in a 100-lap race on an oval track that is 4.75 miles long. Carlos made 4 complete laps during the time that Bobby was out of the race for repairs to his car. Carlos maintains an average speed of 95 miles per hour for the entire race. What average speed will Bobby have to achieve in order to win the race? (Round your answer to the nearest whole number.) Bobby needs an average speed of 99 miles per hour or greater.

47. During one point in the season in 1994, the Boston Red Sox had won 45 baseball games and lost 55 games. Tricia and Mindy were discussing the potential for the Red Sox to win 50 percent of all their games for the season. If the team had 62 games remaining at that point, how many of the future games would they need to win in order to achieve that percent? They would need to win 36 of the next 62 games.

48. Factor $75x^2 - 3$ completely.
$3(5x - 1)(5x + 1)$

49. Factor. $6x^2 - 25x + 25$
$(2x - 5)(3x - 5)$

Putting Your Skills to Work

USING MATHEMATICS TO MAKE COMPUTERS FASTER

Recently two researchers at the University of Tokyo, Junichiro Makino and Makoto Taiji, announced that they had built a computer that will do 1.08 trillion floating point operations per second. The fastest conventional supercomputer can theoretically only do things at a maximum speed of 280 billion floating point operations per second. These two men designed this special computer to do only one thing: calculate the force of gravity of one heavenly body on another. This specialized computer will perform tasks involving the analysis of stars in 3 months that would take 5 years on an ordinary supercomputer. In computer design, it is not usually possible to measure how long the computer takes to do one specific task, but rather how long it takes to do several tasks, because the speeds are too amazingly fast. See if you can answer the following questions.

PROBLEMS FOR INDIVIDUAL INVESTIGATION

Express your answers in scientific notation.

1. How long in seconds does it take to do one floating point operation with the new computer produced by Makino and Taiji? 9.259×10^{-13} second

2. How long in seconds does it take to do one floating point operation at the maximum speed of the fastest conventional supercomputer? 3.571×10^{-12} second

PROBLEMS FOR COOPERATIVE INVESTIGATION AND ANALYSIS

Together with other members of your class see if you can determine the following.

3. How many floating point operations (of the type designed by Makino and Taiji) can be done by a new experimental computer in the time it takes the fastest conventional supercomputer? more than 3 operations

INTERNET CONNECTIONS: Go to ``http://www.prenhall.com/tobey'' to be connected

Site: *Enabling Technologies for Petaflops Computing* or a related site

This site gives a review of a book describing the computers of the future, which will be even faster than the one designed by Makino and Taiji. Use the information from the review to answer the following questions.

4. A petaFLOP computer would be how many times as fast as Makino and Taiji's computer?

5. How long in seconds will it take to do one floating point operation with a petaFLOP computer? Express your answer in scientific notation.

8.4 Review of Methods for Solving Systems of Equations

After studying this section, you will be able to:

1 *Choose an appropriate method to solve a system of equations algebraically.*
2 *Identify inconsistent and dependent systems by algebraic methods.*

1 *Choosing an Appropriate Method to Solve a System of Equations Algebraically*

At this point we will review the algebraic methods for solving systems of linear equations and discuss the advantages and disadvantages of each method.

Method	Advantage	Disadvantage
Substitution	Works well if one or more variables have a coefficient of 1 or -1.	Often becomes difficult to use if no variable has a coefficient of 1 or -1.
Addition	Works well if equations have fractional or decimal coefficients. Works well if no variable has a coefficient of 1 or -1.	None

EXAMPLE 1 Select a method and solve each system of equations.

(a) $x + y = 3080$
$2x + 3y = 8740$

(b) $5x - 2y = 19$
$-3x + 7y = 35$

(a) Since both x and y have coefficients of 1, we will select the substitution method.

$y = 3080 - x$ *Solve the first equation for y.*

$2x + 3(3080 - x) = 8740$ *Substitute the expression into the second equation.*

$2x + 9240 - 3x = 8740$ *Remove parentheses.*

$-1x = -500$ *Simplify.*

$x = 500$ *Divide each side by -1.*

$y = 3080 - x$

$= 3080 - 500$ *Substitute 500 for x.*

$= 2580$ *Simplify.*

The solution is $x = 500$ and $y = 2580$.

(b) Because neither x nor y has a coefficient of 1 or -1, we select the addition method. We choose to eliminate the y variable. Thus, we would like the coefficients of y to be -14 and 14.

$7(5x) - 7(2y) = 7(19)$ *Multiply each term of the first equation by 7.*

$2(-3x) + 2(7y) = 2(35)$ *Multiply each term of the second equation by 2.*

$35x - 14y = 133$ *We now have an equivalent system of equations.*
$-6x + 14y = 70$

$29x = 203$ *Add the two equations.*

$x = 7$ *Divide each side by 29.*

Substitute $x = 7$ into one of the original equations.

$$5(7) - 2y = 19$$

$$35 - 2y = 19 \qquad \textit{Solve for } y.$$

$$-2y = -16$$

$$y = 8$$

The solution is $x = 7$ and $y = 8$.

Practice Problem 1 Select a method and solve each system of equations.

(a) $3x + 5y = 1485$
$x + 2y = 564$

(b) $7x + 6y = 45$
$6x - 5y = -2$ ■

2 *Identifying Inconsistent and Dependent Systems by Algebraic Methods*

Recall that an inconsistent system of linear equations is a system of parallel lines. Since parallel lines never intersect, the system has no solution. Could we determine this algebraically?

EXAMPLE 2 Solve the following system algebraically.

$$3x - y = -1$$

$$3x - y = -7$$

Clearly, the addition method would be very convenient in this case.

TEACHING TIP Note that when all *letters* vanish but numbers don't vanish and the result is *not true,* we have parallel lines and *no solution.*

$3x - y = -1$	*Keep the first equation unchanged.*
$-3x + y = +7$	*Multiply each term in the second equation by -1.*
$0 = 6$	*Add the two equations.*

Notice we have $0 = 6$, which we know is not true. The statement $0 = 6$ is inconsistent with known mathematical facts. No possible x and y values can make that equation true. There is no solution to this system of equations.

This example shows us that, if we obtain a mathematical statement that is not true (inconsistent with known facts), we can identify the entire system as inconsistent. If graphed, these lines would be parallel.

Practice Problem 2 Solve this system algebraically.

$$4x + 2y = 2$$

$$-6x - 3y = 6$$ ■

What happens if we try to solve a dependent system of equations algebraically? Recall that a dependent system consists of two lines that coincide (are the same line).

EXAMPLE 3 Solve the following system algebraically.

$$x + y = 4$$
$$3x + 3y = 12$$

Let us use the substitution method.

$y = 4 - x$ *Solve the first equation for y.*

$3x + 3(4 - x) = 12$ *Substitute 4 − x for y in the second equation.*

$3x + 12 - 3x = 12$ *Remove parentheses.*

$12 = 12$

Notice we have $12 = 12$, which is always true. Thus we obtain an equation that is true for any value of x. An equation that is always true is called an **identity.** All the points that lie on the line will satisfy both equations. There are an infinite number of solutions.

This example shows us that, if we obtain a mathematical statement that is always true (an identity), we can identify the equations as dependent. There are an unlimited number of solutions to the system that has dependent equations.

TEACHING TIP Note that when *all letters* vanish and the result is a *true statement* (including $0 = 0$), the two equations represent the *same line* and have *infinite solutions.*

Practice Problem 3 Solve this system algebraically.

$$3x - 9y = 18$$
$$-4x + 12y = -24$$ ■

TEACHING TIP After discussing Example 3, ask students to do the following as an Additional Class Exercise.

Find the solution for each of the following systems if possible. If there is no solution, so state. If the system has an infinite number of solutions, so state.

A. $-4x + 3y = 0$
$5x - 6y = 9$

B. $6x - 2y = 8$
$-9x + 3y = -12$

C. $-15x - 18y = 10$
$5x + 6y = -2$

Answers:

A. One solution when $x = -3$ and $y = -4$

B. Infinite number of solutions

C. No solution

TWO LINEAR EQUATIONS WITH TWO VARIABLES

Graph	Number of Solutions	Algebraic Interpretation
(6, −3) Two lines intersect at one point	One unique solution	You obtain one value for x and one value for y. For example, $x = 6, y = -3$
Parallel lines	No solution	You obtain an equation that is inconsistent with known facts. For example, $0 = 6$
Lines coincide	Infinite number of solutions	You obtain an equation that is always true. For example, $8 = 8$

8.4 Exercises

Verbal and Writing Skills

1. If there is no solution to a system of linear equations, the graphs of the equations are __parallel lines__. Solving the system algebraically, you will obtain an equation that is __inconsistent__ with known facts.

2. If an algebraic attempt at solving a system of linear equations results in an identity, the system is said to be __dependent__. There are an __infinite__ number of solutions, and the graphs of the lines __coincide__.

3. If there is exactly one solution, the graphs of the equations __intersect__. This system is said to be __independent__ and __consistent__.

4. A student solved a system of equations, but obtained the equation $-36x = 0$. He was unsure of what to do next. What should he do next? What should he conclude about the system of equations? He should solve for x. $x = 0$ then substitute back into one of the original equations and solve for y. The system is independent and consistent.

If possible, solve each system by an algebraic method (substitution or addition method, without the use of graphing). Otherwise, state that the problem has no solution or an infinite number of solutions.

5. $x - 4y = 6$
$-x + 2y = 4$
$-2y = 10$
$x = -14,\ y = -5$

6. $-2x + 2y = 7$
$4x - 2y = 1$
$2x = 8$
$x = 4,\ y = \frac{15}{2}$

7. $4x - 6y = 10$
$-10x + 15y = -25$
infinite number of solutions

8. $3x + 2y = -2$
$-4x + 5y = 18$
$12x + 8y = -8$
$-12x + 15y = 54$
$y = 2,\ x = -2$

9. $-2x - 3y = 15$
$5x + 2y = 1$
$-4x - 6y = 30$
$15x + 6y = 3$
$x = 3,\ y = -7$

10. $2x - 4y = 5$
$-4x + 8y = 9$
no solution

11. $5x + 10y = 15$
$2x + 4y = -1$
no solution

12. $3x - y = 4$
$-9x + 3y = -12$
infinite number of solutions

13. $5x - 3y = 9$
$7x + 2y = -6$
$10x - 6y = 18$
$21x + 6y = -18$
$x = 0,\ y = -3$

14. $4x + 5y = 16$
$6x - 7y = 24$
$24x + 30y = 96$
$24x - 28y = 96$
$x = 4,\ y = 0$

15. $3x - 2y = 70$
$0.6x + 0.5y = 50$
$3x - 2y = 70$
$6x + 5y = 500$
$y = 40,\ x = 50$

16. $5x - 3y = 10$
$0.9x + 0.4y = 30$
$5x - 3y = 10$
$9x + 4y = 300$
$x = 20,\ y = 30$

17. $0.3x - 0.5y = 0.4$
$0.6x + 1.0y = 2.8$
$-6x + 10y = -8$
$6x + 10y = 28$
$y = 1, x = 3$

18. $-0.05x + 0.02y = -0.01$
$0.10x + 0.04y = -0.22$
$-10x + 4y = -2$
$10x + 4y = -22$
$y = -3, x = -1$

19. $\frac{4}{3}x + \frac{1}{2}y = 1$
$\frac{1}{3}x - y = -\frac{1}{2}$
$8x + 3y = 6$
$2x - 6y = -3$
$y = \frac{2}{3}, x = \frac{1}{2}$

20. $6(x + y) = 7 + y$
$7(2 + y) = 3x + 1$
$6x + 5y = 7$
$-6x + 14y = -26$
$x = 2, y = -1$

21. $3(b - 1) = 2(a + b)$
$4(a - b) = -b + 1$
$-4a + 2b = 6$
$4a - 3b = 1$
$a = -5, b = -7$

22. $a + 2(b + 1) = 1$
$4(a - b) = 22 + b$
$-4a - 8b = 4$
$4a - 5b = 22$
$a = 3, b = -2$

Solve for x and y. Round your answer to three decimal places.

23. $26x - 39y = 208$ $x = 5, y = -2$
$51x + 68y = 119$

24. $0.625x + 0.375y = 0.875$ $x = 1.688$
$0.118x + 0.294y = 0.058$ $y = -0.480$

Cumulative Review Problems

Solve for x in each of the following equations.

25. $4(x + 3) + 3(x - 4) = -3 + 2(x - 1)$ $x = -1$

26. $\frac{1}{6}x - \frac{3}{2}x = -\frac{2}{3}$ $x = \frac{1}{2}$

27. $20x + 12\left(\frac{9}{2} - x\right) = 70$ $x = 2$

28. $0.3(150) + x = 0.4(x + 150)$ $x = 25$

In 1995 the income tax rate for Federal income tax of 15% applied to single people who had a taxable income of $23,350 or less and to married people filing jointly who had a combined taxable income of $39,000 or less.

29. What was Robert's taxable income if he paid an income tax of $3225 for 1995 and he filed as a single person? $21,500

30. What was the combined taxable income of Alberto and Maria Lazano if they paid an income tax of $5775 for 1995 and they filed as a married couple? $38,500

8.5 Solve Word Problems Using a System of Equations

MathPro

Video 21

SSM

After studying this section, you will be able to:

1 *Solve a variety of applied word problems by using a system of two linear equations containing two variables.*

1 *Using a System of Equations to Solve Word Problems*

Word problems can be solved in a variety of ways. In Chapter 3 and throughout the text, you have used one equation with one unknown to describe situations found in real-life applications. Some of these same problems can be solved using two equations in two unknowns. The method you use will depend on what you find works best for you. Become familiar with both methods before you decide. Generally, you will want to use an equation to represent each fact provided in the problem situation.

TEACHING TIP After Example 1, show the students this:
Tickets for a local concert were priced differently for weekday evenings and for weekend evenings. For the Thursday concert, advance tickets were \$8 and tickets at the door were \$10. For the Friday concert, advance tickets were \$11 and tickets at the door were \$15. On Thursday the revenue from ticket sales was \$6800. On Friday the revenue from ticket sales was \$9600. On both nights exactly the same number of people purchased advance tickets. On both nights exactly the same number of people purchased tickets at the door. How many people bought tickets in advance, and how many bought tickets at the door? (Let a = no. of advance tickets, d = no. of tickets sold at door)

Thursday $\quad 8a + 10d = 6800 \quad (1)$

Friday $\quad 11a + 15d = 9600 \quad (2)$

Multiply (1) by -3; multiply (2) by 2.

$$\begin{aligned} -24a - 30d &= -20{,}400 \\ +22a + 30d &= 19{,}200 \\ \hline -2a &= -1200 \\ a &= 600 \\ 8(600) + 10d &= 6800 \\ d &= 200 \end{aligned}$$

So, each night the sales were 600 advance tickets, 200 tickets at the door.

EXAMPLE 1 The student concert committee on campus sold tickets for a concert given by a popular music group. Tickets purchased in advance were \$4. Tickets purchased at the door were \$5. A total of 480 students bought tickets and attended the concert. The revenue from the ticket sales was \$2100. How many students bought tickets in advance and how many bought tickets at the door?

1. *Understand the problem.*
We are provided with two basic pieces of information
(1) The number of tickets bought
(2) The total revenue
This will be the basis for our two equations.

There are two different prices for the tickets. These will represent the two unknowns.

Let a = the number of tickets purchased in advance

d = the number of tickets purchased at the door

2. *Write an equation(s).*
Pick out the information you need to represent (1) total bought and (2) total revenue.

$$\underbrace{a}_{\text{number bought in advance}} + \underbrace{d}_{\text{number bought at door}} = \underbrace{480}_{\text{total bought}} \qquad \textbf{(1)}$$

$$\underbrace{4a}_{\substack{\text{revenue from \$4 tickets} \\ \text{\$4 (number bought)}}} + \underbrace{5d}_{\substack{\text{revenue from \$5 tickets} \\ \text{\$5(number bought)}}} = \underbrace{2100}_{\text{total revenue}} \qquad \textbf{(2)}$$

3. *Solve the system of equations and state the answer.*
We will solve the system by the substitution method.

$$a + d = 480 \quad \textbf{(1)}$$
$$4a + 5d = 2100 \quad \textbf{(2)}$$

$d = 480 - a$ *Solve equation* **(1)** *for variable d.*

$4a + 5(480 - a) = 2100$ *Use equation* **(2)** *and replace d by* $480 - a$.

$4a + 2400 - 5a = 2100$ *Solve for a.*

$$-1a + 2400 = 2100$$
$$-1a = -300$$
$$a = 300$$

$d = 480 - a$ *Substitute* 300 *for a to find d.*

$d = 480 - 300$

$d = 180$

Thus 300 tickets were sold in advance and 180 tickets were sold at the door.

4. *Check.*

Do these answers seem reasonable? *Yes.*

Does the number of tickets bought in advance and at the door total 480?

$$300 + 180 = 480$$

Does revenue from advance sales and from sales at the door total $2100?

$$4(300) + 5(180) \stackrel{?}{=} 2100$$

$$1200 + 900 = 2100$$

Practice Problem 1 During the first week of college, Alicia took math and verbal placement tests. She just received her test scores. The combined score of the math test and the verbal test was 1153 points. She shared that figure with her friends, but she did not reveal the score for the separate tests. Later, she reluctantly admitted that the difference between the two scores was 409. Can you determine her actual score on each test if you know her math test score was the higher of the two? ■

EXAMPLE 2 A technician in a data processing office is trying to verify the rate at which two electronic card-sorting machines operate. Yesterday the first machine sorted for 3 minutes and the second machine sorted for 4 minutes. The total work load both machines processed during that time period was 10,300 cards. Two days ago the first machine sorted for 2 minutes and the second machine for 3 minutes. The total work load both machines processed during that time period was 7400 cards. Can you determine the number of cards per minute sorted by each machine?

1. *Understand the problem.*

The number of cards processed by two different machines on two different days provides the basis for two linear equations with two unknowns.

The two equations will represent what occurred on two different days.

(1) yesterday

(2) two days ago

The number of cards sorted by each machine is what we need to find.

Let x = the number of cards per minute sorted by the first machine

y = the number of cards per minute sorted by the second machine

2. *Write the equation(s).*

Yesterday the first machine sorted for 3 minutes and the second machine sorted for 4 minutes, processing 10,300 cards.

first machine		second machine		total work	
3(no. of cards per minute)		4(no. of cards per minute)			
$3x$	$+$	$4y$	$=$	10,300	**(1)**

Two days ago the first machine sorted for 2 minutes and the second machine sorted for 3 minutes, processing 7400 cards.

first machine		second machine		total work	
2(no. of cards per minute)		3(no. of cards per minute)			
$2x$	$+$	$3y$	$=$	7400	**(2)**

TEACHING TIP After explaining Example 2, ask students to do the following as an Additional Class Exercise.
The West Chop lighthouse has two diesel generators. Yesterday the large one was run for 7 hours and the smaller one for 3 hours. Together they consumed 41 gallons of fuel. Today the large one was run for 5 hours and the smaller one for 4 hours. Together they consumed 33 gallons of fuel. How many gallons per hour does the large generator use? How many gallons per hour does the small generator use?

Sol. The large generator uses 5 gallons per hour. The small one uses 2 gallons per hour.

3. *Solve the system of equations and state the answer.*
We will use the addition method to solve this system.

$$3x + 4y = 10{,}300 \quad \textbf{(1)}$$

$$2x + 3y = 7{,}400 \quad \textbf{(2)}$$

$$6x + 8y = 20{,}600$$ *Multiply equation* **(1)** *by* 2.

$$-6x - 9y = -22{,}200$$ *Multiply equation* **(2)** *by* -3.

$$-1y = -1600$$ *Add the two equations to eliminate x and solve for y.*

$$y = 1600$$

$$2x + 3(1600) = 7400$$ *Substitute this value of y into equation* **(2)**.

$$2x + 4800 = 7400$$ *Simplify and solve for x.*

$$2x = 2600$$

$$x = 1300$$

Thus, the first machine sorts at the rate of 1300 cards per minute and the second machine at the rate of 1600 cards per minute.

4. *Check:* The numbers appear to be reasonable. Just by guessing, we would probably estimate that each machine would sort somewhere between 1000 and 2000 cards per minute. We can verify each statement in the problem:

Yesterday: Were 10,300 cards sorted?

$$3(1300) + 4(1600) \stackrel{?}{=} 10{,}300$$

$$3900 + 6400 \stackrel{?}{=} 10{,}300$$

$$10{,}300 = 10{,}300 \checkmark$$

Two days ago: Were 7400 cards sorted?

$$2(1300) + 3(1600) \stackrel{?}{=} 7400$$

$$2600 + 4800 \stackrel{?}{=} 7400$$

$$7400 = 7400 \checkmark$$

Practice Problem 2 After a recent disaster, the Red Cross brought in a pump to evacuate water from a basement. After 3 hours a larger pump was brought in and the two ran for an additional 5 hours, removing 49,000 gallons of water. The next day, after running both pumps for 3 hours, the larger one was taken to another site. The smaller pump finished the job after another 2 hours. If on the second day 30,000 gallons were pumped, how much water does each pump remove per hour? ■

EXAMPLE 3 Fred is considering working for a company that sells encyclopedias. During his job interview the company was not willing to release the specific salary structure for beginning sales representatives. Fred found out that all starting representatives receive the same annual base salary and a standard commission of a certain percentage of the sales they make during the first year. He has been told that one representative sold \$50,000 worth of encyclopedias her first year and that she earned \$14,000. He was able to find out that another representative sold \$80,000 worth of encyclopedias and that he earned \$17,600. Determine the base salary and the commission rate of a beginning sales representative.

The earnings of the two different sales representatives will represent the two equations. Earnings are based on

$$\text{base salary} + \text{commission} = \text{total earnings}$$

What is unknown? The base salary and the commission rate are what we need to find.

Let b = the amount of base salary

c = the commission rate

$b + 50{,}000c = 14{,}000$ (1) earnings for the first sales representative

$b + 80{,}000c = 17{,}600$ (2) earnings for the second sales representative

We will use the addition method to solve these equations.

$-b - 50{,}000c = -14{,}000$ *Multiply equation* **(1)** *by* -1.

$b + 80{,}000c = 17{,}600$ *Leave equation* **(2)** *unchanged.*

$30{,}000c = 3600$ *Add the two equations and solve for c.*

$$c = \frac{3600}{30{,}000}$$

$c = 0.12$ *The commission rate is 12% or 0.12.*

$b + (50{,}000)(0.12) = 14{,}000$ *Substitute this value into equation* **(1)**.

$b + 6000 = 14{,}000$ *Simplify and solve for b.*

$b = 8000$ *The base salary is \$8000 per year.*

Thus the base salary is \$8000 per year, and the commission rate is 12%.

Check: You should verify that these values are reasonable and that they satisfy the statements in the original problem.

Practice Problem 3 Last week Rick bought 7 gallons of unleaded premium and 8 gallons of unleaded regular gasoline and paid \$27.22. This week he purchased 8 gallons of unleaded premium and 4 gallons of unleaded regular and paid \$22.16. He forgot to record how much each type of fuel cost per gallon. Can you determine these values? ■

The use of two variables in two equations is very helpful in solving a class of problems often referred to as mixture problems.

EXAMPLE 4 A lab technician is required to prepare 200 liters of a solution. The prepared solution must contain 42% fungicide. The technician wishes to combine a solution that contains 30% fungicide with a solution that contains 50% fungicide. How much of each solution should he use?

Understand the problem. The technician needs to make one solution from two different solutions. We need to find the amount (number of liters) of each solution that will give us 200 liters of a 42% solution. The two unknowns are easy to identify.

Let x = the number of liters of the 30% solution needed

y = the number of liters of the 50% solution needed

Now, what are the two equations? We know that the 200 liters of the final solution will be made by combining x and y.

$$x + y = 200 \qquad \textbf{(1)}$$

The second piece of information is about the percent of fungicide in each solution. See the picture at the left.

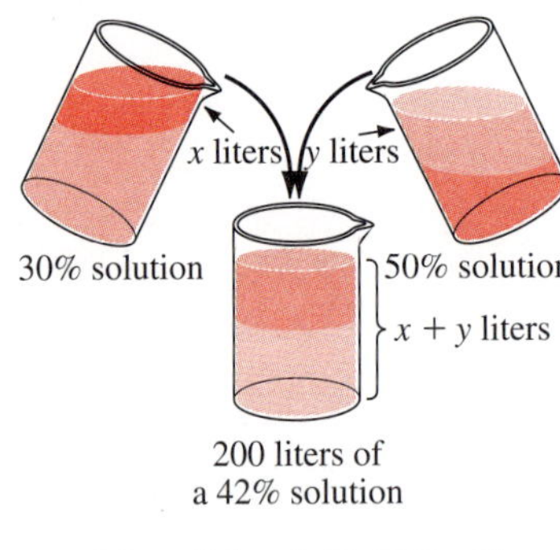

$0.3x + 0.5y = 0.42\,(200)$ **(2)**

Thus our system of equations is

$$\begin{aligned} x + \quad y &= 200 \qquad \textbf{(1)} \\ 0.3x + 0.5y &= 84 \qquad \textbf{(2)} \end{aligned}$$

We will solve this system by the addition method.

$1.0x + 1.0y = 200$ *Rewrite equation* **(1)** *in an equivalent form, but write the coefficients as decimals.*

$-0.6x - 1.0y = -168$ *Multiply equation* **(2)** *by* -2.

$0.4x = 32$ *Solve for x.*

$$\frac{0.4x}{0.4} = \frac{32}{0.4}$$

$$x = 80$$

$x + y = 200$ *Substitute the value* $x = 80$ *into one of the original equations and solve for y.*

$$80 + y = 200$$

$$y = 120$$

The technician will use 80 liters of the 30% fungicide and 120 liters of the 50% fungicide to obtain the required solution.

Check. Do the amounts total 200 liters?

$$80 + 120 = 200 \checkmark$$

Do the amounts yield a 42% strength mixture?

$$0.30x + 0.50y \stackrel{?}{=} 0.42(200)$$

$$0.30(80) + 0.50(120) \stackrel{?}{=} 84$$

$$24 + 60 = 84 \checkmark$$

Practice Problem 4 Dwayne Laboratories needs to ship 4000 liters of H_2SO_4 (sulfuric acid). Dwayne keeps in stock solutions that are 20% H_2SO_4 and 80% H_2SO_4. If the shipment is to be 65% strength, how should Dwayne's stock be mixed? ■

TEACHING TIP After covering mixture problems in Example 4, you may wish to do an Additional Class Exercise of this type.
A chemist has a 40% acid solution and an 80% acid solution. She wishes to mix together some of each to obtain 12 deciliters of a 50% solution. How much of each should she use?
Sol.

Let x = the number of deciliters of the 40% solution

Let y = the number of deciliters of the 80% solution

$$x + y = 12$$

$$0.40x + 0.80y = (0.50)(12)$$

We have 9 deciliters of 40% solution, 3 deciliters of 80% solution.

EXAMPLE 5 Mike has a boat on Lazy River. He would like to find the speed of his boat and the speed of the current on the river. He took a 48-mile trip up the river traveling against the current in exactly 3 hours. He refueled and made the return trip in exactly 2 hours. What is the speed of his boat in still water and the speed of the current on the river?

Let x = the speed of the boat in mph

y = the speed of the river current in mph

You may want to draw a picture to show the boat traveling in each direction.

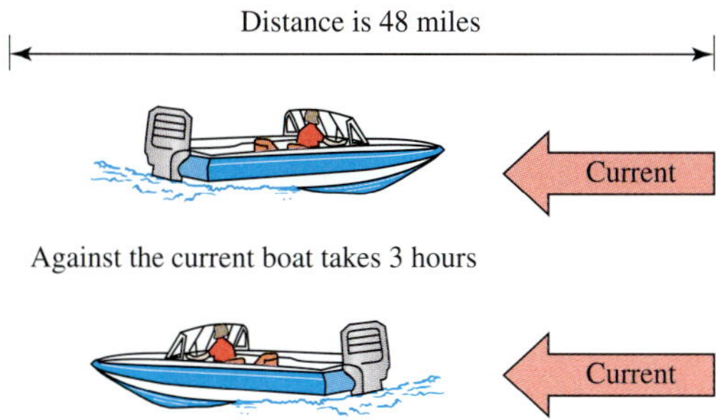

Traveling *against* the current it took 3 hours to travel 48 miles. We can calculate the rate (speed) the boat travels upstream. Use $d = r \cdot t$.

$$48 = 3 \cdot r$$

$$r = 16 \text{ mph}$$

The boat is going *against* the current.

$$x - y = 16 \qquad \textbf{(1)}$$

Traveling with the current it took 2 hours to travel 48 miles. We can calculate the rate (speed) the boat travels downstream. Use $d = r \cdot t$.

$$48 = 2 \cdot r$$

$$r = 24 \text{ mph}$$

The boat is going *with* the current.

$$x + y = 24 \qquad \textbf{(2)}$$

This system is solved most easily by the addition method.

$$x - y = 16 \qquad \textbf{(1)}$$

$$\underline{x + y = 24} \qquad \textbf{(2)}$$

$2x \quad = 40$ *Add the two equations.*

$x = 20$ *Solve for x by dividing both sides by 2.*

$x + y = 24$ **(2)** *Substitute the value of $x = 20$ into one of the original equations and solve for y.*

$20 + y = 24$ *Substitute $x = 20$ into equation* **(2)**.

$y = 4$

Thus the speed of Mike's boat is 20 mph and the speed of the current in Lazy River is 4 mph.

Check. Can you verify these answers?

Practice Problem 5 Mr. Caminetti traveled downstream in his boat a distance of 72 miles in a time of 3 hours. His return trip up the river against the current took 4 hours. Find the speed of the boat in still water. Find the speed of the current in the river. ■

8.5 Exercises

Applications

Solve each problem by using two equations with two variables.

1. One number added to twice another number gives a sum of 42. The difference between the two numbers is −3. Find the numbers.

$x + 2y = 42$
$x - y = -3$
$3y = 45$
$y = 15, x = 12$
The numbers are 12 and 15.

2. Twice a number added to another number gives a sum of 29. The difference between the numbers is −2. Find each number.

$2x + y = 29$
$x - y = -2$
$3x = 27$
$x = 9, y = 11$
The numbers are 9 and 11.

3. A restaurant chef is shopping around for the best price for sourdough bread. The Meilleur Pain company has bread priced \$0.12 per loaf more than the Corner Loaf company. If you average the two prices for sourdough bread, you obtain \$0.36 per loaf. Find the price of bread for both companies.

$\frac{x + y}{2} = 0.36$ A loaf from Meilleur Pain is \$0.42.
$x = 0.12 + y$ A loaf from Corner Loaf is \$0.30.

4. In all right triangles the sum of the two smaller interior angles is 90°. In a certain right triangle the difference between those two angles is 28°. Find the number of degrees in each of these two angles.

$x + y = 90$
$x - y = 28$
$2x = 118$ $x = 59, y = 31$
Smaller angle is 31°.
Larger angle is 59°.

5. There are 46 people working in the special effects department of a major motion picture studio. The majority of the employees have worked for free as interns for special effects production companies. The difference between the number of staff who interned and those who did not is 18. How many employees interned? How many did not?

$x + y = 46$ 32 worked as interns.
$x - y = 18$ 18 did not work as interns.

6. A large department store starts its salespersons at a salary of \$4.50 per hour. The starting assistant managers earn \$6.50 per hour. Two roommates work at the same store. The person who was an assistant manager worked 4 more hours than the salesperson. The first seven days of the job they earned a total of \$532.00. How many hours did each work?

$\left.\begin{array}{l}4.50x + 6.50y = 532\\ y = x + 4\end{array}\right\}$ $450x + 650(x + 4) = 53{,}200$
$1100x = 50{,}600$
$x = 46$
sales = 46 hours
asst. man. = 50 hours

7. The latest matinee prices for a Broadway musical in New York are \$65.00 for an orchestra seat and \$55.00 for a mezzanine seat. Today, 900 paying customers saw the show, and the box office took in \$55,800.00. How many of each type of seat were sold?

$x + y = 900$
$55x + 65y = 55{,}800$
630 orchestra seats; 270 mezzanine seats

8. Woodman's Catering Company pays cooks \$105 per shift and wait staff \$75 per shift. Yesterday 29 people worked for this company at a wedding reception. If the payroll was \$2415, how many cooks and how many wait staff were employed?

$x + y = 29$
$105x + 75y = 2415$
8 cooks and 21 wait staff were employed.

9. KidTime Daycare has plans to expand its playground. Presently it takes 1050 feet of fencing to enclose its rectangular playground. The expansion plans call for the daycare to triple the width of the playground and to double the length. Now the center will require 2550 feet of fencing. What was the length and width of the original playground?

$\left.\begin{array}{l}2W + 2L = 1050\\ 6W + 4L = 2550\end{array}\right\}$ The original width was 225 feet. The original length was 300 feet.

10. Bill is the busiest psychotherapist at a local clinic. Last week he treated 50 patients. The head of the clinic told him that the rest of the staff (eight other psychotherapists) saw four times as many men and three times as many women as Bill saw during the week. This amounted to 179 patients. How many men and how many women did Bill treat?

$M + W = 50$
$4M + 3W = 179$
He treated 29 men and 21 women.

11. Rose learned, during her beginning mechanics course, that her leaking radiator holds 16 quarts of water. When the radiator was partially full, she temporarily patched the leak. At that time it was 50% antifreeze. After she added a mixture of 80% antifreeze and waited a little while, she found that it was 65% antifreeze. How many quarts of solution were in the radiator just before Rose added antifreeze? How many quarts of 80% solution antifreeze did she add?

$x + y = 16$
$0.50x + 0.80y = 0.65(16)$
8 quarts before she added. She add 8 quarts.

12. A candy company wants to use 20 kilograms of a special 45% fat content chocolate to conform with customer dietary demands. To obtain this, a 50% fat content Hawaiian chocolate is combined with a 30% fat content domestic chocolate in a special melting process. How many kilograms of the 50% fat chocolate and how many kilograms of the 30% fat chocolate are used to make the required 20 kilograms?

$x + y = 20$
$0.50x + 0.30y = 0.45(20)$
15 kg of 50% fat content chocolate; 5 kg of 30% fat content chocolate

13. The hiking club wants to sell 50 pounds of an energy snack food that will cost them only \$1.80 per pound. The mixture will be made up of nuts that cost \$2.00 per pound and raisins that cost \$1.50 per pound. How much of each should they buy to obtain 50 pounds of the desired mixture?

$x + y = 50$
$2.00x + 1.50y = 50(1.80)$
30 lb of nuts
20 lb of raisins

14. How many bunches of local daisies that cost \$2.50 per bunch should be mixed with imported daisies that cost \$3.50 per bunch to obtain 100 bunches of daisies that will cost \$2.90 per bunch?

$x + y = 100$
$2.50x + 3.50y = 2.90(100)$
40 bunches of imported daisies
60 bunches of local daisies

15. An airplane flies between Boston and Cleveland. Due to the trip routing the plane flies a distance of 630 miles for each trip. The trip with a tailwind (the wind traveling the same direction as the plane) took 3 hours. The return trip traveling against the wind took 3.5 hours. Find the speed of the wind. Find the speed of the airplane in still air.

$3(p + w) = 630$
$3.5(p - w) = 630$
The plane flies 195 mph. The wind speed is 15 mph.

16. An airplane travels between two cities that are 3000 kilometers apart. The trip against the wind took 6 hours. The return trip with the benefit of the wind was 5 hours. What is the wind speed in kilometers per hour? What is the speed of the plane in still air?

$\frac{3000}{6} = r - w$
$\frac{3000}{5} = r + w$
$2r = 500 + 600$
$r = 550$
$w = 50$

wind speed = 50 kilometers per hour
plane's speed in still air = 550 kilometers per hour

17. Walter and Dianne Meibaum went on their first overnight sailing trip, 45 miles each way. The trip against the current took 4 hours. The return trip with the assistance of the current took only 3 hours. Find the speed of the sailboat in still water. Find the speed of the current.

$3(b + c) = 45$
$4(b - c) = 45$
The boat travels 13.125 miles per hour. The current travels 1.875 miles per hour.

18. Ginny, a beginning rower, can row a boat with the current at 5.5 miles per hour on the Saco River. Against the current, she rows at 4.2 miles per hour. How fast is the current? How fast can she row the boat in still water?

$b + c = 5.5$
$b - c = 4.2$
The current is 0.65 miles per hour. Ginny rows 4.85 miles per hour.

★ **19.** Sam's commission is a given percent of the total sales cost of the surgical transplant equipment he sells. He receives an 8% commission for selling heart transplant kits and a 6% commission for selling kidney transplant kits. Last month he earned \$4480 in commissions. This month he sold the same quantity of heart and kidney kits; however, his company gave him a raise and he now earns 10% commission for selling heart kits and 8% for kidney kits. This month he earned \$5780. What is the value of the heart kits he sold last month? What is the value of the kidney kits?

$0.08h + 0.06k = 4480$
$0.10h + 0.08k = 5780$
\$29,000 for a heart kit
\$36,000 for a kidney kit

Cumulative Review Problems

20. Add. $\frac{1}{x-3} - \frac{1}{x+3} + 2$ $\quad \frac{2x^2 - 12}{x^2 - 9}$

21. Evaluate when $x = -2$ and $y = 3$.
$2x^2 - 3y + 4y^2$ $\quad 35$

22. Simplify. $(-4x^3y^2)^3$ $\quad -64x^9y^6$

Putting Your Skills to Work

USING MATHEMATICS TO PREDICT THE NUMBER OF PETS

More U.S. households have dogs than have cats. However, many experts in the field say that with increased urbanization the number of households with cats is increasing at a greater rate than the number of households with dogs.

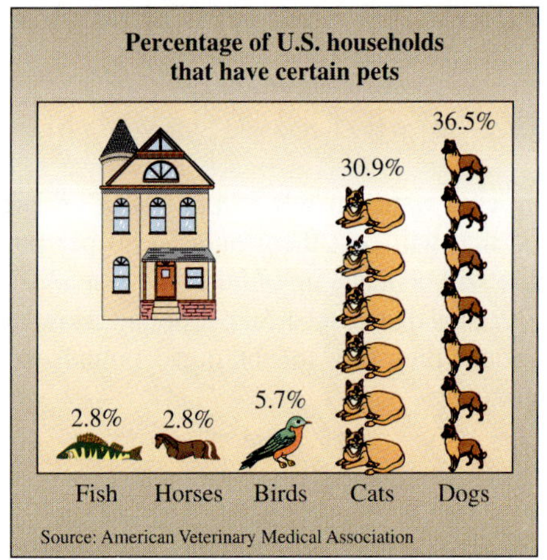

PROBLEMS FOR INDIVIDUAL INVESTIGATION AND ANALYSIS

Let P = the percentage of households that have one or more pets and x = the number of years since 1995.

1. If since 1995 the percentage of households that have one or more dogs as pets is increasing at the rate of 0.2% per year, write an equation of the form $P = 36.5 + ax$ that will estimate the future percentage x years from 1995. What is the value of a?
$P = 36.5 + 0.2x$ $\quad a = 0.2$

2. If since 1995 the percentage of households that have one or more cats as pets is increasing at the rate of 0.3% per year, write an equation of the form $P = 30.9 + bx$ that will estimate the future percentage x years from 1995. What is the value of b?
$P = 30.9 + 0.3x$ $\quad b = 0.3$

PROBLEMS FOR COOPERATIVE INVESTIGATION AND GROUP ACTIVITIES

Together with some other members of your class see if you can answer the following.

3. Solve the two equations from (1) and (2) simultaneously to determine the number of years from 1995 it will be until the percentage is the same for cats and dogs. 56 years

4. What will be the percentage of households when this equality is achieved? (Round your answer to the nearest tenth of a percent.) 47.7%

INTERNET CONNECTIONS: Go to ``http://www.prenhall.com/tobey'' to be connected

Site: Companion Animal Overpopulation: The Facts & Figures (Doris Day Animal League) or a related site

This site gives facts and figures related to pet overpopulation. Use the information to answer the following questions.

5. Suppose that a neighborhood has 12 cats and their offspring. How many kittens can these cats produce in 7 years?

6. Let x be the number of dogs that enter animal shelters in a year. Write an equation of the form $K = ax$ that will estimate the number of these dogs that are purebreds. What is the value of a?

7. Let x be the number of cats that enter animal shelters in a year. Write an equation of the form $K = bx$ that will estimate the number of these cats that are killed. What is the value of b?

Chapter Organizer

Topic	Procedure	Examples
Solving a system of equations: substitution method, p. 464.	**1.** Solve for one variable in terms of the other variable in one of the two equations. **2.** Substitute the expression you obtain for this variable into the other equation. **3.** Solve this equation to find the value of the variable. **4.** Substitute this value for the variable into one of the equations with two variables. **5.** Check each solution in both equations.	Solve: $2x + 3y = -6$ $3x - y = 13$ Solve second equation for y: $-y = -3x + 13$ $y = 3x - 13$ Substitute this expression into the first equation: $2x + 3(3x - 13) = -6$ $2x + 9x - 39 = -6$ $11x = -6 + 39$ $x = 3$ Substitute $x = 3$ into $y = 3x - 13$: $y = 3(3) - 13 = 9 - 13 = -4$ *Check:* $2(3) + 3(-4) \stackrel{?}{=} -6$ $3(3) - (-4) \stackrel{?}{=} 13$ $6 - 12 \stackrel{?}{=} -6$ $9 - (-4) \stackrel{?}{=} 13$ $-6 = -6$ ✓ $9 + 4 = 13$ ✓ The solution is $(3, -4)$.
Solving a system of equations: addition method, p. 470.	**1.** Multiply each term of one or both equations by some nonzero integer so that the coefficients of one of the variables are opposites. **2.** Add the equations of this new system. **3.** Solve the resulting equation for the variable. **4.** Substitute the value found into either of the original equations to find the value of the second variable. **5.** Check your solution in both of the original equations.	Solve: $3x - 2y = 0$ $-4x + 3y = 1$ Eliminate the x's: a common multiple of 3 and 4 is 12. Multiply the first equation by 4 and the second by 3, then add. $12x - 8y = 0$ $-12x + 9y = 3$ $y = 3$ Substitute $y = 3$ into either equation and solve for x. Let us pick $3x - 2y = 0$: $3x - 2(3) = 0$ $3x - 6 = 0,$ $x = 2$ *Check:* $3(2) - 2(3) \stackrel{?}{=} 0$ $-4(2) + 3(3) \stackrel{?}{=} 1$ $6 - 6 \stackrel{?}{=} 0$ $-8 + 9 \stackrel{?}{=} 1$ $0 = 0$ ✓ $1 = 1$ ✓ The solution is $(2, 3)$.
Solving a system of equations by graphing, p. 459.	Graph each equation on the same coordinate system. One of the following will be true. **1.** The lines intersect. The point of intersection is the solution to the system. Such a system is consistent. **2.** The lines are parallel. There is no point of intersection. The system has no solution. Such a system is inconsistent. **3.** The lines coincide. Every point on the line represents a solution. There are an infinite number of solutions. Such a system is dependent.	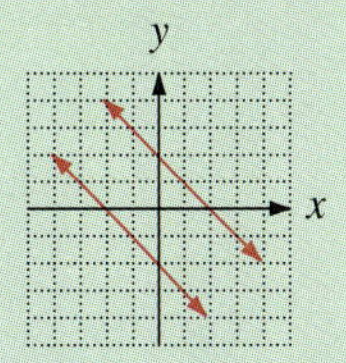 **1.** 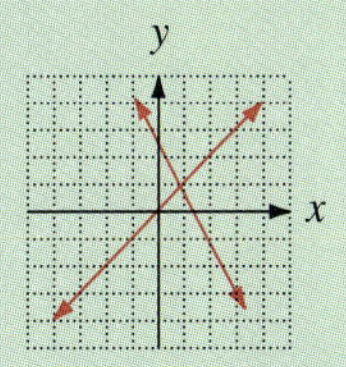 y x **2.** y x **3.** y x 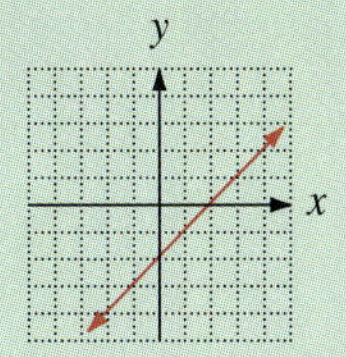

Chapter Organizer

<table>
<tr><th>Topic</th><th>Procedure</th><th>Examples</th></tr>
<tr><td>Choosing a method to solve a system of equations, p. 479.</td><td>1. Substitution works well if one or more variables have a coefficient of 1 or -1.
2. Addition works well for integer, fractional, or decimal coefficients.
3. Graphing gives you a picture of the system. It works well if the system has integer solutions or if the system is inconsistent or dependent.</td><td>Use substitution.
$x + y = 8$
$2x - 3y = 9$
Use addition.
$\frac{3}{5}x - \frac{1}{4}y = 4$
$\frac{1}{5}x + \frac{3}{4}y = 8$
Graph to get a picture of the system. If the system is consistent and if it is difficult to determine the solutions because they are fractions, use an algebraic method.</td></tr>
<tr><td>To identify inconsistent and dependent systems algebraically, p. 480.</td><td>1. If you obtain an equation that is inconsistent with known facts, such as $0 = 2$, the system is inconsistent. There is no solution.
2. If you obtain an equation that is always true, such as $0 = 0$, the system is dependent. There are an infinite number of solutions.</td><td>Solve: $x + 2y = 1$
$3x + 6y = 12$
Multiply the first equation by -3:
$-3x - 6y = -3$
$3x + 6y = 12$
$0 = 9$
0 is not equal to 9, so there is no solution. The system is inconsistent.
Solve: $2x + y = 1$
$6x + 3y = 3$
Multiply the first equation by -3:
$-6x - 3y = -3$
$6x + 3y = 3$
$0 = 0$
Since 0 is always equal to 0, the system is dependent. There are an infinite number of solutions.</td></tr>
<tr><td>Solving word problems with a system of equations, p. 484.</td><td>1. Understand the problem. Choose a variable to represent each unknown quantity.
2. Write a system of equations in two variables.
3. Solve the system of equations and state the answer.
4. Check the answer.</td><td>Apples sell for $0.35 per pound. Oranges sell for $0.40 per pound. Nancy bought 12 pounds of apples and oranges for $4.45. How many pounds of each fruit did she buy?
Let x = number of pounds of apples
y = number of pounds of oranges
12 pounds in all were purchased. $x + y = 12$
The purchase cost $4.45. $0.35x + 0.40y = 4.45$
Multiply the second equation by 100 and we have the system
$x + y = 12$
$35x + 40y = 445$
Multiply the top equation by -35 and add the equations.
$-35x - 35y = -420$
$35x + 40y = 445$
$5y = 25$
$y = 5$
Substituting into $x + y = 12$:
$x + 5 = 12$
$x = 7$
Nancy purchased 7 pounds of apples and 5 pounds of oranges.
Check: Are there 12 pounds of apples and oranges?
$7 + 5 = 12$ ✓
Would this purchase cost $4.45?
$0.35(7) + 0.40(5) \stackrel{?}{=} 4.45$
$2.45 + 2.00 = 4.45$ ✓</td></tr>
</table>

Chapter 8 Review Problems

8.1 *Solve each system of equations by graphing each line and finding the point of intersection.*

1. $3x + y = 4$
$3x - 2y = 10$

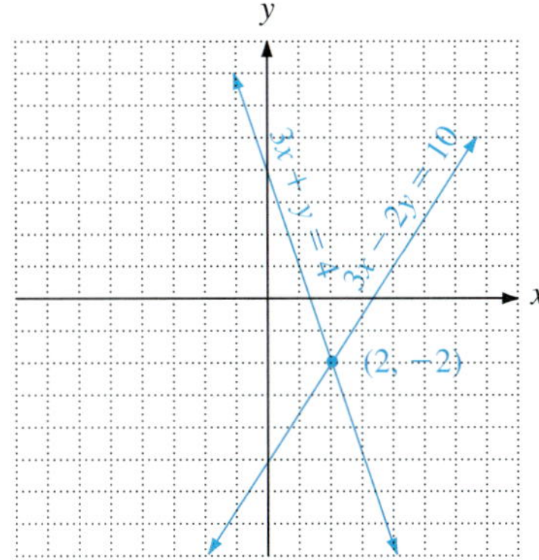

2. $-3x + y = -2$
$-2x - y = -8$

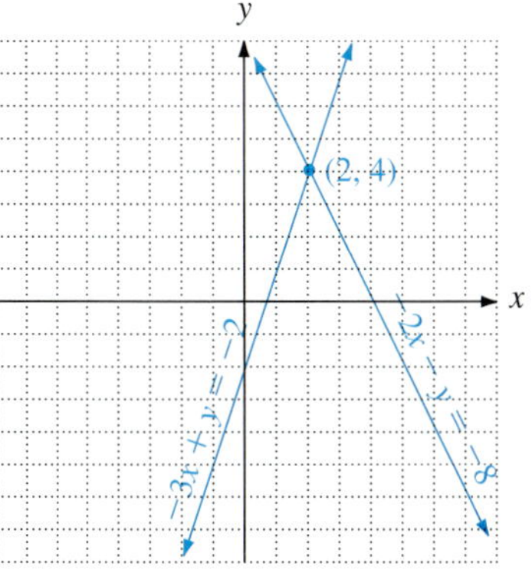

3. $-4x + 3y = 15$
$2x + y = -5$

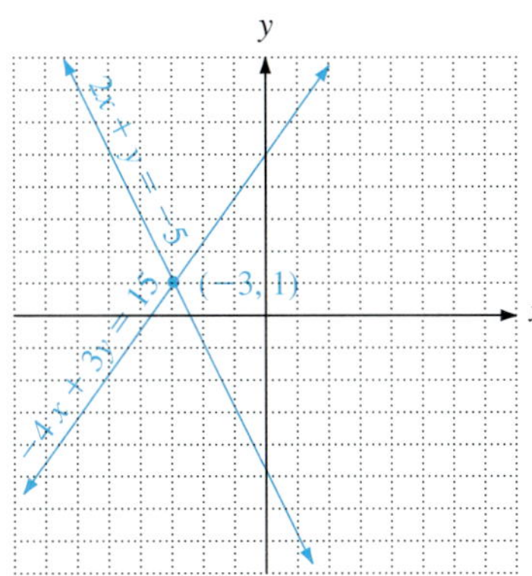

4. $2x - y = 1$
$3x + y = -6$

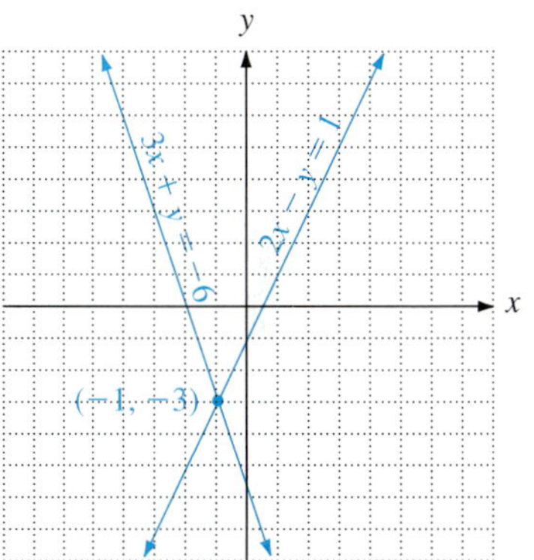

8.2 *Solve by the substitution method. Check your solution.*

5. $x + 3y = 18$
$2x + y = 11$
$x = 3,\ y = 5$

6. $x + y = 6$
$-2x + y = -3$
$x = 3,\ y = 3$

7. $3x + 2y = 0$
$2x - y = 7$
$x = 2,\ y = -3$

8. $5x - y = 4$
$2x - \frac{1}{3}y = 6$
$x = 14,\ y = 66$

8.3 *Solve by the addition method. Check your solution.*

9. $6x - 2y = 10$
$2x + 3y = 7$
$x = 2,\ y = 1$

10. $4x - 5y = -4$
$-3x + 2y = 3$
$x = -1,\ y = 0$

11. $-5x + 8y = 4$
$2x - 7y = 6$
$x = -4,\ y = -2$

12. $2x + 4y = 10$
$4x - 3y = 9$
$x = 3,\ y = 1$

8.4 *If possible, solve for both variables by any appropriate method to obtain one solution. If it is not possible to obtain one solution, state the reason.*

13. $3x - 2y = 17$
$x + y = 9$
$(7, 2)$

14. $x - 2y = 2$
$2x + 3y = 11$
$(4, 1)$

15. $11x + 3y = 12$
$4x + y = 5$
$(3, -7)$

16. $7x - 5y = 8$
$-x + y = -2$
$(-1, -3)$

17. $5x - 3y = 32$
$5x + 7y = -8$
$(4, -4)$

18. $4x + 3y = 46$
$-2x + 3y = -14$
$(10, 2)$

19. $7x + 3y = 2$
$-8x - 7y = 2$
$\left(\frac{4}{5}, -\frac{6}{5}\right)$

20. $x + 5y = 7$
$-3x - 4y = 1$
$(-3, 2)$

21. $5x - 11y = -4$
$6x - 8y = -10$
$(-3, -1)$

22. $3x + 5y = -9$
$4x - 3y = 17$
$(2, -3)$

23. $2x - 4y = 6$
$-3x + 6y = 7$
No solution
Inconsistent system

24. $4x - 7y = 8$
$5x + 9y = 81$
$(9, 4)$

25. $2x - 9y = 0$
$3x + 5 = 6y$
$\left(-3, -\frac{2}{3}\right)$

26. $2x + 10y = 1$
$-4x - 20y = -2$
Infinite number of solutions, dependent equations

27. $5x - 8y = 3x + 12$
$7x + y = 6y - 4$
$(-2, -2)$

28. $1 + x - y = y + 4$
$4(x - y) = 3 - x$
$(-1, -2)$

29. $3x + y = 9$
$x - 2y = 10$
$(4, -3)$

30. $x + 12y = 0$
$3x - 5y = -2$
$\left(-\frac{24}{41}, \frac{2}{41}\right)$

31. $2(x + 3) = y + 4$
$4x - 2y = -4$
Infinite number of solutions. Dependent equations.

32. $x + y = 3000$
$x - 2y = -120$
$(1960, 1040)$

33. $4x - 3y + 1 = 6$
$5x + 8y + 2 = -74$
$(-4, -7)$

34. $5x + 4y + 3 = 23$
$8x - 3y - 4 = 75$
$(8, -5)$

35. $\frac{2x}{3} - \frac{3y}{4} = \frac{7}{12}$
$8x + 5y = 9$
$\left(\frac{29}{28}, \frac{1}{7}\right)$

36. $\frac{x}{2} + \frac{y}{5} = 4$
$\frac{x}{3} + \frac{y}{5} = \frac{10}{3}$
$(4, 10)$

37. $\frac{1}{5}a + \frac{1}{2}b = 6$
$\frac{3}{5}a - \frac{1}{2}b = 2$
$(10, 8)$

38. $\frac{2}{3}a + \frac{3}{5}b = -17$
$\frac{1}{2}a - \frac{1}{3}b = -1$
$(-12, -15)$

39. $0.2s - 0.3t = 0.3$
$0.4s + 0.6t = -0.2$
$\left(\frac{1}{2}, -\frac{2}{3}\right)$

40. $0.3s + 0.2t = 0.9$
$0.2s - 0.3t = -0.6$
$\left(\frac{15}{13}, \frac{36}{13}\right)$

41. $3m + 2n = 5$
$4n = 10 - 6m$
Infinite number of solutions, dependent equations

42. $12n - 8m = 6$
$4m + 3 = 6n$
Infinite number of solutions, dependent equations

43. $3(x + 2) = -2 - (x + 3y)$
$3(x + y) = 3 - 2(y - 1)$
$(-5, 4)$

44. $13 - x = 3(x + y) + 1$
$14 + 2x = 5(x + y) + 3x$
$(9, -8)$

45. $0.2b = 1.4 - 0.3a$
$0.1b + 0.6 = 0.5a$
$(2, 4)$

46. $0.3a = 1.1 - 0.2b$
$0.3b = 0.4a - 0.9$
$(3, 1)$

47. $\frac{b}{5} = \frac{2}{5} - \frac{a - 3}{2}$
$4(a - b) = 3b - 2(a - 2)$
$(3, 2)$

48. $9(b + 4) = 2(2a + 5b)$
$\frac{b}{5} + \frac{a}{2} = \frac{18}{5}$
$(12, -12)$

Solve for x and y. Assume that a and b are nonzero constants.

★**49.** $ax - by = ab + b^2$
$ax + by = 2a^2 + ab - b^2$
$\left(a + b, \frac{a^2 - b^2}{b}\right)$

★**50.** $5 + ax + by = 2a + 5$
$2b + by = ax$
$\left(\frac{a + b}{a}, \frac{a - b}{b}\right)$

8.5 *Solve each word problem.*

51. During a concert, \$3280 was received. A total of 760 students attended the concert. Reserved seats were \$6 and general admission seats were \$4. How many of each kind of ticket was sold?
120 reserved seats, 640 general admission seats

52. In a local fruit market, apples are on sale for \$0.25 per pound. Oranges are on sale for \$0.39 per pound. Coach Smith purchased 12 pounds of a mixture of apples and oranges before the basketball team left on a bus trip to play at Center City. If the coach paid \$3.70, how many pounds of each fruit did he get? 7 pounds of apples, 5 pounds of oranges

53. A plane travels 1500 miles in 5 hours with the benefit of a tailwind. On the return trip it requires 6 hours to fly against the wind. Can you find the speed of the plane in still air and the wind speed? 275 mph is the speed of the plane in still air; 25 mph is the speed of the wind

54. A chemist needs to combine two solutions of a certain acid. How many liters of a 30% acid solution should be added to 40 liters of a 12% acid solution to obtain a final mixture that is 20% acid? 32 liters

55. The girls and boys sponsoring the Scouts Car Wash had 86 customers on Saturday. They charged \$4 to wash regular-sized cars and \$3.50 to wash compact cars. The gross receipts for the day were \$323. How many cars of each type were washed?
44 regular sized cars, 42 compact cars

56. Carl manages a highway department garage in Maine. A mixture of salt and sand is stored in the garage for use during the winter. Carl needs 24 tons of a salt/sand mixture that is 25% salt. He will combine shipments of 15% salt and shipments of 30% salt to achieve the desired 24 tons. How much of each type should he use?
8 tons of 15% salt, 16 tons of 30% salt

57. A water ski boat traveled 23 kilometers per hour going with the current. It returned and went back in the opposite direction against the current and traveled only 15 kilometers per hour. What was the speed of the boat? What was the speed of the current? Speed of the boat is 19 kilometers per hour, speed of the current is 4 kilometers per hour.

58. Fred is analyzing the cost of producing two different items at an electronics company. An electrical sensing device requires 5 grams of copper and requires 3 hours to assemble. A smaller sensing device made by the same company requires 4 grams of copper but requires 5 hours to assemble. The first device has a production cost of \$27. The second device has a production cost of \$32. How much does it cost the company for a gram of this type of copper? What is the hourly labor cost at this company? (Assume that production cost is obtained by adding copper cost and labor cost.) It costs \$3 for a gram of copper. The hourly labor rate is \$4.

Putting Your Skills to Work

THE MATHEMATICS OF THE BREAK-EVEN POINT

Systems of equations are used in the business world to determine the **break-even point.** Below the break-even point, a business is operating at a loss. Dave's Candy Company has fixed costs from salaries, rent, light, heat, and telephone of $300 per day. Each pound of candy produced costs $2 per pound for ingredients. Each pound of candy sold yields $3 in revenue.

Let $x =$ the number of pounds of candy
$C =$ cost per day, $R =$ revenue per day

$$C = \underbrace{\$300}_{\text{fixed expenses}} + \underbrace{2x}_{\$2\text{ per pound}} \qquad R = \underbrace{3x}_{\$3\text{ per pound}}$$

The break-even point is where cost equals revenue, the intersection of the two lines. This occurs when $x = 300$ pounds of candy.

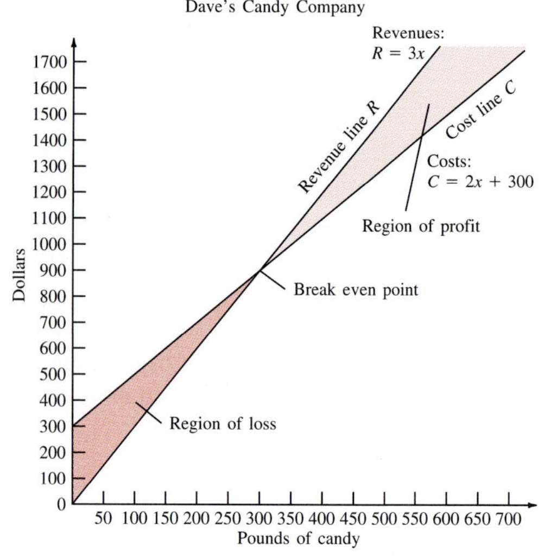

PROBLEMS FOR INDIVIDUAL INVESTIGATION

1. 200 pounds of candy are made and sold in one day. Is this a loss or a profit? Why? Loss because it costs $100 more than you make.
2. If you manufacture and sell daily 625 pounds of candy, how much is the profit or loss? $325 profit

PROBLEMS FOR COOPERATIVE GROUP ACTIVITY AND JOINT INVESTIGATION

Together with other members of your class answer these questions. Campus T-Shirts, Inc., is a local student-run enterprise. It has daily fixed costs of $400. Every T-shirt costs the company $5 to make. Every T-shirt sold yields $7 in revenue.

3. Write a system of equations to describe the cost and revenue for Campus T-shirts, Inc. and draw a graph to illustrate profit and loss for the company. $C = 400 + 5x \quad R = 7x$
4. Determine the break-even point by solving the system of equations. Break-even point is at (200, 1400). 200 T-shirts will cost $1400 to make and will bring in $1400.
5. What is the profit or loss if 320 shirts are made and sold each day? $240 profit

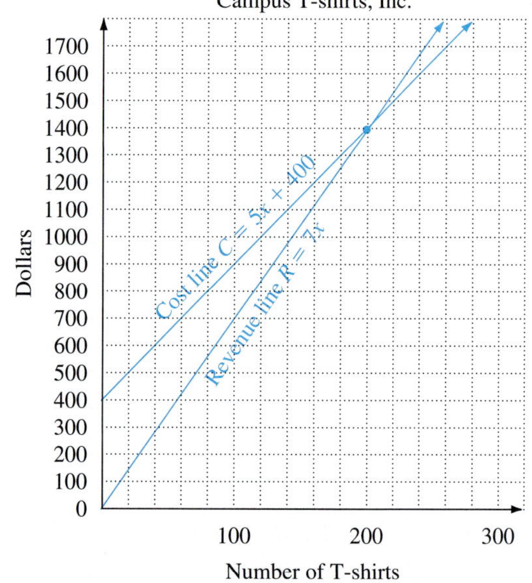

INTERNET CONNECTIONS: Go to ''http://www.prenhall.com/tobey'' to be connected

Site: The Manufacturing Game: ''Thingamajig'' or a related site

Suppose you are a member of a team that plans to manufacture and sell thingamajigs according to the information on this web page. Assume that the fixed costs are $325 and that your team can assemble two thingamajigs every minute.

6. Write a system of equations to describe your cost and revenue. (When finding the revenue equation, assume that you are to make and sell more than 10 thingamajigs.) Draw a graph to illustrate profit and loss for your team.
7. Determine the break-even point by solving the system of equations. What is your profit or loss if you make and sell 320 thingamajigs?

Chapter 8 Test

Solve by the method specified.

1. Substitution method.

$$\begin{aligned} 3x - y &= -5 \\ -2x + 5y &= -14 \end{aligned}$$

2. Addition method.

$$\begin{aligned} 3x + 4y &= 7 \\ 2x + 3y &= 6 \end{aligned}$$

3. Graphing method.

$$\begin{aligned} 2x - y &= 4 \\ 4x + y &= 2 \end{aligned}$$

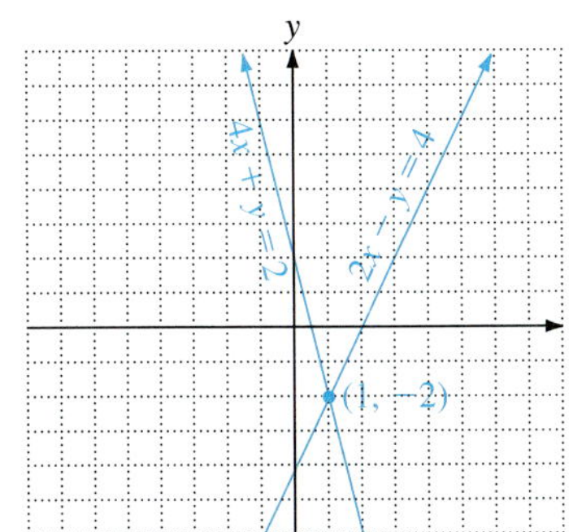

4. Any method.

$$\begin{aligned} x - 3y &= 0 \\ 2x - 3y &= -6 \end{aligned}$$

Solve by any method to find one solution if this is possible. If there is not one solution to a system, state why.

5. $$\begin{aligned} 2x - y &= 5 \\ -x + 3y &= 5 \end{aligned}$$

6. $$\begin{aligned} 2x + y &= 4 \\ 3x + 7y &= 17 \end{aligned}$$

7. $$\begin{aligned} \frac{2}{3}x - \frac{1}{5}y &= 2 \\ \frac{4}{3}x + 4y &= 4 \end{aligned}$$

8. $$\begin{aligned} -2x + 4y &= 14 \\ 3x - 6y &= 5 \end{aligned}$$

9. $$\begin{aligned} 5x - 2 &= y \\ 10x &= 4 + 2y \end{aligned}$$

10. $$\begin{aligned} 0.3x + 0.2y &= 0 \\ 1.0x + 0.5y &= -0.5 \end{aligned}$$

11. $$\begin{aligned} 2(x + y) &= 2(1 - y) \\ 5(-x + y) &= 2(23 - y) \end{aligned}$$

12. $$\begin{aligned} 3(x - y) &= 11 - y \\ 8(y - 1) &= 12x - 5 \end{aligned}$$

1. $(-3, -4)$

2. $(-3, 4)$

3. See graph

4. $(-6, -2)$

5. $(4, 3)$

6. $(1, 2)$

7. $(3, 0)$

8. no solution, inconsistent system

9. dependent equations, infinite number of solutions

10. $(-2, 3)$

11. $(-5, 3)$

12. no solution, inconsistent system

13. First number is 5, Second number is −3

14. 12,896 of the $12 tickets, 17,894 of the $8 tickets

15. A shirt costs $20 and a pair of slacks costs $24.

16. 500 per hr by 1st machine; 400 per hr by 2nd machine

17. 450 km/hr speed of jet; 50 km/hr speed of wind

Solve each word problem.

13. Twice one number plus three times a second is one. Twice the second plus three times the first is nine. Find the numbers.

14. Both $8 and $12 tickets were sold for the game. In all, 30,790 paid for admission, resulting in a "gate" of $297,904. How many of each type of ticket was sold?

15. Five shirts and three pairs of slacks cost $172. Three of the same shirts and four pairs of the same slacks cost $156. How much does each shirt cost? How much is a pair of slacks?

16. Two machines place address labels on business envelopes. The first machine operated for 8 hours and the second machine for 3 hours. That day 5200 envelopes were affixed with address labels. The next day 4400 envelopes were addressed when the first machine was operated for 4 hours and the second machine for 6 hours. How many envelopes are addressed per hour by each machine?

17. A jet plane flew 2000 kilometers against the wind in a time of 5 hours. It refueled and flew back with the wind in 4 hours. Find the wind speed in kilometers per hour. Find the speed of the jet in still air in kilometers per hour.

Cumulative Test for Chapters 0–8

Approximately one-half of this test covers the content of Chapters 0–7. The remainder covers the content of Chapter 8.

Simplify.

1. 11% of \$37.20 (Round your answer to the nearest cent)

2. $\dfrac{-2^4 \cdot (2^2)}{2^3}$

3. $(3x^2 - 2x + 1)(5x - 3)$

4. $\dfrac{x^2 - 7x + 12}{x^2 - 9} \div (x^2 - 8x + 16)$

Solve.

5. $5(3 - x) \geq 6$

6. $x^2 - 4x - 45 = 0$

7. 12% of what number is 360?

8. $\dfrac{3x - 7}{5x^2 - 7x - 1} = 0$

9. Factor. $5a^3 - 5a^2 - 60a$

10. Solve for x. $4x = 2(12 - 2x)$

1. \$4.09

2. -8

3. $15x^3 - 19x^2 + 11x - 3$

4. $\dfrac{1}{(x + 3)(x - 4)}$

5. $x \leq \frac{9}{5}$

6. $x = -5, x = 9$

7. 3000

8. $x = \frac{7}{3}$

9. $5a(a - 4)(a + 3)$

10. $x = 3$

11. $x = 8; y = -8$

12. $x = 24; y = 8$

13. infinite number of solutions, dependent equations

14. (1, 1)

15. inconsistent system of equations, no solution

16. 4000 liters of 8%; 8000 liters of 20%

17. 1 mph

18. 1200 and 1800 labels per hour

Solve for x and y, if possible.

11. $$\begin{aligned} 2x + y &= 8 \\ 3x + 4y &= -8 \end{aligned}$$

12. $$\begin{aligned} \frac{1}{3}x - \frac{1}{2}y &= 4 \\ \frac{3}{8}x - \frac{1}{4}y &= 7 \end{aligned}$$

13. $$\begin{aligned} 1.3x - 0.7y &= 0.4 \\ -3.9x + 2.1y &= -1.2 \end{aligned}$$

14. $$\begin{aligned} x + 2y &= 3 \\ 5x - 6y &= -1 \end{aligned}$$

Solve each word problem.

15. The equations in a system are graphed and parallel lines are obtained. What is this system called? What does that mean?

16. A mixture of 8% and 20% solutions yields 12,000 liters of a 16% solution. What quantities of the original solutions were used?

17. A boat trip 6 miles upstream takes 3 hours. The return takes 90 minutes. How fast is the stream?

18. Two printers were used for a 5-hour job of printing 15,000 labels. The next day a second run of the same labels took 7 hours because one printer broke down after 2 hours. How many labels can each printer process per hour?

CHAPTER 9

Radicals

Mathematical formulas involving radicals serve a number of practical purposes. For example, there is a formula that is commonly used by law enforcement officials that allows the prediction of the rate of speed of a car prior to an accident based on the length of the skid marks on the pavement. Would you like to find out how much you can learn just from the length of a skid mark? Turn to the Putting Your Skills to Work on page 552.

Pretest Chapter 9

1. 11
2. $\frac{5}{8}$
3. 13
4. -10
5. 2.236
6. not real
7. x^2
8. $5x^2y^3$
9. $7\sqrt{2}$
10. $x^2\sqrt{x}$
11. $6x\sqrt{x}$
12. $2a^2b\sqrt{2b}$
13. $4\sqrt{2}$
14. $\sqrt{2} + \sqrt{3}$
15. $4\sqrt{2}$
16. $13\sqrt{7} - 7$
17. $11\sqrt{2} - 18\sqrt{3}$
18. $18\sqrt{2}$
19. $4 + 2\sqrt{6}$
20. $23 + 8\sqrt{7}$
21. 1
22. -1
23. $17 - 4\sqrt{15}$
24. $\frac{6}{x}$
25. $\frac{5\sqrt{3}}{3}$
26. $3\sqrt{5} - 6$
27. $13 - 2\sqrt{42}$
28. $2\sqrt{13}$
29. $\sqrt{11}$
30. 7
31. $\frac{35}{3}$
32. $\frac{1}{2}$
33. 7

If you are familiar with the topics in this chapter, take this test now. Check your answers with those in the back of the book. If an answer was wrong or you couldn't do a problem, study the appropriate section of the chapter.

If you are not familiar with the topics in this chapter, don't take this test now. Instead, study the examples, work the practice problems, and then take the test. This test will help you to identify those concepts that you have mastered and those that need more study.

Assume all variables in the radicand represent positive real numbers.

Section 9.1

Find the square roots, if possible. (If it is impossible to obtain a real number for an answer, say so.) Approximate to the nearest thousandth.

1. $\sqrt{121}$ **2.** $\sqrt{\dfrac{25}{64}}$ **3.** $\sqrt{169}$

4. $-\sqrt{100}$ **5.** $\sqrt{5}$ **6.** $-\sqrt{-1}$

Section 9.2

Simplify.

7. $\sqrt{x^4}$ **8.** $\sqrt{25x^4y^6}$ **9.** $\sqrt{98}$

10. $\sqrt{x^5}$ **11.** $\sqrt{36x^3}$ **12.** $\sqrt{8a^4b^3}$

Section 9.3

Combine, if possible.

13. $5\sqrt{2} - \sqrt{2}$

14. $\sqrt{2} + \sqrt{3}$

15. $3\sqrt{2} - \sqrt{8} + \sqrt{18}$

16. $2\sqrt{28} + 3\sqrt{63} - \sqrt{49}$

17. $3\sqrt{8} + \sqrt{12} + \sqrt{50} - 4\sqrt{75}$

Section 9.4

Multiply. Simplify your answer.

18. $(2\sqrt{6})(3\sqrt{3})$

19. $\sqrt{2}(\sqrt{8} + 2\sqrt{3})$

20. $(\sqrt{7} + 4)^2$

21. $(\sqrt{11} - \sqrt{10})(\sqrt{11} + \sqrt{10})$

22. $(\sqrt{m} + \sqrt{m+1})(\sqrt{m} - \sqrt{m+1})$

23. $(2\sqrt{3} - \sqrt{5})^2$

Section 9.5

Simplify. Rationalize all denominators.

24. $\sqrt{\dfrac{36}{x^2}}$ **25.** $\dfrac{5}{\sqrt{3}}$ **26.** $\dfrac{3}{\sqrt{5} + 2}$ **27.** $\dfrac{\sqrt{7} - \sqrt{6}}{\sqrt{7} + \sqrt{6}}$

Section 9.6

Use the Pythagorean theorem *to find the length of the third side of the right triangle labeled x.*

28.

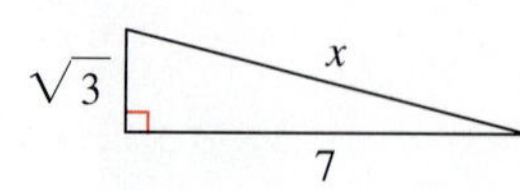

29.

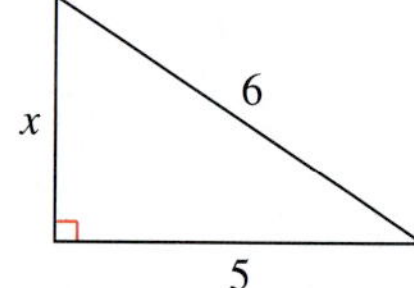

Solve each radical equation and verify your answer.

30. $2 + \sqrt{3x + 4} = x$ **31.** $\sqrt{3x + 1} = 6$ **32.** $\sqrt{11x - 5} = \sqrt{4 - 7x}$

Section 9.7

33. If y varies inversely as x and $y = \frac{2}{3}$ when $x = 21$, find the value of y if $x = 2$.

9.1 Square Roots

MathPro

Video 21

SSM

After studying this section, you will be able to:

1 *Evaluate a square root of a perfect square.*
2 *Approximate a square root to the nearest thousandth.*

1 *Evaluating the Square Root of a Perfect Square*

How long is the side of a square whose area is 4? Recall that the formula for the area of a square is

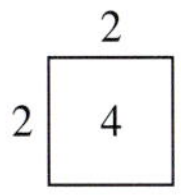

area of a square $= s^2$ — *Our question then becomes, what number times itself is* 4?

$s^2 = 4$

$s = 2$ — *Because* $(2)(2) = 4$.

$s = -2$ — *Because* $(-2)(-2) = 4$.

(Square with side 2 and area 4.)

We say that 2 is the square root of 4 because $(2)(2) = 4$. We can also say that -2 is the square root of 4 because $(-2)(-2) = 4$. Note that 4 is a **perfect square.** The square root of a perfect square is an integer.

The symbol $\sqrt{\ }$ is used in mathematics for the **square root** of a number. The symbol itself $\sqrt{\ }$ is called the *radical sign*. When we use this symbol, it is understood to be the nonnegative square root, or principal square root. If we want to find the negative square root of a number, we use the symbol $-\sqrt{\ }$.

TEACHING TIP It is fairly important to take the time to clear up some possible confusion about negative signs and square roots. After presenting Example 1 to the students, you might want to offer the following contrasts.

A. If someone asks what are the square roots of 121, the answers are $+11$ and -11.

B. If someone asks us to evaluate $\sqrt{121}$, this is the principal or positive square root and thus the answer is $+11$.

C. If someone asks us to evaluate $\sqrt{-121}$, at this point we would just state that this expression does not yield a real number.

D. Finally, if the negative sign is in front of the radical, such as $-\sqrt{121}$, then the answer is -11.

> **Definition of Square Root**
>
> For all nonnegative numbers N, the square root of N (written $\sqrt{N}$) is defined to be the nonnegative number a if and only if $a^2 = N$.

Notice in the definition that we did not use the words "for all *positive* numbers N" because we also want to include the square root of 0. $\sqrt{0} = 0$ since $0^2 = 0$. So, for positive numbers *and* for zero, the definition of a square root is $\sqrt{N} = a$ if and only if $a^2 = N$.

Let us now examine the use of this definition with a few examples.

EXAMPLE 1 Find. **(a)** $\sqrt{144}$ **(b)** $-\sqrt{9}$

(a) $\sqrt{144} = 12$ since $(12)^2 = (12)(12) = 144$.

(b) The symbol $-\sqrt{9}$ is read "the negative square root of 9."

$$-\sqrt{9} = -3 \text{ since } (-3)^2 = (-3)(-3) = 9.$$

Practice Problem 1 Find. **(a)** $\sqrt{64}$ **(b)** $-\sqrt{121}$ ■

TEACHING TIP Remind students that if they can quickly identify and evaluate such expressions as $\sqrt{4}$ and $\sqrt{9}$ it will greatly help them throughout this chapter.

The number beneath the radical sign is called the **radicand.** The radicand will not always be an integer.

EXAMPLE 2 Find. **(a)** $\sqrt{\frac{1}{4}}$ **(b)** $-\sqrt{\frac{4}{9}}$

(a) $\sqrt{\frac{1}{4}} = \frac{1}{2}$ since $\left(\frac{1}{2}\right)^2 = \left(\frac{1}{2}\right)\left(\frac{1}{2}\right) = \frac{1}{4}$.

(b) $-\sqrt{\frac{4}{9}} = -\frac{2}{3}$ since $\left(-\frac{2}{3}\right)\left(-\frac{2}{3}\right) = \frac{4}{9}$.

Practice Problem 2 Find. **(a)** $\sqrt{\frac{9}{25}}$ **(b)** $-\sqrt{\frac{121}{169}}$ ■

EXAMPLE 3 Find. **(a)** $\sqrt{0.09}$ **(b)** $\sqrt{1600}$

(a) $\sqrt{0.09} = 0.3$ since $(0.3)(0.3) = 0.09$.
Notice that $\sqrt{0.09}$ is *not* 0.03 because $(0.03)(0.03) = 0.0009$. Remember to count the decimal places when you multiply decimals. When finding the square root of a decimal, you should do the multiplication as a check.

(b) $\sqrt{1600} = 40$ since $(40)(40) = 1600$.

Practice Problem 3 Find. **(a)** $-\sqrt{0.0036}$ **(b)** $\sqrt{2500}$ ■

2 *Approximating a Square Root*

Not all numbers are perfect squares. How can we find the square root of 2? That is, what number times itself equals 2? $\sqrt{2}$ is an irrational number. It cannot be written as $\frac{p}{q}$, where p, q are integers, $q \neq 0$. $\sqrt{2}$ is a nonterminating, nonrepeating decimal and the best we can do is to approximate its value. To do so, you can use the square root key on your

To use a calculator, enter 2 and push the $\boxed{\sqrt{x}}$ or $\boxed{\sqrt{\ }}$ key.

$\boxed{2}\ \boxed{\sqrt{x}}\ \boxed{1.4142136}$

Remember, this is only an approximation. Most calculators are limited to an eight-digit display.

To use the table, look under the x column until you find the 2 row. Then look under the $\sqrt{x}$ column in that row. The value is rounded to the nearest thousandth.

$$\sqrt{2} \approx 1.414$$

x	$\sqrt{x}$
1	1.000
2	1.414
3	1.732
4	2.000
5	2.236
6	2.449
7	2.646

EXAMPLE 4 Use a calculator or a table to approximate each square root. Round to the nearest thousandth.

(a) $\sqrt{28}$ ***(b)*** $\sqrt{191}$

Using a calculator, we have

(a) 28 $\boxed{\sqrt{x}}$ 5.291502622, thus $\sqrt{28} \approx 5.292$

(b) 191 $\boxed{\sqrt{x}}$ 13.82027496, thus $\sqrt{191} \approx 13.820$

TEACHING TIP It is helpful to point out to students some example of why we want to approximate square roots. The distance between consecutive bases in a baseball diamond is exactly 90 feet. However, the distance from home plate to second base is exactly $\sqrt{16{,}200}$ feet. To approximate this distance to the nearest thousandth, we would use a calculator and obtain 127.279 feet, rounded to the nearest thousandth of a foot. If we wanted to measure to see if a baseball diamond was accurately constructed, we might need that approximate value.

Practice Problem 4 Approximate each square root.

(a) $\sqrt{3}$ **(b)** $\sqrt{35}$ **(c)** $\sqrt{127}$ ■

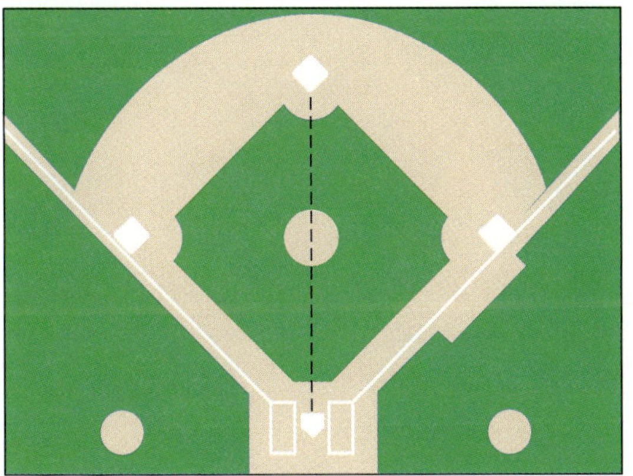

On a professional softball diamond, the distance from home plate to second base is exactly $\sqrt{7200}$ feet. Try to approximate this value using a calculator.

A Topic for Further Thought

Let's look at the radicand carefully. Does the radicand need to be a nonnegative number? What if we write $\sqrt{-4}$? Is there a number that you can square to get -4? Obviously, there is *no real number* that you can square to get -4. We know that $(2)^2 = 4$ and $(-2)^2 = 4$. Any number that is squared will be nonnegative. We therefore conclude that $\sqrt{-4}$ does not represent a real number. We want to work with real numbers, so our definition of the square root $\sqrt{n}$ requires that n be nonnegative. Thus $\sqrt{-4}$ is *not a real number.*

You may encounter a more sophisticated number system called **complex numbers** in a higher level mathematics course. In the complex number system negative numbers will have square roots.

9.1 Exercises

Verbal and Writing Skills

1. Define a square root in your own words.
Answers will vary.

2. Is $\sqrt{576} = 24$? Why or why not?
Yes, because $(24)(24) = 576$.

3. Is $\sqrt{0.9} = 0.3$? Why or why not?
No, because $(0.3)(0.3) = 0.09$.

4. How would you find $\sqrt{90}$?
Use a calculator or square root table.

Find the two square roots of each number.

5. 9 ± 3

6. 4 ± 2

7. 64 ± 8

8. 25 ± 5

9. 81 ± 9

10. 100 ± 10

11. 121 ± 11

12. 144 ± 12

Find the square root. Do not *use a calculator or a table of square roots.*

13. $\sqrt{25}$ 5

14. $\sqrt{36}$ 6

15. $\sqrt{0}$ 0

16. $\sqrt{1}$ 1

17. $\sqrt{49}$ 7

18. $\sqrt{16}$ 4

19. $\sqrt{100}$ 10

20. $\sqrt{9}$ 3

21. $-\sqrt{36}$ -6

22. $-\sqrt{25}$ -5

23. $\sqrt{0.81}$ 0.9

24. $\sqrt{0.64}$ 0.8

25. $\sqrt{1.96}$ 1.4

26. $\sqrt{1.21}$ 1.1

27. $\sqrt{\frac{16}{25}}$ $\frac{4}{5}$

28. $\sqrt{\frac{9}{64}}$ $\frac{3}{8}$

29. $\sqrt{\frac{49}{64}}$ $\frac{7}{8}$

30. $\sqrt{\frac{49}{100}}$ $\frac{7}{10}$

31. $\sqrt{400}$ 20

32. $\sqrt{3600}$ 60

33. $-\sqrt{10{,}000}$ -100

34. $\sqrt{40{,}000}$ 200

35. $\sqrt{169}$ 13

36. $\sqrt{225}$ 15

37. $-\sqrt{\frac{1}{64}}$ $-\frac{1}{8}$

38. $\sqrt{\frac{81}{100}}$ $\frac{9}{10}$

39. $\sqrt{\frac{25}{49}}$ $\frac{5}{7}$

40. $-\sqrt{\frac{9}{64}}$ $-\frac{3}{8}$

41. $\sqrt{0.36}$ 0.6

42. $\sqrt{0.25}$ 0.5

43. $\sqrt{14{,}400}$ 120

44. $-\sqrt{19{,}600}$ -140

★**45.** $\sqrt{441}$ 21

★**46.** $\sqrt{961}$ 31

47. $\sqrt{0.0169}$ 0.13

48. $\sqrt{0.000225}$ 0.015

Use a calculator or the square root table to approximate each square root to the nearest thousandth.

49. $\sqrt{42}$ 6.481

50. $\sqrt{46}$ 6.782

51. $\sqrt{59}$ 7.681

52. $\sqrt{52}$ 7.211

53. $-\sqrt{133}$ -11.533

54. $-\sqrt{146}$ -12.083

55. $-\sqrt{195}$ -13.964

56. $-\sqrt{182}$ -13.491

First estimate the answer. Then use a calculator to find each answer. Round your results to two decimal places.

57. $-\sqrt{360}$ -18.97

58. $\sqrt{12.5}$ 3.54

59. $\sqrt{45.5}$ 6.75

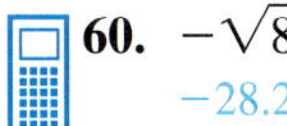

60. $-\sqrt{800}$ -28.28

Applications

The time in seconds it takes for an object to fall s feet is given by $t = \frac{1}{4}\sqrt{s}$. *Thus an object falling 100 feet would take* $t = \frac{1}{4}\sqrt{100} = \frac{1}{4}(10) = 2.5$ *seconds.*

61. How long would it take an object dropped off a building to fall 900 feet? 7.5 seconds

62. How long would it take an object dropped out of a tower to fall 1600 feet? 10 seconds

63. The Sears Tower in Chicago is the world's tallest building. It is 1454 feet tall. If a metal frame fell off of the building from a height of 1444 feet, how long would it take the metal frame to hit the ground?
9.5 seconds

64. The CN Tower in Toronto, Ontario, Canada is the world's tallest self-supporting structure. It is 1821 feet tall. If a worker installing an antenna for cellular phone reception drops a bolt from a height of 1764 feet, how long would it take the bolt to hit the ground?
10.5 seconds

To Think About

Without the aid of a calculator, see if you can evaluate the following. (You may need to make several educated guesses!)

65. $\sqrt{\sqrt{256}}$ 4

66. $\sqrt{\sqrt{\sqrt{100{,}000{,}000}}}$ 10

Cumulative Review Problems

67. Solve. $3x + 2y = 8$
$7x - 3y = 11$

$9x + 6y = 24$
$14x - 6y = 22$ $x = 2; y = 1$

68. Solve if possible. $2x - 3y = 1$
$-8x + 12y = 4$

no solution, inconsistent system of equations

69. A snowboard racer bought 3 new snowboards and 2 pairs of racing goggles last month for \$850. This month, she bought 4 snowboards and 3 pairs of goggles for \$1150. How much did each snowboard cost? How much did each pair of goggles cost?
Each snowboard is \$250.
Each pair of goggles is \$50.

70. With a tail wind, the Qantas flight from Los Angeles to Sydney, Australia, covered 7280 miles in 14 hours, nonstop. The return flight of 7140 miles took 17 hours because of a head wind. If the airspeed of the plane remained constant, and if the return flight encountered a constant head wind equal to the earlier tail wind, what was the plane's airspeed?
The plane's airspeed is 470 mph.
The wind is 50 mph.

9.2 Simplify Radicals

MathPro

Video 22

SSM

After studying this section, you will be able to:

1. *Simplify radical expressions with a radicand that is a perfect square.*
2. *Simplify radical expressions with a radicand that is not a perfect square.*

1 *Simplifying a Radical Expression with a Radicand That Is a Perfect Square*

We know that $\sqrt{25} = \sqrt{5^2} = 5$. $\sqrt{9} = \sqrt{3^2} = 3$ and $\sqrt{36} = \sqrt{6^2} = 6$. If we write the radicand as a square, the square root is the base. That is $\sqrt{x^2} = x$ when x is nonnegative. We can use this idea to simplify radicals.

EXAMPLE 1 Find. **(a)** $\sqrt{9^4}$ **(b)** $\sqrt{126^2}$ **(c)** $\sqrt{17^6}$

(a) Using the law of exponents, we rewrite 9^4 as a square.

$\sqrt{9^4} = \sqrt{(9^2)^2}$ *Raise a power to a power* $(9^2)^2 = 9^4$.

$= 9^2$ *To check, multiply* $(9^2)(9^2) = 9^4$.

(b) $\sqrt{126^2} = 126$

(c) $\sqrt{17^6} = \sqrt{(17^3)^2}$ *Raise a power to a power* $(17^3)^2 = 17^6$.

$= 17^3$ *To check, multiply* $(17^3)(17^3) = 17^6$.

Practice Problem 1 Find. **(a)** $\sqrt{6^2}$ **(b)** $\sqrt{13^4}$ **(c)** $\sqrt{18^{12}}$ ■

This same concept can be used with variable expressions. We will restrict the values of each variable to be nonnegative if the variable is under the radical sign. Thus *all variable radicands in this chapter* are assumed to *represent positive numbers or zero.*

EXAMPLE 2 Find. **(a)** $\sqrt{x^6}$ **(b)** $\sqrt{x^{16}}$ **(c)** $\sqrt{y^{24}}$

(a) $\sqrt{x^6} = \sqrt{(x^3)^2}$ *By the law of exponents,* $(x^3)^2 = x^6$.

$= x^3$

(b) $\sqrt{x^{16}} = x^8$ **(c)** $\sqrt{y^{24}} = y^{12}$

Practice Problem 2 Find. **(a)** $\sqrt{x^4}$ **(b)** $\sqrt{y^{18}}$ **(c)** $\sqrt{x^{30}}$ ■

In each of these examples we found the square root exactly. It is as if we ''removed the radical sign'' when we found the square root. When you find a square root exactly, do *not* leave a radical sign in your answer.

Often the coefficient of the variable will not be written in exponent form. You will want to be able to recognize the squares of numbers from 1 to 15, 20, and 25. If you do, you will be able to find the square root faster than by using a calculator or a table of square roots.

Before you begin, you need to know the multiplication rule for square roots.

Multiplication Rule for Square Roots

For all nonnegative numbers a, b,

$$\sqrt{a} \cdot \sqrt{b} = \sqrt{ab} \quad \text{and} \quad \sqrt{ab} = \sqrt{a} \cdot \sqrt{b}$$

EXAMPLE 3 Find. **(a)** $\sqrt{225x^2}$ **(b)** $\sqrt{x^6y^{14}}$ **(c)** $\sqrt{169x^8y^{10}}$

a) $\sqrt{225x^2} = 15x$ *Using the multiplication rule for square roots,* $\sqrt{225x^2} = \sqrt{225}\,\sqrt{x^2} = 15x.$

(b) $\sqrt{x^6y^{14}} = x^3y^7$ **(c)** $\sqrt{169x^8y^{10}} = 13x^4y^5$

To check, square each answer. The square of the answer will be the expression under the radical sign.

Practice Problem 3 Find.

(a) $\sqrt{625y^4}$ **(b)** $\sqrt{x^{16}y^{22}}$ **(c)** $\sqrt{121x^{12}y^6}$ ■

2 *Simplifying a Radical Expression with a Radicand That Is Not a Perfect Square*

Most of the time when we encounter square roots the radicand is not a perfect square. Thus, we cannot remove the radical sign and find the square root. It may, however, be possible to simplify the expression. We will want to do so, so that we can combine these expressions, as we will be doing in Section 9.3.

EXAMPLE 4 Simplify. **(a)** $\sqrt{20}$ **(b)** $\sqrt{50}$ **(c)** $\sqrt{48}$

To begin, when you look at a radicand, look for perfect squares.

(a) $\sqrt{20} = \sqrt{4 \cdot 5}$ *Since 4 is a perfect square, we can write 20 as* $4 \cdot 5$.

$= \sqrt{4}\sqrt{5}$ *Recall* $\sqrt{4} = 2$

$= 2\sqrt{5}$

(b) $\sqrt{50} = \sqrt{25 \cdot 2}$
$= \sqrt{25}\,\sqrt{2}$
$= 5\sqrt{2}$

(c) $\sqrt{48} = \sqrt{16 \cdot 3}$
$= \sqrt{16}\,\sqrt{3}$
$= 4\sqrt{3}$

TEACHING TIP After discussing Example 4, ask students to simplify the following three radical expressions as an Additional Class Exercise.

A. $\sqrt{45}$
B. $\sqrt{52}$
C. $\sqrt{325}$

Sol.

A. $3\sqrt{5}$
B. $2\sqrt{13}$
C. $5\sqrt{13}$

Practice Problem 4 Simplify. **(a)** $\sqrt{98}$ **(b)** $\sqrt{12}$ **(c)** $\sqrt{75}$ ■

The same procedure can be used with expressions containing variables. The key is to think squares. That is, think, ''How can I rewrite the expression so that it contains a perfect square.''

TEACHING TIP After discussing Example 5, ask students to simplify the following three radical expressions as an Additional Class Exercise.

A. $\sqrt{x^{11}}$
B. $\sqrt{w^{14}}$
C. $\sqrt{z^{23}}$

Sol.

A. $x^5\sqrt{x}$
B. w^7
C. $z^{11}\sqrt{z}$

EXAMPLE 5 Simplify. **(a)** $\sqrt{x^3}$ **(b)** $\sqrt{y^5}$ **(c)** $\sqrt{x^7y^9}$

(a) Recall that the law of exponents tells us to add the exponents when multiplying two numbers that have the same base. Think of rewriting 3 as a sum that contains an even number. $3 = 2 + 1$

$\sqrt{x^3} = \sqrt{x^2}\sqrt{x}$ *Because* $(x^2)(x) = x^3$ *by the law of exponents.*

$= x\sqrt{x}$

(b) $\sqrt{y^5} = \sqrt{y^4}\sqrt{y}$ *Notice that* y^4 *is the largest possible even exponent.*

$= y^2\sqrt{y}$

In general, to simplify a square root that has a variable with an exponent in the radicand, first write it as a product of the form $\sqrt{x^n}\sqrt{x}$, where n is the largest possible even exponent.

(c) $\sqrt{x^7y^9} = \sqrt{x^6}\sqrt{x}\sqrt{y^8}\sqrt{y}$

$= x^3\sqrt{x}\,y^4\sqrt{y}$

$= x^3y^4\sqrt{xy}$

In the final form, we place the factors with exponents first and the radical factor second.

Practice Problem 5 Simplify. **(a)** $\sqrt{x^{11}}$ **(b)** $\sqrt{x^5y^3}$ ■

Putting this all together, to simplify a radical, you factor each part of the expression under the radical sign and take out the perfect squares.

TEACHING TIP After discussing Example 6, ask students to simplify the following two radical expressions as an Additional Class Exercise.

A. $\sqrt{32x^3y^8}$
B. $\sqrt{108b^{10}c^{13}}$

Sol.

A. $4xy^4\sqrt{2x}$ B. $6b^5c^6\sqrt{3c}$

EXAMPLE 6 Simplify. **(a)** $\sqrt{12y^5}$ **(b)** $\sqrt{27x^5y^6}$ **(c)** $\sqrt{18x^3y^7w^{10}}$

(a) $\sqrt{12y^5} = \sqrt{4 \cdot 3 \cdot y^4 \cdot y}$

$= 2y^2\sqrt{3y}$ *Remember,* $y^4 = (y^2)^2$ *and* $\sqrt{(y^2)^2} = y^2$.

(b) $\sqrt{27x^5y^6} = \sqrt{9 \cdot 3 \cdot x^4 \cdot x \cdot y^6}$

$= 3x^2y^3\sqrt{3x}$

Notice that since the exponent of y is even, y^6 is a perfect square. y will not appear under the radical sign in the answer.

(c) $\sqrt{18x^3y^7w^{10}} = \sqrt{9 \cdot 2 \cdot x^2 \cdot x \cdot y^6 \cdot y \cdot w^{10}}$

$= 3xy^3w^5\sqrt{2xy}$

Practice Problem 6 Simplify.

(a) $\sqrt{48x^{11}}$ **(b)** $\sqrt{72x^8y^9}$ **(c)** $\sqrt{121x^6y^7z^8}$ ■

9.2 Exercises

Simplify. Leave the answer in exponent form.

1. $\sqrt{7^2}$ 7
2. $\sqrt{6^2}$ 6
3. $\sqrt{10^4}$ 10^2
4. $\sqrt{8^4}$ 8^2
5. $\sqrt{9^6}$ 9^3

6. $\sqrt{5^8}$ 5^4
7. $\sqrt{56^6}$ 56^3
8. $\sqrt{39^8}$ 39^4
9. $\sqrt{3^{150}}$ 3^{75}
10. $\sqrt{2^{180}}$ 2^{90}

Simplify. Assume that all variables represent positive numbers.

11. $\sqrt{w^{12}}$ w^6
12. $\sqrt{y^6}$ y^3
13. $\sqrt{t^{18}}$ t^9
14. $\sqrt{w^{22}}$ w^{11}
15. $\sqrt{y^{26}}$ y^{13}

16. $\sqrt{w^{14}}$ w^7
17. $\sqrt{49x^6}$ $7x^3$
18. $\sqrt{81y^4}$ $9y^2$
19. $\sqrt{144x^2}$ $12x$
20. $\sqrt{196y^{10}}$ $14y^5$

21. $\sqrt{400x^8}$ $20x^4$
22. $\sqrt{2500y^{12}}$ $50y^6$
23. $\sqrt{x^2y^6}$ xy^3
24. $\sqrt{x^4y^{12}}$ x^2y^6
25. $\sqrt{25x^8y^4}$ $5x^4y^2$

26. $\sqrt{16x^2y^{20}}$ $4xy^{10}$
27. $\sqrt{64x^2y^4}$ $8xy^2$
28. $\sqrt{100x^{12}y^8}$ $10x^6y^4$
29. $\sqrt{256a^2b^8c^4}$ $16ab^4c^2$
30. $\sqrt{400a^{20}b^6z^4}$ $20a^{10}b^3z^2$

Simplify. Assume that all variables represent positive numbers.

31. $\sqrt{20}$ $2\sqrt{5}$
32. $\sqrt{12}$ $2\sqrt{3}$
33. $\sqrt{24}$ $2\sqrt{6}$
34. $\sqrt{27}$ $3\sqrt{3}$
35. $\sqrt{18}$ $3\sqrt{2}$

36. $\sqrt{32}$ $4\sqrt{2}$
37. $\sqrt{50}$ $5\sqrt{2}$
38. $\sqrt{48}$ $4\sqrt{3}$
39. $\sqrt{72}$ $6\sqrt{2}$
40. $\sqrt{60}$ $2\sqrt{15}$

41. $\sqrt{90}$ $3\sqrt{10}$
42. $\sqrt{125}$ $5\sqrt{5}$
43. $\sqrt{98}$ $7\sqrt{2}$
44. $\sqrt{54}$ $3\sqrt{6}$
45. $\sqrt{108}$ $6\sqrt{3}$

46. $\sqrt{8x^3}$
$2x\sqrt{2x}$

47. $\sqrt{12y^5}$
$2y^2\sqrt{3y}$

48. $\sqrt{27w^5}$
$3w^2\sqrt{3w}$

49. $\sqrt{18y^{13}}$
$3y^6\sqrt{2y}$

50. $\sqrt{16x^7}$
$4x^3\sqrt{x}$

51. $\sqrt{49x^5}$
$7x^2\sqrt{x}$

52. $\sqrt{75x^3y^5}$
$5xy^2\sqrt{3xy}$

53. $\sqrt{50x^7y}$
$5x^3\sqrt{2xy}$

54. $\sqrt{48y^3w}$
$4y\sqrt{3yw}$

55. $\sqrt{12x^2y^3}$
$2xy\sqrt{3y}$

56. $\sqrt{75x^2y^3}$
$5xy\sqrt{3y}$

57. $\sqrt{27a^8y^7}$
$3a^4y^3\sqrt{3y}$

58. $\sqrt{135x^5y^7}$
$3x^2y^3\sqrt{15xy}$

59. $\sqrt{140x^7y^{11}}$
$2x^3y^5\sqrt{35xy}$

60. $\sqrt{150a^3b^{13}}$
$5ab^6\sqrt{6ab}$

61. $\sqrt{180a^5bc^2}$
$6a^2c\sqrt{5ab}$

62. $\sqrt{80a^3b^3c^6}$
$4abc^3\sqrt{5ab}$

63. $\sqrt{63a^4b^6c^3}$
$3a^2b^3c\sqrt{7c}$

64. $\sqrt{28a^2b^5c^8}$
$2ab^2c^4\sqrt{7b}$

65. $\sqrt{81x^{12}y^{11}w^5}$
$9x^6y^5w^2\sqrt{yw}$

To Think About

Simplify. Assume that all variables represent positive numbers.

66. $\sqrt{(x+3)^2}$
$x + 3$

67. $\sqrt{x^2 + 12x + 36}$
$x + 6$

68. $\sqrt{16x^2 + 8x + 1}$
$4x + 1$

69. $\sqrt{4x^2 + 20x + 25}$
$2x + 5$

70. $\sqrt{x^2y^2 + 14xy^2 + 49y^2}$
$y(x + 7)$

71. $\sqrt{16x^2 + 16x + 4}$
$2(2x + 1)$

Cumulative Review Problems

72. Solve. $3x - 2y = 14$
$y = x - 5$
$3x - 2(x - 5) = 14 \Rightarrow x = 4;\ y = -1$

73. Solve. $\frac{1}{3}(2x - 4) = \frac{1}{4}x - 3$
$x = -4$

74. Surprisingly, Asia and Australia lead the world in nuclear energy consumption. In 1984 they used 310 million tons of equivalent energy. In 1993 they used 560 million tons. Estimate how much energy will be used in the year 2000. 754 million tons of equivalent energy

75. In 1993 the United States consumed 84 quadrillion Btu of energy. This was 265% of the energy consumed by China in that same year. How many quadrillion Btu of energy were consumed by China? Approx. 31.7 quadrillion Btu of energy

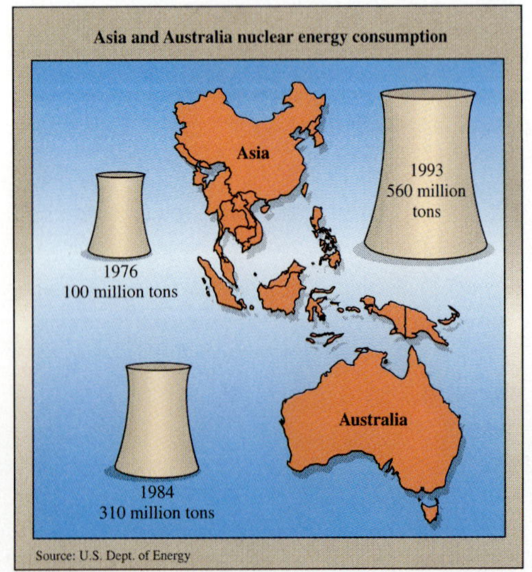

9.3 Addition and Subtraction of Radicals

After studying this section, you will be able to:

1 *Add or subtract radical expressions with common radicals.*

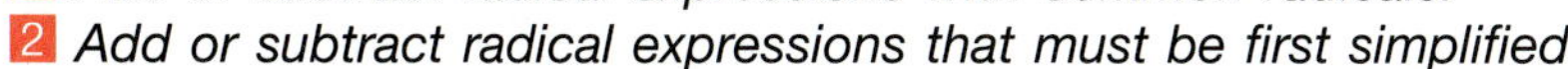

2 *Add or subtract radical expressions that must be first simplified.*

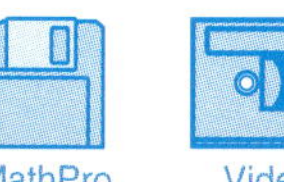

MathPro

Video 22

SSM

1 *Adding and Subtracting Radicals with Common Radicals*

Recall that when you simplify an algebraic expression you combine like terms. $8b + 5b = 13b$ because both terms contain the same variable b, and you add the coefficients. This same idea is used when combining radicals. To add or subtract radicals, the number or the algebraic expression under the radical sign must be the same.

EXAMPLE 1 Combine. **(a)** $5\sqrt{2} - 8\sqrt{2}$ **(b)** $7\sqrt{a} + 3\sqrt{a} - 5\sqrt{a}$

First check to see if the radicands are the same. If they are, you can combine the radicals.

(a) $5\sqrt{2} - 8\sqrt{2} = -3\sqrt{2}$ **(b)** $7\sqrt{a} + 3\sqrt{a} - 5\sqrt{a} = 5\sqrt{a}$

Practice Problem 1 Combine. **(a)** $7\sqrt{11} + 4\sqrt{11}$
(b) $4\sqrt{t} - 7\sqrt{t} + 6\sqrt{t} - 2\sqrt{t}$ ■

Be careful. Not all the terms in an expression will have common radicals.

EXAMPLE 2 Combine. $5\sqrt{2a} + 3\sqrt{a} - 7\sqrt{2} + 3\sqrt{2a}$

The only terms that have the same radicand are $5\sqrt{2a}$ and $3\sqrt{2a}$. These terms may be combined. All other terms stay the same.

$$5\sqrt{2a} + 3\sqrt{a} - 7\sqrt{2} + 3\sqrt{2a} = 8\sqrt{2a} + 3\sqrt{a} - 7\sqrt{2}$$

TEACHING TIP After discussing Example 2, ask students to combine like terms in the following expression as an Additional Class Exercise.

$7\sqrt{3xy} - 5\sqrt{3y} - 9\sqrt{3xy} - 12\sqrt{3y}$

Sol. $-2\sqrt{3xy} - 17\sqrt{3y}$

Practice Problem 2 Combine. $3\sqrt{x} - 2\sqrt{xy} - 5\sqrt{y} + 7\sqrt{xy}$ ■

2 *Adding and Subtracting Radical Expressions That Must First Be Simplified*

Sometimes it is necessary to simplify one or more radicals before their terms can be combined. Be alert to combine only like radicals.

EXAMPLE 3 Combine. **(a)** $2\sqrt{3} + \sqrt{12}$ **(b)** $\sqrt{12} - \sqrt{27} + \sqrt{50}$

(a) $2\sqrt{3} + \sqrt{12} = 2\sqrt{3} + \sqrt{4 \cdot 3}$ *Look for perfect square factors.*

$= 2\sqrt{3} + 2\sqrt{3}$ *Simplify and combine like radicals.*

$= 4\sqrt{3}$

(b) $\sqrt{12} - \sqrt{27} + \sqrt{50} = \sqrt{4 \cdot 3} - \sqrt{9 \cdot 3} + \sqrt{25 \cdot 2}$

$= 2\sqrt{3} - 3\sqrt{3} + 5\sqrt{2}$ *Only $2\sqrt{3}$ and $-3\sqrt{3}$ can be combined.*

$= -\sqrt{3} + 5\sqrt{2}$ or $5\sqrt{2} - \sqrt{3}$

Practice Problem 3 Combine. **(a)** $\sqrt{50} - \sqrt{18} + \sqrt{98}$
(b) $\sqrt{12} + \sqrt{18} - \sqrt{50} + \sqrt{27}$ ■

EXAMPLE 4 Combine. $\sqrt{2a} + \sqrt{8a} + \sqrt{27a}$

$$\sqrt{2a} + \sqrt{8a} + \sqrt{27a} = \sqrt{2a} + \sqrt{4 \cdot 2a} + \sqrt{9 \cdot 3a}$$

Look for perfect square factors.

$$= \sqrt{2a} + 2\sqrt{2a} + 3\sqrt{3a}$$

Be careful. $\sqrt{2a}$ *and* $\sqrt{3a}$ *are* ***not*** *like radicals. The expression under the radical sign must be the same.*

$$= 3\sqrt{2a} + 3\sqrt{3a}$$

Practice Problem 4 Combine. $\sqrt{9x} + \sqrt{8x} - \sqrt{4x} + \sqrt{50x}$ ■

Special care should be taken if the original radical has a numerical coefficient. If the radical can be simplified, the two resulting numerical coefficients should be multiplied.

TEACHING TIP Students tend to have much more difficulty adding radical terms if the square root has a numerical coefficient prior to the simplification step. After you work out Example 5 for students, you may want to do the following as an Additional Class Exercise.
Combine.

$$5\sqrt{28} - 3\sqrt{7} + 6\sqrt{63}$$

Sol.

$$5\sqrt{4}\sqrt{7} - 3\sqrt{7} + 6\sqrt{9}\sqrt{7}$$
$$= 5(2)\sqrt{7} - 3\sqrt{7} + 6(3)\sqrt{7}$$
$$= 10\sqrt{7} - 3\sqrt{7} + 18\sqrt{7}$$
$$= 25\sqrt{7}$$

EXAMPLE 5 Combine. **(a)** $3\sqrt{18} - 5\sqrt{2}$ **(b)** $2\sqrt{20} + 3\sqrt{45} - 4\sqrt{80}$

(a) $3\sqrt{18} - 5\sqrt{2} = 3 \cdot \sqrt{9} \cdot \sqrt{2} - 5\sqrt{2}$

$$= 3 \cdot 3\sqrt{2} - 5\sqrt{2}$$

Remember to multiply $3 \cdot 3$ here.

$$= 9\sqrt{2} - 5\sqrt{2}$$
$$= 4\sqrt{2}$$

(b) $2\sqrt{20} + 3\sqrt{45} - 4\sqrt{80} = 2 \cdot \sqrt{4} \cdot \sqrt{5} + 3 \cdot \sqrt{9} \cdot \sqrt{5} - 4 \cdot \sqrt{16} \cdot \sqrt{5}$

$$= 2 \cdot 2 \cdot \sqrt{5} + 3 \cdot 3 \cdot \sqrt{5} - 4 \cdot 4 \cdot \sqrt{5}$$
$$= 4\sqrt{5} + 9\sqrt{5} - 16\sqrt{5}$$
$$= -3\sqrt{5}$$

Practice Problem 5 Combine. **(a)** $2\sqrt{18} + 3\sqrt{8}$
(b) $\sqrt{27} - 4\sqrt{3} + 2\sqrt{75}$ ■

TEACHING TIP When you have finished explaining Example 6, you may wish to do the following Additional Class Exercise for your students.
Combine.

$$2\sqrt{3x^3} - 5x\sqrt{27x} + 4\sqrt{3x}$$

Sol.

$$2\sqrt{x^2}\sqrt{3x} - 5x\sqrt{9}\sqrt{3x} + 4\sqrt{3x}$$
$$= 2x\sqrt{3x} - 15x\sqrt{3x} + 4\sqrt{3x}$$
$$= -13x\sqrt{3x} + 4\sqrt{3x}$$

Note that $-13x\sqrt{3x}$ and $4\sqrt{3x}$ are **not like terms.**

EXAMPLE 6 Combine. $3a\sqrt{8a} + 2\sqrt{50a^3}$

$3a\sqrt{8a} + 2\sqrt{50a^3}$ *Simplify each radical.*

$$= 3a\sqrt{4}\sqrt{2a} + 2\sqrt{25a^2}\sqrt{2a}$$
$$= 3a \cdot 2 \cdot \sqrt{2a} + 2 \cdot 5a \cdot \sqrt{2a}$$
$$= 6a\sqrt{2a} + 10a\sqrt{2a}$$
$$= 16a\sqrt{2a}$$

(*Note*: If you are unsure of the last step, show the use of the distributive property in performing the addition.)

$$6a\sqrt{2a} + 10a\sqrt{2a} = (6 + 10)a\sqrt{2a}$$
$$= 16a\sqrt{2a}$$

Practice Problem 6 Combine. $2\sqrt{12x} - 3\sqrt{45x} - 3\sqrt{27x} + \sqrt{20}$ ■

9.3 Exercises

Verbal and Writing Skills

1. List the steps involved in simplifying an expression such as $\sqrt{75x} + \sqrt{48x} + \sqrt{16x^2}$.
1. Simplify each radical term.
2. Combine like radicals.

2. Julio said that, since $\sqrt{8}$ and $\sqrt{2}$ are not similar, $\sqrt{8} + \sqrt{2}$ could not be simplified. What do you think? Simplify $\sqrt{8}$ as $2\sqrt{2}$ and then combine. That is, $\sqrt{8} + \sqrt{2} = 2\sqrt{2} + \sqrt{2} = 3\sqrt{2}$.

Combine each of the following, if possible. Do not use a calculator or a table of square roots.

3. $\sqrt{3} - 5\sqrt{3} + 2\sqrt{3}$
$-2\sqrt{3}$

4. $\sqrt{6} + 4\sqrt{6} - 7\sqrt{6}$
$-2\sqrt{6}$

5. $\sqrt{2} + 8\sqrt{3} - 5\sqrt{3} + 4\sqrt{2}$
$5\sqrt{2} + 3\sqrt{3}$

6. $\sqrt{5} - \sqrt{6} + 3\sqrt{5} - 2\sqrt{6}$
$4\sqrt{5} - 3\sqrt{6}$

7. $3\sqrt{x} - 2\sqrt{xy} - 8\sqrt{x}$
$-5\sqrt{x} - 2\sqrt{xy}$

8. $\sqrt{ab} - 2\sqrt{b} + 3\sqrt{ab}$
$4\sqrt{ab} - 2\sqrt{b}$

9. $\sqrt{3} - \sqrt{12}$
$\sqrt{3} - 2\sqrt{3} = -\sqrt{3}$

10. $\sqrt{5} - \sqrt{20}$
$\sqrt{5} - 2\sqrt{5} = -\sqrt{5}$

11. $\sqrt{50} + 3\sqrt{32}$
$5\sqrt{2} + 12\sqrt{2} = 17\sqrt{2}$

12. $2\sqrt{12} + \sqrt{48}$
$4\sqrt{3} + 4\sqrt{3} = 8\sqrt{3}$

13. $2\sqrt{8} - 3\sqrt{2}$
$4\sqrt{2} - 3\sqrt{2} = \sqrt{2}$

14. $2\sqrt{27} - 4\sqrt{3}$
$6\sqrt{3} - 4\sqrt{3} = 2\sqrt{3}$

15. $\sqrt{75} + \sqrt{3} - 2\sqrt{27}$
$5\sqrt{3} + \sqrt{3} - 6\sqrt{3} = 0$

16. $\sqrt{28} - 3\sqrt{7} + 2\sqrt{63}$
$2\sqrt{7} - 3\sqrt{7} + 6\sqrt{7} = 5\sqrt{7}$

17. $2\sqrt{12} + \sqrt{20} + \sqrt{36}$
$4\sqrt{3} + 2\sqrt{5} + 6$

18. $\sqrt{24} - 2\sqrt{54} + 2\sqrt{18}$
$2\sqrt{6} - 6\sqrt{6} + 6\sqrt{2} = -4\sqrt{6} + 6\sqrt{2}$

19. $2\sqrt{48} - 3\sqrt{8} + \sqrt{50}$
$8\sqrt{3} - 6\sqrt{2} + 5\sqrt{2} = 8\sqrt{3} - \sqrt{2}$

20. $\sqrt{50} + \sqrt{28} - \sqrt{9}$
$5\sqrt{2} + 2\sqrt{7} - 3$

21. $3\sqrt{3x} + \sqrt{12x}$
$3\sqrt{3x} + 2\sqrt{3x} = 5\sqrt{3x}$

22. $5\sqrt{2y} - \sqrt{18y}$
$5\sqrt{2y} - 3\sqrt{2y} = 2\sqrt{2y}$

23. $3\sqrt{8xy} - \sqrt{50xy}$
$6\sqrt{2xy} - 5\sqrt{2xy} = \sqrt{2xy}$

24. $2\sqrt{45ab} - \sqrt{20ab}$
$6\sqrt{5ab} - 2\sqrt{5ab} = 4\sqrt{5ab}$

25. $1.2\sqrt{3x} - 0.5\sqrt{12x}$
$1.2\sqrt{3x} - \sqrt{3x} = 0.2\sqrt{3x}$

26. $-1.5\sqrt{2a} + 0.2\sqrt{8a}$
$-1.5\sqrt{2a} + 0.4\sqrt{2a} = -1.1\sqrt{2a}$

27. $\sqrt{20y} + 2\sqrt{45y} - \sqrt{5y}$
$2\sqrt{5y} + 6\sqrt{5y} - \sqrt{5y} = 7\sqrt{5y}$

28. $\sqrt{72w} - 3\sqrt{2w} - \sqrt{50w}$
$6\sqrt{2w} - 3\sqrt{2w} - 5\sqrt{2w} = -2\sqrt{2w}$

29. $3\sqrt{y^3} + 2y\sqrt{y}$
$3y\sqrt{y} + 2y\sqrt{y} = 5y\sqrt{y}$

30. $4x\sqrt{x} - \sqrt{16x^3}$
$4x\sqrt{x} - 4x\sqrt{x} = 0$

31. $2x\sqrt{8x} + 5\sqrt{18x}$
$4x\sqrt{2x} + 15\sqrt{2x}$

32. $-3a\sqrt{12a} + \sqrt{75a}$
$-6a\sqrt{3a} + 5\sqrt{3a}$

33. $5\sqrt{8x^3} - 3x\sqrt{50x}$
$10x\sqrt{2x} - 15x\sqrt{2x} = -5x\sqrt{2x}$

34. $-2\sqrt{27y^3} + y\sqrt{12y}$
$-6y\sqrt{3y} + 2y\sqrt{3y} = -4y\sqrt{3y}$

35. $3\sqrt{27x^2} - 2\sqrt{48x^2}$
$9x\sqrt{3} - 8x\sqrt{3} = x\sqrt{3}$

36. $-5\sqrt{8y^2} + 2\sqrt{32y^2}$
$-10y\sqrt{2} + 8y\sqrt{2} = -2y\sqrt{2}$

37. $2\sqrt{6y^3} - 2y\sqrt{54}$
$2y\sqrt{6y} - 6y\sqrt{6}$

38. $x^2\sqrt{12x^3} - 2\sqrt{48x^7}$
$2x^3\sqrt{3x} - 8x^3\sqrt{3x} = -6x^3\sqrt{3x}$

39. $-y\sqrt{175y^2} + \sqrt{700y^4}$
$-5y^2\sqrt{7} + 10y^2\sqrt{7} = 5y^2\sqrt{7}$

40. $3\sqrt{20x^3} - 4x\sqrt{5x^2}$
$6x\sqrt{5x} - 4x^2\sqrt{5}$

41. $5x\sqrt{8x} - 24\sqrt{50x^3}$
$-110x\sqrt{2x}$

Applications

To estimate the cost of erecting a radio tower, we need to know how many feet of support cable are needed.

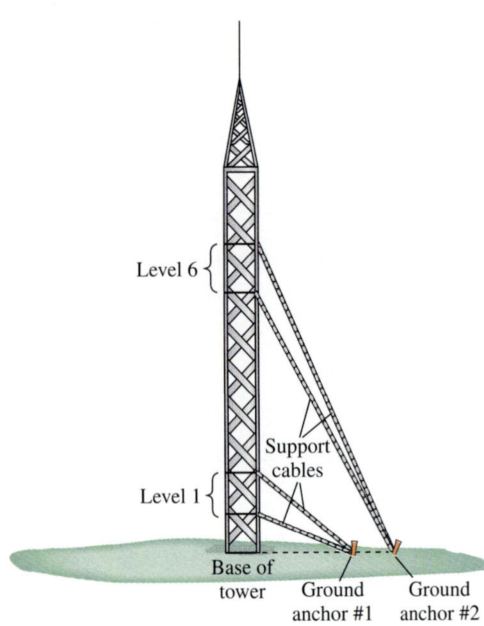

42. Ground anchor #2 is exactly 54 feet from the base of the tower (dashed line in diagram). The bottom ground support cable from level 6 to ground anchor #2 is exactly 130 feet higher than the base of the tower. The top ground support cable from level 6 to ground anchor #2 is exactly 135 feet higher than the base of the tower. How long is each support cable to level 6? (Round your answer to the nearest hundredth of a foot.)
145.40 ft to top; 140.77 ft to bottom

43. Ground anchor #1 is exactly 28 feet from the base of the tower (dashed line in diagram). The bottom ground support cable from level 1 to ground anchor #1 is exactly 42 feet higher than the base of the tower. The top ground support cable from level 1 to ground anchor #1 is exactly 47 feet higher than the base of the tower. How long is each support cable to level 1? (Round your answer to the nearest hundredth of a foot.) 54.71 ft to top; 50.48 ft to bottom

Cumulative Review Problems

44. Solve and graph on a number line. $7 - 3x \leq 11$

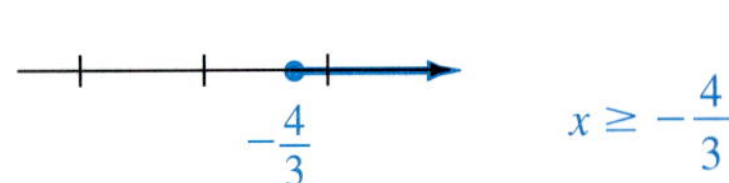

$x \geq -\frac{4}{3}$

45. Graph on a coordinate system. $3y - 2x \leq 6$

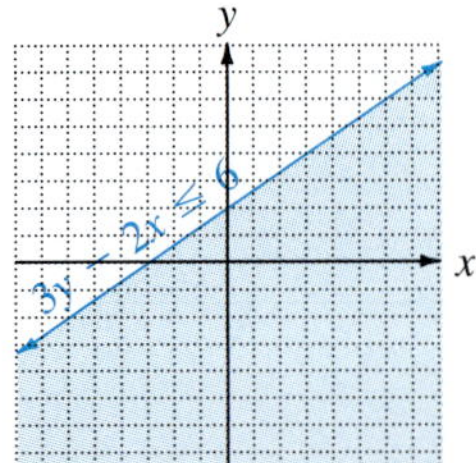

46. Enrico went to the local music store and bought 10 CDs for \$107.90. Some of the CDs were \$7.99 and some were \$11.99. How many of each did Enrico buy?
3 CDs at \$7.99 and 7 CDs at \$11.99.

47. Shelly invested \$8000 in two places. She bought a certificate of deposit with a return of 7% and she invested the rest of the money at 5% in a money market fund. At the end of the year she earned a total of \$432 from these two places. How much did she invest in each place?
\$1600 at 7% to purchase a certificate of deposit; \$6400 at 5% invested in a money market fund

9.4 Multiplication of Radicals

MathPro

Video 22

SSM

After studying this section, you will be able to:

1. *Multiply monomial radical expressions.*
2. *Multiply a monomial radical expression by a polynomial.*
3. *Multiply two polynomial radical expressions.*

1 *Multiplying Monomial Radical Expressions*

Recall the basic rule for multiplication of radicals

$$\sqrt{a}\sqrt{b} = \sqrt{ab}$$

We will use this concept to multiply radical expressions. Note that the direction ''multiply'' means express the product in simplest form.

EXAMPLE 1 Multiply. $\sqrt{7}\sqrt{14x}$

$$\sqrt{7}\sqrt{14x} = \sqrt{98x}$$

We do *not* stop here, because the radical $\sqrt{98x}$ can be simplified.

$\sqrt{98x} = \sqrt{49 \cdot 2x} = \sqrt{49}\sqrt{2x} = 7\sqrt{2x}$

Practice Problem 1 Multiply. $\sqrt{3a}\sqrt{6a}$ ■

EXAMPLE 2 Multiply.

(a) $(2\sqrt{3})(5\sqrt{7})$ **(b)** $(x\sqrt{2})(4\sqrt{10})$ **(c)** $(2a\sqrt{3a})(3\sqrt{6a})$

(a) $(2\sqrt{3})(5\sqrt{7}) = 10\sqrt{21}$ *Multiply the coefficients* $(2)(5) = 10$. *Multiply the radicals* $\sqrt{3}\sqrt{7} = \sqrt{21}$.

(b) $(x\sqrt{2})(4\sqrt{10}) = 4x\sqrt{20}$ *Simplify* $\sqrt{20}$.

$= 4x\sqrt{4}\sqrt{5}$

$= 8x\sqrt{5}$

(c) $(2a\sqrt{3a})(3\sqrt{6a}) = 6a\sqrt{18a^2}$ *Simplify* $\sqrt{18a^2}$.

$= 6a\sqrt{9}\sqrt{2}\sqrt{a^2}$ *Multiply* $(6a)(3)(a)$.

$= 18a^2\sqrt{2}$

Practice Problem 2 Multiply.

(a) $(2\sqrt{3})(5\sqrt{5})$ **(b)** $(2x\sqrt{6})(3\sqrt{6})$ **(c)** $(4\sqrt{3x})(2x\sqrt{6x})$ ■

2 *Multiplying a Monomial Radical Expression by a Polynomial*

Recall that a binomial consists of two terms. $\sqrt{2} + \sqrt{5}$ is a binomial. We use the distributive property when multiplying a binomial.

EXAMPLE 3 Multiply. $\sqrt{5}(\sqrt{2} + 3\sqrt{7})$

$$= \sqrt{5}\sqrt{2} + 3\sqrt{5}\sqrt{7}$$
$$= \sqrt{10} + 3\sqrt{35}$$

In similar fashion, we can multiply by a polynomial of three terms.

Practice Problem 3 Simplify. $2\sqrt{3}(\sqrt{3} + \sqrt{5} - \sqrt{12})$ ■

Note that $\sqrt{3} \cdot \sqrt{3} = 3$. This is a specific case of the multiplication rule for radicals.

> For any nonnegative real number a,
> $$\sqrt{a} \cdot \sqrt{a} = a$$

TEACHING TIP After explaining to students how to do Examples 3 and 4, ask them to multiply the following expressions and simplify their answers as an Additional Class Exercise.
Multiply. $\sqrt{w}(5\sqrt{w} - 2\sqrt{3})$
Sol. $5w - 2\sqrt{3w}$

EXAMPLE 4 Multiply. $\sqrt{a}(3\sqrt{a} - 2\sqrt{5})$

$= 3\sqrt{a}\sqrt{a} - 2\sqrt{5}\sqrt{a}$ *Use the distributive property.*
$= 3a - 2\sqrt{5a}$

Practice Problem 4 Multiply. $2\sqrt{x}(4\sqrt{x} - x\sqrt{2})$ ■

Be sure to simplify all radicals after multiplying.

EXAMPLE 5 Multiply. $2\sqrt{2x}(3\sqrt{10x} - 2\sqrt{6})$

$= 6\sqrt{20x^2} - 4\sqrt{12x}$ *Use the distributive property.*
$= 6\sqrt{4x^2}\sqrt{5} - 4\sqrt{4}\sqrt{3x}$ *Simplify $\sqrt{20x^2}$ and $\sqrt{12x}$.*
$= 6(2x)\sqrt{5} - 4(2)\sqrt{3x}$
$= 12x\sqrt{5} - 8\sqrt{3x}$

Practice Problem 5 Multiply. $\sqrt{6x}(\sqrt{15x} + \sqrt{10x^2})$ ■

3 *Multiplying Two Polynomial Radical Expressions*

Recall the FOIL method used for algebraic expressions to multiply two binomials. The same method can be used for radical expressions.

Algebraic Expressions

$$(2x + y)(x - 2y) = 2x^2 - 4xy + xy - 2y^2$$
$$= 2x^2 - 3xy - 2y^2$$

Radical Expressions

$$(2\sqrt{3} + \sqrt{5})(\sqrt{3} - 2\sqrt{5}) = 2\sqrt{9} - 4\sqrt{15} + \sqrt{15} - 2\sqrt{25}$$
$$= 2(3) - 3\sqrt{15} - 2(5)$$
$$= 6 - 3\sqrt{15} - 10$$
$$= -4 - 3\sqrt{15}$$

Let's look at this procedure more closely.

EXAMPLE 6 Multiply. $(\sqrt{2} + 5)(\sqrt{2} - 3)$

$$(\sqrt{2} + 5)(\sqrt{2} - 3)$$

$= \sqrt{4} - 3\sqrt{2} + 5\sqrt{2} - 15$ *Multiply to obtain the four products.*

$= 2 + 2\sqrt{2} - 15$ *Combine the middle terms.*

$= -13 + 2\sqrt{2}$

TEACHING TIP After showing students how to do Example 6, you may want to show them the following similar, but more difficult, problem as an Additional Class Exercise. Multiply. $(3\sqrt{5} + 9)(3\sqrt{5} - 2)$

Sol.

$$9\sqrt{25} - 6\sqrt{5} + 27\sqrt{5} - 18$$
$$= 45 + 21\sqrt{5} - 18$$
$$= 27 + 21\sqrt{5}$$

Practice Problem 6 Multiply. $(\sqrt{2} + \sqrt{6})(2\sqrt{2} - \sqrt{6})$ ■

EXAMPLE 7 Multiply. $(\sqrt{2} - 3\sqrt{6})(\sqrt{2} + \sqrt{6})$

$$(\sqrt{2} - 3\sqrt{6})(\sqrt{2} + \sqrt{6})$$

$= \sqrt{4} + \sqrt{12} - 3\sqrt{12} - 3\sqrt{36}$ *Multiply to obtain the four products.*

$= 2 - 2\sqrt{12} - 18$ *Combine the middle terms.*

$= -16 - 4\sqrt{3}$ *Simplify* $-2\sqrt{12}$.

Practice Problem 7 Multiply. $(\sqrt{6} + \sqrt{5})(\sqrt{2} + 2\sqrt{5})$ ■

If an expression with radicals is squared, write the expression as the product of two binomials and multiply.

EXAMPLE 8 Multiply. $(2\sqrt{3} - \sqrt{6})^2$

$$= (2\sqrt{3} - \sqrt{6})(2\sqrt{3} - \sqrt{6})$$
$$= 4\sqrt{9} - 2\sqrt{18} - 2\sqrt{18} + \sqrt{36}$$
$$= 12 - 4\sqrt{18} + 6$$
$$= 18 - 12\sqrt{2}$$

TEACHING TIP After showing students how to do Example 8, ask them to do the following multiplication problem as an Additional Class Exercise.
Multiply. $(3\sqrt{5} - 2\sqrt{10})^2$

Sol. $85 - 60\sqrt{2}$

Practice Problem 8 Multiply. $(3\sqrt{5} - \sqrt{10})^2$ ■

9.4 Exercises

Multiply. Be sure to simplify any radicals in your answer. Do not use a calculator or a table of square roots.

1. $\sqrt{7}\sqrt{5}$ $\sqrt{35}$

2. $\sqrt{3}\sqrt{11}$ $\sqrt{33}$

3. $\sqrt{2}\sqrt{22}$ $2\sqrt{11}$

4. $\sqrt{5}\sqrt{15}$ $5\sqrt{3}$

5. $\sqrt{2a}\sqrt{6a}$ $2a\sqrt{3}$

6. $\sqrt{8}\sqrt{6b}$ $4\sqrt{3b}$

7. $\sqrt{5x}\sqrt{10}$ $5\sqrt{2x}$

8. $\sqrt{12b}\sqrt{2b}$ $2b\sqrt{6}$

9. $(3\sqrt{5})(2\sqrt{6})$ $6\sqrt{30}$

10. $(-3a\sqrt{2})(5a\sqrt{3a})$ $-15a^2\sqrt{6a}$

11. $(2x\sqrt{x})(3x\sqrt{5x})$ $6x^3\sqrt{5}$

12. $(4\sqrt{12})(2\sqrt{6})$ $48\sqrt{2}$

13. $(-3\sqrt{ab})(2\sqrt{b})$ $-6b\sqrt{a}$

14. $(2\sqrt{x})(5\sqrt{xy})$ $10x\sqrt{y}$

15. $\sqrt{3}(\sqrt{7} + 5\sqrt{3})$ $\sqrt{21} + 15$

16. $\sqrt{7}(2\sqrt{3} - 3\sqrt{2})$ $2\sqrt{21} - 3\sqrt{14}$

17. $\sqrt{2}(\sqrt{10} - 3\sqrt{6})$ $2\sqrt{5} - 6\sqrt{3}$

18. $\sqrt{5}(\sqrt{15} + 3\sqrt{5})$ $5\sqrt{3} + 15$

19. $2\sqrt{x}(\sqrt{x} - 8\sqrt{5})$ $2x - 16\sqrt{5x}$

20. $-3\sqrt{b}(2\sqrt{a} + 3\sqrt{b})$ $-6\sqrt{ab} - 9b$

21. $\sqrt{6}(\sqrt{2} - 3\sqrt{6} + 2\sqrt{10})$ $2\sqrt{3} - 18 + 4\sqrt{15}$

22. $\sqrt{10}(\sqrt{5} - 3\sqrt{10} + 5\sqrt{2})$ $5\sqrt{2} - 30 + 10\sqrt{5}$

23. $2\sqrt{a}(3\sqrt{b} + \sqrt{ab} - 2\sqrt{a})$ $6\sqrt{ab} + 2a\sqrt{b} - 4a$

24. $3\sqrt{x}(\sqrt{y} + 2\sqrt{xy} - 4\sqrt{x})$ $3\sqrt{xy} + 6x\sqrt{y} - 12x$

25. $(2\sqrt{3} + \sqrt{6})(\sqrt{3} - 2\sqrt{6})$ $6 - 3\sqrt{18} - 12 = -6 - 9\sqrt{2}$

26. $(3\sqrt{5} + \sqrt{3})(\sqrt{5} + \sqrt{3})$ $15 + 4\sqrt{15} + 3 = 18 + 4\sqrt{15}$

27. $(8 + \sqrt{3})(2 + 2\sqrt{3})$ $16 + 18\sqrt{3} + 6 = 22 + 18\sqrt{3}$

28. $(5 + \sqrt{2})(6 + \sqrt{2})$ $30 + 11\sqrt{2} + 2 = 32 + 11\sqrt{2}$

29. $(2\sqrt{7} - 3\sqrt{3})(\sqrt{7} + \sqrt{3})$ $14 - \sqrt{21} - 9 = 5 - \sqrt{21}$

30. $(5\sqrt{6} - \sqrt{2})(\sqrt{6} + 3\sqrt{2})$ $30 + 14\sqrt{12} - 6 = 24 + 28\sqrt{3}$

31. $(\sqrt{3} + 2\sqrt{6})(2\sqrt{3} - \sqrt{6})$ $6 + 3\sqrt{18} - 12 = -6 + 9\sqrt{2}$

32. $(\sqrt{2} + 3\sqrt{10})(2\sqrt{2} - \sqrt{10})$ $4 + 5\sqrt{20} - 30 = -26 + 10\sqrt{5}$

33. $(3\sqrt{7} - \sqrt{8})(\sqrt{8} + 2\sqrt{7})$ $3\sqrt{56} + 42 - 8 - 2\sqrt{56} = 2\sqrt{14} + 34$

34. $(\sqrt{12} - \sqrt{5})(\sqrt{5} + 2\sqrt{12})$ $\sqrt{60} + 24 - 5 - 2\sqrt{60} = -2\sqrt{15} + 19$

35. $(2\sqrt{5} - 3)^2$
$20 - 12\sqrt{5} + 9 = 29 - 12\sqrt{5}$

36. $(3\sqrt{2} + 5)^2$
$18 + 30\sqrt{2} + 25 = 43 + 30\sqrt{2}$

37. $(\sqrt{3} + 5\sqrt{2})^2$
$3 + 10\sqrt{6} + 50 = 53 + 10\sqrt{6}$

38. $(\sqrt{5} - 2\sqrt{6})^2$
$5 - 4\sqrt{30} + 24 = 29 - 4\sqrt{30}$

39. $(\sqrt{6} - 2\sqrt{3})^2$
$6 - 4\sqrt{18} + 12 = 18 - 12\sqrt{2}$

40. $(2\sqrt{3} + 5\sqrt{5})^2$
$12 + 20\sqrt{15} + 125 = 137 + 20\sqrt{15}$

41. $(\sqrt{a} - \sqrt{ab})^2$
$a - 2a\sqrt{b} + ab$

42. $(\sqrt{xy} - 2\sqrt{x})^2$
$xy - 4x\sqrt{y} + 4x$

43. $(3x\sqrt{y} + \sqrt{5})(3x\sqrt{y} - \sqrt{5})$
$9x^2y - 5$

44. $(5a\sqrt{3} - 3\sqrt{2})(5a\sqrt{3} + 3\sqrt{2})$
$75a^2 - 18$

45. $(5x\sqrt{x} - 2\sqrt{5x})(\sqrt{5} - \sqrt{x})$
$5x\sqrt{5x} - 5x^2 - 10\sqrt{x} + 2x\sqrt{5}$

46. $(4a\sqrt{a} - 2\sqrt{3a})(\sqrt{a} - 6\sqrt{3})$
$4a^2 - 24a\sqrt{3a} - 2a\sqrt{3} + 36\sqrt{a}$

To Think About

47. (a) If a can be any number, can you think of an example of when it is not true that $\sqrt{a} \cdot \sqrt{a} = a$? It is not true when a is negative.
(b) Why is it necessary to state the restriction "for any nonnegative real number a"?
(a) It is not true when a is negative.
(b) In order to deal with real numbers.

48. Hank was trying to evaluate the expression $(\sqrt{a} + \sqrt{b})^2$. He wrote $(\sqrt{a} + \sqrt{b})^2 = (\sqrt{a})^2 + (\sqrt{b})^2 = a + b$. This was a serious error! What is wrong with his work? How should he have done the problem?
$(\sqrt{a} + \sqrt{b})^2 = (\sqrt{a} + \sqrt{b})(\sqrt{a} + \sqrt{b})$. Hank's final answer does not have a middle term.
$(\sqrt{a} + \sqrt{b})^2 = a + 2\sqrt{ab} + b$.

Cumulative Review Problems

49. Factor. $36x^2 - 49y^2$. $(6x + 7y)(6x - 7y)$

50. Factor. $x^8 - 1$ $(x^4 + 1)(x^2 + 1)(x + 1)(x - 1)$

51. Ships often measure their speed in knots (or nautical miles per hour). The regular mile (statute mile) that we are more familiar with is 5280 feet. A nautical mile is about 6076 feet. Write a formula of the form $m = ak$ where m is the number of statute miles, k is the number of nautical miles, and a is a constant number (in this case a fraction). Use the formula to find out how fast a destroyer is going in miles per hour if it is making 35 knots.
$m = \left(\frac{1519}{1320}\right)k$ Destroyer is going 40.3 mph.

52. Phil and Melissa Label found that the collision and theft portion of their car insurance could be reduced by 15% if they purchased an auto security device. They purchased and had such a device installed for \$350. The yearly collision and theft portion of their car insurance is \$280. How many years will it take for the security device to pay for itself? It will take $8\frac{1}{2}$ years.

9.5 Division of Radicals

MathPro

Video 23

SSM

After studying this section, you will be able to:

1. *Simplify a fraction involving radicals by using the quotient rule for square roots.*
2. *Rationalize the denominator of a fraction with a square root in the denominator.*
3. *Rationalize the denominator of a fraction with a binomial in the denominator containing at least one square root.*

1 *Simplifying a Fraction Involving Radicals by Using the Quotient Rule*

Just as there is a multiplication rule for square roots, there is a quotient rule for square roots. The rules are similar.

Quotient Rule for Square Roots

For all positive numbers a and b,

$$\frac{\sqrt{a}}{\sqrt{b}} = \sqrt{\frac{a}{b}} \quad \text{and} \quad \sqrt{\frac{a}{b}} = \frac{\sqrt{a}}{\sqrt{b}}$$

This can be used to divide or to simplify radical expressions involving division. We will be using both parts of the quotient rule interchangeably.

EXAMPLE 1 Simplify. **(a)** $\frac{\sqrt{75}}{\sqrt{3}}$ **(b)** $\sqrt{\frac{25}{36}}$

(a) We notice that 3 is a factor of 75. We use the quotient rule to rewrite the expression.

$$\frac{\sqrt{75}}{\sqrt{3}} = \sqrt{\frac{75}{3}} \quad \textit{Divide.}$$

$$= \sqrt{25} \quad \textit{Simplify.}$$

$$= 5$$

(b) Since both 25 and 36 are perfect squares, we will rewrite this as the quotient of square roots.

$$\sqrt{\frac{25}{36}} = \frac{\sqrt{25}}{\sqrt{36}}$$

$$= \frac{5}{6}$$

Practice Problem 1 Simplify. **(a)** $\sqrt{\frac{x^3}{x}}$ **(b)** $\sqrt{\frac{49}{x^2}}$ ■

TEACHING TIP After showing students Example 2, ask them to do the following simplification problem as an Additional Class Exercise.

Simplify. $\sqrt{\frac{w^8x^3}{16}}$

Sol. $\frac{\sqrt{w^8x^3}}{\sqrt{16}} = \frac{w^4x\sqrt{x}}{4}$

EXAMPLE 2 Simplify. $\sqrt{\frac{20}{x^6}}$

$$\sqrt{\frac{20}{x^6}} = \frac{\sqrt{20}}{\sqrt{x^6}} = \frac{\sqrt{4}\sqrt{5}}{x^3} = \frac{2\sqrt{5}}{x^3}$$

Don't forget to simplify $\sqrt{20}$ *to* $2\sqrt{5}$.

Practice Problem 2 Simplify. $\sqrt{\frac{50}{a^4}}$ ■

2 *Rationalizing the Denominator of a Fraction with a Square Root in the Denominator*

Sometimes, when calculating with fractions that contain radicals, it is advantageous to have an integer in the denominator. If a fraction has a radical in the denominator, we rationalize the denominator. That is, we change the radical in the denominator into an integer. Remember, we do not want to change the value of the fraction. Thus we will multiply the numerator and the denominator by the same number.

EXAMPLE 3 Rationalize the denominator. **(a)** $\frac{3}{\sqrt{2}}$ **(b)** $\frac{a}{\sqrt{5}}$

(a) $\frac{3}{\sqrt{2}} = \frac{3}{\sqrt{2}} \times 1 = \frac{3}{\sqrt{2}} \times \frac{\sqrt{2}}{\sqrt{2}} = \frac{3\sqrt{2}}{\sqrt{4}} = \frac{3\sqrt{2}}{2}$

(b) $\frac{a}{\sqrt{5}} = \frac{a}{\sqrt{5}} \times 1 = \frac{a}{\sqrt{5}} \times \frac{\sqrt{5}}{\sqrt{5}} = \frac{a\sqrt{5}}{5}$

Practice Problem 3 Simplify. **(a)** $\frac{9}{\sqrt{7}}$ **(b)** $\frac{\sqrt{5}}{\sqrt{6}}$

(Note that a fraction is not simplified unless the denominator is rationalized.) ■

We will want to use the smallest possible radical that will yield the square root of a perfect square. Very often we will not use the radical that is in the denominator.

EXAMPLE 4 Simplify. **(a)** $\frac{\sqrt{7}}{\sqrt{8}}$ **(b)** $\frac{3}{\sqrt{x^3}}$

TEACHING TIP After demonstrating Example 4, ask students to do the following simplification problem as an Additional Class Exercise. Rationalize the denominator.

$\frac{2}{\sqrt{ab^5}}$

Sol. $\frac{2\sqrt{ab}}{ab^3}$

(a) Think, "What times 8 will make a perfect square?"

$$\frac{\sqrt{7}}{\sqrt{8}} = \frac{\sqrt{7}}{\sqrt{8}} \times \frac{\sqrt{2}}{\sqrt{2}} = \frac{\sqrt{14}}{\sqrt{16}} = \frac{\sqrt{14}}{4}$$

(b) Think, "What times x^3 will give an even exponent?"

$$\frac{3}{\sqrt{x^3}} \times \frac{\sqrt{x}}{\sqrt{x}} = \frac{3\sqrt{x}}{\sqrt{x^4}} = \frac{3\sqrt{x}}{x^2}$$

You may want to simplify $\sqrt{x^3}$ before you rationalize the denominator.

$$\frac{3}{\sqrt{x^3}} = \frac{3}{x\sqrt{x}} \times \frac{\sqrt{x}}{\sqrt{x}} = \frac{3\sqrt{x}}{x\sqrt{x^2}} = \frac{3\sqrt{x}}{(x)(x)} = \frac{3\sqrt{x}}{x^2}$$

Practice Problem 4 Simplify. **(a)** $\frac{\sqrt{2}}{\sqrt{12}}$ **(b)** $\frac{6a}{\sqrt{a^7}}$ ■

EXAMPLE 5 Simplify. $\dfrac{\sqrt{2}}{\sqrt{27x}}$

Since "what multiplied by 27 will be a perfect square" is not apparent, we will begin by simplifying the denominator.

$$\frac{\sqrt{2}}{\sqrt{27x}} = \frac{\sqrt{2}}{3\sqrt{3x}}$$

Now it is easy to see that we multiply by $\frac{\sqrt{3x}}{\sqrt{3x}}$ *to rationalize the denominator.*

$$= \frac{\sqrt{2}}{3\sqrt{3x}} \times \frac{\sqrt{3x}}{\sqrt{3x}}$$

$$= \frac{\sqrt{6x}}{3\sqrt{9x^2}}$$

$$= \frac{\sqrt{6x}}{9x}$$

Practice Problem 5 Simplify. $\dfrac{\sqrt{2x}}{\sqrt{8x}}$ ■

3 *Rationalizing the Denominator of a Fraction with a Binomial Denominator*

Sometimes the denominator of a fraction is a binomial with a radical term. $\frac{1}{5 - 3\sqrt{2}}$ is just such a fraction. How can we eliminate the radical term in the denominator? Recall that when we multiply $(x + 3)(x - 3)$ the product does not have a middle term. That is, $(x + 3)(x - 3) = x^2 - 9$. Both terms are perfect squares. We can use this idea to eliminate the radical in $5 - 3\sqrt{2}$.

$$(5 - 3\sqrt{2})(5 + 3\sqrt{2}) = 5^2 - (3\sqrt{2})^2 = 25 - 18 = 7$$

$(5 - 3\sqrt{2})$ and $(5 + 3\sqrt{2})$ are called *conjugates*.

TEACHING TIP Take the time to explain what a conjugate is. Then quiz students in class by asking them to identify the conjugate of each expression.

A. $7 + \sqrt{3}$
Conjugate: $7 - \sqrt{3}$

B. $2\sqrt{5} + 7\sqrt{11}$
Conjugate: $2\sqrt{5} - 7\sqrt{11}$

C. $\sqrt{x} - 5$
Conjugate: $\sqrt{x} + 5$

D. $4\sqrt{x} - 5\sqrt{2}$
Conjugate: $4\sqrt{x} + 5\sqrt{2}$

EXAMPLE 6 Rationalize the denominator. **(a)** $\dfrac{2}{\sqrt{3} - 4}$ **(b)** $\dfrac{\sqrt{x}}{\sqrt{5} + \sqrt{3}}$

(a) The conjugate of $\sqrt{3} - 4$ is $\sqrt{3} + 4$.

$$\frac{2}{(\sqrt{3} - 4)} \cdot \frac{(\sqrt{3} + 4)}{(\sqrt{3} + 4)} = \frac{2\sqrt{3} + 8}{\sqrt{9} + 4\sqrt{3} - 4\sqrt{3} - 16}$$

$$= \frac{2\sqrt{3} + 8}{3 - 16}$$

$$= \frac{2\sqrt{3} + 8}{-13}$$

$$= -\frac{2\sqrt{3} + 8}{13}$$

(b) The conjugate of $\sqrt{5} + \sqrt{3}$ is $\sqrt{5} - \sqrt{3}$.

$$\frac{\sqrt{x}}{(\sqrt{5}+\sqrt{3})} \cdot \frac{(\sqrt{5}-\sqrt{3})}{(\sqrt{5}-\sqrt{3})} = \frac{\sqrt{5x}-\sqrt{3x}}{\sqrt{25}-\sqrt{15}+\sqrt{15}-\sqrt{9}}$$

$$= \frac{\sqrt{5x}-\sqrt{3x}}{5-3}$$

$$= \frac{\sqrt{5x}-\sqrt{3x}}{2}$$

Be careful not to combine $\sqrt{5x}$ and $\sqrt{3x}$ in the numerator. They are not like square roots.

Practice Problem 6 Simplify. **(a)** $\dfrac{4}{\sqrt{3}+\sqrt{5}}$ **(b)** $\dfrac{\sqrt{a}}{\sqrt{10}-3}$ ■

EXAMPLE 7 Rationalize the denominator. $\dfrac{\sqrt{3}+\sqrt{2}}{\sqrt{3}-\sqrt{2}}$

The conjugate of $\sqrt{3} - \sqrt{2}$ is $\sqrt{3} + \sqrt{2}$.

$$\frac{(\sqrt{3}+\sqrt{2})}{(\sqrt{3}-\sqrt{2})} \cdot \frac{(\sqrt{3}+\sqrt{2})}{(\sqrt{3}+\sqrt{2})}$$

$$= \frac{\sqrt{9}+\sqrt{6}+\sqrt{6}+\sqrt{4}}{(\sqrt{3})^2-(\sqrt{2})^2} \quad \textit{Multiply.}$$

$$= \frac{3+2\sqrt{6}+2}{3-2} \quad \textit{Simplify and combine like terms.}$$

$$= \frac{5+2\sqrt{6}}{1}$$

$$= 5 + 2\sqrt{6}$$

TEACHING TIP In addition to Example 7, you may want to show them the following as an Additional Class Exercise. Be sure to emphasize the steps in reducing the fraction.

Rationalize the following denominator.

$$\frac{4\sqrt{3}+3\sqrt{2}}{5\sqrt{3}-6\sqrt{2}}$$

$$\frac{4\sqrt{3}+3\sqrt{2}}{5\sqrt{3}-6\sqrt{2}} \cdot \frac{5\sqrt{3}+6\sqrt{2}}{5\sqrt{3}+6\sqrt{2}}$$

$$= \frac{60+39\sqrt{6}+36}{75-72}$$

$$= \frac{96+39\sqrt{6}}{3}$$

$$= 32 + 13\sqrt{6}$$

Practice Problem 7 Rationalize the denominator. $\dfrac{\sqrt{7}-\sqrt{x}}{\sqrt{7}+\sqrt{x}}$ ■

9.5 Exercises

Simplify. Be sure to rationalize all denominators. Do not use a calculator or a table of square roots.

1. $\dfrac{\sqrt{32}}{\sqrt{2}}$ 4
2. $\dfrac{\sqrt{7}}{\sqrt{28}}$ $\dfrac{1}{2}$
3. $\dfrac{\sqrt{5}}{\sqrt{20}}$ $\dfrac{1}{2}$
4. $\dfrac{\sqrt{24}}{\sqrt{6}}$ 2
5. $\dfrac{\sqrt{343}}{\sqrt{7}}$ 7
6. $\dfrac{\sqrt{5}}{\sqrt{125}}$ $\dfrac{1}{5}$
7. $\dfrac{\sqrt{6}}{\sqrt{x^4}}$ $\dfrac{\sqrt{6}}{x^2}$
8. $\dfrac{\sqrt{12}}{\sqrt{x^2}}$ $\dfrac{2\sqrt{3}}{x}$
9. $\dfrac{\sqrt{75}}{\sqrt{3}}$ 5
10. $\dfrac{\sqrt{3}}{\sqrt{48}}$ $\dfrac{1}{4}$
11. $\dfrac{\sqrt{8}}{\sqrt{x^6}}$ $\dfrac{2\sqrt{2}}{x^3}$
12. $\dfrac{\sqrt{20}}{\sqrt{x^8}}$ $\dfrac{2\sqrt{5}}{x^4}$
13. $\dfrac{3}{\sqrt{7}}$ $\dfrac{3\sqrt{7}}{7}$
14. $\dfrac{2}{\sqrt{10}}$ $\dfrac{\sqrt{10}}{5}$
15. $\dfrac{5}{\sqrt{6}}$ $\dfrac{5\sqrt{6}}{6}$
16. $\dfrac{3}{\sqrt{5}}$ $\dfrac{3\sqrt{5}}{5}$
17. $\dfrac{x}{\sqrt{3}}$ $\dfrac{x\sqrt{3}}{3}$
18. $\dfrac{\sqrt{y}}{\sqrt{6}}$ $\dfrac{\sqrt{6y}}{6}$
19. $\dfrac{\sqrt{8}}{\sqrt{x}}$ $\dfrac{2\sqrt{2x}}{x}$
20. $\dfrac{\sqrt{18}}{\sqrt{y}}$ $\dfrac{3\sqrt{2y}}{y}$
21. $\dfrac{3}{\sqrt{12}}$ $\dfrac{\sqrt{3}}{2}$
22. $\dfrac{4}{\sqrt{20}}$ $\dfrac{2\sqrt{5}}{5}$
23. $\dfrac{7}{\sqrt{a^3}}$ $\dfrac{7\sqrt{a}}{a^2}$
24. $\dfrac{5}{\sqrt{b^5}}$ $\dfrac{5\sqrt{b}}{b^3}$
25. $\dfrac{x}{\sqrt{2x^5}}$ $\dfrac{\sqrt{2x}}{2x^2}$
26. $\dfrac{y}{\sqrt{5x^3}}$ $\dfrac{y\sqrt{5x}}{5x^2}$
27. $\dfrac{\sqrt{18}}{\sqrt{2x^3}}$ $\dfrac{3\sqrt{x}}{x^2}$
28. $\dfrac{\sqrt{24}}{\sqrt{6x^7}}$ $\dfrac{2\sqrt{x}}{x^4}$
29. $\sqrt{\dfrac{2}{7}}$ $\dfrac{\sqrt{14}}{7}$
30. $\sqrt{\dfrac{3}{10}}$ $\dfrac{\sqrt{30}}{10}$
31. $\sqrt{\dfrac{3}{ab^2}}$ $\dfrac{\sqrt{3a}}{ab}$
32. $\sqrt{\dfrac{5}{2a^2b}}$ $\dfrac{\sqrt{10b}}{2ab}$
33. $\dfrac{9}{\sqrt{32x}}$ $\dfrac{9\sqrt{2x}}{8x}$
34. $\dfrac{3}{\sqrt{50x}}$ $\dfrac{3\sqrt{2x}}{10x}$
35. $\dfrac{3}{\sqrt{2}-1}$ $3\sqrt{2}+3$
36. $\dfrac{2}{\sqrt{3}+1}$ $\sqrt{3}-1$
37. $\dfrac{3}{\sqrt{2}+\sqrt{5}}$ $\sqrt{5}-\sqrt{2}$
38. $\dfrac{4}{\sqrt{7}-\sqrt{2}}$ $\dfrac{4(\sqrt{7}+\sqrt{2})}{5}$
39. $\dfrac{\sqrt{6}}{\sqrt{6}-\sqrt{3}}$ $2+\sqrt{2}$
40. $\dfrac{\sqrt{3}}{\sqrt{5}+\sqrt{3}}$ $\dfrac{\sqrt{15}-3}{2}$
41. $\dfrac{3x}{2\sqrt{2}-\sqrt{5}}$ $x(2\sqrt{2}+\sqrt{5})$
42. $\dfrac{4x}{2\sqrt{7}+2\sqrt{6}}$ $x(2\sqrt{7}-2\sqrt{6})$
43. $\dfrac{x}{\sqrt{3}-2x}$ $\dfrac{x(\sqrt{3}+2x)}{3-4x^2}$
44. $\dfrac{y}{\sqrt{5}-2y}$ $\dfrac{y(\sqrt{5}+2y)}{5-4y^2}$
45. $\dfrac{\sqrt{7}}{\sqrt{8}+\sqrt{7}}$ $2\sqrt{14}-7$
46. $\dfrac{\sqrt{5}}{\sqrt{5}+\sqrt{6}}$ $\sqrt{30}-5$
47. $\dfrac{\sqrt{5}-\sqrt{2}}{\sqrt{5}+\sqrt{2}}$ $\dfrac{7-2\sqrt{10}}{3}$
48. $\dfrac{\sqrt{7}+\sqrt{3}}{\sqrt{7}-\sqrt{3}}$ $\dfrac{5+\sqrt{21}}{2}$

49. $\dfrac{2\sqrt{3}+\sqrt{6}}{\sqrt{6}-\sqrt{3}}$ $3\sqrt{2}+4$

50. $\dfrac{2\sqrt{5}+3}{\sqrt{5}+\sqrt{3}}$ $\dfrac{10+3\sqrt{5}-2\sqrt{15}-3\sqrt{3}}{2}$

51. $\dfrac{4\sqrt{3}+2}{\sqrt{8}-\sqrt{6}}$ $5\sqrt{6}+8\sqrt{2}$

52. $\dfrac{3\sqrt{5}+4}{\sqrt{15}-\sqrt{3}}$ $\dfrac{19\sqrt{3}+7\sqrt{15}}{12}$

53. $\dfrac{x-4}{\sqrt{x}+2}$ $\sqrt{x}-2$

54. $\dfrac{x-9}{\sqrt{x}-3}$ $\sqrt{x}+3$

★55. $\dfrac{\sqrt{x}+2}{\sqrt{x}-\sqrt{2}}$ $\dfrac{x+\sqrt{2x}+2\sqrt{2}+2\sqrt{x}}{x-2}$

★56. $\dfrac{\sqrt{x}-4}{\sqrt{x}+\sqrt{3}}$ $\dfrac{x-\sqrt{3x}-4\sqrt{x}+4\sqrt{3}}{x-3}$

To Think About

57. **(a)** Multiply $(x+\sqrt{2})$ by its conjugate. x^2-2

(b) Use the results of part (a) to show that, while x^2-2 cannot be factored by earlier methods, it can be factored using radicals. $x^2-2=(x+\sqrt{2})(x-\sqrt{2})$

(c) Factor using radicals. x^2-12 $(x+2\sqrt{3})(x-2\sqrt{3})$

58. $\frac{3}{\sqrt{11}}$ when rationalized is $\frac{3\sqrt{11}}{11}$. Evaluate each expression with a calculator. What do you notice? Why? 0.904534034 The values are the same. They are equivalent expressions.

59. $\frac{7}{\sqrt{35}}$ when rationalized is $\frac{\sqrt{35}}{5}$. Evaluate each expression with a calculator. What do you notice? Why? 1.183215957 The values are the same. They are equivalent expressions.

Cumulative Review Problems

60. Solve for x (round to two decimal places). $1.06(0.80x)=32.64$ $x=38.49$

61. Evaluate $3x^2-x\div y+y^2-y$ for $x=12$ and $y=-3$.
$3(12)^2-(12)\div(-3)+(-3)^2-(-3)=432+16=448$

62. Dr. Frank Day, a chemist, has determined that the types of percolated coffee he buys have 70% more caffeine per cup than the types of instant coffee he usually buys. He determined that the percolated coffee has 120 mg of caffeine per cup. Approximately how much caffeine per cup is in the instant coffee?
approximately 71 mg per cup

63. Wally and Mary Coyle are considering two family medical insurance plans that are quite similar. The Gold Plan costs \$3000 per year and covers all doctor visits and all prescription drugs after a \$50 deductible per doctor visit and a \$30 deductible per prescription. The Master Plan costs \$4000 per year and covers all doctor visits and all prescription drugs. Wally and Mary have discovered that half of all their doctor visits result in a written prescription. How many doctor visits per year would they need to average for the Master Plan to be more economical? They would need more than 15 doctor visits per year.

9.6 The Pythagorean Theorem and Radical Equations

After studying this section, you will be able to:

1 *Use the Pythagorean theorem.*
2 *Solve radical equations.*

1 *Using the Pythagorean Theorem*

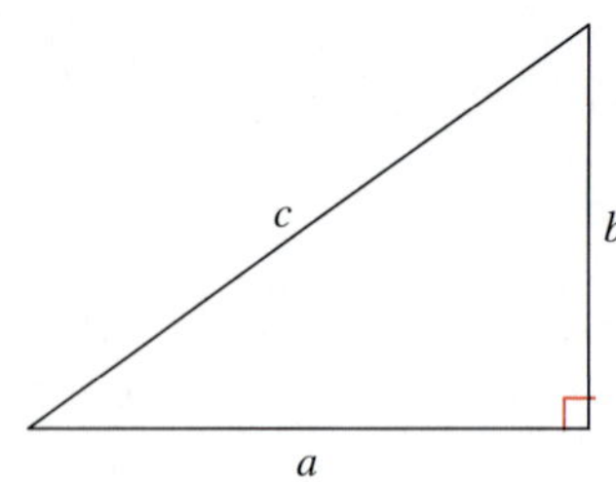

In ancient Greece, mathematicians studied a number of properties of right triangles. They were able to prove that the square of the longest side of a right triangle is equal to the sum of the squares of the other two sides. This property is called the Pythagorean theorem, in honor of the Greek mathematician Pythagoras (ca. 590 B.C.). The shorter sides, a and b, are referred to as the **legs** of a right triangle. The longest side c is called the **hypotenuse** of a right triangle. We state the theorem thus:

> **Pythagorean Theorem**
>
> In any right triangle, if c is the length of the hypotenuse and a and b are the lengths of the two legs, then
>
> $$c^2 = a^2 + b^2$$

If we know any two sides of a right triangle, we can find the third side using this theorem.

Recall that in Section 9.1 we showed that if $s^2 = 4$ then $s = 2$ or -2. In general, if $x^2 = a$, then $x = \pm\sqrt{a}$. This is sometimes called "taking the square root of each side of an equation." The abbreviation $\pm\sqrt{a}$ means $+\sqrt{a}$ or $-\sqrt{a}$. It is read "plus or minus $\sqrt{a}$."

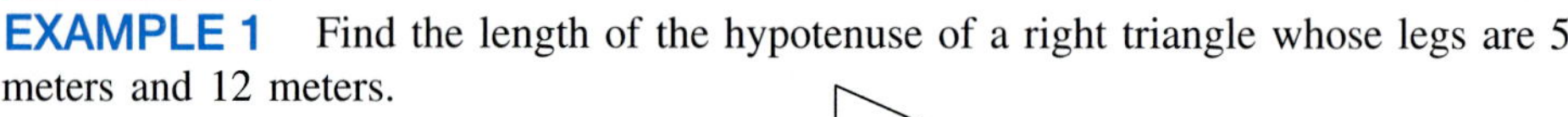

EXAMPLE 1 Find the length of the hypotenuse of a right triangle whose legs are 5 meters and 12 meters.

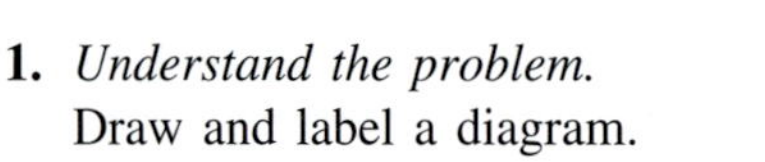

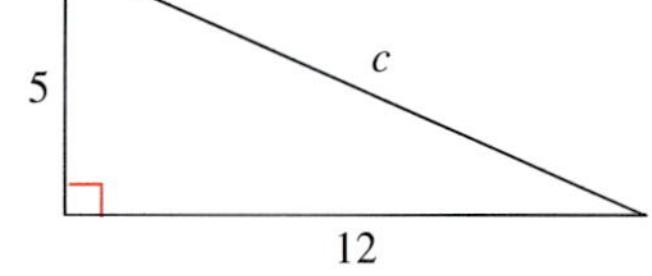

1. *Understand the problem.*
   Draw and label a diagram.
2. *Write an equation.*
   Write the Pythagorean theorem. Substitute the known values into the equation.

$$c^2 = a^2 + b^2$$
$$c^2 = 5^2 + 12^2$$

3. *Solve and state the answer.*

$$c^2 = 25 + 144$$
$$c^2 = 169 \qquad \textit{Take the square root of each side of the equation.}$$
$$c = \pm\sqrt{169}$$
$$= \pm 13$$

   The hypotenuse is 13 meters. We do not use -13 because length is not negative.

4. *Check.*
   Substitute the values for a, b, and c in the Pythagorean theorem and evaluate.
   Does $13^2 = 5^2 + 12^2$? Yes. The solution checks. ✓

Practice Problem 1 Find the length of the hypotenuse of a right triangle with legs of 9 centimeters and 12 centimeters. ■

Sometimes you will need to find one of the legs of a right triangle given the hypotenuse and the other leg.

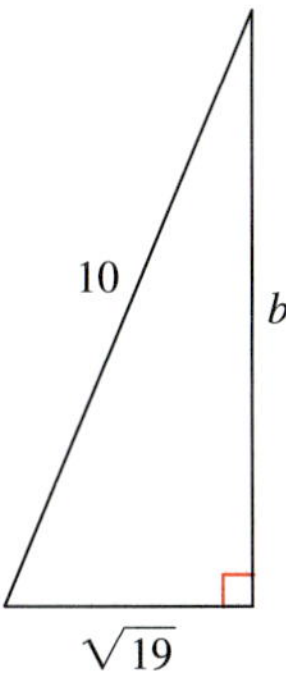

EXAMPLE 2 Find the leg of a right triangle that has a hypotenuse of 10 yards and one leg of $\sqrt{19}$ yards.

Draw a diagram. The diagram is in the margin on the right,

$c^2 = a^2 + b^2$ *Write the Pythagorean theorem.*

$10^2 = (\sqrt{19})^2 + b^2$ *Substitute the known values into the equation.*

$100 = 19 + b^2$

$81 = b^2$

$\pm 9 = b$ *Take the square root of each side of the equation.*

The leg is 9 yards.

TEACHING TIP After explaining Example 2, ask your students to find the missing side of each of the following right triangles as an Additional Class Exercise.

A. Find a if $b = 7$ and $c = 8$.
B. Find c if $a = \sqrt{3}$ and $b = 2$.
C. Find b if $a = 5$ and $c = 13$.

Sol.

A. $a = \sqrt{15}$
B. $c = \sqrt{7}$
C. $b = 12$

Practice Problem 2 The hypotenuse of a right triangle is $\sqrt{17}$ meters. One leg is 1 meter long. Find the length of the other leg. ■

Sometimes the answer will be an irrational number such as $\sqrt{33}$. It is best to leave the answer in radical form unless you are asked for an approximate answer. To find the approximate value, use a calculator or the square root table. $\sqrt{33}$ is an exact answer.

EXAMPLE 3 A 25-foot ladder is placed against a building. The foot of the ladder is 8 feet from the wall. Approximately how high on the building will the ladder reach? Round your answer to the nearest tenth.

$$c^2 = a^2 + b^2$$

$$25^2 = 8^2 + b^2$$

$$625 = 64 + b^2$$

$$561 = b^2$$

$$\pm \sqrt{561} = b$$

We want only positive value for the distance, so $b = \sqrt{561}$. Using a calculator, we have

561 [√] 23.68543856 Rounding, we obtain $b \approx 23.7$

If you do not have a calculator, you will need to use the square root table in Appendix A. However, this square root table only goes to number 200! To approximate from the square root table, you will have to write the radical in a different fashion.

$b = \sqrt{3}\sqrt{187}$ *Since* $\sqrt{ab} = \sqrt{a}\sqrt{b}$ *and* $3 \cdot 187 = 561$.

$b \approx (1.732)(13.675)$ *Replace each radical by a decimal approximation from the table.*

$b \approx 23.7$ *Multiply and round to the nearest tenth.*

The ladder reaches approximately 23.7 feet up on the building.

TEACHING TIP After explaining Example 3, ask students to solve the following applied problem as an Additional Class Exercise.
Topsfield is exactly 15 miles north of Allentown. Bridgeton is exactly 8 miles east of Allentown. How long is a road that travels in a straight line from Topsfield to Bridgeton?

Sol. Exactly 17 miles.

Practice Problem 3 A support line is placed 3 meters away from the base of an 8-meter pole. If the support line is attached to the top of the pole and pulled tight (assume that it is a straight line), how long is the support line from the ground to the pole? Approximate your answer to the nearest tenth. ■

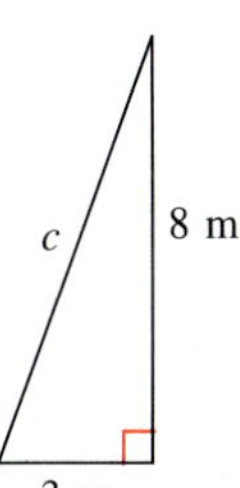

2 *Solving Radical Equations*

A **radical equation** is an equation with an unknown letter in one or more of the radicands. This type of equation can be simplified by squaring each side of the equation. The apparent solution to a radical equation *must* be checked by substitution into the *original* equation.

EXAMPLE 4 Solve and check your answer. $\sqrt{x+4} = 11$

$$\sqrt{x+4} = 11$$

$$(\sqrt{x+4})^2 = (11)^2 \quad \textit{Square each side of the equation.}$$

$$x + 4 = 121 \quad \textit{Simplify and solve for } x.$$

$$x = 117$$

Check: $\sqrt{117+4} \stackrel{?}{=} 11$

$$\sqrt{121} \stackrel{?}{=} 11$$

$$11 = 11 \quad ✓$$

Thus $x = 117$ is the solution.

Practice Problem 4 Solve and check. $\sqrt{2x-5} = 6$ ■

For some equations, you will need to *isolate the radical before* you square each side of the equation.

EXAMPLE 5 Solve and check your answer. $1 + \sqrt{5x-4} = 5$

$$1 + \sqrt{5x-4} = 5 \quad \textit{We want to isolate the radical first.}$$

$$\sqrt{5x-4} = 4 \quad \textit{Subtract 1 from each side to isolate the radical.}$$

$$(\sqrt{5x-4})^2 = (4)^2 \quad \textit{Square each side.}$$

$$5x - 4 = 16 \quad \textit{Simplify and solve for } x.$$

$$5x = 20$$

$$x = 4$$

Check: $1 + \sqrt{5(4)-4} \stackrel{?}{=} 5$

$$1 + \sqrt{20-4} \stackrel{?}{=} 5$$

$$1 + \sqrt{16} \stackrel{?}{=} 5$$

$$1 + 4 = 5 \quad ✓$$

Thus $x = 4$ is the solution.

Practice Problem 5 Solve and check. $\sqrt{3x-2} - 7 = 0$ ■

Some problems will have an apparent solution, but that solution does not always check. An obtained solution that does not satisfy the original equation is called an **extraneous root.**

EXAMPLE 6 Solve and check. $\sqrt{x+3} = -7$

$$\sqrt{x+3} = -7$$

$$(\sqrt{x+3})^2 = (-7)^2 \qquad \textit{Square each side.}$$

$$x + 3 = 49 \qquad \textit{Simplify and solve for } x.$$

$$x = 46$$

Check: $\sqrt{46+3} \stackrel{?}{=} -7$

$$\sqrt{49} \stackrel{?}{=} -7$$

$$7 \neq -7!$$ *Does not check! The apparent solution cannot be verified.*

There is no solution to Example 6.

Practice Problem 6 Solve and check. $\sqrt{5x+4} + 2 = 0$ ■

Wait a minute! Is that original equation in Example 6 possible? No. $\sqrt{x+3} = -7$ is an impossible statement. By our definition of the $\sqrt{\ }$ symbol, we defined it to mean the positive square root of the radicand. If you noticed this in the first step, you could write down "no solution" immediately.

When you square both sides of the equation you may obtain a quadratic equation. In such cases, you must check both apparent solutions.

EXAMPLE 7 Solve and check. $\sqrt{3x+1} = x + 1$.

TEACHING TIP Spend some extra time going over Example 7. This is a very important example for students to study.

$$\sqrt{3x+1} = x + 1$$

$$(\sqrt{3x+1})^2 = (x+1)^2 \qquad \textit{Square each side.}$$

$$3x + 1 = x^2 + 2x + 1 \qquad \textit{Simplify and set the equation equal to } 0.$$

$$0 = x^2 - x$$

$$0 = x(x-1) \qquad \textit{Factor the quadratic equation.}$$

$$x = 0 \qquad x - 1 = 0 \qquad \textit{Set each factor equal to } 0 \textit{ and solve.}$$

$$x = 1$$

Check: If $x = 0$: $\sqrt{3(0)+1} \stackrel{?}{=} 0 + 1$ If $x = 1$: $\sqrt{3(1)+1} \stackrel{?}{=} 1 + 1$

$$\sqrt{0+1} \stackrel{?}{=} 0 + 1 \qquad \sqrt{3+1} \stackrel{?}{=} 2$$

$$\sqrt{1} \stackrel{?}{=} 1 \qquad \sqrt{4} \stackrel{?}{=} 2$$

$$1 = 1 \ \checkmark \qquad 2 = 2 \ \checkmark$$

Thus $x = 0$ and $x = 1$ are both solutions.

It is possible in these cases for both apparent solutions to check, for both to be extraneous, or for one to check and the other to be extraneous.

Practice Problem 7 Solve and check. $-2 + \sqrt{6x-1} = 3x - 2$ ■

EXAMPLE 8 Solve and check. $\sqrt{2x-1} = x - 2$

$$\sqrt{2x-1} = x - 2$$

$$(\sqrt{2x-1})^2 = (x-2)^2 \qquad \textit{Square each side.}$$

$$2x - 1 = x^2 - 4x + 4 \qquad \textit{Simplify and set the equation to 0.}$$

$$0 = x^2 - 6x + 5 \qquad \textit{Solve for } x.$$

$$0 = (x-5)(x-1)$$

$$x - 5 = 0 \qquad x - 1 = 0$$

$$x = 5 \qquad x = 1$$

Check: If $x = 5$: $\sqrt{2(5)-1} \stackrel{?}{=} 5 - 2$

$$\sqrt{10-1} \stackrel{?}{=} 3$$

$$\sqrt{9} \stackrel{?}{=} 3$$

$$3 = 3 \checkmark$$

If $x = 1$: $\sqrt{2(1)-1} \stackrel{?}{=} 1 - 2$

$$\sqrt{2-1} \stackrel{?}{=} -1$$

$$\sqrt{1} \stackrel{?}{=} -1$$

$$1 \neq -1$$

It does not *check. In this case* $x = 1$ *is an* extraneous root.

Thus only $x = 5$ is a solution to this equation.

Practice Problem 8 Solve and check. $2 - x + \sqrt{x+4} = 0$ ■

EXAMPLE 9 Solve and check. $\sqrt{3x+3} = \sqrt{5x-1}$

Here there are two radicals. Each radical is already isolated.

$$(\sqrt{3x+3})^2 = (\sqrt{5x-1})^2 \qquad \textit{Square each side.}$$

$$3x + 3 = 5x - 1 \qquad \textit{Simplify and solve for } x.$$

$$3 = 2x - 1$$

$$4 = 2x$$

$$2 = x$$

Check: $\sqrt{3(2)+3} \stackrel{?}{=} \sqrt{5(2)-1}$

$$\sqrt{6+3} \stackrel{?}{=} \sqrt{10-1}$$

$$\sqrt{9} = \sqrt{9} \checkmark$$

Thus $x = 2$ is a solution.

Practice Problem 9 Solve and check. $\sqrt{2x+1} = \sqrt{x-10}$ ■

9.6 Exercises

Use the Pythagorean theorem to find the length of the third side of each right triangle. Leave any irrational answers in radical form.

1. 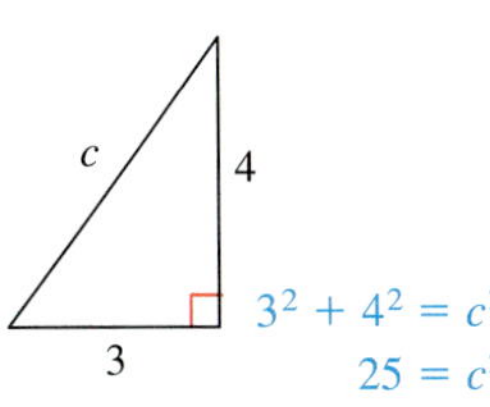

$3^2 + 4^2 = c^2$
$25 = c^2$
$c = 5$

2. 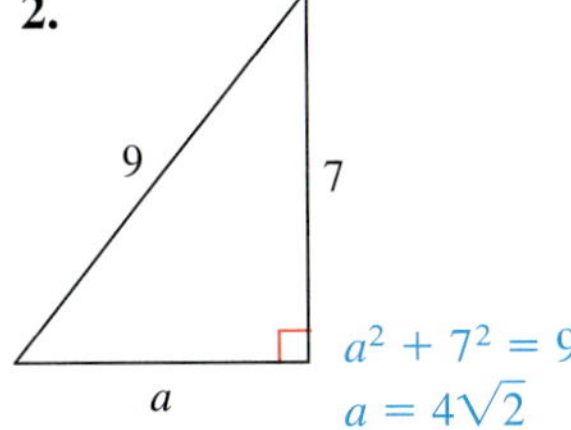

$a^2 + 7^2 = 9^2$
$a = 4\sqrt{2}$

3. 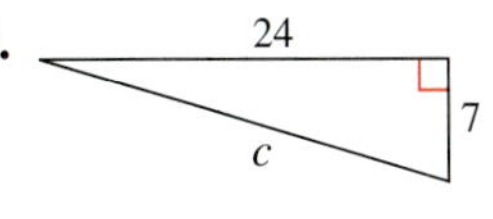

$24^2 + 7^2 = c^2$
$625 = c^2$
$c = 25$

4. 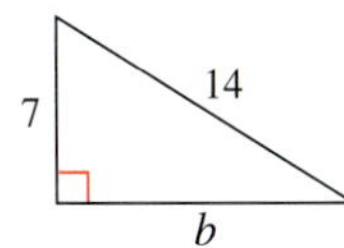

$7^2 + b^2 = 14^2$
$b^2 = 147$
$b = 7\sqrt{3}$

Problems 5 through 14 refer to a right triangle. Find the exact length of the missing side.

5. $a = 5$, $b = 7$
Find c.
$c^2 = 5^2 + 7^2 = 25 + 49 = 74$
$c = \sqrt{74}$

6. $a = 6$, $b = 9$
Find c.
$c^2 = 36 + 81 = 117$
$c = 3\sqrt{13}$

7. $a = \sqrt{18}$, $b = 4$
Find c.
$c^2 = 18 + 16 = 34$
$c = \sqrt{34}$

8. $a = \sqrt{19}$, $b = 7$
Find c.
$c^2 = 19 + 49 = 68$
$c = 2\sqrt{17}$

9. $c = 20$, $b = 18$
Find a.
$a^2 + 18^2 = 20^2$
$a^2 = 400 - 324 = 76$
$a = 2\sqrt{19}$

10. $c = 17$, $b = 16$
Find a.
$a^2 + 16^2 = 17^2$
$a^2 = 289 - 256 = 33$
$a = \sqrt{33}$

11. $c = \sqrt{23}$, $a = 4$
Find b.
$16 + b^2 = (\sqrt{23})^2$
$b^2 = 23 - 16 = 7$
$b = \sqrt{7}$

12. $c = \sqrt{27}$, $a = 5$
Find b.
$25 + b^2 = (\sqrt{27})^2$
$b^2 = 27 - 25 = 2$
$b = \sqrt{2}$

13. $c = 12.96$, $b = 8.35$ Find a to the nearest hundredth.
9.91

14. $a = 7.61$, $b = 5.38$ Find c to the nearest hundredth.
9.32

Applications

Draw a diagram and use the Pythagorean theorem to solve each word problem. Approximate the answer to the nearest tenth.

15. A ladder is 18 feet long. It reaches 15 feet up on a wall. How far is the base of the ladder from the wall?
$x^2 + 15^2 = 18^2$
$x^2 + 225 = 324$
$x = \sqrt{99} = 3\sqrt{11} \approx 9.9$ ft

16. A kite is flying on a string 100 feet long. Assume that the string is a straight line. How high is the kite if it is flying above a point 20 feet away along the ground?
$x^2 + 20^2 = 100^2$
$x^2 = 10{,}000 - 400$
$x \approx 98$ ft

17. A baseball diamond is really constructed from a square. Each side of the square is 90 feet long. How far is it from home plate to second base? *Hint:* Draw the diagonal.
$x^2 = 90^2 + 90^2$
$x^2 = 16{,}200$
$x \approx 127.3$ ft

18. A boat must be moored 20 feet away from a dock. The boat will be 10 feet below the level of the dock at low tide. What is the minimum length of rope needed to go from the boat to the dock at low tide?
$x^2 = 100 + 400$
$x = 10\sqrt{5} \approx 22.4$ ft

19. A man is standing 6 miles from the center of a mountain. He is at the same level as the base of the mountain. The mountain is 3 miles high. How far is the man from the top of the mountain?
$x^2 = 36 + 9$
$x \approx 6.7$ mi

20. A plane flies 7 miles north of the airport. A helicopter flies 5 miles east of the airport. How far apart are the plane and helicopter?
$x^2 = 49 + 25$
$x^2 = 74$
$x = \sqrt{74} \approx 8.6$ mi

Solve for the variable. Check your solutions.

21. $\sqrt{x} = 11$ $x = 121$

22. $\sqrt{x} = 3.5$ $x = 12.25$

23. $\sqrt{x+8} = 5$
$x + 8 = 25$
$x = 17$

24. $\sqrt{x+5} = 7$
$x + 5 = 49$
$x = 44$

25. $\sqrt{3x+6} = 2$
$3x + 6 = 4$
$x = -\frac{2}{3}$

26. $\sqrt{3x-8} = 4$
$3x - 8 = 16$
$x = 8$

27. $\sqrt{2x+2} = \sqrt{3x-5}$
$2x + 2 = 3x - 5$
$7 = x$

28. $\sqrt{5x-5} = \sqrt{4x+1}$
$5x - 5 = 4x + 1$
$x = 6$

29. $\sqrt{2x} - 5 = 4$
$\sqrt{2x} = 9$
$2x = 81$ $x = \frac{81}{2}$

30. $\sqrt{13x} - 2 = 5$
$13x = 49$
$x = \frac{49}{13}$

31. $\sqrt{3x+10} = x$
$3x + 10 = x^2$
$x^2 - 3x - 10 = 0$
$(x-5)(x+2) = 0$
$x = 5$ only

32. $\sqrt{5x-6} = x$
$5x - 6 = x^2$
$x^2 - 5x + 6 = 0$
$(x-2)(x-3) = 0$
$x = 2, x = 3$

33. $\sqrt{5y+1} = y + 1$
$5y + 1 = y^2 + 2y + 1$
$y^2 - 3y = y(y-3) = 0$
$y = 0, y = 3$

34. $\sqrt{2y+9} = y + 3$
$2y + 9 = y^2 + 6y + 9$
$y^2 + 4y = 0$
$y(y+4) = 0$
$y = 0$ only

35. $x - 3 = \sqrt{2x-3}$
$x^2 - 6x + 9 = 2x - 3$
$x^2 - 8x + 12 = 0$
$(x-6)(x-2) = 0$
$x = 6$ only

36. $x - 2 = \sqrt{2x-1}$
$x^2 - 4x + 4 = 2x - 1$
$x^2 - 6x + 5 = 0$
$(x-1)(x-5) = 0$
$x = 5$ only

37. $\sqrt{3y+1} - y = 1$
$3y + 1 = y^2 + 2y + 1$
$y^2 - y = 0$
$y = 0, y = 1$

38. $\sqrt{3y-8} + 2 = y$
$3y - 8 = y^2 - 4y + 4$
$y^2 - 7y + 12 = 0$
$(y-3)(y-4) = 0$
$y = 3, y = 4$

39. $\sqrt{5+2x} - 1 - x = 0$
$5 + 2x = x^2 + 2x + 1$
$4 = x^2$
$2 = x$ only

40. $\sqrt{4x+13} + 1 - 2x = 0$
$4x + 13 = 4x^2 - 4x + 1$
$4x^2 - 8x - 12 = 0$
$(x-3)(x+1) = 0$
$x = 3$ only

41. $\sqrt{6y+1} - 3y = y$
$6y + 1 = 16y^2$
$0 = 16y^2 - 6y - 1$
$0 = (8y+1)(2y-1)$
$y = \frac{1}{2}$ only

42. $\sqrt{2x+3} - 2 = x$
$2x + 3 = x^2 + 4x + 4$
$0 = x^2 + 2x + 1$
$0 = (x+1)^2$
$x = -1$

To Think About

Solve for x.

43. $\sqrt{2x+5} = 2\sqrt{2x} + 1$
$2x + 5 = 8x + 4\sqrt{2x} + 1$
$-6x + 4 = 4\sqrt{2x}$
$36x^2 - 48x + 16 = 32x$
$36x^2 - 80x + 16 = 0$
$(9x-2)(x-2) = 0$
$x = \frac{2}{9}$ only

44. $\sqrt{x-2} + 3 = \sqrt{4x+1}$
$x - 2 + 6\sqrt{x-2} + 9 = 4x + 1$
$6\sqrt{x-2} = 3x - 6$
$2\sqrt{x-2} = x - 2$
$4x - 8 = x^2 - 4x + 4$
$x^2 - 8x + 12 = 0$
$(x-2)(x-6) = 0$
$x = 2, x = 6$

Cumulative Review Problems

45. Solve if possible. $\frac{2x}{x-7} + 3 = \frac{14}{x-7}$
$2x + 3x - 21 = 14$
$x = 7$ no solution

46. Multiply mentally. $(3x-1)^2$ $9x^2 - 6x + 1$

47. $f(x) = 2x^2 - 3x + 6$
Find $f(-2)$. 20

48. $x^2 + y^2 = 9$ Is this a relation or a function?
relation

9.7 Word Problems Involving Radicals: Direct and Inverse Variation

After studying this section, you will be able to:

1 *Solve problems involving direct variation.*
2 *Solve problems involving inverse variation.*

1 *Solving Problems Involving Direct Variation*

Often in daily life there is a relationship between two measurable quantities. For example, we turn up the thermostat and the heating bill increases. We say that the heating bill varies directly as the temperature on the thermostat. If y varies directly as x, then $y = kx$, where k is a constant. The constant is often called the *constant of variation.* Consider the following example.

EXAMPLE 1 Cliff works part-time in a local supermarket while going to college. His salary varies directly as the number of hours worked. Last week he earned \$33.60 for working 7 hours. This week he earned \$52.80. How many hours did he work?

Let S = his salary
h = the number of hours he worked
k = the constant of variation

Since his salary varies directly as the number of hours he worked, we write

$$S = k \cdot h$$

We can find the constant k by substituting the known values of $S = 33.60$ when $h = 7$.

$$33.60 = k \cdot 7 = 7k$$

$$\frac{33.60}{7} = \frac{7k}{7} \qquad \textit{Solve for } k.$$

$$4.80 = k \qquad \textit{The constant of variation is } 4.80.$$

$$S = 4.80h \qquad \textit{Replace } k \textit{ in the variation equation by } 4.80.$$

How many hours did he work to earn \$52.80?

$$S = 4.80h \qquad \textit{The direct variation equation with } k = 4.80.$$

$$52.80 = 4.80h \qquad \textit{Substitute } 52.80 \textit{ for } S \textit{ and solve for } h.$$

$$\frac{52.80}{4.80} = \frac{4.80h}{4.80}$$

$$11 = h$$

Cliff worked 11 hours this week.

Practice Problem 1 The *change* in temperature measured in Celsius varies directly as the change measured in Fahrenheit. The change from freezing water to boiling water is 100° in Celsius and 180° in Fahrenheit. If the Fahrenheit temperature drops 20°, what will be the change in temperature on the Celsius scale? ■

Solving a Direct Variation Problem

1. Write the direct variation equation.
2. Solve for the constant k by substituting in known values.
3. In the direct variation equation replace k by the value obtained in step 2.
4. Solve for the desired value.

TEACHING TIP After explaining Example 2 to students, ask them to do the following as an Additional Class Exercise.
If x varies directly with y, and $x = 9$ when $y = 2$, find x when $y = 7$.

Sol. 31.5

EXAMPLE 2 If y varies directly as x and $y = 20$ when $x = 3$, find the value of y when $x = 21$.

Step 1 $y = kx$

Step 2 To find k, we substitute in $y = 20$ and $x = 3$.

$$20 = k(3)$$

$$20 = 3k$$

$$\frac{20}{3} = k$$

Step 3 We now write the variation equation with k replaced by $\frac{20}{3}$.

$$y = \frac{20}{3}x$$

Step 4 We replace x by 21. Now we find y.

$$y = \left(\frac{20}{\cancel{3}}\right)(\overset{7}{\cancel{21}})$$

$$= (20)(7) = 140$$

Thus $y = 140$ when $x = 21$.

Practice Problem 2 If y varies directly as x and $y = 18$ when $x = 5$, find the value of y when $x = \frac{20}{23}$. ■

Direct variation equations apply to a number of cases. The following table shows some of the more common forms of direct variation. In each case, $k =$ the constant of variation.

Sample Direct Variation Situations

VERBAL DESCRIPTION	VARIATION EQUATION
y varies directly as x	$y = kx$
b varies directly as the square of c	$b = kc^2$
l varies directly as the cube of m	$l = km^3$
d varies directly as the square root of h	$d = k\sqrt{h}$

EXAMPLE 3 In a certain class of racing cars the maximum speed varies directly as the square root of the horsepower of the engine. If a car with 225 horsepower is able to achieve a maximum speed of 120 mph, what speed could it achieve if the engine developed 256 horsepower?

Let V = the maximum speed
h = the horsepower of the engine
k = the constant of variation

Step 1 Since the maximum speed (V) varies directly as the square root of the horsepower of the engine,

$$V = k\sqrt{h}$$

Step 2 $120 = k\sqrt{225}$ *Substitute known values of V and h.*

$120 = k \cdot 15$ *Solve for k.*

$8 = k$

Step 3 Now we can write the direct variation equation with a known value for k.

$$V = 8\sqrt{h}$$

Step 4 $V = 8\sqrt{256}$ *Substitute the value of $h = 256$.*

$= (8)(16)$ *Solve for V.*

$= 128$

Thus a maximum speed of 128 mph could be achieved if the engine developed 256 horsepower.

Practice Problem 3 For a certain class of automobiles, the distance to stop the car varies directly as the square of its speed. On ice-covered roads a car that is traveling at a speed of 20 mph can stop in 60 feet. If the car is traveling at 40 mph, how many feet will be necessary for it to come to a stop? ■

2 *Solving Problems Involving Inverse Variation*

If one variable is a constant multiple of the reciprocal of the other, the two variables are said to *vary inversely*. If y varies inversely as x, we express this by the equation $y = \frac{k}{x}$, where k is the constant of variation. Inverse variation problems can be solved by a four-step procedure similar to the direct variation problems.

Solving an Inverse Variation Problem

1. Write the inverse variation equation.
2. Solve for the constant k by substituting in known values.
3. In the inverse variation equation, replace k by the value obtained in step 2.
4. Solve for the desired value.

TEACHING TIP After explaining Example 4 to students, ask them to do the following as an Additional Class Exercise.
If a varies inversely as b, and $b = 10$ when $a = 3$, find b when $a = 12$.

Sol. 2.5

EXAMPLE 4 If y varies inversely as x and $y = 12$ when $x = 7$, find the value of y when $x = \frac{2}{3}$.

Step 1 $y = \frac{k}{x}$

Step 2 $12 = \frac{k}{7}$ *Substitute known values of x and y to find k.*

$84 = k$

Step 3 $y = \frac{84}{x}$

Step 4 To find y when $x = \frac{2}{3}$, we substitute: $y = \frac{84}{\frac{2}{3}}$

On the right side we have $\frac{84}{1} \div \frac{2}{3} = \frac{\overset{42}{\cancel{84}}}{1} \cdot \frac{3}{\underset{1}{\cancel{2}}} = 42 \cdot 3 = 126$

Thus $y = 126$ when $x = \frac{2}{3}$.

Practice Problem 4 If y varies inversely as x and $y = 8$ when $x = 15$, find the value of y when $x = \frac{3}{5}$. ■

Test distance

Test distance

EXAMPLE 5 A car manufacturer is thinking of reducing the size of the wheel used in a subcompact car. The number of times a car wheel must turn to cover a given distance varies inversely as the radius of the wheel. (Notice that this says that the smaller the wheel, the more times it must turn to cover a given distance.) A wheel with a radius of 0.35 meter must turn 400 times to cover a specified distance on a test track. How many times would it have to turn if the radius is reduced to 0.30 meter (see the sketch)?

Let n = the number of times the car wheel will turn
r = the radius of the wheel
k = the constant of variation

Step 1 Since the number of turns varies inversely as the radius, we can write

$n = \frac{k}{r}$ *Write the variation equation.*

Step 2 $400 = \frac{k}{0.35}$ *Substitute known values of n and r to find k.*

$140 = k$

Step 3 $n = \frac{140}{r}$ *Use the variation equation where k is known.*

How many times will the wheel turn if the radius is 0.30 meter?

Step 4 $n = \frac{140}{0.30}$ *Substitute* 0.30 *for r.*

$n = 466\frac{2}{3}$

The wheel would have to turn $466\frac{2}{3}$ times to cover the same distance if the radius were only 0.30 meter.

Practice Problem 5 Over the last three years the market research division of a calculator company found the volume of sales of scientific calculators varies inversely as the price of the calculator. One year 120,000 calculators were sold at $30 each. How many calculators were sold the next year when the price was $24 for each calculator? ■

The following table demonstrates some situations involving inverse variation. In each case $k =$ the constant of variation.

Sample Inverse Variation Situations

VERBAL DESCRIPTION	VARIATION EQUATION
y varies inversely as x	$y = \dfrac{k}{x}$
b varies inversely as the square of c	$b = \dfrac{k}{c^2}$
l varies inversely as the cube of m	$l = \dfrac{k}{m^3}$
d varies inversely as the square root of t	$d = \dfrac{k}{\sqrt{t}}$

EXAMPLE 6 The illumination of a light source varies inversely as the square of the distance from the source. The illumination measures 25 candlepower when a certain light is 4 meters away. Find the illumination when the light is 8 meters away (see figure).

TEACHING TIP After explaining Example 6 to students, ask them to do the following inverse variation as an Additional Class Exercise.
The resistance to the flow of electric current in the wire of fixed length varies inversely as the square of the diameter of the wire. When the resistance measured is 0.5 ohm, the diameter of the wire is 0.01 centimeter. If we use wire that is 0.02 centimeter in diameter, what will be the resistance?

Sol. 0.125 ohm of resistance

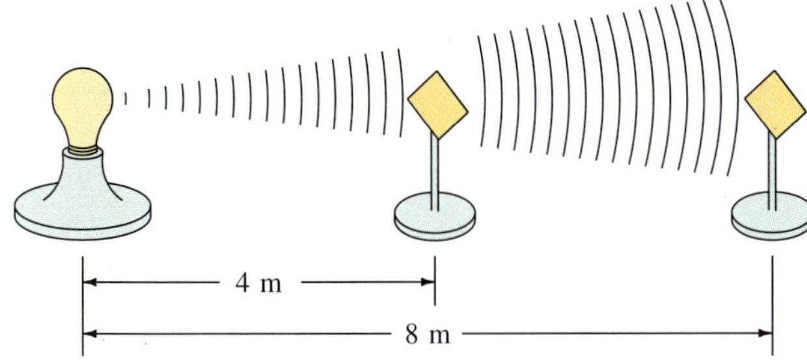

Let $I =$ the measurement of illumination
$d =$ the distance from the light source
$k =$ the constant of variation

Step 1 Since the illumination varies inversely as the square of the distance,

$$I = \frac{k}{d^2}$$

Step 2 We evaluate the constant by substituting the given values.

$$25 = \frac{k}{4^2} \quad \textit{Substitute } I = 25 \textit{ and } d = 4.$$

$$25 = \frac{k}{16} \quad \textit{Simplify and solve for } k.$$

$$400 = k$$

Step 3 We may now write the variation equation with the constant evaluated.

$$I = \frac{400}{d^2}$$

Step 4

$$I = \frac{400}{8^2} \quad \textit{Substitute a distance of 8 meters.}$$

$$= \frac{400}{64} \quad \textit{Square 8.}$$

$$= \frac{25}{4}$$

$$= 6\frac{1}{4}$$

The illumination is $6\frac{1}{4}$ candlepower when the light source is 8 meters away.

Practice Problem 6 If the amount of power in an electrical circuit is held constant, the resistance in the circuit varies inversely as the square of the amount of current. If the amount of current is 0.01 ampere, the resistance is 800 ohms. What is the resistance if the amount of current is 0.02 ampere? ■

Graphs of Variation Equations

The graphs of direct variation equations are shown below. Notice that as x increases y also increases.

Variation Statement	Equation	Graph
1. y varies directly as x	$y = kx$	y x
2. y varies directly as x^2	$y = kx^2$	y x
3. y varies directly as x^3	$y = kx^3$	y x
4. y varies directly as the square root of x	$y = k\sqrt{x}$,	y x

The graphs of inverse variation equations are shown below. Notice that as x increases y decreases.

Variation Statement	Equation	Graph
5. y varies inversely as x	$y = \frac{k}{x}$	y x
6. y varies inversely as x^2	$y = \frac{k}{x^2}$	y x

9.7 Exercises

1. If y varies directly as x, and $y = 7$ when $x = 4$, find y when $x = 9$.
$y = kx$ $\quad 7 = 4k \quad k = \frac{7}{4} \Rightarrow y = \frac{7}{4}x \quad y = \frac{63}{4}$

2. If y varies directly as x, and $y = 9$ when $x = 12$, find y when $x = 5$.
$y = kx$ $\quad 9 = 12k \quad k = \frac{3}{4} \Rightarrow y = \frac{3}{4}x \quad y = \frac{15}{4}$

3. If y varies directly as x and $y = 150$ when $x = 60$, find y when $x = 36$.
$y = kx$ $\quad 150 = k60 \quad k = \frac{150}{60} = \frac{5}{2} \Rightarrow y = \frac{5}{2}(36) \quad y = 90$

4. If y varies directly as x and $y = 900$ when $x = 200$, find y when $x = 30$.
$y = kx$ $\quad 900 = 200k \quad k = \frac{9}{2} \Rightarrow y = \frac{9}{2}x \quad y = 135$

5. If y varies directly as the cube of x, and $y = 12$ when $x = 2$, find y when $x = 7$.
$y = kx^3$ $\quad 12 = 8k \quad k = \frac{3}{2} \Rightarrow y = \frac{3}{2}x^3 \quad y = \frac{1029}{2}$

6. If y varies directly as the square root of x, and $y = 7$ when $x = 4$, find y when $x = 25$.
$y = k\sqrt{x}$ $\quad 7 = k\sqrt{4} \quad k = \frac{7}{2} \Rightarrow y = \frac{7}{2}\sqrt{x} \quad y = \frac{35}{2}$

7. If y varies directly as the square of x, and $y = 800$ when $x = 40$, find y when $x = 25$.
$y = kx^2$ $\quad 800 = 1600k \quad k = \frac{1}{2} \Rightarrow y = \frac{1}{2}x^2 \quad y = \frac{625}{2}$

8. If y varies directly as the square of x, and $y = 15{,}000$ when $x = 100$, find y when $x = 8$.
$y = kx^2$ $\quad 15{,}000 = 10{,}000k \quad k = \frac{3}{2} \Rightarrow y = \frac{3}{2}x^2 \quad y = 96$

9. If y varies directly as x^2, and $y = 1146.88$ when $x = 12.80$, find y when $x = 7.30$.
$y = kx^2$ $\quad 1146.88 = 163.84k \quad k = 7 \Rightarrow y = 7x^2 \quad y = 373.03$

10. If y varies directly as x^3, and $y = 214.375$ when $x = 3.5$, find y when $x = 8.2$.
$y = kx^3$ $\quad 214.375 = 42.875k \quad k = 5 \Rightarrow y = 5x^3 \quad y = 2756.84$

Applications

11. The pressure of water on an object submerged beneath the surface varies directly as the distance beneath the surface. A submarine experiences a pressure of 26 pounds per square inch at 60 feet below the surface. How much pressure will the submarine experience at 390 feet below the surface?
$P = kd \Rightarrow 26 = k \cdot 60 \Rightarrow k = \frac{13}{30} \Rightarrow P = \frac{13}{30} \cdot 390$
$= 169$ lb/sq in.

12. The weight of an object on the surface of the moon varies directly as the weight of an object on the surface of Earth. An astronaut with his protective suit weighs 80 kilograms on Earth's surface; on the moon and wearing the same suit, he will weigh 12.8 kilograms. If his moon rover vehicle weighs 900 kilograms on Earth, how much will it weigh on the moon?
$M = kw \Rightarrow 12.8 = k(80) \Rightarrow M = \frac{12.8}{80} \cdot 900 = 144$ kg

13. The time it takes to fill a storage cube with sand varies directly as the cube of the side of the box. A storage cube that is 2.0 meters on each side can be filled in 7 minutes by a sand loader. How long will it take to fill a storage cube that is 4.0 meters on each side?

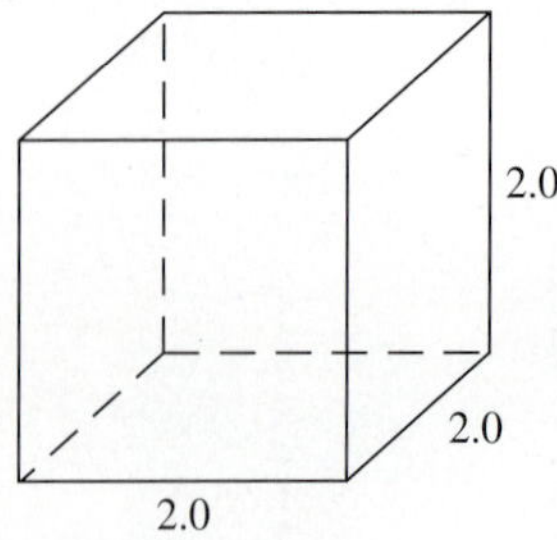

$T = ks^3$
$7 = k(2)^3$
$= \frac{7}{8}(4)^3$
$= 56$ min

14. The revenue from sales of pizza at College Pizza is directly proportional to the advertising budget. When the owners spent \$3000 per month to advertise on local radio and television stations, the monthly gross revenue at College Pizza was \$120,000. If the owners increase their advertising budget to \$5000 per month, what can they expect for a monthly gross revenue?
$S = kA$
$S = 40A$ $\quad$ \$200,000 per month revenue

15. If y varies inversely as x, and $y = 12$ when $x = 4$, find y when $x = 7$.

$y = \frac{k}{x}$ $\quad 12 = \frac{k}{4}$ $\quad k = 48 \Rightarrow y = \frac{48}{x}$ $\quad y = \frac{48}{7}$

16. If y varies inversely as x, and $y = 39$ when $x = 3$, find y when $x = 11$.

$y = \frac{k}{x}$ $\quad 39 = \frac{k}{3}$ $\quad k = 117 \Rightarrow y = \frac{117}{x}$ $\quad y = \frac{117}{11}$

17. If y varies inversely as x, and $y = \frac{1}{4}$ when $x = 8$, find y when $x = 1$.

$y = \frac{k}{x}$ $\quad \frac{1}{4} = \frac{k}{8}$ $\quad k = 2 \Rightarrow y = \frac{2}{x}$ $\quad y = 2$

18. If y varies inversely as x, and $y = \frac{1}{5}$ when $x = 20$, find y when $x = 2$.

$y = \frac{k}{x}$ $\quad \frac{1}{5} = \frac{k}{20}$ $\quad k = 4 \Rightarrow y = \frac{4}{x}$ $\quad y = 2$

19. If y varies inversely as the square of x, and $y = 30$ when $x = 2$, find y when $x = 9$.

$y = \frac{k}{x^2}$

$k = x^2 y$

$k = 120 \Rightarrow y = \frac{120}{81} = \frac{40}{27}$

20. If y varies inversely as the cube of x, and $y = \frac{1}{9}$ when $x = 3$, find y when $x = \frac{1}{2}$.

$y = \frac{k}{x^3}$

$k = x^3 y$

$k = 3 \Rightarrow y = \frac{3}{(\frac{1}{2})^3}$

$y = 24$

21. If y varies inversely as the square root of x, and $y = 2$ when $x = 4$, find y when $x = 81$.

$y = \frac{k}{\sqrt{x}}$ $\quad 2 = \frac{k}{2}$ $\quad k = 4 \Rightarrow y = \frac{4}{\sqrt{x}}$ $\quad y = \frac{4}{9}$

22. If y varies inversely as the square root of x and $y = 2$ when $x = 25$, find y when $x = 121$.

$y = \frac{k}{\sqrt{x}}$ $\quad 2 = \frac{k}{5}$ $\quad k = 10 \Rightarrow y = \frac{10}{\sqrt{x}}$ $\quad y = \frac{10}{11}$

23. If y varies inversely as $\sqrt{x}$, and $y = 5.00$ when $x = 73.96$, find y when $x = 18.49$.

$y = \frac{k}{\sqrt{x}}$ $\quad 5 = \frac{k}{\sqrt{73.96}}$ $\quad k = 43 \Rightarrow y = \frac{43}{\sqrt{18.49}}$ $\quad y = 10$

24. If y varies inversely as $\sqrt{x}$, and $y = 10.0$ when $x = 47.61$, find y when $x = 5.29$.

$y = \frac{k}{\sqrt{x}}$ $\quad 10 = \frac{k}{\sqrt{47.61}}$ $\quad k = 69 \Rightarrow y = \frac{69}{\sqrt{x}}$ $\quad y = 30$

25. The amount of time in minutes that it takes for an ice cube to melt varies inversely as the temperature of the water that the ice cube is placed in. When an ice cube is placed in 60°F water, it takes 2.3 minutes to melt. How long would it take for this size of ice cube to melt if it were placed in 40°F water?

$\text{time} = \frac{k}{\text{temperature}}$

$\text{time} = \frac{138}{\text{temperature}}$ $\quad$ 3.45 minutes

26. For a biosphere, the engineers have decided that part of the structure housing certain experiments will use 14 inches of fiberglass insulation. You know that (a) the thickness of fiberglass insulation is an indicator of its thermal effectiveness and (b) the heat loss through a certain type of fiberglass insulation varies inversely as the thickness of the fiberglass. If the heat loss through 6 inches of the type of fiberglass is 3000 Btu/hour, how much heat will you lose with the 14 inches on the biosphere? (Round to the nearest tenth.)

$L = \frac{k}{T}$ $\quad k = 18{,}000$

Heat loss is approx. 1285.7 Btu/hour.

27. The weight of an object near Earth's surface varies inversely as the square of its distance from the center of Earth. An object weighs 1000 pounds on Earth's surface. This is approximately 4000 miles from the center of Earth. How much will the object weigh 6000 miles from the center of Earth?

$W = \frac{k}{D^2} \Rightarrow 1000 = \frac{k}{4000^2} \Rightarrow W = \frac{(1000)(4000)(4000)}{(6000)^2}$

$= \frac{4000}{9}$ or $444\frac{4}{9}$ pounds

28. The pressure exerted by a given amount of gas on a cubical storage container varies inversely as the cube of the length of the side of the container. A certain amount of gas exerts a pressure of 60 pounds per square inch when placed in a cubical storage container that measures 0.2 meter on each side. If the gas is stored in a box 0.4 meter on each side, what will be the pressure?

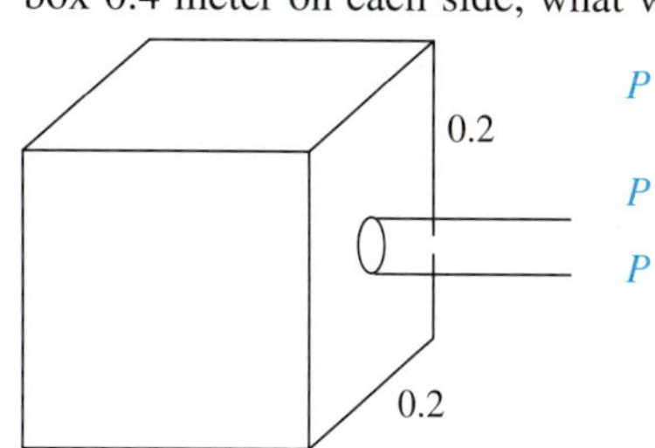

$P = \frac{k}{s^3} \Rightarrow 60 = \frac{k}{(0.2)^3}$

$P = \frac{60(0.2)(0.2)(0.2)}{(0.4)^3}$

$P = 7.5 \text{ lb/in.}^2$

Verbal and Writing Skills

29. Variable y varies directly as variable x. Thus $y = kx$. Describe the graphs of all such direct variations. The graph is a straight line through the origin.

30. Variable y varies inversely as variable x. Thus $y = \frac{k}{x}$. Describe the graphs of all such inverse variations for positive values of k and positive values of x. The graph is similar to that part of the graph of $y = \frac{1}{x}$ that falls in the first quadrant. It is asymptotic to the x-axis and the y-axis.

To Think About

Write an equation to describe the variation and solve each word problem.

31. The amount of heat that passes through a wall can be related to the factors of the temperature difference on the two sides of the wall, the area of the wall, and the thickness of the wall. The heat transfer varies directly as the product of the area of the wall and the temperature difference between the two sides of the wall and inversely as the thickness of the wall. In a certain house 6000 Btu/hour was transferred by a wall measuring 80 square feet. The wall is 0.5 foot thick. The outside temperature was 40°F, the inside temperature was 70°F. Suppose that the wall was exactly 1.0 foot thick and the outside temperature was 20°F. Assuming the same wall area and inside temperature, how much heat would be transferred?

$$H = \frac{K \cdot A \cdot D}{T} = \frac{(0.5)(6000)(50)(80)}{(80)(30)(1)}$$
$$= 5000 \text{ Btu/hr}$$

32. An important element of sailboat design is a calculation of the force of wind on the sail. The force varies directly with the square of the wind speed and directly with the area of the sail. On a certain class of racing boat the force on 100 square feet of sail is 280 pounds when the wind speed is 6 mph. What would be the force on the sail if it is rigged for a storm (the amount of sail exposed to the wind is reduced by lashing part of the sail to the boom) so that only 60 square feet of sail is exposed but the windspeed is 40 mph?

$$F = KS^2A$$
$$280 = K(6)^2(100)$$
$$F = \frac{(280)(60)(40)(40)}{3600}$$
$$= 7467 \text{ lb}$$

Cumulative Review Problems

33. Solve this equation for a. $\frac{80{,}000}{320} = \frac{120{,}000}{320 + a}$

$$\text{LCD} = 320(320 + a)$$
$$480 = 320 + a$$
$$a = 160$$

34. Graph the equation $S = kh$, where $k = 4.80$. Use h for the horizontal axis, S for the vertical. Assume h and S are nonnegative.

$S = 4.80h$

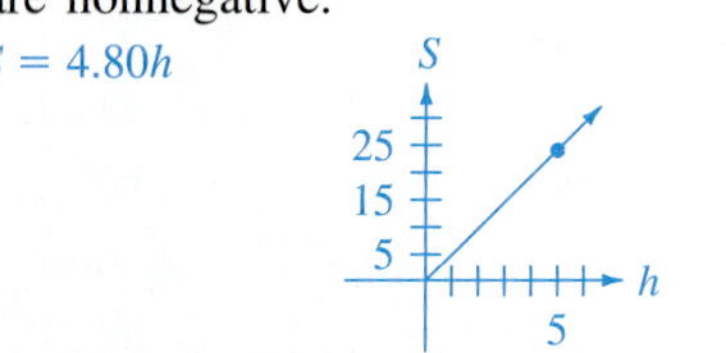

35. Factor. $12x^2 - 20x - 48$ $4(3x + 4)(x - 3)$

36. Solve. $2x - (5 - 3x) = 6(x - 2) - 3(2 - x)$

$$2x - 5 + 3x = 6x - 12 - 6 + 3x$$
$$5x - 5 = 9x - 18$$
$$x = \frac{13}{4}$$

Chapter Organizer

Topic	Procedure	Examples
Square roots, p. 505.	1. The square roots of perfect squares up to $\sqrt{225} = 15$ (or so) should be memorized. Approximate values of others can be found in tables or by calculator. 2. Negative numbers do not have real number square roots.	$\sqrt{196} = 14$ $\quad \sqrt{169} = 13$ $\sqrt{200} \approx 14.142$ $\quad \sqrt{13} \approx 3.606$ $\sqrt{-5}$ does not exist.
Simplifying radicals, p. 510.	$\sqrt{ab} = \sqrt{a}\sqrt{b}$ Radicals are simplified by taking the square roots of all perfect squares and leaving other factors under the radical sign.	$\sqrt{48} = \sqrt{16}\sqrt{3} = 4\sqrt{3}$ $\sqrt{12x^2y^3z} = \sqrt{2^2x^2y^2}\sqrt{3yz}$ $= 2xy\sqrt{3yz}$ $\sqrt{a^{11}} = \sqrt{a^{10} \cdot a} = \sqrt{a^{10}}\sqrt{a} = a^5\sqrt{a}$
Adding and subtracting radicals, p. 515.	Simplify all radicals and then add or subtract like radicals.	$\sqrt{12} + \sqrt{18} + \sqrt{27} + \sqrt{50}$ $= \sqrt{4}\sqrt{3} + \sqrt{9}\sqrt{2} + \sqrt{9}\sqrt{3} + \sqrt{25}\sqrt{2}$ $= 2\sqrt{3} + 3\sqrt{2} + 3\sqrt{3} + 5\sqrt{2}$ $= 5\sqrt{3} + 8\sqrt{2}$
Multiplying radicals, p. 519.	Radicals are multiplied in a way analogous to the multiplication of polynomials. Apply the rule $\sqrt{a}\sqrt{b} = \sqrt{ab}$ and simplify all results.	$(2\sqrt{6} + \sqrt{3})(\sqrt{6} - 3\sqrt{3})$ $= 2\sqrt{36} - 6\sqrt{18} + \sqrt{18} - 3\sqrt{9}$ $= 2(6) - 5\sqrt{18} - 3(3)$ $= 12 - 5\sqrt{9}\sqrt{2} - 9$ $= 12 - 5(3)\sqrt{2} - 9$ $= 3 - 15\sqrt{2}$
Dividing radicals, p. 524.	Division of radical expressions is accomplished by writing the division as a fraction and rationalizing the denominator.	Divide. $\sqrt{x} \div \sqrt{3}$ $\frac{\sqrt{x}}{\sqrt{3}} \cdot \frac{\sqrt{3}}{\sqrt{3}} = \frac{\sqrt{3x}}{\sqrt{9}} = \frac{\sqrt{3x}}{3}$
Conjugates, p. 526.	Note that the sum and difference of two radical expressions multiply to the difference of their squares—eliminating the radicals. These are called conjugates.	$(3\sqrt{6} + \sqrt{10})$ and $(3\sqrt{6} - \sqrt{10})$ are conjugates. $(3\sqrt{6} + \sqrt{10})(3\sqrt{6} - \sqrt{10})$ $= 9\sqrt{6^2} - \sqrt{10^2}$ $= 9 \cdot 6 - 10 = 54 - 10 = 44$
Rationalizing denominators, p. 525.	To simplify a fraction with a radical in the denominator, the radical (in the denominator) must be removed. 1. Multiply the denominator by itself if it is a monomial. 2. Multiply the denominator by its conjugate if it is a binomial. In each case the numerator is multiplied by the same quantity as the denominator.	$\frac{\sqrt{10}}{\sqrt{6}} = \frac{\sqrt{10} \cdot \sqrt{6}}{\sqrt{6} \cdot \sqrt{6}} = \frac{2\sqrt{15}}{6} = \frac{\sqrt{15}}{3}$ $\frac{\sqrt{3}}{\sqrt{3} - \sqrt{2}} = \frac{\sqrt{3}(\sqrt{3} + \sqrt{2})}{(\sqrt{3} - \sqrt{2})(\sqrt{3} + \sqrt{2})}$ $= \frac{\sqrt{9} + \sqrt{6}}{3 - 2} = 3 + \sqrt{6}$
Radicals with fractions, p. 524.	Note that $\sqrt{\frac{5}{3}}$ is the same as $\frac{\sqrt{5}}{\sqrt{3}}$ and so must be rationalized.	$\sqrt{\frac{5}{3}} = \frac{\sqrt{5}}{\sqrt{3}} \cdot \frac{\sqrt{3}}{\sqrt{3}} = \frac{\sqrt{15}}{3}$
Pythagorean theorem, p. 530.	In any right triangle with hypotenuse of length c and legs of lengths a and b, $c^2 = a^2 + b^2$ c b a	Find c to the nearest tenth if $a = 7$ and $b = 9$. $c^2 = 7^2 + 9^2 = 49 + 81 = 130$ $c = \sqrt{130} \approx 11.4$

Chapter Organizer

Topic	Procedure	Examples
Radical equations, p. 532.	To solve an equation containing a square root radical: **1.** Isolate one radical by itself on one side of the equation. **2.** Square both sides. **3.** Solve the resulting equation. **4.** Check all apparent solutions. Extraneous roots may have been introduced in step 2.	Solve. $\sqrt{3x+2} - 4 = 8$ $\sqrt{3x+2} = 12$ $(\sqrt{3x+2})^2 = 12^2$ $3x + 2 = 144$ $3x = 142$ $x = \frac{142}{3}$ $\sqrt{3 \cdot \frac{142}{3} + 2} \stackrel{?}{=} 12$ $\sqrt{144} = 12$ ✓ So $x = \frac{142}{3}$.
Variation: direct, p. 537.	If y varies directly with x, there is a constant of variation, k, such that $y = kx$. Once k is determined, other values of y or x can easily be computed.	y varies directly with x. When $x = 2$, $y = 7$. $y = kx$ $7 = k(2)$ Substituting. $k = \frac{7}{2}$ Solving. $y = \frac{7}{2}x$ What is y when $x = 8$? $y = \frac{7}{2}x = \frac{7}{2} \cdot 8 = 28$
Variation: inverse, p. 539.	If y varies inversely with x, the constant k is such that $y = \frac{k}{x}$	y varies inversely with x. When x is 5, y is 12. What is y when x is 15? $y = \frac{k}{x}$ $12 = \frac{k}{5}$ Substituting. $k = 60$ Solving. $y = \frac{60}{x}$ Substituting. $x = 15$ $y = \frac{60}{15} = 4$

Chapter 9 Review Problems

9.1 *Simplify or evaluate, if possible.*

1. $\sqrt{9}$ 3 **2.** $\sqrt{25}$ 5 **3.** $\sqrt{36}$ 6 **4.** $\sqrt{49}$ 7 **5.** $\sqrt{100}$ 10

6. $\sqrt{4}$ 2 **7.** $\sqrt{-81}$ not a real number **8.** $\sqrt{-64}$ not a real number **9.** $\sqrt{16}$ 4 **10.** $\sqrt{121}$ 11

11. $\sqrt{225}$ 15 **12.** $\sqrt{169}$ 13 **13.** $\sqrt{0.81}$ 0.9 **14.** $\sqrt{0.36}$ 0.6 **15.** $\sqrt{\frac{1}{25}}$ $\frac{1}{5}$

16. $\sqrt{\frac{36}{49}}$ $\frac{6}{7}$ **17.** $-\sqrt{0.0004}$ -0.02 **18.** $\sqrt{0.0009}$ 0.03 **19.** $\sqrt{-0.0016}$ not a real number **20.** $-\sqrt{0.0081}$ -0.09

Approximate, using the square root table in Appendix A or a calculator. Round to nearest thousandth, if necessary.

21. $\sqrt{105}$ 10.247 **22.** $\sqrt{198}$ 14.071 **23.** $\sqrt{77}$ 8.775 **24.** $\sqrt{88}$ 9.381

9.2 *Simplify.*

25. $\sqrt{50}$ $5\sqrt{2}$ **26.** $\sqrt{48}$ $4\sqrt{3}$ **27.** $\sqrt{98}$ $7\sqrt{2}$ **28.** $\sqrt{72}$ $6\sqrt{2}$ **29.** $\sqrt{40}$ $2\sqrt{10}$ **30.** $\sqrt{80}$ $4\sqrt{5}$

31. $\sqrt{x^8}$ x^4 **32.** $\sqrt{y^{10}}$ y^5 **33.** $\sqrt{x^5y^6}$ $x^2y^3\sqrt{x}$ **34.** $\sqrt{a^3b^4}$ $ab^2\sqrt{a}$ **35.** $\sqrt{16x^3y^5}$ $4xy^2\sqrt{xy}$ **36.** $\sqrt{98x^4y^6}$ $7x^2y^3\sqrt{2}$

37. $\sqrt{12x^5}$ $2x^2\sqrt{3x}$ **38.** $\sqrt{27x^7}$ $3x^3\sqrt{3x}$ **39.** $\sqrt{75x^{10}}$ $5x^5\sqrt{3}$ **40.** $\sqrt{125x^{12}}$ $5x^6\sqrt{5}$

41. $\sqrt{120a^3b^4c^5}$ $2ab^2c^2\sqrt{30ac}$ **42.** $\sqrt{121a^6b^4c}$ $11a^3b^2\sqrt{c}$ **43.** $\sqrt{56x^7y^9}$ $2x^3y^4\sqrt{14xy}$ **44.** $\sqrt{99x^{13}y^7}$ $3x^6y^3\sqrt{11xy}$

9.3 *Simplify.*

45. $\sqrt{2} - \sqrt{8} + \sqrt{32}$ $3\sqrt{2}$

46. $3\sqrt{5} - 2\sqrt{20} - \sqrt{5}$ $-2\sqrt{5}$

47. $x\sqrt{3} + 3x\sqrt{3} + \sqrt{27x^2}$ $7x\sqrt{3}$

48. $a\sqrt{2} + \sqrt{12a^2} + a\sqrt{98}$ $8a\sqrt{2} + 2a\sqrt{3}$

49. $5\sqrt{5} - 6\sqrt{20} + 2\sqrt{10}$ $-7\sqrt{5} + 2\sqrt{10}$

50. $3\sqrt{6} - 5\sqrt{18} + 3\sqrt{24}$ $9\sqrt{6} - 15\sqrt{2}$

51. $2\sqrt{28} - 3\sqrt{63} + 2x\sqrt{7}$ $-5\sqrt{7} + 2x\sqrt{7}$

52. $5a\sqrt{3} - 12\sqrt{27} + 2\sqrt{75}$ $5a\sqrt{3} - 26\sqrt{3}$

9.4 *Simplify.*

53. $(2\sqrt{x})(3\sqrt{x^3})$ $6x^2$

54. $(-5\sqrt{a})(2\sqrt{ab})$ $-10a\sqrt{b}$

55. $(\sqrt{2a^3})(\sqrt{8b^2})$ $4ab\sqrt{a}$

56. $(5x\sqrt{x})(-3x^2\sqrt{x})$ $-15x^4$

57. $\sqrt{5}(3\sqrt{5} - \sqrt{20})$ 5

58. $2\sqrt{3}(\sqrt{27} - 6\sqrt{3})$ -18

59. $\sqrt{2}(\sqrt{5} - \sqrt{3} - 2\sqrt{2})$ $\sqrt{10} - \sqrt{6} - 4$

60. $\sqrt{5}(\sqrt{6} - 2\sqrt{5} + \sqrt{10})$ $\sqrt{30} - 10 + 5\sqrt{2}$

61. $(\sqrt{11} + 2)(2\sqrt{11} - 1)$ $20 + 3\sqrt{11}$

62. $(\sqrt{10} + 3)(3\sqrt{10} - 1)$ $27 + 8\sqrt{10}$

63. $(2 + 3\sqrt{6})(4 - 2\sqrt{3})$ $8 - 4\sqrt{3} + 12\sqrt{6} - 18\sqrt{2}$

64. $(5 - \sqrt{2})(3 - \sqrt{12})$ $15 - 3\sqrt{2} - 10\sqrt{3} + 2\sqrt{6}$

65. $(2\sqrt{3} + 3\sqrt{6})^2$ $66 + 36\sqrt{2}$

66. $(5\sqrt{2} - 2\sqrt{6})^2$ $74 - 40\sqrt{3}$

67. $(a\sqrt{b} - 2\sqrt{a})(\sqrt{b} - 3\sqrt{a})$ $ab - 3a\sqrt{ab} - 2\sqrt{ab} + 6a$

68. $(x\sqrt{y} - 2\sqrt{x})(3\sqrt{y} + \sqrt{x})$ $3xy + x\sqrt{xy} - 6\sqrt{xy} - 2x$

9.5 *Rationalize the denominator.*

69. $\dfrac{1}{\sqrt{3x}}$ $\dfrac{\sqrt{3x}}{3x}$

70. $\dfrac{2y}{\sqrt{5}}$ $\dfrac{2y\sqrt{5}}{5}$

71. $\dfrac{x^2y}{\sqrt{8}}$ $\dfrac{x^2y\sqrt{2}}{4}$

72. $\dfrac{3ab}{\sqrt{2b}}$ $\dfrac{3a\sqrt{2b}}{2}$

73. $\sqrt{\dfrac{3}{7}}$ $\dfrac{\sqrt{21}}{7}$

74. $\sqrt{\dfrac{2}{9}}$ $\dfrac{\sqrt{2}}{3}$

75. $\dfrac{15\sqrt{60}}{5\sqrt{15}}$ 6

76. $\dfrac{\sqrt{120}}{2\sqrt{5}}$ $\sqrt{6}$

77. $\dfrac{\sqrt{a^5}}{\sqrt{2a}}$ $\dfrac{a^2\sqrt{2}}{2}$

78. $\dfrac{\sqrt{x^3}}{\sqrt{3x}}$ $\dfrac{x\sqrt{3}}{3}$

79. $\dfrac{a\sqrt{a^2b}}{ab\sqrt{ab^2}}$ $\dfrac{\sqrt{ab}}{b^2}$

80. $\dfrac{2xy\sqrt{x^3}}{2y\sqrt{y^3}}$ $\dfrac{x^2\sqrt{xy}}{y^2}$

81. $\dfrac{3}{\sqrt{5}+\sqrt{2}}$ $\sqrt{5}-\sqrt{2}$

82. $\dfrac{2}{\sqrt{6}-\sqrt{3}}$ $\dfrac{2(\sqrt{6}+\sqrt{3})}{3}$

83. $\dfrac{1-\sqrt{5}}{2+\sqrt{5}}$ $-7+3\sqrt{5}$

84. $\dfrac{1-\sqrt{3}}{3+\sqrt{3}}$ $\dfrac{3-2\sqrt{3}}{3}$

9.6

Use the Pythagorean theorem to find the length of the missing side of each right triangle.

85. $a = 5$, $b = 8$ $c = \sqrt{89}$

86. $c = \sqrt{11}$, $b = 3$ $a = \sqrt{2}$

87. $c = 5$, $a = 3.5$ $b = \dfrac{\sqrt{51}}{2}$

88. Find c and round your answer to the nearest thousandth, if $a = 2.400$ and $b = 2.000$. $c = 3.124$

89. A flagpole is 24 meters tall. A man stands 18 meters from the base of the pole. How far is it from the feet of the man to the top of the pole? 30 meters

90. A city is 20.0 miles east of a major airport. A small town is directly south of the city. The town is 50.0 miles from the airport. How far is it from the town to the city? (Round your answer to the nearest tenth of a mile.) 45.8 miles

91. The diagonal measurement of a small color television screen is 10 inches. The screen is square. How long is each side of the screen? (Round your answer to the nearest tenth of an inch.) 7.1 inches

Solve each radical equation. Be sure to verify your answers.

92. $\sqrt{x+3} = 4$ $x = 13$

93. $\sqrt{2x-3} = 9$ $x = 42$

94. $\sqrt{1-3x} = \sqrt{5+x}$ $x = -1$

95. $\sqrt{-5+2x} = \sqrt{1+x}$ $x = 6$

96. $\sqrt{10x+9} = -1 + 2x$ $x = 4$

97. $\sqrt{2x-5} = 10 - x$ $x = 7$

98. $6 - \sqrt{5x-1} = x + 1$ $x = 2$

99. $4x + \sqrt{x+2} = 5x - 4$ $x = 7$

9.7

100. If y varies directly with the square root of x, and $y = 35$ when $x = 25$, find the value of y when $x = 121$. $y = 77$

101. If y varies inversely with the square of x, and $y = \frac{6}{5}$ when $x = 5$, find the value of y with $x = 15$. $y = \dfrac{2}{15}$

102. If y varies inversely with the cube of x, and $y = 4$ when $x = 2$, find the value of y when $x = 4$. $y = \frac{1}{2}$

103. The insect population in a potato field varies inversely with the amount of pesticide used. When 40 pounds of pesticide was used, the insect population of the field was estimated to be 1000 bugs. How many pounds of pesticide should be required to reduce the number of bugs in the field to 100? 400 lb

104. The length of skid marks on the road when a car slams on the brakes varies directly with the square of the speed of the car. At 30 mph a certain car had skid marks 40 feet long. How long will the skid marks be if the car is traveling at 55 mph? 134 ft

105. The horsepower that is needed to drive a racing boat through water varies directly with the cube of the speed of the boat. What will happen to the horsepower requirement if someone wants to double the maximum speed of a given boat? 8 times as much

Putting Your Skills to Work

CAR ACCIDENT INVESTIGATION

Police use the formula $r = 2\sqrt{5L}$

to estimate the speed a car has been traveling by measuring the skid marks. In the formula, r is the speed in miles per hour and L is the length of the skid marks in feet.

PROBLEMS FOR INDIVIDUAL INVESTIGATION

1. Estimate the speed of a car that left skid marks of 125 feet.
   $r = 2\sqrt{5(125)}$
   $r = 50$ miles per hour

2. How long will the skid marks be at 40 miles per hour?
   $40 = 2\sqrt{5L}$
   $L = 80$ feet

PROBLEMS FOR GROUP INVESTIGATION AND ANALYSIS

Together with other members of your class, complete the following.

3. Make a table of values and graph the equation $r = 2\sqrt{5L}$.

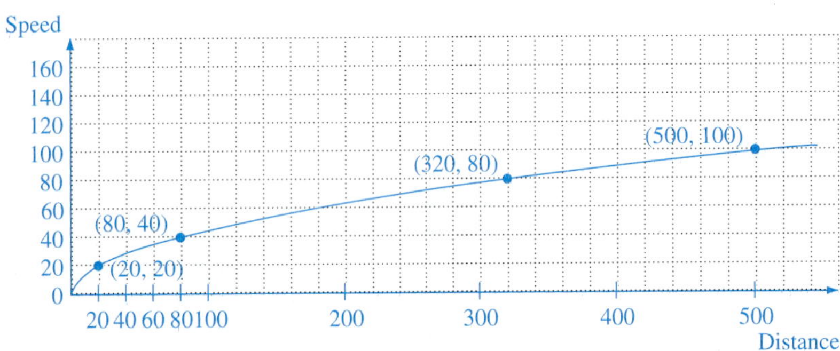

4. Describe what happens when the speed of the car increases from 20 miles per hour to 40 miles per hour. How much longer are the skid marks?
   Between 20 miles per hour and 40 miles per hour the 20-mile-per-hour increase in speed results in a 60-foot increase in the length of the skid marks.

5. Describe what happens when the speed of the car increases from 80 miles per hour to 100 miles per hour. How much longer are the skid marks?
   Between 80 miles per hour and 100 miles per hour the 20-mile-per-hour increase in the speed results in a 180-foot increase in the length of the skid marks. This is 3 times the increase in the length of the skid marks at the lower speed.

INTERNET CONNECTIONS: Go to ''http://www.prenhall.com/tobey'' to be connected

Site: Fatal Crashes, Causes and Costs (Advocates for Highway and Auto Safety) or a related site

This site gives statistics regarding fatal automobile accidents. Use the data for your home state to answer the following questions.

6. How many of the accidents were speed related?

7. How many of the accidents were alcohol related?

8. What was the average cost per fatal accident?

Chapter 9 Test

Evaluate.

1. $\sqrt{81}$

2. $\sqrt{\frac{9}{100}}$

Simplify.

3. $\sqrt{48x^2y^7}$

4. $\sqrt{100x^3yz^4}$

Combine and simplify.

5. $\sqrt{5} + 3\sqrt{20} - 2\sqrt{45}$

6. $\sqrt{4a} + \sqrt{8a} + \sqrt{36a} + \sqrt{18a}$

Multiply.

7. $(2\sqrt{a})(3\sqrt{b})(2\sqrt{ab})$

8. $\sqrt{3}(\sqrt{6} - \sqrt{2} + 5\sqrt{27})$

9. $(2 - \sqrt{6})^2$

10. $(4\sqrt{2} - \sqrt{5})(3\sqrt{2} + \sqrt{5})$

Simplify.

11. $\sqrt{\frac{x}{5}}$

12. $\frac{3}{\sqrt{12}}$

1. 9

2. $\frac{3}{10}$

3. $4xy^3\sqrt{3y}$

4. $10xz^2\sqrt{xy}$

5. $\sqrt{5}$

6. $8\sqrt{a} + 5\sqrt{2a}$

7. $12ab$

8. $3\sqrt{2} - \sqrt{6} + 45$

9. $10 - 4\sqrt{6}$

10. $19 + \sqrt{10}$

11. $\frac{\sqrt{5x}}{5}$

12. $\frac{\sqrt{3}}{2}$

Simplify.

13. $\dfrac{3 - \sqrt{2}}{\sqrt{2} - 6}$

14. $\dfrac{3a}{\sqrt{5} + \sqrt{2}}$

15. Use the table of square roots or a calculator to approximate $\sqrt{156}$. Round your answer to nearest hundredth.

Find the missing side by using the Pythagorean theorem.

16.

x

7

12

17.

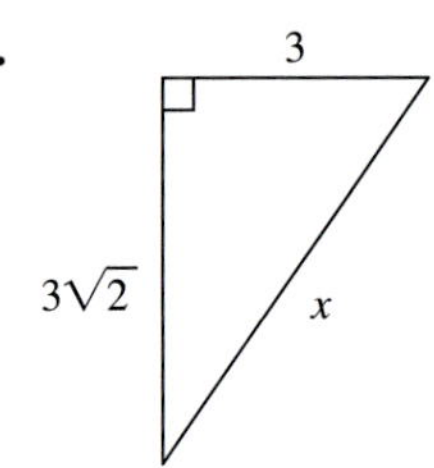

Solve each radical equation. Verify your solutions.

18. $6 - \sqrt{2x + 1} = 0$

19. $x = 5 + \sqrt{x + 7}$

Solve each word problem.

20. The intensity of illumination from a light source varies inversely with the square of the distance from the light source. A photoelectric cell is placed 8 inches from a light source. Then it is moved so that it only receives one-fourth as much illumination. How far is it from the light source?

21. Sales tax varies directly with the cost of the sale. If \$26.10 is the tax on a purchase of \$360, what will the tax be on a purchase of \$12,580?

22. The area of an equilateral triangle varies directly as the square of its perimeter. If the perimeter is 12 centimeters, the area is 6.93 square centimeters. What is the area of an equilateral triangle whose perimeter is 21 centimeters? Round to the nearest tenth.

13. $\dfrac{16 - 3\sqrt{2}}{-34}$ or $\dfrac{3\sqrt{2} - 16}{34}$

14. $a(\sqrt{5} - \sqrt{2})$

15. 12.49

16. 

17. $x = 3\sqrt{3}$

18. $x = \frac{35}{2}$

19. $x = 9$

20. 16 inches

21. \$912.05

22. $A = 21.22\ \text{cm}^2$

Cumulative Test for Chapters 0–9

Approximately one-half of this test covers the content of Chapters 0–8. The remainder covers the content of Chapter 9.

Simplify.

1. $\dfrac{12}{15}$

2. $3\dfrac{1}{4} + 5\dfrac{2}{3}$

3. $6\dfrac{1}{4} \div 6\dfrac{2}{3}$

4. Write 0.07% as a decimal.

Simplify.

5. $-11 - 16 + 8 + 4 - 13 + 31$

6. $(-3)^2 \cdot 4 - 8 \div 2 - (3 - 2)^3$

7. $3x^3yz^2 + 4xyz - 5x^3yz^2$

8. $(3x - 2)^3$

9. $(2x - 3)^2$

10. $(3x - 11)(3x + 11)$

11. $\dfrac{4x}{x + 2} + \dfrac{4}{x + 2}$

12. $\dfrac{x^2 - 5x + 6}{x^2 - 5x - 6} \div \dfrac{x^2 - 4}{x^2 + 2x + 1}$

1. $\frac{4}{5}$

2. $8\frac{11}{12}$

3. $\frac{15}{16}$

4. 0.0007

5. 3

6. 31

7. $-2x^3yz^2 + 4xyz$

8. $27x^3 - 54x^2 + 36x - 8$

9. $4x^2 - 12x + 9$

10. $9x^2 - 121$

11. $\dfrac{4x + 4}{x + 2}$

12. $\dfrac{(x - 3)(x + 1)}{(x - 6)(x + 2)}$

13.	$7x^2y^3\sqrt{2x}$
14.	$14\sqrt{3}$
15.	$9 - 6\sqrt{2}$
16.	$5 + 2\sqrt{6}$
17.	$10 + 22\sqrt{3}$
18.	$\frac{3 + \sqrt{2}}{7}$
19.	-4
20.	not a real number
21.	$c = \sqrt{23}$
22.	$b = 4\sqrt{5}$
23.	$x = 5$
24.	$y = 6$
25.	$y = 10$
26.	$78.5\ \text{m}^2$

Simplify, if possible.

13. $\sqrt{98x^5y^6}$

14. $\sqrt{12} - \sqrt{27} + 3\sqrt{75}$

15. $(\sqrt{6} - \sqrt{3})^2$

16. $\dfrac{\sqrt{3} + \sqrt{2}}{\sqrt{3} - \sqrt{2}}$

17. $(3\sqrt{6} - \sqrt{2})(\sqrt{6} + 4\sqrt{2})$

18. $\dfrac{1}{3 - \sqrt{2}}$

19. $-\sqrt{16}$

20. $\sqrt{-16}$

Use the Pythagorean theorem to find the missing side of each triangle. Be sure to simplify your answer.

21. $a = 4,\ b = \sqrt{7}$

22. $a = 19,\ c = 21$

Solve for the variable. Be sure to verify your answer(s).

23. $\sqrt{4x + 5} = x$

24. $\sqrt{3y - 2} + 2 = y$

25. If y varies inversely as x, and $y = 4$ when $x = 5$, find the value of y when $x = 2$.

26. The surface area of a balloon varies directly with the square of the cross-sectional radius. When the radius is 2 meters, the surface area is 1256 square meters. Find the surface area if the radius is 0.5 meter.

CHAPTER 10

Quadratic Equations

Meteorologists are continuing to use more and more mathematics to predict weather data and to record weather information. The amount of snowfall and rain can sometimes be approximated by quadratic equations. The equation then leads to other useful information about the weather that can be determined by using mathematical procedures. Turn to page 599 and read through the Putting Your Skills to Work. See if you can answer the questions raised about the weather conditions.

Pretest Chapter 10

1. $x = 14, x = -2$

2. $x = 0, x = \frac{4}{3}$

3. $x = \frac{2}{5}, x = 4$

4. $x = \frac{7}{3}, x = -2$

5. $x = \pm 2\sqrt{3}$

6. $x = \pm 5$

7. $x = -4 \pm \sqrt{11}$

8. $x = \frac{-3 \pm \sqrt{65}}{4}$

9. $x = \frac{3 \pm \sqrt{3}}{2}$

10. $x = \frac{1}{4}, x = -2$

11. $x = \frac{-4 \pm \sqrt{13}}{3}$

12. no real solution

13. See graph

14. See graph

15. base = 6 cm, altitude = 13 cm

If you are familiar with the topics in this chapter, take this test now. Check your answers with those in the back of the book. If an answer was wrong or you couldn't do a problem, study the appropriate section of the chapter.

If you are not familiar with the topics in this chapter, don't take this test now. Instead, study the examples, work the practice problems, and then take the test.

This test will help you to identify those concepts that you have mastered and those that need more study.

Section 10.1

Solve each quadratic equation by the factoring method.

1. $x^2 - 12x - 28 = 0$

2. $3x^2 + 2x = 6x$

3. $5x^2 = 22x - 8$

4. $2x + 28 = 6x^2$

Section 10.2

Solve each quadratic equation by taking the square root of each side of the equation.

5. $x^2 - 12 = 0$

6. $3x^2 + 1 = 76$

Solve each quadratic equation by the completing the square method.

7. $x^2 + 8x + 5 = 0$

8. $2x^2 + 3x - 7 = 0$

Section 10.3

Solve each quadratic equation if possible by using the quadratic formula.

9. $2x^2 - 6x + 3 = 0$

10. $4x^2 = -7x + 2$

11. $3x^2 + 8x + 1 = 0$

12. $5x^2 + 3 = 4x$

Section 10.4

Graph the quadratic equations. Locate the vertex.

13. $y = 5 - 3x^2$

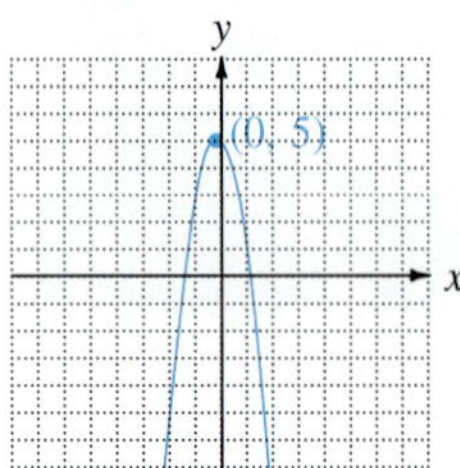

14. $y = 2x^2 - 4x - 1$

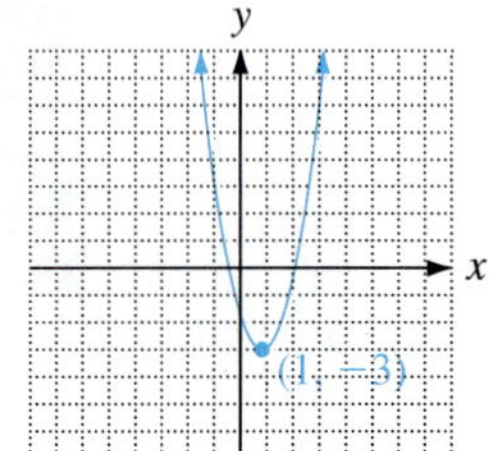

Section 10.5

15. A triangle has an area of 39 square centimeters. The altitude of the triangle is 1 centimeter longer than double the length of the base. Find the dimensions of the triangle.

10.1 Introduction to Quadratic Equations

After studying this section, you will be able to:

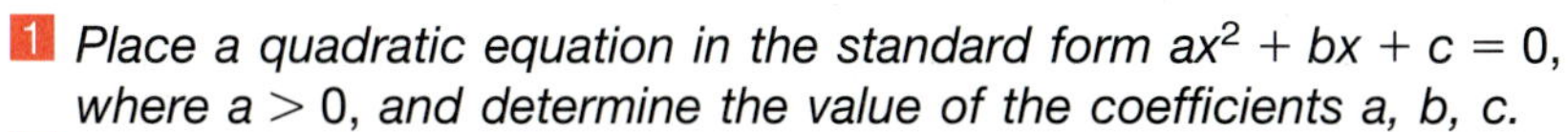

1. *Place a quadratic equation in the standard form $ax^2 + bx + c = 0$, where $a > 0$, and determine the value of the coefficients a, b, c.*
2. *Solve quadratic equations of the form $ax^2 + bx = 0$ by factoring.*
3. *Solve quadratic equations of the form $ax^2 + bx + c = 0$ by factoring.*
4. *Solve applied problems that require the use of quadratic equations.*

MathPro Video 24 SSM

1 *Placing a Quadratic Equation in Standard Form*

In Section 5.7 we introduced quadratic equations. All quadratic equations contain polynomials of the second degree. For example,

$$3x^2 + 5x - 7 = 0, \quad \tfrac{1}{2}x^2 - 5x = 0, \quad 2x^2 = 5x + 9, \quad 3x^2 = 8$$

are all quadratic equations. It is particularly useful to place quadratic equations in standard form.

> The **standard form of a quadratic equation** is $ax^2 + bx + c = 0$, where a, b, and c are real numbers, and a is a positive real number.

You will need to be able to recognize the real numbers that represent a, b, and c in specific equations. It will be easier to do so if these equations are placed in standard form.

EXAMPLE 1 Place each of the following quadratic equations in standard form and identify the real numbers a, b, and c.

(a) $5x^2 - 6x + 3 = 0$ — *This equation is in standard form.*
$ax^2 + bx + c = 0$ — *Match each term to the standard form.*
$a = 5, \quad b = -6, \quad c = 3$

(b) $2x^2 + 5x = 4$ — *The right-hand side is not zero. It is not in standard form.*
$2x^2 + 5x - 4 = 0$ — *Add -4 to each side of the equation.*
$ax^2 + bx + c = 0$ — *Match each term to the standard form.*
$a = 2, \quad b = 5, \quad c = -4$

(c) $-2x^2 + 15x + 4 = 0$ — *This is not in standard form since the coefficient of x^2 is negative.*
$2x^2 - 15x - 4 = 0$ — *Multiply each term on both sides of the equation by -1.*
$ax^2 + bx + c = 0$
$a = 2, \quad b = -15, \quad c = -4$

(d) $7x^2 - 9x = 0$ — *This equation is in standard form.*
$ax^2 + bx + c = 0$ — *Note that the constant term is missing, so we know $c = 0$.*
$a = 7, \quad b = -9, \quad c = 0$

(e) $x^2 + 3x + 5 = 10x - 6$ — *This equation is not in standard form.*
$x^2 + 3x - 10x + 5 + 6 = 0$ — *Add $-10x + 6$ to each side.*
$x^2 - 7x + 11 = 0$ — *Simplify.*
$ax^2 + bx + c = 0$ — *Observe that the coefficient of x^2 is 1.*
$a = 1, \quad b = -7, \quad c = 11$

Practice Problem 1 Place each quadratic equation in standard form and identify the real numbers a, b, and c. Standard form requires that a is a positive number. If $a < 0$, you can multiply each term of the equation by -1 to obtain an equivalent equation.

(a) $2x^2 + 12x - 9 = 0$ **(b)** $7x^2 = 6x - 8$ **(c)** $-x^2 - 6x + 3 = 0$

(d) $10x^2 - 12x = 0$ **(e)** $5x^2 - 4x + 20 = 4x^2 - 3x - 10$ ■

2 *Solving Quadratic Equations of the Form $ax^2 + bx = 0$ by Factoring*

Notice that the terms in quadratic equations of the form $ax^2 + bx = 0$ both have x as a factor. To solve such equations, begin by factoring out the x and any common numerical factor. Thus you may be able to factor out an x or a $5x$. Remember you want to remove the *greatest common factor*. So if $5x$ is a common factor of each term, be sure to remove the factor $5x$.

EXAMPLE 2 Solve. **(a)** $25x^2 - 30x = 0$ **(b)** $7x^2 + 9x - 2 = -8x - 2$

(a) $25x^2 - 30x = 0$

$5x(5x - 6) = 0$ *Factor.*

$5x = 0 \qquad 5x - 6 = 0$ *Set each factor equal to zero.*

$\frac{5x}{5} = \frac{0}{5} \qquad 5x = 6$

$x = 0 \qquad \frac{5x}{5} = \frac{6}{5}$

$x = \frac{6}{5}$

(b) $7x^2 + 9x - 2 = -8x - 2$ *The equation is not in standard form.*

$7x^2 + 9x - 2 + 8x + 2 = 0$ *Add $8x + 2$ to each side.*

$7x^2 + 17x = 0$ *Collect like terms.*

$x(7x + 17) = 0$ *Factor.*

$x = 0 \qquad 7x + 17 = 0$ *Set each factor equal to zero.*

$7x = -17$

$x = -\frac{17}{7}$

One of the solutions to each of the above equations is $x = 0$. This will always be true of equations of this form. One root will be zero. The other root will be a nonzero real number.

Practice Problem 2 Solve. **(a)** $3x^2 + 18x = 0$ **(b)** $2x^2 - 7x - 6 = 4x - 6$ ■

3 *Solving Quadratic Equations of the Form $ax^2 + bx + c = 0$ by Factoring*

If the equation has fractions, clear the fractions by multiplying each term by the least common denominator of all the fractions in the equation. Sometimes the original equation does not look like a quadratic equation. However, it takes the quadratic form once the fractions have been cleared. A note of caution: Always check a possible solution in the original equation. Any solution that would make the denominator of any fraction in the original equation 0 is not a valid solution.

EXAMPLE 3 Solve and check. $8x - 6 + \frac{1}{x} = 0$

$8x - 6 + \frac{1}{x} = 0$	*The equation has a fractional term.*
$x(8x) - (x)(6) + x\left(\frac{1}{x}\right) = x(0)$	*Multiply each term by the LCD = x.*
$8x^2 - 6x + 1 = 0$	*Simplify. The equation is now in standard form.*
$(4x - 1)(2x - 1) = 0$	*Factor.*
$4x - 1 = 0 \quad 2x - 1 = 0$	*Set each factor equal to 0.*
$4x = 1 \quad 2x = 1$	*Solve each equation for x.*
$x = \frac{1}{4} \quad x = \frac{1}{2}$	

Check. Checking fractional roots is more difficult, but you should be able to do it if you work carefully.

$$\text{If } x = \frac{1}{4}: \quad 8\left(\frac{1}{4}\right) - 6 + \frac{1}{\frac{1}{4}} = 2 - 6 + 4 = 0 \quad \left(\text{since } 1 \div \frac{1}{4} = 1 \cdot \frac{4}{1} = 4\right)$$

$$0 = 0 \checkmark$$

$$\text{If } x = \frac{1}{2}: \quad 8\left(\frac{1}{2}\right) - 6 + \frac{1}{\frac{1}{2}} = 4 - 6 + 2 = 0 \quad \left(\text{since } 1 \div \frac{1}{2} = 1 \cdot \frac{2}{1} = 2\right)$$

$$0 = 0 \checkmark$$

Both roots check, so $x = \frac{1}{2}$ and $x = \frac{1}{4}$ are the two roots that satisfy the original equation.

$$8x - 6 + \frac{1}{x} = 0$$

It is also correct to write the answers $x = 0.5$ and $x = 0.25$ in decimal form. This will speed up the checking process, especially if you use a scientific calculator when performing the check.

Practice Problem 3 Solve and check. $3x + 10 - \frac{8}{x} = 0$ ■

TEACHING TIP After explaining Example 3, you may want to do the following problem as an Additional Class Exercise.
Solve for x.

$$\frac{1}{2} - \frac{6}{x} + \frac{10}{x^2} = 0$$

LCD $= 2x^2$

$$2x^2\left(\frac{1}{2}\right) - 2x^2\left(\frac{6}{x}\right) + 2x^2\left(\frac{10}{x^2}\right) = 2x^2(0)$$

$$x^2 - 12x + 20 = 0$$

$$(x - 10)(x - 2) = 0$$

$$x = 10,\ x = 2$$

If the quadratic equation is already in standard form, we examine it to see if it can be factored. If it is possible to factor a quadratic equation, the zero property allows us to find a solution. Recall that if $a \cdot b = 0$, then either $a = 0$ or $b = 0$. Thus we can set each factor of the quadratic equation equal to 0 and solve for x.

EXAMPLE 4 Solve and check. $6x^2 + 7x - 10 = 0$

$(6x - 5)(x + 2) = 0$ *Factor the quadratic equation.*

$6x - 5 = 0 \qquad x + 2 = 0$ *Set each factor equal to zero and solve for x.*

$6x = 5 \qquad x = -2$

$x = \dfrac{5}{6}$

Thus the solutions to the equation are $x = \frac{5}{6}$ and $x = -2$.

Check each solution in the original equation.

$$6\left(\frac{5}{6}\right)^2 + 7\left(\frac{5}{6}\right) - 10 \stackrel{?}{=} 0 \qquad 6(-2)^2 + 7(-2) - 10 \stackrel{?}{=} 0$$

$$6\left(\frac{25}{36}\right) + 7\left(\frac{5}{6}\right) - 10 \stackrel{?}{=} 0 \qquad 6(4) + 7(-2) - 10 \stackrel{?}{=} 0$$

$$\frac{25}{6} + \frac{35}{6} - 10 \stackrel{?}{=} 0 \qquad 24 + (-14) - 10 \stackrel{?}{=} 0$$

$$\frac{60}{6} - 10 \stackrel{?}{=} 0 \qquad 10 - 10 = 0 \checkmark$$

$$10 - 10 = 0 \checkmark$$

Practice Problem 4 Solve and check. $2x^2 + 9x - 18 = 0$ ■

It is important to place the quadratic equation in standard form before factoring. This will sometimes involve removing parentheses and collecting like terms.

EXAMPLE 5 Solve. $x + (x - 6)(x - 2) = 2(x - 1)$

$x + x^2 - 6x - 2x + 12 = 2x - 2$ *Remove parentheses.*

$x^2 - 7x + 12 = 2x - 2$ *Collect like terms.*

$x^2 - 9x + 14 = 0$ *Add $-2x + 2$ to each side.*

$(x - 7)(x - 2) = 0$ *Factor.*

$x = 7, \qquad x = 2$ *Use the zero property.*

The solutions to the equation are $x = 7$ and $x = 2$.

Practice Problem 5 Solve. $-4x + (x - 5)(x + 1) = 5(1 - x)$ ■

EXAMPLE 6 Solve. $\frac{8}{x+1} - \frac{x}{x-1} = \frac{2}{x^2-1}$

The LCD $= (x+1)(x-1)$. We multiply each term by the LCD.

$$(x+1)(x-1)\left(\frac{8}{x+1}\right) - (x+1)(x-1)\left(\frac{x}{x-1}\right) = (x+1)(x-1)\left[\frac{2}{(x+1)(x-1)}\right]$$

$8(x-1) - x(x+1) = 2$ *Simplify.*

$8x - 8 - x^2 - x = 2$ *Remove parentheses.*

$-x^2 + 7x - 8 = 2$ *Collect like terms.*

$-x^2 + 7x - 8 - 2 = 0$ *Add -2 to each side.*

$-x^2 + 7x - 10 = 0$ *Simplify.*

$x^2 - 7x + 10 = 0$ *Multiply each term by -1.*

$(x-5)(x-2) = 0$ *Factor and solve for x.*

$x = 5, \quad x = 2$ *Use the zero property.*

Check. The check is left to the student.

Practice Problem 6 Solve. $\frac{10x+18}{x^2+x-2} = \frac{3x}{x+2} + \frac{2}{x-1}$ ■

4 *Solving Applied Problems That Require the Use of Quadratic Equations*

A manager of a truck delivery service is determining the possible routes he may have to send trucks on during a given day. His delivery service must have the capability to send trucks between any two of the following Texas cities: Austin, Dallas, Tyler, Lufkin, and Houston. He made a rough map of the cities.

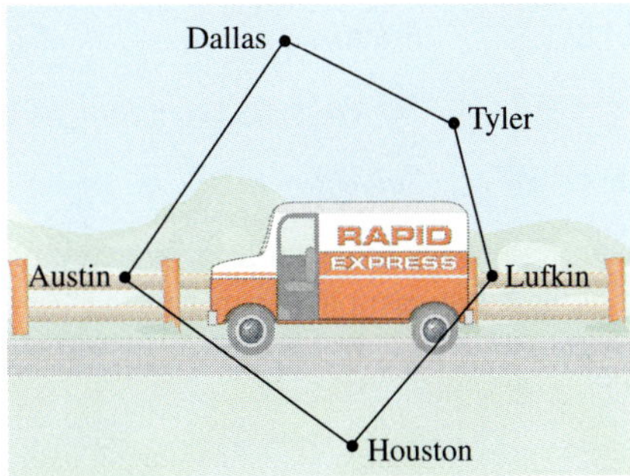

How many possible trips may he have to send trucks on in a given day, if he must be able to send a truck between any two cities? Of course, one possible way is to draw a straight line between every two cities. Let us assume that no three cities lie on a straight line. Therefore, there are always separate lines to connect any two cities. If we draw all these lines, we see there are exactly 10 lines, so the answer is 10 possible truck routes. What if he has to service 18 cities? How many truck routes will there be then?

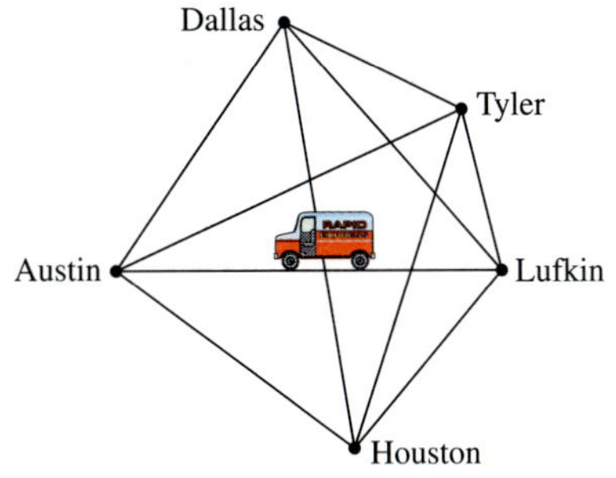

It gets quite difficult (and time consuming) to determine the number of lines using a drawing. In a more advanced math course, it can be proved that the maximum number of truck routes (t) can be found by using the number of cities (n) in the following equation.

$$t = \frac{n^2 - n}{2}$$

Thus, to find the number of truck routes for 18 cities, we merely substitute.

$$\begin{aligned} t &= \frac{(18)^2 - (18)}{2} \\ &= \frac{324 - 18}{2} \\ &= \frac{306}{2} \\ &= 153 \end{aligned}$$

There are a possible 153 truck routes to service 18 cities if no three of those cities lie on one straight line.

EXAMPLE 7 A truck delivery company can handle a maximum of 36 truck routes in one day. How many separate cities can the truck company service?

$$t = \frac{n^2 - n}{2}$$

$$36 = \frac{n^2 - n}{2}$$ *Substitute 36 for the number of truck routes.*

$$72 = n^2 - n$$ *Multiply both sides of the equation by 2.*

$$0 = n^2 - n - 72$$ *Add −72 to each side to obtain standard form.*

$$0 = (n - 9)(n + 8)$$ *Factor.*

$$n - 9 = 0 \qquad n + 8 = 0$$ *Set each factor equal to zero.*

$$n = 9 \qquad n = -8$$ *Solve for n.*

We reject $n = -8$ as meaningless for this problem. We cannot have a negative number of cities. Thus our answer is that the company can service 9 cities.

Check. If we service $n = 9$ cities, does that really result in a maximum number of $t = 36$ truck routes?

$$t = \frac{n^2 - n}{2}$$ *The equation in which t = the number of truck routes and n = the number of cities to be connected.*

$$36 \stackrel{?}{=} \frac{(9)^2 - (9)}{2}$$ *Substitute $t = 36$ and $n = 9$.*

$$36 \stackrel{?}{=} \frac{81 - 9}{2}$$

$$36 \stackrel{?}{=} \frac{72}{2}$$

$$36 = 36 \checkmark$$

Practice Problem 7 How many cities can they service with 28 truck routes? ■

10.1 Exercises

Write each quadratic equation in standard form. Determine the values of a, b, and c.

1. $x^2 + 5x + 6 = 0$
$a = 1, b = 5, c = 6$

2. $x^2 - 7x + 2 = 0$
$a = 1, b = -7, c = 2$

3. $5 - 6x + 8x^2 = 0$
$8x^2 - 6x + 5 = 0$
$a = 8, b = -6, c = 5$

4. $17 + 3x + 5x^2 = 0$
$5x^2 + 3x + 17 = 0$
$a = 5, b = 3, c = 17$

5. $2 = -12x^2 + 3x$
$12x^2 - 3x + 2 = 0$
$a = 12, b = -3, c = 2$

6. $8 = -14x^2 - 9x$
$14x^2 + 9x + 8 = 0$
$a = 14, b = 9, c = 8$

7. $3x^2 - 8x = 0$
$a = 3, b = -8, c = 0$

8. $5x^2 + 15x = 0$
$a = 5, b = 15, c = 0$

9. $x^2 + 15x - 7 = 12x + 8$
$x^2 + 3x - 15 = 0$
$a = 1, b = 3, c = -15$

10. $3x^2 + 10x + 8 = -17x - 2$
$3x^2 + 27x + 10 = 0$
$a = 3, b = 27, c = 10$

11. $2x(x - 3) = (x + 5)(x - 2)$
$x^2 - 9x + 10 = 0$
$a = 1, b = -9, c = 10$

12. $6x(x + 2) = (x + 7)(x - 3)$
$5x^2 + 8x + 21 = 0$
$a = 5, b = 8, c = 21$

Using the factoring method, solve for the roots of each quadratic equation. Be sure to place your equation in standard form before factoring.

13. $18x^2 + 9x = 0$
$9x(2x + 1) = 0$
$x = 0, x = -\frac{1}{2}$

14. $8x^2 - 4x = 0$
$4x(2x - 1) = 0$
$x = 0, x = \frac{1}{2}$

15. $11x^2 = 14x$
$x(11x - 14) = 0$
$x = 0, x = \frac{14}{11}$

16. $8x^2 = 20x$
$4x(2x - 5) = 0$
$x = 0, x = \frac{5}{2}$

17. $7x^2 - 3x = 5x$
$7x^2 - 8x = 0$
$x(7x - 8) = 0$
$x = 0, x = \frac{8}{7}$

18. $9x^2 + 2x = -3x$
$9x^2 + 5x = 0$
$x(9x + 5) = 0$
$x = 0, x = -\frac{5}{9}$

19. $11x^2 - 13x = 8x - 3x^2$
$14x^2 - 21x = 0$
$7x(2x - 3) = 0$
$x = 0, x = \frac{3}{2}$

20. $60x^2 + 14x = 9x^2 - 20x$
$51x^2 + 34x = 0$
$17x(3x + 2) = 0$
$x = 0, x = -\frac{2}{3}$

21. $x^2 + 5x - 24 = 0$
$(x + 8)(x - 3) = 0$
$x = -8, x = 3$

22. $x^2 - 14x + 40 = 0$
$(x - 4)(x - 10) = 0$
$x = 4, x = 10$

23. $2x^2 - 15x - 8 = 0$
$(2x + 1)(x - 8) = 0$
$x = -\frac{1}{2}, x = 8$

24. $2x^2 + 13x - 7 = 0$
$(2x - 1)(x + 7) = 0$
$x = \frac{1}{2}, x = -7$

25. $3x^2 - 13x + 4 = 0$
$(3x - 1)(x - 4) = 0$
$x = \frac{1}{3}, x = 4$

26. $5x^2 - 16x + 3 = 0$
$(5x - 1)(x - 3) = 0$
$x = \frac{1}{5}, x = 3$

27. $x^2 = 9x - 14$
$x^2 - 9x + 14 = 0$
$(x - 2)(x - 7) = 0$
$x = 2, x = 7$

28. $x^2 = 12x - 27$
$x^2 - 12x + 27 = 0$
$(x - 9)(x - 3) = 0$
$x = 9, x = 3$

29. $0 = x^2 + 15x + 50$
$(x + 10)(x + 5) = 0$
$x = -10, x = -5$

30. $0 = x^2 + 14x + 48$
$(x + 6)(x + 8) = 0$
$x = -6, x = -8$

31. $10x^2 + 31x + 15 = 0$
$(5x + 3)(2x + 5) = 0$
$x = -\frac{3}{5}, x = -\frac{5}{2}$

32. $15x^2 + 31x + 10 = 0$
$(5x + 2)(3x + 5) = 0$
$x = -\frac{2}{5}, x = -\frac{5}{3}$

33. $y^2 + 8y + 13 = y + 1$
$y^2 + 7y + 12 = 0$
$(y + 3)(y + 4) = 0$
$y = -3, y = -4$

34. $y^2 + 10y + 17 = 2y + 2$
$y^2 + 8y + 15 = 0$
$(y + 5)(y + 3) = 0$
$y = -5, y = -3$

35. $6y^2 = 11y - 3$
$6y^2 - 11y + 3 = 0$
$(3y - 1)(2y - 3) = 0$
$y = \frac{1}{3}, y = \frac{3}{2}$

36. $8y^2 = 14y - 3$
$8y^2 - 14y + 3 = 0$
$(4y - 1)(2y - 3) = 0$
$y = \frac{1}{4}, y = \frac{3}{2}$

37. $9y^2 - 24y + 16 = 0$
$(3y - 4)(3y - 4) = 0$
$y = \frac{4}{3}$

38. $4y^2 - 12y + 9 = 0$
$(2y - 3)(2y - 3) = 0$
$y = \frac{3}{2}$

39. $n^2 - 10n + 25 = 0$
$(n - 5)(n - 5) = 0$
$n = 5$

40. $n^2 - 12n + 36 = 0$
$(n - 6)(n - 6) = 0$
$n = 6$

41. $n^2 + 5n - 17 = 6n + 3$
$n^2 - n - 20 = 0$
$(n + 4)(n - 5) = 0$
$n = -4, n = 5$

42. $(x - 6)(x + 1) = 8$
$x^2 - 5x - 14 = 0$
$(x + 2)(x - 7) = 0$
$x = -2, x = 7$

43. $3x^2 + 8x - 10 = -6x - 10$
$3x^2 + 14x = 0$
$x(3x + 14) = 0$
$x = 0, x = -\frac{14}{3}$

44. $5x^2 - 7x + 12 = -8x + 12$
$5x^2 + x = 0$
$x(5x + 1) = 0$
$x = 0, x = -\frac{1}{5}$

45. $3x(x - 3) = 6 - 2x$
$3x^2 - 7x - 6 = 0$
$(3x + 2)(x - 3) = 0$
$x = -\frac{2}{3}, x = 3$

46. $2x(x + 6) = 9x + 20$
$2x^2 + 3x - 20 = 0$
$(2x - 5)(x + 4) = 0$
$x = \frac{5}{2}, x = -4$

47. $n^2 + 2n - 11 = 5n + 7$
$n^2 - 3n - 18 = 0$
$(n + 3)(n - 6) = 0$
$n = -3, n = 6$

48. $(x - 3)(x + 1) = 12$
$x^2 - 2x - 15 = 0$
$(x - 5)(x + 3) = 0$
$x = 5, x = -3$

49. $x(x + 8) = 16(x - 1)$
$x^2 - 8x + 16 = 0$
$(x - 4)(x - 4) = 0$
$x = 4$

50. $x(x + 9) = 4(x + 6)$
$x^2 + 5x - 24 = 0$
$(x + 8)(x - 3) = 0$
$x = -8, x = 3$

★**51.** $2x(x + 3) = (3x + 1)(x + 1)$
$x^2 - 2x + 1 = 0$
$(x - 1)(x - 1) = 0$
$x = 1$

★**52.** $2(2x + 1)(x + 1) = x(7x + 1)$
$3x^2 - 5x - 2 = 0$
$(3x + 1)(x - 2) = 0$
$x = -\frac{1}{3}, x = 2$

★**53.** $-3x + (x + 6)(x + 2) = 12 - 3x$
$x^2 + 8x = 0$
$x(x + 8) = 0$
$x = 0, x = -8$

★**54.** $-4x + (x - 6)(x + 2) = 20 - 4x$
$x^2 - 4x - 32 = 0$
$(x - 8)(x + 4) = 0$
$x = 8, x = -4$

Solve each equation. Check your solutions.

55. $\dfrac{4}{x} + \dfrac{3}{x+5} = 2$

$2x^2 + 3x - 20 = 0$
$(2x - 5)(x + 4) = 0$
$x = \dfrac{5}{2}, x = -4$

56. $\dfrac{5}{x+2} = \dfrac{2x-1}{5}$

$2x^2 + 3x - 27 = 0$
$(2x + 9)(x - 3) = 0$
$x = -\dfrac{9}{2}, x = 3$

57. $\dfrac{12}{x-2} = \dfrac{2x-3}{3}$

$2x^2 - 7x - 30 = 0$
$(2x + 5)(x - 6) = 0$
$x = -\dfrac{5}{2}, x = 6$

58. $\dfrac{7}{x} + \dfrac{6}{x-1} = 2$

$2x^2 - 15x + 7 = 0$
$(x - 7)(2x - 1) = 0$
$x = \dfrac{1}{2}, x = 7$

59. $\dfrac{x}{3} + \dfrac{2}{3} = \dfrac{5}{x}$

$x^2 + 2x - 15 = 0$
$(x + 5)(x - 3) = 0$
$x = -5, x = 3$

60. $\dfrac{x}{4} - \dfrac{7}{x} = -\dfrac{3}{4}$

$x^2 + 3x - 28 = 0$
$(x + 7)(x - 4) = 0$
$x = -7, x = 4$

61. $x - 8 = -\dfrac{15}{x}$

$x^2 - 8x + 15 = 0$
$(x - 3)(x - 5) = 0$
$x = 3, x = 5$

62. $x + \dfrac{12}{x} = 8$

$x^2 - 8x + 12 = 0$
$(x - 2)(x - 6) = 0$
$x = 2, x = 6$

63. $\dfrac{24}{x^2 - 4} = 1 + \dfrac{2x-6}{x-2}$

$3x^2 - 2x - 40 = 0$
$(3x + 10)(x - 4) = 0$
$x = -\dfrac{10}{3}, x = 4$

64. $\dfrac{5x+7}{x^2-9} = 1 + \dfrac{2x-8}{x-3}$

$3x^2 - 7x - 40 = 0$
$(3x + 8)(x - 5) = 0$
$x = -\dfrac{8}{3}, x = 5$

To Think About

65. Why can an equation in standard form with $c = 0$ always be solved?
You can always factor out x.

66. Martha solved $(x + 3)(x - 2) = 14$ as follows:

$$x + 3 = 14 \qquad x - 2 = 14$$
$$x = 11 \qquad x = 16$$

Josette said this had to be wrong because these values did not check. Explain what is wrong with Martha's method. You must have the product of two or more factors equal to 0, not 14. Remember the zero factor property.

67. The equation $ax^2 - 7x + c = 0$ has the roots $x = -\frac{3}{2}$ and $x = 6$. Find the values of a and c.

$36a - 42 + c = 0$
$\dfrac{9}{4}a + \dfrac{21}{2} + c = 0$
$36a - \dfrac{9}{4}a - 42 - \dfrac{21}{2} = 0$
$a = \dfrac{14}{9}, c = -14$

Applications

The maximum number of truck routes, t, that are needed to provide direct service to n cities, where no three cities lie in a straight line, is given by the equation

$$t = \frac{n^2 - n}{2}$$

Use this equation in finding the solutions to Problems 68 through 73.

68. How many truck routes will be needed to run service between 20 cities? 190 truck routes

69. How many truck routes will be needed to run service between 22 cities? 231 truck routes

70. A company can handle a maximum of 21 truck routes in one day. How many separate cities can be serviced? 7 cities

71. A company can handle a maximum of 15 truck routes in one day. How many separate cities can be serviced? 6 cities

72. A company can handle a maximum of 45 truck routes in one day. How many separate cities can be serviced? 10 cities

73. A company can handle a maximum of 55 truck routes in one day. How many separate cities can be serviced? 11 cities

The cost in dollars, c, to produce x digital telephone answering machines is given by the equation $c = 0.2(x^2 - 300x + 45{,}000)$. *Use this equation in finding the solutions to Problems 74 to 79.*

74. If the cost to produce the machines is \$4680, what are the two values of x (the two different numbers of machines) that may be produced? 120 or 180 machines may be produced.

75. If the cost to produce the machines is \$4520, what are the two values of x (the two different numbers of machines) that may be produced? 140 or 160 machines may be produced.

76. Take the average of the two answers you obtained in Problem 74. How many machines will be produced? Now use the cost equation to find the cost of producing that many machines. What is significant about your answer? The average is 150 machines. The cost of producing 150 machines is \$4500, which is less than the cost of producing 120 or 180 machines.

77. Take the average of the two answers you obtained in Problem 75. How many machines will be produced? Now use the cost equation to find the cost of producing that many machines. What is significant about your answer? The average is 150 machines. The cost of producing 150 machines is \$4500, which is less than the cost of producing 140 or 160 machines.

78. Find the cost to produce 149 machines and the cost to produce 151 machines. Compare these two costs with your results for Problems 74 and 76. What can you conclude? The cost of producing 149 machines or 151 machines is \$4500.20. This result is less than the results of Problem 74 but more than the results of Problem 76. It appears that the minimum cost of \$4500 occurs when 150 machines are manufactured.

79. Find the cost to produce 148 machines and the cost to produce 152 machines. Compare these two costs with your results for Problems 75 and 77. What can you conclude? The cost of producing 148 machines or 152 machines is \$4500.80. This result is less than the results of Problem 75 but more than the results of Problem 77. It appears that the minimum cost of \$4500 occurs when 150 machines are manufactured.

Cumulative Review Problems

Simplify the following complex fractions.

80. $\dfrac{\dfrac{1}{x} + \dfrac{3}{x-2}}{\dfrac{4}{x^2-4}}$ $\quad \dfrac{(2x-1)(x+2)}{2x}$

81. $\dfrac{\dfrac{1}{2x} + \dfrac{3}{4x}}{\dfrac{7}{x} - \dfrac{3}{2x}}$ $\quad \dfrac{5}{22}$

Simplify.

82. $\dfrac{2x-8}{x^3-16x}$ $\quad \dfrac{2}{x(x+4)}$

Add the fractions.

83. $\dfrac{2x}{3x-5} + \dfrac{2}{3x^2-11x+10}$ $\quad \dfrac{2x^2-4x+2}{3x^2-11x+10}$

10.2 Solutions by the Square Root Property and by Completing the Square

After studying this section, you will be able to:

1. *Solve quadratic equations by the square root property.*
2. *Solve quadratic equations by completing the square.*

1 *Solving Quadratic Equations by the Square Root Property*

Recall that the quadratic equation $x^2 = a$ has two possible solutions, $x = \sqrt{a}$ or $x = -\sqrt{a}$, where a is a nonnegative real number. That is, $x = \pm\sqrt{a}$. This basic idea is called the *square root property*.

> **Square Root Property**
>
> If $x^2 = a$, then $x = \sqrt{a}$ or $x = -\sqrt{a}$, for all nonnegative real numbers a.

EXAMPLE 1 Solve. **(a)** $x^2 = 49$ **(b)** $x^2 = 20$ **(c)** $5x^2 = 125$

(a) $x^2 = 49$
$$x = \pm\sqrt{49}$$
$$x = \pm 7$$

(b) $x^2 = 20$
$$x = \pm\sqrt{20}$$
$$x = \pm 2\sqrt{5}$$

(c) $5x^2 = 125$
$$x^2 = 25$$
$$x = \pm\sqrt{25}$$
$$x = \pm 5$$

Divide both sides by 5 before taking the square root.

Practice Problem 1 Solve. **(a)** $x^2 = 1$ **(b)** $x^2 - 50 = 0$ **(c)** $3x^2 = 81$ ■

EXAMPLE 2 Solve. $3x^2 + 5x = 18 + 5x + x^2$

Simplify the equation by placing all the variable terms on the left and the numerical terms on the right.

$$2x^2 = 18$$
$$x^2 = 9$$
$$x = \pm 3$$

Practice Problem 2 Solve. $4x^2 - 5 = 319$ ■

This technique can also be used if the term that is squared is a binomial.

EXAMPLE 3 Solve. $(x - 2)^2 = 36$

$$(x - 2)^2 = 36$$
$$x - 2 = \pm\sqrt{36} \quad \textit{Take the square root of both sides.}$$
$$x - 2 = \pm 6$$

Note that we now have two equations since $\pm$ indicates plus or minus. We will solve each equation.

$$x - 2 = +6 \qquad x - 2 = -6$$
$$x = 8 \qquad x = -4$$

The two roots of the equation are $x = 8$ and $x = -4$.

Practice Problem 3 Solve. $(2x + 3)^2 = 25$ ■

Sometimes when we use this method we obtain an irrational number. This will occur if the number on the right side of the equation is not a perfect square.

TEACHING TIP After explaining Example 4, ask students as an Additional Class Exercise to solve for x and simplify the answer for the following equation.

$$(4x - 3)^2 = 50$$

Sol. $x = \dfrac{3 \pm 5\sqrt{2}}{4}$

EXAMPLE 4 Solve. $(3x + 1)^2 = 8$

$$(3x + 1)^2 = 8$$

$$3x + 1 = \pm\sqrt{8} \qquad \textit{Take the square root of both sides.}$$

$$3x + 1 = \pm 2\sqrt{2} \qquad \textit{Simplify as much as possible.}$$

Now we must solve the two equations expressed by the plus or minus statement.

$$3x + 1 = +2\sqrt{2} \qquad 3x + 1 = -2\sqrt{2}$$

$$3x = -1 + 2\sqrt{2} \qquad 3x = -1 - 2\sqrt{2}$$

$$x = \frac{-1 + 2\sqrt{2}}{3} \qquad x = \frac{-1 - 2\sqrt{2}}{3}$$

The roots of this quadratic equation are irrational numbers. They are

$$x = \frac{-1 + 2\sqrt{2}}{3} \quad \text{and} \quad x = \frac{-1 - 2\sqrt{2}}{3}$$

We cannot simplify these roots further, so we have them in this form.

Practice Problem 4 Solve. $(2x - 3)^2 = 12$ ■

2 *Solving Quadratic Equations by Completing the Square*

If a quadratic equation is not in a form where we can use the square root property, we can rewrite the equation by a method called *completing the square*. That is, we try to rewrite the equation with a perfect square on the left side so that we have $(ax + b)^2 = c$.

EXAMPLE 5 Solve. **(a)** $x^2 - 2x = 8$ **(b)** $x^2 + 12x = 4$

(a) Think, what can we add to $x^2 - 2x$ to make a perfect square. Let's draw a picture of $x^2 - 2x$. Begin with $(x \quad)(x \quad)$. To have a $-2x$ as a middle term, the binomial must be $(x - 1)(x - 1)$. Now the picture is complete and we can complete the square.

$$x^2 - 2x = 8$$

$$(x - 1)(x - 1) \qquad \textit{We need a } +1 \textit{ to complete the square.}$$

$$x^2 - 2x + 1 = 8 + 1 \qquad \textit{Be sure to add } +1 \textit{ to both sides of the equation.}$$

$$(x - 1)^2 = 9 \qquad \textit{Now we have a perfect square on the left.}$$

$$x - 1 = \pm 3$$

$$x - 1 = 3 \qquad x - 1 = -3$$

$$x = 4 \qquad x = -2$$

The two roots are $x = 4$ and $x = -2$.

(b) Draw a picture of $x^2 + 12x$. Fill in the pieces you know.

$$x^2 + 12x \qquad = 4$$

$$(x + 6)(x + 6) = 4 \qquad \textit{We need a } +36 \textit{ to complete the square.}$$

$$x^2 + 12x + 36 = 4 + 36 \qquad \textit{Add } +36 \textit{ to both sides of the equation.}$$

$$(x + 6)^2 = 40 \qquad \textit{Write the left side as a perfect square.}$$

$$x + 6 = \pm\sqrt{40} \qquad \textit{Solve for } x.$$

$$x + 6 = +2\sqrt{10} \qquad\qquad x + 6 = -2\sqrt{10}$$

$$x = -6 + 2\sqrt{10} \qquad\qquad x = -6 - 2\sqrt{10}$$

The two roots are $x = -6 + 2\sqrt{10}$ and $x = -6 - 2\sqrt{10}$.

Practice Problem 5 Solve. **(a)** $x^2 - 8x = 20$ **(b)** $x^2 + 10x = 3$ ■

Completing the square is a little more difficult when the coefficient of x is an odd number. Let's see what happens.

EXAMPLE 6 Solve by completing the square. $x^2 - 3x - 1 = 0$

Write the equation so that all the x terms are on the left and the numbers are on the right.

$$x^2 - 3x \qquad = 1$$

$$(x \quad)(x \quad)$$

The missing number when added to itself must be -3.

Since $-\frac{3}{2} + (-\frac{3}{2}) = -\frac{6}{2} = -3$, *the missing number is* $-\frac{3}{2}$.

Complete the picture.

$$\left(x - \frac{3}{2}\right)\left(x - \frac{3}{2}\right)$$

$$x^2 - 3x + \frac{9}{4} = 1 + \frac{9}{4} \qquad \textit{Complete the square } (-\tfrac{3}{2})(-\tfrac{3}{2}) = \tfrac{9}{4}.$$

$$\left(x - \frac{3}{2}\right)^2 = \frac{13}{4} \qquad \textit{Solve for } x.$$

$$x - \frac{3}{2} = \pm\sqrt{\frac{13}{4}}$$

$$x - \frac{3}{2} = \pm\frac{\sqrt{13}}{2}$$

$$x - \frac{3}{2} = +\frac{\sqrt{13}}{2} \qquad\qquad x - \frac{3}{2} = -\frac{\sqrt{13}}{2}$$

$$x = \frac{3}{2} + \frac{\sqrt{13}}{2} \qquad\qquad x = \frac{3}{2} - \frac{\sqrt{13}}{2}$$

$$x = \frac{3 + \sqrt{13}}{2} \qquad\qquad x = \frac{3 - \sqrt{13}}{2}$$

The two roots are $x = \dfrac{3 \pm \sqrt{13}}{2}$.

Practice Problem 6 Solve by completing the square. $x^2 - 5x = 7$ ■

If the coefficient a is not 1, we divide all terms of the equation by a so that the coefficient of the squared variable will be 1.

TEACHING TIP After explaining Example 7, provide the following Additional Class Exercise.
Solve for x. $2x^2 - 4x - 5 = 0$

Sol. $x = \dfrac{2 \pm \sqrt{14}}{2}$

EXAMPLE 7 Solve by completing the square. $4y^2 + 4y - 3 = 0$

$$4y^2 + 4y = 3 \quad \textit{Place the numerical term on the right.}$$

$$y^2 + y = \frac{3}{4} \quad \textit{Divide all terms by 4.}$$

$$\left(y + \frac{1}{2}\right)\left(y + \frac{1}{2}\right) \quad \textit{Think } \tfrac{1}{2} + \tfrac{1}{2} = 1. \textit{ The middle term is } 1y.$$

$$y^2 + y + \frac{1}{4} = \frac{3}{4} + \frac{1}{4} \quad \textit{Complete the square } (\tfrac{1}{2})^2 = \tfrac{1}{4}.$$

$$\left(y + \frac{1}{2}\right)^2 = 1 \quad \textit{Solve for } y.$$

$$y + \frac{1}{2} = \pm\sqrt{1}$$

$$y + \frac{1}{2} = \pm 1$$

$$y + \frac{1}{2} = 1 \qquad y + \frac{1}{2} = -1$$

$$y = \frac{2}{2} - \frac{1}{2} \qquad y = -\frac{2}{2} - \frac{1}{2}$$

$$y = \frac{1}{2} \qquad y = -\frac{3}{2}$$

Thus the two roots are $y = \frac{1}{2}$ and $y = -\frac{3}{2}$.

Check.

$$4\left(\frac{1}{2}\right)^2 + 4\left(\frac{1}{2}\right) - 3 \stackrel{?}{=} 0$$

$$4\left(\frac{1}{4}\right) + 4\left(\frac{1}{2}\right) - 3 \stackrel{?}{=} 0$$

$$1 + 2 - 3 = 0 \ \checkmark$$

$$4\left(-\frac{3}{2}\right)^2 + 4\left(-\frac{3}{2}\right) - 3 \stackrel{?}{=} 0$$

$$4\left(\frac{9}{4}\right) + 4\left(-\frac{3}{2}\right) - 3 \stackrel{?}{=} 0$$

$$9 - 6 - 3 = 0 \ \checkmark$$

Both values check.

This method of completing the square will enable us to solve any quadratic equation that has real roots. It is usually faster, however, to factor the quadratic equation if the polynomial is factorable.

Practice Problem 7 Solve by factoring. $4y^2 + 4y - 3 = 0$ ■

10.2 Exercises

Solve by using the square root property.

1. $x^2 = 49$ $x = \pm 7$

2. $x^2 = 100$ $x = \pm 10$

3. $x^2 = 125$ $x = \pm 5\sqrt{5}$

4. $x^2 = 48$ $x = \pm 4\sqrt{3}$

5. $x^2 - 75 = 0$ $x = \pm 5\sqrt{3}$

6. $x^2 - 40 = 0$ $x = \pm 2\sqrt{10}$

7. $3x^2 = 75$ $x = \pm 5$

8. $4x^2 = 64$ $x = \pm 4$

9. $6x^2 = 120$ $x = \pm 2\sqrt{5}$

10. $3x^2 = 150$ $x = \pm 5\sqrt{2}$

11. $4x^2 - 12 = 0$ $x = \pm\sqrt{3}$

12. $5x^2 - 15 = 0$ $x = \pm\sqrt{3}$

13. $2x^2 - 196 = 0$
$x = \pm 7\sqrt{2}$

14. $5x^2 - 140 = 0$
$x = \pm 2\sqrt{7}$

15. $x^2 - 13 = 36$
$x^2 = 49$
$x = \pm 7$

16. $x^2 - 27 = 54$
$x^2 = 81$
$x = \pm 9$

17. $5x^2 + 13 = 73$
$5x^2 = 60$
$x = \pm 2\sqrt{3}$

18. $6x^2 + 5 = 53$
$x^2 = 8$
$x = \pm 2\sqrt{2}$

19. $13x^2 + 17 = 82$
$13x^2 = 65$
$x = \pm\sqrt{5}$

20. $11x^2 + 14 = 47$
$x^2 = 3$
$x = \pm\sqrt{3}$

21. $(x - 3)^2 = 5$
$x - 3 = \pm\sqrt{5}$
$x = 3 \pm \sqrt{5}$

22. $(x + 2)^2 = 3$
$x + 2 = \pm\sqrt{3}$
$x = -2 \pm \sqrt{3}$

23. $(x + 6)^2 = 4$
$x + 6 = \pm 2$
$x = -8, x = -4$

24. $(x - 5)^2 = 36$
$x - 5 = \pm 6$
$x = 11, x = -1$

25. $(2x + 5)^2 = 2$
$2x + 5 = \pm\sqrt{2}$
$x = \dfrac{-5 \pm \sqrt{2}}{2}$

26. $(3x - 4)^2 = 6$
$3x - 4 = \pm\sqrt{6}$
$x = \dfrac{4 \pm \sqrt{6}}{3}$

27. $(3x - 1)^2 = 7$
$3x - 1 = \pm\sqrt{7}$
$x = \dfrac{1 \pm \sqrt{7}}{3}$

28. $(2x + 3)^2 = 10$
$2x + 3 = \pm\sqrt{10}$
$x = \dfrac{-3 \pm \sqrt{10}}{2}$

29. $(7x + 2)^2 = 12$
$7x + 2 = \pm 2\sqrt{3}$
$x = \dfrac{-2 \pm 2\sqrt{3}}{7}$

30. $(8x + 5)^2 = 20$
$8x + 5 = \pm 2\sqrt{5}$
$x = \dfrac{-5 \pm 2\sqrt{5}}{8}$

31. $(5x - 3)^2 = 75$
$5x - 3 = \pm 5\sqrt{3}$
$x = \dfrac{3 \pm 5\sqrt{3}}{5}$

32. $(7x - 2)^2 = 98$
$7x - 2 = \pm 7\sqrt{2}$
$x = \dfrac{2 \pm 7\sqrt{2}}{7}$

Solve for x. Round your answer to the nearest thousandth.

33. $x^2 - 19.263 = 0$
$x = \pm 4.389$

34. $x^2 - 22.986 = 0$
$x = \pm 4.794$

35. $3.120x^2 - 7.986 = 0$
$x = \pm 1.600$

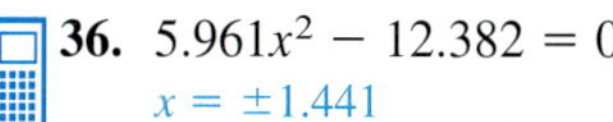
36. $5.961x^2 - 12.382 = 0$
$x = \pm 1.441$

Solve each quadratic equation by completing the square.

37. $x^2 + 14x = 15$
$(x + 7)^2 = 64$
$x = 1, x = -15$

38. $x^2 - 2x = 5$
$(x - 1)^2 = 6$
$x = 1 \pm \sqrt{6}$

39. $x^2 - 4x = 1$
$(x - 2)^2 = 5$
$x = 2 \pm \sqrt{5}$

40. $x^2 + 4x = 12$
$x^2 + 4x + 4 = 12 + 4$
$(x + 2)^2 = 16$
$x = 2, x = -6$

41. $x^2 + 6x + 7 = 0$
$x^2 + 6x + 9 = -7 + 9$
$(x + 3)^2 = 2$
$x = -3 \pm \sqrt{2}$

42. $x^2 - 12x - 4 = 0$
$x^2 - 12x + 36 = 4 + 36$
$(x - 6)^2 = 40$
$x = 6 \pm 2\sqrt{10}$

43. $x^2 - 10x - 4 = 0$

$x^2 - 10x + 25 = 4 + 25$
$(x - 5)^2 = 29$
$x = 5 \pm \sqrt{29}$

44. $x^2 + 20x + 72 = 0$

$x^2 + 20x + 100 = -72 + 100$
$(x + 10)^2 = 28$
$x = -10 \pm 2\sqrt{7}$

45. $x^2 + 3x = 0$

$x + 3x + \frac{9}{4} = \frac{9}{4}$
$\left(x + \frac{3}{2}\right)^2 = \frac{9}{4}$
$x = 0, x = -3$

46. $x^2 + 5x = 3$

$x^2 + 5x + \frac{25}{4} = 3 + \frac{25}{4}$
$\left(x + \frac{5}{2}\right)^2 = \frac{37}{4}$
$x = \frac{-5 \pm \sqrt{37}}{2}$

47. $x^2 + x - 30 = 0$

$x^2 + x + \frac{1}{4} = 30 + \frac{1}{4}$
$\left(x + \frac{1}{2}\right)^2 = \frac{121}{4}$
$x = 5, x = -6$

48. $x^2 - x - 56 = 0$

$x^2 - x + \frac{1}{4} = 56 + \frac{1}{4}$
$\left(x - \frac{1}{2}\right)^2 = \frac{225}{4}$
$x = 8, x = -7$

49. $4x^2 - 8x + 3 = 0$

$x^2 - 2x + 1 = -\frac{3}{4} + 1$
$(x - 1)^2 = \frac{1}{4}$
$x = \frac{3}{2}, x = \frac{1}{2}$

50. $2x^2 - 3x = 9$

$x^2 - \frac{3}{2}x + \frac{9}{16} = \frac{9}{2} + \frac{9}{16}$
$\left(x - \frac{3}{4}\right)^2 = \frac{81}{16}$
$x = 3, x = -\frac{3}{2}$

51. $2x^2 + 10x + 11 = 0$

$x^2 + 5x + \frac{25}{4} = -\frac{11}{2} + \frac{25}{4}$
$\left(x + \frac{5}{2}\right)^2 = \frac{3}{4}$
$x = \frac{-5 \pm \sqrt{3}}{2}$

To Think About

52. If an equation in standard form has $b = 0$, it reduces to $ax^2 - c = 0$. It is easy to find the solutions, if they exist. Explain how.

$ax^2 = c$
$x^2 = \frac{c}{a}, x = \pm\sqrt{\frac{c}{a}}$

53. Let b represent a constant value. Solve for x by completing the square: $x^2 + bx - 7 = 0$.

$x^2 + bx + \frac{b^2}{4} = 7 + \frac{b^2}{4}$ $\quad \left(x + \frac{b}{2}\right)^2 = \frac{28 + b^2}{4}$ $\quad x = \frac{-b \pm \sqrt{28 + b^2}}{2}$

54. Solve for x in terms of b and c. $x^2 + bx + c = 0$

$x^2 + bx + \frac{b^2}{4} = \frac{b^2}{4} - c$ $\quad \left(x + \frac{b}{2}\right)^2 = \frac{b^2 - 4c}{4}$ $\quad x = \frac{-b \pm \sqrt{b^2 - 4c}}{2}$

Cumulative Review Problems

55. Solve. $3a - 5b = 8$

$5a - 7b = 8$

a = −4, b = −4

56. Solve. $\frac{x - 4}{x - 2} - \frac{1}{x} = \frac{x - 3}{x^2 - 2x}$

$x = 1, x = 5$

57. While hiking in the Grand Canyon, Will had to use his hunting knife to clear a fallen branch from the path. Find the pressure on the blade of his knife if the knife is 8 in. long and sharpened to an edge 0.01 in. wide and he applies a force of 16 pounds. Use the formula $P = F/A$ where P is the pressure on the blade of the knife, F is the force in pounds placed on the knife, and A is the area exposed to an object by the sharpened edge of the knife.

The pressure is 200 pounds per square inch.

58. Darlene inspected a sample of 100 new steel belted radial tires. 13 of the tires had defects in workmanship, while 9 of the tires had defects in materials. Of those defective tires it was found that 6 had both defects in workmanship and in materials. In all how many of the 100 tires had any kind of defect? (Assume that the sample consists of only tires with defects in materials or workmanship.) A total of 16 tires were defective. 6 had both kinds of defects, 7 had only defects in workmanship, and 3 had only defects in materials.

10.3 Solutions by the Quadratic Formula

After studying this section, you will be able to:

1. *Solve a quadratic equation with real roots using the quadratic formula.*
2. *Find a decimal approximation to the real roots of a quadratic equation.*
3. *Determine if a given quadratic equation has no real solution.*

1 *Solving a Quadratic Equation by the Quadratic Formula*

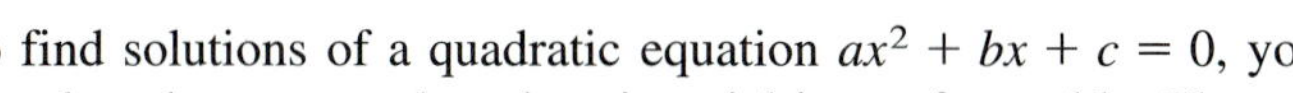

To find solutions of a quadratic equation $ax^2 + bx + c = 0$, you can factor or you can complete the square when the trinomial is not factorable. The method of completing the square can be used on the general equation, and we can derive a general formula for the solution of any quadratic equation. This is called the *quadratic formula.*

> **Quadratic Formula**
>
> The roots of any quadratic equation in the form $ax^2 + bx + c = 0$, where a, b, and c are real numbers and $a \neq 0$, are
>
> $$x = \frac{-b \pm \sqrt{b^2 - 4ac}}{2a}$$

All you will need to know are the values of a, b, and c to solve a quadratic equation using the quadratic formula.

TEACHING TIP Remind students to memorize the quadratic formula perfectly. Suggest that they memorize it over three separate days and write it down each day until they have it ''absolutely perfect.'' A lot of students ''think'' they have it memorized but they will mix up one little detail—which of course makes every answer they obtain by the quadratic formula incorrect.

EXAMPLE 1 Solve by the quadratic formula. $3x^2 + 10x + 7 = 0$

In our given equation, $3x^2 + 10x + 7 = 0$, we have

$a = 3$ (the coefficient of x^2) $b = 10$ (the coefficient of x)

$c = 7$ (the constant term)

$$x = \frac{-b \pm \sqrt{b^2 - 4ac}}{2a} = \frac{-10 \pm \sqrt{(10)^2 - 4(3)(7)}}{2(3)}$$ *Write the quadratic formula and substitute values for a, b, and c.*

$$= \frac{-10 \pm \sqrt{100 - 84}}{6}$$ *Simplify.*

$$= \frac{-10 \pm \sqrt{16}}{6} = \frac{-10 \pm 4}{6}$$

$$x = \frac{-10 + 4}{6} = \frac{-6}{6} = -1$$ *Using the positive sign.*

$$x = \frac{-10 - 4}{6} = \frac{-14}{6} = -\frac{7}{3}$$ *Using the negative sign.*

Thus the two solutions are $x = -1$ and $x = -\frac{7}{3}$.

[*Note:* Here we obtain rational roots. We would obtain the same answer by factoring $3x^2 + 10x + 7 = 0$ into $(3x + 7)(x + 1) = 0$ and setting each factor = 0.]

Practice Problem 1 Solve by using the quadratic formula. $x^2 - 7x + 6 = 0$ ■

Often the roots will be irrational numbers. Sometimes these roots can be simplified. You should always leave your answer in simplest form.

TEACHING TIP Take some time to go over carefully how the fraction is simplified and reduced in Example 2. Then ask students to do the following problem as an Additional Class Exercise.

$$3x^2 - 2x - 4 = 0$$

Sol. $x = \dfrac{1 \pm \sqrt{13}}{3}$

EXAMPLE 2 Solve by the quadratic formula. $2x^2 - 4x - 9 = 0$

$$a = 2, \quad b = -4, \quad c = -9$$

$$x = \frac{-b \pm \sqrt{b^2 - 4ac}}{2a} = \frac{-(-4) \pm \sqrt{(-4)^2 - 4(2)(-9)}}{2(2)}$$

$$= \frac{4 \pm \sqrt{16 + 72}}{4} = \frac{4 \pm \sqrt{88}}{4}$$

Notice that we can simplify $\sqrt{88}$, so

$$= \frac{4 \pm 2\sqrt{22}}{4} = \frac{2(2 \pm \sqrt{22})}{4} = \frac{\overset{1}{\cancel{2}}(2 \pm \sqrt{22})}{\underset{2}{\cancel{4}}}$$

Be careful. Here we were able to divide out by 2 because 2 was a factor of every term.

$$x = \frac{2 \pm \sqrt{22}}{2}$$

Practice Problem 2 Solve. $3x^2 - 8x + 3 = 0$ ■

The quadratic equation *must* be written in the standard form $ax^2 + bx + c = 0$ *before* the quadratic equation can be used. Several algebraic steps may be needed to accomplish that objective.

EXAMPLE 3 Solve by the quadratic formula. $2x^2 = 1 - \frac{7}{3}x$

$2x^2 = 1 - \dfrac{7}{3}x$ *The equation is not in standard form.*

$3(2x^2) = 3(1) - 3\left(\dfrac{7}{3}x\right)$ *Multiply each term by the LCD of 3.*

$6x^2 = 3 - 7x$ *Simplify.*

$6x^2 + 7x - 3 = 0$ *Add 7x and −3 to both sides.*

$a = 6, \quad b = 7, \quad c = -3$

$x = \dfrac{-7 \pm \sqrt{(7)^2 - 4(6)(-3)}}{2(6)}$ *Substitute $a = 6$, $b = 7$, $c = -3$ into the quadratic formula.*

$$= \frac{-7 \pm \sqrt{49 + 72}}{12}$$

$$= \frac{-7 \pm \sqrt{121}}{12} = \frac{-7 \pm 11}{12}$$

Thus we have $x = -\frac{18}{12} = -\frac{3}{2}$ or $x = \frac{4}{12} = \frac{1}{3}$.

Practice Problem 3 Solve. $2x^2 = 3x + 1$ ■

2 *Finding a Decimal Approximation to the Real Roots of a Quadratic Equation*

EXAMPLE 4 Find the roots of $3x^2 - 5x = 7$. Approximate your answer to the nearest thousandth.

$3x^2 - 5x - 7 = 0$ *Place the equation in standard form.*

$$a = 3, \quad b = -5, \quad c = -7$$

$$x = \frac{-(-5) \pm \sqrt{(-5)^2 - 4(3)(-7)}}{2(3)} = \frac{5 \pm \sqrt{25 + 84}}{6}$$

$$= \frac{5 \pm \sqrt{109}}{6}$$

Using most scientific calculators, we can find $\frac{5 + \sqrt{109}}{6}$ with these keystrokes.

5 [+] 109 [√] [=] [÷] 6 [=] 2.5733844

Rounding to the nearest thousandth, we have $x \approx 2.573$.

To find $\frac{5 - \sqrt{109}}{6}$ we have

5 [−] 109 [√] [=] [÷] 6 [=] -0.9067178

Rounding to the nearest thousandth, we have $x \approx -0.907$.

If you do not have a calculator with a square root key, look up $\sqrt{109}$ in the square root table. $\sqrt{109} \approx 10.440$. Using this result, we have

$$x \approx \frac{5 + 10.440}{6} = \frac{15.440}{6} = 2.573 \quad \text{(rounded to the nearest thousandth)}$$

and

$$x \approx \frac{5 - 10.440}{6} = \frac{-5.440}{6} = -0.907 \quad \text{(rounded to nearest thousandth)}$$

If you have a graphing calculator, there is an alternate method to finding the decimal approximation of the roots of a quadratic equation. In Section 10.4 you will cover graphing quadratic equations. You will discover that finding the roots of a quadratic equation of the form $ax^2 + bx + c = 0$ is equivalent to finding the values of x on the graph of $y = ax^2 + bx + c$ where $y = 0$. These may be approximated quite quickly with a graphing calculator.

Practice Problem 4 Find the roots of $2x^2 = 13x + 5$. Approximate your answer to the nearest thousandth. ■

TEACHING TIP It is actually possible to use the quadratic formula if a, b, and c are fractions, that is, rational numbers. However, the resulting calculations are usually more difficult. Therefore, we stress as very important that students should always multiply each term by the LCD to eliminate the fractions.

3 Quadratic Equations with No Real Solutions

EXAMPLE 5 Solve by the quadratic formula. $2x^2 + 5 = -3x$

The equation is not in standard form.

$2x^2 + 5 = -3x$ *Add 3x to both sides.*

$2x^2 + 3x + 5 = 0$ *We can now find a, b, and c.*

$a = 2, \quad b = 3, \quad c = 5$ *Substitute these values into the quadratic formula.*

$$x = \frac{-3 \pm \sqrt{(3)^2 - 4(2)(5)}}{2(2)} = \frac{-3 \pm \sqrt{9 - 40}}{4} = \frac{-3 \pm \sqrt{-31}}{4}$$

There is no real number that is $\sqrt{-31}$. Since we are using only real numbers in this book, we say there is no solution to the problem. (*Note:* In more advanced math courses, these types of numbers, called complex numbers, will be studied.)

Practice Problem 5 Solve. $5x^2 + 2x = -3$ ■

In any given quadratic equation, we can tell if the roots are not real numbers. Look at the quadratic formula.

$$\frac{-b \pm \sqrt{b^2 - 4ac}}{2a}$$

The expression under the radical sign is called the *discriminant.*

$$b^2 - 4ac \text{ is the discriminant.}$$

If the discriminant is a negative number, the roots are not real numbers. There is no real number solution to the equation.

EXAMPLE 6 Determine whether or not $3x^2 = 5x - 4$ has real number solution(s).

First, we place the equation in standard form. Then we need only check the discriminant.

$3x^2 = 5x - 4$

$3x^2 - 5x + 4 = 0$

$a = 3, \quad b = -5, \quad c = 4$ *Substitute these values for a, b, and c into the discriminant and evaluate.*

$$b^2 - 4ac = (-5)^2 - 4(3)(4) = 25 - 48 = -23$$

The discriminant is negative. Thus $3x^2 = 5x - 4$ has no real number solution(s).

Practice Problem 6 Determine whether or not each equation has real number solution(s).

(a) $5x^2 = 3x + 2$ **(b)** $2x^2 - 5 = 4x$ ■

10.3 Exercises

Solve by the quadratic formula. If no real roots are possible, say so.

1. $x^2 + 4x + 1 = 0$

$x = \dfrac{-4 \pm \sqrt{16 - 4}}{2} = -2 \pm \sqrt{3}$

2. $x^2 - 2x - 4 = 0$

$x = \dfrac{2 \pm \sqrt{4 + 16}}{2} = 1 \pm \sqrt{5}$

3. $x^2 - 3x - 8 = 0$

$x = \dfrac{3 \pm \sqrt{9 + 32}}{2} = \dfrac{3 \pm \sqrt{41}}{2}$

4. $x^2 + 9x + 7 = 0$

$x = \dfrac{-9 \pm \sqrt{81 - 28}}{2} = \dfrac{-9 \pm \sqrt{53}}{2}$

5. $2x^2 - 7x - 9 = 0$

$x = \dfrac{7 \pm \sqrt{49 + 72}}{4}; x = -1, x = \dfrac{9}{2}$

6. $2x^2 - 6x + 1 = 0$

$x = \dfrac{6 \pm \sqrt{36 - 8}}{4} = \dfrac{3 \pm \sqrt{7}}{2}$

7. $5x^2 + x - 1 = 0$

$x = \dfrac{-1 \pm \sqrt{1 + 20}}{10}$

$x = \dfrac{-1 \pm \sqrt{21}}{10}$

8. $2x^2 + 12x - 5 = 0$

$x = \dfrac{-12 \pm \sqrt{144 + 40}}{4}$

$x = \dfrac{-6 \pm \sqrt{46}}{2}$

9. $5x - 1 = 2x^2$

$x = \dfrac{5 \pm \sqrt{25 - 8}}{4} = \dfrac{5 \pm \sqrt{17}}{4}$

10. $5x^2 = -3x + 2$

$x = \dfrac{-3 \pm \sqrt{9 + 40}}{10}$

$x = \dfrac{2}{5}, x = -1$

11. $6x^2 - 3x = 1$

$x = \dfrac{3 \pm \sqrt{9 + 24}}{12}$

$x = \dfrac{3 \pm \sqrt{33}}{12}$

12. $3 = 7x - 4x^2$

$x = \dfrac{7 \pm \sqrt{49 - 48}}{8}$

$x = 1, x = \dfrac{3}{4}$

13. $x + \dfrac{3}{2} = 3x^2$

$0 = 6x^2 - 2x - 3$

$x = \dfrac{2 \pm \sqrt{4 + 72}}{12}$

$x = \dfrac{1 \pm \sqrt{19}}{6}$

14. $\dfrac{2}{3}x^2 - x = \dfrac{5}{2}$

$4x^2 - 6x - 15 = 0$

$x = \dfrac{6 \pm \sqrt{36 + 240}}{8}$

$x = \dfrac{3 \pm \sqrt{69}}{4}$

15. $9x^2 - 6x = 2$

$9x^2 - 6x - 2 = 0$

$x = \dfrac{6 \pm \sqrt{36 + 72}}{18}$

$x = \dfrac{1 \pm \sqrt{3}}{3}$

16. $6x^2 = 13x + 5$

$6x^2 - 13x - 5 = 0$

$x = \dfrac{13 \pm \sqrt{169 + 120}}{12}$

$x = \dfrac{5}{2}, x = -\dfrac{1}{3}$

17. $4x^2 + 3x + 2 = 0$

$x = \dfrac{-3 \pm \sqrt{9 - 32}}{8}$

no real solution

18. $\dfrac{x}{2} + \dfrac{2}{x} = \dfrac{5}{2}$

$x^2 - 5x + 4 = 0$

$x = \dfrac{5 \pm \sqrt{25 - 16}}{2}$

$x = 4, x = 1$

19. $\dfrac{x}{3} + \dfrac{2}{x} = \dfrac{7}{3}$

$x^2 - 7x + 6 = 0$

$x = \dfrac{7 \pm \sqrt{49 - 24}}{2}$

$x = 6, x = 1$

20. $3x^2 = 2x - 5$

$3x^2 - 2x + 5 = 0$

$x = \dfrac{2 \pm \sqrt{4 - 60}}{6}$

no real solution

21. $5y^2 = 3 - 7y$

$5y^2 + 7y - 3 = 0$

$y = \dfrac{-7 \pm \sqrt{49 + 60}}{10}$

$y = \dfrac{-7 \pm \sqrt{109}}{10}$

22. $5y - 6 = -2y^2$

$2y^2 + 5y - 6 = 0$

$y = \dfrac{-5 \pm \sqrt{25 + 48}}{4}$

$y = \dfrac{-5 \pm \sqrt{73}}{4}$

23. $\dfrac{1}{3}m^2 = \dfrac{1}{2}m + \dfrac{3}{2}$

$2m^2 - 3m - 9 = 0$

$m = \dfrac{3 \pm \sqrt{9 + 72}}{4}$

$m = 3, m = -\dfrac{3}{2}$

24. $\dfrac{3}{5}m^2 - m = \dfrac{2}{5}$

$3m^2 - 5m - 2 = 0$

$m = \dfrac{5 \pm \sqrt{25 + 24}}{6}$

$m = 2, m = -\dfrac{1}{3}$

Solve for x by using the quadratic formula. Round your answer to the nearest thousandth.

25. $1.04x^2 - 17.982x + 74.108 = 0$ $x = 10.511, x = 6.779$

26. $0.986x^2 + 9.685x - 4.231 = 0$ $x = 0.419, x = -10.241$

Use the quadratic formula to find the roots of the following equations. Find a decimal approximation to the nearest thousandth.

27. $x^2 + 3x - 7 = 0$

$x = \dfrac{-3 \pm \sqrt{9 + 28}}{2}$

$x = 1.541, x = -4.541$

28. $x^2 + 5x - 2 = 0$

$x = \dfrac{-5 \pm \sqrt{25 + 8}}{2}$

$x = 0.372, x = -5.372$

29. $3x^2 - 2x - 3 = 0$

$x = \dfrac{2 \pm \sqrt{4 + 36}}{6}$

$x = 1.387, x = -0.721$

30. $2x^2 + 3x - 3 = 0$

$x = \dfrac{-3 \pm \sqrt{9 + 24}}{4}$

$x = 0.686, x = -2.186$

31. $5x^2 + 10x + 1 = 0$

$x = \dfrac{-5 \pm 2\sqrt{5}}{5}$

$x = -1.894, x = -0.106$

32. $2x^2 + 9x + 2 = 0$

$x = \dfrac{-9 \pm \sqrt{65}}{4}$

$x = -4.266, x = -0.234$

Mixed Practice

Solve for x. Choose the method you feel works best.

33. $t^2 + 1 = \dfrac{13}{6}t$

$6t^2 - 13t + 6 = 0$

$t = \dfrac{3}{2}, t = \dfrac{2}{3}$

34. $2t^2 - 1 = \dfrac{7}{3}t$

$6t^2 - 7t - 3 = 0$

$t = \dfrac{3}{2}, t = -\dfrac{1}{3}$

35. $3(s^2 + 1) = 10s$

$3s^2 - 10s + 3 = 0$

$s = 3, s = \dfrac{1}{3}$

36. $3 + p(p + 2) = 18$

$p^2 + 2p - 15 = 0$

$p = 3, p = -5$

37. $(p - 2)(p + 1) = 3$

$p^2 - p - 5 = 0$

$p = \dfrac{1 \pm \sqrt{21}}{2}$

38. $(s - 8)(s + 7) = 5$

$s^2 - s - 61 = 0$

$s = \dfrac{1 \pm 7\sqrt{5}}{2}$

39. $y^2 - \dfrac{2}{5}y = 2$

$5y^2 - 2y - 10 = 0$

$y = \dfrac{1 \pm \sqrt{51}}{5}$

40. $y^2 - 3 = \dfrac{8}{3}y$

$3y^2 - 8y - 9 = 0$

$y = \dfrac{4 \pm \sqrt{43}}{3}$

41. $3x^2 - 7 = 0$

$x = \pm\dfrac{\sqrt{21}}{3}$

42. $7x^2 + 5x = 0$

$x = 0, x = -\dfrac{5}{7}$

43. $x(x - 2) = 7$

$x = 1 \pm 2\sqrt{2}$

44. $(3x + 1)^2 = 12$

$x = \dfrac{-1 \pm 2\sqrt{3}}{3}$

Cumulative Review Problems

45. *Simplify.* $\dfrac{3x}{x - 2} + \dfrac{4}{x + 2} - \dfrac{x + 22}{x^2 - 4}$

$\dfrac{3(x + 5)}{x + 2}$

46. Give equations in **(a)** slope-intercept form and **(b)** standard form with integer coefficients for the line through $(-8, 2)$ and $(5, -3)$.

(a) $y = -\dfrac{5x}{13} - \dfrac{14}{13}$ **(b)** $5x + 13y = -14$

47. *Divide.*

$(x^3 + 8x^2 + 17x + 6) \div (x + 3)$

$x^2 + 5x + 2$

48. *Solve for g.*

$2p = 3gf - 7p + 20$

$\dfrac{9p - 20}{3f} = g$

10.4 Graphing Quadratic Equations

After studying this section, you will be able to:

1. *Graph equations of the form $y = ax^2 + bx + c$ by graphing ordered pairs of numbers.*
2. *Graph equations of the form $y = ax^2 + bx + c$ by using the vertex.*

1 *Graphing $y = ax^2 + bx + c$ Using Ordered Pairs*

Recall that the graph of a linear equation is always a straight line. What is the graph of a quadratic equation? What happens to the graph if one of the variable terms is squared? We will use point plotting to look for a pattern. We begin with a table of values.

EXAMPLE 1 Graph. **(a)** $y = x^2$ **(b)** $y = -x^2$

(a) $y = x^2$

x	y
−3	9
−2	4
−1	1
0	0
1	1
2	4
3	9

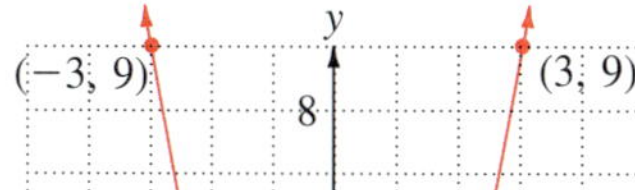

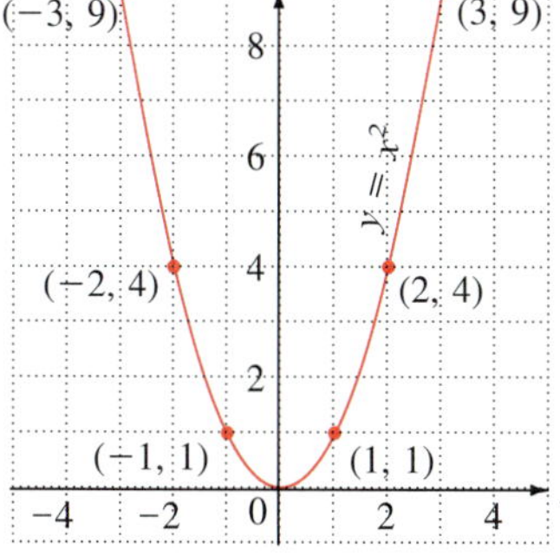

(b) $y = -x^2$

x	y
−3	−9
−2	−4
−1	−1
0	0
1	−1
2	−4
3	−9

Practice Problem 1 Graph. **(a)** $y = 2x^2$ **(b)** $y = -2x^2$

$y = 2x^2$

x	y
−2	8
−1	2
0	0
1	2
2	8

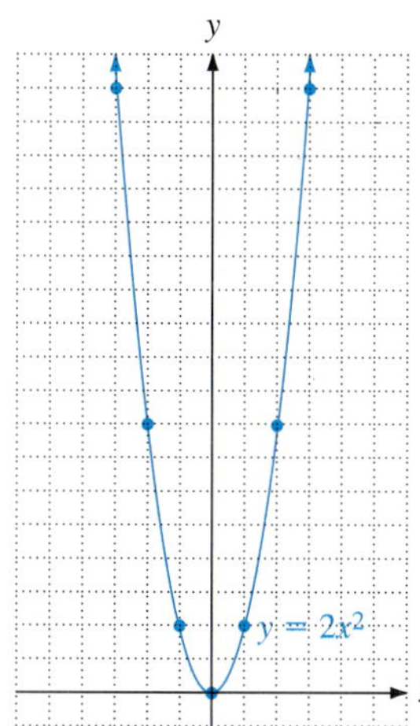

$y = -2x^2$

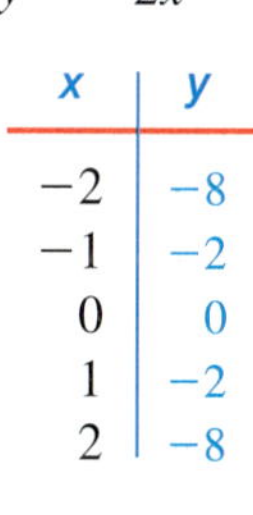

x	y
−2	−8
−1	−2
0	0
1	−2
2	−8

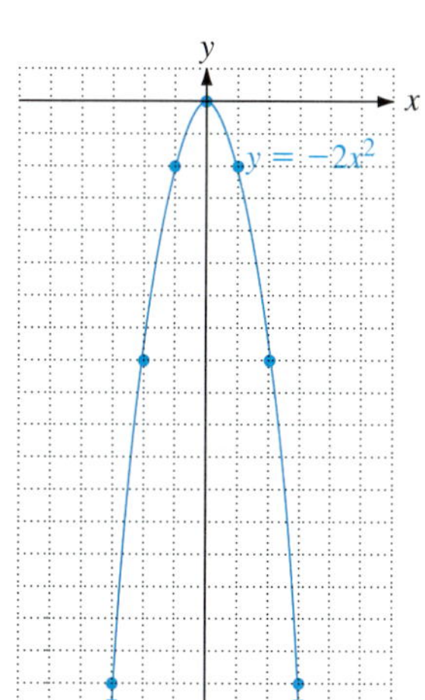

■

To Think About What happens to the graph when the coefficient of x^2 is negative? What happens when the coefficient of x^2 is an integer greater than 1 or less than −1? Graph $y = 3x^2$ and $y = -3x^2$. What might happen if $-1 < a < 1$. Graph $y = \frac{1}{3}x^2$ and $y = -\frac{1}{3}x^2$.

The curves that you have been graphing are called *parabolas.* What happens to the graph of a parabola as the equation changes?

EXAMPLE 2 Graph. **(a)** $y = x^2 - 3$ **(b)** $y = x^2 + 2$

(a) $y = x^2 - 3$

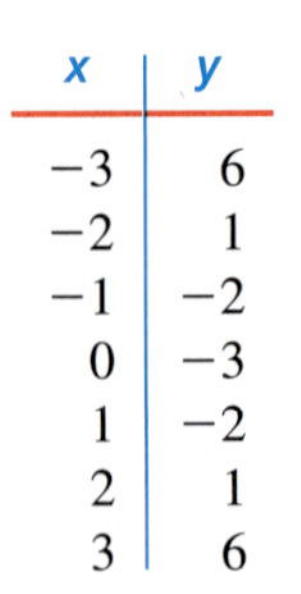

x	y
−3	6
−2	1
−1	−2
0	−3
1	−2
2	1
3	6

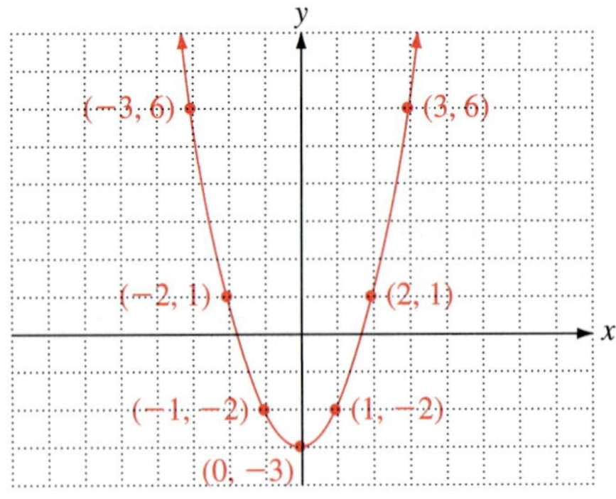

(b) $y = x^2 + 2$

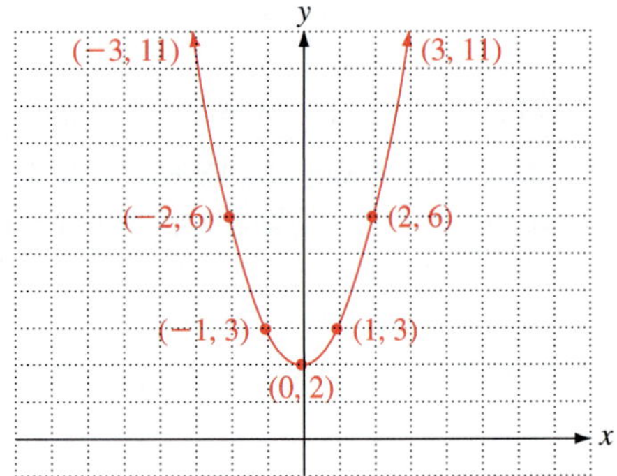

x	y
−3	11
−2	6
−1	3
0	2
1	3
2	6
3	11

Practice Problem 2 Graph. **(a)** $y = (x - 1)^2$ **(b)** $y = (x + 2)^2$

$y = (x - 1)^2$

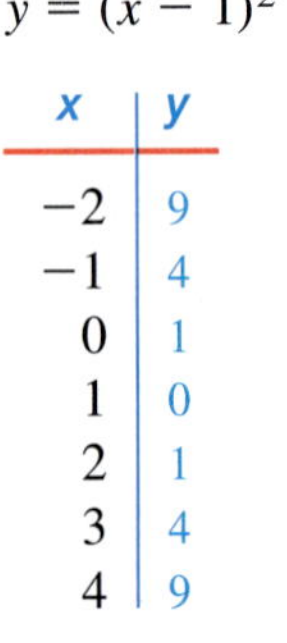

x	y
−2	9
−1	4
0	1
1	0
2	1
3	4
4	9

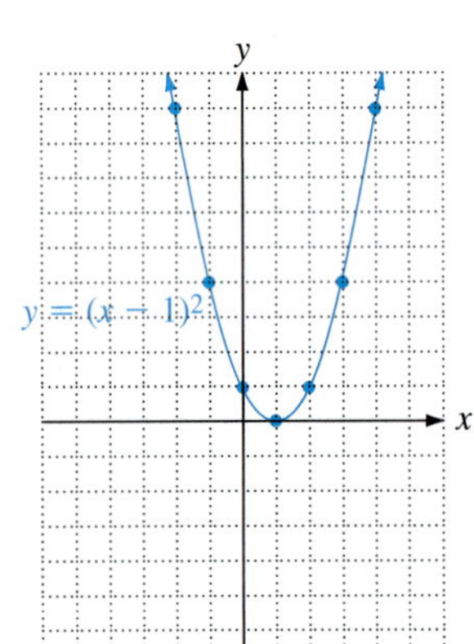

$y = (x + 2)^2$

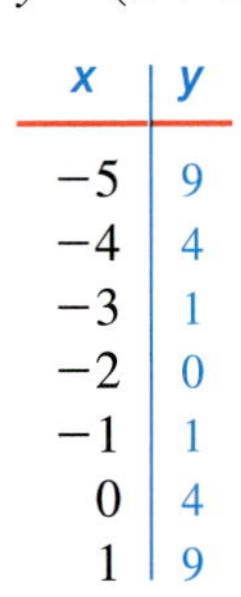

x	y
−5	9
−4	4
−3	1
−2	0
−1	1
0	4
1	9

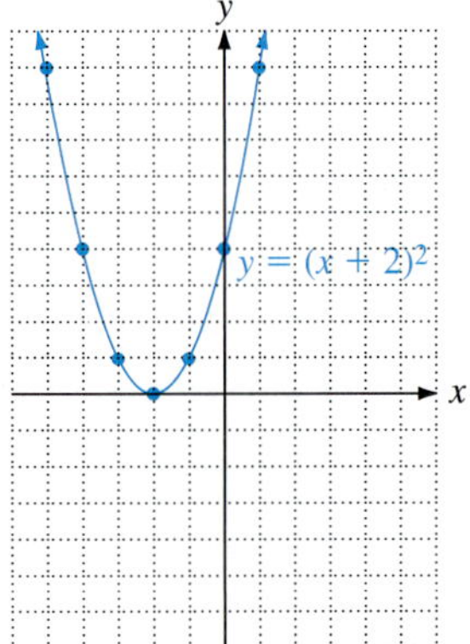

■

To Think About Look at the graphs for Example 2. Describe how the constant c in the equation $y = ax^2 + c$ affects the graph of the parabola. Look at the graphs for Practice Problem 2. Describe how the constant c in the equation $y = (x + c)^2$ affects the graph of the parabola.

In the above examples the equations we worked with were relatively simple. They provided us with an understanding of the general shape of the graph of a quadratic equation. We have seen how changing the equation affects the graph of that equation. Let's look at a slightly more difficult equation to graph.

We introduce a new word, vertex. The **vertex** is the lowest point on a parabola that opens upward. The **vertex** is the highest point on a parabola that opens downward.

EXAMPLE 3 Graph $y = x^2 - 2x$. Identify the coordinates of the vertex.

x	y	
-2	8	
-1	3	
0	0	
1	-1	vertex
2	0	
3	3	
4	8	

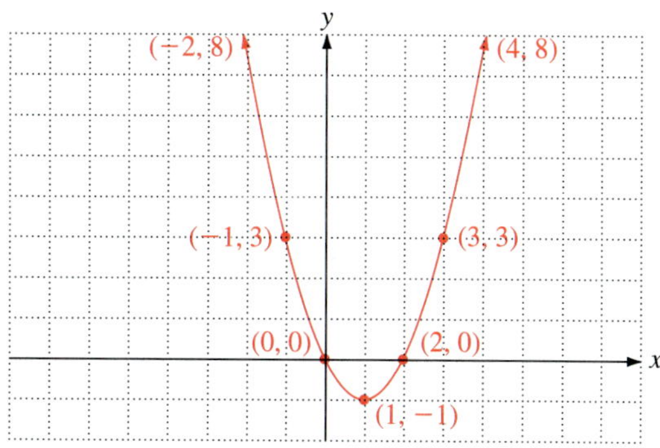

The vertex is at $(1, -1)$.

Notice the x-intercepts are at $(0, 0)$ and $(2, 0)$. The x-coordinates are the solutions to the equation.

$$x^2 - 2x = 0$$

$$x(x - 2) = 0$$

$$x = 0, \qquad x = 2$$

Practice Problem 3 Graph. **(a)** $y = x^2 + 2x$ **(b)** $y = -x^2 - 2x$ For each table of values, use x from -4 to 2. Identify the coordinates of the vertex.

$y = x^2 + 2x$

x	y
-4	8
-3	3
-2	0
-1	-1
0	0
1	3
2	8

$y = -x^2 - 2x$

x	y
-4	-8
-3	-3
-2	0
-1	1
0	0
1	-3
2	-8

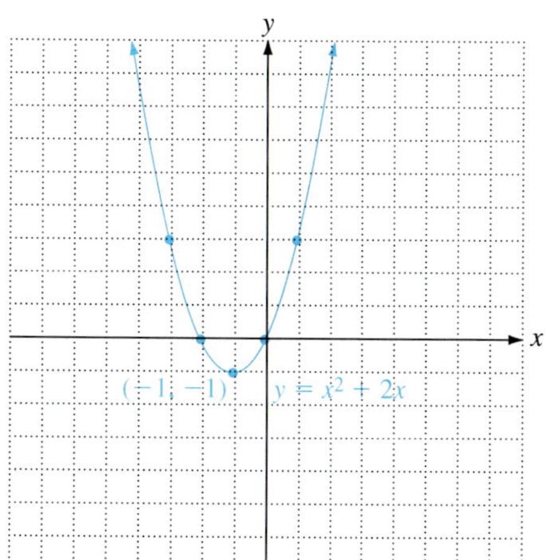

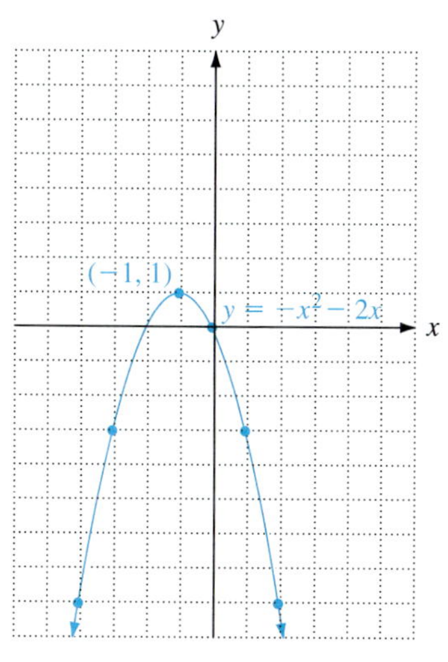

GRAPHING CALCULATOR

Graphing Quadratic Equations

Graph $y = x^2 - 4x$ on a graphing calculator using an appropriate window.

Display:

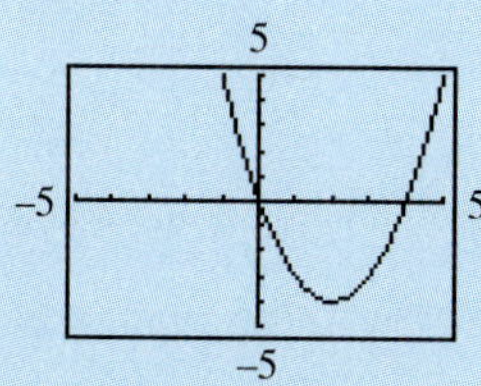

Next you can use the Trace and Zoom features of your calculator to determine the vertex.

Some calculators have a feature that will calculate the maximum or minimum point on the graph. Use the feature that calculates the minimum point to find the vertex.

Display:

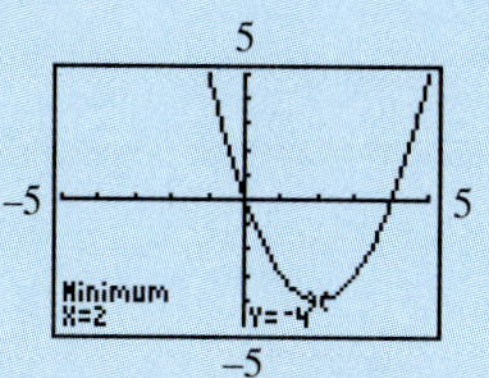

Thus the vertex is at $(2, -4)$.

2 *Graphing Equations of the Form $y = ax^2 + bx + c$ by Using the Vertex*

Since the general shape of the parabola will be the same if the coefficient of the squared term is 1, the vertex is the key to graphing parabolas. Let's take a look at the previous graphs and identify the vertex of each parabola.

Equation:	$y = x^2$	$y = x^2 - 3$	$y = (x - 1)^2$ $y = x^2 - 2x + 2$	$y = (x + 2)^2$ $y = x^2 + 4x + 4$	$y = x^2 - 2x$
Vertex:	$(0, 0)$	$(0, -3)$	$(1, 0)$	$(-2, 0)$	$(1, -1)$

Notice that the x-coordinate of the vertex of the parabola for the first two equations is 0. You may also notice that these two equations do not have a middle term. That is, for $y = ax^2 + bx + c$, $b = 0$. This is not true of the last three equations. As you may have already guessed, there is a relationship between the equation and the x-coordinate of the vertex of the parabola.

Vertex of a Parabola

The x-coordinate of the vertex of $y = ax^2 + bx + c$ is

$$x = \frac{-b}{2a}$$

Once we have the x-coordinate, we can find the y-coordinate using the equation. We can plot the point and draw a general sketch of the curve. The sign of the squared term will tell us if the graph opens upward or downward.

EXAMPLE 4 $y = -x^2 - 2x + 3$. Determine the vertex and the x-intercepts. Then sketch the graph.

We will first determine the coordinates of the vertex. We begin by finding the x-coordinate.

$$x = \frac{-b}{2a} = \frac{-(-2)}{2(-1)} = \frac{2}{-2} = -1$$

$y = -(-1)^2 - 2(-1) + 3$ *Determine y when x is -1.*

$= 4$ *The point $(-1, 4)$ is the vertex of the parabola.*

Now we will determine the x-intercepts, those points where the graph crosses the x-axis. This occurs when $y = 0$. These are the solutions to the equation. Note that this equation is factorable.

$0 = -x^2 - 2x + 3$ *Find the solutions to the equation.*

$0 = (-x + 1)(x + 3)$ *Factor.*

$-x + 1 = 0 \qquad x + 3 = 0$ *Solve for x.*

$x = 1 \qquad x = -3$ *The x-intercepts are at $(1, 0)$ and at $(-3, 0)$.*

We now have enough information to sketch the graph. Since $a = -1$, the parabola opens downward and the vertex $(-1, 4)$ is the highest point. The graph crosses the x-axis at $(-3, 0)$ and at $(1, 0)$.

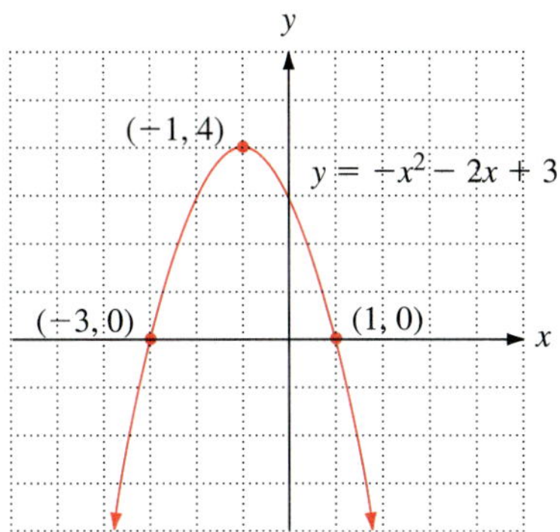

Practice Problem 4 $y = x^2 - 4x - 5$. Determine the vertex and the x-intercepts. Then sketch the graph.

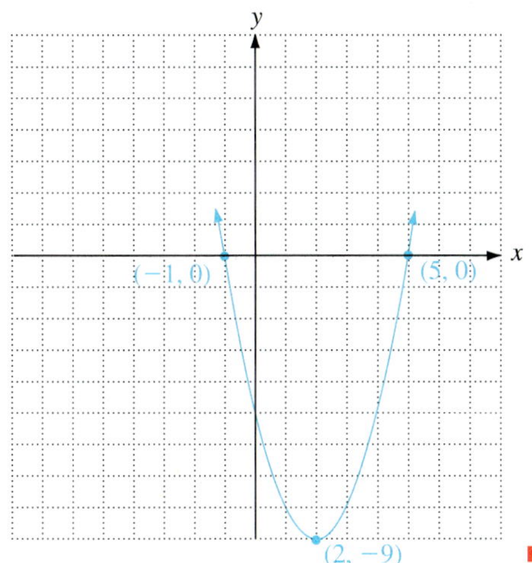

EXAMPLE 5 The height of a ball in feet is given by the equation $h = -16t^2 + 64t + 5$, where t is measured in seconds.

(a) Graph the equation $h = -16t^2 + 64t + 5$.

(b) What physical significance does the vertex of the graph have?

(c) After how many seconds will the ball hit the ground?

(a) Since the equation is a quadratic equation, the graph is a parabola. a is negative. Thus the parabola opens downward and the vertex is the highest point. We begin by finding the coordinates of the vertex. Our equation has ordered pairs (t, h) instead of (x, y). The t-coordinate of the vertex is

$$t = \frac{-b}{2a} = \frac{-64}{2(-16)} = \frac{-64}{-32} = 2$$

Substitute $t = 2$ into the equation to find the h-coordinate of the vertex.

$$h = -16(2)^2 + 64(2) + 5 = -64 + 128 + 5 = 69$$

The vertex (t, h) is $(2, 69)$.

To sketch this graph, it would be helpful to know the h-intercept, that is, the point where $t = 0$.

$$\begin{aligned} h &= -16(0)^2 + 64(0) + 5 \\ &= 5 \end{aligned}$$

The h-intercept is at $(0, 5)$.

The t-intercept is where $h = 0$. We will use the quadratic formula to find t when h is 0, that is, to find the roots of the quadratic equation.

$$0 = -16t^2 + 64t + 5$$

$$t = \frac{-64 \pm \sqrt{(64)^2 - 4(-16)(5)}}{2(-16)}$$

$$t = -\frac{64 \pm \sqrt{4096 + 320}}{-32} = -\frac{64 \pm \sqrt{4416}}{-32}$$

Using a calculator or a square root table, we have

$$t = -\frac{64 + 66.453}{-32} \approx 4.077, \qquad t \approx 4.1 \quad \text{(to the nearest tenth)}$$

$$t = -\frac{64 - 66.453}{-32} \approx -0.077$$

We will not use negative t values. We are studying the height of the ball from the time of throwing ($t = 0$ seconds) to the time that it hits the ground (approximately $t = 4.1$ seconds). In this situation it is not useful to find points for negative values of t because they have no meaning in this situation.

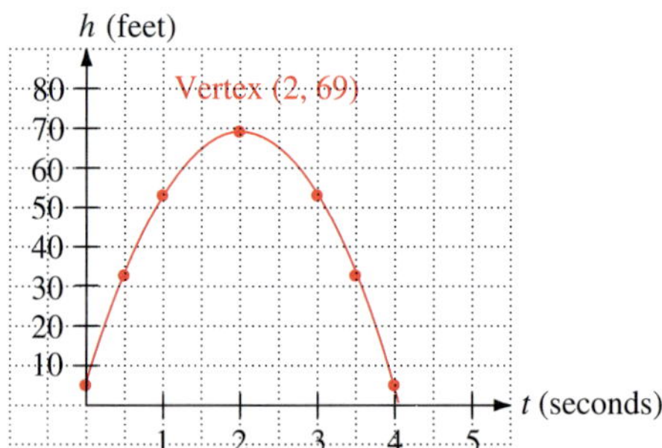

(b) The vertex represents the greatest height of the ball. The ball is 69 feet above the ground 2 seconds after it is thrown.

(c) The ball will hit the ground at approximately 4.1 seconds after it is thrown. This is the point at which the graph intersects the t-axis, the positive root of the equation.

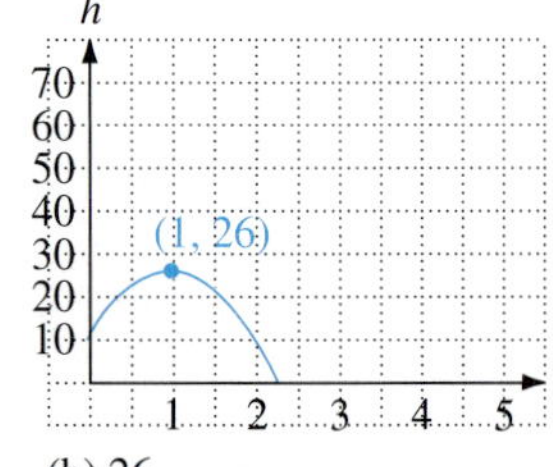

(b) 26
(c) 2.3 seconds

Practice Problem 5 A ball is thrown upward with a speed of 32 feet per second from a distance 10 feet above the ground. The height of the ball in feet is given by the equation $h = -16t^2 + 32t + 10$, where t is measured in seconds.

(a) Graph the equation $h = -16t^2 + 32t + 10$.

(b) What is the greatest height of the ball?

(c) After how many seconds will the ball hit the ground? ■

10.4 Exercises

Graph each equation.

1. $y = 2x^2 - 1$

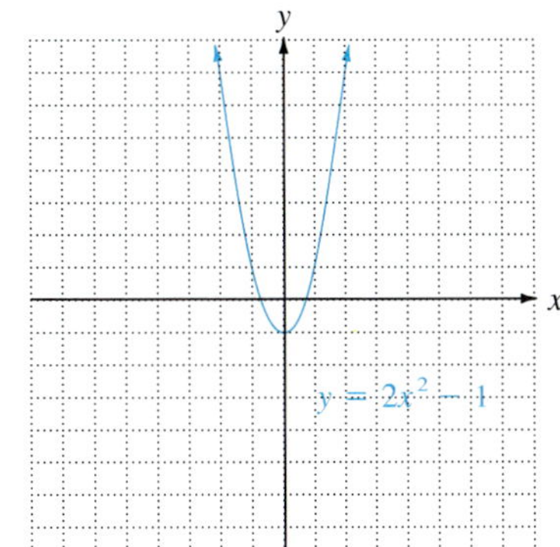

2. $y = x^2 + 2$

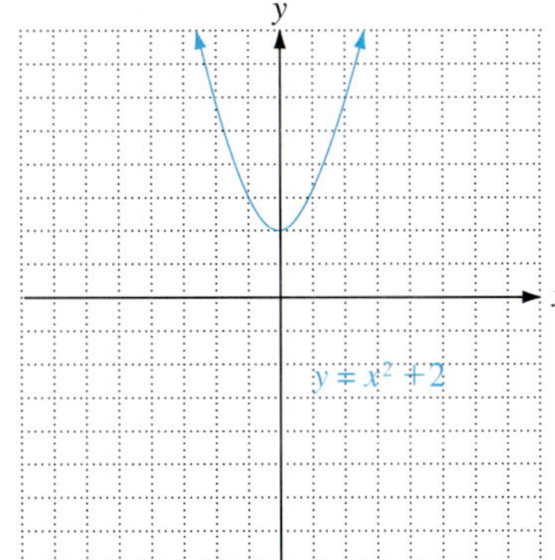

3. $y = -\frac{1}{3}x^2$

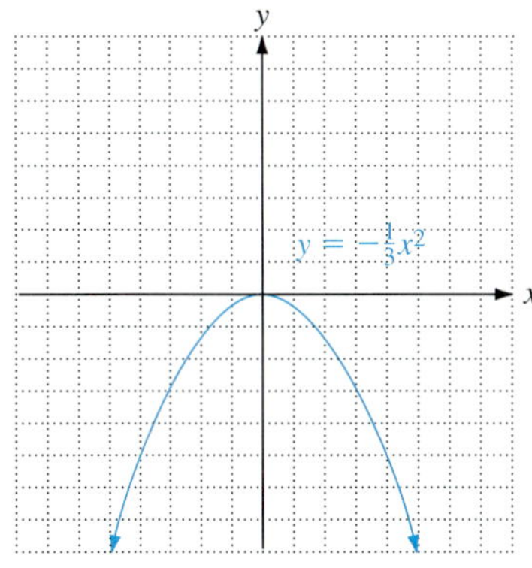

4. $y = \frac{1}{2}x^2$

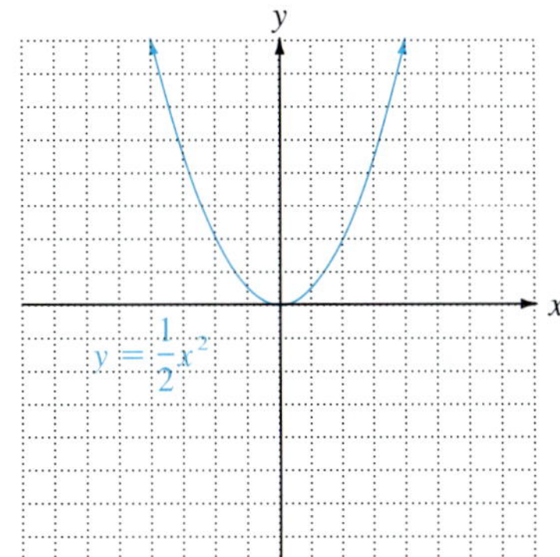

5. $y = x^2 - 5$

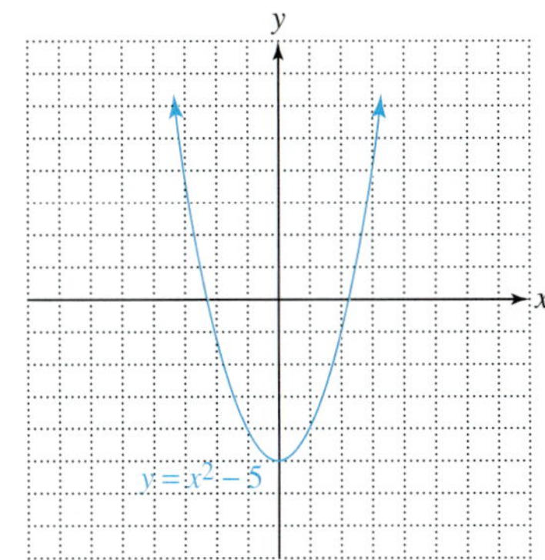

6. $y = 3x^2 + 2$

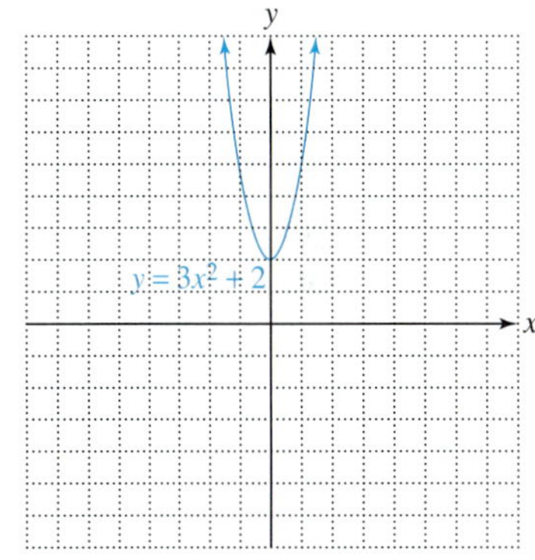

7. $y = (x - 2)^2$

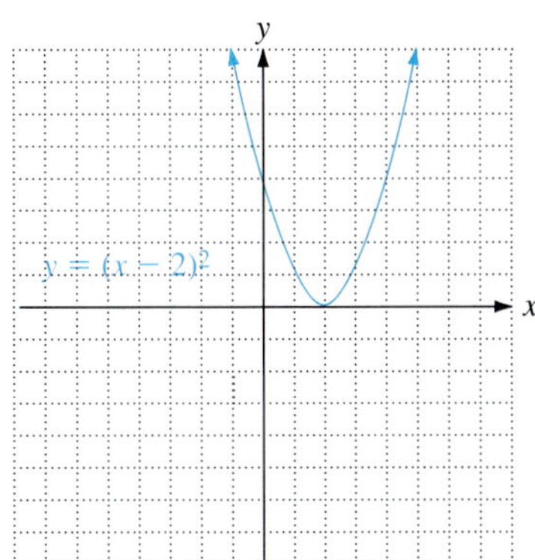

8. $y = (x + 1)^2$

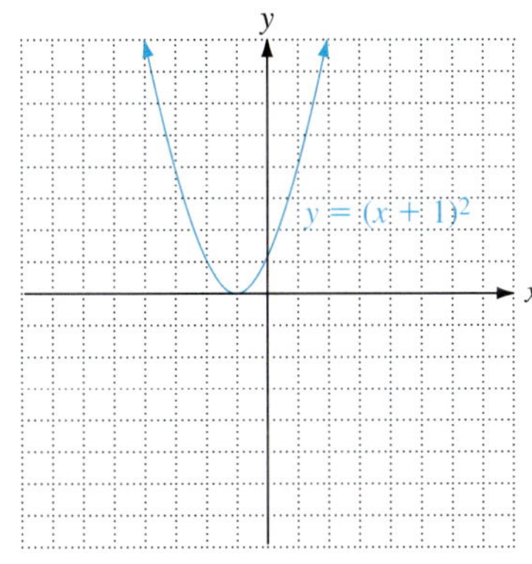

9. $y = -\frac{1}{2}x^2 + 4$

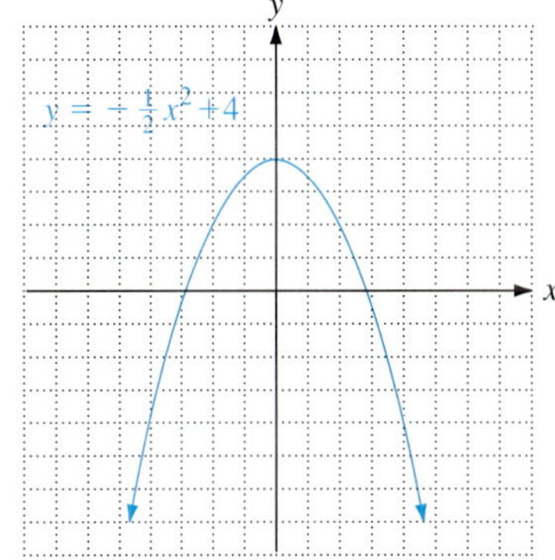

10. $y = -\frac{1}{2}x^2 + 4$

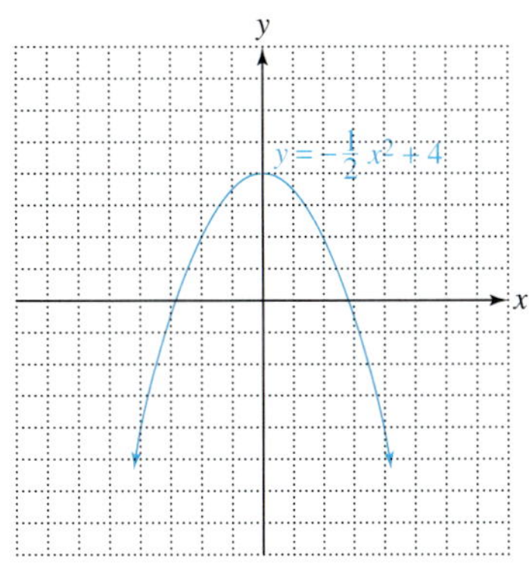

11. $y = \frac{1}{2}(x - 3)^2$

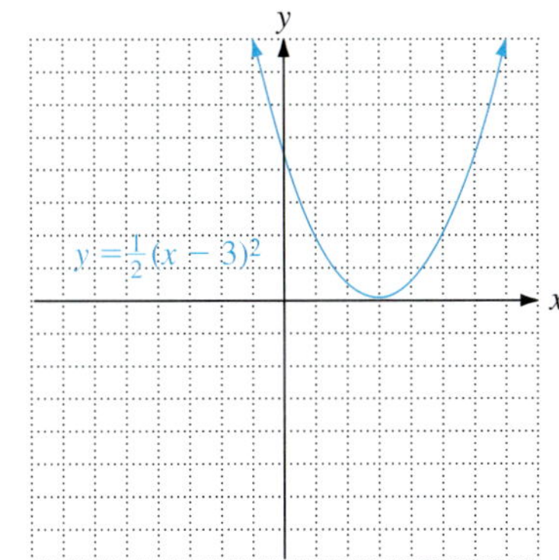

12. $y = \frac{1}{2}(x + 2)^2$

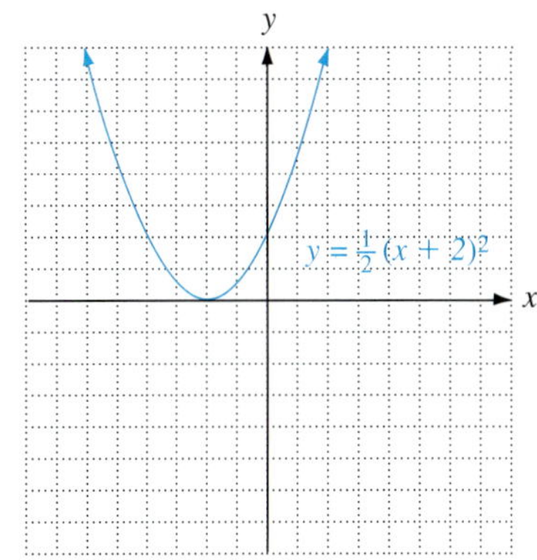

Graph each equation. Identify the coordinates of the vertex.

13. $y = x^2 + 4x$

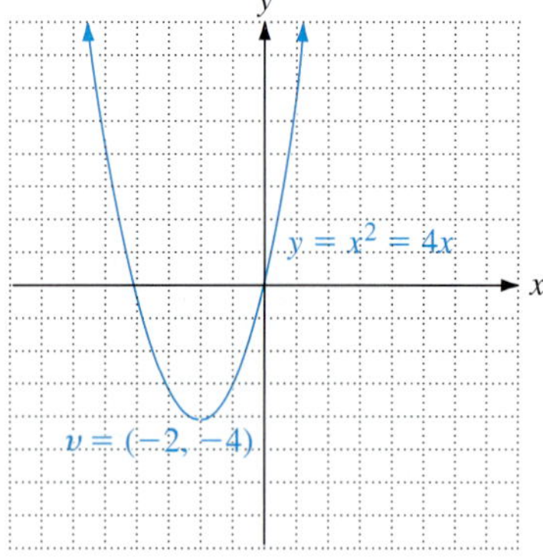

14. $y = -x^2 + 4x$

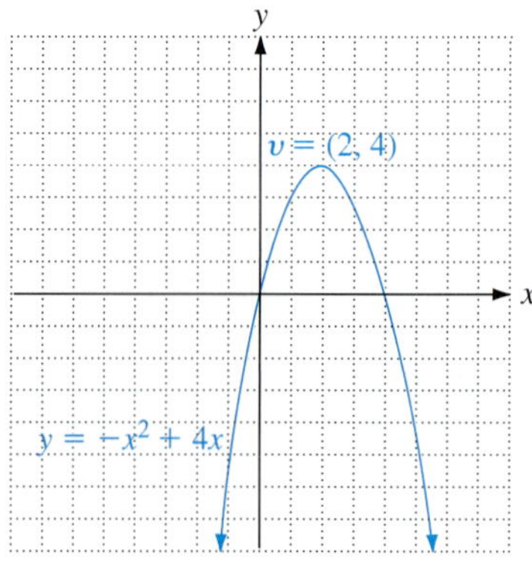

15. $y = x^2 - 4x$

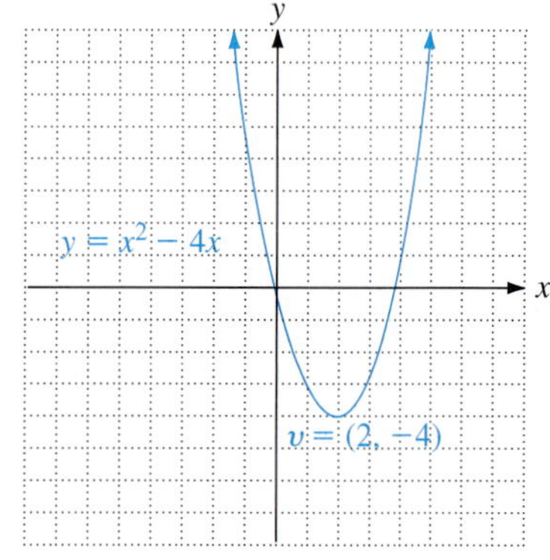

16. $y = -x^2 - 4x$

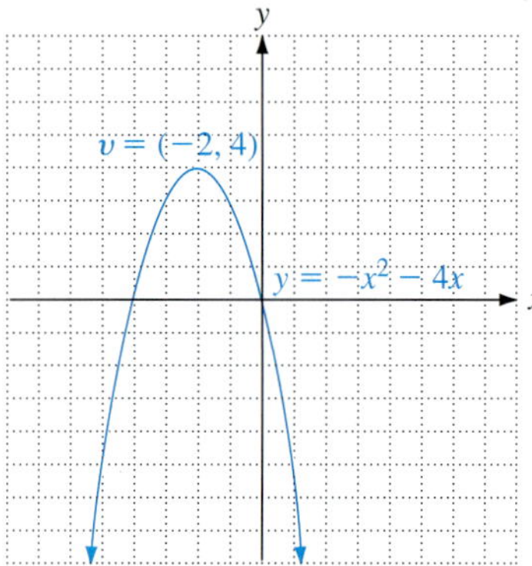

17. $y = -2x^2 + 8x$

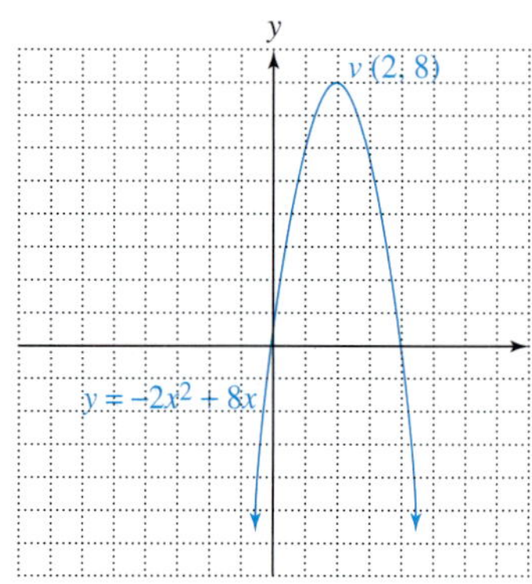

18. $y = x^2 + 6x$

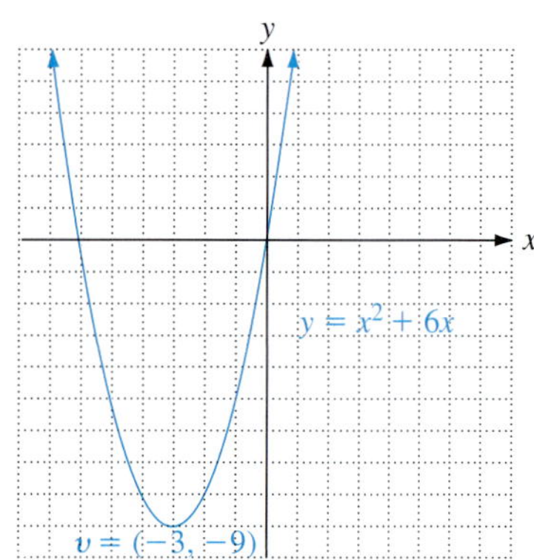

Determine the vertex and the x-intercepts. Then sketch the graph.

19. $y = x^2 - 4x + 3$

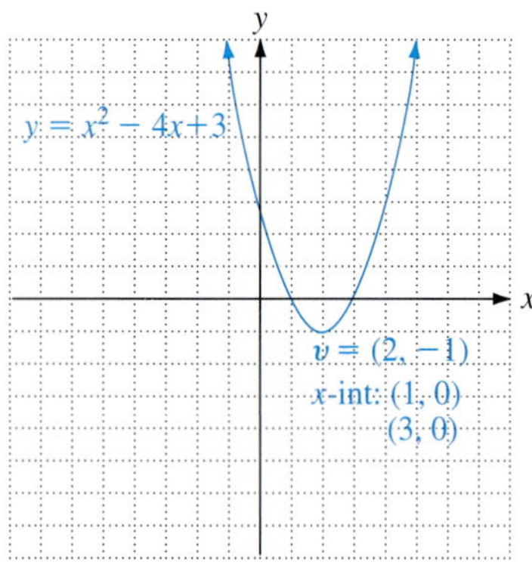

20. $y = x^2 + 2x - 3$

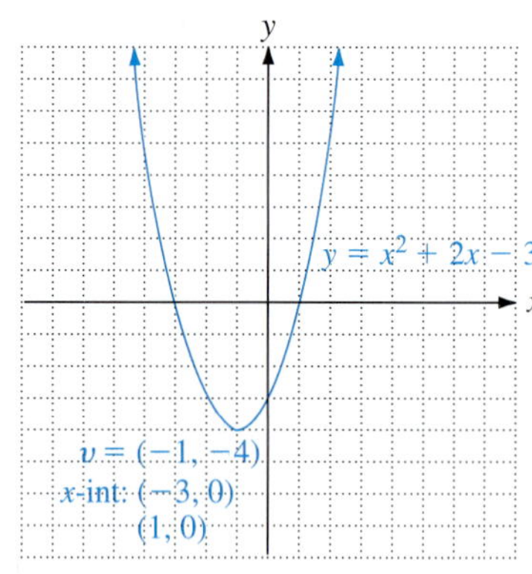

21. $y = -x^2 - 6x - 5$

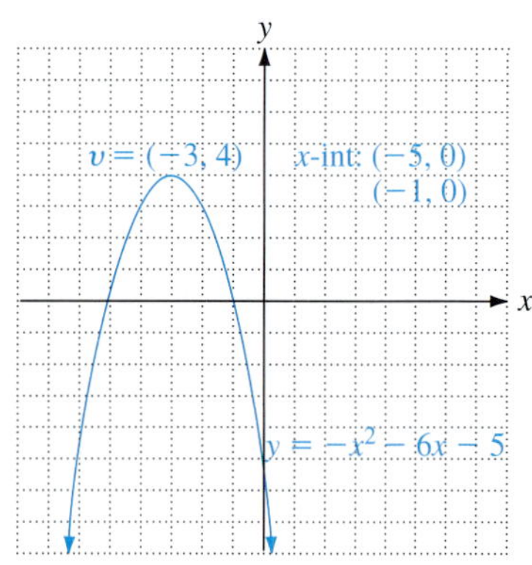

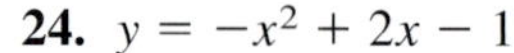

22. $y = -x^2 + 2x + 3$

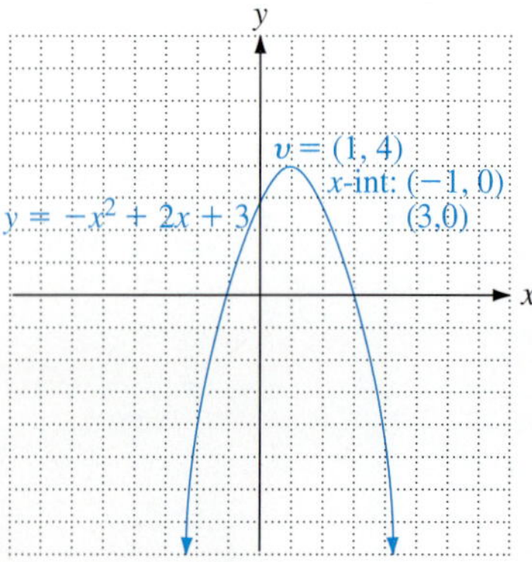

23. $y = x^2 + 6x + 9$

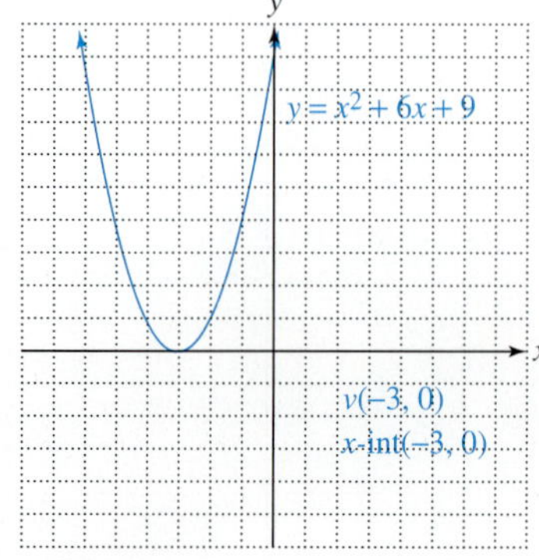

24. $y = -x^2 + 2x - 1$

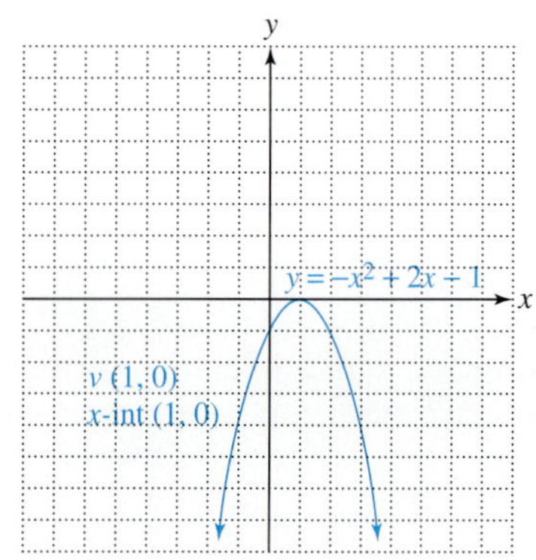

Applications

25. A miniature rocket is launched so that its height h in meters after t seconds is given by the equation $h = -4.9t^2 + 19.6t + 10$.

(a) Draw a graph of the equation.

(b) How high is the rocket after 3 seconds? 24.7 meters

(c) What is the highest the rocket will travel? 29.6 meters

(d) How long after the launch will it take for the rocket to strike the earth? 4.46 seconds

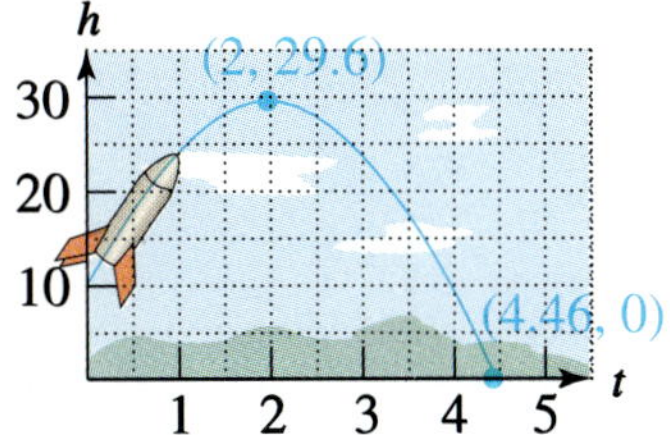

26. A miniature rocket is launched so that its height h in meters after t seconds is given by the equation $h = -4.9t^2 + 39.2t + 4$.

(a) Draw a graph of the equation.

(b) How high is the rocket after 2 seconds? 62.8 meters

(c) What is the highest the rocket will travel? 82.4 meters

(d) How long after the launch will it take for the rocket to strike the earth? 8.1 seconds

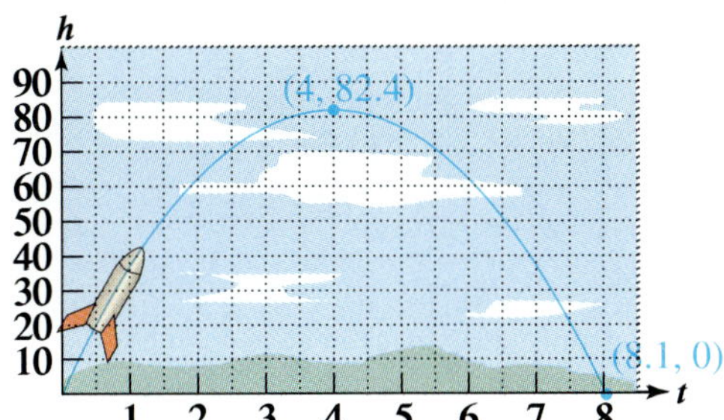

Optional Graphing Calculator Problems

Use your graphing calculator to find the vertex, the y-intercept, and the x-intercepts of the following. Round your answer to the nearest thousandth when necessary.

27. $y = 3x^2 + 7x - 9$
vertex $(-1.167, -13.083)$
y-intercept $(0, -9)$
x-intercepts $(0.922, 0)$ and $(-3.255, 0)$

28. $y = x^2 + 10.6x - 212.16$
vertex $(-5.3, -240.25)$
y-intercept $(0, -212.16)$
x-intercepts $(10.2, 0)$ and $(-20.8, 0)$

29. $y = -156x^2 - 289x + 1133$
vertex $(-0.926, 1266.848)$
y-intercept $(0, 1133)$
x-intercepts $(1.923, 0)$ and $(-3.776, 0)$

30. $y = -301x^2 - 167x + 1528$
vertex $(-0.277, 1551.164)$
y-intercept $(0, 1528)$
x-intercepts $(1.993, 0)$ and $(-2.548, 0)$

Cumulative Review Problems

31. If y varies inversely as x and $y = \frac{1}{3}$ when $x = 21$, find y when $x = 7$. $y = \frac{7}{x}$ $y = 1$

32. If y varies directly as x^2, and $y = 12$ when $x = 2$, find y when $x = 5$. $y = 3x^2$ $y = 75$

Putting Your Skills to Work

HELPING THE FOOTBALL COACH PLAN THE DEFENSE

A football is thrown a distance of 40 yards by a quarterback. The ball is thrown and caught 5 feet above ground. The football is 15 feet above ground at the peak of the throw. The ball follows a path that is part of a parabola. The vertex of the parabola is the highest point the ball reaches. The height h of the football in feet at a distance of x yards from the quarterback is given by the equation

$$h = -\tfrac{1}{40}(x - 20)^2 + 15$$

A graph of the equation is shown below.

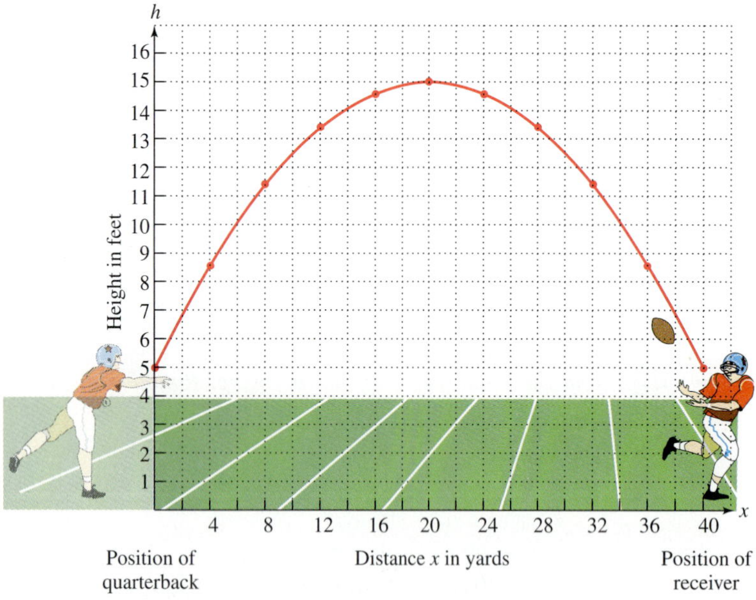

PROBLEMS FOR INDIVIDUAL INVESTIGATION

Approximate the answers to the following questions based on the graph.

1. How high is the football above ground when it is 8 yards from the quarterback? 11.5 feet

2. How high is the football above ground when it is 32 yards from the quarterback? 11.5 feet

3. If a member of the opposing team is rushing toward the quarterback and can leap up a distance of 8 feet to intercept or deflect the ball, how close does he have to be to impede the throw of the ball? 3 yards

4. If a member of the opposing team is covering the receiver and he can leap up a distance of 9 feet to intercept or deflect the ball, how close does he need to be to the receiver to keep the receiver from catching the ball? 5 yards

PROBLEMS FOR GROUP INVESTIGATION AND ANALYSIS

Together with some of your classmates, complete the following.

Suppose that the equation $h = -\frac{1}{35}(x - 15)^2 + 12$ describes the most commonly used throw by an opposing quarterback. Graph the equation from $x = 0$ to $x = 30$.

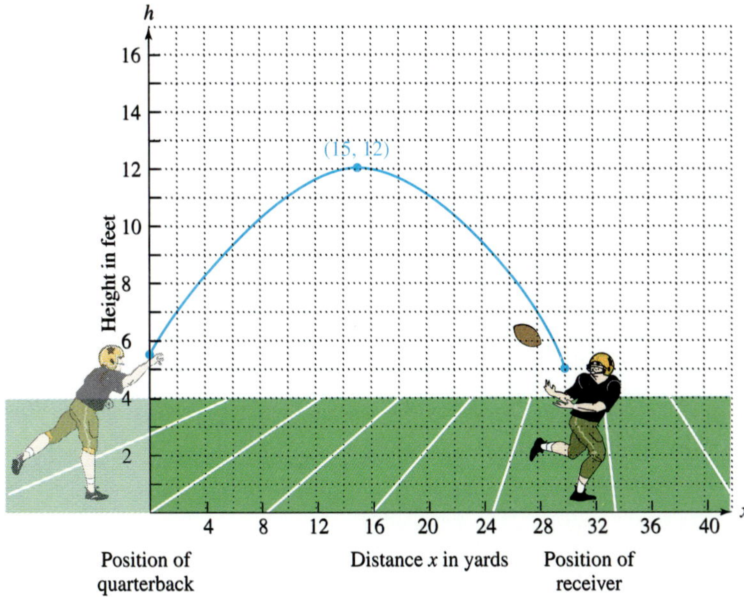

Answer the following questions based on your graph.

5. What is the maximum height of the football when it is thrown? 12 feet

6. How far will the football travel to the receiver if the receiver catches it at a height of 5 feet above ground? approximately 31 yards

7. If a person can leap 7 feet up to intercept or deflect the ball, how close does he have to be to the quarterback to impede the throw of the ball? 2 yards

8. If a person can leap 9 feet up to intercept or deflect the ball, how close does he have to be to the receiver to keep the receiver from catching the ball? 5 yards

Note: Using videocameras and computer technology, it is possible to find an equation for the path of a football when thrown by a quarterback.

INTERNET CONNECTIONS: Go to ''http://www.prenhall.com/tobey'' to be connected

Site: Annual Records and Team Accolades (New York Giants) or a related site

8. This site shows the number of games that were won, lost, and tied by the New York Giants each season. Choose any 15-year period of the Giants' history. Tell what time period you chose and give the percent of games won, percent of games lost, and percent of games tied during this period.

10.5 Formulas and Applied Problems

MathPro

Video 25

SSM

After studying this section, you will be able to:

1 *Solve applied problems requiring the use of a quadratic equation.*

1 *Solving Applied Problems Using a Quadratic Equation*

We can use what we have learned about quadratic equations to solve word problems. Some word problems will involve geometry. Recall that the Pythagorean theorem states that in a right triangle $a^2 + b^2 = c^2$, where a and b are legs and c is the hypotenuse. We will use this theorem to solve word problems involving right triangles.

EXAMPLE 1 The hypotenuse of a right triangle is 25 meters in length. One leg is 17 meters longer than the other. Find the length of each leg.

1. *Understand the problem.*
Draw a picture.

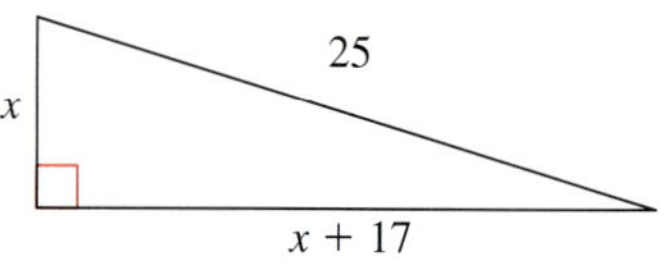

2. *Write an equation.*
The problem involves a right triangle.
Use the Pythagorean theorem.

$$a^2 + b^2 = c^2$$

$$x^2 + (x + 17)^2 = 25^2$$

3. *Solve the equation and state the answer.*

$$x^2 + (x + 17)^2 = 25^2$$

$$x^2 + x^2 + 34x + 289 = 625$$

$$2x^2 + 34x - 336 = 0$$

$$x^2 + 17x - 168 = 0$$

$$(x + 24)(x - 7) = 0$$

$$x + 24 = 0 \qquad x - 7 = 0$$

$$x = -24 \qquad x = 7$$

(Note that $x = -24$ is not a valid solution.)
One leg is 7 meters in length. The other leg is $x + 17$ or $7 + 17 = 24$ meters in length.

4. *Check.*
Check the conditions in the original problem. Do the two legs differ by 17 meters?

$$24 - 7 \stackrel{?}{=} 17 \qquad 17 = 17 \checkmark$$

Do the sides of the triangle satisfy the Pythagorean theorem?

$$7^2 + 24^2 \stackrel{?}{=} 25^2 \qquad 49 + 576 \stackrel{?}{=} 625 \qquad 625 = 625 \checkmark$$

Practice Problem 1 The hypotenuse of a right triangle is 30 meters in length. One leg is 6 meters shorter than the other leg. Find the length of each leg. ■

Sometimes it may help to use two variables in the initial part of the problem. Then one variable can be eliminated by substitution.

EXAMPLE 2 The Ski Club is renting a bus to travel to Mount Snow. The members agreed to share the cost of $180 equally. On the day of the ski trip three members were sick with the flu and could not go. This raised by $10 the share of each person going on the trip. How many people originally planned to attend?

1. *Understand the problem.*
Let s = the number of students in the ski club
c = the cost for each student in the original group

2. *Write an equation(s).*

$$\text{number of students} \times \text{cost per student} = \text{total cost}$$

$$s \cdot c = 180 \qquad \textbf{(1)}$$

Now if 3 people are sick, the number of students drops by 3, but the cost for each increases by $10. The total is still $180. Therefore,

$$(s - 3)(c + 10) = 180 \qquad \textbf{(2)}$$

3. *Solve the equation(s) and state the answer.*

$$c = \frac{180}{s} \qquad \textit{Solve equation } \textbf{(1)} \textit{ for } c.$$

$$(s - 3)\left(\frac{180}{s} + 10\right) = 180 \qquad \textit{Substitute } \tfrac{180}{s} \textit{ for } c \textit{ in equation } \textbf{(2)}.$$

$$(s - 3)\left(\frac{180 + 10s}{s}\right) = 180 \qquad \textit{Add } \tfrac{180}{s} + 10. \textit{ Use the LCD} = s.$$

$$(s - 3)(180 + 10s) = 180s \qquad \textit{Multiply both sides of the equation by } s.$$

$$10s^2 + 150s - 540 = 180s \qquad \textit{Use FOIL to multiply the binomials.}$$

$$10s^2 - 30s - 540 = 0$$

$$s^2 - 3s - 54 = 0$$

$$(s - 9)(s + 6) = 0$$

$$s = 9 \qquad s = -6$$

The number of students originally in the ski club was 9.

4. *Check.*
Would the cost increase by $10 if the number dropped from 9 students to 6 students? 9 people in the club would mean that each would pay $20 (180 ÷ 9 = 20). If the number is reduced by 3, there were 6 people who took the trip. Their cost was $30 (180 ÷ 6 = 30). The increase is $10.

$$20 + 10 \stackrel{?}{=} 30 \qquad 30 = 30 \quad \checkmark$$

Practice Problem 2 Several friends from the University of New Mexico decided to charter a sport fishing boat for the afternoon while on a vacation at the coast. They agreed to share the rental cost of $240 equally. At the last minute, two of them could not go on the trip. This raised the cost by $4 for each person in the group. How many people originally planned to charter the boat? ■

EXAMPLE 3 Minette is fencing a garden that borders the back of a large barn. She has 120 feet of fencing. She would like a rectangular garden that measures 1350 square feet in area. She wants to use the back of the barn, so she only needs to use fencing on three sides. What dimensions should she use for her garden?

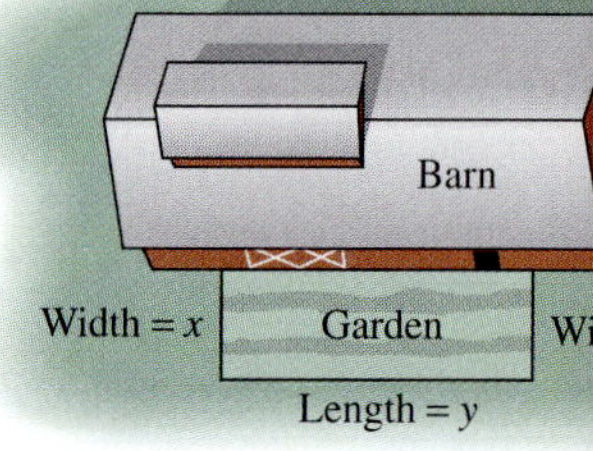

1. *Understand the problem.*
   Draw a picture.

   Let x = the width of the garden in feet

   y = the length of the garden in feet

2. *Write an equation(s).*
   Look at the drawing. The 120 feet of fencing will be needed for the width twice (two sides) and the length once (one side).

$$120 = 2x + y \qquad \textbf{(1)}$$

   The area formula is $A = \text{(width)(length)}$.

$$1350 = x(y) \qquad \textbf{(2)}$$

3. *Solve the equation(s) and state the answer.*

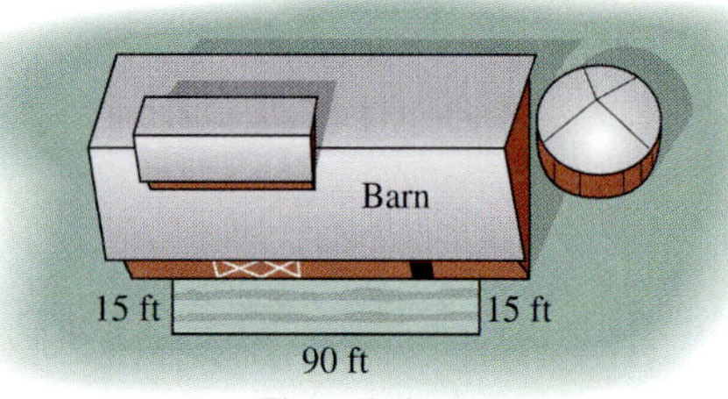

First solution

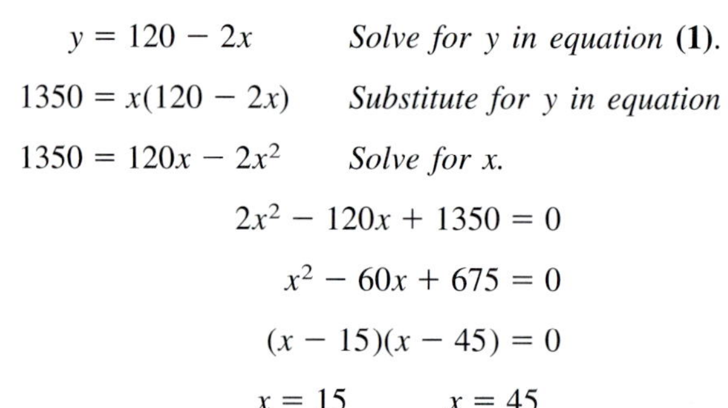

$y = 120 - 2x$ *Solve for y in equation* **(1)**.

$1350 = x(120 - 2x)$ *Substitute for y in equation* **(2)**.

$1350 = 120x - 2x^2$ *Solve for x.*

$$2x^2 - 120x + 1350 = 0$$

$$x^2 - 60x + 675 = 0$$

$$(x - 15)(x - 45) = 0$$

$$x = 15 \qquad x = 45$$

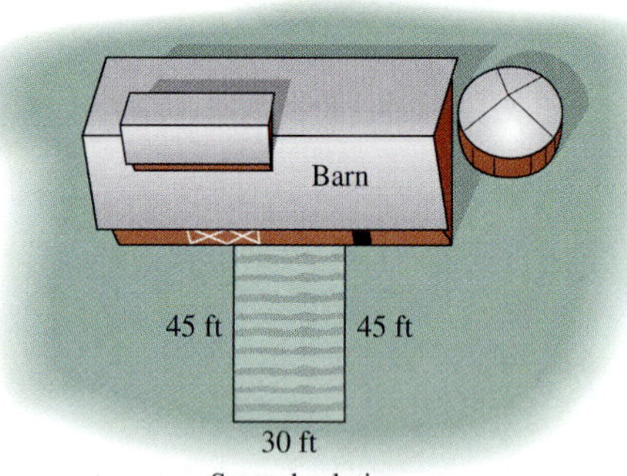

Second solution

First solution: If the width is 15 feet, the length is 90 feet.

Check $15 + 15 + 90 = 120$ ✓

$(15)(90) = 1350$ ✓

Second solution: If the width is 45 feet, the length is 30 feet.

Check $45 + 45 + 30 = 120$ ✓

$(45)(30) = 1350$ ✓

Thus we see that Minette has two choices that satisfy the given requirements (see the figure). She would probably choose the shape that is the most practical for her. (If her barn is not 90 feet long, she could not use this first solution!)

Remark: Isn't there another way to do this problem? Yes. Use only the variable x to represent the length and width. If you let x = the width and $120 - 2x$ = the length, you can immediately write equation (3), which is

$$1350 = x(120 - 2x) \qquad \textbf{(3)}$$

However, many students find that this is difficult to do, and they would prefer to use two variables. Try to work these problems in a way that seems fairly easy to you.

Practice Problem 3 An Arizona rancher has a small fenced-in area against a canyon wall. Therefore, he only needs three sides of fencing to make a rectangular holding area for cattle. He has 160 feet of fencing available. He wants to make the rectangular area with dimensions that provide him with an area of 3150 square feet. What dimensions should he use for the rectangular-shaped holding area? ■

10.5 Exercises

Applications

Solve each word problem.

1. The hypotenuse of a right triangle is 40 yards long. One leg of the triangle is 8 yards longer than the other leg. Find the length of each leg of the triangle.
$(x + 8)^2 + x^2 = 1600$
$x^2 + 8x - 768 = 0$
$(x + 32)(x - 24) = 0$
32 yards, 24 yards

2. The hypotenuse of a right triangle is 15 yards long. One leg of the triangle is 3 yards shorter than the other leg. Find the length of each leg of the triangle.
$(x - 3)^2 + x^2 = 225$
$x^2 - 3x - 108 = 0$
$(x - 12)(x + 9) = 0$
12 yards, 9 yards

3. The sporting goods store at the mall is a perfect rectangle. Its diagonal is 13 meters and the width of the rectangle is 7 meters shorter than its length. Find the length and the width of the store.
length = 12 meters
width = 5 meters

4. Manuel placed a ladder 20 feet long against the house. The distance from the top of the ladder to the bottom of the house is 4 feet greater than the distance from the bottom of the house to the foot of the ladder. How far is the foot of the ladder from the house? How far up on the house does the top of the ladder touch the building?

$x^2 + (x + 4)^2 = 20^2$
$2x^2 + 8x - 384 = 0$
$x^2 + 4x - 192 = 0$
$(x + 16)(x - 12) = 0$
Foot of ladder is 12 ft.
Top is 16 ft.

5. To help out Nina and Tom with their new twins, the employees at Tom's office decided to give the new parents a gift certificate worth \$420 to furnish the babies' room, sharing the cost equally. At the last minute, 7 people from the executive staff chipped in, lowering the cost per person by \$5. How many people originally planned to give the gift?
21 people originally planned to give a gift.

6. Everyone in the School Street Townhouse Association agreed to equally share the cost of a new \$6000 septic system. However, before the system was installed, 5 new units were built and the 5 families were asked to contribute their share. This lowered the contribution of each family by \$200. How many families originally agreed to share the cost of the new septic system?
Originally there were 10 families.

7. The yacht club need to raise \$18,000 in upkeep fees. The fees were assigned equally to each member. Due to an increase of 90 new members, the cost of the upkeep fee went down \$10 for each member. How many members were originally in the club? After the increase, how many members are there now? There were originally 360 members. After the increase there were 450 members.

8. The members of the varsity football team decided to give the coach a gift that cost \$80. On the day that they collected money, 10 people were absent. They were in a hurry, so they collected an equal amount from each member and bought the gift. Later the team captain said that, if everyone had been present to chip in, each person would have contributed \$4 less. How many people actually contributed? How many people in total were on the football team?
$xy = 80$
$(x + 4)(y - 10) = 80$
$4y^2 - 800 - 40y = 0$
$(y - 20)(y + 10) = 0$
Ten people contributed.
Twenty people were on the football team.

9. The new housing development is planting a cooperative organic garden. The garden is in the shape of a rectangle, and one side rests against the community sports facility, so the co-op needs fencing on only 3 sides. If they have 140 feet of fencing available and the garden is 2000 square feet in area, what are the possible dimensions for the garden? If the width is 20 feet, the length is 100 feet. If the width is 50 feet, the length is 40 feet.

10. A security fence encloses a rectangular area on one side of Fenway Park in Boston. Three sides of fencing are used, since the fourth side of the area is formed by a building. The enclosed area measures 1250 square feet. Exactly 100 feet of fencing is used to fence in three sides of this rectangle. What are the possible dimensions that could have been used to construct this area? The width is 25 feet, the length along the building is 50 feet.

★ **11.** A pilot is testing a new experimental craft. The cruising speed is classified information. In a test the jet traveled 2400 miles. The pilot revealed that if he had increased his speed 200 mph the total trip would have taken 1 hour less. Can you determine the cruising speed of the jet? The cruising speed is 600 mph.

★ **12.** A gymnasium floor is being covered by square shock-absorbing tiles. The old gym floor required 864 square tiles. The new tiles are 2 inches larger in both length and width than the old tiles. The new flooring will require only 600 tiles. What is the length of a side of one of the new shock-absorbing tiles? (*Hint:* Since the area is the same in each case, write an expression for area with old tiles and one with new tiles. Set the two expressions equal to each other.)
The length of the new tile is 12 inches on each side.

To Think About

The sum of all the counting numbers from 1 to 6 is $1 + 2 + 3 + 4 + 5 + 6 = 21$. A shortcut to the answer is $6(7)(\frac{1}{2}) = 21$. In general the sum of the numbers from 1 up to n, which we can write as $1 + 2 + \cdots + n$, can be computed by the formula

$$\frac{n(n+1)}{2} = s \qquad \text{where } s \text{ is the sum}$$

13. If the sum of the first n counting numbers is 91, find n.

$$\frac{n(n+1)}{2} = 91$$
$$n^2 + n = 182$$
$$(n + 14)(n - 13) = 0$$
$$n = 13$$

14. If the sum of the first n counting numbers is 136, find n.

$$\frac{n(n+1)}{2} = 136$$
$$n^2 + n = 272$$
$$(n + 17)(n - 16) = 0$$
$$n = 16$$

The Springfield Boot Company has determined that the daily profit of the company can be predicted by the equation $P = -2x^2 + 360x - 14{,}400$, where x is the number of pairs of boots produced each day.

15. What is the daily profit if 85 pairs of boots are produced each day? $1750

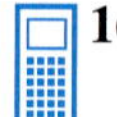
16. **(a)** How many pairs of boots each day need to be produced to obtain the maximum daily profit?
90 boots

(b) What is the maximum daily profit? $1800

Cumulative Review Problems

17. Graph on a number line the solution to $3x + 11 \geq 9x - 4$.

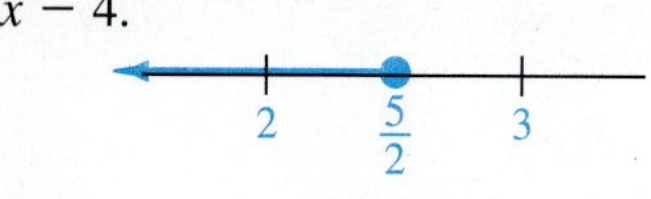

18. Rationalize the denominator. $\dfrac{\sqrt{3}}{\sqrt{5} + 2}$

$\sqrt{15} - 2\sqrt{3}$

Chapter Organizer

Topic	Procedure	Examples
Solve quadratic equations by factoring, p. 560.	1. Clear the equation of fractions, if necessary. 2. Write as $ax^2 + bx + c = 0$. 3. Factor. 4. Set each factor equal to 0. 5. Solve the resulting equations.	Solve. $x + \frac{1}{2} = \frac{3}{2x}$ Multiply each term by LCD $= 2x$. $2x(x) + 2x\left(\frac{1}{2}\right) = 2x\left(\frac{3}{2x}\right)$ $2x^2 + x = 3$ $2x^2 + x - 3 = 0$ $(2x + 3)(x - 1) = 0$ $2x + 3 = 0 \qquad x - 1 = 0$ $2x = -3 \qquad x = 1$ $x = -\frac{3}{2}$
Solving quadratic equations: taking the square root of each side of the equation, p. 569.	A quadratic equation of the form $ax^2 - c = 0$ can be solved by solving for x^2. Then use the property that if $x^2 = a$ then $x = \pm\sqrt{a}$. This amounts to taking the square root of each side of the equation.	$2x^2 + 1 = 99$ $2x^2 = 99 - 1$ $2x^2 = 98$ $\frac{2x^2}{2} = \frac{98}{2}$ $x^2 = 49$ $x = \pm\sqrt{49} = \pm 7$
Solving quadratic equations: completing the square, p. 570.	We can write a quadratic equation in the form $(ax + b)^2 = c$ by completing the square. 1. Write the equation as $ax^2 + bx = -c$. 2. Divide by a. 3. Take half the x coefficient, square it, and add this to both sides. 4. Factor the perfect square on the left. 5. Take the square root of both sides. 6. Solve for x.	Solve. $5x^2 - 3x - 7 = 0$ $5x^2 - 3x = 7$ $x^2 - \frac{3}{5}x = \frac{7}{5}$ $\left[\frac{1}{2} \cdot \left(-\frac{3}{5}\right)\right]^2 = \frac{9}{100}$ $x^2 - \frac{3}{5}x + \frac{9}{100} = \frac{7}{5} + \frac{9}{100}$ $\left(x - \frac{3}{10}\right)^2 = \frac{149}{100}$ $x - \frac{3}{10} = \pm\frac{\sqrt{149}}{10}$ $x = \frac{3}{10} \pm \frac{\sqrt{149}}{10}$ $x = \frac{3 \pm \sqrt{149}}{10}$
Solving quadratic equations: quadratic formula, p. 575.	1. Put the equation in standard form: $ax^2 + bx + c = 0$. 2. Carefully determine the values of a, b, and c. 3. Substitute these into the formula: $x = \frac{-b \pm \sqrt{b^2 - 4ac}}{2a}$ 4. Simplify.	Solve. $3x^2 - 4x - 8 = 0$ $a = 3, b = -4, c = -8$ $x = \frac{-(-4) \pm \sqrt{(-4)^2 - 4(3)(-8)}}{2(3)}$ $= \frac{4 \pm \sqrt{16 + 96}}{6}$ $= \frac{4 \pm \sqrt{112}}{6} = \frac{4 \pm \sqrt{16 \cdot 7}}{6}$ $= \frac{4 \pm 4\sqrt{7}}{6} = \frac{\cancel{2}(2 \pm 2\sqrt{7})}{\cancel{2} \cdot 3} = \frac{2 \pm 2\sqrt{7}}{3}$
The discriminant, p. 578.	The *discriminant* is $b^2 - 4ac$. If negative, there are no real solutions (since it is not possible to take the square root of a negative number and get a real number).	What kinds of solutions does this equation have? $3x^2 + 2x + 1 = 0$ $b^2 - 4ac = 2^2 - 4 \cdot 3 \cdot 1$ $= 4 - 12 = -8$ This equation has no real solutions.

Chapter Organizer

Topic	Procedure	Examples
Properties of parabolas, p. 581.	**1.** The graph of $y = ax^2 + bx + c$ is a parabola. It opens up if $a > 0$ and down if $a < 0$. **2.** The vertex of a parabola is its lowest ($a > 0$) or highest ($a < 0$) point. Its x-coordinate is $x = \frac{-b}{2a}$.	$y = 3x^2 + 4x - 11$ has vertex at $x = \frac{-b}{2a} = \frac{-4}{6} = -\frac{2}{3}$ The parabola opens upward.
Graphing parabolas, p. 581.	Graph the vertex of a parabola and several other points to get a good idea of its graph. Usually, we find the y-intercept and the x-intercepts if any exist. The following procedure is helpful. **1.** Determine if $a < 0$ or $a > 0$. **2.** Find the y-intercept. **3.** Find the x-intercepts. **4.** Find the vertex. **5.** Plot one or two extra points if necessary. **6.** Draw the graph.	Graph. $y = x^2 - 2x - 8$ $a = 1, b = -2, c = -8$; $a > 0$, so the parabola opens upward. Let $x = 0$. $y = 0^2 - 2(0) - 8 = -8$ The y-intercept is $(0, -8)$. $0 = x^2 - 2x - 8$ $0 = (x - 4)(x + 2)$ $x = 4, \quad x = -2$ The x-intercepts are $(4, 0)$ and $(-2, 0)$. The vertex is at $x = \frac{-b}{2a} = -\frac{-(-2)}{2} = 1$. If $x = 1$, $y = (1)^2 - 2(1) - 8 = -9$. The vertex is at $(1, -9)$. y x $(-2, 0)$ $(4, 0)$ $(0, -8)$ $(1, -9)$ $y = x^2 - 2x - 8$
Formulas and applied problems, p. 592.	Some word problems require the use of a quadratic equation. You may want to follow this four-step procedure for problem solving. **1.** Understand the problem. **2.** Write an equation. **3.** Solve the equation and state the answer. **4.** Check.	The area of a rectangle is 48 square inches. The length is three times the width. Find the dimensions of the rectangle. **1.** Draw a picture. $3w$ w **2.** $(3w)w = 48$ **3.** $3w^2 = 48$ $w^2 = 16$ $w = 4, l = 12$ **4.** $(4)(12) = 48$

Putting Your Skills to Work

MEASURING THE DEPTH OF SNOWFALL

During the 1995–1996 winter, eastern Massachusetts received record levels of snowfall. During one storm, a heavy snowfall changed to a rain storm. Thus the depth of snow first increased and then decreased as the warm rains washed away the snow. By the end of the storm there was only rain and no snow remaining on the ground. The snowfall depth d (measured in inches at the Fowler Observatory in Wenham, Massachusetts) during this storm could be approximated by using the equation $d = -0.1(t^2 - 34t + 120)$ when t is the number of hours from midnight on Sunday night just before the storm hit. (Since amount of snowfall is always positive or zero, this model is only valid when d is greater than or equal to zero.) Use this equation to answer the following questions.

PROBLEMS FOR INDIVIDUAL INVESTIGATION

1. At what time on what day did the snowstorm begin? The storm began at 4:00 A.M. Monday.

2. At what time and on what day did the storm end? The storm ended at 6:00 A.M. Tuesday.

PROBLEMS FOR GROUP INVESTIGATION AND ANALYSIS

Together with some other members of your class determine an answer to the following.

3. At what time and on what did the depth of snow reach a maximum? The maximum depth occurred at 5:00 P.M. Monday.

4. What was the maximum depth of the snow? The maximum depth of snow was 16.9 inches.

5. During what 2-hour period was the depth of the snow increasing at the greatest rate? What was the increase in depth during that 2-hour period? On Monday from 4:00 A.M. to 6:00 A.M. the greatest increase occurred, which was 4.8 inches of snow.

INTERNET CONNECTIONS: Go to ''http://www.prenhall.com/tobey'' to be connected

Site: Fairbanks, Alaska Climatology (Snow) or a related site

This site includes graphs showing typical snowfall in Fairbanks. Note that the red curve on the Daily Average Snow Depth graph indicates the average amount of snowfall each day.

6. On the average, what month has the most snowfall in Fairbanks?

7. On the average, about how much snow falls on April 14?

8. On the average, about how much more snow falls on January 7 than on October 31?

Chapter 10 Review Problems

10.1 *Write the quadratic equation in standard form. Determine the values of a, b, and c.* **Do not solve.**

1. $5x^2 = -2x + 3$
$5x^2 + 2x - 3 = 0$
$a = 5, b = 2, c = -3$

2. $7x^2 + 6x = -8x^2 - 4$
$15x^2 + 6x + 4 = 0$
$a = 15, b = 6, c = 4$

3. $(x - 4)(2x - 1) = x^2 + 7$
$x^2 - 9x - 3 = 0$
$a = 1, b = -9, c = -3$

4. $x(3x + 1) = -5x + 10$
$3x^2 + 6x - 10 = 0$
$a = 3, b = 6, c = -10$

5. $\frac{1}{x^2} - \frac{3}{x} + 5 = 0$
$5x^2 - 3x + 1 = 0$
$a = 5, b = -3, c = 1$

6. $\frac{x}{x+2} - \frac{3}{x-2} = 5$
$4x^2 + 5x - 14 = 0$
$a = 4, b = 5, c = -14$

7. $x^2 + 7x + 8 = 4x^2 + x + 8$
$3x^2 - 6x = 0$
$a = 3, b = -6, c = 0$

8. $12x^2 - 9x + 1 = 15 - 9x$
$12x^2 - 14 = 0$
$a = 12, b = 0, c = -14$

Place each equation in standard form. Solve by factoring.

9. $x^2 + 21x + 20 = 0$
$(x + 20)(x + 1) = 0$
$x = -20, x = -1$

10. $x^2 + 10x + 25 = 0$
$(x + 5)(x + 5) = 0$
$x = -5$

11. $x^2 + 6x - 27 = 0$
$(x - 3)(x + 9) = 0$
$x = 3, x = -9$

12. $4x^2 - 12x + 9 = 0$
$(2x - 3)(2x - 3) = 0$
$x = \frac{3}{2}$

13. $6x^2 + 23x = -15$
$6x^2 + 23x + 15 = 0$
$(6x + 5)(x + 3) = 0$
$x = -\frac{5}{6}, x = -3$

14. $4x^2 = 5x - 1$
$4x^2 - 5x + 1 = 0$
$(4x - 1)(x - 1) = 0$
$x = \frac{1}{4}, x = 1$

15. $12x^2 = 4 - 13x$
$12x^2 + 13x - 4 = 0$
$(3x + 4)(4x - 1) = 0$
$x = -\frac{4}{3}, x = \frac{1}{4}$

16. $9x^2 = -27x - 8$
$9x^2 + 27x + 8 = 0$
$(3x + 1)(3x + 8) = 0$
$x = -\frac{1}{3}, x = -\frac{8}{3}$

17. $x^2 + \frac{5}{6}x - 1 = 0$
$6x^2 + 5x - 6 = 0$
$(3x - 2)(2x + 3) = 0$
$x = \frac{2}{3}, x = -\frac{3}{2}$

18. $x^2 + \frac{7}{6}x - \frac{1}{2} = 0$
$6x^2 + 7x - 3 = 0$
$(3x - 1)(2x + 3) = 0$
$x = \frac{1}{3}, x = -\frac{3}{2}$

19. $x^2 + \frac{2}{15}x = \frac{1}{15}$
$15x^2 + 2x - 1 = 0$
$(5x - 1)(3x + 1) = 0$
$x = \frac{1}{5}, x = -\frac{1}{3}$

20. $1 + \frac{13}{12x} - \frac{1}{3x^2} = 0$
$12x^2 + 13x - 4 = 0$
$(4x - 1)(3x + 4) = 0$
$x = \frac{1}{4}, x = -\frac{4}{3}$

21. $5x^2 + 16x = 0$
$x(5x + 16) = 0$
$x = 0, x = -\frac{16}{5}$

22. $-2x^2 + 9x = 0$
$-x(2x - 9) = 0$
$x = 0, x = \frac{9}{2}$

23. $x^2 + 12x - 3 = 8x - 3$
$x^2 + 4x = 0$
$x(x + 4) = 0$
$x = 0, x = -4$

24. $6x^2 - 13x + 2 = 5x + 2$
$6x^2 - 18x = 0$
$6x(x - 3) = 0$
$x = 0, x = 3$

25. $1 + \frac{2}{3x+4} = \frac{3}{3x+2}$
$9x^2 + 15x = 0$
$3x(3x + 5) = 0$
$x = 0, x = -\frac{5}{3}$

26. $2 - \frac{5}{x+1} = \frac{3}{x-1}$
$2x^2 - 8x = 0$
$2x(x - 4) = 0$
$x = 0, x = 4$

27. $5 + \frac{24}{2 - x} = \frac{24}{2 + x}$

$5x^2 - 48x - 20 = 0$

$(5x + 2)(x - 10) = 0$

$x = -\frac{2}{5}, x = 10$

28. $1 + \frac{14}{x^2} = \frac{9}{x}$

$x^2 - 9x + 14 = 0$

$(x - 7)(x - 2) = 0$

$x = 7, x = 2$

29. $3 = \frac{10}{x} + \frac{8}{x^2}$

$3x^2 - 10x - 8 = 0$

$(3x + 2)(x - 4) = 0$

$x = -\frac{2}{3}, x = 4$

30. $\frac{4}{9} = \frac{5x}{3} - x^2$

$9x^2 - 15x + 4 = 0$

$(3x - 1)(3x - 4) = 0$

$x = \frac{1}{3}, x = \frac{4}{3}$

10.2 *Solve by taking the square root of each side.*

31. $x^2 = 100$

$x = \pm 10$

32. $x^2 = 121$

$x = \pm 11$

33. $x^2 + 3 = 28$

$x^2 = 25$

$x = \pm 5$

34. $x^2 - 4 = 32$

$x^2 = 36$

$x = \pm 6$

35. $x^2 - 5 = 17$

$x^2 = 22$

$x = \pm\sqrt{22}$

36. $x^2 + 11 = 50$

$x^2 = 39$

$x = \pm\sqrt{39}$

37. $2x^2 - 1 = 15$

$x^2 = 8$

$x = \pm 2\sqrt{2}$

38. $3x^2 + 1 = 61$

$x^2 = 20$

$x = \pm 2\sqrt{5}$

39. $3x^2 + 6 = 60$

$x^2 = 18$

$x = \pm 3\sqrt{2}$

40. $2x^2 - 5 = 43$

$x^2 = 24$

$x = \pm 2\sqrt{6}$

41. $(x - 4)^2 = 7$

$x - 4 = \pm\sqrt{7}$

$x = 4 \pm \sqrt{7}$

42. $(x - 2)^2 = 3$

$x - 2 = \pm\sqrt{3}$

$x = 2 \pm \sqrt{3}$

43. $(5x + 6)^2 = 20$

$5x + 6 = \pm\sqrt{20}$

$x = \frac{-6 \pm 2\sqrt{5}}{5}$

44. $(3x + 8)^2 = 12$

$3x + 8 = \pm\sqrt{12}$

$x = \frac{-8 \pm 2\sqrt{3}}{3}$

Solve by completing the square.

45. $x^2 + 6x + 5 = 0$

$x^2 + 6x + 9 = 4$

$(x + 3)^2 = 4$

$x = -1, x = -5$

46. $x^2 + 8x + 15 = 0$

$x^2 + 8x + 16 = 1$

$(x + 4)^2 = 1$

$x = -5, x = -3$

47. $x^2 - 4x + 2 = 0$

$x^2 - 4x + 4 = 2$

$(x - 2)^2 = 2$

$x = 2 \pm \sqrt{2}$

48. $x^2 + 2x - 6 = 0$

$x^2 + 2x + 1 = 7$

$(x + 1)^2 = 7$

$x = -1 \pm \sqrt{7}$

49. $2x^2 - 8x - 90 = 0$

$x^2 - 4x + 4 = 49$

$(x - 2)^2 = 49$

$x = 9, x = -5$

50. $3x^2 - 36x - 60 = 0$

$x^2 - 12x + 36 = 56$

$(x - 6)^2 = 56$

$x = 6 \pm 2\sqrt{14}$

51. $3x^2 + 6x - 6 = 0$

$x^2 + 2x + 1 = 3$

$(x + 1)^2 = 3$

$x = -1 \pm \sqrt{3}$

52. $2x^2 + 10x - 3 = 0$

$x^2 + 5x + \frac{25}{4} = \frac{31}{4}$

$\left(x + \frac{5}{2}\right)^2 = \frac{31}{4}$

$x = \frac{-5 \pm \sqrt{31}}{2}$

10.3 *Solve by the quadratic formula.*

53. $x^2 + 3x - 40 = 0$

$x = \dfrac{-3 \pm \sqrt{169}}{2}$

$x = -8,\ x = 5$

54. $x^2 - 9x + 18 = 0$

$x = \dfrac{9 \pm \sqrt{81 - 72}}{2}$

$x = 6,\ x = 3$

55. $x^2 + 4x - 6 = 0$

$x = \dfrac{-4 \pm \sqrt{16 + 24}}{2}$

$x = \dfrac{-4 \pm \sqrt{40}}{2}$

$x = -2 \pm \sqrt{10}$

56. $x^2 + 4x - 8 = 0$

$x = \dfrac{-4 \pm \sqrt{16 + 32}}{2}$

$x = -2 \pm 2\sqrt{3}$

57. $2x^2 - 7x + 4 = 0$

$x = \dfrac{7 \pm \sqrt{49 - 32}}{4}$

$x = \dfrac{7 \pm \sqrt{17}}{4}$

58. $2x^2 + 5x - 6 = 0$

$x = \dfrac{-5 \pm \sqrt{25 + 48}}{4}$

$x = \dfrac{-5 \pm \sqrt{73}}{4}$

59. $2x^2 + 4x - 5 = 0$

$x = \dfrac{-4 \pm \sqrt{16 + 40}}{4}$

$x = \dfrac{-2 \pm \sqrt{14}}{2}$

60. $2x^2 + 6x - 5 = 0$

$x = \dfrac{-6 \pm \sqrt{36 + 40}}{4}$

$x = \dfrac{-3 \pm \sqrt{19}}{2}$

61. $3x^2 - 5x = 4$

$x = \dfrac{5 \pm \sqrt{25 + 48}}{6}$

$x = \dfrac{5 \pm \sqrt{73}}{6}$

62. $4x^2 + 3x = 2$

$x = \dfrac{-3 \pm \sqrt{9 + 32}}{8}$

$x = \dfrac{-3 \pm \sqrt{41}}{8}$

63. $4x^2 = 1 - 4x$

$x = \dfrac{-4 \pm \sqrt{16 + 16}}{8}$

$x = \dfrac{-1 \pm \sqrt{2}}{2}$

64. $3x^2 - 8x = -2$

$x = \dfrac{8 \pm \sqrt{64 - 24}}{6}$

$x = \dfrac{4 \pm \sqrt{10}}{3}$

Solve by any method. If there is no real number solution, so state.

65. $2x^2 - 9x + 10 = 0$

$x = \dfrac{5}{2},\ x = 2$

66. $4x^2 - 4x - 3 = 0$

$x = \dfrac{3}{2},\ x = -\dfrac{1}{2}$

67. $9x^2 - 6x + 1 = 0$

$x = \dfrac{1}{3}$

68. $2x^2 - 11x + 12 = 0$

$x = 4,\ x = \dfrac{3}{2}$

69. $3x^2 - 6x + 2 = 0$

$x = \dfrac{3 \pm \sqrt{3}}{3}$

70. $5x^2 - 7x = 8$

$x = \dfrac{7 \pm \sqrt{209}}{10}$

71. $4x^2 + 4x = x^2 + 5$

$x = \dfrac{-2 \pm \sqrt{19}}{3}$

72. $5x^2 + 7x + 1 = 0$

$x = \dfrac{-7 \pm \sqrt{29}}{10}$

73. $x^2 + 5x = -1 + 2x$

$x = \dfrac{-3 \pm \sqrt{5}}{2}$

74. $3x^2 = 6 - 7x$

$x = -3,\ x = \dfrac{2}{3}$

75. $2y^2 = -3y - 1$

$y = -\dfrac{1}{2},\ y = -1$

76. $4y^2 = 5y - 1$

$y = 1,\ y = \dfrac{1}{4}$

77. $3x^2 + 1 = 73 + x^2$

$x = \pm 6$

78. $2x^2 - 1 = 35$

$x = \pm 3\sqrt{2}$

79. $\dfrac{(y - 2)^2}{20} + 3 + y = 0$

$y = -8$ only

80. $\dfrac{(y + 2)^2}{5} + 2y = -9$

$y = -7$ only

81. $3x^2 + 1 = 6 - 8x$

$x = \dfrac{-4 \pm \sqrt{31}}{3}$

82. $2x^2 + 10x = 2x - 7$

$x = \dfrac{-4 \pm \sqrt{2}}{2}$

83. $2y - 10 = 10y(y - 2)$

$y = \dfrac{11 \pm \sqrt{21}}{10}$

84. $\dfrac{3y - 2}{4} = \dfrac{y^2 - 2}{y}$

$y = -4,\ y = 2$

85. $\dfrac{y^2 + 5}{2y} = \dfrac{2y - 1}{3}$

$y = -3,\ y = 5$

86. $\dfrac{5x^2}{2} = x - \dfrac{7x^2}{2}$

$x = 0,\ x = \dfrac{1}{6}$

10.4 *Graph each quadratic equation. Label the vertex.*

87. $y = 2x^2$

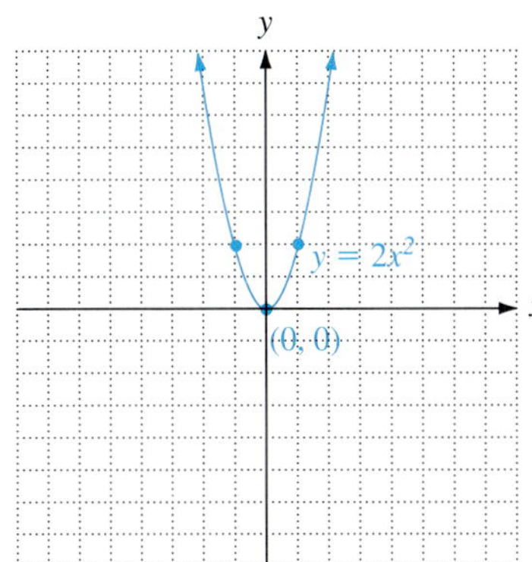

88. $y = x^2 + 4$

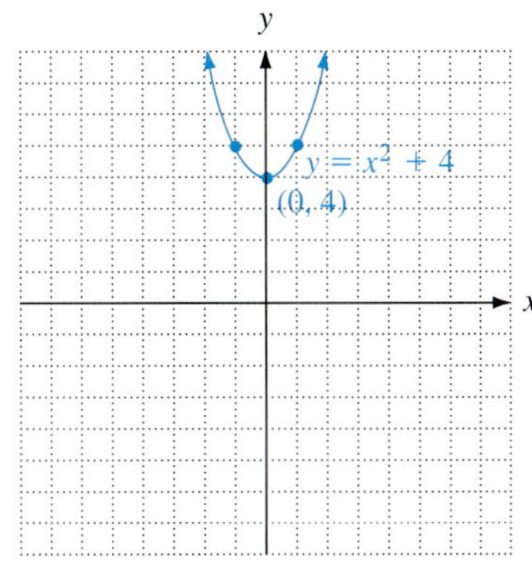

89. $y = x^2 - 3$

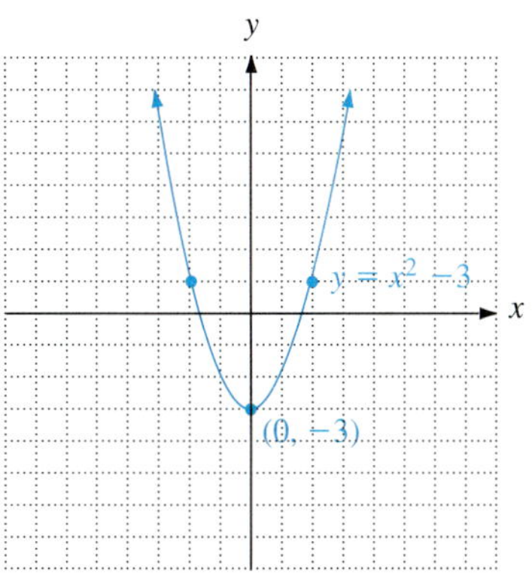

90. $y = -\frac{1}{2}x^2$

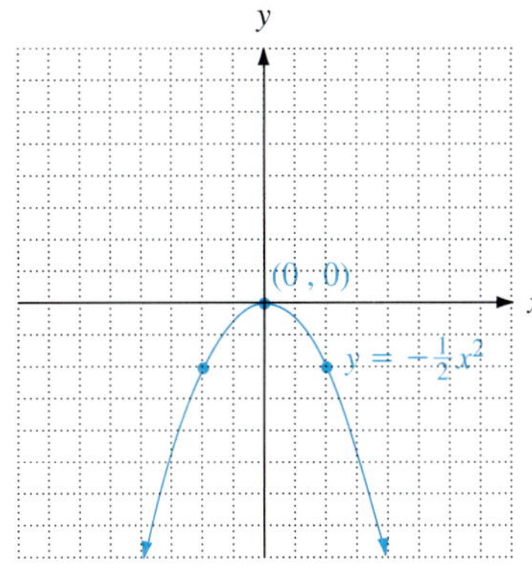

91. $y = x^2 - 3x - 4$

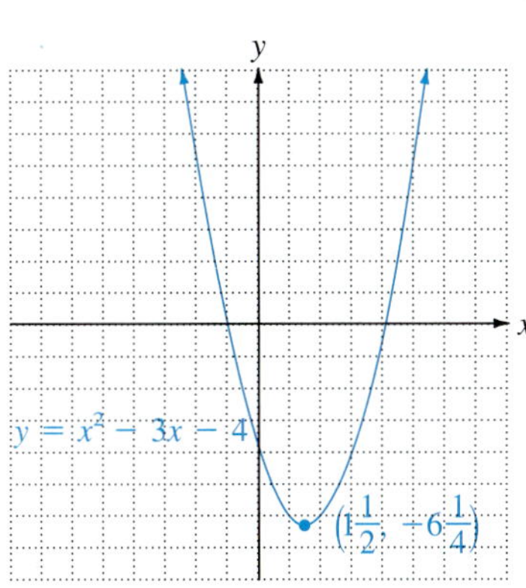

92. $y = \frac{1}{2}x^2 - 2$

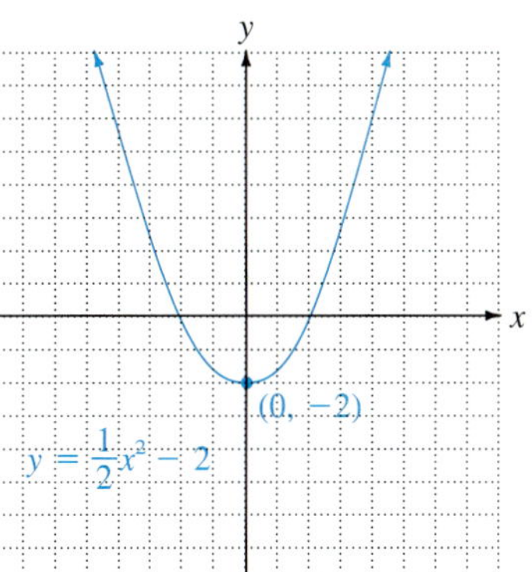

93. $y = -2x^2 + 12x - 17$

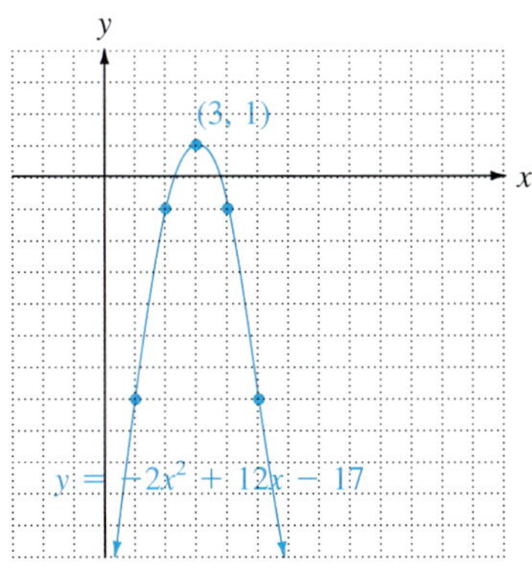

94. $y = -3x^2 - 2x + 4$

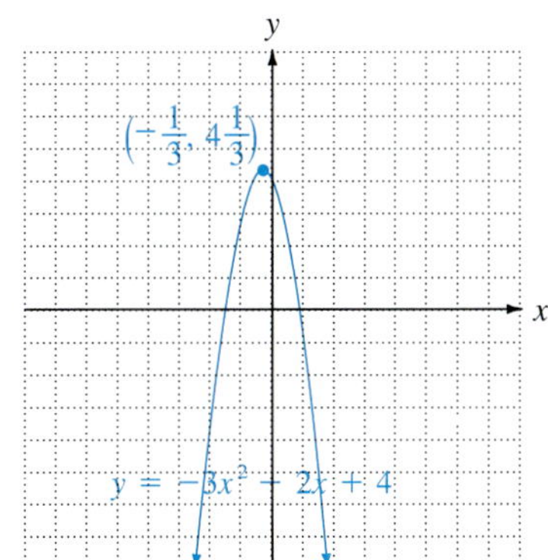

10.5 *Solve each word problem.*

95. Jon is building a dog pen against the house. He has 11 feet of fencing and will pen in an area of 15 square feet. What are the dimensions of the pen?

or

$l = 6$ feet, $w = 2.5$ feet

96. The hypotenuse of a right triangle is 20 centimeters. One leg of the right triangle is 4 centimeters less than the other. What is the length of each leg of the right triangle? 16 centimeters, 12 centimeters

97. Last year the Golf Club on campus raised \$720 through dues assigned equally to each member. This year there were four fewer members. As a result, the dues went up by \$6 for each member. How many members were in the Golf Club last year?

24 members were in the club last year.

98. One positive number is 2 more than twice another positive number. The difference of the squares of these numbers is 119. Find the numbers that satisfy these conditions.

$(2x + 2)^2 - x^2 = 119$

$3x^2 + 8x - 115 = 0$

$(3x + 23)(x - 5) = 0$

5 and 12

99. The length of a room exceeds its width by 4 feet. A rug covers the floor area except for a border 2 feet wide all around it. The area of the border is 68 square feet. What is the area of the rug? 38.25 ft^2

★ **100.** Alice drove 90 miles to visit her cousin. Her average speed on the trip home was 15 mph faster than her speed on the trip going. Her total travel time for both going and returning was 3.5 hours. What was her average rate of speed on each trip?

going 45 mph, returning 60 mph

Chapter 10 Test

Solve by any desired method. If there is no real number solution, say so.

1. $5x^2 + 7x = 4$

2. $3x^2 + 13x = 10$

3. $2x^2 = 2x - 5$

4. $6x^2 - 7x - 5 = 0$

5. $12x^2 + 11x = 5$

6. $18x^2 + 32 = 48x$

7. $2x^2 - 11x + 3 = 5x + 3$

8. $2x^2 + 5 = 37$

9. $2x(x - 6) = 6 - x$

10. $x^2 - x = \dfrac{3}{4}$

1. $x = \dfrac{-7 \pm \sqrt{129}}{10}$

2. $x = -5,\ x = \frac{2}{3}$

3. no real solution

4. $x = -\frac{1}{2},\ x = \frac{5}{3}$

5. $x = -\frac{5}{4},\ x = \frac{1}{3}$

6. $x = \frac{4}{3}$

7. $x = 0,\ x = 8$

8. $x = \pm 4$

9. $x = 6,\ x = -\frac{1}{2}$

10. $x = \frac{3}{2},\ x = -\frac{1}{2}$

11.	See graph
12.	See graph
13.	9 meters and 12 meters
14.	7 seconds

Graph each quadratic equation. Locate the vertex.

11. $y = 3x^2 - 6x$

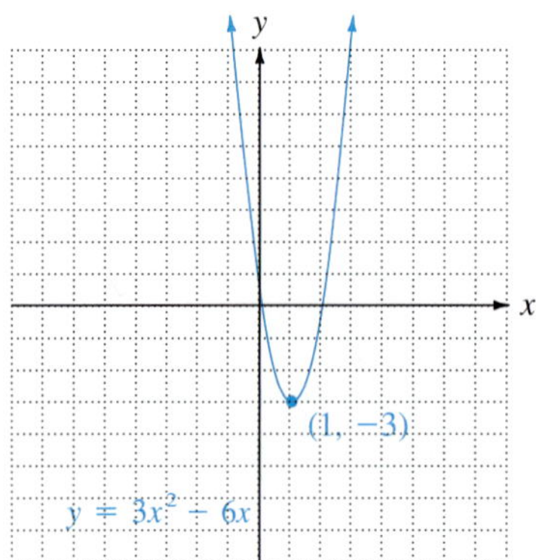

12. $y = -x^2 + 8x - 12$

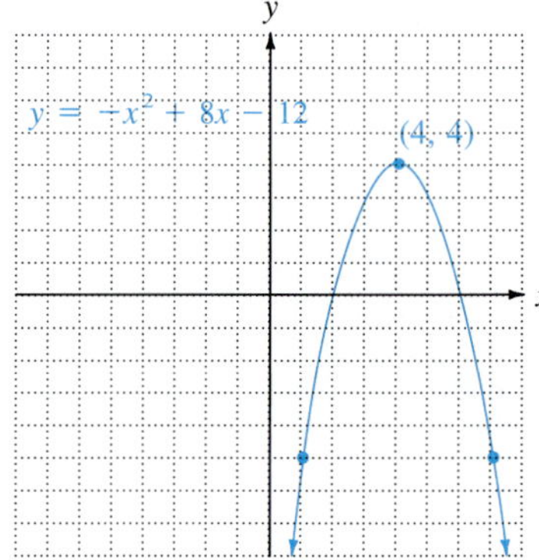

Solve each word problem.

13. The hypotenuse of a right triangle is 15 meters in length. One leg is 3 meters longer than the other leg. Find the length of each leg.

14. When an object is thrown upward, its height (S) in meters is given (approximately) by the quadratic equation

$$S = -5t^2 + vt + h$$

where v = the initial upward velocity in meters per second

t = the time of flight in seconds

h = the height above level ground from which the object is thrown

Suppose that a ball is thrown upward with a velocity of 33 meters/second at a height of 14 meters above the ground. When will it strike the ground?

Cumulative Test for Chapters 1–10

Approximately one-half of this test covers Chapters 0–9. The remainder covers the content of Chapter 10.

1. Simplify. $3x\{2y - 3[x + 2(x + 2y)]\}$

2. Evaluate $2x^2 - 3xy + y^2$ when $x = 2$ and $y = -3$.

3. Solve for x. $\frac{1}{2}(x - 2) = \frac{1}{3}(x + 10) - 2x$

4. Factor completely. $16x^4 - 1$

5. Multiply. $(3x + 1)(2x - 3)(x + 4)$

6. Solve for x. $\frac{3x}{x^2 - 4} = \frac{2}{x + 2} + \frac{4}{2 - x}$

7. Graph the line $y = -\frac{3}{4}x + 2$. Identify the slope and the y-intercept.

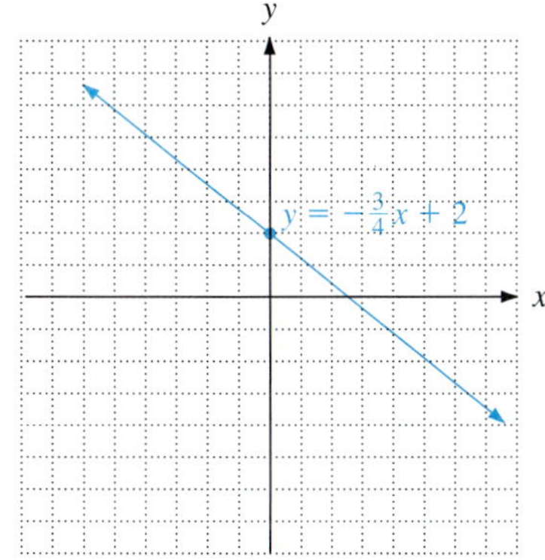

8. Find the equation of a line with slope $= -3$ that passes through the point $(2, -1)$.

9. Solve for (x, y).

$$3x + 2y = 5$$
$$7x + y = 19$$

10. Simplify. $(-3x^3y^2)^4$

11. Simplify. $\sqrt{18x^5y^6z^3}$

12. Multiply. $(\sqrt{2} + \sqrt{3})(2\sqrt{2} - 4\sqrt{3})$

1. $-27x^2 - 30xy$

2. 35

3. 2

4. $(2x - 1)(2x + 1)(4x^2 + 1)$

5. $6x^3 + 17x^2 - 31x - 12$

6. $x = -\frac{12}{5}$

7. slope $-\frac{3}{4}$; y-intercept $= +2$

8. $y = -3x + 5$

9. $(3, -2)$

10. $81x^{12}y^8$

11. $3x^2y^3z\sqrt{2xz}$

12. $-2\sqrt{6} - 8$

13. $-5\sqrt{5} + 11$

14. $r = \dfrac{A \pm \sqrt{A^2 + 3A}}{3}$

15. $b = \dfrac{\pm\sqrt{H + 6}}{5}$

16. $x = \frac{7}{2}, x = -5$

17. $x = \dfrac{-1 \pm 2\sqrt{5}}{2}$

18. $R = -12, R = 4$

19. $x = \dfrac{-11 \pm \sqrt{97}}{6}$

20. $x = \dfrac{-1}{7}, x = -5$

21. $x = \pm 7\sqrt{2}$

22. $x = \pm 5$

23. See graph

24. 6 meters, 16 meters

13. Rationalize the denominator. $\dfrac{\sqrt{5} - 3}{\sqrt{5} + 2}$

14. Solve for r. $A = 3r^2 - 2Ar$

15. Solve for b. $H = 25b^2 - 6$

Solve.

16. $2x^2 + 3x = 35$

17. $(2x + 1)^2 = 20$

18. $\dfrac{60}{R} = \dfrac{60}{R + 8} + 10$

19. $3x^2 + 11x + 2 = 0$

20. $7x^2 + 36x + 5 = 0$

21. $x^2 = 98$

22. $3x^2 + 4 = 79$

23. Graph $y = x^2 + 6x + 10$. Locate the vertex.

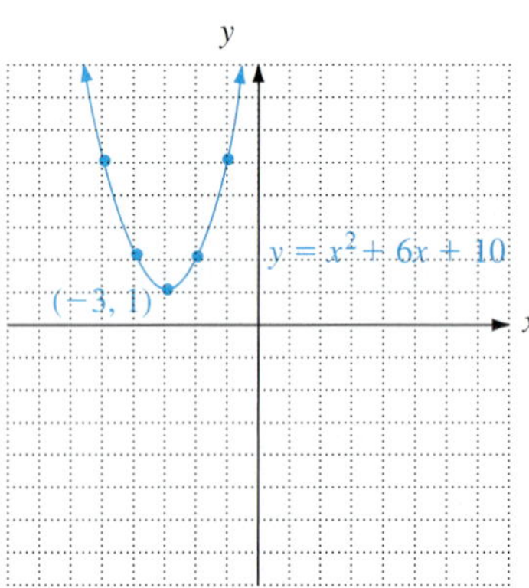

24. The length of a rectangle is 2 meters less than three times its width. The area is 96 square meters. Find its length and width.

Practice Final Examination

The following problems cover the content of Chapters 0–10. Follow the directions for each problem and simplify your answers.

1. Simplify.
$-2x + 3y\{7 - 2[x - (4x + y)]\}$.

2. Evaluate if $x = -2$ and $y = 3$.
$2x^2 - 3xy - 4y$

3. Simplify. $(-3x^2y)(-6x^3y^4)$

4. Combine like terms.
$5x^2y - 6xy + 8xy - 3x^2y - 10xy$

5. Solve for x.
$$\frac{1}{2}(x + 4) - \frac{2}{3}(x - 7) = 4x$$

6. Solve for b. $p = 2a + 2b$

7. Solve for x and graph the resulting inequality on a number line.
$5x + 3 - (4x - 2) \leq 6x - 8$

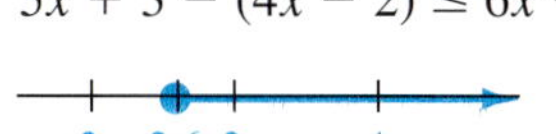

8. Multiply. $(2x + y)(x^2 - 3xy + 2y^2)$

9. Factor completely. $4x^2 - 18x - 10$

10. Factor completely. $25x^4 - 25$

11. Combine.
$$\frac{2}{x - 3} - \frac{3}{x^2 - x - 6} + \frac{4}{x + 2}$$

12. Simplify. $$\frac{\dfrac{3}{x} + \dfrac{5}{2x}}{1 + \dfrac{2}{x + 2}}$$

13. Solve for x.
$$\frac{2}{x + 2} = \frac{4}{x - 2} + \frac{3x}{x^2 - 4}$$

14. Find the slope of the line and then graph the line. $5x - 2y - 3 = 0$

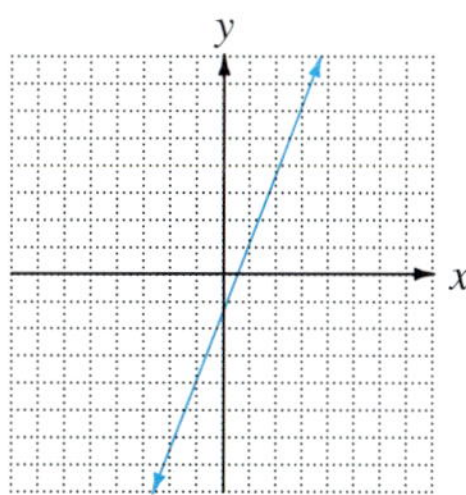

15. Find an equation of the line that has a slope of $-\frac{3}{4}$ and passes through the point $(-2, 5)$.

16. Two quarter-circles are attached to a rectangle. The dimensions are indicated in the figure. Find the area of the region. Use 3.14 as an approximation for π.

8 in.
3 in. 3 in.
3 in. 3 in.
14 in.

Answers:

1. $-2x + 21y + 18xy + 6y^2$
2. 14
3. $18x^5y^5$
4. $2x^2y - 8xy$
5. $x = \frac{8}{5}$
6. $\frac{p - 2a}{2} = b$
7. $x \geq 2.6$
8. $2x^3 - 5x^2y + xy^2 + 2y^3$
9. $2(2x + 1)(x - 5)$
10. $25(x^2 + 1)(x + 1)(x - 1)$
11. $\frac{6x - 11}{(x + 2)(x - 3)}$
12. $\frac{11(x + 2)}{2x(x + 4)}$
13. $x = -\frac{12}{5}$
14. Slope $= \frac{5}{2}$
15. $3x + 4y = 14$
16. 24 sq. in. + 14.13 sq. in. = 38.13 sq. in.

17. $x = 7, y = -2$

18. $a = -2, b = -3$

19. $8x\sqrt{2}$

20. $6\sqrt{3} - 12 + 12\sqrt{2}$

21. $-\dfrac{\sqrt{15} + \sqrt{35} + \sqrt{21} + 7}{2}$

22. $x = \dfrac{5}{3}, x = -1$

23. $y = \dfrac{3 \pm \sqrt{7}}{2}$

24. $x = \pm 2$

25. $\sqrt{39}$

26. The number is 5.

27. Width is 7 meters. Length is 12 meters.

28. \$3000 at 10%, \$4000 at 14%

29. 200 reserved seat tickets and 160 general admission tickets

30. The base is 8 meters. The altitude is 17 meters.

17. Solve for x and y.

$$2x + 3y = 8$$
$$3x - 5y = 31$$

18. Solve for a and b.

$$a - \frac{3}{4}b = \frac{1}{4}$$
$$\frac{3}{2}a + \frac{1}{2}b = -\frac{9}{2}$$

19. Simplify and combine.
$2x\sqrt{50} + \sqrt{98x^2} - 3x\sqrt{18}$

20. Multiply and simplify.
$\sqrt{6}(3\sqrt{2} - 2\sqrt{6} + 4\sqrt{3})$

21. Simplify by rationalizing the denominator.
$\dfrac{\sqrt{3} + \sqrt{7}}{\sqrt{5} - \sqrt{7}}$

22. Solve for x. $3x^2 - 2x - 5 = 0$

23. Solve for y. $2y^2 = 6y - 1$

24. Solve for x. $4x^2 + 3 = 19$

25. In the right triangle shown, sides b and c are given. Find the length of side a.

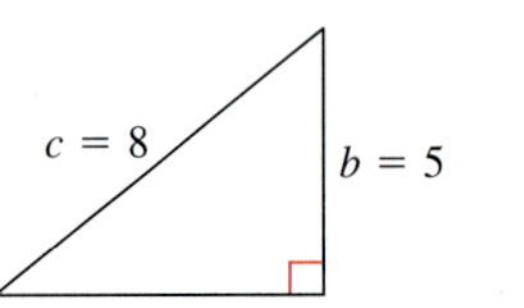

26. A number is tripled and then increased by 6. The result is 21. What is the original number?

27. A rectangular region has a perimeter of 38 meters. The length of the rectangle is 2 meters shorter than double the width. What are the dimensions of the rectangle?

28. Margaret invested \$7000 for one year. She placed part of the money in a tax-free bond earning 10% interest. She placed the rest in a money market account earning 14% interest. At the end of one year she had earned \$860 in interest. How much did she invest at each interest rate?

29. A benefit concert was held on the campus to raise scholarship money. A total of 360 tickets were sold. Admission prices were \$5 for reserved seats and \$3 for general admission. Total receipts were \$1480. How many reserved seat tickets were sold? How many general admission tickets were sold?

30. The area of a triangle is 68 square meters. The altitude of the triangle is one meter longer than double the base of the triangle. Find the altitude and base of the triangle.

Glossary

Absolute Value of a Number (1.1) The absolute value of a number x is the distance between 0 and the number x on the number line. It is written as $|x|$. $|x| = x$ if $x \geq 0$, but $|x| = -x$ if $x < 0$.

Altitude of a Geometric Figure (3.5) The height of the geometric figure. In the three figures shown the altitude is labeled a.

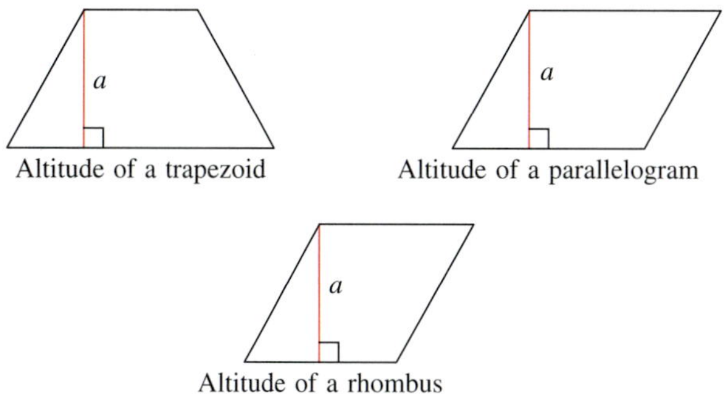

Altitude of a trapezoid
Altitude of a parallelogram
Altitude of a rhombus

Altitude of a Triangle (3.5) The height of any given triangle. In the three triangles shown the altitude is labeled a.

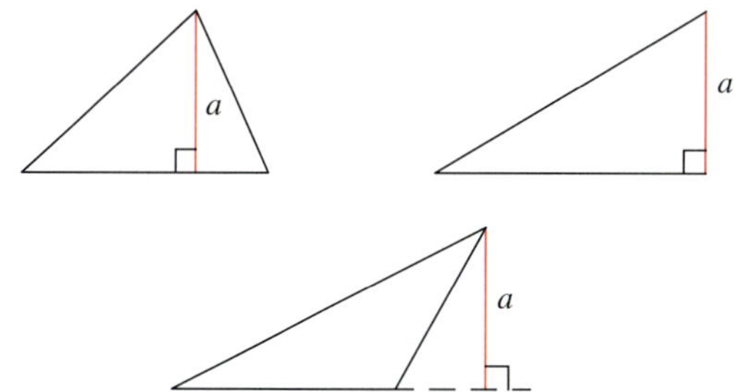

Associative Property of Addition (1.1) If a, b, and c are real numbers, then

$$a + (b + c) = (a + b) + c$$

This property states that if three numbers are added it does not matter *which two numbers* are added first, the result will be the same.

Associative Property of Multiplication (1.3) If a, b, and c are real numbers, then

$$a \times (b \times c) = (a \times b) \times c$$

This property states that if three numbers are multiplied it does not matter *which two numbers* are multiplied first, the result will be the same.

Base (1.4) The number or variable that is raised to a power. In the expression 2^6, the number 2 is the base.

Base of a triangle (3.5) The side of a triangle that is perpendicular to the altitude.

Binomial (1.5) A polynomial of two terms. The expressions $a + 2b$, $6x^3 + 1$, and $5a^3b^2 + 6ab$ are all binomials.

Circumference of a Circle (3.5) The distance around a circle. The circumference of a circle is given by the formula $C = \pi d$ or $C = 2\pi r$, where d is the diameter of the circle and r is the radius of the circle.

Coefficient (4.1) A coefficient is a factor or a group of factors in a product. In the term $4xy$ the coefficient of y is $4x$, but the coefficient of xy is 4. In the term $-5x^3y$ the coefficient of x^3y is -5.

Commutative Property for Addition (1.1) If a and b are any real numbers, then $a + b = b + a$.

Commutative Property for Multiplication (1.3) If a and b are any real numbers, then $ab = ba$.

Complex Fraction (6.4) A fraction that contains at least one fraction in the numerator or in the denominator or both. These three fractions are complex fractions:

$$\frac{7 + \frac{1}{x}}{x^2 + 2}, \quad \frac{1 + \frac{1}{5}}{2 - \frac{1}{7}}, \quad \text{and} \quad \frac{1}{\frac{3}{4}}$$

Constant (2.3) Symbol or letter that is used to represent exactly one single quantity during a particular problem or discussion.

Coordinates of a Point (7.1) An ordered pair of numbers (x, y) that specifies the location of a point in a rectangular coordinate system.

Degree of a Polynomial (4.3) The degree of the highest-degree term of a polynomial. The degree of the polynomial $5x^3 + 2x^2 - 6x + 8$ is 3. The degree of the polynomial $5x^2y^2 + 3xy + 8$ is 4.

Degree of a Term of a Polynomial (4.3) The sum of the exponents of the variables in the term. The degree of $3x^3$ is 3. The degree of $4x^5y^2$ is 7.

Denominator (0.1) and (6.1) The bottom number or algebraic expression in a fraction. The denominator of

$$\frac{3x - 2}{x + 4}$$

is $x + 4$. The denominator of $\frac{3}{7}$ is 7. The denominator of a fraction may not be zero.

Dependent Equations (8.1) Two equations are dependent if every value that satisfies one equation satisfies the other. A system of two dependent equations in two variables will not have a unique solution.

Diagonal of a Four-sided Figure (3.5) A line connecting two nonadjacent corners of the figure. In each of the figures shown, line AC is a diagonal.

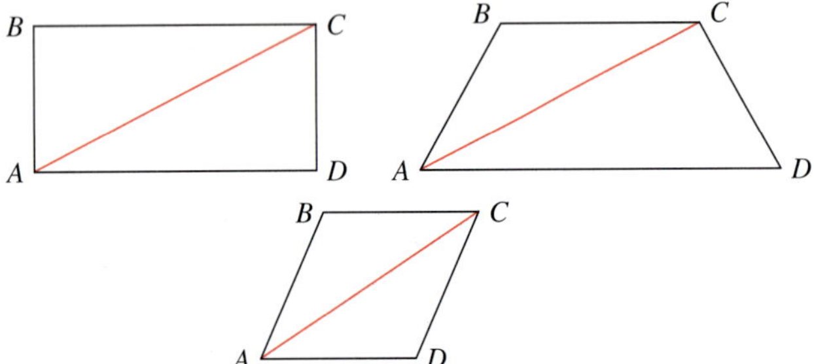

Difference (3.1) The result of subtracting one number or expression from another. The mathematical expression $x - 6$ can be written in words as the difference between x and 6.

Difference of Two Squares Polynomial (5.5) A polynomial of the form $a^2 - b^2$ that may be factored by using the formula

$$a^2 - b^2 = (a + b)(a - b).$$

Distributive Property (1.5) For all real numbers a, b, and c, it is true that $a(b + c) = ab + ac$.

Dividend (0.4) The number that is to be divided by another. In the problem $30 \div 5 = 6$, the three parts are as follows:

5 is the divisor → $5\overline{)30}$ with quotient 6

30 is the dividend

6 is the quotient

Divisor (0.4) The number you divide into another.

Domain of a Relation (7.6) In any relation, the set of values that can be used for the independent variable is called its domain. This is the set of all the first coordinates of the ordered pairs that define the relation.

Equilateral Triangle (3.5) A triangle with three sides equal in length and three angles that measure 60°. Triangle ABC is an equilateral triangle.

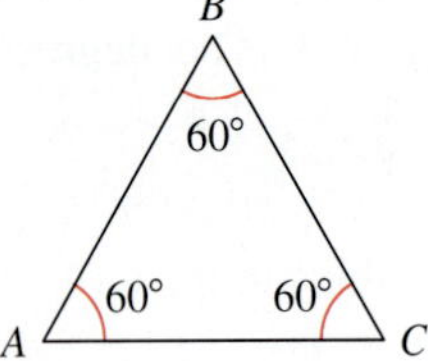

Even Integers (1.3) Integers that are exactly divisible by 2, such as $\ldots, -4, -2, 0, 2, 4, 6, \ldots$.

Exponent (1.4) The number that indicates the power of a base. If the number is a positive integer it indicates how many times the base is multiplied. In the expression 2^6, the exponent is 6.

Expression (3.1) A mathematic expression is any quantity using numbers and variables. Therefore, $2x$, $7x + 3$, and $5x^2 + 6x$ are all mathematical expressions.

Extraneous Solution (6.5) and (9.6) An obtained solution to an equation that when substituted back into the original equation, does *not* yield an identity. $x = 2$ is an extraneous solution to the equation

$$\frac{x}{x - 2} - 4 = \frac{2}{x - 2}$$

An extraneous solution is also called an extraneous root.

Factor (0.1) and (5.1) When two or more numbers, variables, or algebraic expressions are multiplied, each is called a factor. If we write $3 \cdot 5 \cdot 2$, the factors are 3, 5, and 2. If we write $2xy$, the factors are 2, x, and y. In the expression $(x - 6)(x + 2)$, the factors are $(x - 6)$ and $(x + 2)$.

Fractions

Algebraic Fractions (6.1) The indicated quotient of two algebraic expressions.

$$\frac{x^2 + 3x + 2}{x - 4} \quad \text{and} \quad \frac{y - 6}{y + 8}$$

are algebraic fractions. In these fractions the value of the denominator cannot be zero.

Numerical Fractions (0.1) A set of numbers used to describe parts of whole quantities. A numerical fraction can be represented by the quotient of two integers for which the denominator is not zero. The numbers $\frac{1}{5}$, $-\frac{2}{3}$, $\frac{8}{2}$, $-\frac{4}{31}$, $\frac{8}{1}$, and $-\frac{12}{1}$ are all numerical fractions. The set of rational numbers can be represented by numerical fractions.

Function (7.6) A relation in which no two different ordered pairs have the same first coordinate.

Hypotenuse of a Right Triangle (9.6) The side opposite the right angle in any right triangle. The hypotenuse is always the longest side of a right triangle. Side AB is the hypotenuse of triangle ABC.

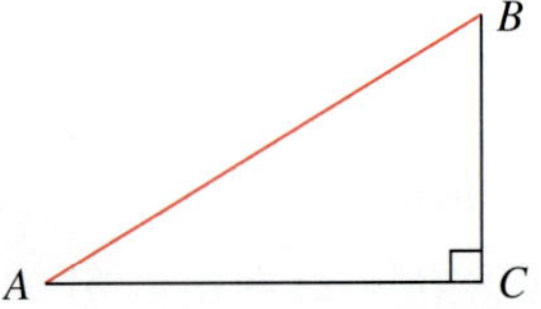

Identity (2.1) A statement that is always true. The equations $5 = 5$, $7 + 4 = 7 + 4$, and $x + 8 = x + 8$ are examples of identities.

Imaginary Number (9.1) A complex number that is not a real number. Imaginary numbers can be created by taking the square root of a negative real number. The numbers $\sqrt{-9}$, $\sqrt{-7}$, and $\sqrt{-12}$ are all imaginary numbers.

Improper Fraction (0.1) A numerical fraction whose numerator is larger than or equal to its denominator. $\frac{8}{3}$, $\frac{5}{2}$, and $\frac{7}{7}$ are improper fractions.

Inconsistent System of Equations (8.1) A system of equations that does not have a solution.

Independent Equations (8.1) Two equations that are not dependent are said to be independent.

Inequality (2.6), (2.7), and (7.5) A mathematical relationship between quantities that are not equal. $x \leq -3$, $w > 5$, and $x < 2y + 1$ are mathematical inequalities.

Integers (1.1) The set of numbers . . . , $-5, -4, -3, -2, -1, 0, 1, 2, 3, 4, 5$, ...

Intercepts of an Equation (7.2) The point or points where the graph of the equation crosses the x-axis or the y-axis or both. (*See* x-intercept.)

Irrational Number (9.1) A real number that cannot be expressed in the form $\frac{a}{b}$, where a and b are integers and $b \neq 0$. $\sqrt{2}$, π, $5 + 3\sqrt{2}$, and $-4\sqrt{7}$ are irrational numbers.

Isosceles Triangle (3.5) A triangle with two equal sides and two equal angles. Triangle ABC is an isosceles triangle. Angle BAC is equal to angle ACB. Side AB is equal in length to side BC.

Least Common Denominator of Numerical Fractions (0.2) The smallest whole number that is exactly divisible by all denominators of a group of fractions. The least common denominator (LCD) of $\frac{1}{6}$, $\frac{2}{3}$, and $\frac{3}{5}$ is 30. The least common denominator is also called the lowest common denominator.

Leg of a Right Triangle (9.6) One of the two shorter sides of a right triangle. Side AC and side BC are legs of triangle ABC.

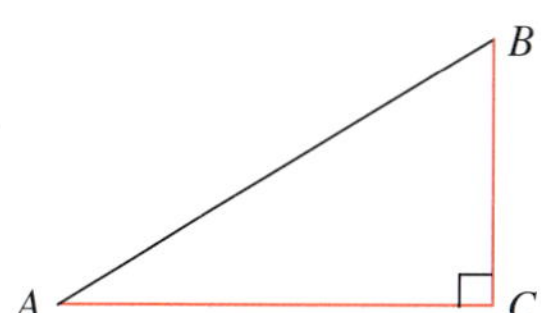

Like Terms (1.6) Terms that have identical variables and exponents. In the expression $5x^3 + 2xy^2 + 6x^2 - 3xy^2$, the term $2xy^2$ and the term $-3xy^2$ are like terms.

Linear Equation in Two Variables (7.2) An equation of the form $Ax + By = C$, where A, B, and C are real numbers. The graph of a linear equation in two variables is a straight line.

Line of Symmetry of a Parabola (10.4) A line that can be drawn through a parabola such that, if the graph were folded on this line, the two halves of the curve would correspond exactly. The line of symmetry through the parabola formed by $y = ax^2 + bx + c$ is given by $x = \frac{-b}{2a}$. The line of symmetry of a parabola always passes through the vertex of a parabola.

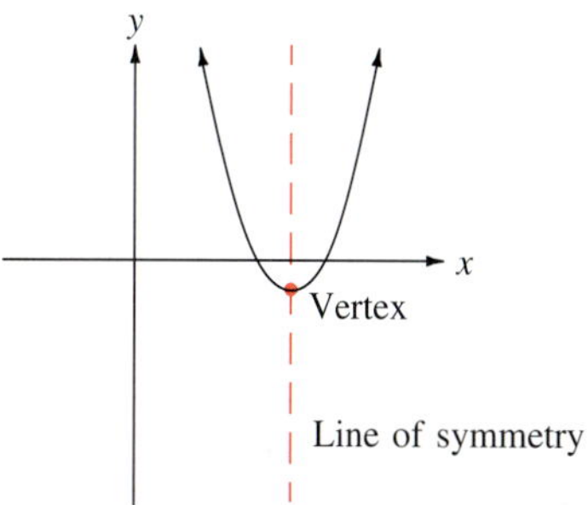

Mixed Number (0.1) A number that consists of an integer written next to a proper fraction. $2\frac{1}{3}$, $4\frac{6}{7}$, and $3\frac{3}{8}$ are all mixed numbers. Mixed numbers are sometimes called mixed fractions or mixed numerals.

Monomial (1.5) A polynomial of one term. The expressions $3xy$, $5a^2b^3cd$, and -6 are all monomials.

Natural Numbers (0.1) The set of numbers 1, 2, 3, 4, 5, ... This set is also called the set of counting numbers.

Numeral (0.1) The symbol used to describe a number.

Numerator (0.1) The top number or algebraic expression in a fraction. The numerator of

$$\frac{x + 3}{5x - 2}$$

is $x + 3$. The numerator of $\frac{12}{13}$ is 12.

Numerical Coefficient (4.1) The number that is multiplied by a variable or a group of variables. The numerical coefficient in $5x^3y^2$ is 5. The numerical coefficient in $-6abc$ is -6. The numerical coefficient in x^2y is 1. A numerical coefficient of 1 is not usually written.

Odd Integers (1.3) Integers that are not exactly divisible by 2, such as $-3, -1, 1, 3, 5, 7, 9$, ...

Opposite of a Number (1.1) Two numbers that are the same distance from zero on the number line but lie on different sides of it are considered opposites. The opposite of -6 is 6. The opposite of $\frac{22}{7}$ is $-\frac{22}{7}$.

Ordered Pair (7.1) A pair of numbers presented in a specified order. An ordered pair is often used to specify a location on a graph. Every point in a rectangular coordinate system can be represented by an ordered pair (x, y).

Origin (7.1) The point $(0, 0)$ in a rectangular coordinate system.

Parabola (10.4) A curve created by graphing a quadratic equation. The curves shown are all parabolas. The graph of the equation $y = ax^2 + bx + c$, where $a \neq 0$, will always be a parabola.

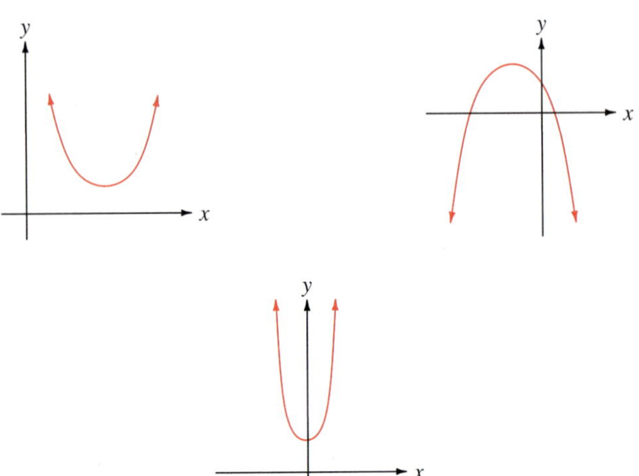

Parallel Lines (7.3) and (8.1) Two straight lines that never intersect. The graph of an inconsistent system of two linear equations in two variables will result in parallel lines.

Parallelogram (3.5) A four-sided figure with opposite sides parallel. Figure $ABCD$ is a parallelogram.

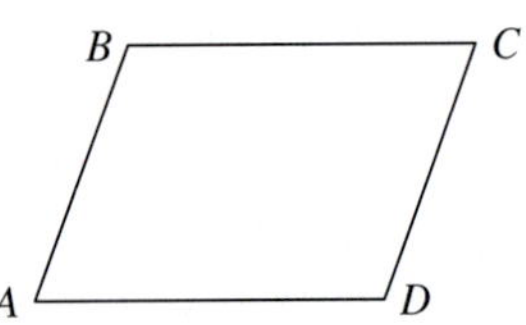

Percent (0.5) Hundredths or "per one hundred"; indicated by the % symbol. Thirty-seven hundredths $\left(\frac{37}{100}\right) = 37\%$ (thirty-seven percent).

Perfect Square Number (5.5) A number that is the square of an integer. The numbers 1, 4, 9, 16, 25, 36, 49, 64, 81, 100, 121, 144, . . . are perfect square numbers.

Perfect Square Trinomial (5.5) A polynomial of the form $a^2 + 2ab + b^2$ or $a^2 - 2ab + b^2$ that may be factored using one of the following formulas:

$$a^2 + 2ab + b^2 = (a + b)^2$$

or

$$a^2 - 2ab + b^2 = (a - b)^2$$

Perimeter (3.5) The distance around any plane figure. The perimeter of this triangle is 13. The perimeter of this rectangle is 20.

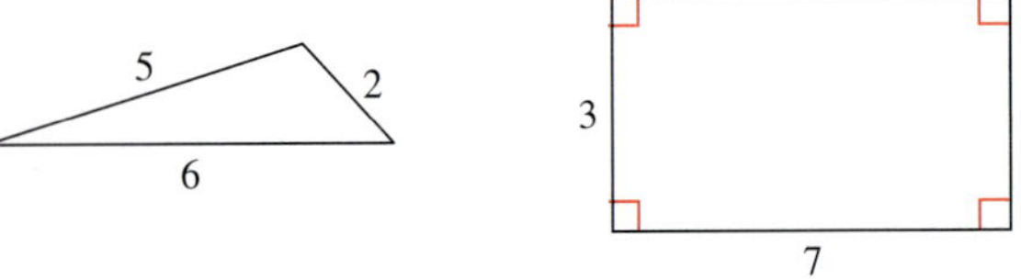

Pi (3.5) An irrational number, denoted by the symbol π, that is approximately equal to 3.141592654. In most cases 3.14 can be used as a sufficiently accurate approximation for π.

Polynomial (1.5) Expressions that contain terms with nonnegative integer exponents. The expressions $5ab + 6$, $x^3 + 6x^2 + 3$, -12, and $x + 3y - 2$ *are all polynomials.* The expressions $x^{-2} + 2x^{-1}$, $2\sqrt{x} + 6$, and $\frac{5}{x} + 2x^2$ *are not polynomials.*

Prime Number (0.1) Any natural number greater than 1 whose only natural number factors are 1 and itself. The first eight prime numbers are 2, 3, 5, 7, 11, 13, 17, and 19.

Prime Polynomial (5.6) A prime polynomial is a polynomial that cannot be factored by the methods of elementary algebra. $x^2 + x + 1$ is a prime polynomial.

Principal (3.4) In monetary problems, the principal is the original amount of money invested or borrowed.

Principal Square Root (9.1) For any given nonnegative number N, the principal square root of N (written $\sqrt{N}$) is the nonnegative number a if and only if $a^2 = N$. The principal square root of 25 ($\sqrt{25}$) is 5.

Proper Fraction (0.1) A numerical fraction whose numerator is less than its denominator; $\frac{3}{7}$, $\frac{2}{5}$, and $\frac{8}{9}$ are proper fractions.

Proportion (6.6) A proportion is an equation stating that two ratios are equal.

$$\frac{a}{b} = \frac{c}{d}$$

is a proportion.

Pythagorean Theorem (9.6) In any right triangle, if c is the length of the hypotenuse and a and b are the lengths of the two legs, then $c^2 = a^2 + b^2$.

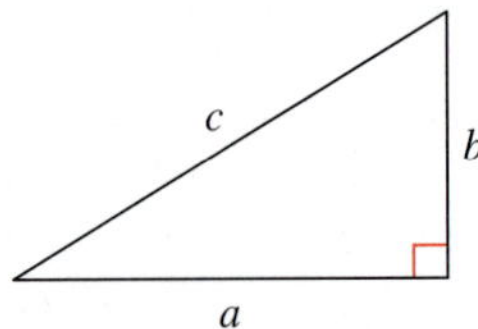

Quadratic Equation (5.7) A quadratic equation is a polynomial equation with one variable that contains at least one term with the variable squared, but no term with the variable raised to a higher power. $5x^2 + 6x - 3 = 0$, $x^2 = 7$, and $5x^2 = 2x$ are all quadratic equations.

Quadratic Formula (10.3) If $ax^2 + bx + c = 0$ and $a \neq 0$, then the solutions to the quadratic equation are given by

$$x = \frac{-b \pm \sqrt{b^2 - 4ac}}{2a}$$

Quotient (0.4) The result of dividing one number or expression by another. In the problem $12 \div 4 = 3$, the quotient is 3.

Radical (9.1) An expression composed of a radical sign and a radicand. The expressions $\sqrt{5x}$, $\sqrt{\frac{3}{5}}$, $\sqrt{5x + b}$, and $\sqrt{10}$ are called radicals.

Radical Equation (9.6) An equation that contains one or more radicals. $\sqrt{x + 4} = 12$, $\sqrt{x} = 3$, and $\sqrt{3x} + 4 = x + 2$ are all radical equations.

Radical Sign (9.1) The symbol $\sqrt{\ }$ used to indicate the square root of a number.

Radicand (9.1) The expression beneath the radical sign. The radicand in $\sqrt{5ab}$ is $5ab$.

Range of a Relation (7.6) In any relation, the set of values that represents the dependent variable is called its range. This is the set of all the second coordinates of the ordered pairs that define the relation.

Ratio (6.6) The ratio of one number a to another number b is the quotient $a \div b$ or $\frac{a}{b}$.

Rational Numbers (1.1) and (9.1) A number that can be expressed in the form $\frac{a}{b}$, where a and b are integers and $b \neq 0$. $\frac{7}{3}, -\frac{2}{5}, \frac{7}{-8}, \frac{5}{1}$, 1.62, and 2.7156 are rational numbers.

Rationalizing the Denominator (9.5) The process of transforming a fraction that contains a radical in the denominator to an equivalent fraction that does not contain a radical in the denominator.

Original Fraction	*Equivalent Fraction With Rationalized Denominator*
$\frac{3}{\sqrt{2}}$	$\frac{3\sqrt{2}}{2}$

Rationalizing the Expression (9.5) The process of transforming a radical that contains a fraction to an equivalent expression that does not contain a fraction inside a radical.

Original Radical	*Equivalent Expression after It Has Been Rationalized*
$\sqrt{\frac{3}{5}}$	$\frac{\sqrt{15}}{5}$

Real Number (9.1) Any number that is rational or irrational. 2, 7, $\sqrt{5}$, $\frac{3}{8}$, π, $-\frac{7}{5}$, and $-3\sqrt{5}$ are all real numbers.

Rectangle (3.5) A four-sided figure with opposite sides parallel and all interior angles measuring 90°. The opposite sides of a rectangle are equal.

Relation (7.6) A relation is any set of ordered pairs.

Rhombus (3.5) A parallelogram with four equal sides. Figure *ABCD* is a rhombus.

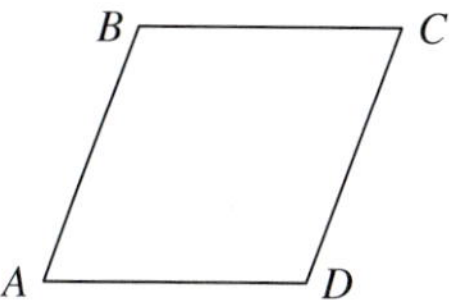

Right Angle (3.5) An angle that measures 90°. Right angles are usually labeled in a sketch by using a small square to indicate that it is a right angle. Here angle *ABC* is a right angle.

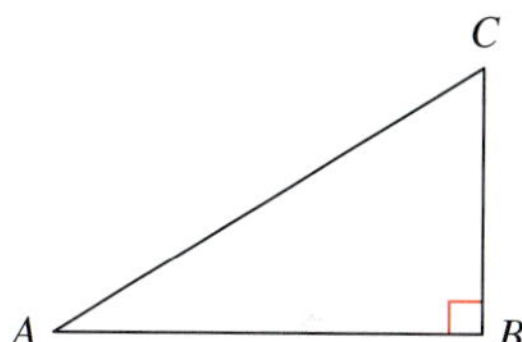

Right Triangle (3.5) A triangle that contains a right angle.

Root of an Equation (2.1) and (5.7) A value of the variable that makes an equation into a true statement. The root of an equation is also called the solution of an equation.

Scientific Notation (4.2) A positive number is written in scientific notation if it is in the form $a \times 10^n$, where $1 \leq a < 10$ and n is an integer.

Slope-Intercept Form (7.3) The equation of a line that has slope m and the y-intercept at $(0, b)$ is given by $y = mx + b$.

Slope of a Line (7.3) The ratio of change in y over the change in x for any two different points on a nonvertical line. The slope m is determined by

$$m = \frac{y_2 - y_1}{x_2 - x_1}$$

where $x_2 \neq x_1$ for any two points (x_1, y_1) and (x_2, y_2) on a nonvertical line.

Solution of an Equation (2.1) A number that, when substituted into a given equation, yields an identity. The solution of an equation is also called the root of an equation.

Solution of a Linear Inequality (2.7) The possible values that make a linear inequality true.

Solution of an Inequality in Two Variables (7.5) The set of all possible ordered pairs that when substituted into the inequality will yield a true statement.

Solution to a System of Two Equations in Two Variables (8.2) An ordered pair that can be substituted into each equation and an identity will be obtained in each case.

Square (3.5) A rectangle with four equal sides.

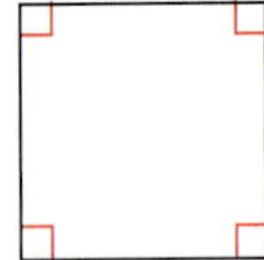

Square Root (9.1) For any given nonnegative number N, the square root of N is the number a if $a^2 = N$. One square root of 16 is 4 since $(4)^2 = 16$. Another square root of 16 is -4 since $(-4)^2 = 16$. When we write $\sqrt{16}$, we want only the positive root, which is 4.

Standard Form of a Quadratic Equation (5.7) A quadratic equation that is in the form $ax^2 + bx + c = 0$.

Subscript of a Variable (3.2) A small number or letter written slightly below and to the right of a variable. In the expressions $5 = 2(x - x_0)$, the subscript of x is 0. In the expression $t_f = 5(t_a - b)$ the subscript of the first t is f. The subscript of the second t is a. A subscript is used to indicate a different value of the variable.

System of Equations (8.1) A set of two or more equations that must be considered together.

Term (1.5) A number, a variable, or a product of numbers and variables. For example, in the expression $a^3 - 3a^2b + 4ab^2 + 6b^3 + 8$, there are five terms. They are a^3, $-3a^2b$, $4ab^2$, $6b^3$, and 8. The terms of a polynomial are separated by plus and minus signs.

Trapezoid (3.5) A four-sided figure with two sides parallel. The parallel sides are called the bases of the trapezoid. Figure $ABCD$ is a trapezoid.

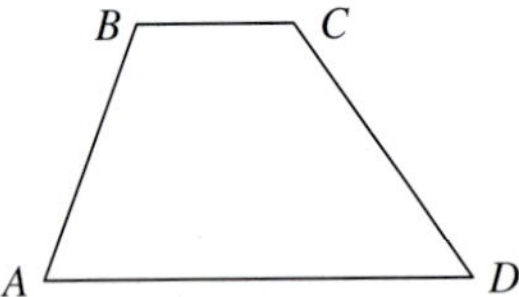

Trinomial (1.5) A polynomial of three terms. The expressions $x^2 + 6x - 8$ and $a + 2b - 3c$ are trinomials.

Variable (1.4) A letter that is used to represent a number or a set of numbers.

Variation (9.7) An equation relating values of one variable to those of other variables. An equation of the form $y = kx$, where k is a constant, indicates *direct variation*. An equation of the form $y = \frac{k}{x}$, where k is a constant, indicates *inverse variation*. In both cases, k is called the *constant of variation*.

Vertex of a Parabola (10.4) The lowest point on a parabola opening upward or the highest point on a parabola opening downward. The x-coordinate of the vertex of the parabola formed by the equation $y = ax^2 + bx + c$ is given by $x = \frac{-b}{2a}$.

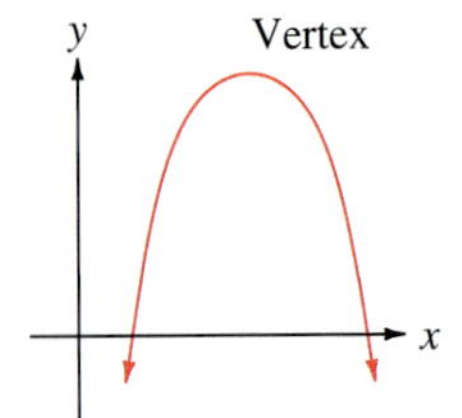

Parabola opening downward

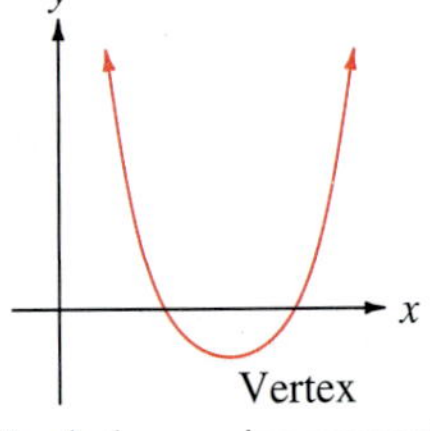

Parabola opening upward

Vertical Line Test (7.6) If a vertical line can intersect the graph of a relation more than once, the relation is not a function.

Whole Numbers (0.1) The set of numbers 0, 1, 2, 3, 4, 5, ...

x-Intercept (7.2) The number a is the x-intercept of a line if the line crosses the x-axis at $(a, 0)$. The x-intercept of line l on the graph below is 4.

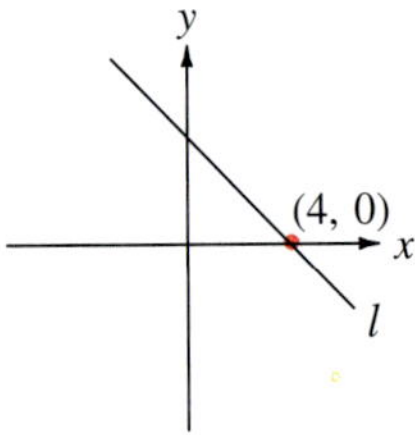

y-Intercept (7.2) The number b is the y-intercept of a line if the line crosses the y-axis at $(0, b)$. The y-intercept of line p on the graph below is 3.

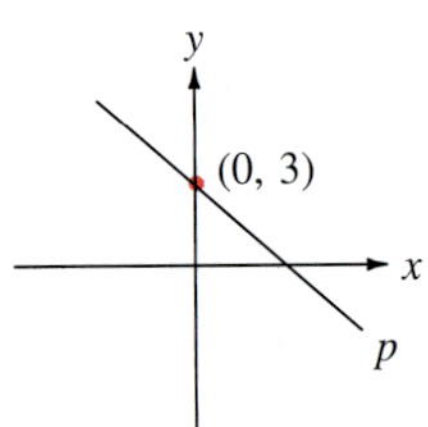

APPENDIX A
Table of Square Roots

x	$\sqrt{x}$	x	$\sqrt{x}$	x	$\sqrt{x}$	x	$\sqrt{x}$	x	$\sqrt{x}$
1	1.000	41	6.403	81	9.000	121	11.000	161	12.689
2	1.414	42	6.481	82	9.055	122	11.045	162	12.728
3	1.732	43	6.557	83	9.110	123	11.091	163	12.767
4	2.000	44	6.633	84	9.165	124	11.136	164	12.806
5	2.236	45	6.708	85	9.220	125	11.180	165	12.845
6	2.449	46	6.782	86	9.274	126	11.225	166	12.884
7	2.646	47	6.856	87	9.327	127	11.269	167	12.923
8	2.828	48	6.928	88	9.381	128	11.314	168	12.961
9	3.000	49	7.000	89	9.434	129	11.358	169	13.000
10	3.162	50	7.071	90	9.487	130	11.402	170	13.038
11	3.317	51	7.141	91	9.539	131	11.446	171	13.077
12	3.464	52	7.211	92	9.592	132	11.489	172	13.115
13	3.606	53	7.280	93	9.644	133	11.533	173	13.153
14	3.742	54	7.348	94	9.695	134	11.576	174	13.191
15	3.873	55	7.416	95	9.747	135	11.619	175	13.229
16	4.000	56	7.483	96	9.798	136	11.662	176	13.266
17	4.123	57	7.550	97	9.849	137	11.705	177	13.304
18	4.243	58	7.616	98	9.899	138	11.747	178	13.342
19	4.359	59	7.681	99	9.950	139	11.790	179	13.379
20	4.472	60	7.746	100	10.000	140	11.832	180	13.416
21	4.583	61	7.810	101	10.050	141	11.874	181	13.454
22	4.690	62	7.874	102	10.100	142	11.916	182	13.491
23	4.796	63	7.937	103	10.149	143	11.958	183	13.528
24	4.899	64	8.000	104	10.198	144	12.000	184	13.565
25	5.000	65	8.062	105	10.247	145	12.042	185	13.601
26	5.099	66	8.124	106	10.296	146	12.083	186	13.638
27	5.196	67	8.185	107	10.344	147	12.124	187	13.675
28	5.292	68	8.246	108	10.392	148	12.166	188	13.711
29	5.385	69	8.307	109	10.440	149	12.207	189	13.748
30	5.477	70	8.367	110	10.488	150	12.247	190	13.784
31	5.568	71	8.426	111	10.536	151	12.288	191	13.820
32	5.657	72	8.485	112	10.583	152	12.329	192	13.856
33	5.745	73	8.544	113	10.630	153	12.369	193	13.892
34	5.831	74	8.602	114	10.677	154	12.410	194	13.928
35	5.916	75	8.660	115	10.724	155	12.450	195	13.964
36	6.000	76	8.718	116	10.770	156	12.490	196	14.000
37	6.083	77	8.775	117	10.817	157	12.530	197	14.036
38	6.164	78	8.832	118	10.863	158	12.570	198	14.071
39	6.245	79	8.888	119	10.909	159	12.610	199	14.107
40	6.325	80	8.944	120	10.954	160	12.649	200	14.142

Unless the value of $\sqrt{x}$ ends in 000, all values are rounded to the nearest thousandth.

APPENDIX B

Metric Measurement and Conversion of Units

After studying this section, you will be able to:

1. *Convert from one metric unit of measurement to another.*
2. *Convert between metric and U.S. units of measure.*
3. *Convert from one type of unit of measure to another.*

1 *Converting from One Metric Unit of Measurement to Another*

Metric measurements are becoming more common in the United States. The metric system is used in many parts of the world and in the sciences. The basic unit of length in the metric system is the meter. For smaller measurements, centimeters are commonly used. There are 100 centimeters in 1 meter. For larger measurements, kilometers are commonly used. There are 1000 meters in 1 kilometer. The following tables give the metric units of measurement for length.

Metric Length

*1 kilometer (km)	=	1000 meters
1 hectometer (hm)	=	100 meters
1 dekameter (dam)	=	10 meters
*1 meter (m)	=	1 meter
1 decimeter (dm)	=	0.1 meter
*1 centimeter (cm)	=	0.01 meter
*1 millimeter (mm)	=	0.001 meter

The four most common units of metric measurement are indicated by an asterisk.

Metric Weight

1 kilogram (kg)	=	1000 grams
1 gram (g)	=	1 gram
1 milligram (mg)	=	0.001 gram

Metric Volume

1 kiloliter (kL)	=	1000 liters
1 liter (L)	=	1 liter
1 milliliter (mL)	=	0.001 liter

One way to convert from one metric unit to another is to multiply by 1. We know mathematically that to multiply by 1 will yield an equivalent expression since $(a)(1) = a$. In our calculations when solving word problems, we treat dimension symbols much like we treat variables in algebra.

EXAMPLE 1 Fred drove a distance of 5.4 kilometers. How many meters is that?

$$5.4 \text{ km} \cdot 1 = 5.4 \text{ km} \cdot \frac{1000 \text{ m}}{1 \text{ km}} = \frac{(5.4)(1000)}{1} \cdot \frac{\text{km}}{\text{km}} \cdot \text{m}$$

$$= 5400 \text{ m}$$

The distance is 5400 meters.

Practice Problem 1 June's room is 3.4 meters wide. How many centimeters is that? ■

EXAMPLE 2 A chemist measured 67 milliliters of a solution. How many liters is that?

$$67 \text{ mL} \cdot \frac{0.001 \text{ L}}{1 \text{ mL}} = 0.067 \text{ L}$$

The chemist measured 0.067 liters of the solution.

Practice Problem 2 The container has 125 liters of water. How many kiloliters is that? ■

2 *Converting between Metric and U.S. Units of Measure*

The most common relationships between the U.S. and metric systems are listed in the following table. Most of these values are approximate.

Metric Conversion Ratios

LENGTH:	1 inch = 2.54 centimeters
	39.37 inches = 1 meter
	1 mile = 1.61 kilometers
	0.62 mile = 1 kilometer
WEIGHT:	1 pound = 454 grams
	2.20 pounds = 1 kilogram
	1 ounce = 28.35 grams
	0.0353 ounce = 1 gram
LIQUID CAPACITY:	1 quart = 946 milliliters
	1.06 quarts = 1 liter
	1 gallon = 3.785 liters

EXAMPLE 3 A box weighs 190 grams. How many ounces is that? Round your answer to the nearest hundredth.

$$190 \text{ g} \cdot \frac{0.0353 \text{ oz}}{1 \text{ g}} = 6.707 \text{ oz}$$

$$\approx 6.71 \text{ oz} \quad \textit{rounding to the nearest hundredth}$$

The box weighs approximately 6.71 ounces.

Practice Problem 3 A bag of groceries weighs 5.72 pounds. How many kilograms is that? ■

EXAMPLE 4 Juanita drives 23 kilometers to work each day. How many miles is the trip? Round your answer to the nearest mile.

$$23 \text{ km} \cdot \frac{0.62 \text{ mi}}{1 \text{ km}} = 14.26 \text{ mi} \approx 14 \text{ mi} \quad \text{(rounded to the nearest mile)}$$

Juanita drives approximately 14 miles to work each day.

Practice Problem 4 Carlos installed an electrical connection that is 8.00 centimeters long. How many inches long is the connection? Round your answer to the nearest hundredth. ■

EXAMPLE 5 Anita purchased 42.0 gallons of gasoline for her car last month. How many liters did she purchase? Round your answer to the nearest liter.

$$42.0 \text{ gal} \cdot \frac{3.785 \text{ L}}{1 \text{ gal}} = 158.97 \text{ L} \approx 159 \text{ L} \quad \text{(rounded to the nearest liter)}$$

Anita purchased approximately 159 liters of gasoline last month.

Practice Problem 5 Warren purchased a 3.00-liter bottle of Coca-Cola. How many quarts of Coca-Cola is that? Round your answer to the nearest hundredth of a quart. ■

3 *Converting from One Type of Unit of Measure to Another Type*

Sometimes you will need to convert from one type of unit of measure to another. For example you may need to convert days to minutes or miles per hour to feet per second. Recall the U.S. units of measure.

Length	Time
12 inches = 1 foot	60 seconds = 1 minute
3 feet = 1 yard	60 minutes = 1 hour
5280 feet = 1 mile	24 hours = 1 day
1760 yards = 1 mile	7 days = 1 week
Weight	**Volume**
16 ounces = 1 pound	2 cups = 1 pint
2000 pounds = 1 ton	2 pints = 1 quart
	4 quarts = 1 gallon

EXAMPLE 6 A car was traveling at 50.0 miles per hour. How many feet per second was the car traveling? Approximate your answer to the nearest tenth of a foot per second.

$$\frac{50 \text{ mi}}{\text{hr}} \cdot \frac{5280 \text{ ft}}{1 \text{ mi}} \cdot \frac{1 \text{ hr}}{60 \text{ min}} \cdot \frac{1 \text{ min}}{60 \text{ sec}} = \frac{(50)(5280) \text{ ft}}{(60)(60) \text{ sec}} = \frac{73.333\ldots \text{ ft}}{\text{sec}}$$

The car was traveling at 73.3 feet per second, rounded to the nearest tenth of a foot per second.

Practice Problem 6 A speeding car was traveling at 70.0 miles per hour. How many feet per second was the car traveling? Approximate your answer to the nearest tenth of a foot per second. ■

B Exercises

1. How many meters are in 34 km? 34,000 m
2. How many meters are in 128 km? 128,000 m
3. How many centimeters are in 57 m? 5700 cm
4. How many centimeters are in 46 m? 4600 cm
5. How many millimeters are in 25 cm? 250 mm
6. How many millimeters are in 63 cm? 630 mm
7. How many meters are in 563 mm? 0.563 m
8. How many meters are in 831 mm? 0.831 m
9. How many milligrams are in 29.4 g? 29,400 mg
10. How many milligrams are in 75.2 g? 75,200 mg
11. How many kilograms are in 98.4 g? 0.0984 kg
12. How many kilograms are in 62.7 g? 0.0627 kg
13. How many milliliters are in 7 L? 7000 mL
14. How many milliliters are in 12 L? 12,000 mL
15. How many kiloliters are in 4 mL? 0.000004 kL
16. How many kiloliters are in 3 mL? 0.000003 kL

Use the table of metric conversion ratios to find each of the following. Round all answers to the nearest tenth.

17. How many inches are in 4.2 cm?

$$4.2 \text{ cm} \times \frac{1 \text{ in.}}{2.54 \text{ cm}} \approx 1.7 \text{ in.}$$

18. How many inches are in 3.8 cm?

$$3.8 \text{ cm} \times \frac{1 \text{ in.}}{2.54 \text{ cm}} \approx 1.5 \text{ in.}$$

19. How many kilometers are in 14 mi?

$$14 \text{ mi} \times \frac{1.61 \text{ km}}{1 \text{ mi}} \approx 22.5 \text{ km}$$

20. How many kilometers are in 13 mi?

$$13 \text{ mi} \times \frac{1.61 \text{ km}}{1 \text{ mi}} \approx 20.9 \text{ km}$$

21. How many meters are in 110 in.?

$$110 \text{ in.} \times \frac{2.54 \text{ cm}}{1 \text{ in.}} = 279.4 \text{ cm} \approx 2.8 \text{ m}$$

22. How many meters are in 150 in?

$$150 \text{ in.} \times \frac{2.54 \text{ cm}}{1 \text{ in.}} = 381 \text{ cm} \approx 3.8 \text{ m}$$

23. How many centimeters are in 7 in.?

$$7 \text{ in.} \times \frac{2.54 \text{ cm}}{1 \text{ in.}} \approx 17.8 \text{ cm}$$

24. How many centimeters are in 9 in.?

$$9 \text{ in.} \times \frac{2.54 \text{ cm}}{1 \text{ in.}} \approx 22.9 \text{ cm}$$

25. A box weighing 2.4 lb would weigh how many grams?

$$2.4 \text{ lb} \times \frac{454 \text{ g}}{1 \text{ lb}} = 1089.6 \text{ g}$$

26. A box weighing 1.6 lb would weigh how many grams?

$$1.6 \text{ lb} \times \frac{454 \text{ g}}{1 \text{ lb}} = 726.4 \text{ g}$$

27. A man weighs 78 kg. How many pounds does he weigh?

$$78 \text{ kg} \times \frac{2.2 \text{ lb}}{1 \text{ kg}} = 171.6 \text{ lb}$$

28. A woman weighs 52 kg. How many pounds does she weigh?

$$52 \text{ kg} \times \frac{2.2 \text{ lb}}{1 \text{ kg}} = 114.4 \text{ lb}$$

29. Ferrante purchased 3 qt of milk. How many liters is that?

$$3 \text{ qt} \times \frac{1 \text{ L}}{1.06 \text{ qt}} \approx 2.8 \text{ L}$$

30. Wong Tin purchased 1 gal of milk. How many liters is that? 3.8 L

Answer the following questions. Round all answers to the nearest hundredth.

31. How many inches are in 3050 miles?

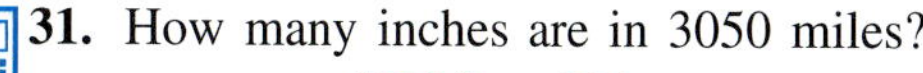

$$3050 \text{ mi} \times \frac{5280 \text{ ft}}{1 \text{ mi}} \times \frac{12 \text{ in.}}{1 \text{ ft}} = 193{,}248{,}000 \text{ in.}$$

32. How many inches are in 4500 miles?

$$4500 \text{ mi} \times \frac{5280 \text{ ft}}{1 \text{ mi}} \times \frac{12 \text{ in.}}{1 \text{ ft}} = 285{,}120{,}000 \text{ in.}$$

33. A truck traveled at a speed of 40 feet per second. How many miles per hour is that?

$$\frac{40 \text{ ft}}{1 \text{ sec}} \times \frac{1 \text{ mi}}{5280 \text{ ft}} \times \frac{3600 \text{ sec}}{1 \text{ hr}} \approx 27.27 \text{ miles per hour}$$

34. A car traveled at a speed of 55 feet per second. How many miles per hour is that?

$$\frac{55 \text{ ft}}{1 \text{ sec}} \times \frac{1 \text{ mi}}{5280 \text{ ft}} \times \frac{3600 \text{ sec}}{1 \text{ hr}} \approx 37.5 \text{ miles per hour}$$

35. How many years are in 3,500,000 seconds? (Use the approximate value that 365 days = 1 year.)

$$3{,}500{,}000 \text{ sec} \times \frac{1 \text{ min}}{60 \text{ sec}} \times \frac{1 \text{ hr}}{60 \text{ min}} \times \frac{1 \text{ da}}{24 \text{ hr}} \times \frac{1 \text{ yr}}{365 \text{ da}} \approx 0.11 \text{ yr}$$

36. How many years are in 2,800,000 seconds? (Use the approximate value that 365 days = 1 year.)

$$2{,}800{,}000 \text{ sec} \times \frac{1 \text{ min}}{60 \text{ sec}} \times \frac{1 \text{ hr}}{60 \text{ min}} \times \frac{1 \text{ da}}{24 \text{ hr}} \times \frac{1 \text{ yr}}{365 \text{ da}} \approx 0.09 \text{ yr}$$

APPENDIX C

Interpreting Data from Tables, Charts, and Graphs

After studying this section, you will be able to interpret data from:

1. *A Table*
2. *A Chart*
3. *A Pictograph*
4. *A Bar Graph*
5. *A Line Graph*
6. *A Pie Graph or Circle Graph*

1 *A table is a device used to organize information into categories. Using it you can readily find details about each category.*

EXAMPLE 1

TABLE OF NUTRITIVE VALUES OF CERTAIN POPULAR "FAST FOODS"

Type of Sandwich	Calories	Protein (g)	Fat (g)	Cholesterol (g)	Sodium (mg)
Burger King Whopper	630	27	38	90	880
McDonald's Big Mac	500	25	26	100	890
Wendy's Bacon Cheeseburger	440	22	25	65	870
Burger King BK Broiler	280	20	10	50	770
McDonald's McChicken	415	19	20	50	830
Wendy's Grilled Chicken	290	24	7	60	670

Source: U.S. Government Agencies and Food Manufacturers

(a) Which food item has the least amount of fat per serving?

(b) Which beef item has the least amount of calories per serving?

(c) How much more protein does a Burger King Whopper have than a McDonald's Big Mac?

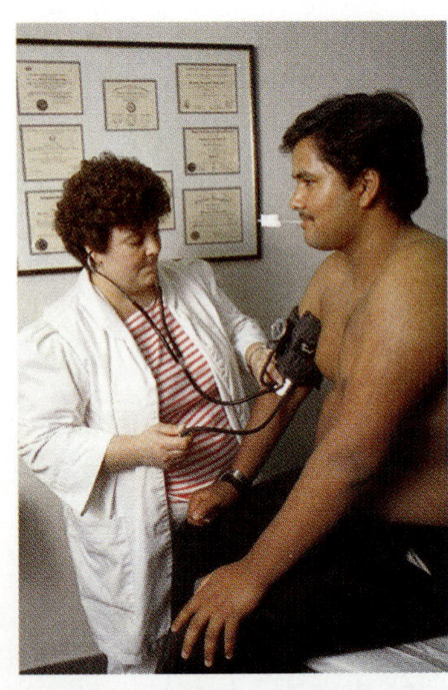

Solution:

(a) The least amount of fat, 7 grams, is in the Wendy's Grilled Chicken Sandwich.

(b) The least amount of calories, 440, for a beef sandwich is the Wendy's Bacon Cheeseburger.

(c) The Burger King Whopper has 2 grams of protein more than the McDonald's Big Mac.

Practice Problem 1

(a) Which sandwich has the lowest level of sodium?

(b) Which sandwich has the highest level of cholesterol?

2 *A chart is a device used to organize information in which not each category is the same. Example 2 illustrates a chart containing different types of data.*

EXAMPLE 2 The following chart shows how people in Topsfield indicated they spent their free time.

Survey of use of leisure time chart

Category	Activity	Hours spent per week
Single Men	Gym Outdoor sports Dating Watching pro sports Reading & TV	6 4 7 12 3
Single Women	Gym Outdoor sports Dating Time with friends Reading & TV	4 2 7 10 9
Couples	Time with family Time as a couple Time with friends Reading & TV	21 8 4 9
Children	Watching TV Playing outside Reading	28 8 1

Use the chart to answer the following questions about people in Topsfield.

(a) What is the average amount of time a couple spends together as a family during the week?

(b) How much more time do children spend watching TV than playing outside?

(c) What activity do single men spend most of their time doing?

Solution:

(a) The average amount of time a couple spends together as a family is 21 hours per week.

(b) Children spend 20 more hours per week watching TV than playing outside.

(c) Single men spend more time per week watching pro sports (12 hr) than any other activity.

Practice Problem 2

(a) What two categories do single women spend the most time doing?

(b) What do couples spend most of their time doing?

(c) What is the most significant numerical difference in terms of number of hours per week spent by single women versus single men?

3 Pictographs

A pictograph uses a visually appropriate symbol to represent an amount of items. A pictograph is used in Example 3.

EXAMPLE 3 Consider the pictograph below.

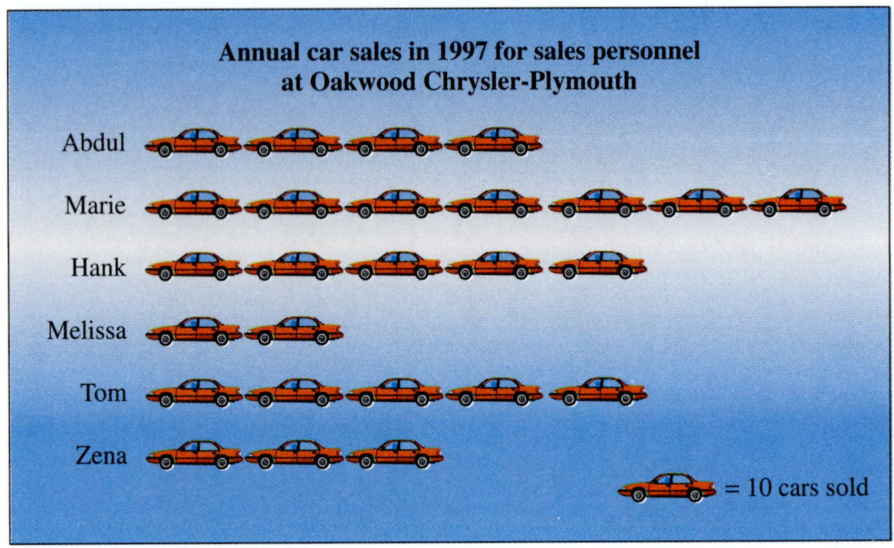

(a) How many cars did Melissa sell in 1997?

(b) Who sold the greatest number of cars?

(c) How many more cars did Tom sell than Zena?

Solution:

(a) Melissa sold $2 \times 10 = 20$ cars.

(b) Maria sold the greatest number of cars.

(c) Tom sold $5 \times 10 = 50$ cars. Zena sold $3 \times 10 = 30$ cars. Now $50 - 30 = 20$. Therefore, Tom sold 20 cars more than Zena.

Practice Problem 3

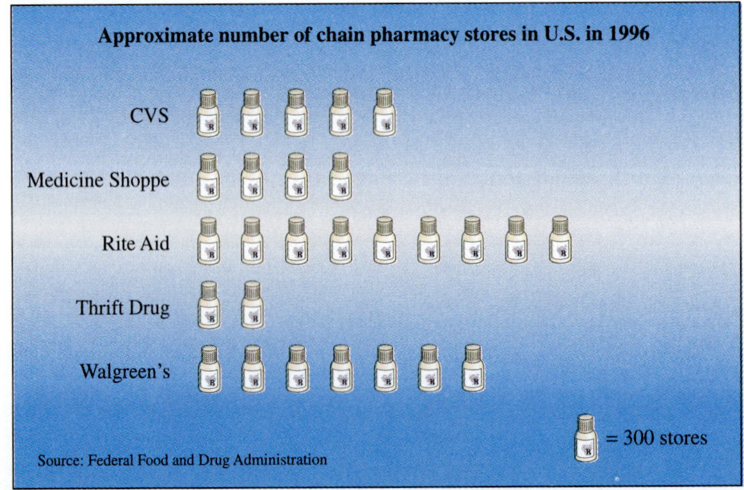

(a) Approximately how many stores does Walgreen's have?

(b) Approximately how many more stores does Rite Aid have than CVS?

(c) What is the combined number of stores of Thrift Drug and the Medicine Shoppes?

4 *Bar Graphs*

A bar graph is helpful for making comparisons and noting changes or trends. A scale is provided so that the height of the bar graph indicates a specific number. A bar graph is displayed in Example 4. A bar graph may be represented horizontally or vertically. In either case the basic concepts of interpreting a bar graph are the same.

EXAMPLE 4 The approximate population of California by year is given in this bar graph.

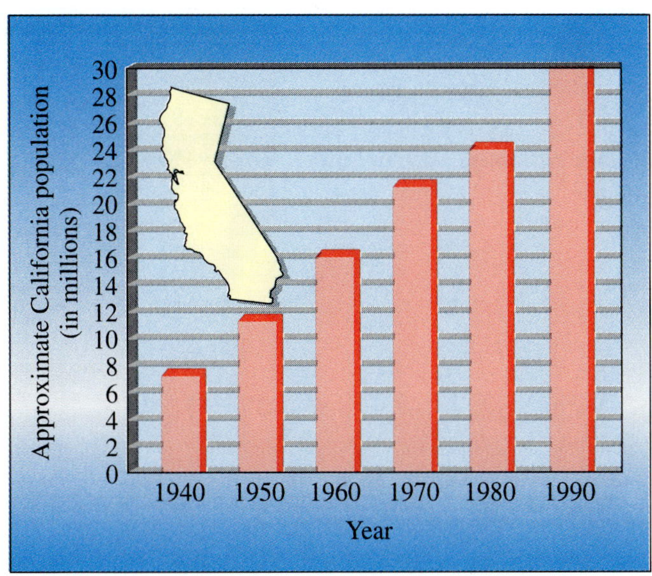

(a) What was the approximate population of California in 1990?

(b) How much greater was the population of California in 1980 than in 1970?

Solution:

(a) The approximate population of California in 1990 was about 30 million.

(b) In 1980 it was 24 million. In 1970 it was 21 million. The population was approximately 3 million people more in California in 1980 than in 1970.

Practice Problem 4 The following bar graph depicts the number of fatal accidents for U.S. air carriers for scheduled flight service for aircraft with 30 seats or more.

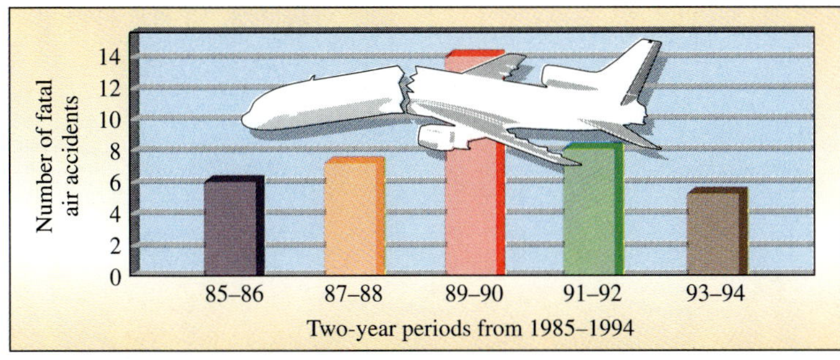

Source: National Transportation Safety Board

(a) What two-year period had the greatest number of fatal accidents?

(b) What was the increase in the number of fatal accidents from the 1987–1988 period to the 1989–1990 period?

(c) What was the decrease in the number of fatal accidents from the 1991–1992 period to the 1993–1994 period?

5 *Line Graphs*

A line graph is often used to display data when significant changes or trends are present. In a line graph only a few points are actually plotted from measured values. The points are then connected by straight lines in order to show a trend. The intervening values between points may not exactly lie on the line. A line graph is displayed in Example 5.

EXAMPLE 5 The following line graph shows the number of customers per month coming to a restaurant in a tourist vacation community.

(a) What month had the greatest number of customers?

(b) How many customers come to the restaurant in April?

(c) How many more customers came in July than in August?

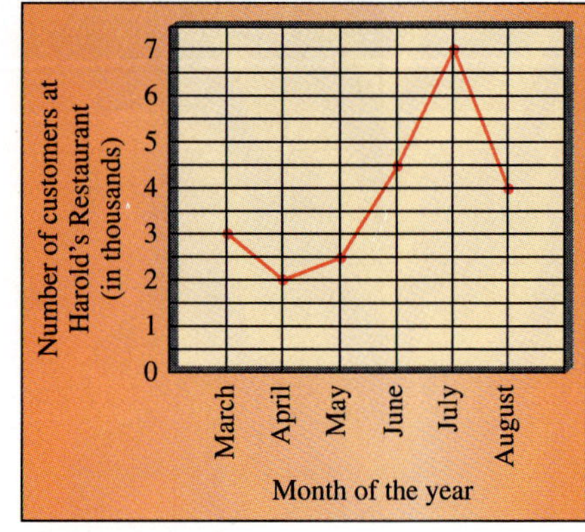

Solution:

(a) More customers came during the month of July.

(b) Approximately 2000 customers came in April.

(c) In July there were 7000 customers, while in August there were 4000 customers. Thus, there were 3000 more customers in July than in August.

Practice Problem 5 The quality of the air is measured by the Pollutant Standards Index (PSI). To meet the national air quality standards set by the U.S. government the air in a city cannot have a PSI greater than 100. The following line graph indicates the number of days that the PSI was greater than 100 in the city of Baltimore during a six-year period.

(a) What was the number of days the PSI exceeded 100 in Baltimore in 1993?

(b) In what year did Baltimore have the fewest days in which the PSI exceeded 100?

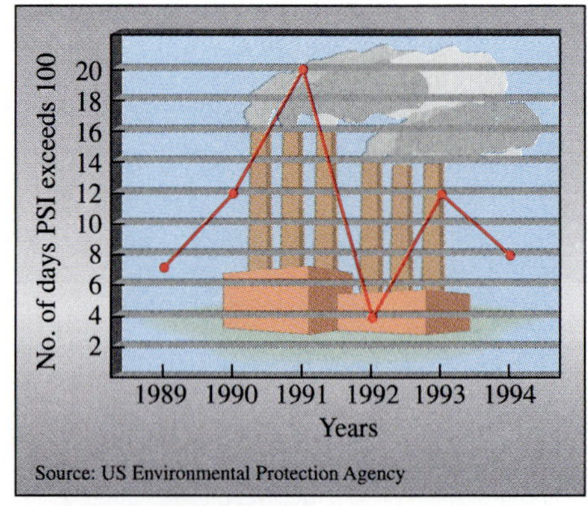

6 Pie Graphs or Circle Graphs

A pie graph or a circle graph indicates how a whole quantity is divided into parts. These graphs help you to visualize the size of the relative proportions of parts. Each piece of the pie or circle is called a sector. Example 6 uses a pie graph.

EXAMPLE 6 Together, the Great Lakes from the largest body of fresh water in the world. The total area of these five lakes is about 290,000 square miles, almost all of which is suitable for boating. The percentage of this total area taken up by each of the Great Lakes is shown in the pie graph.

(a) What percentage of the area is taken up by Lake Michigan?

(b) What lake takes up the largest percentage of area?

(c) How many square miles are taken up by Lake Huron and Lake Michigan together?

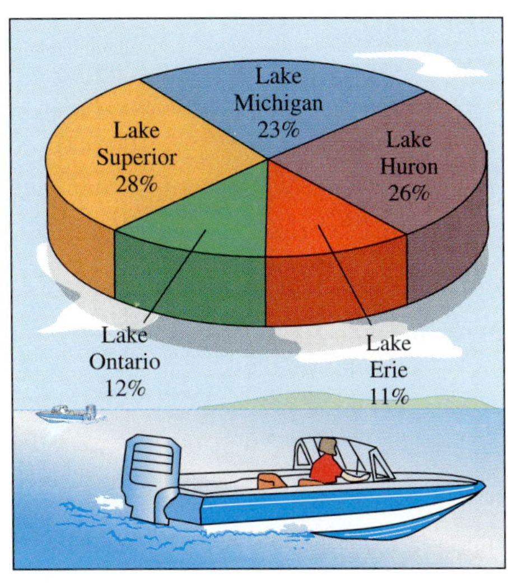

Solution:

(a) Lake Michigan takes up 23% of the area.

(b) Lake Superior takes up the largest percentage.

(c) If we add 26 + 23, we get 49. Thus Lake Huron and Lake Michigan together take up 49% of the total area. 49% of 290,000 = (0.49)(290,000) = 142,100 square miles.

Practice Problem 6 Seattle receives on average about 37 inches of rain per year. However, the amount of rainfall per month varies significantly. The percent of rainfall that occurs during each quarter of the year is shown by the circle graph.

(a) What percent of the rain in Seattle falls between April and June?

(b) Forty percent of the rainfall occurs in what three-month period?

(c) How many inches of rain fall in Seattle from January to March?

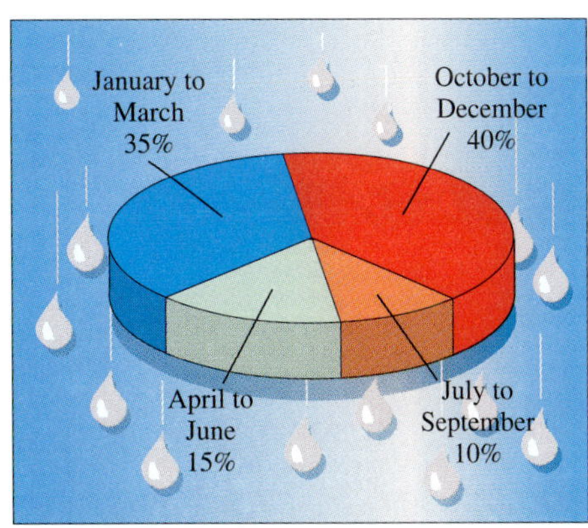

Source: U.S. National Oceanic and Atmospheric Administration

C Exercises

In each case study carefully the appropriate visual display; then answer the following questions. Consider the table below in answering Questions 1–10.

TABLE OF FACTS OF THE ROCKY MOUNTAIN STATES

State	Area in Square Miles	Date Admitted To the Union	Estimated 1990 Population	Number of Representatives in U.S. Congress	Popular Name
Colorado	104,247	1876	3,500,000	5	Centennial State
Idaho	83,557	1890	1,100,000	2	Gem State
Montana	147,138	1889	800,000	2	Treasure State
Nevada	110,540	1864	1,100,000	1	Silver State
Utah	84,916	1896	1,700,000	2	Beehive State
Wyoming	97,914	1890	500,000	1	Equality State

1. What is the area of Utah in square miles?
84,916 square miles

2. What is the area of Montana in square miles?
147,138 square miles

3. What is the 1990 estimated population of Colorado?
3,500,000

4. What is the 1990 estimated population of Nevada?
1,100,000

5. How many representatives in the U.S. Congress come from Idaho? 2

6. How many representatives in the U.S. Congress come from Wyoming? 1

7. What is the popular name for Montana?
Treasure State

8. What is the popular name for Utah?
Beehive State

9. Which of these six states was the first one to be admitted to the Union? Nevada

10. In what year did two of these six states both get admitted to the Union? 1890

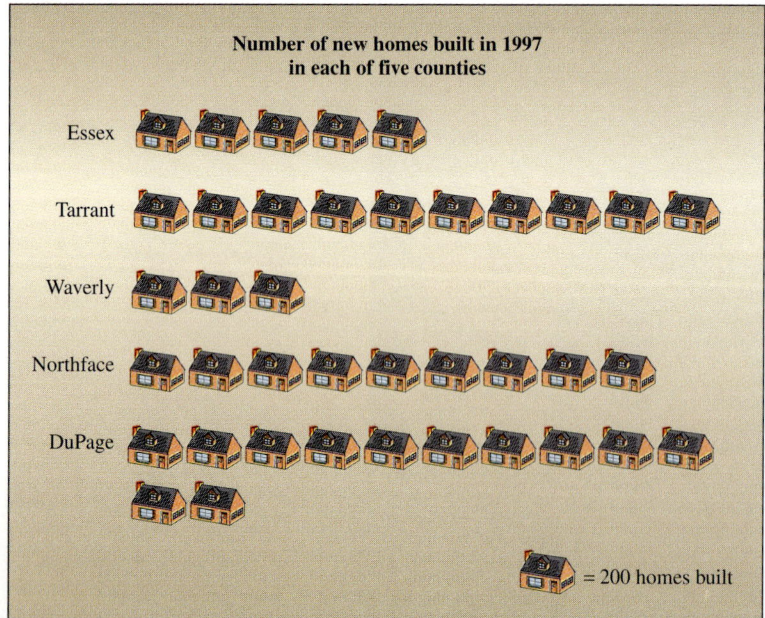

Use this pictograph to answer Questions 11–16.

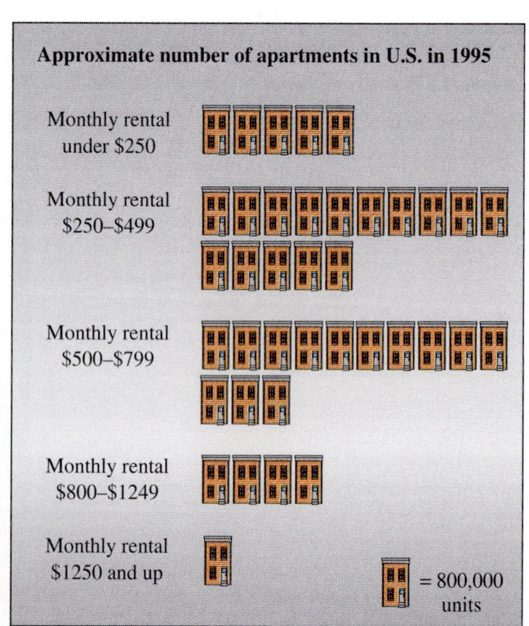

Use this pictograph to answer Questions 17–20.
Source: U.S. Bureau of the Census

11. How many homes were built in Tarrant County in 1997? 2000

12. How many homes were built in Essex County in 1997? 1000

13. In what county were the most homes built? DuPage County

14. How many more homes were built in Tarrant County than Waverly County? 1400 more homes

15. How many homes were built in Essex County and Northface County combined? 2800

16. How many homes were built in DuPage County and Waverly County combined? 3000

17. How many apartment units are rented for under $250 per month? 4,000,000

18. How many apartment units are rented for $800–$1249 per month? 3,200,000

19. How many more apartment units are available in the $500–$799 range than in the $800–$1249 range? 7,200,000 more units

20. How many more apartment units are available in the $800–$1249 range than in the $1250 and up range? 2,400,000 more units

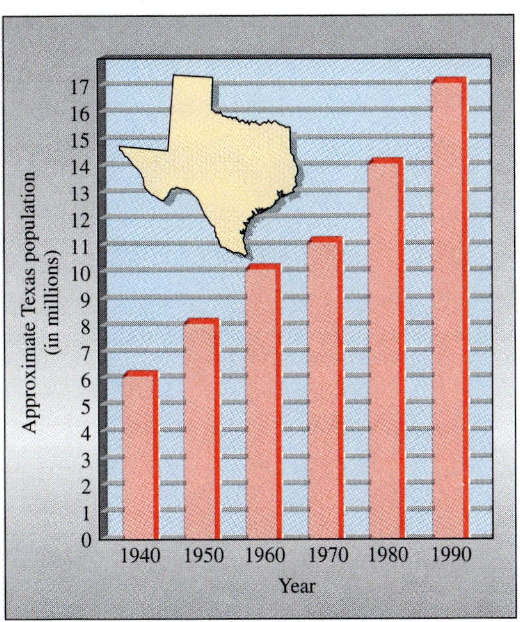

Use this bar graph to answer Questions 21–26.

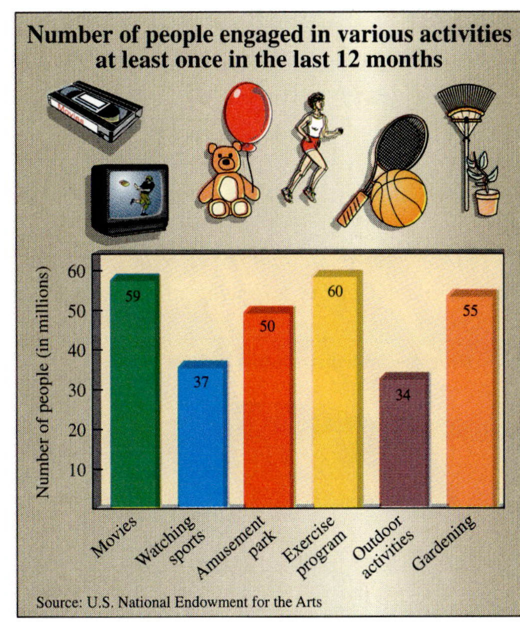

Use this bar graph to answer Questions 27–30.

21. What was the population in Texas in 1950? About 8 million

22. What was the population in Texas in 1990? About 17 million

23. Between what two years was the increase in population the greatest? It is a tie. Between 1970 and 1980 and also between 1980 and 1990 the increase was 3 million.

24. Between what two years was the increase in population the smallest? Between 1960 and 1970. It only increased by one million.

25. How many more people lived in Texas in 1970 than in 1950? 3 million more

26. How many more people lived in Texas in 1980 than in 1960? 4 million more

27. According to the bar graph, how many people watched a sports event at least once in the last 12 months? 37 million

28. According to the bar graph, how many people were involved in gardening at least once in the last 12 months? 55 million

29. What two activities were the most common? Movies and an exercise program

30. What two activities were the least common? Watching sports and outdoor activities

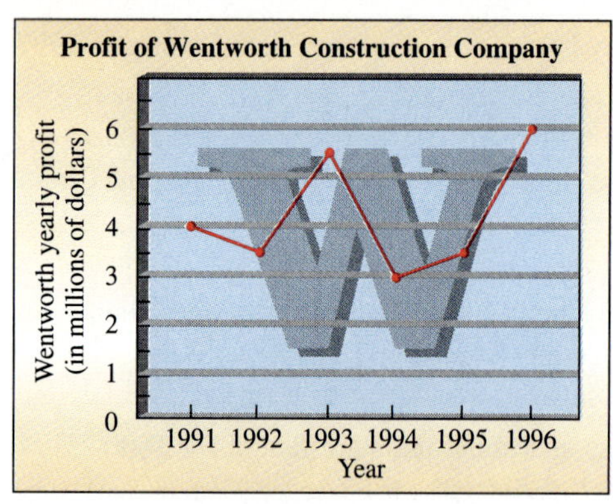

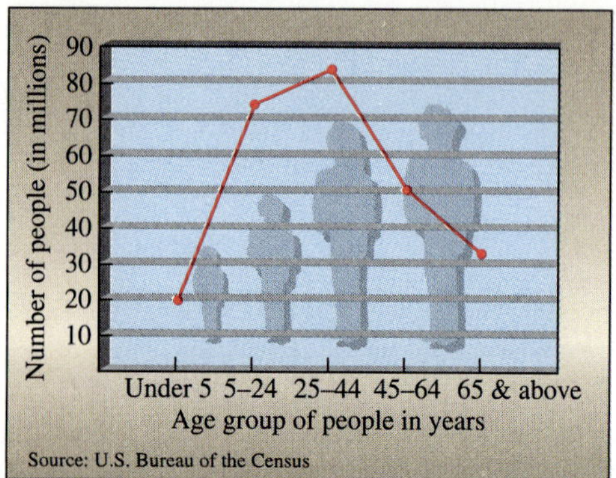

Use this line graph to answer Questions 31–36.

Use this line graph to answer Questions 37–40.

31. What was the profit in 1993? $5.6 million

32. What was the profit in 1992? $3.5 million

33. How much greater was the profit in 1996 than 1995? $2.5 million greater

34. In what year did the smallest profit occur? 1994

35. Between what two years did the profit decrease the most? Between 1993 and 1994

36. Between what two years did the profit increase the most? Between 1995 and 1996

37. How many people in the U.S. are in the age group of 5 to 24 years? 74 million

38. How many people in the U.S. are in the age group of 25 to 44 years? 83 million

39. Thirty-three million people are in what age group? 65 years and above

40. Fifty-one million people are in what age group? Age 45–64 years

Researchers estimate that the religious faith distribution of the 5,000,000,000 people in the world in 1988 was approximately that displayed in the following circle graph. Use this graph to answer Questions 41–46.

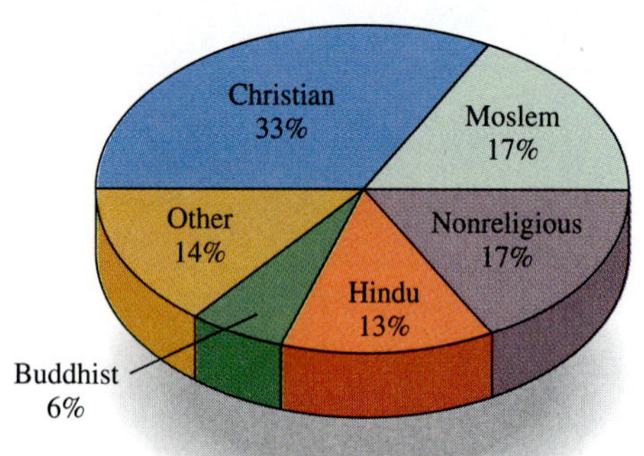

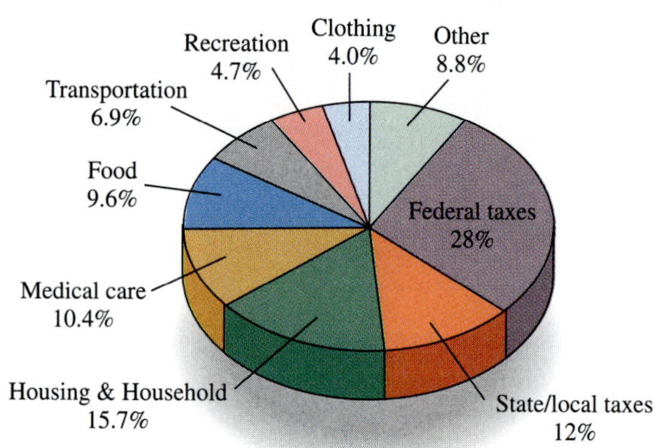

Use this circle graph to answer Questions 41–46.

Use this circle graph to answer Questions 47–50.

41. What percent of the world's population is either Hindu or Buddhist? 19%

42. What percent of the world's population is either Moslem or nonreligious? 34%

43. What percent of the world's population is *not* Moslem? 83%

44. What percent of the world's population is *not* Christian? 67%

45. If there are approximately 5.8 billion people in the world in 1997, how many of them would we expect to be Hindu? 754 million people

46. If there are approximately 6.2 billion people in the world in 1999, how many of them would we expect to be Moslem? 1.054 billion people

47. What percent of the family income is spent for federal, state, and local taxes? 40%

48. What percent of the family income is spent for food, medical care, and housing? 35.7%

49. If the average two-income family earns $52,000 per year, how much is spent on transportation? $3588

50. If the average two-income family earns $52,000 per year, how much is spent on recreation? $2444

APPENDIX D

Inductive and Deductive Reasoning

After studying this section you will be able:

1. *Use inductive reasoning to reach a conclusion.*
2. *Use deductive reasoning to reach a conclusion.*

1 *Inductive Reasoning*

When we reach a conclusion based on specific observations, we are using **inductive reasoning.** Much of our early learning is based on simple cases of inductive reasoning. If a child touches a hot stove or other appliance several times and each time he gets burned, he is likely to conclude, "If I touch something that is hot, I will get burned." This is inductive reasoning. The child has thought about several actions and their outcomes and has made a conclusion or generalization.

The next few examples show how inductive reasoning can be used in mathematics.

EXAMPLE 1 Find the next number in the sequence 10, 13, 16, 19, 22, 25, 28, ...
Solution: We observe a pattern that each number is 3 more than the preceding number. $10 + 3 = 13$; $13 + 3 = 16$, and so on. Therefore, if we add 3 to 28, we conclude that the next number in the sequence is 31.

Practice Problem 1 Find the next number in the sequence 24, 31, 38, 45, 52, 59, 66, ... ■

EXAMPLE 2 Find the next number in the sequence 1, 8, 27, 64, 125, ...
Solution: The sequence can be written as $1^3, 2^3, 3^3, 4^3, 5^3, \ldots$.
Each successive integer is cubed. The next number would be 6^3 or 216.

Practice Problem 2 Find the next number in the sequence.

3, 8, 15, 24, 35, 48, 63, 80, . . . ■

EXAMPLE 3 Guess the next seven digits in the following irrational number:

5.636336333633336 . . .

Solution: Between the 6's there are the digits 3, 33, 333, 3333, and so on. The pattern is that the number of 3's keeps increasing by 1 each time. Thus the next seven digits are 3333363.

Practice Problem 3 Guess the next seven digits in the following irrational number:

6.1213314441 . . . ■

EXAMPLE 4 Find the next two figures that would appear in the sequence.

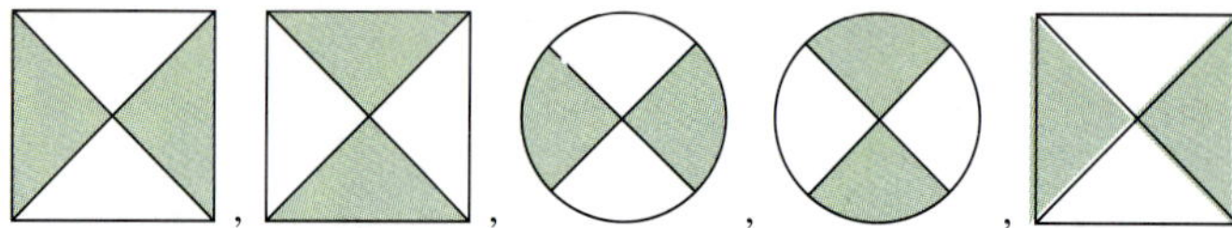

Solution: We notice an alternating pattern: square, square, circle, circle, square . . . . We would next expect a square followed by a circle.

We notice a shading pattern of horizontal, vertical, horizontal, vertical, horizontal . . . . We would next expect vertical, then horizontal. Thus, the next two figures are

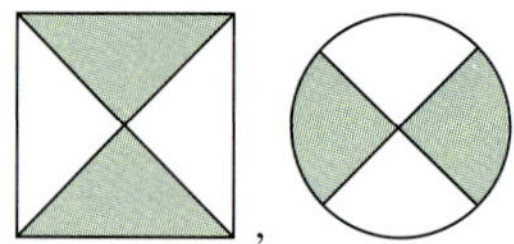

Practice Problem 4 Find the next two figures that would appear in the sequence.

■

How accurate is inductive reasoning? Do we always come to the right conclusion? Conclusions arrived at by inductive reasoning are always tentative. They may require further investigation. When we use inductive reasoning we are using specific data to reach a general conclusion. However, we may have reached the wrong conclusion. To illustrate:

Take the sequence of numbers 20, 10, 5. What is the next number? You might say 2.5. In the sequence 20, 10, 5, each number appears to be one–half of the preceding number. Thus we would predict 20, 10, 5, 2.5, ...

But wait! There is another possibility. Maybe the sequence is 20, 10, 5, −5, −10, −20, −25, −35, −40, ...

To know for sure which answer is correct, we would need more information, such as more numbers in the sequence to verify the pattern. **You should always treat inductive reasoning conclusions as tentative, requiring further verification.**

2 *Deductive Reasoning*

Deductive reasoning requires us to take general facts, postulates, or accepted truths and use them to reach a specific conclusion. Suppose we know the following rules or ''facts'' of algebra:

For all numbers a, b, c and for all numbers $d \neq 0$:

1. *Addition principle of equations:* If $a = b$, then $a + c = b + c$.
2. *Division principle of equations:* If $a = b$, then $\frac{a}{d} = \frac{b}{d}$.
3. *Multiplication Principle of Equations:* If $a = b$, then $ac = bc$.
4. *Distributive Property:* $a(b + c) = ab + ac$.

EXAMPLE 5 Use deductive reasoning and the four properties listed above to justify each step in solving the equation.

$$2(7x - 2) = 38$$

Solution:

STATEMENT	REASON
1. $14x - 4 = 38$	**1.** Distributive property: $a(b + c) = ab + ac$ Here we distributed the 2.
2. $14x = 42$	**2.** Addition principle of equations: If $a = b$ then $a + c = b + c$. Here we added 4 to each side of the equation.
3. $\dfrac{14x}{14} = \dfrac{42}{14}$	**3.** Division principle of equations: If $a = b$ then $\dfrac{a}{d} = \dfrac{b}{d}$
$x = 3$	Here we divided each side by 14.

Practice Problem 5 Use deductive reasoning and the four properties listed above to justify each step in solving the equation.

$$\frac{1}{6}x = \frac{1}{3}x + 4 \blacksquare$$

Sometimes we need to make conclusions about angles and lines in geometry. The following properties are useful. We will refer to angles 1, 2, 3, 4.

1. If two lines intersect, the opposite angles are equal. Here $\angle 1 = \angle 2$ and $\angle 4 = \angle 3$.

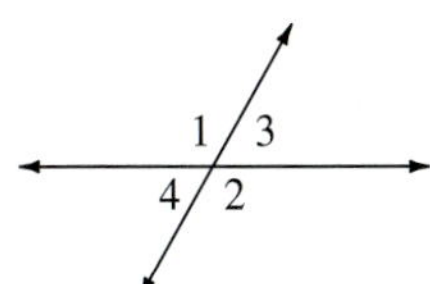

2. If two lines intersect, the adjacent angles are supplementary (they add up to 180°). Here, $\angle 4 + \angle 1 = 180°$, and $\angle 1 + \angle 3 = 180°$, also $\angle 3 + \angle 2 = 180°$, and $\angle 4 + \angle 2 = 180°$.
3. If a transversal (intersecting line) crosses two parallel lines, the alternate interior angles are equal. In the figure at right, line P is a transversal. If line M is parallel to line N, then $\angle 3 = \angle 4$ and $\angle 1 = \angle 2$.

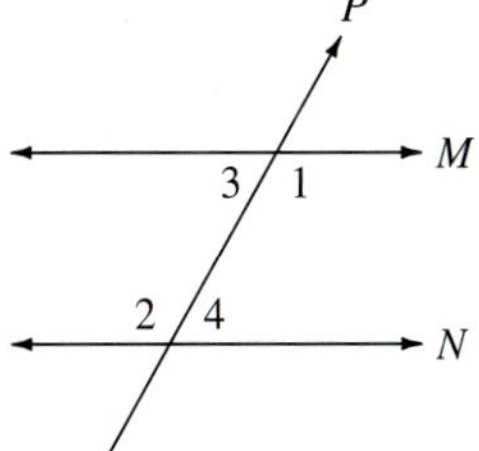

4. If two alternate interior angles on each side of a transversal cutting two straight lines are equal, the two straight lines are parallel. If $\angle 3 = \angle 4$, then line M is parallel to line N.
5. Supplements of equal angles are equal. In the figure at right, if $\angle 1$ and $\angle 2$ are supplementary, and $\angle 3$ and $\angle 4$ are supplementary, then if $\angle 1 = \angle 4$, then $\angle 2 = \angle 3$.

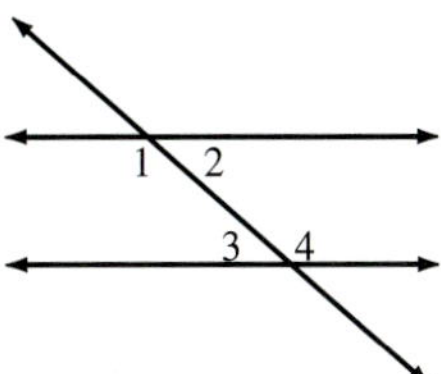

6. Transitive property of equality:

$$\text{If } a = b \text{ and } b = c, \text{ then } a = c.$$

We will now use these facts to prove some geometric conclusions.

EXAMPLE 6 Prove that line M is parallel to line N if $\angle 1 = \angle 3$.

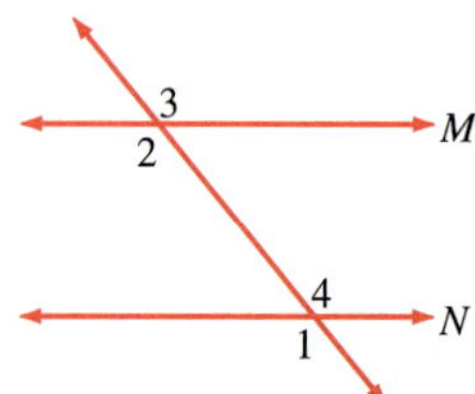

Solution:

STATEMENT	REASON
1. $\angle 3 = \angle 2$ and $\angle 4 = \angle 1$	**1.** If two lines intersect, the opposite angles are equal.
2. $\angle 2 = \angle 4$	**2.** Transitive property of equality.
3. Therefore, line M is parallel to line N. (We write this as $M \parallel N$.)	**3.** If two alternate interior angles on each side of a transversal cutting two straight lines are equal, the straight lines are parallel.

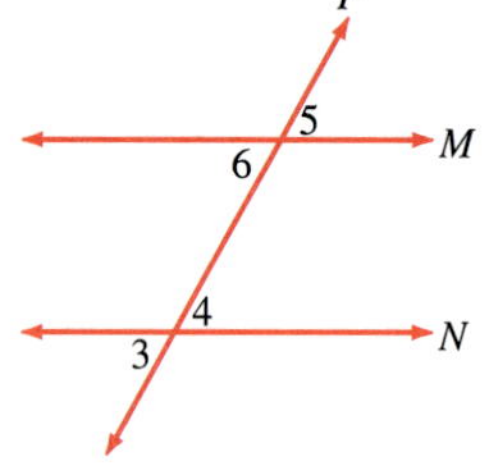

Practice Problem 6 Refer to the diagram in the margin. Line M is parallel to line N. $\angle 5 = 26°$. Prove $\angle 4 = 26°$. ■

Now let us see if we can use deductive reasoning to solve the following problem.

EXAMPLE 7 For next semester four professors from the psychology department have expressed the following desires for freshman courses. Each professor will teach only *one* freshman course next semester. The four freshman courses are General Psychology, Social Psychology, Psychology of Adjustment, and Educational Psychology.

1. Professors A and B don't want to teach General Psychology.
2. Professor C wants to teach Social Psychology.
3. Professor D will be happy to teach any course.
4. Professor B wants to teach Psychology of Adjustment.

Which professor will teach Educational Psychology, if all professors are given the courses they desire?

Solution: Let us organize the facts by listing the four freshman courses and each professor: A, B, C, and D.

General Psychology	Social Psychology	Psychology of Adjustment	Educational Psychology
A	A	A	A
B	B	B	B
C	C	C	C
D	D	D	D

STEP	REASON
1. We cross off Professors *A* and *B* from the General Psychology list.	**1.** Professors *A* and *B* don't want to teach General Psychology.

General Psychology			
~~*A*~~	*A*	*A*	*A*
~~*B*~~	*B*	*B*	*B*
C	*C*	*C*	*C*
D	*D*	*D*	*D*

STEP	REASON
2. We cross Professor *C* off every list (except Social Psychology) and mark that he will teach it.	**2.** Professor *C* wants to teach Social Psychology.

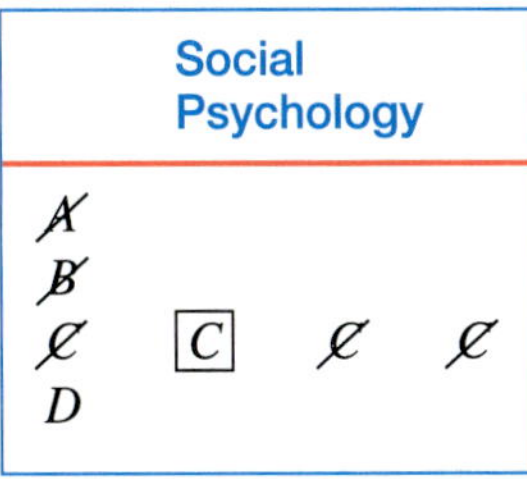

	Social Psychology		
~~*A*~~			
~~*B*~~			
~~*C*~~	[*C*]	~~*C*~~	~~*C*~~
D			

STEP	REASON
3. Professor *D* is thus the only person who can teach General Psychology. We cross him off every other list and mark that he will teach General Psychology.	**3.** Professor *D* is happy to teach any course.

~~*A*~~			
~~*B*~~			
~~*C*~~	[*C*]	~~*C*~~	~~*C*~~
[*D*]	~~*D*~~	~~*D*~~	~~*D*~~

STEP	REASON
4. We cross out all courses for Professor *B* except Psychology of Adjustment.	**4.** Professor *B* wants to teach Psychology of Adjustment.

		Psychology of Adjustment	
~~*A*~~	~~*A*~~	*A*	*A*
~~*B*~~	~~*B*~~	[*B*]	~~*B*~~
~~*C*~~	[*C*]	~~*C*~~	~~*C*~~
[*D*]	~~*D*~~	~~*D*~~	~~*D*~~

STEP	REASON
5. Professor *A* will teach Educational Psychology.	**5.** He is the only professor left. All others are assigned.

Educational Psychology
[*A*]

Practice Problem 7 A Honda, Toyota, Mustang, and Camaro are parked side by side, but not in that order.

1. The Camaro is parked on the right end.
2. The Mustang is between the Honda and the Toyota.
3. The Honda is not next to the Camaro.

Which car is parked on the left end? ■

D Exercises

Find the next number in the sequence.

1. 2, 4, 6, 8, 10, 12, . . . 14
2. 0, 5, 10, 15, 20, 25, . . . 30
3. 7, 16, 25, 34, 43, . . . 52
4. 12, 25, 38, 51, 64, . . . 77
5. 1, 16, 81, 256, 625, . . . 1296
6. 1, 6, 13, 22, 33, 46, . . . 61
7. −7, 3, −6, 4, −5, 5, −4, 6, . . . −3
8. 2, −4, 8, −16, 32, −64, 128, . . . −256
9. $5x$, $6x - 1$, $7x - 2$, $8x - 3$, $9x - 4$, . . . $10x - 5$
10. $60x$, $30x$, $15x$, $7.5x$, $3.75x$, $1.875x$, . . . $0.9375x$

Guess the next seven digits in the following irrational numbers.

11. 8.181181118 . . . 1111811
12. 3.043004300043 . . . 0000430
13. 12.98987987698765 . . . 9876549
14. 7.6574839201102938 . . . 4756657
15. 2.14916253649 . . . 6481100
16. 6.112223333 . . . 4444455
17. Find the next row in this triangular pattern.

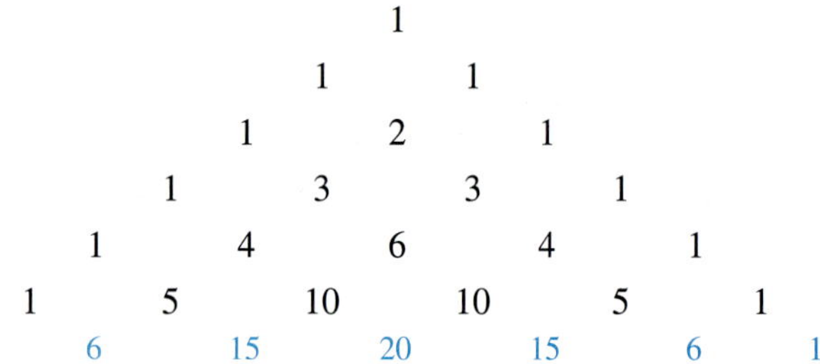

18. Find the next two figures that would appear in the sequence.

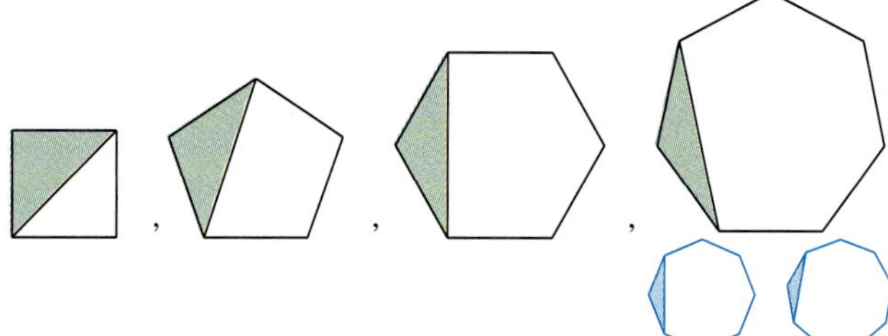

Use deductive reasoning and the four properties discussed in Example 5 to justify each step in solving each equation.

19. $12x - 30 = 6$
    $x = 3$
    Steps will vary.
20. $3x + 7 = 4$
    $x = -1$
    Steps will vary.
21. $4x - 3 = 3x - 5$
    $x = -2$
    Steps will vary.
22. $2x - 9 = 4x + 5$
    $x = -7$
    Steps will vary.
23. $8x - 3(x - 5) = 30$
    $x = 3$
    Steps will vary.
24. $3x - 5(x - 1) = -5$
    $x = 5$
    Steps will vary.
25. $\frac{1}{2}x + 6 = \frac{3}{2} + 6x$
    $x = \frac{9}{11}$
    Steps will vary.
26. $\frac{4}{5}x - 3 = \frac{1}{10}x + \frac{3}{5}$
    $x = \frac{36}{7}$
    Steps will vary.

Use deductive reasoning and the properties of geometry discussed in Example 6 to prove the following.

27. If $\angle 1 = \angle 5$, prove that line P is parallel to line S. Ans. will vary.

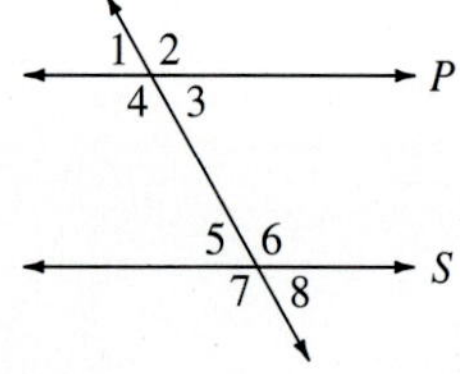

28. If line R is parallel to line S, prove that $\angle 6 = \angle 3$. Ans. will vary.

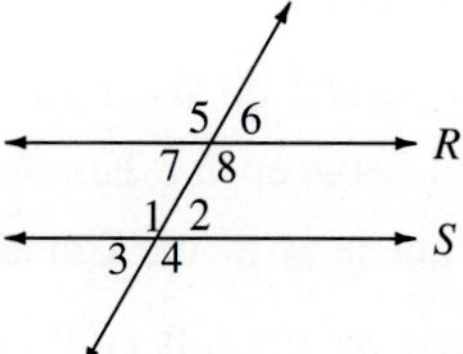

29. William, Brent, Charlie, and Dave competed in the Olympics. These four divers at the Olympics placed first, second, third, and fourth in the competition.

1. James ranked between Brent and Dave.
2. William did better than Brent.
3. Brent did better than Dave.

Who finished in each of the four places?

William was first, Brent was second, James was third, Dave was fourth.

30. The four floors of Hotel Royale are color-coded.

1. The blue floor is directly below the green floor.
2. The red floor is next to the yellow floor.
3. The green floor is above the red floor.
4. The blue floor is above the yellow floor.
5. There is no floor below the red floor.

What is the order of the colors from top to bottom of Hotel Royale?

Top floor is green.
Second floor is blue.
Third floor is yellow.
Bottom floor is red.

31. There are four people named Peter, Michael, Linda and Judy. Each of them has one occupation. They are a teacher, a butcher, a baker, and a candlestick maker.

1. Linda is the baker.
2. The teacher plans to marry the baker in October.
3. Judy and Peter were married last year.

Who is the teacher? The teacher is Michael.

32. A president of a company, a lawyer, a salesman, and a doctor are seated at the head table at a banquet.

1. The lawyer prosecuted the doctor in a malpractice lawsuit and they should not sit together.
2. The salesman should be at one end of the table.
3. The lawyer should be at the far right.

From left to right how should they be seated?

From left to right the correct order is salesman, doctor, president, lawyer.

33. John is considering purchasing a new car when he starts his new full-time job. The car with the best gas mileage is a Honda Civic, the car with the most conveniently located dealer is a Toyota Corolla, the car with the lowest price is the Ford Escort, and the car that has best handling is the Dodge Neon. His next door neighbor has a Honda and he does not want to copy his neighbor. If he does not buy a Toyota then he will definitely not go to graduate school. He told his best friend that he probably will not purchase a car based on its handling. He has definitely decided to go to graduate school. What car is he most likely to purchase?

Toyota Corolla

34. Detective Smith recently found a stolen Corvette abandoned near Skull Rocks. A small empty boat was anchored 1 mile offshore with no person in the boat. Divers later found the body of a 35-year-old man $\frac{1}{2}$ mile further out to sea than the boat. If the man entered the water after 10:00 AM then the current of the outgoing tide was strong enough to take a dead body a distance of $\frac{1}{2}$ mile further out to sea. The coroner determined when the body was found at 4:00 PM that the man had been dead for 7 hours. The Detective determined he died instantly when his head hit the rocks directly beneath the anchored boat. Could the man have acted alone based on the facts revealed so far? No

35. Dr. Parad is deciding if he should do a root canal in order to save Fred's tooth or if he should just remove the tooth and put in a false tooth. If Dr. Parad works alone on the tooth the procedure will take 2 hours and 15 minutes. Fred wants to save the tooth but he does not want the dental bill for Tuesday's work to be over \$200. Dr. Parad charges \$100 per hour for dental procedures. If the bill is under \$150 it is because Dr. Parad's efficient dental assistant is helping with the oral surgery. The Dental procedure is scheduled for Tuesday, when Dr. Parad does not have any dental assistants available. Will Fred's tooth be saved or will he get a false tooth?

Unless Fred is willing to pay over \$200 he will probably get a false tooth.

36. If Marcia passes the Bi-Lingual Teacher's exam, she will stay in the Chicago area. If she devotes at least 50 hours of study time to practice her Spanish she is confident that she can pass the exam. Either Marcia will stay in the Chicago area or she will move back to Massachusetts. If she did not pass the Bi-Lingual Teacher's exam it is because she needed 12 hours of study time per month. Marcia started studying for the exam in April. She will be moving to Massachusetts. When was the Bi-Lingual Teacher's exam given if we know that Marcia took the exam?

The exam took place in August or earlier in the same year.

Solutions to Practice Problems

Chapter 0

0.1 Practice Problems

1. **(a)** $\dfrac{10}{16} = \dfrac{2 \times 5}{2 \times 8} = \dfrac{5}{8}$ **(b)** $\dfrac{24}{36} = \dfrac{12 \times 2}{12 \times 3} = \dfrac{2}{3}$

(c) $\dfrac{42}{36} = \dfrac{6 \times 7}{6 \times 6} = \dfrac{7}{6}$

2. **(a)** $\dfrac{4}{12} = \dfrac{2 \times 2 \times 1}{2 \times 2 \times 3} = \dfrac{1}{3}$ **(b)** $\dfrac{25}{125} = \dfrac{5 \times 5 \times 1}{5 \times 5 \times 5} = \dfrac{1}{5}$

(c) $\dfrac{73}{146} = \dfrac{73 \times 1}{73 \times 2} = \dfrac{1}{2}$

3. **(a)** $\dfrac{18}{6} = \dfrac{3 \times 6}{6} = 3$ **(b)** $\dfrac{146}{73} = \dfrac{73 \times 2}{73} = 2$

(c) $\dfrac{28}{7} = \dfrac{7 \times 4}{7} = 4$

4. 56 out of 154 $= \dfrac{56}{154} = \dfrac{2 \times 7 \times 4}{2 \times 7 \times 11} = \dfrac{4}{11}$

5. **(a)** $\dfrac{12}{7} = 12 \div 7 = 7\overline{)12} = 1\dfrac{5}{7}$

$\begin{array}{r} 1 \\ 7\overline{)12} \\ \underline{7} \\ 5 \end{array}$ Remainder

(b) $\dfrac{20}{5} = 20 \div 5 = 5\overline{)20} = 4$

$\begin{array}{r} 4 \\ 5\overline{)20} \\ \underline{20} \\ 0 \end{array}$ Remainder

6. **(a)** $3\dfrac{2}{5} = \dfrac{(3 \times 5) + 2}{5} = \dfrac{15 + 2}{5} = \dfrac{17}{5}$

(b) $1\dfrac{3}{7} = \dfrac{(1 \times 7) + 3}{7} = \dfrac{7 + 3}{7} = \dfrac{10}{7}$

(c) $2\dfrac{6}{11} = \dfrac{(2 \times 11) + 6}{11} = \dfrac{22 + 6}{11} = \dfrac{28}{11}$

(d) $4\dfrac{2}{3} = \dfrac{(4 \times 3) + 2}{3} = \dfrac{12 + 2}{3} = \dfrac{14}{3}$

7. **(a)** $\dfrac{3}{8} = \dfrac{?}{24}$ Observe $8 \times 3 = 24$

$\dfrac{3 \times}{8 \times}\left|\dfrac{3}{3}\right| = \dfrac{9}{24}$

(b) $\dfrac{5}{6} = \dfrac{?}{30}$

$\dfrac{5 \times}{6 \times}\left|\dfrac{5}{5}\right| = \dfrac{25}{30}$

(c) $\dfrac{12}{13} = \dfrac{?}{26}$

$\dfrac{12 \times}{13 \times}\left|\dfrac{2}{2}\right| = \dfrac{24}{26}$

(d) $\dfrac{2}{7} = \dfrac{?}{56}$

$\dfrac{2 \times}{7 \times}\left|\dfrac{8}{8}\right| = \dfrac{16}{56}$

(e) $\dfrac{5}{9} = \dfrac{?}{27}$

$\dfrac{5 \times}{9 \times}\left|\dfrac{3}{3}\right| = \dfrac{15}{27}$

(f) $\dfrac{3}{10} = \dfrac{?}{60}$

$\dfrac{3 \times}{10 \times}\left|\dfrac{6}{6}\right| = \dfrac{18}{60}$

(g) $\dfrac{3}{4} = \dfrac{?}{28}$

$\dfrac{3 \times}{4 \times}\left|\dfrac{7}{7}\right| = \dfrac{21}{28}$

(h) $\dfrac{8}{11} = \dfrac{?}{55}$

$\dfrac{8 \times}{11 \times}\left|\dfrac{5}{5}\right| = \dfrac{40}{55}$

0.2 Practice Problems

1. **(a)** $\dfrac{3}{6} + \dfrac{2}{6} = \dfrac{3 + 2}{6} = \dfrac{5}{6}$

(b) $\dfrac{3}{11} + \dfrac{2}{11} + \dfrac{6}{11} = \dfrac{3 + 2 + 6}{11} = \dfrac{11}{11} = 1$

(c) $\dfrac{5}{8} + \dfrac{2}{8} + \dfrac{7}{8} = \dfrac{5 + 2 + 7}{8} = \dfrac{14}{8} = 1\dfrac{3}{4}$

2. **(a)** $\dfrac{11}{13} - \dfrac{6}{13} = \dfrac{11 - 6}{13} = \dfrac{5}{13}$

(b) $\dfrac{8}{9} - \dfrac{2}{9} = \dfrac{8 - 2}{9} = \dfrac{6}{9} = \dfrac{2}{3}$

3. Find LCD of $\dfrac{1}{8}$ and $\dfrac{5}{12}$.

$8 = 2 \cdot 2 \cdot 2$
$12 = \downarrow \ 2 \cdot 2 \cdot 3$
$2 \cdot 2 \cdot 2 \cdot 3 = 24$

LCD = 24

4. Find the LCD using Prime Factors.

$\dfrac{8}{35}$ and $\dfrac{6}{15}$

$35 = 7 \cdot 5$
$15 = \ \ 5 \cdot 3$
$7 \cdot 5 \cdot 3$ LCD = 105

5. Find LCD of $\dfrac{5}{12}$ and $\dfrac{7}{30}$.

$12 = 3 \cdot 2 \cdot 2$
$30 = 3 \quad 2 \cdot 5$
$3 \cdot 2 \cdot 2 \cdot 5$ LCD = 60

6. Find LCD of $\dfrac{1}{18}$, $\dfrac{2}{27}$ and $\dfrac{5}{12}$.

$12 = 2 \cdot 2 \cdot 3$
$18 = 2 \quad 3 \cdot 3$
$27 = \quad 3 \cdot 3 \cdot 3$
$2 \cdot 2 \cdot 3 \cdot 3 \cdot 3$ LCD = 108

7. Add $\dfrac{1}{8} + \dfrac{5}{12}$

First find the LCD.

$8 = 2 \cdot 2 \cdot 2$
$12 = 2 \cdot 2 \quad \cdot 3$
$2 \cdot 2 \cdot 2 \cdot 3$ LCD = 24

Then change to equivalent fractions and add.

$\dfrac{1}{8} \times \left|\dfrac{3}{3}\right| + \dfrac{5}{12} \times \left|\dfrac{2}{2}\right| = \dfrac{3}{24} + \dfrac{10}{24} = \dfrac{3 + 10}{24} = \dfrac{13}{24}$

8. $\dfrac{3}{5} + \dfrac{4}{25} + \dfrac{1}{10}$

First find the LCD.

$5 = \quad 5$
$10 = 2 \cdot 5$
$25 = \quad 5 \cdot 5$
$2 \cdot 5 \cdot 5$ LCD = 50

Then change to equivalent fractions and add.

$\dfrac{3}{5} \times \left|\dfrac{10}{10}\right| + \dfrac{4}{25} \times \left|\dfrac{2}{2}\right| + \dfrac{1}{10} \times \left|\dfrac{5}{5}\right| = \dfrac{30}{50} + \dfrac{8}{50} + \dfrac{5}{50}$

$= \dfrac{30 + 8 + 5}{50} = \dfrac{43}{50}$

9. $\dfrac{1}{49} + \dfrac{3}{14}$

First find the LCD.

$14 = 2 \cdot 7$
$49 = \quad 7 \cdot 7$
$2 \cdot 7 \cdot 7$ LCD = 98

Then change to equivalent fractions and add.

$\dfrac{1}{49} \times \left|\dfrac{2}{2}\right| + \dfrac{3}{14} \times \left|\dfrac{7}{7}\right| = \dfrac{2}{98} + \dfrac{21}{98} = \dfrac{2 + 21}{98} = \dfrac{23}{98}$

10. $\frac{1}{12} - \frac{1}{30}$

First find the LCD.

$12 = 2 \cdot 2 \cdot 3$

$30 = \quad 2 \cdot 3 \cdot 5$

$2 \cdot 2 \cdot 3 \cdot 5 \qquad \text{LCD} = 60$

Then change to equivalent fractions and subtract.

$$\frac{1}{12} \times \left|\frac{5}{5}\right| - \frac{1}{30} \times \left|\frac{2}{2}\right| = \frac{5}{60} - \frac{2}{60} = \frac{5-2}{60} = \frac{3}{60} = \frac{1}{20}$$

11. $\frac{2}{3} + \frac{3}{4} - \frac{3}{8}$

First find the LCD.

$3 = \qquad 3$

$4 = \quad 2 \cdot 2$

$8 = \quad 2 \cdot 2 \cdot 2$

$2 \cdot 2 \cdot 2 \cdot 3 \qquad \text{LCD} = 24$

Then change to equivalent fractions and add and subtract.

$$\frac{2}{3} \times \left|\frac{8}{8}\right| + \frac{3}{4} \times \left|\frac{6}{6}\right| - \frac{3}{8} \times \left|\frac{3}{3}\right|$$

$$= \frac{16}{24} + \frac{18}{24} - \frac{9}{24} = \frac{16 + 18 - 9}{24} = \frac{25}{24} = 1\frac{1}{24}$$

12. **(a)** $1\frac{2}{3} + 2\frac{4}{5} = \frac{5}{3} + \frac{14}{5} = \frac{5}{3} \times \left|\frac{5}{5}\right| + \frac{14}{5} \times \left|\frac{3}{3}\right| = \frac{25}{15} + \frac{42}{15}$

$$= \frac{25 + 42}{15} = \frac{67}{15} = 4\frac{7}{15}$$

(b) $5\frac{1}{4} - 2\frac{2}{3} = \frac{21}{4} - \frac{8}{3} = \frac{21}{4} \times \left|\frac{3}{3}\right| - \frac{8}{3} \times \left|\frac{4}{4}\right| = \frac{63}{12} - \frac{32}{12}$

$$= \frac{63 - 32}{12} = \frac{31}{12} = 2\frac{7}{12}$$

13.

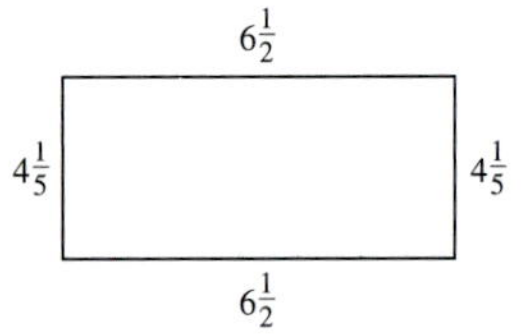

$$4\frac{1}{5} + 4\frac{1}{5} + 6\frac{1}{2} + 6\frac{1}{2}$$

$$= \frac{21}{5} + \frac{21}{5} + \frac{13}{2} + \frac{13}{2}$$

$\text{LCD} = 10$

$$\frac{21}{5} \times \left|\frac{2}{2}\right| + \frac{21}{5} \times \left|\frac{2}{2}\right| + \frac{13}{2} \times \left|\frac{5}{5}\right| + \frac{13}{2} \times \left|\frac{5}{5}\right|$$

$$= \frac{42}{10} + \frac{42}{10} + \frac{65}{10} + \frac{65}{10} = \frac{42 + 42 + 65 + 65}{10} = \frac{214}{10} = 21\frac{2}{5}$$

0.3 Practice Problems

1. **(a)** $\frac{2}{7} \times \frac{5}{11} = \frac{2 \cdot 5}{7 \cdot 11} = \frac{10}{77}$

(b) $\frac{8}{9} \times \frac{3}{10} = \frac{8 \cdot 3}{9 \cdot 10} = \frac{24}{90} = \frac{4}{15}$

2. **(a)** $\frac{3}{5} \times \frac{4}{3} = \frac{3 \cdot 4}{5 \cdot 3} = \frac{4}{5}$

(b) $\frac{9}{10} \times \frac{5}{12} = \frac{3 \cdot 3}{2 \cdot 5} \times \frac{5}{2 \cdot 2 \cdot 3} = \frac{3}{8}$

3. **(a)** $4 \times \frac{2}{7} = \frac{4}{1} \times \frac{2}{7} = \frac{4 \cdot 2}{1 \cdot 7} = \frac{8}{7} = 1\frac{1}{7}$

(b) $12 \times \frac{3}{4} = \frac{12}{1} \times \frac{3}{4} = \frac{2 \cdot 2 \cdot 3}{1} \times \frac{3}{2 \cdot 2} = 9$

4. **(a)** $2\frac{1}{5} \times \frac{3}{7} = \frac{11}{5} \times \frac{3}{7} = \frac{11 \cdot 3}{5 \cdot 7} = \frac{33}{35}$

(b) $3\frac{1}{3} \times 1\frac{2}{5} = \frac{10}{3} \times \frac{7}{5} = \frac{2 \cdot 5}{3} \times \frac{7}{5} = \frac{14}{3} = 4\frac{2}{3}$

5. $3\frac{1}{2} \times \frac{1}{14} \times 4 = \frac{7}{2} \times \frac{1}{14} \times \frac{4}{1} = \frac{7}{2} \times \frac{1}{2 \cdot 7} \times \frac{2 \cdot 2}{1} = 1$

6. **(a)** $\frac{2}{5} \div \frac{1}{3} = \frac{2}{5} \times \frac{3}{1} = \frac{6}{5}$

(b) $\frac{12}{13} \div \frac{4}{3} = \frac{2 \cdot 2 \cdot 3}{13} \times \frac{3}{2 \cdot 2} = \frac{9}{13}$

7. **(a)** $\frac{3}{7} \div 6 = \frac{3}{7} \div \frac{6}{1} = \frac{3}{7} \times \frac{1}{6} = \frac{3}{42} = \frac{1}{14}$

(b) $8 \div \frac{2}{3} = \frac{8}{1} \times \frac{3}{2} = 12$

8. **(a)** $\dfrac{\frac{3}{11}}{\frac{5}{7}} = \frac{3}{11} \div \frac{5}{7} = \frac{3}{11} \times \frac{7}{5} = \frac{21}{55}$

(b) $\dfrac{\frac{12}{5}}{\frac{8}{15}} = \frac{12}{5} \div \frac{8}{15} = \frac{2 \cdot 2 \cdot 3}{5} \times \frac{3 \cdot 5}{2 \cdot 2 \cdot 2} = \frac{9}{2} = 4\frac{1}{2}$

9. **(a)** $1\frac{2}{5} \div 2\frac{1}{3} = \frac{7}{5} \div \frac{7}{3} = \frac{7}{5} \times \frac{3}{7} = \frac{3}{5}$

(b) $4\frac{2}{3} \div 7 = \frac{14}{3} \times \frac{1}{7} = \frac{2 \cdot 7}{3} \times \frac{1}{7} = \frac{2}{3}$

(c) $\dfrac{1\frac{1}{5}}{1\frac{2}{7}} = 1\frac{1}{5} \div 1\frac{2}{7} = \frac{6}{5} \div \frac{9}{7} = \frac{6}{5} \times \frac{7}{9} = \frac{2 \cdot 3}{5} \times \frac{7}{3 \cdot 3} = \frac{14}{15}$

10. $64 \div 5\frac{1}{3} = 64 \div \frac{16}{3} = 64 \times \frac{3}{16} = 12$ jars

11. $25\frac{1}{2} \times 5\frac{1}{4} = \frac{51}{2} \times \frac{21}{4} = \frac{1071}{8} = 133\frac{7}{8}$ miles

0.4 Practice Problems

1. **(a)** 1.371 $\quad 1\frac{371}{1000}$

One and three hundred seventy one thousandths.

(b) $\frac{9}{100}$ = Nine hundredths.

2. **(a)** $\frac{3}{8}$

```
   .375
8)3.000
  2 4
    60
    56
     40
     40
      0
```

(b) $\frac{7}{200}$

```
      0.035
200)7.000
    6 00
    1 000
    1 000
        0
```

(c) $\frac{33}{20}$

```
     1.65
20)33.00
   20
   13 0
   12 0
    1 00
    1 00
       0
```

3. **(a)** $\frac{1}{6}$

```
 .166 = 0.16̄
6)1.000
  6
  40
  36
   40
   36
    4
```

(b) $\frac{5}{11}$

```
  .4545 = .4̄5̄
11)5.0000
   4 4
     60
     55
      50
      44
       60
       55
        5
```

4. **(a)** 0.8 $\quad \frac{8}{10} = \frac{2 \cdot 2 \cdot 2}{2 \cdot 5} = \frac{4}{5}$

(b) 0.88 $\quad \frac{88}{100} = \frac{11 \cdot 2 \cdot 2 \cdot 2}{5 \cdot 5 \cdot 2 \cdot 2} = \frac{22}{25}$

(c) 0.45 $\quad \frac{45}{100} = \frac{5 \cdot 3 \cdot 3}{5 \cdot 5 \cdot 2 \cdot 2} = \frac{9}{20}$

(d) 0.148 $\quad \frac{148}{1000} = \frac{2 \cdot 2 \cdot 37}{5 \cdot 5 \cdot 5 \cdot 2 \cdot 2 \cdot 2} = \frac{37}{250}$

(e) 0.612 $\quad \frac{612}{1000} = \frac{17 \cdot 3 \cdot 3 \cdot 2 \cdot 2}{5 \cdot 5 \cdot 5 \cdot 2 \cdot 2 \cdot 2} = \frac{153}{250}$

(f) 0.016 $\quad \frac{16}{1000} = \frac{2 \cdot 2 \cdot 2 \cdot 2}{5 \cdot 5 \cdot 5 \cdot 2 \cdot 2 \cdot 2} = \frac{2}{125}$

5. **(a)**

$$\begin{array}{r} 3.12 \\ 5.08 \\ 1.42 \\ \hline 9.62 \end{array}$$

(b)

$$\begin{array}{r} 152.003 \\ -\,136.118 \\ \hline 15.885 \end{array}$$

(c)

$$\begin{array}{r} 1.1 \\ 3.16 \\ 5.123 \\ \hline 9.383 \end{array}$$

(d)

$$\begin{array}{r} 1.0052 \\ -\,0.1234 \\ \hline 0.8818 \end{array}$$

6. **(a)**

$$\begin{array}{r} 0.061 \\ 5.0008 \\ 1.3 \\ \hline 6.3618 \end{array}$$

(b)

$$\begin{array}{r} 18.000 \\ -\,0.126 \\ \hline 17.874 \end{array}$$

7.

$$\begin{array}{r} 0.5 \\ \times\,0.3 \\ \hline 0.15 \end{array}$$

8.

$$\begin{array}{r} 0.12 \\ \times\,0.4 \\ \hline 0.048 \end{array}$$

9. **(a)**

$$\begin{array}{r} 1.23 \\ \times\,0.005 \\ \hline 0.00615 \end{array}$$

(b)

$$\begin{array}{r} 0.00002 \\ \times\,0.003 \\ \hline 0.00000006 \end{array}$$

10.

$$\begin{array}{r} 0.62 \\ 0.18\overline{)0.11\,16} \\ 10\,8 \\ \hline 36 \\ 36 \\ \hline 0 \end{array}$$

11.

$$\begin{array}{r} 0.3 \\ 0.3\overline{)0.0\,9} \end{array}$$

12.

$$\begin{array}{r} 30{,}000. \\ .06\overline{)1800.00} \end{array}$$

13. $4.9\overline{)0.0\,1764}$

$$\begin{array}{r} .0036 \\ 49\overline{)0.1764} \\ 147 \\ \hline 294 \\ 294 \\ \hline 0 \end{array}$$

14. **(a)** $0.0016 \times 100 = 0.16$
Move decimal point 2 places to the right.
(b) $2.34 \times 1000 = 2{,}340$
Move decimal point 3 places to the right.
(c) $56.75 \times 10{,}000 = 567{,}500$
Move decimal point 4 places to the right.

15. **(a)** $\dfrac{5.82}{10}$ (Move decimal point 1 place to the left.) 0.582
(b) $\dfrac{123.4}{1000}$ (Move decimal point 3 places to the left.) 0.1234
(c) $\dfrac{0.00614}{10{,}000}$ (Move decimal point 4 places to the left.)
0.000000614

0.5 Practice Problems

1. **(a)** $0.92 = 92\%$ **(b)** $0.418 = 41.8\%$ **(c)** $0.7 = 70\%$
2. **(a)** $0.0019 = 0.19\%$ **(b)** $0.0736 = 7.36\%$
   (c) $0.0003 = 0.03\%$
3. **(a)** 304% **(b)** 518.6% **(c)** 210%
4. **(a)** 0.82 **(b)** 3.47
5. **(a)** 0.07 **(b)** 0.093 **(c)** 0.002
6. **(a)** 1.31 **(b)** 3.016 **(c)** 0.0004
7. Change % to decimal and multiply.
   (a) $0.18 \times 50 = 9$ **(b)** $0.04 \times 64 = 2.56$
   (c) $1.56 \times 35 = 54.6$ **(d)** $0.008 \times 60 = 0.48$
   (e) $0.013 \times 82 = 1.066$ **(f)** $0.00002 \times 564 = 0.01128$
8. $4.2\% \times 18{,}000 = 0.042 \times 18{,}000$
   (a) \$756.00 **(b)** $18{,}000 + 756.00 = \$18{,}756.00$
9. $\dfrac{37}{148}$ reduces to $\dfrac{37 \cdot 1}{37 \cdot 2 \cdot 2} = \dfrac{1}{4} = .25 = 25\%$
10. **(a)** $\dfrac{24}{48} = \dfrac{2 \cdot 2 \cdot 2 \cdot 3}{2 \cdot 2 \cdot 2 \cdot 2 \cdot 3} = \dfrac{1}{2} = 0.5 = 50\%$
    (b) $\dfrac{4}{25} = 0.16 = 16\%$
11. $\dfrac{430}{1256} = \dfrac{215}{628} = 0.342 = 34\%$

0.6 Practice Problems

1. $100{,}000 \times 400 = 40{,}000{,}000$
2. $12\frac{1}{2} \rightarrow 10$; $9\frac{3}{4} \rightarrow 10$; First room $10 \times 10 = 100$ square feet.
   $14\frac{1}{4} \rightarrow 15$; $18\frac{1}{2} \rightarrow 20$; Second room $15 \times 20 = 300$ square feet.
   Both rooms $100 + 300 = 400$ square feet.
3. **(a)** $422.8 \rightarrow 400$ miles $19.3 \rightarrow 20$ gallons of gas
   $\dfrac{400}{20} = 20$ miles/gallon
   (b) $1.29\frac{9}{10} \rightarrow \$1.00$ per gallon $3862 \rightarrow 4{,}000$ miles
   $\dfrac{4{,}000 \text{ miles}}{20 \text{ mi/gallon}} = 200$ gallons
   200 gallons $\times$ \$1.00 per gallon = \$200.00
4. **(a)** 3,580,000,000 estimate $\rightarrow$ $\dfrac{4{,}000{,}000{,}000 \text{ miles}}{40{,}000 \text{ mi/hour}}$
   43,300 estimate $\rightarrow$
   100,000 hours
   (b) 24 hours/day estimate $\rightarrow \dfrac{100{,}000}{20} = 5{,}000$ days
5. 56.93% estimate $\rightarrow 60\% \rightarrow 0.6$
   293,567.12 estimate $\rightarrow$ 300,000
   $0.6 \times 300{,}000 = \$180{,}000$

0.7 Practice Problems

1.

Gather the Facts	What Am I Solving for	What Must I Calculate	Key Points to Remember
Living Room measures $16\frac{1}{2}$ ft $\times$ $10\frac{1}{2}$ ft. The carpet cost \$20.00 per square yard.	Area of room in square feet Area of room in square yards. Cost of the carpet.	Multiply $16\frac{1}{2}$ by $10\frac{1}{2}$ ft. to get the area in square feet. Divide the number of square feet by 9 to get the number of square yards. Multiply the number of square yards by \$20.00.	9 sq. feet = 1 square yard

$16\frac{1}{2} \times 10\frac{1}{2} = 173\frac{1}{4}$ square feet
$173\frac{1}{4} \div 9 = 19\frac{1}{4}$ square yards
$19\frac{1}{4} \times 20 = \$385.00$ total cost of carpet
CHECK:
Estimate area of room $16 \times 10 = 160$ square feet.
Estimate area in square yards $160 \div 10 = 16$
Estimate the cost $16 \times 20 = \$320.00$
This is close to our answer of \$385.00. Our answer seems reasonable.

2. **(a)** $\dfrac{10}{55} = 0.181 = 18\%$ **(b)** $\dfrac{3{,}660{,}000}{13{,}240{,}000} = 0.276 = 28\%$
(c) $\dfrac{15}{55} = 0.273 = 27\%$ **(d)** $\dfrac{3{,}720{,}000}{13{,}240{,}000} = 0.281 = 28\%$
(e) We notice that 18% of the company's sales force is located in the Northwest, and they were responsible for 28% of the sales volume. The percent of sales compared to the percent of sales force is about 150%. 27% of the company's sales force is located in the Southwest, and they were responsible for 28% of the sales volume. The percent of sales compared to the percent of sales force is approximately 100%. It would appear that the Northwest sales force is more effective.

Chapter 1

1.1 Practice Problems

1.

	Number	Integer	Rational Number	Irrational Number
(a)	$-\frac{2}{5}$		X	
(b)	1.515151...		X	
(c)	-8	X	X	
(d)	π			X

2. (a) Population growth of 1,259 is +1,259.
   (b) Depreciation of \$763 is −\$763.00.
   (c) Wind chill factor of minus 10 is −10.
3. (a) The opposite of $+\frac{2}{5}$ is $-\frac{2}{5}$.
   (b) The opposite of −1.92 is +1.92.
4. $(-23) + (-35) = -58$
5. $\left(\frac{-3}{5}\right) + \left(\frac{-4}{7}\right) = \frac{-21}{35} + \frac{-20}{35} = -\frac{41}{35}$
6. $(-12.7) + (-9.38) = -22.08$
7. $(-7) + (-11) + (-33) = -51$
8. $(-9) + (+3) = -6$
9. $(+37) + (-40) = -3$
10. $\left(-\frac{5}{12}\right) + \left(+\frac{7}{12}\right) + \left(-\frac{11}{12}\right)$
    $= \frac{2}{12} + \left(-\frac{11}{12}\right) = -\frac{9}{12} = -\frac{3}{4}$
11. $(-6) + (-8) + (+2) = -14 + 2 = -12$
12. $(-6) + (+5) + (-7) + (-2) + (+5) + (+3)$

$$\begin{array}{rr} -6 & +5 \\ -7 & +5 \\ -2 & +3 \\ \hline -15 & 13 \end{array} \qquad -15 + (+13) = -2$$

13. (a) $-2.9 + (-5.7) = -8.6$
    (b) $\frac{2}{3} + \left(-\frac{1}{4}\right)$
    $= \frac{8}{12} + \left(-\frac{3}{12}\right) = \frac{5}{12}$
    (c) $-10 + (-3) + 15 + 4$
    $= -13 + 19 = 6$

1.2 Practice Problems

1. $(-9) - (+3)$
   $= -9 + (-3)$
   $= -12$
2. $(+9) - (+12)$
   $= +9 + (-12)$
   $= -3$
3. $\left(-\frac{1}{5}\right) - \left(+\frac{1}{4}\right) = -\frac{1}{5} + \left(-\frac{1}{4}\right) = -\frac{9}{20}$
4. $\left(-\frac{3}{17}\right) - \left(-\frac{3}{17}\right) = -\frac{3}{17} + \left(+\frac{3}{17}\right) = 0$
5. Subtract −18 from 30
   $30 - (-18)$
   $= 30 + (+18)$
   $= 48$
6. (a) $-21 - 9$
   $= -21 + (-9)$
   $= -30$
   (b) $17 - 36$
   $= 17 + (-36)$
   $= -19$
7. $350 - (-186)$
   $= 350 + (186)$
   $= 536$ The helicopter is 536 feet from the sunken vessel.

1.3 Practice Problems

1. (a) $(-6)(-2) = +12$ (b) $(7)(9) = +63$
   (c) $\left(-\frac{3}{5}\right)\left(\frac{2}{7}\right) = -\frac{6}{35}$ (d) $(40)(-20) = -800$
2. $(-3)(-4)(-3)$
   $= (+12)(-3) = -36$
3. (a) $(-2)(-3)(-4)(-1)$
   $= (2)(3)(4)(1)$
   $= (6)(4)(1)$
   $= (24)(1) = +24$
   (b) $(-1)(-3)(-2)$
   $= -(1)(3)(2)$
   $= -(3)(2)$
   $= -6$
   (c) $(-4)(-\frac{1}{4})(-2)(-6)$
   $= (4)(\frac{1}{4})(2)(6)$
   $= (1)(2)(6)$
   $= (2)(6) = +12$
4. $(-2)(3)(-4)(-1)(-2)$
   $= (2)(3)(4)(1)(2)$
   $= (6)(4)(1)(2)$
   $= (24)(1)(2)$
   $= (24)(2) = +48$
5. (a) $(-36) \div (-2) = +18$ (b) $18 \div 9 = +2$
   (c) $\frac{50}{-10} = -5$ (d) $(-49) \div 7 = -7$
6. (a) $(-1.242) \div (-1.8)$

$$\begin{array}{r} .69 \\ 18.\overline{)12.42} \\ \underline{10\ 8} \\ 1\ 62 \\ \underline{1\ 62} \\ 0 \end{array}$$

   Thus $(-1.242) \div (-1.8) = 0.69$

   (b) $(0.235) \div (-0.0025)$

$$\begin{array}{r} 94. \\ 25\overline{)2350.} \\ \underline{225} \\ 100 \\ \underline{100} \\ 0 \end{array}$$

   Thus $(0.235) \div (-0.0025) = -94$

7. $\left(-\frac{5}{16}\right) \div \left(-\frac{10}{13}\right) = \left(-\frac{5}{16}\right) \times \left(-\frac{13}{10}\right) = \frac{13}{32}$
8. (a) $\dfrac{-12}{-\frac{4}{5}} = -12 \div \left(-\frac{4}{5}\right) = -12 \times \left(-\frac{5}{4}\right) = 15$
   (b) $\dfrac{-\frac{2}{9}}{\frac{8}{13}} = \left(-\frac{2}{9}\right) \times \left(\frac{13}{8}\right) = -\frac{13}{36}$

1.4 Practice Problems

1. (a) 6^4 (b) $(-2)^5$ (c) 108^3 (d) $(-11)^6$
2. (a) $5^3 = (5)(5)(5) = 125$
   (b) $3^5 = (3)(3)(3)(3)(3) = 243$
   (c) $9^4 = (9)(9)(9)(9) = 6561$
3. (a) $(-3)^3 = -27$
   (b) $-2^4 = -(2^4) = -16$
   (c) $(-2)^6 = 64$
   (d) $(-1)^{10} = 1$
4. (a) $(\frac{1}{3})^3 = (\frac{1}{3})(\frac{1}{3})(\frac{1}{3}) = \frac{1}{27}$
   (b) $(0.3)^4 = (0.3)(0.3)(0.3)(0.3) = 0.0081$
   (c) $(\frac{3}{2})^4 = (\frac{3}{2})(\frac{3}{2})(\frac{3}{2})(\frac{3}{2}) = \frac{81}{16} = 5\frac{1}{16}$

1.5 Practice Problems

1. (a) $-3(x + 2y) = -3x - 6y$
   (b) $-a(a - 3b) = -a^2 + 3ab$
2. (a) $-(-3x + y) = 3x - y$
   (b) $-(5r - 3s - 1) = -5r + 3s + 1$

3. (a) $\frac{3}{5}(a^2 - 5a + 25) = \frac{3}{5}a^2 - 3a + 15$
(b) $2.5(x^2 - 3.5x + 1.2) = 2.5x^2 - 8.75x + 3$
4. (a) $(3)(x)(3)(x) + (3)(x)(y) = 9x^2 + 3xy$
(b) $(-4)(x)(x) - (-4)(x)(2)(y) + (-4)(x)(3) = -4x^2 + 8xy - 12x$
5. (a) $(3x^2)(-4) - (2x)(-4) = -12x^2 + 8x$
(b) $(4y)(4y) - (7x^2)(4y) + (1)(4y) = 16y^2 - 28x^2y + 4y$

1.6 Practice Problems

1. (a) $5a$ and $8a$ are like terms.
$2b$ and $-4b$ are like terms.
(b) y^2 and $-7y^2$ are like terms.
2. (a) They are $5x^2$, $-6x$, 8, $-4x$, $-9x^2$, and 2.
(b) $5x^2$ and $-9x^2$ are like terms.
$-6x$ and $-4x$ are like terms.
8 and 2 are like terms.
3. (a) $5a + 7a + 4a = (5 + 7 + 4)a = 16a$
(b) $16y^3 + 9y^3 = (16 + 9)y^3 = 25y^3$
4. $-8y^2 - 9y^2 + 4y^2 = (-8 - 9 + 4)y^2 = -13y^2$
5. (a) $-x + 3a - 9x + 2a = -x - 9x + 3a + 2a = -10x + 5a$
(b) $5ab - 2ab^2 - 3a^2b + 6ab = 5ab + 6ab - 2ab^2 - 3a^2b$
$= 11ab - 2ab^2 - 3a^2b$
(c) $7x^2y - 2xy^2 - 3x^2y - 4xy^2 + 5x^2y$
$= 7x^2y - 3x^2y + 5x^2y - 2xy^2 - 4xy^2 = 9x^2y - 6xy^2$
6. $5xy - 2x^2y + 6xy^2 - xy - 3xy^2 - 7x^2y$
$= 5xy - xy - 7x^2y - 2x^2y + 6xy^2 - 3xy^2$
$= 4xy - 9x^2y + 3xy^2$
7. $5a(2 - 3b) - 4(6a + 2ab) = 10a - 15ab - 24a - 8ab$
$= 10a - 24a - 15ab - 8ab = -14a - 23ab$
8. $\frac{1}{7}a^2 - \frac{5}{12}b + 2a^2 - \frac{1}{3}b$
$= \left(\frac{1}{7} + 2\right)a^2 + \left[\left(-\frac{5}{12}\right) + \left(-\frac{1}{3}\right)\right]b$
$= \frac{15}{7}a^2 + \left[\left(-\frac{5}{12}\right) + \frac{(-1)(4)}{(3)(4)}\right]b = \frac{15}{7}a^2 - \frac{9}{12}b = \frac{15}{7}a^2 - \frac{3}{4}b$

1.7 Practice Problems

1. $18 - (-3)^3$
$= 18 - (-27)$
$= 18 + (+27)$
$= 45$

2. $12 \div (-2)(-3) - 2^4$
$= 12 \div (-2)(-3) - 16$
$= (-6)(-3) - 16$
$= 18 - 16 = 2$

3. $6 - (8 - 12)^2 + 8 \div 2$
$= 6 - (-4)^2 + 8 \div 2$
$= 6 - (16) + 4$
$= -10 + 4$
$= -6$

4. $\left(-\frac{1}{7}\right)\left(-\frac{14}{5}\right) + \left(-\frac{1}{2}\right) \div \left(\frac{3}{4}\right)$
$= \left(-\frac{1}{7}\right)\left(-\frac{14}{5}\right) + \left(-\frac{1}{2}\right) \times \left(\frac{4}{3}\right)$
$= \left(\frac{2}{5}\right) + \left(-\frac{2}{3}\right)$
$= \left(\frac{6}{15}\right) + \left(-\frac{10}{15}\right) = -\frac{4}{15}$

1.8 Practice Problems

1. Evaluate $4 - \frac{1}{2}x$, for $x = -8$.
$4 - \frac{1}{2}(-8)$
$= 4 + 4$
$= 8$

2. Evaluate for $x = -3$.
(a) $-x^4 = -(-3)^4$
$= -(81) = -81$
(b) $(-x)^4 = [-(-3)]^4$
$= (3)^4 = 81$

3. Evaluate $(5x)^3 + 2x$, for $x = -2$.
$[5(-2)]^3 + 2(-2)$
$= (-10)^3 + (-4)$
$= (-1000) + (-4)$
$= -1004$

4. Area of a triangle is
$A = \frac{1}{2}bh$
height = 3 meters (m)
base = 7 meters (m)
$A = \frac{1}{2}(7\text{m})(3\text{m})$
$A = \frac{21}{2}(\text{m})^2$
$A = 10.5$ square meters

5. Area of a circle is
$A = \pi r^2 \quad r = 3$ meters
$A = \pi(3 \text{ m})^2$
$A = 3.14(9)(\text{m})^2$
$A = 28.26$ square meters

6. Formula $\quad C = \frac{5}{9}(F - 32)$
$F = 68°$
$C = \frac{5}{9}(68 - 32)$
$C = \frac{5}{9}(36)$
$C = 20°$ Celsius

7. Use the formula.
$k = 1.61$ (m) $\quad$ Replace m by 35
$k = 1.61$ (35)
$k = 56.35$ $\quad$ The truck is violating the speed limit.

1.9 Practice Problems

1. $5[4x - 3(x - 2)]$
$= 5[4x - 3x + 6]$
$= 5[x + 6]$
$= 5x + 30$

2. $3ab - [2ab - (2 - a)]$
$= 3ab - [2ab - 2 + a]$
$= 3ab - 2ab + 2 - a$
$= ab + 2 - a$

3. $3[4x - 2(1 - x)] - [3x + (x - 2)]$
$3[4x - 2 + 2x] - [3x + x - 2]$
$12x - 6 + 6x - 3x - x + 2 = 14x - 4$
4. $-2\{5x - 3x[2x - (x^2 - 4x)]\}$
$-2\{5x - 3x[2x - x^2 + 4x]\}$
$-2\{5x - 6x^2 + 3x^3 - 12x^2\}$
$-2\{3x^3 - 12x^2 - 6x^2 + 5x\}$
$-2\{3x^3 - 18x^2 + 5x\} = -6x^3 + 36x^2 - 10x$

Chapter 2

2.1 Practice Problems

1. $x + 0.3 = 1.2$
$\quad -0.3 \;\; -0.3$
$x \quad = 0.9$
Check. $0.9 + 0.3 \stackrel{?}{=} 1.2$
$1.2 = 1.2$

2. $17 = x - 5$
$+5 \quad +5$
$22 = x$
Check. $17 \stackrel{?}{=} 22 - 5$
$17 = 17$

3. $5 - 12 = x - 3$
$-7 = x - 3$
$+3 \quad +3$
$-4 = x$
Check. $5 - 12 \stackrel{?}{=} -4 - 3$
$-7 = -7$

4. $x + 8 = -22 + 6 \qquad x = -2$
$-2 + 8 \stackrel{?}{=} -22 + 6$
$6 \neq -16$ $\quad$ This is not true.
Thus $x = -2$ is not a solution. Solve to find the solution.
$x + 8 = -22 + 6 = -16$
$x = -16 - 8$
$x = -24$

5. $\frac{1}{20} - \frac{1}{2} = x + \frac{3}{5}$
$\frac{1}{20} - \frac{1 \cdot 10}{2 \cdot 10} = x + \frac{3 \cdot 4}{5 \cdot 4}$
$\frac{1}{20} - \frac{10}{20} = x + \frac{12}{20}$
$x = \frac{1}{20} - \frac{10}{20} - \frac{12}{20} = -\frac{21}{20} = -1\frac{1}{20}$

Check.
$\frac{1}{20} - \frac{1}{2} \stackrel{?}{=} -1\frac{1}{20} + \frac{3}{5}$
$\frac{1}{20} - \frac{10}{20} \stackrel{?}{=} -\frac{21}{20} + \frac{12}{20}$
$-\frac{9}{20} = -\frac{9}{20}$

2.2 Practice Problems

1. $(\not{8})\frac{1}{\not{8}}x = -2(8)$
$x = -16$

2. $\frac{\not{9}x}{\not{9}} = \frac{72}{9}$
$x = 8$

3. $\frac{\not{6}x}{\not{6}} = \frac{50}{6}$
$x = \frac{25}{3}$

4. $\frac{-\not{27}x}{-\not{27}} = \frac{54}{-27}$
$x = -2$

5. $\frac{-x}{-1} = \frac{36}{-1}$
$x = -36$

6. $\frac{-51}{-6} = \frac{-6x}{-6}$
$\frac{17}{2} = x$

7. $\dfrac{21}{4.2} = \dfrac{\cancel{4.2}x}{\cancel{4.2}}$

$5 = x$

2.3 Practice Problems

1.
$$\begin{aligned} 9x + 2 &= 38 \\ -2 \quad & \;\; -2 \\ \hline \frac{9x}{9} &= \frac{36}{9} \\ x &= 4 \end{aligned}$$
Check. $9(4) + 2 \stackrel{?}{=} 38$; $36 + 2 \stackrel{?}{=} 38$; $38 = 38$ ✓

2.
$$\begin{aligned} 13x &= 2x - 66 \\ -2x \quad & \;\; -2x \\ \hline \frac{11x}{11} &= \frac{-66}{11} \\ x &= -6 \end{aligned}$$
Check. $13(-6) \stackrel{?}{=} 2(-6) - 66$; $-78 \stackrel{?}{=} -12 - 66$; $-78 = -78$ ✓

3.
$$\begin{aligned} 3x + 2 &= 5x + 2 \\ -2 \quad & \;\; -2 \\ \hline 3x &= 5x \\ -5x \quad & \;\; -5x \\ \hline -2x &= 0 \\ x &= 0 \end{aligned}$$
Check. $3(0) + 2 \stackrel{?}{=} 5(0) + 2$; $2 = 2$ ✓

4.
$$\begin{aligned} -z + 8 - z &= 3z + 10 - 3 \\ -2z + 8 &= 3z + 7 \\ -3z \quad & \;\; -3z \\ \hline -5z + 8 &= 7 \\ -8 &= -8 \\ \hline \frac{-5z}{-5} &= \frac{-1}{-5} \\ z &= \frac{1}{5} \end{aligned}$$

5.
$$\begin{aligned} 2x^2 - 6x + 3 &= -4x - 7 + 2x^2 \\ -2x^2 \quad & \;\; -2x^2 \\ \hline -6x + 3 &= -4x - 7 \\ +4x \quad & \;\; +4x \\ \hline -2x + 3 &= -7 \\ -3 &= -3 \\ \hline \frac{-2x}{-2} &= \frac{-10}{-2} \\ x &= 5 \end{aligned}$$

6.
$$\begin{aligned} 4x - (x + 3) &= 12 - 3(x - 2) \\ 4x - x - 3 &= 12 - 3x + 6 \\ 3x - 3 &= -3x + 18 \\ +3x \quad & \;\; +3x \\ \hline 6x - 3 &= 18 \\ +3 \quad & \;\; +3 \\ \hline \frac{6x}{6} &= \frac{21}{6} \\ x &= \frac{21}{6} = \frac{7}{2} \end{aligned}$$

Check.
$$4\left(\frac{7}{2}\right) - \left(\frac{7}{2} + 3\right) \stackrel{?}{=} 12 - 3\left(\frac{7}{2} - 2\right)$$
$$14 - \left(\frac{13}{2}\right) \stackrel{?}{=} 12 - 3\left(\frac{3}{2}\right)$$
$$\frac{28}{2} - \frac{13}{2} \stackrel{?}{=} \frac{24}{2} - \frac{9}{2}; \quad \frac{15}{2} = \frac{15}{2}$$

7.
$$\begin{aligned} 4(-2x - 3) &= -5(x - 2) + 2 \\ -8x - 12 &= -5x + 10 + 2 \\ -8x - 12 &= -5x + 12 \\ +5x \quad & \;\; +5x \\ \hline -3x - 12 &= 12 \\ +12 &= +12 \\ \hline \frac{-3x}{-3} &= \frac{24}{-3} \\ x &= -8 \end{aligned}$$

8.
$$\begin{aligned} 0.3x - 2(x + 0.1) &= 0.4(x - 3) - 1.1 \\ 0.3x - 2x - 0.2 &= 0.4x - 1.2 - 1.1 \\ -1.7x - 0.2 &= 0.4x - 2.3 \\ -0.4x \quad & \;\; -0.4x \\ \hline -2.1x - 0.2 &= -2.3 \\ +0.2 \quad & \;\; +0.2 \\ \hline \frac{-2.1x}{-2.1} &= \frac{-2.1}{-2.1} \\ x &= 1 \end{aligned}$$

9.
$$\begin{aligned} 5(2z - 1) + 7 &= 7z - 4(z + 3) \\ 10z - 5 + 7 &= 7z - 4z - 12 \\ 10z + 2 &= 3z - 12 \\ -3z \quad & \;\; -3z \\ \hline 7z + 2 &= -12 \\ -2 \quad & \;\; -2 \\ \hline \frac{7z}{7} &= -\frac{14}{7} \\ z &= -2 \end{aligned}$$

2.4 Practice Problems

1.
$$\begin{aligned} \frac{3}{8}x - \frac{3}{2} &= \frac{1}{4}x \\ 3x - 12 &= 2x \\ -2x \quad & \;\; -2x \\ \hline x - 12 &= 0 \\ +12 \quad & \;\; +12 \\ \hline x &= 12 \end{aligned}$$

2.
$$\begin{aligned} \frac{5x}{4} - 1 &= \frac{3x}{4} + \frac{1}{2} \\ 5x - 4 &= 3x + 2 \\ -3x \quad & \;\; -3x \\ \hline 2x - 4 &= +2 \\ +4 &= +4 \\ \hline \frac{2x}{2} &= \frac{6}{2} \\ x &= 3 \end{aligned}$$
Check.
$$\frac{5(3)}{4} - 1 \stackrel{?}{=} \frac{3(3)}{4} + \frac{1}{2}$$
$$\frac{15}{4} - 1 \stackrel{?}{=} \frac{9}{4} + \frac{1}{2}$$
$$\frac{15}{4} - \frac{4}{4} \stackrel{?}{=} \frac{9}{4} + \frac{2}{4}$$
$$\frac{11}{4} = \frac{11}{4}$$

3.
$$\begin{aligned} \frac{5x}{6} - \frac{5}{8} &= \frac{3x}{4} - \frac{1}{3} \\ 20x - 15 &= 18x - 8 \\ -18x \quad & \;\; -18x \\ \hline 2x - 15 &= -8 \\ +15 &= +15 \\ \hline \frac{2x}{2} &= \frac{7}{2} \\ x &= \frac{7}{2} \end{aligned}$$

4.
$$\begin{aligned} \frac{1}{3}(x - 2) &= \frac{1}{4}(x + 5) - \frac{5}{3} \\ \frac{1}{3}x - \frac{2}{3} &= \frac{1}{4}x + \frac{5}{4} - \frac{5}{3} \\ 4x - 8 &= 3x + 15 - 20 \\ 4x - 8 &= 3x - 5 \\ -3x \quad & \;\; -3x \\ \hline x - 8 &= -5 \\ +8 \quad & \;\; +8 \\ \hline x &= 3 \end{aligned}$$

5.
$$\begin{aligned} 2.8 &= 0.3(x - 2) + 2(0.1x - 0.3) \\ 2.8 &= 0.3x - 0.6 + 0.2x - 0.6 \\ 10(2.8) &= 10(0.3x) - 10(0.6) + 10(0.2x) - 10(0.6) \\ 28 &= 3x - 6 + 2x - 6 \\ 28 &= 5x - 12 \\ +12 \quad & \;\; +12 \\ \hline 40 &= 5x \\ 8 &= x \end{aligned}$$

2.5 Practice Problems

1.
$$\begin{aligned} d &= rt \\ 240 &= r(5) \\ \frac{240}{5} &= r \\ 48 &= r \end{aligned}$$

2. Solve for m. $\dfrac{E}{c^2} = \dfrac{mc^2}{c^2} \qquad \dfrac{E}{c^2} = m$

3. Solve for y.

$$\begin{array}{rcl} 8 - 2y + 3x & = & 0 \\ -3x & = & -3x \\ \hline 8 - 2y & = & -3x \\ -8 & & -8 \\ \hline \dfrac{-2y}{-2} & = & \dfrac{-3x-8}{-2} \\ y & = & \dfrac{3x+8}{2} \end{array}$$

4. Solve for d.

$$c = \pi d$$
$$\frac{c}{\pi} = \frac{\pi d}{\pi}$$
$$\frac{c}{\pi} = d$$

2.6 Practice Problems

1. **(a)** $7 > 2$ **(b)** $-3 > -4$ **(c)** $-1 < 2$
(d) $-8 < -5$ **(e)** $0 > -2$ **(f)** $\frac{2}{5} > \frac{3}{8}$

2. **(a)** $x > 5$; x is greater than 5

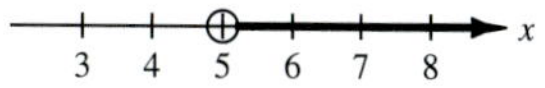
3 4 5 6 7 8 x

(b) $x \le -2$; x is less than or equal to -2

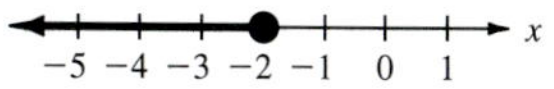
−5 −4 −3 −2 −1 0 1 x

(c) $3 > x$; 3 is greater than x (or x is less than 3)

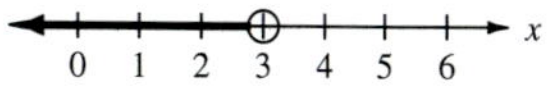
0 1 2 3 4 5 6 x

(d) $x \ge -\frac{3}{2}$; x is greater than or equal to $-\frac{3}{2}$

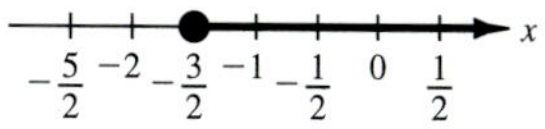
$-\frac{5}{2}$ −2 $-\frac{3}{2}$ −1 $-\frac{1}{2}$ 0 $\frac{1}{2}$ x

3. **(a)** $t \le 180$ **(b)** $d < 15{,}000$

2.7 Practice Problems

1. **(a)** $9 > 6$, $3 > 2$ **(b)** $2 < 5$, $4 < 10$ **(c)** $-8 < -3$, $-4 < 1$ **(d)** $-3 < 7$, $-10 < 0$

2. **(a)** $7 > 2$, $-14 < -4$ **(b)** $-3 < -1$, $3 > 1$
(c) $-10 > -20$, $1 < 2$ **(d)** $-15 < -5$, $3 > 1$

3.

$$\begin{array}{rcl} 8x - 2 & < & 3 \\ +2 & & +2 \\ \hline \dfrac{8x}{8} & < & \dfrac{5}{8} \\ x & < & \dfrac{5}{8} \end{array}$$

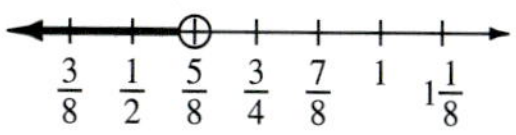
$\frac{3}{8}$ $\frac{1}{2}$ $\frac{5}{8}$ $\frac{3}{4}$ $\frac{7}{8}$ 1 $1\frac{1}{8}$

4.

$$\begin{array}{rcl} 7 + x & < & 28 \\ -7 & & -7 \\ \hline x & < & 21 \end{array}$$

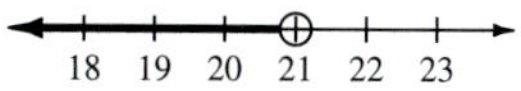
18 19 20 21 22 23

5.

$$\begin{array}{rcl} \frac{1}{2}x + 3 & < & \frac{2}{3}x \\ 3x + 18 & < & 4x \\ -4x & & -4x \\ \hline -x + 18 & < & 0 \\ -18 & & -18 \\ \hline \dfrac{-x}{-1} & > & \dfrac{-18}{-1} \\ x & > & 18 \end{array}$$

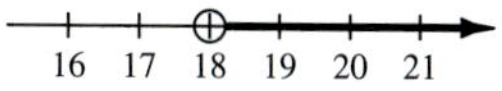
16 17 18 19 20 21

6.

$$\frac{1}{2}(3 - x) \le 2x + 5$$
$$\frac{3}{2} - \frac{1}{2}x \le 2x + 5$$
$$3 - x \le 4x + 10$$
$$-5x \le 7$$
$$x \ge -\frac{7}{5}$$

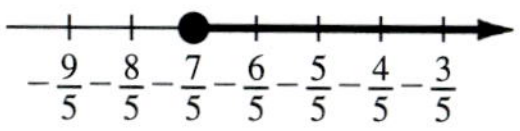
$-\frac{9}{5}$ $-\frac{8}{5}$ $-\frac{7}{5}$ $-\frac{6}{5}$ $-\frac{5}{5}$ $-\frac{4}{5}$ $-\frac{3}{5}$

7.

$$2000\text{N} - 700{,}000 \ge 2{,}500{,}000$$
$$2{,}000\text{N} \ge 3{,}200{,}000$$
$$\text{N} \ge 1{,}600$$

Chapter 3

3.1 Practice Problems

1. **(a)** $x + 4$ **(b)** $3x$
(c) $x - 8$ **(d)** $\frac{1}{4}x$

2. **(a)** $3x + 8$ **(b)** $3(x + 8)$

3. **(a)** $3x - 4$ **(b)** $\frac{2}{3}(x + 5)$

4. Let a = Ann's hours per week
Then $a - 17$ = Marie's hours per week

5. width $= w$
length $= 2w + 5$

6. 1st angle $= s - 16$
2nd angle $= s$
3rd angle $= 2s$

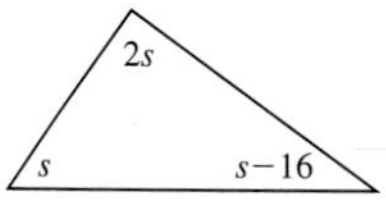
2s
s
s−16

7. Let x = the number of students in the fall
$\frac{2}{3}x$ = the number of students in the spring
$\frac{1}{5}x$ = the number of students in the summer

3.2 Practice Problems

1.

$$\frac{3}{4}x = -81$$
$$x = -108$$

2.

$$3x - 2 = 49$$
$$3x = 51$$
$$x = 17$$

3. $(x) + (3x - 12) = 24$
First number $x = 9$, Second number $= 3(9) - 12 = 15$

4. Let the precipitation for S. Dakota $= x$
then Utah $= x - 2$

$$x + x - 2 = 14$$
$$2x - 2 = 14$$
$$2x = 16$$
$$x = 8 \quad \text{therefore}$$

S. Dakota had 8 inches of rain
Utah had 6 inches

5. **(a)** $\frac{220}{4} = 55$ mph **(b)** $\frac{225}{4.5} = 50$ mph
(c) The trip leaving the city by 5 mph.

6.

$$\frac{x + x + 78 + 80 + 100 + 96}{6} = 90$$
$$354 + 2x = 540$$
$$2x = 186$$
$$x = 93$$

She needs a 93 on the final exam.

3.3 Practice Problems

1. short piece $= x$
long piece $= x + 17$

$$(x) + (x + 17) = 89$$
$$2x + 17 = 89$$
$$2x = 72$$
$$x = 36 \text{ feet}$$

Therefore Short piece $x = 36$
Long piece $(36) + 17 = 53$ feet

2. Family 1 $= x + 360$
Family 2 $= x$
Family 3 $= 2x - 200$

$$(x) + (x + 360) + (2x - 200) = 3960$$
$$4x + 160 = 3960$$
$$x = 950$$

Therefore Family #2 = \$950.00
Family #3 = 2(950) − 200 = \$1700.00
Family #1 = 950 + 360 = \$1310.00

3. Let the longest side $= x$

$$\begin{aligned} \text{other side} &= x - 30 \\ \text{shorter side} &= \tfrac{1}{2}x \\ x + x - 30 + \tfrac{1}{2}x &= 720 \\ 5x - 60 &= 1440 \\ 5x &= 1500 \\ x &= 300 \end{aligned}$$

Therefore the

longest side = 300 meters

other side = 270 meters

shorter side = 150 meters

3.4 Practice Problems

1. $$\begin{aligned} 3(22) + (0.18)m &= 363 \\ 66 + 0.18m &= 363 \\ 0.18m &= 297 \\ m &= 1650 \text{ miles} \end{aligned}$$

2. $$\begin{aligned} 0.38x &= 4560 \\ x &= \$12{,}000.00 \end{aligned}$$

3. $$\begin{aligned} x + 0.07x &= 13{,}910 \\ 1.07x &= 13{,}910 \\ x &= \$13{,}000.00 \end{aligned}$$

4. $$\begin{aligned} x &= 7000(0.12)(1) \\ x &= \$840.00 \end{aligned}$$

5. $$\begin{aligned} 0.15x + 0.12(8000 - x) &= 1155 \\ 0.15x + 960 - 0.12x &= 1155 \\ 0.03x + 960 &= 1155 \\ 0.03x &= 195 \\ x &= \$6{,}500.00 \end{aligned}$$

Therefore she invested \$6,500.00 at 15%
(8,000 − 6,500) = \$1,500.00 at 12%

6. $$\begin{aligned} x &= \text{Dimes} \\ x + 5 &= \text{Quarters} \\ 0.10x + 0.25(x + 5) &= 5.10 \\ 0.10x + 0.25x + 1.25 &= 5.10 \\ 0.35x + 1.25 &= 5.10 \\ 0.35x &= 3.85 \\ x &= 11 \end{aligned}$$

Therefore she has 11 dimes, 16 quarters

7. $$\begin{aligned} \text{Nickels} &= 2x \\ \text{Dimes} &= x \\ \text{Quarters} &= x + 4 \\ 0.05(2x) + 0.10(x) + 0.25(x + 4) &= 2.35 \\ 0.1x + 0.10x + 0.25x + 1 &= 2.35 \\ 0.45x + 1 &= 2.35 \\ 0.45x &= 1.35 \\ x &= 3 \end{aligned}$$

Therefore the boy has 3 dimes, 6 nickels, 7 quarters

3.5 Practice Problems

1. $$\begin{aligned} A &= \frac{1}{2}ab \\ &= \frac{1}{2}(20)(14) \\ &= (10)(14) \\ &= 140 \text{ square inches} \end{aligned}$$

2. $$\begin{aligned} A &= LW \\ 120 &= L(8) \\ \frac{120}{8} &= L \\ L &= 15 \text{ yards} \end{aligned}$$

3. $$\begin{aligned} A &= \frac{1}{2}a(b_1 + b_2) \\ 256 &= \frac{1}{2}a(12 + 20) \\ 256 &= \frac{1}{2}(a)(32) \\ 256 &= 16a \\ \frac{256}{16} &= a \\ a &= 16 \text{ feet} \end{aligned}$$

4. $$\begin{aligned} C &= 2\pi r \\ C &= 2(3.14)(15) \\ C &= 94.2 \\ C &= 94 \text{ meters} \end{aligned}$$

5. $$\begin{aligned} P &= a + b + c \\ P &= 15 + 15 + 15 \\ P &= 45 \text{ cm} \end{aligned}$$

6. $$\begin{aligned} \text{Surface Area} &= 4\pi r^2 \\ &= 4(3.14)(5)^2 \\ &= 314 \text{ m}^2 \end{aligned}$$

7. $$\begin{aligned} V &= \pi r^2 h \\ V &= 3.14(3)^2(4) \approx 113 \text{ ft}^3 \end{aligned}$$

8. Calculate the area of the pool.
$A = L \times W$
$A = 8 \times 12 = 96$ sq. ft.
Now add 6 feet to the length and 6 feet to the width of the pool and calculate the area.
$A = L \times W$
$A = 18 \times 14 = 252$ sq. ft.
Now subtract the areas.
$252 - 96 = 156$ sq. ft.
at \$12.00 a square foot.
$156 \times 12 = \$1{,}872.00$

9. $$\begin{aligned} V &= LWH \\ V &= (6)(5)(8) = 240 \text{ ft}^3 \end{aligned}$$
$$\text{Weight} = 240 \text{ ft}^3 \times \frac{62.4 \text{ lb}}{1 \text{ ft}^3} = 14{,}976 \text{ lb} \approx 15{,}000 \text{ lb}$$

3.6 Practice Problems

1. **(a)** height ≤ 6 **(b)** speed > 65
(c) area ≥ 560 **(d)** profit margin ≥ 50

2. $$\begin{aligned} \frac{1050 + 1250 + 950 + x}{4} &\geq 1{,}100 \\ \frac{x + 3250}{4} &\geq 1{,}100 \\ x + 3250 &\geq 4{,}400 \\ x &\geq 1{,}150 \end{aligned}$$
The rope must hold at least 1,150 pounds.

3. $$\begin{aligned} 1400 + 0.02x &> 2{,}200 \\ 0.02x &> 800 \\ x &> \$40{,}000.00 \end{aligned}$$

Chapter 4

4.1 Practice Problems

1. **(a)** $a^7 \cdot a^5 = a^{12}$ **(b)** $w^{10} \cdot w = w^{11}$
2. **(a)** $x^3 \cdot x^9 = x^{12}$ **(b)** $3^7 \cdot 3^4 = 3^{11}$
(c) $a^3 \cdot b^2 = a^3 \cdot b^2$ (cannot be simplified)
3. **(a)** $(-a^8)(a^4) = -a^{12}$ **(b)** $(3y^2)(-2y^3) = -6y^5$
(c) $(-4x^3)(-5x^2) = 20x^5$
4. **(a)** $(-5a^3b)(-2ab^4) = 10a^4b^5$
(b) $(2xy)\left(-\frac{1}{4}x^2y\right)(6xy^3) = -3x^4y^5$
5. **(a)** $\frac{10^{13}}{10^7} = 10^6$ **(b)** $\frac{x^{11}}{x} = x^{10}$
6. **(a)** $\frac{c^3}{c^4} = \frac{1}{c}$ **(b)** $\frac{10^{31}}{10^{56}} = \frac{1}{10^{25}}$
7. **(a)** $\frac{-7x^7}{-21x^9} = \frac{1}{3x^2}$ **(b)** $\frac{15x^{11}}{-3x^4} = -5x^7$
8. **(a)** $\frac{x^7y^9}{y^{10}} = \frac{x^7}{y}$ **(b)** $\frac{12x^5y^6}{-24x^3y^8} = -\frac{x^2}{2y^2}$
9. **(a)** $\frac{10^7}{10^7} = 1$ **(b)** $\frac{12a^4}{15a^4} = \frac{4}{5}$
10. $\frac{-20a^3b^8c^4}{28a^3b^7c^5} = -\frac{5b}{7c}$
11. $\frac{(-6ab^5)(3a^2b^4)}{16a^5b^7} = -\frac{9b^2}{8a^2}$
12. **(a)** $(a^4)^3 = a^{12}$ **(b)** $(10^5)^2 = 10^{10}$
13. **(a)** $(3xy)^3 = 27x^3y^3$ **(b)** $(yz)^{37} = y^{37}z^{37}$
14. $\left(\frac{4a}{b}\right)^6 = \frac{4^6a^6}{b^6} = \frac{4096a^6}{b^6}$
15. **(a)** $\left(\frac{-2x^3y^0z}{4xz^2}\right)^5 = \left(\frac{-x^2}{2z}\right)^5 = \frac{(-1)^5(x^2)^5}{2^5z^5} = -\frac{x^{10}}{32z^5}$
(b) $\left(\frac{4x^8y^5}{-3x^4z}\right)^3 = \left(\frac{4x^4y^5}{-3z}\right)^3 = \frac{4^3x^{12}y^{15}}{(-3)^3z^3} = -\frac{64x^{12}y^{15}}{27z^3}$

4.2 Practice Problems

1. **(a)** $x^{-12} = \frac{1}{x^{12}}$ **(b)** $w^{-5} = \frac{1}{w^5}$ **(c)** $z^{-2} = \frac{1}{z^2}$
(d) $y^{-7} = \frac{1}{y^7}$
2. **(a)** $4^{-3} = \frac{1}{4^3} = \frac{1}{64}$ **(b)** $6^{-2} = \frac{1}{6^2} = \frac{1}{36}$
3. **(a)** $\frac{3}{w^{-4}} = 3w^4$ **(b)** $\frac{x^{-6}y^4}{z^{-2}} = \frac{y^4z^2}{x^6}$ **(c)** $x^{-6}y^{-5} = \frac{1}{x^6y^5}$
4. **(a)** $(2x^4y^{-5})^{-2} = 2^{-2}x^{-8}y^{10} = \frac{y^{10}}{4x^8}$
(b) $\frac{y^{-3}z^{-4}}{y^2z^{-6}} = \frac{z^6}{y^2y^3z^4} = \frac{z^6}{y^5z^4} = \frac{z^2}{y^5}$
5. **(a)** 7.82×10^4 **(b)** 4.786×10^6
6. **(a)** 9.8×10^{-1} **(b)** 7.63×10^{-4} **(c)** 9.2×10^{-5}
7. **(a)** 2960.0 **(b)** 1,930,000.0 **(c)** 0.0543 **(d)** 0.00008562

8. 3.09×10^{16} m

9. (a) $(5.6 \times 10^4)(1.4 \times 10^9)$
$= (5.6)(1.4)(10^4)(10^9)$
$= 7.84 \times 10^{13}$

(b) $\dfrac{1.11 \times 10^{-4}}{3.7 \times 10^{-7}} = \dfrac{1.11}{3.7} \times \dfrac{10^7}{10^4} = 0.3 \times 10^3 = 3.0 \times 10^2$

10. $(159 \text{ parsecs})\dfrac{(3.09 \times 10^{13} \text{ kilometers})}{1 \text{ parsec}} = 491.31 \times 10^{13}$ kilometers

$d = r \times t$

$491.31 \times 10^{13} \text{ km} = \dfrac{50{,}000 \text{ km}}{1 \text{ hr}} \times t$

$\dfrac{491.31 \times 10^{13} \text{ km}(1 \text{ hr})}{50{,}000 \text{ km}} = t$

$\dfrac{4.9131 \times 10^{15} \text{ km}(1 \text{ hr})}{5.0 \times 10^4 \text{ km}} = t$

$0.9826 \times 10^{11} \text{ hr} = t$

9.83×10^{10} hours

4.3 Practice Problems

1. $(-8x^3 + 3x^2 + 6) + (2x^3 - 7x^2 - 3)$
$= [-8x^3 + 2x^3] + [3x^2 - 7x^2] + [6 - 3] = -6x^3 - 4x^2 + 3$

2. $\left(-\dfrac{1}{3}x^2 - 6x - \dfrac{1}{12}\right) + \left(\dfrac{1}{4}x^2 + 5x - \dfrac{1}{3}\right)$
$= \left[-\dfrac{1}{3}x^2 + \dfrac{1}{4}x^2\right] + [-6x + 5x] + \left[-\dfrac{1}{12} - \dfrac{1}{3}\right]$
$= -\dfrac{1}{12}x^2 - x - \dfrac{5}{12}$

3. $(3.5x^3 - 0.02x^2 + 1.56x - 3.5) + (-0.08x^2 - 1.98x + 4)$
$= 3.5x^3 + [-0.02x^2 - 0.08x^2] + [1.56x - 1.98x] + [-3.5 + 4]$
$= 3.5x^3 - 0.10x^2 - 0.42x + 0.5$

4. $(-3a^3 + 2ab + 7b) + (-4a^3 - 6ab + 3b^3)$
$= [-3a^3 - 4a^3] + [2ab - 6ab] + 7b + 3b^3$
$= -7a^3 - 4ab + 7b + 3b^3$

5. $(5x^3 - 15x^2 + 6x - 3) - (-4x^3 - 10x^2 + 5x + 13)$
$= [5x^3 + 4x^3] + [-15x^2 + 10x^2] + [6x - 5x] + [-3 - 13]$
$= 9x^3 - 5x^2 + x - 16$

6. $(x^3 - 7x^2y + 3xy^2 - 2y^3) - (2x^3 + 4xy - 6y^3)$
$= [x^3 - 2x^3] - 7x^2y + 3xy^2 - 4xy + [-2y^3 + 6y^3]$
$= -x^3 - 7x^2y + 3xy^2 - 4xy + 4y^3$

4.4 Practice Problems

1. $-8x^5 + 12x^4$
2. (a) $21 - 14x$ (b) $-3x^3 - 6x^2 + 12x$
(c) $6x^4y + 12x^3y^2 - 6xy^3$
3. (a) $12x^3 - 28x^2 + 4x$ (b) $18x^4y - 12x^3y + 6x^2y$
4. $5x^2 - 10x - x + 2 = 5x^2 - 11x + 2$
5. $24ac - 8ad - 15bc + 5bd$
6. $3x^2 - 3x + 2x - 2 = 3x^2 - x - 2$
7. $6a^2 - 9ab + 4ab - 6b^2 = 6a^2 - 5ab - 6b^2$
8. $(3x - 2y)(3x - 2y) = 9x^2 - 6xy - 6xy + 4y^2 = 9x^2 - 12xy + 4y^2$
9. $10x^4 + 12x^2y^2 + 15x^2y^2 + 18y^4 = 10x^4 + 27x^2y^2 + 18y^4$
10. $(2x - 1)(7x + 3)$
$= 14x^2 + 6x - 7x - 3$
$= 14x^2 - x - 3$ square feet

4.5 Practice Problems

1. $(6x)^2 - (7)^2 = 36x^2 - 49$
2. $(3x)^2 - (5y)^2 = 9x^2 - 25y^2$
3. $(4x^2)^2 - (1)^2 = 16x^4 - 1$
4. (a) $(5x)^2 + 2(5x)(4) + (4)^2 = 25x^2 + 40x + 16$
(b) $(4a)^2 - 2(4a)(9b) + (9b)^2 = 16a^2 - 72ab + 81b^2$
5.
$$\begin{array}{r} 3x^2 - 2xy + 4y^2 \\ x - 2y \\ \hline -6x^2y + 4xy^2 - 8y^3 \\ +3x^3 - 2x^2y + 4xy^2 \\ \hline 3x^3 - 8x^2y + 8xy^2 - 8y^3 \end{array}$$

6. $(2x^2 - 5x + 3)(x^2 + 3x - 4)$
$= 2x^2(x^2 + 3x - 4) - 5x(x^2 + 3x - 4) + 3(x^2 + 3x - 4)$
$= 2x^4 + 6x^3 - 8x^2 - 5x^3 - 15x^2 + 20x + 3x^2 + 9x - 12$
$= 2x^4 + x^3 - 20x^2 + 29x - 12$

7. $(3x - 2)(3x + 2)(2x + 3)$
$= (9x^2 - 4)(2x + 3)$
$$\begin{array}{r} 9x^2 - 4 \\ 2x + 3 \\ \hline 27x^2 + 0x - 12 \\ 18x^3 + 0x^2 - 8x \\ \hline 18x^3 + 27x^2 - 8x - 12 \end{array}$$

4.6 Practice Problems

1. $\dfrac{14y^2 - 49y}{7y} = \dfrac{14y^2}{7y} - \dfrac{49y}{7y} = 2y - 7$

2. $\dfrac{15y^4 - 27y^3 - 21y^2}{3y^2} = 5y^2 - 9y - 7$

3.
$$\begin{array}{r} x^2 + 6x + 7 \\ x + 4\overline{)x^3 + 10x^2 + 31x + 35} \\ \underline{x^3 + 4x^2} \\ 6x^2 + 31x \\ \underline{6x^2 + 24x} \\ 7x + 35 \\ \underline{7x + 28} \\ 7 \end{array}$$
Ans. $x^2 + 6x + 7 + \dfrac{7}{x + 4}$

4.
$$\begin{array}{r} 2x^2 + x + 1 \\ x - 1\overline{)2x^3 - x^2 + 0x + 1} \\ \underline{2x^3 - 2x^2} \\ x^2 + 0x \\ \underline{x^2 - x} \\ x + 1 \\ \underline{x - 1} \\ 2 \end{array}$$
Ans. $2x^2 + x + 1 + \dfrac{2}{x - 1}$

5.
$$\begin{array}{r} 5x^2 + x - 2 \\ 4x - 3\overline{)20x^3 - 11x^2 - 11x + 6} \\ \underline{20x^3 - 15x^2} \\ 4x^2 - 11x \\ \underline{4x^2 - 3x} \\ -8x + 6 \\ \underline{-8x + 6} \\ 0 \end{array}$$
Ans. $5x^2 + x - 2$

Check. $(4x - 3)(5x^2 + x - 2)$
$= 20x^3 + 4x^2 - 8x - 15x^2 - 3x + 6$
$= 20x^3 - 11x^2 - 11x + 6$

Chapter 5

5.1 Practice Problems

1. (a) $7(3a - b)$ (b) $p(1 + rt)$
2. $4a(3a + 4b^2 - 3ab)$
3. (a) $8(2a^3 - 3b^3)$ (b) $r^3(s^2 - 4rs + 7r^2)$
4. $9ab^2c(2a^2 - 3bc - 5ac)$
5. $6xy^2(5x^2 - 4x + 1) = 30x^3y^2 - 24x^2y^2 + 6xy^2$
6. $(a + b)(3 + x)$ 7. $(8y - 1)(9y^2 - 2)$
8. $\pi b^2 - \pi a^2 = \pi(b^2 - a^2)$

5.2 Practice Problems

1. $(2x - 7)(3y - 8)$

2. $3x(2x - 5) + 2(2x - 5)$
$= (2x - 5)(3x + 2)$

3. $a(x + 2) + 4b(x + 2)$
$= (x + 2)(a + 4b)$

4. $4(3x + y) + 7a(3x + y)$
$= (3x + y)(4 + 7a)$

5. $6a^2 + 3ac + 10ab + 5bc$
$= 3a(2a + c) + 5b(2a + c)$
$= (2a + c)(3a + 5b)$

6. $6xy + 14x - 15y - 35$
$= 2x(3y + 7) - 5(3y + 7)$
$= (3y + 7)(2x - 5)$

7. $3x + 6y - 5ax - 10ay$
$= 3(x + 2y) - 5a(x + 2y)$
$= (3 - 5a)(x + 2y)$

8. $10ad + 27bc - 6bd - 45ac$
$= 10ad - 6bd - 45ac + 27bc$
$= 2d(5a - 3b) - 9c(5a - 3b)$
$= (2d - 9c)(5a - 3b)$
Check: $10ad - 6bd - 45ac + 27bc$

5.3 Practice Problems

1. $(x + 6)(x + 2)$
2. $(x + 2)(x + 15)$
3. $(x + 1)(x + 6)$
4. $(x - 9)(x - 2)$
5. $(y - 3)(y - 8)$
6. $(x + 6)(x - 1)$
7. $(a + 3)(a - 8)$
8. $(x + 20)(x - 3)$
Check: $x^2 - 3x + 20x - 60$
$= x^2 + 17x - 60$
9. $(x + 5)(x - 12)$
10. $(a^2 + 7)(a^2 - 6)$
11. $3x^2 + 45x + 150$
$= 3(x^2 + 15x + 50)$
$= 3(x + 5)(x + 10)$
12. $4x^2 - 8x - 140$
$= 4(x^2 - 2x - 35)$
$= 4(x + 5)(x - 7)$

5.4 Practice Problems

1. $(2x - 5)(x - 1)$
2. $(9x - 1)(x - 7)$
3. $(3x - 2)(x + 2)$
4. $(3x - 7)(x + 2)$
5. $2x^2 - 2x - 5x + 5$
$= 2x(x - 1) - 5(x - 1)$
$= (2x - 5)(x - 1)$
6. $9x^2 - 63x - x + 7$
$= 9x(x - 7) - 1(x - 7)$
$= (x - 7)(9x - 1)$
7. $3x^2 + 6x - 2x - 4$
$= 3x(x + 2) - 2(x + 2)$
$= (x + 2)(3x - 2)$
8. $8x^2 - 8x - 6$
$= 2(4x^2 - 4x - 3)$
$= 2(2x + 1)(2x - 3)$
9. $24x^2 - 38x + 10$
$= 2(12x^2 - 19x + 5)$
$= 2(12x^2 - 4x - 15x + 5)$
$= 2[4x(3x - 1) - 5(3x - 1)]$
$= 2(4x - 5)(3x - 1)$

5.5 Practice Problems

1. $(1 + 8x)(1 - 8x)$
2. $(6x + 7)(6x - 7)$
3. $(10x + 9y)(10x - 9y)$
4. $(x^4 + 1)(x^4 - 1)$
$= (x^4 + 1)(x^2 + 1)(x^2 - 1)$
$= (x^4 + 1)(x^2 + 1)(x + 1)(x - 1)$
5. $(4x + 1)(4x + 1) = (4x + 1)^2$
6. $(5x - 3)(5x - 3) = (5x - 3)^2$
7. (a) $(5x - 6y)(5x - 6y) = (5x - 6y)^2$
(b) $(8x^3 - 3)(8x^3 - 3) = (8x^3 - 3)^2$
8. $9x^2 - 12x - 3x + 4$
$= 3x(3x - 4) - 1(3x - 4)$
$= (3x - 1)(3x - 4)$
9. $20x^2 - 45$
$= 5(4x^2 - 9)$
$= 5(2x - 3)(2x + 3)$
10. $3(25x^2 - 20x + 4)$
$= 3(5x - 2)(5x - 2)$
$= 3(5x - 2)^2$

5.6 Practice Problems

1. (a) $9y^2(x^4 - 1)$
$= 9y^2(x^2 + 1)(x^2 - 1)$
$= 9y^2(x^2 + 1)(x + 1)(x - 1)$
(b) $-(4x^2 - 12x + 9)$
$= -(2x - 3)(2x - 3)$
$= -(2x - 3)^2$
(c) $3(x^2 - 12x + 36)$
$= 3(x - 6)(x - 6)$
$= 3(x - 6)^2$
(d) $5x(x^2 - 3xy + 2x - 6y)$
$= 5x[x(x - 3y) + 2(x - 3y)]$
$= 5x(x + 2)(x - 3y)$
2. $x^2 - 9x - 8$
The factors of -8 are
$(-2)(4) = -8$
$(2)(-4) = -8$
$(-8)(1) = -8$
$(-1)(8) = -8$
None of these pairs will add up to be the coefficient of the middle term. Thus the polynomial cannot be factored.
It is prime.
3. $25x^2 + 82x + 4$
Check to see if this is a perfect square trinomial.
$2[(5)(2)] = 2(10) = 20$ This is not the coefficient of the middle term. The grouping number equals 100. No factors add to 82.
It is prime.

5.7 Practice Problems

1. $10x^2 - x - 2 = 0$
$(5x + 2)(2x - 1) = 0$

$5x + 2 = 0$
$5x = -2$
$x = -\frac{2}{5}$

$2x - 1 = 0$
$2x = 1$
$x = \frac{1}{2}$

Check.
$10\left(-\frac{2}{5}\right)^2 - \left(-\frac{2}{5}\right) - 2 \stackrel{?}{=} 0$
$10\left(\frac{4}{25}\right) + \frac{2}{5} - 2 \stackrel{?}{=} 0$
$\frac{8}{5} + \frac{2}{5} - 2 \stackrel{?}{=} 0$
$\frac{10}{5} - \frac{10}{5} \stackrel{?}{=} 0$
$0 = 0$

$10\left(\frac{1}{2}\right)^2 - \frac{1}{2} - 2 \stackrel{?}{=} 0$
$10\left(\frac{1}{4}\right) - \frac{1}{2} - 2 \stackrel{?}{=} 0$
$\frac{5}{2} - \frac{1}{2} - 2 \stackrel{?}{=} 0$
$\frac{4}{2} - 2 \stackrel{?}{=} 0$
$2 - 2 \stackrel{?}{=} 0$
$0 = 0$

Thus $x = -\frac{2}{5}$ and $x = \frac{1}{2}$ are both roots for the equation.

2. $3x^2 - 5x + 2 = 0$
$(3x - 2)(x - 1) = 0$
$3x - 2 = 0$
$3x = 2$
$x = \frac{2}{3}$
$x - 1 = 0$
$x = 1$

3. $7x^2 + 11x = 0$
$x(7x + 11) = 0$
$x = 0$
$7x + 11 = 0$
$7x = -11$
$x = \frac{-11}{7}$

4. $x^2 - 6x + 4 = -8 + x$
$x^2 - 7x + 12 = 0$
$(x - 3)(x - 4) = 0$
$x - 3 = 0$
$x = 3$
$x - 4 = 0$
$x = 4$

5. $x(3x - 7) = 20$
$3x^2 - 7x = 20$
$3x^2 - 7x - 20 = 0$
$(3x + 5)(x - 4) = 0$
$3x + 5 = 0$
$x = -\frac{5}{3}$
$x - 4 = 0$
$x = 4$

6. $\frac{2x^2 - 7x}{3} = 5$
$2x^2 - 7x = 15$
$2x^2 - 7x - 15 = 0$
$(2x + 3)(x - 5) = 0$
$2x + 3 = 0$
$x = -\frac{3}{2}$
$x - 5 = 0$
$x = 5$

7. Let w = width, then
$3w + 2$ = length
$(3w + 2)w = 85$
$3w^2 + 2w = 85$
$3w^2 + 2w - 85 = 0$
$(3w + 17)(w - 5) = 0$
$3w + 17 = 0$
$w = -\frac{17}{3}$
$w - 5 = 0$
$w = 5$
The only valid answer is width = 5 meters
length = $3(5) + 2 = 17$ meters

8. Let b = base
$b - 3$ = altitude
$\frac{b(b - 3)}{2} = 35$
$b^2 - 3b = 70$
$b^2 - 3b - 70 = 0$
$(b + 7)(b - 10) = 0$
$b + 7 = 0$
$b = -7$
This is not a valid answer.
$b - 10 = 0$
$b = 10$
Thus the base = 10 centimeters
altitude = $10 - 3 = 7$ centimeters

9. $-5t^2 + 45 = 0$
$-5(t^2 - 9) = 0$
$t^2 - 9 = 0$
$(t + 3)(t - 3) = 0$
$t = 3 \qquad t = -3 \qquad t = -3$ is not a valid answer.
Thus it will be 3 seconds before she breaks the water's surface.

Chapter 6

6.1 Practice Problems

1. (a) $\frac{7}{3} = \frac{?}{18}$ (b) $\frac{28}{63} = \frac{7 \times 2 \times 2}{7 \times 3 \times 3} = \frac{4}{9}$
$\frac{7 \cdot 6}{3 \cdot 6} = \frac{42}{18}$

2. $\frac{12x - 6}{14x - 7} = \frac{6(2x - 1)}{7(2x - 1)} = \frac{6}{7}$

3. $\frac{4x - 6}{2x^2 - x - 3} = \frac{2(2x - 3)}{(2x - 3)(x + 1)} = \frac{2}{x + 1}$

4. $\frac{7x^2 - 14xy}{14x^2 - 28xy} = \frac{7x(x - 2y)}{14x(x - 2y)} = \frac{7x}{14x} = \frac{1}{2}$

5. $\frac{x^3 + 11x^2 + 30x}{3x^3 + 17x^2 - 6x} = \frac{x(x^2 + 11x + 30)}{x(3x^2 + 17x - 6)} = \frac{(x + 5)(x + 6)}{(3x - 1)(x + 6)}$
$= \frac{x + 5}{3x - 1}$

6. $\frac{2x - 5}{5 - 2x} = \frac{-1(-2x + 5)}{(5 - 2x)} = \frac{-1(5 - 2x)}{(5 - 2x)} = -1$

7. $\frac{4x^2 + 3x - 10}{25 - 16x^2} = \frac{(4x - 5)(x + 2)}{(5 + 4x)(5 - 4x)} = \frac{(4x - 5)(x + 2)}{-1(4x - 5)(5 + 4x)}$
$= \frac{x + 2}{-1(5 + 4x)} = -\frac{x + 2}{5 + 4x}$

8. $\frac{4x^2 - 9y^2}{4x^2 + 12xy + 9y^2} = \frac{(2x + 3y)(2x - 3y)}{(2x + 3y)(2x + 3y)} = \frac{2x - 3y}{2x + 3y}$

9. $\frac{12x^3 - 48x}{6x - 3x^2} = \frac{12x(x^2 - 4)}{3x(2 - x)} = \frac{12x(x + 2)(x - 2)}{3x(2 - x)} = -4(x + 2)$

6.2 Practice Problems

1. $\frac{10x}{x^2 - 7x + 10} \cdot \frac{x^2 + 3x - 10}{25x} = \frac{10x}{(x - 5)(x - 2)} \cdot \frac{(x - 2)(x + 5)}{25x}$
$= \frac{2(x + 5)}{5(x - 5)}$

2. $\frac{2y^2 - 6y - 8}{y^2 - y - 2} \cdot \frac{y^2 - 5y + 6}{2y^2 - 4y - 6}$
$= \frac{2(y + 1)(y - 4)}{(y + 1)(y - 2)} \cdot \frac{(y - 2)(y - 3)}{2(y + 1)(y - 3)} = \frac{y - 4}{y + 1}$

3. $\frac{x^2 + 5x + 6}{x^2 + 8x} \div \frac{2x^2 + 5x + 2}{2x^2 + x}$
$= \frac{(x + 2)(x + 3)}{x(x + 8)} \cdot \frac{x(2x + 1)}{(2x + 1)(x + 2)} = \frac{x + 3}{x + 8}$

4. $\frac{x + 3}{x - 3} \div (9 - x^2) = \frac{x + 3}{x - 3} \cdot \frac{1}{(3 + x)(3 - x)} = \frac{1}{(x - 3)(3 - x)}$

6.3 Practice Problems

1. $\frac{2s + t}{2s - t} + \frac{s - t}{2s - t} = \frac{2s + t + s - t}{2s - t} = \frac{3s}{2s - t}$

2. $\frac{b}{(a - 2b)(a + b)} - \frac{2b}{(a - 2b)(a + b)} = \frac{b - 2b}{(a - 2b)(a + b)}$
$= \frac{-b}{(a - 2b)(a + b)}$

3. $\frac{7}{6x + 21}, \frac{13}{10x + 35}$ $\quad 6x + 21 = 3(2x + 7)$
$10x + 35 = 5(2x + 7)$
$\text{LCD} = 3 \cdot 5 \cdot (2x + 7) = 15(2x + 7)$

4. (a) $\frac{3}{50xy^2z}, \frac{19}{40x^3yz}$ $\quad 50xy^2z = 2 \cdot 5^2 \cdot x \cdot y^2 \cdot z$
$40x^3yz = 2^3 \cdot 5 \cdot x^3 \cdot y \cdot z$
$\text{LCD} = 2^3 \cdot 5^2 \cdot x^3 \cdot y^2 \cdot z$
$\text{LCD} = 200x^3y^2z$

(b) $\frac{12}{x^2 + 5x}, \frac{16}{x^2 + 10x + 25}$ $\quad x^2 + 5x = x(x + 5)$
$x^2 + 10x + 25 = (x + 5)(x + 5)$
$\text{LCD} = x(x + 5)^2$

(c) $\frac{2}{x^2 + 5x + 6}, \frac{6}{3x^2 + 5x - 2}$
$x^2 + 5x + 6 = (x + 3)(x + 2)$
$3x^2 + 5x - 2 = (3x - 1)(x + 2)$
$\text{LCD} = (x + 2)(x + 3)(3x - 1)$

5. $\frac{7}{a} + \frac{3}{abc} = \frac{7bc + 3}{abc}$ $\quad \text{LCD} = abc$

6. $\frac{2a - b}{(a + 2b)(a - 2b)} + \frac{2}{(a + 2b)}$ $\quad \text{LCD} = (a + 2b)(a - 2b)$
$= \frac{2a - b}{(a + 2b)(a - 2b)} + \frac{2(a - 2b)}{(a + 2b)(a - 2b)} = \frac{2a - b + (2a - 4b)}{(a + 2b)(a - 2b)}$
$= \frac{4a - 5b}{(a + 2b)(a - 2b)}$

7. $\frac{3}{x^2} + \frac{4x}{xy} + \frac{7}{y}$ $\quad \text{LCD} = x^2y$
$= \frac{3y}{x^2y} + \frac{4x^2}{x^2y} + \frac{7x^2}{x^2y} = \frac{3y + 4x^2 + 7x^2}{x^2y} = \frac{11x^2 + 3y}{x^2y}$

8. $\frac{7a}{a^2 + 2ab + b^2} + \frac{4}{a^2 + ab}$
$= \frac{7a}{(a + b)(a + b)} + \frac{4}{a(a + b)}$ $\quad \text{LCD} = a(a + b)^2$
$= \frac{7a^2}{a(a + b)(a + b)} + \frac{4(a + b)}{a(a + b)(a + b)} = \frac{7a^2 + 4a + 4b}{a(a + b)^2}$

9. $\frac{x + 7}{3x - 9} - \frac{x - 6}{x - 3} = \frac{x + 7}{3(x - 3)} - \frac{x - 6}{x - 3}$ $\quad \text{LCD} = 3(x - 3)$
$= \frac{x + 7 - 3(x - 6)}{3(x - 3)} = \frac{x + 7 - 3x + 18}{3(x - 3)} = \frac{-2x + 25}{3(x - 3)}$

10. $\frac{x - 2}{x^2 - 4} - \frac{x + 1}{2x^2 + 4x} = \frac{x - 2}{(x + 2)(x - 2)} - \frac{x + 1}{2x(x + 2)}$
$\text{LCD} = 2x(x + 2)(x - 2)$
$= \frac{2x(x - 2)}{2x(x + 2)(x - 2)} - \frac{(x - 2)(x + 1)}{2x(x + 2)(x - 2)}$
$= \frac{2x^2 - 4x - (x^2 - x - 2)}{2x(x + 2)(x - 2)} = \frac{2x^2 - 4x - x^2 + x + 2}{2x(x + 2)(x - 2)}$
$= \frac{x^2 - 3x + 2}{2x(x + 2)(x - 2)} = \frac{\cancel{(x - 2)}(x - 1)}{2x(x + 2)\cancel{(x - 2)}} = \frac{x - 1}{2x(x + 2)}$

6.4 Practice Problems

1. $\frac{\frac{1}{a} + \frac{1}{b}}{\frac{2}{ab^2}} = \frac{\frac{b + a}{ab}}{\frac{2}{ab^2}}$
$= \frac{b + a}{ab} \div \frac{2}{ab^2} = \frac{b + a}{ab} \cdot \frac{ab^2}{2} = \frac{\cancel{ab^2}^{\,b}(a + b)}{2\cancel{ab}} = \frac{b(a + b)}{2}$

2. $\frac{\frac{1}{a} + \frac{1}{b}}{\frac{1}{a} - \frac{1}{b}} = \frac{\frac{b + a}{ab}}{\frac{b - a}{ab}} = \frac{b + a}{\cancel{ab}} \cdot \frac{\cancel{ab}}{b - a} = \frac{b + a}{b - a}$

3. $\frac{\frac{x}{x^2 + 4x + 3} + \frac{2}{x + 1}}{x + 1} = \frac{\frac{x}{(x + 1)(x + 3)} + \frac{2}{x + 1}}{x + 1}$
$= \frac{\frac{x + 2(x + 3)}{(x + 1)(x + 3)}}{x + 1} = \frac{\frac{x + 2x + 6}{(x + 1)(x + 3)}}{(x + 1)} = \frac{3x + 6}{(x + 1)(x + 3)} \cdot \frac{1}{(x + 1)}$
$= \frac{3(x + 2)}{(x + 1)^2(x + 3)}$

4. $\dfrac{\dfrac{6}{x^2-y^2}}{\dfrac{1}{x-y}+\dfrac{3}{x+y}} = \dfrac{\dfrac{6}{(x+y)(x-y)}}{\dfrac{(x+y)+3(x-y)}{(x+y)(x-y)}} = \dfrac{6}{(x+y)(x-y)}$

$\div \dfrac{x+y+3x-3y}{(x+y)(x-y)} = \dfrac{6}{(x+y)(x-y)} \cdot \dfrac{(x+y)(x-y)}{4x-2y}$

$= \dfrac{6}{\cancel{(x+y)(x-y)}} \cdot \dfrac{\cancel{(x+y)(x-y)}}{2(2x-y)} = \dfrac{3}{2x-y}$

5. $\dfrac{\dfrac{2}{3x^2}-\dfrac{3}{y}}{\dfrac{5}{xy}-4}$ LCD $= 3x^2y$

$\dfrac{3x^2y\left(\dfrac{2}{3x^2}\right) - 3x^2y\left(\dfrac{3}{y}\right)}{3x^2y\left(\dfrac{5}{xy}\right) - 3x^2y(4)} = \dfrac{2y-3x^2(3)}{3x(5)-12x^2y} = \dfrac{2y-9x^2}{15x-12x^2y}$

6. $\dfrac{\dfrac{6}{x^2-y^2}}{\dfrac{7}{x-y}+\dfrac{3}{x+y}} = \dfrac{\dfrac{6}{(x+y)(x-y)}}{\dfrac{7}{x-y}+\dfrac{3}{x+y}}$ LCD $= (x+y)(x-y)$

$\dfrac{(x+y)(x-y)\left(\dfrac{6}{(x+y)(x-y)}\right)}{(x+y)(x-y)\left(\dfrac{7}{x-y}\right) + (x+y)(x-y)\left(\dfrac{3}{x+y}\right)}$

$= \dfrac{6}{7(x+y)+3(x-y)} = \dfrac{6}{7x+7y+3x-3y} = \dfrac{6}{10x+4y}$

$= \dfrac{3}{5x+2y}$

6.5 Practice Problems

1. $\dfrac{2}{x}+\dfrac{4}{x} = 3 - \dfrac{1}{x} - \dfrac{17}{8}$

LCD $= 8x$

$16 + 32 = 24x - 8 - 17x$

$48 = 7x - 8$

$56 = 7x$

$x = 8$

Check. $\dfrac{2}{8}+\dfrac{4}{8} \stackrel{?}{=} 3 - \dfrac{1}{8} - \dfrac{17}{8}$

$\dfrac{6}{8} \stackrel{?}{=} 3 - \dfrac{18}{8}$

$\dfrac{6}{8} \stackrel{?}{=} \dfrac{24}{8} - \dfrac{18}{8}$

$\dfrac{6}{8} = \dfrac{6}{8}$ ✓

2. $\dfrac{4}{2x+1} = \dfrac{6}{2x-1}$ LCD $= (2x+1)(2x-1)$

$(2x+1)(2x-1)\left[\dfrac{4}{2x+1}\right] = (2x+1)(2x-1)\left[\dfrac{6}{2x-1}\right]$

$4(2x-1) = 6(2x+1)$

$8x - 4 = 12x + 6$

$-4x = 10$

$x = -\dfrac{5}{2}$

Check. $\dfrac{4}{2\left(-\dfrac{5}{2}\right)+1} \stackrel{?}{=} \dfrac{6}{2\left(-\dfrac{5}{2}\right)-1}$

$\dfrac{4}{-5+1} \stackrel{?}{=} \dfrac{6}{-5-1}$

$\dfrac{4}{-4} \stackrel{?}{=} \dfrac{6}{-6}$

$-1 = -1$ ✓

3. $\dfrac{x-1}{x^2-4} = \dfrac{2}{x+2} + \dfrac{4}{x-2}$

$\dfrac{x-1}{(x+2)(x-2)} = \dfrac{2}{x+2} + \dfrac{4}{x-2}$ LCD $= (x+2)(x-2)$

$\cancel{(x+2)(x-2)}\left[\dfrac{x-1}{\cancel{(x+2)(x-2)}}\right] = \cancel{(x+2)}(x-2)\left[\dfrac{2}{\cancel{x+2}}\right]$

$+ (x+2)\cancel{(x-2)}\left[\dfrac{4}{\cancel{x-2}}\right]$

$x - 1 = 2(x-2) + 4(x+2)$

$x - 1 = 2x - 4 + 4x + 8$

$x - 1 = 6x + 4$

$-5x = 5$

$x = -1$

Check. $\dfrac{-1-1}{(-1)^2-4} \stackrel{?}{=} \dfrac{2}{-1+2} + \dfrac{4}{-1-2}$

$\dfrac{-2}{-3} \stackrel{?}{=} \dfrac{2}{1} + \dfrac{4}{-3}$

$\dfrac{2}{3} \stackrel{?}{=} \dfrac{6}{3} - \dfrac{4}{3}$

$\dfrac{2}{3} = \dfrac{2}{3}$

4. $\dfrac{2x}{x+1} = \dfrac{-2}{x+1} + 1$ LCD $= (x+1)$

$(x+1)\left[\dfrac{2x}{x+1}\right] = (x+1)\left[\dfrac{-2}{x+1}\right] + (x+1)[1]$

$2x = -2 + x + 1$

$2x = x - 1$

$x = -1$

Check. $\dfrac{2(-1)}{-1+1} \stackrel{?}{=} \dfrac{-2}{-1+1} + 1$

$\dfrac{-2}{0} \stackrel{?}{=} \dfrac{-2}{0} + 1$

These expressions are not defined, therefore, there is **no solution** to this problem.

6.6 Practice Problems

1. $\dfrac{8}{420} = \dfrac{x}{315}$

$8(315) = 420x$

$2{,}520 = 420x$

$x = 6$

It would take Brenda 6 hours to drive 315 miles.

2. $\dfrac{\dfrac{5}{8}}{30} = \dfrac{2\dfrac{1}{2}}{x}$

$\dfrac{5}{8}x = 30\left(2\dfrac{1}{2}\right)$

$\dfrac{5}{8}x = 75$

$x = 120$

Therefore $2\frac{1}{2}$ inches would represent 120 miles.

3. $\dfrac{13}{x} = \dfrac{16}{18}$

$13(18) = 16x$

$234 = 16x$

$x = 14\dfrac{5}{8}$ cm

4. $\dfrac{6}{7} = \dfrac{x}{38.5}$

$6(38.5) = 7x$

$231 = 7x$

$x = 33$ feet

5. Train A time $= \dfrac{180}{x+10}$ Train B time $= \dfrac{150}{x}$

$\dfrac{180}{x+10} = \dfrac{150}{x}$

$180x = 150(x+10)$

$180x = 150x + 1500$

$30x = 1500$

$x = 50$

Train B travels 50 kilometers per hour. Train A travels $50 + 10 = 60$ kilometers per hour.

6.

	Number of Hours	Part of the Job Done in One Hour
John	6 hours	$\frac{1}{6}$
Dave	7 hours	$\frac{1}{7}$
John & Dave Together	x	$\frac{1}{x}$

$$\frac{1}{6} + \frac{1}{7} = \frac{1}{x} \qquad \text{LCD} = 42x$$

$7x + 6x = 42$

$13x = 42$

$$x = 3\frac{3}{13} \qquad \frac{3}{13}\text{ hour} \times \frac{60 \text{ min}}{1 \text{ hour}} = \frac{180}{13}\text{ min} = 13.846 \text{ min}$$

Thus, doing the job together will take 3 hours and 14 minutes.

Chapter 7

7.1 Practice Problems

1. Point B is three units to the right of the origin and 4 units up.
2. Point D is one unit to the right and 4 units up.
   Point E is four units to the right and one unit up.
3. 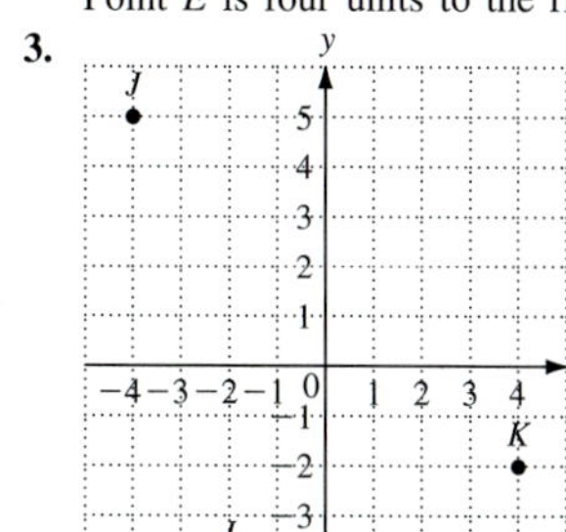

4. 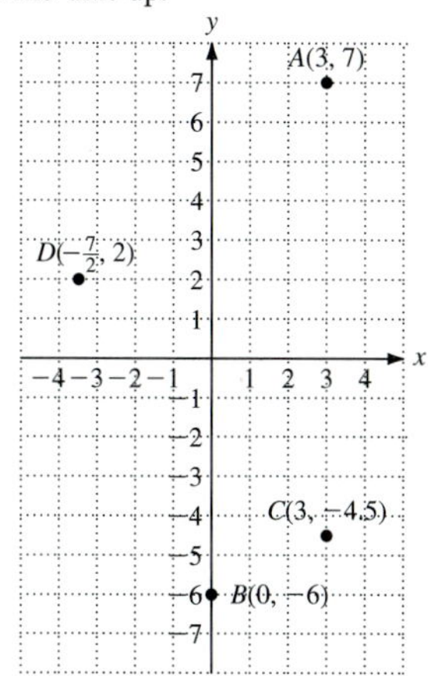

5. $B = (2, 7)$
6. $A = (-2, -1)$; $B = (-1, 3)$; $C = (0, 0)$; $D = (2, -1)$; $E = (3, 1)$

7. $3x - 4y = 12$

   (a) $3(0) - 4y = 12$
   $-4y = 12$
   $y = -3 \qquad (0, -3)$

   (b) $3x - 4(3) = 12$
   $3x - 12 = 12$
   $3x = 24$
   $x = 8 \qquad (8, 3)$

   (c) $3x - 4(-6) = 12$
   $3x + 24 = 12$
   $3x = -12$
   $x = -4 \qquad (-4, -6)$

8. $y = -2x - 5$

   (a) $y = -2(3) - 5$
   $y = -6 - 5$
   $y = -11 \qquad (3, -11)$

   (b) $y = -2(-4) - 5$
   $y = 8 - 5$
   $y = 3 \qquad (-4, 3)$

7.2 Practice Problems

1. Graph $x + y = 10$.
   Let $x = 0$ then
   $0 + y = 10$
   $y = 10$

   Let $x = 5$ then
   $5 + y = 10$
   $y = 5$

   Let $x = 2$ then
   $2 + y = 10$
   $y = 8$

   Plot the ordered pairs.
   (0, 10), (5, 5) and (2, 8)

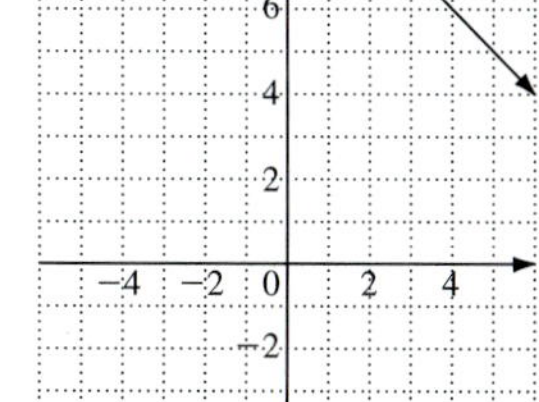

2. $7x + 3 = -2y + 3$
   $7x + 2y = 0$
   Let $x = 0$ then
   $7(0) + 2y = 0$
   $2y = 0$
   $y = 0$
   Let $x = -2$ then
   $7(-2) + 2y = 0$
   $-14 + 2y = 0$
   $2y = 14$
   $y = 7$

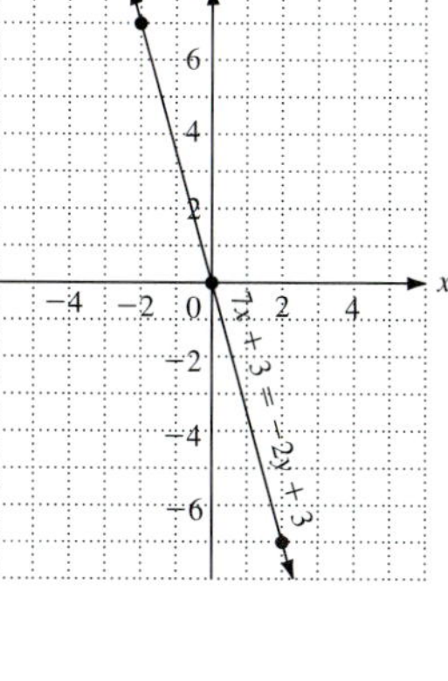

   Let $x = 2$ then
   $7(2) + 2y = 0$
   $14 + 2y = 0$
   $2y = -14$
   $y = -7$
   Graph the ordered pairs. (0, 0), (2, −7) and (−2, 7)

3. $3x - 2y = 6$
   Let $y = 0$ then
   $3x - 2(0) = 6$
   $3x = 6$
   $x = 2$
   Let $y = 3$ then
   $3x - 2(3) = 6$
   $3x - 6 = 6$
   $3x = 12$
   $x = 4$
   Let $y = -3$ then
   $3x - 2(-3) = 6$
   $3x + 6 = 6$
   $3x = 0$
   $x = 0$
   Graph the ordered pairs. (2, 0), (4, 3) and (0, −3)

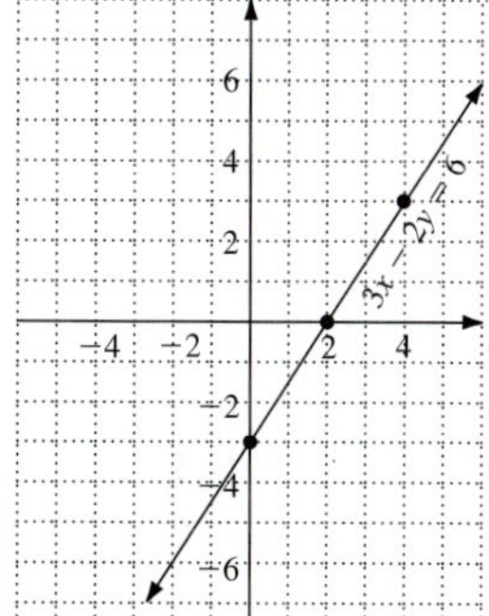

4. $2y - x = 6$
   Find the two intercepts.

Let $x = 0$	Let $y = 0$
$2y - 0 = 6$	$2(0) - x = 6$
$2y = 6$	$-x = 6$
$y = 3$	$x = -6$

   Third point
   Let $y = 1$
   $2(1) - x = 6$
   $2 - x = 6$
   $-x = 4$
   $x = -4$
   Graph the ordered pairs.
   (0, 3), (−6, 0), and (−4, 1)

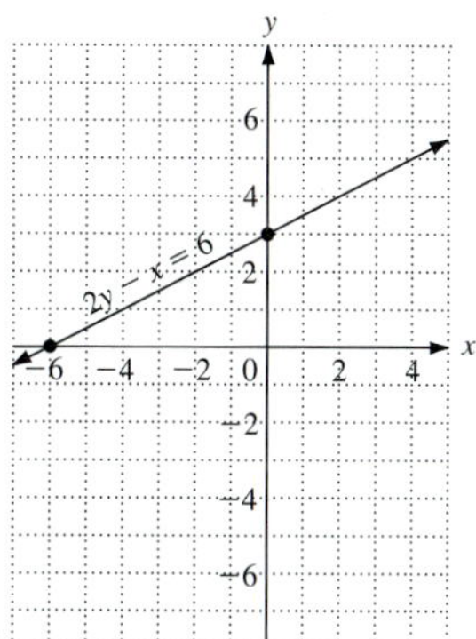

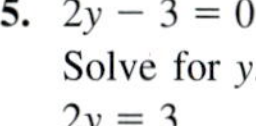

5. $2y - 3 = 0$
   Solve for y.
   $2y = 3$
   $$y = \frac{3}{2}$$
   This line is parallel to the x-axis. It is a horizontal line $1\frac{1}{2}$ units above the x-axis.

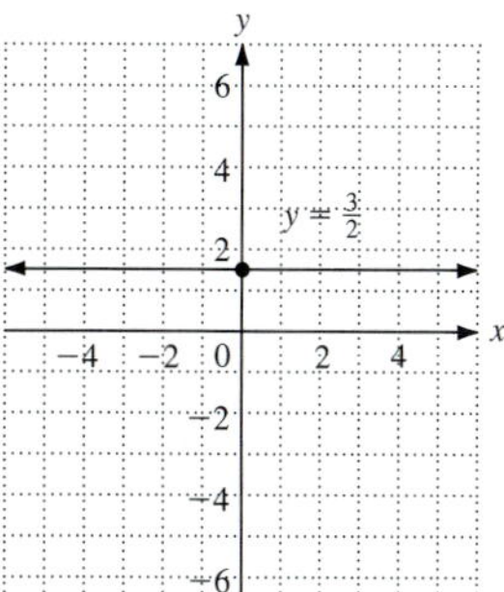

6. $x + 3 = 0$
Solve for x.
$x = -3$
This line is parallel to the y-axis. It is a vertical line 3 units to the left of the y-axis.

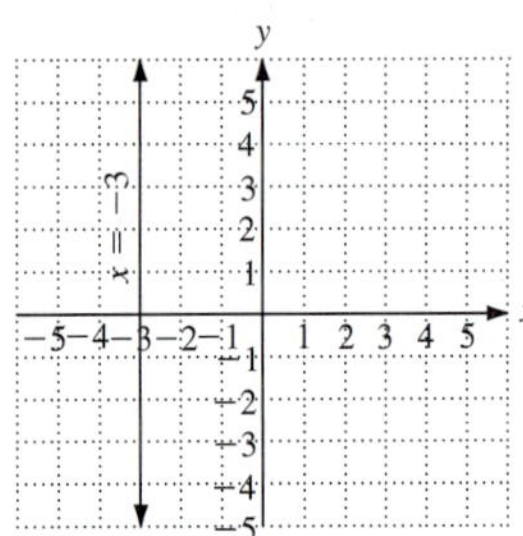

7.3 Practice Problems

1. $m = \dfrac{-1-(1)}{-4-(6)} = \dfrac{-2}{-10} = \dfrac{1}{5}$ $\quad m = \dfrac{1}{5}$

2. $m = \dfrac{6-(-4)}{1-(-1)} = \dfrac{10}{2} = 5$ $\quad m = 5$

3. $m = \dfrac{1-0}{-1-2} = \dfrac{1}{-3}$ $\quad m = -\dfrac{1}{3}$

4. **(a)** $m = \dfrac{4-0}{-4-(-4)} = \dfrac{4}{0}$

$\dfrac{4}{0}$ is undefined. Therefore there is no slope and the line is a vertical line through $x = -4$.

(b) $m = \dfrac{-11-(-11)}{3-(-7)} = \dfrac{0}{10} = 0$

$m = 0$. The line is a horizontal line through $y = -11$.

5. $y = -2x + 6$
The slope is -2 and the y-intercept is 6.

6. $4x - 2y = -5$ Solve for y.
$-2y = -4x - 5$
$y = 2x + \dfrac{5}{2}$ Slope $= 2$ y-intercept $= \dfrac{5}{2}$

7. **(a)** $y = mx + b$
$m = -\dfrac{3}{7}$ y-intercept $= \dfrac{2}{7}$
$y = -\dfrac{3}{7}x + \dfrac{2}{7}$

(b) $y = -\dfrac{3}{7}x + \dfrac{2}{7}$
$7y = -3x + 2$
$7y + 3x = 2$
$3x + 7y = 2$

8. y-intercept $= -1$. Thus the coordinates of the y-intercept for this line are $(0, -1)$. Plot the point. Slope is $\dfrac{\text{rise}}{\text{run}}$ since the slope for this line is $\dfrac{3}{4}$, we will go up(rise) 3 units and go over(run) 4 units to the right. This is the point $(4, 2)$.

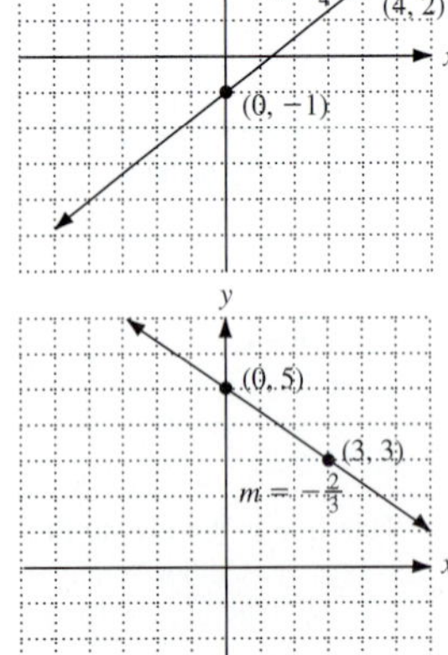

9. $y = -\dfrac{2}{3}x + 5$
y-intercept is 5, plot the point $(0, 5)$. The slope is $-\dfrac{2}{3}$. Begin at $(0, 5)$, go down 2 units and to the right 3 units. Draw a line that connects the points $(0, 5)$ and $(3, 3)$.

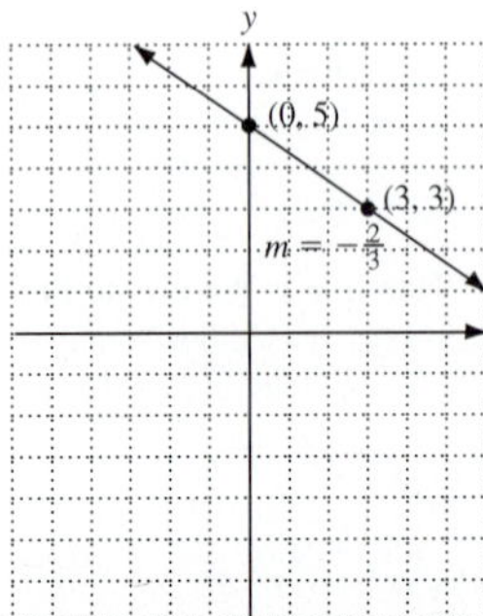

10. **(a)** $m = \dfrac{1}{4}$ **(b)** $m = -4$ **11.** **(a)** $m = \dfrac{1}{4}$ **(b)** $m = -4$

7.4 Practice Problems

1. $y = mx + b$
$4 = \dfrac{1}{2}(-2) + b$ Solve for b.
$4 = -1 + b$
$5 = b$ Thus y-intercept is 5.
We can now write the equation of the line. $y = \dfrac{1}{2}x + 5$

2. $y = mx + b$
$12 = -\dfrac{3}{4}(-8) + b$ Solve for b.
$12 = 6 + b$
$6 = b$ Thus y-intercept is 6.
We can now write the equation of the line. $y = -\dfrac{3}{4}x + 6$

3. Find the slope.
$m = \dfrac{1-5}{-1-3} = \dfrac{-4}{-4} = 1$
Using either of the two points given, substitute x and y values into the equation.
$m = 1$ $x = 3$ and $y = 5$.
$y = mx + b$
$5 = 1(3) + b$ Solve for b.
$5 = 3 + b$
$2 = b$ Thus y-intercept is 2.
$y = x + 2$

4. y-intercept is $(0, 1)$. Look for another point in the line. We choose $(6, 2)$. Count the number of units from 1 to 2 (rise). Count the number of units from 0 to 6 (run).
$m = \dfrac{1}{6}$ Now we can write the equation of the line.
$y = mx + b$
$y = \dfrac{1}{6}x + 1$

7.5 Practice Problems

1. Graph $x - y \geq -10$.
Begin by graphing the line $x - y = -10$. You may use any method discussed previously. The easiest test point is $(0, 0)$. Substitute $x = 0$, $y = 0$ in the inequality.
$x - y \geq -10$
$0 - 0 \geq -10$
$0 \geq -10$ True
Therefore shade the side of the line that includes the point $(0, 0)$.

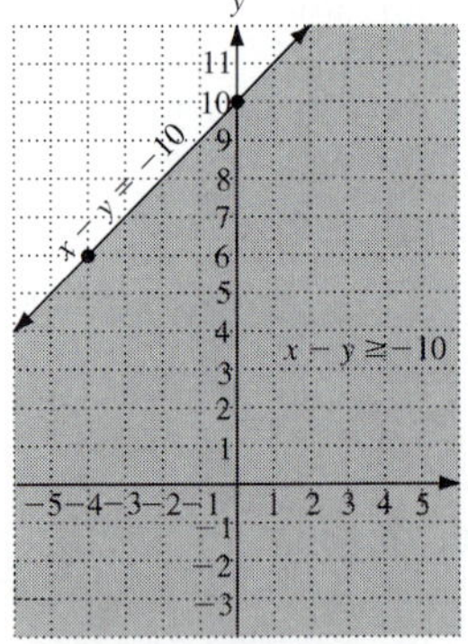

2. $y > \dfrac{1}{2}x$
Since the line goes through the origin $(0, 0)$ choose another test point. We will choose $(-1, 1)$.
$y > \dfrac{1}{2}x$ $\quad 1 > \dfrac{1}{2}(-1)$
$1 > -\dfrac{1}{2}$ True
Therefore shade the side of the line that includes the point $(-1, 1)$.

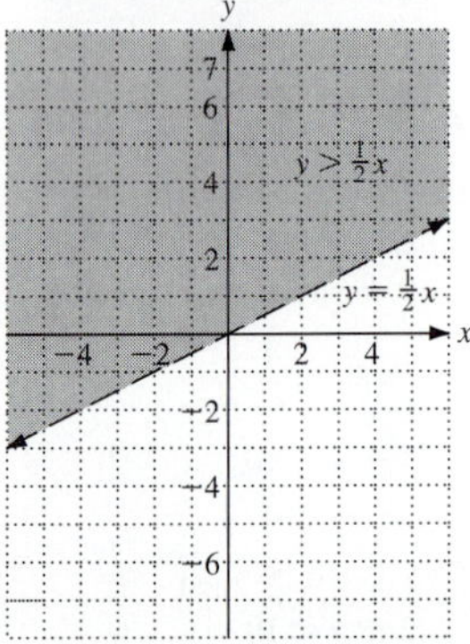

3. $y \geq -3$
A test point is not needed in the case of horizontal and vertical lines. Usually the region can be determined by inspection.

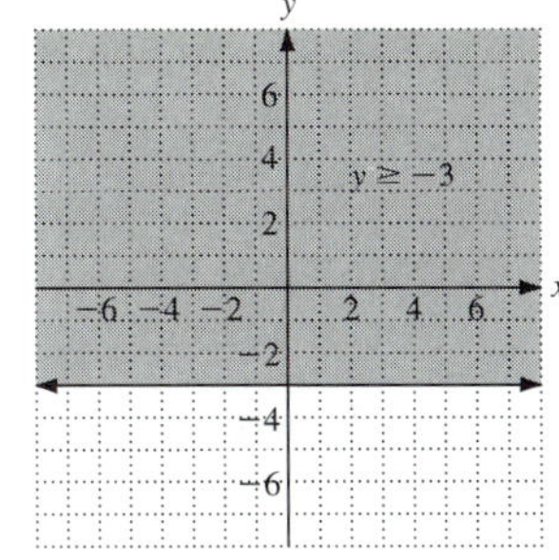

7.6 Practice Problems

1. The domain is $\{-3, 3, 0, 20\}$. The range is $\{-5, 5\}$.
2. **(a)** Function **(b)** Not a function
3. **(a)** Function
   The domain is {all nonnegative real numbers}.
   The range is {all nonnegative real numbers}.
   (b) Not a function
4. $y = x^2 - 2$

x	$y = x^2 - 2$	y
-2	$y = (-2)^2 - 2 = 2$	2
-1	$y = (-1)^2 - 2 = -1$	-1
0	$y = 0 - 2 = -2$	-2
1	$y = (1)^2 - 2 = -1$	-1
2	$y = (2)^2 - 2 = 2$	2

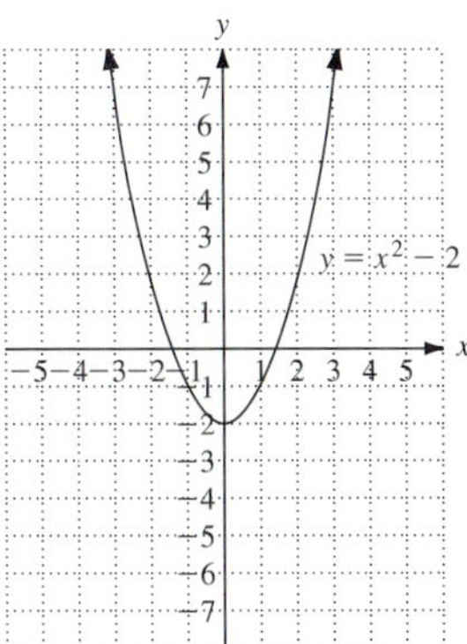

5. $x = y^2 - 1$

y	$x = y^2 - 1$	x	y
-2	$x = (-2)^2 - 1 = 3$	3	-2
-1	$x = (-1)^2 - 1 = 0$	0	-1
0	$x = 0^2 - 1 = -1$	-1	0
1	$x = 1^2 - 1 = 0$	0	1
2	$x = 2^2 - 1 = 3$	3	2

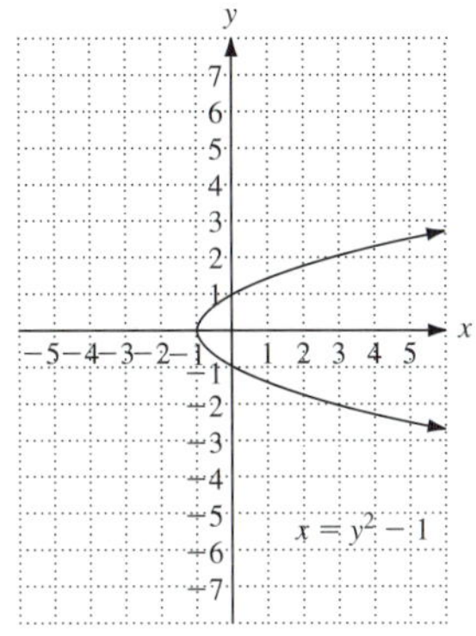

6. $y = \frac{6}{x}$

x	$y = \frac{6}{x}$	y
-3	$y = \frac{6}{-3} = -2$	-2
-2	$y = \frac{6}{-2} = -3$	-3
-1	$y = \frac{6}{-1} = -6$	-6
0	We cannot divide by 0.	
1	$y = \frac{6}{1} = 6$	6
2	$y = \frac{6}{2} = 3$	3
3	$y = \frac{6}{3} = 2$	2

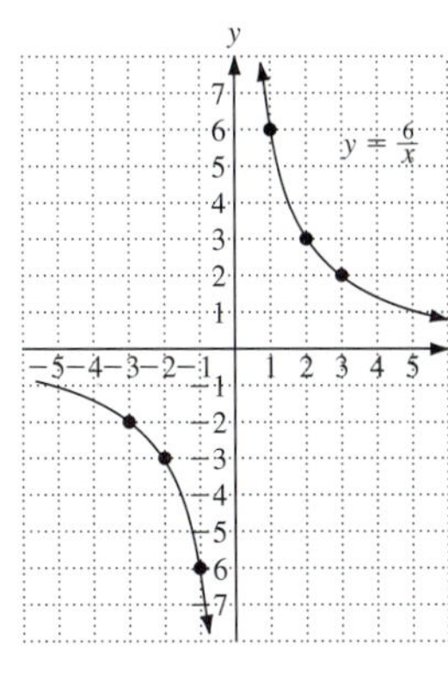

7. **(a)** Not a function **(b)** Function **(c)** Not a function
8. $f(x) = -3x + 4$
   $f(5) = -3(5) + 4$
   $f(5) = -15 + 4$
   $f(5) = -11$
9. $f(x) = -2x^2 + 3x - 8$
   (a) $f(-1) = -2(-1)^2 + 3(-1) - 8$
   $f(-1) = -2 - 3 - 8$
   $f(-1) = -13$
   (b) $f(2) = -2(2)^2 + 3(2) - 8$
   $f(2) = -8 + 6 - 8$
   $f(2) = -10$
   (c) $f(-3) = -2(-3)^2 + 3(-3) - 8$
   $f(-3) = -18 - 9 - 8$
   $f(-3) = -35$
   (d) $f(0) = -2(0)^2 + 3(0) - 8$
   $f(0) = 0 + 0 - 8$
   $f(0) = -8$

Chapter 8

8.1 Practice Problems

1. $x + y = 12$
   $-x + y = 4$
   The lines intersect at the point (4, 8). Thus the solution to the system of equation is $x = 4$, $y = 8$.

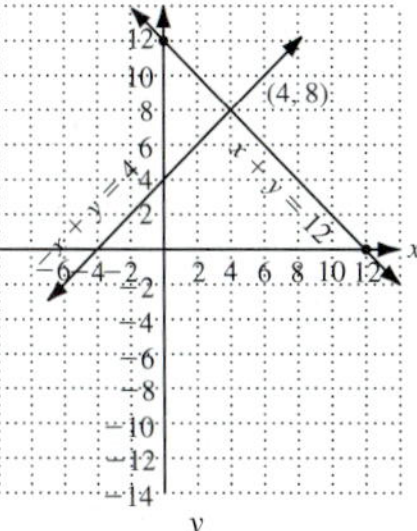

2. $4x + 2y = 8$
   $-6x - 3y = 6$
   These lines are parallel. They do not intersect. Hence there is no solution to this system of equations.

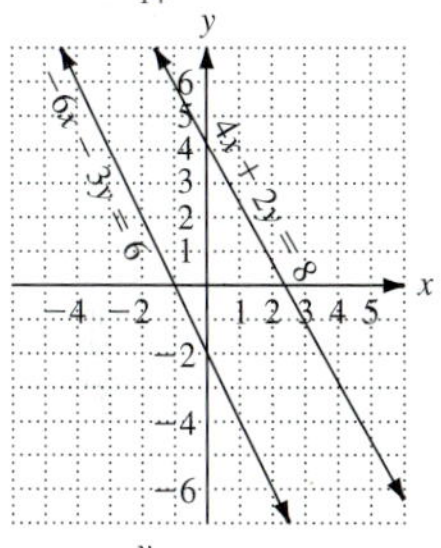

3. $3x - 9y = 18$
   $-4x + 12y = -24$
   Notice both equations represent the same line. Thus there is an infinite number of solutions to this system. The equations are said to be dependent.

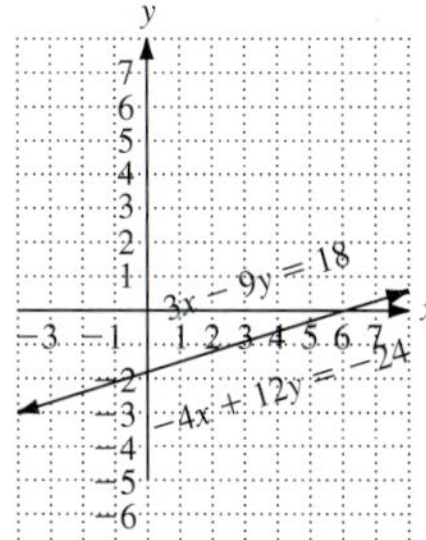

8.2 Practice Problems

1. $y = 3x + 10$
   $2x - 3y = -16$
   Substitute $y = 3x + 10$ into the second equation.
   $2x - 3(3x + 10) = -16$
   $2x - 9x - 30 = -16$
   $-7x = 14$
   $x = -2$
   Now we obtain the value of the second variable.

$y = 3(-2) + 10$
$y = -6 + 10$
$y = 4$
Check the solution $x = -2$ and $y = 4$ in both original equations.

$y = 3x + 10 \qquad 2x - 3y = -16$
$4 \stackrel{?}{=} 3(-2) + 10 \qquad 2(-2) - 3(4) \stackrel{?}{=} -16$
$4 \stackrel{?}{=} -6 + 10 \qquad -4 - 12 \stackrel{?}{=} -16$
$4 = 4 \qquad -16 = -16$

2. $5x + 3y = 19$
$2x - y = 12$
Solve the second equation for y.
$-y = -2x + 12$
$y = 2x - 12$
Now substitute $y = 2x - 12$ into the first equation.
$5x + 3(2x - 12) = 19$
$5x + 6x - 36 = 19$
$11x - 36 = 19$
$11x = 55$
$x = 5$
Now obtain the value for the second variable.
$2x - y = 12$
$2(5) - y = 12$
$10 - y = 12$
$-y = 2$
$y = -2$
The solution is $x = 5$, $y = -2$. Check this result.

3. $\frac{x}{3} - \frac{y}{2} = 1$ Multiply by the LCD. $\rightarrow 2x - 3y = 6$
$x + 4y = -8$
Solve the second equation for x.
$x = -4y - 8$
Substitute $x = -4y - 8$ into the equation $2x - 3y = 6$.
$2(-4y - 8) - 3y = 6$
$-8y - 16 - 3y = 6$
$-11y - 16 = 6$
$-11y = 22$
$y = -2$
Obtain the value of the second variable.
$x + 4y = -8$
$x + 4(-2) = -8$
$x - 8 = -8$
$x = 0$
The solution is $x = 0$, $y = -2$. Remember to check the solution in both original equations.

8.3 Practice Problems

1. $3x + 5y = 1$
$\underline{-3x + y = 11}$
$6y = 12$
$y = 2$ Substitute for y in the second equation.
$-3x + 2 = 11$
$-3x = 9$
$x = -3$
The solution is $x = -3$, $y = 2$. Check this result.

2. $3x + y = 7$ Multiply by 2. $\rightarrow 6x + 2y = 14$
$5x - 2y = 8 \qquad\qquad \underline{+\ 5x - 2y = 8}$
$11x = 22$
$x = 2$
Substitute $x = 2$ into one of the original equations.
$5(2) - 2y = 8$
$10 - 2y = 8$
$-2y = -2$
$y = 1$
The solution is $x = 2$, $y = 1$. Check this result.

3. $4x + 5y = 7$ Multiply by -3. $\rightarrow -12x - 15y = -21$
$3x + 7y = 2$ Multiply by 4. $\rightarrow \underline{12x + 28y = 8}$
$13y = -13$
$y = -1$
Substitute $y = -1$ into one of the original equations.
$4x + 5(-1) = 7$
$4x - 5 = 7$
$4x = 12$
$x = 3$
The solution is $x = 3$, $y = -1$. Check this result.

4. $\frac{2}{3}x - \frac{3}{4}y = 3$ Multiply by the LCD. $\rightarrow 8x - 9y = 36$
Use the addition method to solve the system.
$8x - 9y = 36 \qquad\qquad 8x - 9y = 36$
$-2x + y = 6$ Multiply by 4. $\rightarrow \underline{-8x + 4y = 24}$
$-5y = 60$
$y = -12$
Substitute $y = -12$ into one of the original equations.
$-2x - 12 = 6$
$-2x = 18$
$x = -9$
The solution is $x = -9$, $y = -12$. Check this result.

5. Multiply each term of each equation by 10.
$0.2x + 0.3y = -0.1 \Rightarrow 2x + 3y = -1$
$0.5x - 0.1y = -1.1 \Rightarrow 5x - y = -11$
$2x + 3y = -1 \qquad\qquad 2x + 3y = -1$
$5x - y = -11$ Multiply by 3. $\rightarrow \underline{15x - 3y = -33}$
$17x = -34$
$x = -2$
Substitute $x = -2$ into one of the equations.
$2(-2) + 3y = -1$
$-4 + 3y = -1$
$3y = 3$
$y = 1$
The solution is $x = -2$, $y = 1$. Check this result. Be sure to substitute $x = -2$ and $y = 1$ into the original equations.

8.4 Practice Problems

1. (a) $3x + 5y = 1485$
$x + 2y = 564$
Solve for x in the second equation and solve using the substitution method.
$x = -2y + 564$
$3(-2y + 564) + 5y = 1485$
$-6y + 1692 + 5y = 1485$
$-y = -207$
$y = 207$
Substitute $y = 207$ into the second equation and solve for x.
$x + 2(207) = 564$
$x + 414 = 564$
$x = 150$
The solution is $x = 150$, $y = 207$. Check this result.

(b) $7x + 6y = 45$
$6x - 5y = -2$
Using the addition method, multiply the first equation by -6 and the second equation by 7.
$-42x - 36y = -270$
$\underline{42x - 35y = -14}$
$-71y = -284$
$y = 4$
Substitute $y = 4$ into one of the original equations and solve for x.
$7x + 6(4) = 45$
$7x + 24 = 45$
$7x = 21$
$x = 3$
The solution is $x = 3$, $y = 4$. Check this result.

2. $4x + 2y = 2$
$-6x - 3y = 6$
Use the addition method. Multiply the first equation by 3 and the second equation by 2.
$12x + 6y = 6$
$-12x - 6y = 12$
$0 = 18$
The statement $0 = 18$ is inconsistent. There is no solution to this system of equations.

3. $3x - 9y = 18$
$-4x + 12y = -24$
Use the addition method. Multiply the first equation by 4 and the second equation by 3.
$12x - 36y = 72$
$-12x + 36y = -72$
$0 = 0$
The statement $0 = 0$ is true and can identify the equations as dependent. There are an unlimited number of solutions to the system.

8.5 Practice Problems

1. Let x = Math Score
y = Verbal Score
$x + y = 1153$
$x - y = 409$
$2x = 1562$
$x = 781$
Substitute $x = 781$ into one of the original equations.
$781 + y = 1153$
$y = 372$
Therefore she scored 781 on Math and 372 on Verbal.

2. Let x = No. of gal/hr pumped by smaller pump
y = No. of gal/hr pumped by larger pump
1st day $8x + 5y = 49{,}000$
2nd day $5x + 3y = 30{,}000$
Multiply the first equation by 5 and the second equation by -8.
$40x + 25y = 245{,}000$
$-40x - 24y = -240{,}000$
$y = 5{,}000$
Substitute $y = 5{,}000$ into one of the original equations and solve for x.
$8x + 5(5000) = 49{,}000$
$8x + 25{,}000 = 49{,}000$
$8x = 24{,}000$
$x = 3{,}000$
The smaller pump removes 3000 gallons per hour.
The larger pump removes 5000 gallons per hour.

3. Let x = cost per gallon of unleaded premium gasoline.
y = cost per gallon of unleaded regular gasoline.
Last week's purchase $7x + 8y = 27.22$
This week's purchase $8x + 4y = 22.16$
Use the addition method. Multiply the first equation by 8 and the second equation by -7.
$56x + 64y = 217.76$
$-56x - 28y = -155.12$
$36y = 62.64$
$y = 1.74$
Substitute $y = 1.74$ into one of the original equations and solve for x.
$7x + 8(1.74) = 27.22$
$7x + 13.92 = 27.22$
$7x = 13.30$
$x = 1.90$
Unleaded premium gasoline costs \$1.90 per gallon.
Unleaded regular gasoline costs \$1.74 per gallon.

4. Let x = the amount of 20% solution
y = the amount of 80% solution
$x + y = 4000$
$.20x + .80y = 2600$
Solve for x in the first equation.
$x = 4000 - y$
Substitute $x = 4000 - y$ into the second equation.
$.20(4000 - y) + .80y = 2600$
$800 - .2y + .8y = 2600$
$800 + 0.6y = 2600$
$0.6y = 1800$
$y = 3000$
Now substitute $y = 3000$ into the first equation.
$x + y = 4000$
$x + 3000 = 4000$
$x = 1000$
Therefore he will need 1,000 liters of 20% solution and 3,000 liters of 80% solution.

5. To go downstream with the current we have as the rate of travel
$r = \frac{d}{t} = \frac{72}{3} = 24$ mph
To go upstream against the current we have as the rate of travel
$r = \frac{d}{t} = \frac{72}{4} = 18$ mph
Let x = the speed of boat in mph.
y = the speed of the current in mph.
$x + y = 24$
$x - y = 18$
$2x = 42$
$x = 21$
Substitute $x = 21$ into one of the original equations and solve for y.
$21 + y = 24$
$y = 3$
Thus the boat speed is 21 mph and the speed of the current is 3 mph.

Chapter 9

9.1 Practice Problems

1. **(a)** $\sqrt{64} = 8$ because $8^2 = 64$
(b) $-\sqrt{121} = -11$ because $(-11)^2 = 121$

2. **(a)** $\sqrt{\frac{9}{25}} = \frac{\sqrt{9}}{\sqrt{25}} = \frac{3}{5}$ **(b)** $-\sqrt{\frac{121}{169}} = \frac{-\sqrt{121}}{\sqrt{169}} = -\frac{11}{13}$

3. **(a)** $-\sqrt{0.0036} = -0.06$ **(b)** $\sqrt{2500} = 50$

4. **(a)** 1.732 **(b)** 5.916 **(c)** 11.269

9.2 Practice Problems

1. **(a)** 6
(b) 13^2
(c) 18^6

2. **(a)** x^2
(b) y^9
(c) x^{15}

3. **(a)** $25y^2$
(b) x^8y^{11}
(c) $11x^6y^3$

4. **(a)** $\sqrt{49} \cdot \sqrt{2} = 7\sqrt{2}$
(b) $\sqrt{4} \cdot \sqrt{3} = 2\sqrt{3}$
(c) $\sqrt{25} \cdot \sqrt{3} = 5\sqrt{3}$

5. **(a)** $\sqrt{x^{11}} = \sqrt{x^{10}} \cdot \sqrt{x} = x^5\sqrt{x}$
(b) $\sqrt{x^5y^3} = \sqrt{x^4y^2} \cdot \sqrt{xy} = x^2y\sqrt{xy}$

6. **(a)** $\sqrt{48x^{11}} = \sqrt{16x^{10}} \cdot \sqrt{3x} = 4x^5\sqrt{3x}$
(b) $\sqrt{72x^8y^9} = \sqrt{36x^8y^8} \cdot \sqrt{2y} = 6x^4y^4\sqrt{2y}$
(c) $\sqrt{121x^6y^7z^8} = \sqrt{121x^6y^6z^8} \cdot \sqrt{y} = 11x^3y^3z^4\sqrt{y}$

9.3 Practice Problems

1. **(a)** $11\sqrt{11}$ **(b)** $\sqrt{t}$

2. $3\sqrt{x} - 2\sqrt{xy} + 7\sqrt{xy} - 5\sqrt{y}$
$= 3\sqrt{x} + 5\sqrt{xy} - 5\sqrt{y}$

3. (a) $\sqrt{50} - \sqrt{18} + \sqrt{98}$
$= \sqrt{25} \cdot \sqrt{2} - \sqrt{9} \cdot \sqrt{2} + \sqrt{49} \cdot \sqrt{2}$
$= 5\sqrt{2} - 3\sqrt{2} + 7\sqrt{2} = 2\sqrt{2} + 7\sqrt{2} = 9\sqrt{2}$

(b) $\sqrt{12} + \sqrt{18} - \sqrt{50} + \sqrt{27}$
$= \sqrt{4} \cdot \sqrt{3} + \sqrt{9} \cdot \sqrt{2} - \sqrt{25} \cdot \sqrt{2} + \sqrt{9} \cdot \sqrt{3}$
$= 2\sqrt{3} + 3\sqrt{3} + 3\sqrt{2} - 5\sqrt{2} = 5\sqrt{3} - 2\sqrt{2}$

4. $\sqrt{9x} + \sqrt{8x} - \sqrt{4x} + \sqrt{50x}$
$= 3\sqrt{x} + 2\sqrt{2x} - 2\sqrt{x} + 5\sqrt{2x}$
$= \sqrt{x} + 7\sqrt{2x}$

5. (a) $2\sqrt{18} + 3\sqrt{8}$
$= 2\sqrt{9}\sqrt{2} + 3\sqrt{4}\sqrt{2}$
$= 2(3)\sqrt{2} + 3(2)\sqrt{2}$
$= 6\sqrt{2} + 6\sqrt{2} = 12\sqrt{2}$

(b) $\sqrt{27} - 4\sqrt{3} + 2\sqrt{75}$
$= \sqrt{9}\sqrt{3} - 4\sqrt{3} + 2\sqrt{25}\sqrt{3}$
$= 3\sqrt{3} - 4\sqrt{3} + 10\sqrt{3} = 9\sqrt{3}$

6. $2\sqrt{12x} - 3\sqrt{45x} - 3\sqrt{27x} + \sqrt{20}$
$= 2\sqrt{4} \cdot \sqrt{3x} - 3\sqrt{9} \cdot \sqrt{5x} - 3\sqrt{9} \cdot \sqrt{3x} + \sqrt{4}\sqrt{5}$
$= 2(2)\sqrt{3x} - 3(3)\sqrt{5x} - 3(3)\sqrt{3x} + 2\sqrt{5}$
$= 4\sqrt{3x} - 9\sqrt{3x} - 9\sqrt{5x} + 2\sqrt{5}$
$= -5\sqrt{3x} - 9\sqrt{5x} + 2\sqrt{5}$

9.4 Practice Problems

1. $\sqrt{3a}\sqrt{6a} = \sqrt{18a^2} = 3a\sqrt{2}$

2. (a) $(2\sqrt{3})(5\sqrt{5}) = 10\sqrt{15}$

(b) $(2x\sqrt{6})(3\sqrt{6}) = 6x\sqrt{36} = 36x$

(c) $(4\sqrt{3x})(2x\sqrt{6x}) = 8x\sqrt{18x^2} = 24x^2\sqrt{2}$

3. $2\sqrt{3}(\sqrt{3} + \sqrt{5} - \sqrt{12})$
$= 2\sqrt{9} + 2\sqrt{15} - 2\sqrt{36}$
$= 2(3) + 2\sqrt{15} - 2(6) = 6 + 2\sqrt{15} - 12 = -6 + 2\sqrt{15}$

4. $2\sqrt{x}(4\sqrt{x} - x\sqrt{2})$
$= 8\sqrt{x^2} - 2x\sqrt{2x}$
$= 8x - 2x\sqrt{2x}$

5. $\sqrt{6x}(\sqrt{15x} + \sqrt{10x^2})$
$= \sqrt{90x^2} + \sqrt{60x^3}$
$= 3x\sqrt{10} + 2x\sqrt{15x}$

6. $(\sqrt{2} + \sqrt{6})(2\sqrt{2} - \sqrt{6})$
$= 2\sqrt{4} - \sqrt{12} + 2\sqrt{12} - \sqrt{36}$
$= 2(2) + \sqrt{12} - 6 = 4 + 2\sqrt{3} - 6 = -2 + 2\sqrt{3}$

7. $(\sqrt{6} + \sqrt{5})(\sqrt{2} + 2\sqrt{5})$
$= \sqrt{12} + 2\sqrt{30} + \sqrt{10} + 2\sqrt{25}$
$= 2\sqrt{3} + 2\sqrt{30} + \sqrt{10} + 2(5)$
$= 2\sqrt{3} + 2\sqrt{30} + \sqrt{10} + 10$

8. $(3\sqrt{5} - \sqrt{10})^2$
$= (3\sqrt{5} - \sqrt{10})(3\sqrt{5} - \sqrt{10})$
$= 9\sqrt{25} - 3\sqrt{50} - 3\sqrt{50} + \sqrt{100}$
$= 45 - 6\sqrt{25}\sqrt{2} + 10 = 55 - 30\sqrt{2}$

9.5 Practice Problems

1. (a) $\sqrt{\frac{x^3}{x}} = \sqrt{x^2} = x$ (b) $\sqrt{\frac{49}{x^2}} = \frac{7}{x}$

2. $\sqrt{\frac{50}{a^4}} = \frac{5\sqrt{2}}{a^2}$

3. (a) $\frac{9}{\sqrt{7}} \cdot \frac{\sqrt{7}}{\sqrt{7}} = \frac{9\sqrt{7}}{7}$ (b) $\frac{\sqrt{5}}{\sqrt{6}} \cdot \frac{\sqrt{6}}{\sqrt{6}} = \frac{\sqrt{30}}{6}$

4. (a) $\frac{\sqrt{2}}{\sqrt{12}} \cdot \frac{\sqrt{3}}{\sqrt{3}} = \frac{\sqrt{6}}{\sqrt{36}} = \frac{\sqrt{6}}{6}$

(b) $\frac{6a}{\sqrt{a^7}} \cdot \frac{\sqrt{a}}{\sqrt{a}} = \frac{6a\sqrt{a}}{\sqrt{a^8}} = \frac{6a\sqrt{a}}{a^4} = \frac{6\sqrt{a}}{a^3}$

5. $\frac{\sqrt{2x}}{\sqrt{8x}} = \frac{\sqrt{2x}}{2\sqrt{2x}} \cdot \frac{\sqrt{2x}}{\sqrt{2x}} = \frac{\sqrt{4x^2}}{2\sqrt{4x^2}} = \frac{2x}{4x} = \frac{1}{2}$

6. (a) $\frac{4}{\sqrt{3} + \sqrt{5}} \cdot \frac{\sqrt{3} - \sqrt{5}}{\sqrt{3} - \sqrt{5}}$
$= \frac{4(\sqrt{3} - \sqrt{5})}{3 - 5} = \frac{4(\sqrt{3} - \sqrt{5})}{-2}$
$= \frac{2(\sqrt{3} - \sqrt{5})}{-1} = -2(\sqrt{3} - \sqrt{5})$

(b) $\frac{\sqrt{a}}{(\sqrt{10} - 3)} \cdot \frac{(\sqrt{10} + 3)}{(\sqrt{10} + 3)}$
$= \frac{\sqrt{a}(\sqrt{10} + 3)}{\sqrt{100} + 3\sqrt{10} - 3\sqrt{10} - 9}$
$= \frac{\sqrt{a}(\sqrt{10} + 3)}{10 - 9} = \sqrt{a}(\sqrt{10} + 3)$

7. $\frac{\sqrt{7} - \sqrt{x}}{\sqrt{7} + \sqrt{x}} \cdot \frac{\sqrt{7} - \sqrt{x}}{\sqrt{7} - \sqrt{x}}$
$= \frac{7 - 2\sqrt{7x} + x}{7 - x} = \frac{7 + x - 2\sqrt{7x}}{7 - x}$

9.6 Practice Problems

1. $c^2 = a^2 + b^2$
$c^2 = 9^2 + 12^2$
$c^2 = 81 + 144$
$c^2 = 225$
$c = \pm\sqrt{225}$
$c = \pm 15$
$c = +15$ because length is not negative.

2. $(\sqrt{17})^2 = (1)^2 + b^2$
$17 = 1 + b^2$
$16 = b^2$
$\sqrt{16} = b$
$4 = b$
The leg is 4 meters long.

3. $c^2 = 3^2 + 8^2$
$c^2 = 9 + 64$
$c^2 = 73$
$c = \sqrt{73}$
$c = 8.5$ meters

4. $\sqrt{2x - 5} = 6$
$(\sqrt{2x - 5})^2 = (6)^2$
$2x - 5 = 36$
$2x = 41$
$x = 20.5$

Check.
$\sqrt{2(20.5) - 5} \stackrel{?}{=} 6$
$\sqrt{41 - 5} \stackrel{?}{=} 6$
$\sqrt{36} \stackrel{?}{=} 6$
$6 = 6$ ✓

5. $\sqrt{3x - 2} - 7 = 0$
$\sqrt{3x - 2} = 7$
$(\sqrt{3x - 2})^2 = (7)^2$
$3x - 2 = 49$
$3x = 51$
$x = 17$

Check.
$\sqrt{3(17) - 2} - 7 \stackrel{?}{=} 0$
$\sqrt{51 - 2} - 7 \stackrel{?}{=} 0$
$\sqrt{49} - 7 \stackrel{?}{=} 0$
$7 - 7 \stackrel{?}{=} 0$
$0 = 0$ ✓

6. $\sqrt{5x + 4} + 2 = 0$
$\sqrt{5x + 4} = -2$
$(\sqrt{5x + 4})^2 = (-2)^2$
$5x + 4 = 4$
$5x = 0$
$x = 0$

Check.
$\sqrt{5(0) + 4} + 2 \stackrel{?}{=} 0$
$\sqrt{4} + 2 \stackrel{?}{=} 0$
$2 + 2 \stackrel{?}{=} 0$
$4 \neq 0$

No, this does not check. There is no solution.

7. $-2 + \sqrt{6x - 1} = 3x - 2$
$\sqrt{6x - 1} = 3x$
$(\sqrt{6x - 1})^2 = (3x)^2$
$6x - 1 = 9x^2$
$0 = 9x^2 - 6x + 1$ *Factor.*
$0 = (3x - 1)^2$
$0 = 3x - 1$
$x = \frac{1}{3}$ is a double root.

Check.
$-2 + \sqrt{6\left(\frac{1}{3}\right) - 1} \stackrel{?}{=} 3\left(\frac{1}{3}\right) - 2$
$-2 + \sqrt{1} \stackrel{?}{=} 1 - 2$
$-2 + 1 \stackrel{?}{=} 1 - 2$
$-1 = -1$ ✓

8. $2 - x + \sqrt{x+4} = 0$

$$\sqrt{x+4} = x - 2$$
$$(\sqrt{x+4})^2 = (x-2)^2$$
$$x + 4 = x^2 - 4x + 4$$
$$0 = x^2 - 5x$$
$$0 = x(x-5)$$
$$x = 0 \qquad x - 5 = 0$$
$$x = 5$$

Check.

$$2 - 0 + \sqrt{0+4} \stackrel{?}{=} 0$$
$$2 + \sqrt{4} \stackrel{?}{=} 0$$
$$2 + 2 \stackrel{?}{=} 0$$
$$4 \neq 0$$

$x = 0$ does not check

$$2 - 5 + \sqrt{5+4} \stackrel{?}{=} 0$$
$$-3 + \sqrt{9} \stackrel{?}{=} 0$$
$$-3 + 3 \stackrel{?}{=} 0$$
$$0 = 0 \quad ✓$$

The only solution is $x = 5$.

9. $\sqrt{2x+1} = \sqrt{x-10}$

$$(\sqrt{2x+1})^2 = (\sqrt{x-10})^2$$
$$2x + 1 = x - 10$$
$$x = -11$$

$x = -11$ produces a negative radicand and thus is extraneous. No solution.

9.7 Practice Problems

1. Change of 100°C = change of 180°F.

Let C = the change in C

f = the change in F

k = the constant of variation

$$C = kf$$
$$100 = k \cdot 180$$
$$k = \frac{5}{9}$$

if $C = kf$ then

$$C = \frac{5}{9}f$$

if $f = -20$ then

$$C = \frac{5}{9}(-20)$$
$$C = -11\frac{1}{9}$$

The temperature drops $11\frac{1}{9}$ degrees Celsius.

2. $y = kx$

To find k we substitute $y = 18$ and $x = 5$.

$$18 = k(5)$$
$$\frac{18}{5} = k$$

We now write the variation equation with k replaced by $\frac{18}{5}$.

$$y = \frac{18}{5}x$$

Replace x by $\frac{20}{23}$ and solve for y.

$$y = \frac{18}{5} \cdot \frac{20}{23}$$
$$y = \frac{72}{23}$$

Thus $y = \frac{72}{23}$ when $x = \frac{20}{23}$.

3. Let d = the distance to stop the car

s = speed of the car

k = the constant of variation

Since the distance varies directly as the square of the speed, we have

$$d = ks^2$$

To evaluate k we substitute the known distance and speed.

$$60 = k(20)^2$$
$$60 = k(400)$$
$$k = \frac{3}{20}$$

Now write the variation equation with the constant evaluated.

$$d = \frac{3}{20}s^2$$
$$d = \frac{3}{20}(40)^2 \quad \textit{Substitute } s = 40.$$
$$d = \frac{3}{20}(1600)$$
$$d = 240$$

The distance to stop the car going 40 mph on an ice-covered road is 240 feet.

4. $y = \frac{k}{x}$

$$8 = \frac{k}{15} \quad \textit{Substitute known values of } x \textit{ and } y \textit{ to find } k.$$
$$120 = k$$
$$y = \frac{120}{x} \quad \textit{Write the variation equation with } k \textit{ replaced by } 120.$$
$$y = \frac{120}{\frac{3}{5}} \quad \textit{Find } y \textit{ when } x = \frac{3}{5}.$$
$$y = 200 \quad \text{Thus } y = 200 \text{ when } x = \frac{3}{5}.$$

5. Let V = the volume of sales

p = price of calculators

k = the constant of variation

Since volume varies inversely as the price then

$$V = \frac{k}{p}$$ *Evaluate k by substituting known values for V and p.*

$$120{,}000 = \frac{k}{30}$$
$$3{,}600{,}000 = k$$
$$V = \frac{3{,}600{,}000}{24}$$ *Write the variation equation with the evaluated constant and substitute $p = 24.00$.*

$$V = 150{,}000$$

Thus the volume increased to 150,000 calculators sold per year when the price was reduced to $24 per calculator.

6. Let R = the resistance in the circuit

i = the amount of current

k = the constant of variation

Since the resistance in the circuit varies inversely as the square of the amount of current

$$R = \frac{k}{i^2}$$

To evaluate k, substitute the known values for R and i.

$$800 = \frac{k}{(0.01)^2}$$
$$k = 0.08$$
$$R = \frac{0.08}{(0.02)^2}$$ *Write the variation equation with the evaluated constant of $k = 0.08$ and substitute $i = 0.02$.*

$$R = \frac{0.08}{0.0004}$$
$$R = 200$$

Thus the resistance is 200 ohms if the amount of current is increased to 0.02 ampere.

Chapter 10

10.1 Practice Problems

1. (a) $2x^2 + 12x - 9 = 0$

$a = 2,\ b = 12,\ c = -9$

(b) $7x^2 = 6x - 8$
$7x^2 - 6x + 8 = 0$
$a = 7,\ b = -6,\ c = 8$

(c) $-x^2 - 6x + 3 = 0$
$x + 6x - 3 = 0$
$a = 1,\ b = 6,\ c = -3$

(d) $10x^2 - 12x = 0$
$a = 10,\ b = -12,\ c = 0$

(e) $5x^2 - 4x + 20 = 4x^2 - 3x - 10$
$x^2 - x + 30 = 0$
$a = 1,\ b = -1,\ c = 30$

2. **(a)** $3x^2 + 18x = 0$
$3x(x + 6) = 0$
$3x = 0 \qquad x + 6 = 0$
$x = 0 \qquad x = -6$
The two roots are 0 and -6.

(b) $2x^2 - 7x - 6 = 4x - 6$
$2x^2 - 11x = 0$
$x(2x - 11) = 0$
$x = 0 \qquad 2x - 11 = 0$
$x = \dfrac{11}{2}$
The two roots are 0 and $\dfrac{11}{2}$.

3. $3x + 10 - \dfrac{8}{x} = 0$
$3x^2 + 10x - 8 = 0$
$(3x - 2)(x + 4) = 0$
$3x - 2 = 0 \qquad x + 4 = 0$
$x = \dfrac{2}{3} \qquad x = -4$

Check. $3\left(\dfrac{2}{3}\right) + 10 - \dfrac{8}{\frac{2}{3}} \stackrel{?}{=} 0 \qquad 3(-4) + 10 - \dfrac{8}{-4} \stackrel{?}{=} 0$
$2 + 10 - 12 \stackrel{?}{=} 0 \qquad -12 + 10 + 2 \stackrel{?}{=} 0$
$2 - 2 \stackrel{?}{=} 0 \qquad -2 + 2 \stackrel{?}{=} 0$
$0 = 0$ ✓ $\qquad 0 = 0$ ✓

Both roots check so $x = \dfrac{2}{3}$ and $x = -4$ are the two roots that satisfy the equation $3x + 10 - \dfrac{8}{x} = 0$.

4. $2x^2 + 9x - 18 = 0$
$(2x - 3)(x + 6) = 0$
$2x - 3 = 0 \qquad x + 6 = 0$
$x = \dfrac{3}{2} \qquad x = -6$

Check. $2\left(\dfrac{3}{2}\right)^2 + 9\left(\dfrac{3}{2}\right) - 18 \stackrel{?}{=} 0 \qquad 2(-6)^2 + 9(-6) - 18 \stackrel{?}{=} 0$
$2\left(\dfrac{9}{4}\right) + \dfrac{27}{2} - 18 \stackrel{?}{=} 0 \qquad 2(36) - 54 - 18 \stackrel{?}{=} 0$
$\dfrac{9}{2} + \dfrac{27}{2} - 18 \stackrel{?}{=} 0 \qquad 72 - 54 - 18 \stackrel{?}{=} 0$
$\dfrac{36}{2} - 18 \stackrel{?}{=} 0 \qquad 18 - 18 \stackrel{?}{=} 0$
$18 - 18 \stackrel{?}{=} 0 \qquad 0 = 0$ ✓
$0 = 0$ ✓

Both roots check so $x = \dfrac{3}{2}$ and $x = -6$ are the two roots that satisfy the equation $2x^2 + 9x - 18 = 0$.

5. $-4x + (x - 5)(x + 1) = 5(1 - x)$
$-4x + x^2 - 4x - 5 = 5 - 5x$
$x^2 - 8x - 5 = 5 - 5x$
$x^2 - 3x - 10 = 0$
$(x + 2)(x - 5) = 0$
$x + 2 = 0 \qquad x - 5 = 0$
$x = -2 \qquad x = 5$

Check. $-4(-2) + (-2 - 5)(-2 + 1) \stackrel{?}{=} 5(1 + 2)$
$8 + (-7)(-1) \stackrel{?}{=} 5(3)$
$8 + 7 \stackrel{?}{=} 15$
$15 = 15$ ✓
$-4(5) + (5 - 5)(5 + 1) \stackrel{?}{=} 5(-4)$
$-20 + 0 \stackrel{?}{=} -20$
$-20 = -20$ ✓

Both roots check so $x = -2$ and $x = 5$ are the two roots that satisfy the equation $-4x + (x - 5)(x + 1) = 5(1 - x)$.

6. $\dfrac{10x + 18}{x^2 + x - 2} = \dfrac{3x}{x + 2} + \dfrac{2}{x - 1}$
$\dfrac{10x + 18}{(x - 1)(x + 2)} = \dfrac{3x}{x + 2} + \dfrac{2}{x - 1}$ *Multiply by the LCD.*
$10x + 18 = 3x(x - 1) + 2(x + 2)$
$10x + 18 = 3x^2 - 3x + 2x + 4$
$0 = 3x^2 - 11x - 14$
$0 = (3x - 14)(x + 1)$
$3x - 14 = 0 \qquad x + 1 = 0$
$x = \dfrac{14}{3} \qquad x = -1$

7. t = the number of truck routes
n = the number of cities they can service
$t = \dfrac{n^2 - n}{2}$
$28 = \dfrac{n^2 - n}{2}$
$56 = n^2 - n$
$0 = n^2 - n - 56$
$0 = (n + 7)(n - 8)$
$n + 7 = 0 \qquad n - 8 = 0$
$n = -7 \qquad n = 8$
We reject $n = -7$. There cannot be a negative number of cities. Thus our answer is 8 cities.

10.2 Practice Problems

1. **(a)** $x^2 = 1$
$x = \pm\sqrt{1}$
$x = \pm 1$

(b) $x^2 - 50 = 0$
$x^2 = 50$
$x = \pm\sqrt{50}$
$x = \pm 5\sqrt{2}$

(c) $3x^2 = 81$
$x^2 = 27$
$x = \pm\sqrt{27} = \pm 3\sqrt{3}$

2. $4x^2 - 5 = 319$
$4x^2 = 324$
$x^2 = 81$
$x = \pm\sqrt{81} = \pm 9$

3. $(2x + 3)^2 = 25$
$\sqrt{(2x + 3)^2} = \pm\sqrt{25}$
$2x + 3 = \pm 5$
$2x + 3 = 5 \qquad 2x + 3 = -5$
$2x = 2 \qquad 2x = -8$
$x = 1 \qquad x = -4$
The two roots of the equation are $x = 1$ and $x = -4$.

4. $(2x - 3)^2 = 12$
$\sqrt{(2x - 3)^2} = \pm\sqrt{12}$
$2x - 3 = \pm 2\sqrt{3}$
$2x = 3 \pm 2\sqrt{3}$
$x = \dfrac{3 \pm 2\sqrt{3}}{2}$

5. **(a)** $x^2 - 8x = 20$
$x^2 - 8x + (4)^2 = 20 + 16$
$x^2 - 8x + 16 = 36$
$(x - 4)^2 = 36$
$\sqrt{(x - 4)^2} = \sqrt{36}$
$x - 4 = \pm 6$
$x = 4 \pm 6$
$x = 10 \qquad x = -2$

(b) $x^2 + 10x = 3$

$x^2 + 10x + (5)^2 = 3 + 25$

$(x + 5)^2 = 28$

$\sqrt{(x + 5)^2} = \pm\sqrt{28}$

$x + 5 = \pm\sqrt{28}$

$x = -5 \pm 2\sqrt{7}$

6. $x^2 - 5x = 7$

$x^2 - 5x + \left(\frac{5}{2}\right)^2 = 7 + \left(\frac{5}{2}\right)^2$

$\left(x - \frac{5}{2}\right)^2 = 7 + \frac{25}{4}$

$\left(x - \frac{5}{2}\right)^2 = \frac{28}{4} + \frac{25}{4}$

$\left(x - \frac{5}{2}\right)^2 = \frac{53}{4}$

$x - \frac{5}{2} = \pm\sqrt{\frac{53}{4}}$

$x = \frac{5}{2} \pm \frac{\sqrt{53}}{2}$

$x = \frac{5 \pm \sqrt{53}}{2}$

7. $4y^2 + 4y - 3 = 0$

$(2y + 3)(2y - 1) = 0$

$2y + 3 = 0 \qquad 2y - 1 = 0$

$y = -\frac{3}{2} \qquad y = \frac{1}{2}$

10.3 Practice Problems

1. $x^2 - 7x + 6 = 0$

$x = \frac{-(-7) \pm \sqrt{(-7)^2 - 4(1)(6)}}{2(1)}$

$x = \frac{7 \pm \sqrt{49 - 24}}{2}$

$x = \frac{7 \pm \sqrt{25}}{2}$

$x = \frac{7 \pm 5}{2}$

$x = \frac{7 + 5}{2} \qquad x = \frac{7 - 5}{2}$

$x = 6 \qquad x = 1$

2. $3x^2 - 8x + 3 = 0$

$x = \frac{-(-8) \pm \sqrt{(-8)^2 - 4(3)(3)}}{2(3)}$

$x = \frac{8 \pm \sqrt{64 - 36}}{6}$

$x = \frac{8 \pm \sqrt{28}}{6}$

$x = \frac{8 \pm 2\sqrt{7}}{6}$

$x = \frac{2(4 \pm \sqrt{7})}{6} = \frac{4 \pm \sqrt{7}}{3}$

3. $2x^2 = 3x + 1$

$2x^2 - 3x - 1 = 0$

$x = \frac{-(-3) \pm \sqrt{(-3)^2 - 4(2)(-1)}}{2(2)}$

$x = \frac{3 \pm \sqrt{9 + 8}}{4}$

$x = \frac{3 \pm \sqrt{17}}{4}$

4. $2x^2 = 13x + 5$

$2x^2 - 13x - 5 = 0$

$x = \frac{-(-13) \pm \sqrt{(-13)^2 - 4(2)(-5)}}{2(2)}$

$x = \frac{13 \pm \sqrt{169 + 40}}{4}$

$x = \frac{13 \pm \sqrt{209}}{4}$

$x = \frac{13 + \sqrt{209}}{4} \qquad x = \frac{13 - \sqrt{209}}{4}$

$x \approx 6.864$ and $x \approx -0.364$

5. $5x^2 + 2x = -3$

$5x^2 + 2x + 3 = 0$

$x = \frac{-2 \pm \sqrt{(2)^2 - 4(5)(3)}}{2(5)}$

$x = \frac{-2 \pm \sqrt{4 - 60}}{10}$

$x = \frac{-2 \pm \sqrt{-56}}{10}$

There is no real number that is $\sqrt{-56}$.
There is no solution to this problem.

6. (a) $5x^2 - 3x - 2 = 0$

$b^2 - 4ac = (-3)^2 - 4(5)(-2)$

$= 9 + 40 = 49$

Because $49 > 0$ there will be two real roots.

(b) $2x^2 - 4x - 5 = 0$

$b^2 - 4ac = (-4)^2 - 4(2)(-5)$

$= 16 + 40 = 56$

Because $56 > 0$ there will be two real roots.

10.4 Practice Problems

1. (a) $y = 2x^2$

x	y
−2	8
−1	2
0	0
1	2
2	8

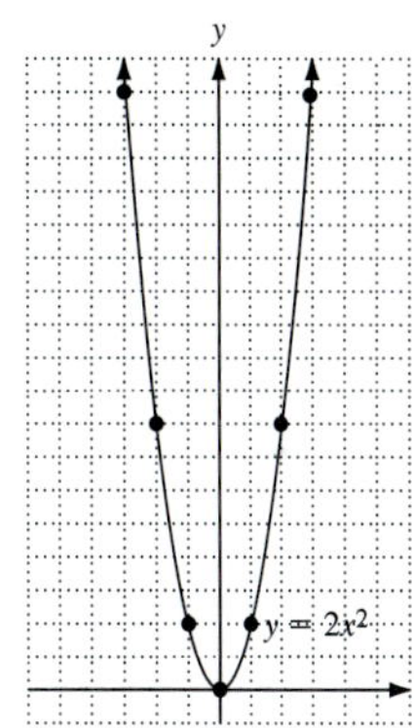

(b) $y = -2x^2$

x	y
−2	−8
−1	−2
0	0
1	−2
2	−8

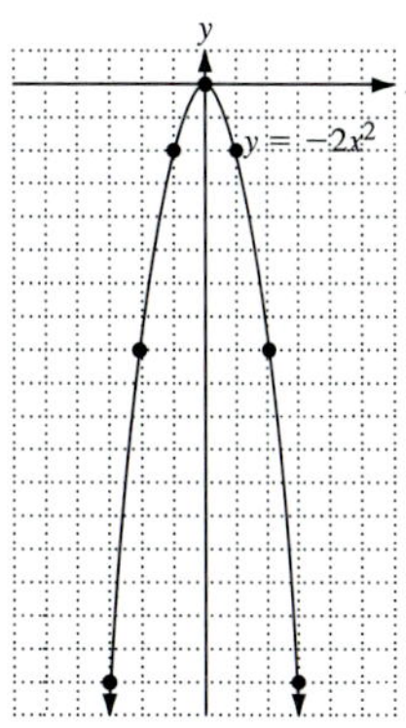

2. (a) $y = (x - 1)^2$

x	y
−2	9
−1	4
0	1
1	0
2	1
3	4
4	9

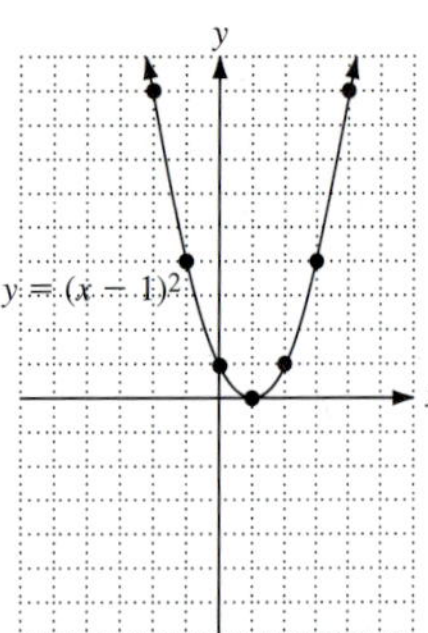

(b) $y = (x + 2)^2$

x	y
−5	9
−4	4
−3	1
−2	0
−1	1
0	4
1	9

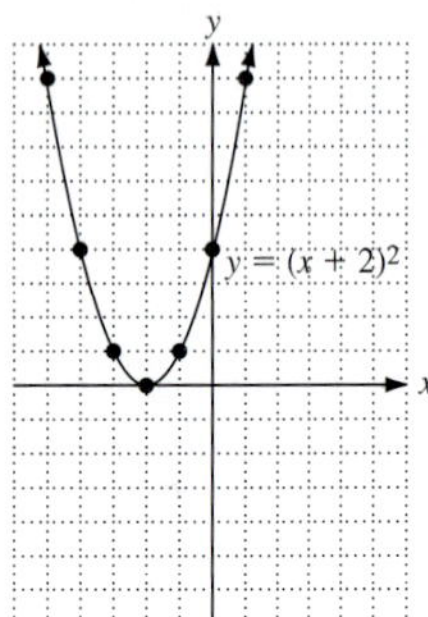

3. (a) $y = x^2 + 2x$

x	y	
−4	8	
−3	3	
−2	0	
−1	−1	vertex
0	0	
1	3	
2	8	

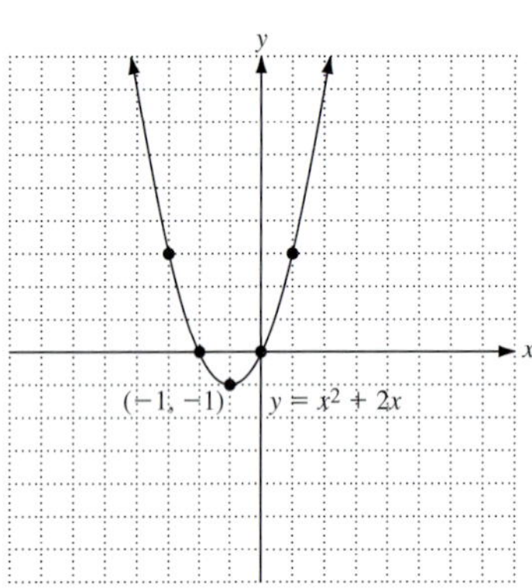

(b) $y = -x^2 - 2x$

x	y	
−4	−8	
−3	−3	
−2	0	
−1	1	vertex
0	0	
1	−3	
2	−8	

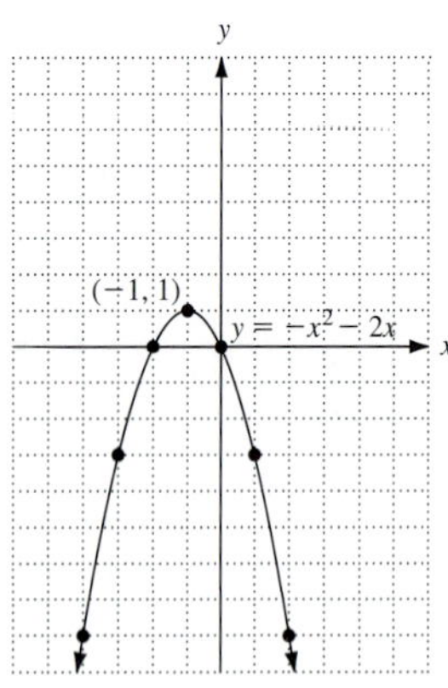

4. $y = x^2 - 4x - 5$

$x = \dfrac{-b}{2a} = \dfrac{-(-4)}{2(1)} = 2$

if $x = 2$ then

$y = (2)^2 - 4(2) - 5$

$y = 4 - 8 - 5$

$y = -9$

The point $(2, -9)$ is the vertex.

x-intercepts:

$x^2 - 4x - 5 = 0$

$(x + 1)(x - 5) = 0$

$x + 1 = 0 \qquad x - 5 = 0$

$x = -1 \qquad x = 5$

The x-intercepts are the points $(-1, 0)$ and $(5, 0)$.

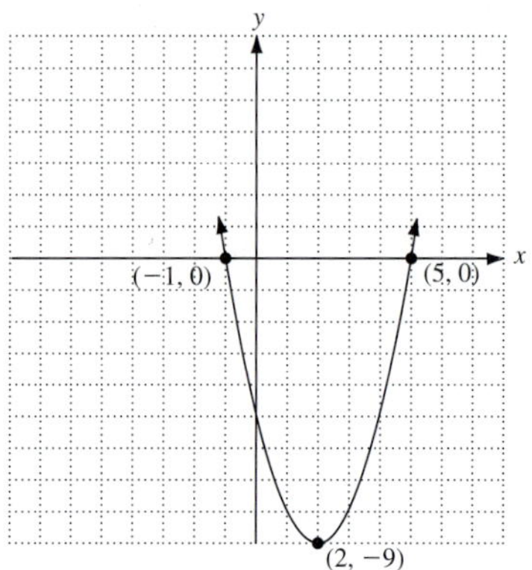

5. $h = -16t^2 + 32t + 10$

(a) Since the equation is a quadratic equation, the graph is a parabola. a is negative, therefore the parabola opens downward and the vertex is the highest point.

$$\text{Vertex} = t = \frac{-b}{2a} = \frac{-32}{2(-16)} = \frac{-32}{-32} = 1$$

if $t = 1$ then $h = -16(1)^2 + 32(1) + 10$

$h = -16 + 32 + 10$

$h = 26$

The vertex (t, h) is $(1, 26)$.

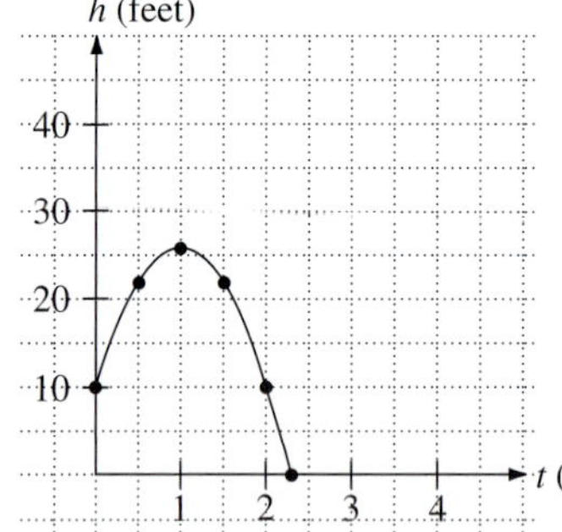

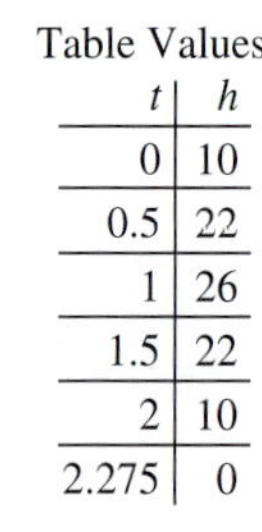
Table Values

t	h
0	10
0.5	22
1	26
1.5	22
2	10
2.275	0

Now find the h-intercept. The point where $t = 0$.

$h = -16t^2 + 32t + 10$

$h = 10$

The h-intercept is at $(0, 10)$.

(b) The ball is 26 feet above the ground one second after it is thrown. This is the greatest height of the ball.

(c) The ball hits the ground when $h = 0$.

$0 = -16t^2 + 32t + 10$

$t = \dfrac{-32 \pm \sqrt{(32)^2 - 4(-16)(10)}}{2(-16)}$

$t = \dfrac{-32 \pm \sqrt{1024 + 640}}{-32}$

$t = \dfrac{-32 \pm \sqrt{1664}}{-32}$

$t = \dfrac{-32 \pm 40.792}{-32}$

$t = \dfrac{-32 + 40.792}{-32}$ and $t = \dfrac{-32 - 40.792}{-32}$

$t = \dfrac{8.792}{-32}$ $\qquad t = \dfrac{-72.792}{-32}$

$t = -0.275$ $\qquad t = 2.275$

In this situation it is not useful to find points for negative values of t. Therefore $t = 2.3$ seconds. The ball will hit the ground approximately 2.3 seconds after it is thrown.

10.5 Practice Problems

1. $a^2 + b^2 = c^2$ $\qquad$ Let $a = x$

$b = x - 6$

$x^2 + (x - 6)^2 = 30^2$

$x^2 + x^2 - 12x + 36 = 900$

$2x^2 - 12x + 36 = 900$

$2x^2 - 12x - 864 = 0$

$x^2 - 6x - 432 = 0$

$(x + 18)(x - 24) = 0$

$x + 18 = 0 \qquad x - 24 = 0$

$x = -18 \qquad x = 24$

One leg is 24 meters, the other leg is $x - 6$ or $24 - 6 = 18$ meters.

2. Let c = cost for each student in the original group

s = number of students

Number of students × cost per student = total cost

$s \times c = 240.00$

If 2 people cannot go, then the number of students drops by 2 but the cost of each increases by \$4.00.

$(s - 2)(c + 4) = 240$

If $s \times c = 240$ then

$c = \dfrac{240}{s}$

$(s - 2)\left(\dfrac{240}{s} + 4\right) = 240$ *Multiply by the LCD = s.*

$(s - 2)(240 + 4s) = 240s$

$4s^2 + 232s - 480 = 240s$

$4s^2 - 8s - 480 = 0$

$s^2 - 2s - 120 = 0$

$(s - 12)(s + 10) = 0$

$s - 12 = 0 \qquad s + 10 = 0$

$s = 12 \qquad s = -10$

The number of students originally going on the trip was 12.

Check. If 12 students originally planned the trip then

$12 \times c = 240$

$c = \$20.00$ per student

If 2 students dropped out, does this mean they must increase the cost of the trip by \$4.00?

$10 \times c = 240$

$c = \$24.00$ YES!

3. Let x = width

y = length

$160 = 2x + y$

The area of a rectangle is (width)(length).

$A = x(y)$ *Substitute $y = 160 - 2x$.*

$3150 = x(160 - 2x)$

$3150 = 160x - 2x^2$

$2x^2 - 160x + 3150 = 0$

$x^2 - 80x + 1575 = 0$

$(x - 45)(x - 35) = 0$

$x = 45 \qquad x = 35$

First Solution

If the width is 45 feet then the length is

$160 = 2(45) + y$

$160 = 90 + y$

$y = 70$ feet

Check. $45 + 45 + 70 \stackrel{?}{=} 160$

$90 + 70 \stackrel{?}{=} 160$

$160 = 160$ ✓

Second Solution

If the width is 35 feet then the length is

$160 = 2(35) + y$

$160 = 70 + y$

$y = 90$ feet

Check. $35 + 35 + 90 \stackrel{?}{=} 160$

$70 + 90 \stackrel{?}{=} 160$

$160 = 160$ ✓

Selected Answers

Chapter 0

Pretest Chapter 0

1. $\frac{7}{9}$ **2.** $\frac{2}{3}$ **3.** $3\frac{3}{4}$ **4.** $\frac{33}{7}$ **5.** 6 **6.** 35 **7.** 120 **8.** $\frac{13}{9}$ **9.** $\frac{5}{12}$ **10.** $8\frac{5}{12}$ **11.** $1\frac{33}{40}$ **12.** $\frac{10}{9}$ or $1\frac{1}{9}$
13. $\frac{21}{2}$ or $10\frac{1}{2}$ **14.** $\frac{7}{2}$ or $3\frac{1}{2}$ **15.** $\frac{28}{39}$ **16.** 0.625 **17.** $0.\overline{5}$ **18.** 19.651 **19.** 2.0664 **20.** 0.45 **21.** 2.246 **22.** 70%
23. 0.6% **24.** 3.27 **25.** 0.02 **26.** 407.5 **27.** 7.2% **28.** 10,000 **29.** 21,000,000 **30.** 20 miles per gallon
31. 28.4 miles per gallon

0.1 Exercises

7. $\frac{4}{7}$ **9.** $\frac{1}{5}$ **11.** 5 **13.** $3\frac{4}{5}$ **15.** $31\frac{1}{4}$ **17.** $14\frac{5}{6}$ **19.** $\frac{20}{7}$ **21.** $\frac{43}{8}$ **23.** $\frac{79}{6}$ **25.** $\frac{20}{28}$ **27.** $\frac{6}{21}$ **29.** $\frac{39}{51}$
31. $\frac{5}{7}$ **33.** $\frac{43}{120}$ **35.** 119 **37.** $\frac{3}{5}$ **39.** Aaron Dunsay $\frac{3}{5}$, Paul Banks $\frac{2}{5}$, Tom Re $\frac{4}{5}$

0.2 Exercises

3. 36 **5.** 20 **7.** 63 **9.** 90 **11.** $\frac{9}{11}$ **13.** $\frac{3}{17}$ **15.** $\frac{23}{24}$ **17.** $\frac{3}{28}$ **19.** $\frac{19}{18}$ or $1\frac{1}{18}$ **21.** $\frac{17}{40}$ **23.** $\frac{41}{24}$ or $1\frac{17}{24}$
25. $\frac{17}{36}$ **27.** $\frac{5}{4}$ or $1\frac{1}{4}$ **29.** $\frac{13}{20}$ **31.** $\frac{11}{36}$ **33.** $8\frac{5}{12}$ **35.** $7\frac{5}{12}$ **37.** $3\frac{13}{20}$ **39.** $2\frac{3}{4}$ **41.** $7\frac{10}{21}$ **43.** $1\frac{5}{9}$
45. $10\frac{7}{24}$ miles **47.** $4\frac{1}{12}$ hours **49.** $A = 12$ inches, $B = 15\frac{7}{8}$ inches **51.** $1\frac{5}{8}$ inches **53.** $\frac{9}{11}$ **54.** $\frac{133}{5}$ **55.** $19\frac{3}{7}$ **56.** 12

Putting Your Skills to Work

1. $10\frac{1}{8}$ in. **2.** from 1952 to 1972

0.3 Exercises

1. $\frac{6}{55}$ **3.** 6 **5.** $\frac{24}{7}$ or $3\frac{3}{7}$ **7.** $\frac{10}{3}$ or $3\frac{1}{3}$ **9.** 5 **11.** $\frac{2}{3}$ **13.** $\frac{1}{7}$ **15.** 14 **17.** $\frac{15}{4}$ or $3\frac{3}{4}$ **19.** $\frac{8}{35}$ **21.** $\frac{7}{12}$
23. 1 **25.** $\frac{5}{2}$ or $2\frac{1}{2}$ **27.** $\frac{219}{4}$ or $54\frac{3}{4}$ **29.** $\frac{17}{2}$ or $8\frac{1}{2}$ **31.** 28 **33.** $\frac{3}{16}$ **35.** $71\frac{1}{2}$ yds

Putting Your Skills to Work

1. $\$3287\frac{1}{2}$ **2.** $\$2431\frac{1}{4}$ **3.** $\$113\frac{3}{4}$

0.4 Exercises

5. 0.15 **7.** 0.625 **9.** 0.3 **11.** $\frac{3}{20}$ **13.** $\frac{2}{25}$ **15.** $\frac{9}{8}$ **17.** 7.72 **19.** 9.38 **21.** 208.791 **23.** 87.2 **25.** 18.512
27. 2.834 **29.** 0.001428 **31.** 0.35 **33.** 77.28 **35.** 3.18 **37.** 6.2 **39.** 6.3 **41.** 0.0076 **43.** 2 **45.** 243
47. 0.73892 **49.** 0.005808 **51.** 24.36 **53.** 25,712 **55.** 0.003472 **57.** 666.9 miles **58.** $\frac{2}{3}$ **59.** $\frac{1}{6}$ **60.** $\frac{93}{100}$
61. $\frac{11}{10}$ or $1\frac{1}{10}$

Putting Your Skills to Work

1. 468,000,000 people **2.** 249,000,000 people

0.5 Exercises

3. 62.4% **5.** 30.4% **7.** 156% **9.** 0.04 **11.** 0.002 **13.** 2.50 **15.** 358 **17.** 0.52 **19.** 296.8 **21.** 176 **23.** 5%
25. 5% **27.** 150% **29.** 0.4% **31.** 137.5% **33.** 14,100 **35.** 76.11% **37.** 648 CDs were defective. **39.** 77%
41. **(a)** 12,090 miles traveled on business. **(b)** $3747.90 for travel expenses.

Putting Your Skills to Work

1. 1991 **2.** It differed by 6.3% between 1984 and 1990 or between 1984 and 1993.

0.6 Exercises

1. 180,000 **3.** 160,000,000 **5.** 290 **7.** 30,000 **9.** 0.1 **11.** \$4000 **13.** 800 square feet **15.** 20 mpg **17.** \$6400 **19.** \$20 **21.** \$1200 **23.** \$1,600,000,000 **25.** \$240 **26.** \$180

0.7 Exercises

1. \$462.50 **3.** 95 cubic yards **5.** Jog $2\frac{2}{15}$ miles; walk $3\frac{1}{9}$ miles; rest $4\frac{4}{9}$ minutes; walk $1\frac{7}{9}$ miles **7.** Betty; Melinda increases each activity by $\frac{2}{3}$ by day 3 and Betty increases each activity by $\frac{7}{9}$ by day 3. **8.** $3\frac{107}{135}$ **9.** $3\frac{3}{5}$ **10.** Answers will vary. **11.** 27 miles per gallon **13.** 142,560; Answers will vary. Samples include: improved schools, new job opportunities, availability of affordable housing. **15.** 49%; 61%; business travelers; Answers may vary; A sample is: Men 21 or older spend more of their travel time on business than for pleasure. **17.** 18% **19.** 69%

Chapter 0 Review Problems

1. $\frac{3}{4}$ **2.** $\frac{3}{10}$ **3.** $\frac{1}{3}$ **4.** $\frac{3}{5}$ **5.** $\frac{19}{7}$ **6.** $6\frac{4}{5}$ **7.** $6\frac{3}{4}$ **8.** 15 **9.** 15 **10.** 12 **11.** 22 **12.** $\frac{11}{12}$ **13.** $\frac{11}{24}$ **14.** $\frac{4}{15}$ **15.** $\frac{17}{48}$ **16.** $\frac{173}{30}$ or $5\frac{23}{30}$ **17.** $\frac{79}{20}$ or $3\frac{19}{20}$ **18.** $\frac{8}{15}$ **19.** $\frac{23}{12}$ or $1\frac{11}{12}$ **20.** $\frac{3}{5}$ **21.** $\frac{2}{7}$ **22.** $\frac{30}{11}$ or $2\frac{8}{11}$ **23.** $\frac{21}{2}$ or $10\frac{1}{2}$ **24.** $\frac{19}{8}$ or $2\frac{3}{8}$ **25.** $\frac{20}{7}$ or $2\frac{6}{7}$ **26.** 2 **27.** 4 **28.** $\frac{3}{10}$ **29.** $\frac{2}{3}$ **30.** $\frac{3}{20}$ **31.** 6 **32.** 4.462 **33.** 7.737 **34.** 11.427 **35.** 4.814 **36.** 13.7 **37.** 111.1121 **38.** 0.00862 **39.** 362,341 **40.** 0.07956 **41.** 10.368 **42.** 0.00186 **43.** 0.07132 **44.** 0.002 **45.** 90 **46.** 0.07 **47.** 0.5 **48.** 37.5% **49.** $\frac{6}{25}$ **50.** 0.014 **51.** 0.361 **52.** 0.0002 **53.** 1.253 **54.** 260 **55.** 90 **56.** 18 **57.** 12.5% **58.** 60% **59.** 97,200,000 **60.** 75% **61.** 400,000,000,000 **62.** 2500 **63.** 600,000 **64.** 19 **65.** \$12,000 **66.** 30 **67.** \$480 or \$640 **68.** \$300 **69.** \$349.07 **70.** 6%

Chapter 0 Test

1. $\frac{8}{9}$ **2.** $\frac{7}{8}$ **3.** $\frac{45}{7}$ **4.** $3\frac{3}{11}$ **5.** 33 **6.** $\frac{13}{18}$ **7.** $\frac{17}{12}$ or $1\frac{5}{12}$ **8.** $\frac{1}{2}$ **9.** $\frac{39}{8}$ or $4\frac{7}{8}$ **10.** $\frac{5}{6}$ **11.** $\frac{5}{4}$ or $1\frac{1}{4}$ **12.** $\frac{4}{3}$ or $1\frac{1}{3}$ **13.** $\frac{21}{55}$ **14.** $\frac{7}{2}$ or $3\frac{1}{2}$ **15.** $\frac{65}{8}$ or $8\frac{1}{8}$ **16.** $\frac{159}{74}$ or $2\frac{11}{74}$ **17.** $\frac{18}{25}$ **18.** 0.4375 **19.** 14.64 **20.** 1806.62 **21.** 1.312 **22.** 16.32 **23.** 230 **24.** 19.3658 **25.** 7.3% **26.** 1.965 **27.** 63 **28.** 0.336 **29.** 12.5% **30.** 30% **31.** 52% **32.** 18 **33.** 100 **34.** 99 **35.** \$2700 **36.** 65%

Chapter 1

Pretest Chapter 1

1. 3 **2.** −4 **3.** −18 **4.** 20 **5.** −42 **6.** $\frac{8}{3}$ **7.** 28 **8.** −4 **9.** 16 **10.** 64 **11.** $-\frac{8}{27}$ **12.** −256 **13.** $-6x^2 + 4x^2y - 2xz$ **14.** $-3x + 4y + 12$ **15.** $5x^2 - 6x^2y - 11xy$ **16.** $7x - 2y + 7$ **17.** $5x - 7y$ **18.** $-x^2y + 17y^2$ **19.** 28 **20.** 197 **21.** $\frac{107}{12}$ **22.** 6.37 **23.** 18 **24.** 29 **25.** 25°C **26.** $-9x - 3y$ **27.** $-x^2 + 6xy$

1.1 Exercises

11. −53 **13.** $-2\frac{3}{8}$ **15.** +7 **17.** $-\frac{3}{4}$ **19.** $+\frac{5}{13}$ **21.** −4 **23.** −13 **25.** $\frac{1}{3}$ **27.** 55 **29.** −31 **31.** $-\frac{7}{13}$ **33.** −3.8 **35.** 0.4 **37.** −25 **39.** $\frac{1}{35}$ **41.** −11 **43.** 3 **45.** −2 **47.** 13 **49.** −3.2 **51.** 17 **53.** −28 **55.** $-\frac{3}{8}$ **57.** $-\frac{11}{12}$ **59.** −3.4 **61.** −32 **63.** 1 **65.** 16.39 **67.** −5°F **69.** −265 **71.** 57°F **73.** \$140 **74.** 18 **75.** 28 **76.** 0 **77.** Under one of the following cases: (a) $|x| > |y|$ with x negative and y positive, (b) $|y| > |x|$ with x positive and y negative, or (c) x and y both negative. **78.** $x = y$ **79.** \$39.00

Putting Your Skills to Work

1. \$36,335 **2.** \$112,286.40

1.2 Exercises

1. 3 **3.** 10 **5.** −6 **7.** −26 **9.** −26 **11.** 8 **13.** 5 **15.** −5 **17.** 0 **19.** −4 **21.** −0.9 **23.** 4.47 **25.** 2 **27.** $1\frac{7}{20}$ **29.** $-1\frac{7}{12}$ **31.** −53 **33.** −99 **35.** 7.1 **37.** $\frac{14}{5}$ or $2\frac{4}{5}$ **39.** $8\frac{3}{5}$ or $\frac{43}{5}$ **41.** $-\frac{21}{20}$ or $-1\frac{1}{20}$ **43.** −8.5

45. 0.0499 **47.** $-5\frac{4}{5}$ **49.** $-\frac{3}{10}$ **51.** 7 **53.** -48 **55.** -2 **57.** 11 **59.** -62 **61.** 7 **63.** -41 **65.** 38
67. \$149 **69.** A drop of 16°F or -16°F **70.** -21 **71.** -51 **72.** -19 **73.** -8°C

1.3 Exercises

1. -36 **3.** 30 **5.** 60 **7.** -24 **9.** 0 **11.** 24 **13.** 0.264 **15.** -1.75 **17.** -24 **19.** $\frac{5}{16}$ **21.** -3 **23.** 0
25. -6 **27.** 36 **29.** -16 **31.** -2 **33.** 4 **35.** -0.9 **37.** $-\frac{3}{10}$ **39.** $-\frac{20}{3}$ or $-6\frac{2}{3}$ **41.** -30 **43.** $\frac{9}{16}$ **45.** $-\frac{3}{4}$
47. -24 **49.** 36 **51.** -16 **53.** -18 **55.** $-\frac{8}{35}$ **57.** $\frac{2}{27}$ **59.** -4 **61.** -2 **63.** 17 **65.** -72 **67.** -1
69. \$30 **71.** 150 or 150 m right of 0 **72.** -6.69 **73.** -72 **74.** 4.63 **75.** -88

1.4 Exercises

7. 8 **9.** 81 **11.** 343 **13.** -27 **15.** 64 **17.** -125 **19.** $\frac{1}{16}$ **21.** $\frac{8}{125}$ **23.** 0.81 **25.** 0.125 **27.** 4096
29. -4096 **31.** 6^5 **33.** w^2 **35.** x^4 **37.** y^5 **39.** 89 **41.** 161 **43.** -91 **45.** -282 **47.** -576 **49.** -512
51. 6.094744099 **53.** 16,777,216 **55.** -19 **56.** $-\frac{5}{3}$ or $-1\frac{2}{3}$ **57.** -8 **58.** 2.52

1.5 Exercises

7. $6x + 24y$ **9.** $-6a + 15b$ **11.** $8x + 2y - 4$ **13.** $6x^2 + 18xy$ **15.** $8x^2 - 2xy - 12x$ **17.** $-15x - 45 + 35y$ **19.** $\frac{x^2}{4} + \frac{x}{2} - 2$
21. $4x^2 - 12x - \frac{12}{7}$ **23.** $y^2 - \frac{4xy}{3} - 2y$ **25.** $6a^2 + 3ab - 3ac - 12a$ **27.** $-12 + 6x$ **29.** $25x^3 - 5x$ **31.** $12x^2 - 15xy - 18x$
33. $-4x^2 - 4xy + 20x$ **35.** $-4a^2b + 2ab^2 + ab$ **37.** $-8x^2y + 4xy^2 - 12xy$ **39.** $3.75a^2 - 8.75a + 5$
41. $-0.12a^2 + 0.08ab - 0.008ab^2$ **43.** -16 **44.** 64 **45.** 1 **46.** 14

1.6 Exercises

7. $10x$ **9.** $-28x^3$ **11.** $18x^4 + 7x^2$ **13.** $5a + 2b - 7a^2$ **15.** $-4ab - 7$ **17.** $7.1x - 3.5y$ **19.** $-2x - 8.7y$ **21.** $\frac{3}{4}x^2 - \frac{10}{3}y$
23. $-\frac{1}{15}x - \frac{2}{21}y$ **25.** $-5x - 7y - 20$ **27.** $5x^2y - 10xy^6 - xy^2$ **29.** $5bc - 6ac$ **31.** $x^2 - 10x + 3$ **33.** $-10y^2 - 16y + 12$
35. $5a - 3ab - 8b$ **37.** $-12x + 13y$ **39.** $3a^2 + 27ab$ **41.** $-17xy + 27y^2$ **43.** $71x - 27$ **44.** $-\frac{2}{15}$ **45.** $-\frac{5}{6}$ **46.** $\frac{23}{50}$
47. $-\frac{15}{98}$

1.7 Exercises

5. 24 **7.** 7 **9.** 6 **11.** -29 **13.** 14 **15.** 145 **17.** -6 **19.** 42 **21.** $\frac{9}{4}$ or $2\frac{1}{4}$ **23.** 0.848 **25.** $\frac{3}{10}$
27. 7.56 **29.** $\frac{1}{4}$ **31.** x **32.** $-\frac{19}{12}$ or $-1\frac{7}{12}$ **33.** -1 **34.** $\frac{72}{125}$

1.8 Exercises

1. -5 **3.** -2 **5.** $12\frac{1}{2}$ **7.** -26 **9.** 10 **11.** 3 **13.** -50 **15.** -5 **17.** 9 **19.** 39 **21.** -2 **23.** 15
25. 20 **27.** 29 **29.** **(a)** 8 **(b)** 4 **(c)** 14 **31.** **(a)** -19 **(b)** 8 **(c)** 6 **33.** 42 **35.** 29 **37.** 84 square inches
39. 78.5 square meters **41.** 14°F **43.** \$53,851 **45.** -76°F to -22°F **47.** 12.4 miles **48.** 16 **49.** $-x^2 + 2x - 4y$
50. -34 **51.** -0.72

Putting Your Skills to Work

1. BMI = 21.3 **2.** Yes, just slightly. BMI = 25.6

1.9 Exercises

5. $-3x - y$ **7.** $5a + 3b$ **9.** $-x^4 - x^2y^2$ **11.** $-245 + 25x$ **13.** $4x^2 - 2x + 6$ **15.** $-12y^3 - 8y^2 + 32y$ **17.** $4x - 6y - 3$
19. $2a - 6ab - 8b^2$ **21.** $-x^3 + 2x^2 - 3x - 12$ **23.** $3a^2 + 16b + 12b^2$ **25.** $15x - 10y$ **27.** $4x^2 + 18x + 42$ **29.** $12a^2 - 8b$
31. 97.52°F **32.** 465,426.5 sq ft **33.** 35 **34.** 50

Putting Your Skills to Work

1. \$300.20 **2.** \$20,140

Chapter 1 Review Problems

1. -8 **2.** -4.2 **3.** -9 **4.** 1.9 **5.** $-\frac{1}{3}$ **6.** $-\frac{7}{22}$ **7.** $\frac{1}{6}$ **8.** $\frac{22}{15}$ **9.** -13 **10.** -2 **11.** 8 **12.** 13

13. -33 **14.** 9.2 **15.** $-\frac{13}{8}$ **16.** $\frac{1}{2}$ **17.** -22.7 **18.** -27 **19.** -4 **20.** 16 **21.** -29 **22.** 1 **23.** -42

24. -3 **25.** -2 **26.** 27 **27.** -5 **28.** 2 **29.** $-\frac{2}{3}$ **30.** $-\frac{25}{7}$ **31.** 10 **32.** -72 **33.** 30 **34.** -30 **35.** $-\frac{1}{2}$

36. $-\frac{4}{7}$ **37.** -30 **38.** -5 **39.** -9.1 **40.** 0.9 **41.** 10.1 **42.** -1.2 **43.** -0.4 **44.** -2.3 **45.** 24 yards

46. $-22°$ **47.** 7363 feet **48.** $2\frac{1}{4}$ point loss **49.** 81 **50.** 64 **51.** -125 **52.** $\frac{1}{8}$ **53.** -64 **54.** 0.49 **55.** $\frac{25}{36}$

56. $\frac{27}{64}$ **57.** $15x - 35y$ **58.** $6x^2 - 14xy + 8x$ **59.** $-7x^2 + 3x - 11$ **60.** $-6xy^2 - 3xy + 3y^2$ **61.** $-5a^2b + 3bc$

62. $14x - 2y$ **63.** $-5x^2 - 35x - 9$ **64.** $10x^2 - 8x - \frac{1}{2}$ **65.** -44 **66.** 30 **67.** 1 **68.** -19 **69.** 1 **70.** -4

71. -15 **72.** 10 **73.** -16 **74.** $\frac{32}{5}$ **75.** \$810 **76.** 86°F **77.** \$2119.50 **78.** \$8580.00 **79.** $x - 8$ **80.** $-19x - 30$

81. $-2 + 10x$ **82.** $-12x^2 + 63x$ **83.** $5xy^3 - 6x^3y - 13x^2y^2 - 6x^2y$ **84.** $-14y^2 - 4xy + x^2$ **85.** $x - 10y + 35 - 15xy$

86. $10x - 22y - 36$ **87.** $-4a - 24b + 32ab + 18a^2 - 6a^2b$ **88.** $-10a + 25ab - 15b^2 - 10ab^2$ **89.** $-3x - 9xy + 18y^2$

Chapter 1 Test

1. 4 **2.** $\frac{1}{3}$ **3.** 0.2 **4.** 96 **5.** -70 **6.** 4 **7.** -3 **8.** $-\frac{10}{3}$ or $-3\frac{1}{3}$ **9.** -64 **10.** 1.44 **11.** -125 **12.** $\frac{16}{81}$

13. $-5x^2 - 10xy + 35x$ **14.** $6a^2b^2 + 4ab^3 - 14a^2b^3$ **15.** $-6xy - 5x^2y - 4xy^2$ **16.** $2a^2b + \frac{15}{2}ab$ **17.** $8a^2 + 20ab$ **18.** $5a + 30$

19. $14x - 16y$ **20.** 8 **21.** -25 **22.** 122 **23.** 37 **24.** 1 **25.** 96.6 **26.** $-3a - 9ab + 3b^2 - 3ab^2$

27. $-69x + 90y - 51$

Chapter 2

Pretest Chapter 2

1. $x = 58$ **2.** $x = 48$ **3.** $x = -4$ **4.** $x = 4$ **5.** $x = -\frac{7}{2}$ **6.** $x = -\frac{1}{7}$ **7.** $x = \frac{5}{2}$ **8.** $x = \frac{17}{10}$ **9.** **(a)** $F = \frac{9C + 160}{5}$ **(b)** $F = 5°$

10. **(a)** $r = \frac{I}{Pt}$ **(b)** 6% **11.** $<$ **12.** $>$ **13.** $>$ **14.** $<$ **15.** -2 x **16.** 6 x

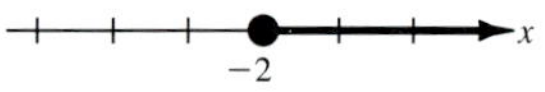

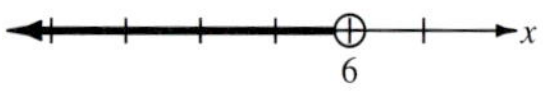

2.1 Exercises

7. 3 **9.** -1 **11.** 20 **13.** 7 **15.** 1 **17.** 0 **19.** -2 **21.** 66 **23.** 27 **25.** -14 **27.** no, $x = 26$

29. no, $x = 7$ **31.** no, $x = 8$ **33.** yes **35.** 4 **37.** 22 **39.** 3.5 **41.** -1.2 **43.** $\frac{1}{3}$ **45.** $\frac{4}{5}$ **47.** $\frac{7}{6}$ or $1\frac{1}{6}$

49. $27\frac{1}{8}$ **51.** 7.2 **53.** -27.4976 **54.** $7x - 20$ **55.** $-2x - 4y$ **56.** $-7x^2 - 4x - 7$ **57.** $-2y^2 - 4y + 4$ **58.** 76.5% did not

2.2 Exercises

5. 63 **7.** -15 **9.** 80 **11.** -48 **13.** 7 **15.** $-\frac{8}{3}$ **17.** 50 **19.** 15 **21.** -7 **23.** $\frac{2}{5}$ **25.** no, $x = -7$ **27.** yes

29. -0.8 **31.** $\frac{9}{4}$ **33.** -0.7 **35.** 3 **37.** -4 **39.** $\frac{15}{11}$ **41.** $\frac{7}{6}$ **43.** -5.26 **45.** -1 **46.** 42 **47.** -37 **48.** 21

49. $93\frac{1}{3}$% is acceptable.

2.3 Exercises

1. 6 **3.** 11 **5.** -17 **7.** 13 **9.** 15 **11.** 28 **13.** -36 **15.** 8 **17.** 3 **19.** 5 **21.** -9 **22.** yes **23.** yes

24. no, $x = -11$ **25.** no, $x = 12$ **26.** 3; 3 **27.** 7; 7 **29.** $-\frac{5}{6}$ **31.** 5 **33.** -4 **35.** 6 **37.** 4 **39.** $\frac{1}{2}$ **41.** 1

43. 4 **45.** -4 **47.** 0 **49.** 2 **51.** 6 **53.** 5 **55.** -7 **57.** $-\frac{2}{3}$ **59.** $-\frac{1}{4}$ **61.** -0.25 or $-\frac{1}{4}$ **63.** 8 **65.** -6

67. $\frac{1}{2}$ **69.** -4.23 **71.** $14x^2 - 14xy$ **72.** $5x + 25y$ **73.** $28x + 54$ **74.** $-10x - 60$

2.4 Exercises

1. 2 **3.** $2\frac{1}{2}$ or $\frac{5}{2}$ **5.** 1 **7.** 24 **9.** 20 **11.** 7 **13.** -13 **15.** 3 **17.** yes **19.** no **21.** 1 **23.** 0 **25.** -7

27. -3 **29.** 4 **31.** -22 **33.** 3 **35.** -12 **37.** $-\frac{13}{12}$ or $-1\frac{1}{12}$ **39.** $\frac{4}{113}$ **41.** $\frac{18}{17}$ **43.** $\frac{2}{3}$ **45.** $\frac{27}{14}$ or $1\frac{13}{14}$

46. $\frac{19}{20}$ **47.** $\frac{539}{40}$ or $13\frac{19}{40}$ **48.** $\frac{22}{5}$ or $4\frac{2}{5}$

47. $\frac{539}{40}$ or $13\frac{19}{40}$ 48. $\frac{22}{5}$ or $4\frac{2}{5}$

2.5 Exercises

1. 50 miles per hour **3.** **(a)** 10 meters **(b)** 16 meters **5.** **(a)** $y = \frac{3}{5}x - 3$; -6 **7.** $h = \frac{2A}{b}$ **9.** $r = \frac{I}{Pt}$ **11.** $b = y - mx$

13. $\frac{2A - ab_2}{a} = b_1$ **15.** $y = -\frac{5}{9}x - 2$ **17.** $x = \frac{7}{6}y + 14$ **19.** $x = \frac{c - by}{a}$ **21.** $r^2 = \frac{s}{4\pi}$ **23.** $m = \frac{E}{c^2}$ **25.** $t^2 = \frac{2S}{g}$

27. $I = \frac{E}{R}$ **29.** $H = \frac{V}{LW}$ **31.** $h = \frac{3V}{\pi r^2}$ **33.** $L = \frac{P - 2W}{2}$ **35.** $b^2 = c^2 - a^2$ **37.** $F = \frac{9}{5}C + 32$ **39.** $V = \frac{kT}{P}$

41. I doubles **42.** I increases four times **43.** A increases four times **44.** A becomes $\frac{1}{4}A$ **45.** 31.2 **46.** 0.096 **47.** 40%

48. 2.5%

Putting Your Skills to Work

1. 26 mpg **2.** 20 mpg if he were traveling at 70 miles per hour. He would be traveling at approx. 65 miles per hour if he obtained 23 mpg.

2.6 Exercises

3. > **5.** < **7.** < **9.** < **11.** < **13.** < **15.** 0.0247 **17.** −7 −6 −5 −4 −3 **19.** 2 3 4 5 6

21. $\frac{1}{4}$ $\frac{1}{2}$ $\frac{3}{4}$ 1 $\frac{5}{4}$ **23.** −3.8 −3.6 −3.4 **25.** 10 15 20 25 30 **27.** $x \geq -1$ **29.** $x < -20$

31. $x > 6.5$ **33.** $c < 56$ **35.** $h \geq 37$ **37.** $-\frac{5}{2}$ −3 −2 −1 0 1 2

38. −4 0 $\frac{7}{2}$ **39.** 6.08 **40.** 16.25 **41.** 2% **42.** 37.5%

2.7 Exercises

1. $x > 8$ 4 5 6 7 8 9 10

3. $x > -9$ −11 −10 −9 −8 −7

5. $x \geq 8$ 5 6 7 8 9 10 11

7. $x < \frac{7}{2}$ 1 $\frac{3}{2}$ 2 $\frac{5}{2}$ 3 $\frac{7}{2}$ 4

9. $x > -\frac{5}{4}$ −2 $-\frac{7}{4}$ $-\frac{6}{4}$ $-\frac{5}{4}$ −1 $-\frac{3}{4}$

11. $x > -6$ −10 −9 −8 −7 −6 −5 −4

13. $x \leq -2$ −6 −5 −4 −3 −2 −1 0

17. $x > 2$ **19.** $x \geq 4$ **21.** $x < -1$ **23.** $x \geq -6$ **25.** $x < -3$ **27.** $x > 0$ **29.** $x > -\frac{4}{11}$ **31.** $x > -1.46$ **33.** $x \geq 76$

35. **(a)** 24,000 chips **(b)** 76,000 chips; 61,600 chips

Putting Your Skills to Work

1. \$800 **2.** \$1580

Chapter 2 Review Problems

1. −10 **2.** −1 **3.** 5 **4.** 4 **5.** 10 **6.** −13 **7.** 40.4 **8.** −7 **9.** −2 **10.** −64 **11.** −11 **12.** $\frac{22}{3}$ or $7\frac{1}{3}$

13. −3 **14.** −1 **15.** 4 **16.** 3 **17.** 3 **18.** −6 **19.** 3 **20.** $-\frac{7}{2}$ or $-3\frac{1}{2}$ **21.** −3 **22.** 2.1 **23.** $-\frac{7}{3}$ or $-2\frac{1}{3}$

24. $-\frac{2}{7}$ **25.** 5 **26.** 0 **27.** 1 **28.** 20 **29.** $\frac{2}{3}$ **30.** 5 **31.** $\frac{35}{11}$ **32.** 4 **33.** −17 **34.** $\frac{2}{5}$ **35.** 32 **36.** $\frac{26}{7}$

37. −1 **38.** $-\frac{5}{2}$ or $-2\frac{1}{2}$ **39.** 4 **40.** −17 **41.** 4 **42.** 0 **43.** 4 **44.** −5 **45.** 3 **46.** −28 **47.** $-\frac{17}{5}$ **48.** 9

49. 0 **50.** 3 **51.** $y = 3x - 10$ **52.** $y = \frac{-5x - 7}{2}$ **53.** $r = \frac{A - P}{Pt}$ **54.** $h = \frac{A - 4\pi r^2}{2\pi r}$

55. $p = \frac{3H - a - 3}{2}$ **56.** **(a)** $T = \frac{1000\ C}{WR}$ **(b)** $T = 6000$ **57.** **(a)** $y = \frac{5}{3}x - 4$ **(b)** $y = 11$ **58.** **(a)** $R = \frac{E}{I}$ **(b)** $R = 5$

59. $x \le 2$

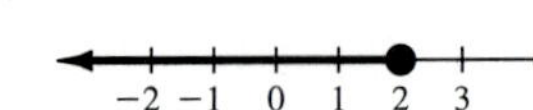

60. $x \ge 1$

−2 −1 0 1 2 3

61. $x < -4$

−5 −4 −3 −2 −1 0

62. $x > -3$

−5 −4 −3 −2 −1 0

63. $x \ge 3$

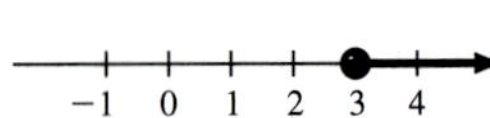

64. $x < 2$

−1 0 1 2 3 4

65. $x \le -8$

−10 −9 −8 −7 −6 −5

66. $x \ge 3$

0 1 2 3 4 5

67. $x < 10$

7 8 9 10 11 12

68. $x > -3$

−5 −4 −3 −2 −1 0

69. $x > \frac{17}{2}$

$\frac{15}{2}$ 8 $\frac{17}{2}$ 9 $\frac{19}{2}$ 10

70. $x \ge \frac{19}{7}$

$\frac{15}{7}$ $\frac{17}{7}$ $\frac{19}{7}$

71. $x \ge -15$

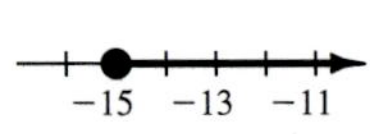

72. $x > \frac{7}{5}$

$\frac{7}{5}$ $\frac{8}{5}$ $\frac{9}{5}$ 2 $\frac{11}{5}$ $\frac{12}{5}$

73. $n \le 46$ **74.** $n \le 17$

Chapter 2 Test

1. 2.1 **2.** 2 **3.** $-\frac{7}{2}$ **4.** −30 **5.** 2 **6.** −1.2 **7.** 7 **8.** $\frac{7}{3}$ **9.** 13 **10.** 125 **11.** 10 **12.** −4 **13.** 12 **14.** $-\frac{1}{5}$ **15.** 3 **16.** 2 **17.** 18 **18.** $w = \frac{A - 2P}{3}$ **19.** $w = \frac{6 - 3x}{4}$ **20.** $a = \frac{2A - hb}{h}$ **21.** $y = \frac{10ax - 5}{8ax}$

22. $x \le -\frac{1}{2}$ **23.** $x > -\frac{5}{4}$ **24.** $x < 2$ **25.** $x \ge \frac{1}{2}$

$-\frac{3}{2}$ $-\frac{1}{2}$

$-\frac{7}{4}$ $-\frac{5}{4}$ $-\frac{3}{4}$

0 1 2 3 4

$-\frac{1}{2}$ 0 $\frac{1}{2}$ 1 $\frac{3}{2}$

Cumulative Test for Chapters 0–2

1. $\frac{4}{21}$ **2.** $\frac{79}{20}$ or $3\frac{19}{20}$ **3.** $\frac{32}{15}$ or $2\frac{2}{15}$ **4.** 0.6888 **5.** 0.12 **6.** 61.2 **7.** 30 **8.** $12ab - 28ab^2$ **9.** $25x^2$ **10.** $2x + 6y - 4xy + 16xy^2$ **11.** 5 **12.** 4 **13.** 1 **14.** $y = \frac{3x + 2}{7}$ **15.** $b = \frac{3H - 8a}{2}$ **16.** $t = \frac{I}{Pr}$ **17.** $a = \frac{2A}{h} - b$

18. $x < 5$ **19.** $x \le 3$ **20.** $x > -15$ **21.** $x \le -1$ **22.** $x \ge -12$

2 3 4 5

1 2 3 4 5

−17 −16 −15 −14 −13

−2 −1 0 1 2

−13 −12 −11 −10

23. Minimum score of 92

Chapter 3

Pretest Chapter 3

1. $2x - 30$ **2.** $0.40x + 150$ **3.** 5 **4.** 1st number = 12; 2nd number = 4; 3rd number = 7 **5.** $x = 8.6$ m; $2x - 3 = 14.2$ m; $2x + 1 = 18.2$ m **6.** $x = 7.5$ pounds; $x - 3.5 = 4$ pounds; $x - 2 = 5.5$ pounds **7.** width = 8 m; length = 29 m **8.** \$400 at 12%; \$600 at 9% **9.** 22,000 people **10.** 3 nickels; 7 dimes; 10 quarters **11.** 72 square inches **12.** 339.12 cubic meters **13.** 25.12 centimeters **14.** \$6859.80 **15.** $h \le 72$ **16.** $r \ge 450$ **17.** $a < 15$ **18.** $c > 8{,}900$ **19.** More than \$30,000 **20.** Length ≤ 16 feet

3.1 Exercises

1. $x + 5$ **3.** $x - 5$ **5.** $\frac{x}{3}$ **7.** $3x$ **9.** $\frac{x}{8}$ **11.** $2x + 5$ **13.** $\frac{2}{7}x - 3$ **15.** $5(x + 12)$ **17.** $\frac{1}{2}x - 5$ **19.** $3x + \frac{1}{2}x$ **21.** $4x - \frac{1}{2}x$ **23.** $b + 15$ = no. of hours Alicia works; b = no. of hours Barbara works **25.** x = income from retirement fund; $x - 833$ = income from mutual fund **27.** $3w + 3$ = length of rectangle; w = width of rectangle **29.** s = attendance on Saturday; $s + 1600$ = attendance on Friday; $s - 783$ = attendance on Thursday **31.** 1st angle = $t + 19$; 2nd angle = $3t$; 3rd angle = t **33.** $3t - 200$ = attendance on Friday; t = attendance on Thursday **35.** x = the cost of the history book in dollars; $x + 13$ = the cost of the biology book in dollars; $x - 27$ = the cost of the English book in dollars **37.** February = $\frac{5}{6}x$; March = x; April = $\frac{3}{4}x$ **39.** x = the atomic weight of chromium; $3x + 51.2$ = the atomic weight of lead **40.** 17 **41.** 10 **42.** 12 **43.** 7

3.2 Exercises

1. 45 **3.** 2368 **5.** 290 **7.** 39 **9.** 51 **11.** −2 **13.** 72; 84 **15.** −5; 3 **17.** 80 **19.** 77.0973; 118.9605 **21.** 15 used bikes **23.** 25 months **25.** 6 CDs **27.** 4 French fries **29.** 6 hours **31.** 6 miles per hour **33.** 5 miles **35.** He traveled 64 mph on the highway and 52 mph on the mountain road. It was 12 mph faster on the highway. **37.** 96 **39.** 9 mpg **41. (a)** $F - 40 = \frac{x}{4}$; **(b)** 200 chirps; **(c)** 77° **42.** $10x^3 - 30x^2 - 15x$ **43.** $-2a^2b + 6ab - 10a^2$ **44.** $-5x - 6y$ **45.** $-4x^2y - 7xy^2 - 8xy$ **46.** 3000 apples

3.3 Exercises

1. Older computer has 0.34 gigabyte; new computer has 2.8 gigabytes. **3.** Long piece is 32 m long; short piece is 15 m long. **5.** The main span of the Verrazano Narrows bridge is 4260 feet; the main span of the George Washington bridge is 3500 feet; the main span of the Golden Gate Bridge is 4200 feet. **7.** Magaret lost 66 lb; Tony lost 46 lb; Jan lost 44 lb **9.** The first flight is 2.375 hours; the second flight is 4.75 hours; the third flight is 7.125 hours. **11.** Width is 6 cm; length is 15 cm. **13.** Width is 4 m, length is 8 m. **15.** First side is 36 cm; second side is 18 cm; third side is 32 cm. **17.** Longest side = 18 in.; Shortest side = 13 in.; Third side = 15 in. **19.** Length = 16.858 cm; Width = 7.762 cm **21.** Original square was 11 m × 11 m **23.** $-8x^3 + 12x^2 - 32x$ **24.** $5a^2b + 30ab - 10a^2$ **25.** $-19x + 2y - 2$ **26.** $9x^2y - 6xy^2 + 7xy$

3.4 Exercises

1. 56 pages **3.** 18 tours **5.** 7 hours **7.** $360 **9.** 800 crimes **11.** $14,500 **13.** $5000 **15.** $2600 at 12%; $1400 at 14% **17.** $200,000 at 8%; $120,000 at 12% **19.** $6500 at 14%; $2500 at 9% **21.** 13 quarters; 9 nickels **23.** 18 nickels; 6 dimes; 9 quarters **25.** eight $10 bills; sixteen $20 bills; eleven $100 bills **27.** First angle measures 70°; Second angle measures 35°; Third angle measures 75° **29.** $1328 at 7%, $1641 at 11% **31.** 225 miles **33.** 12 **34.** 15 **35.** −28 **36.** −25

Putting Your Skills to Work

1. Monthly payments will be $248.90. He will pay $1947.20 in interest. **2.** Monthly payments will be $141.21. She will pay $389.04 in interest.

3.5 Exercises

7. 98 in.^{2} **9.** 10 feet **11.** Length = 15 in. It is a square. Perimeter = 60 in. **13.** 28.26 m^2 **15.** 24 inches **17.** 73 meters **19.** 62.8 meters **21.** 9 inches **23.** equal sides = 12 feet; base = 8 feet **25.** 13° **27.** First angle = 100°; Second angle = 50°; Third angle = 30° **29.** 2512 in.^{3} **31.** 3 feet **33.** 552 square inches **35. (a)** 113.04 cm^3 **(b)** 113.04 cm^2 **37.** $23.85 **39. (a)** 285.74 sq. in. **(b)** $2143.05 **41. (a)** 36 cubic yd. **(b)** $126 **42.** No. The perimeter may be 19 feet or it may be 15.5 feet. **43. (a)** 36π in.^{2} **(b)** 10π in.^{2}; the sphere **44.** 6.28 ft **45.** $3x^2 + 8x - 24$ **46.** $-11x - 17$ **47.** $-4x - 9$ **48.** $22x - 84$

Putting Your Skills to Work

1. 60% **2.** 60 days

3.6 Exercises

1. $x > 67{,}000$ **3.** $x \leq 120$ **5.** $x < 34$ **7.** $x \geq 3500$ **9.** $x \leq 350$ **11.** The third side must be less than or equal to 224 ft. **13.** She must score 90 or more. **15.** He can take a maximum of 17 clients to lunch. **17.** The length must be 17 ft or longer. **19.** He must sell 15 cars or more. **21.** She must sell more than $60,000 per month of medical supplies. **23.** The Fahrenheit temperature must be less than 230°. **25.** He must purchase 12 chairs or less. **27.** More than 454 discs must be manufactured and sold. **29.** $x < -4$ **30.** $x \leq 3$ **31.** $x \geq 4\frac{2}{3}$ **32.** $x > -4$

Chapter 3 Review Problems

1. $x + 19$ **2.** $\frac{2}{3}x$ **3.** $x - 56$ **4.** $3x$ **5.** $2x + 7$ **6.** $2x - 3$ **7.** $5x - \frac{x}{3}$ **8.** $\frac{x+6}{4}$ **9.** $h + 370$ = the speed of the jet; h = the speed of the helicopter **10.** $g - 2000$ = the cost of the truck; g = the cost of the garage **11.** $3w + 5$ = the length of the rectangle; w = the width of the rectangle **12.** r = the number of retired people; $4r$ = the number of working people; $0.5r$ = the number of unemployed people **13.** b = the number of degrees in angle B; $2b$ = the number of degrees in angle A; $b - 17$ = the number of degrees in angle C **14.** a = the number of students in algebra; $a + 29$ = the number of students in biology; $0.5a$ = the number of students in geology **15.** 12 **16.** 18 **17.** 44 **18.** 1 **19.** 16 years old **20.** $40 **21.** 6.6 hours; 6 hours **22.** 90 **23.** $W = 10$ m; $L = 23$ m **24.** $W = 12$ ft; $L = 30$ ft **25.** 1st side = 8 yd; 2nd side = 15 yd; 3rd side = 17 yd **26.** 1st angle = 32°; 2nd angle = 96°; 3rd angle = 52° **27.** 31.25 yd; 18.75 yd **28.** Jon = $30,000; Lauren = $18,000 **29.** 280 miles **30.** 310 kilowatt-hours **31.** $200.00 **32.** $22,500 **33.** $7000 at 12%; $2000 at 8% **34.** $2000 at $4\frac{1}{2}$%; $3000 at 6% **35.** 18 nickels; 6 dimes; 9 quarters **36.** 7 nickels; 8 dimes; 10 quarters **37.** 8 m **38.** $16\frac{1}{2}$ in. **39.** 71° **40.** 42 sq. mi. **41.** 56 in. **42.** 168 square centimeters **43.** 1920 ft^3 **44.** 113.04 cm^3 **45.** 12 in. **46.** 254.34 cm^2 **47.** $23,864 **48.** $3440 **49.** 84 **50.** 80 dimes; 8 quarters **51.** 200 miles **52.** more than $300,000 **53.** $16.15 **54.** 3 hours **55.** $24\frac{4}{5}$ min **56.** 4 ounces **57.** 16 **58.** 14 games

Putting Your Skills to Work

1. $d = 5t$; $d = 120\left(t - \frac{1}{12}\right)$ **2.** $5t + 120\left(t - \frac{1}{12}\right) = 100$; $t = \frac{22}{25}$. Time is approx. 52.8 minutes. Yes.

Chapter 3 Test

1. 35 **2.** 36 **3.** -12 **4.** first side = 20 m; second side = 30 m; third side = 16 m **5.** width = 20 m; length = 47 m
6. 1st pollutant = 8 ppm; 2nd pollutant = 4 ppm; 3rd pollutant = 3 ppm **7.** 15 months **8.** \$8500 **9.** \$1400 at 14%; \$2600 at 11%
10. 16 nickels; 7 dimes; 8 quarters **11.** 138.16 in. **12.** 192 square inches **13.** 4187 in.^{3} **14.** 70 cm^2 **15.** \$450
16. at least an 82 **17.** more than \$110,000

Cumulative Test for Chapters 0–3

1. 3.69 **2.** $\frac{31}{24}$ **3.** $-23y + 18$ **4.** 28 **5.** $\frac{2H - 3a}{5}$ **6.** $x \geq -3$; 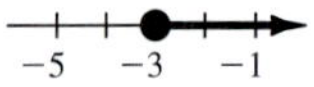 **7.** 40
8. Psychology students = 50; World Hist. students = 84 **9.** width = 7 cm; length = 32 cm **10.** \$135,000 **11.** \$3000 at 15%; \$4000 at 7%
12. 7 nickels; 12 dimes; 4 quarters **13.** A = 162.5 m^2; \$731.25 **14.** V = 113.04 cubic inches; 169.56 pounds

Chapter 4

Pretest Chapter 4

1. 3^{16} **2.** $-10x^7$ **3.** $-24a^5b^4$ **4.** x^{23} **5.** $\frac{-2}{x^2y^2}$ **6.** $\frac{5bc}{3a}$ **7.** x^{50} **8.** $-8x^6y^3$ **9.** $\frac{9a^2b^4}{c^6}$ **10.** $\frac{3x^2}{y^3z^4}$ **11.** $\frac{-2b^2}{a^3c^4}$
12. $\frac{1}{9}$ **13.** 6.38×10^{-4} **14.** 1,894,000,000,000 **15.** $-2x^2 + 7x - 29$ **16.** $x^3 - 5x^2 + 11x - 12$ **17.** $-6x + 21x^2$
18. $2x^4 + 8x^3 - 2x^2$ **19.** $5x^4y^2 - 15x^2y^3 + 5xy^4$ **20.** $x^2 + 14x + 45$ **21.** $12x^2 + xy - 6y^2$ **22.** $2x^4 - 7x^2y^2 + 3y^4$
23. $64x^2 - 121y^2$ **24.** $36x^2 + 36xy + 9y^2$ **25.** $25a^2b^2 - 60ab + 36$ **26.** $4x^3 - 13x^2 + 11x - 2$ **27.** $16x^4 - 81$
28. $7x^3 - 6x^2 + 11x$ **29.** $5x^2 - 6x + 2 - \frac{3}{3x - 2}$

4.1 Exercises

5. 6; x, y; 11, and 1 **7.** $3x^3y^2$ **9.** $-3a^2b^2c^3$ **11.** 3^{15} **13.** 5^{26} **15.** $3^5 \cdot 8^2$ **17.** $-60x^3$ **19.** $12x^5$ **21.** $-12x^{11}$
23. $-10x^4y^3$ **25.** $\frac{10}{21}x^4y^5$ **27.** $-8.05x^5wy^4$ **29.** 0 **31.** $-24x^4y$ **33.** $72x^6y^{11}$ **35.** $-28a^5b^6$ **37.** $-3x^3y$ **39.** $12a^7b^6$
41. 0 **43.** $-24w^5xyz^6$ **45.** $\frac{1}{x^2}$ **47.** y^7 **49.** 3^{28} **51.** $\frac{1}{13^{10}}$ **53.** 3^4 **55.** $\frac{a^8}{4}$ **57.** $\frac{x^7}{y^9}$ **59.** $\frac{1}{2b}$ **61.** $\frac{2x^2}{y^2}$
63. $\frac{-1}{2a^2b}$ **65.** 1 **67.** $-\frac{2x^4}{y^5}$ **69.** $\frac{y^5}{10x^4}$ **71.** $6x^2$ **73.** $\frac{y^2}{16x^3}$ **75.** $\frac{3a}{4}$ **77.** $\frac{5x^6}{7y^8}$ **79.** $-\frac{1}{17x^5y^2z}$ **81.** $2x$
83. $\frac{-2}{3a^3}$ **84.** $-27x^5yz^3$ **85.** $-7ab^5$ **86.–88.** Answers will vary. **89.** a^{5b} **90.** b^{3x} **91.** c^{y+2} **93.** w^{40} **95.** $a^{12}b^4$
97. $a^{24}b^{16}c^8$ **99.** $3^8x^4y^8$ **101.** $16a^{20}$ **103.** $\frac{12^5x^5}{y^{10}}$ **105.** $\frac{36x^2}{25y^6}$ **107.** $-32a^{25}b^{10}c^5$ **109.** $-64x^3z^{12}$ **111.** $\frac{a}{4b^4}$
113. $-8a^7b^{11}$ **115.** $\frac{16y^8}{9z^6}$ **117.** $\frac{y^{18}}{8}$ **119.** $\frac{a^{15}b^5}{c^{25}d^5}$ **121.** $16x^9y^3$ **123.** $\frac{b^{35}}{a^{20}c^{15}}$ **124.** $-3x^3y^4z^7$ **125.** $\pm 2x^5y^4z^7$
126.–128. Answers will vary. **129.** 2^{17} **131.** -11 **132.** -46 **133.** $-\frac{7}{4}$ **134.** 5 **135.** Approximately 27.9 million

4.2 Exercises

1. $\frac{3}{x^2}$ **3.** $\frac{1}{2xy}$ **5.** $\frac{3xz^3}{y^2}$ **7.** $3x$ **9.** $\frac{x^6}{y^2z^3w^{10}}$ **11.** $\frac{1}{8}$ **13.** $\frac{z^8}{9x^2y^4}$ **15.** $\frac{1}{x^6y}$ **17.** $-\frac{y^6}{8x^9}$ **19.** 1.2378×10^5
21. 7.42×10^{-4} **23.** 7.652×10^9 **25.** 56,300 **27.** 0.000033 **29.** 983,000 **31.** 300,000,000 meters per second
33. 1.695×10^{-23} cubic centimeters **35.** 1.0×10^1 **37.** 3.2×10^{-19} **39.** 4.5×10^5 **41.** 1.81×10^4 dollars **43.** 1.9×10^{11} hours
45. 1.15×10^{11} hours **47.** 4.175×10^{-16} gram **49.** 4.717×10^{26} joules **50.** -0.8 **51.** -1 **52.** $-\frac{1}{28}$

Putting Your Skills to Work

1. 3.14×10^8 acres **2.** 2.246×10^9 acres

4.3 Exercises

1. $2x - 17$ **3.** $-3x^2 - 10x + 10$ **5.** $x^3 - 2x^2 + 5x - 1$ **7.** $\frac{5}{6}x^2 + \frac{1}{2}x - 9$ **9.** $1.7x^3 - 3.4x^2 - 13.2x - 5.4$ **11.** $-11x - 2$
13. $2x^4 + x^2$ **15.** $-\frac{1}{6}x^2 + 7x - \frac{3}{10}$ **17.** $-2x^3 + 5x - 10$ **19.** $-4.7x^4 - 0.7x^2 - 1.6x + 0.4$ **21.** $10x + 10$ **23.** $-3x^2y + 6xy^2 - 4$
25. $-11a^4 - 5a^3 - 2$ **27.** $a^3 - 8a^2b + 5ab^2 - 7b^2$ **29.** $-58x^2 + 27xy - 8y^2$ **30.** $-x^2y^2 + 7xy$ **31.** $7x^3 + 41x^2 + 22$
32. $-25x^3 - 20x + 21$ **33.** $\frac{3y - 2}{8}$ **34.** $\frac{3A}{2CD}$ **35.** $b = \frac{2A}{h} - c$ or $b = \frac{2A - ch}{h}$ **36.** $\frac{5xy}{B}$ **37.** Approximately \$696.8 billion dollars

4.4 Exercises

1. $15x^3 - 3x^2$ **3.** $-15x^3 + 10x^2 - 25x$ **5.** $2x^4 - 6x^3 + 2x^2 - 12x$ **7.** $-5x^3y^3 + 15x^2y^2 - 25xy$ **9.** $3x^7 + 6x^5 - 9x^4 + 21x^3$
11. $6ab^2 - 4a^2b^3 - 10ab^5$ **13.** $-15x^4y^2 + 6x^3y^2 - 18x^2y^2$ **15.** $4b^4 + 6b^3 - 8b^2$ **17.** $3x^4 - 9x^3 + 15x^2 - 6x$

19. $-2x^3y^3 + 12x^2y^2 - 16xy$ **21.** $-28x^5y + 12x^4y + 8x^3y - 4x^2y$ **23.** $-15a^4b^2 - 3a^3b^2 + 21ab^2$ **25.** $12x^7 - 6x^5 + 18x^4 + 54x^3$ **27.** $-24x^7 + 12x^5 - 8x^3$ **29.** $x^2 + 13x + 30$ **31.** $x^2 + 8x + 12$ **33.** $x^2 - 3x - 18$ **35.** $x^2 - 11x + 30$ **37.** $-5x^2 - 16x - 3$ **39.** $3x^2 + 4xy - 27x - 36y$ **41.** $12y^2 - y - 6$ **43.** $20y^2 - 22y + 6$ **44.** The signs are incorrect. The result should be $-3x + 6$. **45.** The last term is incorrect. The result should be $-3x + 7$. **46.** $12x$ **47.** $49x^2 - 9$, difference of 2 perfect squares **49.** $-28x^2 + 2xy + 6y^2$ **51.** $15a^2 - 28ab^2 + 12b^4$ **53.** $4x^4 - 25y^4$ **55.** $64x^2 - 32x + 4$ **57.** $25x^4 + 20x^2y^2 + 4y^4$ **59.** $21m^2 - gm - 10g^2$ **61.** $6a^2b^2 - 31abd + 35d^2$ **63.** $30x^2 + 43xy - 8y^2$ **65.** $35x^2 + 21xz - 10xy - 6yz$ **67.** $4ac - 3ab - 32bc + 24b^2$ **69.** $\frac{1}{6}x^6y^8 + \frac{5}{126}x^4y^{10} - \frac{1}{21}x^2y^{12}$ **71.** -10 **72.** $-\frac{25}{7}$ **73.** 8 dimes; 11 quarters **74.** -5 **75.** 40% of people in China, 97% of people in America, 12% of people in India **76.** Nine \$20 bills, eight \$10 bills, twenty-three \$5 bills

4.5 Exercises

5. $x^2 - 16$ **7.** $x^2 - 81$ **9.** $64x^2 - 9$ **11.** $4x^2 - 49$ **13.** $4x^2 - 25y^2$ **15.** $49s^2t^2 - 1$ **17.** $144x^4 - 49$ **19.** $9x^4 - 16y^6$ **21.** $16x^2 - 8x + 1$ **23.** $64x^2 + 48x + 9$ **25.** $25x^2 - 70x + 49$ **27.** $4x^2 + 12xy + 9y^2$ **29.** $64x^2 - 80xy + 25y^2$ **31.** $121 + 44y + 4y^2$ **33.** $36w^2 + 60wz + 25z^2$ **35.** $49x^2 - 9y^2$ **37.** $16x^4 - 56x^2y + 49y^2$ **39.** $x^3 - 11x + 6$ **41.** $9y^3 + 3y^2z - 8yz^2 - 4z^3$ **43.** $4x^4 - 7x^3 + 2x^2 - 3x - 1$ **45.** $2x^3 + 3x^2 - 17x - 30$ **47.** $3x^3 - 13x^2 - 6x + 40$ **49.** $2x^3 - 7x^2 - 32x + 112$ **51.** $a^4 + a^3 - 13a^2 + 17a - 6$ **53.** $2x^6 + 5x^5y - 6x^4y^2 - 18x^2y^2 - 45xy^3 + 54y^4$ **55.** \$11,000 at 7%; \$7,000 at 11% **56.** width = 7 m; length = 10 m **57.** Approximately −28.12°F **58.** 2.0625×10^5 meters

Putting Your Skills to Work

1. 1.53×10^6 nanometers **2.** 1.53 millimeters

4.6 Exercises

1. $4x^2 - 10x + 2$ **3.** $2y^2 - 3y - 1$ **5.** $7x^4 - 3x^2 + 8$ **7.** $8x^4 - 9x + 6$ **9.** $10x^3 - 7x^2 - x + 5$ **11.** $3x + 5$ **13.** $x - 2 + \frac{-20}{x - 7}$ **15.** $3x^2 - 4x + 8 + \frac{-10}{x + 1}$ **17.** $x^2 - x + 2$ **19.** $2x^2 + x - 2$ **21.** $6y^2 + 3y - 8 + \frac{7}{2y - 3}$ **23.** $y^2 - 4y - 1 + \frac{-9}{y + 3}$ **25.** $2y^2 + y + 2 + \frac{1}{2y - 1}$ **27.** $y^3 - 3y^2 + 6y - 16 + \frac{47}{y + 3}$ **29.** $y^3 + 2y^2 - 5y - 10 + \frac{-25}{y - 2}$ **31.** 77,000 gallons **32.** Approximately 3.4 million

Chapter 4 Review Problems

1. $-18a^7$ **2.** 5^{23} **3.** $6x^4y^6$ **4.** $40x^5y^5$ **5.** 8^{17} **6.** $\frac{1}{7^{12}}$ **7.** $\frac{1}{x^4}$ **8.** w^5 **9.** $\frac{x^4}{3}$ **10.** $-\frac{3}{5x^5y^4}$ **11.** $-\frac{2a}{3b^6}$ **12.** $\frac{5a^3b}{3c}$ **13.** x^{24} **14.** $125x^3y^6$ **15.** $9a^6b^4$ **16.** $27a^6b^6$ **17.** $\frac{2x^4}{3y^2}$ **18.** $\frac{25a^2b^4}{c^6}$ **19.** $\frac{1}{2xy^2}$ **20.** $\frac{y^9}{64w^{15}z^6}$ **21.** $\frac{1}{x^3}$ **22.** $\frac{1}{x^6y^8}$ **23.** $\frac{2y^3}{x^6}$ **24.** $\frac{x^5}{2y^6}$ **25.** $\frac{1}{4x^6}$ **26.** $\frac{3y^2}{x^3}$ **27.** $\frac{4w^2}{x^5y^6z^8}$ **28.** $\frac{b^5c^3d^4}{27a^2}$ **29.** 1.563402×10^{11} **30.** 8.8367×10^4 **31.** 1.79632×10^5 **32.** 7.8×10^{-3} **33.** 6.173×10^{-5} **34.** 1.3×10^{-1} **35.** 120,000 **36.** 83,670,000,000 **37.** 3,000,000 **38.** 0.25 **39.** 0.00000005708 **40.** 0.000000006 **41.** 2.0×10^{13} **42.** 9.36×10^{19} **43.** 9.6×10^{-10} **44.** 7.8×10^{-11} **45.** 3.504×10^8 kilometers **46.** 7.94×10^{14} cycles **47.** 6×10^9 **48.** $-5x^2 - 11x - 18$ **49.** $-5x^2 - 16x - 12$ **50.** $-x^3 + 2x^2 - x + 8$ **51.** $7x^3 - 3x^2 - 6x + 4$ **52.** $-11x^2 + 7xy - 20y^2$ **53.** $3x^2 - 6y + 20xy$ **54.** $5x^3y^3 - 2x^2y^2 + 10xy - 4$ **55.** $11x^2y^2 + xy - 2y - 3y^2$ **56.** $-x^2 - 2x + 10$ **57.** $-6x^2 - 6x - 3$ **58.** $15x^2 + 2x - 1$ **59.** $32x^2 - 32x + 6$ **60.** $20x^2 + 48x + 27$ **61.** $10x^3 - 30x^2 + 15x$ **62.** $-6x^3y^3 + 10x^2y^2 - 12xy$ **63.** $4x^4 - 12x^3 + 20x^2 - 8x$ **64.** $5a^2 - 8ab - 21b^2$ **65.** $8x^4 - 12x^2 - 10x^2y + 15y$ **66.** $-15x^6y^2 - 9x^4y + 6x^2y$ **67.** $9x^2 - 12x + 4$ **68.** $25x^2 - 9$ **69.** $49x^2 - 36y^2$ **70.** $25a^2 - 20ab + 4b^2$ **71.** $64x^2 + 144xy + 81y^2$ **72.** $4x^3 + 27x^2 + 5x - 3$ **73.** $2x^4 + x^3 - 8x^2 - 5x + 12$ **74.** $2x^3 - 7x^2 - 42x + 72$ **75.** $8x^3 + 36x^2 + 54x + 27$ **76.** $2y^2 + 3y + 4$ **77.** $6x^3 + 7x^2 - 18x$ **78.** $4x^2y - 6x + 8y$ **79.** $53x^3 - 12x^2 + 19x + 13$ **80.** $3x + 7$ **81.** $4x + 1$ **82.** $2x^2 - 5x + 13 + \frac{-27}{x + 2}$ **83.** $3x^2 + 2x + 4 + \frac{9}{2x - 1}$ **84.** $4x - 5 + \frac{1}{2x + 1}$ **85.** $4x + 1$ **86.** $x^2 + 3x + 8$ **87.** $2x^2 + 4x + 5 + \frac{11}{x - 2}$ **88.** $3x^3 - x^2 + x - 5$ **89.** $4x^2 + 5x + 1$

Chapter 4 Test

1. 3^{34} **2.** 3^{28} **3.** 8^{24} **4.** $6x^7y^7$ **5.** $\frac{-7x^3}{5}$ **6.** $-125x^3y^{18}$ **7.** $\frac{49a^{14}b^4}{9}$ **8.** $\frac{1}{4a^5b^6}$ **9.** $\frac{1}{125}$ **10.** $\frac{6c^5}{a^4b^3}$ **11.** $\frac{2w^6}{x^3y^4z^8}$ **12.** 1.765×10^{-7} **13.** 582,000,000 **14.** 2.4×10^{-6} **15.** $-2x^2 + 5x$ **16.** $3x^2 + 3xy - 6y - 7y^2$ **17.** $-21x^5 + 28x^4 - 42x^3 + 14x^2$ **18.** $15x^4y^3 - 18x^3y^2 + 6x^2y$ **19.** $10a^2 + 7ab - 12b^2$ **20.** $6x^3 - 11x^2 - 19x - 6$ **21.** $49x^4 + 28x^2y^2 + 4y^4$ **22.** $81x^2 - 4y^2$ **23.** $12x^4 - 14x^3 + 25x^2 - 29x + 10$ **24.** $3x^4 + 4x^3y - 15x^2y^2$ **25.** $3x^3 - x + 5$ **26.** $2y^2 + 3y - 1$

Cumulative Test for Chapters 0–4

1. $-\frac{11}{24}$ **2.** -0.74 **3.** $-\frac{6}{7}$ **4.** \$8.96 **5.** $2x^2 - 13x$ **6.** 35 **7.** $x = -\frac{9}{2}$ **8.** $x = 44$ **9.** $x > -1$ **10.** $f = \frac{2B}{3a} - \frac{c}{3}$ or $f = \frac{2B - ac}{3a}$ **11.** 12,400 employees **12.** \$199.20 **13.** $15x^2 - 47x + 28$ **14.** $9x^2 - 30x + 25$ **15.** $8x^3 + 12x^2 + 6x + 1$ **16.** $-20x^5y^8$ **17.** $\frac{-2x^3}{3y^9}$ **18.** $-27x^3y^{12}z^6$ **19.** $\frac{9z^8}{w^2x^3y^4}$ **20.** 1.36×10^{15} **21.** 4.0×10^{-35} **22.** $5x^3 + 7x^2 - 6x + 50$ **23.** $-36x^3y^2 + 18x^2y^3 - 48xy^4$ **24.** $2x^4 - 15x^3 + 28x^2 - 33x + 12$

Chapter 5

Pretest Chapter 5

1. $2x(x - 3y + 6y^2)$ **2.** $(3x + 4y)(a - 2b)$ **3.** $18ab(2b - 1)$ **4.** $(a - 2b)(5 - 3x)$ **5.** $(3x - 4)(x + y)$ **6.** $(7x - 3)(3x - 2)$ **7.** $(x - 24)(x + 2)$ **8.** $(x - 5)(x - 3)$ **9.** $(x + 5)(x + 1)$ **10.** $2(x + 6)(x - 2)$ **11.** $3(x - 9)(x + 7)$ **12.** $(5x - 2)(3x - 2)$ **13.** $(3y - 2z)(2y + 3z)$ **14.** $4(3x + 5)(x + 2)$ **15.** $(9x^2 + 4)(3x + 2)(3x - 2)$ **16.** $(7x - 2y)^2$ **17.** $(5x + 8)^2$ **18.** $3x(2x - 1)(x + 3)$ **19.** $2y^2(4x - 3)^2$ **20.** Cannot be factored **21.** $x = 1, x = -\frac{3}{2}$ **22.** $x = 5, x = 6$ **23.** $x = -\frac{2}{3}, x = 3$

5.1 Exercises

5. $5b(a - 1)$ **7.** $9wz(2 - 3w)$ **9.** $2x(4x^2 - 5x - 7)$ **11.** $6(2xy - 3yz - 6xz)$ **13.** $a(a^3y - a^2y^2 + ay^3 - 2)$ **15.** $2x^5(3x^4 - 4x^2 + 2)$ **17.** $9ab(ab - 4)$ **19.** $8a(5a - 2b - 3)$ **21.** $17x(2x - 3y - 1)$ **23.** $6xy(3y + x + 4 - 2xy)$ **25.** $(x - 2y)(5 - z)$ **27.** $(x - 7)(5x + 3)$ **29.** $(3y + 5z)(7x - 6t)$ **31.** $(bc - 1)(5a + b + c)$ **33.** $(bc - 3a)(3c - 2 - 6b)$ **35.** $(x - 2y)(3x^2 - 1)$ **37.** $(9x + 5y)(2ab + 1)$ **39.** $y(5a - 3b + 1)(3x^2 - 19x - 1)$ **41.** $57{,}982(3x^2 - 6y^2 - 11z^2)$ **43.** $C = 25.95(a + b + c + d)$ **44.** $4x^2 - 9$ **45.** $16a^2 - 8ab + b^2$ **46.** 17, 19, 21 **47.** \$26,000 **48.** \$165 billion **49.** 2.24 ft/min

5.2 Exercises

1. $(x - 2)(3 + y)$ **3.** $(a + 3b)(2x - y)$ **5.** $(x - 4)(x^2 + 3)$ **7.** $(3a + b)(x - 2)$ **9.** $(a + 2b)(5 + 6c)$ **11.** $(a - b)(5 - 2x)$ **13.** $(y - 2)(y - 3)$ **15.** $(x + y)(3a - 1)$ **17.** $(3x + y)(2a - 1)$ **19.** $(x + 6)(x - 8)$ **21.** $(4x + 3w)(7x + 2y^2)$ **23.** $(4a - 3b)(2a + 5e)$ **25.** $(4x + 7w)(11x^2 - 9y^2)$ **28.** $121x^2 - 49y^2$ **29.** trinomial **30.** 8 seconds **31.** \$420 profit per car

5.3 Exercises

3. $(x + 1)(x + 8)$ **5.** $(x + 7)(x + 5)$ **7.** $(x - 3)(x - 1)$ **9.** $(x - 11)(x - 2)$ **11.** $(x + 4)(x - 3)$ **13.** $(x - 14)(x + 1)$ **15.** $(x + 5)(x - 4)$ **17.** $(x - 8)(x + 3)$ **19.** $(x - 2)(x - 5)$ **21.** $(x - 7)(x + 2)$ **23.** $(x + 7)(x - 2)$ **25.** $(x - 4)(x - 5)$ **27.** $(x + 15)(x + 2)$ **29.** $(y - 5)(y + 1)$ **31.** $(a + 8)(a - 2)$ **33.** $(x - 4)(x - 8)$ **35.** $(x + 7)(x - 3)$ **37.** $(x + 5)(x + 8)$ **39.** $(x - 22)(x + 1)$ **41.** $(x + 12)(x - 3)$ **43.** $(x - 7y)(x + 6y)$ **45.** $(x - 7y)(x - 9y)$ **47.** $(x^2 - 8)(x^2 + 5)$ **49.** $(x^2 - 15)(x^2 + 3)$ **51.** $2(x - 2)(x - 4)$ **53.** $3(x + 4)(x - 6)$ **55.** $4(x + 5)(x + 1)$ **57.** $7(x + 5)(x - 2)$ **59.** $6(x + 1)(x + 2)$ **61.** $2(x - 7)(x - 9)$ **63.** $\frac{A - P}{Pr}$ **64.** $x \geq -\frac{5}{3}$ **65.** 120 miles **66.** Sales must be at least \$130,000 **67.** 25°C **68.** June

$-\frac{5}{3}$

5.4 Exercises

1. $(3x + 1)(x + 2)$ **3.** $(2x - 1)(x - 2)$ **5.** $(3x - 7)(x + 1)$ **7.** $(2x + 1)(x - 3)$ **9.** $(5x - 2)(x + 1)$ **11.** $(3x - 2)(2x - 3)$ **13.** $(2x - 5)(x + 4)$ **15.** $(3x + 1)(3x + 2)$ **17.** $(3x + 2)(2x - 3)$ **19.** $(3x - 2)(2x - 5)$ **21.** $(2x - 1)(2x + 9)$ **23.** $(9y - 4)(y - 1)$ **25.** $(5a + 2)(a - 3)$ **27.** $(6x - 1)(2x - 3)$ **29.** $(5x - 2)(3x + 2)$ **31.** $(5x + 3)(2x + 3)$ **33.** $(6x + 1)(2x - 3)$ **35.** $(2x^2 - 1)(x^2 + 8)$ **37.** $(2x - y)(2x + 5y)$ **39.** $(5x - 4y)(x + 4y)$ **41.** $2(5x + 6)(x + 1)$ **43.** $3(2x - 1)(2x - 3)$ **45.** $3(3x - 1)(2x + 5)$ **47.** $2x(3x + 1)(x - 3)$ **49.** $(2x + 5)(6x - 7)$ **51.** $(4x - 3)(5x - 3)$ **53.** $\frac{1}{10}$ **54.** $(x + 9)(x - 4)$ **55.** 680 minutes **56.** 18.8 million children

5.5 Exercises

1. $(3x + 4)(3x - 4)$ **3.** $(4 - 3x)(4 + 3x)$ **5.** $(5x - 4)(5x + 4)$ **7.** $(2x - 5)(2x + 5)$ **9.** $(6x - 5)(6x + 5)$ **11.** $(1 - 7x)(1 + 7x)$ **13.** $(9x - 7y)(9x + 7y)$ **15.** $(5 + 11x)(5 - 11x)$ **17.** $(9x + 10y)(9x - 10y)$ **19.** $(5a + 1)(5a - 1)$ **21.** $(9x^2 - 2y)(9x^2 + 2y)$ **23.** $(7x + 1)^2$ **25.** $(y - 3)^2$ **27.** $(3x - 4)^2$ **29.** $(7x + 2)^2$ **31.** $(x + 5)^2$ **33.** $(7 - 5x)^2$ **35.** $(5x - 4)^2$ **37.** $(9x + 2y)^2$ **39.** $(3x - 7y)^2$ **41.** $(4a + 9b)^2$ **43.** $(3x^2 - y)^2$ **45.** $(7x + 1)(7x + 9)$ **47.** $(4x^2 + 1)(2x + 1)(2x - 1)$ **49.** $(x^5 + 6y^5)(x^5 - 6y^5)$ **51.** $(2x^4 + 3)^2$ **53.** Because no matter what combination you try, you cannot multiply two binomials to equal $9x^2 + 1$. **55.** There is one answer. It is 49. **57.** $3(3x - 1)(3x + 1)$ **59.** $3(7x - y)(7x + y)$ **61.** $3(2x - 3)(2x - 3)$ **63.** $2(7x + 3)(7x + 3)$ **65.** $(x - 2)(x - 7)$ **67.** $(2x - 1)(x + 3)$ **69.** $(4x - 11)(4x + 11)$ **71.** $(3x + 7)^2$ **73.** $3(x + 5)(x - 3)$ **75.** $7(x - 3)(x + 3)$ **77.** $5(x + 2)^2$ **79.** $2(x - 9)(x - 7)$ **81.** $2x^3 + 7x^2 - 10x - 24$ **82.** $x > \frac{5}{2}$ **83.** $x^2 + 3x + 4 + \frac{-3}{x - 2}$ **84.** $2x^2 + x - 5$ **85.** 3838 ft above sea level **86.** 5 miles

5.6 Exercises

1. $3(3x - 2y + 5)$ **3.** $9(2x - y)(2x + y)$ **5.** $(3x - 2y)^2$ **7.** $(x + 5)(x + 3)$ **9.** $(5x - 3)(x - 2)$ **11.** $(x - 3y)(a + 2b)$ **13.** $3(x^2 + 2)(x^2 - 2)$ **15.** $(2x - 3)^2$ **17.** $(2x - 3)(x - 4)$ **19.** $(x - 10y)(x + 7y)$ **21.** $(a - 4)(x - 5)$ **23.** $5x(3 - x)(3 + x)$ **25.** $5xy^3(x - 1)^2$ **27.** $3xy(3z + 2)(3z - 2)$ **29.** $3(x + 7)(x - 5)$ **31.** $5(x - 2)(x - 4)$ **33.** $-1(2x^2 + 1)(x + 2)(x - 2)$ **35.** prime **37.** $3(x - 2y)(x - 3y)$ **39.** $3x(2x + y)(5x - 2y)$ **41.** $4(2x - 1)(x + 4)$ **43.** $11x^2(3x - 1)(3x + 1)$ **45.** $a(x + 2y)(a - 3)$ **47.** prime **48.** The number d must be a perfect square such as 1, 4, 9, 16, 25, 36, 49, 64, 81, 100, . . . **49.** \$28,000 **50.** 372 live strains

5.7 Exercises

1. 5, −2 **3.** −5, −8 **5.** $-\frac{1}{2}, 3$ **7.** $\frac{3}{2}, 2$ **9.** 8, −2 **11.** 2, 5 **13.** $\frac{2}{3}, \frac{3}{2}$ **15.** 0, 8 **17.** 3, −3 **19.** 0, 1 **21.** −5, 4 **23.** 6 **25.** $-\frac{1}{2}$ **27.** 0, −2 **29.** 9, −6 **31.** $-\frac{5}{3}, 3$ **33.** You can always factor out x. **34.** You must have the product of two or more factors equal to zero, not 14. Remember the zero factor property. **35.** $L = 14$ m; $W = 10$ m **37.** 4; 14

39. 12 meters above ground after 2 sec **41.** 5 additional helicopters **43.** $-12x^3y^7$ **44.** $12a^{10}b^{13}$ **45.** $-\frac{3a^4}{2b^2}$ **46.** $\frac{1}{3x^5y^4}$

Putting Your Skills to Work

1. \$765,000 **2.** \$3,160,000

Chapter 5 Review Problems

1. $3x^2y(5x - 3y)$ **2.** $7x(3x - 2)$ **3.** $7xy(x - 2y - 3x^2y^2)$ **4.** $25a^4b^4(2b - 1 + 3ab)$ **5.** $9x^2(3x - 1)$ **6.** $2(x - 2y + 3z + 6)$
7. $(a + 3b)(2a - 5)$ **8.** $3xy(5x^2 + 2y + 1)$ **9.** $(3x - 7)(a - 2)$ **10.** $(a + 5b)(a - 4)$ **11.** $(x^2 + 3)(y - 2)$ **12.** $(4a - 5)(d + 5c)$
13. $(5x - 1)(3x + 2)$ **14.** $(5w - 3)(6w + z)$ **15.** $(x - 6)(x + 3)$ **16.** $(x - 4)(x - 6)$ **17.** $(x + 6)(x + 8)$ **18.** $(x + 3y)(x + 5y)$
19. $(x^2 + 7)(x^2 + 6)$ **20.** $(x + 3)(x - 13)$ **21.** $5(x + 1)(x + 3)$ **22.** $3(x + 1)(x + 12)$ **23.** $2(x - 6)(x - 8)$ **24.** $4(x - 5)(x - 6)$
25. $(4x - 5)(x + 3)$ **26.** $(3x - 1)(4x + 5)$ **27.** $(5x + 4)(3x - 1)$ **28.** $(3x - 2)(2x - 3)$ **29.** $(2x - 3)(x + 1)$ **30.** $(3x - 1)(x + 5)$
31. $(10x - 1)(2x + 5)$ **32.** $(5x - 1)(4x + 5)$ **33.** $(4a + 1)(a - 3)$ **34.** $(4a + 3)(a - 1)$ **35.** $2(x - 1)(3x + 5)$ **36.** $2(x + 1)(3x - 5)$
37. $2(2x - 3)(x - 5)$ **38.** $4(x - 9)(x + 4)$ **39.** $2(x - 1)(5x - 6)$ **40.** $(2x + 11)(x - 4)$ **41.** $(3x - 1)(4x + 3)$ **42.** $2(x - 1)(8x + 15)$
43. $(3x - 2y)(2x - 5y)$ **44.** $2(3x - y)(x - 5y)$ **45.** $(3x^2 - 8)(x^2 + 1)$ **46.** $(3x^2 - 2)(x^2 + 4)$ **47.** $(7x + y)(7x - y)$
48. $4(2x - 3y)(2x + 3y)$ **49.** $(3x - 2)^2$ **50.** $(4x - 1)(4x + 1)$ **51.** $(5x - 6)(5x + 6)$ **52.** $(10x - 3)(10x + 3)$ **53.** $(1 - 7x)(1 + 7x)$
54. $(2 - 7x)(2 + 7x)$ **55.** $(6x + 1)^2$ **56.** $(5x - 2)^2$ **57.** $(4x - 3y)^2$ **58.** $(7x - 2y)^2$ **59.** $2(x - 3)(x + 3)$ **60.** $3(x - 5)(x + 5)$
61. $2(2x + 5)^2$ **62.** $2(5x + 6)^2$ **63.** $(2x + 3y)(2x - 3y)$ **64.** $(x + 3)^2$ **65.** $(x - 3)(x - 6)$ **66.** $(x + 15)(x - 2)$
67. $(x - 1)(6x + 7)$ **68.** $(5x - 2)(2x + 1)$ **69.** $4(3x + 4)$ **70.** $4xy(2xy - 1)$ **71.** $10x^2y^2(5x + 2)$ **72.** $13ab(2a^2 - b^2 + 4ab^3)$
73. $x(x - 8)^2$ **74.** $2(x + 10)^2$ **75.** $3(x - 3)^2$ **76.** $x(5x - 6)^2$ **77.** $(7x + 10)(x - 1)$ **78.** $(5x + 2)(x - 2)$
79. $xy(3x + 2y)(3x - 2y)$ **80.** $x^3a(3a + 4x)(a - 5x)$ **81.** $2(3a + 5b)(2a - b)$ **82.** $(4a - 5b)^2$ **83.** $(a - 1)(7 - b)$ **84.** $(3d - 4)(1 - c)$
85. $(2x - 1)(1 + b)$ **83.** $(b - 7)(5x + 4y)$ **87.** $x(2a - 1)(a - 7)$ **88.** $x(x + 4)(x - 4)(x + 1)(x - 1)$ **89.** $(x^2 + 9y^6)(x + 3y^3)(x - 3y^3)$
90. $(3x^2 - 5)(2x^2 + 3)$ **91.** $yz(14 - x)(2 - x)$ **92.** $x(3x + 2)(4x + 3)$ **93.** $(2w + 1)(8w - 5)$ **94.** $3(2w - 1)^2$
95. $2y(2y - 1)(y + 3)$ **96.** $(5y - 1)(2y + 7)$ **97.** $8y^8(y^2 - 2)$ **98.** $49(x^2 + 1)(x + 1)(x - 1)$ **99.** prime **100.** prime
101. $4y(2y^2 - 5)(y^2 + 3)$ **102.** $3x(3y + 7)(y - 2)$ **103.** $(4x^2y - 7)^2$ **104.** $2xy(8x + 1)(8x - 1)$ **105.** $(2x + 5)(a - 2b)$
106. $(2x + 1)(x + 3)(x - 3)$ **107.** $-3, 6$ **108.** $2, -5$ **109.** $0, \frac{1}{6}$ **110.** $0, -\frac{11}{6}$ **111.** $-5, \frac{1}{2}$ **112.** $-8, -3$ **113.** $-5, -9$
114. $-\frac{3}{5}, 2$ **115.** -3 **116.** $-3, \frac{3}{4}$ **117.** $\frac{1}{5}, 2$ **118.** base = 10 cm, altitude = 7 cm **119.** width = 7 feet, length = 15 feet
120. 6 seconds **121.** 8 amperes, 12 amperes

Putting Your Skills to Work

1. 2.4 million students **2.** Approximately 2.8%

Chapter 5 Test

1. $(x + 14)(x - 2)$ **2.** $(3x + 10y)(3x - 10y)$ **3.** $(5x + 1)(2x + 5)$ **4.** $(3a - 5b)^2$ **5.** $x(7 - 9x + 14y)$ **6.** $(x + 2y)(3x - 2w)$
7. $2x(3x - 4)(x - 2)$ **8.** $c(5a - b)(a - 2b)$ **9.** $4(5x^2 + 2y^2)(5x^2 - 2y^2)$ **10.** $(3x - y)(3x - 4y)$ **11.** $7x(x - 6)$ **12.** prime
13. prime **14.** $-5y(2x - 3y)^2$ **15.** $(4x + 1)(4x - 1)$ **16.** $(x^8 + 1)(x^4 + 1)(x^2 + 1)(x + 1)(x - 1)$ **17.** $(x + 3)(2a - 5)$
18. $(a + 2b)(w + 2)(w - 2)$ **19.** $3(x - 6)(x + 5)$ **20.** $x(2x + 5)(x - 3)$ **21.** $-5, -9$ **22.** $-\frac{7}{3}, -2$ **23.** $-\frac{5}{2}, 2$
24. width = 7 miles, length = 13 miles

Cumulative Test for Chapters 0–5

1. 15% **2.** 0.494 **3.** -10.36 **4.** $8x^4y^{10}$ **5.** -64 **6.** $14x^2 + x - 3$ **7.** $2x^3 - 12x^2 + 19x - 3$ **8.** $x \leq -3$ **9.** 2
10. -15 **11.** $t = \frac{2s - 2a}{3}$ **12.** $(3x - 1)(2x - 1)$ **13.** $(6x + 1)(x - 1)$ **14.** $(3x + 2)(3x - 1)$ **15.** $(11x + 8y)(11x - 8y)$
16. $-4(5x + 6)(4x - 5)$ **17.** prime **18.** $x(4x + 5)^2$ **19.** $(9x^2 + 4b^2)(3x + 2b)(3x - 2b)$ **20.** $(2x + 3)(a - 2b)$
21. $(x^2 + 5)(x^2 + 3)$ **22.** $-8, 3$ **23.** $\frac{5}{3}, 2$ **24.** base = 6 miles, altitude = 19 miles

Chapter 6

Pretest Chapter 6

1. -2 **2.** $\frac{3x - 2}{4x + 3}$ **3.** $\frac{b}{2a - b}$ **4.** $\frac{2a - b}{6}$ **5.** $\frac{x}{x + 5}$ **6.** $\frac{1}{x + 3}$ **7.** $\frac{y}{x - 1}$ **8.** 3 **9.** $\frac{-2y - 7}{(y - 1)(2y + 3)}$ **10.** $\frac{1}{x + 5}$
11. $\frac{5x^2 + 6x - 9}{3x(x - 3)^2}$ **12.** $\frac{2a - 3}{5a^2 + a}$ **13.** $\frac{a^2 - 2a - 2}{3a^2(a + 1)}$ **14.** $\frac{-x^2 - y^2}{x^2 + 2xy - y^2}$ **15.** $x = -5$ **16.** $x = 7$ **17.** 9.3 **18.** \$263.50

6.1 Exercises

1. 3 **3.** $\frac{6}{x}$ **5.** $\frac{2}{x - 4}$ **7.** $x - 3y$ **9.** $\frac{3x}{x - 3y}$ **11.** $\frac{x + 2}{x}$ **13.** $\frac{x + 3}{4x - 1}$ **15.** $\frac{x(x - 4)}{x + 6}$ **17.** $\frac{3x - 2}{x + 4}$ **19.** $\frac{3x - 5}{4x - 1}$
21. $\frac{x - 4}{x - 2}$ **23.** $\frac{x + 5}{x - 3}$ **25.** $\frac{-3}{2}$ **27.** $\frac{-2x - 3}{x + 5}$ **29.** $\frac{4x + 5}{2x - 1}$ **31.** $\frac{-2y - 3}{y + 2}$ **33.** $\frac{a - b}{2a - b}$ **35.** $\frac{3x - 2y}{3x + 2y}$ **37.** $\frac{x(x - 2)}{2(x + 3)}$
39. $9x^2 - 42x + 49$ **40.** $100x^2 - 81y^2$ **41.** $2x^3 - 9x^2 - 2x + 24$ **42.** $2x^4 - 5x^3 - 10x^2 - 14x - 15$ **43.** $1\frac{5}{8}$ acre per lot
44. 6 hours, 25 minutes

6.2 Exercises

3. $\frac{x+5}{x+7}$ **5.** $\frac{x^2}{2(x-2)}$ **7.** $\frac{x-2}{x+3}$ **9.** $\frac{(x+8)(x+7)}{7x(x-3)}$ **11.** $\frac{x+1}{(x+2)(x+3)}$ **13.** $\frac{3(x+2y)}{4(x+3y)}$ **15.** $\frac{y(x+y)}{(x+1)^2}$ **17.** $\frac{-3(x-2)}{(x+5)}$
19. $\frac{(x+5)(x-2)}{3x-1}$ **21.** 1 **23.** $\frac{(2x+1)(4x-1)}{(x+1)(x-6)}$ **25.** $\frac{3}{2}$ **26.** $7x^3 - 22x^2 + 2x + 3$ **27.** 5 milligrams of medication
28. \$2.38

6.3 Exercises

1. $\frac{2}{3x^3}$ **3.** $\frac{x+8}{x+2}$ **5.** $\frac{-1}{x-4}$ **7.** $\frac{2x-7}{5x+7}$ **9.** $\frac{2x+7}{2x-7}$ **11.** $5a^3$ **13.** $120x^4y^5$ **15.** $x^2 - 16$ **17.** $(x+2)(4x-1)^2$
19. $\frac{8y+2}{xy^2}$ **21.** $\frac{2x+9}{(x+3)(x+2)}$ **23.** $\frac{4y}{(y+1)(y-1)}$ **25.** $\frac{10a+5b+2ab}{2ab(2a+b)}$ **27.** $\frac{x^2-3x+24}{4x^2}$ **29.** $\frac{6x+20}{(x+5)^2(x-5)}$
31. $\frac{2x}{(x+1)(x-1)}$ **33.** $\frac{a+5}{6}$ **35.** $\frac{-8z-7}{30z}$ **37.** $\frac{9x+2}{(3x-4)(4x-3)}$ **39.** $\frac{-4x-2}{x(x-2)^2}$ **41.** $\frac{x^2-5x-2}{(x+2)(x+3)(x-1)}$
43. $\frac{3a-3b+2ab+a^3-a^2b}{ab(a-b)}$ **45.** $\frac{9y}{y-3}$ **47.** $\frac{3}{y+4}$ **49.** $\frac{3x^2+3x+2}{(x+3)(x-3)}$ **51.** $\frac{(4y+1)(y-1)}{2y(3y+1)(y-3)}$ **53.** $\frac{9}{2(5x-1)}$
55. $\frac{6x}{y-7}$ **57.** $\frac{7}{3a-2}$ **58.** $x = -7$ **59.** $y = \frac{5ax+6bc}{2a}$ **60.** $x < \frac{3}{2}$ **61.** $81x^{12}y^{16}$ **62.** at least 17 days per month
63. Approx. 377,400 more people were unemployed.

Putting Your Skills to Work

1. Approx. 138.6 miles **2.** between 2000 rpm and 3000 rpm

6.4 Exercises

1. $\frac{3y}{2x}$ **3.** $y + x$ **5.** $\frac{y}{x}$ **7.** $\frac{x-3}{x}$ **9.** $\frac{-2x}{3(x+1)}$ **11.** $\frac{3a^2+9}{a^2+2}$ **13.** $\frac{a^2-a-14}{a^2-3a}$ or $\frac{a^2-a-14}{a(a-3)}$ **15.** $\frac{x(x-2)}{x^2+4}$
17. $\frac{2x-5}{3(x+3)}$ **19.** $\frac{y+1}{-y+1}$ **21.** $\frac{y-1}{2y^3+2y^2-y-1}$ **23.** $w = \frac{P-2l}{2}$ **24.** $x > -1$

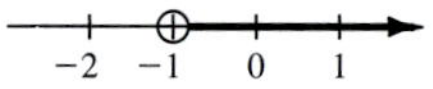

6.5 Exercises

1. 3 **3.** 2 **5.** $-\frac{3}{4}$ **7.** −2 **9.** 4 **11.** $\frac{-14}{3}$ **13.** There is no solution. **15.** −3 **17.** −5 **19.** 12
23. There is no solution. **25.** There is no solution. **27.** 3 **29.** $\frac{12}{5}$ **31.** 7 **33.** 2, 4 **34.** −2, −3 **35.** $(3x+4)(2x-3)$
36. $\frac{-10}{3}$ **37.** width = 7 m, length = 20 m **38.** Monthly payment is \$946.45.

6.6 Exercises

1. 15 **3.** $40\frac{4}{5}$ **5.** $\frac{40}{3}$ **7.** 22.75 or $22\frac{3}{4}$ **9.** 126 books **11.** 19.25 gallons **13.** **(a)** 522.88 British pounds **(b)** \$214.20
15. 56 miles per hour **17.** 46 gallons **19.** 29 miles **21.** $18\frac{17}{20}$ inches **23.** $114\frac{2}{7}$ meters **25.** 48 inches **27.** 86 meters
29. 61.5 miles per hour **31.** The commuter airline flies at 250 kilometers per hour. The helicopter flies at 210 kilometers per hour.
33. $2\frac{2}{9}$ hours or 2 hours 13 minutes **35.** $8\frac{4}{7}$ hours or 8 hours 34 minutes **37.** 6.316×10^{-7} **38.** 582,000,000 **39.** $\frac{w^8}{x^3y^2z^4}$
40. $\frac{27}{8}$ or $3\frac{3}{8}$

Putting Your Skills to Work

1. 462 days more **2.** approx. 21 orbits

Chapter 6 Review Problems

1. $-\frac{4}{5}$ **2.** $\frac{x}{x-y}$ **3.** $\frac{x+3}{x-4}$ **4.** $\frac{2x+1}{x+1}$ **5.** $\frac{x+3}{x-7}$ **6.** $\frac{2(x+4)}{3}$ **7.** $\frac{2x+3}{2x}$ **8.** $\frac{x}{x-5}$ **9.** $\frac{2(x-4y)}{2x-y)}$ **10.** $\frac{2-y}{3y-1}$
11. $\frac{x-2}{5x^2+x-6}$ **12.** $4x + 2y$ **13.** $\frac{x-25}{x+5}$ **14.** $\frac{x-4}{4}$ **15.** $\frac{2(x+3)(x-3)}{3y(y+3)}$ **16.** $\frac{(y+2)(2y+1)(y-2)}{(y+1)(2y-1)(y+1)}$ **17.** $\frac{3(x+2)}{x(x+5)}$
18. $\frac{4}{5(x-5)}$ **19.** $\frac{4y(3y-1)}{(3y+1)(2y+5)}$ **20.** $\frac{3y(4x+3)}{2(x-5)}$ **21.** $\frac{(x+2)}{2}$ **22.** $\frac{3}{16}$ **23.** $\frac{1}{x(4x-3)}$ **24.** $\frac{2(x+3y)}{x-4y}$
25. $\frac{9x+2}{x(x+1)}$ **26.** $\frac{5x^2+7x+1}{x(x+1)}$ **27.** $\frac{(x-1)(x-2)}{(x+3)(x-3)}$ **28.** $\frac{10x-22}{(x+2)(x-4)}$ **29.** $\frac{2xy+4x+5y+6}{2y(y+2)}$ **30.** $\frac{(2a+b)(a+4b)}{ab(a+b)}$
31. $\frac{3x-2}{3x}$ **32.** $\frac{2x^2+7x-2}{2x(x+2)}$ **33.** $\frac{1-2x-x^2}{(x+5)(x+2)}$ **34.** $\frac{7+15x-3x^2}{(x-4)(x-5)}$ **35.** $\frac{3(2x-1)(x+1)}{(x+2)(x+3)(x-3)}$ **36.** $\frac{x^2+13x+1}{(x+2)(x-1)(x+4)}$

37. $\frac{1}{11}$ **38.** $\frac{5}{3x^2}$ **39.** $w - 2$ **40.** 1 **41.** $-\frac{y^2}{2}$ **42.** $\frac{x + 2y}{y(x + y + 2)}$ **43.** $\frac{-1}{a(a + b)}$ **44.** $\frac{-3a - b}{b}$ **45.** $\frac{-3y}{2(x + 2y)}$
46. $\frac{5y(x + 5y)^2}{x(x - 6y)}$ **47.** 15 **48.** 2 **49.** $\frac{2}{7}$ **50.** -8 **51.** -4 **52.** -2 **53.** $\frac{1}{5}$ **54.** $\frac{1}{2}$ **55.** $-4, 6$ **56.** no solution
57. 2 **58.** 6 **59.** no solution **60.** -2 **61.** 9 **62.** 0 **63.** 1.3 **64.** 2.8 **65.** 26.4 **66.** 2 **67.** 11 hours
68. 40 miles per hour **69.** 8.3 gallons **70.** 167 cookies **71.** 46 gallons **72.** 91.5 miles
73. train: 60 miles per hour; car: 40 miles per hour **74.** 3 hours 5 minutes **75.** 1200 feet tall **76.** 182 feet tall

Chapter 6 Test

1. $\frac{2}{3a}$ **2.** $\frac{2x^2(2 - y)}{(y + 2)}$ **3.** $\frac{5}{12}$ **4.** $\frac{1}{3b^2}$ **5.** $\frac{(a - 2)(4a + 7)}{(a + 2)^2}$ **6.** $\frac{23x - 64}{2(x - 2)(x - 3)}$ **7.** $\frac{x - a}{ax}$ **8.** $-\frac{x + 2}{x + 3}$ **9.** $\frac{x}{4}$
10. $\frac{x + 1}{x(x + 3)^2}$ **11.** $\frac{2x - 3y}{4x + y}$ **12.** $\frac{x}{(x + 2)(x + 4)}$ **13.** $x = 3$ **14.** $x = 4$ **15.** no solution **16.** $x = \frac{47}{6}$ **17.** $x = \frac{45}{13}$
18. 6 hours, 25 minutes **19.** \$368 **20.** 102 feet

Cumulative Test for Chapters 0–6

1. 0.006 cm **2.** \$250 **3.** \$49.98 **4.** 6 **5.** $h = \frac{A}{\pi r^2}$ **6.** $x > 1.25$ **7.** $x > 5$

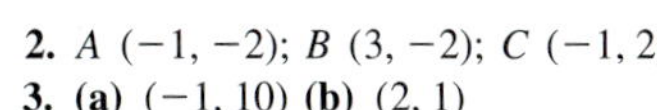

8. $(a + b)(3x - 2y)$ **9.** $2a(4a + b)(a - 5b)$ **10.** $-12x^5y^7$ **11.** $\frac{2x + 5}{x + 7}$ **12.** $3x + 1$ **13.** $\frac{11x - 3}{2(x + 2)(x - 3)}$
14. $\frac{x^2 + 2x + 6}{(x + 4)(x - 3)(x + 2)}$ **15.** -2 **16.** $\frac{-9}{2}$ **17.** $\frac{x + 8}{6x(x - 3)}$ **18.** $\frac{3ab^2 + 2a^2b}{5b^2 - 2a^2}$ **19.** $\frac{98}{5}$ **20.** 208 miles
21. 484 phone calls

Chapter 7

Pretest Chapter 7

1.

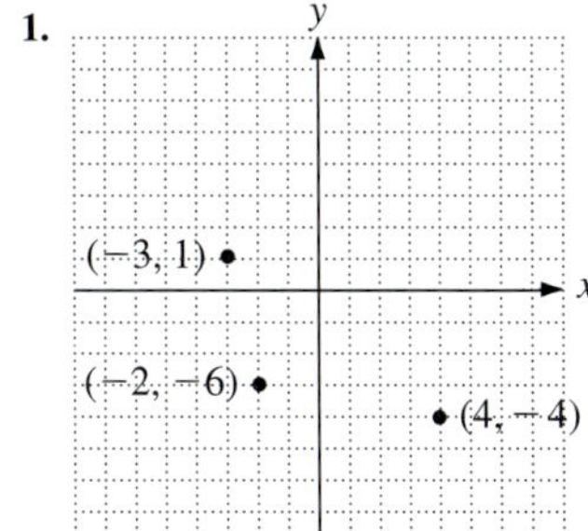

2. $A\ (-1, -2)$; $B\ (3, -2)$; $C\ (-1, 2)$
3. **(a)** $(-1, 10)$ **(b)** $(2, 1)$

4.

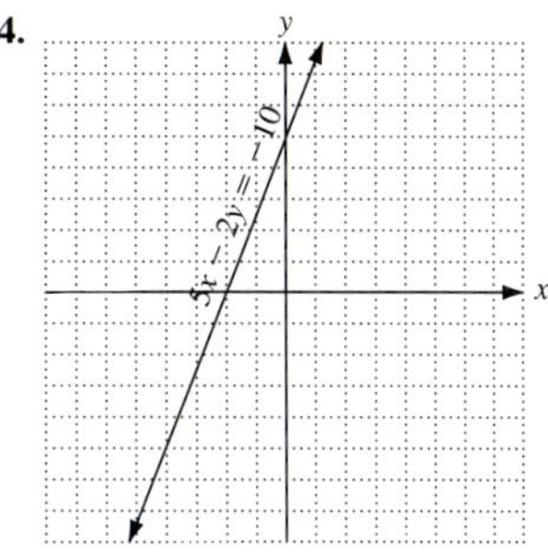

5.

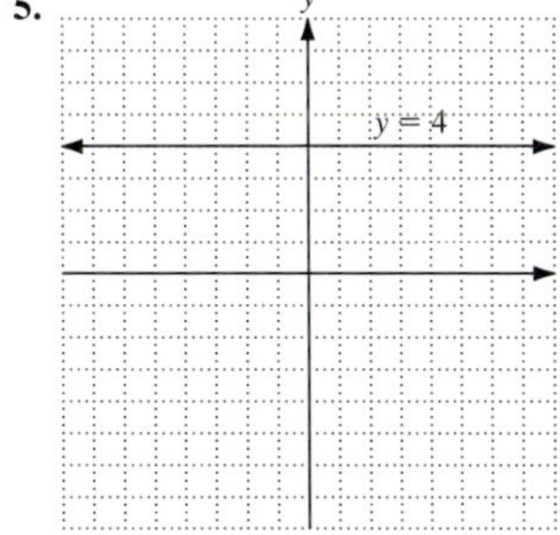

6. $\frac{2}{7}$ **7.** $\frac{4}{3}$
8. **(a)** $y = 4x - 5$ **(b)** $4x - y = 5$

9.

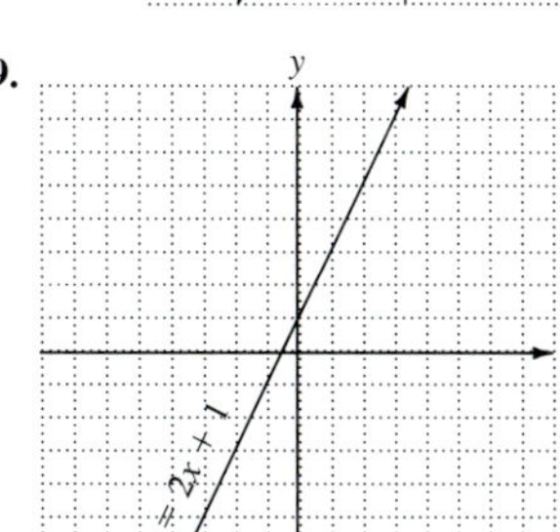

10. $y = -\frac{3}{2}x + 2$
11. **(a)** $-\frac{1}{4}$ **(b)** 4
12. $y = \frac{2}{3}x - 1$
13. $y = -x + 5$

14.

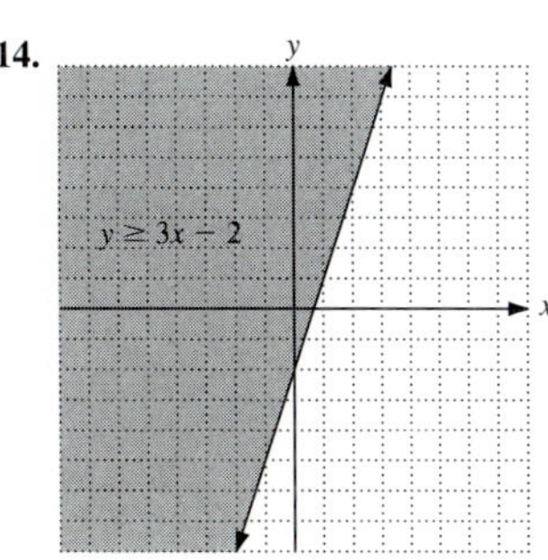

15.

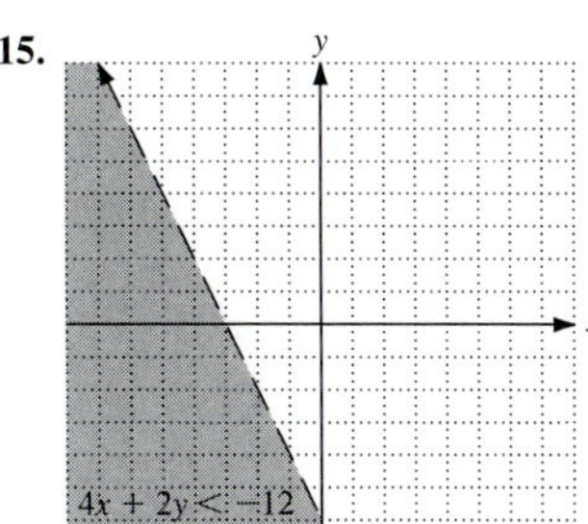

16. yes **17.** no **18.** **(a)** -13 **(b)** 5 **19.** **(a)** 72 **(b)** $\frac{1}{2}$

7.1 Exercises

1.

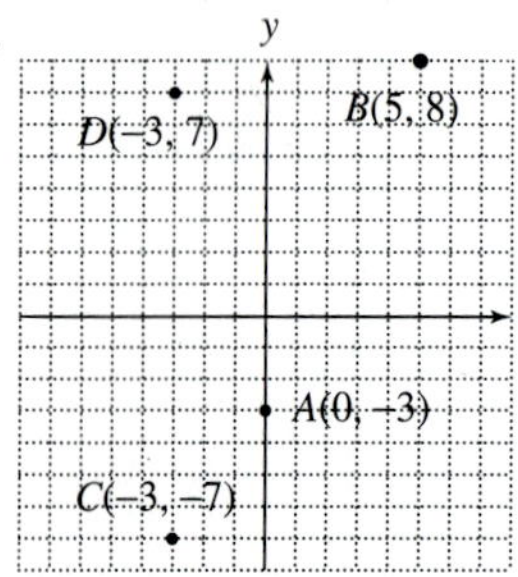

3.

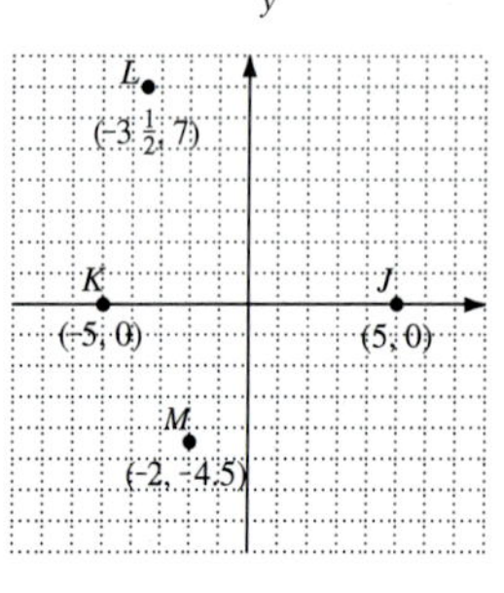

5. R: $(-3, -5)$; S: $\left(-4\frac{1}{2}, 0\right)$; X: $(3, -5)$, Y: $\left(2\frac{1}{2}, 6\right)$

13. **(a)** $(0, 8)$ **(b)** $(4, 20)$

15. **(a)** $(-6, 15)$ **(b)** $(3, -3)$

17. **(a)** $(8, 0)$ **(b)** $(-12, -5)$

19. **(a)** $(-7, 3)$ **(b)** $(-2, -5)$

21. $\dfrac{8x + 8y}{x + y} = 8$ **22.** $\dfrac{(x + y)(x - y)}{(x + y)(x + 3y)} \cdot \dfrac{(x + 3y)(x - 3y)}{(x - 3y)(x - 3y)} = \dfrac{x - y}{x - 3y}$ **23.** Use $\pi \approx 3.14$ Area is approx. 1133.54 square yards.

24. 922 members

7.2 Exercises

5. $(0, 1)$, $(-2, 5)$, $(1, -1)$

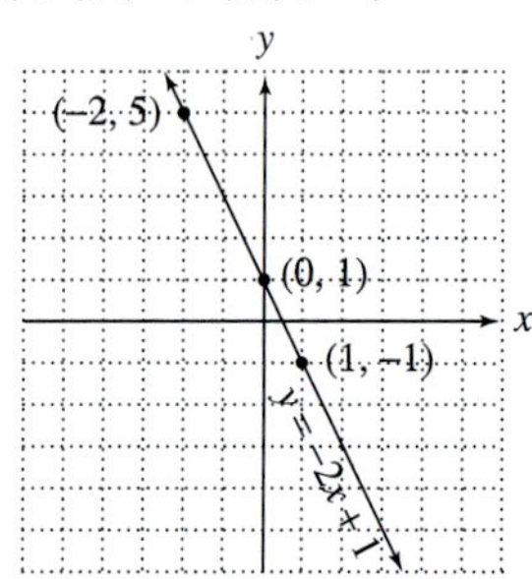

7. $(0, -4)$, $(2, -2)$, $(4, 0)$

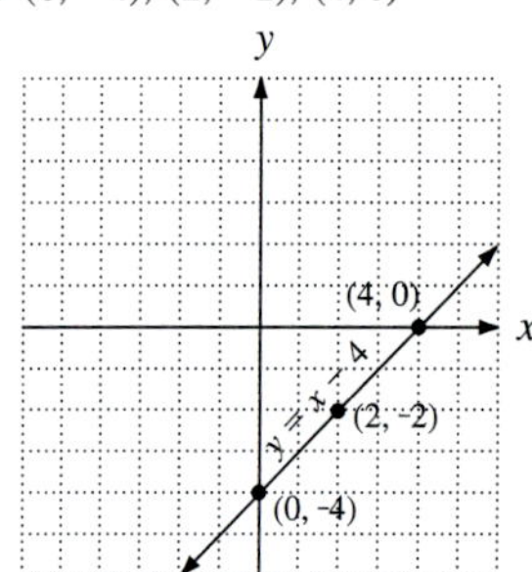

9.

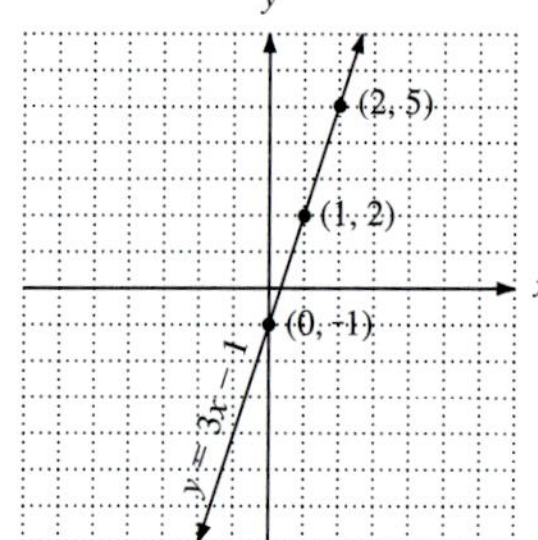

11.

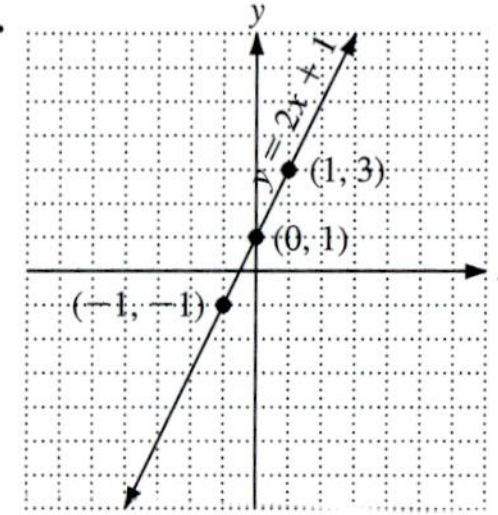

13.

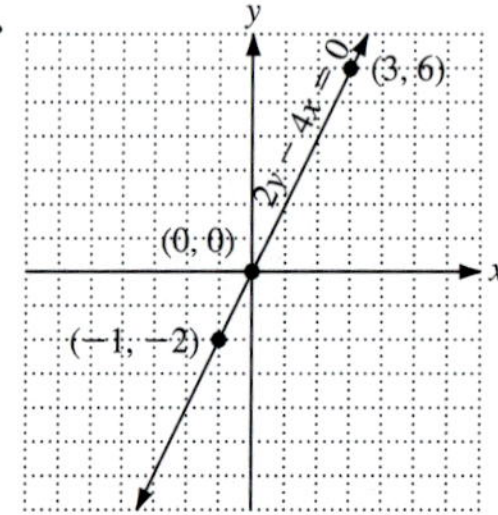

15.

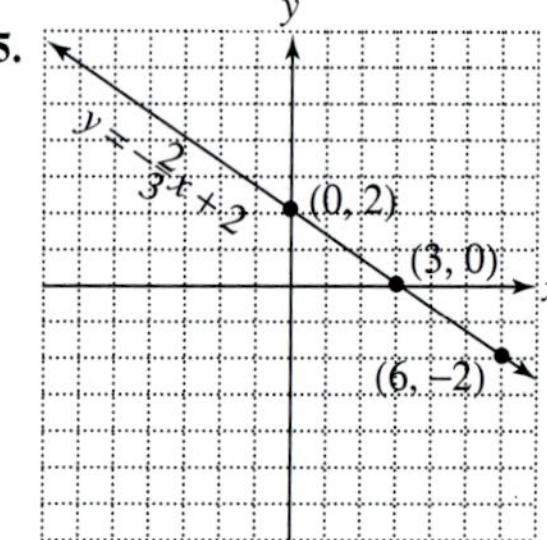

17.

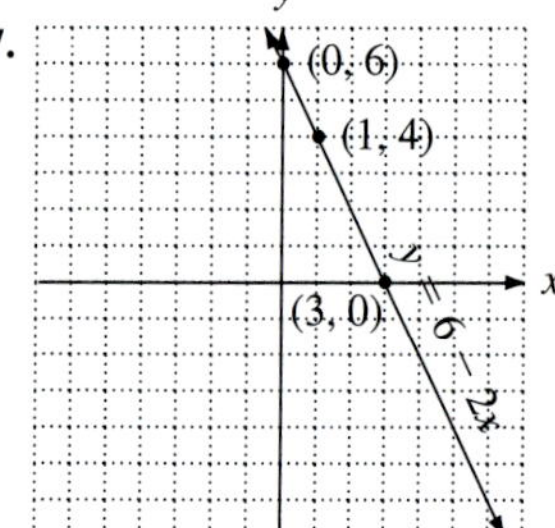

19.

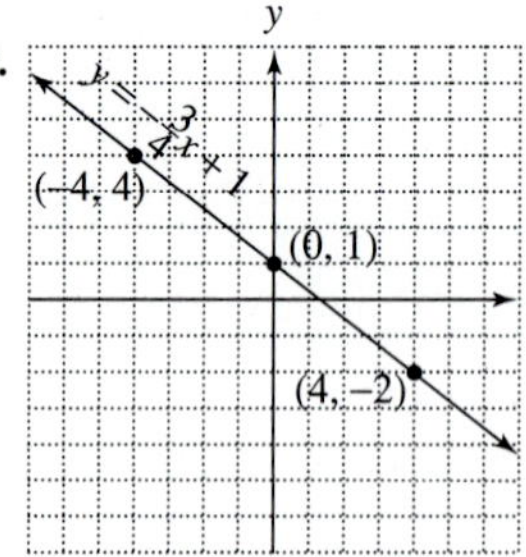

21.

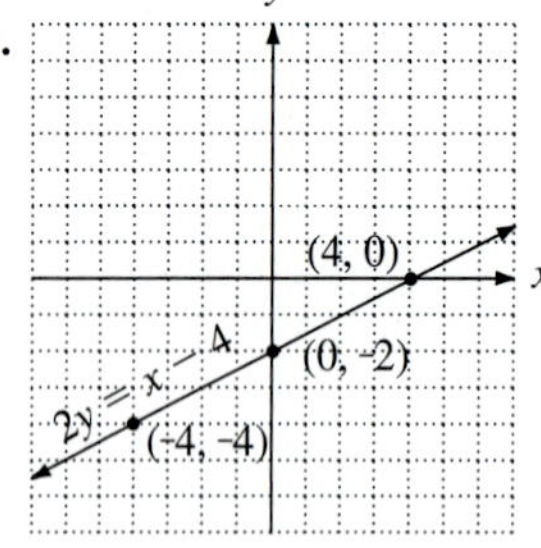

23.

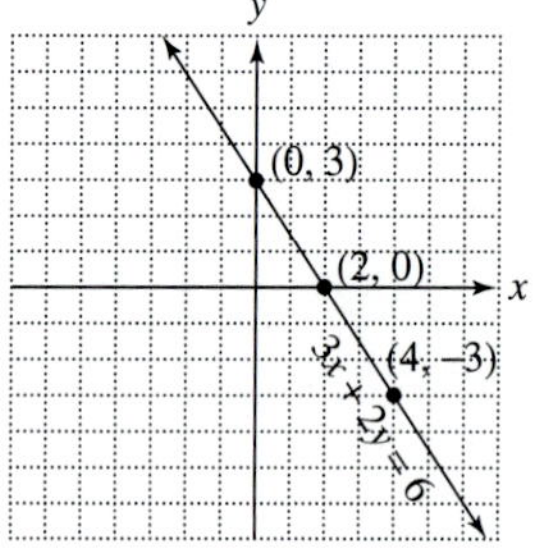

25.

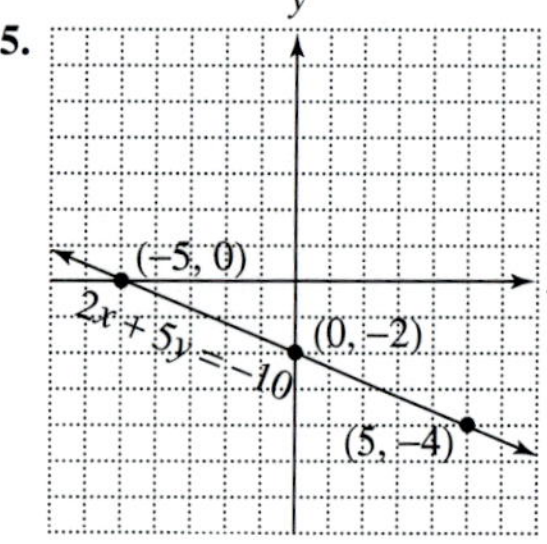

27.

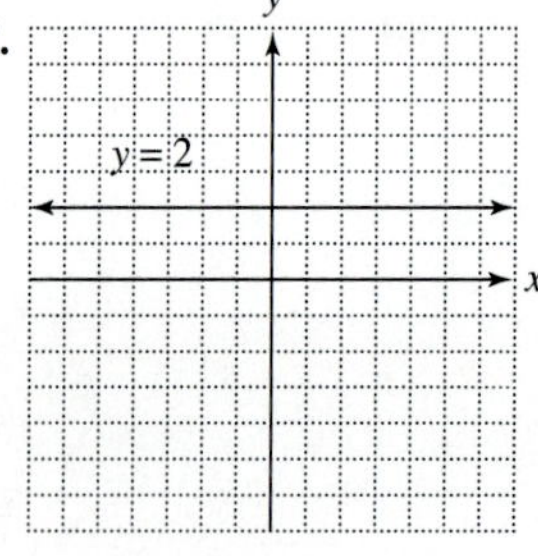

29.

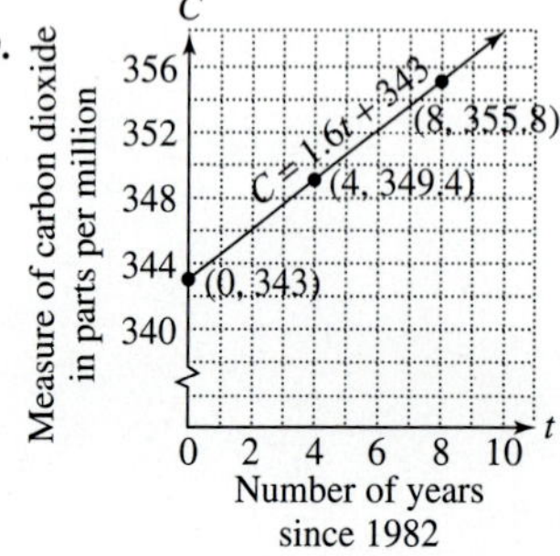

31. $x = 28$

32. $x \geq -\dfrac{14}{3}$

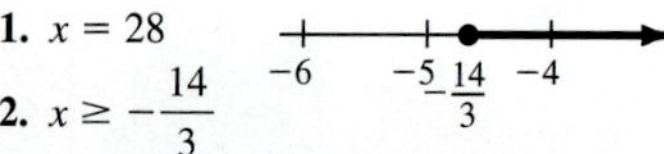

33. $L = 17$ meters; $W = 9.5$ meters

34. 212 faculty members

7.3 Exercises

1. -6 **3.** $\frac{3}{5}$ **5.** $-\frac{2}{5}$ **7.** 0 **9.** $-\frac{4}{3}$ **11.** -9 **15.** $m = 8; b = 9$ **17.** $m = 3; b = -7$ **19.** $m = \frac{5}{6}; b = -\frac{2}{9}$ **21.** $m = -6; b = 0$ **23.** $m = -4; b = -\frac{1}{3}$ **25.** $m = 3; b = 2$ **27.** $m = -\frac{5}{2}; b = \frac{3}{2}$ **29.** $m = \frac{7}{3}; b = -\frac{4}{3}$ **31. (a)** $y = \frac{3}{4}x + 2$ **(b)** $3x - 4y = -8$ **33. (a)** $y = 6x - 3$ **(b)** $6x - y = 3$ **35. (a)** $y = \frac{-5}{2}x + 6$ **(b)** $5x + 2y = 12$ **37. (a)** $y = -2x + \frac{1}{2}$ **(b)** $4x + 2y = 1$

39.

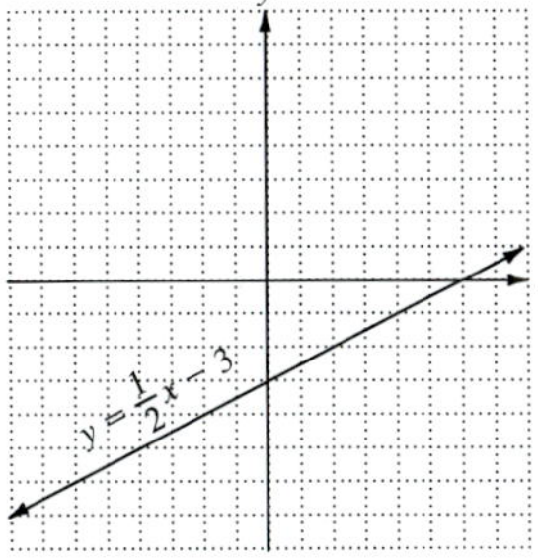

41.

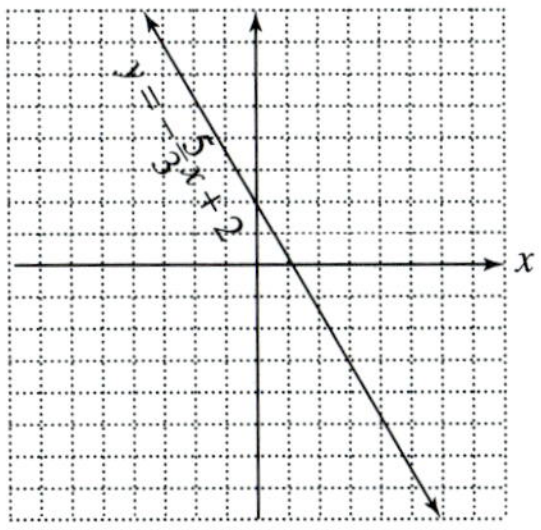

43.

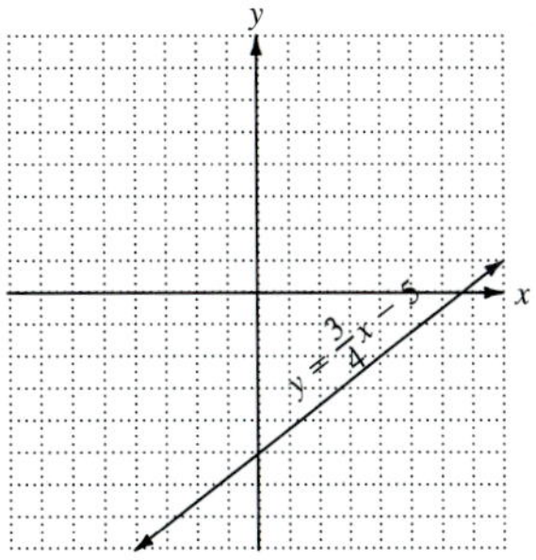

45.

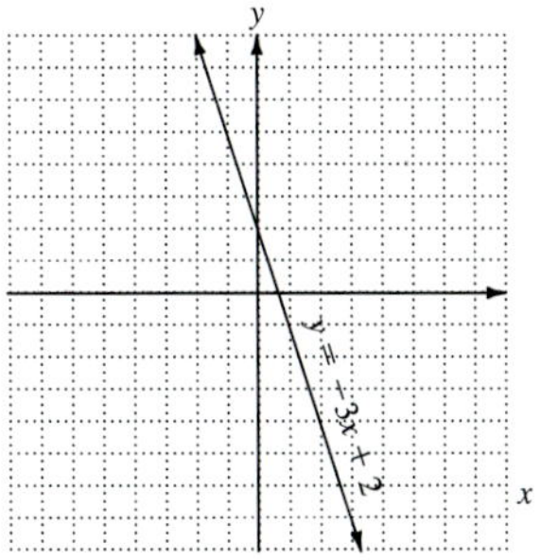

47.

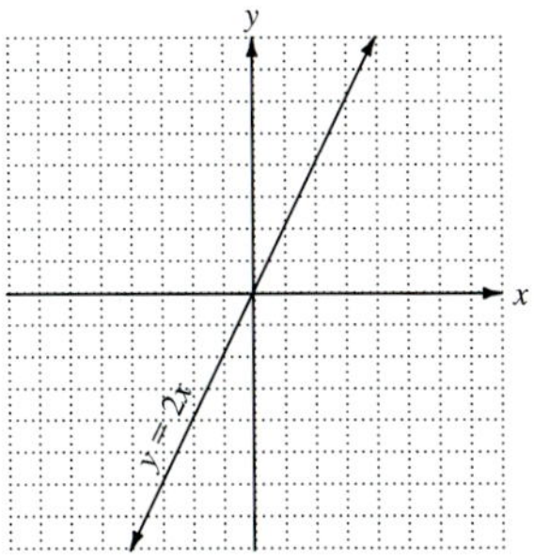

49. (a) $\frac{13}{5}$ **(b)** $-\frac{5}{13}$ **51. (a)** $\frac{1}{8}$ **(b)** -8 **53. (a)** 2 **(b)** $-\frac{1}{2}$ **55. (a)** $-\frac{1}{2}$ **(b)** 2

56. $x > 4$

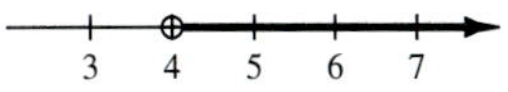

57. $x < \frac{12}{5}$

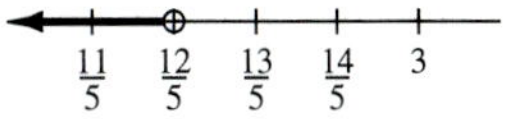

58. $x \le 24$

59. $x \le -22$

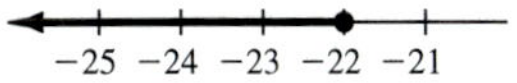

60. 148 miles at maximum speed

61. A minimum of 23 counselors

Putting Your Skills to Work

1. 25 pounds per square inch

2. 44 feet

3. $\frac{5}{11}$

4.

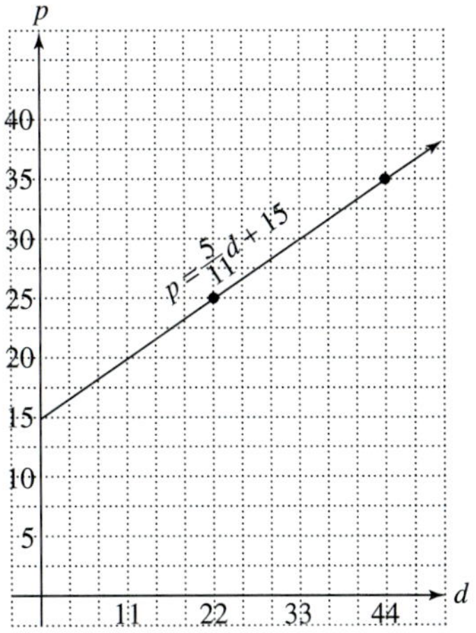

5. $\frac{5}{11}$; yes. You can determine the slope of a line using the coordinates of two points, from the equation in slope-intercept form, from the graph by counting.

6.

d	0	11	22	33	44
p	15	20	25	30	35

(a) 5 **(b)** 11 **(c)** Slope is the difference between the p values over the difference between the d values.

7.4 Exercises

1. $y = 6x + 20$ **3.** $y = -2x + 11$ **5.** $y = \frac{2}{3}x$ **7.** $y = -3x + \frac{7}{2}$ **9.** $y = -\frac{2}{5}x - 1$ **11.** $y = -4$ **13.** $y = 2x + 3$ **15.** $y = -3x + 1$ **17.** $y = -3x + 3$ **19.** $y = 5x - 10$ **21.** $y = 9x - 12$ **23.** $y = \frac{1}{3}x + \frac{1}{2}$ **25.** $y = -\frac{2}{3}x + 1$ **27.** $y = \frac{2}{3}x - 4$ **29.** $y = 3x$ **31.** $y = -2$ **33.** $x = 6$ **34.** $y = 5$ **35.** $x = -5$ **36.** $y = -3$ **37.** $y = 2$ **38.** $x = -1$ **39.** $y = 2x + 3$ **40.** $y = -\frac{1}{2}x + 3$ **41.** $y = -\frac{1}{2}x + 4$ **42.** $y = 5x - 1$ **43.** $t = \frac{24}{5}$ **44.** $\frac{1}{(2x - 5)^2}$ **45.** 4.95% **46.** \$12,640

7.5 Exercises

3. 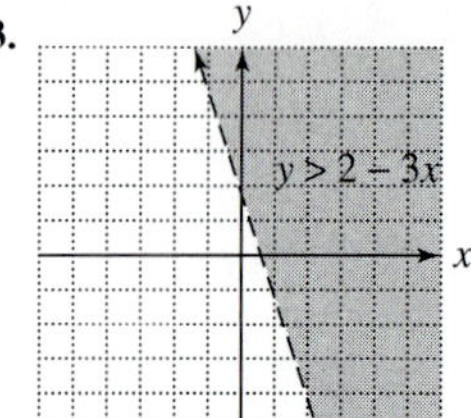

5. 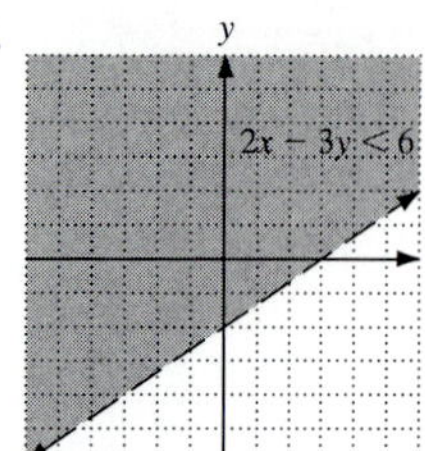

7. 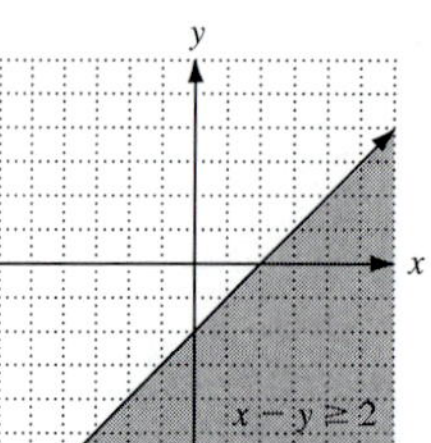

9. 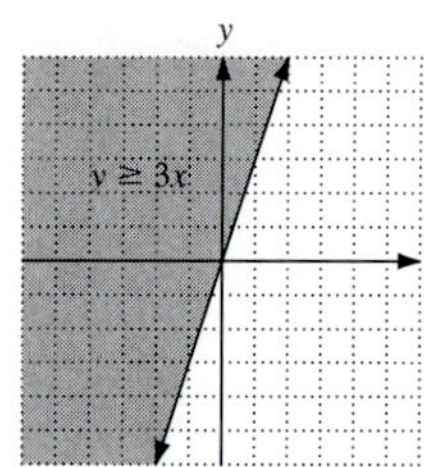

11. 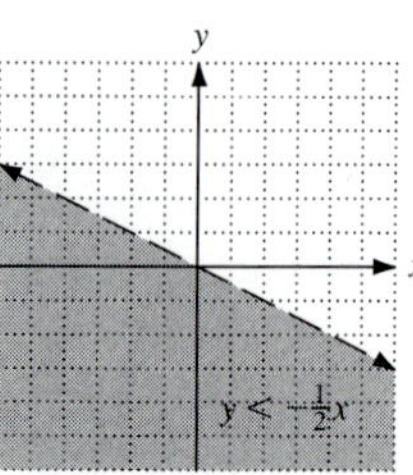

13. 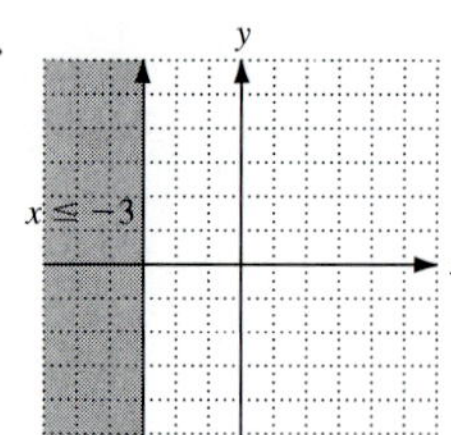

15. 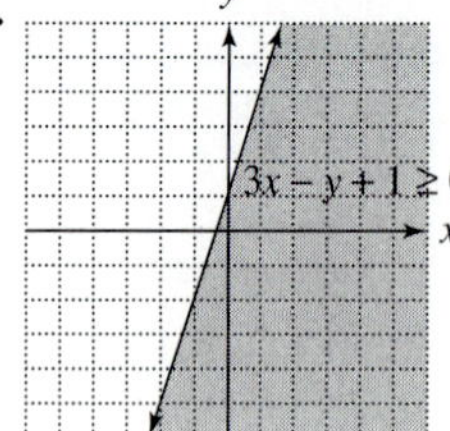

17. 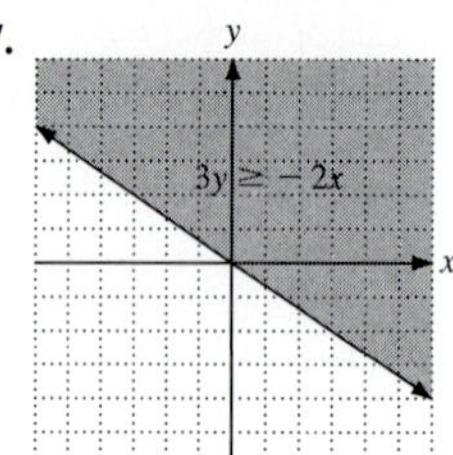

19. 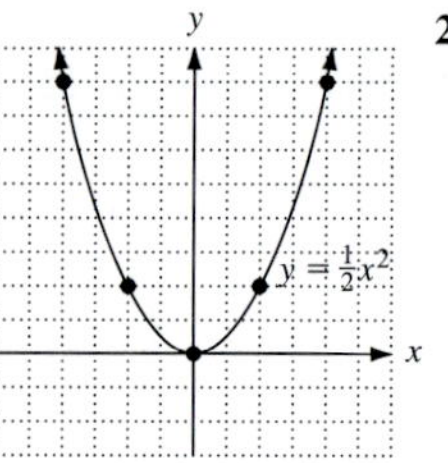

21. 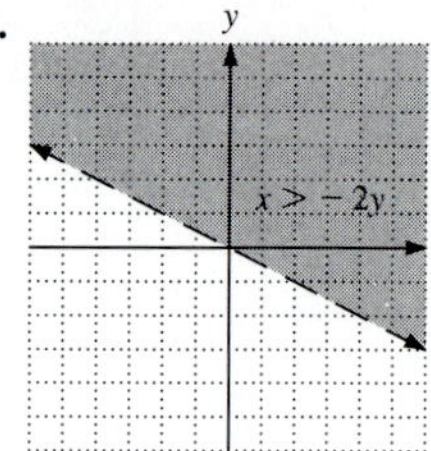

23. **(a)** $x < 3$ **(b)** $x < 3$ **(c)** $3x + 3$ will be greater than $5x - 3$ for those values of x where the graph of $y = 3x + 3$ lies *above* the graph of $y = 5x - 3$. **24.** $y = -\frac{7}{3}x + 7$ **25.** $x = -\frac{3}{2}, x = -2$ **26.** She will need 16 more calls. **27.** She can drive 9,200 miles above the 36,000 mile limit.

7.6 Exercises

7. Domain = {1, 3, 5, 7}, Range = {−2, 7, 8}, Function **9.** Domain = $\{\frac{1}{4}, \frac{1}{2}, \frac{3}{4}\}$, Range = {5, 6, 10}, Not a function
11. Domain = {1, 3, 8, 10}, Range = {1, 3, 8, 10}, Function **13.** Domain = {5, 5.6, 5.8, 6}, Range = {5.8, 6, 8}, Function
15. Domain = {30, 40, 50}, Range = {2, 30, 60, 80}, Not a function

17. 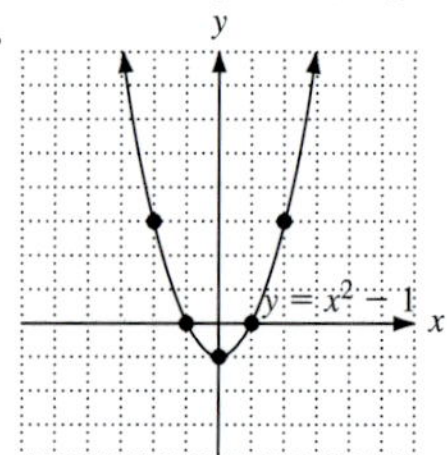

19. 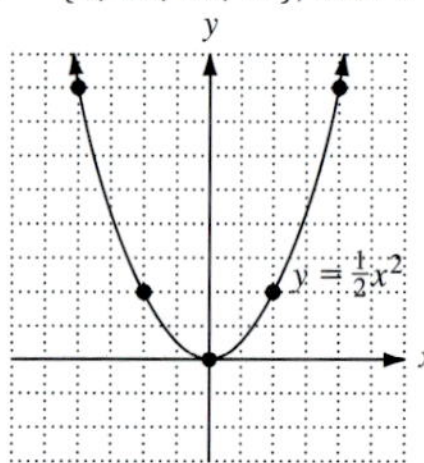

21. 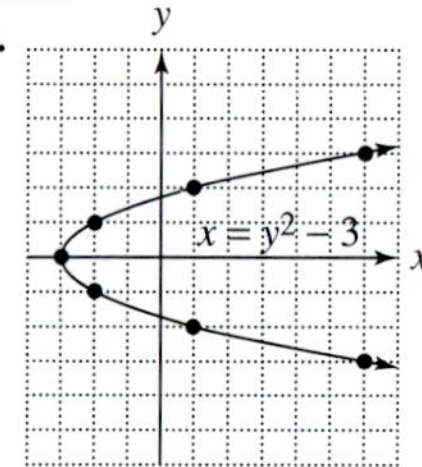

23. 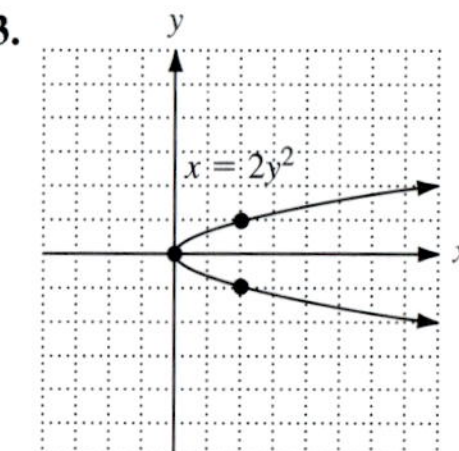

25. 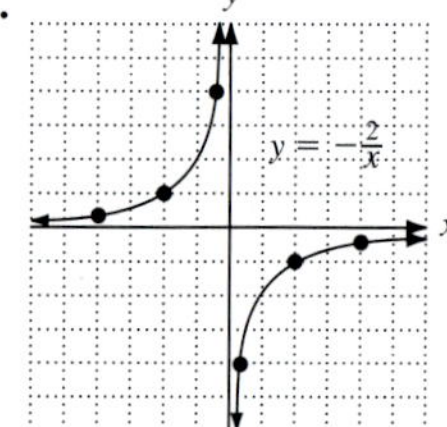

27. 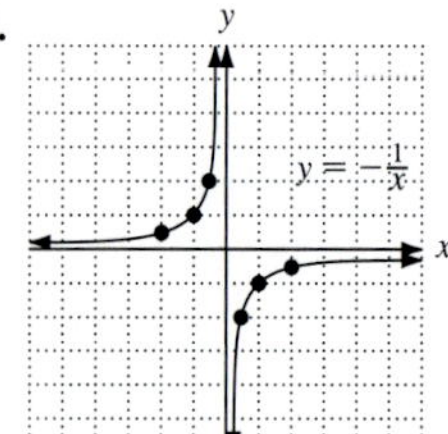

29. 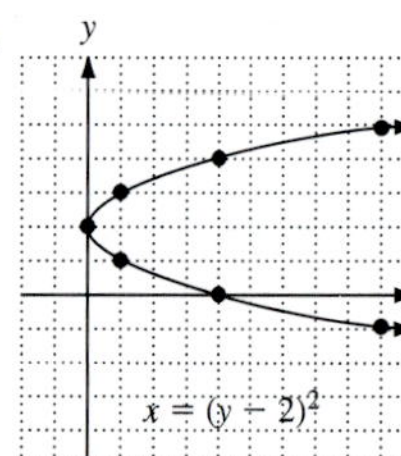

31. 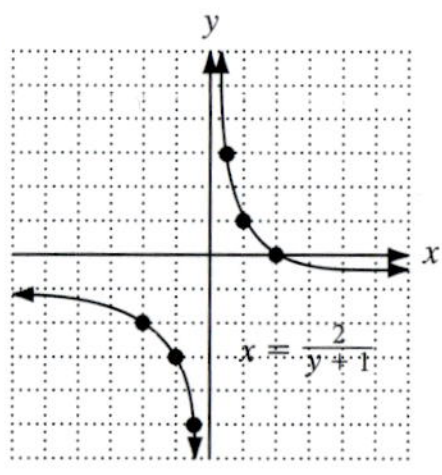

33. Function **35.** Not a function **37.** Function **39.** Not a function **41.** **(a)** −11 **(b)** 4 **(c)** 24 **43.** **(a)** −4 **(b)** 2 **(c)** 29
45. **(a)** −4 **(b)** −3 **(c)** 12 **47.** **(a)** 3 **(b)** −13 **(c)** −33 **49.** **(a)** −8 **(b)** 7 **(c)** 4 **50.** $-2x^2 - 14x + 6$ **51.** $-3x^2 - 7x + 11$
52. $2x^2 - x + 3$ **53.** $2x + 4 + \frac{-1}{3x - 1}$

Putting Your Skills to Work

1. Approximately 138 dollars **2.** Yes, about 50 dollars

Chapter 7 Review Problems

1.

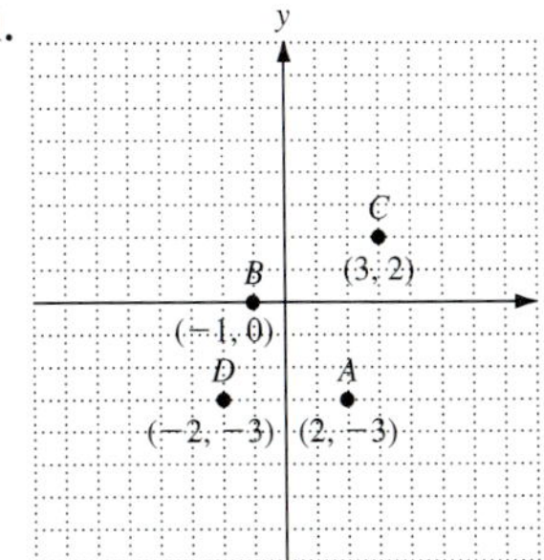

2.

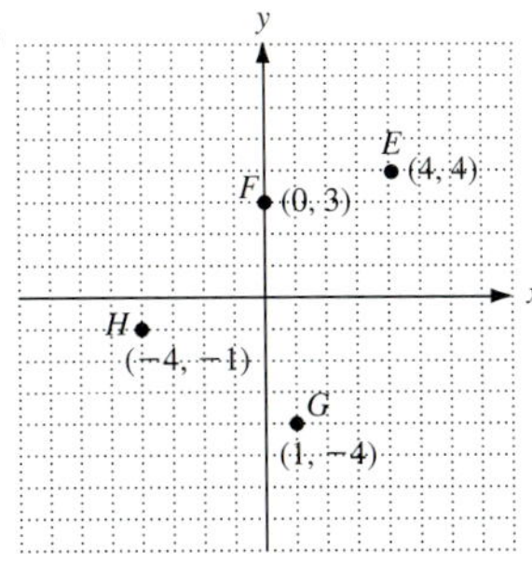

3. **(a)** $(0, -5)$ **(b)** $(3, 4)$
4. **(a)** $(1, 2)$ **(b)** $(-4, 4)$
5. **(a)** $(6, -1)$ **(b)** $(6, 3)$

6.

7.

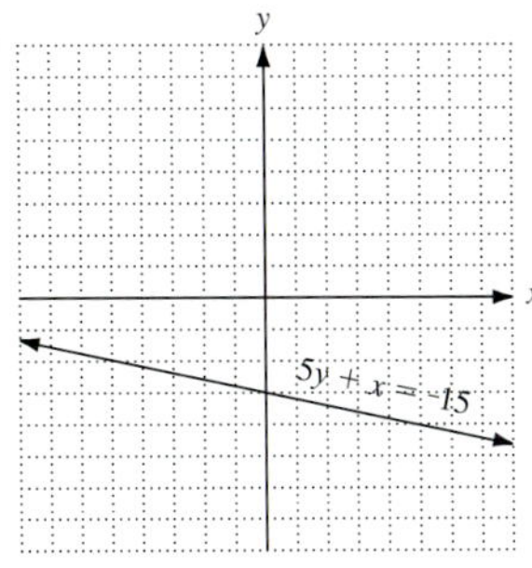

8.

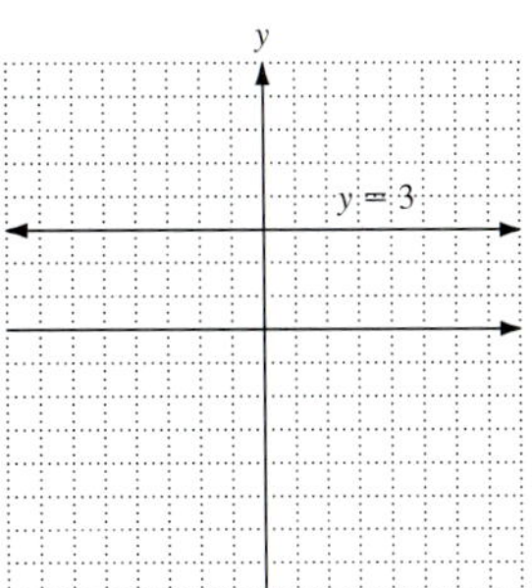

9. $-\frac{5}{6}$ **10.** $m = \frac{5}{8}, b = \frac{3}{2}$ **11.** $y = -\frac{1}{2}x + 3$ **12.** $y = 7$ **13.** $\frac{3}{2}$ **14.** $\frac{1}{7}$

15.

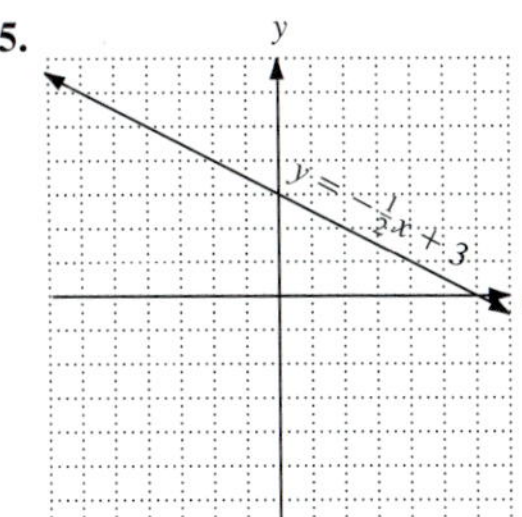

16.

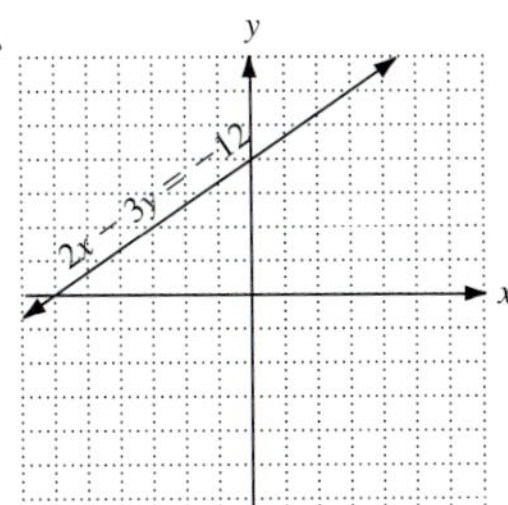

17.

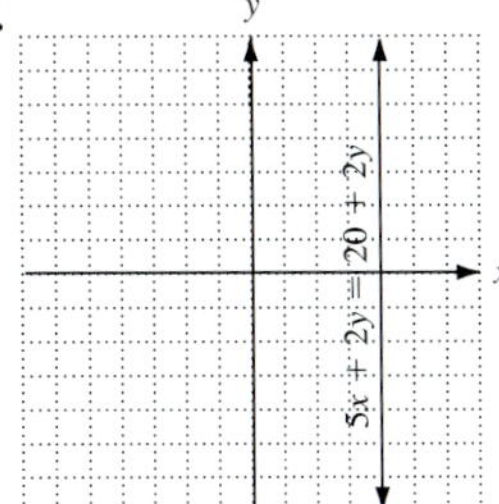

18. $y = 2x - 4$ **19.** $y = -6x + 14$ **20.** $y = \frac{1}{3}x - 5$ **21.** $y = -\frac{1}{2}x + \frac{9}{2}$ **22.** $y = x - 5$ **23.** $y = 5$ **24.** $y = \frac{2}{3}x - 3$
25. $y = -3x + 1$ **26.** $x = -5$

27.

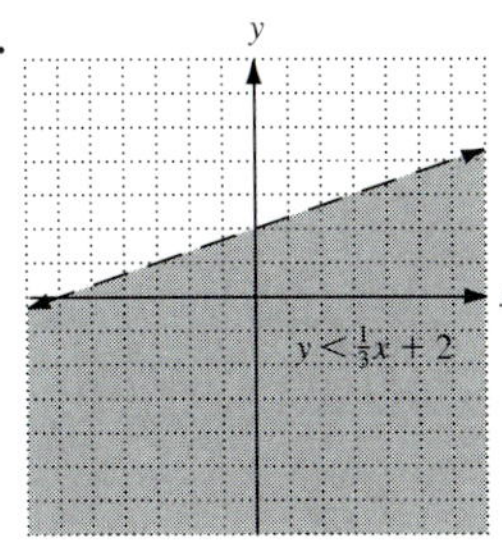

28.

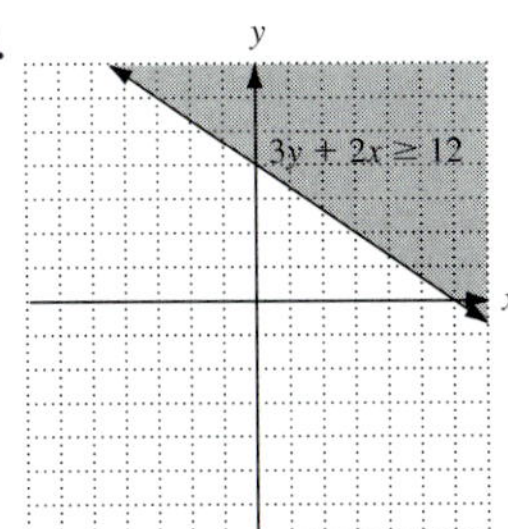

29.

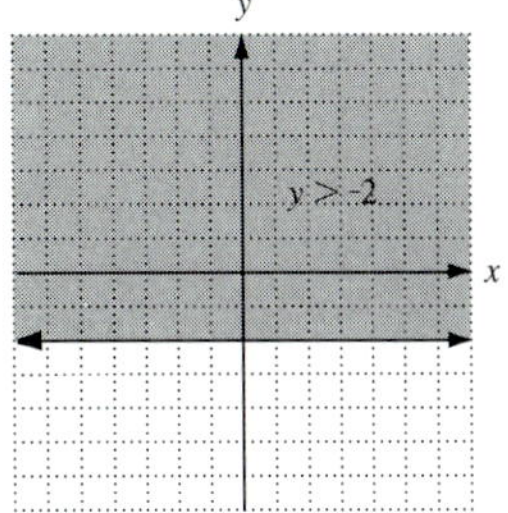

30. Domain: $\{-3, 0, 1, 6\}$, Range: $\{-2, 4, 7\}$, function **31.** Domain: $\{0, 1, 2\}$, Range: $\{6, 4, 8\}$, not a function **32.** Domain: $\{5, -6, -5\}$, Range: $\{-6, 5\}$, not a function **33.** Domain: $\{3, -7, -3, 7\}$, Range: $\{-7, -3, 3, 7\}$, function **34.** function **35.** not a function **36.** function

37.

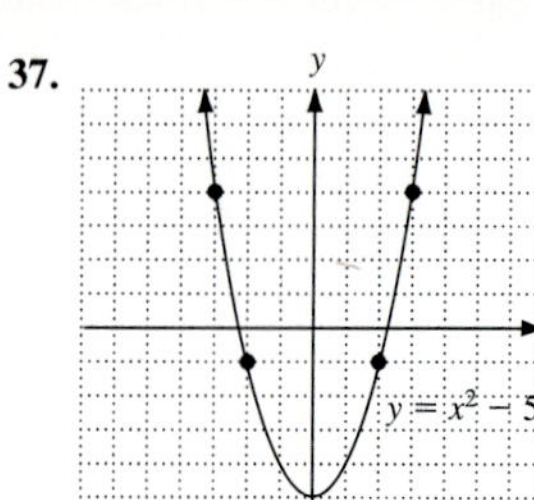

38.

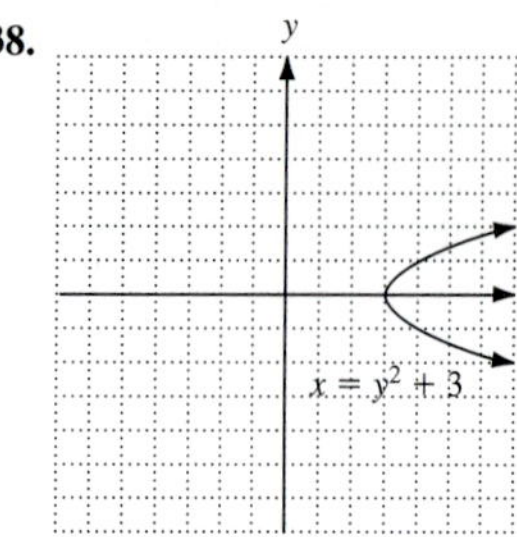

39.

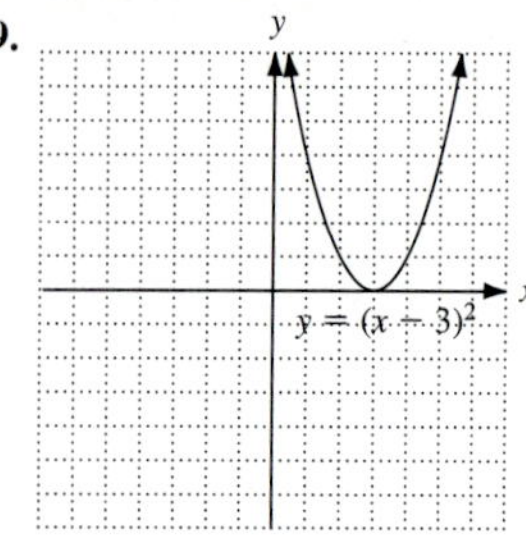

40. **(a)** 7 **(b)** 19 **41.** **(a)** 9 **(b)** 24 **42.** **(a)** -1 **(b)** 2 **43.** **(a)** 31 **(b)** 59 **44.** **(a)** 1 **(b)** -6 **45.** **(a)** 1 **(b)** $\frac{1}{5}$

Chapter 7 Test

1.

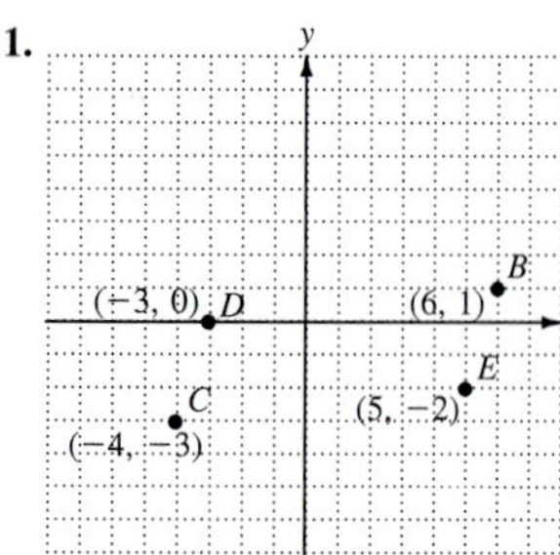

2.

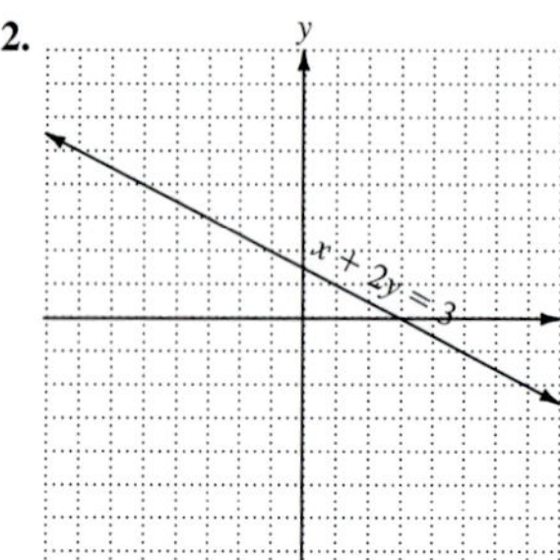

3.

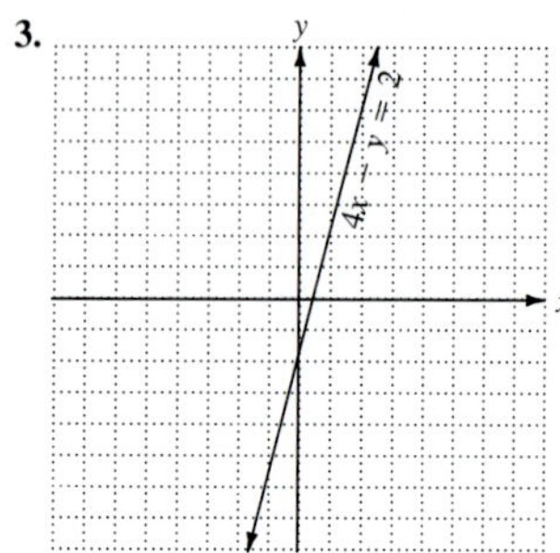

4.

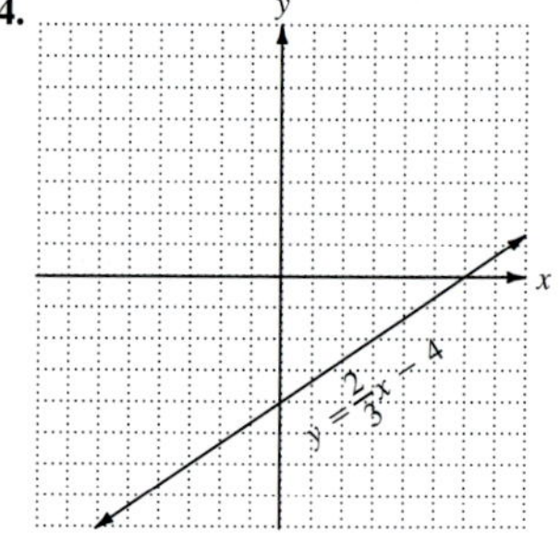

5. $m = -\frac{1}{2}, b = \frac{5}{2}$ **6.** $\frac{1}{2}$ **7.** $x - 2y = 8$ **8.** 0 **9.** $x + 3y = 17$ **10.** $x = 2$, no slope

11.

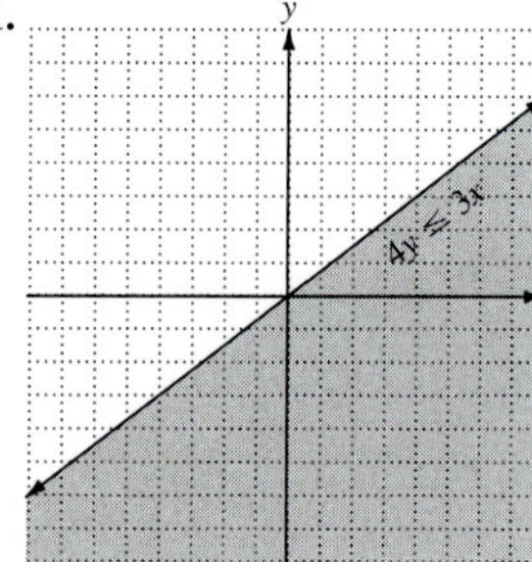

12.

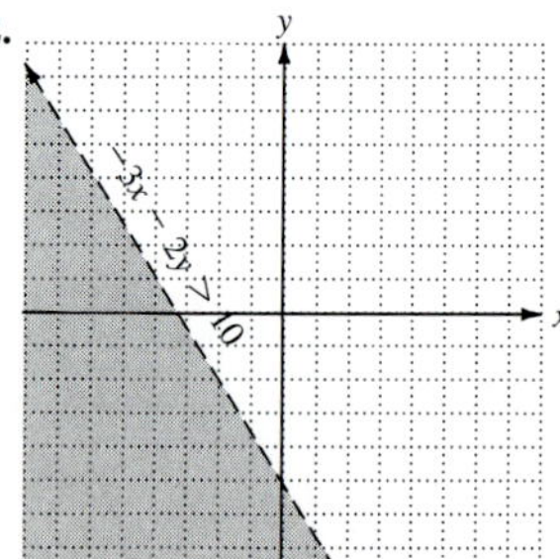

13. Yes. No two different ordered pairs have the same first coordinate. **14.** Yes. It passes the vertical line test.

15.

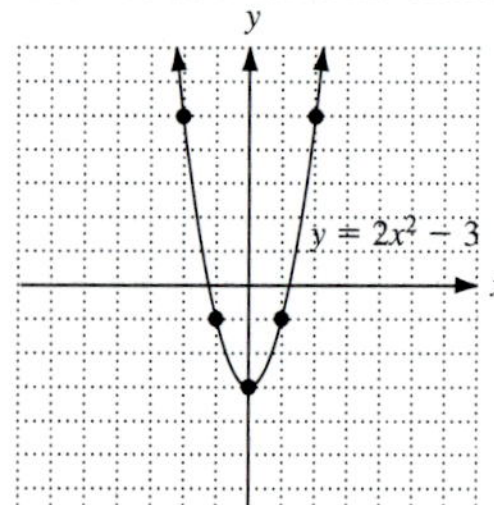

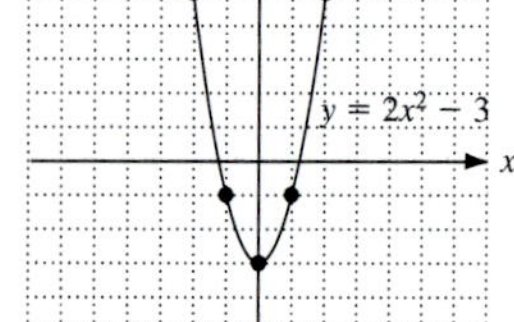

16. **(a)** -4 **(b)** -2 **17.** **(a)** -3 **(b)** $-\frac{3}{4}$

Cumulative Test for Chapters 0–7

1. $-27x^9y^{12}z^3$ **2.** $-\dfrac{3a^2}{4b^2}$ **3.** 7 **4.** 40 **5.** 5 **6.** $w = \dfrac{A - lh}{l}$ **7.** $2(5a + 7b)(5a - 7b)$ **8.** $(2x + 5)(x + 4)$ **9.** $2(x - 1)$

10. $3x - y = 10$ **11.** $x = 7$ **12.** $x - 3y = -11$ **13.** 0 **14.** $\frac{3}{7}$

15.

16.

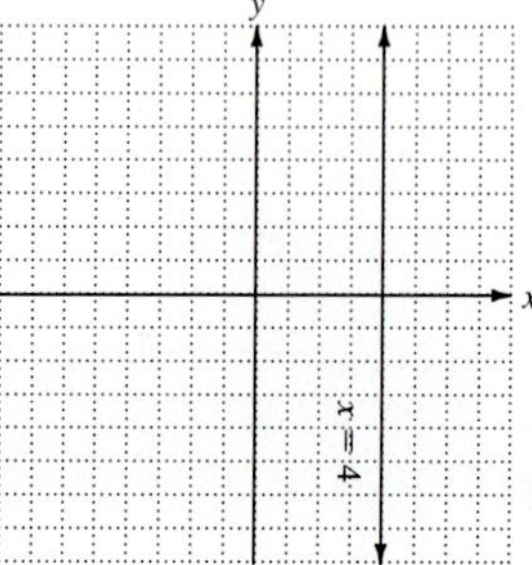

17.

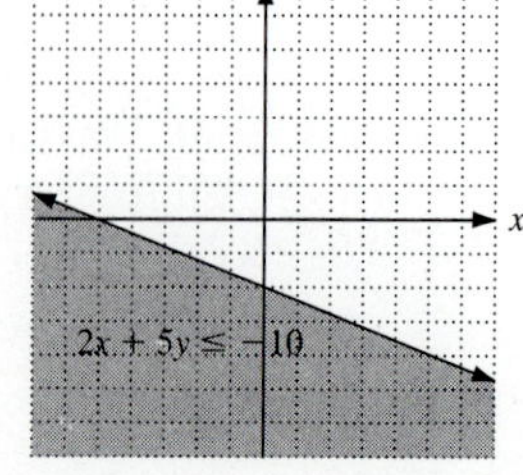

18. No

19. Yes **20.** **(a)** 5 **(b)** 3

Chapter 8

Pretest Chapter 8

1. $x = -3,\ y = 2$

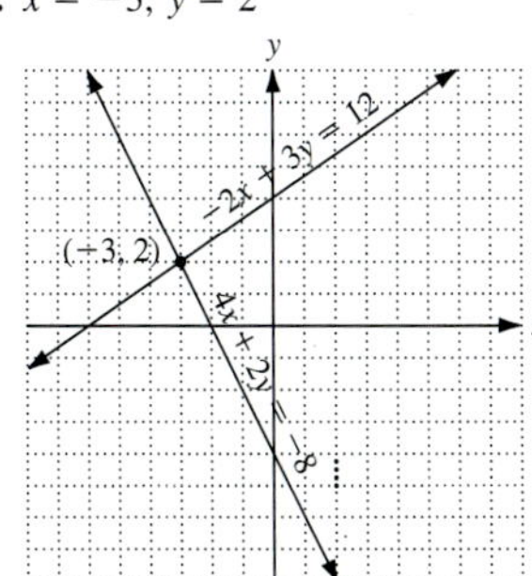

2. no solution

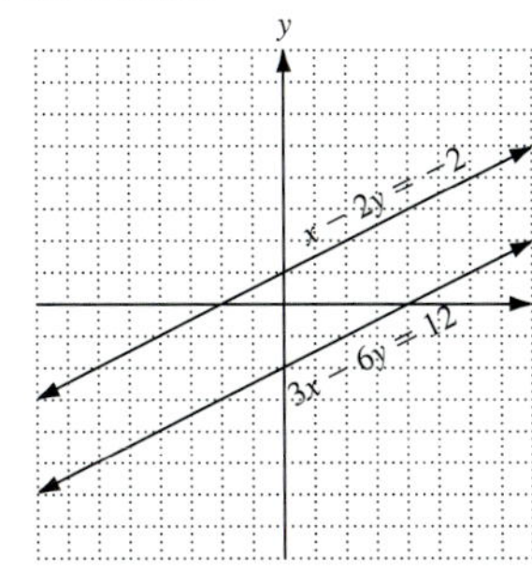

3. $x = -3,\ y = -2$ 4. $x = \frac{3}{5},\ y = \frac{3}{5}$ 5. $x = 2,\ y = 0$ 6. $x = 6,\ y = 2$ 7. $x = 2,\ y = 3$ 8. intersect 9. are parallel 10. coincide 11. $x = 3,\ y = \frac{1}{2}$ 12. $x = 0,\ y = -5$ 13. $x = 2,\ y = -1$ 14. James can sort 290 letters per hour. Michael can sort 690 letters per hour.

8.1 Exercises

1.

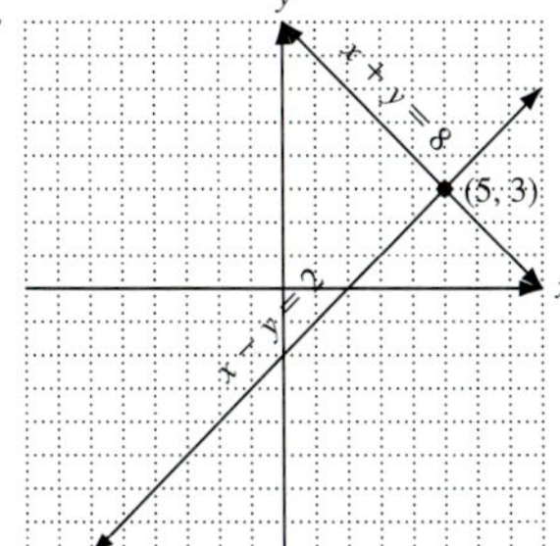

3.

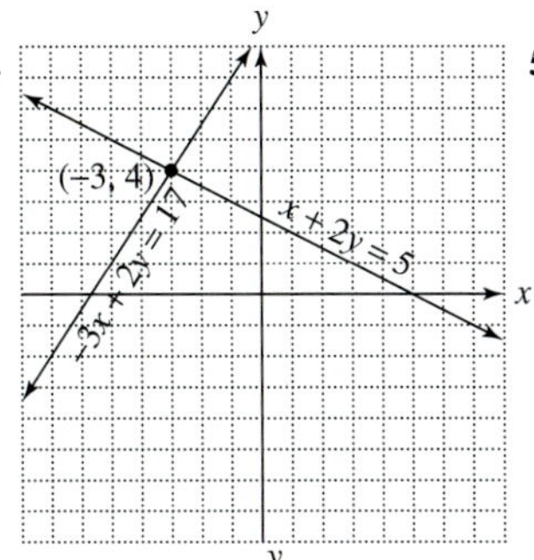

5.

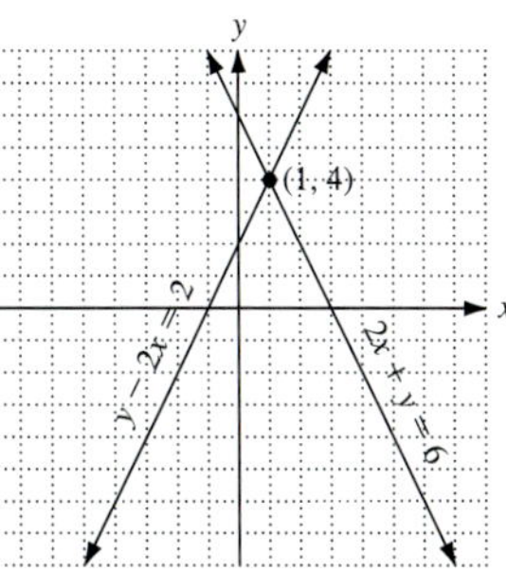

7.

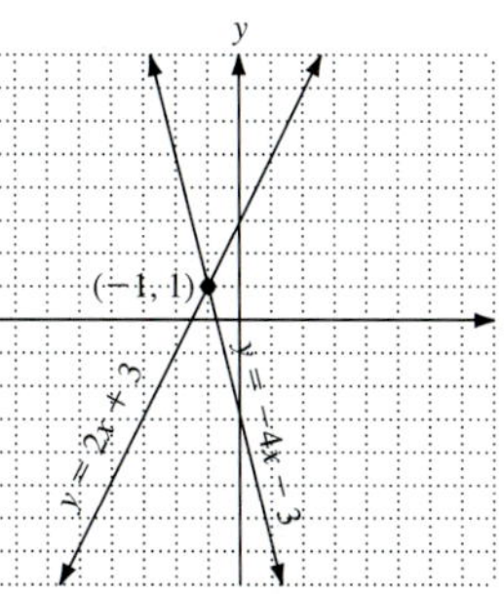

9.

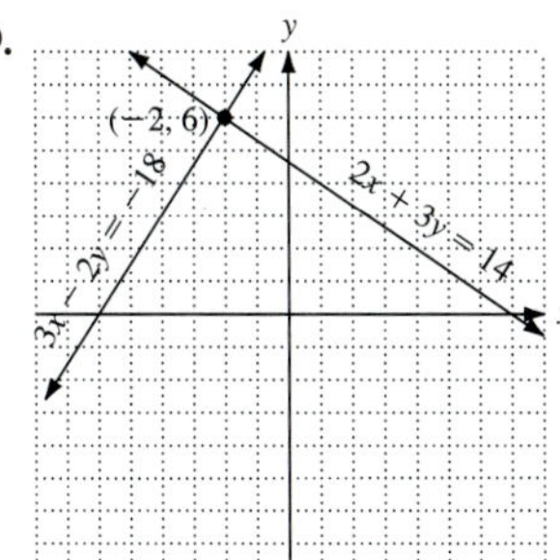

11.

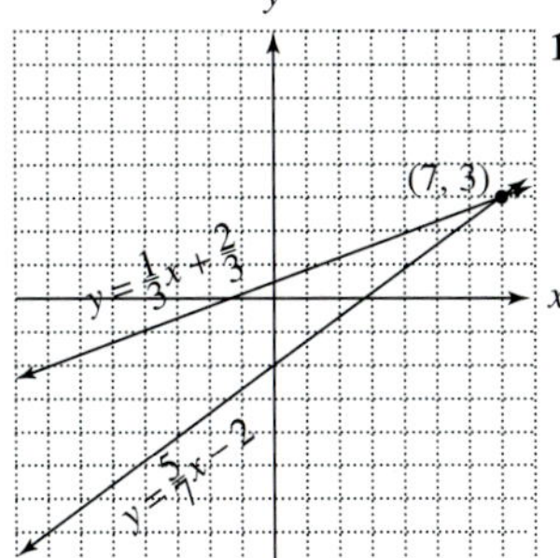

13.

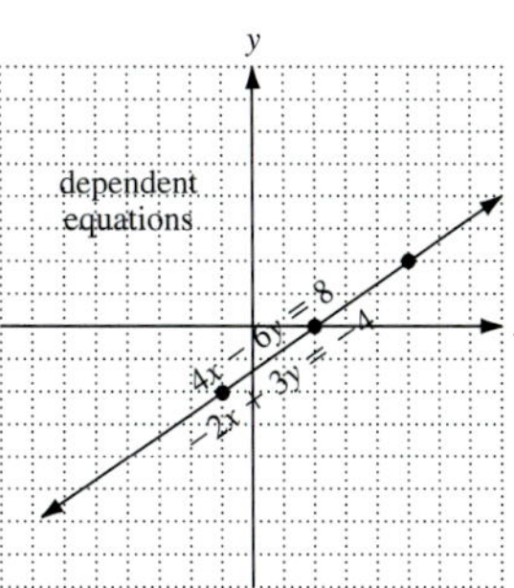

15.

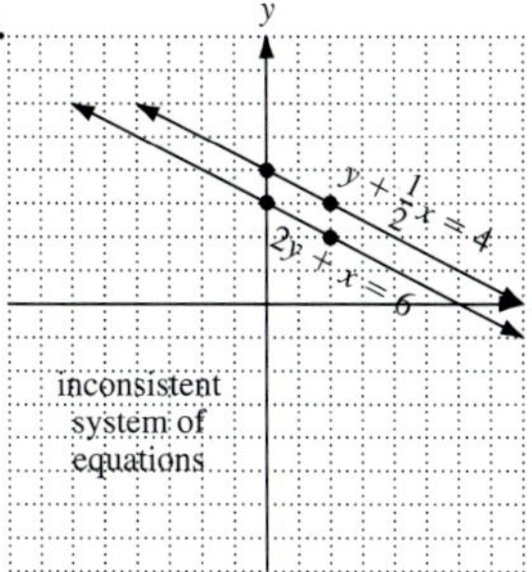

17.

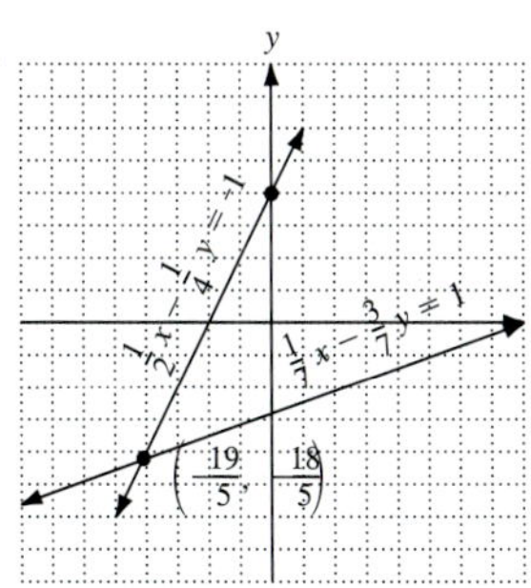

25. $(21, 27)$ 27. $(-29.61, -71.52)$ 28. $\frac{1}{8} \le x$ 29. $d = \dfrac{5y + 16}{x}$ 30. $9x^2 - 42x + 49$ 31. $(5x - 6)^2$

8.2 Exercises

1. $x = 1;\ y = 1$ 3. $x = 3;\ y = -1$ 5. $x = 1;\ y = 2$ 7. $x = \dfrac{8}{5};\ y = \dfrac{16}{5}$ 9. $x = 3;\ y = -5$ 11. $x = 3;\ y = 0$ 13. $a = 9;\ b = 5$ 15. $x = 2;\ y = 0$ 17. $p = \dfrac{2}{3};\ q = \dfrac{5}{3}$ 19. $x = 2;\ y = -3$ 21. $x = -3;\ y = 6$ 23. $x = -3;\ y = 1$ 25. $x = 7;\ y = -2$ 27. $x = -3;\ y = 3$ 29. $x = 2;\ y = 1$ 31. $x = 3;\ y = \dfrac{1}{4}$ 33. $x = 2.9346;\ y = 1.0993$ 40. $m = -\dfrac{7}{11};\ b = \dfrac{19}{11}$

41.

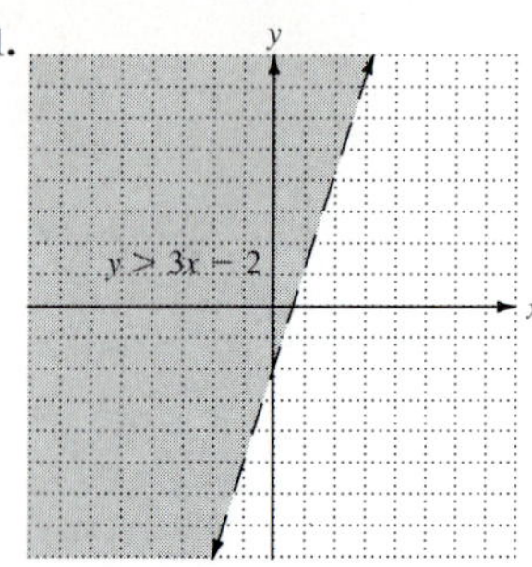

42. Approximately 398 meters per minute **43.** $3,914,000

8.3 Exercises

3. $x = -1; y = 4$ **5.** $x = 1; y = 2$ **7.** $x = 0; y = 5$ **9.** $x = -4; y = -3$ **11.** $x = 8; y = -1$ **13.** $a = -6; b = 2$ **15.** $x = -\frac{9}{4}; y = 1$ **17.** $x = 2; y = 2$ **19.** $x = 1; y = -2$ **21.** $x = 6; y = 3$ **23.** $x = 2; y = \frac{1}{5}$ **25.** $x = -3; y = 5$ **27.** $x = -2; y = \frac{4}{3}$ **29.** $x = -4; y = -2$ **33.** $x = 1; y = 0$ **35.** $x = -4; y = -7$ **37.** $x = 8; y = -10$ **39.** $x = 7; y = -2$ **40.** $x = 5; y = -3$ **41.** $x = -3; y = -4$ **42.** $x = 3; y = 7$ **43.** $x = 2; y = 5$ **44.** $x = a - b; y = a + b$ **45.** $x = 4a; y = 2b$ **46.** 99 miles per hour or faster **47.** They will need to win 36 of the next 62 games. **48.** $3(5x - 1)(5x + 1)$ **49.** $(2x - 5)(3x - 5)$

Putting Your Skills to Work

1. 9.259×10^{-13} second **2.** 3.571×10^{-12} second **3.** More than 3 operations

8.4 Exercises

5. $x = -14, y = -5$ **7.** infinite number of solutions **9.** $x = 3, y = -7$ **11.** no solution **13.** $x = 0, y = -3$ **15.** $x = 50, y = 40$ **17.** $y = 1, x = 3$ **19.** $x = \frac{1}{2}, y = \frac{2}{3}$ **21.** $a = -5, b = -7$ **23.** $x = 5, y = -2$ **25.** $x = -1$ **26.** $x = \frac{1}{2}$ **27.** $x = 2$ **28.** $x = 25$ **29.** $21,500 **30.** $38,500

8.5 Exercises

1. The numbers are 12 and 15. **3.** A loaf from Meilleur Pain is $0.42. A loaf from Corner Loaf is $0.30. **5.** 32 worked as interns. 18 did not work as interns. **7.** 630 orchestra seats, 270 mezzanine seats **9.** Width was 225 feet, length was 300 feet. **11.** 8 qts before, 8 qts added **13.** 30 lb of nuts, 20 lb of raisins **15.** The plane flies 195 mph. The wind speed is 15 mph. **17.** The boat travels 13.125 mph in still water, the current travels 1.875 mph **19.** $29,000 for a heart kit; $36,000 for a kidney kit **20.** $\frac{2x^2 - 12}{x^2 - 9}$ **21.** 35 **22.** $-64x^9y^6$

Putting Your Skills to Work

1. $P = 36.5 + 0.2x,\ a = 0.2$ **2.** $P = 30.9 + 0.3x,\ b = 0.3$

Chapter 8 Review Problems

1.

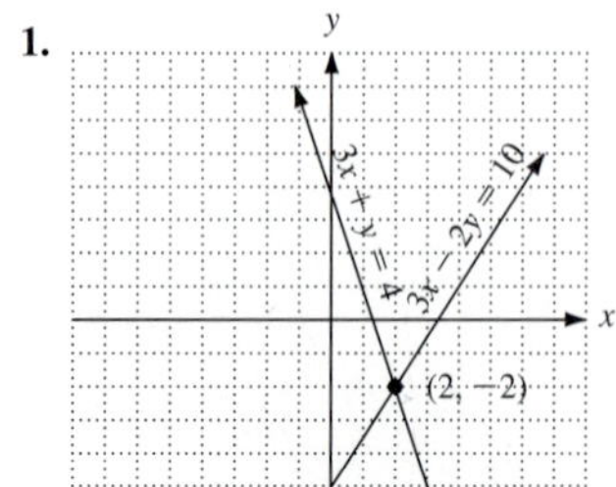

2.

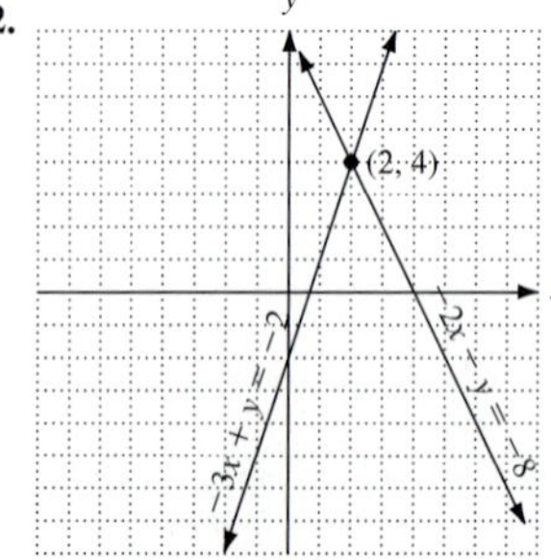

3.

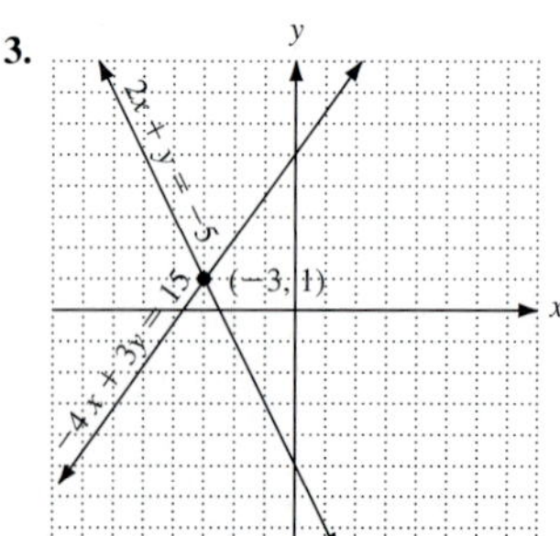

4.

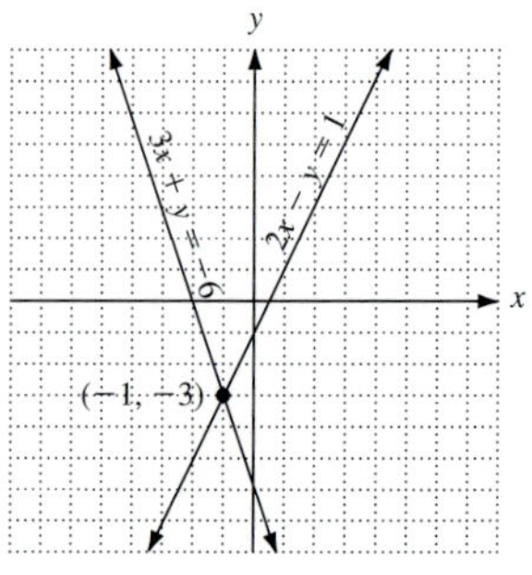

5. $x = 3, y = 5$ **6.** $x = 3, y = 3$ **7.** $x = 2, y = -3$ **8.** $x = 14, y = 66$ **9.** $x = 2, y = 1$ **10.** $x = -1, y = 0$ **11.** $x = -4, y = -2$ **12.** $x = 3, y = 1$ **13.** $(7, 2)$ **14.** $(4, 1)$ **15.** $(3, -7)$ **16.** $(-1, -3)$ **17.** $(4, -4)$ **18.** $(10, 2)$ **19.** $\left(\frac{4}{5}, -\frac{6}{5}\right)$ **20.** $(-3, 2)$ **21.** $(-3, -1)$ **22.** $(2, -3)$ **23.** No solution, Inconsistent system **24.** $(9, 4)$ **25.** $\left(-3, -\frac{2}{3}\right)$ **26.** Infinite number of solutions, dependent equations **27.** $(-2, -2)$ **28.** $(-1, -2)$ **29.** $(4, -3)$ **30.** $\left(-\frac{24}{41}, \frac{2}{41}\right)$

31. Infinite number of solutions. Dependent equations. **32.** (1960, 1040) **33.** $(-4, -7)$ **34.** $(8, -5)$ **35.** $\left(\frac{29}{28}, \frac{1}{7}\right)$ **36.** (4, 10)
37. (10, 8) **38.** $(-12, -15)$ **39.** $\left(\frac{1}{2}, -\frac{2}{3}\right)$ **40.** $\left(\frac{15}{13}, \frac{36}{13}\right)$ **41.** Infinite number of solutions, dependent equations
42. Infinite number of solutions, dependent equations **43.** $(-5, 4)$ **44.** $(9, -8)$ **45.** (2, 4) **46.** (3, 1) **47.** (3, 2)
48. $(12, -12)$ **49.** $\left(a + b, \frac{a^2 - b^2}{b}\right)$ **50.** $\left(\frac{a + b}{a}, \frac{a - b}{b}\right)$ **51.** 120 reserved seats, 640 general admission seats
52. 7 pounds of apples, 5 pounds of oranges **53.** 275 mph is the speed of the plane in still air; 25 mph is the speed of the wind **54.** 32 liters
55. 44 regular sized cars, 42 compact cars **56.** 8 tons of 15% salt, 16 tons of 30% salt
57. Speed of the boat is 19 kilometers per hour, speed of the current is 4 kilometers per hour.
58. It costs $3 for a gram of copper. The hourly labor rate is $4.

Putting Your Skills to Work

1. Loss because it costs $100 more than you make when 200 pounds of candy are made and sold in one day. **2.** $325 profit

Chapter 8 Test

1. $(-3, -4)$
2. $x = -3$; $y = 4$

3.

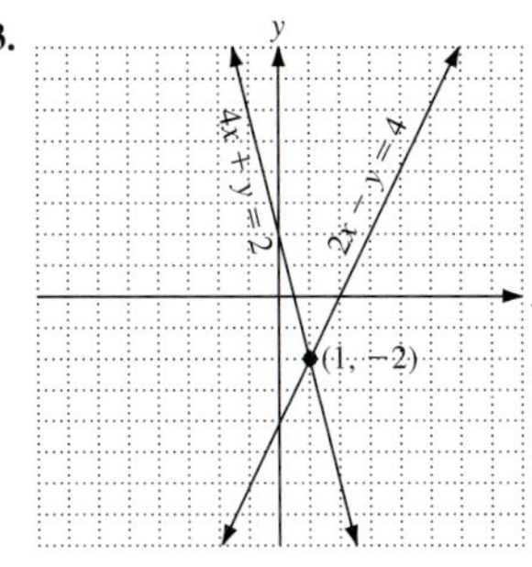

4. $(-6, -2)$ **5.** (4, 3) **6.** (1, 2) **7.** (3, 0) **8.** no solution, inconsistent system **9.** dependent equations, infinite number of solutions
10. $(-2, 3)$ **11.** $(-5, 3)$ **12.** no solution, inconsistent system **13.** $(5, -3)$
14. 12,896 of the $12 tickets, 17,894 of the $8 tickets **15.** A shirt costs $20 and a pair of slacks costs $24. **16.** 500 per hr by 1st machine; 400 per hr by 2nd machine **17.** 450 km/hr speed of jet; 50 km/hr speed of wind

Cumulative Test for Chapters 0–8

1. $4.09 **2.** -8 **3.** $15x^3 - 19x^2 + 11x - 3$ **4.** $\frac{1}{(x + 3)(x - 4)}$ **5.** $x \leq \frac{9}{5}$ **6.** $x = -5, x = 9$ **7.** 3000 **8.** $x = \frac{7}{3}$
9. $5a(a - 4)(a + 3)$ **10.** $x = 3$ **11.** $x = 8, y = -8$ **11.** $x = 24, y = 8$ **13.** infinite number of solutions, dependent equations
14. (1, 1) **15.** inconsistent system of equations, no solution **16.** 4000 liters of 8%; 8000 liters of 20% **17.** 1 mph
18. 1200 and 1800 labels per hour

Chapter 9

Pretest Chapter 9

1. 11 **2.** $\frac{5}{8}$ **3.** 13 **4.** -10 **5.** 2.236 **6.** not real **7.** x^2 **8.** $5x^2y^3$ **9.** $7\sqrt{2}$ **10.** $x^2\sqrt{x}$ **11.** $6x\sqrt{x}$
12. $2a^2b\sqrt{2b}$ **13.** $4\sqrt{2}$ **14.** $\sqrt{2} + \sqrt{3}$ **15.** $4\sqrt{2}$ **16.** $13\sqrt{7} - 7$ **17.** $11\sqrt{2} - 18\sqrt{3}$ **18.** $18\sqrt{2}$ **19.** $4 + 2\sqrt{6}$
20. $23 + 8\sqrt{7}$ **21.** 1 **22.** -1 **23.** $17 - 4\sqrt{15}$ **24.** $\frac{6}{x}$ **25.** $\frac{5\sqrt{3}}{3}$ **26.** $3\sqrt{5} - 6$ **27.** $13 - 2\sqrt{42}$ **28.** $2\sqrt{13}$
29. $\sqrt{11}$ **30.** 7 **31.** $\frac{35}{3}$ **32.** $\frac{1}{2}$ **33.** 7

9.1 Exercises

5. ± 3 **7.** ± 8 **9.** ± 9 **11.** ± 11 **13.** 5 **15.** 0 **17.** 7 **19.** 10 **21.** -6 **23.** 0.9 **25.** 1.4 **27.** $\frac{4}{5}$ **29.** $\frac{7}{8}$
31. 20 **33.** -100 **35.** 13 **37.** $-\frac{1}{8}$ **39.** $\frac{5}{7}$ **41.** 0.6 **43.** 120 **45.** 21 **47.** 0.13 **49.** 6.481 **51.** 7.681
53. -11.533 **55.** -13.964 **57.** -18.97 **59.** 6.75 **61.** 7.5 seconds **63.** 9.5 seconds **65.** 4 **66.** 10 **67.** $x = 2$; $y = 1$
68. no solution, inconsistent system of equations **69.** Each snowboard is $250. Each pair of goggles is $50.
70. The plane's airspeed is 470 mph in still air. The wind blows at 50 mph.

9.2 Exercises

1. 7 **3.** 10^2 **5.** 9^3 **7.** 56^3 **9.** 3^{75} **11.** w^6 **13.** t^9 **15.** y^{13} **17.** $7x^3$ **19.** $12x$ **21.** $20x^4$ **23.** xy^3
25. $5x^4y^2$ **27.** $8xy^2$ **29.** $16ab^4c^2$ **31.** $2\sqrt{5}$ **33.** $2\sqrt{6}$ **35.** $3\sqrt{2}$ **37.** $5\sqrt{2}$ **39.** $6\sqrt{2}$ **41.** $3\sqrt{10}$ **43.** $7\sqrt{2}$
45. $6\sqrt{3}$ **47.** $2y^2\sqrt{3y}$ **49.** $3y^6\sqrt{2y}$ **51.** $7x^2\sqrt{x}$ **53.** $5x^3\sqrt{2xy}$ **55.** $2xy\sqrt{3y}$ **57.** $3a^4y^3\sqrt{3y}$ **59.** $2x^3y^5\sqrt{35xy}$
61. $6a^2c\sqrt{5ab}$ **63.** $3a^2b^3c\sqrt{7c}$ **65.** $9x^6y^5w^2\sqrt{yw}$ **66.** $x + 3$ **67.** $x + 6$ **68.** $4x + 1$ **69.** $2x + 5$ **70.** $y(x + 7)$
71. $2(2x + 1)$ **72.** $x = 4$; $y = -1$ **73.** -4 **74.** Between 743 tons and 756 tones of equivalent energy
75. Approx. 31.7 quadrillion Btu of energy

9.3 Exercises

3. $-2\sqrt{3}$ **5.** $5\sqrt{2} + 3\sqrt{3}$ **7.** $-5\sqrt{x} - 2\sqrt{xy}$ **9.** $-\sqrt{3}$ **11.** $17\sqrt{2}$ **13.** $\sqrt{2}$ **15.** 0 **17.** $4\sqrt{3} + 2\sqrt{5} + 6$
19. $8\sqrt{3} - \sqrt{2}$ **21.** $5\sqrt{3x}$ **23.** $\sqrt{2xy}$ **25.** $0.2\sqrt{3x}$ **27.** $7\sqrt{5y}$ **29.** $5y\sqrt{y}$ **31.** $4x\sqrt{2x} + 15\sqrt{2x}$ **33.** $-5x\sqrt{2x}$
35. $x\sqrt{3}$ **37.** $2y\sqrt{6y} - 6y\sqrt{6}$ **39.** $5y^2\sqrt{7}$ **41.** $-110x\sqrt{2x}$ **43.** 54.71 ft to top; 50.48 ft to bottom

44. $x \geq -\frac{4}{3}$

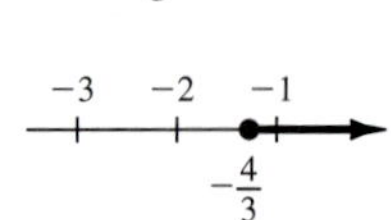

45.

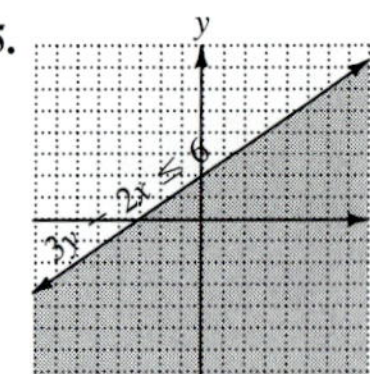

46. 3 CDs at \$7.99 and 7 CDs at \$11.99 **47.** \$1600 at 7% and \$6400 at 5%

9.4 Exercises

1. $\sqrt{35}$ **3.** $2\sqrt{11}$ **5.** $2a\sqrt{3}$ **7.** $5\sqrt{2x}$ **9.** $6\sqrt{30}$ **11.** $6x^3\sqrt{5}$ **13.** $-6b\sqrt{a}$ **15.** $\sqrt{21} + 15$ **17.** $2\sqrt{5} - 6\sqrt{3}$ **19.** $2x - 16\sqrt{5x}$ **21.** $2\sqrt{3} - 18 + 4\sqrt{15}$ **23.** $6\sqrt{ab} + 2a\sqrt{b} - 4a$ **25.** $-6 - 9\sqrt{2}$ **27.** $22 + 18\sqrt{3}$ **29.** $5 - \sqrt{21}$ **31.** $-6 + 9\sqrt{2}$ **33.** $2\sqrt{14} + 34$ **35.** $29 - 12\sqrt{5}$ **37.** $53 + 10\sqrt{6}$ **39.** $18 - 12\sqrt{2}$ **41.** $a - 2a\sqrt{b} + ab$ **43.** $9x^2y - 5$ **45.** $5x\sqrt{5x} - 5x^2 - 10\sqrt{x} + 2x\sqrt{5}$ **47.** **(a)** It is not true when a is negative. **(b)** In order to deal with real numbers. **48.** Hank's final answer does not have a middle term. $(\sqrt{a} + \sqrt{b})^2 = a + 2\sqrt{ab} + b$. **49.** $(6x + 7y)(6x - 7y)$ **50.** $(x^4 + 1)(x^2 + 1)(x + 1)(x - 1)$ **51.** $m = \left(\frac{1519}{1320}\right)k$, destroyer is traveling at 40.3 miles per hour. **52.** 6 years

9.5 Exercises

1. 4 **3.** $\frac{1}{2}$ **5.** 7 **7.** $\frac{\sqrt{6}}{x^2}$ **9.** 5 **11.** $\frac{2\sqrt{2}}{x^3}$ **13.** $\frac{3\sqrt{7}}{7}$ **15.** $\frac{5\sqrt{6}}{6}$ **17.** $\frac{x\sqrt{3}}{3}$ **19.** $\frac{2\sqrt{2x}}{x}$ **21.** $\frac{\sqrt{3}}{2}$ **23.** $\frac{7\sqrt{a}}{a^2}$ **25.** $\frac{\sqrt{2x}}{2x^2}$ **27.** $\frac{3\sqrt{x}}{x^2}$ **29.** $\frac{\sqrt{14}}{7}$ **31.** $\frac{\sqrt{3a}}{ab}$ **33.** $\frac{9\sqrt{2x}}{8x}$ **35.** $3\sqrt{2} + 3$ **37.** $\sqrt{5} - \sqrt{2}$ **39.** $2 + \sqrt{2}$ **41.** $x(2\sqrt{2} + \sqrt{5})$ **43.** $\frac{x(\sqrt{3} + 2x)}{3 - 4x^2}$ **45.** $2\sqrt{14} - 7$ **47.** $\frac{7 - 2\sqrt{10}}{3}$ **49.** $3\sqrt{2} + 4$ **51.** $5\sqrt{6} + 8\sqrt{2}$ **53.** $\sqrt{x} - 2$ **55.** $\frac{x + \sqrt{2x} + 2\sqrt{2} + 2\sqrt{x}}{x - 2}$ **57.** **(a)** $x^2 - 2$ **(b)** $(x + \sqrt{2})(x - \sqrt{2})$ **(c)** $(x + 2\sqrt{3})(x - 2\sqrt{3})$ **58.** 0.904534033 The values are the same. They are equivalent expressions. **59.** 1.183215957 The values are the same. They are equivalent expressions. **60.** 38.49 **61.** 448 **62.** Approx. 71 mg per cup **63.** They would need more than 15 doctor visits per year.

9.6 Exercises

1. 5 **3.** 25 **5.** $\sqrt{74}$ **7.** $\sqrt{34}$ **9.** $2\sqrt{19}$ **11.** $\sqrt{7}$ **13.** 9.91 **15.** 9.9 ft **17.** 127.3 ft **19.** 6.7 mi **21.** 121 **23.** 17 **25.** $-\frac{2}{3}$ **27.** 7 **29.** $\frac{81}{2}$ **31.** 5 only **33.** $y = 0, y = 3$ **35.** 6 only **37.** $y = 0, y = 1$ **39.** $x = 2$ **41.** $\frac{1}{2}$ only **43.** $\frac{2}{9}$ only **44.** $x = 2, x = 6$ **45.** no solution **46.** $9x^2 - 6x + 1$ **47.** 20 **48.** relation

9.7 Exercises

1. $y = \frac{63}{4}$ **3.** $y = 90$ **5.** $y = \frac{1029}{2}$ **7.** $y = \frac{625}{2}$ **9.** $y = 373.03$ **11.** 169 lb/sq in. **13.** 56 min **15.** $y = \frac{48}{7}$ **17.** $y = 2$ **19.** $y = \frac{40}{27}$ **21.** $y = \frac{4}{9}$ **23.** $y = 10$ **25.** 3.45 minutes **27.** $444\frac{4}{9}$ pounds **31.** 5000 Btu/hr **32.** 7467 lb **33.** $a = 160$

34. S; 25, 15, 5; h; 5

35. $4(3x + 4)(x - 3)$ **36.** $x = \frac{13}{4}$

Chapter 9 Review Problems

1. 3 **2.** 5 **3.** 6 **4.** 7 **5.** 10 **6.** 2 **7.** not a real number **8.** not a real number **9.** 4 **10.** 11 **11.** 15 **12.** 13 **13.** 0.9 **14.** 0.6 **15.** $\frac{1}{5}$ **16.** $\frac{6}{7}$ **17.** −0.02 **18.** 0.03 **19.** not a real number **20.** −0.09 **21.** 10.247 **22.** 14.071 **23.** 8.775 **24.** 9.381 **25.** $5\sqrt{2}$ **26.** $4\sqrt{3}$ **27.** $7\sqrt{2}$ **28.** $6\sqrt{2}$ **29.** $2\sqrt{10}$ **30.** $4\sqrt{5}$ **31.** x^4 **32.** y^5 **33.** $x^2y^3\sqrt{x}$ **34.** $ab^2\sqrt{a}$ **35.** $4xy^2\sqrt{xy}$ **36.** $7x^2y^3\sqrt{2}$ **37.** $2x^2\sqrt{3x}$ **38.** $3x^3\sqrt{3x}$ **39.** $5x^5\sqrt{3}$ **40.** $5x^6\sqrt{5}$ **41.** $2ab^2c^2\sqrt{30ac}$ **42.** $11a^3b^2\sqrt{c}$ **43.** $2x^3y^4\sqrt{14xy}$ **44.** $3x^6y^3\sqrt{11xy}$ **45.** $3\sqrt{2}$ **46.** $-2\sqrt{5}$ **47.** $7x\sqrt{3}$ **48.** $8a\sqrt{2} + 2a\sqrt{3}$ **49.** $-7\sqrt{5} + 2\sqrt{10}$ **50.** $9\sqrt{6} - 15\sqrt{2}$ **51.** $-5\sqrt{7} + 2x\sqrt{7}$ **52.** $5a\sqrt{3} - 26\sqrt{3}$ **53.** $6x^2$ **54.** $-10a\sqrt{b}$ **55.** $4ab\sqrt{a}$ **56.** $-15x^4$ **57.** 5 **58.** −18 **59.** $\sqrt{10} - \sqrt{6} - 4$ **60.** $\sqrt{30} - 10 + 5\sqrt{2}$ **61.** $20 + 3\sqrt{11}$ **62.** $27 + 8\sqrt{10}$ **63.** $8 - 4\sqrt{3} + 12\sqrt{6} - 18\sqrt{2}$ **64.** $15 - 3\sqrt{2} - 10\sqrt{3} + 2\sqrt{6}$ **65.** $66 + 36\sqrt{2}$ **66.** $74 - 40\sqrt{3}$ **67.** $ab - 3a\sqrt{ab} - 2\sqrt{ab} + 6a$ **68.** $3xy + x\sqrt{xy} - 6\sqrt{xy} - 2x$ **69.** $\frac{\sqrt{3x}}{3x}$ **70.** $\frac{2y\sqrt{5}}{5}$ **71.** $\frac{x^2y\sqrt{2}}{4}$ **72.** $\frac{3a\sqrt{2b}}{2}$ **73.** $\frac{\sqrt{21}}{7}$ **74.** $\frac{\sqrt{2}}{3}$ **75.** 6 **76.** $\sqrt{6}$ **77.** $\frac{a^2\sqrt{2}}{2}$ **78.** $\frac{x\sqrt{3}}{3}$ **79.** $\frac{\sqrt{ab}}{b^2}$ **80.** $\frac{x^2\sqrt{xy}}{y^2}$ **81.** $\sqrt{5} - \sqrt{2}$ **82.** $\frac{2(\sqrt{6} + \sqrt{3})}{3}$ **83.** $-7 + 3\sqrt{5}$ **84.** $\frac{3 - 2\sqrt{3}}{3}$ **85.** $c = \sqrt{89}$ **86.** $a = \sqrt{2}$ **87.** $b = \frac{\sqrt{51}}{2}$ **88.** $c = 3.124$ **89.** 30 meters **90.** 45.8 miles **91.** 7.1 inches **92.** $x = 13$ **93.** $x = 42$ **94.** $x = -1$ **95.** $x = 6$ **96.** $x = 4$ **97.** $x = 7$ **98.** $x = 2$ **99.** $x = 7$ **100.** $y = 77$ **101.** $y = \frac{2}{15}$ **102.** $y = \frac{1}{2}$ **103.** 400 lb **104.** 134 ft **105.** 8 times as much

Putting Your Skills to Work

1. 50 miles per hour **2.** 80 feet

Chapter 9 Test

1. 9 **2.** $\frac{3}{10}$ **3.** $4xy^3\sqrt{3y}$ **4.** $10xz^2\sqrt{xy}$ **5.** $\sqrt{5}$ **6.** $8\sqrt{a} + 5\sqrt{2a}$ **7.** $12ab$ **8.** $3\sqrt{2} - \sqrt{6} + 45$ **9.** $10 - 4\sqrt{6}$ **10.** $19 + \sqrt{10}$ **11.** $\dfrac{\sqrt{5x}}{5}$ **12.** $\dfrac{\sqrt{3}}{2}$ **13.** $\dfrac{16-3\sqrt{2}}{-34}$ or $\dfrac{3\sqrt{2}-16}{34}$ **14.** $a(\sqrt{5} - \sqrt{2})$ **15.** 12.49 **16.** $x = \sqrt{95}$ **17.** $x = 3\sqrt{3}$ **18.** $x = \frac{35}{2}$ **19.** $x = 9$ **20.** 16 inches **21.** \$912.05 **22.** $A = 21.22 \text{ cm}^2$

Cumulative Test for Chapters 0–9

1. $\frac{4}{5}$ **2.** $8\frac{11}{12}$ **3.** $\frac{15}{16}$ **4.** 0.0007 **5.** 3 **6.** 31 **7.** $-2x^3yz^2 + 4xyz$ **8.** $27x^3 - 54x^2 + 36x - 8$ **9.** $4x^2 - 12x + 9$ **10.** $9x^2 - 121$ **11.** $\dfrac{4x+4}{x+2}$ **12.** $\dfrac{(x-3)(x+1)}{(x-6)(x+2)}$ **13.** $7x^2y^3\sqrt{2x}$ **14.** $14\sqrt{3}$ **15.** $9 - 6\sqrt{2}$ **16.** $5 + 2\sqrt{6}$ **17.** $10 + 22\sqrt{3}$ **18.** $\dfrac{3+\sqrt{2}}{7}$ **19.** -4 **20.** not a real number **21.** $c = \sqrt{23}$ **22.** $b = 4\sqrt{5}$ **23.** $x = 5$ **24.** $y = 6$ **25.** $y = 10$ **26.** 78.5 m^2

Chapter 10

Pretest Chapter 10

1. $x = 14, -2$ **2.** $x = 0, \frac{4}{3}$ **3.** $x = \frac{2}{5}, 4$ **4.** $x = \frac{7}{3}, -2$ **5.** $x = \pm 2\sqrt{3}$ **6.** $x = \pm 5$ **7.** $x = -4 \pm \sqrt{11}$ **8.** $x = \dfrac{-3 \pm \sqrt{65}}{4}$ **9.** $x = \dfrac{3 \pm \sqrt{3}}{2}$ **10.** $x = \frac{1}{4}, x = -2$ **11.** $x = \dfrac{-4 \pm \sqrt{13}}{3}$ **12.** no real solution

13. (0, 5) **14.** (1, −3) **15.** base = 6 cm; altitude = 13 cm

10.1 Exercises

1. $a = 1, b = 5, c = 6$ **3.** $a = 8, b = -6, c = 5$ **5.** $a = 12, b = -3, c = 2$ **7.** $a = 3, b = -8, c = 0$ **9.** $a = 1, b = 3, c = -15$ **11.** $a = 1, b = -9, c = 10$ **13.** $x = 0, x = -\dfrac{1}{2}$ **15.** $x = 0, x = \dfrac{14}{11}$ **17.** $x = 0, x = \dfrac{8}{7}$ **19.** $x = 0, x = \dfrac{3}{2}$ **21.** $x = -8, x = 3$ **23.** $x = -\dfrac{1}{2}, x = 8$ **25.** $x = \dfrac{1}{3}, x = 4$ **27.** $x = 2, x = 7$ **29.** $x = -10, x = -5$ **31.** $x = -\dfrac{3}{5}, x = -\dfrac{5}{2}$ **33.** $y = -3, y = -4$ **35.** $y = \dfrac{1}{3}, y = \dfrac{3}{2}$ **37.** $y = \dfrac{4}{3}$ **39.** $n = 5$ **41.** $n = -4, n = 5$ **43.** $x = 0, x = -\dfrac{14}{3}$ **45.** $x = -\dfrac{2}{3}, x = 3$ **47.** $n = -3, n = 6$ **49.** $x = 4$ **51.** $x = 1$ **53.** $x = 0, x = -8$ **55.** $x = \dfrac{5}{2}, x = -4$ **57.** $x = -\dfrac{5}{2}, x = 6$ **59.** $x = -5, x = 3$ **61.** $x = 3, x = 5$ **63.** $x = -\dfrac{10}{3}, x = 4$ **65.** You can always factor out x. **66.** You must have the product of two or more factors equal to 0, not 14. Remember the zero factor property. **67.** $a = \dfrac{14}{9}, c = -14$ **69.** 231 truck routes **71.** 6 cities **73.** 11 cities **75.** 140 or 160 machines may be produced. **77.** The average is 150 machines. The cost of producing 150 machines is \$4500, which is less than the cost of producing 140 or 160 machines. **79.** The cost of producing 148 machines or 152 machines is \$4500.80. This result is less than the result of problem 75 but more than the result of problem 77. The minimum cost of \$4500 occurs when 150 machines are manufactured. **80.** $\dfrac{(2x-1)(x+2)}{2x}$ **81.** $\dfrac{5}{22}$ **82.** $\dfrac{2}{x(x+4)}$ **83.** $\dfrac{2x^2 - 4x + 2}{3x^2 - 11x + 10}$

10.2 Exercises

1. $x = \pm 7$ **3.** $x = \pm 5\sqrt{5}$ **5.** $x = \pm 5\sqrt{3}$ **7.** $x = \pm 5$ **9.** $x = \pm 2\sqrt{5}$ **11.** $x = \pm\sqrt{3}$ **13.** $x = \pm 7\sqrt{2}$ **15.** $x = \pm 7$ **17.** $x = \pm 2\sqrt{3}$ **19.** $x = \pm\sqrt{5}$ **21.** $x = 3 \pm \sqrt{5}$ **23.** $x = -8, -4$ **25.** $x = \dfrac{-5 \pm \sqrt{2}}{2}$ **27.** $x = \dfrac{1 \pm \sqrt{7}}{3}$ **29.** $x = \dfrac{-2 \pm 2\sqrt{3}}{7}$ **31.** $x = \dfrac{3 \pm 5\sqrt{3}}{5}$ **33.** $x = \pm 4.389$ **35.** $x = \pm 1.600$ **37.** $x = 1, x = -15$ **39.** $x = 2 \pm \sqrt{5}$ **41.** $x = -3 \pm \sqrt{2}$ **43.** $x = 5 \pm \sqrt{29}$ **45.** $x = 0, x = -3$ **47.** $x = 5, x = -6$ **49.** $x = \dfrac{3}{2}, x = \dfrac{1}{2}$ **51.** $x = \dfrac{-5 \pm \sqrt{3}}{2}$ **52.** $x = \pm\sqrt{\dfrac{c}{a}}$ **53.** $x = \dfrac{-b \pm \sqrt{28 + b^2}}{2}$ **54.** $x = \dfrac{-b \pm \sqrt{b^2 - 4c}}{2}$ **55.** $a = -4, b = -4$ **56.** $x = 1, x = 5$ **57.** The pressure is 200 pounds per square inch. **58.** A total of 16 tires were defective 6 had both kinds of defects. 7 had only defects in workmanship, and 3 had only defects in materials.

10.3 Exercises

1. $x = -2 \pm \sqrt{3}$ **3.** $x = \frac{3 \pm \sqrt{41}}{2}$ **5.** $x = -1, x = \frac{9}{2}$ **7.** $x = \frac{-1 \pm \sqrt{21}}{10}$ **9.** $x = \frac{5 \pm \sqrt{17}}{4}$ **11.** $x = \frac{3 \pm \sqrt{33}}{12}$ **13.** $x = \frac{1 \pm \sqrt{19}}{6}$ **15.** $x = \frac{1 \pm \sqrt{3}}{3}$ **17.** no real solution **19.** $x = 6, x = 1$ **21.** $y = \frac{-7 \pm \sqrt{109}}{10}$ **23.** $m = 3, m = -\frac{3}{2}$ **25.** $x = 10.511, x = 6.779$ **27.** $x = 1.541, x = -4.541$ **29.** $x = 1.387, x = -0.721$ **31.** $x = -1.894, x = -0.106$ **33.** $t = \frac{3}{2}, t = \frac{2}{3}$ **35.** $s = 3, s = \frac{1}{3}$ **37.** $p = \frac{1 \pm \sqrt{21}}{2}$ **39.** $y = \frac{1 \pm \sqrt{51}}{5}$ **41.** $x = \pm\frac{\sqrt{21}}{3}$ **43.** $x = 1 \pm 2\sqrt{2}$ **45.** $\frac{3(x + 5)}{x + 2}$ **46. (a)** $y = -\frac{5x}{13} - \frac{14}{13}$ **(b)** $5x + 13y = -14$ **47.** $x^2 + 5x + 2$ **48.** $\frac{9p - 20}{3f} = g$

10.4 Exercises

1.

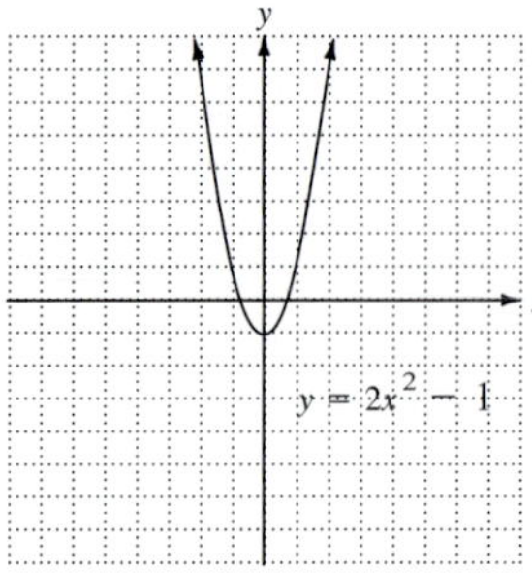

3.

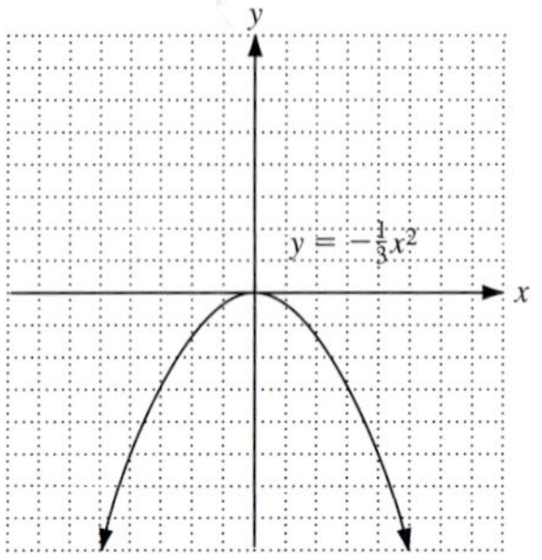

5.

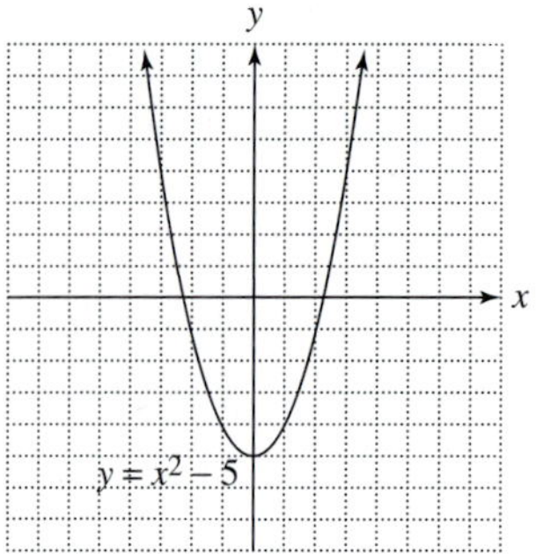

7.

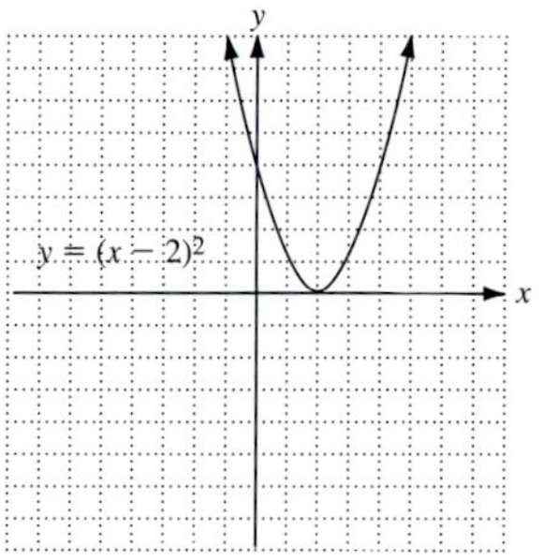

9.

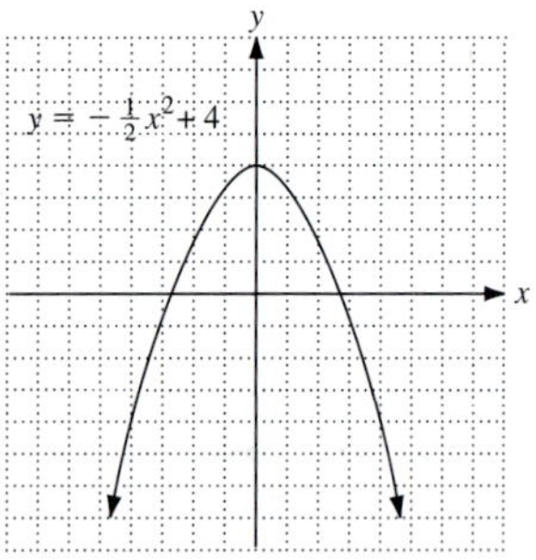

11.

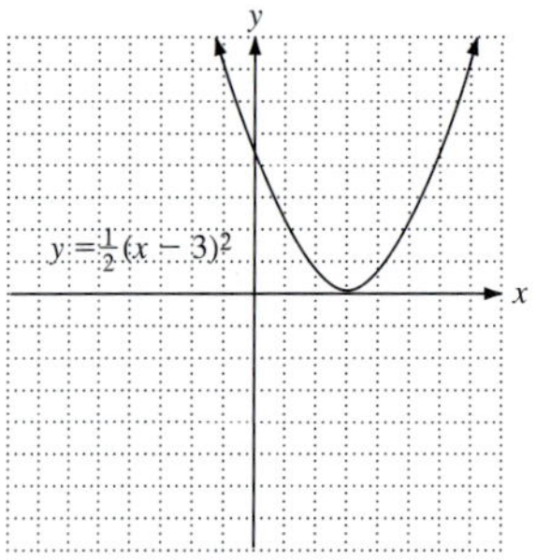

13.

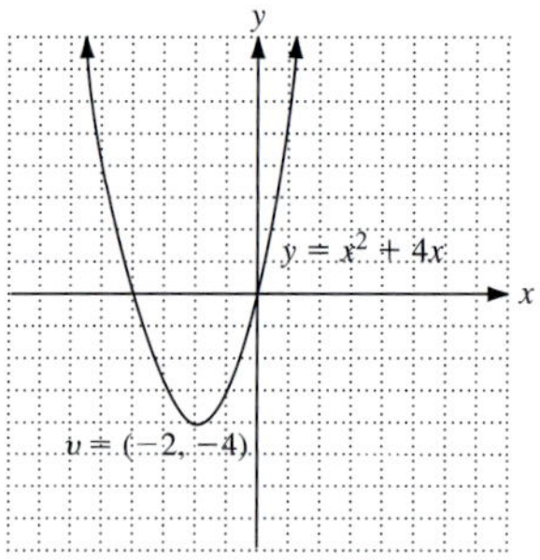

15.

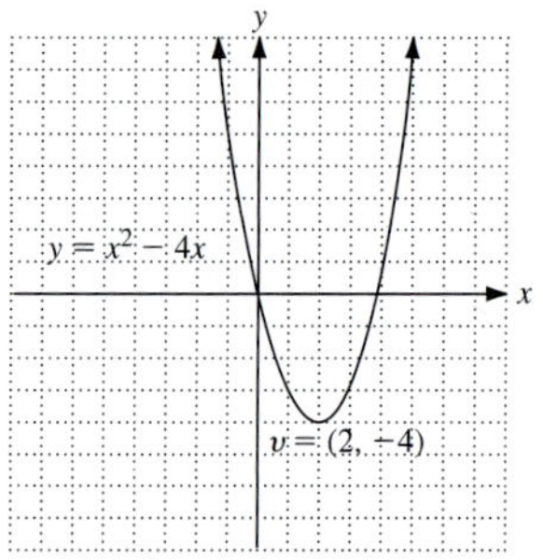

17.

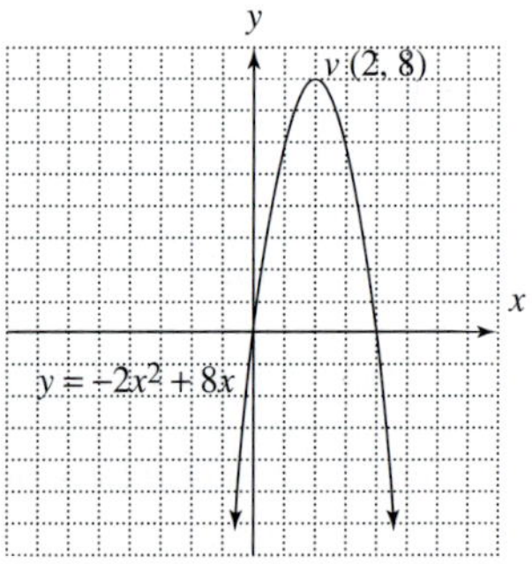

19.

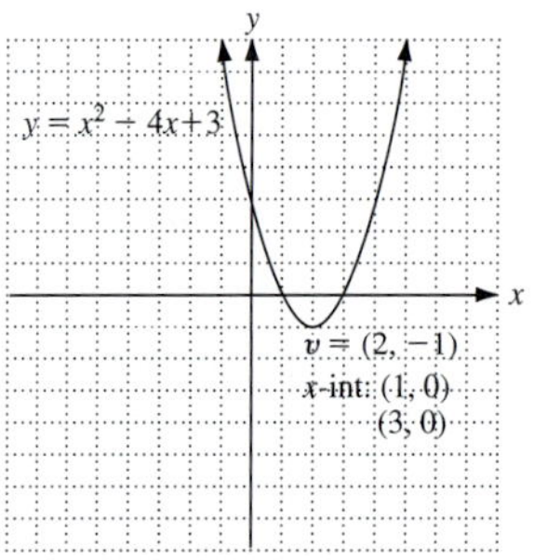

21.

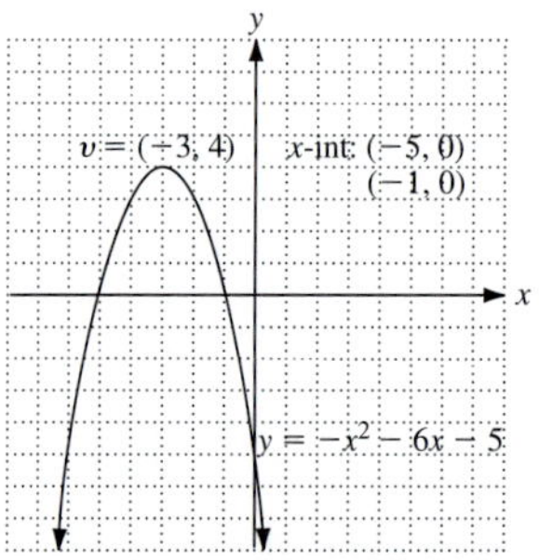

23.

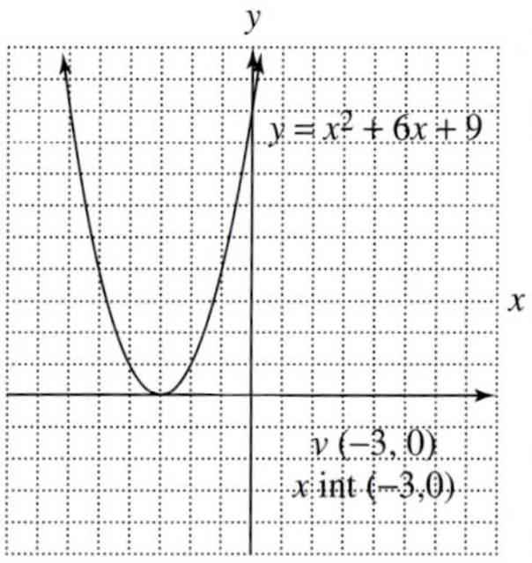

25. (a)

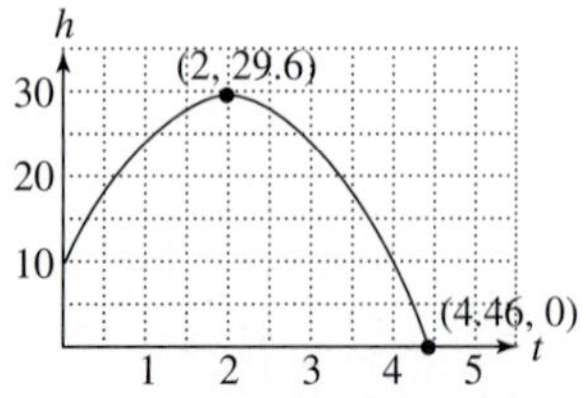

(b) 24.7 meters **(c)** 29.6 meters **(d)** 4.46 seconds **27.** vertex $(-1.167, -13.083)$, y-intercept $(0, -9)$, x-intercepts $(0.922, 0)$ and $(-3.255, 0)$ **29.** vertex $(-0.926, 1266.848)$, y-intercept $(0, 1133)$, x-intercepts $(1.923, 0)$ and $(-3.776, 0)$ **31.** $y = 1$ **32.** $y = 75$

Putting Your Skills to Work

1. 11.5 feet **2.** 11.5 feet **3.** 3 yards **4.** 5 yards

10.5 Exercises

1. 32 yards, 24 yards **3.** $L = 12$ m, $W = 5$ m **5.** 21 people originally planned to give the gift. **7.** There were originally 360 members. After the increase there were 450 members. **9.** If the width is 20 feet, the length is 100 feet. If the width is 50 feet, the length is 40 feet. **11.** The cruising speed is 600 mph **13.** 13 **14.** 16 **15.** \$1750 **16. (a)** 90 boots **(b)** \$1800 **17.**

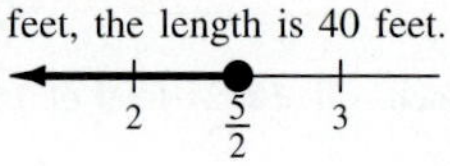

18. $\sqrt{15} - 2\sqrt{3}$

Putting Your Skills to Work

1. The storm began at 4:00 A.M. Monday. **2.** The storm ended at 6:00 A.M. Tuesday.

Chapter 10 Review Problems

1. $a = 5, b = 2, c = -3$ **2.** $a = 15, b = 6, c = 4$ **3.** $a = 1, b = -9, c = -3$ **4.** $a = 3, b = 6, c = -10$ **5.** $a = 5, b = -3, c = 1$ **6.** $a = 4, b = 5, c = -14$ **7.** $a = 3, b = -6, c = 0$ **8.** $a = 12, b = 0, c = -14$ **9.** $-20, -1$ **10.** -5 **11.** $3, -9$ **12.** $\frac{3}{2}$ **13.** $-\frac{5}{6}, -3$ **14.** $\frac{1}{4}, 1$ **15.** $-\frac{4}{3}, \frac{1}{4}$ **16.** $-\frac{1}{3}, -\frac{8}{3}$ **17.** $\frac{2}{3}, -\frac{3}{2}$ **18.** $\frac{1}{3}, -\frac{3}{2}$ **19.** $\frac{1}{5}, -\frac{1}{3}$ **20.** $\frac{1}{4}, -\frac{4}{3}$ **21.** $0, -\frac{16}{5}$ **22.** $0, \frac{9}{2}$ **23.** $0, -4$ **24.** $0, 3$ **25.** $0, -\frac{5}{3}$ **26.** $0, 4$ **27.** $-\frac{2}{5}, 10$ **28.** $7, 2$ **29.** $-\frac{2}{3}, 4$ **30.** $\frac{1}{3}, \frac{4}{3}$ **31.** ± 10 **32.** ± 11 **33.** ± 5 **34.** ± 6 **35.** $\pm\sqrt{22}$ **36.** $\pm\sqrt{39}$ **37.** $\pm 2\sqrt{2}$ **38.** $\pm 2\sqrt{5}$ **39.** $\pm 3\sqrt{2}$ **40.** $\pm 2\sqrt{6}$ **41.** $4 \pm \sqrt{7}$ **42.** $2 \pm \sqrt{3}$ **43.** $\frac{-6 \pm 2\sqrt{5}}{5}$ **44.** $\frac{-8 \pm 2\sqrt{3}}{3}$ **45.** $-1, -5$ **46.** $-5, -3$ **47.** $2 \pm \sqrt{2}$ **48.** $-1 \pm \sqrt{7}$ **49.** $9, -5$ **50.** $6 \pm 2\sqrt{14}$ **51.** $-1 \pm \sqrt{3}$ **52.** $\frac{-5 \pm \sqrt{31}}{2}$ **53.** $-8, 5$ **54.** $6, 3$ **55.** $-2 \pm \sqrt{10}$ **56.** $-2 \pm 2\sqrt{3}$ **57.** $\frac{7 \pm \sqrt{17}}{4}$ **58.** $\frac{-5 \pm \sqrt{73}}{4}$ **59.** $\frac{-2 \pm \sqrt{14}}{2}$ **60.** $\frac{-3 \pm \sqrt{19}}{2}$ **61.** $\frac{5 \pm \sqrt{73}}{6}$ **62.** $\frac{-3 \pm \sqrt{41}}{8}$ **63.** $\frac{-1 \pm \sqrt{2}}{2}$ **64.** $\frac{4 \pm \sqrt{10}}{3}$ **65.** $\frac{5}{2}, 2$ **66.** $\frac{3}{2}, -\frac{1}{2}$ **67.** $\frac{1}{3}$ **68.** $4, \frac{3}{2}$ **69.** $\frac{3 \pm \sqrt{3}}{3}$ **70.** $\frac{7 \pm \sqrt{209}}{10}$ **71.** $\frac{-2 \pm \sqrt{19}}{3}$ **72.** $\frac{-7 \pm \sqrt{29}}{10}$ **73.** $\frac{-3 \pm \sqrt{5}}{2}$ **74.** $-3, \frac{2}{3}$ **75.** $-\frac{1}{2}, -1$ **76.** $1, \frac{1}{4}$ **77.** ± 6 **78.** $\pm 3\sqrt{2}$ **79.** -8 only **80.** -7 only **81.** $\frac{-4 \pm \sqrt{31}}{3}$ **82.** $\frac{-4 \pm \sqrt{2}}{2}$ **83.** $\frac{11 \pm \sqrt{21}}{10}$ **84.** $-4, 2$ **85.** $-3, 5$ **86.** $0, \frac{1}{6}$

87. $y = 2x^2$, $(0, 0)$

88. $y = x^2 + 4$, $(0, 4)$

89. $y = x^2 - 3$, $(0, -3)$

90.

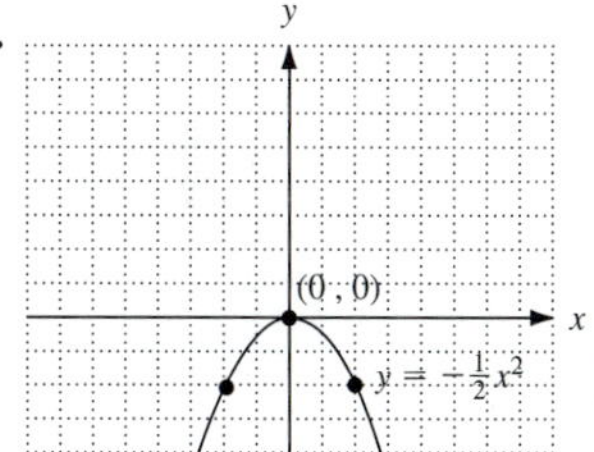

91.

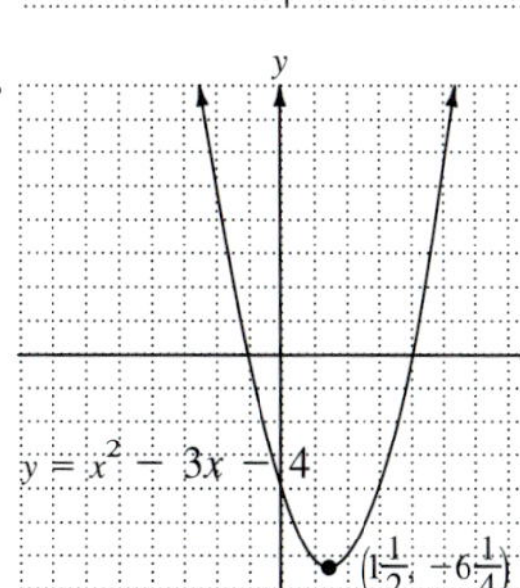

92.

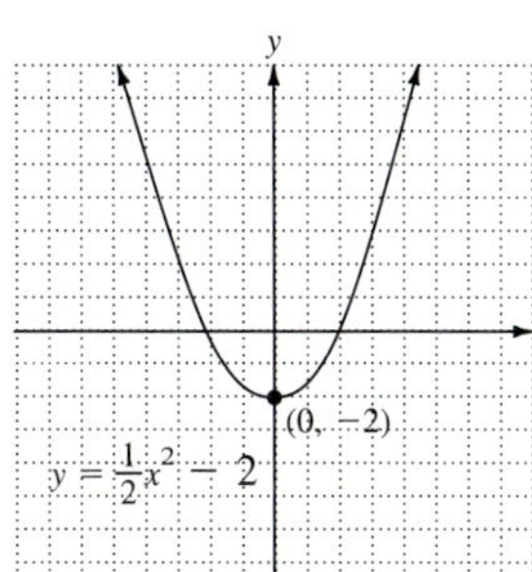

93.

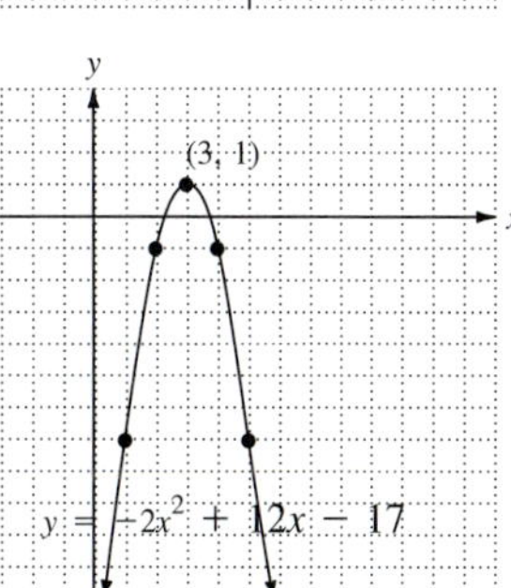

94.

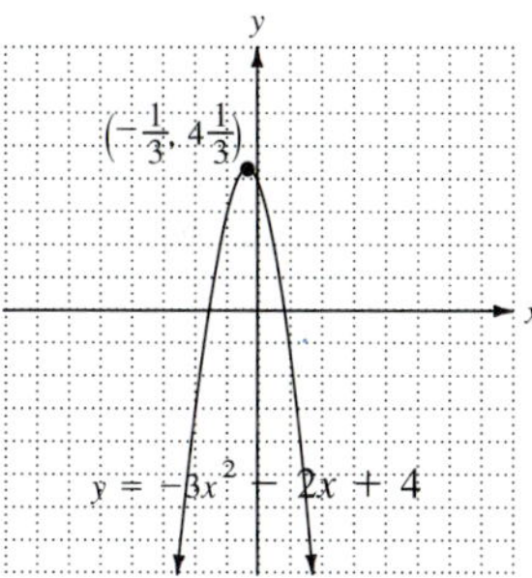

95. $l = 5$ feet, $w = 3$ feet or $l = 6$ feet, $w = 2.5$ feet **96.** 16 centimeters, 12 centimeters **97.** 24 members were in the club *last year.* **98.** 5 and 12 **99.** 38.25 ft^2 **100.** going 45 mph, returning 60 mph

Chapter 10 Test

1. $\frac{-7 \pm \sqrt{129}}{10}$ **2.** $-5, \frac{2}{3}$ **3.** no real solution **4.** $-\frac{1}{2}, \frac{5}{3}$ **5.** $-\frac{5}{4}, \frac{1}{3}$ **6.** $\frac{4}{3}$ **7.** $0, 8$ **8.** ± 4 **9.** $-\frac{1}{2}, 6$ **10.** $\frac{3}{2}, -\frac{1}{2}$

11.

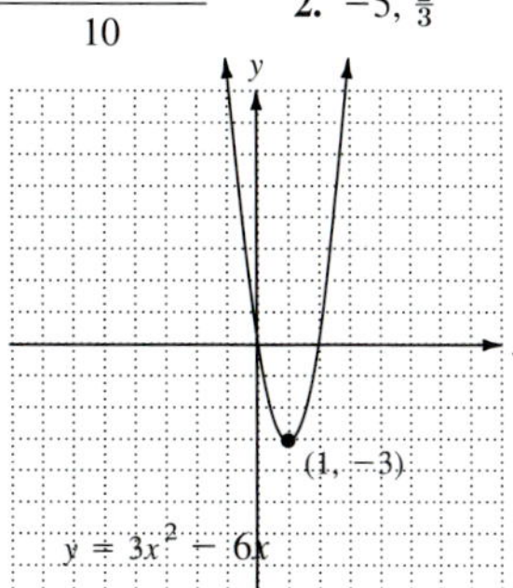

12.

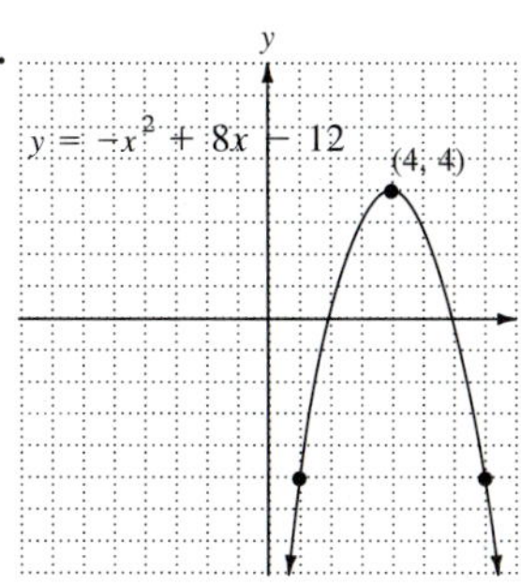

13. 9 meters and 12 meters **14.** 7 seconds

Cumulative Test for Chapters 1–10

1. $-27x^2 - 30xy$ **2.** 35 **3.** 2 **4.** $(2x - 1)(2x + 1)(4x^2 + 1)$ **5.** $6x^3 + 17x^2 - 31x - 12$ **6.** $-\frac{12}{5}$

7.

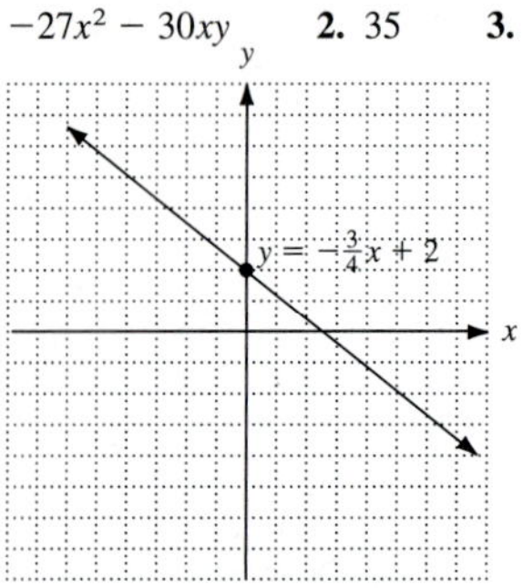

slope $-\frac{3}{4}$; y-intercept $= +2$

8. $y = -3x + 5$ **9.** $(3, -2)$ **10.** $81x^{12}y^8$ **11.** $3x^2y^3z\sqrt{2xz}$ **12.** $-2\sqrt{6} - 8$ **13.** $-5\sqrt{5} + 11$ **14.** $r = \dfrac{A \pm \sqrt{A^2 + 3A}}{3}$ **15.** $b = \pm\dfrac{\sqrt{H + 6}}{5}$ **16.** $\frac{7}{2}$, -5 **17.** $-\frac{1}{2} \pm \sqrt{5}$ **18.** -12, 4 **19.** $\dfrac{-11 \pm \sqrt{97}}{6}$ **20.** $\dfrac{-1}{7}$, -5 **21.** $\pm 7\sqrt{2}$ **22.** ± 5

23.

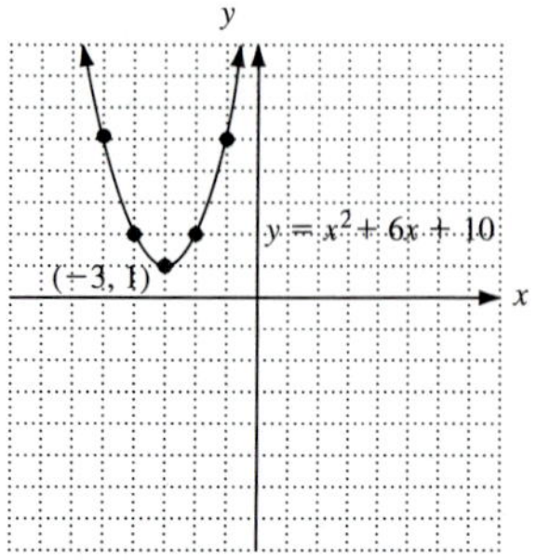

24. width = 6 meters, length = 16 meters

Practice Final Examination

1. $-2x + 21y + 18xy + 6y^2$ **2.** 14 **3.** $18x^5y^5$ **4.** $2x^2y - 8xy$ **5.** $\frac{8}{5}$ **6.** $b = \dfrac{p - 2a}{2}$ **7.** $x \geq 2.6$

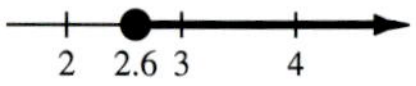

8. $2x^3 - 5x^2y + xy^2 + 2y^3$ **9.** $2(2x + 1)(x - 5)$ **10.** $25(x^2 + 1)(x + 1)(x - 1)$ **11.** $\dfrac{6x - 11}{(x + 2)(x - 3)}$ **12.** $\dfrac{11(x + 2)}{2x(x + 4)}$ **13.** $-\frac{12}{5}$

14. $m = \frac{5}{2}$

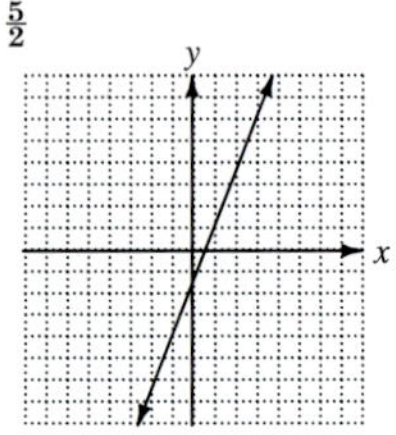

15. $3x + 4y = 14$ **16.** 38.13 sq. in. **17.** 7, -2 **18.** -2, -3 **19.** $8x\sqrt{2}$ **20.** $6\sqrt{3} - 12 + 12\sqrt{2}$ **21.** $-\dfrac{\sqrt{15} + \sqrt{35} + \sqrt{21} + 7}{2}$ **22.** $\dfrac{5}{3}$, -1 **23.** $\dfrac{3 \pm \sqrt{7}}{2}$ **24.** ± 2 **25.** $\sqrt{39}$ **26.** The number is 5. **27.** Width is 7 meters. Length is 12 meters. **28.** \$3000 at 10%, \$4000 at 14% **29.** 200 reserved seat tickets and 160 general admission tickets **30.** The base is 8 meters. The altitude is 17 meters.

Appendix B

Exercises

1. 34,000 m **3.** 5700 cm **5.** 250 mm **7.** 0.563 m **9.** 29,400 mg **11.** 0.0984 kg **13.** 7000 mL **15.** 0.000004 kL **17.** 1.7 in. **19.** 22.5 km **21.** 2.8 m **23.** 17.8 cm **25.** 1089.6 g **27.** 171.6 lb **29.** 2.8 L **31.** 193,248,000 in. **33.** 0.45 miles per hour **35.** 0.11 yr

Appendix C

Exercises

1. 84,916 square miles **3.** 3,500,000 **5.** 2 **7.** Treasure State **9.** Nevada **11.** 2000 homes **13.** Dupage County **15.** 2800 homes **17.** 4,000,000 apartment units **19.** 7,200,000 more apartment units **21.** 8,000,000 **23.** The greatest increase of 3 million people occurred twice: first, in the period between 1970 and 1980 and also in the period between 1980 and 1990 **25.** 3 million more **27.** 37 million **29.** Movies and an Exercise Program **31.** \$5.6 million **33.** \$2.5 million greater **35.** Between 1993 and 1994 **37.** 74 million **39.** an age group of 65 years and above **41.** 19% **43.** 83% **45.** 754 million people **47.** 40% **49.** \$3588

Appendix D

Exercises

1. 14 **3.** 52 **5.** 1296 **7.** -3 **9.** $10x - 5$ **11.** 1111811 **13.** 9876549 **15.** 6481100 **17.** 1 6 15 20 15 6 1 **19.** $x = 3$, steps will vary **21.** $x = -2$, steps will vary **23.** $x = 3$, steps will vary **25.** $x = \dfrac{9}{11}$, steps will vary **27.** answers will vary **29.** William was first, Brent was second, James was third, Dave was fourth **31.** The teacher is Michael **33.** Toyota Corolla **35.** Unless Fred is willing to pay over \$200 for Tuesday's procedure, he will probably get a false tooth.

Applications Index

Index

Photo Credits

CHAPTER 0 **CO** Kathy Ferguson/Photo Edit **p. 48** Robert Clay/Monkmeyer Press

CHAPTER 1 **CO** Richard Pasley/Stock Boston **p. 110** John Coletti/Stock Boston **p. 113** Tom McCarthy/PhotoEdit **p. 117** Zigy Kaluzny/Tony Stone Images

CHAPTER 2 **CO** Richard R. Hansen/Photo Researchers, Inc. **p. 157** Jeffrey Sylvester/FPG International **p. 166** Richard Pasley/Stock Boston

CHAPTER 3 **CO** Bachmann/Photo Researchers, Inc. **p. 191** Siteman/Monkmeyer Press **p. 196** Gilbert Grant/Photo Researchers, Inc. **p. 213** Laima Druskis/Stock Boston **p. 219** Donald C. Johnson/The Stock Market **p. 227** Bill Ross/Woodfin Camp & Associates **p. 234** Rosanne Olson/Tony Stone Images

CHAPTER 4 **CO** Peter Menzel/Stock Boston

CHAPTER 5 **CO** Steven Dunwell/The Image Bank **p. 330** B. Daemmrich/The Image Works

CHAPTER 6 **CO** Porterfield/Chickering/Photo Researchers, Inc. **p. 363** Eric Schweikardt/The Image Bank

CHAPTER 7 **CO** Bachmann/Stock Boston **p. 426** Ron Church/Photo Researchers, Inc.

CHAPTER 8 **CO** Roger Tully/Tony Stone Images **p. 478** David Parker/Science Photo Library/Photo Researchers, Inc.

CHAPTER 9 **CO** Goodwin/Monkmeyer Press **p. 552** Tony Freeman/PhotoEdit

CHAPTER 10 **CO** Paul Chesley/Tony Stone Images **p. 590** All-Sport Photographic Ltd. **p. 599** Susan Van Etten/Monkmeyer Press

APPENDIX C **p. A–6** Michael Newman/PhotoEdit **p. A–7** Stock Boston **p. A–10** Lawrence Migdale/Stock Boston **p. A–11** Ellis Herwig/Stock Boston